MEDICAL MYCOLOGY

The Pathogenic Fungi and The Pathogenic Actinomycetes

John Willard Rippon, Ph.D.

Associate Professor of Medicine
The Pritzker School of Medicine
The University of Chicago
Chicago, Illinois

W. B. SAUNDERS COMPANY Philadelphia, London, Toronto

W. B. Saunders Company: West Washington Square
Philadelphia, PA 19105

1 St. Anne's Road
Eastbourne, East Sussex BN21 3UN, England

1 Goldthorne Avenue
Toronto, Ontario M8Z 5T9, Canada

Library of Congress Cataloging in Publication Data

Rippon, John Willard.

Medical mycology the pathogenic fungi and the pathogenic actinomycetes.

Includes bibliographies.

1. Medical mycology. I. Title. [DNLM: 1. Actinomycetales. 2. Actinomycosis 3. Fungi. 4. Mycoses. WC450 R593m 1974]

RC117.R5 616.9'69 73-80980

ISBN 0-7216-7585-9

Medical Mycology ISBN 0-7216-7585-9

Print No.: 9 8 7 6 5 4

Dedicated to Antonio González Ochoa

When I was a student, my late teacher and friend, Professor Ester Meyer, praised *el maestro* as an exemplary investigator, teacher, clinician, and gentleman. In my later professional association with him, I found these qualities not merely true but magnified.

Note to the Reader

A set of three color filmstrips on *Medical Mycology* is available from the publisher. The filmstrips are for individual use and teaching purposes, and nicely complement this textbook.

In this text the color filmstrip frames are referred to in the legends of various illustrations by the notation FS A 1, as an example. The three filmstrips (A, B, and C) consist of a total of 220 separate frames.

PREFACE

The purpose of this volume is threefold: (1) to provide a basic presentation of the subject matter of medical mycology; (2) to present an overview of research efforts concerning both the fungal pathogen and its host; and (3) to serve as a reference source for both historical and current literature of importance in the development of the field. In this way it is hoped that this text will be found useful to students at all levels—the professional medical mycologist, the investigator in this and related disciplines, and the clinician and pathologist who may be concerned with fungous diseases only occasionally or whose practice involves frequent encounters with the mycoses. There is presently a renewed interest in mycotic diseases, particularly in the so-called opportunistic infections, and in basic research on host-parasite interactions. For this reason, it was felt that a text should be designed to include these recent developments in the field of medical mycology and to provide a fundamental understanding of the diseases and the organisms that cause them. It is hoped that this text will provide such an expanded view of the subject.

The history and development of medical mycology are characterized by several stages. In the days of Gruby, Malmsten, and Schoenlein, around 1840, there was a wave of excitement in this new field, and investigators were determining the fungal etiology of several dermatologic diseases, such as the tineas and thrush. A few years later, the sister field of bacteriology began to overshadow clinical mycology because of the trail-blazing work of Koch and Pasteur, and the study of fungal disease went into its first decline. The critical investigator was replaced by the dilettante, and in that period many papers appeared describing dozens of new fungal pathogens (generally contaminants) and attributing a fungal etiology to everything from warts and acne to psoriasis and pemphigus.

In the early 1900's, this dermatologic and mycologic chaos was brought into order temporarily by a dermatologist who was also a scientist. Sabouraud's monumental work on the tineas cleared the air, and he set forth the principles of medical mycology: careful observation of the disease, critical evaluation of its etiology, and a cautious, unbiased approach to therapy. Soon after this, however, there was another period of decline and confusion. There were by then some 130 synonyms for *Candida albicans* and descriptions of more than 300 dermatophyte species in 40 genera. Beginning in the late 1930's, however, trained mycologists, such as Conant and Emmons, began working in the field and placed the nomenclature of the etiologic agent and evaluation of the disease syndrome on a scientific basis. Conant's lucid *Manual of Clinical Mycology* became the bible of the discipline and was later joined by the works of Emmons and Lacaz. These works clearly defined and described the pathogenic fungi as well as the diseases they produce and remain standard texts in the field.

Beginning in the early 1960's, a new aspect of medical mycology has emerged and gained increasing importance. This is the growing incidence of opportunistic fungous infections. To a large extent, this has paralleled advances in medicine involving the treatment of neoplasms, collagen diseases, and other debilitations as well as the progress made in organ transplantation. Such techniques prolong the life of the patient but abrogate normal host defenses and frequently allow saprophytic species to proliferate and invade, thereby causing disease.

In this volume I have generally approached the subject matter along traditional lines used in previous textbooks of medical mycology. In addition, I have tried to emphasize the growing importance of opportunistic infections and to delineate the differences between pathogenic and opportunistic infecting fungi and the diseases they elicit.

Another area of great importance in medical mycology is the increased basic research aimed at elucidating the physiology, biochemistry, and mechanisms of pathogenicity of the fungi in addition to the responses of the host when challenged by infection. An overview of some of this work is included in the appropriate chapters. This is not intended to be a review of all investigations in a particular field but only an indication of some areas in which active research is being carried out. Sections are also included concerning both natural and experimental infections in animals. I have tried to compare and contrast the diseases as they occur in man and animals, and in this way I hope that the text will be useful in the study of both human and veterinary mycology. In addition, there are chapters on other important aspects of medical mycology, such as the pharmacology of antimycotic agents, allergic diseases, intoxications, the genetics of pathogenic fungi, and taxonomy.

In choosing the references to be included in the various chapters, I have elected to pick those that illustrate two aspects of the subject. The first group includes the historically significant papers that first described a disease or some important facet of it. It has been very rewarding and sometimes amusing to read original literature on a particular disease. Papers up to 1960 that are still pertinent or whose content has not been updated by more recent investigations or reviews have been included. The second criterion for choice of a paper was its recentness and significance in the field. Most of the references cited have appeared since 1960. These papers not only include new information but also review previous publications and thus act as a bibliography for those interested in a particular subject.

I am profoundly and deeply grateful for the assistance of Dr. Josephine Morello. She has patiently transformed (translated) the language of my manuscript into readable English. This has been done after many hours of writing, rewriting, and debating, leading to deletion and compromise. The reader should be aware that if he finds this text at all readable it is because of her arduous efforts.

I also wish to thank my board of consultants, Libero Ajello, E. S. McDonough, Angela Restrepo, Shirley McMillen, and Howard Larsh, mycologists; Sharon Thomsen, Francis Straus, and Philip Graff, pathologists; Francisco Battistini and Allan L. Lorincz, dermatologists; John Fennessy, radiologist; and Nicholas J. Gross and Frederick C. Kittle, specialists in diseases of the chest. I would also like to thank Martha Berliner, P. Kulkavni, C. Satyanarayana, F. Mariet, S. Banerjee, and F. Pifano, among others, for the use of illustrations.

Special thanks go to my friend and associate in many research projects, Edward D. Garber, who has contributed the chapter on genetics and taxonomy.

The maps were prepared by Robert Williams and the drawings by Robert Williams, Charles Wellek, and John Rippon. A final thank you is given to Daila Shefner for proofreading this manuscript.

JOHN WILLARD RIPPON

CONTENTS

The Dimorphic Pathogenic Fungi

The Opportunistic Fungi

Part Three
Allergic Diseases Mycetismus, and Mycotoxicosis

Part Four
Taxonomy and Fungal Genetics

Part Five
Pharmacology of Actinomycotic Drugs

INTRODUCTION TO MEDICAL MYCOLOGY

The ability of fungi to cause disease appears to be an accidental phenomenon. With the exception of a few dermatophytes, pathogenicity among the molds is not necessary for the maintenance of dissemination of the species. Further, the fungi that are able to cause disease seem to do so because of some peculiar trait of their metabolism not shared by taxonomically related species. Thus the survival and growth of fungi at the elevated temperature of the body, the reduced oxidation-reduction environment of tissue, and the ability to overcome the host's defense mechanisms set apart these few species from the great numbers of saprophytic and plant pathogenic fungi.

Among the best examples of this transient adaptation to invasion and growth within tissue are the dimorphic fungi. In nature they grow as soil saprophytes, usually in a restricted ecologic niche, producing mycelium and spores similar to other fungi. However, when their spores are inhaled or gain entrance by other means into man or other animals, the organisms are able to adapt and grow in this unnatural environment. In so doing, drastic changes occur in their morphology, metabolism, cell wall content and structure, enzyme systems, and methods of reproduction. If the host's defenses are unable to counteract the organism—and this is rare—the infection leads to the death of both host and parasite. Infection for the fungus is a blind alley and is not contagious to other hosts or generally of use in disseminating the species. Such diseases include histoplasmosis, blastomycosis, coccidioidomycosis, and, to a degree, sporotrichosis. The previously mentioned also represent primary pathogens among the fungi in that they are able to cause disease in a normal healthy host, provided that sufficient numbers are present in the infecting dose. In debilitated hosts, their course of infection and disease is exaggerated.

The second major group of fungus diseases are the dermatophytoses. This is a closely related group of organisms with the ability to utilize keratin and to establish a kind of equilibrium, albeit transitory, with the host. In the soil are numerous related keratolytic species, most of which seldom, if ever, are recovered from clinical disease. Others, especially the anthropophilic species, are universal agents of ringworm and probably no longer have a significant soil reservoir, depending on human or fomite contact for transmission. These are among the commonest infectious agents of man. The dermatophytes are in a sense specialized saprophytes, as they do not invade living tissue, utilizing only the dead cornified appendages of the host, such as hair, skin, nails, fur, and feathers. The clinical disease caused by the dermatophytes is, for the most part, a toxic and allergic reaction to the presence of the organisms and its by-products.

A group of diseases referred to as the subcutaneous mycoses include chromomycosis, mycetoma, sporotrichosis, and subcutaneous phycomycosis. These are organisms of limited invasive ability, gaining entrance to the body by traumatic implantation, and may take years to produce a noticeable disease. The clinical course is a chronic progressive one and very slow to develop. Mycetoma or madura foot and chromomycosis are clinical entities that may have a number of different species as etiologic agents. Mycetoma in particular may be caused by a diverse group of bacterial and fungal organisms that are totally unrelated to one another, but the clinical disease elicited is similar. The organisms are all soil saprophytes of regional epidemiology.

There remains a large category of opportunistic fungous infections. Included are diseases which are manifest almost exclusively in patients debilitated by some other cause and whose normal defense mechanisms are impaired. Formerly, some of these diseases and the fungal infections were in almost constant association such as mucormycosis with diabetes, candidosis with hypothyroidism and other endocrine disturbances, and aspergillosis with chronic lung disease. In present-day medicine, the advent of cytotoxic drugs, long-term steroid treatment, and immunosuppressive agents has markedly increased the number and severity of diseases in this category. The diverse array of organisms being isolated from these cases emphasizes that probably all fungi may be considered potential pathogens when normal defenses are abrogated. Fungi are particularly remarkable for their ability to adapt and propagate in a wide variety of environmental situations; thus their invasion of debilitated patients is not surprising.

Also included in this book will be a brief discussion of conditions arising from ingestion or inhalation of fungal products. These include allergic manifestations to spores, such as farmer's lung and conidiosporosis; toxic reactions due to consumption of infected food products, such as aflatoxin; and, of universal interest, mushroom poisoning.

HISTORY

Medical mycology is a science with traditions like any other specialty of human endeavor. For this reason, some bacterial diseases are included in this discussion. For may years actinomycetes were considered a "link" between bacteria and fungi. The diseases they produce were chronic granulomatous diseases similar to true mycotic infections. Morphologically, physiologically, and biochemically the actinomycetes are bacteria. Further, they are susceptible to antibacterial antibiotics; fungi are not, neither are they susceptible to antifungal drugs. However, because of similarities in clinical disease, the pathogenic actinomycetes are treated along with eumycotic infections.

The discovery of the causal relation of certain fungi to infectious disease precedes by several years the pioneer work of Pasteur and Koch with pathogenic bacteria. Schoenlein and Gruby studied the fungus of the scalp infection favus (*Trichophyton schoenleinii*) in 1839, and in the same year Lagenbeck described the yeastlike organism of thrush (*Candida albicans*). Robert Remak had described favus earlier, but his work was ignored. Gruby even isolated the fungus of favus and produced the disease by inoculating a healthy head, thus fulfilling Koch's postulates before Koch formulated them. Prior to this, Bassi described the fungal etiology of muscardine of silkworms (*Beauveria bassiana*). In spite of its earlier beginnings, medical mycology was soon overshadowed by bacteriology and has never received as much attention, though some of the fungous diseases (dermatomycoses) are among the more common infections of man. This is perhaps attributable to the relatively benign nature of the common mycoses, the rarity of the more serious ones, and the difficulty of differentiation of these structurally complex forms which, in a practical sense, sets them off sharply from the bacteria.

A great impetus was given the study of fungous diseases by the careful work of perhaps the most famous name in the field of medical mycology—Raymond Sabouraud. The publication of the classic work on dermatophyte infections, "Les Teignes," was a model of scientific observation. Unfortunately, his followers were not as careful, and the literature became cluttered with numerous synonyms for almost every fungus infection and with a fungal etiology for almost every human disease. There are over 100 names given for the yeastlike organism, *Candida albicans.* Some true fungous infections (histoplasmosis and coccidioidomycosis) were at first described as caused by protozoan parasites, but the works of Ophuls, Brumpt, Gilchrist, and Smith (most of them dermatologists) delineated their true nature and the extent of their epidemiology. A group of Latin American scientist-clinicians is responsible for a large portion of our present knowledge. This group includes González Ochoa, Almeida, Mackinnon, and others. The terrible confusion in nomenclature was finally brought into order by the work of the outstanding mycologists, Norman Conant and Chester Emmons. Present research is aimed toward improved diagnostic techniques, specific serologic tests for fungous diseases, accurate taxonomy, and new and improved chemotherapeutic agents. Moreover, efforts are being made to elucidate the mechanisms of fungal pathogenicity. The field of medical mycology is indeed fortunate, now, to have a number of competent investigators directing their attention to the problems of this discipline.

PATHOGENIC FUNGI

Today we recognize some 50 "pathogens" among the approximately 100,000 species of fungi—about 20 that may cause systemic infections, about 20 which are regularly isolated from cutaneous infections, and a dozen that are associated with severe, localized, subcutaneous disease. In addition, there is a long list of opportunistic organisms that may cause disease in the debilitated patient. A few of the diseases discussed in this book, e.g., actinomycosis, candidosis, and pityriasis versicolor, are caused by endogenous organisms, i.e., species that are part of the normal flora of man; all other fungous and actinomycetous infections are exogenous in origin.

The criterion of pathogenicity is one of the poorest that can be used in differentiation of microorganisms. Because pathogenicity is variable and difficult to determine, by its use parasitic microorganisms are grouped together that are, in fact, much more closely related to certain free-living forms than they are to one another. The superficial nature of pathogenicity as a differential characteristic is nowhere better illustrated than among the fungi, for the pathogenic forms which constitute the subject matter of medical mycology form a heterogeneous group that includes some of the actinomycetes, certain molds and moldlike fungi, and a number of yeasts and yeastlike organisms. As stated before, very few are primary pathogens, i.e., able to produce disease in a healthy host, and infection is a blind alley for the organism. With the possible exception of two or three dermatophytes, none of this group is an obligate parasite, and most are misplaced soil saprophytes. From a general biological point of view, then, the pathogenicity of certain fungi is of very minor significance; from the point of view of the parasitized host, man, it is of considerably greater interest. For lack of a better definition, then, a fungal pathogen will denote an organism regularly isolated from a given disease process; rarely, isolated organisms from a variety of clinical circumstances will be considered opportunistic organisms (Table 1).

The fungi are structurally complex, showing a bewildering variety of reproductive structures associated with sexual and asexual processes, in addition to vegetative nonreproductive elements. Their differentiation into genera, species, and varieties is made, in large part, on a morphologic basis—especially the morphology of the reproductive structures. In contrast to bacteria, physiologic and immunologic characteristics of fungi are usually of minor or no importance for purposes of differentiation or identification. The biochemistry of the fungi has been extensively investigated, with the elucidation of decomposition occurring in nature, as well as the application of fungi to industrial processes, e.g., antibiotic production, alcohol and citric acid formation, and so on. Present knowledge of the respiratory mechanisms of the cell as well as of catabolic and anabolic pathways of metabolism is derived in large part from the studies of yeasts. However, identification of the fungi is still an exercise in contemplative observation.

IMMUNITY AND SEROLOGY

Natural immunity to fungous diseases is very high. Infection therefore depends on exposure to a sufficient inoculum size of the organism, and the general resistance of the host. This is well demonstrated by the two general categories of systemic fungous diseases. In the first, infection occurs in the patient who is in a particular endemic area and who inhales sufficient numbers of spores. The majority of such infections are asymptomatic or resolve quickly. This is followed by a specific resistance to reinfection. Only rarely does the infection become serious. On the other hand, opportunistic infections are caused by universally present organisms of low virulence. Establishment of disease depends entirely on lowered resistance of the host. If the patient recovers from such an infection, no specific resistance is noted.

Table 1. *Clinical Types of Fungous Infections**

Type	*Disease*	*Causative Organism*
Superficial infections	Pityriasis versicolor	*Pityrosporum orbiculare (Malassezia furfur)*
	Piedra	*Trichosporon cutaneum* (white)
		Piedraia hortai (black)
Cutaneous infections	Ringworm of scalp, glabrous skin, nails	Dermatophytes (*Microsporum* sp., *Trichophyton* sp., *Epidermophyton* sp.)
	Candidosis of skin, mucous membranes, and nails	*Candida albicans* and related species
Subcutaneous infections	Chromomycosis	*Fonsecaea pedrosoi* and related forms
	Mycotic mycetoma	*Allescheria boydii, Madurella mycetomii,* etc.
	Subcutaneous phycomycosis	*Basidiobolus haptosporus*
		Entomophthora coronata
	Rhinosporidiosis	*Rhinosporidium seeberi*
	Lobomycosis	*Loboa loboi*
	Sporotrichosis	*Sporothrix schenckii*
Systemic infections	Pathogenic fungous infections	
	Histoplasmosis	*Histoplasma capsulatum*
	Blastomycosis	*Blastomyces dermatitidis*
	Paracoccidioidomycosis	*Paracoccidioides brasiliensis*
	Coccidioidomycosis	*Coccidioides immitis*
	Opportunistic fungous infections	
	Cryptococcosis	*Cryptococcus neoformans*
	Aspergillosis	*Aspergillus fumigatus,* etc.
	Mucormycosis	*Mucor* sp., *Absidia* sp., *Rhizopus* sp.
	Candidosis, systemic	*Candida albicans*

*From Rippon, J. W. *In* Burrows, W. 1973. *Textbook of Microbiology*, 20th ed. Philadelphia, W. B. Saunders Company, p. 683.

Humoral antibodies play little or no part in our ability to contain a fungous infection. Cellular defenses are the only efficacious resistance to invasion. This is well illustrated by the types of infections exhibited by patients with the various forms of lymphomatous disease or with genetic defects of leukocytic function. In patients who have defects or dyscrasias of their T-cell lymphocytes, opportunistic fungous infections are common. If the defect is only in the B cells, fungous disease is uncommon, whereas bacterial infections, particularly those caused by gram-positive cocci, are frequent. The importance of cellular defense mechanisms is also reflected in the pathologic manifestations evoked by infection by the organism. The pathogenic fungi in the normal host induce a pyogenic reaction, followed by a granulomatous reaction. The response elicited by invasion by opportunistic fungi is necrotic and suppurative. The host is deficient and cannot contain the organism. To a certain extent, all fungous infections are opportunistic. The normal, healthy, well-nourished adult is resistant to all fungi, except in cases of overwhelming exposure. Slight debilitation, in some cases transient, may afford the opportunity for establishment of disease. These subtle differences in host defenses are a fascinating and as yet unexploited area of research. Such factors, and they appear to be numerous, determine who will have a fungous infection, either pathogenic or opportunistic, the severity of the disease, and the final outcome of the host-parasite interaction. Though some differences in the virulence of different strains of fungi have been found, these appear to be of minor importance.

Fungi are poor antigens. For this reason, in most cases, serology is of little importance as a diagnostic or prognostic aid. Complement-fixing antibodies, precipitins, and so forth, when present, may be used to evaluate the status of disease in a few cases. Often, however, there are many cross-reactions, conflicting opinions on interpretation, and no standardization of antigens, so that the usefulness of these procedures is very limited. Standardization of antigens and the use of immunodiffusion procedures in a few diseases are the subjects of much research presently and offer many advantages over the older complement-fixation procedures. Specific fluorescent antibody procedures are also useful in most diseases.

Hypersensitivity and allergy to fungi are

manifested in various ways. Frank allergy may simply be due to inhalation of the spores of fungi. This may induce asthma, asthmalike diseases, and even fibrosis and consolidation of lungs. Allergies may also manifest as "ids" or eruptions associated with cutaneous infections due to dermatophytes or *Candida* sp. Allergic reactions, such as erythema nodosum, may be part of the primary infection of a systemic disease, as in coccidioidomycosis. Hypersensitivity, as demonstrated by a positive skin reaction (e.g., to coccidioidin or histoplasmin), is often associated with good resistance to infection or reinfection.

PATHOLOGY

The tissue response of the host to the offending agent varies widely with the variety of organism. In dermatophyte infections, erythema is generally produced and is probably a response to the irritation caused by the organism or by its products of metabolism. Occasionally, severe inflammation, followed by scar tissue and keloid formation, will occur. In organisms that invade living tissue, such as those responsible for subcutaneous and systemic disease, a rather uniform pyogenic response is generally elicited. This usually gives way to a variety of chronic disease processes, which are listed in Table 2.

THERAPY

In general, antibacterial antibiotics are not effective in treating fungous diseases, and the few widely used antifungal agents are similarly ineffective against bacteria. For this

Table 2. *Tissue Reactions in Fungous Diseases**

Chronic inflammatory reactions
- Lymphocytes, plasma cells, neutrophils, and fibroblasts, occasionally giant cells
- *Rhinosporidium seeberi*
- Subcutaneous phycomycosis

Pyogenic reactions
- Acute or chronic, suppurative neutrophilic infiltrate
- *Actinomyces israelii:* sulphur granules, also lipid-laden peripheral histiocytes
- *Nocardia asteroides*
- Acute aspergillosis
- Acute candidosis

Mixed pyogenic and granulomatous reactions
- Neutrophilic infiltration and granulomatous reaction, lymphocytes, plasma cells
- *Blastomyces dermatitidis*
- *Paracoccidioides brasiliensis*
- *Coccidioides immitis:* neutrophils, especially at broken spherule
- *Sporothrix schenckii:* organism rarely seen in tissue
- Chromomycosis: chronic pyogenic inflammation; epithelioid cell nodules and giant cells
- Mycetoma; in addition, may be large foamy giant cells similar to xanthoma

Pseudoepitheliomatous hyperplasia
- Following chronic inflammation in skin (hyperplasia of epidermal cells, hyperkeratosis, elongation of rete ridges)
- *B. dermatitidis*
- *P. brasiliensis*
- Chromomycosis
- *Coccidioides immitis*

Histiocytic granuloma
- Histiocytes frequently with intracellular organisms, sometimes becoming multinucleate giant cells
- *Histoplasma capsulatum*
- Meningeal *Cryptococcus neoformans*

Granuloma with caseation
- Granulomatous reaction, Langhans' giant cells (L.G.C.), central necrosis
- *Histoplasma capsulatum*
- *Coccidioides immitis*

Granuloma "sarcoid" type
- Nonnecrotizing
- *Cryptococcus neoformans*
- Occasionally *Histoplasma capsulatum*

Fibrocaseous pulmonary granuloma; "tuberculoma"
- *H. capsulatum:* thick fibrous wall surrounding epithelioid and L.G.C. organisms in soft center, often calcification
- *Coccidioides immitis:* Thin fibrous wall, rarely calcified
- *Cryptococcus neoformans:* poorly defined

Thrombotic arteritis
- Thrombosis, purulent coagulative necrosis, invasion of vessels
- Aspergillosis
- Mucormycosis

Fibrosis
- Proliferating fibroblasts, deposition of collagen; may resemble keloid
- *Loboa loboi*

Sclerosing foreign body granuloma
- In paranasal sinuses or following viral infection
- *Aspergillus* sp., bizarre hyphae in giant cells

*A Gram stain is used for actinomycosis, nocardiosis, actinomycotic mycetoma, and candidosis; otherwise a periodic acid-Schiff (PAS) stain is recommended. Methenamine silver stains are useful for both actinomycotic and eumycotic organisms.

reason it is very important to establish a diagnosis of the fungal agent early in the course of the disease. It is not unusual to find patients on long-term antitubercular therapy, when, at last, it was found that a fungus was responsible for their disease.

In the preantibiotic era, about the only treatment for systemic mycoses was supportive therapy of the patient. Clinical cures can now be obtained with some agents. Cutaneous infections were treated with various combinations of keratolytic and antimicrobial agents, such as Castellani's paint and Whitfield's ointment. Modifications of these are still useful today in dermatologic practice. Sulfur ointments and salicylic acid remain the treatments of choice for some of these diseases.

Only a few useful antifungal agents are presently available for the therapy of fungous diseases. They generally have the twin drawbacks of being insoluble or only partially soluble in water and variably toxic in therapeutic dosage. One group of such substances, formed by various species of *Streptomyces*, is the polyene antibiotics, which are effective against certain pathogenic fungi. They have the general structure $(CH{=}CH)_n$, and, of the dozen or so which have been described, nystatin and amphotericin B have proved to be the most useful. Nystatin is a diene-tetraene polyene derived from *Streptomyces noursei*, and amphotericin B is a heptene derived from *S. nodosus*. Both contain the diaminomethyl deoxypentose and mycosamine. Nystatin also contains a lactone structure. Both appear to act against the cell membrane of the fungus, causing leakage of potassium and other metabolites. Both are relatively insoluble in water or saline. Nystatin is effective in topical *Candida* infections and occasionally in keratitis and otomycosis of other fungal origin, but not in deep mycoses or dermatophyte infections. Amphotericin B is effective against the deep mycoses, e.g., blastomycosis, cryptococcosis, histoplasmosis, coccidioidomycosis, and systemic candidosis, but its usefulness is somewhat limited by toxicity. However, this agent, given in dextrose solution to prevent precipitation, offers about the only effective treatment for these diseases. New drugs, such as pimaricin from *S. natalensis*, hamycin, and saramycetin (×5079C), are now receiving clinical trials. Some have the drawbacks of toxicity (pimaricin), high relapse rate (hamycin, dihydroxystilbamidine), and difficulty of production (saramycetin). Only continued clinical trial will determine if they will achieve the wide acceptance gained for nystatin and amphotericin B. Other new drugs for specific diseases are regularly appearing, as 5-fluorocytosine for cryptococcosis, and these will be discussed in separate chapters. Potassium iodide, an old treatment for deep mycoses, is still the treatment of choice for sporotrichosis and subcutaneous phycomycosis.

Another widely used antibiotic, griseofulvin, produced by *Penicillium griseofulvum*, is an effective chemotherapeutic agent in the dermatomycoses, but not in the deep mycoses. It has been found to be 7-chloro-2′,4,6-trimethyoxy-6′-methylspirobenzofuran-2 (3HO, 1′-[2]-cyclohexane)-3,4′-dione. It may be applied topically but is of questionable efficacy by this route. When given orally, it is absorbed and later concentrated in the keratin-containing structures. It causes abnormal multiple branching of the fungus and interferes with protein and nucleic acid synthesis. The latter probably is the most important mode of action. This antibiotic was isolated in 1939 by Raistrick but was of no interest at that time because it had no antibacterial activity. It was used for the treatment of some fungous infections of plants, but its chemotherapeutic activity in mammalian dermatophyte infection was not exploited until the work of Blank and Roth in 1959, as discussed in the chapters on dermatophytes and pharmacology. Since then, it has been increasingly more widely and successfully used for the treatment of the dermatomycoses. Again, its usefulness is limited by solubility problems and difficulty of absorption. The latter is partially alleviated by administration following a fatty meal. Long periods of treatment are often necessary with this drug, and the relapse rate is very high. A recently introduced antidermatophyte drug that has gained wide acceptance is tolnaftate. It is used topically and sometimes combined with orally administered griseofulvin. It is not effective in hair and nail infections.

A great deal of research is presently underway to develop more effective and less toxic antifungal agents. Because they are bacteria, the pathogenic actinomycetes are susceptible to antibacterial chemotherapy.

LABORATORY MYCOLOGY

The isolation of fungi is a relatively easy procedure; the identification and determination of significance are much more difficult. Unlike pathogenic bacteria, fungi are not nutritionally fastidious, literally growing on wet cement. A simple source of organic nitrogen and carbohydrate will suffice. The most commonly employed media in medical mycology is Sabouraud's agar. This contains beef extract and dextrose at a pH of 5.6, which discourages bacterial growth. It is not necessarily true that fungi grow best on this media (they do not), but it is traditional to use it. Further, the colonial morphology noted, studied, described, and taught is that of the fungus growing on Sabouraud's media. Colony morphology varies greatly with the media on which the organism is grown. Therefore, it is recommended that, for standardized gross morphologic appraisal, this media continue to be used. In present-day use, the media contains added antimicrobial agents to make it more selective for fungal pathogens. An antibacterial antibiotic, such as Aureomycin or Chloromycetin, to reduce bacterial overgrowth and Actidione (cycloheximide) to retard the growth of nonpathogenic fungi are added. The latter antibiotic, derived from *Streptomyces griseus,* is a general inhibitor of eukaryotic protein synthesis, having no effect on prokaryotic cells. For some unknown reason (probably relative uptake), most pathogenic fungi, dermatophytes and deep-infecting fungi, are not inhibited by it, though most soil saprophytes, such as *Pencillium, Aspergillus,* and *Alternaria,* are inhibited by it. However, some pathogenic fungi, e.g., *Cryptococcus,* for which it was once used in therapy, are also inhibited, and antibiotic-free media must be used when these are suspected. The opportunistic fungal pathogens, such as *Aspergillus, Mucor,* and *Fusarium,* are also sensitive to Actidione. Since the antibacterial antibiotics inhibit actinomycetes, nonselective bacteriologic media must be used in their isolation.

Clinical specimens of all types may be planted on the media and incubated for growth. Since most fungi grow better at room temperature, this is the temperature most commonly utilized for primary isolation. Generally, tubes should be observed from four to six weeks, as most fungal pathogens grow very slowly.

The identification of the isolated fungus is an exercise in contemplative observation. Classification in mycology is based on morphologic attributes of the fungus. Only familiarity by repeated study of the various organisms will lead to confidence in identification. There is no easy way. The procedure for identification is to accumulate the set of characteristics of an unknown and then determine its identity. This involves (1) the gross morphology of the colony or thallus (the obverse), its color, texture, topography, and rate of growth; (2) the reverse, for presence or absence of characteristic pigment; and (3) the microscopic morphology, its size, shape, topography and arrangement of spores, types of hyphal appendages, and hyphal modifications. In addition to these, one must be familiar with strain variations which are common and numerous, and variations influenced by culture media, age of growth, temperature, and perhaps the phase of the moon.

The handling of the isolated growth for identification differs depending on the category of disease from which the specimen is obtained. Dermatophytes are planted on slide cultures, as described in the Appendix, to observe their microscopic morphology. Scraping the culture for a wet mount is to be discouraged, as spore arrangement is an important characteristic for identification. Yeasts are planted on special media for spore production, assimilation, and fermentation profiles. Growth from deep-infecting fungi in the mycelial stage is handled with great care, as the spores are infectious. *Histoplasma* and *Coccidioides* are quite dangerous and should be examined and manipulated only in a sterile hood. The yeast or parasitic stages of dimorphic fungi are not infectious and may be examined with the usual precautions.

After the organism has been isolated and identified, its significance must be determined. Of course, a single colony of a known pathogen, such as *Histoplasma, Blastomyces,* and *Coccidioides,* is to be considered significant. With other organisms, significance may depend on where they were isolated. *Candida albicans* from skin lesions or internal organs is significant but, from the buccal cavity, rectum, or vagina, may represent normal flora. In the latter instances, the numbers of the organisms isolated are important. In the case of opportunistic infections, such as *Aspergillus, Fusarium,* and *Mucor,* the problem of significance is greater. Any fungus isolated from a sterilely obtained specimen from a closed body cavity or organ should be regarded as a possible etiologic agent of disease. On the other hand, a single colony of *Aspergillus* from a sputum culture, wound swab, or other nonsterile specimen usually represents a contaminant. Here again, numbers isolated are important. If the same organism is repeatedly isolated and in some numbers, then the organism should be considered as an etiologic agent of disease. This is particularly true in debilitated patients in whom infection with opportunistic fungi, such as *Aspergillus, Nocardia,* and *Mucor,* is increasingly common.

There are three other procedures which may be used as adjuncts to the diagnosis of disease of fungal origin. The first is demonstration of the organism in biopsy or tissue sections. Presently many specific stains exist for the histopathologic examination and demonstration of fungi. The staining is often difficult to carry out, and control slides must always be run. The presence of a few fungal organisms in tissue is most easily seen in the Gomori-Grocott chromic acid methenamine silver method. Detail of structure for ease of identification is best seen using the Gridley or Hotchkiss-McManus modification of the periodic acid-Schiff stain. These stains are useful for fungi but do not generally stain actinomycetes. Demonstration of *Nocardia, Streptomyces, Actinomyces,* and so forth, since they are bacteria, is easy using the Brown-Brenn modification of the Gram stain or the methenamine stain.

Serologic procedures are often very useful in diagnosis and prognosis of patients with a systemic fungous disease. They are of little value in dermatophyte infection or subcutaneous infections. Precipitin titers usually appear first, followed by complement-fixing titers. The disappearance or persistence of the latter is correlated with the progress of the patient, and serial tests are very useful in determining prognosis. These procedures are somewhat complicated and are best done by an experienced serologist. Much expertise and special equipment are also required for application of the fluorescent antibody techniques. These may be very useful for the demonstration of specific fungi in tissue, sputum, and exudate and may be the only procedures to establish a diagnosis in the event cultures fail or were not taken.

A relatively simple procedure coming into greater use in the laboratory is immunodiffusion. Through the use of an Ouchterlony plate, patients' sera are placed in one well and antigen in the other. The presence or absence of precipitin lines and the numbers of lines may indicate the etiologic agent. In some diseases, precipitin patterns have been correlated with presence or absence of active disease in contrast to past infection. As more specific antigens are produced, this procedure will become increasingly useful in the diagnosis and prognosis of fungal diseases.

Animal inoculation is sometimes helpful in establishing an etiology. Often clinical material will fail to grow on culture and stains may not show the organism, but experimental animals will become infected, and the etiology will be established. Animals are particularly useful in the survey of soils for the presence of pathogenic fungi. In present-day laboratory practice with the advent of newer cultural, serologic, and histologic techniques, animal inoculation is less often used and is reserved for particularly difficult cases.

CLINICAL MYCOLOGY

As previously outlined, the isolation and identification of fungal pathogens is a long, arduous, time-consuming procedure. There are a few simple diagnostic aids that may be done in the examining room or laboratory which will help to establish a fungal etiology for a disease process more rapidly.

For many years in dermatology the potassium hydroxide slide for examining skin scales has been used. This rapid procedure will demonstrate the presence of mycelial elements of dermatophytes and establish a diagnosis. It is also useful to visualize the short, hyphal strands and yeast forms of pityriasis versicolor and yeast and mycelial elements of dermatocandidosis. In addition, this procedure has many applications with other infections. Aspirates of the micropustules in a blastomycosis lesion, treated with KOH and examined, will show the large yeast cells and establish a diagnosis. Cultures may require four to six weeks to grow. Spherules in coccidiomycosis may also be demonstrated by this method. Concentrated or unconcentrated sputum, wound scraping, skin scraping, biopsy material, and specimens of all sorts can be examined by the KOH slide method, and often fungal elements can be seen. This procedure usually will not establish a specific etiology but only indicate the category of etiologic agent. In mycetoma, in which either a bacterial or mycotic etiology may be present, an examination of sinus tract exudate will usually reveal whether the infection is actinomycotic or eumycotic, and appropriate therapy may be instituted.

Another device that is useful in the diagnosis of cutaneous infections of fungal origin is the Wood's lamp. With a peak of 3650 Å, the lamp will demonstrate a characteristic fluorescence produced by some microbial agents. Classically, this procedure was used to demonstrate hair infection by *Microsporum* species in children. Surveys of large populations could be done rapidly and easily. The infected hairs show a bright green fluorescence. Today this is less useful because of the large increase in nonfluorescing tinea capitis due to *Trichophyton tonsurans*. The Wood's light will also show a pink fluorescence in skin lesions of erythrasma, a gold fluorescence in areas of pityriasis versicolor, and a bright green in toe webs, lesions, and wounds infected by *Pseudomonas*.

Specific diagnosis of fungal disease requires the isolation and identification of the agent. The previously noted procedures are only useful in indicating a possible etiology.

STANDARD REFERENCES

TEXTBOOKS ON MEDICAL MYCOLOGY

Conant, N. H., D. T. Smith, et al. 1971. *Manual of Clinical Mycology*. 3rd ed. W. B. Saunders Company, Philadelphia.

Emmons, C. W., C. H. Binford, and J. P. Utz. 1970. *Medical Mycology*. 2nd ed. Lea and Febiger, Philadelphia.

Lacaz, Carlos da Silva. 1967. *Compendio de Micologia Medica*. Savier. São Paulo, Brazil.

LABORATORY MANUALS ON MEDICAL MYCOLOGY

Ajello, L., L. K. Georg, et al. 1963. *Laboratory Manual for Medical Mycology*. Public Health Service Publication No. 994. U.S. Government Printing Office. Washington, D.C.

Hazen, E. L., M. A. Gordon, et al. 1970. *Laboratory Identification of Pathogenic Fungi, Simplified*. 3rd ed. Charles C Thomas. Springfield, Ill.

Beneke, E. S., and A. Rogers. 1971. *Medical Mycology Laboratory Manual*. 3rd ed. Burgess Publishing Company. Minneapolis, Minn.

Rebell, G., and D. Taplin. 1970. *Dermatophytes, Their Recognition and Identification*. 2nd ed. University of Miami Press. Coral Gables, Fla.

TEXTBOOKS ON GENERAL MYCOLOGY

Alexopoulos, C. J. 1962. *Introductory Mycology*. 2nd ed. John Wiley and Sons. London.

Bessey, E. A. 1961. *Morphology and Taxonomy of Fungi*. Hafner Publishing Company. New York.

Burnett, J. H. 1968. *Fundamentals of Mycology*. St. Martin's Press. New York.

Ainsworth, G. C., and A. S. Sussman. 1966. *The Fungi*. 3 vols. Academic Press. New York.

DICTIONARY

Ainsworth, G. C. 1971. *Dictionary of the Fungi* Commonwealth Mycological Institute, Kew, Surrey, England.

REVIEWS AND ABSTRACTS OF CURRENT LITERATURE

Review of Medical and Veterinary Mycology. Commonwealth Mycological Institute, Kew, Surrey, England.

THERAPY OF FUNGOUS INFECTION

Hildick-Smith, G., H. Blank, et al. 1964. *Fungus Diseases and Their Treatment*. Little, Brown and Company. Boston.

Part One

The Pathogenic Actinomycetes

Chapter 1

THE PATHOGENIC ACTINOMYCETES

The human pathogenic actinomycetes are so-called "higher" bacteria and are classified in the order Actinomycetales of the class Schizomycetes. This order includes some chronic, disease-producing organisms, such as the etiologic agents of tuberculosis and Hansen's disease (leprosy). By tradition, these two organisms have been studied along with other bacteria. The remaining pathogenic actinomycetes, however, were thought to be transitional forms between bacteria and fungi and were included in the sphere of medical mycology. The etiologic agents of lumpy jaw and actinomycotic mycetoma show some funguslike characteristics, such as the branching of the organism in tissue, the formation of an extensive mycelian network that may occur in tissue or in culture and the production of chronic disease states. However, cell wall analysis shows the presence of the characteristic bacterial muramic acid which, along with the lack of a structural membrane-bound nucleus, lack of mitochondria, typical bacterial size, and susceptibility to antibacterial antibiotics, defines these organisms as bacteria and not fungi. As far as their role as a phylogenetic "link" to the fungi is concerned, the presence in fungi of eukaryotic nuclei and mitochondria, and fungal conformity to Mendelian genetics make their derivation from prokaryotic bacteria, independent of other eukaryotic organisms, extremely improbable.

MORPHOLOGY

The actinomycetes grow in the form of fine, straight, or wavy filaments or hyphae, 0.5 to 0.8 μ in diameter. These hyphae show both lateral and dichotomous branching and may grow out from the medium to form an aerial mycelium. On solid media, the filaments occur in tangled masses, while in liquid media and in tissue, there is a tendency for them to grow in clusters of radiating dendritic clumps, which are sometimes lobulated. There are four genera of medical interest: the anaerobic *Actinomyces* and the aerobic *Nocardia*, *Actinomadura*, and *Streptomyces*[18] (Table 1–1). Classification of species within the genera and even separation of the genera themselves is controversial.[11] The following characteristics are generally accepted by workers in the field. *Actinomyces* includes organisms which are anaerobic or microaerophilic and nonacid-fast and in which the vegetative mycelium breaks up into bacillary or coccoid elements. The *Nocardia* are aerobic and sometimes partially acid-fast. They may fragment into bacillary or coccoid forms and produce chains of squared spores (arthrospores), 1 to 2 μ long, by simple fragmentation of hyphal branches. Segmentation of the filaments occurs in some species as early as 24 hours, while in others it is delayed three weeks or more; the segmented filaments fragment to form bacillary forms, 4 to 6 μ in length, which are morphologically indistinguishable from many other bacteria. In most smear preparations of the pathogenic forms, the filaments are broken up, and the appearance is that of ordinary bacilli. In stained smears of tissue or sputum, *Actinomyces* tend to be very small bacilli with one or two short bifurcations, while *Nocardia*, particularly *N. asteroides*, may show long sinuous threads with occasional branches, in

Table 1–1. *The Pathogenic Actinomycetes, Related Organisms, and Their Diseases**

Disease	*Organism*	*Geographic Distribution*
Actinomycosis	*Actinomyces israelii* (man) *Actinomyces bovis* (cattle) *Bifidobacterium eriksonii* *Arachnia propionicus*	Ubiquitous
Nocardiosis (pulmonary and systemic)	*Nocardia asteroides* *Nocardia brasiliensis*	Ubiquitous Mexico, South America, Africa, India
Mycetoma (actinomycotic)	*Actinomadura madurae* *Actinomadura pelletieri* *Nocardia caviae* *Streptomyces somaliensis* *Streptomyces paraguayensis*	Ubiquitous Africa, South America Rare Africa, South America, United States, Arabia South America
Erythrasma	*Corynebacterium minutissimum*	Ubiquitous
Cracked heel	*Actinomyces keratolytica (?)*	India, United States
Trichomycosis axillaris	*Corynebacterium tenuis*	Ubiquitous
Epidemic eczema	*Dermatophilus congolensis*	Australia, Africa, United States

*From Rippon, J. W. *In* Burrows, W. 1973. *Textbook of Microbiology*. 20th ed. Philadelphia, W. B. Saunders Company, p. 687.

addition to bacillary forms. Grains, or granules which are microcolonies, are commonly found in visceral actinomycosis, in visceral and subcutaneous infections of *Nocardia brasiliensis*, and in subcutaneous infections of *Nocardia* species other than *N. asteroides*. *N. asteroides* does not form grains in visceral infections and does so rarely, if at all, in subcutaneous disease.

In *Streptomyces*, there is generally more aerial mycelium, no fragmentation of the mycelium into bacillary or coccoid forms, and no acid-fastness of mycelial elements. Chains of round-to-oval spores are produced consecutively within a specialized hyphal element. These spores are partially acid-fast. The maturation of the spore-bearing hyphae is often associated with the formation of spirals, which range from open, barely perceptible coils to those which are so compressed that adjacent turns are in contact. The spirals may be dextrose or sinistrose; both direction and tightness of coiling are constant within species. The spores are more resistant than the filaments and will survive at 60°C for as long as three hours, but they are less resistant than bacterial spores to heat and toxic agents. The tips of some filaments may become swollen and club-shaped.

All, or practically all, the actinomycetes are gram-positive, and some of the *Nocardia* are acid-fast. Grains will be stained by hematoxylin and eosin, but single filaments or bacilli will not. All morphologic forms are well demonstrated by methenamine silver stains.

PHYSIOLOGY AND COMPOSITION

Recent work on the chemical composition of the actinomycetes has modified the previous concepts of classification of the aerobic actinomycetes. Mordarski[26] has found a particular lipid fraction in the *Nocardia* which she calls "lipid characteristic of *Nocardia*" (LCNA). It is present in the *Nocardia* but not in the *Actinomadura* or *Streptomyces*. Becker et al.[3] describe a simple paper chromatographic technique for determining the cell wall composition of *Streptomyces* and *Nocardia*. LL-diaminopimelic acid (DAP) and glycine are found in *Streptomyces; meso*-DAP, arabinose, and galactose are found in *Nocardia*. Two species considered to be *Nocardia* or *Streptomyces*, e.g., *N. madurae* and *N. pelletieri*, do not fit either pattern, as they contain *meso*-DAP but no arabinose or LCNA. This supports the contention of Lechavalier that they should be placed in a separate genus, *Actinomadura*. This would include *A. madurae*, *A. pelletieri*, and *A. dassonvillea*, the latter having been reported so far only from soil and animal infection. Goodfellow, applying the adansonian cluster technique to all the criteria used for identification, has delineated the three genera and their species.[18] The physiologic characteristics of the pathogenic *Nocardia* and *Streptomyces* species have been summarized by Gordon.[4] The morphologic and physiologic constants of Actinomyces and related anaerobic forms are given by Moore and Holdeman[4] and Ajello.[2]

CULTIVATION

Both spores and mycelia of actinomycetes grow in subculture. On solid media, the growth of aerobic forms is dry, tough and leathery, sometimes wrinkled, adherent to and piled above the medium; in many instances, it resembles the growth of *Mycobacterium*. *Nocardia* species, especially *N. asteroides*, grow well on the usual *Mycobacterium* media. In some cases, especially among the *Streptomyces*, the growth appears powdery or chalky, owing to the formation of aerial mycelia. In liquid media, growth occurs in the form of a dry, wrinkled surface film or, more often, as flakes or aggregates which adhere to the sides of the flask, especially at the surface, or sink to the bottom.

Pigment formation, with colors ranging over the entire spectrum, is common among the actinomycetes. Differentiation is usually made between pigmentation of the vegetative mycelium and that of the spore-bearing aerial mycelium, as well as between pigment diffusing into the medium. Soluble purple and brown pigments are often observed on protein-containing media. These characteristics may be strain-variable, media-variable, and temperature-variable. The actinomycetes, especially the saprophytic forms, are physiologically active, utilizing a variety of nitrogen and carbon compounds, and many are actively proteolytic. The important differences for species identification are discussed in later chapters. An earthy-to-musty odor, like that of freshly turned soil or of a damp basement, is produced by many species, especially of *Streptomyces*. The optimum temperature for growth is usually 20° to 30°C, though some of the pathogenic species grow best at 37°C, and thermophilic species, analogous to thermophilic bacteria, are known. The great majority of actinomycetes are aerobic and nonfastidious nutritionally, but a few are anaerobic or microaerophilic and are nutritionally exacting.

Differentiation of the actinomycetes is done in part on a morphologic basis and in part on a compositional and physiologic basis, the former including pigmentation and saccharolytic and proteolytic activity.

IMMUNOLOGY AND SEROLOGY

Immunologic investigations on actinomycetes have been numerous, confusing, controversial, and generally of limited value. Mostly this has been because of examination of only a few strains, misidentification, and lack of a standardized technique. Perhaps this situation has been best summarized by an early investigator, Bretey, who stated that *Nocardia* and *Streptothrix* are "bad antigens." Some of the more recent work has shown stable antigenic differences in these groups. González Ochoa and Vasquez-Hoyos[17] could serologically separate four groups of actinomycetes: (1) bovine and certain *Nocardia*, (2) the somaliensis group, (3) the madurae group and (4) the paraguayensis group. Genera could be delineated by a slide agglutination test of Schneidau and Shaefer,[34] separating *Nocardia* and *Streptomyces*. Kwapinski,[22] using complement fixation, has found patterns of antigens common to and specific for several genera, including *Mycobacterium*, *Streptomyces*, *Nocardia*, and *Actinomyces*. A complement-fixation test for the diagnosis of *N. asteroides* infection in cows has been devised.[30] Dyson and Slack[10] produced a specific skin test for *N. asteroides* infection in experimental animals. González Ochoa and Baranda,[15] with a carbohydrate fraction, developed a specific skin test for patients infected with *N. brasiliensis*, and Bojalil and Zamora[5] have a protein fraction specific for *N. asteroides* and *N. brasiliensis* infections.

Presumptive diagnosis of actinomycosis could not be reliably demonstrated by Georg et al.[13] using an immunodiffusion technique. Serologic procedures for antigenic grouping of strains of actinomyces have been investigated, and a polyvalent diagnostic reagent has been developed, using fluorescent antibody procedures.[6]

PATHOGENICITY

Three species of anaerobic actinomycetes are capable of producing disease in man. They are *Actinomyces israelii*, *A. naeslundii* (which may be a variant of *A. israelii*), and

Bifidobacterium eriksonii. These are all normal flora of the human buccal cavity and tonsillar crypts and are of limited invasive ability. They probably cannot elicit disease without the aid of trauma to the tissue and the presence of associated bacteria. *A. bovis* is found in lumpy jaw of cattle. Among the *Nocardia, N. brasiliensis* is probably the most virulent, while *N. asteroides* is considered an opportunist. *N. caviae, Actinomadura madurae, A. pelletieri, Streptomyces paraguayensis,* and *S. somaliensis* are all soil saprophytes which may cause mycetoma in man upon traumatic implantation into tissue. Other species of *Nocardia* and *Streptomyces* have occasionally been reported in human infection but are rare or of questionable identification. Infection by all actinomycetes initially elicits an acute inflammatory pyogenic response, followed by chronic inflammatory reaction and sometimes generalized spread. The degree and severity of disease depends on the physical status of the patient and the virulence of the organism.

Actinomycosis

DEFINITION

Actinomycosis is a chronic suppurative and granulomatous disease characterized by peripheral spread and extension to contiguous tissue, rare hematogenous spread, and the formation of multiple draining sinus tracts. These sinuses drain from suppurative pyogenic lesions. The exudate contains firm, lobulated grains (sulfur granules) or microcolonies of the etiologic agent, which may reach 1 to 2 mm in diameter, and adherent cellular debris, associated microorganisms and coccoid or bacillary forms of the etiologic agent. The disease is caused in man most commonly by *Actinomyces israelii* and in animals by *Actinomyces bovis.* Other rarely implicated species include *A. naeslundii, Bifidobacterium eriksonii,* and *Arachnia propionicus.*

The disease affects the cervicofacial, thoracic, and abdominal regions in man and may result in severe disfigurement and disability. Localized lesions are amenable to therapy, but most cases of extensive involvement or hematogenous spread are fatal. Its ability to invade and destroy bone is notorious.

Synonymy. Lumpy jaw, leptothricosis, streptotricosis.

HISTORY

Lumpy jaw or actinomycosis was once a fairly common disease of man and is still frequent in cattle. Today it is uncommon in general medical practice and most often diagnosed in retrospect. This change is largely due to the widespread practice of giving a "shot" of antibiotic indiscriminately for almost any symptom. The etiologic agents, endogenous *Actinomyces,* are quite sensitive to most antibacterial antibiotics, including penicillin and sulfas. Formerly, infection was often associated with tooth extraction or dental surgery, which provided traumatized tissue in which these organisms and associated (synergistic) microorganisms could grow. Prophylactic use of antibiotics following oral procedures has largely eliminated this hazard. Poor oral hygiene and carious teeth are still the most common predisposing factors. An important aspect in the decline of infection is the high level of oral hygiene achieved in the "developed" nations of the world. That this is not the case in "emerging" nations is attested to by the continued numerous and increasing case reports of actinomycosis and related disease in these areas. Presently, the more common types of serious disease are those associated with abdominal injury or disease of the alimentary tract, and those in the thoracic area, the result of aspiration of buccal material, or extension from focal cervical lesions associated with poor oral hygiene and general health.

Actinomycosis was undoubtedly observed early in the nineteenth century, as actinomycotic tumors were described erroneously in 1826 by Leblanc as osteosarcomas. It was first recognized as a specific parasitic disease in 1876 by Bollinger, who named the disease "lumpy jaw." In 1877, Harz described the disease in cattle and called the etiologic agent *Actinomyces bovis* (ray fungus of cattle), not because of its cultural characteristics but because of its appearance in tissue. Previously, the syndrome of human actinomycosis was recognized by Lebert in 1857, and in 1874 Cohn, who described what must have been authentic disease in the lachrymal canal, called the agent *Streptothrix foersteri.* Bollinger's study material included other similar diseases of cattle, such as "woody tongue," which is caused by *Actinobacillus lignieresi.* In 1878, Israel and Ponfick recognized the disease in humans and isolated the organism, and its anaerobic nature was described by Israel and Wolff in 1891. They remarked that it could not be isolated from exogenous sources. In 1890, Boestrum isolated an aerobic organism, probably a contaminating *Streptomyces,* from lesion material and assumed that *A. bovis* was carried to the mouth on straw from exogenous sources. This erroneous information was carried in textbooks for some 40 years and was used to explain why the disease was more common among rural inhabitants. Boestrum's organism, sometimes called *A. hominis* or *A. graminis,* has proved to be a contaminant, and it is definitely established that the organism isolated by Wolff and Israel was that observed by Bollinger and Harz and the true etiologic agent of the disease. In 1910, Lord's experiments *in vitro* demonstrated the presence of the organism in the normal mouth, in tonsillar crypts, and about carious teeth. These findings were confirmed by Emmons[11] and Rosebury,[33] and the modern concepts of the organism and disease were summarized by Erikson.[12] The disease is very easy to diagnose and, as observed by Cope,[9] "exists wherever there is a microscope and a laboratory." Before the advent of antibiotics, the disease was diagnosed very frequently, though accurate mortality rates are not known. Following the introduction in 1948 of penicillin therapy for the disease by Nichols and Herrell,[28] there was a sharp decrease in the number of deaths and severity of disease, especially of the cervicofacial type. Recently there has been some concern that the incidence is again on the rise.[1]

ETIOLOGY, ECOLOGY, AND DISTRIBUTION

Presently, most workers agree that there are two commonly encountered species which cause the disease syndrome, "lumpy jaw" or actinomycosis. *A. bovis* is the usual cause in cattle, and *A. israelii* the predominant organism in human infection, although it is occasionally found also in cattle. *A. naeslundii,* a facultative anaerobe found in the human mouth, was considered a harmless commensal for years, but recently it has been shown to be involved in authentic cases of the disease.[8] Some workers consider it a variant of *A. israelii. Bifidobacterium eriksonii,* a new species, has been described in five cases of pulmonary actinomycosis without granule formation.[13] *Arachnia propionicus* has been isolated from cervicofacial actinomycosis and lachrymal canaliculitis.[15] It differs from the *Actinomyces* by its production of propionic acid and by a cell wall containing diaminopimelic acid.

Actinomycosis, even in closed lesions, appears to be a cooperative disease of *Actinomyces* and a mixed flora of other bacteria. Holm has referred to these as "associates" and claims that they exert a synergistic activity in the pathogenesis of actinomycosis. Laboratory investigation tends to confirm this. The injection of pure culture of *A. bovis* or *A. israelii* seldom produces lesions in animals without the aid of adjuvants, such as gastric mucin.[24] Fusiform bacilli, anaerobic streptococci, gram-negative bacilli, *Hemophilus aphrophilus,* or related organisms and *Actinobacillus actinomycetemcomitans,* named for its frequent association, are found together with *A. israelii* in cervicofacial disease. *A. actinomycetemcomitans,* the streptococci, and *Hemophilus* are commonly found in the thoracic form of the disease, while various aerobic and anaerobic intestinal flora, including *Escherichia coli,* are found in abdominal or pelvic disease.

Many unsuccessful attempts have been made to isolate the pathogenic *Actinomyces* from vegetation. The investigations of Lord, Emmons, Slack, and others lead to the conclusion that these organisms are endogenous in man and animals, leading a parasitic existence as nonpathogens in the mucous membranes of the oral cavity, in the caries and tartar of teeth, in tonsillar crypts, in the alimentary tract, and perhaps in the respiratory tract. Some authors indicate that granules without adherent neutrophils can be found in undiseased tonsils.

Several literature reviews prior to 1950 cite the case distribution as 50 to 60 per cent cervicofacial, 20 per cent abdominal, 15 per cent thoracic, and the remainder in other organs. Harvey et al.,[19] in a study of 37 cases, found 24 per cent cervicofacial, 13 per cent pulmonary, and 63 per cent abdominal disease, probably reflecting a decrease in focal oral infections associated with the use of antibiotics.

CLINICAL DISEASE

Symptomatology

The clinical picture may vary with the location of the disease, and it is generally classified as cervicofacial, abdominal, or thoracic, depending on the site of primary infection. Rarely the disease may disseminate by hematogenous spread to other organs.

Cervicofacial Actinomycosis. In the cervicofacial type, the organism enters through trauma to the mucous membrane of the mouth or pharynx, by way of carious teeth or through the tonsils. Salivary or lacrimal glands are sometimes involved. The common initial symptoms are pain and swelling along the alveolar ridge and associated areas. The regional lymph nodes are generally not enlarged at first. The swelling becomes firm; nodular or branny masses are described as "wooden" or "lumpy." The skin over the region may discolor (Fig. 1–1 FS A 1). The hard masses soften and become abscesses and, later, multiple granulomas. They break to the surface, forming multiple sinus tracts which discharge purulent material containing sulfur granules (Fig. 1–2 FS A 2). Pain is minimal, but trismus may develop, impairing mastication; otherwise, there is little discomfort to the patient if the disease remains localized. Infections in the maxilla may extend to the cranial bones, giving rise to meningitis, or into the orbit of the eye and the middle ear. Mandibular disease invades the tongue (ligneous phlegmon) and sublingual salivary glands and, by extension, the neck and scapular and upper extremities. Direct extension into the lungs and pleural cavity may also occur. The sinus tracts may heal and reform in the same area later.

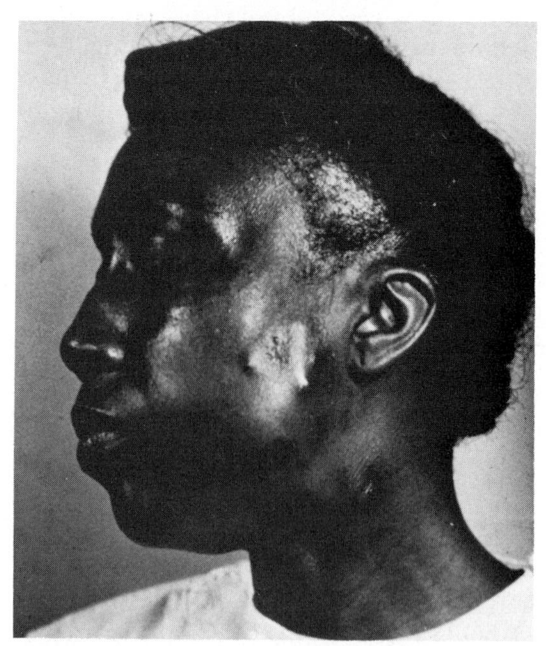

Figure 1–2. FS A 2. Actinomycosis. Multiple sinus tracts opening over ramus and mandible, "lumpy jaw."

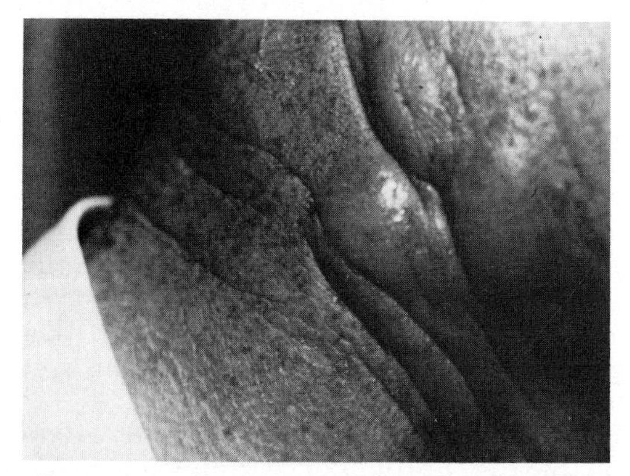

Figure 1–1. FS A 1. Actinomycosis. Erythematous swelling over jaw. (Courtesy of D. Mincey.)

Early in the disease there is no bone involvement as shown by x-ray examination. Later, however, a periostitis develops, followed by destruction of bone cortex, by osteomyelitis, and occasionally by expansion of the cortex with cyst formation (Fig. 1–3). The marked ability of this disease to readily burrow through bony material is a point of contrast to nocardiosis and some eumycotic infections.

The cervicofacial type of actinomycosis has the best prognosis. With surgical debridement and excision as an adjunct to proper antibiotic therapy, the disease may be cured without much difficulty, if it has remained localized.

Differential diagnosis includes other granulomatous lesions, especially tuberculosis. Also to be considered are other mycotic diseases, such as blastomycosis and coccidioidomycosis, nocardiosis, glanders, gumma, tularemia, osteomyelitis, and neoplasm.

Thoracic Actinomycosis. Thoracic actinomycosis may result from direct extension of the disease from the neck, from extension from abdominal or hepatic infection through the diaphragm, or as a primary infection from aspiration of the organism through the mouth. The most common sites of infection are the hilar region and the basal parenchymal areas; this latter area is in contrast to tuberculosis, which is generally apical only. The initial symptoms are those of a subacute pulmonary infection, with a mild fever, cough, and production of purulent sputum without hemoptysis. As the disease progresses, small abscesses develop in the lung, and the sputum may become blood-streaked, suggesting lung destruction. In time, the infection spreads to include the pleura and thoracic wall and then penetrates to the surface to form typical discharging sinuses (Fig. 1–4 FS A 3). Increasing dyspnea, fever, night sweats, anemia, and general wasting are seen.

The early physical signs of actinomycosis resemble those of tuberculosis. X-rays show massive smooth consolidations, usually bilateral, in the lower half of the lung field. Consolidations may project from the hilar area, suggesting neoplasm (Fig. 1–5). Sometimes rarefactions are seen in these masses, which indicates actinomycosis rather than tumor. Pleural effusion is frequent. Lesions usually are nonencapsulated, and adhesions are common. Ribs are involved and show destructive and proliferative changes as previously described.

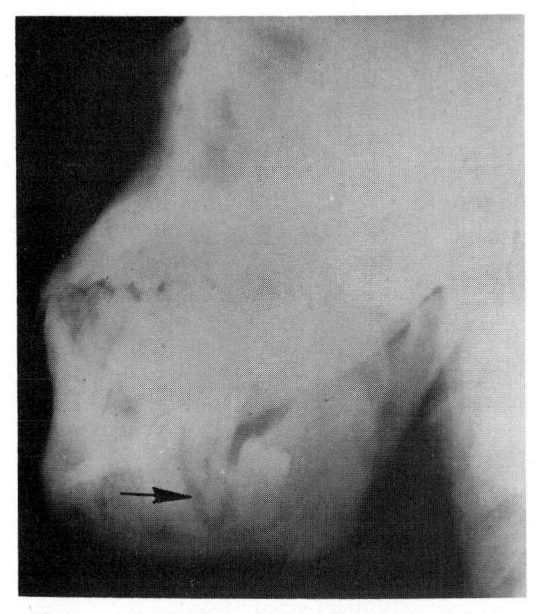

Figure 1–3. Actinomycosis. Destruction of bone around impacted tooth. (Courtesy of D. Mincey.)

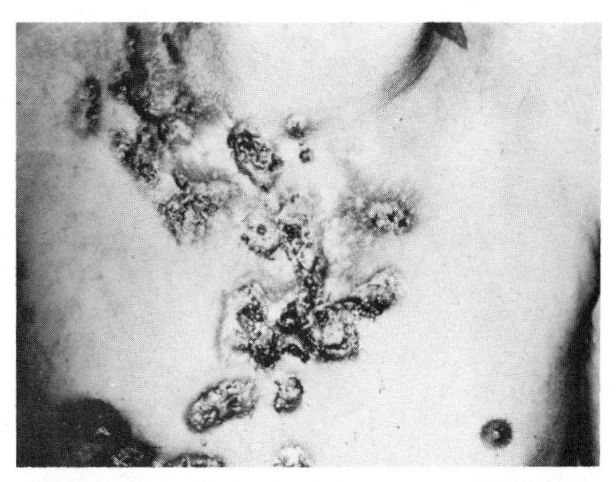

Figure 1–4. FS A 3. Actinomycosis. Multiple-discharging sinuses developing from hilar and cervical region. The disease is extended to the chest from the cervicofacial area (scrofuloderma type).

The disease must be differentiated from tuberculosis, the distribution of lesions being helpful, and from nocardiosis, lung abscesses, bronchiectasis, and tumor.

Abdominal Actinomycosis. This form of the disease may result from perforation of the intestinal wall by such objects as fish and chicken bones or from knife and gunshot wounds. Perforating ulcers may also initiate abdominal actinomycosis, but most frequently the primary source is the diseased appendix. Such organs as the fallopian tubes, gallbladder, and liver, may appear to be the primary sites, but thorough examination and history usually reveal another primary source. Rare primary infections of each of these organs probably occurs.

The initial symptoms of abdominal actinomycosis are insidious and related to the involved organ. In primary colon disease, an indistinct irregular palpable mass may be present and must be differentiated from carcinoma. In this type of infection, extension to the liver is frequent, and jaundice may occur. Involvement of the gallbladder and urinary tract with accompanying dysfunction has been noted, but other causes of cystitis, pyelonephritis, and cholecystitis must be

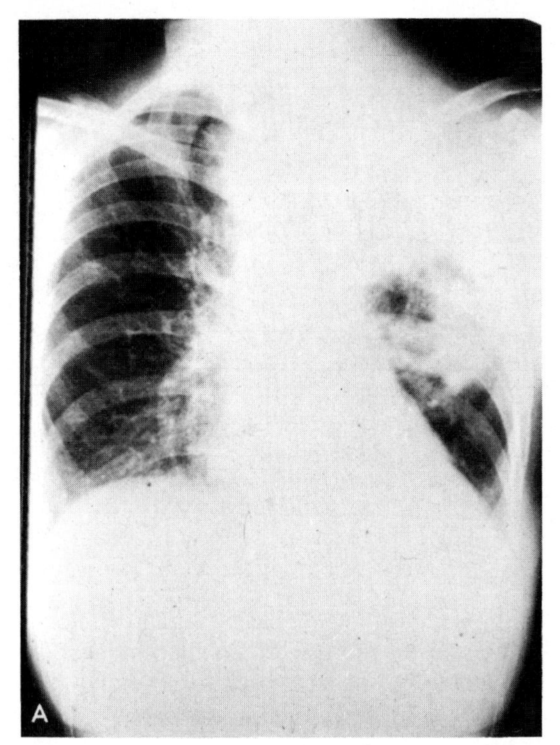

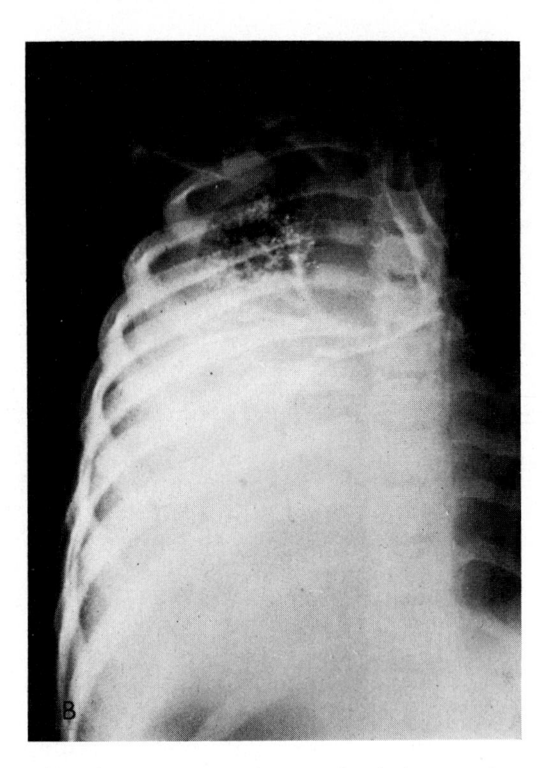

Figure 1–5. Actinomycosis. *A,* Pulmonary involvement resulting from extension of cervicofacial disease. There is diffuse, dense infiltrate of the upper lobe at the left lung and pleural thickening over the lateral upper lobe. Soft tissue swelling is seen in the left superclavicular region. There was a scrofula at this point. (Courtesy of J. Fennessey.) *B,* Basilar involvement of the right lung. This bronchogram demonstrates a large pleural effusion or empyema. There is incomplete filling of the lower lobe with no bronchiectasis. (Courtesy of J. Fennessey.)

ruled out. Extension to the spinal column produces destructive bone lesions, with collapse and compression of the spinal cord and psoas abscess. After extensive spread to contiguous tissue, penetration of the abdominal wall and sinus formation may occur, with extrusion of pus and sulfur granules. Unless draining sinuses develop, however, actinomycosis is rarely diagnosed before exploratory laparotomy.

X-rays of the abdomen show indistinct masses, enlargement of liver or spleen, or an abnormal process involving the vertebrae. The change in the latter is that of periostitis, with erosion of the articular facets, lamina, and transverse and spinous process. This is in contrast to tuberculous disease, which favors the vertebral bodies. Signs and symptoms include localized pain, with a palpable process, jaundice, weight loss, chills, spiking fever, vomiting, and nocturnal sweats. Differential diagnoses include neoplasms, amoebiasis, liver abscess, psoas abscess of other origin, chronic appendicitis, salpingitis, and pyelonephritis.

Other areas of primary actinomycotic infection include bladder, kidney, humerus, and central nervous system. Though hematogenous spread is rare, this must always be considered when lesions are found outside the usual areas. Direct inoculation into the skin, resulting in mycetoma, has been reported several times, and *A. israelii* is considered an etiologic agent of actinomycotic mycetoma. This disease and its pathology are discussed in a later chapter.

Actinomycosis presents such a variety of clinical signs and symptoms that it must be differentiated from a number of chronic infections and neoplastic diseases. In addition to those already mentioned, granuloma inguinale, staphylococcic actinophytosis (botryomycosis), sarcoidosis, carcinoma of the retroperitoneal tissues or iliac bones, typhoid fever, and the various deep mycotic infections should be ruled out.

Prognosis and Therapy. Prior to the use of antibiotics and sulfas, the prognosis of actinomycosis was poor. Therapy included surgical management of the involved area, along with irradiation, vaccination, and various types of nonspecific chemotherapy, such as thymol. The use of iodides probably resulted from their success in treatment of actinobacillus infection in cattle. However, they are of questionable value, their beneficial effect being derived from their direct action on the involved tissue rather than from their antimicrobial efficacy. The sulfonamides were introduced in 1938, and successful treatments with blood levels of 6 mg per 100 ml have been reported. Since the work of Nichols and Herrell,[28] penicillin has been considered the treatment of choice. Regimens vary from 1 to 6 million units to 10 to 20 million units IV per day, depending on the severity of disease. Therapy is administered for 30 to 45 days prior to surgical incision, drainage, or excision, or the lesions are allowed to heal by intention. Following surgical manipulation, treatment is continued for 12 to 18 months, with an oral dosage of 2 to 5 million units penicillin V per day. Chlortetracycline, chloramphenicol, oxytetracycline, streptomycin, Terramycin, sulfadiazine, and other sulfonamides are useful as supplements to penicillin therapy. Successful treatment has been reported in cervicofacial actinomycosis using tetracycline alone administered in dosages of 250 mg, four times per day, for three weeks.[29] However, most authors agree that tetracycline is of limited value, especially in extensive disease. In cases in which penicillin allergy precludes its use, lincomycin has been found to be effective. Doses of 600 mg IM every 12 hours and 500 mg six times a day by mouth brought the disease under control. The dose was then reduced to 500 mg four times a day by mouth for a year.[25]

The antimycotic agents, amphotericin B, nystatin, griseofulvin, and saramycetin are not effective.

Pathology

Gross Pathology. Systemic actinomycosis often is not diagnosed until autopsy. Although the gross pathology may vary, depending on anatomic location, the usual picture is of a chronic, extending, suppurative, scarring, inflammatory process. Primary sites of infection are marked by a dense, cellular infiltrate, producing firm, nodular lesions of varying sizes which may be painfully swollen. This is followed by a softening of the tissues as granulomas form, and finally, suppuration within the granulomas to form multiple interconnecting abscesses and sinus tracts. This may be followed by cicatricial healing and spreading by burrowing, usually along facial planes or muscle or tissue, leaving deep, communicating, scarred sinus tracts which may heal and later reopen. Debris and purulent material may be present within the lumen.

Autopsy. Cervicofacial actinomycosis is rarely encountered at autopsy, although extension from this site to the cervical spine and cranial cavity may result in fatal meningitis. An occasional complication is thrombosis of the jugular vein. Extension downward to the superior portion of the thoracic cage also occurs. Abdominal actinomycosis, on the other hand, is not uncommonly diagnosed at autopsy. Abdominal lesions are multiple abscesses which most often are found in the abdominal wall. If the infection began in the large intestine or appendix, minute abscesses of the mucosa and submucosa or ulcerations may be found, along with rectal fistulae with multiple tracts. Extension from the retrocecal region along the iliopsoas muscle to the kidney may produce draining sinuses to the inguinal region, and may be accompanied by chronic active interstitial nephritis. The liver is frequently involved. Multilocular abscesses and pylephlebitis are seen. In cut section, foamy macrophages with large accumulations of lipids are seen at the edges of the abscesses and are responsible for the yellowish color observed on gross examination. Grey patches of scar tissue may also be present. Extension from the liver through the diaphragm to the pleural cavity is not uncommon. Involvement of the retroperitoneal tissues, psoas muscle, cecum, kidneys, spleen, fallopian tubes, and ovaries has been observed, with rare hematogenous spread to lungs, brain, meninges, or other tissues.

Primary actinomycosis of the lungs is manifested by multiple abscesses, bronchiectasis, pleural and pulmonary adhesions, and pleural thickening. Bronchial fistulae, emphysematous blebs in both lungs and pleura, dense fibrosis, empyema, and rarely extension to the pericardium may occur. As discussed earlier, in about one quarter of reviewed cases, involvement of the bone has been described. Pathology of the scapula, humerus, lumbar vertebrae, ribs, sacrum, and long bones occurs. Occasional lesions involving arm, mastoid, orbit, axilla, and buttock have been described. The latter was a primary lesion, apparently the result of a bite.

Histopathology. The histopathologic appearance of actinomycosis is similar in the various organs involved. The pathologic process consists of an acute pyogenic response, which evolves into a chronic granulomatous lesion, suppuration, and abscess formation. These may later heal, with fibrosis and scar formation. The characteristic granules, or drusen of German writers, have an adherent mass of polymorphonuclear neutrophils which are attached to the radially arranged eosinophilic "clubs" of the granule. This mass of leukocytes makes it easy to pick out granules in hematoxylin and eosin–stained preparations (Fig. 1–6 FS A 4). Enclosing the area is a band of cellular infiltrate, consisting of lymphocytes, plasma cells, epithelioid cells, and histiocytes, forming a huge abscess (Figs. 1–7 and 1–8). Peripherally, there is an area of granulation tissue containing lipoid-laden histiocytes (giving the area a yellowish color) and proliferating vascular structures. Giant cells are rare but may occur in the area. Since all lesions contain a mixed bacterial flora, some of the histopathologic changes may be due to associated bacteria. In many cases, abscesses are found to contain only such organisms and no granules of *Actinomyces israelii*. In these cases it is often necessary to search through many sections to find the typical gram-positive branching rods. Grains are usually found in the purulent areas of these sections.

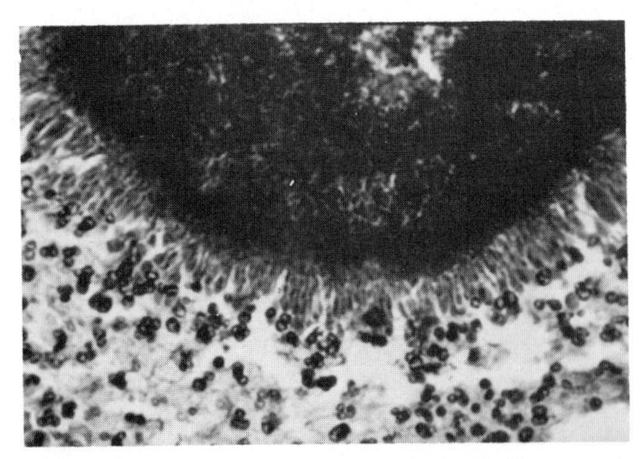

Figure 1–6. FS A 4. Actinomycosis. Sulfur granule showing eosinophilic "clubs" with attached neutrophils. ×400.

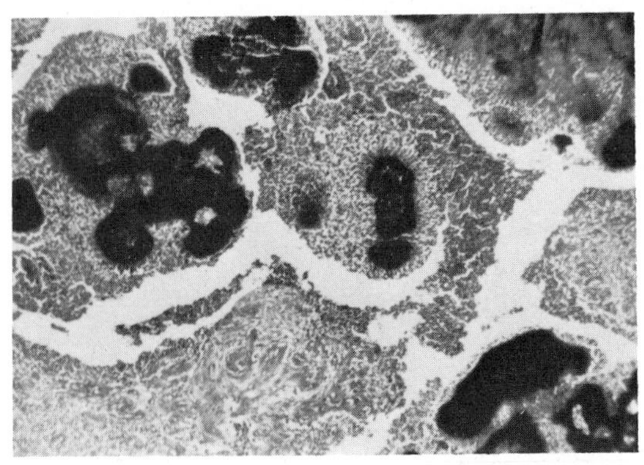

Figure 1–7. Actinomycosis. Multiple granules in sinus tracts.

The histologic picture of other bacterial and mycotic granulomas may resemble that of actinomycosis. This is particularly true of staphylococcal actinophytosis (botryomycosis), which may present with a similar gross and microscopic appearance. The grains are of similar size and architecture, again with adherent neutrophils, but there are no eosinophilic "rays" radiating from the granule. Though the grain is basophilic on hematoxylin and eosin stain, it will be seen that there is central degeneration and a loss of basophilia. High-power examination shows that the organisms present in the granule are cocci. The histologic picture usually does not show the sharp demarcation between the leukocytic center and the peripheral zone. The lipid-laden histiocytes, a hallmark of actinomycosis, are missing.

The grains in actinomycosis, when examined microscopically, are seen to consist of dense rosettes of club-shaped filaments in radial arrangement. The individual rosettes vary from 30 to 400 μ, with the average between 100 and 300 μ. The large sulfur granules are visible to the naked eye and can often be seen on gauze dressings over an area of drainage. A single actinomyces filament has a diameter of 1 μ or less; thus it is easily differentiated from eumycotic

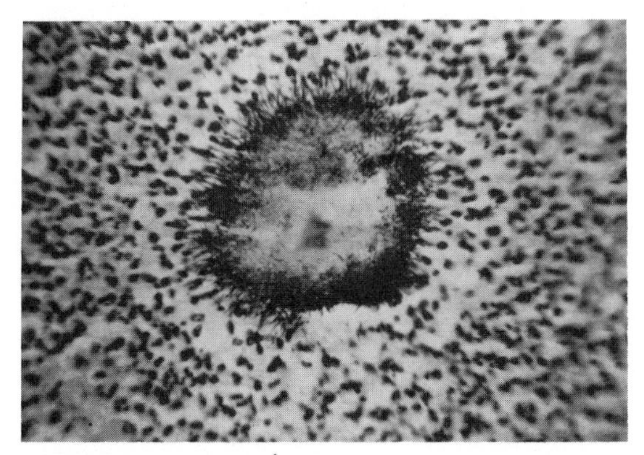

Figure 1–8. Actinomycosis. Gram stain of granule. The radiating rays (ray fungus) are gram-positive. ×400.

filaments, which are from 3 to 4 μ. *Nocardia* filaments are also 1 μ in diameter. The rosette itself is made up of three kinds of structures: a central core of branching filaments, which are irregularly disposed but with a general radial arrangement; refringent, club-shaped bodies at the periphery, which are also radially arranged; and spherical coccuslike bodies interspersed throughout. The clubs may be plainly seen when the granules are crushed and examined in fresh preparation, or a stain such as eosin, which colors the sheaths of the clubs, may be used. The filaments are gram-positive; therefore, the Gram stain (Fig. 1–8) is useful for tissue sections in which grains are scarce and only unorganized filaments or coccoid forms are found. The Brown-Brenn or the MacCallum-Goodpasture modification of the Gram stain is useful. The methenamine silver stains are excellent because they stain all filaments regardless of age; however, they do not distinguish gram-positive actinomyces from gram-negative associates. In hematoxylin and eosin–stained sections, the core of the granules are basophilic with an eosinophilic periphery. Although quite useful when the grains are numerous, hematoxylin and eosin stain will not show single filaments. The organism is not acid-fast or stained by special stains for fungi. *Actinomyces* filaments are only occasionally seen not organized into grains. In such cases, they must be distinguished from *Nocardia asteroides*, which does not usually form grains. In general, *Nocardia* will show long, sinuous threads, in addition to short bacillary and coccoid elements, whereas *Actinomyces* consist of a fine meshwork of short, branching bacillary and coccoid forms.

The club-shaped bodies at the margin of the granules are conspicuous for their high refringency and generally structureless homogeneous appearance. They are pear-shaped swellings of the terminal ends of the filaments, which arise as distinct transformations of the filaments. In young colonies the hyaline substance which comprises the clubs is soft and water-soluble, but as the colony ages, the clubs attain a firmer consistency. Pine and Overman[32] have shown the clubs to have a different chemical composition from the cell walls and to be set down in layers which become firm with impregnation of $Ca_3(PO_4)_2$. Their formation appears to be associated with the resistance of tissues: when resistance to invasion is slight, they are absent; filaments alone are found. It is probable that the clubs are the result of antigen-antibody complexing. Clubs, as a rule, are more common in bovine than in human disease. They are infrequent or absent in grains of *A. naeslundii* and are not found in culture of *A. israelii* or *A. bovis*. Grains have not been reported in infection by *B. eriksonii*.

ANIMAL DISEASE

The majority of actinomycotic lesions in cattle are found in or about the head. In particular, the lower jaw is affected; hence, the common name, "lumpy jaw," for this disease. Among dairy cattle, Guernseys more frequently show disease than other breeds.[3] The organism isolated is generally *A. bovis*, although *A. israelii* has been authenticated in two cases. In addition to the growth in the tongue and maxillary bone, actinomycotic lesions occur in the pharynx, lungs, skin, lymph glands, and subcutaneous tissue, especially of the head and neck, and occasionally in other organs, notably the liver. The growth of the parasite usually leads to the formation of a hard tumor, which gradually increases in size and burrows into the adjacent tissues, softening and disintegrating the bony structure of the head. At the same time, new tissue is forming so that great distortion often ensues. Extension of the disease takes place by gradual invasion of the contiguous tissues, metastases being uncommon. When death occurs, it is not as a rule due to any toxic effect but to the mechanical action of the tumor pressing upon or occluding the respiratory passages or interfering with the taking or mastication of food. Generalized actinomycosis in animals is rare. When it does occur, the bloodstream rather than the lymph seems to be the channel by which the disease is spread. Secondary abscesses are found mainly in the liver.

The characteristic yellow granules, or drusen, are found in the suppurating masses of the tumor. If the pus is shaken up with water in a test tube, the small granules become evident and sink to the bottom of the tube. These structures are so typical, their demonstration by microscopic examination in a case of obscure suppuration is sufficient to suggest a diagnosis. This must be substantiated by crushing the granules and staining for the typical forms. The new growth, granulation tissue, consists chiefly of epithelioid and spindle-shaped connective tissue cells; small giant cells are rarely present. To the naked eye, actinomycotic lesions in the lung and udder often resemble tuberculous nodules and, at times, have undoubtedly been mistaken for them. Microscopic examination suffices to establish their nature.

The incidence of actinomycosis in cattle is uncertain, in part it is readily confused with actinobacillosis. A fair estimate of disease rate might be between 0.2 to 2 per cent in the United States.[3] The "woody tongue" disease caused by the gram-negative *Actinobacillus lignieresi* is similar to actinomycosis in that the tongue, lymph nodes, and soft tissue are invaded, but there is little or no involvement of bone. Of the other domestic animals, swine are most frequently affected. Disease in horses, sheep, dogs, and cats has been reported.

Experimental Animals

Attempts to infect experimental animals with *A. israelii* have in general been disappointing in that only a small proportion of inoculated animals develop disease, and lesions are limited and benign. Some success has been achieved by traumatizing the tissue first and including the associated bacterial flora with *A. israelii.* Experimental infection with pure cultures has been achieved in mice by Meyer[24] using hog gastric mucin to enhance the invasiveness of the organism. Others have had less success in rabbits, guinea pigs, and hamsters using repeated injections of the organism.[20] In experimental disease, the essential features of the natural infection are observed, including the formation of tubercle-like nodules and the development of structurally typical granules with clubs.[19]

IMMUNOLOGY AND SEROLOGY

Agglutinins, precipitins, and complement-fixing antibodies have been demonstrated in patients with actinomycosis, but there is little consistency in the reactions or correlation with the disease state. There is a notable lack of evidence that they are involved in combating or protecting the individual against disease. Skin test antigens elicit both immediate and delayed type reactions. Active defense against the organism probably occurs at a cellular level. It has been suggested that the "clubbing" of the organism seen at the periphery of the granule represents a response of the organism to the cellular defense of the host and is an antigen-antibody complex.

By use of the reciprocal agglutination absorption method, and subsequently fluorescent antibody, Slack and Gerencser[35] separated the various pathogenic actinomycetes into four groups: *A. naeslundii, A. bovis, B. eriksonii,* and *A. israelii.* Brock and Georg[6] have found two serotypes of the *A. israelii.* Serotype 1 represents 95 per cent of clinical isolants and differs from serotype 2 in that most stains show the classic molar tooth colony and ferment arabinose. The majority of serotype 2 colonies are smooth and do not ferment the sugar. There is some evidence that *A. naeslundii* is a variant of *A. israelii.* A polyvalent fluorescent antibody is group-specific for the two serotypes of *A. israelii* as well as for *A. naeslundii* and is useful for identification of the organism and its demonstration in clinical material.[6]

LABORATORY DIAGNOSIS

Direct Examination

The isolation and identification of *Actinomyces* is difficult, but a tentative diagnosis can be made by finding the "sulfur granules" in purulent exudate. The pus may be spread out in a Petri dish and examined for the presence of granules. The granules vary in size from barely visible specks to 2.5 mm in diameter. They are yellowish-white to white, firm, and spherical or lobulated. They may be abundant in material from draining sinuses but are rare in sputum. A single granule may be placed in a drop of water, crushed under a microscope slide cover, and examined with the low-power objective. The granule will be seen as an opaque mass with a periphery of clear, gelatinous protrusions or club-shaped bodies (Fig. 1–11,*A* FS A 5,*A*). Clubs are sometimes absent. A Gram stain of the material examined under oil immersion shows delicate, intertwined gram-positive filaments and coccoid and bacillary bodies. This is important in differentiating the granules of *Actinomyces* from those of other bacterial species. Staphylococcal granules are also gram-positive but are composed only of cocci, and gram-negative organisms, such as *Bacteroides* and *Proteus*, may also form loose pseudogranules. If granules are not present, a Gram stain of the material may reveal short-branching filaments, which may be *Actinomyces*. *Nocardia* can usually be ruled out by staining with the modified acid-fast stain.[2] They are acid-fast, as are the spores of *Streptomyces* species.

Culture Methods

Actinomyces may be cultured from purulent exudates, sputum, drainage material, and biopsy and autopsy material. If grains are present, they should be washed in several rinses of sterile saline to remove "associate" bacteria, then crushed in saline with a sterile glass rod against the side of a tube. The material can then be transferred to a brain-heart infusion blood agar plate, and a single loopful streaked onto several plates. Other primary isolation media has been recommended from time to time, but in the author's experience the previously mentioned procedure is sufficient in most cases. The plates are incubated under anaerobic conditions with CO_2 (a candle jar is sufficient). A second set of plates is incubated aerobically to discern the oxygen requirements of the organism and for growth of associated bacteria.

The colonial morphology of *A. israelii, A. bovis*, and *A. naeslundii* grown anaerobically on solid media is sufficiently distinctive that, with experience and the use of selected physiologic characteristics, the organisms can be differentiated from those of contaminating bacteria and from each other. At 48 hours, colonies of *A. israelii* are loose masses of branching filaments with a spiderlike appearance or are white, rough, granular specks resembling sand grains. *A. bovis* colonies are generally small, punctate, moist drops with entire edges, but when examined closely will be seen to have a smooth but grainy surface. *A. naeslundii* and *B. eriksonii* may be similar to either of the above at this stage. After four to six days incubation, the colonies are often less than 1 mm in diameter. At this time all organisms are usually opaque, dead white, or rarely have a slight gray or creamy tinge. *A. israelii* is generally a rough colony (R form), starting as a mass of branching filaments (spider) and developing into a lobulated, glistening "molar tooth" colony (Fig. 1–11,*B* FS A 5,*B*). The rare S variant or serotype 2 may be transparent and regular in form, thus resembling *A. bovis* (Fig. 1–9). Colonies are picked to freshly heated and cooled thioglycollate broth, heart infusion broth, or *Actinomyces* maintenance media.[2] In broth, *A. israelii* grows slowly, forming a hard, granular, fuzzy-edged colony. *A. bovis* generally grows as a smooth (S) form in which colonies are at first like dewdrops, then smooth, convex, and grainy with entire edges. The rare R variant may resemble *A. israelii.* In broth, *A. bovis* produces a soft, lobular colony which is easily dispersed. *A. naeslundii* and *B. eriksonii* resemble the latter. The *A. naeslundii* variant readily becomes microaerophilic or aerobic; *B. eriksonii* is strictly anaerobic. *A. bovis* and *A. israelii* may become microaerophilic. In physiologic tests, *A. israelii* usually ferments xylose and mannitol, reduces nitrate, and does not hydrolyze starch; *A. bovis* hydrolyzes starch, does not reduce nitrate, does not ferment mannitol and usually not xylose; *A. naeslundii* reduces nitrate but does not fer-

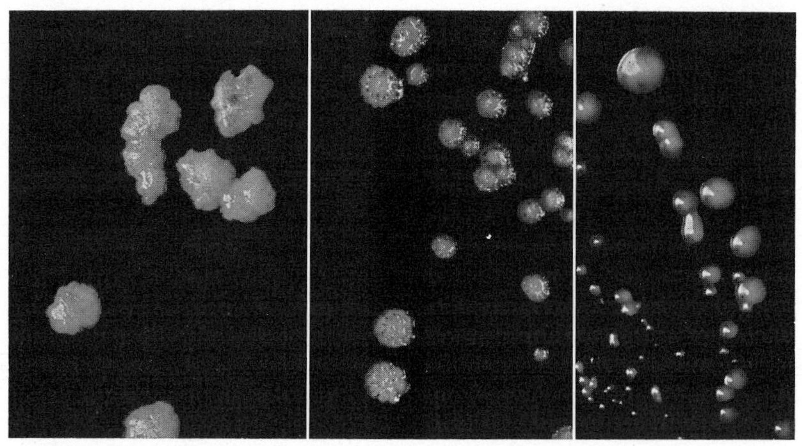

Figure 1–9. Colonies of *Actinomyces israelii* on brain-heart infusion agar after six days' incubation. *Left* and *center,* rough (R) "molar tooth" serotype 1. ×3. *Right,* colonies of the smooth (S) serotype 2. ×6. (Rosebury, Epps, and Clark.)[33]

ment xylose or mannitol. The anaerobic *Corynebacterium* organisms are proteolytic, as determined by gelatin liquefaction, and are catalase-positive. As blood contains catalase, this test cannot be done on blood agars.

MICROBIOLOGY

Genus *Actinomyces* Harz 1877

Synonymy. *Streptothrix* Cohn 1875; *Cladothrix* Cohn 1875; *Discomyces* Rivolta 1878.

Actinomyces israelii (Kruse) Lachner-Sandoval 1898

Synonymy. *Streptothrix israelii* Kruse 1896; *Discomyces israelii* Gedoelst 1902; *Nocardia israeli* Castellani and Chalmers 1913; *Cohnistreptothrix israeli* Pinoy 1913; *Brevistreptothrix israeli* Lignieres 1924.

Colony Morphology. Brain-heart infusion blood agar plates. At 24 to 48 hours anaerobic incubation, serotype 1 *A. israelii* forms a minute, spiderlike colony with branching "mycelial" elements radiating from a central point. This is best observed by holding the plate at an angle to or above a light source. By ten days the colony is hard, lobulated, "molar tooth" gray-white, and glistening. The serotype 2 is small, shiny entire, and "bacterial" at 24 to 48 hours and smooth convex entire and shiny at ten days.

Microscopic Morphology. Short rods with one or more branches and coccoid forms common in young cultures (Fig. 1–10). One μ diameters by 3 to 7 μ long. In old colonies and in tissue they may be long and extensive but less so than *Nocardia.* They are gram-positive and nonacid-fast.

Physiologic Characteristics. Esculin hydrolysis +, catalase 0, indole 0, nitrate-variable, gelatin digestion 0, mannose A, lactose A, sucrose A, maltose A, raffinose A, xylose A, glycerol 0, mannitol variable, starch A, starch hydrolysis 0, microaerophilic.

Actinomyces bovis Harz 1877

Synonymy. *Discomyces bovis* Rivolta 1878; *Bacterium actinocladothrix* Afanassieu 1888; *Nocardia actinomyces* Trevisan 1889; *Cladothrix bovis* Mace 1891; *Streptothrix actinomyces* Rossi-Doria 1892; *Oospora bovis* Sanvageau and Radias 1892; *Nocardia bovis* Blanchard 1895; *Actinomyces bovis sulfureus* Gosperinini 1894; *Cladothrix actinomyces* Mace 1897; *Streptothrix actinomycotica* Foulerton 1899; *Streptothrix bovis communis* Foulerton 1901; *Streptothrix bovis* Chester 1901; *Sphaerotilus bovis* Engler 1907.

Colony Morphology. Brain-heart infusion blood agar plates. At 24 to 48 hours anaerobic incubation, the colony is small, smooth, glistening, entire, white, "bacteria"-like. At ten days it is smooth, flat-to-convex, opaque, white, grainy, similar to *A. israelii* serotype 2.

Microscopic Morphology. Short diphtheroid-like rods; coccoid bodies are found in

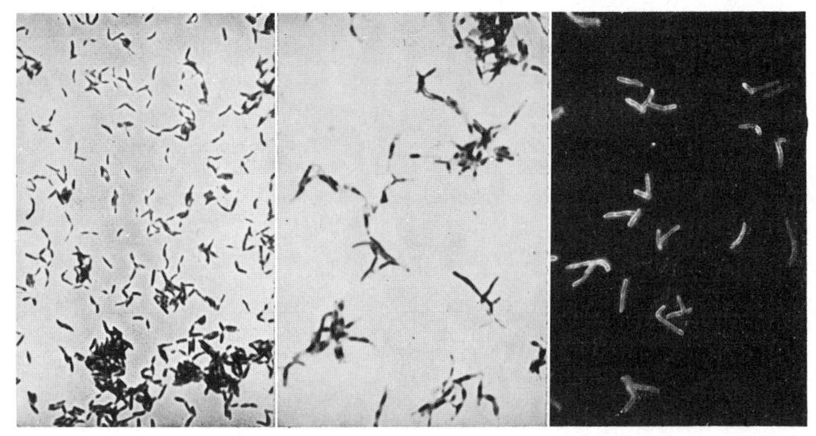

Figure 1–10. *Actinomyces israelii. Left,* darkfield. ×900. *Middle* and *right,* Gram stains of rough and smooth cultures, respectively. ×1200. (Rosebury, Epps, and Clark.)[33]

young colonies. Short, branching rods occur in older colonies. Fragmentation to coccoid bodies occurs in tissue and in culture. Gram-positive, nonacid-fast.

Physiologic Characteristics. Esculin hydrolysis +, catalase 0, indole 0, nitrate +, gelatin hydrolysis 0, mannose variable, lactose A, sucrose A, maltose A, raffinose 0, xylose 0, glycerol 0, mannitol 0, starch A, starch hydrolysis variable, microaerophilic.

Actinomyces naeslundii Thompson and Lovestedt 1951

Colony Morphology. Brain-heart infusion blood agar plates. After 24 to 48 hours

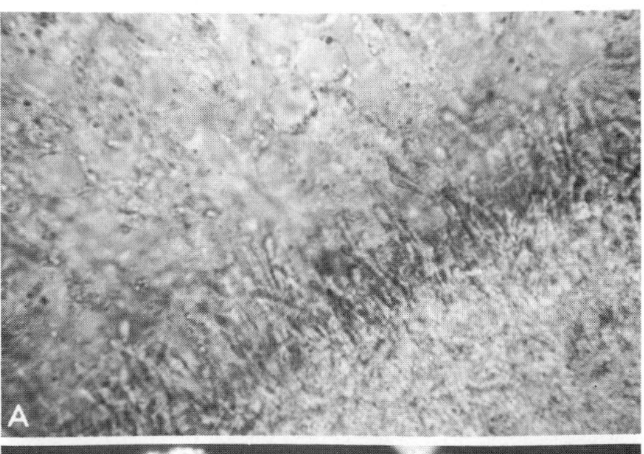

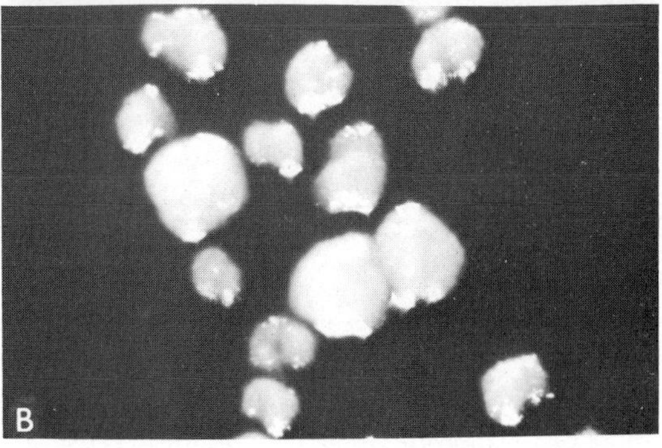

Figure 1–11. FS A 5. *A,* Wet mount of sulfur granule showing gelatinous club-shaped bodies. *B,* Molar tooth colonies of *Actinomyces israelii* growing on blood agar. One week, 37°C, anaerobic culture.

anaerobic incubation, a small, smooth colony appears with mycelian radiation surrounding it. By ten days, the colony is convex-to-flat entire edge, raised, rough, grainy, or similar to *A. bovis.*

Microscopic Morphology. Short rods, branching diphtheroid forms, and small coccuslike cells are seen in young colonies. Fragmentation to coccoid bodies occurs in time. Gram-positive, nonacid-fast.

Physiologic Characteristics. Esculin hydrolysis +, catalase 0, indole 0, nitrate 0, gelatin 0, mannose A, lactose A, sucrose A, maltose A, raffinose A, xylose 0, glycerol 0, mannitol 0, starch A, starch hydrolysis 0, microaerophilic.

*Bifidobacterium eriksonii**

Synonymy. *Actinomyces eriksonii* Georg and Robertstad 1965.

Colony Morphology. Brain heart infusion blood agar plates. At 24 to 48 hours anaerobic incubation, a dense mycelian colony appears, which is flat and granular with a conical center. By ten days, it is conical or convex, soft, entire, and white in color.

Microscopic Morphology. Branching diphtheroid-like rods are seen in smears from young colonies. Fragmentation to coccoid bodies occurs in time. Gram-positive, nonacid-fast.

Physiologic Characteristics. Esculin hydrolysis +, catalase 0, nitrate 0, gelatin hydrolysis 0, glucose A, xylose A, mannitol A, raffinose A, starch hydrolysis +, mannose 0, strict anaerobe.

Fide L. K. Georg.

Arachnia propionica (Buchanan and Pine) Pine and Georg 1969

Colony Morphology. Brain-heart infusion blood agar plates. After 24 to 48 hours anaerobic incubation, a spiderlike colony similar to *A. israelii* appears. However, by ten days the colony becomes smooth, round, convex to flat, and appears more like *A. bovis.*

Microscopic Morphology. Typical *Actinomyces* morphology occurs in culture and tissue. Young colonies contain short-branching, diphtheroid-like forms, and fragmentation to coccoid bodies occurs in time.

Physiologic Characteristics. Esculin hydrolysis +, mannose A, mannitol A, raffinose A, xylose variable, starch hydrolysis 0, nitrate +, gelatin hydrolysis 0, microaerophilic.

Tables 1–2 and 1–3 summarize the preceding material.

Table 1–2. *Summary of Characteristics of the Etiologic Agents of Actinomycosis**

	A. bovis	*A. israelii*	*A. naeslundii*	*Bifidobacterium eriksonii*	*Arachnia propionica*
Aerobic	Variable	Variable	Variable	0	Variable
Anaerobic	+	+	+	+	+
Catalase	0	0	0	0	0
Litmus milk	Reduced	Reduced	Reduced	Reduced with curd	Reduced
Propionic acid production	0	0	0	0	+
Fermentation:					
Glucose	A	A	A	A	A
Mannitol	0	A	0	A	A or variable
Mannose	0 or variable	A	A	A	A
Xylose	A or variable	A	0	A	0
Raffinose	—	A	A	A	A
Starch hydrolysis	+	0	0	+	0

*A compilation from various sources. Media, techniques, and specific instructions are detailed in the C.D.C. manual[2] and the *Manual of Clinical Microbiology*,[4] and in reviews 6, 8, and 13.

Catalase-positive *Actinomyces* have been described,[14] and other organisms, such as *Rothia dentocariosa*,[7] may be confused with pathogenic *Actinomyces*.

Table 1–3. *Colony and Microscopic Morphology of the Actinomyces*

	A. bovis	*A. israelii*	*Bifidobacterium eriksonii*	*A. naeslundii*	*Arachnia propionica*
Microscopic morphology	Little branching, mostly diphtheroid, rarely filamentous	Rods, clubs, and branched forms common; rarely long and filamentous	Mostly diphtheroid, coccoid to awl shaped with bifurcate ends	Many branching rods and thick irregular forms, some diphtheroid	Short branching rods becoming, with age, long filamentous branched cells
Colony morphology	Shiny, grainy, entire-edged, convex colonies; some with scalloped edge	Very hard raised lobulated "molar tooth;" grow into agar and are removed whole; S or serotype 2 similar to *A. bovis*	Grainy lobulated edge, convex colony	Similar to *A. bovis*	Filamentous or cobweb colony; flat to lumpy but less than *A. israelii*
Grains in man or animals	+	+	−	Rare	+

REFERENCES

1. Actinomycosis (editorial). 1962. J.A.M.A., *181*:721–722.
2. Ajello, L., L. K. Georg, et al. 1963. *Laboratory Manual for Medical Mycology.* Public Health Service Publication No. 994. U.S. Government Printing Office. Washington, D.C.
3. Becker, R. B., C. J. Wilcox, et al. 1964. *Genetic Aspects of Actinomycosis and Actinobacillosis in Cattle.* Ohio Res. Bull. No. 938, 24 pp.
4. Blair, J. E., E. H. Lennette, et al. 1970. *Manual of Clinical Microbiology.* American Society for Microbiology. Bethesda, Md.
5. Bojalil, L. F., and A. Zamora. 1963. Precipitin and skin test in the diagnosis of mycetoma due to *N. brasiliensis.* Proc. Soc. Exp. Biol. Med., *113*: 40–43.
6. Brock, D. W., and L. K. Georg. 1969. Characterization of *Actinomyces israelii* serotypes 1 and 2. J. Bacteriol., *97*:589–593.
7. Brown, J. M., and L. K. Georg, et al. 1969. Laboratory identification of *Rothia dentocariosa* and its occurrence in human material. Appl. Microbiol., *17*:150–156.
8. Coleman, R. M., L. K. Georg and A. Rozzell. 1969. *Actinomyces naeslundii* as an agent of human actinomycosis. Appl. Microbiol., *18*:420–426.
9. Cope, V. Z. 1939. *Actinomycosis.* Oxford University Press. London.
10. Dyson, J. E., and J. M. Slack. 1963. Improved antigens for skin testing in nocardiosis. I. Alcohol precipitates of culture supernates. Amer. Rev. Resp. Dis., *88*:80–86.
11. Emmons, C. W. 1935. Actinomyces and actinomycosis. Puerto Rico J. Public Health Trop. Med., *11*:63–76.
12. Erikson, D. 1949. The morphology, cytology and taxonomy of the Actinomyces group. Medical Research Council (Great Britain) Special Report Series No. 240.
13. Georg, L. K., G. W. Robertstad, et al. 1965. A new pathogenic anaerobic Actinomyces species. J. Infect. Dis., *115*:88–99.
14. Georg, L. K., L. Pine, et al. 1969. Actinomyces viscosis comb. nov., a catalase positive, facultative member of the genus Actinomyces. Int. J. Syst. Bacteriol., *19*:291–293.
15. Gerencser, M. A., and J. M. Slack. 1967. Isolation and characterization of *Actinomyces propionicus.* J. Bacteriol., *94*:109–115.
16. González Ochoa, A., and F. Baranda. 1953. Una prueba cutánea para el diagnóstico del micetoma actinomicósico por *Nocardia brasiliensis.* Rev. Inst. Salub. Enferm. Trop., *13*:189–197.
17. González Ochoa, A., and A. Vasquez-Hoyos. 1953. Relaciones serologicas de los Principales actinomycetes pathogenicos. Rev. Inst. Salub. Enferm. Trop., *13*:177–187.
18. Goodfellow, M. 1971. Numerical taxonomy of the genus Nocardioform bacteria. J. Gen. Microbiol., *69*:33–80.
19. Harvey, J. C., J. R. Cantrell, et al. 1957. Actinomycosis: Its recognition and treatment. Ann. Intern. Med., *46*:868–885.
20. Hazen, E. L., and G. N. Little. 1958. *Actinomyces bovis* and "anaerobic diphtheroids." J. Lab. Clin. Med., *51*:968–976.
21. Howell, A., W. C. Murphy, et al. 1959. Oral strains of actinomyces. J. Bacteriol., *78*:82–95.
22. Kwapinski, J. B. 1963. Antigenic structure of actinomycetales. VI. Serological relationship between antigenic fractions of actinomyces and Nocardia. J. Bacteriol., *86*:179–186.
23. Lambert, F. W., J. M. Brown, et al. 1967. Identification of *Actinomyces israelii* and *Actinomyces naeslundii* by fluorescent-antibody and agar-gel diffusion techniques. J. Bacteriol., *94*:1287–1295.
24. Meyer, E., and P. Verges. 1950. Mouse pathogenicity as a diagnostic aid in the identification of *Actinomyces bovis.* J. Lab. Clin. Med., *36*:667–674.
25. Mohr, J., E. R. Rhoades, et al. 1970. Actinomycosis treated with Lincomycin. J.A.M.A., *212*:2260–2261.
26. Mordarska, H., and M. Mordarski. 1969. Comparative studies on the occurrence of lipid A, diaminopimelic acid and arabinose in Nocardia

cells. Arch. Immunol. Ther. Exp. (Warsz), *17*:739–743.

27. Negroni, P. 1936. Datos estadisticos sobre 50 casos de actinomycosis y de su tratamiento vacunoterapica. Rev. Argent. Dermatosif., *20*:458.
28. Nichols, D. R., and W. E. Herrell. 1948. Penicillin in the treatment of actinomycosis. J. Lab. Clin. Med., *33*:521–525.
29. O'Mahoney, J. B. 1966. The use of tetracycline in the treatment of actinomycosis. Br. Dent. J., *121*:23–25.
30. Pier, A. C., and J. B. Enright. 1962. *Nocardia asteroides* as a mammary pathogen of cattle. III. Immunologic reaction of infected animals. Am. J. Vet. Res., *23*:284–292.
31. Pine, L., and L. K. Georg. 1969. Reclassification of *Actinomyces propionicus*. Int. J. Syst. Bacteriol., *19*:267–272.
32. Pine, L., and J. R. Overman. 1966. Differentiation of capsules and hyphae in clubs of bovine sulphur granules. Sabouraudia, *5*:141–143.
33. Rosebury, J., L. S. Epps, et al. 1944. A study of the isolation, cultivation and pathogenicity of *Actinomyces israelii* recovered from the human mouth and from actinomycosis in man. J. Infect. Dis., *74*:131–149.
34. Schneidau, J. O., and M. F. Shaefer. 1960. Studies on Nocardia and the Actinomycetales. II. Antigenic relationships shown by slide agglutination. Am. Rev. Resp. Dis., *82*:64–76.
35. Slack, J. M., and M. A. Gerencser. 1966. Revision of serological group of Actinomyces. J. Bacteriol., *91*:2107.
36. Thompson, L., and S. A. Lovestedt. 1951. An Actinomyces-like organism obtained from the human mouth. Proc. Staff Meet. Mayo Clin., *26*:169–175.

Note: A complete bibliography to 1969 of actinomycosis was published by Yousef, Al-Doory. 1971. Mycopathol. Mycol. Appl., *44*:1–88.

Chapter 2

NOCARDIOSIS

DEFINITION

Nocardiosis is a chronic, suppurative (less commonly granulomatous) disease caused by the soil-inhabiting aerobic actinomycetes, *Nocardia asteroides* and *Nocardia brasiliensis*. The primary infection is pulmonary. The disease may be subclinical or pneumonic, chronic or, rarely, acute, and may become systemic by hematogenous spread. The organism has a predilection for the central nervous system and, less commonly, other organs, such as the kidney. The patients affected are generally debilitated by other diseases or by medication, and *N. asteroides* is an opportunist rather than a primary pathogen. *N. brasiliensis* is more virulent and may be considered to be a primary pathogen.

HISTORY

In 1888, Nocard[26] described an aerobic actinomycete causing *farcin du boeuf* in cattle in Guadeloupe. The disease was confused with farcy or glanders in Europe and the West Indies, and the early reports were so contradictory and equivocal it is difficult to equate the description with the specific disease, nocardiosis. Trevisan erected the genus *Nocardia* in honor of Nocard in 1889, with the type species *Nocardia farcinica.* Eppinger later (1890) described a pseudotuberculosis in a patient with miliary pulmonary involvement, bronchial node disease, and brain abscesses, which showed branching hyphae in pus. The name *Cladothrix asteroides* was given. The organism was renamed *Nocardia asteroides* by Blanchard and Nocard in 1896. Since then, the organism has been described in a number of medical and veterinary reports and called *Streptothrix carnea, S. freeri, S. asteroides, Actinomyces gypsoides,* and *Proactinomyces asteroides.* The early literature did not always differentiate the disease from actinomycosis; therefore, the incidence is difficult to assess. Henrici and Gardner in 1921 accepted 26 reported cases as valid. In 1943, Waksman and Henrici precisely differentiated the etiologic agent, *N. asteroides,* from other actinomycetes, and following this, in 1946 Kirby and McNaught[18] described 32 additional cases. Ballenger and Goldring[2] in 1957 accepted 95 cases as authentic, while Murray et al.[24] in 1961 added more. Since that time, new aspects of nocardiosis have appeared in multiple case reports, particularly in relation to its association with other diseases and with the use of antileukemic drugs, cytotoxins, immune depressants, and corticosteroids.[5,31,39] The opportunistic nature of the infection is now emphasized.[31]

Although nocardiosis can occur in an apparently healthy individual,[2] most cases are now seen in debilitated patients. This is not surprising in view of the enormous increase in opportunistic infections being recorded. In renal transplant patients, 87 per cent of deaths are due to infections.[5] The agents include gram-negative bacteria, *Candida, Aspergillus, Nocardia, Histoplasma,* and *Pneumocystis carinii.* In cancer patients, 14 per cent of all deaths are due to overwhelming fungal and nocardial infection. Formerly rare, nocardiosis is now a frequently encountered, if not common, disease.

Recent studies, which include a more careful examination of the sputum and reevaluation of reported isolations, both in frank disease or as an incidental finding, indicate that the disease is more common than previously thought. Many of these infections may be of a minor transient, benign, or chronic nature, without clinical pulmonary disease. Dissemination may occur in the absence of overt lung pathology, so that care must be taken when diagnosing disease in extrapulmonary areas as primary. There is also evidence that the organism may be a transient member of the normal flora of the tracheal and bronchial tree.[16]

The genus *Nocardia* formerly contained various species isolated from the clinical entity, mycetoma. Most of these have been transferred to the genus *Streptomyces* by Mackinnon[21] and González Ochoa and are discussed in Chapter 4. *N. brasiliensis* is a common agent of mycetoma and can also cause systemic nocardiosis.

ETIOLOGY, ECOLOGY, AND DISTRIBUTION

Two organisms, *Nocardia asteroides* and *N. brasiliensis*, are considered as valid etiologic agents of the clinical disease, nocardiosis, in man (Fig. 2–1 FS A 6). Recent studies by González Ochoa[10] emphasize that *N. brasiliensis* is more virulent than *N. asteroides*—the former being able to cause infection readily in experimental animals as well as systemic disease in normal patients, whereas the latter showed marked strain variation as far as experimental disease was concerned. It has been concluded that *N. asteroides* is rarely a primary pathogen but rather an opportunist. Other investigators have reported varying results.[8]

Gordon and Hagen[13] first isolated *N. asteroides* from soil by the paraffin baiting technique. These findings have been confirmed by other investigators, and it appears that the organism has a worldwide distribution. *N. brasiliensis* has also been recovered from soil. *N. asteroides* has been associated with disease in cattle, small animals, and fish and has also been isolated from the normal skin.[36] Hosty,[16] while surveying the sputa from tuberculosis patients in an Alabama hospital, found *N. asteroides* in 175 out of 85,000 specimens. The patients had no evidence of nocardial infection. Thus the Nocardia might be considered as a saprophyte or minor secondary invader in these cases. As these patients all had some lung disease, it is difficult to extrapolate to the normal population as survey studies have not been adequate.[29] However, it is suggested that *N. asteroides* may be at least a transient member of normal pulmonary flora. Although preexisting lung disease may favor a frank infection by the organism, this is not a necessary prerequisite.

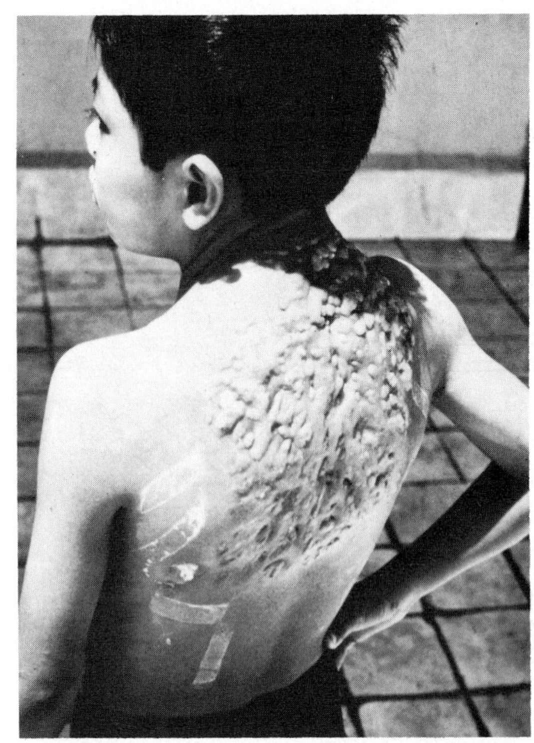

Figure 2–1. FS A 6. *Nocardia brasiliensis.* Pulmonary infection with sinus formation and massive involvement of subcutaneous and cutaneous tissues. (Courtesy of H. Levine.)

Nocardiosis has been reported in all parts of the world. There is no racial or occupational pattern noted. Even though the organism is in the soil, rural incidence is no higher than urban. In reviewed cases, the male to female ratio is given as 2:1 or 3:1. Though infection has been reported in all age groups, the incidence is higher in the 30 to 50 year old category.

CLINICAL DISEASE

Symptomatology

With rare exceptions, nocardiosis is a pulmonary disease of respiratory origin resulting from the inhalation of spores.

Cerebral nocardiosis associated with pulmonary disease is frequently reported, and there is conclusive evidence of hematogenous spread. Primary or secondary lesions in the gastrointestinal tract at a site of preexisting mucosal ulceration and, rarely, appendiceal involvement may result from ingestion of contaminated material or sputa from infected lungs. Nocardiosis has occurred at the site of a bruise where there was no rupture of the integument; therefore caution must be observed in assessing skin lesions as primary. However, authentic primary cutaneous nocardiosis has been described.[38]

In approximately 75 per cent of all reported cases, symptomatic pulmonary involvement has been noted. A pneumonia develops which may be lobular or lobar and is often chronic. The lesion may be a solitary lung abscess, an acute necrotizing pneumonia, scattered infiltrates resembling miliary tuberculosis, a pulmonary mycetoma, or a progressive fibrosis and extension to the pleura and chest wall, with penetration simulating actinomycosis. It is evident, then, that with such a variety of presenting signs the clinical diagnosis is difficult and necessitates laboratory confirmation. The presenting symptoms are not distinctive. Fever is present, ranging from 37.2 to 41°C. Anorexia, malaise, weight loss, pleurisy, nocthidrosis, and, rarely, chills may be found. Cough is usually nonproductive and hacking at first but becomes mucopurulent and blood-streaked. If cavitation occurs, massive hemoptysis may result. Consolidation of one or more lobes of the lung may occur, but caseation and granuloma formation are absent or rare. There is no favored lung site, as apical, hilar, and basilar regions are equally involved. The percussion note may be impaired, with diminished breath sounds over an affected area, and rales may be present.

Though metastatic lesions may occur anywhere in the body, the brain is the most common secondary site of nocardiosis. Brain lesions may be multiple, but a large single abscess may result by extension or coalescence (Fig. 2–2 FS A 7*A*,*B*). Usually the meninges are not involved enough to cause diagnostic changes in the spinal fluid. The symptoms may be primary if lung involvement is minimal. The onset may be sudden or gradual and include, in descending frequency, headache, lethargy, convulsions, peripheral numbness, nuchal rigidity, psychic confusion, aphasia, tremor, and paresis. Rarely, involvement of the spinal column may lead to compression. Though several adjacent vertebrae may be involved, the disks tend to be spared. Mottled osteolytic lesions may occur in the vertebrae or cranial bones. Capsulization of nocardial abscesses is sometimes seen with vascularizations which appear as "tumor blushes" on angiogram.

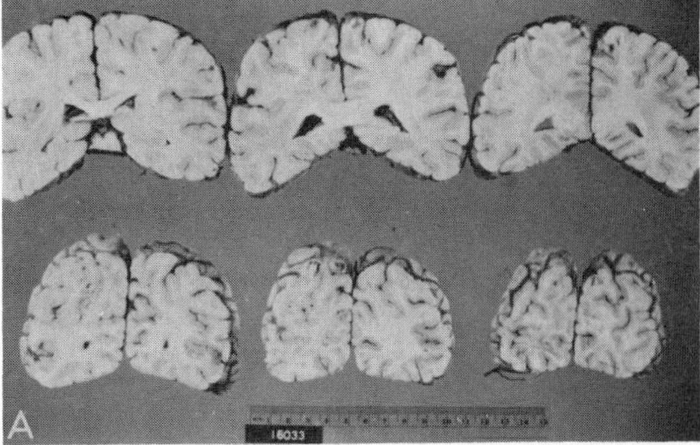

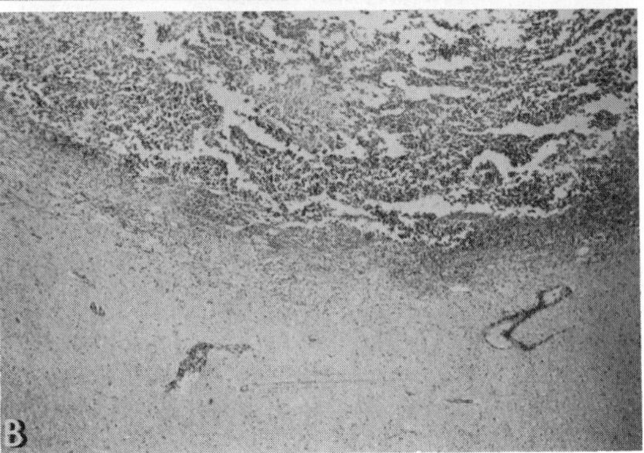

Figure 2–2. FS A 7. *A*, Nocardiosis. Abscess in brain from patient who had primary pulmonary disease. *B*, Histologic picture of brain lesion showing necrosis. ×100.

The kidney is probably the third most commonly involved organ in nocardiosis. Lesions extend from the cortex into the medulla. On occasion, the disease may metastasize to the pericardium, myocardium, spleen, liver, and adrenals. Lymphadenopathy occurs at various sites, but particularly in the cervical and axillary nodes. In contradistinction to actinomycosis, bone involvement is rare but reported. The eyes are rarely infected, and symptoms include papilledema, blurring of visual fields, diplopia, and paralysis of the extrinsic muscle. Again, it is evident that these are vague findings which can be attributed to a number of other conditions. Canaliculitis of the lacrimals in the absence of other symptoms has been attributed to *N. asteroides* as well as to other actinomycetes.

X-ray Findings

Roentgenographic examination of the chest may show pulmonary infiltration, hilar enlargement,[29a] marked pleural thickening (often 2 to 3 cm), cavitation, diffuse mottling, and nodules (Fig. 2–3). Bilateral involvement is common. Most often the abscesses in the lung or parenchyma, though multiple, are too small to be visualized.

Differential Diagnosis

In pulmonary infection, nocardiosis mimics tuberculosis in all stages. Subsequent involvement of pleural and chest walls suggests actinomycosis (Fig. 2–1 FS A 6). As there is no clinical syndrome that characterizes nocardiosis, other conditions must be considered. These include eumycotic infections of the lungs, bacterial brain abscess, carcinomatosis of lung and brain, sarcoma, or late syphilis. Systemic disease is often not diagnosed until surgical intervention or at autopsy.

It is extremely important to differentiate nocardiosis from tuberculosis and actinomycosis as each has a specific treatment.

Nocardiosis due to *N. asteroides* is a disease of the compromised patient. It is therefore a possible, often terminal complication in a number of disease states. Consequently, in patients suffering chronic progressive diseases or diseases causing impairment of normal immune mechanisms, the possibility of this or other facultative pathogens as infecting agents must be constantly considered. In general, both bacterial and eumycotic opportunistic infections are increasing in medical practice. Nocardiosis has been particularly associated with pulmonary alveolar proteinosis,[1] Cushing's syndrome, diabetes mellitus, and the use of corticosteroids, immunosuppressive agents, and perhaps antibiotics.[31]

Prognosis and Therapy

In spite of the sensitivity of the infectious agent to several antibiotics and sulfas, the prognosis of nocardiosis is not good, especially if metastasis has occurred. Several factors contribute to this. Usually the patient is debilitated, his defenses compromised either by disease or by medication, so that

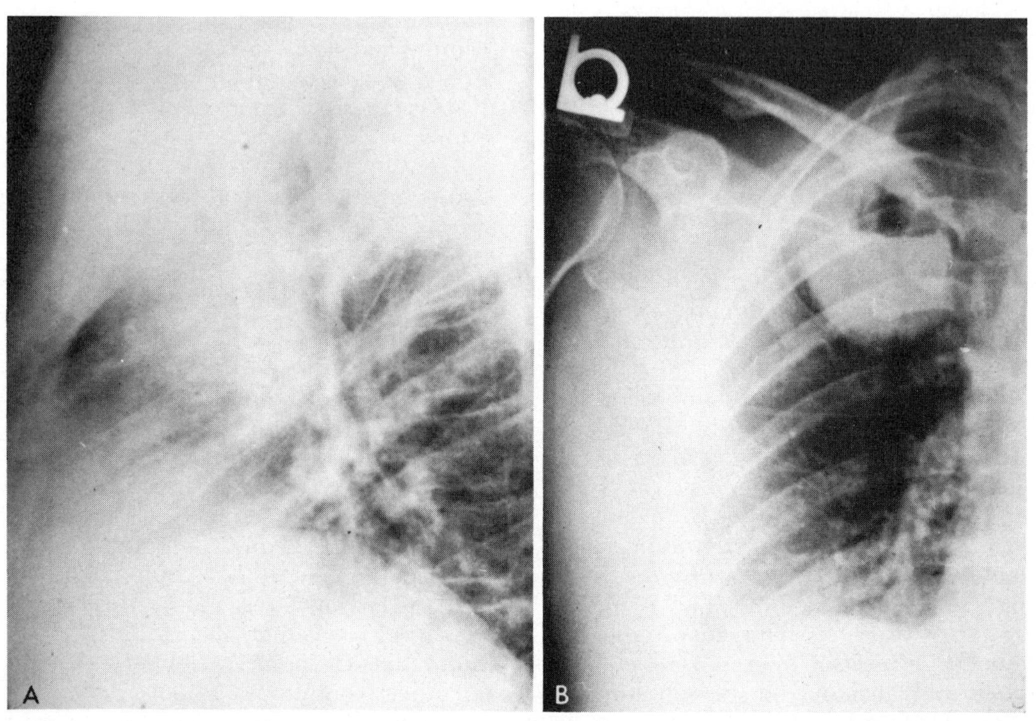

Figure 2–3. Nocardiosis. Pulmonary infection in a patient with lupus erythematosus on steroids. Large cavity in right upper lobe with an air fluid level. The cavity is well circumscribed in contrast to an acute bacterial disease which usually shows more of a reaction around lesions. (Courtesy of J. Fennessey.)

he may be little more than culture media, as viewed by the organism. Adequate and long-term therapy with appropriate drugs is necessary in nocardiosis. Before the advent of sulfonamide therapy, systemic disease was universally fatal. Uncomplicated pulmonary disease can be arrested, but systemic involvement presages difficulty. Central nervous system involvement is still about 87 per cent fatal. The overall mortality is currently estimated at 50 per cent.[37] This may be an exaggeration, as inapparent or mild infection often goes undiagnosed.

The treatment of choice in nocardiosis is sulfadiazine. This has not changed since it was established in 1944. Blood levels of 9 to 12 mg per 100 ml should be maintained and, in severe cases, 10 to 20 mg per 100 ml. This requires daily administration of 4 to 6 g of sulfadiazine or as high as 10 g.[1a] Therapy should be continued for three to six months, or longer. Black and McNellis[4] recently studied the *in vitro* sensitivity of the sulfas and newer antimicrobial agents. None were superior to sulfadiazine and sulfamethoxazole, although fusidic acid was very effective. Clinically, the two sulfas are essentially equally effective. Antibiotics may be used as adjunct therapy. In addition to sulfas, streptomycin, at a dose of 2 g per day, is often used during the acute disease. Aureomycin has shown good activity *in vitro* and warrants trial. Penicillin, chloramphenicol, and tetracycline, as well as other drugs, have been used but alone have failed to arrest the disease. Cephaloridine has been used successfully (1 g every six hours), as well as a combination of sulfamethoxazole and trimethoprim.[2,7] The latter is a bacterial folate inhibitor at a point other than that involved with sulfa drugs.[2] The synergism exhibited by these drugs *in vivo* cannot be demonstrated *in vitro*.[3] The converse is also true in that sensitivity *in vitro* does not correlate with clinical response.[1b,34] The usual dose is 2.5 g per day of sulfamethoxazole and 480 mg per day of trimethoprim.[2,7] Needless to say, the antimycotic agents, such as amphotericin B and nystatin, are useless. Sulfones have been tried but are of questionable value. In cerebral disease, cycloserine is added to the regimen (250 mg every six hours),[15] as well as other antibiotics, such as tetracycline. Both cycloserine and the sulfas penetrate well into the central nervous system. Ampicillin is currently being investigated as an adjunct to sulfa therapy. The results of testing *in vitro* vary considerably with strains and techniques.[20a]

With adequate drug cover, surgical intervention is a valuable adjunct. Abscesses of the brain and lungs and empyema require drainage.[15] Excision or debridement of diseased tissue is also useful.

PATHOLOGY

Gross Pathology

Nocardiosis presents a similar picture in almost all involved organs. Multiple abscess formation with central necrosis, and little or no peripheral fibrosis are the usual findings. Only rarely is a tuberculoma or organized granuloma with giant cell infiltration seen. Caseation is rare. The picture, then, cannot be distinguished from infections caused by pyogenic bacteria. Grossly, the abscesses may appear homogeneous and shiny and may resemble lymphosarcoma. These lesions are generally soft and friable and on cut-section may contain granular, greyish-white material resembling liquefactive necrosis of tumor tissue. There is less host response elicited in nocardiosis than in actinomycosis and, as a consequence, less fibrosis and scarring. The characteristic burrowing and sinus formation of the latter disease is generally absent in nocardiosis due to *N. asteroides* as are true sulfur granules. Both, however, may be seen in infections due to *N. brasiliensis*.

The lung changes may be a combination of pneumonic consolidation, abscesses, diffuse miliary and nodular lesions, and cavities with adhesions. Pleural involvement gives rise to fibrinous pleurisy, empyema, multiple abscesses, and effusion with some fibrosis. Tracheobronchial adenopathy is found, but infarction and atelectasis are rare. Other organs present with suppurative abscesses of the general pyogenic cellular type.

Histopathology

Suppuration is the classic histologic picture of nocardiosis, regardless of the anatomic site of infection. There is a dense, polymorphonuclear leukocyte infiltrate within the nodules, as well as fibrin, lymphocytes, and plasma cells. Nodules may necrotize when they attain a large size, followed by cavitation. Fibrosis, Langhans' giant cells, and caseation necrosis rarely occur, thus allowing differentiation of the lesion from that generally seen in tuberculosis. Encapsulated lesions are sometimes found, especially in the brain.

The organism is easily missed in histologic examination, because it is not stained when the usual hematoxylin and eosin method is used. With the modified Gram's stain (Brown-Breen, MacCallum-Goodpasture), delicate, multiple-branched filaments are seen, along with beaded chains of bacillary bodies. These may aggregate into a pseudogranule but are of loose consistency compared to the grains in actinomycosis. A modified acid-fast stain will demonstrate the organism, but its tendency to fragment into short, bacillary forms may cause it to be confused with the tubercle bacillus. The filaments are well demonstrated by the methenamine silver stains if staining time is extended. The special stains for fungi (PAS, Gridley) are not of use. Methods for staining *Nocardia* and *Actinomyces* have been reviewed by Robboy and Vickery.[30]

ANIMAL DISEASE

Natural infections by *N. asteroides* have been recorded in cats, cattle, dogs, horses, goats, pigs, rabbits, and fish. Considering that the organism is abundant in soil, it is not surprising that a variety of animals are infected. The original isolate from farcy of cattle was named *N. farcinica,* but it is probable that this organism was identical to the later described *N. asteroides.*[11,14] Infection in cattle can be epidemic. *N. asteroides* commonly infects the udder, causing mastitis; it may also be found in the jaw, cervical lymph nodes, gastrointestinal tract, and other areas commonly invaded by *Actinomyces bovis.* There is less bone involvement in nocardiosis, and granules are absent or rare. The disease entity, farcy, is characterized by tumefaction and granulomatous inflammation of the superficial lymph nodes, vessels, and subcutis, leading to suppuration. The organism which is still called *N. farcinica* by some[25] is considered by others to be within the species limits of *N. asteroides.*[14] Epidemics among hogs and in trout hatcheries have also been recorded.[33] The dog is commonly infected, and similar types of pathology are seen in both animal and human infections.

Another aerobic actinomycete, *Nocardia dassonvillei,*[32] has been described in cutaneous lesions similar to those in farcy, and in granulomatous lesions of the intestine in cattle. Infection in man has thus far not been reported. The cell wall analysis and lipid content of this organism are similar to those of *Streptomyces madurae* and *Streptomyces pelletierii* and differ from other *Nocardia* and *Streptomyces.* It has been proposed that the three species may be included in a new genus, *Actinomadura* (Lechevalier, 1970).

Experimental infection in animals with *N. asteroides* has been investigated numerous times with varying results. It does not appear to be of use in establishing identification of the species. The first successful report of infection was described by Strauss and Kligman,[35] using gastric mucin as an adjuvant in white mice. In a large survey of strains, Georg[8] found variable pathogenicity for guinea pigs. Macotela-Ruiz and Mariat[22] demonstrated granules in lesions produced in guinea pigs and mice inoculated with *N. asteroides* or *N. brasiliensis.* The grains from the former were soft, lobulated, easily dispersed masses, whereas firm, hard, multiple-rayed granules were found with *N. brasiliensis.* González Ochoa[10] studied the relative pathogenicity of *N. asteroides, N. brasiliensis,* and *N. caviae.* Grains were found in infections with the latter two species but not with *N. asteroides.* He concluded that the so-called *N. asteroides* isolated from mycetoma was in reality *N. caviae.* He emphasized the virulent nature of *N. brasiliensis* compared to *N. asteroides* and *N. caviae.* Other workers have reported regular and reproducible infections in mice with *N. asteroides.*[20,23]

IMMUNOLOGY AND SEROLOGY

Natural resistance to infection by *N. asteroides* is probably high, as inhalation of spores from this common soil organism is undoubtedly frequent. This again emphasizes the opportunistic nature of the organism in eliciting disease. A skin test (asteroidin), described as not cross-reacting with tuberculin, was reported by Drake and Henrici.[6] test antigen.[19] Pier and Enright[28] have developed a specific skin test for nocardiosis in cattle.

Vaccine prophylaxis was utilized without concrete evidence of efficacy before the advent of specific therapy. In experimental infections, Macotela-Ruiz and Mariat[22] could find no difference in susceptibility between animals sensitized with killed *Nocardia* cells and controls. Circulating antibodies have been reported in patients, but their role in defense has not been assessed.

Serologic procedures have been tried on sera from patients with nocardiosis, but no consistency of response has been found. Complement-fixing and agglutinating antibodies have been reported.[28] The fluorescent antibody technique for specific identification of *N. asteroides* has been attempted, but results were equivocal.[19] At the present time, there is no reliable diagnostic or prognostic serologic procedure for the disease.

LABORATORY DIAGNOSIS

Direct Examination

Sputum, pus, tissue material, and so forth can be examined for *N. asteroides.* The materials to be examined may be digested, then concentrated by centrifugation. Gram's strain of the material will show long, sinuous, branching, gram-positive filaments and fragmented bacillary bodies (Fig. 2–4). The branching tends to be at long intervals and at right angles to the main axis of the mycelium. A modified acid-fast stain will show beaded or fragmented acid-fast bacillary forms. This distinguishes the organism from *Actinomyces,* but it may be confused with the tubercle bacillus. Both *N. asteroides* and *N. brasiliensis* are acid-fast; most other actinomycetes are not. Rarely, *N. asteroides* aggregates into a soft pseudogranule, whereas *N. brasiliensis* regularly forms true granules.

Culture Methods

Unlike Actinomyces, *N. asteroides* is not fastidious and grows readily on ordinary laboratory media without antibiotics. It is aerobic, and its optimal growth temperature is 37°C. Both *N. asteroides* and *N. brasiliensis* develop slowly on routine media. By two to three weeks, they attain a diameter of 5 to 10 mm. The colonies are waxy, folded, and heaped at first. They may later develop areas of downy or tufted aerial mycelia. The whole surface may become dry and powdery.

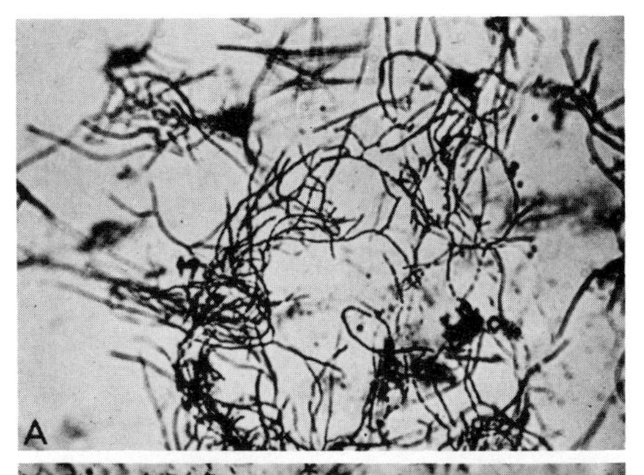

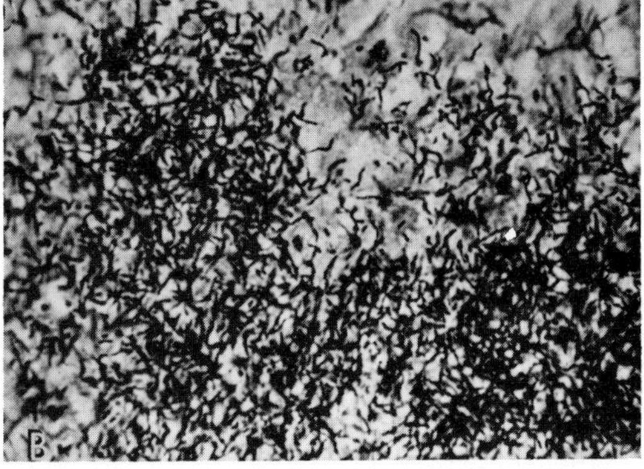

Figure 2–4. *A,* Nocardiosis. Gram stain of *Nocardia asteroides* sputum. Note the long, sinuous, branching rods. ×1000. (From Rippon, J. W. *In* Burrows, W. 1973. *Textbook of Microbiology.* 20th ed. Philadelphia, W. B. Saunders Company, p. 693.) *B,* Actinomycosis. Gram stain of *Actinomyces israelii* in tissue debris. The bacilli are short, branching rods and tend to aggregate. ×1000.

A musty, dirtlike odor is sometimes present. Though the classic colony of *N. asteroides* is glabrous, wrinkled, folded, and orange, the color range includes pink, white, buff, brown, lavender, and salmon (Fig. 2–5 FS A 8). The definitive work of Gordon and Mihm is recommended[14] as an aid in identification. The color range for colonies of *N. brasiliensis* is similar to that for those of *N. asteroides.*

Stained smears from growth on laboratory media will show coccoid and bacillary elements. The long, branched filaments are usually not seen from agar preparations but are easily obtained from growth in broth. Acid fastness is quite variable[30] from growth on agar but is enhanced if the organism is grown in litmus milk. Sporulation occurs by fragmentation into arthrospores. This can best be seen by the slide culture technique.[8] Branching also may be observed, thus differentiating the organism from various mycobacteria.

Colonies on Löwenstein-Jensen or similar isolation media for tubercle bacilli appear

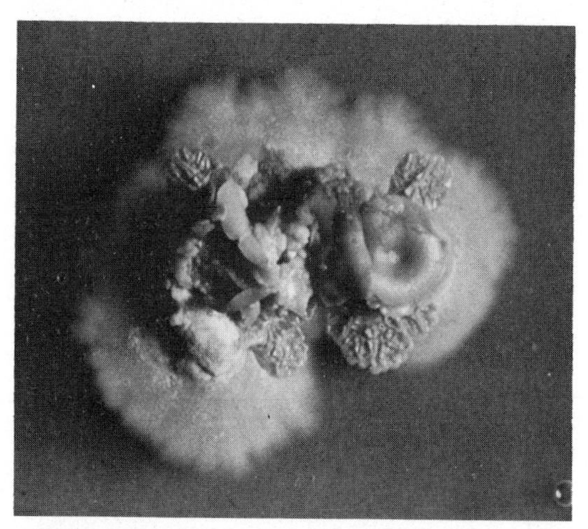

Figure 2–5. FS A 8. Nocardiosis. Colony of *Nocardia asteroides*. It is wrinkled, folded, glabrous, and bright orange. In one area, a white overgrowth is beginning where sporulation will occur.

within one to two weeks. They are folded, moist, and glabrous and indistinguishable from so-called "atypical" mycobacteria. The latter are strongly acid-fast and do not show branching.

The differentiation of *N. asteroides* from *N. brasiliensis* and other actinomycetes is detailed in the section on mycetoma. *N. asteroides* does not grow in gelatin test media, produce gelatinase,[5] digest casein or tyrosine, or peptonize bromcresol purple milk.[12] *N. brasiliensis* will have a positive reaction in all these previously mentioned areas. Both are urease-positive and partially acid-fast when the Kenyoun technique is modified by the use of 1 per cent H_2SO_4 to decolorize.

MICROBIOLOGY

Genus *Nocardia* Trevisan 1889

Synonymy. *Proactinomyces* Jensen 1931; *Asteroides* Puroni and Leonardi 1935.

Nocardia asteroides (Eppinger 1891) Blanchard 1895

Synonymy. *Cladothrix asteroides* Eppinger 1891; *Streptothrix carnea* Rossi-Doria 1891; *Streptothrix freeri* Musgrave and Clegg 1907; *Actinomyces gypsoides* Henrici and Gardner 1921; *Discomyces asteroides* Gedoelst 1902; *Oospora asteroides* Sauvageau and Radais 1892; *Actinomyces asteroides* Gasparini 1894; *Actinomyces asteroides* var. *serratus* Sartory, Meyer, and Meyer 1930; *Proactinomyces asteroides* var. *crateriforme* Baldacci 1938; *Proactinomyces asteroides* var. *decolor* Baldacci 1938.

Colony Morphology. Sabouraud's dextrose agar (SDA) 25°C or 37°C. The young colony is glabrous, folded, wrinkled, or granular, sometimes mycobacterialike. The color is commonly orange but varies from yellow to pink to brown. In time, a fuzzy or chalky overgrowth of aerial mycelium occurs. A damp, soil-like odor may occur in time.

Microscopic Morphology. Delicate, sinuous, irregularly branching mycelium is found in culture. The diameter averages 1 μ. In time, the mycelian strands fragment unevenly to form bacillary or coccoid bodies. The aerial mycelium also fragments, but more regularly, and forms a series of squarish conidia or spores. Branching is less common and the mycelian strands are longer in tissue and culture as compared to *Actinomyces* species. It is gram-positive and irregularly acid-fast.

Physiologic Characteristics. Milk not coagulated, casein hydrolysis 0, tyrosine hydrolysis 0, xanthine hydrolysis 0, growth in 0.4 per cent gelatin 0, thermotolerant to 45°C for growth.[2a]

Nocardia brasiliensis (Lindenberg 1909) Castellani and Chalmers 1913

Synonymy. *Discomyces brasiliensis* Lindenberg 1909; *Streptothrix brasiliensis* Grecco 1916; *Actinomyces mexicanus* Boyd and Crutchfield 1921; *Oospora brasiliensis* Sartory 1923; *Actinomyces brasiliensis* Gomes 1923; *Nocardia pretoria* Pijper and Pullinger 1927; *Nocardia transvalensis* Pijper and Pullinger 1927; *Nocardia mexicanus* Ota 1928; *Actinomyces transvalensis* Nannizzi 1934; *Actinomyces pretoria* Nannizzi 1934; *Proactinomyces brasiliensis* Negroni 1954; *Proactinomyces mexicanus*. *Proactinomyces transvalensis*, and *Proactinomyces pretoria* Negroni 1954.

Colony morphology and cultural characteristics are covered in Chapter 4, Mycetoma.

REFERENCES

1. Andriole, V. T., M. Ballas, et al. 1964. The association of nocardiosis and pulmonary alveolar proteinosis. Ann. Intern. Med., *60*:266–275.

1a. Aron, R., and W. Gordon. 1972. Pulmonary nocardiosis, case report and evaluation of current therapy. S. Afr. Med. J., *46*:29–32.

1b. Bach, M. C., L. D. Sabath, et al. 1973. Susceptibility

of *Nocardia asteroides* to 45 antimicrobial agents *in vitro.* Antimicrob. Agents Chemother., *3*:1–8.
2. Ballenger, C. N., Jr., and D. Goldring. 1957. Nocardiosis in childhood. J. Pediatr., *50*:145–169.
2a. Berd, D. 1973. *Nocardia asteroides.* A taxonomic study with clinical correlations. Am. Rev. Resp. Dis., *108*:909–917.
3. Black, W. A., and D. A. McNellis. 1970. Sensitivity of Nocardia to trimethoprim and sulphonamides *in vitro.* J. Clin. Pathol., *23*:423–426.
4. Black, W. A., and D. A. McNellis. 1970. Susceptibility of Nocardia species to modern antimicrobial agents. Antimicrob. Agents Chemother., pp. 346–349.
5. Cohen, M. L., E. B. Weiss, et al. 1971. Successful treatment of *Pneumocystis carinii* and *Nocardia asteroides* in a renal transplant patient. Am. J. Med., *50*:269–276.
6. Drake, C. H., and A. T. Henrici. 1943. *Nocardia asteroides,* its pathogenicity and allergic properties. Am. Rev. Tuberc., *48*:184–198.
7. Evans, R. A., and R. E. Benson. 1971. Complicated Nocardiosis treated with trimethoprim and sulfamethoxazole. Med. J. Aust., *58*:684–685.
8. Georg, L. K., L. Ajello, et al. 1961. The identification of *Nocardia asteroides* and *Nocardia brasiliensis.* Am. Rev. Resp. Dis., *84*:337–347.
9. Gonzalez-Mendoza, A., and F. Mariat. 1964. Sur-l'hydrolyse de la gelatine comme caractere differential entre *Nocardia asteroides* and *Nocardia brasiliensis.* Ann. Inst. Pasteur (Paris), *107*:560–564.
10. González Ochoa, A. 1973. Virulence of Nocardiae. Can. J. Microbiol., *19*:901–904.
11. Goodfellow, M. 1971. Numerical taxonomy of the genus Nocardioform bacteria. J. Gen. Microbiol., *69*:33–80.
12. Gordon, M. A. 1970. Aerobic pathogenic actinomycetes, In *Manual of Clinical Microbiology.* Eds., J. E. Blair, E. H. Lennette, and J. P. Truant. American Society for Microbiol., Bethesda, Md.
13. Gordon, R. E., and W. A. Hagen. 1936. A study of some acid-fast actinomycetes from soil with special reference to pathogenicity for animals. J. Infect. Dis., *59*:200–206.
14. Gordon, R. E., and J. M. Mihm. 1962. The type species of the genus Nocardia. J. Gen. Microbiol., *27*:1–10.
15. Hoeprich, P. D., D. Brandt, et al. 1968. Nocardial brain abscess cured with cycloserine and sulfonamides. Am. J. Med. Sci., *255*:208–216.
16. Hosty, T. S., C. McDurmont, et al. 1961. Prevalence of *Nocardia asteroides* in sputa examined by a tuberculosis diagnostic laboratory. J. Lab. Clin. Med., *99*:90–93.
17. Kingsbury, E. W., and J. M. Slack. 1969. A polypeptide skin test antigen from *Nocardia asteroides.* II. Sabouraudia, 7:85–89.
18. Kirby, W. M. M., and J. B. McNaught. 1946. Actinomycosis due to *Nocardia asteroides.* Arch. Intern. Med., *78*:578–591.
19. Kurup, P. V., H. S. Randhawa, et al. 1970. Nocardiosis: a review. Mycopathology, *40*:194–219.
20. Kurup, P. V., H. S. Randhawa, et al. 1970. Pathogenicity of *Nocardia caviae, N. Asteroides* and *N. brasiliensis.* Mycopathology, *40*:113–130.
20a. Lerner, P. I., and Baum, G. 1973. Antimicrobial susceptibility of *Nocardia* species. Antimicrob. Agents Chemother., *4*:85–93.
21. Mackinnon, J. E., and R. Artagaveytia-Allende. 1956. The main species of pathogenic aerobic actinomycetes causing mycetomas. Trans. R. Soc. Trop. Hyg., *50*:31–40.
22. Macotela-Ruiz, E., and F. Mariat. 1963. Sur la production de mycetoma experimentaux par *Nocardia brasiliensis* et *Nocardia asteroides.* Bull. Soc. Pathol. Exot., *56*:46–54.
23. Mason, K. N., and B. M. Hathaway. 1969. A study of *Nocardia asteroides.* Arch. Pathol., *87*:389–392.
24. Murray, J. F., S. M. Finegold, et al. 1961. The changing spectrum of nocardiosis. A review and presentation of nine cases. Am. Rev. Resp. Dis., *83*:315–330.
25. Musrafa, I. E. 1967. Studies of bovine farcy in the Sudan. I. Pathology of the disease. II. Mycology of the disease. J. Comp. Pathol. Ther., *77*:223–229, 231–236.
26. Nocard, E. 1888. Note sur la maladie des boeufs de la Guadaloupe connue sous le nom de farcin. Ann. Inst. Pasteur (Paris), *2*:293–302.
27. Peabody, J. W., and J. H. Seabury. 1960. Actinomycosis and nocardiosis. A review of basic differences in therapy. Am. J. Med., *28*:99–115.
28. Pier, A. C., and J. B. Enright. 1962. *Nocardia asteroides* as a mammary pathogen of cattle. III. Immunological reaction of infected animals. Am. J. Vet. Res., *23*:284–292.
29. Raich, R. A., F. Casey, et al. 1961. Pulmonary and cutaneous nocardiosis. Am. Rev. Resp. Dis., *83*:505–509.
29a. Rankin, R. S., and J. S. Westcott. 1973. Superior vena cava syndrome caused by *Nocardia* mediastinitis. Am. Rev. Resp. Dis., *108*:361–363.
30. Robboy, S. J., and A. L. Vickery. 1970. Tinctorial and morphologic properties distinguishing actinomycosis and nocardiosis. New Engl. J. Med., *282*:593–596.
31. Saltzman, H. A., E. W. Chick, et al. 1962. Nocardiosis as a complication of other diseases. Lab. Invest., *11*:1110–1117.
32. Simonella, P., and N. R. Brizioli. 1968. Relazioni alla tubercolina nei bovinin portatori di *Nocardia dassonvillei.* Atti. Soc. Ital. Sci. Vet., *22*:918–923.
33. Snieszko, S. F., G. L. Bullock, et al. 1964. Nocardial infection in hatchery fingering rainbow trout. J. Bacteriol., *88*:1809–1810.
34. Strauss, R. E., A. M. Kligman, et al. 1951. The chemotherapy of actinomycosis and nocardiosis. Arch. Dermatol., *98*:489–493.
35. Strauss, R. E., and A. M. Kligman. 1951. The use of gastric mucin to lower resistance of laboratory animals to systemic fungus infection. J. Infect. Dis., *58*:151–155.
36. Stropnik, Z. 1965. Isolation of *Nocardia asteroides* from human skin. Sabouraudia, *4*:41–44.
37. Utz, J. P., and A. Treger. 1959. The current status of chemotherapy of systemic fungal disease. Ann. Intern. Med., *51*:1220–1229.
38. Vasarinish, P. 1968. Primary cutaneous nocardiosis. Arch. Derm., *98*:489–493.
39. Young, L. S., D. Armstrong, et al. 1971. *Nocardia asteroides* infection complicating neoplastic disease. Am. J. Med., *50*:356–367.

Chapter 3

OTHER ACTINOMYCETOUS INFECTIONS

Included in this category are some minor infections caused by actinomycetes and diphtheroid organisms. Though some are common infections, they usually cause only slight discomfort to the patient. They are generally brought to the attention of the physician or dermatologist out of curiosity or concern by the patient about nodose axillary or pubic hair, discolored skin, discolored sweat, or plantar fissures. Rarely serious, they may be considered as nuisance curiosities.

Traditions die hard in science. Actinomycetes were once considered to be the link between bacteria and fungi. As previously discussed, it is now established that they are bacteria. Similarly, in the diseases included here, an actinomycete, generally *Nocardia,* was described as the etiologic agent for each one. More recent studies define the organisms as Corynebacterium, not actinomycetes. For convenience, however, they will be grouped together as actinomycetous infections.

Erythrasma

DEFINITION

Erythrasma (dhobie) is a chronic, mild, localized infection of the stratum corneum. It usually remains localized, most commonly involving the axilla or crural areas, but it may also involve the body folds and clefts and intertriginous areas. It is characterized by brownish-red, punctate, or palmate glistening lesions and dry, smooth, finely creased skin. The etiologic agent, *Corynebacterium minutissimum,* is apparently a common resident of the skin, particularly the toe webs. The condition is readily amenable to systemic antibacterial antibiotics, but recurrence is frequent.

HISTORY

This disease was first described by Buchardt in 1859, and the term "erythrasma" was used by Bärensprung in 1862. Skin scales examined by these authors showed delicate filaments which were believed to be of fungal origin. They named the organism *Microsporum minutissimum.* The contagious nature of the disease was demonstrated by Köbner, who in 1884 was able to transmit the disease by rubbing scales from a patient on the skin of a pupil. Early workers did not accept the disease as a separate entity, and it was thought to be either a variant of pityriasis versicolor, a dermatophyte infection, or eczema marginatum. There was an early association of the disease with hyperhidrosis, and Poehlman in 1928 stressed delicate skin, humidity, site, and secretions as factors predisposing to the disease. Though the disease is caused by a delicate gram-positive rod, it may be complicated by fungal or other bacterial infections, as emphasized by Gougerot in 1936.

ETIOLOGY, ECOLOGY, AND DISTRIBUTION

Though it has received various fungal and bacterial names, and as recently as 1958 was reported by Kalkoff to be gram-negative, the etiologic agent of erythrasma has been identified as gram-positive, filamentous, diphtheroid *Corynebacterium minutissimum*. As a low-grade infection, erythrasma appears to be found throughout the world. The most common site of occurrence is the genital folds, where the mild form of the disease has been found in 4 to 20 per cent of populations examined. The patients are usually unaware of the infection. The incidence is higher among males in warm, humid climates. Another usually symptomless site is the toe web. Up to 25 per cent of students examined in the United States and England[17] demonstrated the presence of the organism. Again, most of the patients were without symptoms. However, some had fissuring and maceration along with other bacteria or a tinea pedis. Actual clinical disease is more common in adult males. This is the classic genitocrural type, and there may be a long history of fluctuating severity. A second, generalized type is predominantly present in middle-aged Negro women.

CLINICAL DISEASE

The lesions are punctate to palm-sized, well-circumscribed, maculopapular rashes. The color varies from pink to reddish to reddish-brown, depending on the age of the lesion. The involved area is glistening or greasy-looking and sometimes covered with small furfuraceous scales. The surface of older lesions may show fine creases. The scaly nature of the disease may require scratching to be revealed. The advancing border of the lesion is serpiginous and erythematous. There is no tendency for it to become elevated or vascularized, and lesions may remain symptomless, except for mild irritation and lichenification.

Erythrasma is most commonly found in areas of the groin and upper thigh in contact with the scrotum. Although both thighs may be involved, the left is more often infected because of more contact with the scrotum. Other common sites are the pubes, the axillae, the intergluteal folds, and under the breasts. Hair in the affected areas remains uninvolved. Occlusion and increased humidity tend to exacerbate mild disease.

In the toe web, there may be some scaling, fissuring, and maceration. In such cases, there is usually a mixed flora of other bacteria also present, including *Staphylococcus*, *Pseudomonas*, and *Proteus*. The disease may coexist with dermatophytosis. Infection is usually between the fourth and fifth, and third and fourth toes. Various studies[15,17] have demonstrated that a mild subclinical form of the disease is very common in the population. Surveys are very easy to perform, taking advantage of the coral-red fluorescence of the lesion when seen under the Wood's lamp. By this method, as many as 25 per cent of the people surveyed were demonstrated to harbor organisms. However, they were usually symptom-free.

Much less commonly encountered than either of the preceding is the generalized form of erythrasma, which is found predominantly among middle-aged Negro women. There are well-defined, scaly, lamellar plaques on large areas of the trunk, proximal parts of the limbs, and the folds of the breasts (Fig. 3–1 FS A 9). Again, this condition is more common in hot, humid climates and is referred to as tropical erythrasma. Usually, there is a long history of chronicity, itching, and lichenification.

Chronic erythrasma that has lichenified may be mistaken for neurodermatitis or for tinea cruris. In contrast to dermatophytosis, however, in erythrasma there is little inflammation, no vesiculation, and no satellite lesions. In the obese patient there is replacement of the dry, scaly lesion with a shiny, smooth, red, wet intertrigo, which may confuse the picture. In the typical form of the disease in other areas, erythrasma is most commonly confused with pityriasis versicolor. The latter lacks the erythematous border and shows no tendency to localize in the body folds. A simple KOH mount readily differentiates the two. At some stages, tinea corporis, eczema, lichen simplex, contact dermatitis, and psoriasis may simulate erythrasma.

PROGNOSIS AND TREATMENT

Since the disease is of minor importance to the patient, many chronic cases are seen.

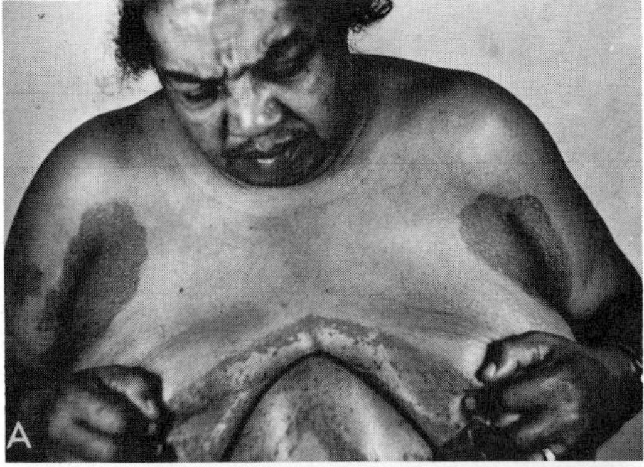

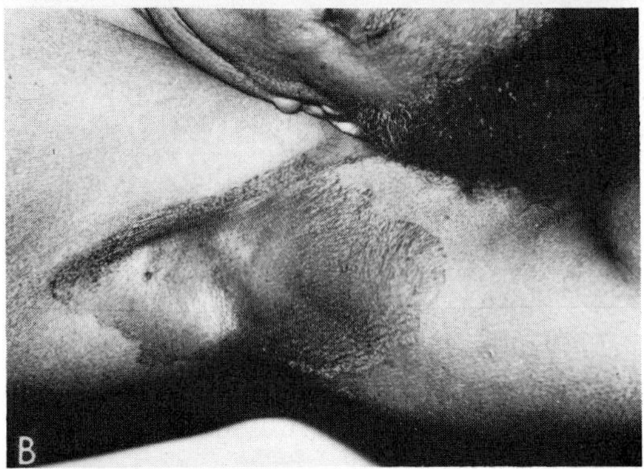

Figure 3–1. FS A 9,*A*. Erythrasma. *A*, Extensive lesions affecting the supramammary folds and the axilla. Lesions were also present on the trunk and legs. *B*, Well-defined, scaly, lamellar plaques in the axilla.

The disease persists indefinitely except for occasional fluctuations in severity. In spite of local or systemic treatment, relapse is common. Local treatments include sodium hyposulfite (20 per cent) or 3 per cent sulfur ointment and the usual keratolytics, such as Whitfield's ointment. Systemic antibiotics, especially erythromycin, in doses of 1 g per day for five days, have evoked complete clinical cure.[15] Other effective antibiotics include tetracycline and Chloromycetin, but erythromycin appears to be the drug of choice. Penicillin and griseofulvin are without effect. Toe web infection is particularly resistant to treatment.

Experimental or natural disease in animals has not been reported.

LABORATORY DIAGNOSIS

Direct examination of scales from erythrasma in KOH mounts under oil will show small coccoid forms, rodlike organisms, and long filaments, 5 to 25 μ in length. These are seen with great difficulty by this procedure, and a much more effective method is to use transparent tape strippings from the area. The tape is pressed on the lesion and removed, carrying with it scales on the sticky side. If cellulose tape is used, a quick rinse with methylene blue can be used before the tape is pressed on a slide and examined by oil immersion. Vinyl tape allows a Gram or Giemsa stain to be done, revealing the organisms more clearly. The coccal forms are 1 μ in diameter, the bacilli 1 to 3 μ, and the filaments average 4 to 7 μ in length.

The Wood's light is of great diagnostic aid in erythrasma. A characteristic coral-red fluorescence can be seen in the active borders and in patches on old lesions (Fig. 3–2 FS A 10,*A*). In contrast, pityriasis versicolor will show some areas of a gold fluorescence. Ulcerated squamous cell carcinoma and carcinoma of the palate have shown some red fluorescence under the Wood's lamp, as have experimental carcinomas of rabbit skin.[17] The fluorescent substance is also produced by the etiologic agent of erythrasma in culture. It appears to be a porphyrin.

CULTURE METHODS

The organism grows readily on a medium containing 20 per cent fetal calf serum and 2 per cent agar in tissue culture media No. 199. The colonies appear in two days, reaching a size of 2 to 3 mm. They are moist, translucent, convex, and nonhemolytic and, while young, will show the characteristic coral-red fluorescence. Gram stains of colonies show typical diphtheroid dimensions and pleomorphism. Culture is not necessary for diagnosis.

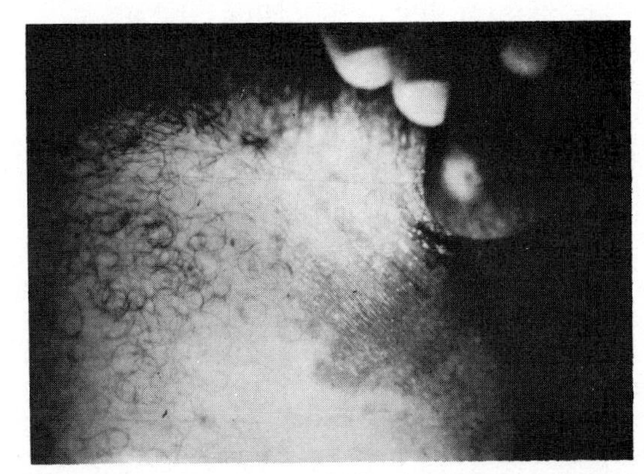

Figure 3–2. FS A 10,*A*. Erythrasma of groin. Wood's lamp demonstration of coral red fluorescence of lesions. (Courtesy of D. Taplin.)

MICROBIOLOGY

***Corynebacterium minutissimum* (Buchardt) Sarkany, Taplin, and Blank 1961**

Synonymy. *Nocardia minutissima* Verdun, 1912; *Microsporon minutissimum* Buchardt 1859; *Sporotrichum minutissimum* Saccardo 1886; *Microsporoides minutissimus* Neveu-Lemaire 1906; *Discomyces minutissimus* Verdun 1907; *Oospora minutissima* Ridet 1911; *Actinomyces minutissimus* Brumpt 1927; *Leptothrix epidermidis; Microsporon gracilis.*

Trichomycosis Axillaris

DEFINITION AND ETIOLOGY

This is a superficial infection of the axillary or pubic hair, characterized by the formation of yellow (flava), red (rubra), or black (nigra) nodules around the hair shaft (Fig. 3–3). The organism involved has been isolated on occasion and named *Corynebacterium tenuis.* The studies of Crissey[2] indicate that the disease is prevalent in the temperate climate but is probably more widespread in the tropics where heat and moisture favor growth of the organism.

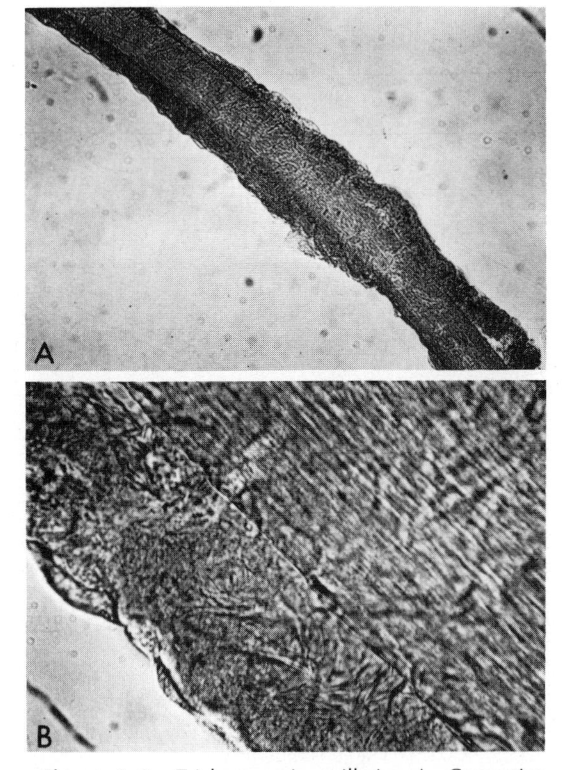

Figure 3–3. Trichomycosis axillaris. *A,* Concretion surrounding hair shaft. ×100. *B,* Higher magnification of concretion. There is no destruction of the hair itself. ×400.

CLINICAL DISEASE

Patients with this condition are usually unaware of it, and the presenting complaint is of discolored, sweat-stained clothes. The concretion is most commonly yellow, least commonly black, and consists of firmly packed coccoid and bacillary forms in a mucinous mass.[11] The consistency is usually firm but may be soft and jellylike (zooglea). The affected hair appears lusterless and brittle and is easily broken. The infection does not extend to the hair base or root, thus differentiating it from ringworm. The flava variety resembles white piedra, but microscopic examination quickly differentiates them. Also included in the differential diagnosis are pediculosis, monilethrix (beaded hair), and trichorrhexis nodosa. Treatment includes depilation and daily application of formalin (2 per cent), bichloride of mercury (1 per cent), or sulfur (3 per cent). Recurrence is common.

LABORATORY DIAGNOSIS

KOH mounts of the hairs will show the nodules to be composed of delicate, short bacilli and coccoidlike diphtheroid organisms enmeshed in mucilaginous material.[11] The flava variety may show a distinct golden fluorescence. The organism has been cultured on enriched media. Recent work by McBride[9] has shown that the organisms readily colonize hair in broth cultures. They then concentrated colored salts, giving the nodules a red or black color. She suggests that there is only one species involved and that it has a unique ability to concentrate colored substances, apparently coming from the patient, giving rise to the variously colored nodules.

MICROBIOLOGY

***Corynebacterium tenuis* (Castellani) Crissey et al. 1952**

Synonymy. *Nocardia tenuis* Castellani 1912; *Discomyces tenuis* Castellani, 1912; *Cohnistreptothrix tenuis* Ota 1928; *Actinomyces tenuis* Dodge 1955.

Pitted Keratolysis

DEFINITION AND ETIOLOGY

Pitted keratolysis, cracked heel, or keratolysis plantare sulcatum is a superficial infection of the stratum corneum, characterized by circular areas of erosion on the plantar surfaces. The condition is much more prevalent than formerly recognized.[8,18] Although the disease is usually asymptomatic, under conditions of heat and dampness it may lead to a severe and debilitating anhidrotic syndrome, with thinning and erosion of the skin. The disease is caused by an organism variously classified as *Actinomyces, Nocardia, Micromonospora, Streptomyces,* or *Corynebacterium.* The latter is probably the true agent, but it has not yet been published as such. The agent readily dissolves keratin and, in histologic section, can be found in the floor of the pits. It appears as coccoid bodies and filaments with diphtheroid morphology.

The disease was first described in India in 1910 by Castellani,[1a] and was largely ignored. The recent studies of Zaias et al.[18] indicate it to be common in the normal population and especially in military personnel assigned to the tropics. Gill and Buckels described the condition in detail.[4] The symptomatic form was described by Lamberg.[8] He found that extreme humidity, intense heat, and poor sanitation exaggerated the disease and produced disability among military personnel, who sometimes required hospitalization. He suggested that prolonged wearing of boots and incubator-like conditions of tropical warfare were responsible for the exacerbation of disease.

CLINICAL FEATURES

The horny layer of the soles, toes, and heels shows numerous areas of erosion (Fig. 3–4 FS A 10,*B*). These are discrete, circular patches that later coalesce to form large areas of denuded stratum corneum. The edges may be discolored with a greenish or brownish outline. Fissures may appear within these areas and lead to secondary bacterial complications with an accompanying foul odor. Hyperhidrosis is noted. In the symptomatic form in which thinned, reddened areas are seen, anhidrosis occurs within the eroded lesions.[8] If the feet are soaked, the accompanying swelling of the horny layer clearly outlines the lesions.

Histologic examination shows a mild inflammatory reaction in the dermis without thickening, and an almost complete absence of the stratum corneum. On methenamine stain or modified Gram stain, filaments may be seen in the walls and floor of the pits. A similar disease, ulcus interdigitale, may occur on the toes.

PROGNOSIS AND THERAPY

The condition, when mild, resolves quickly with adequate personal hygiene and relief

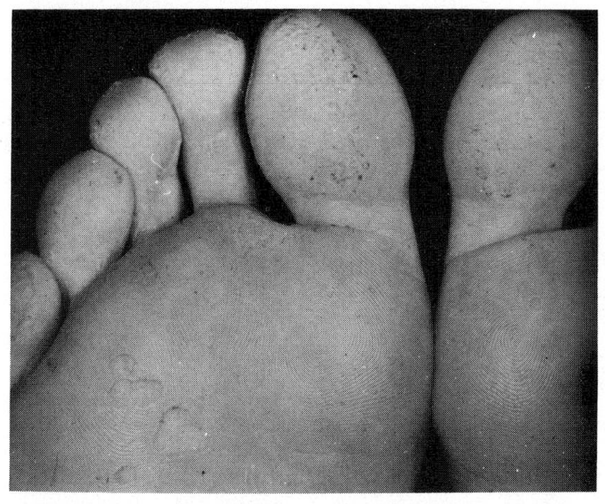

Figure 3–4. FS A 10,*B*. Pitted keratolysis. Plantar surface of toes, showing numerous areas of erosion. (Courtesy of S. Lamberg.)

of the predisposing environment of heat and humidity. For symptomatic disease, effective therapy has been achieved with 20 to 40 per cent formalin in Aquaphor.[7] The normal sweating pattern returns within three days with bed rest, followed by relief of pain and clearing of the erythema.

LABORATORY IDENTIFICATION

KOH mounts of infected scales will show filaments and coccoid forms resembling actinomycetes or diphtheroid organisms. Culture of the agent is difficult and, as yet, controversial. Taplin[16] has isolated a *Corynebacterium* sp. from the lesions, with which he has been able to reproduce the disease on a human volunteer, i.e., Mr. Taplin. However, a published account of the organism has not been made, so that the older designation of Acton and McGuire remains valid, though questionable.

Dermatophilus pedis?

Actinomyces keratolytica **Acton and McGuire 1930**[1]

Dermatophilosis

DEFINITION AND ETIOLOGY

Dermatophilosis (streptotrichosis, epidemic eczema, strawberry foot rot, contagious dermatitis) is a pustular, exudative dermatitis of animals and, rarely, man. The initial lesions are followed by the formation of scales and crusts which heal and fall off, leaving alopecia and scarred areas on the hide. In endemic regions, it may be highly contagious, and epidemics cause severe economic loss among domesticated animals. The etiologic agent is an actinomycete of the Dermatophilaceae, *Dermatophilus congolensis*. The disease has been reported in cattle and sheep primarily, but it also occurs in goats, horses, antelopes, deer, zebras, swine, giraffes, foxes, Colombian ground squirrels, domestic cats, and man.[10] Since the etiologic agent is a *Dermatophilus*, the name for the disease, dermatophilosis, is preferable to other suggested or historical designations.

HISTORY

This disease was first described by Van Saceghem in 1915[17a] as a skin disease in cattle in the Belgian Congo. This work was largely ignored, and a similar disease in Australia received the name "lumpy wool" from Bull in 1929. The literature is confused by reports from various parts of the world, each using a different name, e.g., strawberry foot rot of sheep in Scotland (*Polysepta pedis*), reported by Thompson. The disease was first recognized in the United States in 1961.[3] Three separate outbreaks occurred within that year. It was found in calves from Texas, in horses from New England and New York, and in deer from New York. In the latter instance, the disease was transmitted to man by handling of the infected animals.[3,7] Since then, it has appeared in Iowa, Kansas, and Georgia. It is now considered to be of worldwide distribution. Many names have appeared for the various isolates, but Gordon[5] concludes they are all the same species.

ECOLOGY

Roberts[13] believes the organism is a natural parasite of the epidermis of sheep. Though not recovered from soil, the organism can remain viable in dried crusts for long periods. It dies rapidly if exposed to moisture and probably does not occur in nature as a saprophyte. It is known to be transmitted by direct contact under herd conditions, by flies, and probably by ectoparasites. Epidemics occur when the skin has been damaged by prolonged wetting. The macerated epidermis is less resistant to invasion, and the moisture facilitates release of the infective, motile zoospores and their penetration of susceptible epidermis.

CLINICAL DISEASE

The organism is a parasite of the epidermis and does not invade vascularized tissue. The initial symptom in animals is a pustular dermatitis, followed by formation of scabs and crusts. The hair is matted with exudate (lumpy wool). Healing occurs and the crusts fall off, leaving patches of alopecia. Lesions vary in size and may become confluent. There is some variation of the disease among horses, cattle, and sheep, but the general picture is similar. In man, as reported by Dean,[3] four individuals contracted the disease by handling an infected deer. The onset was characterized by the appearance of multiple, nonpainful pustules on the hands within a week after exposure. The pustules were from 2 to 5 mm in size and contained a white to yellowish serous exudate. A small, shallow, reddish ulcer remained after expression of the fluid. This was followed by formation of a brownish scab that persisted for several days. The lesions healed spontaneously, leaving a purplish-red scar. There was no systemic illness.

Experimental disease has been produced in rabbits.[6] Mice and guinea pigs appear to be resistant.

TREATMENT

In animals, various topical medicaments have been used with irregular results. These include dips of copper sulfate (1:500), 0.5 per cent zinc sulfate, and 1:5000 mercuric chloride or a dip containing arsenic. The organism is susceptible to the usual sulfonamides and antibiotics, but trials have been unrewarding. A long course of penicillin has been effective, but a single dose, regardless of amount, has been unsuccessful.

LABORATORY IDENTIFICATION

Scales and aspirates, stained by methylene blue or Giemsa stain, show branched filaments 2 to 5 μ in diameter, which are divided in longitudinal and transverse planes into packets of up to eight coccoid cells (Fig. 3–5). This picture is pathognomonic for the disease. Pier[12] has shown the immunofluorescent technique to be valuable for diagnosis. The organism is easily cultured from aspirates of unopened pustules at 37°C on blood agar without antibiotics. If there is much contamination, the infected scabs can be rubbed on shaved rabbit skin and cultures made from the resulting

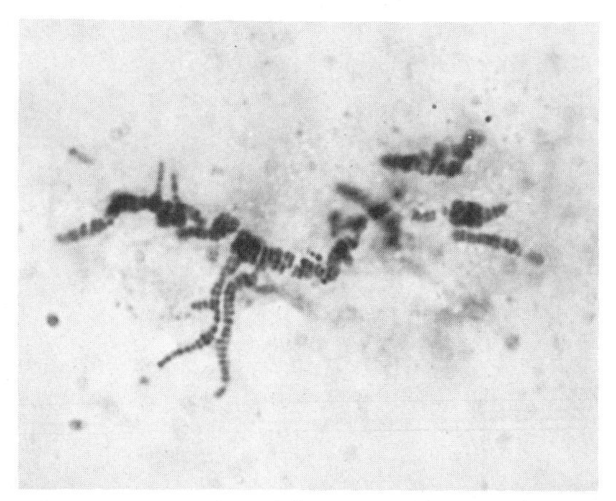

Figure 3–5. Dermatophilosis. Stained smear showing irregular, branched filaments. These are divided both longitudinally and transversely, forming packets of coccoid cells.

pustules after four days. Within 24 hours, small (0.5 to 1.0 mm) colonies appear on agar, which are white to yellow, round or irregular, with a depressed periphery. After five days, a yellow-to-orange pigment appears (Fig. 3–6). The organism is gram-positive, with multiple branching filaments 0.5 to 1.5 μ in diameter. They show the characteristic transverse and longitudinal septation into packets of coccoid spores. Under proper conditions, these hatch and become flagellated and motile. They are not acid-fast (Fig. 3–7).[12a]

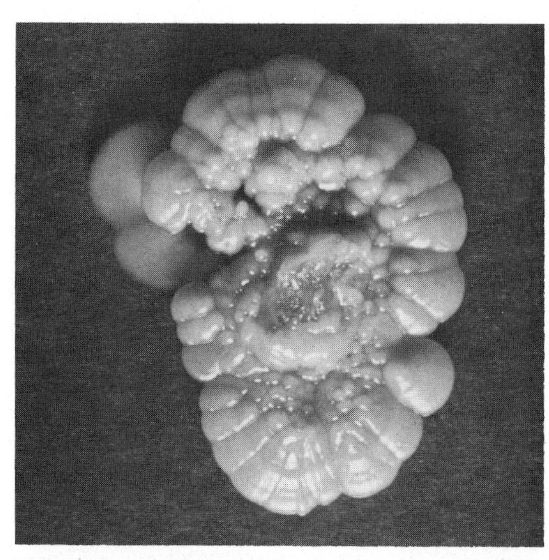

Figure 3–6. Dermatophilosis. Colony of *Dermatophilus congolensis* after five days' growth on brain-heart infusion of glucose blood agar.

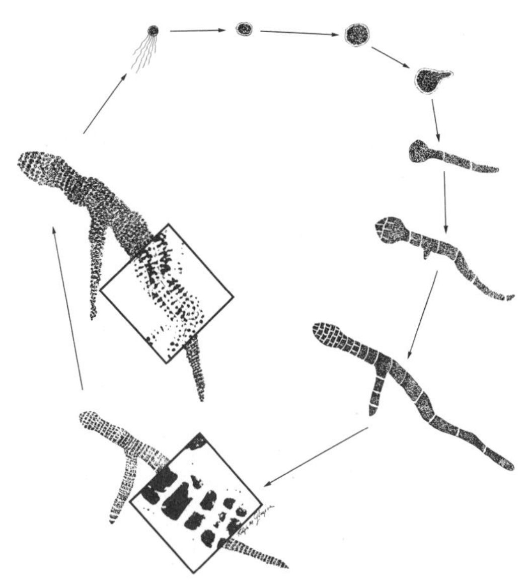

Figure 3–7. Dermatophilus life cycle. (Reprinted by permission from *Veterinary Bacteriology and Virology* by I. A. Merchant and R. A. Packer, 7th edition © 1967, by Iowa State University Press, Ames, Iowa.)

The organism is catalase-positive and urease-positive and is proteolytic for gelatin and casein. It hydrolyzes starch but not xanthine or tyrosine. In enriched broth, it will produce acid but no gas from glucose and fructose and will not ferment sucrose, lactose, xylose, mannitol, dulcitol, salicin, and sorbitol.[5] The cell wall contains *meso*-diaminopimelic acid but no arabinose or galactose.

MICROBIOLOGY

Dermatophilus congolensis Van Saceghem 1915 emend, 1916[17a]

Synonymy. *Actinomyces congolensis* Hudson 1917; *Streptothrix bovis* Snijders and Jensen 1955; *Dermatophilus dermatonomus* (Bull) Austwick 1958; *Nocardia dermatonomus* (Bull) Henry 1952; *Polysepta pedis* Thomson and Bisset 1957; *Dermatophilus pedis* (Thomson and Bisset) Austwick 1958.

REFERENCES

1. Acton, H. W., and C. Maguire. 1930. Keratolysis plantare sulcatum. Indian Med. Gaz., *65*:61–62.

1a. Castellani, A. 1910. Keratoma plantare sulcatum. J. Ceylon Br. Brit. Med. Ass., 7:10–11.

2. Crissey, J. T., G. C. Rebell, et al. 1952. Studies on the causative organism of trichomycosis. J. Invest. Dermatol., *19*:187–198.

3. Dean, D. J., M. A. Gordon, et al. 1961. Streptothricosis: a new zoomatic disease. N.Y. State J. Med., *61*:1283–1287.

4. Gill, K. A., and L. J. Buckels. 1968. Pitted keratolysis. Arch. Dermatol., *98*:7–11.

5. Gordon, M. A. 1964. The genus Dermatophilus. J. Bacteriol., *88*:509–522.

6. Gordon, M. A., and U. Perrin. 1971. Pathogenicity of *Dermatophilus* and *Geodermatophilus.* Infect. Immun., *4*:29–33.

7. Kaplan, W. 1966. Dermatophilosis-A. Recently recognized disease in the United States. Southwest. Vet., *20*:14–19.

8. Lamberg, S. 1969. Symptomatic pitted keratolysis. Arch. Dermatol., *100*:10–11.

9. McBride, M., W. Duncan, et al. 1970. The effects of selenium and tellurium compounds on pigmentation of granules of trichomycosis axillaris. Int. J. Dermatol., *9*:226–231.

10. McClure, H. M., W. Kaplan, et al. 1971. Dermatophilosis in owl monkeys. Sabouraudia, *9*: 185–190.

11. Montes, L. F., C. Vasquez, et al. 1963. Electron microscopic study of infected hair in trichomycosis axillaris. J. Invest. Dermatol., *40*: 273–278.

12. Pier, A. C., J. L. Richard, et al. 1964. Fluorescent antibody and cultural techniques in cutaneous streptothricosis. Am. J. Vet. Res., *25*:1014–1020.

12a. Roberts, D. S. 1961. The life cycle of *Dermatophilus dermatonomus,* the causal agent of ovine mycotic dermatitis. Aust. J. Exp. Biol. Med. Sci., *39*:463–476.

13. Roberts, D. S. 1965. Cutaneous actinomycosis due to the single species *Dermatophilus congolensis Nature (Lond.)*, *206*:1068.

14. Ronchese, F., B. S. Walker, et al. 1954. The reddish-orange fluorescence of necrotic cancerous surfaces under the Wood's light. *Arch. Dermatol.*, *69*:31–42.

15. Sarkany, I., D. Taplin, et al. 1961. The etiology and treatment of erythrasma. *J. Invest.*, *37*:283–290.

16. Taplin, D., and N. Zaias. 1967. Pitted keratolysis. *Report of the Thirteenth International Congress of Dermatology*, pp. 593–595.

17. Temple, D. E., and C. R. Boardman. 1962. The incidence of erythrasma of the toewebs. *Arch. Dermatol.*, *86*:518–519.

17a. Van Saceghem, R. 1916. Etude complementaire sur la dermatose des bovides (impetigo contagieux). Bull. Soc. Pathol. Exot., *10*:290–293.

18. Zaias, N., D. Taplin, et al. 1965. Pitted keratolysis. *Arch. Dermatol.*, *92*:151–154.

Chapter 4

MYCETOMA

DEFINITION

Mycetoma is a clinical syndrome of localized, indolent, deforming, swollen lesions and sinuses, involving cutaneous and subcutaneous tissues, fascia, and bone. It usually occurs on a foot or hand. The disease results from the traumatic implantation of soil organisms into the tissues. The lesions are composed of suppurating abscesses, granulomata, and draining sinuses, with the presence of "grains" which are microcolonies of the etiologic agents. The triad of tumefaction, draining sinuses, and grains is used in a restrictive sense to define the term "mycetoma." The etiologic agents are a wide variety of bacteria (actinomycotic mycetoma) and fungi (eumycotic mycetoma) from plant debris and soil. The involved organisms may also cause other clinical diseases, such as actinomycosis, mycotic granuloma, and chromomycosis, but only when the above criteria are met is the diagnosis of mycetoma valid.

Synonymy. Madura foot, maduromycosis.

HISTORY

The disease was first reported (as Madura foot) by McGill in 1842 from the Madura area of India, and it was confirmed by Colebrook in 1846. The term "mycetoma" was first used by Van Dyke Carter in 1860 to designate all tumors produced by fungi. Further descriptions by Kanthack in 1892 and Vincent in 1894 delineated the disease. The first cases were from India, and the old term "Madura foot" is reflected in the orthographic derivatives: maduromycosis for the disease, *Madurella* and *Indiella* for genera, and *Actinomadura madurae* as a specific etiologic agent. Brumpt in 1905 first stressed that several fungi were capable of eliciting the same clinical disease, and Langeran applied the term "mycetoma" to cases involving *Actinomyces* or *Nocardia* species and filamentous fungi. The division of the disease into two separate categories depending on the etiology was first suggested by Pinoy in 1913, redefined by Chalmers and Archibald in 1916, and again by Lavalle in 1962.[27] They used the terms "actinomycosis" and "maduromycosis" to differentiate the two groups of causative agents. The literature became confused with many ill-defined fungal and bacterial isolants, most of which have been reduced to synonymy with accepted species or discarded. As noted by Emmons, "mycetoma" came to be used as a general term for the presence of any fungal organism in any tissue that was not a specifically defined mycosis. In present usage, the term is applied only when the clinical syndrome described above is found. The two categories are "actinomycotic mycetoma" (actinomycetoma), when the agent is an actinomycete, and "eumycotic mycetoma" (eumycetoma), when a true fungus is involved.[55]

ETIOLOGY, ECOLOGY, AND DISTRIBUTION

The single clinical entity, mycetoma, may be evoked by infection with a diverse group of bacteria and fungi (Table 4–1). About 50 per cent of cases are due to actinomycetes and 50 per cent to true fungi. With the

Table 4–1. *Common Etiologic Agents of Mycetoma in Man and Animals*

Agent	*Grain Color*
Actinomycotic Mycetoma	
Actinomyces israelii	White to yellow
Nocardia asteroides	Grains rare—white
Nocardia brasiliensis	White
Nocardia caviae	White to yellow
Nocardia farcinica	White to yellow*
Actinomadura madurae	White to yellow or pink
Actinomadura pellitieri	Garnet red
Streptomyces somaliensis	Yellow to brown
Streptomyces paraguayensis	Black**
Eumycotic Mycetoma	
Petriellidium (Allescheria) boydii	White
Madurella grisea	Black†
Madurella mycetomii	Black
Acremonium kiliense	White
Leptosphaeria senegalensis	Black
Phialophora jeanselmei	Black
Neotestudina rosatii	White
Pyrenochaeta romeroi	Black†
Curvularia geniculata	Black‡
Drechslera spiciferum	Black‡

*Some authors believe that this organism is within the species limits of *N. asteroides*.

**There is doubt as to the authenticity of this species.

†It is believed that the two species are related, if not identical.

‡From animals.

exception of *Actinomyces israelii*, all are soil saprophytes or plant pathogens which reside on vegetable debris, thorns, sticks, and so forth, and gain entrance to the dermis through abrasion or implantation. In highly endemic regions, there is reason to believe that such factors as continued environmental exposure, inadequate nutrition and hygiene, and general health may play a role in determining the incidence of the disease. Though certain agents show some geographic and ecologic prevalence, most organisms implicated have been found in widely separated areas of the world, and it is probable that they are essentially of worldwide distribution. This again emphasizes occupational and health determinants as predisposing factors for infection.

As in the case with *Sporothrix schenckii*, there is some evidence of inherent differences between soil isolants and their ability to evoke infection. Most strains of *Phialophora jeanselmei* isolated from soil and sewage were unable to grow above 30°C,[49] whereas human isolants readily grew at elevated temperature. Temperature tolerance would be a prime factor for determining potential pathogenicity. As is probable in other mycoses, man acts as a selective agent for strains which can adapt readily to the internal environment and defenses of living tissue.

In some parts of the world, mycetoma occurs commonly. In his classic paper, Abbott[1] noted the admission of 1231 patients with this disease to hospitals in Sudan during two and a half years. This did not include outpatient visits or those patients (particularly women) who for many reasons, including shyness, did not seek medical attention. In 1964 Lynch[29] described an additional several hundred cases and gave a conservative estimate of 300 to 400 new cases per year in that country alone. Several French authors[14,34,48,49] have also emphasized the high endemicity of mycetoma in the equatorial Trans-African Belt extending west from Senegal, Mauritania, Chad, and Nigeria through Algeria and Sudan to the Somali coast. The commonest agents in this region are *Madurella mycetomii* (black grain), *Streptomyces somaliensis* (yellow grain), and *Actinomadura pelletieri* (red grain). A few cases caused by other species are also found. Some agents have been recovered from the thorns, soil and debris of the region.[49]

A second area of high endemicity is Mexico. Although the case rate does not approach that of the African states, mycetoma is a significant disease there. As noted by Lavelle,[27] there are interesting environmental parallels in the two regions. Both lie between the 14° and 33° north latitudes, are transected by the Tropic of Cancer, and have similar climatic conditions. The highly endemic regions of Mexico and Sudan have a rainy season extending from June to October, a dry, cool season from October to March, and hot and dry weather without rainfall from March to June. The amount of precipitation also coincides, varying from 50 to 500 mm. Again, the incidence of disease is highest in rural areas where exposure to the soil, occupation, health, and habits may be contributing factors. The predominant etiologic agents in Mexico differ from those in Sudan. *Nocardia brasiliensis* and *Actinomadura madurae* are most common in the former country. The original region of high incidence, the Madura area of India, also shares many of these climatic conditions.

Other areas of significant incidence include the Mediterranean region, Greece,

Italy, Rumania,[5] Guatemala, the Caribbean Islands, Cambodia, Iran, and areas of South America.[11] Cases are sporadically seen in Europe and the United States.[10] In the latter areas, *Petriellidium (Allescheria) boydii* is the most common agent. Scattered case reports have come from all regions of the world, and the various etiologic agents which are very common in one region are rarely reported from other areas.[5a] It appears, then, that the organisms are universally present and await the proper circumstances for expression.

Most cases are seen in males from rural areas. Although the disease is found in all age groups, clinical attention is generally sought only after many years of development, the decades from 30 to 50 being the most frequently recorded for the disease. All races are susceptible.

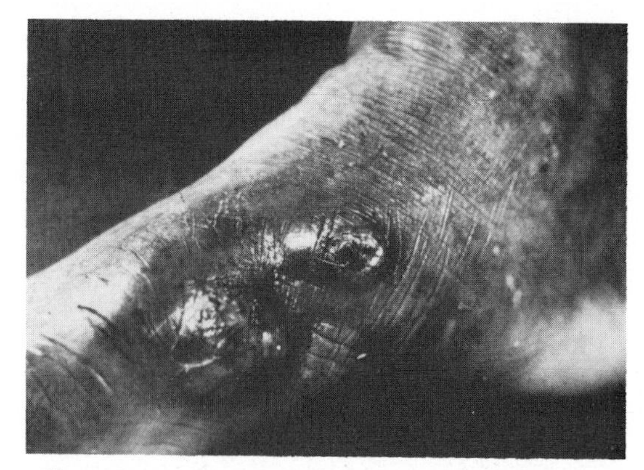

Figure 4–1. FS A 11,*A*. Mycotic mycetoma. Tumorlike process of the foot that has softened and ruptured. There are several draining sinus tracts. *Petriellidium (Allescheria) boydii.* (Courtesy of S. McMillen.)

CLINICAL DISEASE

Symptomatology. Regardless of the etiologic agent, the clinical disease is essentially the same. Variations in the syndrome do occur and are related to the various etiologies, the anatomic site, the duration of the lesion, previous medical management, and the general health of the patient. There is some evidence that, after initial implantation, the organism may remain quiescent for some time, possibly requiring another insult to the area or time to adapt to the host. Almost all case histories record previous injury to the involved region, and noticeable symptoms occur after a lapse of several years. The primary lesion is a locally invasive, indolent, tumorlike process or a small, painless, subcutaneous swelling which slowly enlarges and softens to become phlegmonous (Fig. 4–1 FS A 11,*A*). It ruptures to the surface, forming sinus tracts, then burrows into the deeper tissues, producing swelling and distortion of the foot. The sole often has a convex rather than flat surface (Fig. 4–2 FS A 11,*B, C*). The foot becomes increasingly swollen and distorted, and may be painful. The surface is studded with numerous small eminences, each containing the orifice of a sinus (Fig. 4–3 FS A 12). The organism invades subcutaneous connective tissue, bone (in some types), and ligaments, but tendons, muscles, and nerve tissue are usually spared. Some actinomycetes, however, regularly penetrate muscle tissue. The burrowing follows the fascial planes as in actinomycosis. The abscesses suppurate and drain through multiple sinus tracts. The tracts may remain open for long periods of time, heal, and then reopen. The discharge is a serous exudate, or it may be serosanguinous, seropurulent

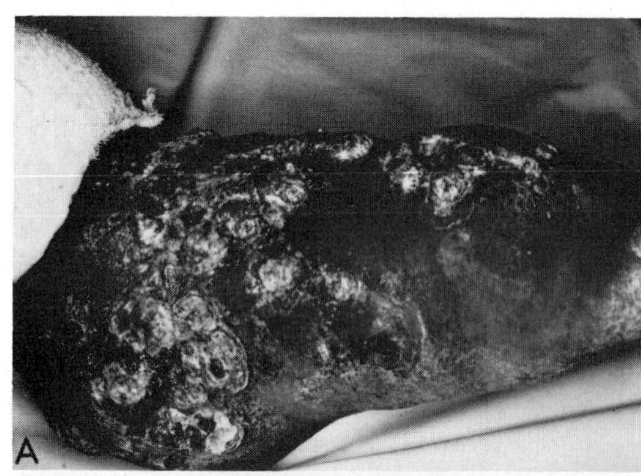

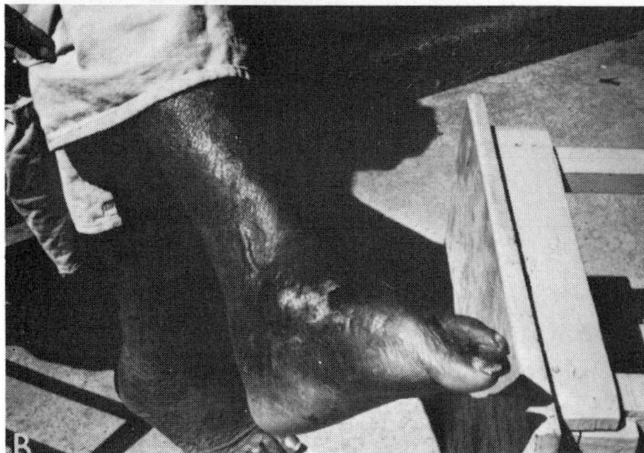

Figure 4–2. *A* (FS A 11,*B*), Actinomycotic mycetoma. Convex sole of foot in a patient with advanced disease. Tumorlike growths are present over the lower leg and foot. There is massive destruction of bone. The hyperplasia indicates an actinomycotic cause of disease. *Actinomadura madurae. B* (FS A 11,*C*), Mycotic mycetoma. Swollen, distorted foot. In this case there is no invasion of muscle or bone and few draining sinus tracts. *Acremonium kiliense.*

or purulent, the latter if there is secondary bacterial infection. Deep tissues are remarkably free of ancillary bacteria. The drainage, when expressed, contains numerous small particles or granules (grains) varying from 300 μ to 5 mm or more in size.

The grains are light-colored (white, yellow, cream), pink or red, or dark-colored (brown or black), depending on the etiologic agent (Fig. 4–4 FS A 13). Their presence serves to separate this disease from pseudomycetomatous diseases, such as yaws and sporotrichosis. Though the color and morphology of the grains are helpful in ascertaining the etiology, several species may produce similar granules, and it is unwise to ascribe a specific cause without further study. Examination of grains alone will permit the differentiation of actinomycotic mycetoma from eumycotic mycetoma and botryomycosis, and, in the hands of an experienced mycologist, a tentative diagnosis of the agent involved. Culture and laboratory identification, however, are always indicated. This is extremely important for the institution of effective therapy. As emphasized by Zaias et al.,[55] the triad of tumefaction, sinus formation, and grains is necessary for the diagnosis of mycetoma.

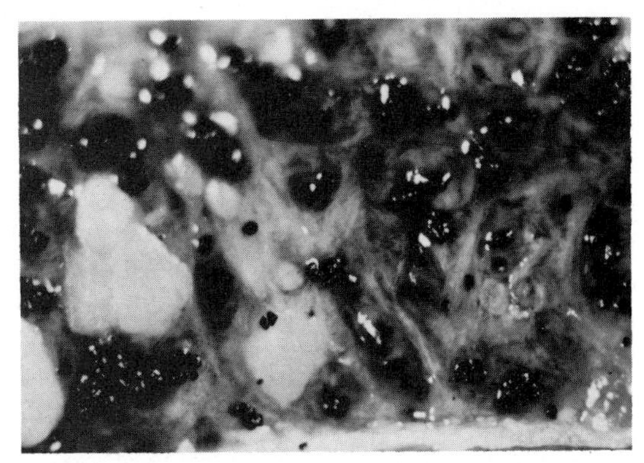

Figure 4–4. FS A 13. Mycotic mycetoma. Cut surface of tissue. Note large black grains surrounded by fibrotic tissue. *Madurella mycetomii.*

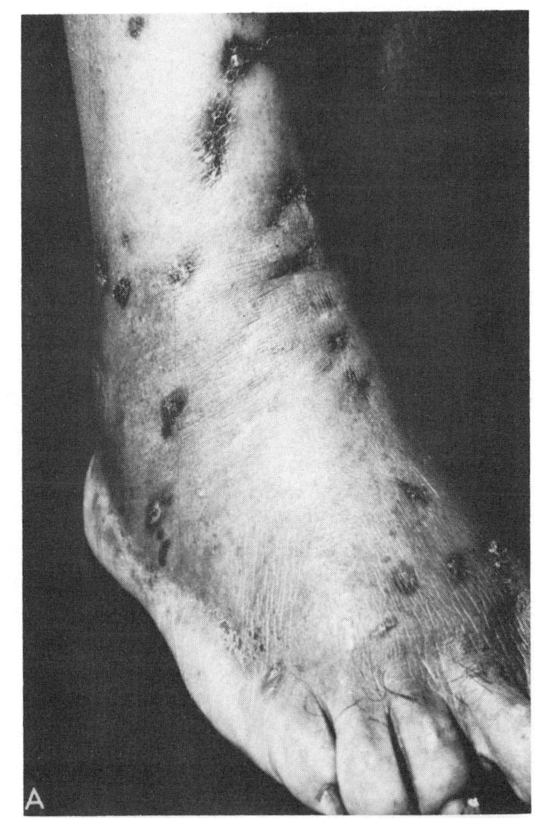

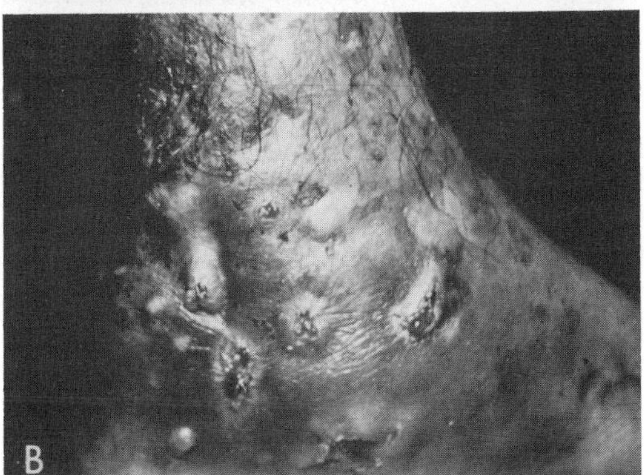

Figure 4–3. FS A 12. *A,* Mycotic mycetoma. Multiple sinus openings in an infection caused by *Madurella grisea*. The surfaces of the lesions are flat and somewhat discolored. The bone is not involved. *B,* Actinomycotic mycetoma. There is destruction of bone, distortion of the foot, and hyperplasia at the openings of the sinus tracts. *Nocardia brasiliensis.* (Courtesy of S. McMillen.)

Following healing of the involved tissue, massive fibrosis occurs, which gives a tumorlike appearance and hard feel to the area. The disease usually remains localized, though it may slowly extend to contiguous tissue and produce severe disfigurement and disability. If the hand is involved, a similar clinical picture is seen, with swelling, sinus tracts, and slow, localized extension. In uncomplicated cases, there is little or no pain, fever, or lymphadenopathy, and review of systems is usually within normal limits.

Hematogenous spread is rare, but recorded. Intraosteal rather than periosteal lesions may indicate that the disease is becoming systemic. When the disease involves the buttocks, chest, or trunk (particularly with *N. brasiliensis*), there is a tendency for it to spread rapidly and widely. With most other agents, visceral involvement is rare, though the isolation of *Petriellidium boydii* with typical granules from lungs has been reported. This agent, along with the other agents, may be involved in meningitis, otitis, ocular infection, cystitis, septicemia, or other disease pictures not within the clinical definition of mycetoma.

The x-ray picture in mycetoma will show necrosis, generalized osteoporosis, and some fusion of the smaller bones. These may be observed if the density of soft tissue is not too great. Areas of osteal hypertrophy and lysis may be present, as well as large, gouged-out lesions occupied by massive granules (Fig. 4–13).

Differential Diagnosis. Mycetoma, as defined, presents so typical a picture as to be readily diagnosed, especially in highly endemic areas. Actinomycosis is excluded from the mycetomas. It gives a similar picture of firm induration, swelling, and sinus tract formation with granules, but the onset, systemic distribution, endogenous nature, and overall clinical appearance are distinctive and form a well-defined entity. However, the etiologic agent, *Actinomyces israelii,* has been reported several times from the clinical syndrome of mycetoma. Systemic nocardiosis with its primary infection in the lungs and quite different clinical course again is a separate disease. Granules are lacking if the agent is *Nocardia asteroides. Nocardia brasiliensis* may be considered an agent of nocardiosis if the primary site of involvement is the chest and lungs.

Pulmonary mycetoma and aspergillomas are also excluded from the accepted definition of mycetoma. In these instances, such an organism, usually *Aspergillus* species, grows in ectatic bronchi or old tuberculous cavities, but lacks inherent qualities of extension or invasion into uninvolved tissue. The lesions may enlarge within the limits of debilitated tissue, and frank invasion of pulmonary parenchyma may ensue, but the pathologic picture is that of a mycotic granuloma rather than a mycetoma.

Botryomycosis (staphylococcic actinophytosis) is a clinical entity that closely resembles mycetoma, with the production of swelling, grains, and draining sinuses (see section on botryomycosis). The agents involved include *Staphylococcus aureus, Pseudomonas aeruginosa, Proteus* sp., and possibly *Escherichia coli.* Examination of the grains quickly identifies this disease.

Several diseases may, at one stage or another, mimic mycetoma. Included are yaws, elephantiasis of the foot (no sinus tracts), cellulitis, and primary or secondary eumycotic infections, such as coccidioidomycosis, sporotrichosis, or ulcerated, hypertrophic chromomycosis. Mycetoma in unusual anatomic sites may mimic progressive angioma of Darier, tuberculosis verrucosa, and acne conglobata.[11]

Prognosis and Therapy. It is imperative that the general class of mycetoma be determined in order to initiate effective therapy. Actinomycotic lesions are more amenable to treatment than are eumycotic mycetomas.

If the lesions remain localized, there is little impairment to general health, although extensive involvement may severely incapacitate the patient. Treatment is made particularly difficult by the inability of drugs to penetrate into cystic and fibrotic areas in sufficient concentration to inhibit the causal agent.[19] Vigorous and prolonged treatment is necessary in order to ensure that an adequate dose is present in the involved tissue. A trial of specific medical treatment combined with conservative surgical management should be considered. Surgical intervention includes exploration and drainage of sinus tracts, debridement of diseased tissue, and removal of cysts from involved bones. Radical surgery and amputation should be considered only as a last resort.

Actinomycotic mycetoma is at least somewhat amenable to treatment. Most success has been obtained by administering high doses (ten million units per day) of penicillin in nonoily preparations,[55] or by following the schedule presented in the chapter on actinomycosis. The newer penicillins, such as phenoxymethylpenicillin, produce higher blood levels than IM procaine penicillin G. The synthetic and penicillinase-resistant penicillins have not as yet received adequate clinical trials. Phenethicillin, ampicillin, cloxacillin, and methicillin show decreasing efficacy, in that order, against *Actinomyces israelii in vitro.* Probenecid (Benemid), 1 gm/day, is advised to decrease renal excretion of penicillin. The standard treatment for *Nocardia asteroides* is given in the chapter on nocardiosis and consists of the administration of 3 to 10 g sulfadiazine per day to achieve blood levels of 8 to 15 mg per 100 ml. This has also been effective in systemic and localized infections due to *N. brasiliensis.*[20] Sulfonamide derivatives and diaminodiphenylsulfone (*p,p′*-sulfonyldianiline, DDS, dapsone) have been used in prolonged treatments with some success. After a starting dose of 50 mg daily, the regimen is increased to 200 mg daily and continued for some time after resolution of the lesions. Sometimes local injections are used as adjunct therapy. As in all cases of mycetoma, medical management must be individualized and treatment regulated by judicious attention to the clinical course of the disease. The newer, longer acting sulfas, such as sulfadimethoxine (Madribon) and sulfamethoxypyridazine (Midicel)[51] have shown some efficacy in clinical trials. The dose varies from 0.5 to 1.5 g per day for six months or longer. When sulfadimethoxine is used, side effects include aplastic anemia, thrombocytopenic purpura, and myocarditis. Perfusion of the affected limb has been used in specialized situations.[55] In infections by *A. madurae, A. pelletieri*, and *S. somaliensis*, the drug most successfully used has been streptomycin. Abbott[1] and Lynch,[29] have noted clinical improvement using an initial dose of 3 g daily for three weeks, decreasing to 2 g daily and then 1 g daily for similar intervals. Bergeron[6] reports the cure of an *A. pelletieri* mycetoma with 2 g of "triple sulfa" four times a day. Other broad-spectrum antibiotics, such as tetracycline, have been used either alone or with streptomycin. Results were less than adequate. Abbott[1] and Lynch[29] stress early diagnosis and conservative surgical measures to prevent extension of disease and the necessity for amputation. In very early lesions, radical excision followed by flooding with tincture of iodine may arrest the process.

Antifungal agents (griseofulvin, amphotericin B, nystatin) have been tried with no success in the treatment of actinomycotic mycetoma.[21]

Eumycotic Mycetoma. Mycetoma of fungal etiology is consistently resistant to chemotherapy. The list of drugs tried without success is long and the results discouraging. The introduction of amphotericin B raised the hope that the disabling and disfiguring lesions could be controlled by chemotherapeutic means. Blood levels of 1.5 to 2 mcg per ml can be achieved safely in patients, but they are usually below the effective range for the agents involved. The *in vitro* sensitivity of *Madurella grisea* varies from 10 to 100 mcg per ml, of *M. mycetomii* from 0.62 to 25 mcg per ml, and of *Leptosphaeria senegalensis* from 0.7 to 100 mcg per ml. Some clinical success in mycetoma due to *M. mycetomii* has been reported with combined therapy of amphotericin B and surgical management. From the literature,[18] it appears that the drug is regularly used in the treatment of mycetoma produced by this agent and possibly by *M. grisea*, but is not very effective in cases caused by other agents. *M. grisea* has been treated successfully by Neuhauser[45] using 20 mg DDS per day. *In vitro* studies by Indira and Sirsi[24] showed partial inhibition of *M. grisea* by 10 mcg per ml of amphotericin B and 20 mcg per ml of nystatin. Hamycin and DDS were without effect. Aureofungin and pimaricin were quite inhibitory at 15 and 10 mcg per ml, respectively. Clinical trials of these drugs have not been reported.

Nielson[43] reports amphotericin B has little or no effect *in vitro* on *Allescheria boydii.* Inhibited strains soon developed resistance. One successful report of *in vitro* sensitivity and clinical cure of *A. boydii* mycetoma is that of Mathews,[35] who used the compound D-25 (2,2′-dihydroxy-5,5′-dichlorodiphenyl sulfide). Treatment consisted of 0.5 to 1 g of the drug in oil injected into the affected area for intervals of six months. Nystatin has been reported as inhibitory, and was used successfully in an ocular infection of *A. boydii.*[17] Most agents of eumycotic mycetoma are inhibited by 28 mcg per ml of nystatin *in vitro*, but there are few clinical trials reported. Mohr and Muchmore[37] report a case of *A. boydii* mycetoma that would resolve when the patient was pregnant. Subsequently, the patient was given norethynodrel (5 to 20 mg per day), and a complete remission occurred. After extensive trials, griseofulvin was found to be of no value.[21,25] Isolated limb perfusion techniques have been utilized to increase the drug level at the affected site. Its use has been of limited value and is appropriate only in specialized cases.[55] Thiabendazole has been tried with some success.[7] Clinical trials of 5-fluorocytosine are now being carried out in Mexico and Latin America.

The historic and perennial use of iodides cannot be ignored, as clinical remissions have occurred with prolonged therapy. As is the case with sporotrichosis, where the agent can grow in media containing 5 per cent KI, the effect is probably on the host tissue rather than on the organism. The drug acts by causing resolution of the granuloma formed around the parasite, thus exposing it to the action of the host's defense mechanisms.

PATHOLOGY

Gross Pathology. The gross pathology of mycetoma depends on the anatomic area affected by the disease. There is distortion and swelling of the involved region. Acute purulent abscesses develop, resolve, and redevelop, burrowing in several directions, undermining and invading the soft tissue and bone and erupting to the surface as draining fistulous tracts. As new areas become active, healing with extensive fibrosis occurs in older areas. If limbs are involved, the process may proceed throughout most of the lifetime of the patient without injury to the general health. Eventually there may be severe incapacitation of the involved appendage. The lesions develop where the organism is implanted. In Africa and most other places, thorns or spines prick the legs or feet; thus the disease is limited to the lower limbs. In Mexico, many cases involve the upper back, neck, and shoulder region, where the area is rubbed and abraded while carrying sugar cane, bundles of firewood, or sisal sacks. Hands may be inoculated by working in the fields, clearing brush, or gathering reeds for making baskets. It can be seen that the mode of infection is the same as with chromomycosis and sporotrichosis. In mycetoma, if the head, chest, or neck regions are involved, complications may arise by extension of the destructive process into vital organs. Lymphatic and hematogenous spread are more common when these are the sites of primary involvement. The gross pathology is identical to that of actinomycosis.

Depending on the etiologic agent involved, there are some subtle differences in the gross appearance of the sinus openings, the tissue reaction, and the tissues invaded. These are discussed in the section on specific organisms. Secondary bacterial infection of the peripheral sinus tracts is common, and may complicate the clinical and pathologic picture. Drainage from the sinuses is usually serous or serosanguinous, but may become purulent and foul-smelling because of

Table 4–2. *Actinomycotic Mycetoma**

Species	*Grain*	*Histology (H and E)*	*Colony and Microscopic Morphology*	*Physiologic Profile*†						
				C	*T*	*X*	*S*	*G*	*U*	*AX*
Nocardia asteroides	Rare; white, soft, irregular, 1 mm	Homogeneous loose clumps of filaments; rare clubs	Rapid growth at 37°C; glabrous, folded, heaped; orange-yellow, tan, etc.; short rods and cocci; rare branched filaments; acid-fast	–	–	–	–‡	–	+	–
Nocardia brasiliensis	White to yellow, lobed, soft, 1 mm	Same as above; clubs common	Rapid growth at 30°C; colony and microscopic morphology same as above; acid-fast	+	+	–	–‡	+	+	–
Nocardia caviae	Same as above	Same as above	Same as above; acid-fast	–	–	+	–‡	–	+	–
Actinomadura madurae	White, soft, oval to lobed, rarely pink, large, 5 mm	Center empty, amorphous; dense mantle peripherally; basophilic wide pink border; loose fringe; clubs	Rapid growth at 37°C; cream-white, rarely clot-red; wrinkled, glabrous; delicate filaments, nonfragmenting, branched arthrospores; non–acid-fast	+	+	–	+	+	–	+
Actinomadura pelletieri	Red, hard, small, oval to lobed, 1 mm	Round homogeneous dark staining; light peripheral band; hard— fractures easily; no clubs	Slow-growing (at 37°C); small, dry, glabrous; light to garnet red; delicate, nonfragmenting, branched filaments; non–acid-fast	+	+	–	–‡	+	–	–
Streptomyces somaliensis	Yellow, hard, round to oval, large, 2 mm	Variable size; amorphous center; light purple with pink patches; dark filaments at edge, entire; no clubs	Slow growth (30°C); cream to brown; wrinkled, glabrous; delicate, branched, nonfragmenting filaments; arthrospores; non–acid-fast	+	+	–	±	+	–	–

Actinomyces israelii is also a cause of light-grained mycetoma. Consult Chapter 1 for identification.

*From Rippon, J. W. *In* Burrows, W. 1973. *Textbook of Microbiology.* 20th ed. Philadelphia, W. B. Saunders Company, p. 695.
†C, casein; T, tyrosine; X, xanthine; S, starch (amylolytic); G, gelatin (proteolytic); U, urea; AX, acid from arabinose and xylose. Details in text and appendix.
‡Some strains are positive.

Table 4–3. *Mycotic Mycetoma (Tissue Dimorphic Fungi)**

Species†	*Grain*	*Histology (H and E)*	*Colony and Microscopic Morphology*	*Physiologic Profile‡*						
				St	*Gel*	*G*	*Gal*	*L*	*M*	*S*
Petriellidium boydii	White, soft, oval to lobed, < 2 mm	Hyaline hyphae, 5 μ; huge swollen cells, < 20 μ; no cement; red border; pink periphery	Rapid-growing; fluffy, mouse-fur grey; large 7-μ unicellular conidia on simple conidiophore; black cleistothecia; 30–37°C; also synnemata in *Graphium* state	+	+	+		0	0	
Madurella grisea	Black, soft to firm, oval to lobed, < 1 mm	Little dark cement in edge; polygonal cells in periphery; center hyaline mycelium	Very slow growing; leathery, tan-grey, later downy; sterile pycnidia; diffusible pigment; 30°C	+	−	+	+	0	+	+
Madurella mycetomii	Black, firm to brittle, oval to lobed, < 2 mm	Compact type with brown-staining cement; vesicular type with brown cement only at edge; swollen cells, < 15 μ; center hyaline mycelium	Very slow growing; downy, velvety, smooth or ridged; cream-apricot to ocher; diffusible brown pigment; black sclerotia, < 2 mm; rare conidia, phialids; 37°C	+	±	+	+	+	+	0
Acremonium kiliense	White, soft, irregular, < 1.5 mm	No cement; hyaline hyphae, > 4 μ; swollen cells, > 12 μ	White glabrous colony, later downy; violet pigment diffusible; curved septate; conidia arranged as head on simple conidiophore; 30°C	0	±					
Phialophora jeanselmei	Black, soft, irregular to vermiscular	Helicoid to serpiginous; center often hollow; no cement; vesicular cells, < 10 μ; brown hyphae	Slow-growing; leathery, black, moist, later velvety; reverse black; toruloid yeast cells, moniliform cells, long tubular phialids; 30°C	0	0	+	+	0	+	+
Leptospharia senegalensis	Black, soft, irregular, ~ 1 mm	Black hyphae; cement in periphery; center hyaline	Rapid-growing; downy grey; reverse black, rare rose pigment, diffusible; black perithecia; < 300 μ; septate ascospores, 25 × 10 μ							

*From Rippon, J. W. *In* Burrows, W. 1973. *Textbook of Microbiology*. 20th ed. Philadelphia, W. B. Saunders Company, p. 696.
†*Pyrenochaeta romeroi* (black grains) and *Neotestudina rosatii* (white grains), and details of above are in text.
‡St, starch; Gel, gelatin; G, glucose; Gal, galactose; L, lactose; M, maltose; S, sucrose.

secondary infections. There is often an oily or fatty content to the drainage related to the necrosis of subcutaneous adipose tissue. Grains may be found in expressed fluid, in purulent exudate, massed and clogging the sinuses internally, or lodged in cystic lesions of bones.

Histopathology. Histologically the lesions are similar regardless of the etiologic agent. The granules are seen in the center of an acute abscess, often coated with a crust of homogeneous material. This is surrounded by a large accumulation of neutrophils in all stages of degeneration. Immediately around the abscess there is an area of dense fibrosis and granulation tissue which is rich in capillaries, epithelioid cells, macrophages, and multinucleated giant cells. True grains in suppurating abscesses and typical clinical symptoms must be present for a diagnosis of mycetoma to be made. This disease is a serious condition which carries with it the associated difficulties in prognosis and therapy, often culminating in amputation. In other fungal diseases, accumulations of hyphae or yeast cells may be present which give the appearance of "pseudogranules." These may be accompanied by fungal elements in giant cells, but the lesions lack the suppuration found around the true granules of mycetoma. A more correct term for such lesions is "mycotic granuloma." These conditions are often more amenable to therapy and carry a better prognosis. Such diseases as coccidioidomycosis, blastomycosis, histoplasmosis, cryptococcosis, cladosporiosis, sporotrichosis, and aspergillosis at some stage may simulate mycotic granuloma. Botryomycosis must also be excluded.

In tissue sections of biopsy or autopsy material from lesions of mycetoma, the grains appear with abscesses (Fig. 4–5 and Fig. 4–6 FS A 14). They are stained by hematoxylin and eosin. The center of actinomycotic grains is usually light-colored and unorganized. There is a strong basophilia surrounding the periphery of the grains and a wide-fringed eosinophilic border. In contrast, eumycotic grains vary and are discussed in the section on mycology. The usual picture of these, however, is broad, light pink mycelium in the periphery, surrounded by a basophilic area. The center may be strongly basophilic or unorganized. The methenamine stains are useful for both types of agents. Thin, dendritic filaments are seen with the actinomycetes, and broad, hyphal units and grossly swollen cells are observed in eumycotic mycetoma. Subtle differences in grain characteristics are discussed later.

In summary, it appears that eumycotic and actinomycotic mycetoma present similar pictures clinically and histologically, but with some minor differences. Both elicit a suppurative response initially. As the disease becomes chronic and established, the eumycotic infection is treated as a foreign body with granuloma formation, epithelioid hyperplasia, and giant cell formation. In actinomycotic mycetoma, the acute suppurative pyogenic reaction persists, and the infection is treated in the same manner as a bacterial invasion. Other points of difference between the two classes of disease include the following: (1) actinomycotic agents produce more extensive and obliterative involvement of bone, with both lytic and hypertrophic changes, and there is late bone involvement with lytic but no hypertrophic effects in the course of disease with eumycotic agents; (2) the actinomycetes invade muscle more readily than eumycetes; and (3) the latter do not cause the cellular proliferation that leads to the raised border around the sinus openings.

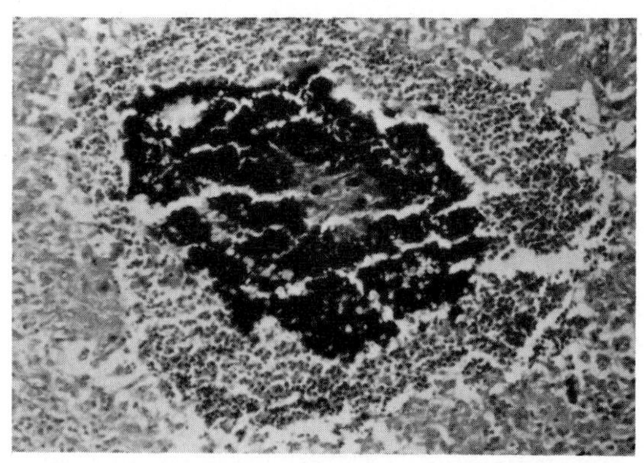

Figure 4–6. FS A 14. Mycotic mycetoma. Black grain of *Phialophora jeanselmei*. The grain is soft and ill-organized. The fungous forms distorted mycelium with balloon cells or cystlike structures. There is a covering of neutrophils attached to the grain, which appears free in the lumen of the sinus tract. The tract walls show histiocytes, giant cells, and a few lymphocytes.

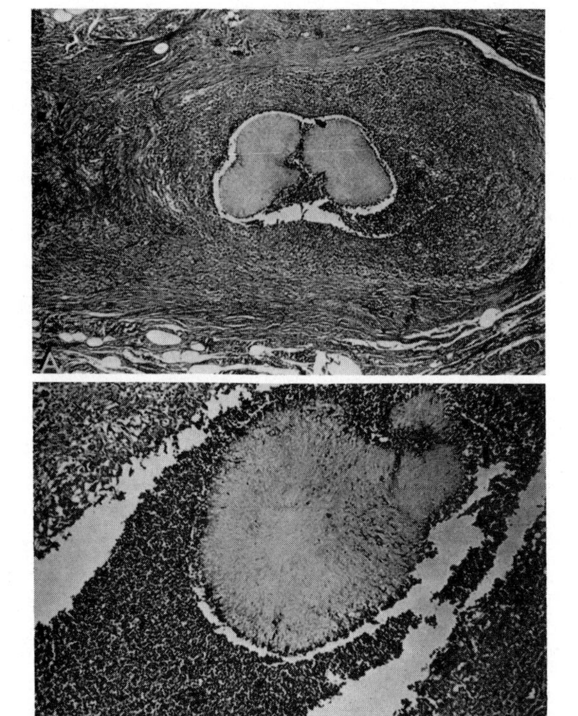

Figure 4–5. Mycotic mycetoma. *A,* Large grain within sinus tract. The periphery of the lesions shows extensive fibrosis followed by a little-organized granulomatous reaction. *Allescheria boydii.* ×100. *B,* Grain of *A. boydii.* There are neutrophils attached to the periphery of the grain. The grain itself has a slight basophilic outer edge and an eosinophilic center. ×400.

ANIMAL DISEASE

Mycetoma is regularly reported in cattle under the name "farcy," and the ascribed agent is called *Nocardia farcinica.* There is much doubt about both the infectious nature of the disease and the validity of the agent. This disease is discussed in the chapter on nocardiosis. Natural infections in lower animals caused by the agents of actinomycotic mycetoma are rare. Ajello[2] described the disease caused by *Nocardia brasiliensis* in a cat. A suppurating granuloma with granules developed following a bite on the hind leg which led to systemic involvement and death. The pathology was similar to that described for the human disease. Eumycotic mycetoma is also rare in lower animals. In 1967, Brodey[9] reviewed twelve cases involving dogs, cats, and horses, and reported a new case in a dog. The published reports all specified the finding of grains (usually dark) in suppurating granulomatous tissue. The disease was usually limited to the legs or feet, and the histology was that described in human disease. The agents when cultured, however, were quite different. So far only two fungi have been implicated—*Drechslera spiciferum* and *Curvularia geniculata.* Neither has been recovered from human mycetoma, although the latter has been found in mycotic keratitis.

Experimental infection in animals has been attempted many times with varying results. Most success has been achieved with the actinomycetes. Macotela-Ruiz and Mariat[32] produced lesions with granules by injecting *N. brasiliensis* or *N. asteroides* into mice, hamsters, and guinea pigs. Some grains had club cells. They concluded that the hamster was the best animal for such studies. Using similar techniques, Rippon and Peck[46] were able to produce mycetoma in the legs of mice with *A. madurae* (Fig. 4–7). Their experiment was designed to show that the proteolytic enzyme, collagenase, was necessary for pathogenicity of this organism. Recent work by González Ochoa[22] has indicated differences in the degree of pathogenicity for three agents of mycetoma. Using several strains of each organism, he found that *N. brasiliensis* was most virulent and always produced grains. Although *N. caviae* was less virulent, it still produced grains, but no grains were found in *N. asteroides* infection, and this organism was significantly less pathogenic than the other agents under his conditions.

Such studies have been less successful with the eumycotic agents. Murray[39] reported infection in mice with *M. mycetomii*; however, his results could not be reproduced by Indira.[24] In a series of papers, Avram[4,5] has reported infection and grain production with *Acremonium kiliense (Cephalosporium falciforme), A. boydii,* and *M. mycetomii.* It appears that true mycetoma can be produced in animals if correct procedures are used. Experimental infection does not carry any diagnostic value in the clinical laboratory, but it can be used for the evaluation of chemotherapeutic agents.

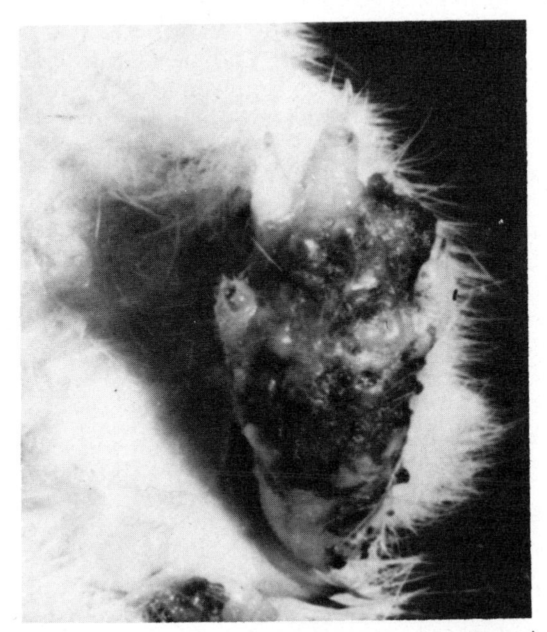

Figure 4–7. Actinomycotic mycetoma. Experimental disease produced in mice by infection with *Actinomadura madurae.* Multiple white grains were extruded from the sinus tracts. There was complete destruction of the bones in the area.[46]

IMMUNOLOGY AND SEROLOGY

Though the etiologic agents of mycetoma are found throughout the world, the overall rarity of the disease attests to a high degree of natural resistance in the normal host, and a low grade of pathogenicity of the organisms. The disease has not been associated with any underlying debilities, as is the case with systemic nocardiosis. If any predisposing factors are to be considered, they would be general health and nutrition, repeated exposure or insult to the affected area, and personal hygiene. The wearing of shoes is a preventative measure, although this increases the chances for ringworm of the foot, a much less hazardous choice.

At present, there are no established serologic procedures for the diagnosis or prognosis of mycetoma. As noted in the chapter on nocardiosis, González Ochoa and Baranda, studying the sera of patients with *N. brasiliensis* infection, found that about half had complement-fixing antibodies, a few had agglutinins, and none had precipitating antibody. There was no correlation between titer and duration, severity, or extent of the disease. Bojalil and Zamora[8] prepared a purified polysaccharide from *N. brasiliensis* with which they were able to demonstrate precipitins in patients' sera, and a specific skin test reaction. In the latter instance, there were no cross-reactions with sera from patients infected by other Nocardia species, leprosy, or tubercle bacilli. Recently, Murray and Mahgoub[41] prepared antigens from several eumycotic and actinomycotic agents of mycetoma using the gel diffusion technique. They found that sera from 69 patients and from mice experimentally infected with *M. mycetomii* gave precipitin lines which were specific for the appropriate antigen. Control sera did not react, nor was there significant cross-reactivity with other fungal antigen. The gel diffusion technique is becoming a useful method in the diagnosis of several mycotic and actinomycotic infections and promises to be an easy and reliable diagnostic procedure in mycetoma infections as well. Since the etiologic agents of this disease are many and of diverse nature, a common or polyvalent antigen seems unfeasible; however, specific reactivity could be useful as an adjunct to cultural techniques when defining the etiology. The appearance of precipitin lines during active disease in histoplasmosis and aspergillosis and their disappearance after cure encourages the investigator that the same may be found for the serodiagnosis of mycetoma.

LABORATORY IDENTIFICATION

Direct Examination. Pus, exudate, or biopsy material from the patient may be examined for the presence of grains. Though the grains vary in size from 0.5 to 2 mm or more, they are usually macroscopically visible, and a study of their morphology, texture, color, and shape may indicate with a fair degree of certainty the identity of the causative organism. As grains are quite similar, especially among the actinomycetes, culture is always necessary for definitive diagnosis. Examination of crushed material in potassium hydroxide preparations will allow delineation of actinomycotic and eumycotic mycetoma, as well as botryomycosis. Actinomycetes produce a grain composed of intertwined filaments, 0.5 to 1 μ in diameter, as well as coccoid and bacillary elements. Eumycotic grains show intertwined, broad (2 to 5 μ) mycelial strands which may have large swollen cells (15 μ or more) at the periphery. A Gram stain will demonstrate the gram-positive filaments of the actinomyces, as well as the gram-positive cocci, or the gram-negative bacilli which produce botryomycosis. These all appear surrounded by a gram-negative matrix. Gram stain, however, is of limited value for observing true fungi. Zaias et al.[55] recommend Albert's stain for studying the fine morphology of fungal grains. Giemsa stain, as recommended by Lacaz, outlines in blue the crust or shell and the cementing substance of eumycotic granules. For preparation of stained smears, the granule should be crushed in its pus. *Nocardia asteroides, N. brasiliensis, N. caviae,* and *N. farcinica* are acid-fast; other species of the genus are not.

Culture Methods. For culture, a deep biopsy is best, as it is usually free from contaminating bacteria and fungi. Otherwise the sample should be handled like the grains of actinomycosis. The granules are washed several times in sterile saline or Eugon broth, crushed with a sterile glass rod or in a Tenbroeck tissue homogenizer, and plated on appropriate media.

Actinomycetes are planted on blood agar or BHI agar. Several plates should be streaked and incubated at 37°C both anaerobically (for *A. israelii*) and aerobically. Although an anaerobic atmosphere including 5 to 10 per cent CO_2 is best for growing *Actinomyces* species, a candle jar will usually suffice. The *Nocardia* species will grow well on Sabouraud's yeast extract agar. The media should contain no antibiotics. Some actinomycetes grow more readily on Löwenstein-Jensen culture medium than on other media. After isolation of the colonies, specific identification can be made by physiologic characteristics.

Specimens containing eumycetes can be washed in antibiotic-containing saline and plated on Sabouraud's yeast extract (0.5 per cent) media without cycloheximide. Several plates should be planted, as the number of a single species present often indicates true etiology. Plates should be incubated at 25°C. As no true fungi are obligate anaerobes, anaerobic procedures are unnecessary. The growth of eumycotic agents is usually very slow, and plates should be kept for six to eight weeks. Identification depends on colony morphology, spore types, and assimilation patterns detailed in the following section. Spore production is often enhanced by growth on low-nutrient media, such as potato-carrot agar or cornmeal agar.

MICROBIOLOGY

The literature on mycetoma contains a long list of species as agents of the disease. Most of these are so poorly described that they are unidentifiable, or the apparent differences were not fully compared with valid species. Continued critical study by several authors, including Emmons, González Ochoa, Ajello, and Mackinnon, has caused this list to be reduced by reclassification and recognition of synonymous organisms. In this section, the more common and valid species of the actinomycetes and eumycetes involved in human mycetoma will be described in detail. In addition the characteristic types of grains, colony morphology (Fig. 4–8 FS A 15), cultural characteristics, physiology, if relevant, and type of disease induced by each will be discussed. Preparation and use of special medias and techniques are included in the Appendix. The microbiology of *Nocardia asteroides* is discussed in Chapter 2.

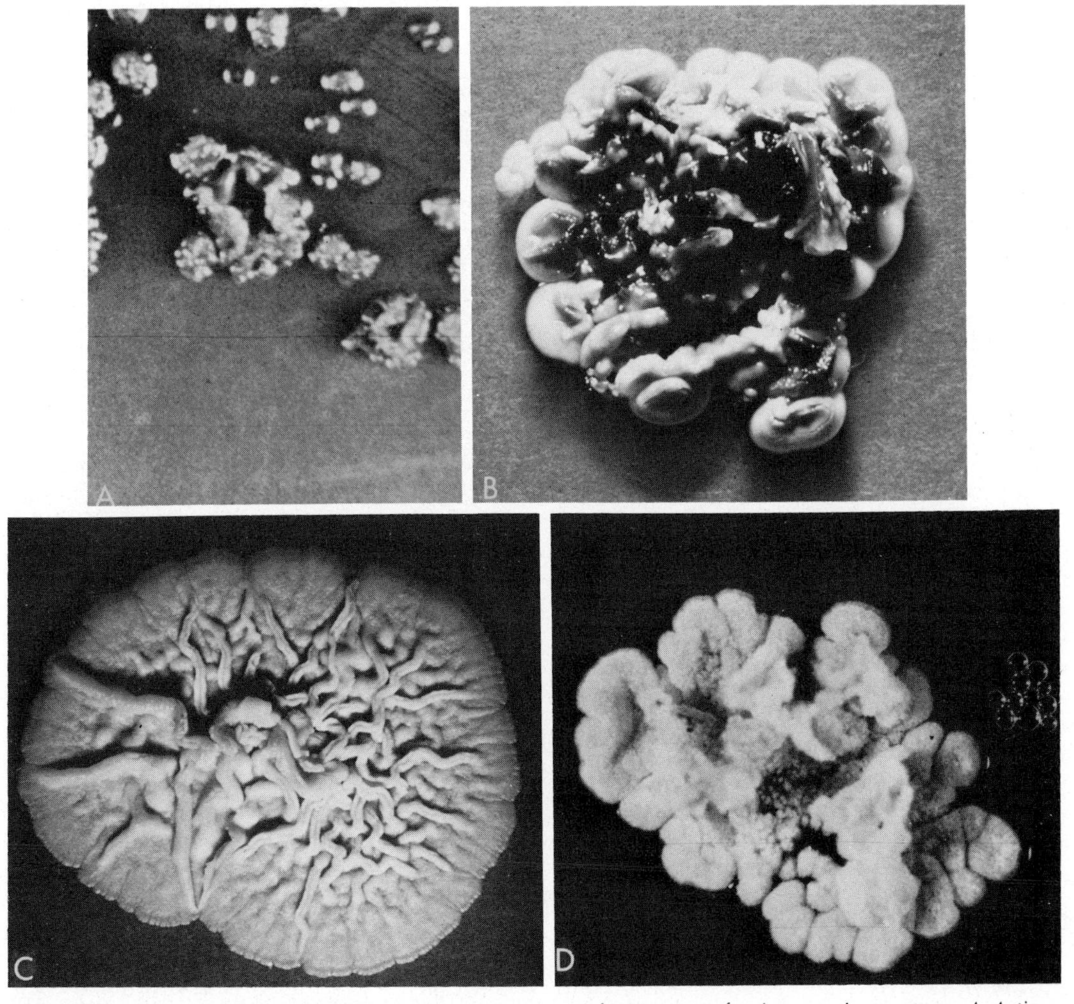

Figure 4–8. FS A 15. Colonial morphology of four common etiologic agents of actinomycotic mycetoma. *A, Actinomadura pelletieri*. This is a moist, red-pink, folded colony. *B, Actinomadura madurae*. This colony is moist, folded, and has areas of blood red-to-pink color. The usual isolant is dull white. *C, Nocardia brasiliensis*. A thin, glabrous, wrinkled, off-white colony. *D, Streptomyces somaliensis*. Creamy, folded colony with tufts of whitish grey mycelium.

Actinomycetes

Nocardia brasiliensis (Lindenberg) Castellani and Chalmers 1913 (for synonymy see Chapter 2)

Clinical Disease. There is some cellular proliferation and local infiltration around the openings of the sinus tracts, often forming a raised border and giving the area a "bumpy" appearance. Punched-out lesions also may occur if there has been secondary bacterial infection. There is frequent and extensive involvement of bone. In conditions that mimic actinomycosis, the slight acid-fastness and aerobic growth of *N. brasiliensis* quickly differentiate the two. Optimum growth is 30° or 37°C.

Grains. The grains are usually less than 1 mm in size, white to yellowish in color, soft, and lobulated. "Clubs" may or may not be seen. They are scattered or may be clumped together in large masses. Dendritic acid-fast filaments are often seen to be fragmenting into bacillary bodies. Some differences in morphology of grains has been noted in different geographic locations.

Colonies. The organism grows rapidly, producing a heaped, wrinkled, and folded colony. The color is yellow to yellow-white, tan, orange, or red-orange and closely resembles the colony of *N. asteroides*. A white, powdery overgrowth may occur, indicative of fragmentation into arthrospores. Rarely, a brownish pigment diffuses into the media.

Stained smears show short, irregular bacilli and coccoid forms. Growth occurs as long filaments in broth cultures. Small, beaded, acid-fast forms may resemble atypical mycobacteria, but slide cultures showing branching filaments differentiate the two.

Culture. Decomposition of tyrosine, gelatin, and casein separate *N. brasiliensis* from *N. asteroides*. *N. brasiliensis* is acid-fast and does not attack xanthine or starch. These tests separate it from other *Nocardia* and *Streptomyces* species.

Nocardia caviae (Erickson) Gordon and Mihm 1962

Synonymy. *Nocardia otitidis cavarum* Snijders 1924; *Actinomyces caviae* Erickson 1935.

Although many references in the literature ascribe mycetoma to *N. asteroides*, it is probable that, in fact, the causative agent is *N. caviae*. The two were distinguished by the critical examination of Gordon and Mihm in 1962.[23] *N. caviae* appears to be an infrequent cause of mycetoma in humans.

Clinical Disease. The few cases recorded resemble *N. brasiliensis* infection. *N. caviae* was first isolated from an ear infection of guinea pigs by Snijders in 1924.

Grains. Similar to *N. brasiliensis*.

Colonies. Growth is described as being moderately rapid, resembling that of both *N. asteroides* and *N. brasiliensis*. On special *Nocardia* media (Bennett's), a cream to peach color is formed. Tufts of short aerial mycelia are seen. The organism is acid-fast.

Culture. *N. caviae* decomposes both xanthine and hydroxanthine but not casein, BCP, milk, tyrosine, or gelatin. Acid is formed from inositol and mannitol, but not from arabinose and xylose. It survives at 50°C for eight hours and grows sparingly at 40°C.

Actinomadura madurae Lechevalier 1970

Synonymy. *Streptomyces madurae* (Vincent) Mackinnon and Artagaveytia-Allende 1956; *Streptothrix madurae* Vincent 1894; *Nocardia madurae* Blanchard 1896; *Actinomyces madurae* Lachner-Sandoval 1898; *Nocardia indica* Chalmers and Christopherson 1916; *Discomyces bahiensis* Piraja da Silva 1919.

Because of differences in cell wall composition, lipid content, and other characteristics, it appears the *Streptomyces madurae, S. pelletierii*, and *S. dassonvillei* will be grouped into a separate genus. The name *Actinomadura* has been suggested by Lechevalier.[28]

Clinical Disease. In infections produced by this organism, a wide raised border is often seen around the sinus openings because of infiltration and hyperplasia of epidermal cells. Bone involvement is notoriously common. Lytic and proliferative changes are seen, and osteophytes or spicules of bone may be found between cavities occupied by large grains. Proteolytic enzymes are elaborated *in vivo* and may affect the type of disease produced.[46] Muscle is readily invaded by the organism.

Grains. The granules produced by *A. madurae* are the largest seen in mycetoma. The average is 1 to 5 mm, and 10 mm has been recorded. Usually they are white to yellowish, irregularly lobulated (mulberry), oval, serpiginous, or angular. The finding of pinkish to light red grains is recorded. The outside edge of the grain has a fringe of filaments which extend radially. These may be up to 50 μ in length. The granules have a shell of homogeneous eosinophilic material surrounding them. The consistency is soft, and the grain is easily crushed to a pasty mass.

Colonies. The usual colonial form is dirty white, glabrous or waxy, wrinkled, and folded, with a flat periphery. The colony is moderately fast growing and so membranous as to peel off the media when subcultures are attempted. A blood-clot red colonial form is not uncommon. The organism grows rapidly on Löwenstein-Jensen media and can be mistaken for atypical mycobacteria.

Stained smears show long, sinuous, branched filaments, which are usually less then 1 μ in diameter. Occasionally, chains of spherical conidia (arthrospores) ranging in size from 0.5 to 1 μ in diameter are seen. The organism is gram-positive and non–acid-fast. It is nonfragmenting. Optimum growth occurs at 37°C.

Culture. *A. madurae* is both proteolytic and amylolytic, digesting casein, gelatin, tyrosine, BCP, milk and starch. It produces acid from arabinose and xylose, It does not digest xanthine. Paraffin is not utilized.

Actinomadura pelletieri Lechevalier 1970

Synonymy. *Streptomyces pelletieri* (Laveran) Mackinnon and Artagaveytia-Allende 1956; *Micrococcus pelletieri* Laveran 1906; *Nocardia africanus* Pijper and Pullinger 1927; *Nocardia pelletieri* Pinoy and Jouenne 1915; *Nocardia genesii* Froes 1931; *Oospora pelletieri* Thiroux and Pelletieri 1912; *Discomyces pelletieri* Neveu-hemaire 1921; *Actinomyces pelletieri* Brumpt 1927; *Aspergillus pelletieri* Smith 1928.

Clinical Disease. Crateriform lobulated nodules are seen around the draining sinuses in *A. pelletieri* infection.[6] There is marked epidermal hyperplasia and an exaggerated granulomatous inflammatory reaction causing the nodular appearance. Bone involvement is similar to that seen with *A. madurae,* though the granules are smaller, as are the lytic lesions and spicules.

Grains. The characteristic granule is small (300 to 500μ) and garnet red. The morphology is oval to spherical, or irregular to lobulated. The edge is entire or slightly denticulate. It is very hard and resists crushing. There is no filamentous fringe as in *A. madurae* grains.

Colonies. Growth is very slow, and small, dry, granular or glabrous adherent colonies are produced. The color is at first pink to peach, developing into a deep cranberry red. Subcultures may in time become more mottled and peach-colored. White overgrowth may occur.

In slide preparations, delicate branched filaments are seen to be 0.5 to 1 μ in diameter. Conidia (arthrospores) are rare. This organism is not acid-fast and is gram-positive. Optimum growth occurs at 37°C.

Culture. *A. pelletieri* is proteolytic, but not amylolytic. Casein, tyrosine, BCP, milk, and gelatin are digested. Starch is usually not digested. Urea, potassium nitrate, and ammonium nitrate are not utilized. There is no acid produced from arabinose, xylose, galactose, mannitol, or maltose. Paraffin is not utilized.

Streptomyces somaliensis (Brumpt) Mackinnon and Artagaveytia-Allende 1956

Synonymy. *Indiella somaliensis* Brumpt 1906; *Nocardia somaliensis* Chalmers and Christopherson 1916; *Actinomyces somaliensis* Brumpt 1927; *Streptothrix somaliensis* Miescher 1917; *Discomyces somaliensis* Neveu-Lemaire 1921.

Clinical Disease. In infections produced by this organism, the openings of the sinus tracts are not raised or indurated as in the infections produced by *Actinomadura* species. Bone is not as extensively involved, but muscle may be invaded.

Grains. Very hard, yellow to brown, large granules up to 2 mm are present. There is a firm, cementlike substance holding the filaments together. When crushed, grains break up into angular fragments. The shape is round to oval, and occasionally lobulate. The grains are rarely a pinkish color.

Colonies. On agar the colony develops rapidly in four to 11 days, and has a creamy, wrinkled, flaky growth. Growth is better on Löwenstein-Jensen media. Sectoring of the colony is common, giving areas of varying consistency and mottled coloring. Areas of brown or black may develop, and the underside often shows a yellow to ocher color. Tufts of brownish, aerial filaments develop, which in time may become grey. A diffusible brown pigment has been reported. The organism is nonfragmenting and may form arthrospores in chains. Stained preparations show delicate branching filaments 0.5 to 1 μ in diameter. Optimum growth occurs at 30°C.

Culture. *S. somaliensis* is both proteolytic and amylolytic. The fermentation pattern is similar to that of *A. madurae,* but no acid from arabinose or xylose is produced.

***Streptomyces paraguayensis* (Almeida) Mackinnon and Artagaveytia-Allende 1956**

Synonymy. *Actinomyces paraguayensis* Almeida 1940; *Nocardia paraguayensis* Conant 1948.

This organism is of doubtful status as a distinct species.[24,26] Mackinnon[30] believes it to be very closely related, if not identical, to *Streptomyces albus.* The latter organism has been reported in mycetoma from Rumania.[5] Dr. R. E. Gordon has found the physiologic and morphologic characteristics to be within this species limits.

The original case described by Almeida was that of a Canadian agricultural worker living in the Paraguayan Chaco, who had lesions on the thorax, shoulders and lung parenchyma, with involvement of the clavicle, achromia, and choroid apophysis. The lesions revealed dark grains with clubs.

The colony of this organism is white to creamy, smooth, and tough, with a raised center. Short, aerial filaments are formed. The usual delicate branching filaments are seen in stained smears. It does not fragment and is not acid-fast.

Eumycetes

***Petriellidium boydii* (Shear) Malloch**

Synonymy. Perfect form *Allescheria boydii* Shear 1921; imperfect form *Monosporium apiospermum* Saccardo 1911; *Scedosporium apiospermum* Saccardo 1914; *Monosporium sclerotiale* Pepere 1914; *Indiella americana* Delamare and Gatti 1929; *Glenospora clapieri* Catanei 1927; *Pseudoallescheria sheari* Negroni and Fisher 1944; *Acromoniella lutei* Leao and Lobo 1940.

Clinical Disease. As with most cases of the eumycotic mycetomas, the swollen appendage is dotted with punctate, flat, fistulous openings and lacks the epidermal hyperplasia seen in many actinomycetous infections. Bone involvement is also less marked. The osteomyelitis is primarily destructive without new bone formation. The organism is the commonest cause of mycetoma in Europe and the United States. It has been isolated from soil and sewage[13] and has been involved in infection of the lung as well as other types of disease.[17,36]

Grains. The granules are large (up to 2 mm), white to yellowish, soft to firm, round to lobulated. The hyphae are broad (up to 5 μ diameter), septate, intertwined, and show numerous swollen cells (15 to 20 μ) at the periphery of the granule. These are sometimes referred to as chlamydospores, but probably reflect a modification of the hyphae to the environment of the tissue (Fig. 4–9). There is no evidence of cement between the hyphal strands.

Colonies. *A. boydii* grows rapidly on all laboratory media. There is an abundance of fluffy or tufted aerial mycelium, which is at first white, but becomes brownish "mouse-fur" grey (Fig. 4–10 FS A 16). The reverse of the colony shows areas of grey to black pigmentation. Rare strains are ivory-colored and membranous.

The hyphae are hyaline and 1 to 3 μ in diameter. The large, pyriform aleuriospores (conidia) are lemon-shaped, 4 to 9 × 6 to 10 μ,

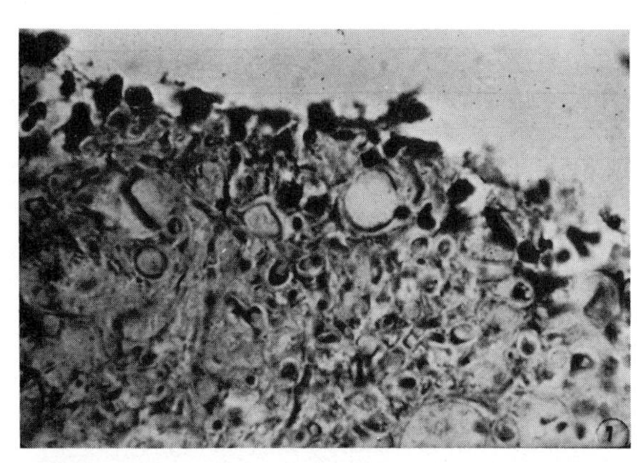

Figure 4–9. Mycotic mycetoma. Edge of grain of *Allescheria boydii.* Note distorted mycelium and large, cystlike chlamydospores. The grain is soft and ill-organized, and there is no cement between the mycelial strands.

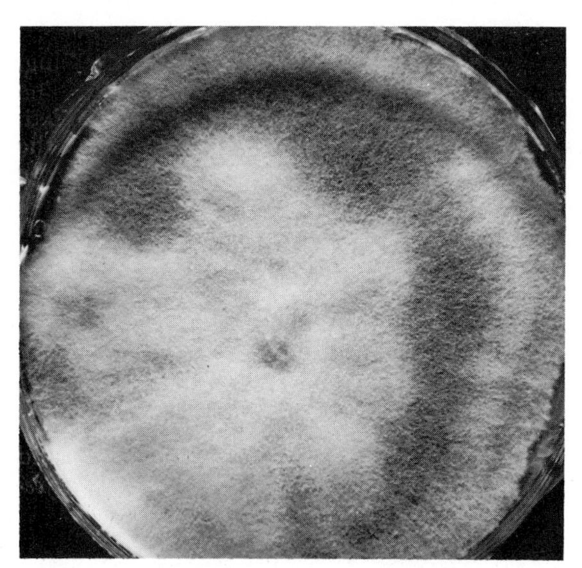

Figure 4–10. FS A 16. Mycotic mycetoma. *Allescheria boydii.* Soft, mouse-grey, furlike colony. Ten days' growth on Sabouraud's glucose agar.

borne singly or in small groups on elongate, simple, or branched conidiophores, or laterally on hyphae. Tufts of conidiophores forming a coremia are also found. Each phore ends in a single conidium. The organism is homothallic, and some isolants produce ascocarps near the agar surface or in the mycelium near the edge of the colony (Fig. 4–11). There is a conspicuous and persistent ascogonium. The cleistothecia measure 100 to 300 μ in diameter and have a yellow-brown to black wall composed of thick-walled polygonal cells. The fruiting body ruptures at maturity. The ovate or clavate asci are evanescent and not found in mature carps. Eight ascospores are produced in each ascus. They are 4 to 5 × 7 to 9 μ, faintly brown, and resemble the asexual aleuriospores. The latter have a truncated scar at the base of the cell where they were attached to the conidiophores. Optimum temperature is 30° to 37°C.

Culture. *A. boydii* is both proteolytic and amylolytic. It assimilates urea, asparagine, potassium nitrate, and ammonium nitrate. Glucose is assimilated but not lactose or maltose.

Madurella mycetomii (Laveran) Brumpt 1905

Synonymy. *Streptothrix mycetomii* Laveran 1902; *Glenospora khartoumensis* Chalmers and Archibald 1916; *Oospora tozeuri* Nicolle and Pinoy 1908; *Madurella tozeuri* Pinoy 1912; *Madurella tabarkae* Blanc and Brun 1919; *Madurella americana* Gammel 1926; *Madurella ikedae* Gammel 1927; *Madurella lackawanna* Hanan and Zurett 1938; *Madurella virido brunnea* Redaelli and Ciferri 1942.

Clinical Disease. The clinical appearance is similar to that of *A. boydii* mycetoma. The infection remains localized and encapsulated at first. Later stages tend to respect anatomic barriers and spread along the fascial planes and the fibrous septa between the muscular bundles. In advanced stages, osteolysis is marked without accompanying osteogenesis.

Grains. *M. mycetomii* grains are black, 0.5 to 1 mm in size, round or lobed. They may be aggregated to form a mass 2 to 4 mm in size. Grains are hard and brittle but can be crushed. They are composed of hyphae which are 2 to 5 μ in diameter, with terminal cells expanded to 12 to 15 μ in diameter and up

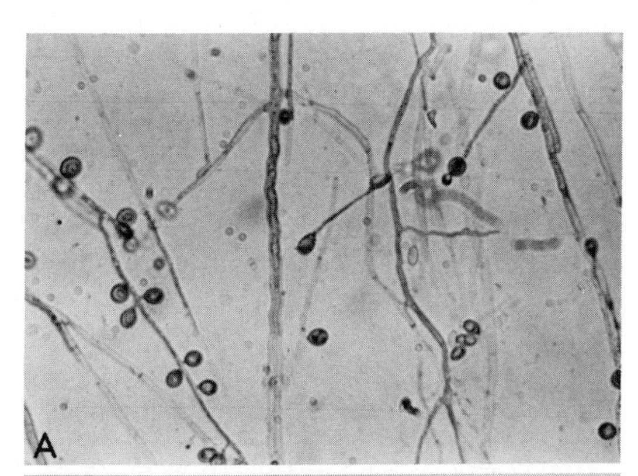

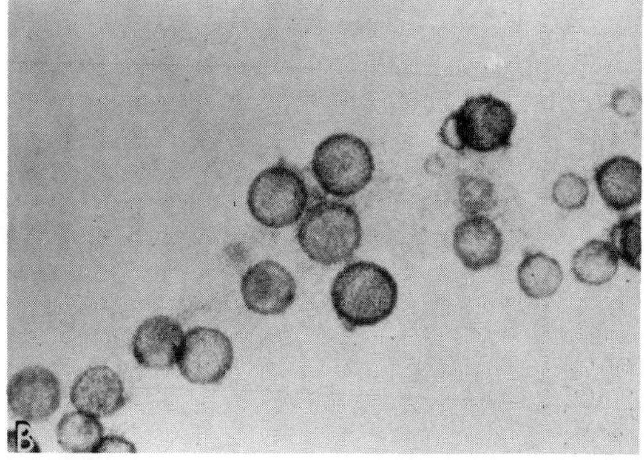

Figure 4–11. *A,* Culture mount in lactophenol cotton blue. Single spores on short projections from mycelium and on elongate condidiophores. ×400. (From Rippon, J. W. *In* Burrows, W. 1973. *Textbook of Microbiology.* 20th ed. Philadelphia, W. B. Saunders Company, p. 697.) *B,* Cleistothecia forming at edge of colony within agar. Obovate ascospores are the same size as asexual conidia but lack an apiculus. The latter is the point of attachment of the conidia to mycelium. ×50.

to 30 μ. The hyphal fragments vary in content of pigment. Two types of grains are described. The first is a compact form filled with dark brown granular cement between the hyphal elements. In hematoxylin and eosin stains, it appears as a uniform rust brown color. A second, vesicular type of grain occurs in which the cement in the periphery is brown and filled with vesicles 6 to 14 μ in diameter. The center is light colored. This latter type resembles the grains of *M. grisea.*

Colonies. The optimum temperature for growth is 37°C. The colony is at first leathery, folded, and heaped, a white to yellow or ocherous brown color, and covered by a greyish down. Later there is an overall growth of bronish aerial mycelia and diffusible pigment. The mycelia average 1 to 5 μ, with moniliform hyphae from 2 to 6 μ. Enlarged chlamydospore-like cells may attain a diameter of 25 μ. In old colonies (two months), black sclerotia, 1 mm in diameter, may be formed, composed of rounded, polygonal-shaped mycelial elements. This is best observed on potato-carrot media. Old colonies take on a reddish-brown hue. Two types of sporulation have been described in growth on cornmeal agar, pyriform aleuriospores (3 to 5 μ), with a truncated base borne on the tips of simple or branched conidiophores, and small, flask-shaped phialides producing rounded spores (3 μ in diameter).

Culture. *M. mycetomii* is amylolytic and weakly proteolytic. Dextrose, galactose, and maltose are assimilated, but not sucrose. This latter reaction separates it from *M. grisea.*

Madurella grisea Mackinnon, Ferrada, and Montemayer 1949

This agent of mycetoma is found chiefly in the Western Hemisphere,[45] although there are rare reports from other parts of the world.[18a]

Clinical Disease. The described clinical picture is similar to that of *A. boydii* mycetoma.[38]

Grains. The granules are black, round to lobed, and similar to those of *M. mycetomii.* The size is up to 1 mm; the consistency is soft at first, becoming hard and brittle. The brown cement is limited to the outer edges of the granule. The center appears hollow in sections and is composed of a loose, hyaline mycelial network. These cells are small (1 to 3 μ) and look like a chain of budding cells.

Colonies. The colonies are slow-growing, with an optimum temperature of 30°C. The appearance is dark, leathery, and folded, with a tan to grey fuzz developing over the surface. Colonies on Sabouraud's dextrose agar are usually sterile except for occasional chlamydospores (Fig. 4–12). On low-nutrient media, pycnidia with pycnidiospores may develop. There is some evidence that this species and *Pyrenochaeta romeroi* are related or identical.[40, 48]

Culture. Like *M. mycetomii,* this species is amylolytic and weakly proteolytic. Dextrose, galactose, sucrose, and maltose are assimilated, but not lactose. The lactose and sucrose reactions separate it from *M. mycetomii.*

Phialophora jeanselmei (Langeron) Emmons 1945

Synonymy. *Torulla jeanselmei* Langeron 1928; *Pullularia jeanselmii* Dodge 1935; *Fonsecaea jeanselmei* Lewis 1948; *Cadophora lignicola* [Nannfeldt] apud Melin and Nannfeldt 1937.

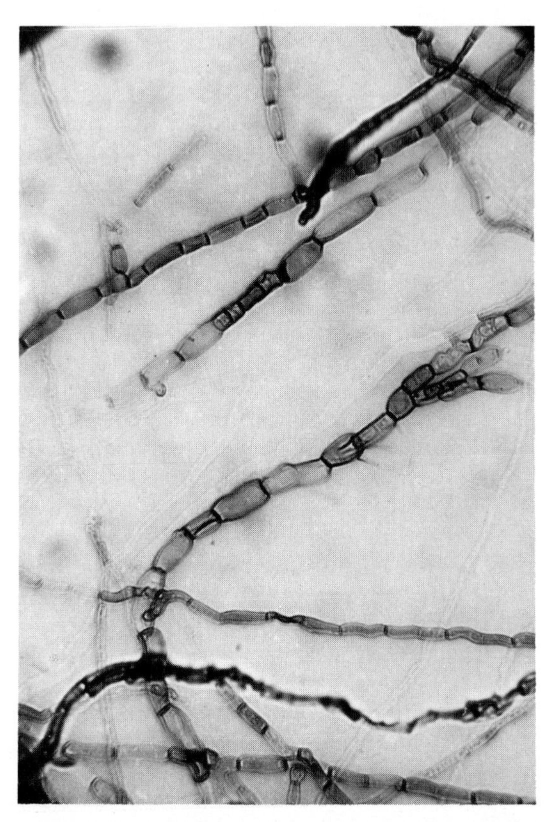

Figure 4–12. *Madurella grisea.* Mycelia of two widths and chains of chlamydospores characteristic of the species.

This organism was first isolated in 1928 from a case of black-grained mycetoma of the foot by Jeanselme on the island of Martinique.[24a] It has occurred infrequently in cases of mycetoma in the United States, and was reclassified by Emmons as a species of *Phialophora* because of its mode of spore production.[16] It was originally designated as a species of *Torula* because of its "toruloid" or yeastlike phase seen in culture.[26a] It is a common inhabitant of soil and has been isolated from wood pulp and named *Cadophora lignicola.* It has been isolated from cases of mycetoma with soft, black, serpiginous granules, from subcutaneous abscesses that show brown chlamydospores and hyphal fragments, and from cases of chromomycosis with sclerotic round bodies in tissue and the hypertrophied verrucous skin lesions characteristic of that disease (Fig. 4–13). Isolants have also been recovered from onychomycosis and tinea pedis, but their significance in these conditions is as yet undetermined. The organism appears to be an opportunist with versatile adaptability.

Clinical Disease. As reported it is similar to that of the other eumycotic mycetomas.

Grains. A vermiform or serpiginous undulating grain is seen that may be hollow in the center. It is black, of a soft consistency, and quite variable in size. It is composed mainly of swollen cells (chlamydospores) which are 5 to 10 μ in size. Brown hyphal elements are also seen. The granule is often helicoid, crescent-shaped, or stringing through a fistula like a boat in a canal. It is quite distinct from other mycetoma grains.

Colonies. To define the cultural characteristics of *P. jeanselmei* is perplexing. The initial slow growing colony is mucoid or leathery, glabrous and dark. Toruloid or yeast cells predominate. Sporulation at this point resembles *Aureobasidium pullulans.* In time, a greyish-black, velvety, aerial mycelium develops with black pigment on the reverse side, giving a dematiaceous appearance. Moniliform hypha and numerous chlamydospores are seen. In older colonies, tubelike phialids are formed with narrow, tapered tips. The flared tip, characteristic of *Phialophora* sp., is lacking. Instead, the phialids vary from minute, barely visible structures to long thin phialophores. The spores are quite variable in size, with an average dimension of 5 to 6 $\mu \times 1$ to 2 μ, and there is no attachment scar. They may be seen clustered about the opening of a phialid. Optimum growth is at 30°C.

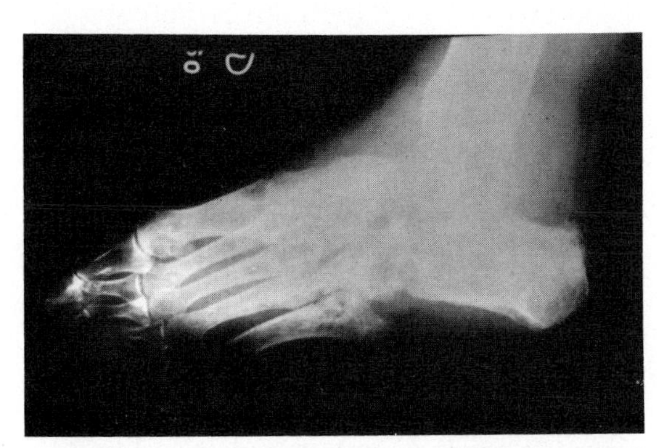

Figure 4–13. Actinomycotic mycetoma. X-ray of foot showing destructive lesions of tarsals and metatarsals. (Courtesy of E. Macotela-Ruiz.)

Culture. *P. jeanselmei* is neither proteolytic nor amylolytic. The former serves to separate it from *Cladosporium* species that may have similar colonial characteristics. Dextrose, galactose, maltose, and sucrose are assimilated. Lactose is not.

Acremonium kiliense Gruetz 1925
Acremonium falciforme Carrión 1951
Acremonium recifei Arêa Leão and Lobo 1934

The first organism was reported under the name *C. falciforme* from Puerto Rico where it was isolated from two cases of white-grained mycetoma. Since then, several cases have been described in America, Africa, and Rumania. *Cephalosporium falciforme* and the *C. recifei* Arêa Leão and Lobo are now considered to be with *Acremonium kiliense* in the genus *Acremonium.*

Clinical Disease. In terms of grains and tissue reaction it is similar to that described for *A. boydii* mycetoma. The *grain* is whitish to yellow in color, soft, irregular, and up to 1.5 mm in diameter. Hyaline hyphae 3 to 5 μ in diameter are found, as well as numerous swollen cells, especially in the periphery of the grain.

Colonies. The growth is relatively slow, and a tufted, downy colony is produced that may be white, buff, pinkish, or lavender. A diffusible currant-red to violet pigment is seen on the reverse of the colony. Hyphae are hyaline and 2 to 4 μ in diameter. Typical *Cephalosporium* conidiophores and conidia are produced. The conidia are crescent- or cycle-shaped, multicellular, and 10×4 μ in diameter. They are borne successively from the tips of conidiophores and held there as a cephalic, "headlike" cluster by mucilaginous exudate. Numerous chlamydospores of variable size are produced, and nodular bodies are regularly seen. Optimum temperature is 30°C.

A. falciforme has multiseptate crescent spores. *A. recifei* has nonseptate crescent spores, and *A. kiliense* has straight spores.

Leptosphaeria senegalensis Segretain, Baylet, Darasses, and Camain 1959

This recently described agent of black-grained mycetoma has been isolated from ten cases in Senegal and Mauritania, and appears to be a common incitant of the disease in that area. So far it is not recorded outside of Africa, although a retrospective study of pathologic material indicates its possible presence in South India.[24,27]

Grains. The grains are soft, black, irregular, and about 1 mm in size. Fragments and aggregates may coalesce to form large masses. These consist of a central core of hyaline hyphae, which are not imbedded in cement. There is a dense periphery of deeply pigmented hyphae interspersed with large vesicular cells in the cement substance. The latter cells also form a hyaline layer just underneath the dark outer periphery. The grains resemble those of *Phialophora jeanselmei, Pyrenochaeta romeroi,* and *Madurella grisea.* There is less lobulation, and the border is less even and more crenated than in the latter two organisms. Also, there are larger vesicles and more irregularly arranged hyphae in *L. senegalensis* grains.

Colonies. The organism grows rapidly, producing a downy grey-brown colony. The reverse is black, and a rose tint may develop in the agar. Ascocarps are usually not produced on Sabouraud's agar. After several months on cornmeal agar, dark brown spherical perithecia are produced, which are 100 to 300 μ in diameter. Asci with interspersed paraphyses are found within the carp. The asci are elongate, being 80 to 110 μ × 20 μ in size. The ascospores are eight in number, elongate to oval, 23 to 30 μ × 8 to 10 μ, and five- to eight-celled. The walls are smooth and hyaline to light brown in color.

Pyrenochaeta romeroi Borelli 1959

This organism was isolated from black-grained mycetoma by Borelli in Venezuela. The granules are soft and black, tubular in shape, and range from 0.5 to 1.5 mm in size. There is a central network of hyphae and a thick band of polygonal inflated cells (chlamydospores) in the peripheral areas. The outer edge is composed of dark, swollen cells. The colony is floccose, woolly, and greyish-black, with a light-colored periphery. The reverse is dematiaceous. The fungus grows rapidly at its optimum temperature of 30°C. Brownish-black osteolate pycnidia are produced, which are 40 to 100 by 50 to 130 μ in size. They are covered with rigid or flexuose, rounded-end, dark setae. The pycnidiospores are elliptical, hyaline to dusky yellow in color, and average 1.5 to 2 × 1 μ or less in size.[14] They are borne in chains on simple conidiophores within the pycnidia. The genus *Pyrenochaeta* occurs commonly in soil and decaying vegetation. Some morphologic studies indicate a close relationship or identity between *P. romeroi* and *M. grisea.*[48] However, other investigators describe serologic differences between the two.[40]

Neotestudina rosatii Segretain and Destombes 1961

This organism has been found in cases of dark-grained mycetoma in the Trans-Africa Belt. The grains are light to brownish, soft, and 0.5 to 1 mm in size. In histologic preparations, a peripheral cement is seen around convoluted hyphae. The center is basophilic in hematoxylin and eosin preparations.

In culture, the organism is slow-growing and produces a leathery, heaped, and folded colony with radial grooves and a flat periphery. The color is tan to brown, with a dark brown dermatiaceous reverse side. Spherical to oval perithecia (about 350 μ) are produced on cornmeal or carrot-potato media after three weeks. The walls are composed of interwoven, dark-colored hyphae, and the interior is filled with colorless mycelia bearing asci. Asci are spherical, 11 to 15 μ long, and contain eight dark, curved, two-celled ascosphores, which are 10 × 4 μ in diameter. Optimum temperature for growth is 30°C.

Two other species have been isolated from mycetoma of animals, *Curvularia geniculata* and *Drechslera spiciferum.* Both are dematiaceous fungi producing a floccose grey-green to black colony. The reverse side is dark. *Curvularia* may produce radiating peripheral filaments and a pink-tinged mycelium in young colonies. The spores are borne on septate, simple or branched, dark conidiophores. There are up to four cells per spore, though more commonly three, with the middle cell enlarged asymmetrically to produce the slight curve. *Drechslera* spores are produced on similar conidiophores, are thick-walled, and contain four or more cells formed by transverse septation. Grains are reported as dark.[9] The disease has been described in cats, dogs, and horses.

Botryomycosis

Botryomycosis or staphylococcic actinophytosis is a chronic, purulent, and granulomatous lesion of the dermis and subdermal tissue. Sulfur granules are produced. This disease mimics actinomycosis and mycetoma. It was originally described as being caused by fungi called "botryomycetes," because clumps of cocci were mistakenly observed as hyphae. The same organism was also wrongly associated with pyogenic granuloma. The disease occurs in animals as chronic, localized, or spreading swollen abscesses that rarely involve lymph nodes or visceral organs. It is a frequent complication of castration in pigs.

Human disease may develop anywhere on the skin, but particularly on the head, hands, and feet.[53,54] The condition is usually localized, but may spread to other organs, such as liver, lungs, kidney, heart, prostate, and lymph nodes. This is usually associated with debilitated patients. Several cases have been reported in children with cystic fibrosis.

The etiologic agent is usually *Staphylococcus aureus,* although there are cases recorded caused by *Pseudomonas aeruginosa,*

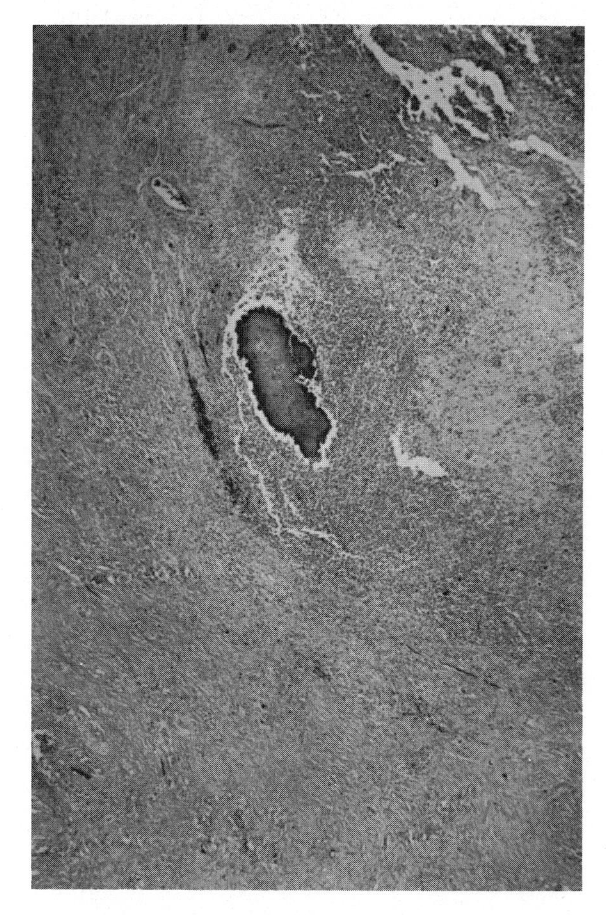

Figure 4–14. Botryomycosis. Small abscess in fibrous scar tissue in neck. Note granule composed of coccoid elements. (Courtesy of S. Thomsen.)

Bacteroides, and *Proteus* species. The general histologic picture is similar to actinomycosis or mycetoma. Neutrophils, lymphocytes, eosinophils, plasma cells, and fibroblasts, as well as histiocytes, spread extensively in cords and sheets, and scattered, foreign-body giant cells are seen. The granules occupy the suppurative center of granulomatous lesions, and are covered by neutrophils. The grains are soft, yellowish-white, lobulated, up to 1 mm in size, and composed of masses of cocci in an eosinophilic milieu (Fig. 4–14). This "cement" is PAS-positive. A "shell" similar to the clubbing of actinomycosis and mycetoma grains has also been described. In patients with this disease, an immune defect or an unusual tissue response for the handling of staphylococci has been postulated. Changes from the normal globulin patterns have not been observed, but the cases are generally associated with other idiosyncrasies of tissue defense and reaction. Little phagocytosis is apparent in the lesions. The organisms isolated are usually of a low order of virulence, but treatment is complicated by the inability of adequate concentrations of antibiotic to penetrate to the organisms sequestered in the grains and granuloma. Treatment is determined by antibiotic sensitivity and necessitates a prolonged course. Surgical exploration and excision are sometimes helpful. The name "granular bacteriosis" has been suggested for this disease.

Mackinnon et al.[31] were able to reproduce the disease in experimental animals. They injected small amounts of *Pseudomonas aeruginosa* intratesticularly into guinea pigs. Granules up to 300 μ in diameter were produced, complete with eosinophilic "rays" and "clubs." In some animals, metastasis to the lung occurred.

REFERENCES

1. Abbott, P. 1956. Mycetoma in the Sudan. Trans. R. Soc. Trop. Med. Hyg., *50*:11–30.
2. Ajello, L., W. W. Walker, et al. 1961. Isolation of Nocardia brasiliensis from a cat. J. Am. Vet. Med. Assoc., *136*:370–376.
3. Ajello, L., and W. C. Basom. 1968. A Mexican case of mycetoma caused by *Streptomyces somaliensis*. Dermatol. Int., 7:17–22.
4. Avram, A. 1967. Grains expérimentaux maduromycosiques et actinomycosiques à *Cephalosporium falciforme, Monosporium apiospermum, Madurella mycetomi*, et *Nocardia asteroides*. Mycopathologia (Den Haag), *32*:319–336.
5. Avram, A., and D. Hatmann. 1968. Quelques aspects concernant l'ostéite mycetomique. A propos d'une etude sur 23 cas roumains. Mykosen, *11*:711–717.

5a. Berd, D. 1973. *Nocardia brasiliensis* infection in the United States. Am. J. Clin. Pathol., *60*:254–258.

6. Bergeron, J., J. F. Mullens, et al. 1969. Mycetoma caused by *Nocardia pelletieri* in the United States. Arch. Dermatol., *99*:564–566.
7. Blank, H., and G. Rebell. 1965. Thiabendazole activity against the fungi of dermatophytosis, mycetomas, and chromomycosis. J. Invest. Dermatol., *44*:219–220.
8. Bojalil, L. F., and A. Zamora. 1963. Precipitin and skin tests in the diagnosis of mycetoma due to *Nocardia brasiliensis*. Proc. Soc. Exp. Biol. Med., *113*:40–43.
9. Brodey, R., H. F. Schryver, et al. 1967. Mycetoma in a dog. J. Am. Vet. Med. Assoc., *151*:442–451.
10. Butz, W. C., and L. Ajello. 1971. Black grain mycetoma. Arch. Dermatol., *104*:197–201.
11. Cardenas, J. V., V. Callev, et al. 1966. Micetomas. Antioquia Med., *16*:117–132.
12. Carrión, A. L. 1956. *Cephalosporium falciforme*, n., sp. a new etiologic agent of maduromycosis. Mycologia, *43*:522–523.
13. Cooke, W. B., and P. Kabler. 1955. Isolation of potentially pathogenic fungi from polluted water and sewage. Public Health. Rep., *70*:684–694.
14. David-Chausse, J., L. Texair, et al. 1968. Mycetoma du pied autochtone a *Pyrenchaeta romeroi*. Bull. Soc. Fr. Dermatol. Syphiligr., *75*:452–453.
15. Destombes, P., F. Mariat, et al. 1961. A propos des mycetomes à Nocardia. Sabouraudia, *1*:161–172.
16. Emmons, C. W. 1945. *Phialophora jeanselmei*, comb. n. from mycetoma of the hand. Arch. Pathol., *39*:364–368.
17. Ernest, J. T., and J. W. Rippon. 1966. Keratitis due to *Allescheria boydii*. Am. J. Ophthalmol., *62*: 1202–1204.
18. Feiger, J. W. 1963. Mycetoma: review of the literature. Milit. Med., *128*:762–765.

18a. Gokhalay, B. B., A. A. Padhye, et al. 1968. Madura foot in India caused by *Madurella grisea*. Sabouraudia, *6*:305–307.

19. Golikov, A. V. 1963. Antibiotic concentrations in actinomycoma and blood of animals with actinomycosis. Antibiotiki, *8*:1045–1048.
20. González Ochoa, A. 1951. Deep tropical mycosis, In Gradwohl, R. B., et al. (eds.), *Clinical Tropical Medicine*. C. V. Mosby, St. Louis, Chapter 50.
21. González Ochoa, A. 1960. Griseofulvin in deep mycoses. Ann. N.Y. Acad. Sci., *89*:254–257.
22. González Ochoa, A. 1973. Virulence of Nocardiae. Can. J. Microbiol., *19*:901–904.
23. Gordon, R. E., and J. M. Mihm. 1962. Identification of *Nocardia caviae* (Erikson) nov. comb. Ann. N.Y. Acad. Sci., *98*:628–636.
24. Indira, P. U., and M. Sirsi. 1968. Studies on maduromycosis. Indian J. Med. Res., *56*:1265–1271.

24a. Jeanselme, M. M., L. Huet, et al. 1928. Nouveau type de mycétome à grains noirs du a une Torulla encore non deceite. Bull. Soc. Fr. Dermatol. Syphiligr., *35*:369–375.

25. Klokke, A. H., G. Swamidasan, et al. 1968. The causal agents of mycetoma in South India. Trans. R. Soc. Trop. Med. Hyg., *62*:509–516.

26. Lacaz, C. da S., and P. S. Minami. 1964. Algunas dados sobre o *Streptomyces paraguayensis.* Hospital (Rio), *65*:41–47.

26a. Langeron, M. 1928. Mycétome à *Torulla jeanselmei* Langeron 1928. Nouveau type de mycétome à grains noirs. Ann. Parasitol. Hum. Comp., *6*:385–390.

27. Lavalle, P. 1962. Agents of mycetoma, In Dalldorf, G. (ed.), *Fungi and Fungus Disease.* Charles C Thomas, Springfield, pp. 50–68.

28. Lechevalier, H. A., and M. P. Lechevalier. 1970. A critical evaluation of the genera of aerobic actinomycetes, In Prauser, H. (ed.), *The Actinomycetales.* Gusta Fisher Verlag, Jena, p. 393–405.

29. Lynch, J. B. 1964. Mycetoma in the Sudan. Ann. R. Coll. Surg. Engl., *35*:319–340.

30. Mackinnon, J. E. 1954. A contribution to the study of the causal organisms of maduromycosis. Trans. R. Soc. Trop. Med. Hyg., *48*:470–480.

31. Mackinnon, J. E., I. A. Conti Diaz, et al. 1969. Experimental botryomycosis produced by *Pseudomonas aeruginosa.* J. Med. Microbiol., *3*:369–372.

32. Macotela-Ruiz, E., and F. Mariat. 1963. Sur la production de mycétomes expérimentaux par *Nocardia brasiliensis* et *Nocardia asteroides.* Bull. Soc. Pathol. Exot., *56*:46–54.

33. Mahgoub, E. S. 1964. The value of gel diffusions in the diagnosis of mycetoma. Trans. R. Soc. Trop. Med. Hyg., *58*:560–563.

34. Mariat, F. Sur la distribution géographique et la répartition des agents de mycétomes. Bull. Soc. Pathol. Exot., *56*:35–45.

35. Mathews, R. S., C. E. Buckley, III, et al. 1968. Local chemotherapy of deep seated fungus infections: Maduromycosis of the foot. Clin. Res., *16*:48.

36. Meyer, E., and R. D. Herrold. 1961. *Allescheria boydii* isolated from a patient with chronic prostatitis. Am. J. Clin. Pathol., *35*:155–159.

37. Mohr, J. A., and H. G. Muchmore. 1968. Maduromycosis due to *Allescheria boydii.* J.A.M.A., *204*:335–336.

38. Montes, L. F., R. G. Freeman, et al. 1969. Maduromycosis due to *Madurella grisea.* Report of the fifth North American case. Arch. Dermatol., *99*:74–79.

39. Murray, I. G., E. T. C. Spooner, et al. 1960. Experimental infection of mice with *Madurella mycetomii.* Trans. R. Soc. Trop. Med. Hyg., *54*: 335–341.

40. Murray, I. G., and H. R. Buckley. 1969. Serological differences between *Pyrenochaeta romeroi* and *Madurella grisea.* Sabouraudia, 7:62–63.

41. Murray, I. G., and E. S. Mahgoub. 1968. Further studies on the diagnosis of mycetoma by double diffusion in agar. Sabouraudia, *6*:106–110.

42. Negroni, P. 1960. La anfotericina B en el tratamiento de los micosis profunda. Rev. Argent. Dermatosif., *40*:204–209.

43. Nielson, H. S., Jr. 1967. Effects of amphotericin B in vitro on perfect and imperfect strains of *Allescheria boydii.* Appl. Microbiol., *15*:86–91.

44. Nielson, H. S., Jr., N. F. Conant, et al. 1968. Report of a mycetoma due to *Phialophora jeanselmei* and undescribed characteristics of the fungus. Sabouraudia, *6*:330–333.

45. Neuhauser, I. 1955. Black grain maduromycosis caused by *Madurella grisea.* Arch. Dermatol., *72*:550–555.

46. Rippon, J. W., and G. Peck. 1967. Experimental infection with *Streptomyces madurae* as a function of collagenase. J. Invest. Dermatol., *49*:371–378.

47. Rosen, F., J. H. Deck, et al. 1965. *Allescheria boydii* unique dissemination to thyroid and brain. Can. Med. Assoc. J., *93*:1125–1127.

48. Segretain, G., and P. Destombes. 1969. Recherche sur les mycetomes à *Madurella grisea* et *Pyrenochaeta romeroi.* Sabouraudia, 7:51–61.

49. Segretain, G., and F. Mariat. 1968. Recherches sur la presence d'agents de mycetomes dans le sol et sur les spineux du Senegal et de la Mauritanie. Bull. Soc. Path. Exot., *61*:194–202.

50. Verghese, A., and A. H. Klokke. 1966. Histologic diagnosis of species of fungus causing mycetoma. Indian J. Med. Res., *54*:524–530.

51. Vipulyasekha, S., and S. Vathanabhuti. 1960. Treatment of nocardial mycetoma with sulphamethoxypyridazine. Br. J. Dermatol., *72*:188–191.

52. Wang, C. J. K., and S. Brownell. 1967. Preliminary studies of some physiological properties of *Torula jeanselmei.* J. Bacteriol., *94*:597–599.

53. Waisman, M. 1962. Staphylococcic actinophytosis (Botryomycosis): Granular bacteriosis of the skin. Arch. Dermatol., *86*:525–529.

54. Winslow, D. J. 1959. Botryomycosis. Am. J. Pathol., *35*:153–167.

55. Zaias, N., D. Taplin, et al. 1969. Mycetoma. Arch. Dermatol., *99*:215–225.

Part Two

The Pathogenic Fungi

Chapter 5

CHARACTERISTICS OF FUNGI

The fungi proper, or Eumycetes, are quite distinct from true bacteria in size, cellular structure, and chemical composition. Differences in cell wall composition and nuclear structure were discussed in the chapter on the actinomycetes. Fungi are eukaryocytes because of their nuclear structure and mode of genetic recombination and, along with algae and protozoa, are protists. The term describes their lack of irreversible differentiation into functional organs and tissues.

Though fungi are essentially single-celled organisms, in some fungal species the cells may show various degrees of specialization, such as in the pseudoparenchymous tissue of the fruiting bodies of higher Ascomycetes and Basidiomycetes. The simplest morphologic form of the fungi is the single-celled, budding *yeast.* As complexity increases, elongation of the cell without separation of newly formed cells results in the threadlike *hypha.* An intertwined mass of hyphae is called a *mycelium.* The popular term *mold* (mould) also refers to the filamentous fungi. The terms "hyphae" and "mycelia" are sometimes used interchangeably. The mycelial mat is known as a *thallus,* but specifically this term refers to the colonial growth derived from a single spore. In the fungi discussed here, it is a loose network of hyphae, also called a colony. *Thallospores* are any of the many types of asexual spores produced on a thallus or colony. In some higher forms, hyphae are cemented together to form large, structurally complex, fruiting bodies, such as mushrooms and puffballs, which may weigh more than 60 pounds. Even though they attain such great size, all fungi are still "primitive" organisms. Any single cell separated from a 60-pound puffball is capable of regrowing the entire structure. This regenerative ability separates the fungi, protozoa, and algae (protists) from the higher multicellular forms such as vascular plants and metazoan animals.

As to the origin of fungi, the debate concerning their evolution from other forms has been long and controversial. Modern scientific analysis has laid to rest the concept of a bacterial ancestor of fungi, where the actinomycetes were the "link" between the two groups. The simplistic notion that fungi are algae without chlorophyll is also untenable at face value. There are several algae without chlorophyll, e.g., *Prototheca* was probably derived from *Chlorella,* but still retains many algaelike characteristics of composition, physiology, and reproduction which are quite distinct from fungi. Some primitive fungi are motile at one stage in their growth cycle as are algae and protozoa, but the chitinous cell wall of most fungi is uncommon in protists. Polyphyletic evolution for the several groups of fungi is still entertained. The presence of cellulose in the cell walls, motility, and the reproductive structures found in the phycomycetes has suggested ancestry with the Rhodophyta (red algae). Sessile protozoa have been proposed as ancestors for the other groups. There is perhaps a clearer relationship between the Ascomycetes and Basidiomycetes than between these latter groups and the Phycomycetes. The Phycomycetes group is divided into several classes by some authors. Further speculation as to the origin of fungi is beyond the scope of this book.

Mycelium. Two main structural types of

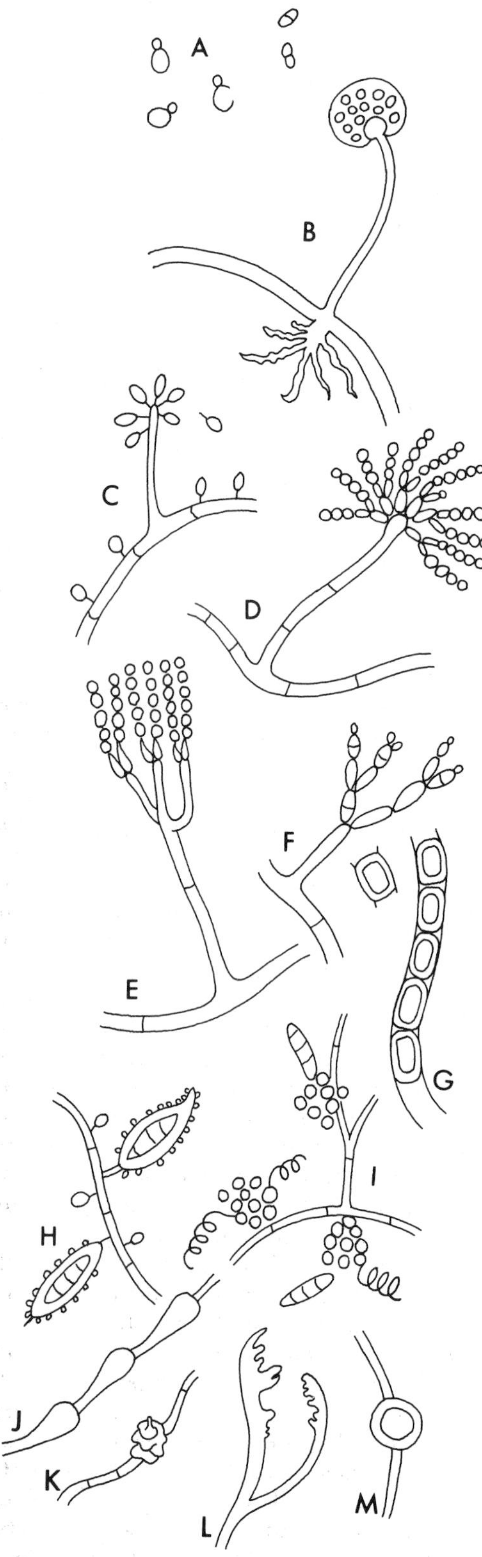

Figure 5–1. Types of asexual spore formation in the fungi. *A,* Budding yeasts, e.g., *Saccharomyces,* and fission yeasts, e.g., *Schizosaccharomyces. B, Rhizopus,* a zygomycete with coenocytic hyphae, sporangiophores, and sporangia. The rootlike processes are rhizoids. *C, Sporothrix,* simple spore formation on a conidiophore and from mycelium. *D, Aspergillus,* deciduous conidia produced by basipetalos budding from raylike projections (sterigmata). The swollen end of the conidiophore bearing the sterigmata is the vesicle. *E, Penicillium,* fingerlike projections (metulae) bear sterigmata that bud off deciduous conidia. *F, Cladosporium,* acrepetalos production of conidia. *G, Coccidioides,* fragmentation of mycelium into arthrospores. *H, Microsporum canis,* microaleuriospores and macroaleuriospores. The thick-walled, echinulate macroaleuriospore is characteristic of the genus *Microsporum. I, Trichophyton mentagrophytes,* microaleuriospores *en grappes,* in grapelike clusters, spiral hyphae. The thin-walled macroaleuriospore is characteristic of the genus *Trichophyton. J,* Racquet mycelium. *K,* Nodular body. *L,* Pectinate body. *M,* Intercalary chlamydospore characteristic of *Microsporum audouinii.*

mycelium may be distinguished. In one of these, the cells making up the hyphae are not marked by cross walls or septa, thereby allowing flow of protoplasm (protoplasmic streaming) throughout the multinucleate structure. Such a structure is said to be nonseptate or coenocytic (Fig. 5–1, *B*). Certain of the algae are of similar structure, and the fungi thus characterized are called Phycomycetes, or algaelike fungi.

In the majority of fungi, however, there are evident septations, each cell being separated from the other by cross walls. The septations have "holes" in them so that there is free flow of cytoplasmic material. Nuclear migration is possible in the Ascomycetes through a large pore in the intracellular septa. In the Basidiomycetes, a special structure comprised of the dolipore and parenthesome prohibits nuclear migration (Fig. 5–2). The dolipore is an expanded, liplike structure around either side of the opening of the septal wall. The parenthesome is a cap or cuplike structure a short distance from either side of the dolipore opening. Held in place by fibrils, it allows cytoplasmic, but not nuclear, migration and covers the dolipore lip when there is a lessening of the pressure on the other side of the septa. Each Basidiomycetes cell contains two dissimilar haploid nuclei (dikaryon), one derived from one parent or mating type, and the other from the other parent. The mycelium of this latter group may also show a special bridge structure, a bumplike eminence connecting one cell to the next. This is called a clamp connection. It is involved in the transfer of one of the two daughter nuclei derived from the

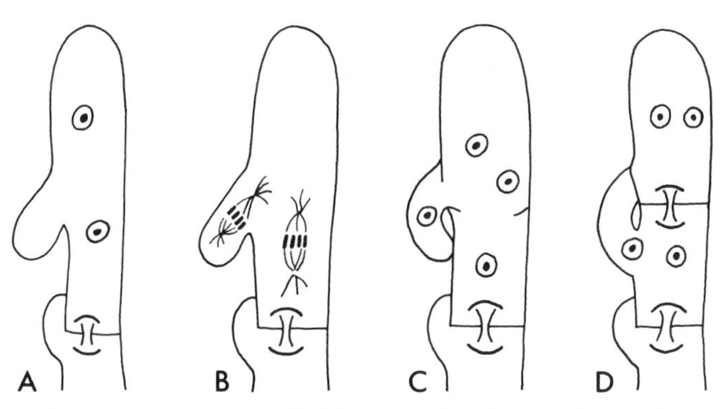

Figure 5–2. Clamp formation in Basidiomycetes. *A,* Dikaryon cell. A backward protrusion develops from the growing end cell. *B,* One nucleus migrates into the neck of the protrusion and the other to the base of the cell. Both divide. *C,* Of the pair produced in the neck, one migrates to apex of cell, and the other migrates backward into growing clamp. Of the pair produced in the base of the cell, one migrates to apex, and one stays in the base. *D,* Septal rearrangement, nuclear migration, and formation of dolipore and parenthesome. Both the new cell and the cell from which it was formed have a pair of nucleii. These nucleii (dikaryon: n + n) represent the haploid nuclei of the original (parental) mating strains. The clamp is shut, and further nuclear migration is prevented.

parental nuclei to a newly formed cell. One nucleus goes through the clamp, which is then closed, and the other is separated by new septum formation. Although this is not the primary criterion for differentiation, the three main classes of fungi may be distinguished in part by the characteristics of their mature nonsporulating hyphae. It is possible, however, to study electron micrographs of hypha in pathologic material or from cultures and to determine the group to which the organism belongs.

The mycelium is further delineated into two general types which differ in function. One of these, the vegetative mycelium, consists of masses of hyphae within the colony, adjacent to and growing into the substrate. They are concerned with the digestion and assimilation of food materials. Fragments of this mycelium will grow and reproduce if transferred. Anastomoses of hyphal elements are commonly seen. The other type, the reproductive or fertile hyphae, usually extends into the air to form aerial mycelium, and gives rise to various types of spores or reproductive bodies. The mode of spore formation and the structure, size, shape, and morphology of spores and spore-bearing elements are the characteristics by which fungi are differentiated, classified, and identified.

In addition to reproducing through sporulation and the vegetative growth of hyphae, many species of fungi reproduce by the simple separation, from any part of the mycelium, of cells known as oidia. These vegetative reproductive forms may give rise to a new mycelium or reproduce themselves by budding like the yeasts, depending on the available nutrients and the environment into which they are placed.

Yeasts. Defined morphologically, a yeast is a single-celled fungus that reproduces by simple budding (Fig. 5–1, *A*). Under ordinary conditions, the life cycle of this fungus is spent as a single-celled, budding structure, although it may elongate to form a pseudomycelium under the influence of certain environmental conditions. In a stricter sense the true yeasts are Ascomycetes, which, under proper conditions of nutrition, temperature, and sexual mating, produce ascospores. Beer yeast, *Saccharomyces cerevisiae,* is one of these. Certain of the imperfect fungi, such as *Candida* and *Cryptococcus,* retain a yeastlike morphology under most circumstances and are referred to as "false yeasts." The *Candida* fungi are probably derived from true yeasts and exhibit pseudomycelium formation under certain nutritional conditions. Cryptococci do not normally show any cellular elongation, though some mycelial variants suggest this genus is related to the Basidiomycetes. Yeasts are not primordial or primitive fungus cells, and it is probable that they developed via evolutionary specialization.

Many filamentous fungi grow and reproduce as yeastlike cells in certain environments. The best example of this is the parasitic yeast stage found in histoplasmosis, blastomycosis, and sporotrichosis. A suitable term for this transition is dimorphism. In the cases cited, the organisms are stimulated primarily by temperature and the tissue environment, and thus are thermal dimorphic organisms. *Coccidioides* also has a parasitic stage, which is not a budding yeast but a spherule with internal spore formation. It is induced by environmental or tissue factors and therefore can be considered dimorphic also. Many other pathogenic, filamentous fungi may be induced into a yeastlike stage by a particular environment. However, under normal conditions of growth, they are mycelial. Induced yeastlike growth may be concomitant with increased pathogenic potential in some species; in others, it is an artifact of growth conditions. In its simplest definition, a yeast is a fungus that is normally a single budding cell under optimal conditions of temperature and growth. If temperature or environmental factors or both are necessary for transition to a yeastlike stage, it is a mold that exhibits dimorphism. A parasitic stage refers to the morphology of the organism found in tissue, if it is different from its normal saprophytic growth. Some fungi that invade tissue (particularly opportunistic fungi) exhibit neither dimorphism nor a parasitic stage. *Aspergillus* and *Mucor* are mycelial in morphology, both as saprophytes and in human tissue. Thermal dimorphism, nutritional dimorphism, and tissue dimorphism are discussed in detail later in this book.

SPORE FORMATION AND CLASSIFICATION

Two classes of spores are distinguished in the fungi: (1) sexual spores, which are produced by the fusion of two nuclei (sexual spores may or may not be derived from different mating strains); and (2) asexual spores, which arise by differentiation of spore-bearing hyphae without nuclear fusion. If a sexual spore is produced only by fusion with a nucleus of another mating type, the fungus is said to be heterothallic. If a nucleus from within the same thallus will serve for fusion with another nucleus of that thallus, the fungus is homothallic. Fungi in which a sexual spore is produced are known as "perfect fungi," and can be assigned to one of the three classes—Phycomycetes,* Ascomycetes, and Basidiomycetes. Many perfect fungi produce numerous asexual spores which serve for species dissemination, and on rare occasions only reproduce sexually with genetic recombination and formation of sexual spores. Many fungi either have lost the ability to form sexual spores or their proper mating type or inducing environment is not known. These latter fungi reproduce by asexual spore-formation only and are called "Fungi Imperfecti" or Deuteromycetes. Some other fungi produce no spores at all and are termed "mycelia sterilia."

For many years most of the pathogenic fungi were known in their asexual stage only, and were named and classed as Deuteromycetes (Fungi Imperfecti). In many of these, recent work has revealed that a sexual stage is produced as a result of contact with an opposite mating type. Since the type and form of sexual reproduction is the basis for biological classification, the new sexual stage is given a descriptive name, allying it with related genera and species. Thus many pathogenic fungi now have two names, an imperfect name (usually the first described) and a perfect name. Although it may be confusing at first, it is taxonomically legal. As an example, the common dermatophyte, *Trichophyton mentagrophytes* (asexual stage name), is also known as *Arthroderma benhamiae*, the descriptive name of its perfect stage. To add to the confusion, the dermatophyte called *Microsporum gypseum* is the imperfect stage of two species of perfect fungi, *Nannizzia gypsea* and *N. incurvata*. Although the asexual spores of the two species are not discernibly different, the sexual fruiting bodies are. This emphasizes the concept that the term "Fungi Imperfecti" is a convenient descriptive filing cabinet for species waiting to be assigned to meaningful categories, if their sexual stages can be discovered. The inherent pathogenicity of fungi is not related to their sexual or asexual stages. These stages merely describe methods of spore production, just as saprophytic stage or parasitic stage describes the morphologic response to the environment in which the fungus is growing.

There are three medically important classes of fungi based on their methods of sexual reproduction: (1) Zygomycetes, in which the fertilized cell becomes a zygote; (2) Ascomycetes, where the sexual spores are contained in an ascus or sac; and (3) Basidiomycetes, which produce sexual spores on the end of a basidium or club-shaped cell.

*The Phycomycetes have now been divided into several classes; of these, only the class Zygomycetes contains fungi of medical importance.

Asexual Reproduction

In many fungi, the primary method of dissemination of the species is by asexual spores. These are usually simple, single-celled bodies, but they may be multicellular and of various shapes, sizes, colors, morphology, and architecture. These characteristics plus the mode of formation, the structure of the spore-bearing mycelium, and the arrangement of the spores *in situ* are the bases used for the identification of species of fungi. In medical mycology these attributes are particularly important, as the perfect state (sexual fruiting body) is seldom found. A somewhat complex system of nomenclature for the different spore types has been devised to aid in their differentiation. Familiarity with these terms is necessary for accurate identification of the isolated pathogen or accidental contaminant from pathologic material.

Media. Since spores and spore production are very important for identification, some attention should be given to media on which the fungus is grown. The common isolation medium used in medical mycology is Sabouraud's agar, which contains peptone and sugar. For most organisms encountered, the type of growth produced on this medium is sufficient for identification, although it tends to promote mycelial growth and suppress sporulation. Spores are generally produced in greater abundance on "deficient" media. These are decoctions of plant material. The most commonly used is cornmeal agar, which is useful in most cases when spores are absent on Sabouraud's agar. Other media include potato-dextrose, potato-carrot, tomato juice, lima bean, and so forth. Three media—Sabouraud's, cornmeal, and potato-dextrose—are sufficient stock for the medical mycology laboratory. In addition, Czapek-Dox (containing salts and glucose) is used in the identification of contaminants, and blood agar is used for induction of dimorphism in systemic pathogens. Primary isolation media have previously been discussed.

Spores and Hyphal Structures. Mycology has been termed an exercise in contemplative observation. One accumulates the characteristics of an isolant, its colonial color and morphology on the obverse, its pigmentation on the reverse, and its spore formation and hyphal structures on microscopic examination. Put together, these details lead to accurate identification, if they are accurately made.

In the common Zygomycetes, the asexual spores are found within a sporangium. For example, in *Rhizopus oryzae*, the white cottony mold, commonly isolated from cases of mucormycosis, soil, compost piles, and so forth, erect, unbranched hyphae (*sporangiophores*) arise from the coenocytic mycelium (Fig. 5–1, *A*). Near the apex of each strand a septum forms. This tip then swells into a globular sporangium. A small evagination of the supporting sporangiophore extends into the sporangium, acting as a sterile supporting structure, the *columella*. Within the sporangium, cleavage furrows form from the peripheral protoplasmic membrane, dividing the internal space into multinucleate protospores. More furrows form and result in the formation of numerous uninucleate sporangiospores. These internally produced asexual spores are released only by rupture of the sporangial wall.

In the higher molds, the asexual spores are not enclosed in a sporangium. Instead they are produced free, either by segmentation, or by budding of the tips of hyphae or from the walls of the hyphae. Such spores are given the general name *conidia*, or conidiospores, and the spore-bearing hyphae, if present, are conidiophores. More specifically, "conidiospore" refers to the spores of an organism such as *Aspergillus* which are successively produced in the conidiophores and are deciduous; that is, they may be shed immediately after production while a new spore is being formed. The term *microconidia* is used to define spores which are single-celled. If the spores have more than one cell, they are a macroconidia. Various spore-bearing structures and modifications may be seen in the fungi, as in the following examples:

(a) In its simplest form conidium production, as found in *Sporothrix schenckii*, is the budding of a spore from the tip of a long slender conidiophore (Fig. 5–1, *C*). The spore is attached to the phore by a tiny neck called an apiculus, which it retains after shedding. There is a bud scar remaining on the phore where the apiculus was attached. Successive spores appear as lateral buds on the conidiophore, segmenting basally from the tip of the phore. Buds may also arise forming conidia anywhere on the thallus mat.

(b) In *Aspergillus* sp. the unbranched conidiophore arises from an enlarged cell of

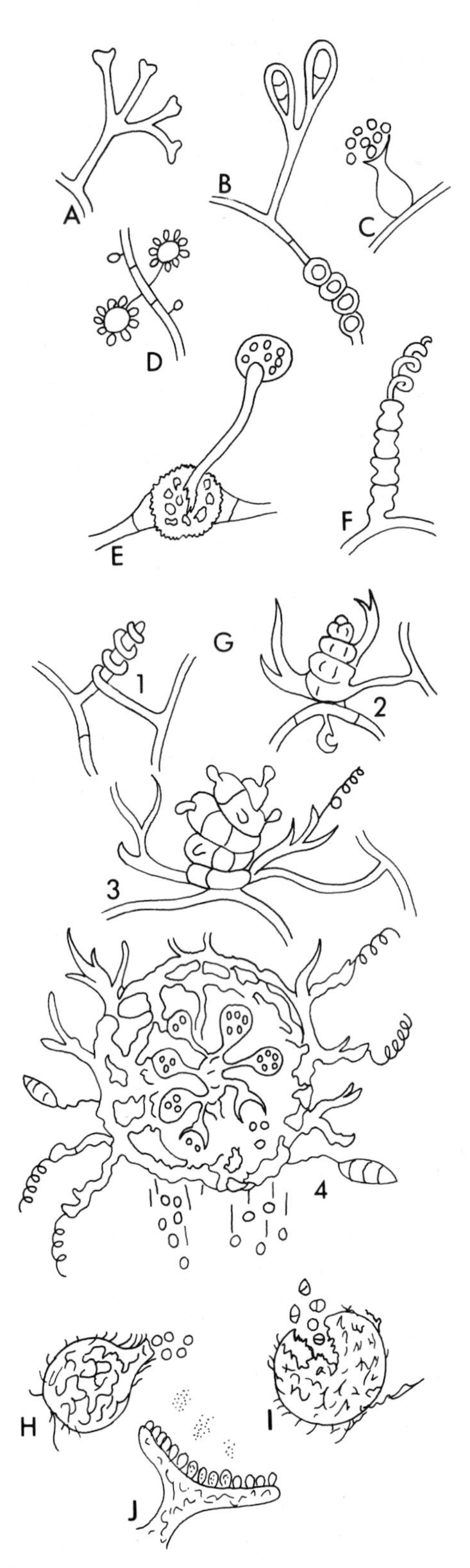

Figure 5–3. Asexual spores, sexual spores, and mycelial forms of fungi. *A,* Favic chandelier of *T. schoenleinii. B, Epidermophyton floccosum,* spatulate (beaver-tail) macroaleuriospore and series of chlamydospores. *C,* Phialide-bearing phialospores. *Phialophora. D, Histoplasma,* echinulate macroaleuriospores. *E,* Zygote (sexual spore) of *Rhizopus,* which opens to produce a sporangium. *F,* Peridial hyphae. *G,* Sexual spore formation in the dermatophytes. (1) The anteridium is entwined by the archigonium. (2) Peridial hyphae begin to surround the ascogenous hyphae. Nuclei from the antheridium have passed into the archegonium, and ascogenous hyphae are formed. (3) Crozier formation and ascus initial are formed. (4) The mature ascocarp contains asci and ascospores. The fruiting body produced by dermatophytes is called a *gymnothecium.* The ascospores can sift through the loose mycelial network. *H,* A perithecium. Mature ascospores are extruded through a pore. *I,* A cleistothecium. The ascospores are released with dissolution of the ascocarp. *J,* An apothecium. The ascospores are usually ejected into the air.

the vegetative mycelium, the foot cell, and terminates in a swollen tip, the vesicle (Fig. 5–1, *D*). A number of bowling pin–shaped stalks, the *sterigmata* or *phialides,* develop from the spherical vesicle. These may be branched or simple, and conidia are abstricted successively from their tip to form chains of conidiospores. This type of sporulation, which has the youngest spore at the base of the chain on the phialide tip, is called basipetalos. Some species of *Aspergillus* also form spores along the vegetative mycelium which are not shed. These are aleuriospores. Phialides with the form of a flower vase are formed by the genus *Phialophora* (Fig. 5–3, *C*).

(c) The common blue-green mold, *Penicillium,* has a similar conidiophore structure, but the terminal portion breaks up into fingerlike, verticillate branches, called *metulae,* and the tips of these (sterigmata) abstrict chains of conidia (Fig. 5–1, *E*).

(d) In *Cladosporium,* the formation of conidia differs from that of many fungi. Instead of successive conidia being formed in a chain in which the terminal spore is the oldest, one spore is formed which buds to form a second, the second a third, and so forth (Fig. 5–1, *F*). This is called *acrepetalos* spore formation. Furthermore, a spore may develop more than one bud, thus producing branching chains of conidia. When branching is extensive, it is called moniliform hyphae.

Aleuriospores resemble conidia in a general way because they are formed on short, lateral branches or directly on the hypha itself rather than on the tips of specialized conidiophores. These spores usually have a thicker point of attachment and are not set free when they mature. Rather they are liberated only when the mycelium that forms them disintegrates. The spores produced by dermatophytes are correctly called *microaleuriospores* and *macroaleuriospores* (Fig. 5–1, *H;* Fig. 5–3, *B, D*). When borne directly on vegetative mycelia, they are referred to as *thyrsi sporiferae, hyphae sporiferae,* or simply as thyrsi. If a number of spores are formed in the same area of a hypha to form a grapelike cluster, they are said to be *en grappes* (Fig. 5–1, *I*). This is seen in *Tricophyton mentagrophytes.* Macroaleuriospores (macroconidia, fuseaux) are large, thick- or thin-walled, septate, and composed of several cells. Macroaleuriospores are formed in large numbers by some fungi, rarely by others, and assume some differential significance. In the dermatophytes, classification into the three imperfect genera (*Trichophyton, Microsporum,* and *Epidermophyton*) is based on the form, size, wall thickness, and presence or absence of echinulations on the macroaleuriospores (Fig. 5–1, *H, I;* Fig. 5–3, *B*).

Fragmentation of preformed mycelium results in the production of a chain of uniformly sized bodies termed *arthrospores* (Fig. 5–1, *G*). This type of spore is common in *Geotrichum, Trichosporon,* and *Coccidioides.* In the latter, the condensation of material around a series of foci results in the formation of arthrospores alternating regularly with empty cells. When freed by fracture of the empty cells, the arthrospores retain irregular bits of the intervening wall. In *Geotrichum* and *Trichosporon,* the barrel-shaped arthrospores are smooth. Many dermatophytes produce arthrospores when they are invading hair or skin, but these are rarely observed in culture. They are not true spores and are perhaps best regarded as oidia in structure and function, though this point is seldom made.

Large, thick-walled structures formed within a hyphae (intercalary) (Fig. 5–1, *M;* Fig. 5–3, *B*) or at the ends of hyphae (terminal) are designated *chlamydospores.* These structures are often irregular in form and size, and some are arranged in a series and assume a spindle form. In such cases, they resemble macroconidia (fuseaux) and, in fact, in much of the older literature are referred to as fuseaux. They are produced abundantly by many fungi along with other spores, or may be produced only under certain conditions and in the absence of other spores. They have little importance in characterizing a fungal species. Sometimes they may become free from the hyphae and bud successively, as in the growth of *Trichophyton verrucosum* at 37°C. It has been suggested that the yeast phase of some fungi, e.g., *Blastomyces dermatitidis,* represents a continually budding chlamydospore. In some dermatophytes, e.g., *Microsporum audouinii,* this may be the only spore seen in culture (Fig. 5–1, *M*).

Vegetative or Hyphal Structures. A number of structures are formed by the vegetative mycelium which have no reproductive significance but are of considerable value in the differentiation of the pathogenic fungi.

Spirals or coiled hyphae, similar to the coiled filaments of actinomycetes, are observed in a number of pathogenic fungi, especially dermatophytes (Fig. 5–1, *I*). They are bedspring-like, helical coils that are found at the ends of peridial hyphae surrounding an ascocarp, but these may be produced in the absence of ascocarps or peridial hyphae. They are very prominent in some strains of *Trichophyton mentagrophytes.*

Nodular organs are formed by some species, e.g., strains of *Microsporum canis* and *Trichophyton mentagrophytes* (Fig. 5–1, *K*). They are enlargements in the mycelium, which consist of closely twisted hyphae. Either side may branch and twine about the parent stem, or different hyphae may entwine together. The resulting structure has the appearance of a knot. This may represent an attempt to initiate a sexual fruiting body.

Racquet mycelium (mycelium *en raquette*) is a term applied to certain hyphae, usually larger than the others, that show a regular enlargement of one end of each segment, large and small ends being in opposition (Fig. 5–1, *J*). The appearance is that of a chain of tennis racquets. This structure is common in many fungi.

The term *pectinate body* is applied to unilateral, short, irregular projections or protuberances forming on one side of a hypha, which give it the appearance of a broken comb (Fig. 5–1, *L*). These are commonly seen in dermatophytes, especially *Microsporum audouinii.*

Favic chandelier (antler hyphae) is the name applied to the structure formed by numerous short, multiple, branches appearing at the end of a hypha. These resemble a reindeer antler or a chandelier (Fig. 5–3, *A*). This structure is not common among the fungi, occurring primarily in *Trichophyton schoenleinii* and occasionally in *Trichophyton violaceum.*

Peridial hypha is a wide, indented, multiseptate hypha which may terminate in a spiral (Fig. 5–3, *F*). This structure is sporadically produced by some strains of fungi. It resembles the ornamentation found around the sexual fruiting body of some perfect dermatophytes or other Ascomycetes, hence its name. It may be numerous in some strains of *Trichophyton mentagrophytes.*

Pycnidia is a mycelial structure resembling the fruiting body (perithecium) of some Ascomycetes. However, the structure is filled with asexually produced conidia. The body is large, up to several millimeters in diameter, and may have a hard wall surrounded by peridial hyphae. Some strains of *T. mentagrophytes* produce numbers of pycnidia, especially when grown on soil-hair agar.

Sexual Reproduction

The sexual stage of fungi is not often observed in the diagnostic laboratory, but an acquaintance with the phenomenon is necessary for the serious student of medical mycology. The fungi are assigned to three general classes, depending on their mode of sexual spore formation. These are Phycomycetes, Ascomycetes, and Basidiomycetes.

Phycomycetes. This is a large and heterologous group of organisms. The organisms are so diverse that some taxonomists divide the Phycomycetes into six separate classes and call them the "Phycomycetes group." Representatives of these classes include the single-celled chytrides; the water molds, such as *Saprolegnia*; the downy mildews, such as *Phytophthora infestans,* the cause of the potato blight; and the white rusts, such as *Albugo,* the etiologic agent of many plant diseases. The Phycomycetes of medical importance are all found in the class, Zygomycetes. These are common fungi of air and soil, including the common bread molds, *Mucor* and *Rhizopus,* which are involved in human mucormycosis. Also included are the fly-attacking fungi, *Entomophthora,* which cause human subcutaneous phycomycosis, and the *Zoopagales* that entrap and eat small animals and worms.

The Zygomycetes found commonly in the laboratory as contaminants or as agents of mucormycosis have a well-developed mycelial thallus. Fertile hyphae of the aerial mycelia produce numerous asexual spores in sporangia atop sporangiophores. There is an extensive vegetative mycelium. In the genus *Rhizopus,* there is also a multiple-branched, short, rootlike extension from the hyphae, called a rhizoid (Fig. 5–1,*B*), which penetrates the nutrient medium. The mycelia are coenocytic and multinucleate. The nuclei appear to divide directly by constriction, without spindle formation or classic mitotic figures.

Sexual reproduction takes place by simple copulation of the tips of multinucleate hyphae. These tips consist of terminal swellings and arise as branches from the mycelial mat. The tips or gametangia are attracted to one another by sex hormones or fuse following chance contact. If compatible mating strains are necessary for initiation of the sexual cycle, the organism is heterothallic; if side branches from the same thallus fuse and form a zygote, the organism is homothallic. After contact, each of the hyphal tips swells, and a septal wall is formed separating the cytoplasm and nuclei in the swollen end from the rest of the hypha. The wall between the adjacent tips then dissolves, and there is mixing of the cytoplasm from the two mating strains, followed by pairing of the nuclei. The new cell (zygote) which is the product of this fusion enlarges, and the walls become thick and pigmented (Fig. 5–3, *E*). When nuclear fusion takes place, a diploid nucleus is formed. In some species, only one nuclear pair survives to form spores; in other species several may survive. After a period of inactivity, the zygote cracks open, a sporangiophore emerges, and a sporangium develops (Fig. 5–3, *E*). At any time during this process, the formation of haploid nuclei may occur by nuclear conjugation followed by reduction division. The timing of this and the mating type of the spores produced within the sporangia are dependent on the species. In some species, there are haploid spores from each of the parental mating strains; in some, only one sex is produced, and in others both sexes and diploid spores are formed. The zygosporangium is similar to the asexual sporangium, and when it is mature, the walls rupture, releasing the spores. This pattern of sexual reproduction is common to the genera of medical importance: *Mucor, Absidia,* and *Rhizopus* sp. of the order Mucorales.

The agents of human subcutaneous phycomycosis, *Basidiobolus* and *Entomophthora* sp., are in a related order, the Entomophthorales. The mycelium is not so extensive as in the Mucorales, but sexual reproduction follows essentially the same pattern. Asexual reproduction differs from the above, because conidia-like structures are produced on the ends of hyphal tips, which are forcibly ejected or "shot" into the surrounding area, seeking insect prey in the case of Entomorphthora, or frog dung in the case of Basidiobolus.

Ascomycetes. The patterns of sexual reproduction and the specialized hyphal mating branches and fruiting bodies produced are too varied among the ascomycetes to be encompassed in a general discussion. However, the end result of all these variations is the formation of the ascus or sac in which the sexual spores (ascospores) are produced. After nuclear conjugation and reduction division have occurred, the ascus usually contains eight haploid nuclei destined to be contained in spores. The spores are produced by a process called free cell formation. Within the ascus, astral rays appear near one pole of the nucleus and radiate around and converge near the other pole, sectoring off a mass of cytoplasm. Cell walls form along the path of the rays, and a spore is produced. The usual spore number is eight, four representing one mating type and four the other. If only four spores are produced, they are binucleate, containing two haploid nuclei of the same mating type. Homothallic sexual reproduction also occurs.

The simplest pattern of sexual reproduction is seen in the true yeasts. Two compatible mating types come in contact and fuse. The nucleus from one yeast cell enters the cytoplasm of the other. The recipient yeast then becomes an ascus, and the nuclei pair, conjugate, and immediately undergo meiosis. Mitotic divisions then occur, and eight ascospores are formed.

In the group of fungi which produce mycelia, the process is more complex. The pattern found among the perfect dermatophytes, *Blastomyces*, and so forth, will be outlined briefly (Fig. 5–3, *G*). A special hyphal structure, the male unit (antheridium), is produced when it is in proximity to a compatible mating type. The antheridium is a rodlike structure around which the female structure (archegonium) wraps tiself. Plasmogamy (fusion of cells by breakdown of the separating cell wall) occurs, and male nuclei migrate into the archegonium. Nuclear pairing occurs, and nuclei then swarm into the many hyphal branches that develop from the archegonium. These branches are called ascogenous hyphae, and a pair of nuclei migrate to their tips. The tip folds over to form a crook or crozier, and a septum is formed, enclosing the nuclear pair in this dikaryotic end cell (Fig. 5–4, *A*). Mitosis occurs (conjugate nuclear division), forming four nuclei (Fig. 5–4, *B*). Septal formation now divides the crozier into three cells: the neck, having one nucleus; the bend, having two nuclei of opposite mating types; and the tip, containing one nucleus of the type opposite the one in the neck (Fig. 5–4, *C*). The nuclei in the bend cell conjugate to form a diploid nucleus. This cell, which is called an ascus mother cell, enlarges and becomes an ascus (Fig. 5–4, *D, E*). Reduction division and ascospore formation occur as previously described (Fig. 5–4, *F, G, H, I*). The tip cell and the neck cell fuse; their nuclei pair, and the fused cell enlarges to form a crozier, whereby the whole process is repeated again. While this is occurring, adjacent mycelium may develop an extensive network around the asci-producing cells.

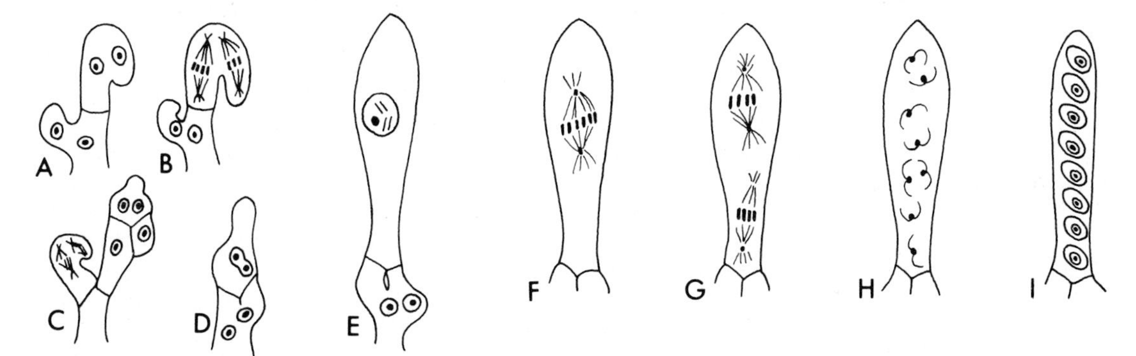

Figure 5–4. Ascospore formation. *A*, Ascogenous hyphae containing both parental nuclei (dikaryon) form a crozier. *B*, The two nuclei divide. One pair of nuclei wander to apex of cell. *C*, Rearrangement of the cell walls of the crozier segregates a pair of nuclei in the apical cell. This is called an *ascus mother cell*. *D*, Nuclear fusion occurs, and a diploid nucleus is produced. *E*, Diploid nucleus in an enlarged ascus mother cell, diplotene formation. *F*, Anaphase I. *G*, Anaphase II and reduction division. *H*, After anaphase III, ascospore delimitation begins. *I*, Mature ascus with eight ascospores.

These are called fruiting bodies or ascocarps. In the dermatophytes, the carp is a loose mesh of hyphae through which the ascospores can sift: a *gymnothecium* (Fig. 5–3, *G4*). If the carp is completely closed and ascospores are released only by disintegration of the carp walls, it is called a *cleistothecium* (Fig. 5–3, *I*). This is the type seen in the perfect *Aspergillus* and *Penicillium* sp. A *perithecium* is a carp with a pore or ostiole through which ascospores may be successively released as they mature (Fig. 5–3, *H*). The common saprophytic fungus, *Neurospora crassa,* which is used for many genetic studies, forms perithecium. When the fruiting body is wide open and cup-shaped, the carp is called an *apothecium.* This group includes cup fungi *(Peziza),* morels, and truffles (Fig. 5–3, *J*).

Basidiomycetes. In general, Basidiomycetes have not been implicated in human disease, but a few instances have been reported of infection produced by the corn smut fungus, *Ustilago,* or the mushrooms, *Schizophyllum* and *Coprinus.* Recently, there has been some evidence that *Cryptococcus, Trichosporon,* and possibly *Candida albicans* may have a sexual stage that belongs in this class. Mushrooms, some of which are toxic and may cause death, and puffballs are also included in this class.

Spores of Basidiomycetes usually germinate to form a mycelial plant (primary mycelium) containing only haploid nuclei. Instead of two sexes, there are often four mating types. When compatible mating strains come in contact, the hyphal tips fuse, but nuclear fusion does not occur. This is called the dikaryon condition (n+n). Both nuclei divide simultaneously in a process called conjugate nuclear division. This is similar to the process that occurs in the crozier cell of the ascomycetes. When the nuclei are about to divide, a short branch from the side of the cell wall forms and arches back away from the hyphal tip (Fig. 5–2, *A*). This branch is called a clamp connection. One nucleus migrates into the neck of the connection and divides. Its spindle fibers point toward the hyphal tip (Fig. 5–2, *B*). One daughter nucleus remains in the clamp, while the other migrates to the hyphal tip (Fig. 5–2,*C*). Meanwhile the other parental nucleus divides in the center of the cell (Fig. 5–2, *B, C*). One daughter migrates to the tip; the other remains more or less in position. The clamp then completes its arch, coming in contact with the cell wall. It then fuses with it, and a septum is formed across the point of origin of the clamp (Fig. 5–2, *D*). Another septum forms through the hypha itself, separating the tip cell. The nucleus that was formerly in the clamp migrates back to the penultimate (original) cell containing the daughter nucleus of the other mating type. The end result is two cells, each of which has two nuclei representing the parental mating strains, and the dikaryon condition continues. This process is repeated each time a new cell is formed, with a clamp pointing back from the growing tip bridging each cell septum. The clamp connection, when present, is a reliable diagnostic characteristic of Basidiomycetes. The dikaryon condition (secondary mycelium) may continue for long periods of time (years) before completion of the sexual cycle occurs. The paired nuclei fuse within a club-shaped terminal cell, the basidium (Fig. 5–5, *A, B, C, D, E, F*). Reduction division occurs, and four spores are produced. These spores are external to the basidium, attached to it by a short neck (Fig. 5–5, *G*). They are forcibly ejected by hydrostatic pressure. In the majority of the Basidiomycetes, a large fruiting body, called a basidiocarp, may form by the massing together of hyphal elements (tertiary mycelium). In mushrooms, the basidia are produced along the gills on the underside of the cap. In the bracken fungi and boletes, they are produced in the lining of pores on the understide of the fruiting body or on the surface of the carp. In the corn smut and wheat rust, a typical carp is not formed.

THE FUNGI IN NATURE

The fungi, though simple organisms, are extremely adaptable to diverse environments. Their chief function in the interplay of biological forms is, to use the popular ecological jargon, to "recycle" organic debris. There is probably no organic substance in the biosphere that is free from attack by fungi. Wood, lignin, vegetation, chitin, keratin, fats, oils, phenols, asphalt, rubber tires, waxes, bones, and so forth, are degradable by fungi. They have been recovered from sulfuric acid–copper plating liquid at pH 0.05, and as fluffy fungus balls, from distilled water bottles, inorganic solutions, acids, and so forth, as "bottle imps" (usually *Penicillium lilacinum*), and they even attack pathology museum specimens contained in

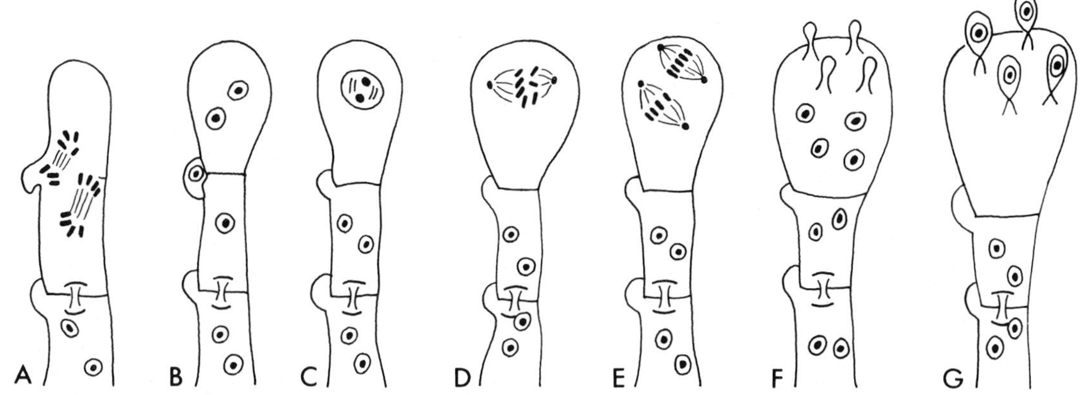

Figure 5–5. Basidiospore formation. *A,* After clamp connection formation and nuclear migration, a probasidium is formed which is still dikaryon (n+n). *B,* Dikaryon probasidium. *C,* Nuclear fusion to form diploid nucleus. *D* and *E,* Meiotic division to form four haploid nuclei (*F*). *G,* Four mature basidiospores on a basidium.

formaldehyde solution. The only requirements for growth are a little organic substrate and moisture. In the process of filling every ecological niche, some one hundred thousand species have evolved. Most are saprobic, but many are plant parasites of great importance. Severe economic loss is caused annually by fungi which attack crops (wheat rust, corn smut, and so forth) and devastate forests (Dutch elm disease) and fruit trees (banana wilt). The history of mankind has been affected by fungi, as witness the potato famine which caused mass migration from Ireland to the new world, or the discovery of a fungal waste product, penicillin, ushering in the new age of chemotherapy and antibiotics. Fungi have adapted to exist as parasites on other fungi as well. Halophilic fungi are known, that appear as white scum or "mycoderma" on brined foodstuffs. Thermophilic fungi are found in compost heaps, where the temperature may reach 45°C (113°F). Some of these temperature-tolerant organisms (as *Aspergillus fumigatus, Rhizopus oryzae,* and *Absidia corymibifera*) readily cause disease in debilitated patients. Fungal products and their metabolites are of great economic importance, for example, the production of alcohol, citric acid, oxalic acid, the production or alteration of steroids, fermentation products, antibiotics, and even the flavoring of coffee and soy sauce.

Few of the fungi are significant parasites of animals. As discussed before, only a small number of fungi have the ability to invade and cause disease in man. However, the remarkable adaptive ability of fungi is nowhere better exemplified than in modern medicine. In order to ameliorate other destructive processes or to treat or prolong life in disease states, therapeutics have often compromised natural defenses of the human host and allowed an ever growing list of saprophytic fungi to gain entrance and invade, usually as a terminal event in a disease. Leukemia, lymphoma, and organ transplant patients on cytotoxins and immunosuppressive agents, diabetics, or those on long-term steroid therapy are particularly vulnerable. Whereas medical mycology was formerly concerned with a few skin-infecting fungi and even fewer pulmonary and systemic infections, the new era in the field is the opportunistic infections of soil and air-borne saprophytes adapting to and taking advantage of the presented organic substrate, the compromised patient. Glibly put, one could say the patient had become a culture media, and these infections might be termed diseases of medical progress.

SUPERFICIAL MYCOSES

Chapter 6

SUPERFICIAL INFECTIONS

The superficial infections include diseases in which a cellular response of the host is generally lacking because the organisms are so remote from living tissue, or infection is so inocuous. There is essentially no pathology elicited by their presence, and patients may be unaware of their condition. These diseases are usually brought to the physician's attention because of their cosmetic effects on the patient.

Pityriasis Versicolor

DEFINITION

Pityriasis versicolor is a chronic, mild, usually asymptomatic infection of the stratum corneum. The lesions are characterized by a branny or furfuraceous consistency; they are discrete or concrescent and appear as discolored or depigmented areas of the skin. The affected areas are principally on the chest, abdomen, upper limbs, and back. The etiologic agent is the lipophilic yeast, *Pityrosporum orbiculare*, formerly known as *Malassezia furfur*.

Synonymy. Tinea versicolor, dermatomycosis furfuracea, chromophytosis, tinea flava, liver spots.

HISTORY

This disease was described very early in the history of medical mycology by Eickstedt (1846) and Sluyter (1847), and termed pityriasis versicolor. They noted the fungal etiology of the lesions. In 1853, Robin named the agent *Microsporum furfur* because he thought it was related to *Microsporum audouini*, and changed the disease epithet to tinea versicolor to relate it to the other ringworm infections. This error was corrected by Baillon (1889), who recognized that the two were not allied, and erected the monotypic genus, *Malassezia furfur*. Gordon[8] first isolated the organism in culture and called it *Pityrosporum orbiculare*. Later work has conclusively shown this organism to be the correct etiology.[13,15,16] The more descriptive term for the disease, pityriasis versicolor, is preferred, and the corrected binomial for the fungus should be *Pityrosporum furfur*.

ETIOLOGY, ECOLOGY, AND DISTRIBUTION

The etiologic agent, *P. orbiculare*, has been shown to be a common endogenous saprophyte of the normal skin. Roberts[16] has isolated it from a normal scalp in association with *P. ovale*, from the chest in 92 per cent of people, and to a lesser extent from the back, trunk, limbs, and, occasionally, the face and other areas. It appears that the organism is universally present as a member of the normal flora of skin, and may be regarded as an opportunist which elicits the disease under special systemic or local conditions. The factors responsible for the overgrowth of the organism in certain people to produce the clinical condition of pityriasis versicolor are not yet known. A relation to squamous cell turnover has been suggested because of the occurrence of the disease in persons with increased endogenous or exogenously administered corticosteroid.[5,10] Under such conditions, the turnover rate of the epithelium is slowed, and many patients develop pityriasis versicolor which disappears when medications cease or endogenous levels are normalized. In the tropics, where a 50 per cent case rate in the population is sometimes seen, this also could be attributed to a slowing of epithelial turnover. Other authors have suggested a genetic predisposition,[15] poor nutritional state, or the accumulation of extracellular glycogen in patients prone to the disease.[12] Poor health, chronic infections, excessive sweating, and at times, pregnancy have been regarded as predisposing factors. The low incidence of conjugal cases (7.5 per cent) also suggests individual predilection.[15] The sexes are about equally affected, and the disease is of worldwide distribution and found in all races. It is quite common in temperate climates and is very prevalent in the tropics and subtropics. Case rates up to 50 per cent have been reported in Mexico, Samoa, Central and South America, India, parts of Africa, Cuba, the West Indies, and the Mediterranean region. The disease is seen in young adults and is usually established in the early twenties. It frequently becomes chronic, and there may be long periods when no lesions are seen. In some climates, a severe form of the disease occurs in infants as a depigmenting diaper rash (achromia parasitica). *P. ovale* is also capable of producing the same clinical condition, especially on the face, eyelids, and ears.

CLINICAL DISEASE

The clinical lesions are very easy to diagnose, and the term versicolor is particularly suitable. The color varies according to the normal pigmentation of the patient, exposure of the area to sunlight, and the severity of the disease. Fawn, yellow-brown, or dark brown patches may be observed, or the lesion may consist of one continuous scaling sheet, involving the entire chest, trunk, or abdomen (Fig. 6–1 FS A 17). The lesions start as tiny, multiple, macular spots that soon scale and enlarge. They may coalesce to form gyrating areas of intermittent scaling of various shades and colors. Papular lesions are sometimes seen and are usually perifolicular

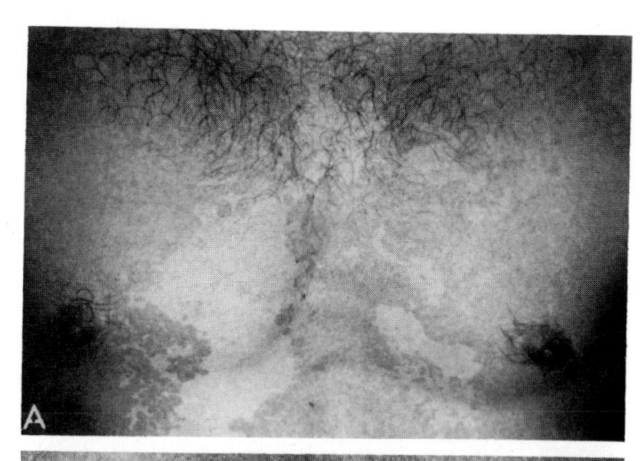

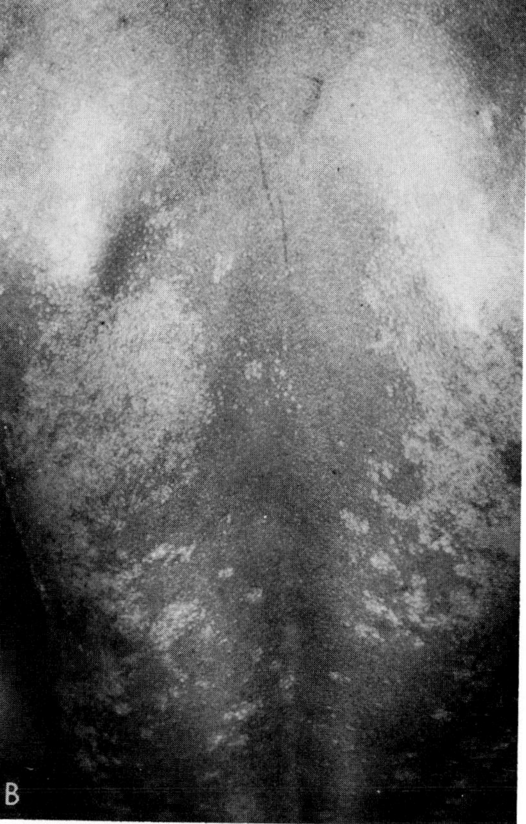

Figure 6–1. FS A 17. Pityriasis versicolor. *A*, Yellow-brown macules with scaling as a continuous sheath, involving the entire chest. Light areas of uninfected normal skin are in the center. *B*, Achromia parasitica. This patient had severe pityriasis as a child and now has a chronic case. In this case the light, unpigmented areas represent the infected sites of the skin.

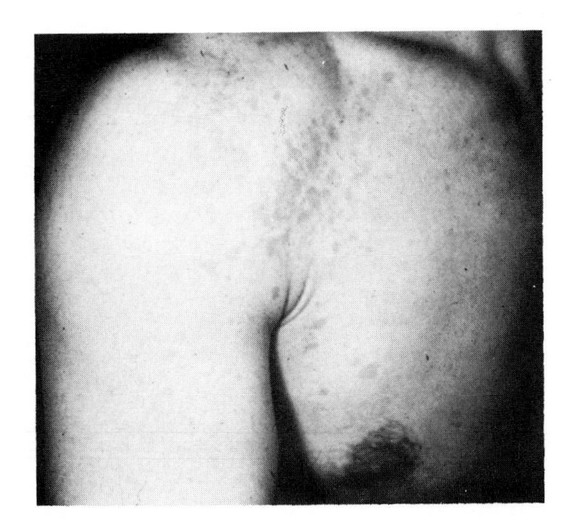

Figure 6–2. Pityriasis versicolor. The common form of the disease, represented by multiple punctate scaling macules with a slight fawn discoloration.

(Fig. 6–2). There is seldom any irritation or inflammatory response, but in some patients a little reddening and some pruritus may occur. Many patients do not become aware of their condition until after exposure to sunlight. The chief complaint is a cosmetic one, in which areas of the skin fail to tan normally. The fungus or its products filter the rays of sunlight and interfere with normal tanning, so that lesions are outlined as lighter areas of color than the surrounding skin.

In dark-skinned infants, particularly in the tropics, a clinical variant is sometimes seen. The infection starts in the diaper-covered areas and spreads rapidly, causing marked depigmentation of the skin. This condition is referred to as pityriasis versicolor alba or achromia parasitica.

Differential Diagnosis. The diagnosis of pityriasis versicolor is quite easy to make by using a potassium hydroxide preparation of skin scrapings or by observing the yellow fluorescence of the affected area under the Wood's lamp. However, in some instances if organisms are sparse, microscopic observation of the fungus may be difficult. In addition, home treatment, especially application of ointments, may interfere with Wood's lamp examination. The infection may be confused with vitiligo or other pigmentary disorders, such as chloasma. In contrast to pityriasis versicolor, tinea circinata, seborrheic dermatitis, pityriasis rosea, erythrasma, secondary syphilis, and pinta are usually somewhat inflammatory. Pityriasis rosea is usually more acute, with rapid spreading after the appearance of the herald spot. Erythrasma is generally restricted to body folds and intertriginous areas. Demonstration of the scaly nature of pityriasis versicolor may require light scratching with a microscope clide, scalpel, or fingernail (the *coup d'ongle* of Besnier). The desquamative lamina are very characteristic of pityriasis versicolor. The squames can be loosened as a single unit encompassing the entire area of the lesion, if small. This can be done with stripping tape, and the squames subsequently stained by crystal violet, methylene blue, or iodine. In this manner the organisms may be readily seen. Most other dermatoses that simulate pityriasis versicolor have irregular or distorted scaling. If observation of the lesion is not hindered by obstructive medications, Wood's lamp examination will show a golden yellow fluorescence, usually in an area which is much greater than the apparent lesion.

Prognosis and Therapy. Numerous topical agents have been used for treating pityriasis versicolor with only qualified success. With or without therapy, the lesions persist, spread, disappear, and reappear or become continuous and chronic. Keratolytic agents, such as Whitfield's ointment or salicylic acid, mild fungicides, as sodium hyposulfite (20 per cent aqueous solution), or 2 per cent sulfur in ointment base will control and cure lesions. Recurrences are common. The treatment of choice appears to be 1 per cent selenium sulfide in a water-miscible ointment base or as shampoo. These agents are applied to the affected areas for ten minutes and then rinsed. If this procedure is repeated three times weekly for several weeks, the lesions should clear. The patient usually remains free of lesions for a year or more. It is necessary to reassure the patient that the disease is not a serious mycosis, and that recurrences are common and to be expected.

PATHOLOGY

The hypha and budding yeasts of *P. orbiculare* are restricted to the outermost layers of the stratum corneum (Fig. 6–3). They may be present in such large quantities that they appear as a continuous layer cemented together and replacing the outer cells of the corneum. Since there is little if any irritation, no pathology of the skin is seen in biopsy sections. The organisms are sometimes seen to mass near the hair follicle and extend into the follicular opening. Cells of the organism may be seen incidentally in biopsies from unrelated dermatoses and cause confusion in diagnosis. The yeast cells are up to 8 μ in diameter, and the hyphal fragments are 3 to 4 μ in width. It is usually easy to distinguish these organisms from other fungi, such as *Candida* or dermatophytes, which may be seen in skin biopsies.

IMMUNOLOGY AND SEROLOGY

It appears that normal skin with a normal epithelial turnover rate precludes the disease process by this member of the normal skin flora. Antibodies are detectable in chronic cases. The indirect immunofluorescent technique may be used to stain the organisms in skin scales and cultures of *Pityrosporum orbiculare*.[13]

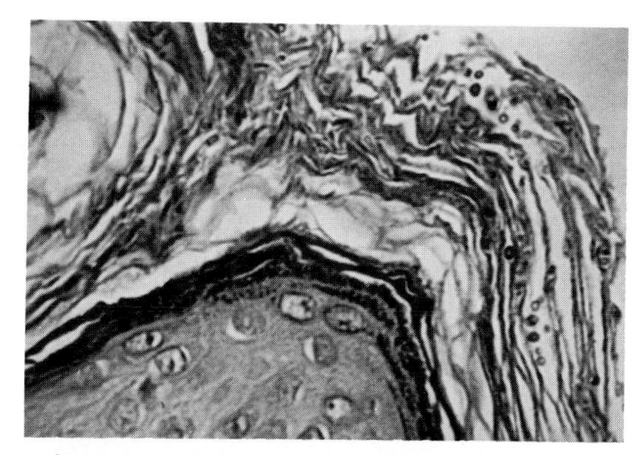

Figure 6–3. Pityriasis versicolor. Yeasts and mycelium in outer layers of stratum corneum. In this case, the lipophilic yeasts were most prevalent at the openings of sebaceous glands. Hematoxylin and eosin stain, ×400.

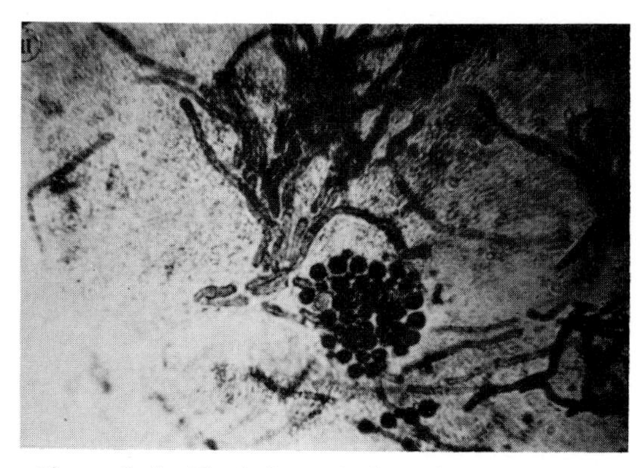

Figure 6–4. Pityriasis versicolor. Skin scales stained with methenamine silver. Short, branched mycelium and grouped, small yeast cell are diagnostic of this disease. ×400. (From Rippon, J. W. *In* Burrows, W. 1973. *Textbook of Microbiology*. 20th ed. Philadelphia, W. B. Saunders Company, p. 702.)

LABORATORY IDENTIFICATION

Direct examination of skin scales in KOH preparations shows the pathognomonic cluster of round, budding yeast cells (up to 8 μ in size) and the short, septate, occasionally branched, hyphal fragments (Fig. 6–4). Under rare circumstances, either the yeast form will predominate or the hyphal form will predominate.

Culture is not necessary for diagnosis, but the organism grows well at 37°C on malt agar or Sabouraud's agar containing streptomycin, penicillin, and Actidione and covered with a layer of olive oil. The colonies consist of yeastlike cells with only a rare elongation to form hyphal elements. Some strains are regularly more hyphal in nature, and there is speculation that these strains may be more likely to produce pityriasis versicolor in susceptible patients.[15,16]

MYCOLOGY

Pityrosporum furfur Rippon 1974

Synonymy. *Malassezia furfur* Robin (Baillon) 1889; *Microsporon furfur* Robin 1853; *Sporotrichum furfur* Saccardo 1886; *Oidium furfur* Zopf 1890; *Malassezia tropica* Castellani 1919; *Malassezia Macfadyeni* Castellani 1908; *Monilia furfur* Vaillemin 1931; *Pityrosporum orbiculare* Gordon 1951.

Tinea Nigra

DEFINITION

Tinea nigra is a superficial, asymptomatic fungus infection of the stratum corneum characterized by brown to black, nonscaly macules (Fig. 6–5 FS A 18). The palmar surfaces are most often affected, but lesions may occur on the plantar and other surfaces of the skin. The etiologic agent is *Cladosporium werneckii* in North and South America. In Asia and Africa the organism is called *C. mansonii.*

Synonymy. Tinea nigra palmaris, keratomycosis nigricans, cladosporiosis epidemica, pityriasis nigra, microsporosis nigra.

HISTORY

Tinea nigra was first described in 1898 by Montoya and Flores in Colombia. However, the description and the illustrations of "Montoyella nigra" by Castellani in 1905 appear to be of pityriasis versicolor.[6a] Castellani claimed that Manson first noted the disease in 1872 and in a later paper called the organism *Cladosporium mansonii.*[6b] The first authentic description of the disease was made by Alexandre Cerqueira (1891) in Bahia, Brazil. He gave it the name keratomycosis nigricans palmaris. However, it was not until 1916 that these findings, along with eight other cases, were published by his son, Castro Cerqueira-Pinto. In 1921, Parreirus Horta discovered the first case in Rio de Janeiro and subsequently isolated the fungus. The black dematiaceous fungus was called *Cladosporium werneckii.* The work of Arêa Leão[2] and Aroeira Neves and Costa[3] in Brazil and Carrion[6] in Puerto Rico have clearly delineated the condition.

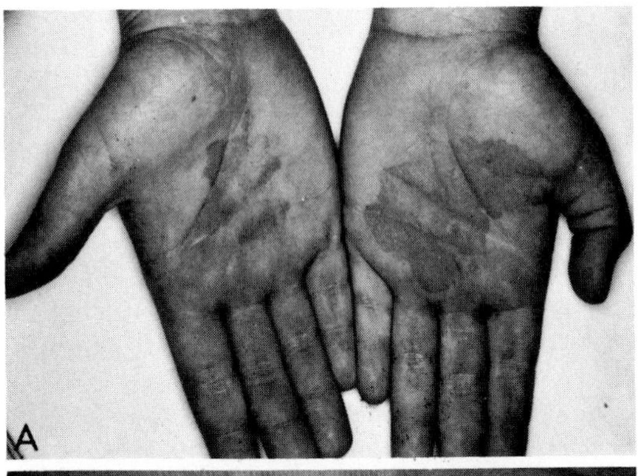

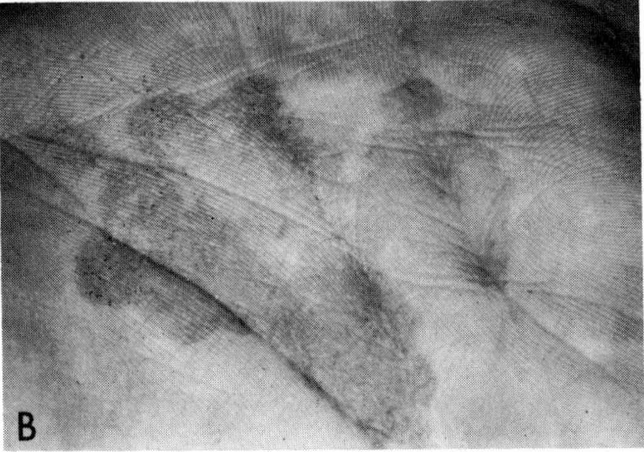

Figure 6–5. FS A 18. *A,* Tinea nigra. Brown-black, nonscaly macules. Bilateral disease. *B,* Enlarged area shows mottled olive-brown macules. The darkest areas are the advancing borders. There is no inflammatory reaction. (Courtesy of S. Lamberg.)

ETIOLOGY, ECOLOGY, AND DISTRIBUTION

The first name given by Castellani was *Cladosporium mansonii* (he termed *Montoyella nigra* a "temporary species"), and for some time this was the organism associated with infections in Asia. It appears that the single species, *Cladosporium werneckii* (Parreiras Horta, 1921),[9a] adequately and correctly defined the etiologic agent of the disease, and that *C. mansonii* is identical to it. Organisms similar to *C. werneckii* are very abundant in soil, sewage, decaying vegetation, and humus. They also grow on wood and paint in humid environments and on shower curtains. The species is extremely variable, and definitive classification is difficult. Some authors maintain that there are two species, *C. werneckii* and *C. mansonii.*[4] Others have suggested that the organism is identical to the black yeast, *Aureobasidium (Pullularia) pullulans.* Cooke[7] has found differences which are sufficient to separate *C. werneckii* from *A. pullulans* but not to separate *C. werneckii* from *C. mansonii.*

The condition tinea nigra is generally considered to be a tropical disease, and is common in Central and South America, Africa, and Asia. However, 15 cases were reported by Van Velsor[18] in North Carolina, and with increased awareness of the condition, new cases are being discovered more frequently in the United States and Europe.[19] Most patients are less than 19 years old, but any age group may be involved since the condition tends to be chronic. In most studies, affected females outnumber males three to one. No predisposing factors have been delineated, and impairment of the immune system does not appear to be relevant. Many patients were noted to be hyperhidrotic.

CLINICAL DISEASE

The lesions of tinea nigra are painless macules which are neither elevated nor scaly. They are sharply marginated and usually single. The condition often begins with the appearance of a light brown macule that spreads centrifugally and darkens. Rarely, eccentric areas of scaling are seen. The color is usually mottled, with deeper pigmentation seen at the advancing periphery. The most common site of infection is the palmar surface and the fingers, but infection of other areas has been reported, including plantar surfaces, neck, and thorax.

The primary importance of tinea nigra is that it is often misdiagnosed as other conditions, particularly malignant melanoma.[19] Such instances have resulted in surgical mutilation and debilitation of the patient, when a simple skin scraping and KOH mount examination would have revealed the true nature of the condition. As has been pointed out repeatedly, diseases which were once considered exotic will continually be appearing in local clinics as a result of jet age, worldwide travel. Not infrequently, vacationers bring home tinea nigra from the beaches of the Caribbean Islands, and the disease is no longer uncommon in the United States. The condition also simulates junctional nevus of the palm, contact dermatitis, pigmentation of Addison's disease, postinflammatory melanosis, melanosis from syphilis, pinta, and staining due to chemicals, dyes, and pigments.

Prognosis and Therapy. Tinea nigra responds readily to keratolytic agents. Daily applications of Whitfield's ointment will clear the condition readily. Tincture of iodine, 2 per cent salicylic acid, or 3 per cent sulfur are also effective. Griseofulvin appears to be ineffective.

There is no tendency for this condition to recur except by reexposure to contaminated material.

PATHOLOGY

There is little or no reaction to the infection. Slight, abnormal thickening of the stratum corneum is sometimes seen with separation of the cells because of the large quantity of fungal elements present. Multiple-branched, brown hyphal filaments appear in masses in the upper layers of the stratum corneum, but the stratum lucidum is spared. Biopsy shows small areas of parakeratosis and a small amount of perivascular infiltrate.

The disease is not known in animals, but experimental infections in man and guinea pigs are easily induced by scarifying the skin, rubbing a culture of *C. werneckii* on the area, and then covering it with a bandage. The incubation period is 10 to 15 days, with most cultures of the *C. mansonii* variant, but *C. werneckii* may require several weeks to produce visible lesions.

No serologic or immunologic studies have been reported. There appears to be no association with other disease states, and no genetic predisposition. The reports of several family members with the condition probably reflects common exposure.

LABORATORY IDENTIFICATION

For direct examination, epidermal scrapings are placed on a microscope slide with a drop of 10 per cent KOH and cleared by heating gently for a minute. Examination with the low power lens reveals brownish to olivaceous, multiple-branched, septate hyphae and budding cells. The hyphal elements are up to 5 μ in diameter. Terminal portions of the hyphae are usually hyaline. Older sections are twisted and tortuous, with numerous septations and thickening of the cell walls that become deeply pigmented. Chlamydospores, swollen cells, yeastlike cells, and fragmented hyphae are also seen. The mycelia differ from dermatophyte hyphae in that the latter are colorless (hyaline), usually are not so branched, and do not show the tapering contour of the terminal branches.

Culture. In culture, the organism grows

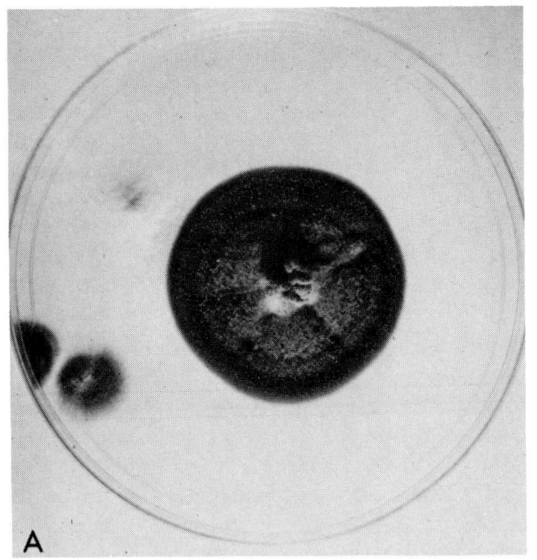

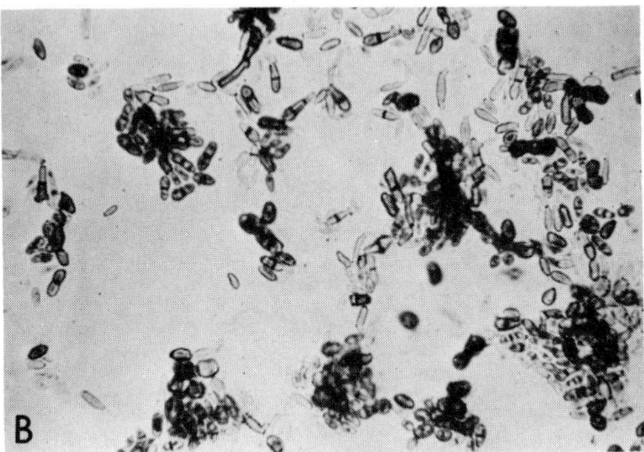

Figure 6–6. FS A 19. *A,* Colony of *Cladosporium werneckii.* Colony after five weeks' growth on Sabouraud's glucose agar. Blackish grey mycelium covers an olive-black, moist colony. *B, Cladosporium* yeast cells. The large, olive-black (*Aureobasidium*-like) yeast cells are large and have thicker walls than the hyaline cells. ×400.

slowly as a shiny, moist, adherent, yeastlike colony which initially is white to grey, but rapidly becomes olive to greenish black (Fig. 6–6,*A* FS A 19,*A*). Three weeks may be necessary before growth is initiated. At this stage (toruloid), microscopic examination shows the colonies to be composed mostly of spherical to oval, budding yeastlike cells, which may occur in chains. Some of the cells have a central cross wall. As the colony ages, an elongate mycelium develops which is tortuous and has numerous septations. It becomes pigmented and develops a sleeve of blastospores. At this stage the mycelium may be very thick (up to 7 μ), with thick-walled, squarish, pigmented cells in a chainlike arrangement (Fig. 6–6,*B* FS A 19,*B*). The terminal portions taper to colorless ends. The buds are numerous and may develop anywhere on the mycelium. They vary from white, elliptical cells (1×3 μ) to olive-colored, two- to three-cell spores (2 to 4×4 to 7 μ). At this stage, the organism resembles *Aureobasidium (Pullularia) pullulans.* Later, a greenish-grey or grey-black fuzz develops over the colony, and the microscopic picture more closely resembles that of a typical *Cladosporium.* The mycelium is now predominant, and the hyphae are more regular and narrow. Sporulation may be absent or sparse. When found, spores are produced on arborescent conidiophores typical of the genus *Cladosporium.* The conidia are acripetalos chains of unicellular spores.

The dimorphism exhibited by *Cladosporium werneckii* can be induced by nutrition and environment. As is the case with *Histoplasma capsulatum,* cysteine and CO_2 will maintain the yeastlike condition. When grown in O_2 or N_2, the growth is mycelial.[9]

MYCOLOGY

***Cladosporium mansonii* Pinoy 1912**

Synonymy. *Cladosporium werneckii* Parreiras Horta 1921; *Caraté noir* Montoya y Florez 1898; *Montoyella nigra* Castellani and Chalmers 1913; *Cladosporium* sp. Reitmann 1930; *Dematium wernecki* Dodge 1935.

Probable Synonymy. *Microsporum mansonii* Castellani 1905; *Foxia mansonii* Castellani 1908; *Torula mansonii* Vuillemin 1929; *Dematium mansonii* Dodge 1935.

There is some question as to whether the organism described by Castellani was from tinea nigra or pityriasis versicolor. If it was from the latter, the organisms described in this list cannot be considered as synonyms.

Piedra

DEFINITION

Piedra is a fungus infection of the hair shaft characterized by the presence of firm, irregular nodules. The nodules are composed of fungal elements cemented together anywhere along the hair shaft, and multiple infections of the same strand are common. Two varieties of piedra are recognized: white piedra, caused by *Trichosporon cutaneum*, and black piedra caused by *Piedraia hortai*.

Synonymy. Tinea nodosa, molestia de Beigel, trichomycosis.

HISTORY

The disease was first described as a fungus infection by Beigel in 1865 in his text, *The Human Hair: Its Growth and Structure*. He isolated the fungus (the "chignon fungus"), and Rabenhorst named it *Pleurococcus beigelii* in 1867. Beigel was probably working with a contaminant fungus, for although his description of the infection in hair is accurate, the organism in his illustrations appears to be *Aspergillus*. The black variety of piedra was described by Malgoi-Hoes in 1901, and the two clinical types were fully differentiated by Horta in 1911.[9b] He called the fungus he isolated *Trichosporon* sp. In 1913, Brumpt named the organism *Trichosporon hortai*. Fonseca and Arêa Leão, in 1928, renamed the organism *Piedraia hortai* because a sexual stage was discovered and its relation to the Ascomycetes became apparent. Some authors contend that the etiologic agents are the same species and that the *Trichosporon* is an immature and imperfect phase of *Piedraia*.

ETIOLOGY, ECOLOGY, AND DISTRIBUTION

Piedraia hortai appears to be related to the Asterineae, a subfamily of loculoascomycetes that form hard masses on plants and trees in tropical climates.[13a] The disease is found commonly throughout the tropical areas of South America, the Far East, and the Pacific Islands, but only sporadically in Africa and the rest of Asia. It is regularly found on monkeys and other primates in these regions. In a study of the pelts of primates in museum collections, Kaplan[11] and Ajello[1] found the infection to be very common. Sometimes the infection is encouraged as a mark of beauty by native peoples. They will purposely not oil their hair (usually done to discourage lice) and will sleep with their heads in a depression in the earth. Under these circumstances, multiple infections are seen, but the greater the number of nobs on the hair shafts, the more pulchritudinous the person is considered. Sometimes these nobs have a religious association.[14] Though scalp hair is most often affected, rare cases are reported of infection on other body regions.

White piedra is common on the temperate periphery of the tropical black piedra belt. The scalp is less frequently involved, and infection is most common in the bearded regions, the axilla, and the groin. The condition occurs sporadically in the United States and Europe, and more commonly in South America and the Orient. Infection in domestic animals, especially the horse, has been noted. *Trichosporon cutaneum* and similar organisms are common in soil, air, sputum, and body surfaces. This group includes

Geotrichum and *Trichosporon*, which have morphologic similarities and are difficult to distinguish.

The specific source of infection is unknown, but swimming in stagnant water has been suggested as a possibility.

CLINICAL DISEASE

There is no discomfort or physical reaction on the part of the patient. The disease is only of cosmetic interest. Its chief importance is in differentiation from pediculosis. The nodules of black piedra are gritty, hard, brown to black encrustations that vary in size from microscopic to a few millimeters and in depth up to 150 μ. The infection starts under the cuticle of the hair shaft. As interpilar growth occurs and continues, the shaft may rupture and be weakened so that breakage may occur. The mass may enlarge and grow on the outside of the hair and completely envelop the hair shaft. In a mature nodule, the periphery is composed of aligned hyphal strands, whereas the cells in the central area are cemented together to form a pseudoparenchymous mass which resembles organized tissue. Within this stroma are locules in which asci are produced (Fig. 6–7 FS A 20). In profile, the nodule appears like a mountain range in the center, with a flat surrounding peripheral plane.

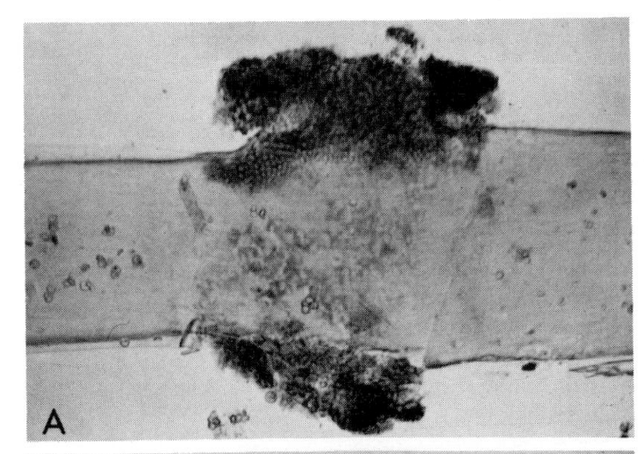

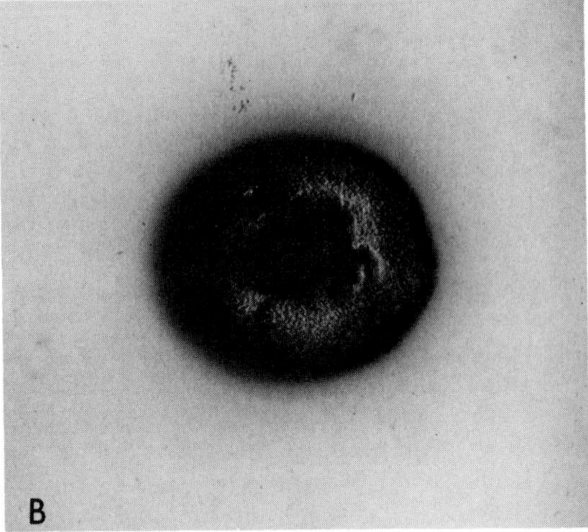

Figure 6–7. FS A 20. *A*, Piedra. Black nodules on hair shaft. Aligned hyphal strands on the periphery of the nodules and a stromalike center containing asci. *B*, *Piedraia hortai.* Dark, brown-black colony, which is elevated and cerebriform in the center and flat at the periphery. Note rusty-red, soluble pigment.

White piedra is characterized by the presence of a softer granule, which is white to light brown. The infection again appears to start beneath the cuticle, possibly following damage. The organism may grow inward and through the shaft to form nodular swellings spaced irregularly along the axis. The hair is weakened at these points and may break. Growth may occur primarily around the hair shaft, forming a soft sleeve of intertwined hyphae which fragment into yeastlike arthrospores (Fig. 6–8). The regular cellular pattern of black piedra is not seen, nor are the nodules as adherent, for they may be stripped off the hair. Several nodules may coalesce to form an extensive mass surrounding the hair shaft. It appears that infection begins soon after emergence of the hair from the follicle, the granule forming and hardening as the hair grows. The most common condition to be considered in the differential diagnosis is the presence of nits and lice. Microscopic examination will quickly rule out this condition. Piedra may simulate such developmental anomalies as monilethrix trichoptilosis and trichorrhexis nodosa. Trichomycosis axillaris may be distinguished by microscopic examination, since hyphal elements are 2 to 4 μ in piedra and 1 μ or less in trichomycosis axillaris. Piedra does not fluoresce under ultraviolet illumination as many trichomycosis axillaris infections do. The ovoid cells of black piedra may resemble the arthrospores of dermatophytes. In piedra, the hair shaft is normal on either side of the cell mass, whereas in tinea capitis the base of the hair shaft and the follicle are involved.

If therapy is desired, it may be achieved

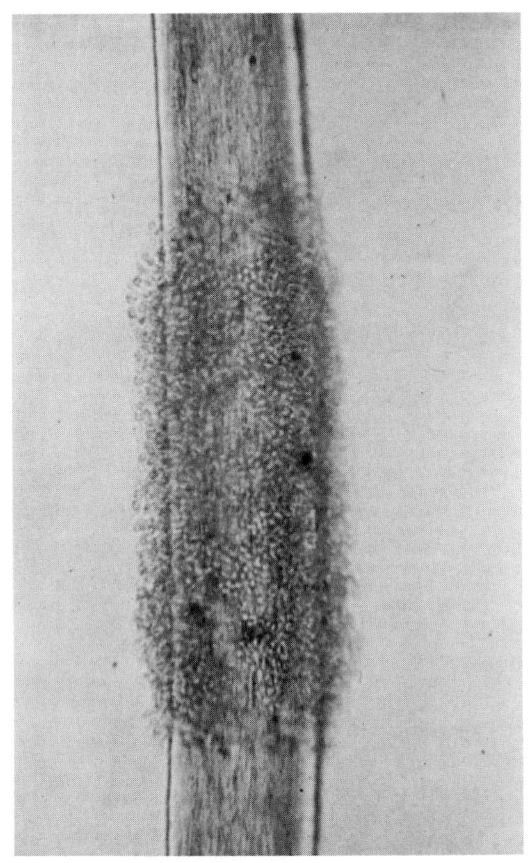

Figure 6–8. White piedra. Soft sleeve of intertwined hyphae and arthrospores around the hair shaft.

simply by shaving or cutting the infected hair. Topical fungicides, such as bichloride of mercury (1:2000), benzoic and salicylic acid combinations, 3 per cent sulfur ointment, and 2 per cent formalin, have been used.

Since, as is the case with the other superficial infections, there is no pathology or inflammatory response, no detectable serologic reaction has been described.

LABORATORY IDENTIFICATION

As previously described, direct microscopic examination of infected material in a 10 per cent potassium hydroxide preparation will differentiate not only the two types of piedra but also the simulating diseases. Black piedra is composed of a tightly packed stroma of regularly arranged, thick-walled, rhomboid cells (resembling arthrospores) and dichotomously branched hyphae held together by a cementlike substance. The hyphae and cells are 4 to 8 μ in diameter, and there is even pigmentation in the walls. A sectioned nodule reveals the asci within locules imbedded in the cellular stroma. There are eight fusiform, curved ascospores ($30 \times 10\ \mu$), the ends of which are prolonged into a spiral filament 10 μ long.

In contrast to black piedra, white piedra granules are softer, more easily detached from hair, and not as discrete. They often form a transparent, greenish, irregular sheath along the hair shaft. In addition, growth may be almost entirely intrapilar and cause only a raised cuticle. The mycelial elements are usually perpendicular to the surface of the hair and are not in an organized structure as in black piedra. The hyphae segment into oval to rectangular cells 2 to 4 μ in diameter, with occasional cells measuring up to 8 μ. Budding blastospores may also be seen, and bacteria may cohabit in the fungal mass to form a zooglea. No asci are seen.

Culture. Both organisms grow on ordinary laboratory media, but *Trichosporon cutaneum* is inhibited by cycloheximide in mycologic selective media (Mycosel). *Piedraia hortai* grow very slowly at 25°C, developing into small, dark brown to black, conical, adherent colonies (Fig. 6–7 FS A 20). The center of the colony is elevated and cerebriform, and the periphery is flat. The colony may be glabrous when young but usually develops a short, greenish brown, aerial mycelium with time. A rusty red pigment may be seen in the media. Microscopic examination reveals thick-walled, closely septate hyphae, chlamydospores, and swollen, irregular cells. In the center of the colony, locules may be found in which asci develop in a manner similar to that seen on the nodules of the hair shaft.[17]

Trichosporon cutaneum grows rapidly on Sabouraud's agar, producing a cream-colored, yeastlike colony that develops radial furrows and irregular folds with age, and often separates the colony from the media (Fig. 6–9). The thallus somewhat resembles a buttercream cake frosting in color and consistency (Fig. 6–10 FS A 21). Old colonies may take on a grey cast. Microscopic examination shows hyaline, septate hyphae which

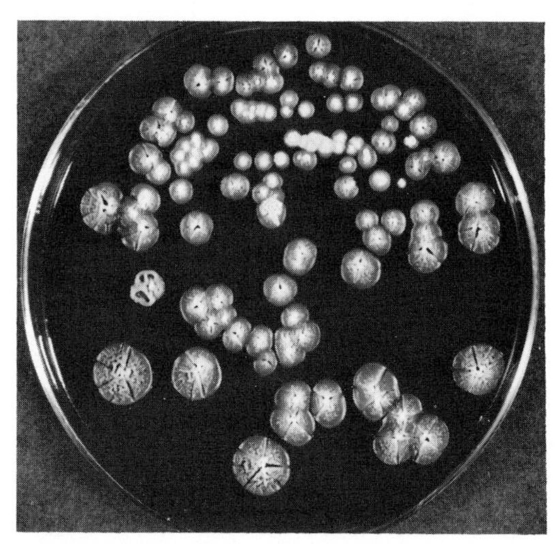

Figure 6–9. White piedra. *Trichosporon cutaneum.* Soft, wrinkled, folded, friable, off-white colonies.

fragment into oval or rectangular arthrospores, 2 to 4 × 3 to 9 μ. Some blastospores are also seen. Electron micrographs of hyphal septa show dolipores and parenthesomes typical of the class Basidiomycetes. It is probable that, if a perfect stage were found, it would be classified as a Basidiomycetes. *T. cutaneum* lacks the ability to ferment carbohydrates. It assimilates glucose, galactose, sucrose, maltose, and lactose, which distinguishes it from other members of the genus *Trichosporon.* Potassium nitrate is not utilized, and arbutin is split.

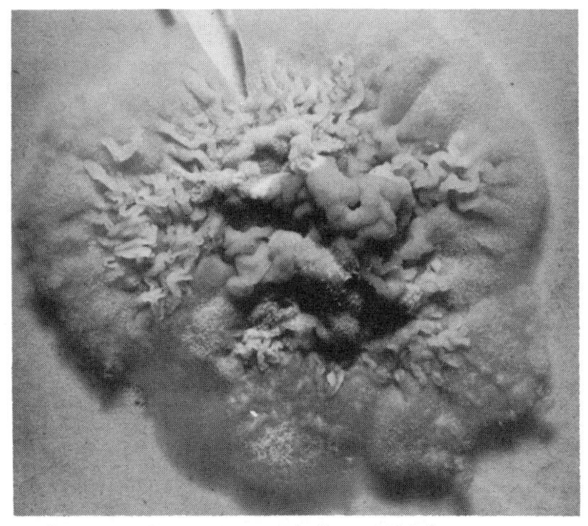

Figure 6–10. FS A 21. Thallus of *Trichosporon cutaneum* of a buttercream color and consistency.

MYCOLOGY

Piedraia hortai (Brumpt) Fonseca and Arêa Leão 1928

Synonymy. *Trichosporon* sp. Horta 1911; *Trichosporon hortai* Brumpt 1913; *Trichosporon guayo* Delamare and Gatti 1928; *Piedraia sarmentoi* Pereira 1930; *Piedraia venezuelensis* Brumpt and Langeron 1934; *Piedraia surinamensis* Dodge 1935; *Piedraia javanica* Boedijn and Verbunt 1938.

Trichosporon beigelii (Rabinhorst) Vuillemin 1902

Synonymy. *Pleurococcus beigelii* Küchenmeister and Rabenhorst 1867; *Trichosporon ovoides* Behrend 1890; *Trichosporon giganteum* Unna 1895; *Trichosporon cerebriforme* (Kambayashi) Ota 1928; *Trichosporon granulosum* (Kambayashi) Ota 1928; *Trichosporon humakuaguensis* Mazza and Nino 1933; *Trichosporon minor* Arêa Leão 1940; *Piedraia colombiana* Dodge 1935; *Trichosporon cutaneum* Ota 1926.

REFERENCES

1. Ajello, L. 1964. Survey of tree shrew pelts for mycotic infections. Mycologia, *56*:455–458.
2. Arêa Leão, A. E., A. Cury, and J. M. Ferreira Filho. 1945. Tinea nigra. Rev. Bras. Biol., *5*:165–177.
3. Aroeira Neves, J., and O. G. Costa. 1947. Tinea nigra. Arch. Dermatol., *55*:67–84.
4. Bastardo de Albornoz, M. C. 1966. Estadio micologica de las 4 primera cepas de *Aureobasidium werneckii* aislatos en Venezuela. Medn. Cutan., *4*:369–378.
5. Boardman, C. R., and F. D. Malkinson. 1962. Tinea versicolor in steroid-treated patients. Arch. Dermatol., *85*:44–52.
6. Carrión, A. L. 1950. Yeastlike dematiaceous fungi infecting the human skin. Arch. Dermatol., *61*:996–1009.
6a. Castellani, A. 1905. Tropical forms of pityriasis versicolor. Br. Med. J., *2*:1271–1272.
6b. Castellani, A., and A. J. Chalmers. 1913. Tinea nigra, In *Manual of Tropical Medicine.* 2nd ed. London, Bailliere, Lindall and Cox.
7. Cooke, W. B. 1959. An ecological life history of *Aureobasidium pullulans* (De Bary) Arnaud. Mycopathologia, *12*:1–45.
8. Gordon, M. 1951. Lipophilic yeastlike organisms associated with tinea versicolor. J. Invest. Dermatol., *17*:267–272.
9. Houston, M. R., and K. H. Meyer, et al. 1969. Dimorphism in *Cladosporium werneckii.* Sabouraudia 7:195–198.
9a. Horta, P. 1921. Tinea nigra. Rev. Med. Cirurg. do Brazil *29*:269–274.

9b. Horta, P. 1911. Sobre una nova forma de piedra. Mem. Inst. Osw. Cruz, *3*:87–88.
10. Jung, E. G., and B. Truniger. 1963. Tinea versicolor and Cushing-syndrome (pityriasis versicolor and Cushing's syndrome). Dermatologica, *127*:18–22.
11. Kaplan, W. 1959. The occurrence of black piedra in primate pelts. Trop. Geogr. Med., *11*:115–126.
12. Keddie, F. M. 1969. A novel reaction caused by tinea versicolor: Extracellular glycogen deposits. J. Invest. Dermatol., *53*:363–372.
13. Keddie, F., and S. Shadomy. 1963. Etiological significance of *Pityrosporum orbiculare* in tinea
13a. Mackinnon, J. E., and G. B. Schouten. 1942. Investigaciones sobre las enfermedades de los cabellos denominadas "piedra" Arch. Soc. Biol. Montevideo, *10*:227–266.
14. Moyer, D. G., and C. Keeler. 1964. Note on culture of black piedra for cosmetic reasons. Arch. Dermatol., *89*:436.
15. Roberts, S. O. B. 1969. Pityriasis: A clinical and mycological investigation. Br. J. Dermatol., *81*:315–326.
16. Roberts, S. O. B. 1969. Pityrosporum orbiculare: Incidence and distribution on clinically normal skin. Br. J. Dermatol., *81*:264–269.
17. Takashio, M., and R. Vanbreuseghem. 1971. Production of ascospores by *Piedraia hortai in vitro.* Mycologia, *63*:612–618.
18. Van Velsor, H., and H. Singletary. 1964. Tinea nigra. Arch. Dermatol., *90*:59–61.
19. Vaffee, A. S. 1970. Tinea nigra palmaris resembling malignant melanoma. New Engl. J. Med., *283*:1112.

CUTANEOUS INFECTIONS

Chapter 7

DERMATOPHYTOSIS AND DERMATOMYCOSIS

INTRODUCTION

The cutaneous infections of man include a wide variety of diseases in which the integument and its appendages, the hair and nails, are involved. Infection is generally restricted to the nonliving cornified layers, but a variety of pathologic changes occur in the host because of the presence of the infectious agent and its metabolic products. The majority of these infections are caused by a homogeneous group of keratinophilic fungi called the dermatophytes. In mycetoma it was seen that a great diversity of unrelated organisms cause essentially a single clinical entity. The dermatophytes, however, are a very similar and closely related group of fungi that cause a wide variety of clinical conditions. A single species may be involved in several disease types, each with its distinctive pathology. These fungi are among the commonest infectious agents of man, and no peoples or geographic areas are without "ringworm." Evolutionary development toward an accommodating host-parasite relationship can be seen among the dermatophytes which is absent among other fungal agents of human disease.

In addition to the dermatophytic fungi, other organisms are sometimes involved in cutaneous infections. These include a wide variety of soil-inhabiting yeasts and molds, and the disease produced by them is called dermatomycosis. Many reports in the literature erroneously cite contaminants and misidentified fungi as the etiologic agents in these infections. However, a few of these cases are authentic, although dermatomycosis occurs less frequently than dermatophytosis.[49,93] Because the disease elicited by these organisms is similar to dermatophyte infections, all of them will be discussed in this chapter.

THE DERMATOPHYTOSES

Definition

In its restricted sense dermatophytosis is an infection produced by a dermatophytic fungus in the keratinized tissues—nails, hair, and stratum corneum of the skin. This term is preferable to dermatomycosis, which would include any fungus infection of the skin, such as secondary spread from a systemic mycosis, or infection by *Candida* species. Dermatophytosis includes several distinct clinical entities, depending on the anatomic site and the etiologic agent involved. The pathology induced in the host initially is an eczemaform response, followed by allergic and inflammatory manifestations. The type and severity of these reactions are related to the immune state of the host, as well as to the strain and species of the organism causing the infection.

History

Because of its visibility, ringworm, like many other dermatologic conditions, has been noted and described from the earliest historical times. Growth of the fungus in skin and scalp is more or less equal in all directions, and the lesions produced tend to creep in a circular or ring form. For this reason, the Greeks named the disease herpes

—a term which still persists, though modified, as *herpes tonsurans, herpes circinatus,* or *herpes desquamans.* The Romans associated the lesions with insects and named the disease tinea, meaning any small insect larva. This name is retained in the clinical terminology of the disease. Tinea also refers to a group of keratinophilic insects, the clothes moths. The English word, ringworm, then, is a combination of the Greek and Latin terms.

One of the most disfiguring of the dermatophyte infections is the disease favus (L. honeycomb). The chronic inflammation, loss of hair, replacement of scalp by scar tissue, and formation of folded, crusted scutula have been recognized in central Europe and the Mediterranean area since classic times. In 1834, Remak examined material from favus and noted the presence of filaments resembling a mold.[124a] He tried to reproduce the disease on his arm by rubbing the organism on his skin, but failed. Schoenlein[145a] in 1839 described the filaments as being molds and concluded that favus was a disease caused by plants. The most remarkable work of that period was done by David Gruby.[64a] In 1841, he published a paper describing the isolation of the fungus of favus (on potato slices) and the production of the disease by inoculation of this fungus onto normal skin. Thus he was the first to establish that a microorganism is responsible for human disease, and Koch's postulates for the criteria of the etiology of infection were fulfilled forty years before Koch formulated them. In addition, Gruby described the yeast, now called *Candida albicans,* from thrush, named the dermatophyte *Microsporum audouinii* from tinea capitis, and recognized the endothrix form of trichophytosis. His work and that of Remak were generally ignored, possibly because of strong antisemitic feelings in medicine at that time. Schoenlein, because of such prejudice, was given most of the credit for the pioneering work on favus and the other dermatophytoses. In 1845 the fungus from favus was named *Oidium schoenleinii* by Lebert and *Achorion schoenleinii* by Remak. Also in 1845, Malmsten erected the genus *Trichophyton* and described *T. tonsurans. T. mentagrophytes* was defined in 1847 by Charles Robin. This author published, in 1853, a compilation of the early works on dermatophytoses in the book, *Histoire Naturelle des Vegetaux Parasites.* It was the first influential work to discuss topical therapy for dermatophyte infections and the importance of epilation in tinea capitis.

Early progress in the field became bogged down in numerous descriptions of fungi from skin infection. Small variations in clinical appearance or colony morphology seemed to warrant a new species. The literature was glutted with incomplete and inaccurate reports. By 1890 one of the great names in dermatology, Sabouraud, began to publish his systematic and scientific studies on the dermatophytoses. He accumulated his work in the volume *Les Teignes,* published in 1910. This book is considered a classic in medical literature. He included a classification system recognizing the three genera of dermatophytes, *Microsporum, Trichophyton,* and *Epidermophyton,* along with the genus *Achorion,* which was based on clinical rather than botanical observation. His basic methodology and astute observations on treatment remained only slightly changed until the advent of griseofulvin. He again stressed epilation of hair, but by this time x-ray was practicable, and he wrote extensively on this subject. Many papers were published by him after *Les Teignes,* but these did little to clarify the taxonomy of dermatophytes. Some of his species have been reduced to synonymy, and the genus *Achorion* has been dropped. In general, as can be seen in Table 7–1, the major tenets of his classification still hold, a tribute to his intuitive insight.

In the succeeding years, the literature again became muddled and confused. Many clinicians described new species based on trivial differences in fungal colony color, morphology, spore size or arrangement, as well as type, extent, topography, and inflammatory and therapeutic response of the disease. This was true for all other types of fungus infection as well. There are at least 172 synonyms for the yeast, *Candida albicans.* A compendium of this early literature can be found in the book, *Medical Mycology,* published by C. W. Dodge in 1935.[38] This volume contains descriptions and references for thousands of species of fungi (including 118 dermatophytes) isolated from all types of human disease. He also theorizes that certain fungi are race specific as, for example, *Trichophyton violaceum,* occurring in Jewish and other semitic peoples, and *Microsporum ferrugineum,* occurring in the Chinese and Japanese. He makes conjectures about many

Table 7–1. *The Classification of Sabouraud and its Present Equivalent*

Genus	Group	Type Species	Synonomy and Present Name
Trichophyton			
	Endothrix	*T. tonsurans*	
	Neoendothrix	*T. flavum*	(*T. tonsurans*)
	Ectothrix		
	Megaspores	*T. roseum*	(*T. megninii*)
	Faviformes	*T. ochraceum*	(*T. verrucosum*)
		T. violaceum	
	Microides		
	Gypseum	*T. mentagrophytes*	
	Niveum	*T. felineum*	(*T. mentagrophytes*)
Microsporum			
Neomicrosporum		*M. canis*	
Eumicrosporum		*M. audouinii*	
Achorion			(*Trichophyton*)
Neoachorion		*A. gallinae*	(*M. gallinae*)
Euachorion		*A. schoenleinii*	(*T. schoenleinii*)
Epidermophyton		*E. floccosum*	

problems involved in immunology, pathology, and distribution of fungous diseases. In addition, many dermatologic conditions were described as having a fungal etiology that do not: namely, pemphigus, epithelioma, eczema, and so forth. Most of these reports resulted after isolation of some contaminant, or from incomplete, inaccurate, and careless observation. All of medicine suffered from such problems at that time; these have largely been eliminated today.

In the 1920's Hopkins and Benham began the scientific study of medical mycology. Their laboratory at Columbia University was one of the first to systematically study fungi involved in disease. Rhoda Benham is considered the founder of modern medical mycology. In 1934, Emmons redefined the dermatophytes according to the botanic rules of nomenclature and taxonomy. He accepted the synonymy of the genus *Achorion* with the genus *Trichophyton*, as proposed by Langeron and Michevitch in 1930, and included all the known dermatophytes in three genera. This was a landmark effort, as it emphasized the care and meticulous scientific observation necessary for accurate identification. Benham proposed a botanic grouping based on colony characteristics and, to a certain extent, clinical disease. Conant modified this and included it in his book, which has become a standard text in the field (Table 7–2). By means of nutritional studies, Georg[54] further clarified the identity of several organisms and established the use of physiologic characteristics and nutritional requirements in identification. She reduced the number of dermatophyte species

Table 7–2. *The Classification of Dermatophyte Groups by Similarity of Colony Morphology As Proposed by Conant, 1954[32]*

Trichophyton Malmsten 1845
- I. Gypseum group
  1. *T. mentagrophytes*
- II. Rubrum group
  1. *T. rubrum*
- III. Crateriform group
  1. *T. tonsurans*
- IV. Faviform group
  1. *T. schoenleinii*
  2. *T. concentricum*
  3. *T. ferrugineum*
  4. *T. violaceum*
  5. *T. verrucosum*
- V. Rosaceum group
  1. *T. megninii*
  2. *T. gallinae*

Microsporum Gruby 1843
- *M. audouinii*
- *M. canis*
- *M. gypseum*

Epidermophyton Sabouraud 1910
- *E. floccosum*

to sixteen. With the description of new, valid species of keratinophilic soil saprophytes and skin pathogens, the number of recognized dermatophytes is now 37. As reviewed by Ajello,[4] these include one species in the genus *Epidermophyton*. 15 in *Microsporum*, and 21 in *Trichophyton*.

The next major development in the taxonomy of dermatophytes came in 1959. In this year Dawson and Gentles[35] described the perfect stage of a keratinophilic soil organism, *Trichophyton ajelloi*. This led to the rapid discovery of the ascomycetous form of many dermatophytes, although in 1927 Nannizzi had described the perfect stage of *M. gypseum*. At present, the use of physiologic characteristics, mating types, and critical antigenic analysis, in addition to the classic methods of descriptive morphology, has placed the taxonomy of dermatophytes and other pathogenic fungi on a firm scientific basis.

Another epochal contribution by Gentles was to revolutionize the therapy of ringworm infections. In 1958 he published a report on the oral administration of griseofulvin, which cured experimental dermatophytosis in a guinea pig.[48] Independently Martin[102] reported similar results. This was the first major change in the treatment of ringworm since Sabouraud plucked the heads of school children in Paris. Williams[171] in 1958 described the first patient to be cured by griseofulvin. The patient, a child, had tinea capitis due to *M. audouinii*. Previously a patient with a life-threatening *T. rubrum* infection had responded dramatically to the drug. This report[24] and the subsequent work of Blank and Roth drew worldwide attention to the efficacy of griseofulvin. Blank and co-workers then established dosage and treatment schedules, which led to general acceptance of the drug as the treatment of choice in almost all forms of dermatophyte infection. Clinical trials were carried out in many parts of the world and included the whole gamut of dermatophytic diseases.[135] These trials, conducted by Riehl, Pardo-Castello, Degos, Esteves, Neves, and others, have been reviewed by Blank and are the standards for therapy.[23,70] Recently several new topical drugs have been introduced. Of these, tolnaftate (Tinactin) has gained wide popularity but is of limited value in some types of infection. Treatment failures and relapses occur with all presently available drugs. The need for better therapeutic agents is apparent.

Dermatophytes remain a major public health problem. The commonness of ringworm infections and the occasional disfiguring severity of the disease have prompted continuing interest and investigation into their etiology, distribution patterns, improved therapy, and mechanisms of pathogenicity.

Etiology, Ecology, and Distribution

The more you study them, the more they look alike.

The group of organisms known as the keratinolytic fungi (families Gymnoascaceae and Onygenaceae) are homogeneous not only in appearance but also in physiology, taxonomy, antigenicity, growth requirements, and limits of infectivity and disease. The ability of these microorganisms to invade and parasitize the cornified tissues is closely associated with, and dependent upon, the utilization of keratin. Keratin is a highly insoluble scleroprotein, and its use as a substrate is rare in nature. Certain insects, including the clothes moth (*Tinea*), the carpet beetles (*Dermestes*), the biting lice (Mallophaga), and possibly some streptomycetes, also utilize keratin.

It had long been recognized that the dermatophytes were very closely related to one another. The asexual stages of all the dermatophyte species were placed in three genera, one monotypic. When the sexual stages were discovered in some dermatophytes, it was again found that only two very similar genera of the ascomycetes could encompass all species. The two perfect genera corresponded closely to the imperfect genera, i.e., all *Microsporum* species with perfect stages belong to the genus *Nannizzia*; all *Trichophyton* species to the genus *Arthroderma*. Antigenically and physiologically, the dermatophytes are also closely related. It has been almost impossible to differentiate genera by an antigenic mosaic. Separation of species also has not been achieved. No serologic tests are available that delineate more than a "group" reaction which includes all dermatophytes. In addition, very few physiologic and nutritional differences between species and genera have been described. The few differences found have been useful in separating similar species.

Table 7–3. *The Currently Recognized Dermatophytes**

Imperfect Genera and Species
Epidermophyton Sabouraud 1910
†*E. floccosum* (Harz 1870) Langeron and Milochevitch 1930
Microsporum Gruby 1843
M. amazonicum Moraes, Borelli, and Feo 1967
†*M. audouinii* Gruby 1843
M. boullardii Dominik and Majchrowicz 1965
†‡*M. canis* Bodin 1902
M. cookei Ajello 1959
‡*M. distortum* DiMenna and Marples 1954
†*M. ferrugineum* Ota 1921
†*M. fulvum* Uriburu 1909
M. gallinae (Megnin 1881) Grigorakis 1929
†*M. gypseum* (Bodin 1907) Guiart and Grigorakis 1928
‡*M. nanum* Fuentes 1956
M. persicolor (Sabouraud 1910) Guiart and Grigorakis 1928
M. praecox Rivalier 1954
M. racemosum Borelli, 1965
M. vanbreuseghemii Georg, Ajello, Friedman, and Brinkman 1962
Trichophyton Malmsten 1845
T. ajelloi (Vanbreuseghem, 1952) Ajello 1968
†*T. concentricum* Blanchard 1895
‡*T. equinum* (Matruchot and Dassonville 1898) Gedoelst 1902
T. erinacei (Smith and Marples) Padhye and Carmichael 1966
T. georgiae Varsavsky and Ajello 1964
T. gloriae Ajello 1967
†*T. gourvilii* Catanei 1933
T. longifusum (Florian and Galgoczy 1964) Ajello 1968
†*T. megninii* Blanchard 1896
†‡*T. mentagrophytes* (Robin 1853) Blanchard 1896
T. phaseoliforme Borelli and Feo 1966
T. proliferans English and Stockdale 1968
†*T. rubrum* (Castellani 1910) Sabouraud 1911
†*T. schoenleinii* (Lebert 1845) Langeron and Milochevitch 1930
‡*T. simii* (Pinoy 1912) Stockdale, Mackenzie, and Austwick 1965
†*T. soudanense* Joyeux 1912
T. terrestre Durie and Frey 1957
T. vanbreuseghemii Rioux, Tarry, and Tuminer 1964
†*T. verrucosum* Bodin 1902
†*T. violaceum* Bodin 1902
†*T. yaoundei* Cochet and Doby Dubois 1957

*Modified from Ajello, L. 1968. A taxonomic review of the dermatophytes and related species. Sabouraudia, *6*:147–159, by permission of E. & S. Livingstone.

†Commonly isolated from human infection.

‡Commonly isolated from animal infection.

The remainder are soil keratinophilic fungi rarely if ever involved in disease. Two other species have recently been described. *T. thuringiense* Koch 1969 is now considered in synonymy with *T. terrestre* (fide Padhye), and *T. fluviomuniense* Miguens 1968 is a granular form of *T. rubrum* (fide Ajello).

At present, there are 37 species of dermatophytes recognized as valid[4] (Table 7–3). Many of these have been found only in soil and as yet have not been reported as causing human or animal disease. The ascigerous (perfect or sexual) stage of 17 dermatophytes has been described. The taxonomy of these fungi as reviewed by Ajello appears to be biologically correct and is the one generally accepted by mycologists. A recent revision by Vanbreuseghem[164] was deemed invalid because he lumped together obviously unrelated species, i.e., some *Trichophyton* and *Microsporum* species in a genus, *Sabouraudites*. An attempt by Benedek to define the "ancestral forms" of dermatophytes apparently was based on unorthodox views of genetics and contaminated cultures.[20]

As far as human disease is concerned, most infections throughout the world are caused by 11 species. Within the continental United States, the list is pared to six. Worldwide travel, however, has increased the host exposure to formerly geographically limited species. Many infections produced by exotic organisms or the rare infections produced by soil keratinophiles present great problems in diagnosis and identification.[130,132] The soil contains many keratinophilic fungi closely related to the dermatophyte genera. Most of these organisms are rarely, if ever, isolated from human infection. However, they may be transient inhabitants of the skin and particularly of animal fur. These genera include *Ctenomyces, Gymnoascus, Chrysosporium, Malbranchea, Aphanoascus,* and so forth. Sometimes they are isolated from the human skin, but there is question of their significance. *Chryosporium* species are frequently cultured from feet. Whether they represent transients or are involved in disease is difficult to determine. However, the clinician and mycologist must be aware of the potential for infection by these soil forms. On occasion, reports can be verified of their evoking clinical disease, as was found recently with *Aphanoascus fulvescens*.[133] The isolant was morphologically quite unlike the usual dermatophyte and might have been considered a contaminant. This report emphasizes the difficulties which arise in identification of species other than those routinely isolated.

For many years, the similarity of dermatophytes to soil-inhabiting Ascomycetes had been recognized. Dassonville and Matruchot (1899) suggested that the asexual dermatophytes were related to the Gymnoascaceae. *T. mentagrophytes* was even transferred to the perfect genus, *Ctenomyces*, by Langeron in 1930. Pycnidia were formed which resembled the cleistothecia of Ascomycetes, but only asexual spores were found. Nannizzi probably was the first to find a sexual fruiting body in a dermatophyte. He published a paper in 1927 describing ascocarps in a culture of *M. gypseum*. He renamed the organism *Gymnoascus gypseus*. This work was not accepted because he used unsterilized soil in his cultures, and his findings were thought to be due to a contaminant. With the work of Dawson and Gentles,[35] it was established that the dermatophytes were Ascomycetes in the family of soil keratinophiles, the Gymnoascaceae. So far, all species described are heterothallic. Interestingly the perfect states of *Blastomyces dermatitidis* and *Histoplasma capsulatum* are also in the same family. It is known that "inducer substrates," such as soil, oatmeal, keratin, iron filings, beetle wings, tomato juice, and chicken feathers, are sometimes necessary to evoke a fruiting body, even when mated pairs are used. By now many dermatophyte species are known to have a perfect state (Table 7–4), and the idiosyncracies of the proper romantic environment are known also. In general, as the dermatophyte becomes more anthropophilic, it gradually loses not only the ability to produce asexual spores but also the ability to form a sexual state. Thus, such species as *M. audouinii, T. rubrum,* and *T. schoenleinii* have so far defied efforts to induce a sexual state. Animal isolants of *T. mentagrophytes* mate easily and produce fertile gymnothecia of *Arthroderma benhamiae* containing asci with ascospores. In man these strains induce a severe inflammatory disease. In contrast, *T. mentagrophytes, var. interdigitale,* usually isolated from noninflammatory lesions, rarely mates with the testor strains of *Arthroderma benhamiae* and can be considered an anthropophile.

The major importance of knowing the perfect stage of dermatophytes is for purposes of identification. New isolants, or isolants which vary in normal colonial morphology, can be mated with testor strains. By this method several previously described species of dermatophytes were found to be variants of recognized species. Another interesting use for the mating phenomenon is in epidemiologic studies. It was found that most isolants of *T. mentagrophytes* from troops in Vietnam were one mating strain of *A. benhamiae*, whereas in the United States another mating strain predominates. This provided evidence that the infections were acquired in Vietnam and not carried from the United States. Certain enzymatic differences have been discovered between the sexes of the same species. In experimental dermatophytic infections, some of these have

Table 7–4. *The Ascigerous Genera and Species of Dermatophytes*

Perfect State	***Imperfect State***
Arthroderma	*Trichophyton*
A. benhamiae Ajello and Cheng 1967	*T. mentagrophytes*
A. ciferrii Varsavsky and Ajello 1964	*T. georgiae*
A. gertleri Bohme 1967	*T. vanbreuseghemii*
A. gloriae Ajello 1967	*T. gloriae*
A. insingulare Padhye and Carmichael 1972	*T. terrestre*
A. lenticularum Pore, Tsao and Plunkett 1965	*T. terrestre*
A. quadrifidum Dawson and Gentles 1961	*T. terrestre*
A. simii Stockdale, Mackenzie, and Austwick 1965	*T. simii*
A. uncinatum Dawson and Gentles 1959	*T. ajelloi*
Nannizzia	*Microsporum*
N. cajetana Ajello 1961	*M. cookei*
N. fulva Stockdale 1963	*M. fulvum*
N. grubia Georg, Ajello, Friedman, and Brinkman 1967	*M. vanbreuseghemii*
N. gypsea Stockdale 1963	*M. gypseum*
N. incurvata Stockdale 1961	*M. gypseum*
N. obtusa Dawson and Gentles 1961	*M. nanum*
N. perisicolor Stockdale 1967	*M. persicolor*
N. racemosa Rush-Munro, Smith, and Borelli 1970	*M. racemosum*

been related to differences in pathogenicity, inflammatory response, and duration of infection.[127,129]

Ecology. By placing hair on soil (hair baiting), Vanbreuseghem and later workers found a rich flora of keratinophilic fungi in this milieu. The significance of this discovery is only now being appreciated. Apparently the soil abounds with many closely related species of keratinophiles. Most of these are seldom if ever involved in dermatophytoses, but they have the potential for it. One can see a natural evolution from keratin-utilizing soil saprophytes (geophilic species) to association with and finally invasion of cornified substrate in living animals (zoophilic species) and man (anthropophilic species). In accepting a primarily parasitic existence, many changes have occurred. Spore production, which is very abundant in soil organisms, is gradually diminished. The more or less strict anthropophiles, *M. audouinii, T. rubrum,* and the members of the faviform group (*T. schoenleinii, T. violaceum*), produce very few asexual spores in culture. Some hair-invading species have evolved a "parasitic" arthrospore, which is produced only during active infection but will remain viable and infectious on fomites for several years. In general, the greater the tendency toward intermittent, chronic, or continuous infection, the lesser the inflammatory response evoked by the organism. The dermatophytes appear to reach a type of equilibrium with the host in matters of growth and lack of irritation. Examples of these are Adamson's fringe and occult tinea pedis. Along with diminished spore production, these organisms are usually physiologically less active, particularly in regard to proteolytic enzymes. This equilibrium may become rather host-specific. Zoophilic strains of *T. mentagrophytes,* which cause a minor or subclinical infection in animals, will evoke a severe inflammatory response in humans. The mild human infection is caused by the downy or var. *interdigitale* form of *T. mentagrophytes.* The importance of defining isolants and species as geophilic, anthropophilic, or zoophilic is in trying to determine source of infection. Tinea capitis due to *M. audouinii* almost invariably means contact with other infected children and becomes a problem in public health and disease control; an infection with *M. canis* usually indicts the family cat (Table 7-5).

Table 7-5. *Ecology of Human Dermatophyte Species*

Anthropophilic	***Zoophilic***	***Geophilic***
Cosmopolitan Species		
E. floccosum	*M. canis*	*M. gypseum*
M. audouinii	*M. gallinae*	*M. fulvum*
T. mentagrophytes var. *interdigitale*	*T. mentagrophytes* var. *mentagrophytes*	*T. ajelloi*
T. rubrum	*T. verrucosum*	*T. terestre*
T. schoenleinii	*T. equinum*	
T. tonsurans	*M. nanum*	
T. violaceum		
Rare and Geographically Limited Species		
M. ferrugineum	*M. distortum*	*Aphanoascus fulvescens*
T. concentricum	*T. erinacei*	
T. gourvillii	*T. simii*	
T. megninii	*M. persicolor*	
T. soudanense		
T. yaoundei		

Distribution. Certain of the dermatophytes are geographically restricted and endemic only in particular parts of the world. *M. ferrugineum* is found in Japan and adjacent areas (Fig. 7-1); *T. concentricum* in the South Seas and small areas of northern South and Central America (Fig. 7-1); *T. yaoundei, T. gourvillii,* and *T. soudanense* in central and west Africa (Fig. 7-2). Other species may be of sporadic but worldwide distribution. Dermatophytes which are endemic within a population are carried by that population to new places. Troop movements, migration of labor, emigration, social habits, and rapid, worldwide travel have all contributed to the changing distribution of ringworm. For example, until recently *T. tonsurans* was rarely isolated in the United States. With the immigration of peoples from Mexico, Puerto Rico, and other Latin American countries, this organism, which is endemic in those regions, has become quite common in several cities (Fig. 7-1a). In New Orleans, Charleston, New York, and Chicago, it is tending to supplant *M. audouinii* as the most frequent cause of tinea capitis[51,64] (Fig. 7-3). Sometimes a new population will show a quite different response to a dermatophyte strain which is endemic in another region. Taplin[27] observed that South Vietnamese troops rarely had the severe disabling form of tinea pedis caused by the endemic strain of *T. mentagrophytes.* American troops, however, in the same region and under the same swamp conditions, developed severe inflammatory

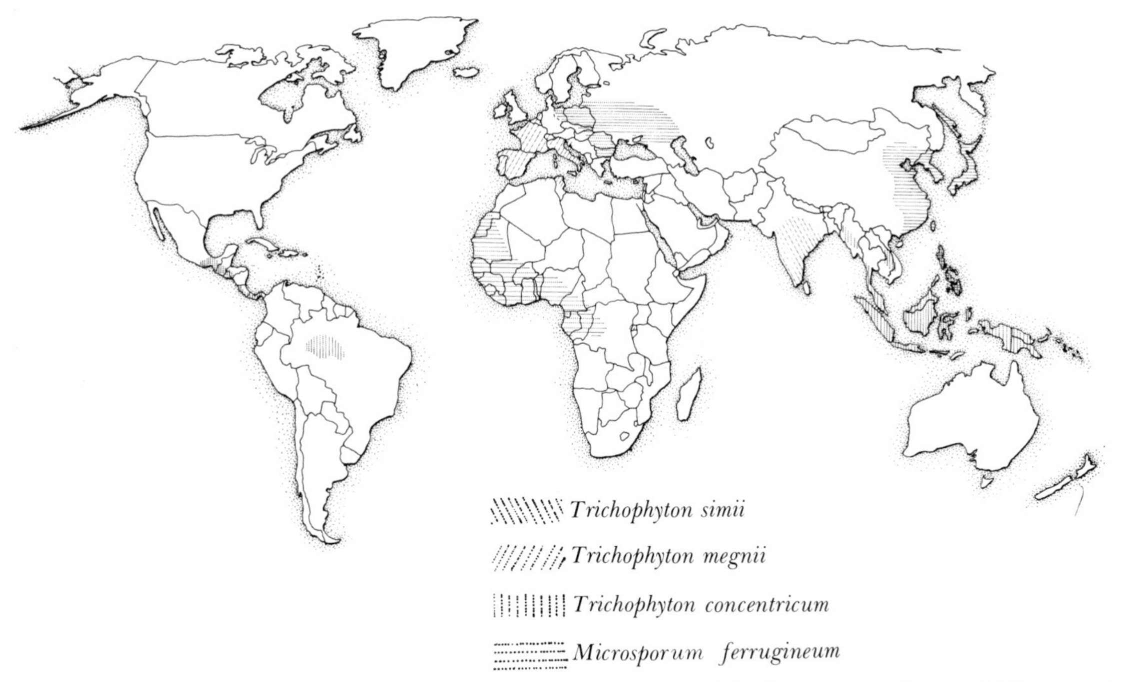

Figure 7–1. Distribution of *Trichophyton simii, Trichophyton megninii, Trichophyton concentricum,* and *Microsporum ferrugineum.*

lesions which sometimes covered the entire lower half of the body. The strain was quite distinct morphologically and was apparently carried by the local water rat. ARVN troops were troubled by *T. rubrum* tinea corporis, which was uncommon among U.S. troops.

Other dermatophytes have a sporadic distribution. *T. schoenleinii* is rare in the United

Figure 7–1a. Distribution of *Trichophyton tonsurans* and *Trichophyton violaceum.*

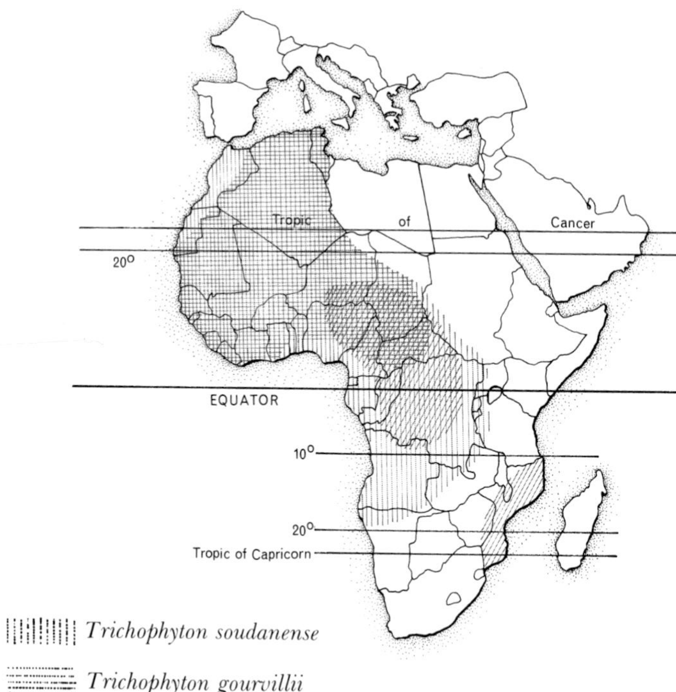

Figure 7–2. Distribution of *Trichophyton soudanense, Trichophyton gourvillii,* and *Trichophyton yaoundei.*

States except in small endemic foci, such as towns in Appalachia. In these areas all residents are descended from immigrants who lived in the same central European village where the disease was endemic (Fig. 7–3). The presence of the African species, *T. soudanense*, in Brazil and the United States may be a legacy of the infamous slave trade that flourished between the fifteenth and nineteenth centuries.[2] The patients involved had not been outside their respective countries. With jet age travel, *T. simii* can be picked

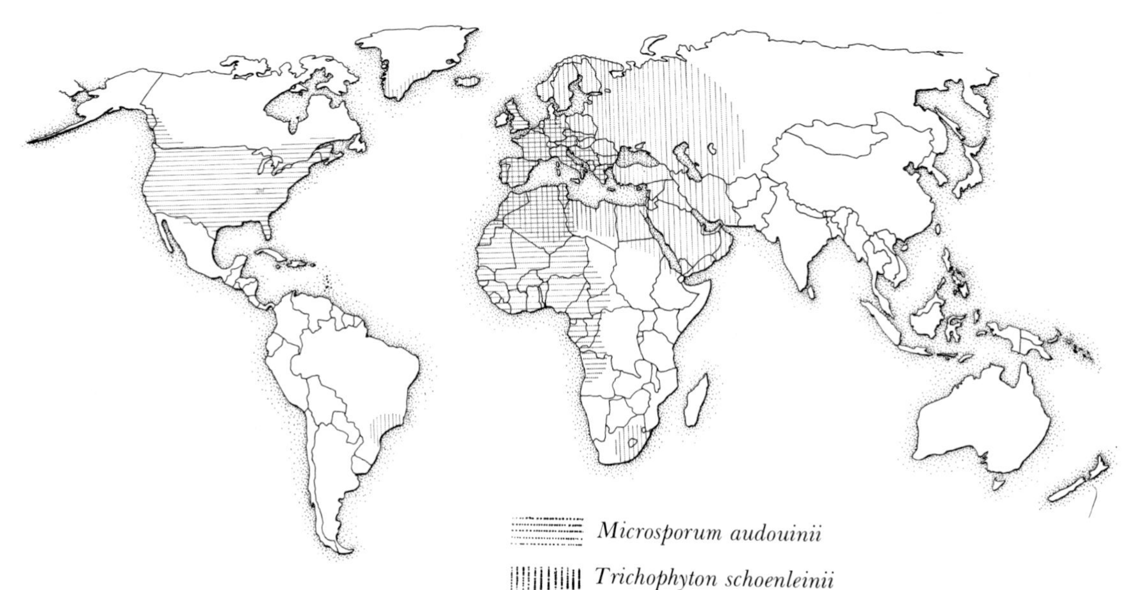

Figure 7–3. Distribution of *Microsporum audouinii* and *Trichophyton schoenleinii.*

Table 7–6. *Dermatophyte Infections—Clinical Diseases and Common Etiologies*

Disease	*Dermatophyte Involved*	*Disease*	*Dermatophyte Involved*
Tinea capitis	*Microsporum,* any species *Trichophyton,* any species except *T. concentricum*	Tinea imbricata	*T. concentricum*
Tinea favosa	*T. schoenleinii* *T. violaceum* (rare) *M. gypseum* (rare)	Tinea cruris	*E. floccosum* *T. rubrum* *T. mentagrophytes (Candida albicans)*
Tinea barbae	*T. mentagrophytes* *T. rubrum* *T. violaceum* *T. verrucosum* *T. megninii* *M. canis*	Tinea pedis	*T. rubrum* *T. mentagrophytes* *E. floccosum (Candida albicans)*
Tinea corporis	*T. rubrum* *T. mentagrophytes* *M. audouinii* *M. canis,* almost any dermatophyte, and *Candida albicans*	Tinea manuum	*T. rubrum* *E. floccosum* *T. mentagrophytes*
		Tinea unguium	*T. rubrum* *T. mentagrophytes* (rare: *T. violaceum, T. schoenleinii, T. tonsurans)* (*C. albicans* and other fungi are involved in a similar clinical disease, onychomycosis)
Tinea of animals (hair, skin, claws, feathers)	*M. canis* *M. nanum* *T. mentagrophytes* *T. verrucosum*	Tina favosa (favus of animals)	*T. equinum* *M. gallinae*

up from monkeys in India (Fig. 7–1), and diagnosed in the midwestern United States.[132] *T. soudanense* acquired in Africa has been identified in Germany, Belgium, and England.[165] These citations emphasize that diagnosis and identification of dermatophytes will be more difficult in the future as endemic species are transported to other areas of the world.

Though they are not debilitating or fatal, dematophytoses are among the most prevalent of human infectious diseases. Tinea pedis affects almost everyone that wears shoes throughout the world. The principal etiologic agents are *T. mentagrophytes* and *T. rubrum.* In contrast, the most frequent causes of the other dermatophytoses (tinea capitis, tinea corporis, and tinea unguium) vary from area to area. Ajello[2] has tabulated these for tinea capitis (Table 7–7). Millions of adults and children in the United States suffer from one or more types of dermatophyte infection. It is estimated that, in the United States alone, the public expenditure for medication for ringworm infections is about $25,000,000 per year.[2]

Source of Infection

The infections transmitted from animal to man (zoophilic) and from soil to man (geophilic) have already been discussed. The methods of transmission of the anthropophilic species remain a complex and controversial problem. *T. schoenleinii* and *T. violaceum* are endemic in certain middle European and Mediterranean regions. They are probably transmitted by direct personal contact, particularly in families. *T. concentricum* is passed from mother to child soon after birth. Infection of entire populations of children in schools and institutions during epidemics of *M. audouinii* has been recorded. Transmission is probably indirect, occurring via fallen hairs and loosened, infected squames rather than by direct contact. The organism has been isolated from combs, brushes, the backs of theater seats, and caps. *T. tonsurans* and *T. violaceum* have been isolated from all of the above and from bed linens. *E. floccosum* has been recovered from towels, undergarments, and jock straps. *T. mentagrophytes* is regularly isolated from locker room floors trod over by "athletes' feet." It is apparent that essentially all persons come into contact with dermatophytes during their lives, yet only a small percentage manifest clinical symptoms of disease.

In light of the prevalence of these organisms, Sulzberger and Baer[157] concluded that dermatophytosis is not a contagious disease. They felt that there must be other factors important for manifestation of disease,

Table 7–7. *Prevalent Agents of Tinea Capitis in the World**

Region / Country	Agent
North America	
United States	*M. audouinii*
Mexico	*T. tonsurans*
Canada	*M. canis*
Caribbean Area	
Cuba, Puerto Rico	*M. canis*
Dominican Republic[31a]	*M. audouinii*
South America	
Argentina, Chile, Uruguay	*M. canis*
Brazil	*T. violaceum*
Peru, Venezuela	*T. tonsurans*
Australasia	
Australia and New Zealand	*M. canis*
Europe	
Denmark, England, Finland, France	*M. canis, M. audouinii*
Greenland	*T. schoenleinii*
Italy, Portugal, Spain, USSR, Yugoslavia	*T. violaceum*
Africa	
Algeria, Egypt, Libya, Tunisia	*T. violaceum*
Angola, Congo, Mozambique, Zaire	*M. ferrugineum*
Morocco, Spanish Sahara	*T. schoenleinii*
Cape Verde Islands	*M. canis*
Nigeria	*M. audouinii*
Asia	
China, India, Israel	*T. violaceum*
Japan	*M. ferrugineum*
Iran, Turkey	*T. schoenleinii*

*Modified from Ajello, L. 1960. Geographic distribution and prevalence of the dermatophytes. Ann. N.Y. Acad. Sci., *89*:30–38.

since we continually wade through a sea of dermatophytes and only a few develop clinical symptoms. Occult tinea pedis is very common, but the difficulties of establishing experimental disease in human volunteers indicates a high degree of natural resistance. It appears that trauma, occlusion, and maceration are necessary for symptomatic infection. Marching, temperature stresses, moisture, and sweaty socks have been cited as necessary factors for exacerbation of previous subclinical disease. An opposing view is that contact with infectious material is the sole factor necessary for disease.[43,44] Gentles[50] and Georg[55] concluded on epidemiologic evidence that tinea pedis is a highly contagious disease. It appears then that there is no simple answer to the mode of spread of anthropophilic dermatophyte infections.

Strain and species differences may account for some of the problem. Infections produced by the granular form of *T. mentagrophytes* are characterized by inflammation and rapid resolution of infection.[129] *T. rubrum* infections often result in chronic disease, characterized by long periods of quiescence and intermittent recrudesence. Once the disease is established in them, individuals may be designated "*T. rubrum* people," for there is a lifelong association with this fungus. The factors involved in this association have been investigated, but no clear-cut determinants have been described. Epithelial turnover has been suggested, as is the case in pityriasis versicolor.[110] Genetic predisposition is possible, but begs the question without defining factors. Hypersensitivity or "immunity," if such exists, is very transient following infection. Reactivity to the trichophytin skin test appears to be of some significance in diagnosis, prognosis, or resistance to infection.[77a] Recent work has indicated that the composition and amount of amino acids found in the sweat of infected patients differs from those found in normal subjects.[120] The authors suggest that this is a factor predisposing to chronic infection. In other studies, certain amino acids, particularly those involved in the urea cycle, were shown to stimulate germination of *M. audouinii* in infected hairs. In contrast to this, other amino acids were inhibitory to germination.[134]

Many other aspects of dermatophyte infections remain undefined. The waxing and waning of epidemics caused by endemic species is known to occur, but is unexplained. *M. audouinii* will infect almost all of the population of a given area, then disappear for long periods of time. The incidence and the severity of *T. violaceum* infection in Yugoslavia and certain other countries varies markedly over a period of years. Simple explanations, such as the presence of a susceptible population, or the appearance of a more virulent strain, do not seem to account for all the vagaries encountered. Seasonal variation in incidence also occurs, and records from clinics and diagnostic laboratories indicate some periodicity. In the midwest, *Candida* infections are most prevalent in the late summer and early fall. *M. audouinii* tinea capitis increases with the opening of the school year. Tinea corporis is common in the midwinter "pale" season, whereas tinea pedis is exacerbated by the sweaty conditions of a humid summer. Kaplan and Ivens examined seasonal incidence of ringworm in dogs and cats.[84] They found that there was a seasonal variation, but the pattern appeared to be different for each dermatophyte.

Pathogenicity and Pathogenesis

Of the many fungi that produce disease in man, only the dermatophytes show evolution toward a parasitic mode of existence. The discovery of the many soil keratinophiles and of their relation to known disease-producing dermatophytes has been mentioned previously. When scrutinized closely, the biologic distance between a "soil dermatophyte" like *Trichophyton ajelloi* and an obligate "skin dermatophyte" like *T. rubrum* does not appear very great. These similarities have led to speculation about the phylogenetic history of skin dermatophytes. As modified from Hildick-Smith and others,[70] these hypotheses include (1) evolution in the soil of a specialized group of fungi with keratinolytic ability; (2) association with furred animals and ability to produce transient infections, e.g., *T. ajelloi* and *M. gypseum*; (3) adaptation to growth in a living keratinizing zone; (4) development of accommodation and equilibrium to the host, e.g., little irritation (Adamson's fringe, occult infection); (5) development of specialized methods of reproduction for successful dissemination of the parasite from host to host, e.g., arthrospores in ectothrix and endothrix hair infections and in infected squames; and (6) adaptation to specific animal hosts, increasing duration of survival, dissemination, and chronicity of infection, e.g., *T. rubrum* in man, *T. erinacei* in hedgehogs, *T. verrucosum* in cattle, *M. canis* in cats, and *T. mentagrophytes* in rodents. As the molecular structure of keratin varies from species to species, it has been suggested that different keratinases have been evolved relatively specific for a given host.[174]

There are several dermatophytes that run the gamut of this active adaptation (evolution). Animal reservoirs of the granular form of *T. mentagrophytes* (var. *mentagrophytes*) have long been recognized. Infection in man by such strains evokes a primary irritant reaction, followed by a severe inflammatory response and rapid termination of infection. This fungus readily invades hair and produces numerous "saprophytic" spores in culture. *T. mentagrophytes* var. *interdigitale* elicits little inflammatory response, produces chronic infections, does not readily infect experimental animals, and in culture produces few "saprophytic" spores. Yet rapid serial transmission of these strains in guinea pigs will transform the *interdigitale* into the granular var. *mentagrophytes* type with a concomitant increase in severity of infection and in spore production.[52] There is some evidence that this occurs in human epidemics also.[143] In contrast to the severe infection described by Sabouraud and produced by the present endogenous strains in Africa, *M. audouinii* has become a fairly benign and widely dispersed anthropophilic species. Most *T. rubrum* infections are chronic and not severe in the normal host, and the isolants produce few spores in culture. However, some rare strains that have been recovered in Miami, Viet Nam, Tunisia, and Nigeria sporulate heavily and are associated with a more inflammatory type of infection.

The dermatophytes show a high degree of specificity as to the tissues attacked. While these fungi are well adapted to parasitize the horny layer of the epidermis and the hair and nails, they appear unable to invade and infect other organs of the body in normal patients. Intravenous injection of *Microsporum* spores or hyphal suspensions of virulent *T. mentagrophytes* does not produce an infection of the internal organs; rather, the fungi become localized in the skin and produce infection only in areas previously dam-

Table 7–8. *The Common Dermatophytes and Their Diseases**

			Species	*Disease in Man*	*Geographical Distribution*
Invading the Hair and Hair Follicles	*Small Spore Varieties*	*Ectothrix Type*	*Microsporum audouini*†	Prepubertal ringworm of the scalp; suppuration rare; child to child	Commonest in Europe, producing about 90 per cent of infections; in U.S. 50 per cent
			Microsporum canis†	Prepubertal ringworm of scalp and glabrous skin; suppuration not infrequent; kerion occasional; from pets	Uncommon in Europe, except England and Scandinavia; responsible for about half the infections in U.S.
			Microsporum gypseum	Ringworm of the scalp and glabrous skin; suppuration and kerion common; from soil	Relatively rare in U.S.; common in South America
			Microsporum fulvum	Ringworm similar to that of *M. gypseum*	Same as above
			Microsporum ferrugineum†	Similar to *M. audouinii*	Africa, India, China, Japan
	Large Spore Varieties	*Endothrix Type*	*Trichophyton tonsurans*	Black-dot ringworm of the scalp; smooth skin; sycosis; tinea unguium; suppuration common; the hair follicles are atrophied	Common in Europe, Russia, Near East, Mexico, Puerto Rico, and South America, but uncommon in U.S. until recently
			Trichophyton violaceum	Black-dot endothrix in both scalp and smooth skin; onychomycosis; suppuration is the rule and kerion frequent	Common in southern Europe, the Balkans, and the Far East; rare in the U.S.
			Trichophyton soudanense *Trichophyton gourvilii* *Trichophyton yaoundi*	Inflammatory, scarring ringworm of scalp	Central and West Africa
		Ectothrix Type	*Trichophyton mentagrophytes*	Commonest cause of intertriginous dermatophytosis of the foot ("athlete's foot"); ringworm of the smooth skin; suppurative folliculitis in scalp and beard	Ubiquitous
			Trichophyton verrucosum	Ringworm of the scalp and smooth skin; suppurative folliculitis in scalp and beard; from cattle	Ubiquitous
			Trichophyton megnini	Sycosis is the most common lesion; infection of smooth skin and nails	Sporadic distribution; Portugal, Sardinia
	No Spores in Hair		*Trichophyton schoenleinii*†	Favus in both scalp and smooth skin; scutulum and kerion	Europe, Near East, Mediterranean region; rare in U.S.
Not Invading the Hair and Hair Follicles			*Epidermophyton floccosum*	Cause of classic eczema marginatum of crural region; causes minority of cases of intertriginous dermatophytosis of foot; not known to infect hair and hair follicles	Ubiquitous, but more common in tropics
			Trichophyton rubrum	Psoriasis-like lesions of smooth skin; tinea unguium; mild suppurative folliculitis in beard; rare invasion of scalp hair endo- and ectothrix described	Ubiquitous
			Trichophyton concentricum	Cause of tinea imbricata; infection of hair and nails uncertain	Common in South Pacific islands, Far East, India, Ceylon; reported in west coast of Central America and northwest coast of South America

*From Rippon, J. W. *In* Burrows, W. 1973. *Textbook of Microbiology.* 20th ed. Philadelphia, W. B. Saunders Company, p. 706.

†Infected hairs show fluorescence by Woods' lamp.

aged by scarification.[28] Dermatophyte growth is quite sensitive to temperature. Normal body temperature (37°C) inhibits growth of most strains and species. Modestly elevated temperatures (41°C), as shown by Lorincz and Sun, kill the organisms and cure experimental infection in animals.[98] However, it has been shown that dermatophytes can be "trained" to grow at elevated temperatures and even assume a yeastlike phase similar to that of several deep infecting fungi.[131] In this transient condition, the organisms can invade deep tissues of experimental animals. As shown by Lorincz et al.[97] and Roth et al.,[139] fresh serum inhibits the growth of dermatophytes. Experimental and clinical evidence of this has been reported. For example, when serum is washed out of tissue cultures of skin, *T. rubrum* invades the living cellular areas,[25] and extensive invasion of epithelium and dermis by this same fungus was seen in a patient with a low level of serum antidermatophyte activity.[26]

The natural history of dermatophyte infection is initially the same in all types of disease. Infection begins in the horny layer of the skin, and the ultimate outcome of disease depends on host, strain, species variation, and anatomic site. On the glabrous skin, the infection spreads centrifugally, showing the classic "ringworm" pattern. The host reaction may be limited to patchy scaling, or proceed to a toxic eczemaform eruption. Later, an inflammatory reaction may occur. In most all cases, apparent resolution occurs, and clinical symptoms disappear. However, organisms may persist for years, and the host becomes a normal carrier. Stress or trauma may exacerbate clinical disease. The natural evolution of dermatophytosis as exemplified by tinea capitis has been delineated by Kligman.[88] He divides the infection into several stages:[89] a period of incubation, enlargement and spread, a refractory period, and a stage of involution. These stages were studied in detail using human volunteers.* Tinea capitis was produced in them using hair infected with *M. audouinii* as inoculum. The following sequence of events was found to occur. In the incubation period (three to four months), hyphae grew in the stratum corneum of the scalp and the follicular orifices. The mycelium entered the follicle and grew downward into it along the surface of the hair shaft. During the period of spread, the fungus grew radially from the point of origin in the scalp stratum corneum. It invaded new follicles as they were encountered. The hyphae penetrated to the upper limits of the zone of active keratinization without invading it. The downward growth at this stage was equal to the upward growth of the forming hair. A ring of extending hyphae was formed just above the keratogenous zone, which is known as Adamson's fringe. As the growing hair carried hyphae to the surface, the specialized parasitic spores (arthrospores) were produced. This sequence occurred only in growing (anagen) hairs. In telogen, or epilated, hairs, the hair is attacked as it would be by a saprophytic keratinophile, with production of penetrating organs and fronds, and macro- and microaleuriospores. If hair growth is interrupted by normal categen or x-ray treatment, the delicate equilibrium is upset. Growth of the fungus is impeded, the hair is lost, and so is the parasite. The refractory period of the infection is characterized by disappearance of hyphae in the scalp, and no new lesions develop. Large quantities of arthrospores are formed in the infected hairs. Gradually a period of involution occurs, and the hair returns to normal. Infected hairs are shed, and the infection resolved. Reinfection and exacerbation may occur. Susceptibility to experimental infection appears to be independent of previous history of dermatophytosis or trichophytin skin test reactivity.

The preceding discussion underlines the delicate host-parasite balance that exists in a dermatophyte infection. In patients treated with oral griseofulvin, this host-parasite balance is abrogated. The drug is laid down in the keratinizing tissues the fungus ceases to penetrate, and is shed with the keratinized debris. On the other hand, if the serum inhibitory factor or factors are absent or diminished, the fungus is no longer impeded in its invasion. As first noted by Majocchi in 1883, granulomatous lesions may develop as the fungus invades the dermis and subcutaneous tissues. Although the Majocchi granuloma, as defined by Pillsbury, refers to small granulomata in hair follicles following shaving of the legs (apparently traumatizing the tissue and implanting the fungus), the

*A review of experimental human infections has been published by Knight, A. G., 1972. J. Invest. Dermatol., 59:354–358.

original description probably referred to disease in patients with an underlying disease. These patients have a lessened ability to contain infection, and widespread granulomatous lesions may develop. In almost all cases the organism is *T. rubrum.* Such lesions have been associated with abnormalities of carbohydrate metabolism, lymphomas, Cushing's syndrome, and impaired immune response (delayed hypersensitivity). Fatal systemic invasion by dermatophytes has been recorded in Japan, Russia, Rumania, and Portugal.[10,113] In these cases the fungus involved was *T. violaceum.*

Clinical Disease

The many species of dermatophytes elicit a number of well-defined clinical syndromes. The same species may be involved in quite different symptomatic disease, depending on the anatomic site. Because these syndromes form distinct entities, they will be discussed separately along with the organism commonly associated with them.

The clinical conditions are (1) tinea capitis (ringworm of the scalp); (2) tinea favosa (favus due to *T. schoenleinii*); (3) tinea corporis (ringworm of the glabrous skin); (4) tinea imbricata (ringworm due to *T. concentricum*); (5) tinea cruris (ringworm of the groin); (6) tinea unguium and onychomycosis (ringworm of the nail); (7) tinea pedis (ringworm of the feet); (8) tinea barbae (ringworm of the beard); and (9) tinea manuum (ringworm of the hand).

With the possible exception of white-spot tinea unguium, all infections begin in the horny layer of the epidermis. Those which infect the hair follicles, hair, and nails soon invade these structures, frequently producing little more than a transient scaling of the epidermis. Infections which are limited to the epidermis may affect the drier parts of the skin, including the palmar and plantar surfaces, or the moister regions, such as the inguinocrural folds and the interdigital spaces. Although a wide variety of clinical types may be observed, the differences are more apparent than real. The histopathologic changes are fundamentally the same in all types of disease. The entire gamut of disease entities produced by dermatophyte infection can be mimicked by *Candida albicans.* The most comprehensive review of the various forms of clinical dermatophytoses is given by Hildick-Smith, Blank, and Sarkany.[70]

Tinea Capitis

DEFINITION

Tinea capitis is a dermatophyte infection of the scalp, eyebrows, and eyelashes caused by species of the genera *Microsporum* and *Trichophyton.* It is characterized by the production of a scaly erythematous lesion and by alopecia that may become severely inflamed, with the formation of deep, ulcerative kerion eruptions. This often results in keloid formation and scarring, with permanent alopecia.

Synonymy. Ringworm of the scalp and hair, tinea tonsurans.

FUNGI INVOLVED

The commoner types of ringworm are classed according to the site of formation of their arthrospores. *Ectothrix* infection is defined as fragmentation of the mycelium into spores around the hair shaft or just beneath the cuticle of the hair (Fig. 7–4). In *Endothrix* infections arthrospore formation occurs by fragmentation of hyphae within the hair shaft (Fig. 7–5). Small-spored ectothrix organisms include *Microsporum audouinii, M. canis,* and *M. ferrugineum.* Masses of spores (1 to 3μ) are produced in a mosaic pattern; all are Woods' lamp–positive. *M. gypseum* and *M. fulvum* produce few spores and are Wood's lamp–negative. *T. mentagrophytes* forms small spores (3 to 4 μ) in chains on the surface of the hair. *T. verrucosum* forms large spores in chains (8 to 12 μ). *T. megninii* also forms medium sized spores in chains. The principal organisms producing endothrix infections are *T. tonsurans, T. violaceum,* and the African species *T. yaoundei, T. soundanense,* and *T.*

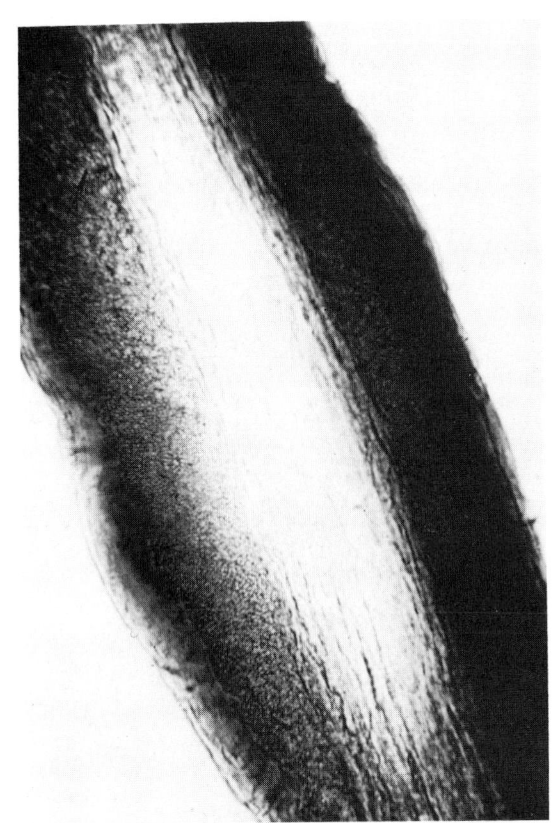

Figure 7–4. Ectothrix infection of hair. Arthrospores form a sheath around the hair shaft. *Microsporum audouinii.* ×500. (From Rippon, J. W. *In* Burrows, W. 1973. *Textbook of Microbiology.* 20th ed. Philadelphia, W. B. Saunders Company, p. 705.)

gourvilii. T. rubrum is not infrequently involved in tinea capitis, but rarely invades hair. Both endothrix and ectothrix sporulation have been described for this species.

SYMPTOMATOLOGY

"Grey-patch ringworm" ectothrix or prepubertal tinea capitis is a very common disease of children which often reaches epidemic proportions. The classic causative agents are *M. canis* and *M. audouinii.* Infection begins as a small erythematous papule around a hair shaft. Within a few days it pales and becomes scaly, and the hair assumes a characteristic grayish, discolored, lusterless appearance. The hair becomes weakened and may break off a few millimeters above the scalp. The lesion spreads, with the formation of more papules in a characteristic ring form, and may coalesce with other infected areas (Fig. 7–6 FS A 22). All hairs within the area are infected. Itching may become severe, and alopecia usually is seen in the infected areas. There may be little inflammatory reaction in *M. audouinii* infections, although on occasion, ulcerations and kerions may form. In *M. canis* and *M. gypseum* infections, inflammatory reactions are more common. Deep ulcerative lesions may develop, and boggy areas of massive inflammatory infiltrate, called kerions, may be produced (Fig. 7–7 FS A 23). This may be followed by scar formation and permanent alopecia. Elevated knobby keloids also develop. Inflammatory reaction usually heralds termination of disease. The tendency for spontaneous cure is most marked in *M. canis* infection, less so with *T. tonsurans,* and is rare with *T. violaceum.* The grey patch seen clinically does not delineate the infection of the scalp. A Woods' lamp examination shows the charac-

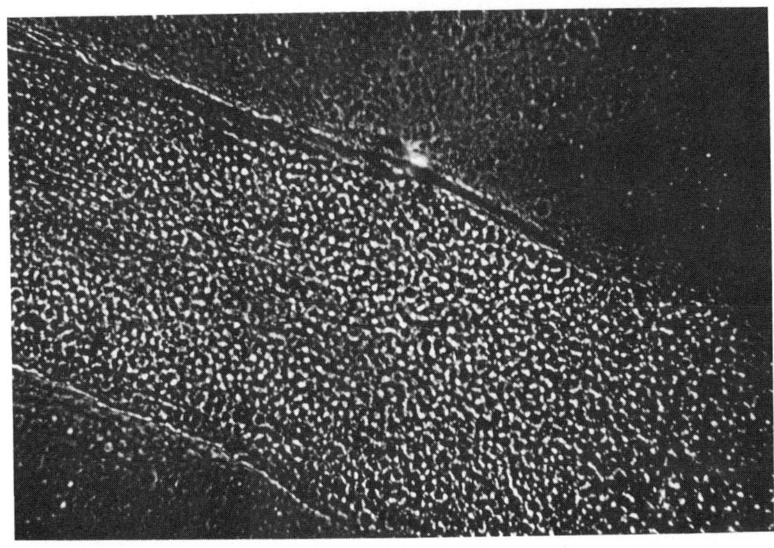

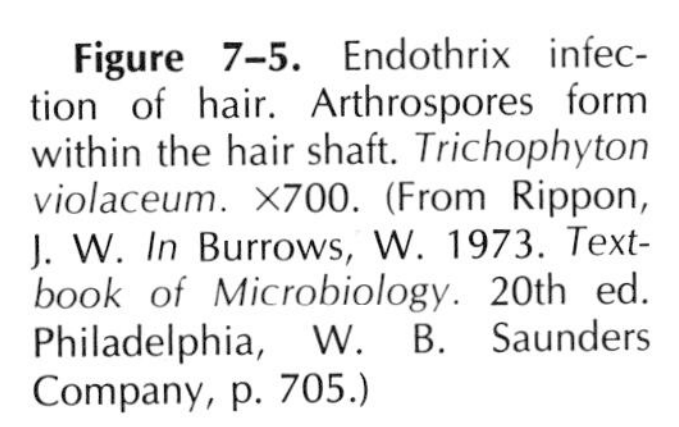

Figure 7–5. Endothrix infection of hair. Arthrospores form within the hair shaft. *Trichophyton violaceum.* ×700. (From Rippon, J. W. *In* Burrows, W. 1973. *Textbook of Microbiology.* 20th ed. Philadelphia, W. B. Saunders Company, p. 705.)

teristic brilliant green fluorescence of infected hairs to extend well beyond the symptomatic areas. Occult infections are also commonly discovered by this examination. Infection rate for boys is up to five times that for girls. However, the reverse is true after puberty.[162]

Uneventful spontaneous cure usually occurs in *Microsporum* infections. This may coincide with the onset of puberty. Rothman[141] considered that the change in composition of sebum with an increase in fungistatic fatty acids might account for this resolution of infection. Even-numbered fatty acids of medium length possess the greatest fungistatic activity. However, some workers have felt that other factors are important. Kligman[88, 89] concluded that infection simply undergoes spontaneous resolution, and that this may occur before, during, or after puberty.

"Black-dot ringworm," endothrix infection produced by *T. tonsurans* and *T. violaceum*, differs from *Microsporum* infection in a number of respects. While initial infection of the hair follicle and hair follows essentially the same course, the lesions are small and often contain only two or three affected hairs in a given area. If the lesion is large, all the hairs within the area are not affected. The lesions are multiple, numerous, and scattered over the scalp. In extensive involvement, the bald patches tend to lose the circumscribed outline of the lesions, and a diffuse alopecia is seen. The infected hairs break off sharply at the follicular orifice, leaving a spore-filled stub or "black dot." The stub sometimes curls as it grows and may be subsurface. In these cases it is necessary to excise the stub with a scalpel in order to culture it. The lesion plaques tend to be angulated, often forming polygons. Sometimes there is moderate inflammation with delayed scarring, which gives an appearance similar to lupus erythematosus. In *Trichophyton* infections, a more severe type of inflammatory reaction is more common than in microsporosis. Kerion formation followed by scarring and permanent alopecia are regularly seen. The var. *sulfureum* of *T. tonsurans* has been associated with the production of multiple kerions and erythema nodosum of the scalp.[46, 167] Multiple infections often occur which involve other dermatophytes. In some cases, three species of dermatophytes, or *Candida* sp. and *Staphylococcus* sp., have been isolated from lesions[136, 167] (Fig. 7–8). There has been an increasing incidence of *T. tonsurans* in the United States. Hairs infected with endothrix

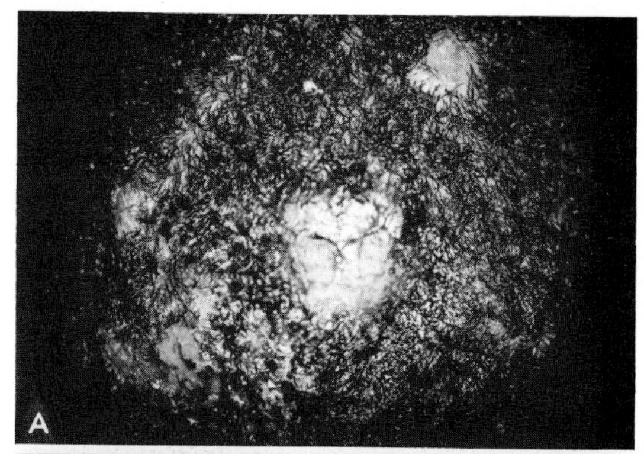

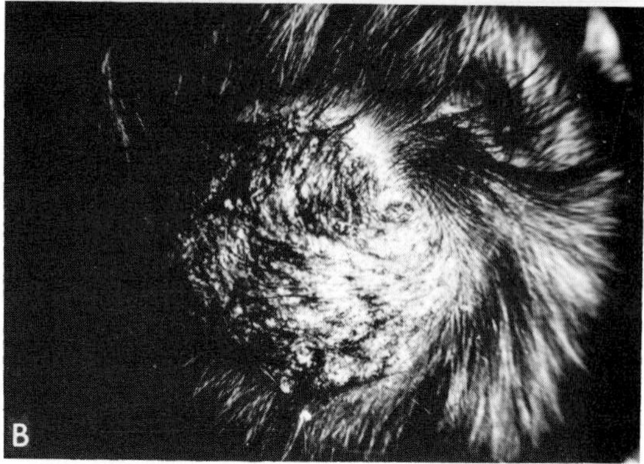

Figure 7–7. FS A 23. Tinea capitis. *A,* Boggy area of inflammatory infiltrate called a *kerion. Microsporum canis. B,* Large kerion consisting of crusts, matted hair, exudate, and scalp debris. *Microsporum canis.*

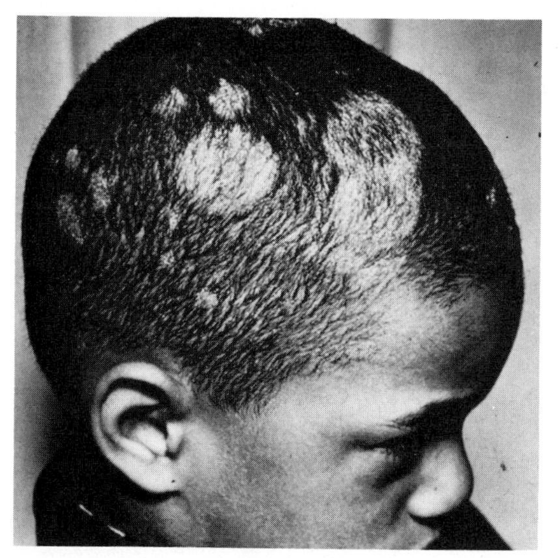

Figure 7–6. FS A 22. "Grey-patch ringworm" of *M. audouinii.* (Courtesy of S. Lamberg.)

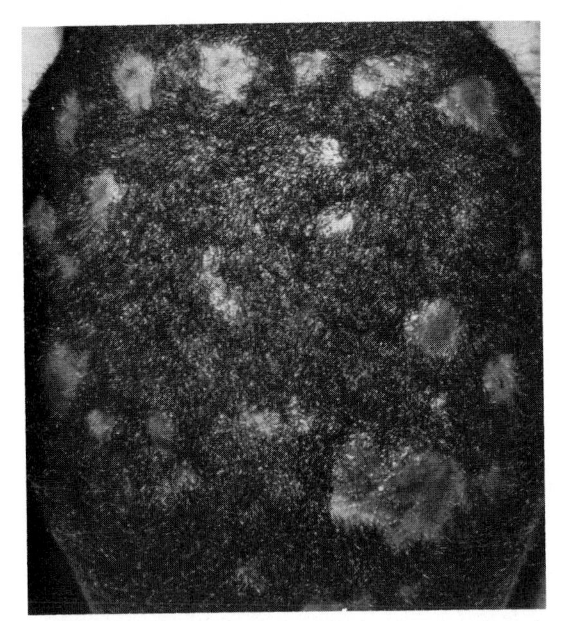

Figure 7–8. Tinea capitis. Multiple kerions and patchy alopecia in scalp infected by *T. tonsurans* var. *sulfureum*, *M. audouinii*, *Candida albicans*, and *Staphylococcus*. (From Varadi, D. P., and J. W. Rippon. 1967. Scalp infection of triple etiology. Arch. Dermatol., 95:299–301.)

Trichophyton do not fluoresce. As a result, in mass examinations for tinea capitis conducted by public health officers, a significant percentage of infections are not detectable. For this reason, dermatophyte test media should be used along with Wood's lamp examination in such surveys.[158,159] Unlike grey-patch ringworm, endothrix infections tend to become chronic and continue into adult life. This is especially true when *T. violaceum* is the etiologic agent.

Infections with ectothrix *Trichophyton (T. mentagrophytes* and *T. verrucosum)* are distinguished clinically by a more marked inflammatory reaction than other forms of tinea capitis. They often produce suppurative folliculitis (Fig. 7–9 FS A 24). These infections are rare and seen mostly in rural areas. The organisms are usually acquired from animals, cattle being the commonest source. Kerion formation is the rule in such infections. There is a massive acute inflammatory infiltrate in the skin of the affected area and deeper tissue. In contiguous areas, a suppurative folliculitis occurs, with indolent cutaneous and subcutaneous infiltration (kerion celsi) (Fig. 7–10 FS A 25). These are boggy on palpation, and pus oozes from the follicles. Several areas of the scalp may be involved simultaneously, and the size of the lesions is variable. The swelling is painful and the hairs easily dislodged. It is important to note that the pus is not due to secondary bacterial invasion but is attributable to the fungus alone, and therefore surgical intervention is not required. On occasion, dull-red, sharply defined plaques studded with pustules may be seen, a condition referred to as agminate folliculitis. Resolution of the infection is accompanied by scarring and patchy permanent alopecia. Individuals with dark, thick hair tend to have more severe inflammatory reactions than those with light, thin hair.

The dermatophytid ("id") reaction may occur in tinea capitis, but less commonly than in tinea pedis. The "id" reaction is an allergic manifestation of infection at a distal site, and the lesions are devoid of organisms. A series of grouped vesicles (pompholyx type) which are tense, itchy, and sometimes painful are found anywhere on the body. However, lesions on the trunk are common in tinea capitis, whereas fingers are more often involved when the primary lesion is a tinea pedis. The lesions may evolve into a scaly eczematoid reaction or a follicular papulovesicular response which covers wide areas of the body and causes great distress and discomfort to the patient.

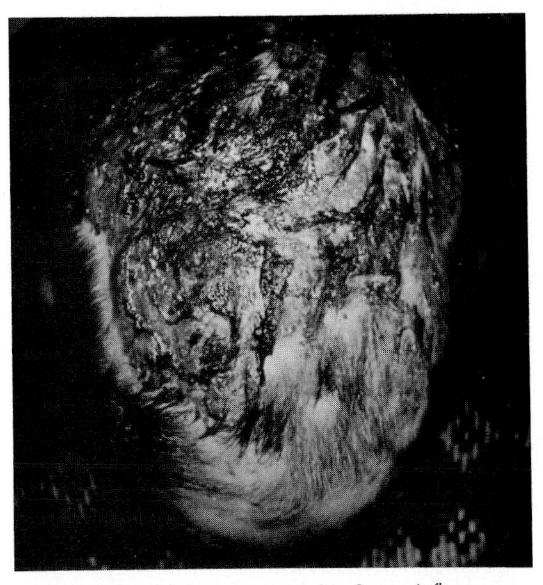

Figure 7–9. FS A 24. Tinea capitis. Severe inflammatory reaction and suppurative folliculitis produced by infection with *T. mentagrophytes*. This strain was mating type a, and produced large quantities of elastase in culture and experimental infections.[126]

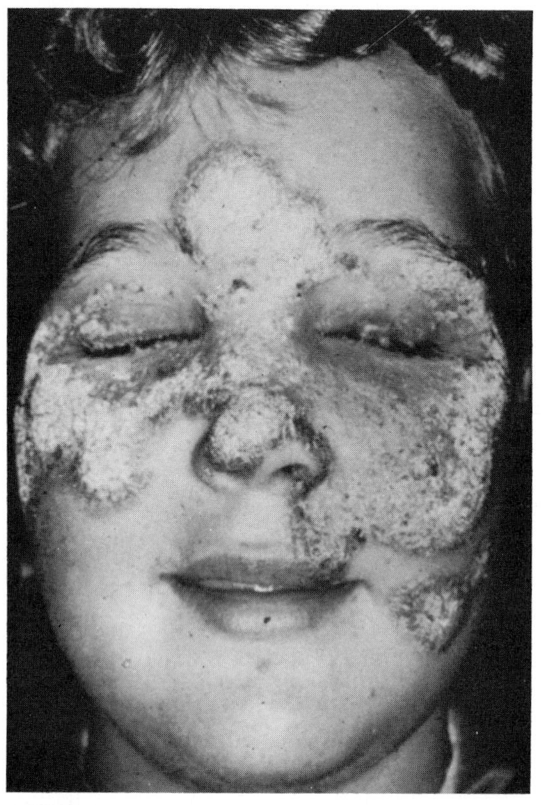

Figure 7–10. FS A 25. Extensive cutaneous and subcutaneous involvement in an infection by *T. verrucosum*. (Courtesy of D. Windhorst.)

A "blue-dot" tinea capitis has also been observed.[93] The infecting organism was a *Penicillium* species. *Aureobasidium pullulans* was also isolated in the same case. The involved hairs showed mycelium and spores around the shaft and in the follicles. A small, crusted area developed around each of the broken hairs, and mechanical removal of the hairs was followed by redevelopment of the lesions. The disease resolved spontaneously after two and one half years. A series of five such cases has been reported.

It was observed by Margarot and Deveze[99a] in 1925 that infected hairs and some fungus cultures fluoresce in ultraviolet light. This "black light" is commonly known as the Wood's lamp. The light is filtered through a Wood's nickel oxide glass (barium silicate with NiO) which passes only the longer ultraviolet rays (peak at 3650 Å). Hair infected with *M. canis*, *M. audouinii*, and *M. ferrugineum* fluoresces a bright green color. Hair infected with *T. schoenleinii* may show a dull green color. *T. verrucosum* exhibits a green fluorescence in infected cow hairs, but infected human hairs do not fluoresce. Hedgehog hairs infected with *T. mentagrophytes* and all *T. simii* infections also fluoresce a bright green.[132] *M. distortum*, *M. nanum*, *M. vanbreuseghemii*, and *M. gypseum* have been reported on occasion to produce infections that fluoresce.

The fluorescent substance appears to be produced by the fungus only in actively growing infected hairs. Saprophytic growth of the organisms on epilated hairs does not produce the active principle. Comparative studies of the water-soluble fluorescent substances show the spectroscopic pattern to be similar in hairs infected by *M. canis*, *M. audouinii*, and *T. schoenleinii*. From infrared spectrophotometric studies, Wolf[173] concluded that these substances were pteridines (pyrimidine-4′,5′:2,3-pyrazine). The exact structure has not as yet been worked out. Infected hairs remain fluorescent for many years after the spores have died.

DIFFERENTIAL DIAGNOSIS

The diagnosis of tinea capitis, especially in children, is suggested by erythematous, scaling patches of alopecia. The presence of brittle lusterless hair or boggy, infiltrative, ulcerative lesions may also be clues. Seborrheic dermatitis, psoriasis, lupus erythematosus, alopecia areata, pseudopelade, impetigo, trichotillomania, pyoderma, folliculitis decalvans, and secondary syphilis are considered in the differential diagnosis. Examination of potassium hydroxide mounts usually determines the proper diagnosis.

In seborrheic dermatitis, the hair involvement is diffuse rather than patchy, the hairs are not broken and the scalp is red, scaly, and itchy. This and other chronic scaling diseases, such as psoriasis, may cause accumulation of scales in matted masses on the scalp. This condition is called "pityriasis amiantacea" (asbestos pityriasis).[19] Scales are more prominent in psoriasis, but again hairs are not broken. Impetigo is difficult to differentiate from inflammatory ringworm, but the pain is usually less severe in the latter. Alopecia areata may have an erythematous border in the early stages of the disease, but reversion to normal skin color occurs. Also in this condition there is a lack of scaling, and the hairs on the border do not break off; however, they pull out easily. These hairs are of the "exclamation point" morphology. The vesicles of the "id" reaction must be differentiated from toxic reactions of many etiologies, including pompholyx, dyshidrosis, and other causes of vesicles and subcorneal bullae.

HISTOPATHOLOGY

There is no specific histologic picture for dermatophyte infections, and hyphae must be shown with special stains. Fungi are seen sparsely in the stratum corneum, penetrating in between and through the squames[68] (Fig. 7–11 FS A 26). Hyphae extend down into the hair shaft and penetrate into the hair lying parallel to it. Hyphal tips grow downward within the hairshaft to the edge of the living keratinizing cells and form "Adamson's fringe." The overall histologic picture in tinea capitis is that of a subacute or chronic dermatitis. In the mildest form, there is intercellular edema of the rete. Parakeratosis, slight vasodilation, and a perivascular infiltrate in the upper dermis is also seen. An exaggerated allergic reaction leads to folliculitis and perifolliculitis, with formation of kerion celsi. This consists initially of an acute leukocytic reaction in the deep dermis and even subcutaneous tissues. The cellular infiltrate gradually becomes that of the chronic inflammatory type. The infiltrate is massive and obliterates much of the normal architecture of the tissue. Generalized follicular keratosis (lichen spinulosus) may be seen as transient eruptions in rapidly developing inflammatory ringworm.

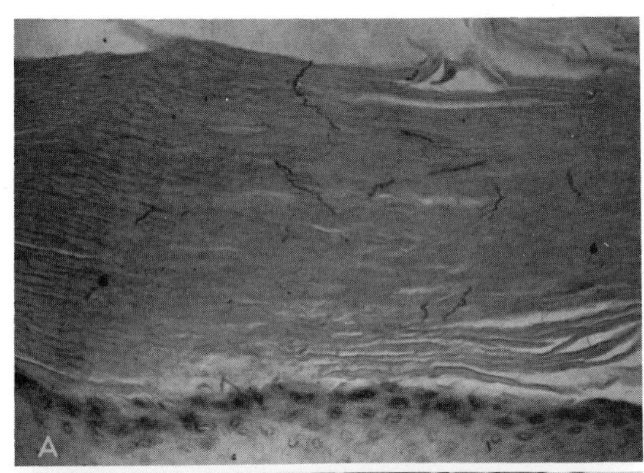

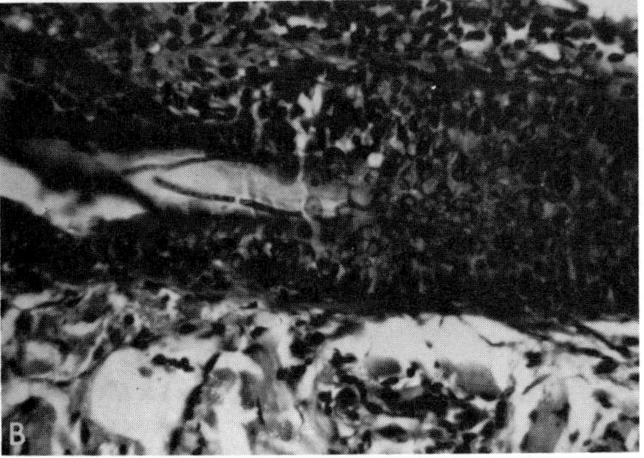

Figure 7–11. FS A 26. *A,* Dermatophyte infection. The hyphal strands penetrate through the stratum corneum but do not invade the living cells of the epidermis. Periodic acid-Schiff stain. ×400. (Courtesy of M. Sulzberger.) *B,* Mycelium growing down a hair shaft to the bulb. ×440.

PROGNOSIS AND THERAPY

Most mild ringworm infections of the microsporosis type (grey-patch) resolve uneventfully with time, usually in early adolescence. The more inflammatory the reaction, the earlier the termination of disease. This is particularly true of animal-acquired ringworm *(M. canis, T. verrucosum, T. mentagrophytes).* Thus most ectothrix infections involute during the normal course of disease and without treatment. However, the patients spread the organisms to others during the infection period. Endothrix infections, on the other hand, tend to become chronic and last far into adult life. *T. violaceum* causes a particularly persistent infection in which patients become vectors for spreading the disease within family groups and the community.[138] Patients should be actively treated to terminate such infections and prevent their spread.

Topical treatment of tinea capitis appears to be without benefit. The infection is unaffected by local applications of fungistatic preparations, keratolytic agents, and antifungal drugs, such as griseofulvin and tolnaftate (Tinactin). However, the addition of topical fungistatic agents (benzoic acid, magenta paint, and so forth) to a more effective treatment regimen (such as systemic administration of griseofulvin, or x-ray epilation) is an important adjunct to therapy. By this means spores are killed and infected debris is removed, thereby preventing spread of infection.

Griseofulvin is the most effective drug presently available for treating tinea capitis.

Dosage schedules vary, but the usual standard treatment (micronized griseofulvin) is 500 mg per day for adults and 250 mg for children in four divided doses. Coarse-particle griseofulvin is given in a dosage of 1 g per day for adults and 500 to 750 mg per day for children. Patients are encouraged to take the drug after a fatty meal to increase absorption from the alimentary tract. After three days, vigorous daily scrubs of the scalp are begun to eliminate infectious debris. Treatment may require several months. Contraindications to this therapy include patients with porphyria or hypersensitivity to griseofulvin.

Though griseofulvin has virtually eliminated the necessity for x-ray epilation as a treatment for tinea capitis, in rare cases it is still advocated by some authors. This procedure is dangerous and can cause severe damage through faults in techniques or inaccurately calibrated x-ray machines. Carefully used, it is said to be a safe, effective procedure. The standard method is known as the Kienböck-Adamson technique. Five preselected areas of the scalp are given a dose of x-ray ranging from 400 to 500 R. The dose is determined by the age of the child and robustness of the hair, i.e., thick, dark hair requires more exposure than thin, light hair. The procedure requires about 40 minutes and is thus difficult to administer to young children. The treatment is rarely given twice to the same patient as this is quite hazardous. Increased incidence of neoplastic diseases has been reported following the use of x-ray epilation.[6] An extensive review of experience with this procedure was reported by Shanks.[147] Most authors agree that this should be reserved as a last resort of therapy. Treatment failures are usually due to technical problems.

Tinea Favosa

DEFINITION

Favus is a clinical entity characterized by the occurrence of dense masses of mycelium and epithelial debris which form yellowish, cup-shaped crusts called scutula. The scutulum develops in a hair follicle with the hair shaft in the center of the raised lesion. Removal of these crusts reveals an oozing, moist red base. After a period of years, atrophy of the skin occurs, leaving a cicatricial alopecia and scarring. Scutula may be formed on the scalp or the glabrous skin. The Latin name favus refers to the similarity of appearance of scutula and honeycombs.

Synonymy. Favus, honeycomb ringworm, erbgrind (Ger.), teigne faveuse (Fr.).

ORGANISMS INVOLVED

The vast majority of cases of favus are caused by *Trichophyton schoenleinii.* The peculiar and characteristic clinical disease evoked by this organism has led mycologists like Remak (1845) and Vanbreuseghem (1962)[164] to place the organism into a separate genus, *Achorion.* However, extensive mycologic investigation of cultural characteristics does not warrant this transfer. Other organisms are capable of producing the same clinical entity. *T. violaceum,* on occasion, may evoke a similar disease, and *Microsporum gypseum* has also been isolated from a few cases. Similar diseases are also produced in animals. *M. gallinae* produces favus of gallinaceous birds, *T. equinum* favus of horses, and *T. mentagrophytes* of the *quinkeanum* variety and *M. persicolor* both produce "mouse favus." *T. schoenleinii* is highly endemic in central and southern Europe, the Middle East, Iran, Kashmir, and Greenland. In North America there are at least two small endemic foci—a mountain region in Kentucky and the Gaspé coast of Quebec. The disease is extremely common among the Bantu in South Africa, where it is called "witkop." Formerly patients with favus were not permitted entry into the United States.

SYMPTOMATOLOGY

Three grades of severity are defined when the infection involves the scalp. The mildest form consists of redness of the scalp in a general follicular distribution and some matting of hair, but without hair loss. In the second grade of severity, the patients show formation of scutula, loss of hair, more redness, and more widespread involvement. Extensive loss of hair (usually over one third of the scalp or more), atrophy of the skin in the bald areas, healing and scarring in the central areas, and formation of new scutula and crusts at the periphery constitute grade three of the disease.

The infection begins as small, yellowish-red, subcuticular puncta. There is an erythematous reaction on the scalp and a variable degree of seborrhea and flaking. At this stage the puncta may develop into the cup-shaped yellowish crusts or the lesions may resemble seborrheic eczema. In the latter type of disease, there is extensive scaling, some matting of hair, and an erythematous base to the lesions (Fig. 7–12 FS A 27). The condition closely resembles seborrheic dermatitis and "tinea amiantacea" and must be differentiated from them. Tinea favosa and tinea capitis have a patchy distribution, whereas seborrheic dermatitis is more generalized and usually affects other parts of the body as well. Soon after infection, the hair becomes lusterless and grey as mycelium penetrates the shaft. On microscopic examination mycelium, "air bubbles," and fat droplets are seen intrapilarly. Arthrospores are very rarely formed and are irregular in shape. As the lesion progresses, the hair is shed and the follicle atrophies (Fig. 7–13 FS A 28). Typically the cup-shaped crusts develop and regress as the lesion advances peripherally. Over a period of years an insidious atrophy of the skin develops, as central clearing occurs. Unlike tinea capitis, there is no tendency for the infection to involute at puberty. The infection may last the lifetime of the patient. The extent and appearance of the lesions depend to a large degree on the general hygiene practiced by the patient. Matted hair, debris, and scutula (sometimes called "godet") along with serous exudate, secondary bacterial involvement, pus, and general filth may be present, and the scalp has an unpleasant "cheesy" or "mousy" odor. Removal of scutula and debris aids in the treatment of the disease.

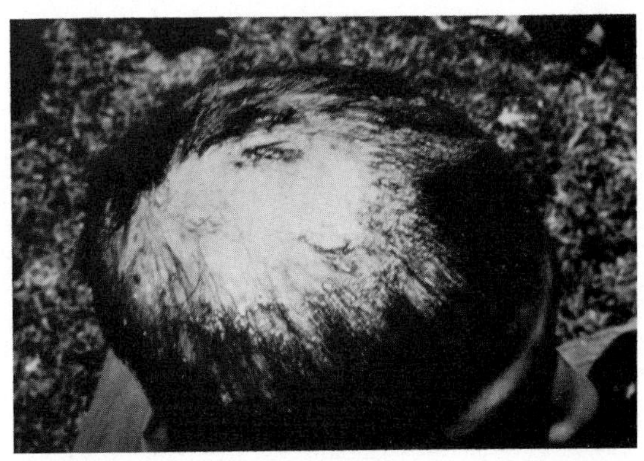

Figure 7–13. FS A 28. Tinea favosa. Advanced disease involving the entire head. There are areas of atrophied skin and alopecia, in addition to some active lesions of infection. (From a small endemic focus in Guatemala, courtesy of R. Mayorga.)

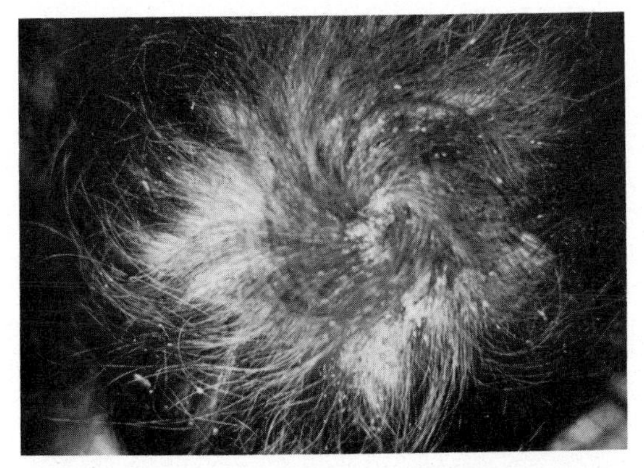

Figure 7–12. FS A 27. Tinea favosa. Seborrheic stage of disease showing matted hair and lesion with erythematous base. The infected hair is grey, whereas the normal hair is pigmented.

Tinea favosa also presents on the glabrous skin. In addition to vesicular, papular, or papulosquamous lesions, typical scutula are formed. The cup-shaped crusts may be very numerous and resemble a range of volcanoes. The skin may atrophy in these areas, a condition not seen in ordinary tinea corporis. *Trichophyton schoenleinii* may also invade the nails. The disease produced in nails is indistinguishable from other forms of tinea unguium.

HISTOPATHOLOGY

Mycelium is present in the horny layer of the scalp, within and around the hairs, and in the scutulum. The scutula consist of intertwined mycelial masses, scales, sebum, and other debris cemented together to form a cup-shaped crust. The periphery is composed of well-preserved hyphae, while in the center, dead and degenerating mycelium and granular debris are found. The scutulum rests on an atrophic epidermis. The dermis shows a mild to moderately severe inflammatory reaction with a round cell infiltrate. The horny layer of the skin often extends over the edge of the scutulum. The etiologic agent, *T. schoenleinii*, elaborates many proteolytic enzymes, including a collagenase and an elastase, which may account for some of the bizarre pathology seen in this disease.[128]

DIFFERENTIAL DIAGNOSIS

When typical, yellowish, cup-shaped scutula are present, along with a mousy or musk-like odor of the scalp and a dull green fluorescence of the hairs under Wood's lamp, the diagnosis is readily apparent. The presence of these symptoms appears to vary geographically. In Iran, less than half of the patients examined showed scutula, whereas they were present in a much higher percentage of people surveyed in Quebec.[22] Marked scalp atrophy is also seen in pseudopelade, lupus erythematosus, and lupus vulgaris. Staphylococcic pyodermas and seborrheic eczema also simulate some of the clinical features of tinea favosa.

PROGNOSIS AND TREATMENT

The infection rarely involutes spontaneously, though the crusts, inflammatory reaction, and debris may gradually subside. The thin, atrophied skin of the scalp is denuded of most hair follicles and sebaceous glands, which are replaced by scar tissue. A cicatricial alopecia may cover essentially the entire scalp. However, within the area, incongruously, a few tufts of normal-appearing hair may remain. Mycelium is usually seen in these hairs and is still sparsely present in the scalp as well. In treated cases, the ultimate prognosis for the scalp is favorable. The extent of sequelae depends on the stage of the disease when arrested. It is important to realize that this is a family-centered infection, in contrast to most cases of tinea capitis which are transmitted by group exposure. Several generations in the same family may be infected and show various stages of the disease.

T. schoenleinii has approximately the same sensitivity to griseofulvin as the other dermatophytes. Resolution of infections has been accomplished with long-term use of the drug. The dosage schedule is the same as for tinea capitis. Cleaning of the debris, removal of crusts and general improvement of scalp hygiene aid in clinical management. All family members should be treated simultaneously. Transmission probably requires long-term association and exposure, so that casual contacts are usually not infected. X-ray epilation was the standard treatment before griseofulvin. It is still widely used as an adjunct to speed results. The MacKee technique, in the hands of a specialist, is popular in some clinics where the disease is endemic.[147]

Tinea Corporis

DEFINITION

Tinea corporis is a dermatophyte infection of the glabrous skin most commonly caused by species of the genera *Trichophyton* and *Microsporum*. The infection is generally restricted to the stratum corneum of the epidermis. The clinical symptoms are a result of the fungal metabolites acting as toxins and allergens. Lesions vary from simple scaling, scaling with erythema, and vesicles to deep granulomata. Lanugo hair in the involved area may be invaded, and the follicle often acts as a reservoir for recrudescence of the disease.

Tinea corporis is a universal affliction of man. The disease is found in all areas of the earth, from the arctic to the equator to the antarctic. However, it is generally more common in the tropics than in temperate climates.

Synonymy. Ringworm of the body, tinea circinata, tinea glabrosa, scherende Flechte (Ger.), herpès circiné trichophytique (Fr.).

ORGANISMS INVOLVED

All species of dermatophytes are able to produce lesions of the glabrous skin, even though some species are more commonly associated with other types of infections. In addition, many species of soil keratinophilic fungi are capable of evoking clinical ringworm. In contrast to tinea capitis, there is little tendency for geographic dominance of a particular species in tinea corporis. Probably the most universally encountered species is *Trichophyton rubrum,* followed in frequency by *T. mentagrophytes.* In areas where there is a heavy infection rate of tinea capitis caused by an endemic species of fungus, there will be a predominance of that species causing tinea corporis.

SYMPTOMATOLOGY

Natural infection begins with the deposition of infected scales, hyphae, or arthrospores on the skin of a susceptible person. Infection may be transmitted by direct contact with an infected individual or animal; by fomites, such as clothing, furniture, and so forth; or by spread from existing, sometimes subclinical, lesions elsewhere (e.g., occult tinea pedis).

Invasion of the horny layer of the skin occurs at the site of inoculation. This is followed by centrifugal spread from the initial site. An incubation period of one to three weeks elapses before clinical signs are evident. The formation of the characteristic rings of inflammatory reaction is more evident in infections of the smooth skin than in hirsute areas. This annular appearance results from the elimination of the fungus and its irritating products from the center of the lesion as the margin spreads peripherally. A second centrifugal spread of the fungus may occur from the original site, with the formation of concentric rings (Fig. 7–14 FS A 29). Hair follicles act as reservoirs of infection, as is commonly seen with *T. rubrum* infections.[37] Some lesions lack any tendency toward spontaneous healing and remain scaly (e.g., *T. tonsurans*) or vesicular (e.g., *T. mentagrophytes*). As with other types of ringworm, the more inflammatory the reaction, the shorter the duration of the infection.

Two types of lesions are most commonly encountered. One is dry and scaly *annulare* (annular patches), and the other *vesiculare* ("Iris" form). The first begins as a small, spreading, elevated area of inflammation. The margin remains red and sometimes slightly swollen, while the central area

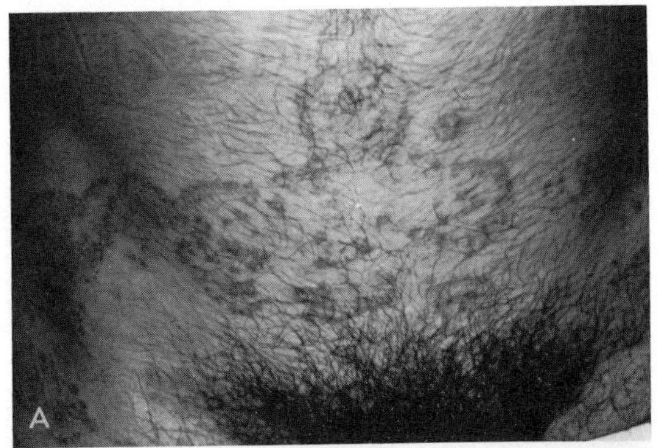

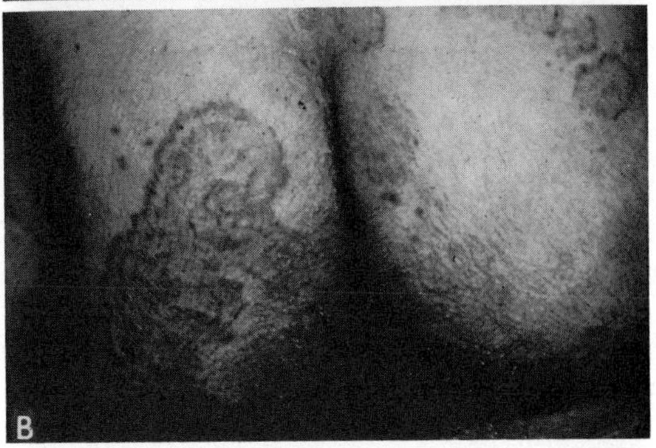

Figure 7–14. FS A 29. Tinea corporis. *A,* Annular appearance of lesions on trunk and (*B*) on gluteal area. (Courtesy of S. Lamberg.)

becomes covered with small scales (Fig. 7–15 FS A 30). Spontaneous healing occurs in the center as the circinate margin advances. Lesions resolve after a few months, or may become chronic and last for the lifetime of the individual. Organisms commonly producing this type of lesion include *T. rubrum* and *Epidermophyton floccosum.*

In the second type of lesion, vesicles appear irregularly or immediately behind the advancing hyperemic and elevated margin. A crust is formed, then healing follows in the center of the lesion to leave a more or less pigmented area (Fig. 7–16 FS A 31). The lesions may become pustular when the fungus invades the hair follicles. For any dermatophyte infection, the severity of response is proportional to the involvement of the hair. In both types of lesions, the fungi are most numerous at the margin of the advancing lesion; therefore scales and vesicles from this region are best suited for study and isolation of the etiologic agent. Lesions usually resolve in a few weeks or months. Chronic infections are uncommon. The usual fungi involved are *T. mentagrophytes* and *T. verrucosum.*

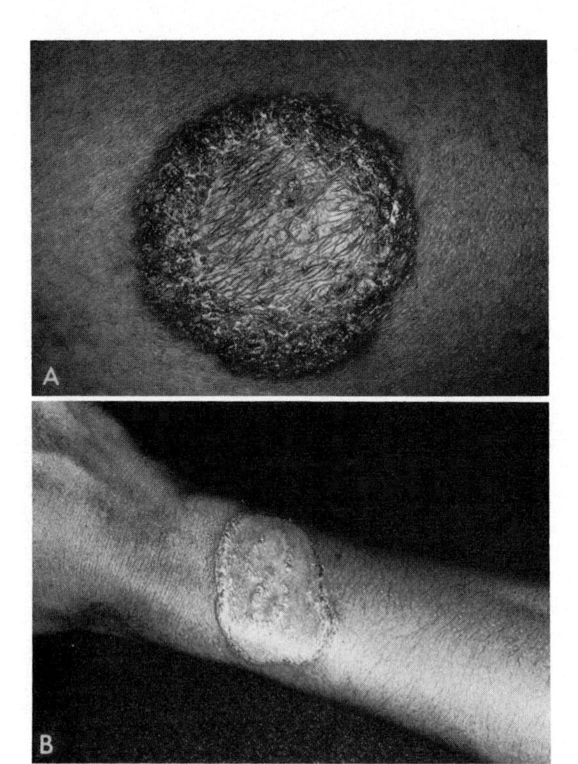

Figure 7–15. FS A 30. Tinea corporis. Vesicular patch form. *A,* Vesicles appear immediately behind active erythematous border. Thigh. *E. floccosum. B,* Similar lesion on arm. *T. rubrum.* (Courtesy of F. Battistini.)

Variations of the above two types of clinical lesions have been described.[70] Several confluent annular patches may converge to form a polycyclic lesion. Extensive hyperkeratosis on a red base is called the *psoriasiform* lesion. The *plaque type* lesion is a stationary, well-defined area with minimal scaling and redness (Fig. 7–17 FS A 32). It is most often caused by *T. rubrum* and commonly becomes chronic and at times barely visible. Pruritus is common in all types of infection. Other more severe types of lesions are sometimes seen. These include *granulomatous lesions, verrucous lesions,* and *tinea profunda.* Small, deep granulomata may be produced around hair follicles. Microscopically these are seen to be perifollicular granulomas caused by fragments of infected hair penetrating the follicle wall (Fig. 7–18 FS A 33). The lesions are most commonly seen on the legs of women after shaving of leg hairs.[172] A more serious condition with a similar histologic picture is Majocchi's granuloma. Involvement is more extensive, and the granulomas are large and sometimes vegetating. The organism producing both types of infection is *T. rubrum.* Patients with Majocchi's granuloma usually have some underlying disease. Blank and Smith[26] describe a patient with an impaired cellular defense mechanism whose infection was accompanied by numerous subcutaneous nodules and abscesses. Biopsy demonstrated invasion of living tissue by mycelium, and the organism was identified as *T. rubrum.* A verrucous type of ringworm of the glabrous skin has been ascribed to *E. floccosum.*[45] The face, forehead, ear, and buttocks were covered with verrucous nodules, and the trunk and legs had large, scaly plaques. In some personally observed cases, vegetating masses were seen on the hands and wrists. No invasion of living tissue was observed in biopsy sections of lesions, and no underlying disease was discovered. In other cases, such as trichophytosis totalis (due to *T. rubrum*), agammaglobulinemia was noted. Patients with this disease who have extensive or unusually severe infection generally respond to griseofulvin treatment but universally relapse when the drug is discontinued. Tinea profunda represents an exaggerated inflammatory response on the glabrous skin and is the equivalent of a kerion of the scalp.

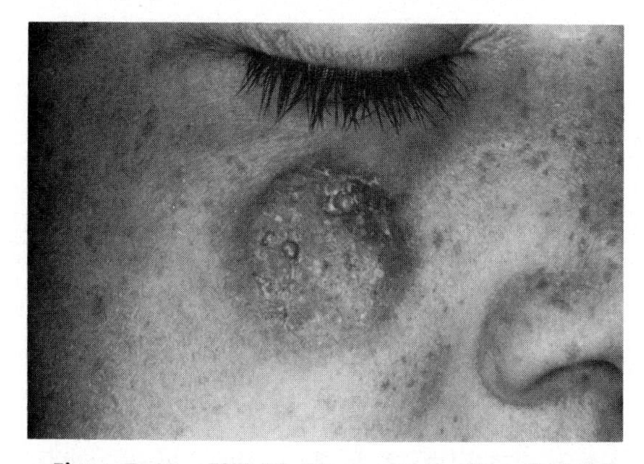

Figure 7–16. FS A 31. Tinea corporis. (tinea faciei[39a]). Vesicular type lesion. Vesicles and crusts are seen over entire area. *T. verrucosum.*

HISTOPATHOLOGY

A toxic reaction in the epidermis is the first tissue response to the presence of the fungus in the stratum corneum. This may subside, become chronic, or develop into an allergic reaction. It presents as a typical eczemaform pattern, although later the

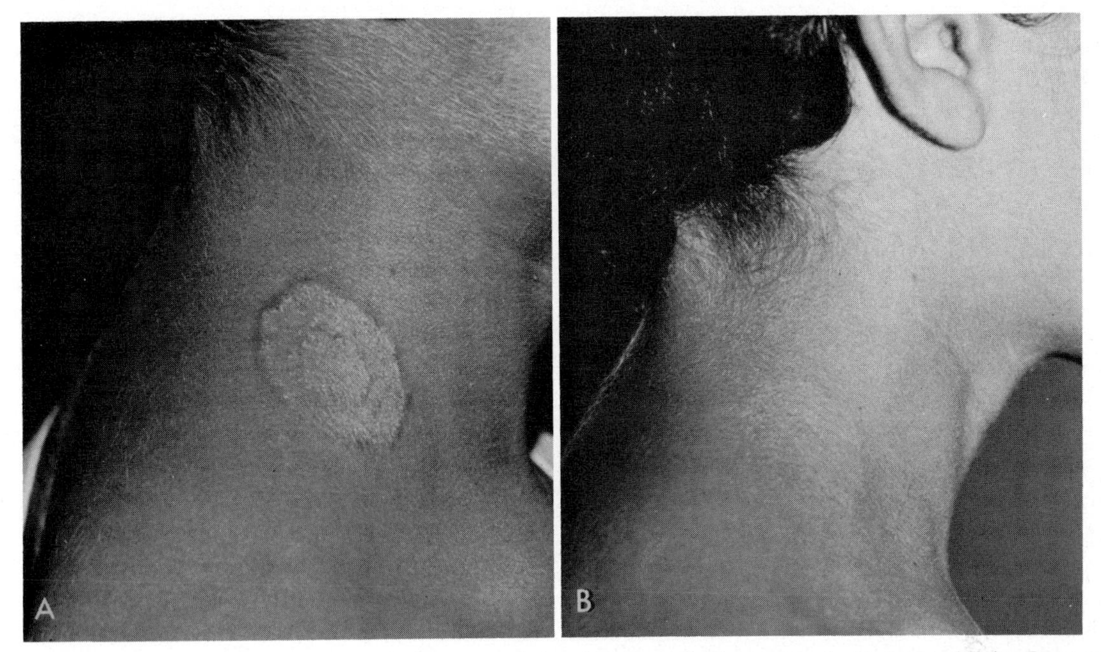

Figure 7–17. FS A 32. Tinea corporis. *A,* Plaque-type lesion. *T. rubrum. B,* After treatment with thiabendazole. (Courtesy of F. Battistini.)

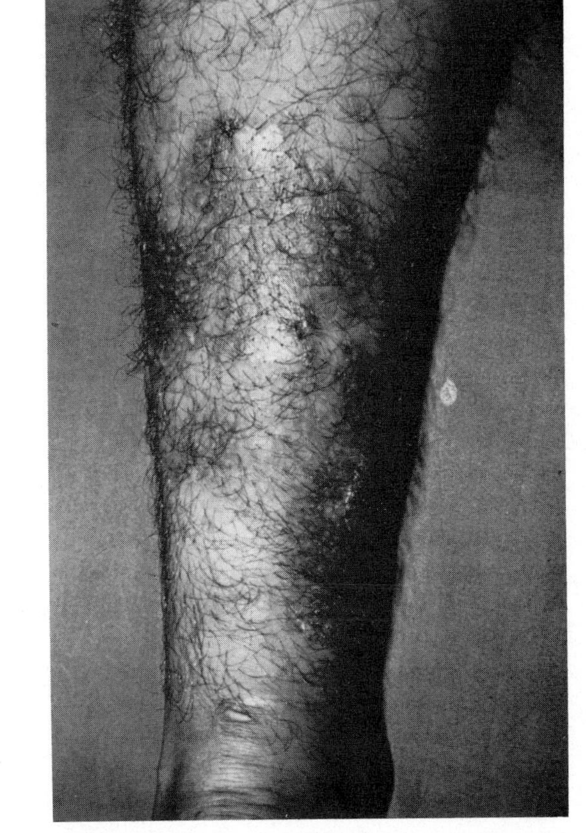

Figure 7–18. FS A 33. Tinea corporis. Granulomatous reactions around hair follicles, resembling Majocchi's granuloma. There is a serous exudate from some follicles. *T. mentagrophytes.*

histologic picture is essentially that of a nonspecific, subacute chronic dermatitis. Special fungus stains demonstrate mycelium in the horny layer of the skin. If the biopsy is from a dry scaly area, hyperkeratosis and parakeratosis are present. Some acanthosis may occur, which, in combination with papillary edema and a perivascular round cell infiltrate, leads to flattening of the rete. The appearance of tinea corporis then is initially one of simple inflammation, and no characteristic features are present. The histologic picture of vesicular lesions is similar, but, in addition, there are subcorneal and intraepidermal vesicles and a more pronounced cellular infiltrate. Intracellular edema affects all parts of the epidermis. Acanthosis is also more marked, and there is often a leukocytic invasion of the epidermis.

In the nodular granulomatous perifolliculitis caused by *T. rubrum,* spores and mycelium are seen in the hair follicle and in the inflammatory infiltrate of the dermis. Irregularly shaped, sporelike structures have been described which may attain 6 μ in diameter. It is the presence of the spores and mycelium in tissue after the rupture of the follicle wall which stimulates the foreign body reaction and granuloma formation. Lymphocytes, histiocytes, epithelioid cells, and some foreign body giant cells are seen.[107]

DIFFERENTIAL DIAGNOSIS

Though many diseases may mimic tinea corporis at some stage, the diagnosis is usually straightforward. Tinea corporis is characterized by the appearance on the glabrous skin of an annular papulosquamous lesion, which is either simply scaly or has a vesiculopustular component at the periphery and is crusted in the center. However, atypical infections are numerous, and the possibility should always be entertained that any red scaly rash on the body may be of fungal etiology. Psoriasis, pityriasis rosea, nummular eczema, granuloma annulare, annular secondary syphilis, lichen planus, seborrheic dermatitis, contact dermatitis, fixed drug eruption, pityriasis versicolor, dermatocandidosis, and erythema annulare are some of the diagnoses to be considered. Nummular eczema is commonly confused with tinea corporis, but the papulovesicular plaques tend to be more symmetrical. The symmetrical pattern of seborrheic dermatitis also distinguishes it from fungal infection, and there usually is involvement of the scalp and intertriginous areas as well. Although the "herald patch" of pityriasis rosea is indistinguishable from tinea corporis, the restriction of the former to the trunk and the symmetrical distribution of lesions delineate this disease. Tinea corporis is readily and rapidly diagnosed by the potassium hydroxide mount. Examination of scales will show septate hyphae and squared or rounded, irregularly arranged arthrospores. Cultures should always be taken. In our experience the number of culture-positive cases when the KOH was negative ranges from 5 to 15 per cent.

PROGNOSIS AND THERAPY

In normal patients, tinea corporis resolves spontaneously after a few months. There is less tendency toward chronicity than in tinea pedis and tinea cruris. Treatment aids in resolution of lesions and effects a clinical cure. Reinfection of the same area may occur within a few weeks to months, if the patient is again exposed to infectious material.

Uncomplicated tinea corporis of the annular, plaque or vesicular type can be treated topically. The drug of choice at present is tolnaftate. It is quite effective, either as a solution or a cream, for lesions of limited size in accessible areas. Tinea corporis due to *T. rubrum, T. mentagrophytes, M. canis, M audouinii,* and generally *T. tonsurans* are amenable to treatment by this drug. More vigorous therapy is sometimes required for infections by *T. verrucosum* and *T. violaceum.* Cleansing of the area to remove scales is a useful adjunct to therapy. Older modes of treatment include 5 per cent ammoniated mercury ointment, Pragmatar ointment, Aquafor containing 3 per cent sulfur and 3 per cent salicylic acid and tincture of iodine. These will also effect clearing of the infection in about two weeks, in uncomplicated cases. Widespread tinea corporis and the more severe types of lesions (granulomatous, verrucous, and tinea profunda) may require systemic griseofulvin therapy. The treatment schedule is usually 1 g administered daily in divided doses. Itching, erythema, and scaling diminish by one week.

The course of therapy is continued for six weeks to two months. When tinea pedis and especially tinea unguium are also present, a longer course of treatment is required. Excellent results have been obtained using thiabendazole (10 per cent cream). Battistini[16] obtained nearly 100 per cent cure in 250 cases. The new agent, haloprogin, is about as effective as tolnaftate and is also effective in dermatocandidosis.

The response to treatment is quite variable regardless of the drug used. Clinical evidence indicates increasing resistance of chronic infections to treatment with griseofulvin, and resistant mutants have been produced in the laboratory.[94]

Tinea Imbricata

DEFINITION

Tinea imbricata is a geographically restricted form of tinea corporis caused by *Trichophyton concentricum.* It is characterized by polycyclic, concentrically arranged rings of papulosquamous patches of scales scattered over and often covering most of the body.

Synonymy. Tokelau, Burmese, Chinese, Indian ringworm, Lofa tokelau, tinea circinata tropical, Gogo.

DISTRIBUTION

The disease occurs in the Pacific, Southeast Asia, and Central and South America. The distribution of the disease may be of anthropological value. It has been suggested that the infection was introduced to the western coasts of South and Central America by pre-Columbian voyagers from Polynesia.[2] This contradicts Thor Heyerdal's theory of population migration, as presumed in Kon Tiki. Tinea imbricata occurs sporadically among the Indians of Mexico, Guatemala, Panama, and Brazil (Fig. 7–1).

SYMPTOMATOLOGY

The disease begins with the appearance of brownish maculopapules which gradually increase in size. The central portions of the lesions become detached, and fissures develop toward the periphery. Around this border, a brownish zone appears. The margins are elevated. By passing a finger over the lesion from the center point to the active edge, a slight resistance can be felt. Layers of the stratum corneum become detached, with the free edges facing the center (Fig. 7–19 FS A 34). This process continues until numerous concentrically arranged, imbricated rings are formed from the initial central lesion. The process by which annular rings are formed in tokelau is quite different from that of annular tinea corporis. Scaling may be profuse and itching severe at first. There is little or no erythema. When the disease is chronic, there is little irritation or

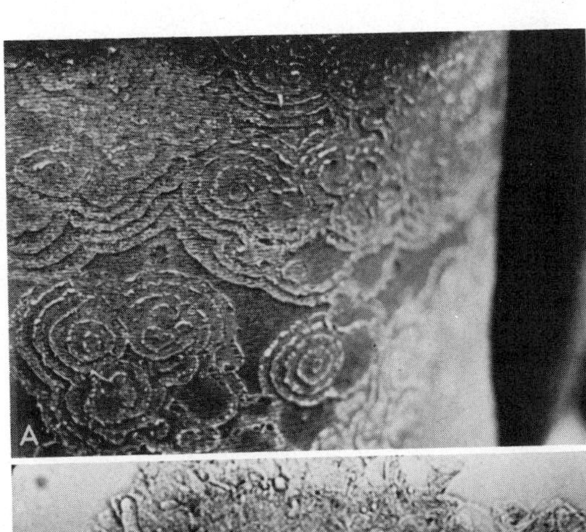

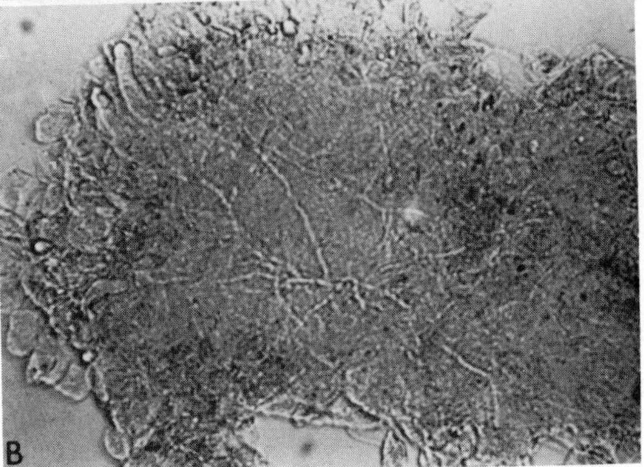

Figure 7–19. FS A 34. Tinea imbricata. *A,* Concentric rings forming over chest. Free edge of scales face center of lesion. *B,* Potassium hydroxide mount of infected skin scale.

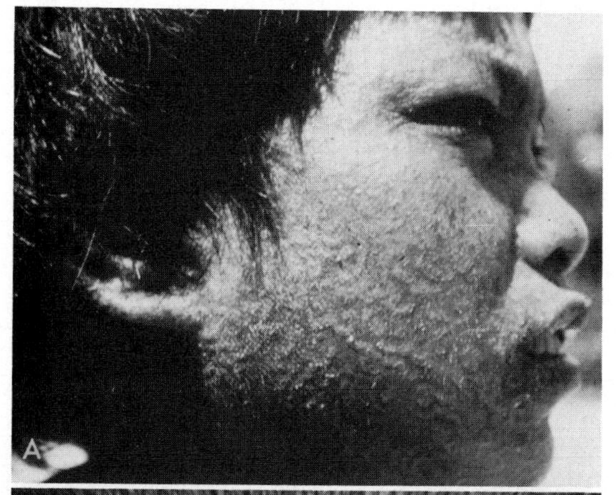

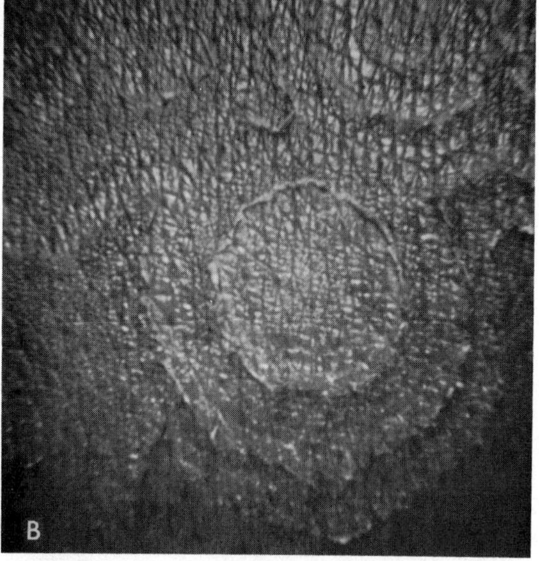

Figure 7–20. Tinea imbricata. *A*, Chronic involvement of face. *B*, Close up of lesion. Note free edge of scale is toward center.

discomfort to the patient (Fig. 7–20). The polymorphic, polycyclic patches are considered marks of beauty by the populace in Thailand and other countries. The face and scalp may be involved, but the hair is spared. If the infection involves the skin over the antecubital fossa, transverse parallel ridges are formed.[99] Nail infection is not uncommon. The etiologic agent, *T. concentricum*, may be accompanied in these lesions by other dermatophytes, e.g., *T. rubrum* and *E. floccosum*. In these cases, the symptoms are usually more severe.

The disease is more common in rural than urban areas. It is assumed that the organism is transmitted by direct, intimate contact, as from a mother to her baby. Once established, the disease becomes chronic and is of lifelong duration. There is no sex or age predilection, and a high percentage of an entire community is often affected. In some studies, it was not uncommon to find some members of a family with long-established disease, while other members were without infection. This emphasizes the low infectivity of *T. concentricum*. In a survey of dermatomycoses of some islands in the Pacific Ocean, *T. imbricata* was found only among Polynesians and not among Caucasians. Some of the latter had lived for generations in the islands and had had direct contact with the Polynesians.[100, 151]

DIFFERENTIAL DIAGNOSIS

The appearance of the polycyclic, polycentric rings with no evidence of erythema is so characteristic that there is little confusion in diagnosis. If the morphology is obscured, the scaling may resemble a form of ichthyosis.

PROGNOSIS AND THERAPY

The disease is chronic and extremely resistant to treatment. It responds readily to griseofulvin, but the required course of therapy is long, and relapse is common. The new lesions are multifocal and inflammatory.[99] Relapse may be accompanied by a rather severe vesicular reaction, with an erythematous base. Itching and irritation are sometimes severe, and the patient is worse off than before therapy was instituted. Complete cures have been recorded with a total dose of 24 g of griseofulvin given over 18 days.[30]

Tinea Cruris

DEFINITION

Tinea cruris is a dermatophyte infection of the groin, perineum, and perianal region, which is acute or chronic and generally severely pruritic. The lesion is characteristically sharply demarcated, with a raised, erythematous margin and thin, dry epidermal scaling. The disease is found in all parts of the world, but is more prevalent in the tropics. It tends to occur when conditions of high humidity lead to maceration of the crural region. A similar condition may involve the axilla or other intertriginous areas.

The disease is most common in men, and rarely involves women except when it is transmitted by intimate contact or fomite. It may reach epidemic proportions in athletic teams, troops, ship crews, and inmates of institutions. In such cases, it is probably most commonly transmitted by towels, linens, and clothing. *E. floccosum,* an etiologic agent of this condition, has been isolated from blankets and sheets.[18]

Synonymy. Ringworm of the groin, Dhobie itch, eczema marginatum, jock itch, gym itch.

ORGANISMS INVOLVED

Infection of the groin usually accompanies dermatophyte disease of the feet, so that the flora involved is commonly the same. In surveys conducted in England, Northern Ireland, Portugal, and Denmark, up to one half of the cases were caused by *Epidermophyton floccosum.* In the United States, *T. rubrum* appears to be the predominant species responsible for tinea cruris. A similar, common disease is caused by *Candida albicans. Trichophyton mentagrophytes* is associated with the more inflammatory pustular type of tinea cruris.

SYMPTOMATOLOGY

The disease begins as a small, round, swollen area of inflammation. This spreads to produce a circinate lesion at first which, because of differences in rate of spread, later becomes serpiginous (Fig. 7–21 FS A 35). In lesions caused by *E. floccosum,* there is a well-marginated, raised border (eczema marginatum) studded with numerous small vesicles or vesiculopustules filled with a serous exudate. Clipping the top of the vesicles for examination aids in the demonstration of the fungus. The central portion is brownish to red in color and covered with thin, branny, furfuraceous scales. The lesions are commonly bilateral, but not necessarily symmetrical. In most instances, the infection begins on the thigh where it is in contact with the scrotum, and spreads rapidly. The disease usually involves the inner thighs and spreads downward further on the left, because of the lower extension of the scrotum on that side. Sometimes the gluteal and pubic regions are involved. If *E. floccosum* is the etiologic agent, infection rarely extends further. In *T. rubrum* infections, the lesions frequently extend over the body, particularly to the waist, buttocks, and thighs (Fig. 7–22 FS A 36). Infections with *T. mentagrophytes* may rapidly involve the chest, back, legs, and feet and cause a severe, incapacitating inflammatory disease.

In acute infections, erythema is present, and there is intense itching. Older lesions are often lichenified, leathery, and plaquelike. Rarely, lesions appear as solitary vesiculopustules without a marginated border and have little tendency to spread. If there is no secondary bacterial involvement, the lesions

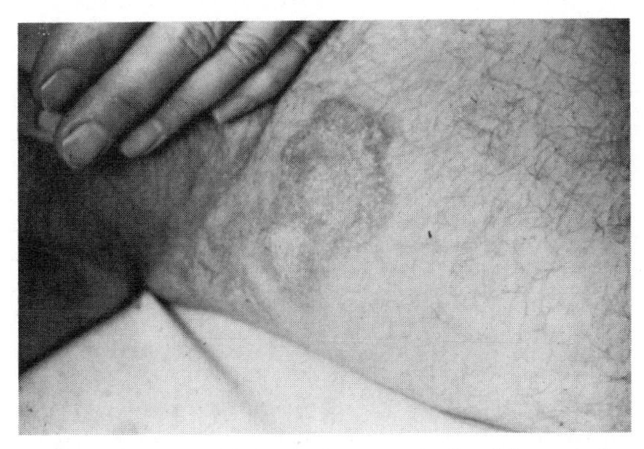

Figure 7–21. FS A 35. Tinea cruris. Serpiginous lesion of left thigh in contact with scrotum. *E. floccosum.*

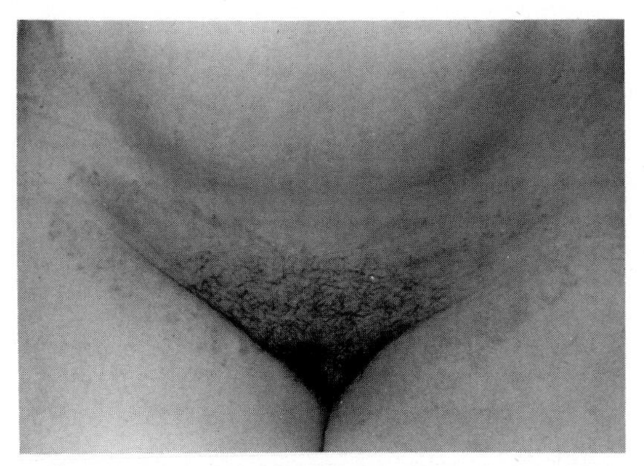

Figure 7–22. FS A 36. Tinea cruris. Extension of lesion from crural area. *T. rubrum.* (Courtesy of F. Battistini.)

are dry and branny. Rarely they may be soggy, whitish, and macerated, more closely resembling a *Candida albicans* infection. In contrast to dermatocandidosis, the scrotum and penis are usually spared in tinea cruris. Experimental inoculations to the scrotum do result in disease, but it is of shorter duration than that of the thigh. La Touche[91] has found that, in a large number of cases of tinea cruris, the scrotum is culturally positive, even in the absence of clinical signs.

PREDISPOSING FACTORS

Concentration of individuals in intimate surroundings, such as barracks, dormitories, and locker rooms, leads to rapid spread of the infection among isolated populations. Perspiration, humidity, irritation from clothes, and other factors which cause maceration of the crural skin increase the susceptibility to infection. Such diseases as diabetes, neurodermatitis, leukorrhea, and friction from skin folds in obese persons frequently are cited as predisposing factors.

DIFFERENTIAL DIAGNOSIS

The typical appearance of eczema marginatum, with its raised border and actively advancing periphery, is diagnostic of tinea cruris. If the lesions are weeping and satellites occur beyond the main lesion, the etiologic agent is probably *Candida albicans*. This is especially true in women, in whom the infection is often associated with vaginal candidosis. Candidal infection is much more common in diabetics than dermatophytosis. Other conditions that may simulate tinea cruris include erythrasma, seborrheic dermatitis, psoriasis, lichen planus chronicus, and contact dermatitis. Seborrheic dermatitis is usually more symmetrical in distribution and rarely restricted to the groin. Psoriasis is more scaly and has a more erythematous base than tinea cruris. In the obese, bacterial infection or simple intertrigo may be common. Lichen planus, contact dermatitis, and eczema are unlikely to be symmetrical and may closely simulate tinea cruris. Use of the Wood's lamp to demonstrate coral red fluorescence is helpful in ruling out erythrasma. In all cases, diagnosis is more frequently made by examination of a potassium hydroxide mount. Cultures are necessary in suspected lesions, even when the KOH is negative.

PROGNOSIS AND THERAPY

With adequate local or systemic therapy and sterilization of clothing, linens, and so forth to prevent reinfection, the prognosis is good unless the etiologic agent is *T. rubrum.* In the latter case a chronic disease may ensue, involving the body, feet, hands, and nails. In all cases of tinea cruris, special attention should be paid to the increased probability of the presence of a clinically evident or occult tinea pedis.[137] Recurrence in uncomplicated cases is prevented by elimination of predisposing factors. Tight-fitting underwear, athletic supporters, sweating, macerating conditions, and obesity are all potential inciters of exacerbation.

Topical treatment with tolnaftate (Tinactin) is usually successful and has few side effects. Other forms of topical therapy often result in a weeping, irritating dermatitis, which is frequently complicated by secondary bacterial infection. Systemic griseofulvin (500 mg per day) is most useful, particularly in *T. rubrum* infections. Symptoms are relieved within three days, and lesions involute within four to six weeks.[135] The course of treatment should be continued until all clinical, microscopic, and cultural signs of the disease disappear. The rapidity with which symptomatic relief is obtained following administration of griseofulvin suggests there is an effect other than an initial direct effect on the fungus. Deposition of griseofulvin in the horny layer with its subsequent inhibition of growth of the dermatophyte would require a longer time lapse to produce relief of symptoms. It has been suggested that, in addition to its effect on fungal metabolism, the initial effect of the drug is one of detoxification or anti-inflammatory action. Thiabendazole (10 per cent cream) is also very effective,[16] as is haloprogin. The latter is also effective in dermatocandidosis.

Tinea Unguium

DEFINITION

Tinea unguium is an invasion of the nail plates by a dermatophyte. For consistency of terminology, this disease is differentiated from onychomycosis, which is an infection of the nails caused by nondermatophytic fungi and yeasts.[18] The disease, tinea unguium, is of at least two types: (1) leukonychia mycotica (superficial white onychomycosis), in which invasion is restricted to patches or pits on the surface of the nail; and (2) invasive, subungual dermatophytosis (ringworm of the nail), in which the lateral or distal edges of the nail are first involved, followed by establishment of the infection beneath the nail plate.

Synonymy. Ringworm of the nail, dermatophytic onychomycosis.

INTRODUCTION

Persistent subungual dermatophytosis was noted by Mahon in the 1860's.[136] The infection was contracted in a fingernail used for epilation of hairs from patients with favus. This classic form of tinea unguium was considered rare until recently. Current surveys indicate the case rate is increasing, and up to 30 per cent of patients with fungous disease of the skin also have tinea unguium.[176] Twenty per cent of all nail disturbances[117] are due to fungi. Some of the apparent increase is probably attributable to better diagnostic methods. The disease and its mycology have recently been reviewed by Zaias.[177]

In the past, leukonychia mycotica was difficult to distinguish from other forms of leukonychia caused by systemic disorders or local insults. The disease was described in 1921 by Ravout and Rabeau and its association with *T. mentagrophytes* established.[122] These investigators were unable to see mycelium in the nail scrapings, but fungi were isolated in cultures. Jessner[76] described several more cases and named the disease leukonychia trichophytica. The term leukonychia mycotica was first used in 1926 by Rost. A comprehensive review of the clinical and mycological aspects of this disease, under the name superficial white onychomycosis, is given in Zaias.[175] Leukonychia mycotica is the preferred term, as it avoids confusion with other conditions and their etiologic agents. The disease is restricted to toenails.

ORGANISMS INVOLVED

Almost all species of dermatophytes have been isolated from ringworm of the nail. The etiologic agents are usually those which are common or endemic in the population. Since invasive ringworm of the nails is generally associated with infection of other cutaneous areas, the causative agents are often the same. Tinea unguium of the fingernail is most commonly due to *T. rubrum*; however, the toenail may be infected by a variety of organisms. Nail involvement associated with tinea corporis and tinea pedis occurs most commonly with *T. rubrum*, *T. mentagrophytes*, and *E. floccosum*; with tinea capitis and tinea favosa, *T. tonsurans*, *T. violaceum*, *T. megninii*, and *T. schoenleinii*; and with tinea imbricata, *T. concentricum*. Rarely encountered species include *M. gypseum*, *M. canis*, *M. audouinii*, *T. soudanense* and *T. gourvilii*.

Leukonychia mycotica is frequently an isolated lesion which is not associated with other forms of dermatophytosis. The most commonly isolated etiologic agent is *T. mentagrophytes*. Both the granular and interdigitale types of this fungus are encountered. Rarely involved organisms include *Cephalosporium roseo-griseum*, members of the *Fusarium oxysporum* group, and *Aspergillus terreus*.

SYMPTOMATOLOGY

Invasive or subungual ringworm of the nail usually begins at the lateral or distal edge of the plate. A minor paronychia usually precedes the infection and may become chronic or resolve. Paronychial inflammation results in production of a pitted or grooved surface on the nail. The initial symptom of nail involvement is a small, well-outlined, yellow or whitish spot which spreads to the base of the nail, or may remain stationary for years (Fig. 7–23 FS A 37). In established infections, the nail plate is brittle, friable, and thickened and may crack because of the piling up of subungual debris. Its color is often brown or black (Fig. 7–24 FS A 38). The accumulation of subungual keratin and debris is considered to be the characteristic feature of tinea unguium.[153,177] Under normal conditions, the nail bed does not contribute to the keratinization of the nail plate, nor does it form keratin under it. External stimulation or irritation due to the presence of the fungus in the nail bed evokes the production of soft, friable keratin. This loosens the nail and, as keratin accumulates, distortion and apparent thickening of the plate occurs. This is in distinct contrast to candidal onychomycosis, in which accumulated debris is absent and the nail is usually not thickened. The cheesy mass of epidermal detritus and keratin provides a fertile milieu for the rapid growth of fungi. Direct invasion of the hard nail plate may occur from below. Irritation of the nail bed stimulates more keratinization, and the nail becomes grossly distorted. A diverse flora consisting of other fungi and bacteria is found as a secondary invader in the debris. The nail matrix is spared. In some cases, minimal architectural changes and discoloration occur, and in rare instances no visible abnormalities are apparent. Another complication that is more common in *T. rubrum* infection is the cracking and separation of the distal part of the plate to leave a thin furrowed base with ragged edges. This occurs more commonly in fingernail infections. Involvement of the whole nail may lead to destruction of the entire appendage.

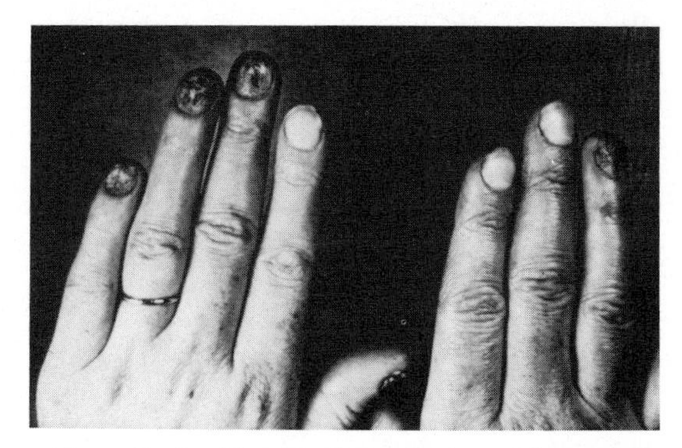

Figure 7–24. FS A 38. Tinea unguium. Advanced disease involving several nails. The infection is chronic and of many years duration, but a few nails remain uninvolved. *T. rubrum.*

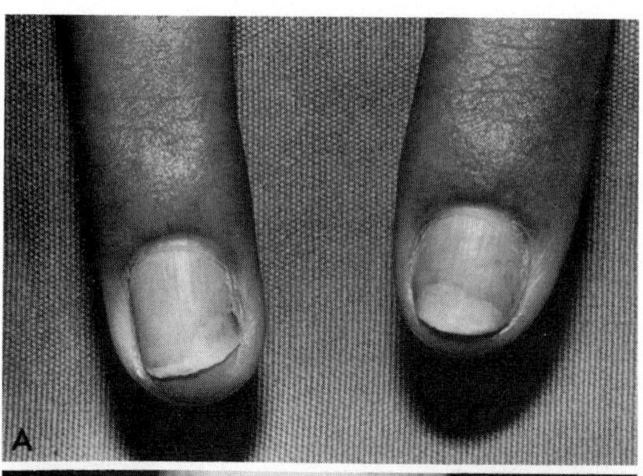

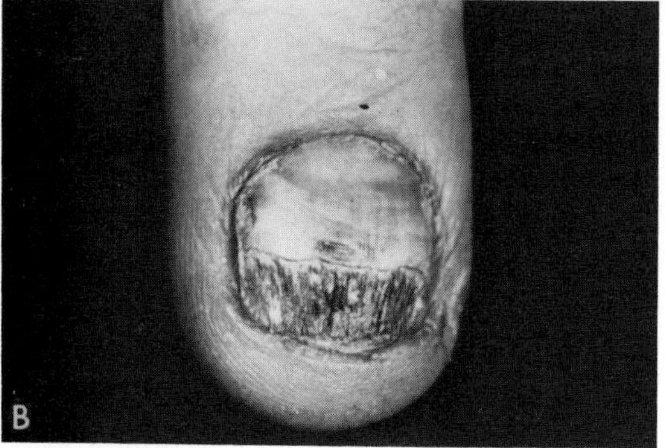

Figure 7–23. FS A 37. *A,* Tinea unguium. Invasive type. Initial infection at distal edge of nail plate. (Courtesy of F. Battistini.) *B,* Advanced disease showing grooved dark brown coloration.

Leukonychia mycotica (superficial white spot tinea unguium) begins as an opaque, circumscribed area on the surface of the nail plate. These lesions are usually punctate at first, irregular in outline, and may be numerous or solitary (Fig. 7–25). They begin at any place on the nail surface—in the center of the nail, near the lunula, in the free edge, or in the lateral folds. An established infection sometimes spreads to involve the entire surface or may remain restricted. The surface of the nail is soft and crumbly. The infection is otherwise asymptomatic and, because of its separation from living tissues,

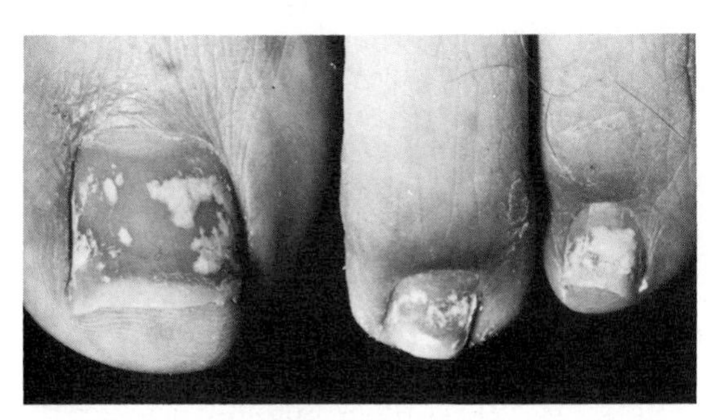

Figure 7–25. Tinea unguium. Leukonychia mycotica. Multiple white, irregular lesions on surface of nail. *T. mentagrophytes*. (Courtesy of N. Zaias.)

does not elicit an inflammatory response. The infection may be chronic and last for many years.

HISTOPATHOLOGY

In the subungual invasive type of tinea unguium, fungi are readily discernible in PAS-stained sections of the infected nail. The hyphal filaments and arthrospores are aligned horizontally between the lamellae of the nail and are generally limited to the lowermost portion of the plate. Onycholysis is not a feature of dermatophytic infection; rather, the lamellae are mechanically separated by the fungi growing between them.[121] Distribution and concentration of the fungus in the nail are quite variable and lead to difficulty when making the diagnosis by potassium hydroxide mounts or histopathologic sections. There is little or no inflammatory response in the underlying tissues.

In contrast to subungual tinea unguium, the mycelial elements in leukonychia mycotica are restricted to the uppermost portion of the nail plate. Fungal invasion rarely involves the deeper layers. Abundant hyphae are seen. As described by Zaias,[175] these are larger and broader than those seen in subungual involvement. They appear to be similar to the "penetrating organs," "eroding fronds," and "carpal bodies" which have been described in soil organisms and dermatophytes during their saprophytic utilization of keratin.[42] Aggregates or masses of distorted hyphae and irregularly shaped arthrospores are often present in sections. In contrast to the more "parasitic" picture seen in subungual tinea unguium, leukonychia mycotica represents an essentially "saprophytic" condition.

DIFFERENTIAL DIAGNOSIS

Subungual tinea unguium is often difficult to diagnose because of the scarcity of fungi and their location in the lowermost sections of the nail plate. Abnormalities of the nail which simulate the condition are congenital, the result of systemic disease, or due to external causes. The congenital conditions include nonmycotic leukonychia, clubbing, Beau's lines (transverse grooves or lines), and pachyonychia congenita. Among the external causes are contact irritants, trauma, onychomycosis due to filamentous fungi and yeasts, onychogryposis, onychophagy, onychotillomania, viral and bacterial infections, neoplasms, ingrowing toenails, subungual exostoses, and fibromas of tuberous sclerosis. Many skin diseases affecting the dorsal skin of the fingers or toes may cause dystrophic nails: eczema, lichen planus, bacterial paronychia, Darier's disease, scleroderma, syringomyelia, Raynaud's disease, hyperthyroidism, keratoderma blennorrhagica, keratoderma palmaris, acrodermatitis perstans, exfoliative dermatitis, and idiopathic onycholysis. Subungual conditions that rarely mimic tinea unguium are hemorrhages due to bacterial endocarditis or trichinosis, the periungual bulla of pemphigus, and argyria. It is apparent that the differential diagnosis covers a wide range of dermatologic and systemic diseases. The condition most closely simulating tinea unguium is psoriasis of the nails. It is extremely difficult to differentiate the two without evidence of psoriasis on other areas of the body. Distorted, deformed, thickened, discolored nails, with an accumulation of debris beneath them, particularly with ragged and furrowed edges, strongly suggest tinea unguium. *Candida* onychomycosis lacks gross distortion and accumulated detritus. Most of the other diseases listed commonly involve several nails and are symmetric in distribution. Inexplicably, tinea unguium may involve a single nail and have an asymmetric distribution. A chronic infection of many years' duration may involve one nail, while another in close proximity remains normal. Leukonychia mycotica is mimicked by leukonychia of other etiology, particularly trauma. Mycologic confirmation is the final proof of diagnosis in all cases.

PROGNOSIS AND TREATMENT

Tinea unguium is the form of dermatophytosis which is most resistant to treatment. It rarely, if ever, resolves spontaneously, and recurrence of the disease in clinically cured nails is common. Topical treatments alone have had a very poor record of cure. Formerly, evulsion of the nail followed by application of fungistatic agents was the only procedure which met with any success. Complete ablation has sometimes been used,[62] but this is an extreme and unwarranted procedure. The condition is chronic, lifelong, resistant to treatment, and may occur in infants[77] as well as adults.

Systemic griseofulvin therapy has resulted in complete remission of the disease in some patients. The course of therapy is long (a year or more), and good results are not assured. In a study by Russell, 80 per cent of patients with fingernail infections were cured after ten months on a dose of 1 g per day, compared with 12 per cent of patients with toenail infections.[142] In another study 8 of 14 patients still had infected toenails after 15 months of griseofulvin therapy, whereas almost total cure of fingernail infections has achieved.[153] Filing down of the nail to paper-thin consistency, followed by soaking in potassium permanganate (1:4000) or painting with phenol, 10 per cent salicylic acid, 1 per cent iodine, or chrysarobin (20 per cent in chloroform), is a useful adjunct to systemic griseofulvin therapy.[75] Some success has been noted with thiabendazole. Glutaraldehyde, 25 per cent diluted 1:1 with phosphate buffer, has effected cure in several cases but was ineffective in others.

Onychomycosis

DEFINITION AND ORGANISMS INVOLVED

As a general term, onychomycosis includes any infection of the nail produced by a fungus. Since infections caused by dermatophyte fungi have a characteristic evolution, pathogenesis, and therapy, they are considered separate and termed tinea unguium. The remaining infections are caused by a heterogeneous group of filamentous fungi and yeasts. Most of these have been found in dystrophic nails, but their significance either in initiation or aggravation of the condition is usually questionable. Primary invasion of the nail plate by *C. albicans* and *Scopulariopsis brevicaulis* is well established. In a recent review, Zaias[176] lists the following fungi as confirmed etiologic agents of onychomycosis: *Aspergillus candidus, A. flavus, A. fumigatus, A. glaucus, A. sydowi, A. terreus, A. ustus, A. versicolor, Cephalosporium* sp., and *Fusarium oxysporum.* Other species sometimes encountered include *Arthroderma tuberculatum* and *Phyllosticta sydowi.* In our clinic, the presence of *Cheatomium globosum* has been confirmed in three cases. Isolation of many other species has been reported, and in several of them, filaments were seen by direct microscopy.[176, 177] Aside from *C. albicans,* the most frequently implicated yeast is *C. parapsilosis.* Dystrophic nails harbor a large flora of saprophytes and secondarily invading fungi, yeasts, and bacteria. The black yeast *Hendersonela tortuloidea* has been involved in infections of the nails as well as the feet.[49]

HISTOPATHOLOGY

Active invasion of the nail plate occurs with *Scopulariopsis brevicaulis.* The characteristic conidia can be seen within the body of the nail. Candidosis of the nails is usually associated with chronic paronychia, which results in distortion of the nail architecture, and a chronic inflammatory response (Fig. 7–26 FS A 39). The yeasts are seen within the nail as well as adnexal tissue. The other fungi are usually located in grooves and cavities of the nail where there is an ac-

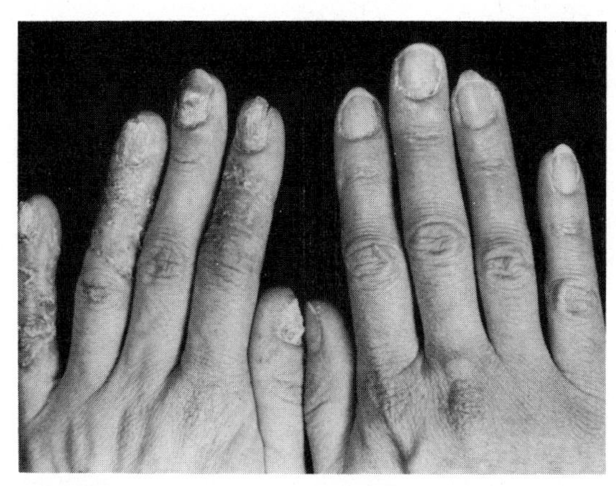

Figure 7–26. FS A 39. Onychomycosis. *Candida albicans.* This is a patient with chronic mucocutaneous candidosis associated with a thymoma. The architecture of the nail is distorted, and there is a chronic paronychia.

cumulation of debris. Active invasion of the nail plate is less frequent and of sporadic distribution. In sections mycelial filaments are seen which resemble those found in tinea unguium.

PROGNOSIS AND TREATMENT

Most cases of onychomycosis occur in abnormal nails. The disease often resolves when the antecedent condition is corrected. In chronic or genetic diseases of the nail in which therapy is not possible, the fungal infection may become chronic. Essentially all of the organisms causing onychomycosis are resistant to griseofulvin. *Candida* infections and their associated paronychia are effectively treated by nystatin ointment (100,000 U per g). Aqueous nystatin is more effective, but the drug is unstable in this milieu. Topical amphotericin B, gentian violet (0.5 per cent), resorcin (10 per cent solution in 70 per cent alcohol), and iodine (1 per cent in chloroform) have been used to treat yeast and filamentous fungal infections of the nail. Some success has been found using thiabendazole (10 per cent in cream base) under occlusive dressing. Glutaraldehyde may prove the treatment of choice after several series have been studied.

Tinea Barbae

DEFINITION

Tinea barbae is a dermatophyte infection of the bearded areas of the face and neck, and therefore is restricted to adult males. Lesions are of two types: a mild superficial type which resembles tinea corporis, and a type in which there is a severe, deep, pustular folliculitis (Fig. 7–27 FS A 40).

Synonymy. Tinea sycosis, Barbers' itch, trichophytie sycosique [Fr.], parasitare Bartifinne [Ger.], ringworm of the beard.

ORGANISMS INVOLVED

Tinea barbae infections are more common in rural areas, and the organisms are usually acquired from animals; therefore they are generally zoophilic dermatophytes. As noted previously, the severity of infection caused by zoophilic dermatophytes is often greater than that produced by anthropophilic fungi. In addition, the severity of the host reaction is also much greater when hair is involved. The combination of these two factors may explain the extremely severe reactions seen in some patients with tinea barbae. The most common organisms involved are *T. mentagrophytes* and *T. verrucosum,* both of which may be acquired from cows. *T. mentagrophytes* is also acquired from horses and dogs. *M. canis* is an uncommon cause of tinea barbae. In areas where *T. schoenleinii* and *T. violaceum* are endemic, they are frequently involved in this disease. Though the latter are anthropophilic fungi, they evoke a severe infection, probably because of hair and follicular involvement. *T. rubrum* is an infrequent cause of tinea barbae and may represent infection acquired from other parts of the body or transmitted as "barbers' itch" from unsanitary barbering practices. A geographically restricted species, *T. megninii,* is not infrequently isolated from barber-transmitted infections in its endemic areas. This organism, not prevalent in any country,

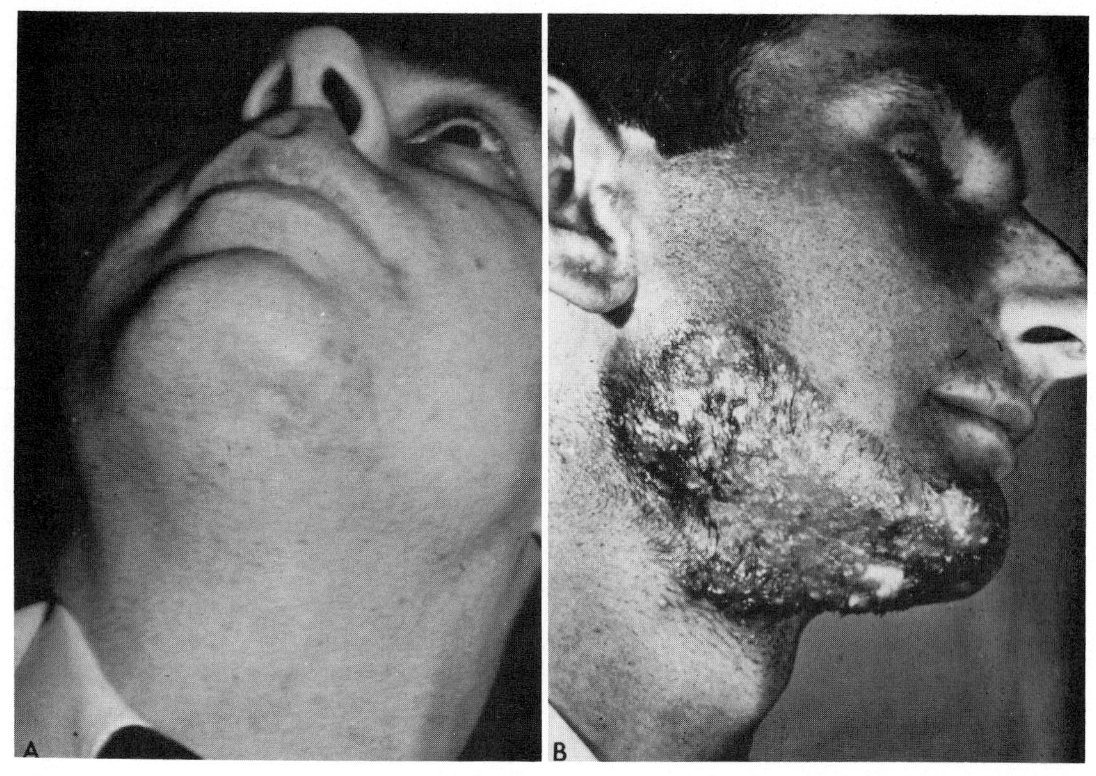

Figure 7–27. FS A 40. Tinea barbae. *A,* Mild superficial type due to *T. mentagrophytes* (JWR). Other isolants of this species may produce a more severe inflammatory disease. *B,* Deep, pustular folliculitis in a patient with *T. verrucosum* infection.

is found in Portugal, Sardinia, Sicily, Africa (as *T. kuryangei*)[4,166] and rarely in other parts of Europe.

SYMPTOMATOLOGY

The superficial type of tinea barbae resembles the lesions of tinea corporis. There is central scaling and a vesiculopustular border. The host reaction is less severe, though alopecia may develop in the center of the lesion. Hair and follicle involvement is less pronounced than in the deep type of infection. *T. rubrum* is generally the causative agent.

The deep or pustular type of tinea barbae is characterized by the presence of deep, follicular pustules that may result in the formation of the nodular, kerion-like lesions seen in tinea capitis. These pustular lesions are initially truly mycotic, and the pus is full of fungal arthrospores. The reaction may be so severe that most of the hair is shed, leading to spontaneous resolution of the disease. Permanent alopecia and scarring are common. The lesions are boggy and edematous. The hairs when epilated are seen to have a pussy whitish mass involving the root and surrounding tissue. Draining sinuses develop and undermine the surrounding tissue. Slight pressure evokes extrusion of purulent material. The lesions are usually solitary and most frequently are found on the maxillary regions. Occasionally the whole bearded area is involved, and extensive reddish-purple verrucose indurations are formed. Enlarged regional lymph nodes, mild pyrexia, and general malaise may accompany severe infections, especially those caused by *T. verrucosum.*[31] The upper lip usually is spared in tinea barbae, in contrast to the bacterial infection, sycosis vulgaris.

HISTOPATHOLOGY

The cellular reaction to tinea barbae is similar to that produced in the more severe types of tinea capitis. Organisms may be seen in the hair shaft and the follicle, and large numbers of arthrospores are present both on the shaft and free in the cellular debris. Sometimes, organisms are absent, and only an acute pyogenic infiltrate is seen. In chronic and resolving lesions, a chronic inflammatory infiltrate with giant cells may be present.

DIFFERENTIAL DIAGNOSIS

A history of contact with animals together with the presence of the severe, inflammatory pustular lesions evoked by *T. verrucosum* or *T. mentagrophytes* suggests a diagnosis of tinea barbae. The follicular pustules, brittle, lusterless, easily-epilated hair, and the presence of actively spreading peripheral borders comprise a classic picture of the disease. If the causative agent is *M. canis*, fluorescence of hairs under a Wood's lamp will be seen. The *Trichophyton* species do not fluoresce. Potassium hydroxide slide mounts readily show the presence of fungal elements and differentiate this disease from sycosis vulgaris. The milder forms of ringworm are less painful and tender than pyodermas caused by staphylococci. Infection by dermatophytes may also involve the eyebrow, but the conjunctiva is spared. Eyebrow infections without other involvement have been noted, particularly in children, and *M. canis* is frequently the etiologic agent (Fig. 7–28 FS A 41). Other conditions that may mimic tinea barbae are contact dermatitis, iododerma, bromoderma, cystic acne, actinomycosis, and pustular syphilids.

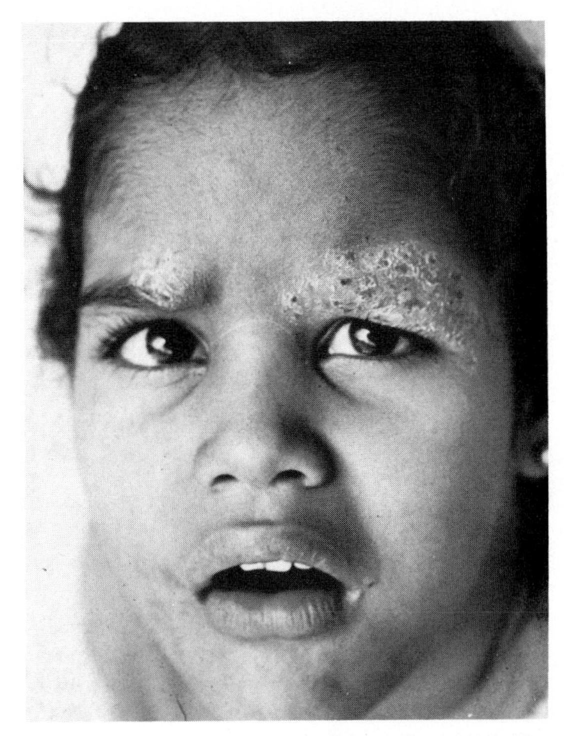

Figure 7–28. FS A 41. Tinea of the eyebrow caused by *M. canis*. (Courtesy of F. Battistini.)

PROGNOSIS AND TREATMENT

Since the majority of cases of tinea barbae are of the inflammatory type, spontaneous resolution usually occurs. Duration of the infection varies with the organism involved. Since *T. verrucosum* and *T. mentagrophytes* are the most virulent organisms, infections produced by them generally resolve in one to three weeks. *M. canis* infection lasts from two to four weeks. Chronic infections lasting more than two months are not uncommon when *T. rubrum* and *T. violaceum* are the etiologic agents.

Griseofulvin may be of some value in treatment of tinea barbae, particularly the chronic type. Rapid disappearance of the general malaise, pain, and discomfort, together with failure to develop satellite lesions and more rapid resolution of the disease, has been reported after treatment of the severe *T. verrucosum* infections.[31] The dose of griseofulvin is 1 or 2 g daily divided in four parts. Therapy should be continued for two to three weeks following disappearance of symptoms.

Formerly, manual or x-ray epilation, together with compression using permanganate soaks (1:4000) or Vleminckx's solution (1:33) were employed. None of these regimens are presently indicated, especially not x-ray epilation. Ammoniated mercury (5 per cent), Quinolor, Desenex, Sopronol, or Asterol was sometimes applied to the lesion. Some of the above may still be useful in resistant cases as an adjunct to griseofulvin therapy. Clipping and shaving of the bearded areas are recommended, along with warm compresses and debridement of diseased tissue.

Tinea Manuum

DEFINITION

Most dermatophyte infections of the hand, particularly of the dorsal aspect, are similar to tinea corporis. Tinea manuum refers to those infections where the interdigital areas and the palmar surfaces are involved and show characteristic pathologic features. The disease was first described by Fox in 1870[45a] and by Pellizzari[118a] in 1888. The subject was reviewed by Mitchell in 1951.[108,114]

ORGANISMS INVOLVED

Although almost all dermatophytes are potential invaders of the hand, the majority of infections are caused by *T. rubrum*, *T. mentagrophytes*, and *E. floccosum*. Tinea manuum is almost always associated with tinea pedis, so that the flora of the latter is usually the etiologic agent of the hand infection.

SYMPTOMATOLOGY

The clinical symptoms displayed by tinea manuum vary considerably, even though the etiologic agent is most often one species, *T. rubrum*. Calnan[29] described five clinical forms. Diffuse hyperkeratosis of the palms and fingers was the most common (Fig. 7–29 FS A 42). This condition is usually unilateral. The second type was crescentic exfoliating skin involvement similar to that seen in tinea pedis. Vesicular, circumscribed patches constituted the third type and were frequently caused by *T. mentagrophytes*. Discrete, red papular and follicular patches were the fourth, and erythematous scaly sheets on the dorsum the fifth type of infection. The latter two types are most commonly caused by *T. rubrum*. No particular predisposing factors are noted except anatomic deformity and occupational compression of the interdigital spaces. The latter leads to maceration and produces a predisposing condition similar to that affecting the feet. *T. mentagrophytes* var. *interdigitale* is a common invader in such cases. *E. floccosum* has been isolated rarely from cases of tinea manuum in which verrucous vegetating processes were seen.

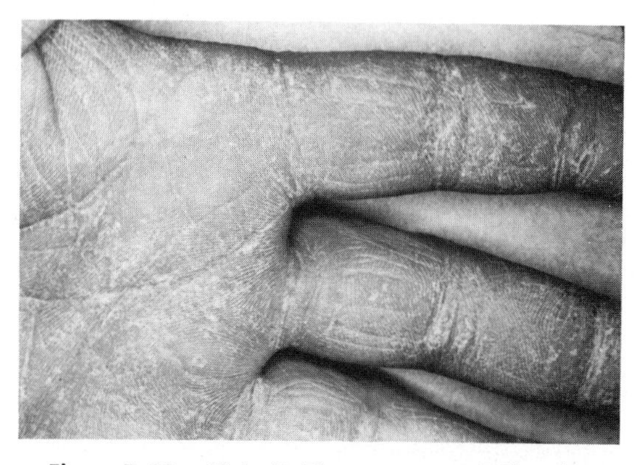

Figure 7–29. FS A 42. Tinea manuum. Diffuse hyperkeratosis of palm and finger. *T. rubrum*. In this chronic infection, only the palmar surface was involved; the dorsum was spared. There were no other lesions on the body.

DIFFERENTIAL DIAGNOSIS

The "id" (dermatophytid) reaction occurs on the hand and is nonspecific in appearance. It results from dermatophyte infection elsewhere on the body, usually the feet. It may resemble dyshidrotic pompholyx or be desquamative and thus resemble tinea manuum. These lesions resolve when the infection at the primary site is cleared. Other conditions that mimic tinea manuum are psoriasis, contact dermatitis, neurodermatitis, chronic pyoderma, secondary syphilis, and dermatocandidosis of the hand. Tinea manuum is usually unilateral.

PROGNOSIS AND TREATMENT

They are the same as for tinea corporis and tinea pedis.

Tinea Pedis

DEFINITION

Tinea pedis is a dermatophyte infection of the feet involving particularly the toe webs and soles. The lesions are of several types, varying from mild, chronic, and scaling to an acute, exfoliative, pustular and bullous disease.

Synonymy. Athlete's foot, ringworm of the foot.

INTRODUCTION

Ringworm of the feet is by far the most common fungus disease of man and is among the most prevalent of all infectious diseases. It is said to be a penalty of civilization and the wearing of shoes. In Western Samoa the disease was found among Europeans who wore shoes but not in barefoot natives.[100] The moisture and warmth of the toe clefts which are induced by shoes and socks provide a humid tropical environment that encourages the growth of fungi. These effects are more pronounced in the space between the fourth and fifth toes, and it is this site that is most frequently involved.

Though tinea pedis is a very common disease, it was not recognized until relatively late. Tilbury Fox[45a] reported on tinea manuum in 1870, but the first recognized case of tinea pedis appears to be that noted by Pellizzari in 1888.[118a, 114] Sabouraud in Paris and Whitfield in Britain wrote on the clinical and mycologic aspects of the disease. Whitfield reported the first British case in 1908. Systematic studies of the mycology and the histologic and clinical appearances of tinea pedis and tinea manuum were published in 1914 by Kaufman-Wolf.[85]

The disease is of worldwide occurrence and is distributed equally among the sexes. In contrast to many other forms of dermatophyte infection, tinea pedis is generally a disease occurring in adult life. Although the infection has been described in children only a few months old,[77] the incidence increases with age. It is probable that infection is related to repeated exposure to dermatophytes, so that people using common bathing facilities such as shower stalls in gymnasiums, barracks, and so forth are more prone to acquire infections earlier in life.[8] Repeated exposure, macerating conditions due to ill-fitting shoes and "sweaty socks," and possibly genetic factors are suggested as the most likely predisposing conditions for the disease. Strauss and Kligman[155] felt that the periodic recurrences of tinea pedis were due to a constant supply of fungi existing in occult lesions. English, on the other hand, feels that these recurrences represent reinfection.[41]

Tinea pedis is one of the most perplexing of infectious diseases. Though estimates of the infection rate for the population range between 30 and 70 per cent, the majority of these are occult or subclinical cases. The reasons why some people contract the disease and others with the same exposure do not are unknown. Deliberate attempts to induce infection both in the feet of normal patients and in those with a past history of infection resulted only in production of acute inflammatory lesions. These resolved spontaneously or with the aid of topical fungistatic drugs. Repeated attempts to produce tinea pedis in volunteers by immersing their feet in water laden with fungi did not result in a single clinical case.[12, 13] This experiment led to the postulation that infection was universal and that tinea pedis was not a contagious disease[157] but rather was dependent on host and environmental factors for expression of the clinical symptoms. Other workers were able to isolate dermatophytes from shower stalls, shoes, floors, and so forth and correlate exposure with infection and species concerned.[41,43,71] In more recent experimental work, Taplin was able to reproduce the severe inflammatory type of tinea pedis seen in United States troops in Southeast Asia.[27] He inoculated socks with strains of *T. mentagrophytes* indigenous to that area. The socks were then worn under occlusive boots, and a severe, persistent tinea pedis was produced that occasionally spread up the leg to involve the groin and buttocks.

ORGANISMS INVOLVED

At present three dermatophytes, *T. mentagrophytes*, *T. rubrum* and *E. floccosum*, cause the majority of cases of tinea pedis. The relative incidence of these varies considerably in different reports.[92] An interesting example of this was provided by English and Gibson,[44] who surveyed the feet of all children in one school. The rate of occurrence of *T. mentagrophytes* was eight times that of other dermatophytes. Yet, in clinical records both in England and the United States, *T. rubrum* was the predominant species. They postulated that *T. mentagrophytes* was more common in occult or inapparent infections, whereas *T. rubrum* caused a chronic and disfiguring type of disease for which patients sought medical advice. *E. floccosum* is a rather constant third. It is particularly prevalent in summer months and may account for 20 per cent of cases in the United States. In other areas (Asia), this species accounts for 44 per cent of cases.[143] In regions where *T. violaceum* is endemic, this organism is also involved in tinea pedis.[92] *T. megninii*, *M. persicolor*,[116b] and *M. canis* are responsible for a small percentage of cases.[112] *T. rubrum* was probably endemic in Japan, China, and the Far East until its recent spread to Asia, Europe, America, and Australia.[29] This worldwide distribution has occurred in the past forty years, and is postulated to have resulted from the vast migration of peoples and troops during the World Wars. *T. rubrum* is now the most cosmopolitan of dermatophytes.

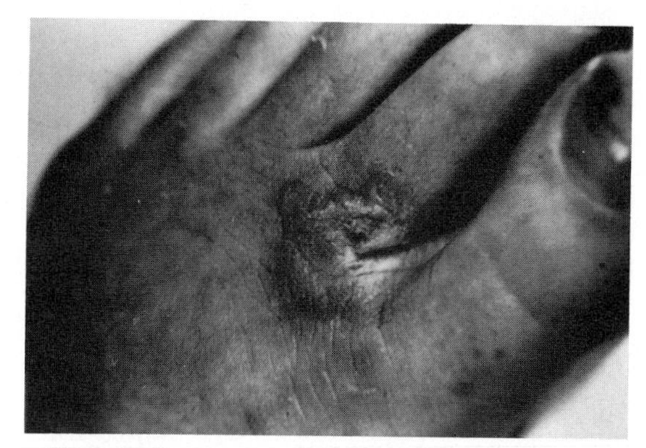

Figure 7–30. FS A 43. Tinea pedis. Intertriginous form between first and second toes. In this lesion, there is an inflammatory response.

SYMPTOMATOLOGY

Tinea pedis occurs in such a variety of clinical types that a satisfactory classification is difficult to devise. The four most commonly seen forms are (1) chronic intertriginous, (2) chronic papulosquamous hyperkeratotic, (3) vesicular or subacute, and (4) acute ulcerative vesiculopustular.

1. The intertriginous form of tinea pedis is the most common type. It appears as a chronic dermatitis, with peeling, maceration, and fissuring of the skin. The areas between the fourth and fifth toes and the third and fourth toes are most often involved. The webs and subdigital and interdigital surfaces are the favored sites. The area is covered with dead, white, macerated epidermis and debris, and there is often a foul odor present. Beneath the debris, the epidermis is erythematous and weeping. The denuded epidermis also harbors the fungus. In successive exacerbations the infection may spread to adjacent areas of the feet, to include the sole, heel arch, and dorsal surface. This condition is very persistent and is associated with hyperhidrosis. The infection may be intensified by hot, humid, summer weather and become severely pruritic. When *E. floccosum* is the fungus involved, marked scaling of the toe and sole, accompanied by numerous punctate satellite lesions, is seen. This may develop into brownish macules (Fig. 7–30 FS A 43). The intertriginous form of tinea pedis is usually amenable to topical treatment.

2. The chronic, papulosquamous, hyperkeratotic type of tinea pedis is very persistent and difficult to treat. It is characterized by the presence of areas of pink skin covered by fine silvery white scales (Fig. 7–31 FS A 44). It is commonly bilateral (Fig. 7–32 FS A 45). Though usually patchy in distribution, the

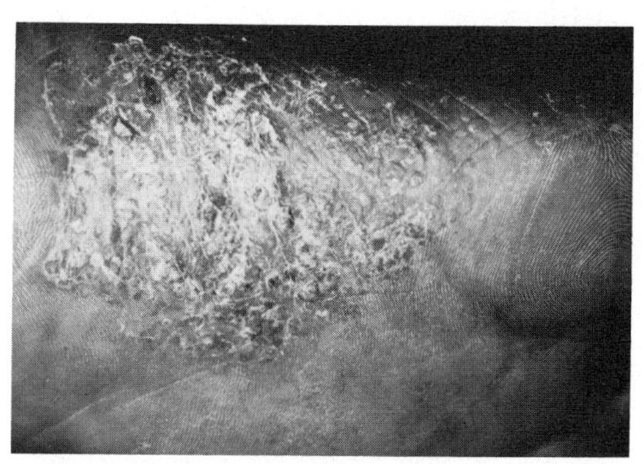

Figure 7–31. FS A 44. Tinea pedis. Chronic papulosquamous hyperkeratotic type. Note areas of pink skin covered with scales. *T. rubrum*.

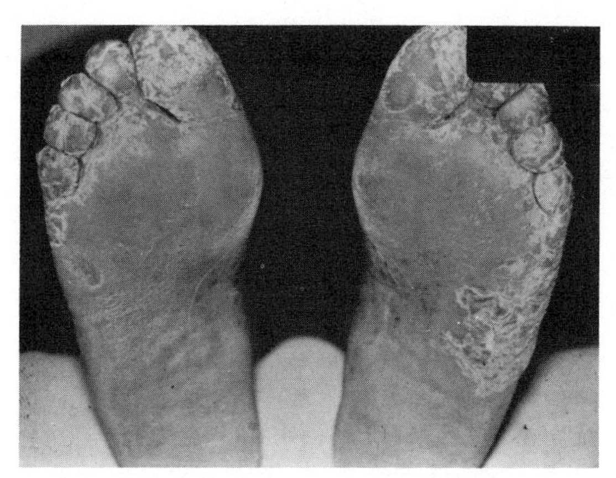

Figure 7–32. FS A 45. Tinea pedis. Chronic papulosquamous type involving both feet. *E. floccosum.*

lesions may involve the whole foot, in which case the disease is termed "moccasin foot." The etiologic agent is usually *T. rubrum.*

3. The vesicular form of tinea pedis is most often caused by *T. mentagrophytes.* The lesion is characterized by the appearance of vesicles, vesiculopustules, and sometimes bullae. The involved area may extend from the intertriginous areas to include the dorsal surface of the foot, the instep, and less frequently the heel and anterior areas (Fig. 7–33 FS A 46). The eruptions vary in size up to 7 to 9 mm, and are isolated or occur in patches. The vesicles are tense and contain a clear, serous exudate. After rupturing, they dry to leave a ragged collarette. The fungus is best demonstrated by clipping off the top of a vesicle or bulla and using this for direct microscopic observation or culture. The fungus is located on the inner top of the vesicle roof. The acute form of the disease frequently resolves spontaneously but often recurs under hot, humid, macerating conditions. In such cases, the disease is often quite inflammatory and may be incapacitating. This form is most often responsible for production of the "id" reaction on other areas of the body (Fig. 7–34). A cellulitis, lymphangitis, and lymphadenitis are occasionally seen, and the disease often resembles erysipelas.

4. In the acute ulcerative form of tinea pedis, the rapid spread of an eczematoid vesiculopustular process is seen. This form of the disease is complicated by secondary bacterial infection. The vesicle fluid is purulent, and ulceration of the epidermis occurs. In rare instances, this process is so fulminating that vast areas or even the entire surface of the sole is shed. In addition, there is pronounced cellulitis, lymphadenitis, lymphangitis, and pyrexia. The "id" reaction is common and may be widespread. Overtreatment with topical preparations often exaggerates this condition (dermatosis medicamentosa).

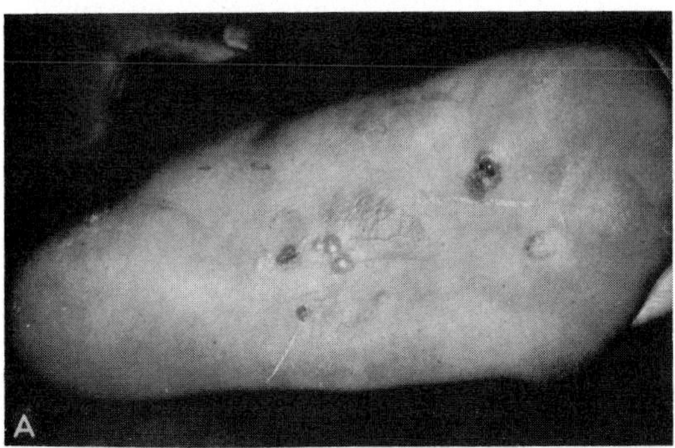

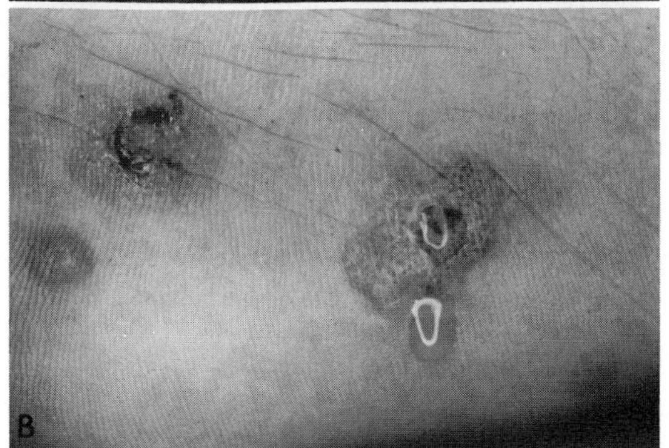

Figure 7–33. FS A 46. Tinea pedis. *A,* Vesicular form involving intertrigo and instep. *B,* Close up of vesicle showing serous exudate. *T. mentagrophytes.* (Courtesy of S. Lamborg.)

HISTOPATHOLOGY

During the acute stage of tinea pedis, intracellular edema and spongiosis with a leukocytic infiltrate are seen in the epidermis. The vesicles are in the upper portion of the epidermis and lie just beneath the stratum corneum. Parakeratosis is present. In chronic lesions, hyperkeratosis, acanthosis, and a chronic inflammatory infiltrate are found.

DIFFERENTIAL DIAGNOSIS

Branny, furfuraceous, scaly patches or groups of vesicles over the soles, together with fissures and a macerated epidermis are

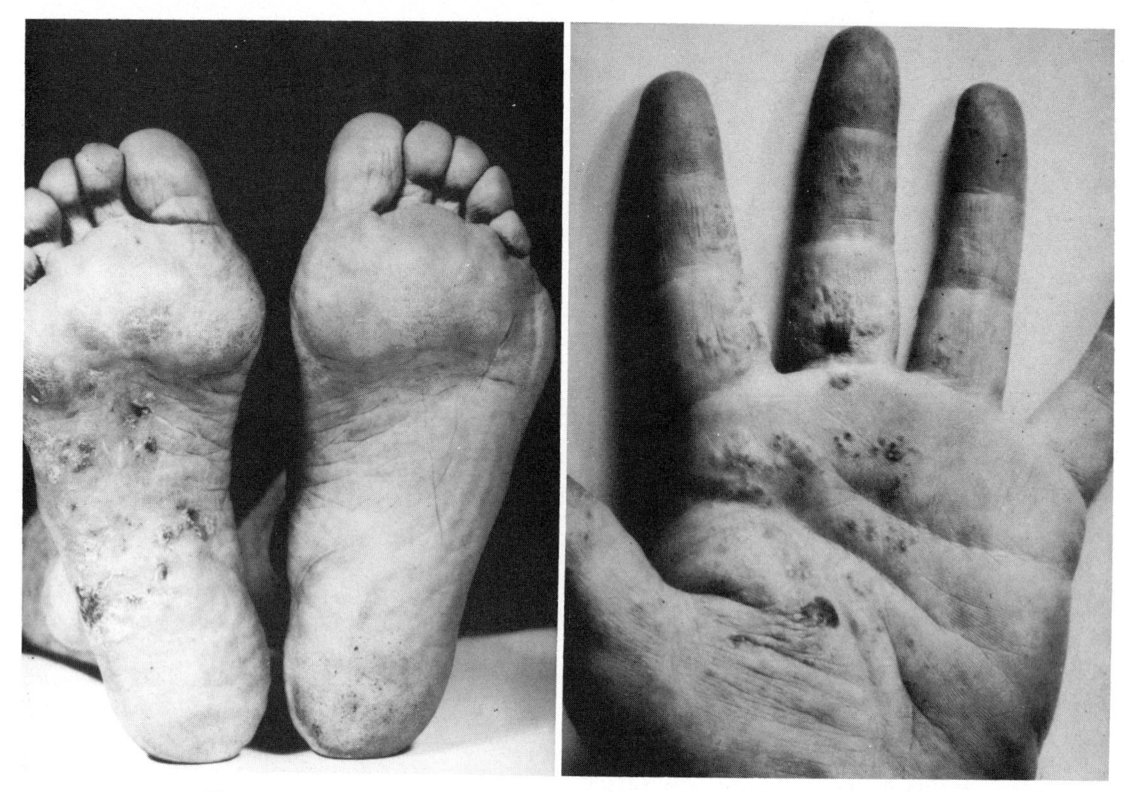

Figure 7–34. Tinea pedis. Vesicular type and associated "id" reaction on hand.

highly suggestive of tinea pedis. Direct microscopy, with a potassium hydroxide mount and culture on selective media, confirms up to 95 per cent of cases. In some instances, several cultures are necessary because of the heavy overgrowth of bacteria and the presence of saprophytic fungi. Multiple or mixed infection with one or more species of dermatophyte and concomitant infection with *Candida albicans* is not uncommon. This may complicate therapy, since most topical and systemic drugs for dermatophytes have no effect on yeasts. Erythrasma of the toe cleft is practically impossible to differentiate from intertriginous tinea pedis. Use of the Wood's lamp to show the coral red fluorescence characteristic of erythrasma can be used to exclude this disease. Bacterial infections caused by *Pseudomonas, Micrococcus,* and *Acinetobacter (Mima)* species also mimic tinea pedis. The clinical condition known as athlete's foot may be caused by many organisms of which fungi are but one.

The differential diagnosis of tinea pedis includes contact dermatitis, pustular psoriasis, idiopathic hyperkeratosis, dyshidrosis, acrodermatitis perstans, dermatitis repens, erysipelas, pyodermias, candidosis, dyshidrosis, secondary syphilis, arsenical keratosis, and fixed drug eruptions.

PROGNOSIS AND THERAPY

Uncomplicated tinea pedis is generally amenable to topical or systemic therapy unless the causative agent is *T. rubrum.* Tolnaftate is widely used and seems to effect a clinical cure in most forms of the disease. The importance of the organism of erythrasma and other bacteria in exaggerating the disease should be recognized. Good hygiene and control of secondary bacterial invaders often lead to remission of symptoms. Systemic griseofulvin therapy (500 mg micronized per day) usually requires two to six weeks before symptomatic improvement occurs, and up to six months or more is needed in resistant cases.[119] Even on such a regimen, some treatment failures are noted. Vascular differences and leaching of the drug from toe webs have been suggested as responsible for the lack of effect of griseofulvin in some cases. Bulla and vesicles usually clear after a short time, but there may be an explosive vesicular reaction in some patients. New sterile blisters may occur on the sole and a severe id on the hands. A mechanism similar to the Herxheimer reaction, which occurs in treated syphilis, has been postulated.

In resistant forms of tinea pedis, combined topical therapy with tolnaftate and oral griseofulvin has been used with success. The older topical agents, such as undecylenic acid, salicylic acid, benzoic acid, and sulfur ointment, are not considered to be of great value in treatment. Undecylenic acid continues to be popular, though rigorous clinical studies fail to confirm it has any effect more significant than simple good foot hygiene.[148] Thymol, phenol-camphor, and strong salicylic acid are irritating and sensitizing. Chronic infections with minor flaking and scaling are not uncommon when *T. rubrum* is the etiologic agent, and these persist despite treatment. Occult or subclinical infections caused by the "downy" or var. *interdigitale* form of *T. mentagrophytes* are also quite persistent, and exacerbations frequently occur. Haloprogin is about as efficaceous as tolnaftate and also clears *Candida* infections.

THE DERMATOPHYTOSES

Animal Disease

Dermatophyte infections of wild and domestic animals have been recognized for many years. It has been pointed out repeatedly that animals act as a reservoir for human dermatophytosis.[56,78,81,149] Ringworm disease in domestic animals constitutes a constant source of infection for persons in contact with them. Thus zoophilic dermatophyte infections are particularly common in rural areas. Fungi from domestic animals, such as dogs and cats, may initiate an epidemic among children. In addition, wild animals also harbor ringworm and may be an indirect source of human infections since the infected hairs shed from these animals may contaminate dwelling places and working areas.[101] An example of this was recently uncovered by Taplin.[27] He discovered that the water rat was the carrier of a particularly virulent strain of *T. mentagrophytes* which infected troops in the Mekong Delta. The animals had no clinical symptoms.

Though long recognized, animal ringworm has only recently been studied in detail. The specific pathologic picture of the infected animal, the dermatophytes involved, and the frequency of transmission of such infections to humans have been reviewed by Georg.[56] The organisms involved are either species-specific dermatophytes, such as *T. equinum* in horses and *M. nanum* in pigs, or universal infecting agents of man and animals, such as *T. mentagrophytes* and *M. canis*. The former species are very rarely isolated from human infection, whereas the latter are as commonly involved in disease of men as true anthropophilic species. Animals also may serve as a vector for animal dermatophytoses and cause epidemics among other animal species. Rodents and cats are probably the chief disseminators of general dermatophytosis. Cases are also on record of transmission from man to animals.[82] Some of the well-recognized dermatophytoses of animals are tabulated in Table 7–9. Complete lists of animal species and reports of disease are given in Ainsworth,[1] Georg,[56] Menges,[104,105,106] and Kaplan.[81]

RINGWORM OF CATS AND DOGS

A more apt name for *Microsporum canis* would be *M. felis*, as cats are the major reservoir for this dermatophyte. The disease in cats is so minimal in the majority of cases that it goes unnoticed. Infected animals spread the disease to children in the family and neighborhood, as well as to dogs and other cats, and leave a trail of infected debris in surrounding areas.[149] The resulting public health problems are obvious.

Clinical Symptoms. Ringworm of the cat caused by *M. canis* is most often subclinical or inconspicuous. In many cases, attention is first drawn to the animal after the development of ringworm by human contacts. The head is the most common site of infection, with areas of hair loss around the nose, eyes, and ears. In clinically apparent lesions, a mild, noninflammatory scaling and patches of alopecia and broken hairs are found. Rarely there is an inflammatory reaction and crust formation. In young animals, lesions are more clearly defined. Discrete circinate lesions are found which show hair

Table 7–9. *Domestic and Wild Animals and the Dermatophytes Causing Natural Infection**

Animal	*Species Recovered and Frequency†*	
Cat	*M. canis*	U
	M. distortum	R
	M. gallinae	R
	M. gypseum	F
	T. mentagrophytes	F
	T. schoenleinii	R
	T. verrucosum	R
	T. violaceum	R
Dog	*M. audouinii*	R
	M. canis	U
	M. cookeii	R
	M. distortum	R
	M. gypseum	F
	M. persicolor	R
	M. vanbreuseghemii	R
	T. equinum	R
	T. megninii	R
	T. mentagrophytes	F
	T. rubrum	R
	T. simii	R
	T. verrucosum	R
	T. violaceum	R
	E. floccosum	R
Cattle	*T. mentagrophytes*	F
	T. verrucosum	U
	M. canis	R
	M. gypseum	R
Sheep	*M. canis*	R
	T. mentagrophytes	R
	T. verrucosum	R
Pigs	*M. canis*	R
	M. nanum	U
	T. mentagrophytes	F
	T. verrucosum	R
Horses	*T. equinum*	U
	M. distortum	R
	M. gypseum	F
	T. mentagrophytes	F
	T. verrucosum	F
Rodents (domestic and wild)	*T. erinacei*	U (hedgehogs)
	T. mentagrophytes	U
	M. canis	F
	M. gallinae	R
	M. gypseum	F
	M. persicolor	U (voles)
	M. vanbreuseghemii	R
Monkeys	*M. audouinii*	R
	M. canis	U
	M. cookeii	R
	M. distortum	R
	M. gypseum	F
	T. mentagrophytes	U
	T. rubrum	R
	T. simii	U
Fowl	*M. gallinae*	U
	T. simii	F

*Modified from Georg,[56] Rebell and Taplin,[119] and Ainsworth and Austwick.[1]
†U = usual; F = frequent; R = rare.

loss, scaling, and a vesicular border. In sickly kittens, the infection often extends to the whole body, forming weeping, crusted lesions, and the disease resembles a severe mange infestation. Favic scutula are sometimes found when the disease is caused by *T. mentagrophytes.* The latter organism evokes a more pronounced inflammatory response. Infection between the paws is common, but whiskers and nails may also be involved.

The disease in dogs is more obvious than in cats. Circular lesions up to 2.5 cm in diameter may appear on any part of the body (Fig. 7–35 FS A 47). Transmission from man to animals is recorded. A boxer contracted a tinea corporis infection caused by *T. rubrum* from his owner, who had the habit of rubbing the dog with his bare feet. The owner had hyperkeratotic tinea pedis.[82] Infections caused by *M. canis* will fluoresce with the Wood's lamp. The characteristic greenish color of the spore sheath around infected hairs is easily seen. Nonfluorescent hairs infected by *T. mentagrophytes* will show arthrospores in chains on direct examination.[53] Infections caused by *M. gypseum* are clearly circular, with alopecia, erythema, scaling, and a peripheral yellowish-white crust.[105]

Topical treatments are the same as those listed for human infections. *M. canis* has a tendency to become chronic, though disease caused by other species may resolve spontaneously with good nutrition and hygiene. Rapid resolution of symptoms follows treatment with oral griseofulvin. The dosage recommended is 60 mg for adult cats and 40 mg for month-old kittens. The drug is given daily until gross symptoms subside (usually two weeks). After this it is given two

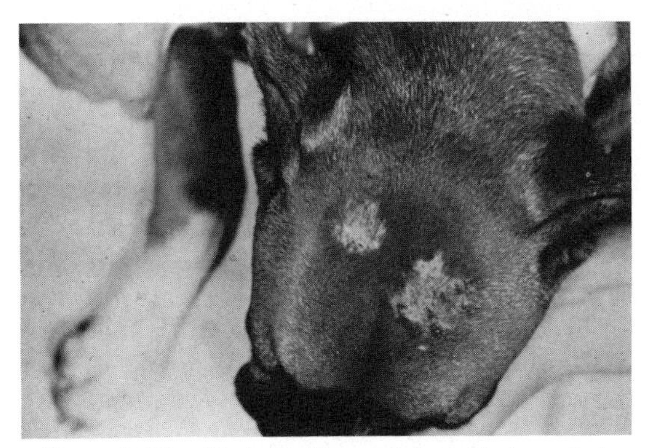

Figure 7–35. FS A 47. Tinea of dog. Circinate lesion on head showing alopecia. *M. canis.* (Courtesy of S. McMillen.)

times a week. In a clinical trial 10 of 18 cats were cured in three weeks; others required up to five weeks.[79] Some seasonal variation in the incidence of ringworm has been found.[84] Individual patterns were found for the several species.

RINGWORM OF HORSES

Clinical Symptoms. The most common clinical picture of dermatophytosis of the horse is that of dry, raised scaling lesions on any part of the animal (Fig. 7–36 FS A 48). Areas that are macerated or rubbed frequently, such as the saddle and girth area, and the hind quarters are the most common sites of infection. Colts and yearlings are most susceptible. The initial lesion is a swelling that can be felt through the hair. The lesions often become small, inflamed ulcers with pussy exudate (girth itch). The hairs appear to be glued together, and the entire mass may be removed as a unit. Lesions enlarge with the loss of peripheral hairs, and a chronic infection is established. As lesions heal, crusts fall off, leaving large bald areas and a "moth-eaten" appearance. Infected debris is found on brushes, saddle gear, and buildings. *T. equinum* is the organism responsible in the great majority of cases, but epizootics of this disease caused by *M. gypseum* are recorded.[83] None of the common horse-infecting dermatophytes are fluorescent.

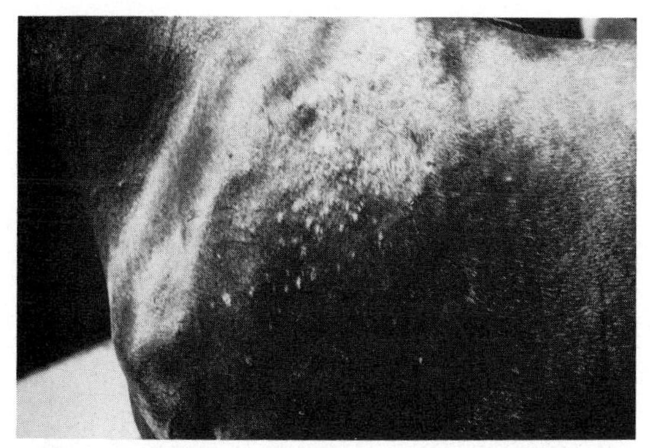

Figure 7–36. FS A 48. Tinea of horse. Multiple, dry, scaly, raised lesion. *T. equinum.*

RINGWORM OF CATTLE

Estimates of the prevalence of ringworm in cattle average 20 per cent, and the vast majority of infections are due to *T. verrucosum.*[104] It is generally considered that calves and yearlings are more susceptible than older animals, for figures of up to 40 per cent of the total number infected have been given for this age group. Although there is a consensus that the infection rate is higher in winter than summer, some authors feel that crowded conditions with increased contact between animals and the presence of infected debris in buildings account for both the higher incidence in calves and the greater infection rate in winter. In support of this, McPherson[103] and Walker[169] found that *T. verrucosum* in infected squames and hair on building woodwork remained viable and infective for from 15 months to 4½ years. Therefore, ample opportunity would be provided for infection of new stock brought into the premises. In addition, this is also a ready source of infection for humans, since children have contracted *T. verrucosum* infection from debris on clothing worn by others in a dairy barn.[169] *T. verrucosum* arthrospores located in a 1.5-mm thickness of skin were able to withstand ultraviolet light equivalent to 437 hours of midday, midsummer sunshine at medial latitudes. Infections may resolve spontaneously, with a subsequent degree of resistance to reinfection evoked. Adult cows with no past history of disease were as susceptible to experimental infection as were calves.[145]

Clinical Symptoms. Ringworm in cattle begins as scattered, discrete, circinate lesions, with slight skin scaling and hair loss. The symptoms may remain stationary and a chronic infection may be established, or, more commonly, the disease develops

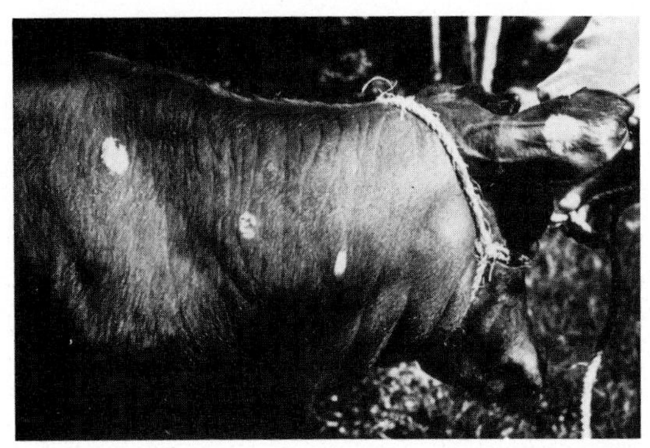

Figure 7–37. FS A 49. Tinea of cattle. Large, circumscribed plaques. Some have an erythematous base, and others are covered with a greyish-white crust. *T. verrucosum.*

acutely into large, circumscribed plaques up to four inches in diameter (Fig. 7–37 FS A 49). These are thickened and covered with a greyish-white crust. Frequently the crusts are large, knoblike bumps that stud large areas of the infected animal. These crusts, which at first are white to grey, become brownish and asbestos-like. They are firmly attached to the animal, and when they are removed, a weeping, bleeding erythematous base is seen. The lesions are often severely inflamed and pruritic. Ulcerations and secondary bacterial infections are frequent. Spontaneous healing follows this stage of the disease, and the lesions become dry, scaly patches, with alopecia and scar formation.

There is at present no satisfactory topical treatment for ringworm of cattle. Systemic griseofulvin has been used successfully,[145] but is prohibitively expensive for treating an entire herd. Good hygiene and sanitation are important for controlling the disease. A fungicidal spray, captan, in a concentration of 0.45 to 0.5 lb per 20 gallons water, used at a rate of 1.0 to 1.5 gallons per animal, has been recommended.[14] This sterilizes infected material and reduces the reservoir of infection. Defungit (bensuldazic acid) used as a wash in a concentration of 0.5 to 1.0 per cent was therapeutically effective under field conditions.[87] As in human dermatophytosis, thiabendazole has been used with some success in cattle ringworm.[111] The mycelium of *Penicillium griseofulvum* has been included in the feed of cattle. This prevented infection and cured those already infected.[39, 69]

RINGWORM OF PIGS

Reports of dermatophyte infection in pigs were rare until 1964. It was believed that these animals were seldom, if ever, subject to dermatophyte disease. A few scattered reports of infection by *T. mentagrophytes* were listed,[58] and Dawson and Gentles mentioned the occurrence of *M. nanum* in pigs in Kenya. Subsequently Ginther et al. in a series of papers described the infection in the United States as being very common in swine, affecting up to 27 per cent of a single herd.[59, 60] Apparently the disease is so benign it has gone unobserved for years, but recently has been shown to have a worldwide distribution. Although disease has been found in all breeds, Yorkshires are the most frequently involved. Infection by *M. nanum* in swine farmers has been recorded.[109]

Clinical Symptoms. The lesions caused by *M. nanum* in pigs are mild and slightly inflamed at first. They then expand centrifugally, and may extend to involve large areas of the body. The initial reaction subsides rapidly, and only inconspicuous scaling and brownish discoloration remain. Lesions behind the ear are most common in chronic infections. No alopecia or systemic disturbances are reported, and the disease becomes chronic and subclinical. Once it is established there is little tendency for spontaneous cure. Infection is extremely rare in piglets or young swine. Occasional transmission to children (as tinea capitis) has been recorded. As infection appears to cause no distress to animals nor is it economically important, treatment of the disease has not been established. In *T. mentagrophytes* infection of swine, a more severe, inflammatory response and pruritus were reported.[58]

FOWL FAVUS

Ringworm (favus) of chickens was first described in 1881 by Megnin and was extensively studied by Sabouraud. In the United States, the first report was by Beach and Halpin in 1918. They called the disease "white comb," and it has appeared sporadically since. Its reported occurrence is worldwide, but it is considered to be a rare infection except in Brazil where it appears to be frequently encountered. The etiologic agent, *M. gallinae,* is one of the few dermatophytes that infects birds. Disease is most common in gallinaceous birds, such as chickens, turkeys, and grouse. Infection of man, dogs and children associated with transmission from fowl has been recorded.[60a,162a] *T. simii* also infects chickens.[161]

Clinical Disease. The disease is characterized by a white, moldy, patchy overgrowth on the comb and wattle. Thick white crusts often develop, and in severe cases the infection becomes generalized to involve the base of feathers. The classic favic scutula commonly described in earlier reports is rarely seen today. Infection of chickens with *T. simii* evokes a more inflammatory reaction, with focal necrosis and crusting.

RINGWORM OF MONKEYS

Dermatophyte disease of monkeys in their native habitat appears to be uncommon. Except for *T. simii* and *M. canis*, few fungal species have been isolated from natural infections. In captivity, however, monkeys acquire infection from a number of sources, such as handlers, rodents, domestic animals, and soil. The list of frequently involved species includes *M. gypseum*, *M. canis*, and *T. mentagrophytes*. All of these evoke a rather severe inflammatory disease, with crusting and hair loss. In contrast, *T. simii* produces silvery scaling lesions with little or no erythema and minimal hair loss. The face is the favored location for infection. *M. gallinae* has also been recovered from infection in a monkey.[63]

OTHER ANIMALS

Many other animal species have been reported to have ringworm. Natural infection in sheep, goats, rabbits,[65] chinchilla (fur slipping), rats, mice, muskrats, foxes, lions, tigers, guinea pigs, and so forth is recorded. The majority of these infections are caused by *T. mentagrophytes* and some by *M. gypseum*. Mouse favus, which is characterized by the presence of numerous white crusted lesions and scutula on the head and body, was ascribed to *T. quinckeanum*. Mating studies, however, have shown that this is not a separate species but a variant of *T. mentagrophytes*. Essentially all rodents carry *T. mentagrophytes* as normal flora, with little or no symptoms of disease. A number of other dermatophytes and keratinophilic fungi are associated with rodent fur – *T. erinacei* in hedgehogs, *M. persicolor* in voles, *Arthroderma curreyi* in wild rabbits, *T. phaseoliforme*, *M. amazonicum*, and *T. terrestre* in mice, and *Chrysosporium* species in many rodents. Griseofulvin has been used in the treatment of natural infections in rabbits and other rodents.[65]

Experimental infection of animals, especially guinea pigs and rabbits, is commonly used to study the pathogenicity, immunity, treatment, and prophylaxis of dermatophyte infection (Fig. 7–38). Disease is most easily established with zoophilic[33] or geophilic species (Fig. 7–39). The area to be infected is plucked, and the prepared site is scarified and inoculated with a suspension of micro-

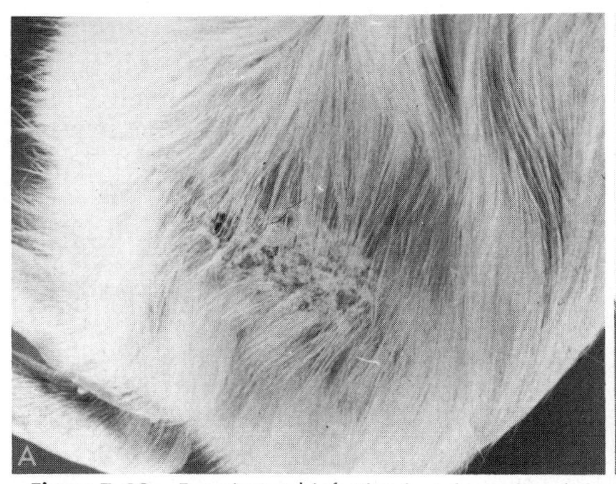

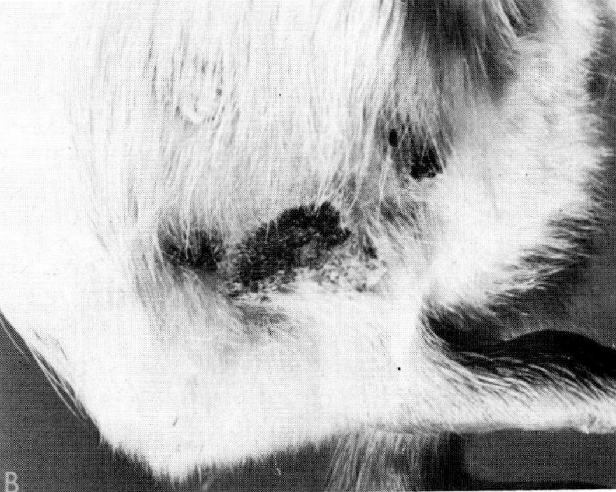

Figure 7–38. Experimental infection in guinea pigs. *A*, Lesion produced by *Arthroderma benhamiae* mating type A. The infection was a chronic, scaling lesion. *B*, Crusted lesion produced by *A. benhamiae* mating type a.

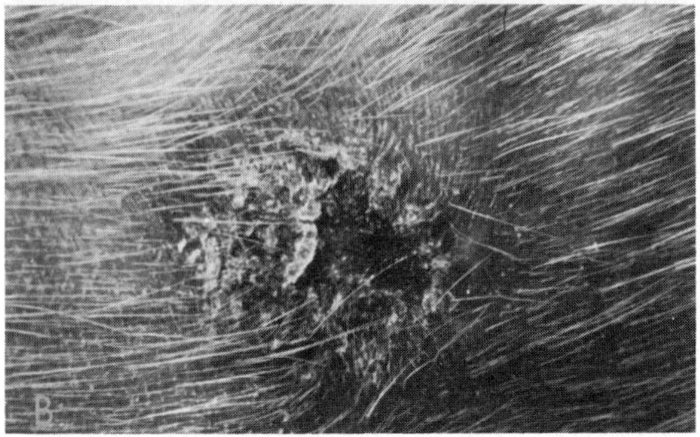

Figure 7–39. *T. simii* infection. *A*, Lesion on arm of human subject who caught it from monkey. *B*, Experimental lesion in guinea pig.

aleuriospores. A most useful device is the Sterneedle multiple puncture gun, which is dipped into the spores and applied to the plucked area. Occlusive dressing is sometimes used. Within one week lesions develop. They are usually inflammatory, and crusting is frequently seen. Resolution of the lesions occurs within three to four weeks, although rarely a chronic, sparsely scaly infection is established which lasts for months or years.[129] Infection involves the stratum corneum of the skin, and hair is frequently invaded with the production of arthrospores. Application of a neck collar prevents licking and chewing of lesions.[80] Experimental infection using anthropophilic species is more difficult to produce. Successful inoculations with *T. rubrum* in rabbits was recorded by Reiss.[124] Castration of the experimental animals augmented the disease, and infections lasting up to nine months were noted. Variations in susceptibility to dermatophyte infection of different laboratory strains of mice and gnotobiotic mice have been examined but no consistent pattern found.[144]

Literature concerning the treatment of experimental infections in animals is sparse. *T. mentagrophytes* of rabbits has been successfully treated with oral griseofulvin. The drug was included in their food at a rate of 0.375 g griseofulvin per pound of feed.

Immunology of Dermatophytosis

No subject in the field of medical mycology has evoked more controversy than "immunity and resistance" in dermatophyte infections. The voluminous literature begins in the early nineteen hundreds and continues to accumulate unabated to the present day.[95] A review of the work on experimental, acquired resistance, natural resistance, hypersensitivity, immunization, and antibody response leads to the following summation. There is no single, clear-cut mechanism which will explain all aspects of susceptibility and immunity to dermatophyte infection. The differences in response to infection, susceptibility to reinfection, and even acquisition of the infection varies considerably with each species of fungus and to a great degree with the individual involved. Very few dermatophyte species evoke the same response in all patients or in experimental animal and human infections.[150] In a given population with equal exposure, only a certain percentage will manifest clinical symptoms of disease. A given individual may have a short, protracted course of infection or no infection with one dermatophyte species, but will develop a mild chronic infection of lifelong duration with another species. The relevant aspects of natural resistance[6a] to dermatophytosis are discussed in the sections on clinical disease. They have been extensively reviewed by Hildick-Smith et al.[70]

The skin is the primary barrier or defense of the body against invasion by microorganisms from the external environment. In dermatophyte infection, this defense is not abrogated. A peculiar relationship exists between the dermatophyte and its host that is unparalleled by other microbial agents of disease. The organism is truly a dermato- or ectophyte, as it does not invade living cells, and its nutritional demands involve no depletion of metabolizable substances from the host. In this sense it is not a parasite. The immune and inflammatory response evoked in the living tissue beneath the site of infection or distal to it are incidental to the disease. They reflect the lack of or low degree of adaptation between the fungus and the host. Attempts to correlate circulating antibody response or specific immune mechanisms to resolution of infection and resistance to reinfection have been unrewarding.

A degree of acquired resistance to the disease has been observed in patients and in experimental infection in animals. Clinical records indicate that, in a large series of children treated for tinea capitis, none returned with a second infection.[47] Similar findings were seen in agricultural workers infected with *T. verrucosum*.[95] These observations are interpreted to mean that there was increased resistance to reinfection as a result of the initial infection. On the other hand, multiple episodes of tinea pedis occur, and reinfection (exacerbation?) is common in patients with this disease. Many of these patients have had a previous history of tinea capitis or tinea corporis. Repeated tinea corporis infection is also common, and each episode differs little from the initial infection.[37] Laboratory animals generally gradually acquire increasing resistance to reinfection with either the same or different species of dermatophytes. In guinea pigs, this resistance is transitory and has been correlated with cutaneous hypersensitivity.[36,170] Both responses are maximal two to three weeks following resolution of primary lesions. Mice and rats appear to develop neither resistance to reinfection nor hypersensitivity.[90]

Hypersensitivity can be demonstrated by skin tests using trichophytin.[67] Intradermal injections of this substance elicit either a delayed (tuberculin) or immediate (urticarial) response. The latter may be passively transferred and is associated with a reaginic antibody tentatively identified as an IgE. Purification and chemical analysis of trichophytin shows it to be a galactomannan peptide.[15] Degradation studies indicate that the immediate reaction is associated with the carbohydrate fraction and delayed reactivity with the peptide moiety. Patients with chronic, widespread infections caused by *T. rubrum* often show only the urticarial reaction when challenged with trichophytin. The same reaction can be elicited in atopic patients without fungous disease. Preliminary impressions from the studies of Lobitz indicate that such patients and those with chronic mucocutaneous candidosis are hyporeactive in their cell-mediated immune systems. This depression is antigen-specific in localized infections but nonspecific in generalized disease. The delayed hypersensitivity reaction is found in either experimental or natural infection by *T. mentagrophytes* and several other dermatophytes. It is associated with the degree of inflammatory response evoked by the primary infection. In guinea pig infection, delayed hypersensitivity correlates with resistance. Attempts to conclusively demonstrate this in human disease have as yet been unsuccessful, although the work of Jones et al.[77a] is suggestive. They concluded that delayed sensitivity is a correlate of immunity, whereas its absence or its coexistence with immediate reactivity is associated with susceptibility to chronic tinea infections. Patients with atopy are particularly prone to chronic infections. Trichophytin, as presently prepared, is not species-specific and is common to almost all dermatophytes from which it is derived. Its relative antigenicity is influenced by the media in which the fungus is grown.[156] Trichophytin-like substances have been isolated from species of *Aspergillus* and *Penicillium*, as well as dermatophytes. Positive skin tests to trichophytin have been elicited in patients with penicillin hypersensitivity and in those with cutaneous tuberculosis. Other studies have detected a high percentage of positive reactors among the normal populace which was free of fungous infection. The demonstration of delayed or immediate-type hypersensitivity to intradermal injection of trichophytin appears to be of limited diagnostic and prognostic value. The development of species-specific antigens would greatly aid in the understanding of dermatophyte disease.

Dermatophytid or "id" reactions are secondary eruptions occurring in sensitized patients as a result of circulation of allergenic products from a primary site of infection. The morphology and site of the "id" lesions vary. The condition, as first described by Jadassohn, resembled lichen scrofulosorum. It is commonly associated with tinea capitis in children. Small, grouped, or diffusely scattered follicular lesions are found on the body. They are symmetrical and central in distribution, but may extend to involve the limbs and face. Horny spines are sometimes observed on top of the involved follicles.

Lesions on the fingers are frequently found in patients with tinea pedis. Generally they are papular, but they may be vesicular (pompholyx type), bullous, or rarely pustular. They appear on the sides of the fingers and wrist, or are grouped on all parts of the body. The dermatophytid reaction responds to desensitization and resolves spontaneously with the elimination of the primary disease. The condition is sometimes exacerbated or exaggerated during treatment with systemic griseofulvin or injection of trichophytin.

The immunologic aspects of dermatophytoses in man and animals have been reviewed by Lepper.[95] Circulating antibodies to natural or experimental dermatophytosis have been demonstrated by a number of techniques.[125] Their significance remains to be established. Precipitating or complement-fixing antibodies and indirect hemagglutination titers were found in rabbits and cattle which were naturally or experimentally infected by several dermatophytes, but no direct relationship to resistance has been found.[1] In man, circulating antibodies that react to antigens from dermatophytes have been detected in patients with and without disease. The C-reactive protein in the β-globulin fraction of human serum reacts with C-reactive substances found in glycopeptide fractions of *E. floccosum*, *Aspergillus fumigatus*, or the galactomannan peptide from *T. mentagrophytes*.[96] Inoculation of living or killed dermatophytes or their extracts results in the formation of circulating antibodies. Just as in other serologic tests, the antisera produced reacts with many species of dermatophytes and other fungi.[86, 125]

Immunization by injection of live or killed fungi, their extracts, or their metabolic products has been attempted many times. In guinea pigs, the injection of acetone-dried powdered mycelium of *T. mentagrophytes* imparted both an immediate and a delayed-type hypersensitivity.[7] Upon challenge with a homologous organism, an increased inflammatory response was noted, and the disease was attenuated. However, resistance was transitory, and complete susceptibility returned after a few months. In the same animals, an immunologically active ribonucleic acid or oligoribonucleotide derived from *T. mentagrophytes* had little skin reactivity, but it did evoke antibodies.[73] Ribonucleic acid fractions have been used successfully by Youmans and others for immunization against tuberculosis or other diseases that evoke a strong hypersensitivity reaction. Topical application of *T. mentagrophytes* extracts to the feet of a group of infection-free persons conferred some resistance when they were subsequently challenged by infection with this organism.[72]

At present, there is no evident correlation between resistance to reinfection and antibody levels as measured by standard *in vitro* techniques. Although a transient resistance to reinfection has been demonstrated, there appears to be little direct relationship between the degree and type of cutaneous hypersensitivity and lasting immunity to dermatophyte disease. The available data necessitate acceptance of the premise that immunity can exist in degrees.[73]

Laboratory Identification

Direct Examination. The diagnosis of dermatophyte infection is most easily confirmed by direct microscopic examination of skin scrapings in a potassium hydroxide slide mount. This is prepared by placing the material to be examined (skin scales, nail scrapings, hair stubs, and so forth) on a microscope slide and adding a drop of 10 per cent KOH, which is mixed well with the specimen and the pieces separated in as thin a layer as possible. A cover slip is placed over the KOH-specimen mixture, and the slide is gently heated. Boiling is avoided, as this precipitates KOH crystals. The slide is allowed to cool and "ripen" a few minutes before examination. The KOH "clears" the specimen by digesting proteinaceous debris, bleaching pigments, and loosening the sclerotic material without damaging the fungus. The hyphae of the fungus are unaffected by this treatment and stand out as highly refractile, long, undulating, branched, septate threads (Fig. 7–40). These threads are seen to course in, around, and through epidermal scales. In a well-cleared mount, one is able to discern nuclei, organelles, and fat droplets within the mycelium. Mature hyphae will show numerous septations. In time, these fragment into rounded or barrel-shaped arthrospores. Examination of hair stubbles often requires more time for clearing, especially if it is dark pigmented hair. The base of the hair shaft and follicular debris are the areas where fungi are most

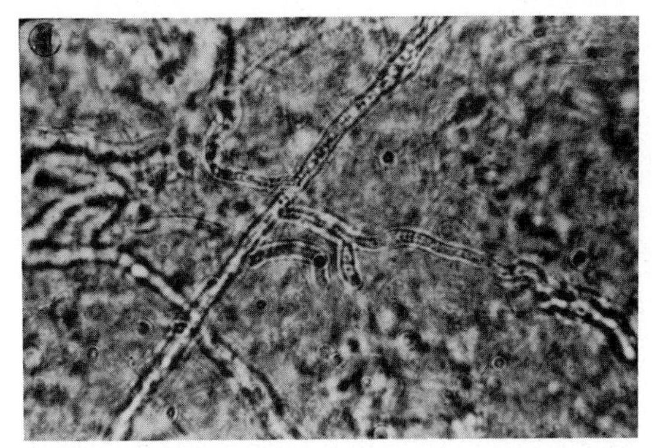

Figure 7–40. Potassium hydroxide mount of tinea corporis. Note refractile, branching, septate hyphae. ×440.

likely to be seen. The arthrospores are outside the hair shaft in chains in a mosaic pattern, or intrapilar, depending on the species involved and whether it is endothrix or ectothrix. Nail scrapings are the most difficult to examine. Fungi are seldom in abundance, and a day or two may be required for good KOH digestion. For all types of specimens, hyphae must be differentiated from other artifacts. These include fibers of cotton, wool, and synthetic materials, starch grains, fat droplets, vegetable detritus, and "mosaic fungus." This last artifact causes the most difficulty for beginners. "Mosaic" is a network of material, including cholesterol crystals, which is deposited around the periphery of keratinized epidermal cells. It can be seen to follow the outline of the cell but not to go through it. This observation, together with the abrupt changes in width, the lack of internal organelles, and the flat crystalline structures with reentrant angles, differentiates it from true hyphae. A host of other debris is seen in nail scrapings. Mycelial elements of *Candida* species in skin and nails are less refractive when seen in KOH mounts and show pseudomycelium, hyphae, and budding yeast cells. *Pityrosporum orbiculare,* the etiologic agent of pityriasis versicolor, has a distinct morphology. Incidental saprophytic fungi and pollen grains may also be found. In experienced hands, the KOH mount is one of the most useful procedures in medical mycology. Besides its use in diagnosis of dermatophyte infections, it can be applied to a number of other specimens, such as sputum, pus, exudates, biopsy material, urine, and fecal material. In such cases, and for the beginner examining skin scrapings, it is useful to add a stain to the KOH. The most convenient of those recommended is the addition of Parker Superchrome Blueblack ink to the KOH solution. The proportions depend on the preference and aesthetic taste of the user. The ink selectively colors the hyphae, making them more pronounced.

The selection of the specimen to be examined determines the success of the procedure to a large extent. In tinea pedis, the outer debris and macerated tissue should be removed from the sole and interdigital spaces. Cleaning the site with an alcohol sponge is useful. As in all fungal lesions, the organisms are most prevalent in the edge or active site; therefore the epidermal scales should be removed from the periphery of the lesion. They may be placed directly on a slide for KOH mounting or into a sterile petri dish or sterile envelope for transport to the examining laboratory. In such containers, the organisms in skin scales remain viable for some time. An air-tight, damp container would promote the growth of saprophytic fungi. Scalpel, scissors, or preferably a double-edged Foman knife are used to obtain scrapings. If bullae or vesicles are present, these are clipped, examined, and cultured. Any antifungal or other topical medicaments must be removed before culturing material from lesions. This often involves vigorous washing. Even if the KOH preparation is negative, culture is warranted.

The procedure used for tinea corporis lesions is basically the same. Scrape the lesion from the center to the outside edge in order to loosen scales. In some infections it is often necessary to scrape until a serous exudate is formed, particularly when the etiologic agent is *T. rubrum.* In tinea capitis and tinea barbae, infected hairs are selected from lesions. In microsporosis this is aided by plucking hairs which fluoresce when a Wood's lamp is held above the head. In trichophytosis the abnormal or distorted hairs near the border of the lesions are picked. In "black-dot" ringworm it is often necessary to use a scalpel to excise the twisted deformed root that may be subsurface. Scutula and crusts from tinea favosa, animal ringworm, and inflammatory ringworm are usually highly contaminated. They should be macerated, washed in alcohol or antibiotic solution, and small, separated sections examined and cultured. Hair and crusts from inflammatory lesions, particularly from the central areas, are often devoid of fungi. Removing the crust and examining the outer edge of the moist red base is usually successful.

Success in finding fungi in tinea unguium and onychomycosis is by far the most difficult to obtain. In white-spot infection the superficial layers are scraped off, cultured, and examined. In tinea unguium most of the upper nail is scraped off and discarded. Only the deepest layers harbor fungi. Sometimes one can remove the debris from under the distal end of the toenail and then vigorously scrape the undersurface of the juncture of the nail with the nail bed. If oozing and slight bleeding occur, the chances of success are greater.

Culture Methods. When the index of suspicion is high, all specimens should be cultured, even when the KOH preparation is negative. Because ringworm can appear quite variable and culture is such an easy procedure, it is advised as a routine part of dermatologic examination. For mycologic examination, cultures are planted by furrowing the specimen into the media with the scraping knife. As much material as the surface of the agar will resonably accommodate should be included. A wide agar slant or specimen jar is recommended. Attention to these details increases the chance of success.

The standard media for primary isolation of dermatophytes is Sabouraud's agar, containing cycloheximide and an antibacterial antibiotic. The cycloheximide (Actidione) in a concentration of 0.1 to 0.4 mg per ml suppresses the growth of most saprophytic fungi without deterring the growth of dermatophytes. The various antibacterial antibiotics used include chloramphenicol (0.05 mg per ml) or Aureomycin (0.1 mg per ml); both are satisfactory. Growth is relatively slow; usually ten days to three weeks are required at the optimum temperatures of 25°C. *T. verrucosum* and rare strains of *T. tonsurans* grow better at 37°C. When growth becomes evident on the primary isolation media, mycelial strands are transferred to a slide culture preparation (see Appendix). On slide culture, two media are advised: cornmeal with 1 per cent glucose to stimulate pigment production of *T. rubrum*, and Sabouraud's agar without antibiotics for visualization of the normal morphology of spores, spore arrangement, and mycelial appendages (Fig. 7–41 FS A 50).

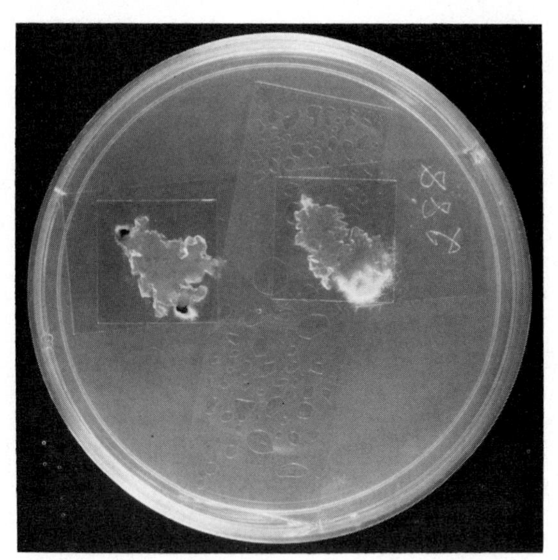

Figure 7–41. FS A 50. Slide culture. The right-hand side is Sabouraud's glucose agar and the left is cornmeal with 1 per cent dextrose. The latter stimulates pigment production in *T. rubrum*.

The slide culture allows observation of the culture while it is growing. When spores are evident, a lacto-phenol cotton blue mounting can be made for accurate observation. Sometimes vitamin-enriched, casein digest media enhance the growth and production of macro- and microaleuriospores. This is particularly true of the faviform *Trichophyton*. L. K. Georg in a series of papers delineated nutritional requirements for some dermatophytes.[54,57] These are very useful in identifying isolates of species closely resembling one another (Table 7–10). The requirement for thiamine can be fulfilled by the pyrimidine moiety in most species.

Table 7–10. *Nutritional Requirements or Growth Enhancement Factors of Dermatophytes**

Species	*Requirement*
T. equinum	Nicotinic acid†
T. megninii	L-histidine
T. tonsurans	Thiamine
T. verrucosum	Inositol and thiamine, 80% of isolants; thiamine only, 20%
T. violaceum	Thiamine

*The tubes inoculated are ammonium nitrate or casein, enriched with the supplement. Most all dermatophytes grow better in vitamin-enriched medias.

†Some nicotinic acid–independent strains have been isolated in New Zealand and Australia.

Some other techniques are useful in separating closely related species. The presence of perforating organs in *in vitro* hair cultures (Fig. 7–48) separates *T. mentagrophytes* from *T. rubrum.*[5] *T. rubrum* produces a red pigment on potato-dextrose slants or cornmeal agar with 1 per cent glucose. *T. mentagrophytes* and *T. tonsurans* do not. Some physiologic characteristics can be used to separate the sexes of the ascigerous stage of dermatophytes as the elaboration of elastase by the plus (+) strain of *Nannizzia fulva* but not the minus (−) strain.[127,129]

For the small office without access to mycologic consultation or in large public health field studies, the routine isolation and identification of dermatophytes is difficult. Recently two new primary medias have been developed which are very helpful in primary isolation of dermatophytes. They are both based on pH change caused by the proteolytic activity of dermatophytes which is lacking in most saprophytic fungi.[61] The first of these to gain popularity was the ink-blue agar of Baxter.[17] The agar is blue, but a colorless area is seen around a growing dermatophyte colony. Some of the drawbacks of this medium are that the mycelium is stained by the dye, thereby masking its natural color, and a number of contaminating bacteria cause color change. A far more satisfactory formulation was devised by Taplin et al.[158,159] The indicator is phenol red (pH 6.8 yellow, pH 8.4 red), and gentamycin, chlortetracycline, and cycloheximide are included to inhibit growth of bacteria and saprophytic fungi. The medium called DTM is commercially available. It has proved very useful in large field surveys because the color change is readily evident and fairly specific for dermatophyte fungi. Some workers feel that the recovery rate is lower on DTM than on Sabouraud's agar. The red color does mask somewhat the colonial morphology and pigmentation. In all cases, the identification of organisms should be confirmed by the use of the slide culture technique in the hands of an experienced mycologist. However, the presence or absence of a dermatophyte is sufficient knowledge for most purposes of therapy.

Mycology

As noted earlier, the number of species of dermatophytes is large; when they are added to the numerous keratinophilic fungi, they amount to a formidable group to describe and identify. Repeated exposure is required in order to become familiar and competent in identifying species and knowing their variability. In the present section some of the clinically important species of dermatophytes will be described, along with their ascigerous state when present. Because so many of these species have had numerous citations in the literature under other names, the synonymy is given for each valid species. The major differential characteristics are given in outline form as obverse and reverse colony morphology, microscopic morphology, and special characteristics if present. The most useful and complete manual covering the identification of dermatophytes and related soil forms is that published by Rebell and Taplin.[123]

Microsporum Gruby 1843

This genus is characterized by the presence of fusiform, obovate to spindle-shaped macroaleuriospores. The walls are thick (to 4 μ in *M. canis*) and pitted, asperulate, echinulate, or spiny. This latter architectural feature may be restricted to the distal end and appears only on a few mature spores. The macroaleuriospores range in size from 7 to 20 × 30 to 160 μ, with from 1 to 15 septations. The form and size of these spores are important in identification. The microaleuriospores, on the other hand, are sessile or stalked, clavate, and 2.5 to 3.5 × 4 to 7μ in size. They are usually not helpful in identification, as they are similar to those found in several other genera. When found, the perfect stage of the members of the genus *Microsporum* is in the genus *Nannizzia* of the family Gymnoascaceae in the class Ascomycetes.

Nannizzia Stockdale 1961

The chief distinguishing features of this genus are the peridial hyphae found around the sexual fruiting body, the gymnothecia. These hyphae are generally verticillately branched, have asperulate to echinulate hyphal cells with one or more constrictions, and terminate in smooth-walled, blunt-ended cells, spikes, or spirals. The individual cells of the peridial hyphae lack the knobby protuberances found in the genus *Arthroderma,* the perfect stage of the genus *Trichophyton.*

Microsporum audouinii Gruby 1843

Synonymy. *Trichophyton decalvans* Malmsten 1848, *Microsporum villosum* Minne 1907, *Microsporum umbonatum* Sabouraud 1907, *Microsporum velvetieum* Sabouraud 1907, *Microsporum tardum* Sabouraud 1910, *Microsporum tomentosum* Sabouraud 1910, *Microsporum depauperatum* Guéguen 1911, *Microsporum rivalieri* Vanbreuseghem 1963, *Closteroaluriosporia audouinii* (Gruby) Grigorakis 1925, *Martensella microspora* Vuillemin 1895, *Sabouraudites audouinii* (Gruby) Ota and Langeron 1923, *Sabouraudites langeronii* Vanbreuseghem 1950.

The correct ending of this genus name should conform to the Greek "-on," as the root word is Greek in origin and would agree with *Trichophyton* and *Epidermophyton*. David Gruby, when naming this genus, made an error in syntax which unfortunately has continued.

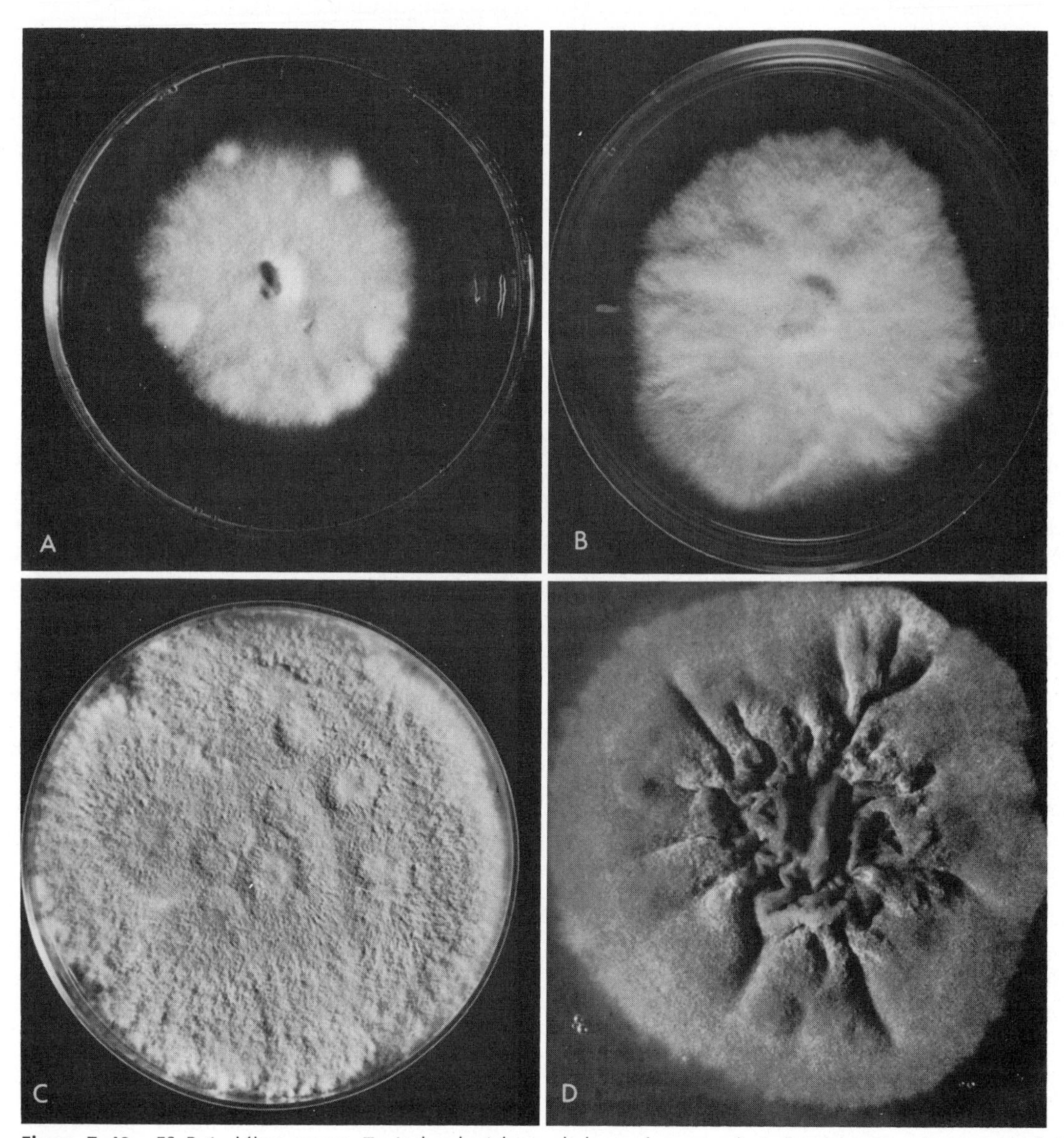

Figure 7–42. FS B 1. *Microsporum.* Typical colonial morphology of commonly isolated species. *A, M. audouinii.* Silky, flat growth with pleomorphic tufts. The reverse is a salmon color. *B,* (FS B 2,*A*), *M. canis.* Floccose growth. The reverse is chrome yellow (FS B 2,*B*). *C* (FS B 3), *M. gypseum (Nannizzia incurvata).* Powdery beige-colored surface. The reverse is variable in color. *D* (FS B 4), *M. ferrugineum* showing the folded, wrinkled, rust-colored colony. The reverse is variable in color.

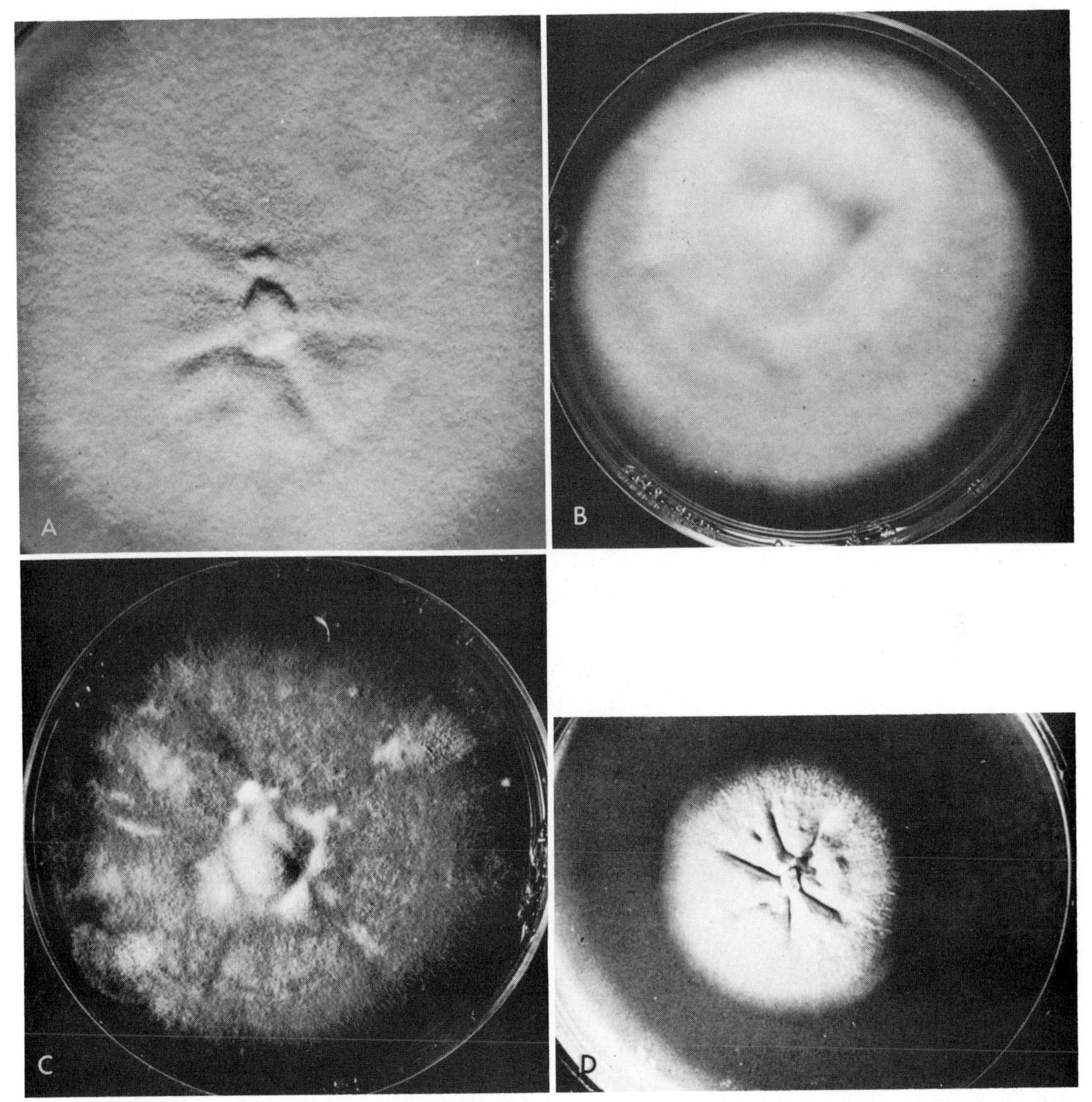

Figure 7–43. FS B 14. *A, M. audouinii* var. *rivalieri. B, M. nanum. C, M. cookei. D, T. verrucosum.*

Colony Obverse (Fig. 7–42,*A* FS B 1). On SDA* the growth is slow, forming a flat, spreading, dense colony with a silky, furry, matted consistency and radiating edges. The color is white or grey to tan or rust-buff. Rarely seen colonial forms include a rust-brown or rose-tan color with radial grooves, as in var. *langeronii.* The var. *rivalieri* has a greyish-white folded thallus with a satiny sheen similar to *T. schoenleinii* (Fig. 7–43,*A* FS B 14,*A*). This latter variety is common in Africa and found in Florida.

Reverse. Salmon, rust, or peach color pigment is characteristically produced by *M. audouinii.* If produced in quantity, it stains the mycelium. Many isolants, however, have little or no pigment.

Microscopic Morphology. Generally few spores are seen (Fig. 7–44,*A*). The usual microscopic examination shows thick-walled terminal or intercalary chlamydospores which are fairly characteristic and permit identification of the culture. Racquet hyphae,

*Sabouraud's dextrose agar is the standard medium on which are based the descriptions of growth characteristics and colony morphology of medically important fungi. Unless otherwise noted, the summary of characteristics is growth on SDA at 25°C for two weeks.

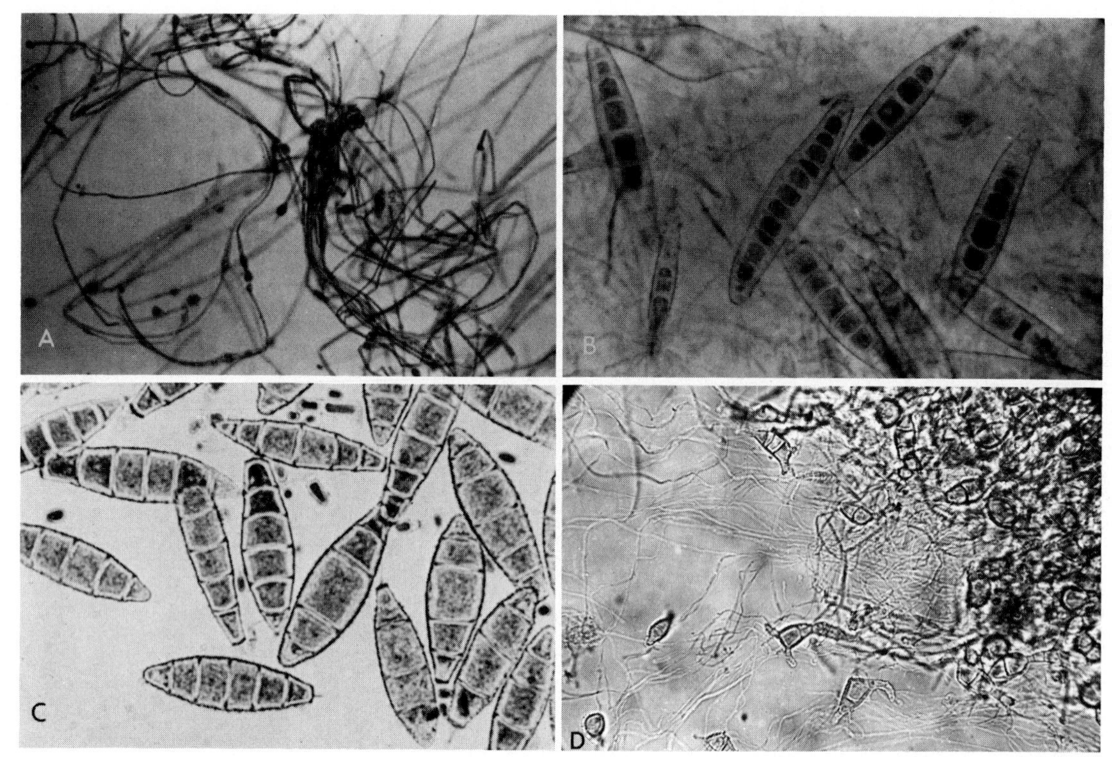

Figure 7–44. *A, M. audouinii.* Microscopic view showing hyphae and a few chlamydospores. *B, M. canis.* Macroaleuriospores. *C, M. gypseum (N. incurvata).* Macroaleuriospores. *D, M. distortum.* Macroaleuriospores.

pectinate bodies, and, rarely, some irregular microaleuriospores may be present. The macroaleuriospores if seen are irregular in shape, elongate, thick-walled, and echinulate (3 × 4 μ).

Differential Characteristics. On polished rice, growth of *M. audouinii* is almost imperceptible in contrast to the luxuriant, pigmented growth of *M. canis.* Growth on yeast extract media or soil-hair agar stimulates the production of macroaleuriospores. Intermediate forms between *M. canis* and *M. audouinii* are not uncommon, and their occurrence suggests a close relationship, if not recent evolutionary derivation, between the two. No perfect stage for these species has yet been found. All strains of *M. audouinii* tested by the *T. simii* sex stimulation procedure show a (+) mating type. *M. audouinii* is anthropophilic, small-spored, ectothrix, and fluorescent in infected hairs. This species probably evolved in Africa, and the var. *rivalieri* may represent the ancestral form. Since only one mating type appears to exist now, it might have had a selective advantage over the other during its adaptation to an anthropophilic organism.

Microsporum canis Bodin 1902

Synonymy. *Microsporum felineum* (Bodin) Mewborn 1902, *Microsporum equinum* Guéguen 1904, *Microsporum lanosum* (Bodin) Sabouraud 1907, *Microsporum caninum* Sabouraud 1908, *Microsporum stillianus* Benedek 1937, *Microsporum peudolanosum* Conant 1937, *Microsporum simiae* Conant 1937, *Microsporum obesum* Conant 1937, *Closterosporia lanosa* Grigorakis 1925. *Closterosporia felinea* Grigorakis 1925.

Colony Obverse. (Fig. 7–42,*B* FS B 2,*A*). The growth is rapid, producing a woolly or cottony, white to yellowish, flat to sparsely grooved colony with radiating edges. It rapidly becomes pleomorphic. Dysgonic isolants are glabrous, heaped, and deeply pigmented.

Reverse. The underside of the colony is characteristically a deep chrome yellow (FS B 2,*B*). This is best viewed in young growth. On potato-dextrose medium, a sparse mycelium and an abundant lemon pigment are seen.

Microscopic Morphology. The large macroaleuriospores (8 to 20 × 40 to 150 μ) are produced in abundance (Fig. 7–44,*B*). They have thick walls (2 μ), up to 15 septa, and are spindle-shaped and echinulate, or pitted. Curved or hooked ends are seen. The dysgonic variant produces elongated spores with few septa, and var. *obesum* produces fat, three-celled spores. The microaleuriospores are slender and clavate, similar to many other species. Racquet hyphae, pectinate bodies, nodular bodies, and chlamydospores are seen.

Differential Characteristics. The growth is abundant on polished rice grains. *M. canis* is zoophilic and native to cats, dogs, and probably horses, apes, and monkeys. Though not considered geophilic, it has been isolated from soil in Hawaii and Rumania. *M. canis* is small-spored, ectothrix, and fluoresces in infected hairs. Colony variants of this organism are common.

Microsporum ferrugineum Ota 1921

Synonymy. *Microsporum japonicum* Dohi and Kambayashi 1921, *Microsporum aureum* Takeya 1925, *Microsporum orientale* Carol 1928, *Trichophyton ferrugineum* (Ota) Langeron and Milochevitch 1930.

Colony Obverse (Fig. 7–42,*D* FS B 4). The growth is slow, forming a heaped, folded, glabrous, reddish-yellow to orange-yellow thallus with a waxy surface. This is the usual strain found in the Far East. A fine, velvety, white overgrowth may occur. Variants without pigment resemble *T. verrucosum* and are common in the Balkans.

Reverse. No characteristic pigment is seen.

Microscopic Morphology. Distorted mycelium without spores is the usual microscopic picture. Faviform, abnormal, mycelial elements and coarse hyphae with prominent crosswalls (bamboo hyphae) are seen.

Differential Characteristics. On deficient media (dilute SDA) and potato-dextrose with charcoal, macroaleuriospores of the *Microsporum* type are occasionally seen. On Löwenstein-Jensen medium, the color is light yellow compared to the dark red-brown of *T. soudanense*. It is anthropophilic, small-spored, ectothrix, and fluorescent in infected hairs.

Microsporum gypseum (Bodin) Guiart and Grigorakis 1928

Synonymy. *Achorion gypseum* Bodin 1907, *Microsporum flavescens* Horta 1911, *Microsporum scorteum* Priestley 1914, *Microsporum xanthodes* Fischer 1918.

Perfect States. *Nannizzia gypsea* (Nannizzi) Stockdale 1963, *Nannizzia incurvata* Stockdale 1961.

Colony Obverse (Fig. 7–42,*C* FS B 3). Colonies grow rapidly, producing a flat, spreading, powdery surface that is rich cinnamon-buff to brown, occasionally with overtones of violet. The powder consists of masses of macroaleuriospores. The edges of the colony are entire to scalloped or ragged. Diffuse pleomorphism rapidly develops.

Reverse. A variety of pigments or none are produced.

Microscopic Morphology. Macroaleuriospores are produced in great abundance. They are thin-walled, 8 to 16 × 20 to 60 μ, roughened, and have 4 to 6 septa (Fig. 7–44,*C*). The usual variety of other spores, including microaleuriospores, is seen. Hair penetration organs are produced.

M. gypseum is geophilic and abundant in soil throughout the world. It is ectothrix but produces few spores. Fluorescence is absent or dull in infected hairs. The complex is made up of two species whose perfect states are: *Nannizzia incurvata* and *Nannizzia gypsea.* The imperfect stage (*M. gypseum*) of these two species is essentially the same and can only be identified by mating with appropriate tester strains. However, *N. gypsea* generally produces a more spreading and a coarser granular colony; the macroaleuriospores are slightly wider, and the surface color is brighter and sometimes a redder color than *N. incurvata. N. incurvata* is pale buff and finely granular and occasionally has a reddish to yellow reverse. Differences in the ornamentation of the macroaleuriospores have been seen by the scanning electron microscope.[167a, 168]

N. gypsea. The gymnothecia are globose, beige to buff, and 300 to 800 μ in diameter. The peridial hyphae are asperulate, with septa constricted 1 to 3 times. They are verticillately branched, and these branches curve away from the main axis. This is in contrast to those of *N. incurvata,* which curve toward the main axis. The branches terminate in elongate, smooth-walled, tapering spikes up to 250 μ in length or, rarely, end in tightly coiled spiral hyphae. Sometimes ellipsoid macroaleuriospores are found at the ends of peridial hyphae. Mutants with smooth-walled macroconidia have been described. The asci are evanescent, thin-walled. 5 to 7 μ in diameter, and eight-spored. Ascospores are smooth, ovoid, 2 × 4 μ in diameter, and appear yellow in mass.

N. incurvata. The gymnothecia, asci, and ascospores are similar to those of *N. gypsea.* The peridial hyphae again are asperulate, septate, with 1 to 3 constrictions and verticillately branched. The branches curve inward toward the main axis and away from the gymnothecial body. They end as blunt tips, spikes, or spiral hyphae.

Microsporum fulvum Uriburu 1909

Perfect State. *Nannizzia fulva* Stockdale 1963.

Until its delineation by Stockdale in 1963, this fungus was treated as a variety of *Microsporum gypseum.* A sexual stage was discovered, and it is now considered to be a separate species.

Colony Obverse. Although it resembles the *M. gypseum* complex, the colony is more floccose and tawny buff in color. The periphery is often white.

Reverse. A dark red undersurface is occasionally seen; otherwise it is colorless to yellow brown.

Microscopic Morphology. The macroaleuriospores of *M. fulvum* are more clavate, cylindrical, or bullet-shaped and are less often in large clusters than those of *M. gypseum.* Microaleuriospores 2 to 3.5 × 3 to 8 μ are produced, but these are indistinguishable from those of other species. Numerous spiral hyphae, which are often branched, are seen. Hair penetration organs are produced.

This is another geophilic species of worldwide distribution. It is sparsely ectothrix and nonfluorescent in infected hairs. Its ascomycetous stage is *N. fulva.*

N. fulva. The gymnothecia are larger (500 to 1300 μ) than *N. gypsea* but otherwise indistinguishable. The asci, ascospores, and peridial hyphae are also similar. Identification depends on mating with tester strains. The sexes are distinguishable by the production of elastase.[127] This enzyme is associated with the (+) mating type and is absent in the minus type. *M. fulvum* is a less pathogenic species than *M. gypseum;* however, its (+) mating type is somewhat more virulent in experimental infection than the (−) type.[129] Others have had inconsistent results in determining elastase production.[169a]

Microsporum nanum Fuentes 1956

Perfect State. *Nannizzia obtusa* Dawson and Gentles 1961.

Colony Obverse (Fig. 7–43,*B* FS B 14,*B*). A rapidly growing, thin, spreading, powdery colony is produced that resembles a variant of *M. gypseum.* It is white to yellow, changing to pinkish-buff in color and has a fringed edge.

Reverse. A red brown pigment is frequently seen.

Microscopic Morphology. The characteristic ovoid to pear-shaped macroaleuriospores are produced in great numbers. They are one- to three-celled, 12 to 18 × 4 to 8 μ in diameter, with thin, verrucous walls. Microaleuriospores are clavate. They occur rarely on SDA but are numerous on soil-hair agar and distinguish this species from *Chrysosporium* sp. The latter have large, ovoid microaleuriospores with a wide attachment base that can be confused with the macroaleuriospores of *M. nanum.* Hair penetration organs are produced by *M. nanum.*

This is a zoophilic species associated with pigs. It has been isolated from soil in pig yards. In man a tinea corporis contracted from pigs has been seen. The organism is sparsely ectothrix and nonfluorescent in infected hair. The perfect stage is *N. obtusa.*

N. obtusa. Gymnothecia are globose, 250 to 450 μ in diameter, and pale buff or yellowish. Peridial hyphae are pale yellow, hyaline, dichotomously branched at an obtuse angle from the main axis, and rarely verticillately branched. The cells of the peridial hyphae are thick-walled, echinulate, cylindrical, and 5 by 13 μ in size, with one to two slight constrictions in the center. The peridial hyphae may end in a long, smooth, tapering spike or tightly coiled spiral hyphae. Asci are subglobose, 5.5 by 5 to 6 μ in size, and evanescent. Eight oblate, smooth-walled, yellowish ascospores are formed. They are 2.7 to 3.2 by 1.2 to 2 μ in size, and yellow in mass. The walls may be somewhat roughened.

Microsporum distortum di Menna and Marples 1954

This species is a rare cause of tinea capitis in New Zealand and Australia. Infection in the U.S. has been traced on occasion to contact with monkeys from South America. This species is similar to and possibly is a variant of *M. canis*. It is small-spored, ectothrix, and fluorescent in infected hairs.

Colony. In general the colonial growth is similar to *M. canis*, but usually has less pigmentation on the underside. The colony is flat, velvety to fuzzy, and white to tan in color.

Microscopic Morphology. The macroaleuriospores are similar to those of *M. canis* but grossly bent and distorted. Sessile clavate microaleuriospores are produced. They are usually more abundant than isolants of *M. canis* (Fig. 7–44, *D*).

Differential Characteristics. *M. distortum* like *M. canis* grows abundantly on rice grains; only the morphology of the macroaleuriospores differentiates the two species.

Microsporum gallinae (Megnin) Grigorakis 1929

Synonymy. *Epidermophyton gallinae* Megnin 1881, *Achorion gallinae* (Megnin) Sabouraud 1910, *Trichophyton gallinae* (Megnin) Silva and Benham 1952.

Colony Obverse. This species grows rapidly, producing a conical, slightly folded, downy to satiny white colony with an entire to slightly scalloped edge. The mycelium may be stained pink (Fig. 7–45,*A* FS B 12,*A*).

Reverse. Within a few weeks a diffusable pigment is formed. Initially it is yellow, but becomes a bright Montmorency sour cherry red or strawberry red (Fig. 7–45,*B* FS B 12,*B*).

Microscopic Morphology. The recent finding of echinulations on the macroaleuriospores has prompted the restoration of this species to the genus *Microsporum*.[63] The macroaleuriospores are 6 to 8 × 15 to 50 μ in size and are often elongate with a blunt tip (spatulate or slipper-shaped). There are two to ten cells, and the walls are usually smooth or sometimes echinulate at the tip. The spores are often curved, with a flared distal end resembling a threatening cobra. They are often attached to dentate, pectinate, or leaf-like hyphae. Clavate and pyriform microaleuriospores are found (Fig. 7–50,*D*). Thiamin or yeast extract added to the media increases sporulation. This is a zoophilic species associated with fowl and is rarely involved in human infection.[60a, 162a] It is ectothrix and nonfluorescent in infected hairs.

Special Characteristics. *M. gallinae* is distinguished from *T. megninii* by good growth on ammonium nitrate media. The latter species requires added L-histidine.

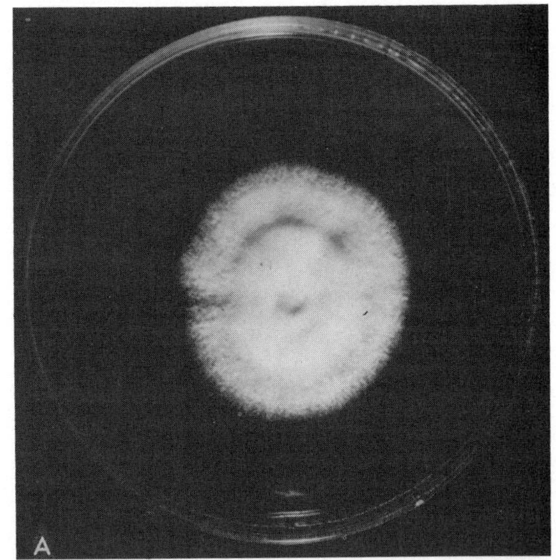

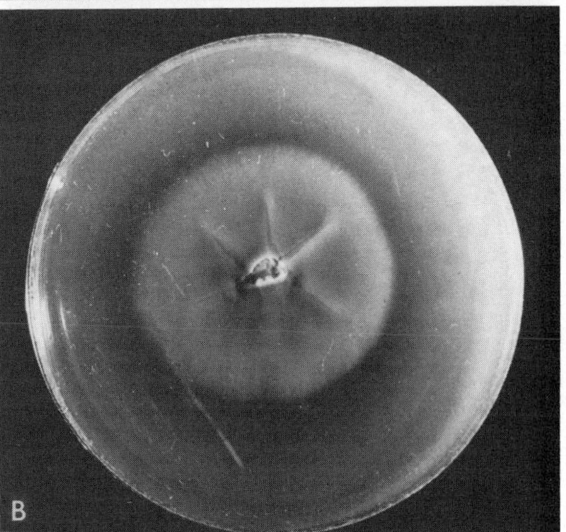

Figure 7–45. FS B 12. *M. gallinae A*, Obverse. *B*, Reverse, showing diffusable cherry-red pigment.

Microsporum persicolor (Sabouraud) Guiart and Grigorakis 1928

Synonymy. *Trichophyton persicolor* Sabouraud 1910.

Perfect State. *Nannizzia persicolor* Stockdale 1967.

For many years there was doubt as to the validity of this species, and most mycologists regarded it as a variety of *T. mentagrophytes*. The discovery of a perfect stage in the genus *Nannizzia* has verified it as a distinct species. This species does not appear to invade hair.

Colony Obverse. A rapidly growing, flat to gently folded, fluffy, yellowish buff to pale pink thallus is produced.

Reverse. Pigmentation of the underside is variable. The color ranges from peach or rose to deep shades of ochre. On sugar-free media, sectors of rose to red to deep wine tints are seen. This distinguishes it from *T. mentagrophytes*, which does not produce color on these media. This red color is also produced on pablum or rice grain media.

Microscopic. Microaleuriospores are usually abundant. They are clavate or fusiform to globose, and arranged in clusters resembling the microscopic morphology of *T. mentagrophytes*. Spiral hyphae are common. The macroaleuriospores are sparsely produced. They are elongate, fusiform to clavate, and usually six-celled. The walls are thin and smooth with some echinulations at the tip. Hair penetration organs are produced.

M. persicolor grows better than *T. mentagrophytes* on ammonium nitrate media, and the predominance of stalked, elongate, clavate microaleuriospores in the former is also a distinguishing feature. This species is zoophilic and is a rare pathogen for man, but it is frequently found in bank voles and field voles. It has been found in soil. The perfect stage is *N. persicolor*.[116b]

N. persicolor. Globose gymnothecia are formed that vary from 250 to 900 μ in diameter and are the usual buff color. The peridial hyphae are branched, asperate to echinulate, and hyaline to pale beige in color. The branches curve outward on the gymnothecia and often end in spirals, which may be dichotomously branched. The cells of the peridial hyphae are symmetrical and have a single central constriction. Asci are evanescent, ovoid, and 4.3 to 6 × 5 to 7 μ in size. Eight ascospores are produced. They are hyaline, lenticular, smooth-walled, yellow in mass, and 2.3 to 3.2 × 1.5 to 2.1 μ in size.

Microsporum cookei Ajello 1959

Perfect State. *Nannizia cajetani* Ajello 1961.

This is a geophilic species reported in rodents, dogs, and rarely in man. It is not known to invade hair.

Colony Morphology. The spreading, rapidly growing colony is coarsely granular and usually a deep pink (Fig. 7–43,*C* FS B 14,*C*). The morphology is similar to strains of *M. gypseum*. The color varies from greenish buff to brown, and a deep vinaceous red is usually found on the reverse.

Microscopic Morphology. The macroaleuriospores are thick-walled, echinulate, 30 to 50 × 10 to 15 μ in size, resembling *M. gypseum* except in wall width. The microaleuriospores are obovoid and abundant. Hair perforating organs are found.

N. cajetani. Gymnothecia are 368 to 686 μ in size, globose, and pale buff to yellow. Peridial hyphae are septate; the cells are elongate, echinulate, and minimally constricted. The ends taper to a long, smooth-walled spike up to 480 μ in length or to tight spiral hyphae. Asci are globose to ovate, 6 to 9 μ in diameter, and contain eight ascospores. These are lenticular, golden in mass, smooth-walled, and 3 to 3.6 × 1.8 μ in size.

Microsporum vanbreuseghemii Georg, Ajello, Friedman and Brinkman 1962

Perfect State. *Nannizzia grubyia* Georg, Ajello, Friedman, and Brinkman 1962.

This is a geophilic species rarely involved in ringworm of man, dog, and squirrel. It is ectothrix and nonfluorescent in infected hair.

Colony Morphology. A fast-growing, flat, spreading colony is produced that is cottony and cream-yellow to lavender pink. The reverse is colorless to yellow. The fungus rapidly becomes pleomorphic.

Microscopic Morphology. Macroaleuriospores are abundant, 59 to 62 × 11 μ in size, echinulate, thick-walled (2 to 3 μ), and cylindro-fusiform in shape, with seven to ten cells. They are similar to those of *T. ajelloi* except that the latter has smooth-walled spores and a purple pigment. *M. precox* and *M. racemosum* have fewer cells in the macroaleuriospores than *M. vanbreuseghemii*. The microaleuriospores are pyriform to obovate. This species readily infects guinea pigs and produces hair perforation organs. The perfect stage is *N. grubyia*.

N. grubyia. The gymnothecia are globose, white to buff in color, and 150 to 600 μ in diameter. Peridial hyphae are hyaline, uncinately or dichotomously branched, curving away from the main axis. The tips end in long, smooth, blunt points or curved, clawlike, phalangiform cells. Spiral hyphae and macroaleuriospores are also found at the tips. Asci are evanescent, globose, 4.8 × 6 μ in diameter, containing eight lenticular, smooth-walled ascospores which are 2.4 × 3 μ in size and pale yellow in mass.

Trichophyton Malmsten 1845

The macroaleuriospores characteristic of this genus are elongate, clavate to fusiform, generally thin-walled (up to 2 μ), smooth, and have zero to ten septa. The size range is 4 to 8 $\times$ 8 to 50 μ. As with other dermatophytes, the microaleuriospores are usually uninucleate, and the cells of the macroaleuriospores multinucleate. Macroaleuriospores are few or absent in many species. Microaleuriospores are globose in shape, 2.5 to 4 μ in diameter, or clavate, pyriform, and 2 to 3 $\times$ 2 to 4 μ in diameter. The perfect stage, when found, is in the genus *Arthroderma.*

Arthroderma Berkeley 1860

The peridial hyphae around the gymnothecia of *Arthroderma* are usually dichotomously branched, with asperulate hyphal cells that are noticeably constricted at the septa. The individual cells have a central constriction, and at either end have a prominent knobby protuberance. They have an overall dumbbell-shaped appearance. This genus is quite similar to *Nannizzia,* and few distinguishing characteristics are constant.[115]

Trichophyton mentagrophytes (Robin) Blanchard 1896

Synonymy. *Microsporon mentagrophytes* Robin 1853, *Achorion quinckeanum* Blanchard 1896, *Trichophyton felineum* Blanchard 1896, *Trichophyton gypseum* Bodin 1902, *Trichophyton granulosum* Sabouraud 1909, *Trichophyton radiolatum* Sabouraud 1910, *Trichophyton lacticolor* Sabouraud 1910, *Trichophyton denticulatum* Sabouraud 1910, *Trichophyton farinulentum* Sabouraud 1910, *Trichophyton asteroides* Sabouraud 1910, *Trichophyton interdigitale* Priestley 1917, *Trichophyton "C"* Hodges 1921, *Trichophyton Kaufmann-Wolf* Ota 1922, *Trichophyton pedis* Ota 1922, *Trichophyton quinckeanum* (Zopf) MacLeod and Muende 1943.

Perfect State. *Arthroderma benhamiae* Ajello and Cheng 1967.

T. mentagrophytes is the commonest dermatophyte of man, animals, and soil. Its morphology is so variable that it has been given a long list of binomials based on colony color, morphology, sporulation, and host association. The characteristics are so inconstant, and so many intermediates are seen that such names are not useful or practical. Almost all these variants mate with the tester strains of *A. benhamiae,* so that a single species is recognized. The dropping of most varietal designations is encouraged. *T. mentagrophytes* is universal, and anthropophilic and zoophilic forms are found. It is usually small-spored, ectothrix, and nonfluorescent in infected hair. Some strains (var. *quinckeanum* type) may produce endothrix infections and show a dull fluorescence. All strains produce hair perforating organs. Guinea pigs are easy to infect with granular strains (var. *mentagrophytes*), but less so with the downy (var. *interdigitale*) type.

Colony Obverse (Fig. 7–46,*B* FS B 6). The anthropophilic form grows as a flat, downy thallus with white edges and a cream-tinted central area. On potato-dextrose agar, a stellate colony with sparse mycelium and numerous spores is seen. Zoophilic isolates produce a flat, rapidly growing, granular colony that is cream, yellowish, buff to tan, and reddish-brown in color. Mycelium is usually sparse, and the powdery appearance is due to quantities of microaleuriospores. The edges are often raylike. Some strains, particularly those from Southeast Asia, show a powdery, lavender-tinged surface. Numerous variations of colony morphology occur.

Reverse. Pigmentation is variable; colorless, yellow-brown, reddish-brown, brown, and a deep wine red resembling *T. rubrum* are seen. In differentiating pigmented colonies from *T. rubrum,* one should note that the latter usually produces a red pigment on potato-dextrose and cornmeal 1 per cent glucose agar, while *T. mentagrophytes* does not.

Microscopic Morphology (Fig. 7–47,*B*). Probably the most consistent feature of *T. mentagrophytes* is production of globose microaleuriospores in grapelike clusters (*en grappe*). These are most abundant in zoophilic-inhabiting strains, and less so in downy strains. In the latter instance the spores are more clavate-shaped and resemble those of *T. rubrum.* Macroaleuriospores are thin-walled, smooth, and variable in shape. The size ranges from 4 to 8 $\times$ 20 to 50 μ and they have three to five cells. They are generally cigar-shaped with a narrow attachment base, and here, in contrast to *T. rubrum,* the end cell of these spores may have a terminal filament or "rat tail" appendage. The typical picture of *T. mentagrophytes* seen under the microscope is massed microaleuriospores, some macroaleuriospores, and several spiral hyphal cells, all in clusters on the vegetative hyphae. Structures resembling peridial hyphae, antlerlike hyphae, arthroaleuriospores, nodular bodies, racquet mycelium, and chlamydospores are also seen. The variety designated as *quinckeanum* (mouse favus) has a gently folded thallus and lateral clavate microaleuriospores. In some strains, isolated from tinea pedis and having deep yellow to orange pigmentation, nodular bodies are numerous and microaleuriospores few in number. Hair perforating organs (Fig. 7–48) and lack of pigment on cornmeal glucose and potato-dextrose agar separate *T. mentagrophytes* from *T. rubrum*.

The perfect state is *Arthroderma benhamiae*. Some granular strains from rodents do not mate with tester strains, suggesting a complex similar to *M. gypseum* or genetic incompatibility.

A. benhamiae. The gymnothecia are globose, white to pale buff, and 250 to 500 μ in diameter (Fig. 7–49 FS B 9). The peridial hyphae are dichotomously branched, curving out from the main axis, and the cells are asperulate and dumbbell-shaped. The cell number in each branch is usually four, in contrast to the two to three cells in terminal branches of *A. simii*. The end cells of *A. benhamiae* end in tapering, smooth-walled spikes or spirals. Asci are evanescent globose to ovate, 4.2 to 7.2 × 3.6 to 6.0 μ in size, and eight spored. The ascospores are smooth, lenticular, and 1.2 to 1.8 × 1.2 to 2.8 μ in size.

Trichophyton rubrum (Castellani) Sabouraud 1911

Synonymy. *Trichophyton purpureum* Bang 1910, *Trichophyton rubidum* Priestley 1917, *Trichophyton "A"* Hodges 1921, *Trichophyton "B"* Hodges 1921, *Trichophyton marginatum* Muijs 1921, *Trichophyton plurizoniforme* MacCarthy 1925, *Trichophyton lanoroseum* MacCarthy 1925, *Trichophyton coccineum* Katoh 1925, *Trichophyton spadix* Katoh 1925, *Trichophyton multicolor* Magalhaes and Neues 1927, *Trichophyton kagawaense* Fujii 1931, *Trichophyton rodhainii* Vanbreuseghem 1949, *Epidermophyton rubrum* Castellani 1910, *Epidermophyton perneti* Castellani 1910, *Epidermophyton salmoneum* Froilano de Mello 1921, *Trichophyton fluviomuniense* Miguens 1968.

T. rubrum is anthropophilic and has recently become the most common and widely distributed dermatophyte of man. It is very rarely isolated from animals and never from soil. It is extremely variable in its morphology and lacks a perfect stage for positive identification. Scalp infections are uncommon, and hair is rarely invaded. In scalp infections involving hair, ecto- and endothrix spores that are nonfluorescent have been described. All strains tested by the *Arthroderma simii* sex stimulation test have been of

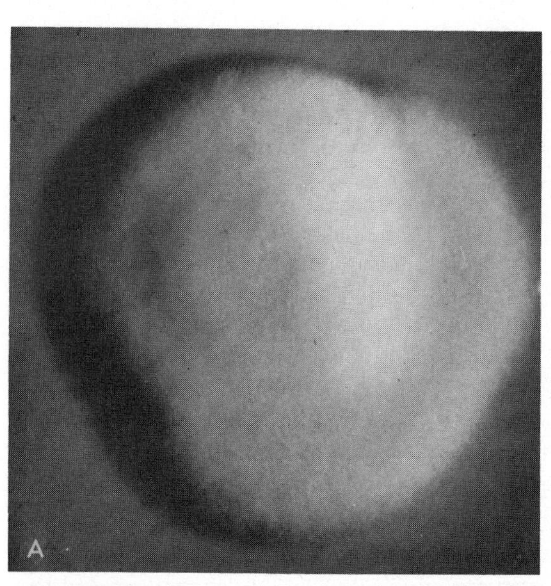

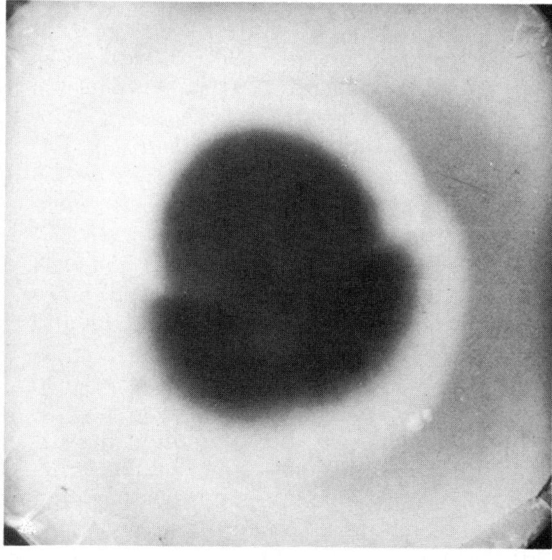

Figure 7–46. *A,* FS B 5, *A, T. rubrum* showing fluffy white mycelium. FS B 5, *B, T. rubrum* showing reverse demonstration of wine-red pigment.

Illustration continued on opposite page.

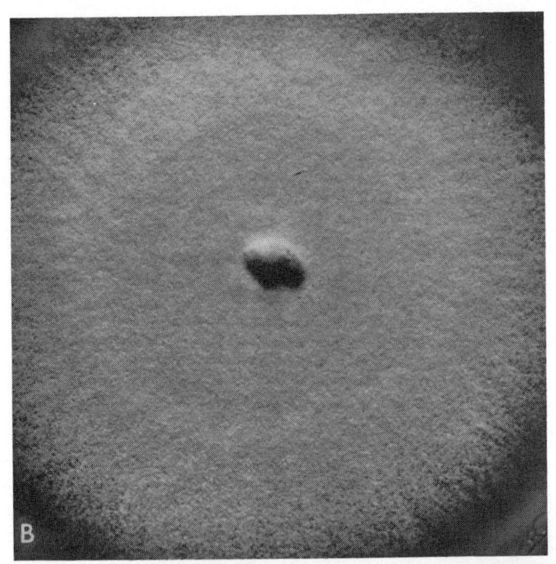

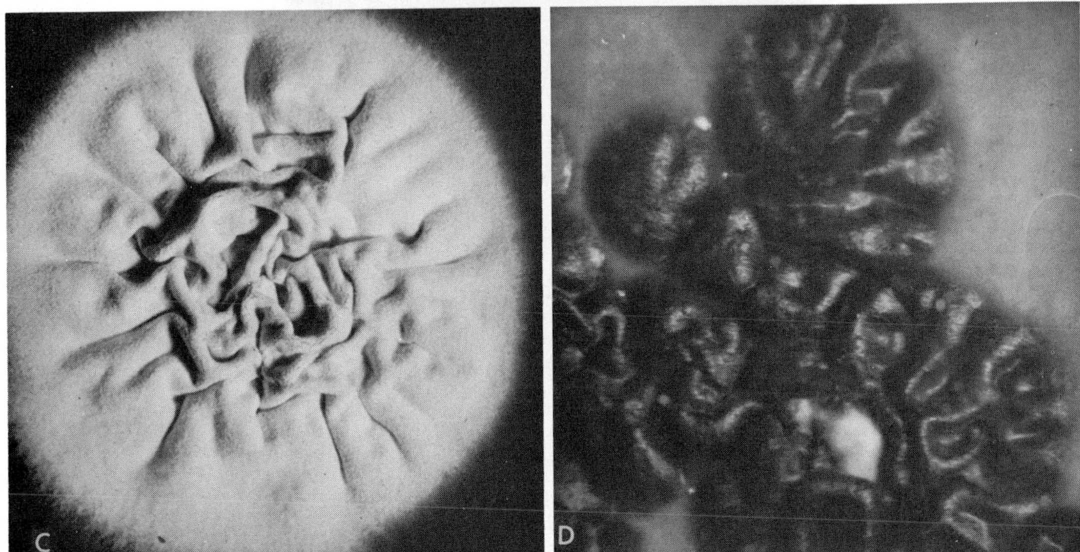

Figure 7–46 *Continued.*
B, FS B 6, *T. mentagrophytes.* Flat granular colony, cream-colored in central area.
C, FS B 7, *T. tonsurans.* Cerebriform colony with suedelike surface. The central convolutions often crack.
D, FS B 8, *T. violaceum.* Glabrous, slightly folded, violet-colored surface. There are white pleomorphic patches.

the (−) mating type. The opposite mating type may have been lost during the evolution to an anthropophilic fungus dependent on man for dissemination of the species. The (−) mating type possibly posessed a selective advantage over the (+) mating type in this regard.

Colony Obverse (Fig. 7–46,*A* FS B 5,*A*). The typical thallus is slow-growing, downy white, generally devoid of spores, and pigmented on the reverse. This is the type commonly isolated from chronic tinea pedis and chronic tinea corporis. Other isolants are less cottony, less pigmented, and produce numerous macroaleuriospores. The sporulating strains usually come from inflammatory tinea corporis, tinea capitis, and granulomatous lesions. Occasional isolants are heaped up, folded, glabrous, intensely pigmented, devoid of spores, and resemble *T. violaceum.* This morphology is seen in the var. *rodhainii* type. Folded colonies of this strain with diminished pigmentation resemble *T. tonsurans.*

Reverse (Fig. 7–46,*A* FS B 5,*B*). The typical pigment is an intense, nondiffusing port wine or venous blood red. If produced in abundance, it not infrequently stains the

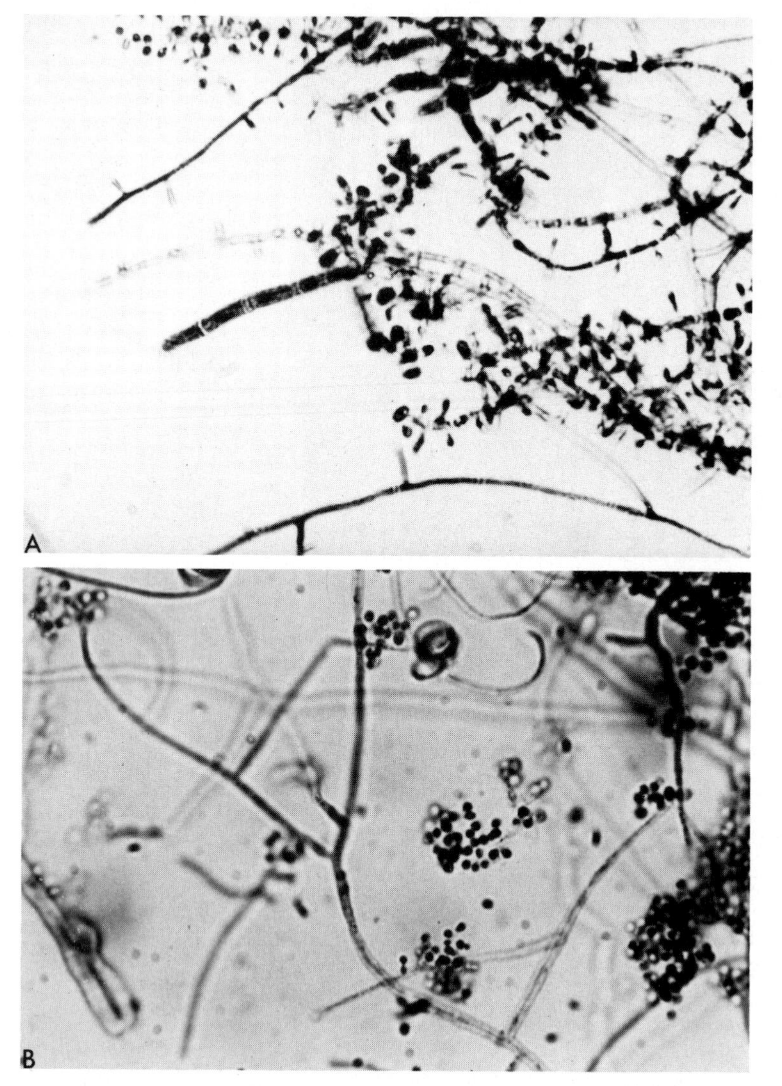

Figure 7–47. *A, T. rubrum* in rare granular form showing macroaleuriospores and "tear drop" microaleuriospores. ×440. *B, T. mentagrophytes.* Microaleuriospores in grapelike clusters and spiral hyphae. ×440. (From Rippon, J. W. *In* Burrows, W. 1973. *Textbook of Microbiology.* 20th ed. Philadelphia, W. B. Saunders Company, pp. 705, 710.)

mycelium on the obverse. Pigment is slow in developing, and is usually first noted on the edge of the colony in agar slants where the media is dried. The color may be yellow initially, developing through a melanoid-green, and finally becoming red. Several pigments are formed by the organism, differing in quantities at different times.[178] A black melanin-like diffusible pigment is sometimes produced. Some strains, particularly those from patients on griseofulvin therapy, fail to show pigment. Growth on potato-dextrose and cornmeal glucose agar enhances pigmentation and separates *T. rubrum* from other red pigmented species.

Microscopic Morphology. The common isolants of *T. rubrum* produce few spores. There are clavate "tear drop" microaleuriospores (2 to 3 × 3 to 5 μ) produced lateral to the hyphae. Clusters of spores in an arborescent or "pine tree" arrangement are seen. Macroaleuriospores are absent or rare except in granular sporulating strains. When present these spores are long, narrow, fusiform (pencil-shaped), usually without a stipe, multicelled, and often develop in groups directly from hyphae. Peridial hyphal-like structures, chlamydospores, pectinate hyphae, and many aberrant structures are also seen (Fig. 7–47,*A*).

Differential Characteristics. Pigment production on special media and lack of *in vitro* hair penetrating organs differentiates *T. rubrum* from *T. mentagrophytes. T. violaceum* grows poorly without thiamine, and *T. megninii* requires L-histidine. Neither the amino acid nor the vitamin is required by *T. rubrum.*

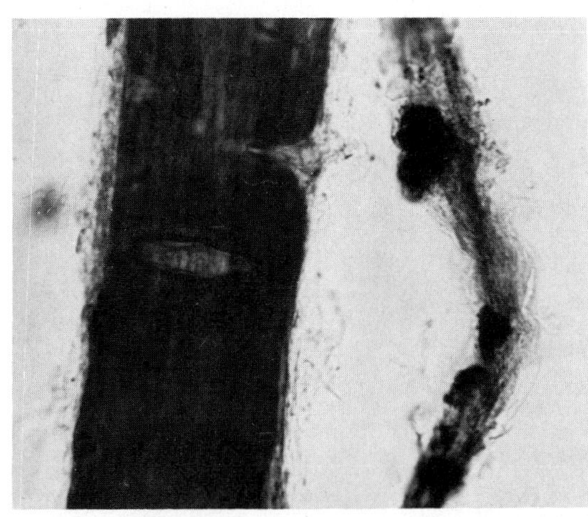

Figure 7–48. Hair perforating organs. These are wedge-shaped indentations produced by hyphae of some dermatophytes while growing saprophytically on hair. ×440.

Trichophyton tonsurans Malmsten 1845

Synonymy. *Trichophyton epilans* Boucher and Megnin 1887, *Trichophyton sabouraudi* Blanchard 1896, *Trichophyton crateriforme* Sabouraud 1902, *Trichophyton flavum* Bodin 1902, *Trichophyton acuminatum* Bodin 1902, *Trichophyton effractum* Sabouraud 1910, *Trichophyton fumatum* Sabouraud 1910, *Trichophyton umbilicalum* Sabouraud 1910, *Trichophyton regulare* Sabouraud 1910, *Trichophyton exsiccatum* Sabouraud 1910, *Trichophyton polygonum* Sabouraud 1910, *Trichophyton plicatile* Sabouraud 1910, *Trichophyton pilosum* Sabouraud 1910, *Trichophyton sulfureum* Sabouraud 1910, *Trichophyton cerebriforme* Sabouraud 1910, *Trichophyton ochropyrraceum* Muijs apud Papengaaji 1924.

This is another species which varies particularly in color, texture, and morphology of the thallus. It is anthropophilic, endothrix, and nonfluorescent in infected hair. The young colonies are sometimes dully fluorescent. *T. tonsurans* was probably originally endemic in central and southern Europe and became prevalent in South America. Recently it has spread to the United States and other countries through the migration of infected peoples. Infection in a horse and a dog is reported.

Figure 7–49. FS B 9. Gymnothecia of *A. benhamiae (T. mentagrophytes)*. They are growing on hair in soil. (Courtesy of G. Rebell.)

Colony Obverse (Fig. 7–46,*C* FS B 7). There are four common colonial forms: crateriforme, cerebriforme, plicatile, and flat. All originally had different specific names. The most frequently isolated strains show a flat growth initially, which is powdery and yellow-tinged. This colony develops into a folded thallus, with a greyish to buff, suedelike surface. Pink tints are sometimes seen, especially if the reverse is heavily pigmented. The var. *sulfureum* type is less powdery, with a deep yellow to chartreuse, suedelike colony resembling *E. floccosum*. A particular colony type may be locally endemic. Pale and white strains are also seen.

Reverse. The pigment on the under surface is commonly yellow-brown to reddish-ochre or deep mahogany red. It sometimes diffuses into the media.

Microscopic Morphology (Fig. 7–50,*A*). The characteristic microaleuriospores are variable in size and shape but usually abundant. Their size is 2 to 5 × 3 to 7 μ. They are clavate, tear-shaped, often borne on elongate stipes (match stick conidia), and formed in clusters on multiple branched, thickened terminal hyphae. Several spores may be expanded and balloon-shaped. Filiform or claw-shaped spores arranged laterally on irregular hyphae resemble a centipede. Macroaleuriospores occur less frequently, are irregular in shape, and are somewhat thick-walled. Chlamydospores, racquet mycelium, and irregular arthrospore-like structures are found.

Differential Characteristics. Growth is minimal or absent in thamine-less media, a characteristic differentiating this species from *T. rubrum* and *T. mentagrophytes*. L-histidine is required for the growth of *T. megninii* but not of *T. tonsurans*.

Trichophyton violaceum Sabouraud apud Bodin 1902

Synonymy. *Trichophyton glabrum* Sabouraud 1910, *Achorion violaceum* Bloch 1911, *Favotrichophyton violaceum* Dodge 1935.

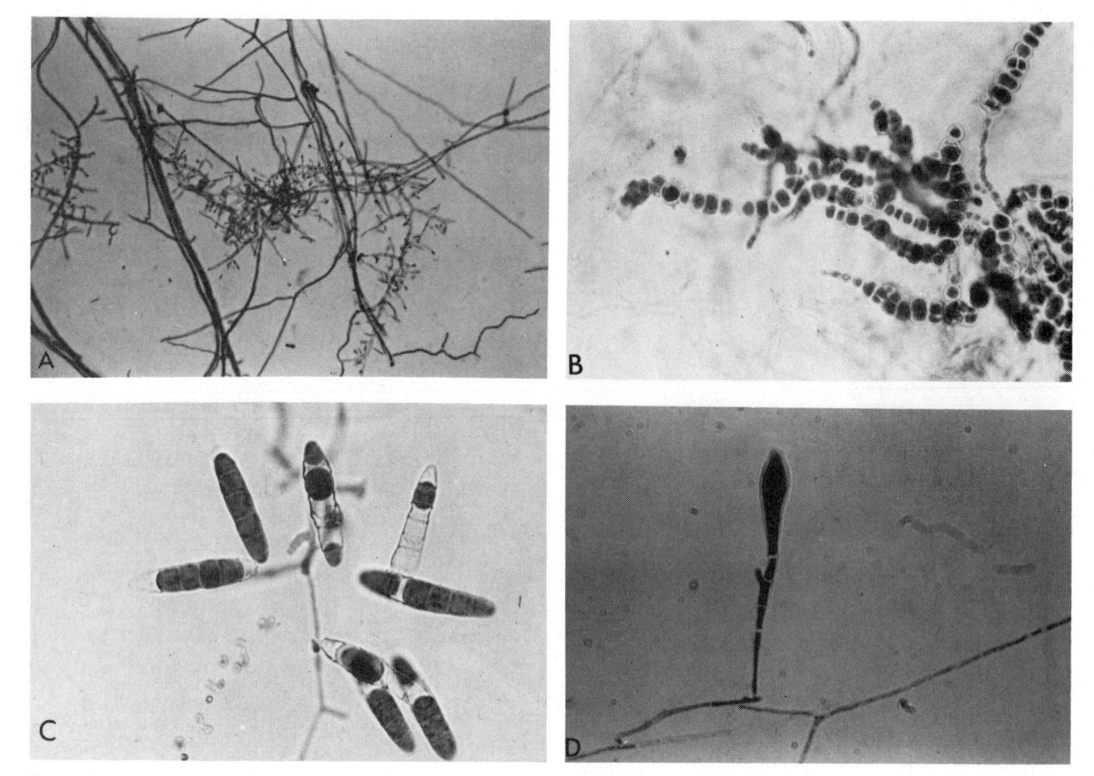

Figure 7–50. *A, T. tonsurans.* Microscopic view showing elongate "match stick" microaleuriospores in groups and at right angles to mycelium. ×150. *B, T. verrucosum.* Chains of chlamydospores produced at 37°C. *C, T. simii.* Macroaleuriospores with endochlamydospores. *D, M. gallinae.* Spatulate macroaleuriospore. ×440.

An anthropophilic species endemic in South America, Mexico, Europe, Asia, and Africa, it is rare in the United States. It is an endothrix organism and nonfluorescent in infected hair. Animal infection is rare, but cases of disease in calf, dog, cat, horse, mouse, and pigeon have been reported.

Colony Obverse (Fig. 7–46,*D* FS B 8). *T. violaceum* is very slow growing, producing a conical or verrucous (faviform) thallus that is heaped up, folded, glabrous or waxy, and deep violet in color. The organism rapidly becomes pleomorphic, sectoring as pale to white mycelia that quickly overgrow the original colony. Old stock cultures become flat, white, and fluffy. Occasional isolants are nonpigmented (*T. glabrum*).

Reverse. The purple pigment that stains the mycelium is also found on the underside.

Microscopic Morphology. Distorted hyphae and the lack of aleuriospores are typical of strains grown on the usual media. The hyphae contain cytoplasmic granules. On media enriched with thiamine, a few clavate microaleuriospores are produced, and rarely an elongate macroaleuriospore is seen.

Differential Characteristics. The partial requirement for thiamin separates this organism from *T. gourvilii, T. rubrum,* and other species that may produce purple pigmented colonies.

Trichophyton verrucosum Bodin 1902

Synonymy. *Trichophyton album* Sabouraud 1908, *Trichophyton ochraceum* Sabouraud 1908, *Trichophyton discoides* Sabouraud 1910, *Trichophyton faviforme* Sabouraud 1892, *Favotrichophyton verrucosum* Neveu-Lemaire 1921.

T. verrucosum is principally associated with cattle and is of worldwide distribution. It is a large-spored ectothrix, and fluorescence of hair in infected cattle has been reported. Infected human hair is not fluorescent.

Colony Obverse. The colony is very slow growing, producing a knoblike or slightly folded, heaped, glabrous, grey-white colony (Fig. 7–43,*D* FS B 14,*D*). This morphology is referred to as the var. *album* type. Two other color variants are described; var. *ochraceum* has a flat, yellow, glabrous colony, and var. *discoides* has a flat, tomentose, grey-white thallus. On blood agar or media enriched with thiamine and inositol, growth is more rapid and spreading. Growth is enhanced by a temperature of 37°C. No characteristic pigment is produced on the reverse.

Microscopic Morphology. On unenriched media, distorted hyphae are seen, with some suggestion of antlerlike branching similar to that of *T. schoenleinii.* No spores are seen. On enriched media, clavate microaleuriospores and rarely elongate fusiform or the rather characteristic rat tail macroaleuriospores are produced. These have from three to five cells and sometimes are shaped like a string bean. At 37°C the fungus grows as a chain of chlamydospores (Fig. 7–50,*B*).

Differential Characteristics. All strains require thiamine, and most require inositol.

Trichophyton schoenleinii (Lebert) Langeron and Milochevitch 1930

Synonymy. *Oidium* (Lebert) *schoenleini* Lebert 1845, *Achorium schoenleini* Remak 1845, *Favuspiltz B* Quincke 1886, *Oospora porriginis* Saccardo 1886, *Schoenleinium achorian* Johan Olsen 1897, *Grubyella* (Lebert) *schoenleini* Ota and Langeron 1923, *Arthrosporia* (Lebert) *schoenleinii* Grigorakis 1925.

The classic cause of favus in man is *T. schoenleinii.* It is endemic throughout Eurasia and Africa but occurs infrequently in the

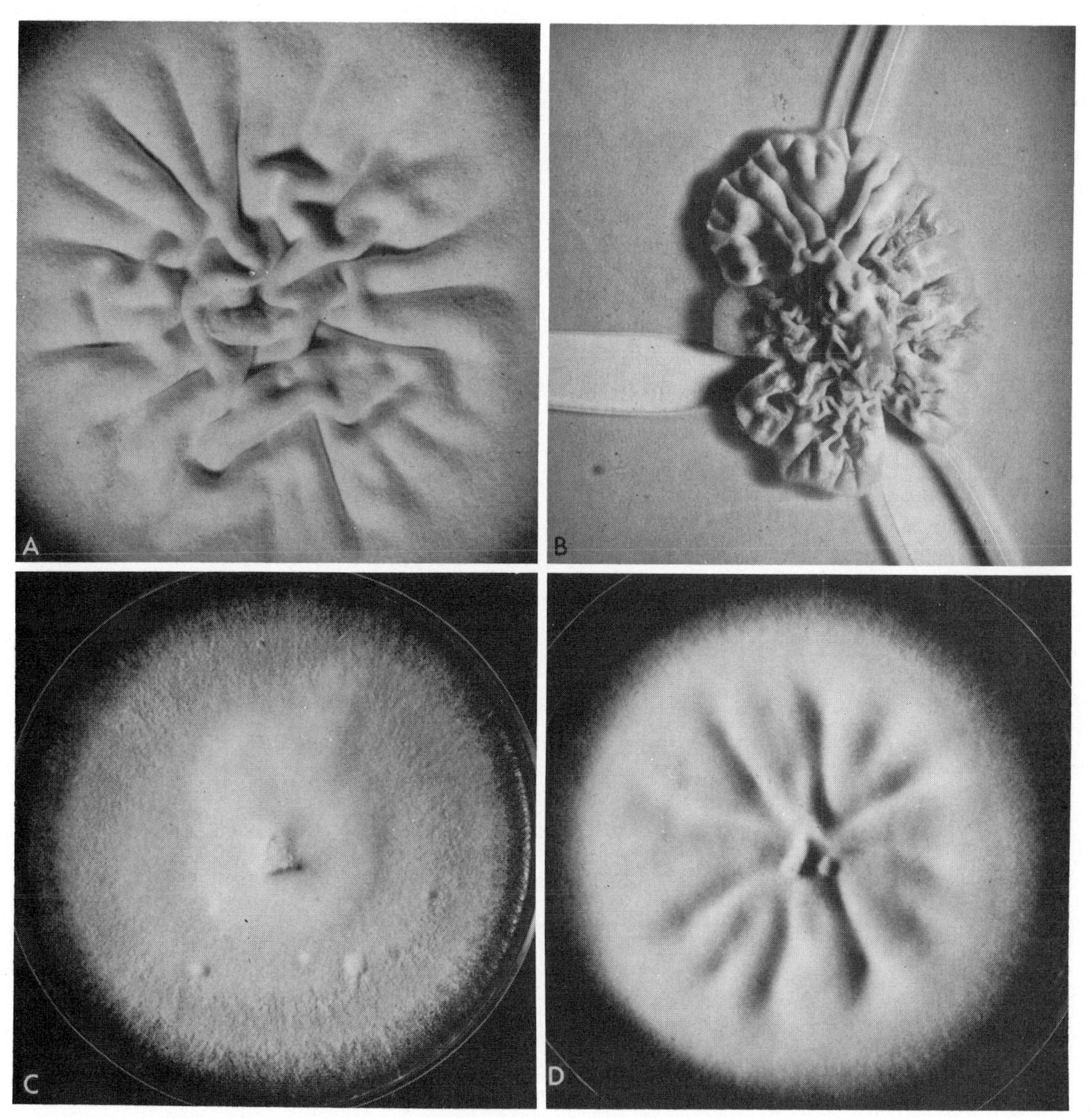

Figure 7–51. FS B 10. *A, T. schoenleinii. B, T. concentricum. C, T. simii. D, T. megninii.*

Western hemisphere. This fungus is rare in animals, but infection of dog, cat, hedgehog, cow, horse, mouse, rabbit, and guinea pig is reported.

Colony Morphology. A slow-growing, glabrous (waxy), or suedelike, off-white colony is produced. There are gentle folds (faviform), but the colony may become very distorted and convoluted, rising off the agar surface or into it, causing cracking and splitting. The growth of some strains is mainly subsurface. In time, old laboratory strains become flat and downy. Some initial isolants are yeastlike in consistency and morphology and grow well at 37°C (Fig. 7–51,*A* FS B 10, *A*).

Microscopic Morphology. Spores are not seen. The hyphae form antlerlike or chandelier-like structures (Fig. 7–52). The branches often have a swollen (nail-head) tip. These are quite characteristic, though a few other species may occasionally form them also. A few distorted clavate microaleuriospores are found in some isolants, particularly when the organism is grown in rice grains. Macroaleuriospores have not been reported.

Differential Characteristics. This fungus grows well at 37°C and does not require thiamine or other special nutrients.

Trichophyton concentricum Blanchard 1896

Synonymy. *Trichophyton mansonii* Castellani 1905, *Trichophyton castellanii* (Perry 1907) Castellani 1908, *Endodermophyton concentricum* Castellani 1910, *Endodermophyton indicum* Castellani 1911, *Endodermophyton mansoni* Castellani 1919, *Endodermophyton tropicale* Castellani 1919, *Endodermophyton roquettei* Fonseca 1925, *Endodermophyton castellanii* Perry 1910, *Lepidophyton* sp. Tribondeau 1899, *Aspergillus lepidophyton* Pinoy 1903, *Aspergillus tokelau* Wehmer 1903.

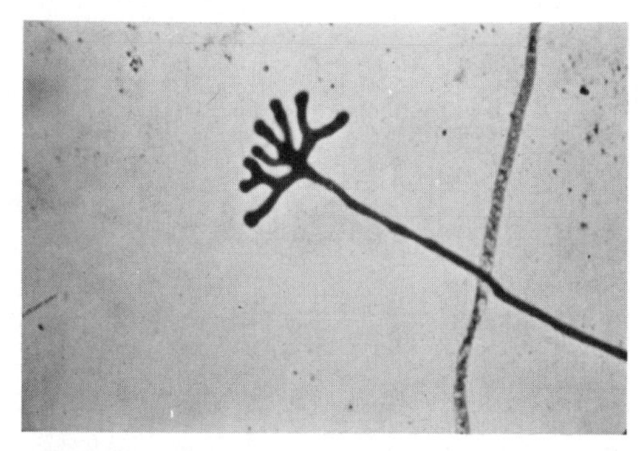

Figure 7–52. "Favic chandelier" of *Trichophyton schoenleinii*. Lactophenol cotton blue. ×400. (From Rippon, J. W. *In* Burrows, W. 1973. *Textbook of Microbiology*. 20th ed. Philadelphia, W. B. Saunders Company, p. 707.)

This fungus causes tinea imbricata and is sporadically distributed in South Asia, the Pacific, and South America. It does not invade hair, and no animal infections have been reported.

Colony Morphology. The thallus is raised, deeply folded, and convoluted. It is glabrous and white at first, becoming cream, amber, brown, or coral red with time. A fuzzy or velvety growth of aerial mycelium may occur. The colony is slow-growing and has a diameter of 5 to 20 mm after ten days (Fig. 7–51,*B* FS B 10,*B*).

Microscopic Morphology. Distorted, convoluted hyphae without spores are seen. The branching mycelium resembles the antlers of *T. schoenleinii,* but the nail head ends are lacking. On bean pod media, lateral irregular macro- and microaleuriospores have been reported to develop. In about 50 per cent of isolants, growth is stimulated by thiamine.

Trichophyton simii (Pinoy) Stockdale, Mackenzie, and Austwick 1965

Synonymy. *Epidermophyton simii* Pinoy 1912, *Pinoyella simii* (Pinoy) Castellani and Chalmers 1919, *Trichophyton mentagrophytes* Emmons 1940.

Perfect State. *Arthroderma simii* Stockdale, Mackenzie, and Austwick 1965.

This organism is a frequent cause of ringworm in monkeys and chickens in endemic areas. It is uncommon in man, but has been isolated regularly from soil. It causes an ecto-endothrix type of hair invasion. Infected guinea pig hair fluoresces a bright blue-green.[132]

Colony Oberse. The fungus grows rapidly, producing a flat, granular colony with a central umbo. Buff is the usual color, but it varies from cream to white (Fig. 7–51,*C* FS B 10,*C*).

Reverse. In time, the colony may have a pigmented undersurface. The color is usually vinaceous (red-brown), but may range from yellow to madder rose. On malt extract agar, quantities of red-brown pigment diffuse into the media. On glucose peptone agar, the colony is yellow, and the reverse straw to salmon.

Microscopic Morphology. Macroaleuriospores are usually produced in great abundance. They are thin-walled, smooth and clavate, cylindriform or fusiform in shape, and have four to ten septa. Spores often occur in clusters. The cells of the macroaleuriospores frequently enlarge, become thick-walled, and are termed endochlamydospores. The cells between the enlarged chlamydospores are often empty and rupture, causing fragmentation of the macroaleuriospore. The resultant free chlamydospores are convex and have a collar left from the adherent portion of the broken cell. Clavate, elongate, pyriform, or peg-shaped microaleuriospores are produced laterally on hyphae (Fig. 7–50,*C*).

A. simii. The gymnothecia are buff, globose, and from 200 to 650 μ in diameter. The morphology is somewhat similar to that of *Arthroderma uncinatum.* Peridial hyphae are hyaline, asperulate, yellow, and dichotomously branched. There are up to three cells per branch, ending in a spike or spiral hypha. Asci are evanescent, subglobose, 5 to 7.7 μ in diameter, with eight lenticular, smooth-walled ascospores. The latter are yellow in mass and 1.7 to 2.1 $\mu \times 2.9 \times 3.3$ μ in size.

The *Arthroderma simii* sexual stimulation test[154] can be utilized to determine the mating types of some other species. In the test, a known mating type of *A. simii* is placed on the plate with a strain of another species. If the mating types are opposite, at the zone of contact dense growth and gymnothecial initials (infertile) are formed. If the mating types are the same, the colonies remain separated by a clear zone. This test has been used to indicate the mating type of species in which the perfect stage has not yet been described, e.g., *T. rubrum* and *M. audouinii.*

Trichophyton equinum (Matruchot and Dassonville) Gedoelst 1902

This species is the cause of infection frequently in horses and rarely in man. It is of worldwide distribution. All strains require nicotinic acid for growth except those from Australia and New Zealand, which are autotrophic (var. *autotrophicum*).

Colony Morphology. The thallus is fluffy, resembling the var. *interdigitale* type of *T. mentagrophytes.* Growth is rapid and flat, but the colony may develop gentle folds. The color is cream-white to yellow. The reverse is bright yellow, changing to dark pink or brown. This pigment may diffuse in the media.

Microscopic Morphology. Thin, elongate to pyriform, stalked microaleuriospores are formed laterally along the hyphae. Rarely they are clustered. Macroaleuriospores are similar to those of *T. mentagrophytes* and are rarely seen in culture. They are fusiform or clavate.

Differential Characteristics. Nicotinic acid (niacin) is required for growth of *T. equinum* except for the strains isolated in New Zealand and Australia. This species is reported to grow on horse hair and not on human hair, but this may not be a reliable characteristic.

Trichophyton megninii Blanchard apud Bouchard 1896

Synonymy. *Trichophyton roseum* Sabouraud apud Bodin 1902, *Trichophyton rosaceum* Sabouraud 1909, *Trichophyton vinosum* Sabouraud 1910, *Ectotrichophyton megninii* Castellani and Chalmers 1919, *Megatrichophyton megninii* Neveu-Lemaire 1921, *Aleuriosporia rosacea* Grigorakis 1925, *Megatrichophyton roseum* (Bodin) Dodge 1935, *Trichophyton kuryangei* Vanbreuseghem and Rosenthal 1961, *Trichophyton à culture rose* Sabouraud 1893.

This is a rare species found in Europe and Africa, particularly in Sardinia and Portugal. Most often this organism causes tinea barbae and is ectothrix, with spores in chains. Animal infection is unknown. An endemic focus has recently been reported from Burundi.[166]

Colony Obverse. The thallus grows moderately rapidly to form a flat or gently folded center with radial furrows. The surface is felt or suedelike, white in color, but often stained pink (Fig. 7–51,*D* FS B 10,*D*).

Reverse. A deep red pigment is elaborated on the underside of the colony. It is less intense than that of *T. rubrum.* The color is more of a bordeaux red than the port wine red color of the latter species.

Differential Characteristics. The requirement for L-histidine separates *T. megninii* from other red pigmented species. Spores are similar to those of *T. rubrum.*

Trichophyton soudanense **Joyeux 1912** (Fig. 7–53, *A* FS B 11,*A*)

Synonymy. *T. sudanense* Bruhns and Alexander 1928.

This is an endothrix species which is endemic in Central and West Africa, with occasional isolations reported in Europe, Brazil, and the United States. It is not known to infect animals.[130, 165]

The thallus is slow-growing, flat to folded in the center of a tough, leathery consistency, and varying from yellow to apricot in color. The colony has a fringed or raylike radiating edge. The reverse is a deep yellow. There is also a violet variant. Round chlamydospores, short, segmented hyphae (arthrospores), and reflexive or right-angle branching mycelia that are bunched into a bushlike bundle are seen on microscopic examination. Pyriform microaleuriospores are formed laterally on the mycelia. On Löwenstein-Jensen media, the colony is dark compared to *M. ferrugineum*, which is a light yellow.

Figure 7–53. FS B 11. *A, T. soudanense. B, T. gourvilii. C, T. yaoundei. D, T. ajelloi.*

***Trichophyton gourvilii* Cantanei 1933** (Fig. 7-53,*B* FS B 11,*B*)

Synonymy. *Trichophyton german II* Gregorio 1931, *Favotrichophyton gourvili* (Catanei) Dodge 1935.

This is an endothrix species endemic in West Africa. The thallus is folded, heaped, and waxy but becomes flat and velvety on transfer. The color is lavender pink to deep garnet red. Typical trichophyton-type micro- and macroaleuriospores develop. Its lack of requirement for histidine and thiamine distinguish it from similar species (Fig. 7-2).

***Trichophyton yaoundei* Cochet and Doby Dubois 1957** (Fig. 7-53,*C* FS B 11,*C*)

This is an endothrix species endemic in equatorial Africa, particularly common in Cameroun, Congo, and Mozambique. It is not known to infect animals. The thallus appears glabrous and moist, and is slow-growing, heaped, folded, and white-cream in color. In time, the colony becomes deep tan to chocolate brown. Some strains grow almost entirely submerged in the medium. Pyriform microaleuriospores are formed. The trichophyton-type macroaleuriospores are rarely seen. Branched, antlerlike mycelial appendages are formed, as well as chlamydospores. It has no specific nutritional requirements. (Fig. 7-2).

***Trichophyton ajelloi* (Vanbreuseghem) Ajello 1967** (Fig. 7-53,*D* FS B 11,*D*)

Synonymy. *Keratinomyces ajelloi* Vanbreuseghem 1952.

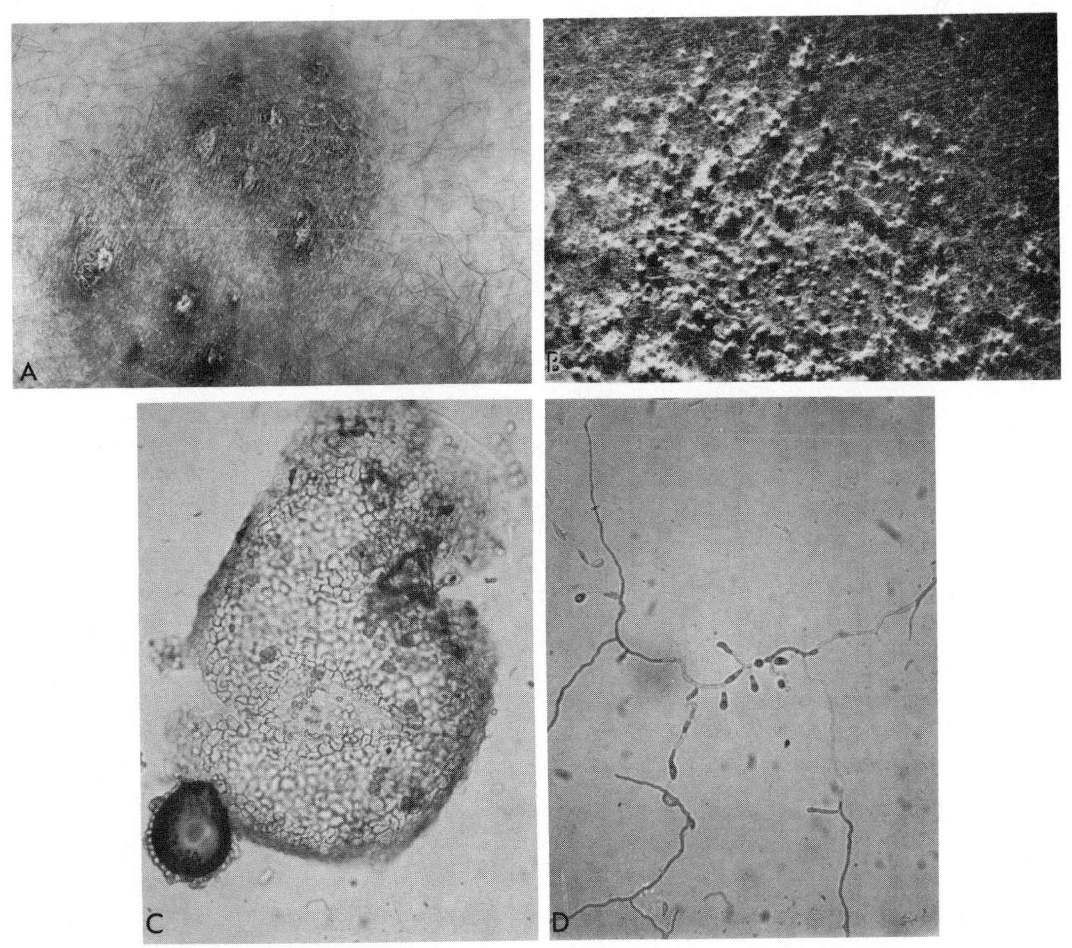

Figure 7-54. Soil keratinophilic fungi that rarely çause disease in man. *Aphanoascus fulvescens*. *A*, Lesion on inner thigh. *B*, Surface of colony covered with numerous brown-black cleistothecia. *C*, Crushed cleistothecium to show thick-walled rhomboid cells of the pseudoparenchymous walls. Note groups of mature ascospores. *D*, Microaleuriospores of the chrysosporium type and intercalary chlamydospores.[133]

Perfect State. *Arthroderma uncinatum* Dawson and Gentles 1961.

This is a very common soil keratinophilic fungus. It has been isolated rarely from tinea corporis of man, and from infections of cattle, dogs, horses and squirrels. It is readily isolated from soil by hair baiting. The colony grows rapidly, producing a flat, thin, powdery to downy, cream, yellow, tan, or orange thallus. The reverse is colorless, red, or a deep blue-black. The thick-walled, smooth macroaleuriospores are numerous. They are fusiform to cylindrical in shape, 20 to 65 × 5 to 10 μ in size, and contain 5 to 12 cells. Pyriform, sessile microaleuriospores are also abundant. Hair perforating organs are produced. Guinea pigs and mice can be infected.

A. uncinatum. Gymnothecia are globose, white to buff, and 300 to 900 μ in diameter. The peridial hyphae are hyaline, uncinately branched away from the main axis, and have knobby, asperulate, centrally-restricted, peridial cells. The tips are blunt to clawlike or have spirals arising either at the end or laterally on the body of the hyphae. Asci and spores are similar to those of *A. benhamiae.*

There are several other *Trichophyton* species and related keratinophils which are abundant in soil (Fig. 7–54). These are rarely involved in human or animal disease. *T. terrestre* Durie and Frey 1957 represents a complex of species with at least three known perfect stages: *Arthroderma quadrifidum* Dawson and Gentles 1961; *A. lenticularum* Pore, Taso, and Plunkett 1965; and *A. insingulare* Padhye and Carmichael 1972.[116,116a] *T. georgiae* Varsavsky and Ajello 1964 has the perfect stage, *Arthroderma ciferrii* Varsavsky and Ajello 1964. *T. gloriae* Ajello 1967 is the imperfect form of *Arthroderma gloriae* Ajello 1967, and was isolated from soil in the southwestern United States. *T. vanbreuseghemii* Rioux, Jarry, and Juminer 1964 was isolated from soil in Tunisia. Its perfect stage is *Athroderma gertleri* Bohme 1967; it has now been found to be of worldwide distribution in soils.

Epidermophyton Sabouraud 1910

This is a monotypic genus characterized by an expanded clavate macroaleuriospore that has a rounded or blunt terminus and smooth walls. Microaleuriospores are absent. Hair is not known to be invaded.

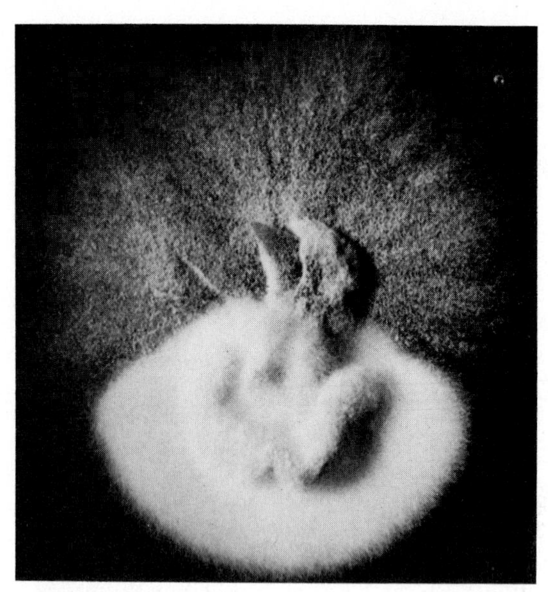

Figure 7–55. FS B 13. *E. floccosum* showing gently folding, suedelike texture. The color is chartreuse. A white pleomorphic overgrowth has begun to cover the surface.

Epidermophyton floccosum (Harz) Langeron and Milochevitch 1930

Synonymy. *Trichothecium floccosum* Harz 1870, *Trichophyton interdigitale* Sabouraud 1905, *Trichophyton inguinale* Sabouraud 1907, *Trichophyton crurus* Castellani 1908, *Epidermophyton inguinale* Sabouraud 1910, *Epidermophyton cruris* Castellani and Chalmers 1910, *Epidermophyton plicarum* Nicolau 1913, *Epidermophyton elypeisoforme* MacCarthy 1925, *Acrothecium floccosum* Harz 1871, *Blastotrichum floccosum* Berlese and Boglino 1886, *Closteriosporia inguinalis* Grigorakis 1925, *Fusoma cruris* Vuillemin 1929.

Colony Morphology (Fig. 7–55 FS B 13). The growth of the thallus is slow and fre-

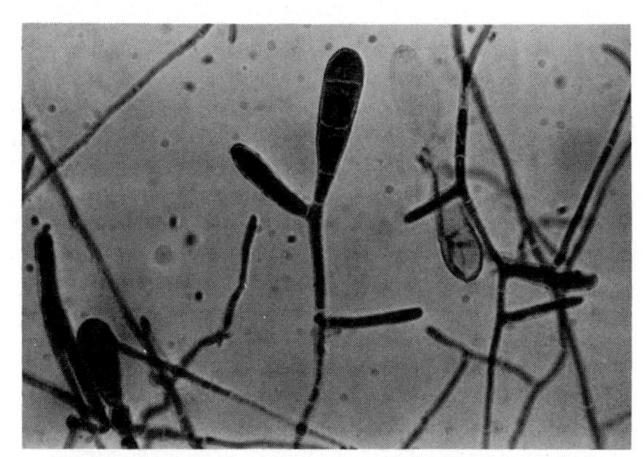

Figure 7–56. *E. floccosum.* Beaver tail–shaped macroaleuriospore. (From Rippon, J. W. *In* Burrows, W. 1973. *Textbook of Microbiology.* 20th ed. Philadelphia, W. B. Saunders Company, p. 709.)

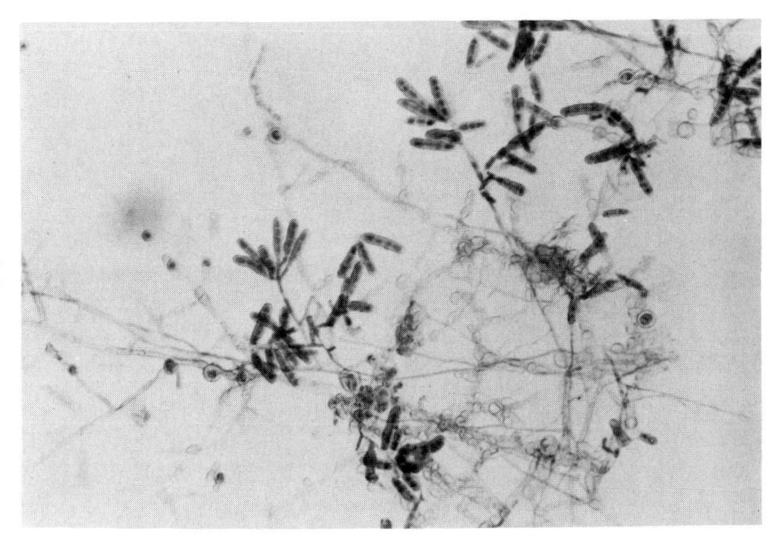

Figure 7–57. *E. floccosum.* Variation in shape of macroaleuriospores and abundant chlamydospores. No microaleuriospores are produced.

quently grainy, lumpy, and sparse on initial isolation. When developed, the colony is gently folded, is fuzzy or suedelike in texture, and is characteristically olive, khaki, or chartreuse in color. Yellow, yellow-brown, and white variants are reported. The underside is colorless to deep yellow-brown.

Microscopic Morphology. The characteristic macroaleuriospores are usually abundant. They are 7 to 12 × 20 to 40 μ in size, beaver tail–shaped or clavate, and have smooth thin walls (Fig. 7–56). These spores often occur in groups. Rarely, pits and knobs have been seen on them. Microaleuriospores are absent. Chlamydospores are frequently very abundant, particularly as the colony ages (Fig. 7–57). They are rounded and thick-walled. Macroaleuriospores sometimes transform into chlamydospores. Racquet mycelium, spirals, and nodular bodies may be found. No specific growth requirements have been reported. Infection in animals has not been described, and experimental infection in laboratory animals is rarely successful.

REFERENCES

1. Ainsworth, G. C., and P. K. C. Austwick. 1959. *Fungal Diseases of Animals.* Bucks, England, Commonwealth Agriculture Bureau Farnham Royal.
2. Ajello, L. 1960. Geographic distribution and prevalence of the dermatophytes. Ann. N.Y. Acad. Sci., *89*:30–38.
3. Ajello, L. 1962. Present day concepts of the dermatophytes. Mycopath., *17*:315–324.
4. Ajello, L. 1968. A taxonomic review of the dermatophytes and related species. Sabouraudia, *6*:147–159.
5. Ajello, L., and L. K. Georg. 1957. *In vitro* hair cultures for differentiating between atypical isolates of *T. mentagrophytes* and *T. rubrum.* Mycopath., *8*:3–17.
6. Alberts, R. E., and A. R. Omram. 1966. Follow-up study of patients treated by x-ray epilation for tinea capitis. Am. J. Public Health, *56*:2114–2120.

6a. Allen, A. M., J. H. Reinhardt, et al. 1973. Griseofulvin in the prevention of experimental human dermatophytosis. Arch. Dermatol., *108*:233–236.

7. Alteras, I. 1969. The therapeutic effect of vaccines extracted from keratinophilic soil fungi. Mycopath., *38*:145–150.
8. Alteras, I., I. Cojocarn, et al. 1967. Occupational mycotic infections in swimming pools and public baths. Derm.-Vener., *12*:409–414.
9. Arêa Leão, A. E., and M. Goto. 1950. O tokelau entré os Indios do Brasil. o hospital (Rio), *37*:225–240.
10. Arauysky, A. N. 1964. Rare mycological findings in pathological material. Mycopath., *22*: 185–200.
11. Baer, R. L., and S. Rosenthal. 1966. The biology of fungus infections of the feet. J.A.M.A., *197*:1017–1020.
12. Baer, R. L., S. A. Rosenthal, et al. 1955. Survival of dermatophytes applied to the feet. J. Invest. Dermatol., *24*:619–622.
13. Baer, R., S. A. Rosenthal, et al. 1956. Experimental investigations on mechanisms producing acute dermatophytosis of feet. J.A.M.A., *160*:184–190.
14. Baker, N. F., and D. W. Davis. 1954. Spray application for treatment of ringworm in cattle. Vet. Med., *49*:275–276.
15. Barker, S. A., and M. D. Trotter. 1960. Isolation of purified trichophytin. Nature (Lond.). *188*:232–233.
16. Battistini, F. Personal communication.

17. Baxter, M. 1965. The use of ink-blue in the identification of dermatophytes. J. Invest. Dermatol., *44*:23–25.
18. Beare, J. M., J. C. Gentles, et al. 1972. Chapter 25 in Rook, A., et al. (eds.), *Textbook of Dermatology*. Philadelphia, F. A. Davis.
19. Becker, S. W., and K. B. Muir. 1929. Tinea amiantacea. Arch. Dermatol., *20*:45–53.
20. Benedek, T. 1967. On the ancestral form of true organ of fructification in the saprophytic stage of the dermatophytes of the faviform group: *Favomicrosporon pinettii* sp. nov. and its perfect form: *Anixiopsis stercoraria* (Hansen) Hansen 1897. Mycopath., *31*:81–143.
21. Beneke, E. S., and A. Rogers. 1971. *Medical Mycology Manual*. Minneapolis, Burgess.
22. Blank, F. 1962. Human favus in Quebec. Dermatologica, *125*:369–381.
23. Blank, H. (ed.). 1960. Griseofulvin and dermatomycoses. An International Symposium sponsored by The University of Miami. Arch. Dermatol., *81*:649–882.
24. Blank, H., F. J. Roth, et al. 1959. The treatment of dermatomycoses with orally administered griseofulvin. Arch. Dermatol., *79*:259–266.
25. Blank, H., S. Sagmi, et al. 1959. The pathogenesis of superficial fungus infections in culture of human skin. Arch. Dermatol., *79*:524–535.
26. Blank, H., and J. G. Smith. 1960. Widespread *Trichophyton rubrum* infection treated with griseofulvin. Arch. Dermatol., *81*:779–789.
27. Blank, H., D. Taplin, et al. 1969. Cutaneous *Trichophyton metagrophytes* infections in Viet Nam. Arch. Dermatol., *99*:135–144.
28. Brocq-Rousseau, D., D. A. Urbain, et al. 1926. Sur l'electivite cutanee des teignes animales, quel que soit leur voie d'introduction dans l'organisme. C. R. Soc. Biol. (Paris), *95*: 966–967.
29. Calnan, C. D. 1958. *Trichophyton rubrum* infections, In R. W. Riddell and G. T. Stewart (eds.), *Fungus Diseases and Their Treatment*. London, Butterworth and Co.
29a. Charney, P., V. M. Torres, et al. 1973. Tolnaftate as a prophylactic agent for tinea pedis. Int. J. Dermatol., *12*:179–185.
30. Chermsirivathana, S., and P. Boonsri. 1961. A case of Tinea imbricata (Hanuman ringworm) treated with fulvicin. Aust. J. Dermatol., *6*:63–66.
31. Cochrane, T., and A. Tullett. 1959. Griseofulvin treatment of acute cattle ringworm infections in man. Br. Med. J., *2*:286–287.
31a. Coiscou, W., A. G. 1967. Algunos aspectas clinicos y mitologices de las tinas del cuero ca bellundo en la Republica Dominica. Rev. Dominicana Dermatol., *1*:4–12.
32. Conant, N. F., D. T. Smith, et al. 1954. *Manual of Clinical Mycology*, 2d ed. Philadelphia, W. B. Saunders.
33. Cox, W. A., and J. A. Moore. 1968. Experimental *Trichophyton verrucosum* infections in laboratory animals. J. Comp. Pathol., *78*:35–41.
34. Cruickshank, C. N. D., M. D. Trotter, et al. 1960. Studies on trichophytin sensitivity. J. Invest. Dermatol., *35*:219–223.
35. Dawson, C. O., and J. C. Gentles, 1959. The perfect stage of *Keratinomyces ajelloi*. Nature (Lond.), *183*:1345–1346.
36. Delameter, E. D. 1941. Experimental studies with dermatophytes. III. Development and duration of immunity and hypersensitivity in guinea pigs. J. Invest. Dermatol., *4*:143–158.
37. Desai, S. C., M. L. A. Bhat, et al. 1963. Biology of *Trichophyton rubrum* infections. Indian J. Med. Res., *51*:233–243.
38. Dodge, C. W. 1935. *Medical Mycology*. St. Louis, C. V. Mosby.
39. Edgson, F. A. 1970. Mass treatment of ringworm of cattle with griseofulvin mycelium. Vet. Rec., *86*:58–59.
39a. Elton, R. F., A. H. Mehregan, et al. 1973. Tinea faciei. Int. J. Dermatol., *12*:394–396.
40. English, M. P. 1957. *Trichophyton rubrum* infection in families. Br. Med. J., *1*:744–746.
41. English, M. P. 1962. Some controversial aspects of tinea pedis. Br. J. Dermatol., *74*:50–56.
42. English, M. P. 1962. The saprophytic growth of keratinophilic fungi on keratin. Sabouraudia, *2*:115–120.
43. English, M. P. 1969. Tinea pedis as a public health problem. Br. J. Dermatol., *81*:705–707.
44. English, M. P., and M. D. Gibson. 1959. Studies in epidemiology of tinea pedis I and II. Br. Med. J., *1*:1442–1445, and 1446–1448.
45. Fisher, B. K., J. G. Smith, et al. 1961. Verrucous epidermophytosis, its response and resistance to griseofulvin. Arch. Dermatol., *84*:375–380.
45a. Fox, T. 1870. Tinea circinata of the hand. Br. Med. J. July 30, p. 116.
46. Franks, A. G., E. M. Rosenbaum, et al. 1952. *T. sulfureum* causing erythema nodosum and multiple kerion formation. Arch. Dermatol., *65*:95–96.
47. Friedman, L., and V. J. Derbes. 1960. The question of immunity in ringworm infections. Ann. N.Y. Acad. Sci., *89*:178–183.
48. Gentles, J. C. 1958. Experimental ringworm in guinea pigs; oral treatment with griseofulvin. Nature (Lond.), *182*:476–477.
49. Gentles, J. C., and E. G. V. Evans. 1970. Infection of the feet and nails with *Hendersonela toruloidea*. Sabouraudia, *8*:72–75.
50. Gentles, J. C., and J. C. Holmes. 1957. Foot ringworm in coal miners. Br. J. Ind. Med., *14*:22–29.
51. Georg, L. K. 1952. *Trichophyton tonsurans* ringworm. A new public health problem. Public Health Rep., *67*:53–56.
52. Georg, L. K. 1954. The relationship between the downy and granular forms of *Trichophyton mentagrophytes*. J. Invest. Dermatol., *23*: 123–141.
53. Georg, L. K., C. S. Roberts, et al. 1957. *Trichophyton mentagrophytes* infection in dogs and cats. J. Am. Vet. Med. Assoc., *130*:427–432.
54. Georg, L. K. 1957. Dermatophytes: New methods in classification. Atlanta, Ga., U.S. Public Health Service.
55. Georg, L. K. 1960. Epidemiology of the dermatophytoses sources of infection, modes of transmission and epidemicity. Ann. N.Y. Acad. of Sci., *89*:69–77.
56. Georg, L. K. 1960. *Animal Ringworm in Public Health*. U.S. Dept. HEW. Public Health Series publication No. 727.

57. Georg, L. K., and L. B. Camp. 1957. Routine nutritional tests for the identification of dermatophytes. J. Bacteriol., *74*:477–490.
58. Ginther, O. J., et al., 1964. First American isolation of *Trichophyton mentagrophytes* in swine. Vet. Med. Small Anim. Clin., *59*:1038–1042.
59. Ginther, O. J., and L. Ajello. 1965. Prevalence of *Microsporum nanum* infection in swine. J. Am. Vet. Med. Assoc., *146*:361–365.
60. Ginther, O. J., G. R. Bubash, et al. 1964. *Microsporum nanum* infection in swine. Vet. Med. Small Anim. Clin., *59*:79–84.
60a. Gip, L. 1964. Isolation of *Trichophyton gallinae* from two patients with tinea cruris. Acta Derm.-venereal. (Stockh.), *44*:251–254.
61. Goldfarb, N. J., and F. Hermann. 1956. A study of pH changes by molds in culture media. J. Invest. Dermatol., *27*:193–201.
62. Goldstein, N., and G. Woodward. 1969, Surgery for tinea pedis. Syndactylization or amputation of toes for chronic, severe fungus infections. Arch. Dermatol., *99*:701–704.
63. Gordon, M. A., and G. N. Little. 1968. *Trichophyton (Microsporum?) gallinae* ringworm in a monkey. Sabouraudia, *6*:207–212.
64. Gray, H. R., J. E. Dalton, et al. 1960. *Trichophyton tonsurans* infection of the scalp in central Indiana State Med. Assoc., *53*:75–80.
64a. Gruby, D. 1841. Sur les mycodermes que constituent la teigne faveus. C. R. Acad. Sci. (Paris), *13*:309–312.
65. Hagen, K. W. 1969. Ringworm in domestic rabbits: Oral treatment with griseofulvin. Lab. Anim. Care., *19*:635–638.
66. Hall, F. R. 1966. Ringworm contracted from cattle in western New York state. Arch. Dermatol., *94*:35–37.
67. Hegyi, E., et al. 1967. The specificity of mycins and their importance in diagnosis of dermatomycosis. Allerg. Asthma (Leipz.), *13*: 164–176.
68. Hetherington, G., R. C. Freeman, et al. 1969. Intracellular location of hyphae in experimental dermatomycosis. Experientia. *25*: 889–890.
69. Hiddleston, W. A. 1970. Antifungal activity of *Penicillium griseofulvin* mycelium. Vet. Rec., *86*:75–76.
70. Hildick-Smith, G., H. Blank, et al. 1963. *Fungus Diseases and Their Treatment.* Boston, Little, Brown and Company.
71. Holmes, J. G. 1958. Tinea pedis in miners, In R. W. Riddell and G. T. Stewart (eds.), *Fungous Diseases and Their Treatment.* Butterworth and Co., London.
72. Huppert, M., and E. L. Keeney. 1959. Immunization against superficial fungous infections. J. Invest. Dermatol., *32*:15–19.
73. Huppert, M. 1962. Immunization against superficial fungous infection. Chapter 17 in C. Dalldorf (ed.), *Fungi and Fungous Diseases.* Springfield, Charles C Thomas.
74. Ito, Y. 1965. On the immunologically active substances of the dermatophytes. J. Invest. Dermatol., *45*:285–294.
75. Jeremiasse, H. P. 1960. Treatment of nail infections with griseofulvin combined with abrasion. Trans. St. John's Hosp. Dermatol. Soc., *45*:92–93.
76. Jessner, M. 1922. Über eine neue Form con Nagelmykosen leukonychia trichophytica. Arch. Dermatol. Syph. (Berlin), *141*:1–8.
77. Jewell, E. W. 1970. *Trichophyton rubrum* onychomycosis in a four month old infant. Cutis, October, p. 1121.
77a. Jones, H. E., J. H. Reinhardt, et al. 1973. A clinical, mycological and immunological survey for dermatophytosis. Arch. Dermatol., *108*:61–65.
78. Kaplan, W. 1967. Epidemiology and public health significance of ringworm in animals. Arch. Dermatol., *96*:404–408.
79. Kaplan, W., and L. Ajello. 1960. Therapy of spontaneous ringworm in cats with orally administered griseofulvin. Arch. Dermatol., *81*:714–723.
80. Kaplan, W., and L. K. Georg. 1957. A device to aid in the development of mycotic and other skin infections in laboratory animals. Mycologia, *49*:604–605.
81. Kaplan, W., L. K. Georg, et al. 1958. Recent developments in animal ringworm and their public health implications. Ann. N.Y. Acad. Sci., *70*:636–649.
82. Kaplan, W., and R. H. Gump. 1958. Ringworm in a dog caused by *Trichophyton rubrum.* Vet. Med., *53*:139–142.
83. Kaplan, W., J. L. Hopping, et al. 1957. Ringworm in horses caused by *Microsporum gypseum.* J. Am. Vet. Med. Assoc., *131*:329–332.
84. Kaplan, W., and M. S. Ivens. 1961. Observations on the seasonal variations in incidence of ringworm in dogs and cats in the United States. Sabouraudia, *1*:91–102.
85. Kaufman-Wolf, M. 1914. Über Pilzerkrankungen der Hände und Füsse. Dermat. Z. (Dermatologica), *21*:385–396.
86. Kielstein, P. 1966. Vergleichende antigen analytishe Untersuchungen einiger Dermatophyten und Schimmel pilze mit Hilfe des Agargelprazipitationtesten. Arch. Exp. Vet. Med., *20*:523–540.
87. Klatt, P. 1969. Treatment of bovine ringworm with two new antimycotics. Blue Book Vet. Prof., *16*:23–26.
88. Kligman, A. M. 1952. The pathogenesis of tinea capitis due to *Microsporum audouinii* and *Microsporum canis.* J. Invest. Dermatol., *18*:231–246.
89. Kligman, A. M. 1955. Tinea capitis due to *M. audouinii* and *M. canis.* Arch. Dermatol., *71*:313–348.
90. Kligman, A. M. 1956. Pathophysiology of ringworm infections in animals with skin cycles. J. Invest. Dermatol., *27*:171–185.
91. La Touche, C. J. 1967. Scrotal dermatophytosis. Br. J. Dermatol., *79*:339–344.
92. Lavalle, P. 1966. Tinea pedis in Mexico. Dermatologia (Mex.), *10*:313–329.
93. Leavell, U. W., E. Tucker, et al. 1966. Blue dot infection of the scalp in two brothers. J. Ky. Med. Assoc., *64*:1107–1110.
94. Lenhart, K. 1970. Griseofulvin-resistant mutants in dermatophytes. Mykosen *13*:139–144.
95. Lepper, A. W. D. 1969. Immunological aspects of dermatomycoses in animals and man. Rev. Med. Vet. Mycol. *6*:435–442.
96. Longbottom, J. L., and J. Pepys. 1964. Pulmonary

aspergillosis: Diagnostic and immunological significance of antigens and C substance in *Aspergillus fumigatus*. J. Pathol. Bacteriol., *88*:141–151.

97. Lorincz, A. L., J. O. Priestly, et al. 1958. Evidence for a humoral mechanism which prevents growth of dermatophytes. J. Invest. Dermatol., *31*:15–17.

98. Lorincz, A. L., and S. H. Sun. 1963. Dermatophyte viability at modestly raised temperatures. Arch. Dermatol., *88*:393–402.

99. MacLennan, R. 1960. A trial of griseofulvin in tinea imbricata. Trans. St. John's Hosp. Dermatol. Soc., *45*:99–100.

99a. Margarot, J., and P. Deveze. 1925. Aspect de quelques dermatosea on lumiere athaparo-violetle note preliminaire. Bull. Soc. Med. Biol. Montpellier, *6*:375–378.

100. Marples, M. J. 1960. Microbiological studies in Western Samoa. II. The isolation of yeast-like organisms from the mouth with a note of some dermatophytes isolated from Tinea. Trans. R. Soc. Trop. Med. Hyg., *54*:166–170.

101. Marples, M. J., and J. M. B. Smith. 1960. The hedgehog as a source of human ringworm. Nature (Lond.), *188*:867–868.

102. Martin, A. R. 1958. The systemic treatment of dermatophytoses. Vet. Rec., *70*:1232.

103. McPherson, E. A. 1957. The influence of physical factors on dermatomycosis in domestic animals. Vet. Rec., *69*:1010–1013.

104. Menges, R. W., and L. K. Georg. 1957. Survey of animal ringworm in the United States. Public Health Rep., *72*:503–509.

105. Menges, R. W., and L. K. Georg. 1957. Canine ringworm caused by *M. gypseum*. Cornell Vet., *47*:90–100.

106. Menges, R. W., G. J. Lane, et al. 1957. Ringworm in wild animals in southwestern Georgia. Am. J. Vet. Res., *18*:672–677.

107. Mikhail, G. R. 1970. *Trichophyton rubrum* granuloma. Int. J. Dermatol., *9*:41–46.

108. Mitchell, J. H. 1951. Ringworm of hands and feet. J.A.M.A., *146*:541–546.

109. Mullins, J. F., C. J. Willis, et al. 1966. *Microsporum nanum*. Arch. Dermatol., *94*:300–303.

109a. Nannizzi, A. 1926. Ricerche sui rapporti mortologici e biologici tra Gymnoascaere e Dermatomice ti. Ann. Mycologici, *24*:85–129.

110. Nelson, L. M., and K. J. McNiece. 1959. Recurrent Cushing's syndrome in *T. rubrum* infection. Arch. Dermatol., *80*:700–704.

111. Neuman, M., and N. Platzner. 1968. Treatment of bovine ringworm with thiabendazole. Refuah Vet., *25*:40–46. (Vet. Inst. Bet Dagan)

112. Neves, H. 1960. Mycological study of 519 cases of ringworm infections in Portugal. Mycopathologia, *13*:121–132.

113. Oliveira, H., R. Trincao, et al. 1960. Tricofitose cutanea general izada com infeccao sistemica por *Trichophyton violaceum*. J. Med. (Pôrta), *41*:629–642.

114. Ormsby, O. S., and J. H. Mitchell. 1916. Ringworm of hands and feet. J.A.M.A., *67*:711–716.

115. Padhye, A. A., and J. W. Carmichael. 1971. The Genus *Arthroderma* Berkeley. Can. J. Bot., *49*:1525–1540.

116. Padhye, A. A., and J. W. Carmichael. 1972. *Arthroderma insingulare* sp. nov. another gymnoascaeceous state of the *Trichophyton terrestre* complex. Sabouraudia, *10*:47–51.

116a. Padhye, A. A., and J. W. Carmichael. 1973. Mating reactions in the *Trichophyton terrestre* complex. Sabouraudia, *11*:64–69.

116b. Padhyde, A. A., F. Blank, et al. 1973. *Microsporum persicolor* infection in the United States. Arch. Dermatol., *108*:561–562.

117. Pardo-Castello, V., and O. A. Pardo. 1960. *Diseases of the Nail*. 3rd ed. Charles C Thomas, Springfield, Illinois.

118. Pankova Ya, V. 1969. On the clinical picture of rubrophytosis in Itsenko-Cushings disease. Vestn. Dermatol. Venerol., *43*:63–69.

118a. Pellizzari, C. 1888. Richorche sul *Trychophyton tonsurans*. G. Ital. Mal. Vener., *29*:8–40.

119. Prazak, G., J. S. Ferguson, et al. 1960. Treatment of tinea pedis with griseofulvin. Arch. Dermatol., *81*:821–826.

120. Pushkarenko, V. I., G. D. Pushkarenko, et al. 1969. Osobennosti aminokislotnogo sostava pota so stop u litz s klinicheski vyrazhennoi epidermofitiei. Vestn. Dermatol. Venerol., *49*:47–49.

121. Raubitschcek, F., and R. Moaz. 1957. Invasion of nails *in vitro* by certain dermatophytes. J. Invest. Dermatol., *28*:261–268.

122. Ravant, P., and H. Rabeau. 1921. Sur une forme speciale de trichophytic ungueale. Ann. Dermatol. Syphiligr. (Paris), *2*:362–364.

123. Rebell, G., and D. Taplin. 1970. Dermatophytes—Their Recognition and Identification. Coral Gables, Fla. University of Miami Press.

124a. Remak, R. 1840. Zur Kenntnis von der planet lichen natur der porrigo lupinosa. Med. Zeitung., *9*:73–74.

124. Reiss, F. 1944. Successful inoculations of animals with *Trichophyton purpureum*. Arch. Dermatol., *49*:242–248.

125. Reyes, A. C., and L. Friedman. 1966. Concerning the specificity of dermatophyte reacting antibody in human and experimental animal sera. J. Invest. Dermatol., *47*:27–34.

126. Rioux, J. A., D. M. Jarry, et al. 1966. Ctenomyces, Arthroderma, ou Trichophyton? Fin d'une controverse et nouvelle acception du terme de dermatophyte. Ann. Parasitol. Hum. Comp., *41*:523–534.

127. Rippon, J. W. 1967. Elastase: Production by ringworm fungi. Science, *157*:947.

128. Rippon, J. W. 1968. Collagenase from *Trichophyton schoenleinii*. J. Bacteriol. *95*:43–46.

129. Rippon, J. W., and D. Garber. 1969. Dermatophyte infection as a function of mating type and associated enzymes. J. Invest. Dermatol., *53*:445–448.

130. Rippon, J. W., and M. Medenica. 1964. Isolation of *Trichophyton soudanense* in the United States. Sabouraudia, *3*:301–304.

131. Rippon, J. W., and G. H. Scherr. 1959. Induced dimorphism in dermatophytes. Mycologia, *51*:902–914.

132. Rippon, J. W., and F. D. Malkinson. 1968. *Trichophyton simii* infection in the United States. Arch. Dermatol., *98*:615–619.

133. Rippon, J., F. C. Lee, et al. 1970. Dermatophyte infection caused by *Aphanoascus fulvescens*. Arch. Dermatol., *102*:552–555.

134. Rippon, J. W., and L. J. Lebeau. 1965. Germination of *Microsporum audouinii* from infected hairs. Mycopath., *26*:273–288.
135. Robinson, H. M., and R. C. V. Robinson. 1959. Griseofulvin, an antifungal antibiotic. Int. Rec. Med., *172*:737–742.
136. Rosenthal, T. 1960. Perspectives in ringworm of the scalp. Arch. Dermatol., *82*:851–856.
137. Rosenthal, S. A., R. L. Baer, et al. 1956. Studies on the dissemination of fungi from the feet of subjects with and without fungus disease of the feet. J. Invest. Dermatol., *26*:41–51.
138. Rosenthal, S., D. Fisher, et al. 1958. A localized outbreak in New York of tinea capitis due to *Trichophyton violaceum*. Arch. Dermatol., *78*:689–691.
139. Roth, F. J., C. C. Boyd, et al. 1959. An evaluation of the fungistatic activity of serum. J. Invest. Dermatol., *32*:549–556.
140. Rothman, S., G. Knox, et al. 1957. Tinea pedis as a source of infection in the family. Arch. Dermatol., *75*:270–271.
141. Rothman, S., A. Smiljanic, et al. 1947. The spontaneous cure of tinea capitis in puberty. J. Invest. Dermatol., *8*:81–98.
142. Russel, B., W. Frain-Bell, et al. 1960. Chronic ringworm of the skin and nails treated with griseofulvin. Report of a therapeutic trial. Lancet, *1*:1141–1147.
143. Sanderson, P. H., and J. C. Sloper. 1953, Skin disease in the British Army in S. E. Asia. III. Relationship between mycotic infections of body and of feet. Br. J. Dermatol., *65*:362–372.
144. Schmitt, J. A., and R. G. Miller. 1967. Variation in susceptibility to experimental dermatomycosis in genetic strains of mice. Mycopath., *32*:306–312.
145. Schulz, J. A., and R. Lippman. 1968. Eradication of bovine ringworm by gricin-vet. Mh. Veterinaermed., *23*:531–534.
145a. Schoenlein, J. L. 1839. Zur pathologie der impetigines. Arch. Anat. Phys. Wiss. Med., p. 82.
146. Sellers, K. C., W. B. V. Sinclair, et al. 1956. Preliminary observations on natural and experimental ringworm in cattle. Vet. Rec., *68*:729–732.
147. Shanks, S. C. 1967. Vale epilatio. X-ray epilation at Goldie Leigh Hospital Woolwich (1922–58). Br. J. Dermatol., *79*:237–238.
148. Shapiro, A. L., and S. Rothman. 1945. Undecylenic acid in the treatment of dermatomycosis. Arch. Dermatol. Syph., *52*:166–171.
149. Simic, L., and S. Perisic. 1969. Microsporosis caused by *M. canis* in humans and a dog transmitted by an imported cat. Mykosen, *12*:699–703.
150. Sloper, J. C. 1955. A study of experimental human infection due to *T. rubrum, T. mentagrophytes* and *E. floccosum*. J. Invest. Dermatol., *25*:21–28.
151. Smith, J. M. B., and M. J. Marples. 1964. Ringworm in the Solomon Islands. Trans. R. Soc. Trop. Med. Hyg., *58*:63–67.
152. Smith, J. M. B., F. M. Rush-Munro, et al. 1969. Animals as a reservoir of human ringworm in New Zealand. Aust. J. Dermatol., *10*: 169–182.
153. Stevenson. C. J., and N. Djavahiszwili. 1961. Chronic ringworm of the nails: Long-term treatment with griseofulvin. Lancet, *1*: 373–374.
154. Stockdale, P. M., D. W. R. Mackenzie, et al. 1965. *Arthroderma simii* sp. nov. The perfect state of *Trichophyton simii* (Pinoy) comb. nov. Sabouraudia, *4*:112–123.
155. Strauss, J. S., and A. M. Kligman. 1957. An experimental study of tinea pedis and onychomycosis of the foot. Arch. Dermatol., *76*:70–79.
156. Stuka, A. J., and R. Burrell. 1967. Factors affecting the antigenicity of *Trichophyton rubrum*. J. Bacteriol., *94*:914–915.
157. Sulzberger, M. B., and R. L. Baer. 1954–1955. Some recent advances in dermatologic mycology. Pages 7–33 in Yearbook of Dermatology and Syphilology. Chicago, Year Book Medical Publishers.
158. Taplin, D., A. M. Allen, et al. 1970. Experience with a new indicator medium (DTM) for isolation of dermatophyte fungi. Proc. Int. Symp. Mycoses. PAHO No. 205, 1970.
159. Taplin, D., N. Zaias, et al. 1969. Isolation and recognition of dermatophytes on a new medium. Arch. Dermatol., *99*:203–209.
160. Taplin, D., and P. Mertz. Human model of dermatophytosis (*Trichophyton rubrum*), in press.
161. Tewari, R. P. 1969. *Trichophyton simii* infections of chickens, dogs and man in India. Mycopath., *39*:293–298.
162. Tritsmans, E., and D. Sano. 1968. Microsporie een zeldzamheid. Arch. Belg. Dermatol. Syphiligr., *24*:13–17.
162a. Torres, G., and L. K. Georg. 1956. A human case of *Trichophyton gallinae* infection. Arch. Dermatol., *74*:191–195.
163. Vanbreuseghem, R. 1949. La culture des dermatophytes in vitro sur des cheveux isoles. Ann. Parasitol., *24*:559–573.
164. Vanbreuseghem, R. 1966. Guide practique de mycologie medicale et veterinarie. Masson et Cie, Paris.
165. Vanbreuseghem, R. 1968. *Trichophyton soudanense* in and outside Africa. Br. J. Dermatol., *80*:140–148.
166. Vanbreuseghem, R., and S. A. Rosenthal. 1961. *Trichophyton kuryangei* sp. noveau Dermatophyte Africaine. Ann. Parasitol. Hum. Comp., *36*:797–803.
167. Varadi, D. P., and J. W. Rippon. 1967. Scalp infection of triple etiology. Arch. Dermatol., *95*:299–301.
167a. Visset, M. F., and C. Vermeil. 1973. Contribution á la connaissance des fungi keratinophiles de l'onest de la France. V. Etude au microscope electronique á balayage. Mycopath., *49*:89–100.
168. Visset, M. 1972. Les formes conidiennes du complexe Microsporum gypseum deservers en microscopie electronique a balayage. Sabouraudia, *10*:191–192.
169. Walker, J. 1955. Possible infection of man by indirect transmission of *Trichophyton discoides*. Br. Med. J., *2*:1430–1433.
169a. Weitzman, I., M. A. Gordon, et al. 1971. Determination of the perfect state, mating complex and elastase of *Microsporum gypseum* complex. J. Invest. Dermatol., *57*:278–282.
170. Wenk, P. 1962. Causes of spontaneous recovery

from trichophytosis in guinea pigs. Z. Tropenmed. Parasitol., *13*:201–215.
171. Williams, D. I., R. H. Marten, et al. 1958. Oral treatment of ringworm with griseofulvin. Lancet, *2*:1212.
172. Wilson. J. W., D. A. Plunkett, et al. 1954. Nodular granulomatous perifolliculitis due to *T. rubrum.* Arch. Dermatol., *69*:258–277.
173. Wolf, F. T., L. A. Jones, et al. 1958. Fluorescent pigment of *Microsporum.* Nature (Lond.), *182*:475.
174. Yu, R. J., S. A. Harmon et al. 1969. Hair digestion by a keratinase of *Trichophyton mentagrophytes.* J. Invest. Dermatol., *53*:166–171.
175. Zaias, N. 1966. Superficial white onychomycosis. Sabouraudia, *5*:99–103.
176. Zaias, N. 1969. Fungi in toe nails. J. Invest. Dermatol., *53*:140–142.
177. Zaias, N. 1972. Onychomycosis. Arch. Dermatol., *105*:263–274.
178. Zussman, R. A., I. Lyon, et al. 1960. Melanoid pigment production in a strain of *T. rubrum.* J. Bacteriol., *80*:708–713.

THE PATHOGENIC YEASTS

Chapter 8

CANDIDOSIS

The terms "yeast" and "yeastlike" are vernacular for a unicellular fungal organism which reproduces by budding. Such a definition is generally recognized as inadequate, in part because some yeasts reproduce by fission, in part because many produce mycelium or pseudomycelium under certain environmental and nutritional conditions, and in part because filamentous fungi (Hyphomycetes) may exist in a unicellular, yeastlike form which reproduces by budding, viz. oidia described in Chapter 5 and the parasitic stage of several pathogenic fungi. On the basis of sexual spore formation, some yeasts are Ascomycetes, others are Basidiomycetes, and still others as yet have not been shown to have a sexual stage and are grouped together as Fungi Imperfecti (Deuteromycetes). Clearly, then, the term "yeast" is of no taxonomic significance and is useful only to describe a morphologic form of a fungus.

The other sections of this text have been grouped primarily on a clinical basis: cutaneous infections, systemic infections, and so forth. It may appear inconsistent to separate a group on purely morphologic grounds, but the most important and most frequently isolated pathogenic yeast, *Candida albicans*, is so protean in its clinical manifestations that it defies categorization on the basis of site of infection. Diseases caused by *Cryptococcus* and *Torulopsis* are less frequently encountered than candidosis. Although the former are usually systemic, cutaneous and mucocutaneous tissues may also be involved. The isolation, characterization, and identification of the yeasts are quite different from those of the mycelial fungi. Morphology is less important, and physiologic characteristics are of greater value for the identification of the yeasts. For their speciation, emphasis is placed on carbohydrate fermentation and assimilation, nitrogen utilization, production of extracellular substances such as capsules, and production of enzymes. The techniques used in zymology, therefore, are similar to those used for the identification of bacteria. For these reasons, the pathogenic yeasts will be considered in a separate section.

The yeasts belong in three classes of fungi: the basidiospore-forming yeasts of the Sporobolomycetaceae of order Ustilaginales (*Leukosporidium* and *Rhodosporidium*), the ascospore-forming yeasts in the family Endomycetaceae of the class Ascomycetes, and the asporogenous yeasts in the form-family Cryptococcaceae in the form-class Deuteromycetes (Fungi Imperfecti). The last group contains most of the human pathogenic species.

The industrial yeasts are, perhaps, the most familiar of the organisms. *Saccharomyces cerevisiae*, an ascomycete, is the common brewers' and bakers' yeast. It occurs as two types: top yeasts which cause a vigorous evolution of carbon dioxide and are found in the froth on the surface of the fermenting mixture, and bottom yeasts which sink to the bottom. Bread yeasts are usually strains of top yeasts. Beer and ale are produced from either type, depending on the variety of beverage. Another species of the genus, *S. ellipsoideus*, is the common wine yeast, occurring naturally on grapes and in the soil of vineyards. Its varieties are named for the various categories of wine which they produce. All these organisms are "perfect" yeasts, the cell body becoming an ascus after sexual union. Other yeasts are lactose fermenters, and are associated with the preparation of fermented milk beverages, such as Kefir and Koumiss, food staples commonly used in southern Europe and Asia. The commonest yeasts encountered as contaminants of bacterial cultures, growing on foods, air borne, or as transient flora of the human skin, are the asporogenous genera *Rhodotorula, Torulopsis, Candida,* and *Cryptococcus.* Besides the frequently encountered pathogenic species of the latter two genera, the saprophytic species, including several *Rhodotorula,* are occasionally isolated from human disease and represent opportunistic infectious agents. Even beer yeast has been encountered in lung infections.

In view of the ubiquitous distribution of yeasts, not only in air, dust, and soil but also on the surface of the body and in the mouth, intestinal tract, and vagina, it is not surprising that these forms have been found in a variety of pathologic processes. A great number of species have been described in this connection, most of them inadequately. In many instances, the yeast had no etiologic relation to the disease, and in others, the same yeast was repeatedly described as a new species, subsequently giving rise to a long list of synonymous names. For these reasons, a very long list of "pathogenic" yeasts has accumulated. Critical examination and consideration has now made it clear that only a few species of yeasts are actually pathogenic for man and animals. The yeasts of medical importance are listed below, along with the generally accepted classification and basic differential characteristics (Table 8–1).

Though most of the above genera form a fairly well-defined group, the genera *Torulopsis* and *Candida* are extremely heterogeneous. Even the separation of the two genera is artificial and tenuous. A number of species of both genera have been shown to be Ascomycetes, belonging in several perfect genera, namely *Candida guilliermondi (Pichia guilliermondi)* and *Torulopis globosa (Citromyces metritensis).* Others have a perfect stage in the class Basidiomycetes, namely *Candida scottii (Leukosporidium scottii).* There is also evidence that *Candida albicans* is a Basidio-

Table 8–1. *Classification of Medically Important Yeasts**

Form-Class: Deuteromycetes

Form-Family: Cryptococcaceae

Genus 1: *Cryptococcus.* Unicellular budding cells only; reproduce by blastospores pinched off the mother cell. Most are urease-positive. Cell surrounded by a heteropolysaccharide capsule and produces starchlike compounds; carotenoid pigments are usually lacking. Inositol is assimilated; sugars are not fermented.
Example: *Cryptococcus neoformans* (cryptococcal meningitis, European blastomycosis)

Genus 2: *Torulopsis.* Same as above, but do not have capsules or produce iodine-positive, starchlike substance; urease-negative, and there is no assimilation of inositol.
Example: *Torulopsis glabrata* (torulopsosis)

Genus 3: *Pityrosporum.* Mostly unicellular budding cells which reproduce by blastospores that cut off from the mother cell by development of a cross-wall. Cells may adhere, forming short hyphal strands. Growth stimulated by lipids. There is no fermentative ability.
Example: *Pityrosporum furfur* (pityriasis versicolor)

Genus 4: *Rhodotorula.* Unicellular budding forms that rarely produce pseudomycelium, are generally encapsulated, but do not produce starchlike substance. They do not assimilate inositol or ferment sugars. Carotenoid pigments are produced.
Example: *Rhodotorula rubra* (rare pulmonary and systemic infections)

Genus 5: *Candida.* Reproduction is by pinched blastospores. They may form pseudomycelium or true mycelium; urease is generally negative; capsules are not formed; starch or carotenoid pigments are not produced; inositol is not assimilated.
Example: *Candida albicans* (candidosis)

Genus 6: *Trichosporon.* Reproduction is by blastospores and arthrospores. Mycelium and pseudomycelium are formed.
Example: *Trichosporan cutaneum* (white piedra and systemic infections)

Genus 7: *Geotrichum.* Reproduction is by arthrospore only. A true mycelium is formed.
Example: *Geotrichum candidum* (rare pulmonary geotrichosis)

*From Rippon, J. W. *In* Burrows, W. 1973. *Textbook of Microbiology.* 20th ed. Philadelphia, W. B. Saunders Company, p. 712. Data from Lodder, J. (ed.). 1970. *The Yeasts.* 2nd ed. Amsterdam, North-Holland Publishing Company.

mycete and, like other *Candida* species, has *Leukosporidium* as a perfect stage. The genus *Rhodotorula* has a number of species with a perfect stage in the Basidiomycete genus, *Rhodosporidium.* Certain of the Cryptococcaceae including *Cryptococcus neoformans* produce clamp connections and have other characteristics which ally them to the Basidiomycetes. Finally, electron micrographs of *Trichosporon cutaneum* have demonstrated Basidiomycete-like septa with dolipores and parenthesomes. The yeasts are a very diverse group of organisms.

The relationship of *C. albicans* to other members of the genus has been investigated by many authors using a variety of techniques. It appears to be the only member of this genus regularly able to evoke fatal disease in man and animals.[62] *C. stellatoidea* and *C. tropicalis,* though less pathogenic than *C. albicans,* are significantly more virulent than other *Candida* species. *C. albicans* is antigenically divided into two groups: A, which it shares with *C. tropicalis,* and B, which is shared by *C. stellatoidea.*[83] Nucleic acid-base composition studies (G-C ratios) indicate that *C. albicans, C. tropicalis, C. clausenii,* and *C. stellatoidea* are related.[82] However, by DNA homology studies, *C. albicans* has close relationship to *C. clausenii* and *C. stellatoidea* but not *C. tropicalis.*[6]

Candidosis

DEFINITION

Candidosis is a primary or secondary infection involving a member of the genus *Candida.* Essentially, however, the disease is an infection caused by *Candida albicans.* The clinical manifestations of disease are extremely varied, ranging from acute, subacute, and chronic to episodic. Involvement may be localized to the mouth, throat, skin, scalp, vagina, fingers, nails, bronchi, lungs, or the gastrointestinal tract, or become systemic as in septicemia, endocarditis, and meningitis. The pathologic processes evoked are also diverse and vary from irritation and inflammation to chronic and acute suppuration or granulomatous response. Since *C. albicans* is an endogenous species, the disease represents an opportunistic infection.

Candidosis and its etiology are so common that it has received a number of names in the past. "Monilia" is perhaps the most popular of the older terms, but it has been designated a *nomen absurdum* and its use is invalid. The term "moniliasis" still persists in some quarters, even though equally archaic designations, such as "phlegms," "ptomaines," and "humors," have long since been abandoned.

Synonymy. Candidiasis, thrush, dermatocandidiasis, bronchomycosis, mycotic vulvovaginitis, muguet, moniliasis.

HISTORY

Hippocrates in his "Epidemics" describes aphthae (white patches) in debilitated patients, and the presence of this clinical condition has been recognized for centuries. The term "thrush" is probably derived from ancient Scandinavian or Anglo-Saxon. "Torsk" is the Swedish equivalent of this word. The disease was noted in Pepys' diary of 1665. It was recognized early as a condition of the newborn, and Veron in 1835 postulated that it was acquired during passage through the womb. Berg (1840) considered it to be transmitted by unhygienic conditions and communal feeding bottles. It was also known to occur in patients with debilitating diseases, confirming Hippocrates' observation. Debilitation was proposed by Bennett in 1844 and Robin in 1853 as the most important prelude to candidal infection. Opposing views were held by Berg (1851) and Valleix (1838), who considered it an unfortunate disease of wide occurrence not related to other diseases or conditions.

Credit for the discovery of the "cryptogamic plant" in a lesion of thrush goes to Lagenbeck.[50b] In 1839 he described a fungus in aphthae, one of the earliest accounts of a parasite in a human disease. He did not consider the fungus to be related to the disease, however, Berg in 1841 and Bennett in

1844 conclusively demonstrated the fungal etiology of thrush. Berg reproduced the disease in healthy babies by inoculating them with aphthous membrane material. One of them died of candidal bronchitis and pneumonia. His colleague David Gruby, who had earlier described the fungus of favus and tinea capitis, studied the thrush fungus and noted its similarities to other fungi. In 1842 he described this fungus as "le vrai Muguet des enfants."[31a] By 1853 Robin, who had great influence on later generations of physicians, recognized that the thrush fungus could become systemic as a terminal event of other illnesses.

The first description of vaginal candidosis was by Wilkinson in 1849. His patient, a 77-year-old woman, had a profuse vaginal discharge. Wilkinson was cautious when he quoted Voget by saying the "epiphytes" could not grow unless a "favorable soil was prepared for them." Previous to that time, and for the next forty years, vaginal discharge as well as thrush were defined in textbooks of medicine as the result of morbid secretions. The relationship of dermatocandidosis to thrush was recognized as early as 1771 by Rosen von Rosenstein, and its relationship to subcutaneous infection by Virchow in the 1850's. Systemic disease by hematogenous spread was described by Zenkey in 1861.

A revival of interest in systemic candidosis and candidal endocarditis occurred after 1940. The occurrence of candidosis as a sequel to the use of antibacterial antibiotics, particularly broad-spectrum antibiotics, evoked a great surge of research. The results have demonstrated the delicate ecosystem of which *Candida* is a member. Many fatal cases of candidosis occurred following abrogation of this balance.[25] In 1940, Joachim and Polayes[41a] described candidal endocarditis as a hazard of heroin injection. About the same time, the association of candidosis and steroid therapy, immunosuppressive drugs, cytotoxic agents, and immune defects became apparent.[91] Presently, *Candida* is recognized as one of the most frequently encountered fungal opportunists.

The classification of the organism *Candida albicans* has been the subject of controversy since its first association with human disease. The term "monilia" with which Candida is often confused was first used by Hill in 1751 to describe fungi from rotting vegetation. This proved to be an invalid description as the organisms were in reality *Aspergillus* sp. The genus *Monilia* was erected by Persoon in 1797 to encompass certain species of fungi isolated from rotting fruit. These are now known to be the imperfect stage of certain ascomycetous genera, for example, *Sclerotinia,* i.e. peach mummies. The original organism isolated by Lagenbeck in 1839 was restudied by Gruby. He considered it a *Sporotrichum.* However, the dimorphic nature of the yeast was recognized as early as 1844 by Bennett, and in 1847 Robin placed it in the genus *Oidium* under the name *Oidium albicans.* This genus name was derived from the morphology of the egg-shaped, oval yeast cell. Robin described in detail the yeast and mycelial forms of the organism. Later, another fungus from rotting wood was termed *Monilia candida* by Bonoden in 1851. Hansen in 1888 described it as a yeast and mycelial fungus. The identity of this organism is unclear. Plaut (1887) found a wood-rotting fungus that produced experimental lesions in the throats of chickens. He concluded that it was the same as Hansen's fungus and, therefore, the etiologic agent of human thrush. This was accepted by Zopf who in 1890 described it as *Monilia albicans.* The name became popular in medical literature though the genus *Monilia* (sensi Persoon) exists as a valid genus for plant pathogens unrelated to *C. albicans.* Much of this confusion is due to the influence of the eminent physician and mycologist, Castellani. He adamantly retained the name *Monilia* and described numerous varieties as new species that have since been shown to be identical to *Candida albicans.* At that time, it was a common practice to publish as new species organisms showing minor variations in morphology on clinical presentation. Lodder lists one hundred synonyms for *C. albicans* in the 1970 edition of *The Yeasts.*[52] To end the confusion, Berkhout in 1923 erected the genus *Candida* to encompass asporgenous yeasts that have "few hyphae, lying flat, falling apart into longer or shorter pieces. Conidia arise by budding on hyphae or each other. Small and colourless."[8] This name was accepted as a *nomen conservandum* by the Eighth Botanical Congress at Paris in 1954.

The terminology for the clinical disease has also been controversial. "Moniliasis" is obviously untenable since the genus that contained the etiologic agents of the disease is invalid. "Candidiasis" came into common use in the United States, but in Canada, England, France, and Italy the term "candidosis" has been accepted as the more desirable descriptive term. Since the other mycotic diseases end with -osis, namely histoplasmosis, cryptococcosis, blastomycosis, sporotrichosis, and so forth, for consistency, the term "candidosis" is preferable. This will preclude mistaking candidiasis for a white worm infection, e.g., schistosomiasis, onchocerciasis, ascariasis, and filariasis.

ETIOLOGY, ECOLOGY, AND DISTRIBUTION

Although the etiologic agent, *Candida albicans*, is usually encountered in most of the clinical forms of candidosis, in some of the less common clinical conditions, such as endocarditis, other species are more frequently isolated. These other species represent normal flora of the cutaneous and mucocutaneous areas and are of very limited pathogenicity. All species may be involved in any form of candidosis, but some are regularly encountered in one particular type. These include *C. parapsilosis* from paronychias, endocarditis, and otitis externa; *C. tropicalis* from vaginitis, bronchopulmonary and systemic infections, and onychomycosis; *C. stellatoidea* from vaginitis; *C. guilliermondi* from endocarditis, dermatocandidosis, and onychomychosis; *C. pseudotropicalis* from vaginitis; *C. krusei* very rarely from endocarditis and vaginitis; and *C. zeylonoides* from onchomycosis.

Candida albicans is a normal inhabitant of the alimentary tract and the mucocutaneous regions.[58] It is regularly present in small numbers in the mouths of normal healthy adults. Poor oral hygiene or even small amounts of antibiotics promote an increase in the number of organisms, though usually without untoward results. In the newborn, however, before an oral ecology is established, even a few organisms presage clinical thrush.[86] The incidence of oral candidosis of the newborn varies in different surveys. Case rates as high as 18 per cent have been recorded, but the average appears to be 4 per cent. It is well established that candidal vaginitis during pregnancy contributes to thrush of the newborn. A small but significant percentage of cases are due to cross-contamination from other infants, mothers, and personnel. This is especially true in nursery epidemics.

The incidence of *C. albicans* in the normal vagina of healthy, nonpregnant women is about 5 per cent.[57] There is a distinct increase in clinical vaginitis in gravid females. Most studies indicate a rate of candidosis of about 18 per cent for nonpregnant women with vaginal discharge, but an average rate of 30 per cent for gravid women.[38] These studies have been carried out in many geographic locations under varied climatic conditions, and such factors do not appear to contribute to the incidence or severity of disease.

The normal alimentary tract has a small but constant population of *C. albicans*. Under normal conditions this is probably influenced by foods, since diet markedly affects the total number of organisms present. In the young, before a balanced flora is established, initial colonization of the intestine is frequently associated with clinical symptoms.[86] Perianal colonization may also occur, followed by diaper rash. In the adult, two extrinsic factors alter the number of *C. albicans* in the intestine. First, it has been established that other members of the intestinal flora exert a control on the population density of the yeasts. Studies have implicated a variety of antimicrobial factors, and probably no single mechanism is totally responsible. However, lactic acid appears to be quite inhibitory to *C. albicans*, and a correlation has been found between the numbers of lactobacilli and other lactic acid–producing organisms present and the number of yeasts. Secreted inhibitory factors, oxidation-reduction potentials, and competition for available nutrients have also been implicated in yeast population control. It was observed quite early that a change in the intestinal flora following orally administered antibiotics greatly influenced the number of *C. albicans*. The overgrowth of organisms may manifest itself only as an irritating pruritis ani or progress to colonization of the intestinal tissue and eventual fatal systemic candidosis. The second factor influencing the population of *C. albicans* is diet. A high-fruit diet appears to favor a rapid increase in the number of intestinal yeasts and probably explains the former postulated association of *Candida* and tropical sprue. In normal adults who maintain this diet, there does not appear to be any clinical symptomatology after acclimatization of the host to the presence of the organism.

The normal skin appears to have a resident yeast flora, but this does not include *C. albicans.* In 2444 scrapings from the skin of 118 normal people, *C. albicans* was recovered only three times.[30] Many other surveys have been carried out which indicate that several factors influence colonization of the skin by *C. albicans.* Whereas normal skin does not harbor a resident flora of *C. albicans*, almost any damage to skin or environmental change leads to rapid colonization. For this reason, *Candida* is not infrequently isolated from a variety of dermatologic conditions. Most of the lesions are situated in moister areas, such as the inframammary folds, the perianal skin, and other intertriginous regions. Fruit pickers, canners, dishwashers, and so forth, are particularly prone to candidal infections of the fingers, since constant contact with a moist environment leads to maceration of skin. In some studies, living in a tropical environment alone or having contact with infected patients increases the recovery of *C. albicans* from the skin.[58] Endocrine balance, the administration of steroids, and other physiologic factors also influence the rapidity and extent of *C. albicans* colonization. Although this organism may not have initiated a particular lesion, once it is established, it contributes to the pathology of the disease.

C. albicans is common to the alimentary tract of almost all mammals and birds, and all such species are susceptible to invasion by this opportunistic fungus. The organism has been recovered from soil rarely, but in these instances it probably represents fecal contamination. It appears unlikely that *C. albicans* normally survives and multiplies in a nonanimal environment. A bacterial parasite of *C. albicans* has been reported[2] and may be important in eliminating it from soil. Unsanitary conditions often lead to contamination of the environment or food with the organism.[24] The other species of *Candida* are regularly isolated from normal skin, particularly between the toes, under the toenails, and from the umbilicus. They also appear to be indigenous to many other animal species as well as to vegetable material and the soil.

Age and sex distribution as well as clinical manifestation of candidosis are markedly affected by varying predisposing factors and the underlying disease of the patient. There are four general conditions in which the normal equilibrium between *Candida* and the host may be sufficiently upset to lead to a pathologic state:

1. Extreme youth. During the normal process of establishing a resident flora, the restricting factors for *Candida* may be absent, and a clinical condition is produced. In normal children this condition resolves rapidly, often without treatment.

2. Physiologic change. Pregnancy appears to affect the carbohydrate content of the vagina and leads to an increase in the population of *Candida.* This overgrowth may be sufficient to cause a clinically apparent vaginitis. The administration of steroids to males or females also leads to proliferation of *Candida.*[55]

3. Prolonged administration of antibiotics. Much evidence has accumulated associating clinical disease with the use of antibacterial antibiotics. The most important effect is the elimination and alteration of the bacterial flora that holds the population of *Candida* in check. Evidence also suggests that there is some effect of the antibiotic on the host tissue which predisposes it to invasion by the organism, and the antibiotic itself may stimulate the growth of the *Candida.* The occurrence of the latter effect has been controversial, and most of the current investigations tend to discount it.[92]

4. General debility and the constitutionally inadequate patient. The list of disease syndromes associated with candidosis is long and varied. The extent and severity of the disease usually correlates with the severity of the underlying illness. The term "debility" encompasses such things as the slight avitaminosis of the aged which leads to perlèche or thrush; diabetes and its associated dermatocandidosis; candidal vegetations of diseased heart valves; and pulmonary or generalized systemic candidosis occurring as a sequela to chronic disease or as a terminal event in the various neoplasias. Debility may be iatrogenic also. Immunosuppressive agents, cytotoxins, and other drugs abrogate the normal defenses of the host and predispose them to candidosis or invasion by other opportunistic organisms. Constitutionally inadequate patients include those with various immune defects and defects associated with abnormal leukocytic function. There are many genetic deficiencies for which *Candida, Aspergillus,* and a long list of other fungal, bacterial, parasitic, and viral opportunistic organisms exploit the defective host as culture milieu.

CLINICAL DISEASE

Candida albicans is perhaps the most protean infectious agent that afflicts man. Only syphilis presents with such a diversity of clinical pictures as are encountered with the various candidoses. All of the tissue and organ systems are subject to invasion, and the pathology evoked is as variable as are the clinical syndromes. In addition to active infection, *Candida albicans* is also involved in several allergic conditions. These various clinical manifestations will be discussed according to the primary organ system involved.

I. Infectious Diseases
  A. Mucocutaneous involvement
    1. Oral: glossitis, stomatitis, chelitis, perlèche
    2. Vaginitis and balanitis
    3. Bronchial and pulmonary
    4. Alimentary: esophagitis, enteric, and perianal disease
    5. Chronic mucocutaneous candidosis
  B. Cutaneous involvement
    1. Intertriginous and generalized candidosis
    2. Paronychia and onychomycosis
    3. Diaper disease (napkin candidosis)
    4. Candidal granuloma
  C. Systemic involvement
    1. Urinary tract
    2. Endocarditis
    3. Meningitis
    4. Septicemia
    5. Iatrogenic candidemia
II. Allergic Diseases
  A. Candidids
  B. Eczema
  C. Asthma
  D. Gastritis

Infectious Diseases

MUCOCUTANEOUS INVOLVEMENT

Oral Candidosis—"Thrush." This is the commonest form of disease produced by overgrowth (colonization) of *Candida albicans.* The mouth of the newborn, similar to the vagina of gravid females, has a low pH which may promote the proliferation of *C. albicans.* It is now well-established that any *C. albicans* in the mouth of the newborn presages the clinical disease until a balanced flora has been established. A cream-white to gray pseudomembrane covers the tongue (Fig. 8–1 FS B 15), soft palate, buccal mucosa, and other oral surfaces. The distribution is discrete, confluent, or patchy. These well-marked signs are usually not evident until the child is one week of age. If *Candida* is absent on the third day of life, the condition rarely develops.[86] The membrane seen on the mucosa is composed of masses of fungi in both the mycelial and yeast growth form. The membranous patches often crumble and have the appearance of milk curds, and, in fact, the latter are sometimes mistaken for signs of thrush. The lesions begin as small focal areas of colonization which enlarge to become a patch. The membrane is rather closely adherent to the underlying mucosa, and its removal reveals a red, oozing base. In severe disease, there may be ulceration and necrosis of the mucous membrane. The pseudomembrane of oral candidosis is neither as firm nor as extensive as a diphtheritic membrane, but it often becomes large enough and the tissue involved sufficiently swollen to impede swallowing and, occasionally, breathing. Extensive involvement may also include the trachea, esophagus, and the angles of the mouth.

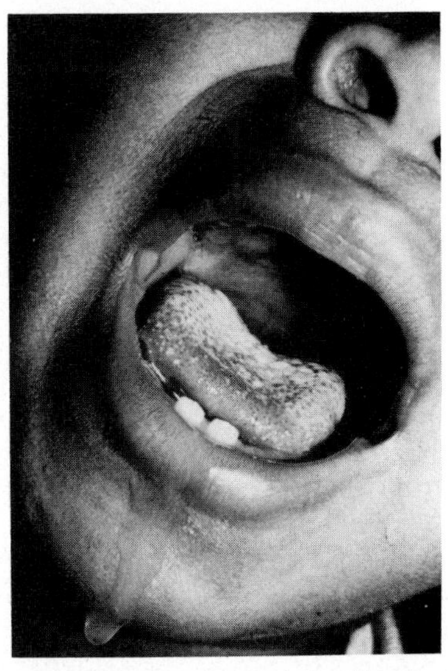

Figure 8–1. FS B 15. Oral candidosis—"thrush." Cream white, curdy patches of pseudomembrane cover the back portion of the tongue.

Oral candidosis also occurs in a condition known as "black hairy tongue." This disease is characterized by hypertrophy of the papillae of the tongue. *Candida* has no etiologic role in this disease, but grows freely in this environment. The same is true in cases of chronic glossitis; the *Candida* proliferate in abundance because of favorable conditions for growth. In both cases, however, the *Candida* contributes to the overall pathology.

Oral thrush in older children and adults is clinically identical to that described for the newborn. In older children, chronic thrush usually indicates polyendocrine disturbances or an underlying defect in natural defenses. In adult patients it may be the result of mild avitaminosis, particularly riboflavin deficiency or a complication of diabetes, advanced neoplasia, or the administration of steroids, antibiotics, or other drugs (Fig. 8–2 FS B 16).

Adult stomatitis which is not related[88] to other diseases is usually the result of badly fitting dentures or poor oral hygiene. Again, *Candida* grows in abundance without having initiated the disease. Deepening and exaggeration of the commissural folds in older patients often leads to a chronic chelitis, the perlèche syndrome (Fig. 8–3). *C. albicans* can often be isolated from the lesions, but it probably has little to do with the pathogenesis of the condition. In true candidal chelitis, there are scattered groups of satellite erosions over the lip, and *Candida* can be isolated from these in abundance.

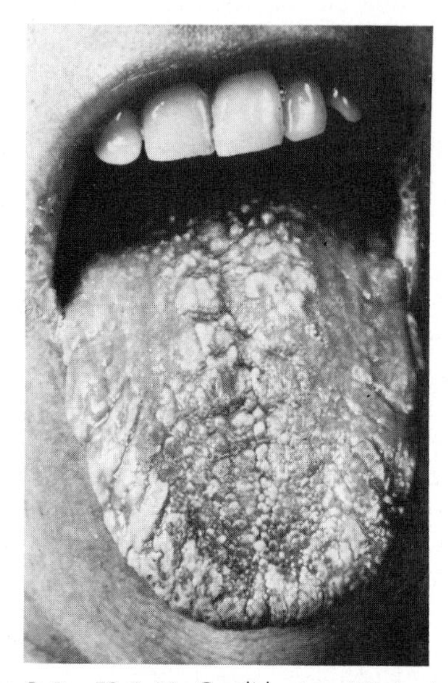

Figure 8–2. FS B 16. *Candida* stomatitis in the adult. The condition was associated with polyendocrine disturbance. Punctate, raised patches are seen involving most of the surface of the tongue.

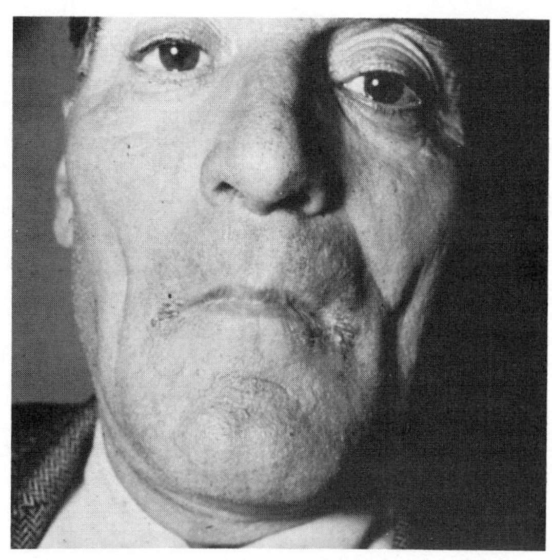

Figure 8–3. *Candida* chelitis—*perlèche.* Exaggeration of commissural folds, satellite lesions on lips, and chronic inflammation of commissures. Angular stomatitis.

Vaginitis and Balanitis. Diabetes, antibiotic therapy, and pregnancy may predispose to vaginal candidosis. The disease is characterized by the presence of a yellow, milky discharge, and patches of grey-white pseudomembranes are seen on the vaginal mucosa. The lesions vary from a slight eczematoid reaction with minimal erythema to a severe disease process with pustules, excoriations, and ulcers. The whole area is greatly inflamed, and pruritis is usually intense. The condition may extend to involve the perineum, the vulva, and the entire inguinal area (Fig. 8–4 FS B 17, *A*).

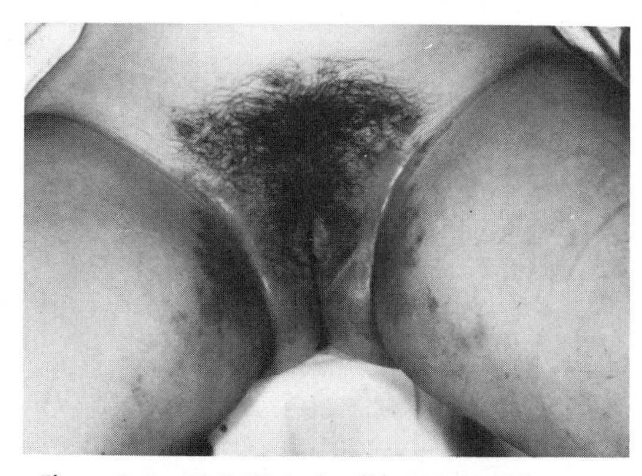

Figure 8–4. FS B 17,*A*. *Candida* vaginitis. Extension to involve perineum and inguinal area. The lesion is extremely pruritic and inflamed.

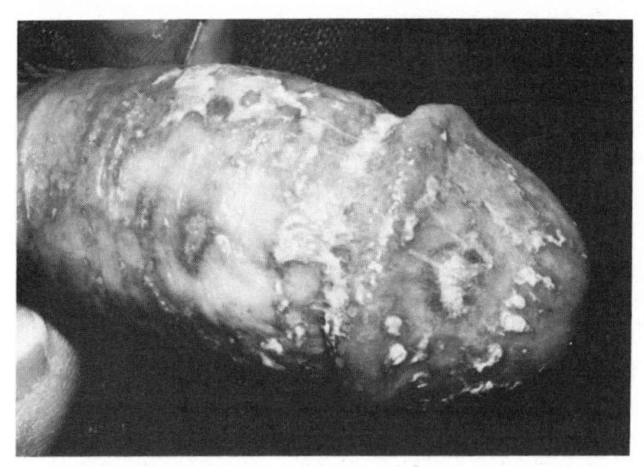

Figure 8–5. FS B 17,*B*. *Candida* balanitis. Superficial erosion and pustules over glans, sulcus, and shaft. (Courtesy of S. Lamberg.)

Candida balanitis or balanoposthitis is a rare condition of males. Often there is a history of vaginitis in the spouse, and the condition is probably a conjugal infection. There are superficial red erosions and thin-walled pustules over the glans and the sulcus corona (Fig. 8–5 FS B 17, *B*). Not infrequently, no yeasts are demonstrable in the lesions, suggesting an allergic rather than an infective disease. Some of these patients have an exaggerated response to the oidiomycin skin test. The condition clears with cure of the vaginitis in the conjugal partner. *Candida* balanitis is also associated with diabetes and, usually, a tinea cruris-like syndrome called dermatocandidosis of the groin.

Bronchial and Pulmonary Candidosis. Bronchial candidosis is a chronic bronchitis with cough, production of sputum, and medium to coarse basilar rales, with linear fibrosis or peribronchial thickening seen on radiologic examination. The etiologic significance of *C. albicans* in this disease is difficult to determine. Many surveys have shown the organism occurs, sometimes in considerable numbers, in essentially all chronic lung conditions.[48] Bronchoscopy may not assess the extent or degree of candidal colonization of the bronchial tree. Castellani first described candidal involvement of the bronchi and lungs. He divided the disease into two categories: the first was a chronic condition called "tea-taster's" cough, which involved only the bronchi. *Candida* may not be related at all to this disease, or it may be a minor colonizer or an allergen. Recently, more attention has been paid to the role of *Candida* in inciting allergic-type diseases. The second type of disease was a progressively fatal pulmonary one. It is a valid candidal infection, and the contribution of the yeast to the pathology is readily established.

Pulmonary candidosis as a primary disease is extremely rare. The diagnosis of this disease must rest on unassailable evidence that rules out all other etiologies of the pathologic process.[18] *Candida* readily colonizes preexisting pathologic conditions attributable to other infectious agents, neoplasms, or chronic functional disorders. In these cases, the organisms may be present in numbers to 10^6 per ml and are easily visible on sputum smears. If their multiplication continues, the *Candida* may contribute to the pathology and cause severe distress and death.

Primary pulmonary candidosis presents with a cough, low-grade fever, night sweats, dyspnea, weight loss, and production of mucoid gelatinous sputum which is often blood-tinged. Roentgenographically there is hilar and peribronchial thickening, and lesions resembling miliary tuberculosis are seen (Fig. 8–6). In severe and extensive infections, the

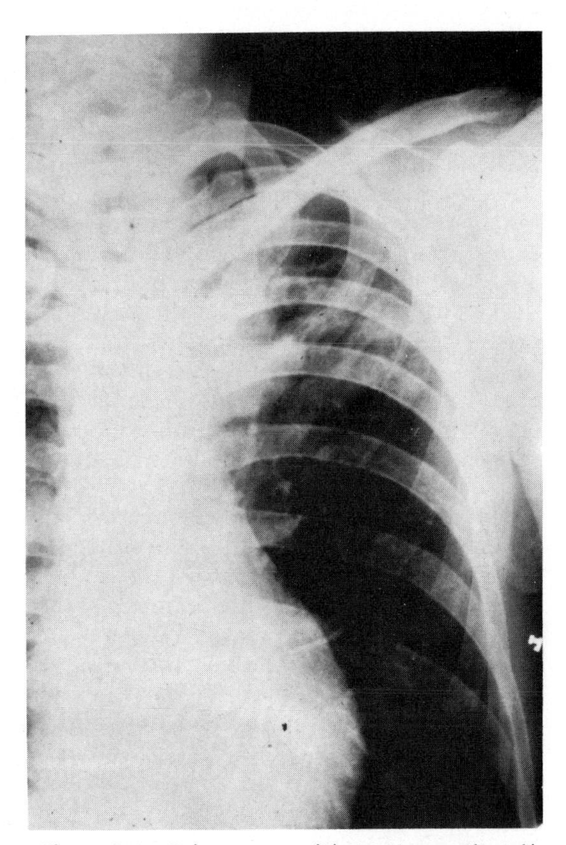

Figure 8–6. Pulmonary candidosis. Nonspecific infiltrate in left apex. There is density immediately above the left hilum. (Courtesy of J. Fennessey.)

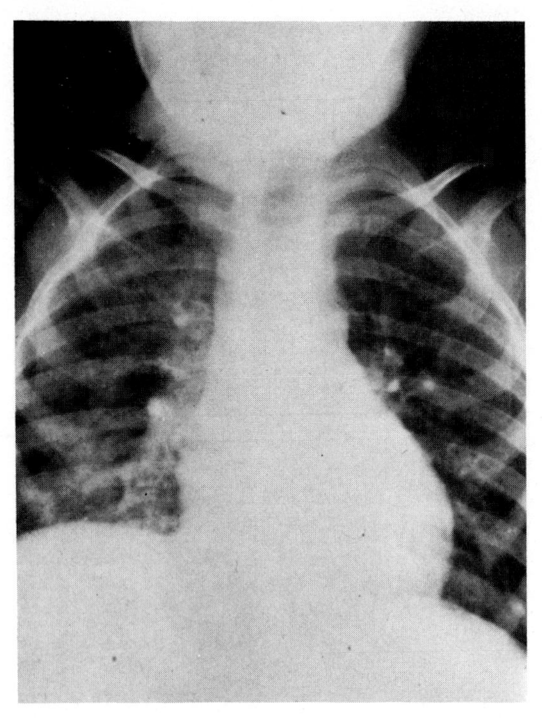

Figure 8–7. Pulmonary candidosis in a child with leukemia. There is diffuse pulmonary infiltrate in many areas of the lung. The lung is underinflated. (Courtesy of J. Fennessey.)

lesions are dense and smooth, and often involve an entire lobe which may show complete consolidation (Fig. 8–7). Usually there are scattered patchy bronchopneumonic lesions, but in severe cases a lobar pneumonia develops. The lesions tend to be labile, and serial x-rays show clearing in some areas with development in others. Medium moist rales occur, but dullness and changes in breath sounds are detectable only in extensive disease. Cavitation of the type seen in bronchiectasis occurs. When two or more lobes are severely involved, death follows from respiratory insufficiency. Pulmonary disease is frequently secondary to septicemia and dissemination of the organism from other loci.

Alimentary Candidosis. Involvement of the esophagus and intestine occurs in several separate disease syndromes. Esophageal candidosis is often an extension of lesions from the oral cavity, especially in thrush of the newborn. Chronic infection in older children is associated with genetic defects and polyendocrine deficiencies and will be discussed in a separate section, Chronic Mucocutaneous Candidosis. In adults, *Candida* infection of the esophagus is associated with antibiotic therapy, corticosteroids, diabetes, and irradiation, as well as various neoplasias, blood dyscrasias, endocrinopathies, and other debilitating conditions.[91] The significance of finding *Candida* in disease of the esophagus is often difficult to assess, especially in children. In adults, however, the diagnosis can be made by culture of material from esophagoscopy and the rather characteristic radiologic findings.[32] Evidence of destruction of the mucous membrane, a ragged and irregular esophageal outline without loss of the longitudinal folds, is seen. The tract is distensible and has a characteristic pattern following barium swallow.[45] In the absence of varices or history of other pathology of the esophagus, a diagnosis of candidosis should be seriously entertained.

Enteric candidosis is one of the most controversial clinical diseases attributed to *Candida*. The diagnosis is very difficult to establish, and the implication of *Candida* as the primary agent of disease is, at the most, tenuous. The clinical condition called candidal enteritis that follows administration of tetracycline and its congeners occurs also when the yeast population is supposedly suppressed by the concomitant use of antifungal antibiotics.[31] On the other hand, there have been many well-documented cases of fatal candidal invasion of the intestine following antibiotic therapy. The final diagnosis of this syndrome requires demonstration of candidal invasion of the intestinal mucosa or repeated isolation of the organism from ulcerative lesions. In biopsy specimens, colonization and invasion of the tissue is accompanied by the downward growth of mycelial elements. These mycelial elements are also seen in fecal smears and indicate invasion of the intestinal wall by *Candida*.[49] At autopsy numerous vegetations and a shaggy mucosal membrane are seen (Fig. 8–18). A quantitative increase in the population of *C. albicans* following antibiotic therapy or change in diet may also cause a chronic "irritable colon" syndrome.[38] This is ascribed to hypersensitivity to the organism or its metabolic products rather than actual infection.

Perianal involvement is common in infants with oral thrush. The condition may exist with or without clinical disease of the intestine. The lesions are initially sharply defined, dull red patches which coalesce and spread with an irregular border. Vesicles form that later break to leave a ragged edge.

Extension of the lesions to include the buttocks and inner thigh is common, and the genitalia, lower abdomen, and entire diaper area are sometimes involved. There is minimal pruritus, and in healthy children the condition resolves quickly following therapy. In adults the pruritus ani which may follow antibiotic therapy has been associated with *Candida* overgrowth. The itching is severe, and the area presents as an intensely inflamed dermatitis. Various studies have implicated other organisms, and *Candida* colonization may be only incidental. Water-soluble toxins from *Candida* applied to the skin are able to elicit a similar clinical picture.

Chronic Mucocutaneous Candidosis. This is a general category describing a clinical manifestation that occurs in persons with various genetic defects. The patients are usually children, and *Candida* is only one of several possible pathogens or opportunistic organisms that may establish disease. The classification of these genetic diseases changes as new defects in leukocyte function or the endocrine systems are discovered.[36] The first major group is associated with dysgenesis of the thymus and subsequent inability to elicit cellular immunity.[9] This category includes (a) dysplasia of the thymus with agammaglobulinemia (Swiss type of agammaglobulinemia); (b) dysplasia of the thymus without agammaglobulinemia (Nezelof-Allibone syndrome); (c) congenital absence of the thymus and parathyroid glands (the third and fourth brachial pouch syndrome, DiGeorge syndrome). All the latter are associated with an increased incidence of candidosis. In contrast, when defects of humoral immunity alone are present (Bruton's hypogammaglobulinemia), there is no increased incidence of candidosis.

The second group which is associated with chronic mucocutaneous candidosis includes polyendocrine deficiencies, such as familial juvenile hypoparathyroidism and hypoadrenocorticism, and thymomas in adults. Patient with defects in leukocyte function are also prone to this type of infection, e.g., children with chronic granulomatous disease whose defective myeloperoxidase system precludes killing of *Candida* following phagocytosis[41, 46] (Fig. 8–8, FS B 18). Idiopathic chronic mucocutaneous candidosis is an additional category in which the underlying defect is as yet undefined.[50a] Pathologically and clinically, it resembles the infection seen in patients with polyendocrine defects.

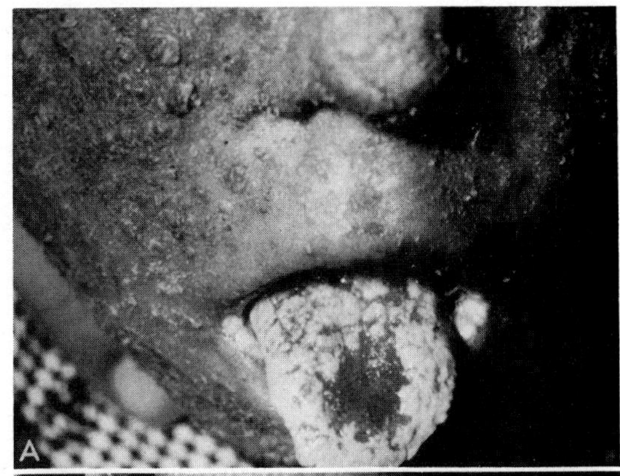

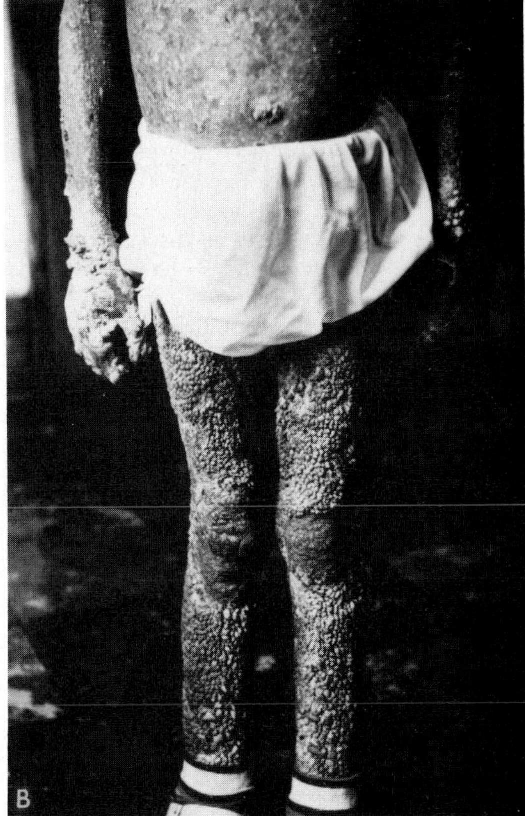

Figure 8–8. FS B 18. *A*, Chronic mucocutaneous candidosis involving face and tongue in a child with chronic granulomatous disease. (Courtesy of D. Windhorst.) *B*, Chronic generalized granulomatous candidosis in a child with an unknown type of leukocyte dysfunction. (Courtesy of S. Lamberg.)

Two final categories involve patients who are unresponsive solely to the *Candida* antigen,[14, 16, 36] or whose sera contain specific IgG which inhibits the clumping effect of normal serum for *Candida*.[17]

Candida granuloma is the final category of disease. This entity is distinguishable clinically and pathologically from the mucocutaneous disease, although they both may exist in the same patient. The granulomas involve the skin or its appendages and are discussed in the following section (Fig. 8–8,*B* FS B 18,*B*).

CUTANEOUS INVOLVEMENT

Intertriginous Candidosis. Cutaneous candidosis involves the intriginous areas of the glabrous skin directly or may occur as a colonization secondary to preexisting lesions on any part of the body caused by varying etiologies. Intertrigo most commonly is seen in the axillae, groin, intermammary folds, intergluteal folds, interdigital spaces, glans penis, and umbilicus. The lesions are quite characteristic and well-defined as weeping, "scalded skin" areas with an erythematous base and a scalloped border (Fig. 8–9 FS B 19). The lesion is surrounded by "satellite" eruptions which develop as discrete vesicles, pustules, or bullae which break to leave a raw surface with eroded ragged edges. These develop and emulate the initial lesions. Clinical variants do occur, and the lesions may appear dry and scaly. Candidosis of the skin is associated with two types of patients. The first of these have metabolic disorders which predispose them to candidal colonization, such as diabetes, obesity, or the sequelae of chronic alcoholism. In the second group, the skin is predisposed to infection by various environmental conditions, including moisture, occlusion, and maceration of the skin under dressings, caused by the wearing of boots and tight clothing in tropical climates and by frequent and continual immersion in water. The latter is common in such occupations as housewife, dishwasher, barmaid, fruit canner, and so forth (Fig. 8–10 FS B 20).[68b] Colonization of the skin under occlusive conditions without invasion of the epidermis produces a contact dermatitis of the primary irritant type.[56] Congenital anatomic defects and poor peripheral circulation also predispose to this type of disease.

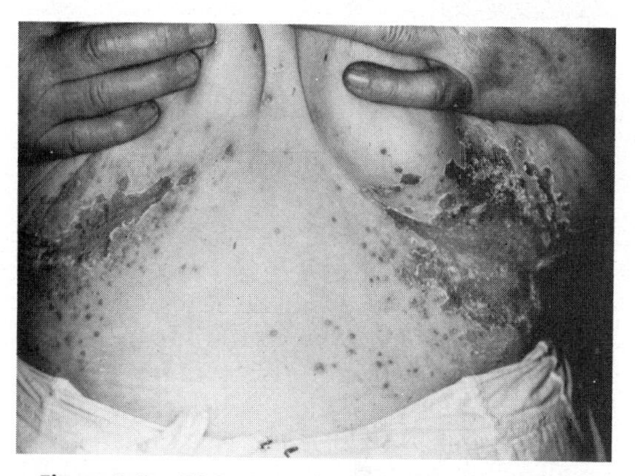

Figure 8–9. FS B 19. Intertriginous candidosis in a diabetic. Note "scalded skin" areas and satellite eruptions.

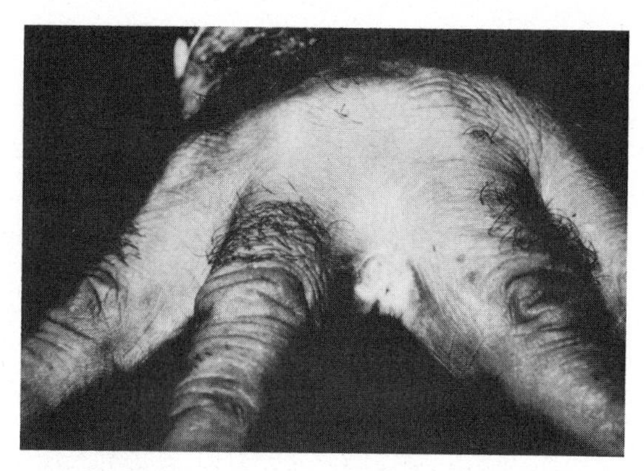

Figure 8–10. FS B 20. Interdigital *Candida* erosion in a bartender.

Generalized dermatocandidosis with widely disseminated lesions occurs in diabetics and in those with a wide variety of ectodermal defects. In the latter group, Candida colonizes damaged tissue but contributes little to the total pathology seen. Another form of candidosis results in the development of purulent follicular papulopustules in intertriginous areas. The pustules have sodden, annular fringes, and folliculitis can occur in the absence of frank intertrigo. The lesions may be potentiated by the use of zinc oxide.

Paronychia and Onychomycosis. These are the commonest forms of cutaneous candidosis. The paronychial folds are readily colonized, particularly in people whose occupations require frequent immersion of appendages in water. The lesions are characterized by the development of painful, reddened swellings extending as far as one centimeter from the paronychial edge. They resemble pyogenic lesions, especially those caused by staphylococci, in the same area. A mixture of bacteria and *Candida* is often present. In chronic paronychia the nail becomes invaded. The resulting onychomycosis appears as a hardened, thickened, brownish-discolored nail plate that is striated or grooved. The nail does not become friable as it does in tinea unguium of dermatophyte etiology. The nail tissue is destroyed in chronic untreated cases.

Diaper Rash. This is not an uncommon sequela of oral and perianal candidosis of the newborn. It also occurs in infants under unhygienic conditions of chronic dampness and irregularly changed, unclean diapers. The initial colonization evokes a primary irritant dermatitis. Invasion of the epidermis by the fungus may ensue, and the condition becomes severe, often spreading to the axillae, face, conjunctiva, and other areas.

Candidal Granuloma. This very rare condition of children was described in detail by Hauser and Rothman, who reviewed 13 cases from the literature.[35] The lesions are quite distinct from other forms of dermatocandidosis and mucocutaneous candidosis. They are described as primary vascularized papules covered with a thick, adherent, yellow-brown crust. These may develop into horns or protrusions up to 2 cm in length (Fig. 8–11). On histologic examination, the lesions appear as poorly organized granulomatous tissue, with giant cells and a chronic inflammatory reaction (Fig. 8–12). The face is most commonly involved, but lesions are also found on the scalp, fingernails, trunk, legs, and pharynx. Defects of the immune system which may predispose to this disease are delayed cutaneous anergy of all types and a lymphopenia.[41] Specific immunologic unresponsiveness to *Candida* antigens has also been postulated.[64] The patients die of the underlying disease in time, often with infection as a terminal event.

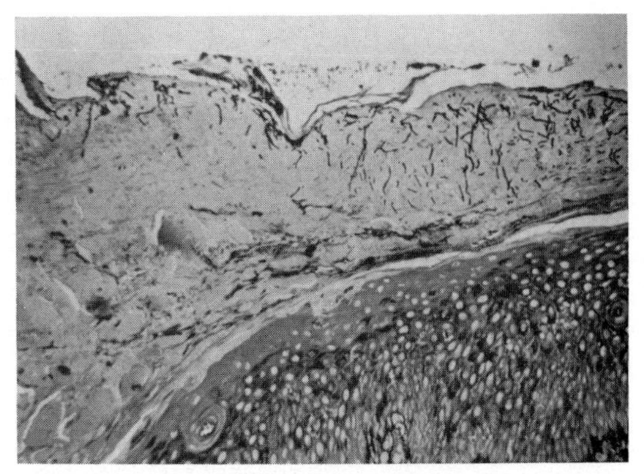

Figure 8–12. Candidal granuloma. Biopsy of hyperkeratic area. The fungus is seen to grow in the stratum corneum but does not invade the epidermis. Gomori-Grocott chromic acid methenamine silver stain (GMS), ×100. (Courtesy of H. Sommers.)

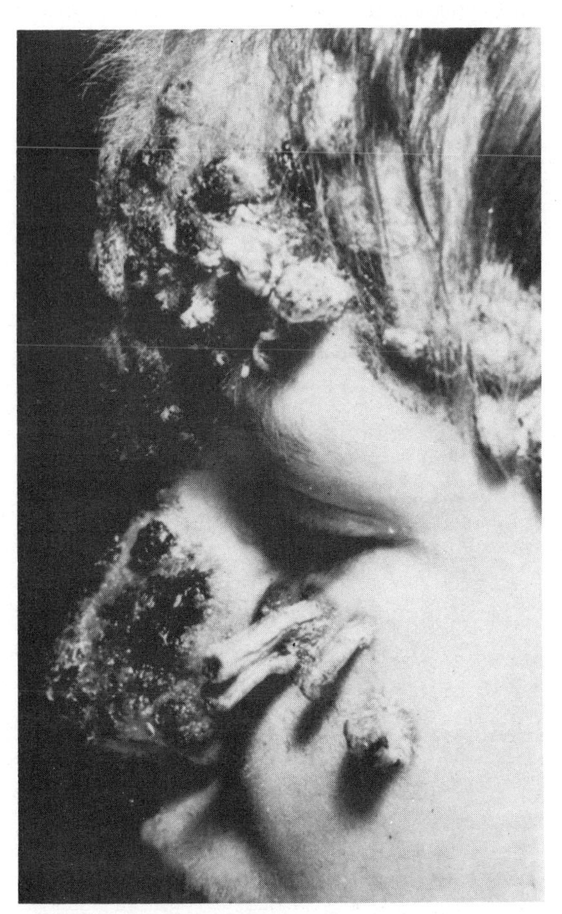

Figure 8–11. Candidal granuloma. Many horns and protrusions are seen over the face of the child.

SYSTEMIC INVOLVEMENT

This is a relatively rare condition except as the terminal event of a debilitating illness. In some cases, it is the result of chronic insult and continued seeding of yeasts into the body because of repeated injections by drug addicts, indwelling catheters, or long-term antibiotic or steroid therapy. The prognosis in these conditions is poor, and the death rate high.

Urinary Tract. Clinical involvement of the urinary tract is reported in association with disseminated candidosis, diabetes, pregnancy, administration of antibiotics, and use of unclean catheters. The condition is more common in women than men by a ratio of four to one. Though bacterial cystitis is not uncommon, candidal invasion of the bladder is infrequent. In the normal patient, the condition clears readily under treatment. The diagnosis is difficult to make, as *Candida* is frequently cultured from the urine. Counts in excess of 1000 colonies per milliliter are

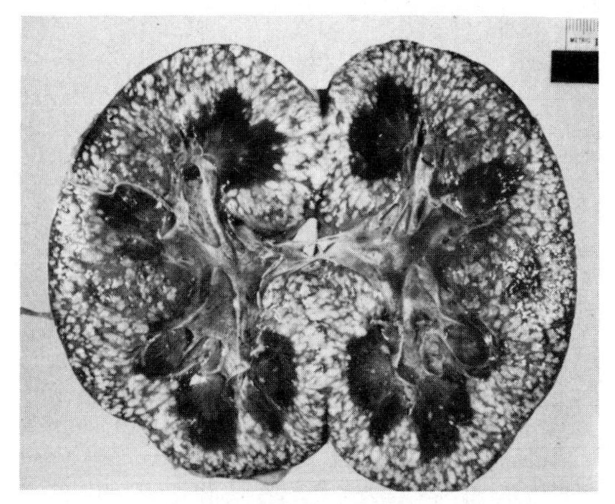

Figure 8–13. FS B 21. Renal candidosis. Cut section of kidney. Miliary solitary and confluent punctate abscesses throughout the cortex and, to a lesser degree, the pyramides. The medulla in general is spared. The patient had acute leukemia. The massive involvement of the cortex indicates a recent, rapid seeding from another focus of infection. In the histologic section, most of the fungus was in the yeast form of growth, again indicating a recent infection.

considered indicative of an active urinary tract infection.[26] On cystoscopy, candidal plaques can be seen on the mucosa of the bladder. These resemble the patches observed in thrush or vaginitis.

Involvement of the kidney usually is the result of generalized dissemination from another focus of infection. Lesions are seen in the medulla and cortex or in the medulla alone. This is in contrast to the disease in experimental animals, which almost always involves the cortex only. In humans where a sudden rapid seeding occurs, numerous lesions of the cortex are seen (Fig. 8–13 FS B 21). Closely examined, these lesions contain more yeastlike than mycelial elements, indicating recent involvement. Invasion of the renal parenchyma in candidal pyelonephritis is reflected by altered renal function.

Endocarditis. This is a rather special form of candidosis, as the etiologic agents are usually species of *Candida* other than *C. albicans*. The clinical symptoms are similar to those of bacterial endocarditis and include fever, murmur, congestive heart failure, anemia, and splenomegaly. Large vegetations are seen on the valves, and there is a high incidence of embolization to the large arteries, both findings being uncommon in bacterial endocarditis. Three groups of patients are susceptible to this condition. The first are those with preexisting valvular disease who also have had treatment with antibacterial antibiotics and an opportunity for entry of normal flora of the skin or intestinal tract into the blood stream. The latter is usually a result of indwelling catheters or prolonged intravenous infusions which predispose to candidemia and subsequent colonization of the valves by skin-inhabiting yeasts. A second group of patients includes drug addicts. Contaminated lots of heroin or needles have been implicated in some cases of candidal endocarditis (Fig. 8–14 FS B 22). The third group falls into a newly recognized category—endocarditis following heart surgery. In one hospital survey, fully 15 per cent of patients receiving a valve prosthesis developed *Candida* endocarditis. Patients without valvular disease are sometimes infected by contaminated blood from oxygenators or by repeated cardiac catheterizations. Some of these infections were cleared with amphotericin B therapy, and the patient received another replacement valve. Most, however, had a fatal outcome. This is true of other forms of candidal endocarditis, but despite therapy

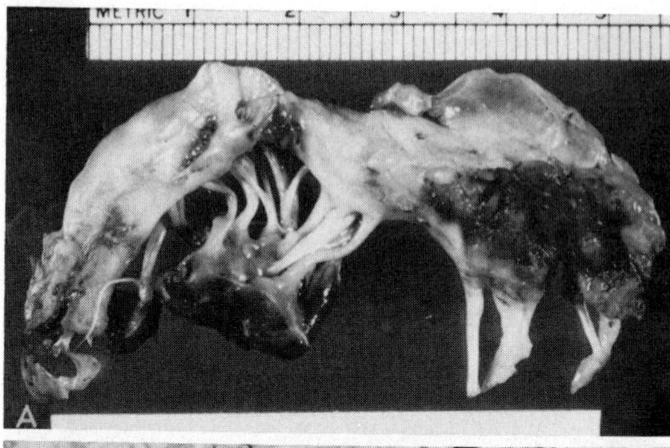

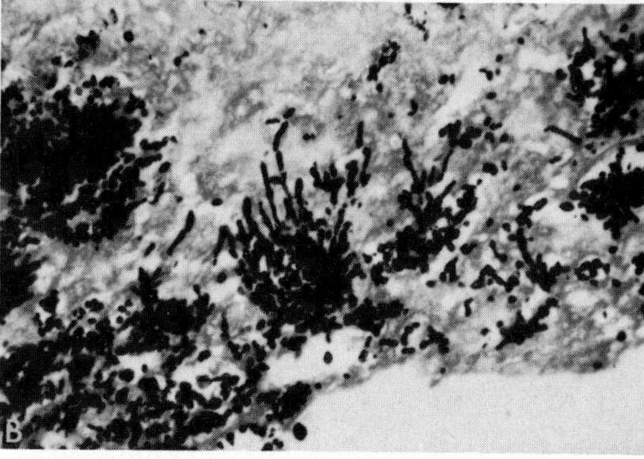

Figure 8–14. FS B 22. *A, Candida* endocarditis. Mitral valve. Acute and chronic endocarditis with prominent vegetations in a long-term drug addict. *B,* Histologic section of vegetation showing mycelial and yeast forms of *C. parapsilosis*. GMS, ×400. (Courtesy of S. Thomsen.)

the immune state of the patient determines the outcome of the disease to a large extent.[67,90] *C. parapsilosis* and *C. guilliermondi* are more common than *C. albicans* as etiologic agents of this syndrome. *C. stellatoidea* endocarditis has also been recorded.

Meningitis. Candidosis of the central nervous system is relatively rare. About 50 cases of this form of the disease have been reported. *Candida* organisms appear to reach the brain by dissemination from foci in the gastrointestinal tract and respiratory system, or they are introduced during intravenous therapy.[22] Most of the patients with this condition have had an underlying disease or were being treated with antibiotics, cytotoxins, or corticosteroids. A few have had no detectable abnormality. The clinical symptoms are those of meningitis, including pain, nuchal rigidity, positive Kernig's and Brudzinski's signs, along with focal neurologic signs, such as aphasia and hemiparesis. Papilledema and increased intracranial pressure are rarely observed, but diplopia, tinnitis, vertigo, stupor, and coma are often present. There appears to be a lack of correlation between the clinical and laboratory signs in candidal meningitis. The spinal fluid is clear, with a low grade pleocytosis composed of mononuclear or polymorphonuclear leukocytes. The protein is generally very high, and the sugar low or normal. These signs may be present when there are no clinical symptoms of meningitis, and the spinal fluid may be normal when clinical symptoms are present. CNS lesions vary from large, solitary abscesses or wide-spread microabscesses to granuloma-like lesions.[72] Invasion of the blood vessel walls initiates thrombosis and secondary infarction. In rare cases, *Candida* meningitis has as fulminant a course as bacterial meningitis. Usually, however, it is more indolent. *C. albicans, C. guilliermondi, C. tropicalis,* and *C. viswanathii* have been isolated from this disease, although most cases are caused by *C. albicans.*

Septicemia. *Candida* septicemia is an increasingly encountered infection, particularly as a terminal event to an underlying disease. Most patients reported have been on antibiotic therapy, and about half were receiving corticosteroids. Approximately three-quarters of the patients had leukemia or other neoplasias.

A transient fungemia may occur following indwelling catheterization, continuous intravenous infusions, or other abrogations to natural barriers. In most patients, the organisms are rapidly cleared, but in debilitated persons a septicemia may be followed by a systemic disease. In normal volunteers positive blood and urine cultures and transient toxemia were observed after oral ingestion of large numbers of *C. albicans,* but the yeasts were rapidly cleared. This demonstrates that overgrowth with large numbers of yeasts can result in some *Candida* passing through the intestinal wall.[50] A positive blood culture with *Candida* should always be considered of serious concern. The clinical signs of septicemia include fever, chills, and impaired renal function. The mortality rate with systemic colonization is about 75 per cent, even with treatment. *C. albicans* is the species involved in most cases, with *C. tropicalis* being recovered about 25 per cent of the time.

Iatrogenic Candidemia. A new category of *Candida* septicemia has only recently been recognized. This is fungemia as a complication of parenteral hyperalimentation. In one series, 6 of 15 infants and children developed *Candida* septicemia while on this regimen.[10] In another study covering an eight-month period, 33 adults with fungal septicemia were seen.[19] In 22 (67 per cent) of these there was a history of receiving hyperalimentation for severe gastrointestinal disease. In no instance was there an association with steroid therapy or immunologic deficiency. If patients were on hyperalimentation for 20 days or more, fully 55 per cent developed candidemia. Ashcraft and Leape[5] advise constant monitoring by blood cultures of patients receiving this procedure. In immunologically competent patients, removal of the catheter alone or in conjunction with low-dosage amphotericin B therapy usually resulted in clearing of the infection. In children the drug can be given in as low a dose as 11 mg over an 11-day period.[59]

ALLERGIC DISEASES INVOLVING CANDIDA

Allergy to the metabolites of *Candida* is a well-established phenomenon, and the clinical condition is called candidids. This reaction is similar to the dermatophytids of ringworm infection in both appearance and genesis. Less well documented is the role of

Candida in other syndromes involving hypersensitivity. These include eczemas, asthma, gastritis, and a condition of the eye similar to the so-called "histoplasma uveitis." A *Candida* gastritis syndrome has been the subject of many conflicting clinical and experimental reports.[92] Extracellular products of *Candida* can induce both delayed and immediate types of hypersensitivity. Most adults without clinical symptoms of infection or allergy react to the intradermal injection of a culture filtrate of *C. albicans* termed "Oidiomycin" (Hollister-Stier). This indicates that antigenic material from the organism or the organism itself can pass through the mucosal wall in sufficient quantity to evoke an immunologic response. It is probable, therefore, that colonization of the bronchi or previously damaged cutaneous tissue by *Candida* or overgrowth of the organism in its normal habitat would be accompanied by manifestation of toxicity and allergy.

Candidids. The lesions of candidids are similar in clinical appearance, morphology, and distribution to those of dermatophytids. They are sterile, grouped, vesicular lesions (pompholyx type) which may occur in the interdigital spaces of the hands or on any part of the body. As with dermatophytids, they disappear following resolution of a *Candida* infection or after desensitization.

Eczema, Asthma, and Gastritis. Cutaneous allergy in the form of urticaria and eczema[78] has been described following candidosis. A skin test may be used to diagnose this condition; however, it must be appreciated that a high percentage of the normal population will give a positive reaction to the test. True allergy is usually indicated if there is an immediate reaction following intradermal injection of the test antigen. In such cases, resolution of the "allergic" diseases has followed desensitization of the patient by the intradermal or subcutaneous injection of extracts from *C. albicans*.[77]

Allergy to *Candida* in the form of urethritis, gastritis, balanitis, rhinosinusitis, and headache has also been described.[77, 78] Hypersensitivity resulting from the overgrowth of *Candida* has been implicated in a condition termed "irritable colon syndrome" and also in skin lesions of *erythema annulare centrifugum* type.[38, 80] Much more study and clinical evaluation are necessary to substantiate the role of *Candida* in these various allergic conditions. The role of *Candida* in allergic uveitis is discussed under mycotic diseases of the eye.

DIFFERENTIAL DIAGNOSIS

In all cases of suspected candidosis, complete cultural confirmation is necessary. A negative mycologic examination is significant, but a positive one is not unassailable proof of candidal involvement. Consideration of the numbers and morphology of the organisms present as well as exclusion of other etiologies are necessary before *Candida* is implicated as the inciting agent of any pathologic process. Isolation of *Candida* from the blood is usually significant and warrants attention. Thrush in the newborn is pathognomonic in its appearance and presents no diagnostic problem. In other patients, leukoplakia, lichen planus, tertiary syphilis, and other lesions resemble cutaneous candidosis. Vaginal candidosis is similar to trichomonas vaginitis and requires laboratory studies for differentiation. Neither are as purulent or severe as acute gonococcal disease. The differential diagnosis of systemic infection must include other mycoses, tuberculosis, neoplasms, or chronic bacterial infections. Since *Candida* may colonize any preexisting cutaneous, mucocutaneous, or respiratory condition, it is very difficult to assess the contribution, if any, of isolated yeasts to the observed pathology.

PROGNOSIS AND THERAPY

As candidosis is primarily an opportunistic infection, prognosis depends almost entirely on the type and severity of the predisposing conditions or diseases. Oral thrush in the newborn healthy child may clear uneventfully, but other forms of *Candida* infection are much more difficult to treat and usually do not clear spontaneously. Control of cutaneous candidosis in the diabetic depends on proper hygiene and regulation of the diabetes. In candidosis associated with macerating conditions, prolonged exposure to moisture, and so forth, elimination of these factors will cause resolution of the disease even without treatment. Chronic disease in the constitutionally inadequate patient can be controlled with therapy, but the condition will return with cessation of therapy. In advanced systemic diseases, candidosis is usually a terminal event which contributes to the ultimate demise of the patient.

Treatment of thrush and cutaneous candidosis was established many years ago and remains relatively unchanged. When compared with other modalities, 1 per cent crystal violet is as effective or better than such drugs as 0.1 per cent hamycin or 1 per cent nystatin.[66] Treatment with crystal violet may cause necrosis of the mucosa, and it is recommended that application be limited to twice daily for three days.[42] Older topical agents, such as sodium caprylate and sodium propionate, are still as effective as Trichomycin or other new drugs. Nystatin can be applied to resistant lesions as a suspension containing 200,000 units per ml. This is applied every two to three hours for several days until clinical improvement is attained. As a viscous suspension, nystatin has been successful in the treatment of chronic esophageal candidosis.[32] It is used in vaginitis as an ointment or as suppositories. As tablets of 500,000 units administered three to four times daily, it is effective in gastrointestinal infection. In combination with steroids in ointment or cream bases, nystatin is effective in treating cutaneous lesions and paronychias. It has been used without much success, as an aerosol in pulmonary candidosis.[4] Natural or laboratory-induced resistance to nystatin apparently is rare.[11]

Amphotericin B (2 or 3 per cent) is sometimes used as a topical agent, though this drug is generally more irritating than nystatin. However, it is the only available effective agent for the treatment of systemic candidosis. The dosage schedule is that recommended for other systemic mycoses (1 mg per kg body weight, total course: 1 to 3 g). Patients should be started on a low dosage of 5.0 mg given on alternate days and increased as can be tolerated. Clinical cures in all forms of systemic disease, endocarditis,[67,90] meningitis,[20] *Candida* granuloma, and chronic mucocutaneous candidosis have been recorded. In *Candida* septicemia resulting from indwelling catheters and hyperalimentation, removal of the needle alone is sometimes sufficient to clear the infection. In immunologically competent patients, low doses of amphotericin B may also be administered. Medoff et al.[59] found that 10 to 355 mg given over a period of 4 to 18 days produced satisfactory blood levels of the drug. No side effects were noted. The patients had both mucocutaneous and systemic disease. The latter was the result of an indwelling catheter.

Candida albicans has a sensitivity to 5-fluorocytosine that ranges from 0.23 to 3.9 μg per ml, although some other species of *Candida* are resistant to 1000 μg per ml.[79] This drug has been used successfully in treating some cases of systemic candidosis. The dosage generally ranges from 8 g daily (oral) to 12 g or more (150 mg/kg/day maximum dose). In recent reports, treatment failures have been noted as well as bone marrow depression leading to death.[69] Many strains of *C. albicans* are resistant to the drug, and others rapidly develop resistance. It appears that the usefulness of this drug in systemic candidosis is limited.[51a]

PATHOLOGY

There has appeared a surfeit of literature concerning the relative pathogenicity of the mycelial and the yeast stage of *Candida albicans*. It was felt that the sprouting of the yeast cell to form mycelium helped the organism to escape macrophage ingestion[54] and was necessary for invasion of tissue.[31] Thus the M (mycelial) phase of *Candida* was considered the pathogenic or parasitic stage, and the Y (yeast) stage the saprophytic form. This is in contrast to the situation with the other dimorphic pathogenic fungi. However, it has now been well established that nutritional and environmental factors influence the Y-M transition of *Candida in vitro*, and that certain enzymes are directly involved.[65,74] Recent investigations conclude that the yeast stage is necessary for initiation of a lesion, and that the mycelium is formed upon exposure to environmental factors which cause inhibition of cell division but not of growth. The result is elongate hyphae.[40,84] The invasive capability of the yeast stage has been well documented.[61] The inflammatory, toxic, and invasive abilities of *C. albicans* are a characteristic of the species and not of a particular growth form. On body surfaces and in fluids, the normal growth form of the organism is as a yeast. When present in large numbers, these organisms cause toxic and inflammatory reactions due to cellular components.[56] Once actual colonization and invasion of a tissue or organ has occurred, the proportion of mycelial elements increases with the age of the lesion. Thus the finding of mycelial strands in sputum, feces, urine, and so forth, connotes either an established *Candida* lesion or significant colonization. This correlation has also been found to hold true in systemic lesions, i.e., a predominance of yeast forms indicates recent dissemination and an early lesion, whereas in old lesions, mycelium, blastospores, and even thick-walled macroconidia (chlamydospores) are found.

Gross Pathology. In systemic involvement, the lesions evoked by *Candida* depend on how recently dissemination occurred. If the condition was a rapidly fatal terminal event to some underlying disease, multiple microabscesses are seen. In chronic conditions, large granulomas are found that simulate tuberculosis. The organ most frequently involved in systemic disease is the kidney. In some studies the cortex alone or the cortex and medulla showed large, white, mycotic abscesses[92] (Fig. 8–13 FS B 21). Other case surveys report large granulomas confined to the medulla. The liver, lungs (Fig. 8–15 FS B 23), blood vessels, spleen, and thyroid are also involved, in that order of frequency. Ulcers of the esophagus and gut are small and usually overlooked unless carefully sought. Rarely, involvement of the eye and brain (Fig. 8–16) have been recorded. In these cases, granulomas or microabscesses were found.[72] In candidal endocarditis, large vegetations are seen on the mitral valves, cusps, and aortic walls (Fig. 8–14 FS B 22). In addition, granulomatous lesions may occur in the myocardium and walls of coronary vessels. A few reports of involvement of other organs have been tabulated.[92]

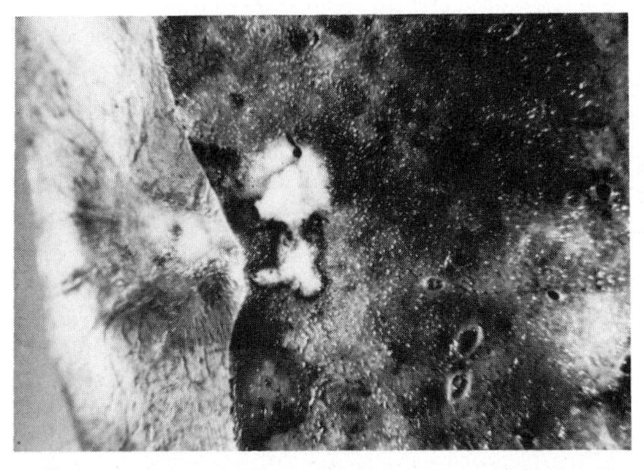

Figure 8–15. FS B 23. Pulmonary candidosis. Cut section of lung. There are firm, white nodules surrounded in some areas by a narrow hyperemic zone. The pleura is indented at this site. (Courtesy of F. Strauss.)

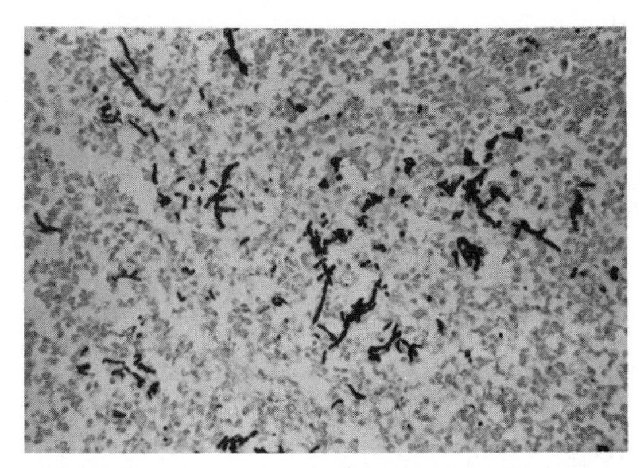

Figure 8–16. Cerebral candidosis. The lesion consisted of a mixture of acute pyogenic and granulomatous responses. The organisms were numerous in the lesions. The patient had been on cytotoxins for treatment of Hodgkin's disease.

Histopathology. In cases of rapid dissemination of disease, the microabscesses contain numerous degenerating neutrophils, and there is an acute pyogenic reaction. Small focal masses of yeast cells surrounded by a zone of histiocytes are scattered throughout the tissues. This is particularly common in cerebral involvement (Fig. 8–17). In very debilitated patients, the cellular reaction is minimal or absent. When granulomas are formed they consist of a chronic inflammatory infiltrate, giant cells, and histiocytes. They are usually not as well organized as tubercular lesions. Lesions of the gut consist of a necrotic reaction and ulcer formation. The fungal elements spread and penetrate downward from the site of the initial colonization (Fig. 8–18). Both yeast and mycelial elements are found, but mycelia predominate (Figs. 8–19 and 8–20).

In superficial cutaneous candidosis, the histologic picture is similar to that of subacute or chronic dermatitis. The fungi are restricted to the stratum corneum (Figs. 8–21 and 8–22). However, on histologic examination, *Candida* granuloma of the skin shows a hyperkeratosis and acanthosis in the epidermis, and a chronic inflammatory reaction with giant cells in the dermis. Very rarely are fungi found in the dermis or epidermis.

Candida in tissue is easily demonstrated. In hematoxylin and eosin–stained sections, the organisms vary in staining intensity and may be missed. The Gridley and GMS stains are

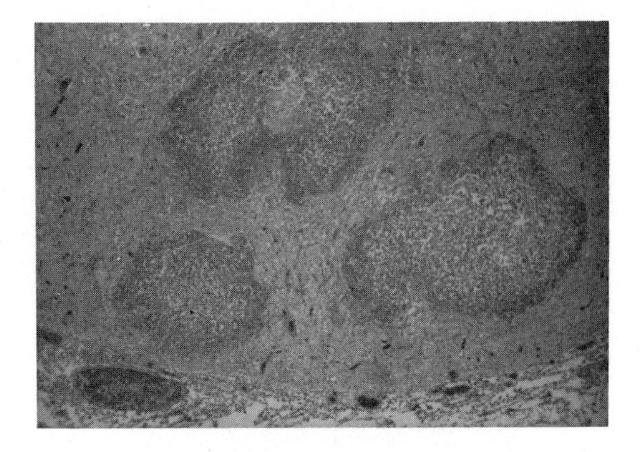

Figure 8–17. Cerebral candidosis. Numerous microabscesses with degenerating neutrophils in central area and a peripheral zone of histiocytes.

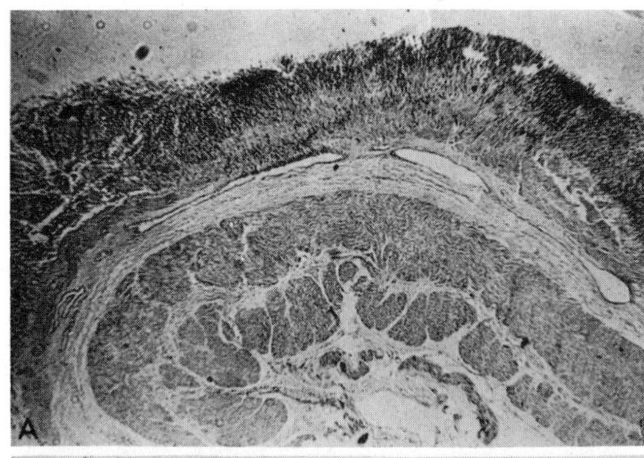

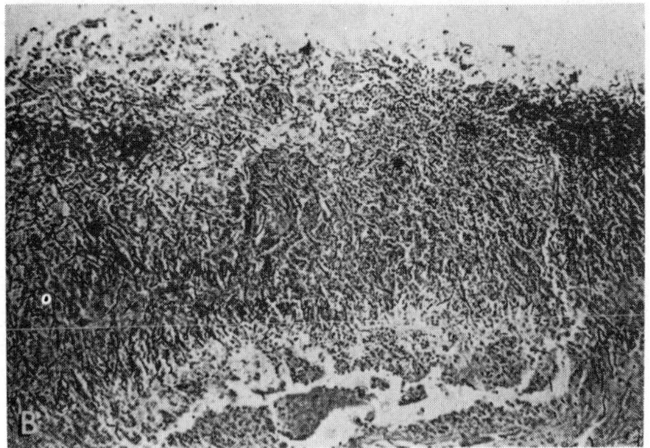

Figure 8–18. *Candida* of intestine. *A,* Areas of necrosis with evidence of *Candida* invasion of ulcers. Gram stain, ×100. *B,* Downward growth of *Candida* mycelium from loci of colonization. Note necrosis without granulomatous response. The patient had terminal leukemia.

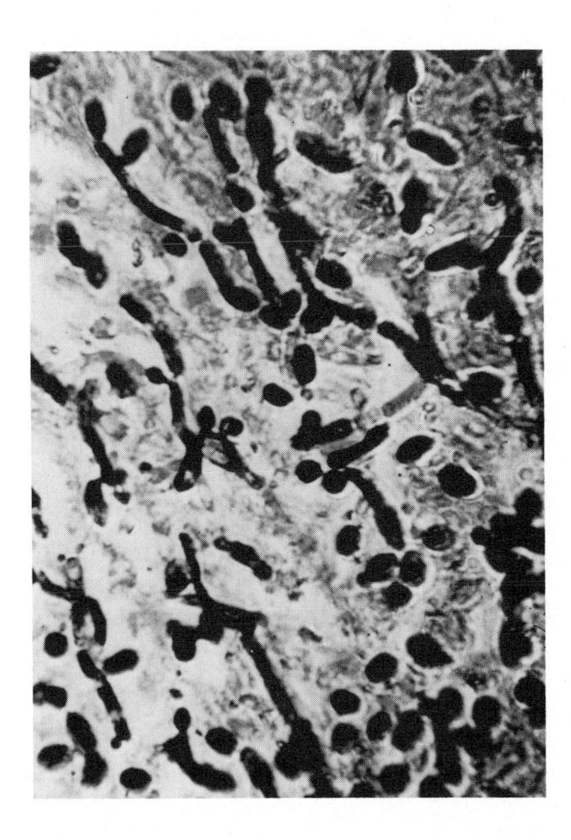

Figure 8–19. Candidosis of kidney. Mixture of yeast and mycelial elements. *C. albicans.* Gram stain, ×600. (From Rippon, J. W. *In* Burrows, W. 1973. *Textbook of Microbiology.* 20th ed. Philadelphia, W. B. Saunders Company, p. 713.)

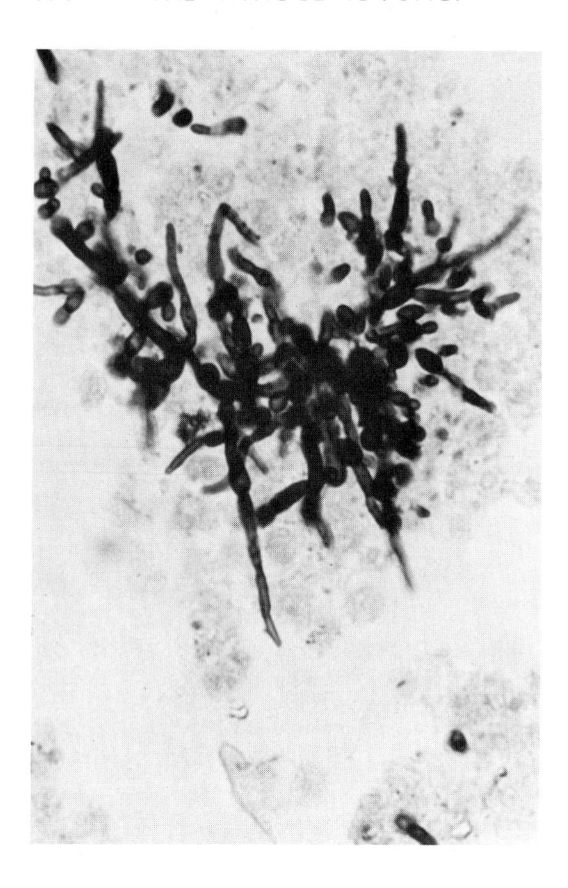

Figure 8–20. C. *tropicalis* in embolus in thigh of a patient with iatrogenic candidosis due to use of contaminated catheter in hyperalimentation. A focus of yeast cells which are sprouting to form pseudohyphae and hyphae is seen. GMS, ×600.

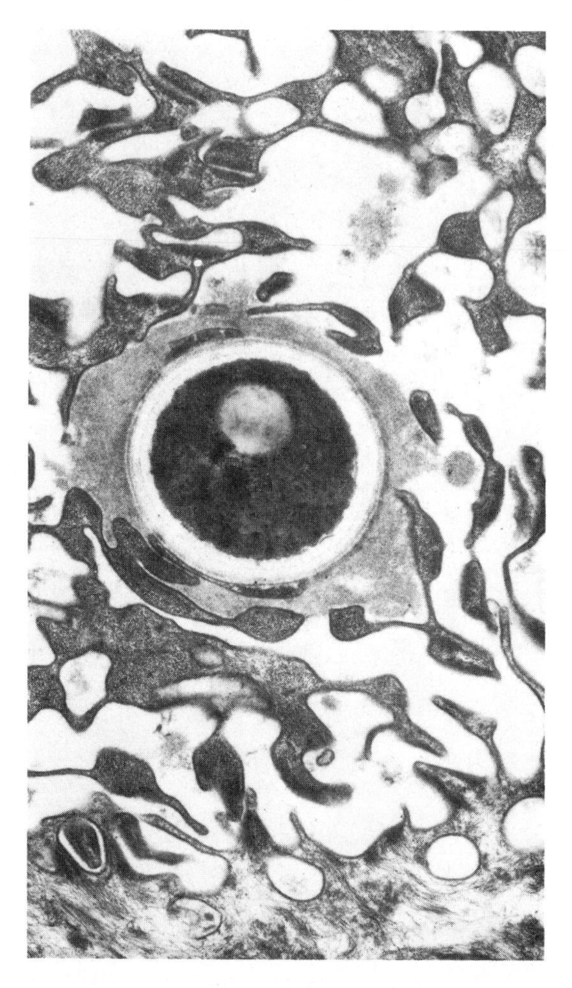

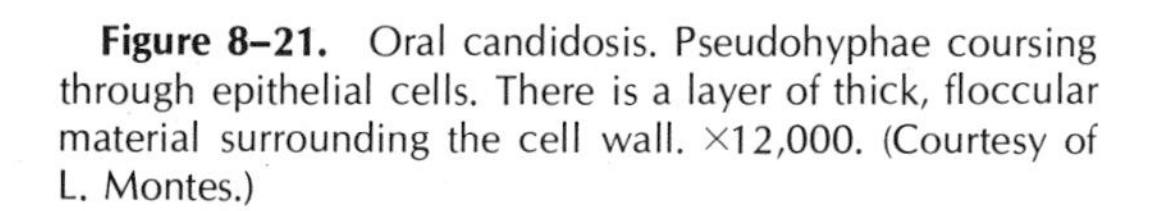

Figure 8–21. Oral candidosis. Pseudohyphae coursing through epithelial cells. There is a layer of thick, floccular material surrounding the cell wall. ×12,000. (Courtesy of L. Montes.)

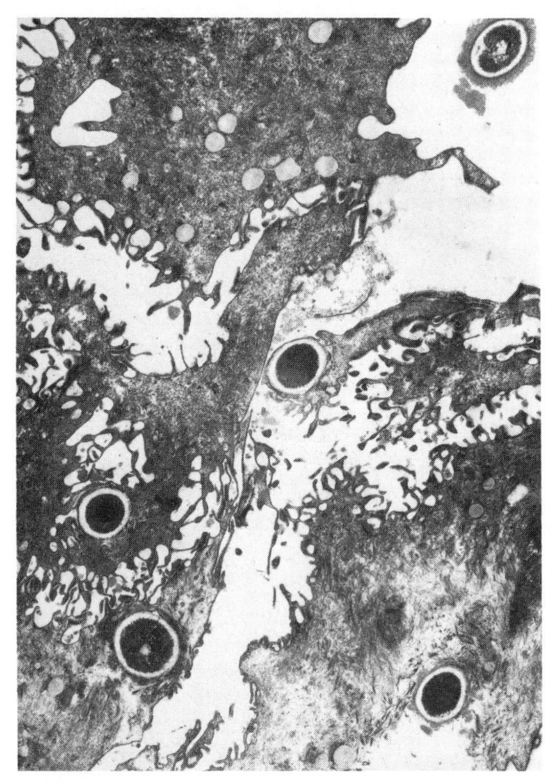

Figure 8–22. Oral candidosis. Intra- and extracellular hyphae of *Candida* in epithelial cells. ×6000. (Courtesy of L. Montes.)

more useful in demonstrating the organisms. The Gram stain is also very helpful. The usual histologic picture consists of masses of intra- and extracellular mycelial and yeast elements. Although the appearance of Candida in invaded tissue is quite characteristic, vacuolated, small, intracellular blastospores may resemble *Histoplasma* yeast cells, and thick-walled spores may emulate the tissue phase of *Blastomyces.* Asteroid bodies in systemic candidosis similar to those seen in sporotrichosis have been reported.[7]

ANIMAL DISEASE

Natural infections of domestic and wild animals are frequently encountered. Birds, particularly domestic fowl, are quite susceptible to buccal, upper alimentary tract, and crop candidosis which clinically and histologically resembles human thrush.[1] *Candida albicans* has been found to be a part of the normal flora of many animal species, and the opportunistic nature of infection is similar to that observed in human disease.[33,81] Most infections occur in the alimentary or respiratory tract, though systemic and chronic cutaneous candidosis in domestic and wild animals have been reported. Cutaneous disease is not uncommon in unhygienically maintained swine.[70]

Experimental cutaneous infections in dogs have been produced under occlusive dressings.[76] Histologically the appearance of the lesion, a contact dermatitis, was the same as was seen in human cases. The mouse and rabbit are equally sensitive (w/w) to intravenous injection of *C. albicans.*[68a] The target organs in normal animals are the kidneys and the heart. The two serologic types (A and B) of *C. albicans* are equally pathogenic, but *C. stellatoidea* (antigenic group B) is relatively avirulent. *C. parapsilosis* and *C. guilliermondi* are also relatively avirulent.[27] *C. tropicalis* is intermediate in virulence, requiring 10 to 1000 times the MLD of *C. albicans.*[34] Some difference in the virulence of various isolants of *C. albicans* has been noted.[62] In the altered host (x-ray, steroids, antibiotics, or concomitant bacterial infection), experimental candidosis is easier to induce, and the pathology more nearly resembles that seen in comparable human disease in which there were predisposing factors to infection.[39] Cortisone, sex hormones, and somatotropic hormones also effect experimental disease. Scherr[75] found that testosterone greatly enhances the rate of dissemination and death in experimental candidosis of mice.

IMMUNOLOGY AND SEROLOGY

The normal human adult has a high innate immunity to infection by *Candida.* The organism lives as a commensal on body surfaces, and disease is very uncommon unless there is alteration of host defenses or predisposing environmental conditions. The factors responsible for this natural resistance appear to be many and varied. Serum components, such as transferrin,[21] clumping factor,[27,54] and so forth, exhibit anticandidal effects *in vitro,* but their role in prevention of disease *in vivo* is unclear. Saturation of transferrin by iron has been noted to predispose to candidal infection in some cases.[15] Presence or absence of circulating antibody appears to be without significance in either clinical disease or experimental infection.[92] Many normal people react to skin-test antigens, but this has not been correlated with past or present disease. Evidence is accumulating from experimental infections[23, 68a] and clinical observations on patients that the major deterrents to candidal invasion consist of at least three components: (a) an intact cellular defense system,[9] (b) opsonic factors in serum, and (c) candidal capability of polymorphonuclear leukocytes.[46,51]

The serology of *Candida* species has been the subject of numerous studies. These have not yet been adequately coordinated or correlated.[92] As would be expected of such a large organism, many antigenic components have been found. A system similar to the Kaufman-White scheme for *Salmonella* has been worked out by Tsuchiya[89a] for the species in the genus *Candida.* This is in general agreement with the nucleic acid studies of Stenderup and Bak.[82] *C. albicans* can be divided into two large serologic groups, A and B. The first group also includes *C. tropicalis*, and the B group includes[12] *C. stellatoidea.*[34]

The serodiagnosis of candidal disease has been the subject of controversy in the literature. It was realized early that patients with candidal infections as well as normal subjects showed hypersensitivity to skin testing with *Candida* antigen, and that many people had agglutinins in their sera. Neither of these findings could be correlated with disease or effective immunity, except that anergy to the skin test was found in disseminated cases of candidosis. Presently, there is increasing evidence that the presence of immunodiffusion bands and perhaps complement fixation titers are of diagnostic and prognostic value.[13,63,84,85,86] All patients with disseminated candidosis, as well as some with *Candida* granuloma and chronic mucocutaneous candidosis, demonstrate precipitin bands in the immunodiffusion test. The test is negative in normal subjects and in those with superficial candidosis. Cytoplasmic antigens containing both a protein fraction and a mannan fraction have been prepared. When reacted with sera from patients with systemic disease,[13,84] these show a band against the protein fraction designated R, and a band against the mannan fraction designated H. The antigen is prepared from disrupted cells, although the skin test antigen Oidiomycin (Hollister-Stier extract diluted 1:10) will give similar results.[85] The presence of multiple bands is thought to correspond to active disease. In the study of Negroni,[63] complement-fixing antibodies were found only in patients with systemic disease. Fluorescent antibody techniques have been developed for the identification of *Candida* species in culture and in tissue.[28] Demonstrable antibodies are absent in severely debilitated patients, such as those who have leukemia or malignant lymphoma. In these and patients on cytotoxins, irradiation, steroids, or other immunosuppressive regimens, tests for circulating antibody are usually negative. However, antigens produced by the fungus are probably present in the patients' sera, but analytic procedures for their detection have not as yet been developed.

LABORATORY IDENTIFICATION

Direct Examination. Scrapings from cutaneous and mucocutaneous lesions can be examined directly either in potassium hydroxide slide mounts or by Gram stain. The rather characteristic mixture of yeast and mycelial phase organisms permits a rapid diagnosis of such infections. Sputum, vaginal discharge, urine specimens, and fecal samples present more difficulty. The mere finding of yeast in such material is of no diagnostic importance; however, mycelial-form organisms usually connote an established colonization of the involved area (Fig. 8–23). Only fresh specimens should be examined in this manner, as *Candida* multiplies rapidly in such milieu and often converts to a mycelial form in time. This would then give an erroneous impression as to number of organisms present and suggest the possibility of established colonization.

In cases of iatrogenic septicemia especially, the organism can be seen in blood smears. Either Wright's stain or Giemsa stain can be used to demonstrate the organism.[3]

Culture Methods. For culture it is imperative that only freshly obtained specimens be examined. At room temperature, the organisms grow rapidly and will give a false impression of initial numbers present. As confirmation of the diagnosis of candidosis often depends on quantitation of yeasts, this precaution is extremely important. Because of the inherent errors in culture methods, multiple specimens should be examined. *Candida* will grow on almost all common laboratory media, although for direct isolation Sabouraud's agar with antibacterial antibiotics is recommended. Most species of

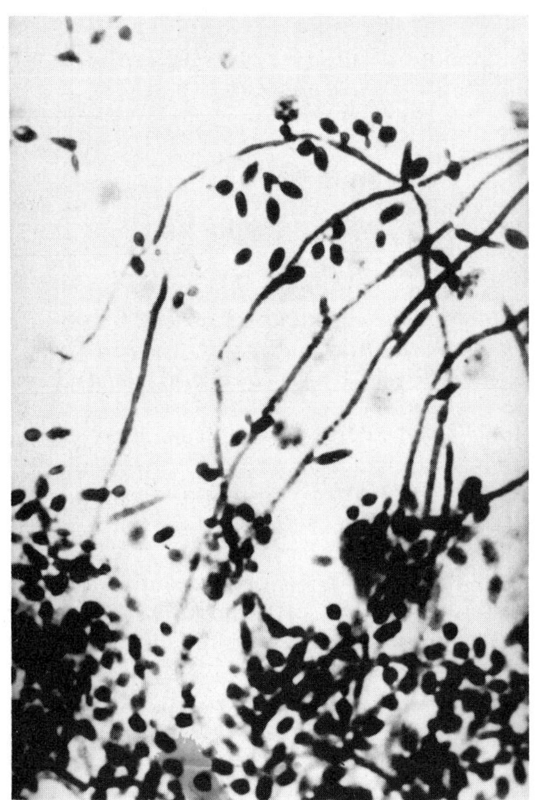

Figure 8–23. Pulmonary candidosis. Gram stain of sputum showing mixture of yeast and mycelial forms. The finding of hyphal units indicates colonization and invasion of tissue. ×440. (From Rippon, J. W. *In* Burrows, W. 1973. *Textbook of Microbiology.* 20th ed. Philadelphia, W. B. Saunders Company, p. 715.)

Candida are unaffected by the cycloheximide used in selective media for pathogenic fungi, but some strains of *C. tropicalis*, *C. krusei*, and *C. parapsilosis* are sensitive to it. Optimal growth of all species occurs at room temperature. A pasty, yeastlike colony appears by 24 to 48 hours.

As *C. albicans* is the most important yeast in human disease, many procedures have been proposed for its rapid identification. One of the most reliable is incubation of the unknown yeast in serum at 37°C for two hours. After such incubation, only *C. albicans* shows sprout mycelium, also called germ tubes (Reynolds-Braude phenomenon), which can be used to identify the organism[71,87] (Fig. 8–24). Furrowing inoculum into the Nickerson-Mankowshi chlamydospore agar (Fig. 8–25) or cornmeal agar (Fig. 8–26) is used to promote chlamydospore production. Among the yeasts encountered in man, only *C. albicans* produces this spore. *C. australis*,[29] which is found in penguin dung, also produces chlamydospores, as does *C. clausenii*, a rarely encountered saprophyte.

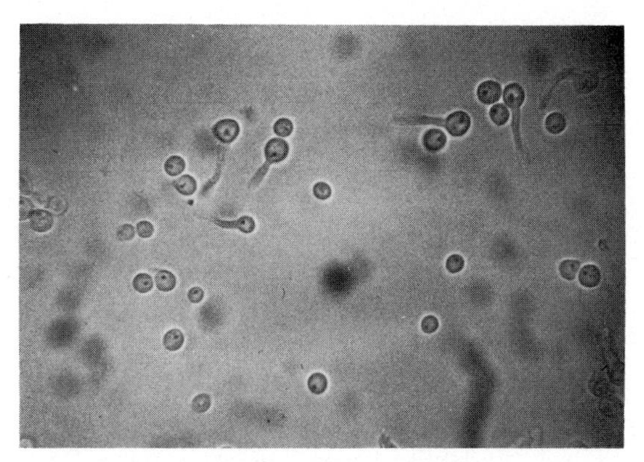

Figure 8–24. Reynolds-Braude phenomenon. *C. albicans.* Sprout mycelium develops after incubation of yeast cells for two hours in serum at 37°C. ×440.

In the event that an isolated yeast cannot be identified as *C. albicans* by chlamydospore production or sprout mycelium, other tests are available. These include mycelial morphology, carbohydrate fermentation, and carbohydrate assimilation. Transition of

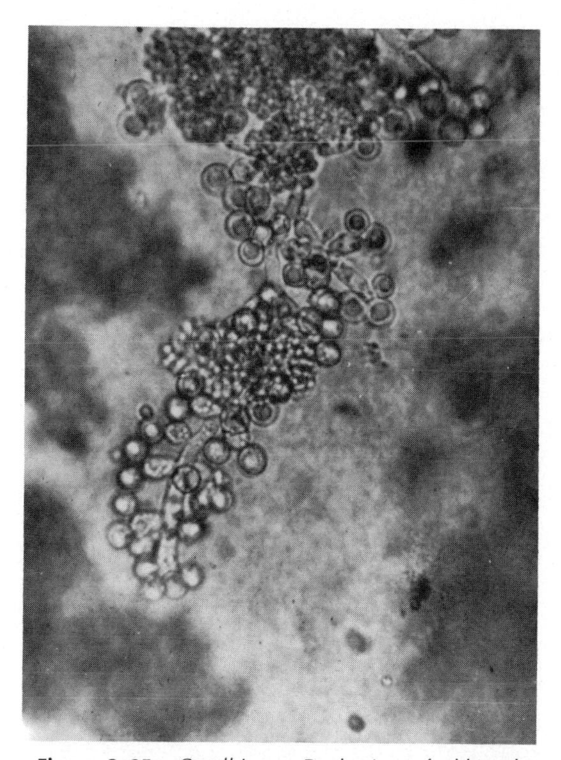

Figure 8–25. *C. albicans.* Production of chlamydospores and internodal blastospores when the fungus is grown on Nickerson-Mankowski medium. ×440. (From Rippon, J. W. *In* Burrows W. 1973. *Textbook of Microbiology.* 20th ed. Philadelphia, W. B. Saunders Company, p. 712. Courtesy of S. McMillen.)

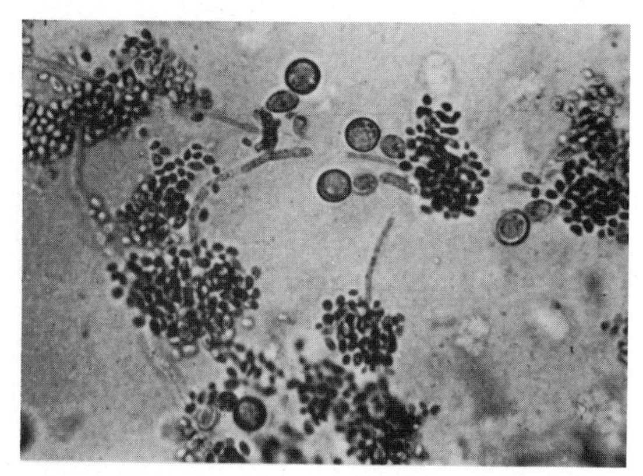

Figure 8–26. *C. albicans.* Large chlamydospores and blastospores produced in furrowed cornmeal agar at 25°C for three days. ×440.

growth to the mycelial stage in the absence of fermentable carbohydrate is a characteristic of the genus *Candida.* On cornmeal or Nickerson-Mankowski agar, the commonly encountered human pathogenic and saprophytic species produce rather characteristic my celial forms and structures (Fig. 8–27). For accurate identification, it is advisable to combine this test with carbohydrate studies. To carry out the latter, the yeast is streaked on a sugar-free medium, such as heart infusion agar, to deplete the organism of fermentable carbohydrates. Growth from such plates is inoculated into a fermentation series and carbohydrate assimilation series (see Appendix). With the combined information, the species can be keyed out in Lodder's *The Yeasts,*[52] or the accompanying chart with its summary of morphologic and physiologic

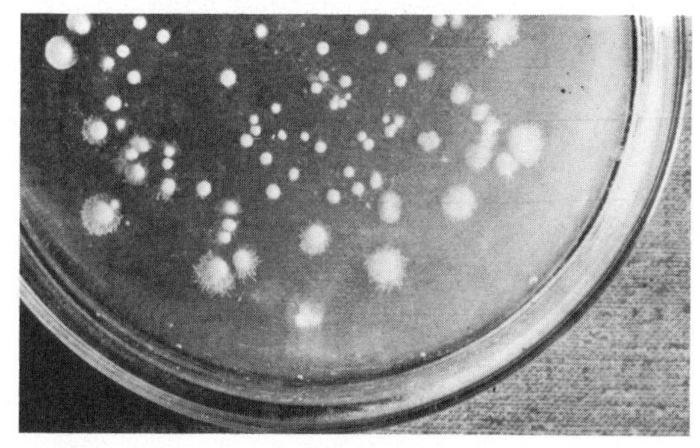

Figure 8–27. *C. albicans.* Colonial growth on Sabouraud's glucose agar at 25°C for two weeks. Hyphae and pseudohyphae can be seen radiating from the large colonies into the agar. (Courtesy of S. McMillen.)

characteristics may be used (Table 8–2). Animal inoculation is not useful in the identification of *Candida* species.

MYCOLOGY

Candida albicans (Robin) Berkhout 1923

Synonymy. *Oidium albicans* Robin 1853; *Monilia albicans* (Robin) Zopf 1890; *Syringospora robinii* Quinquad 1869; *Endomyces pinoyi* Castellani 1912; *Mycotoruloides triadis* Langeron and Talice 1932; *Candida genitalis* Batista and Silveria 1962; *Candida intestinalis* Batista and Silveria 1959; Lodder[52] lists 90 synonyms. The organism was described by Gruby in 1842 as a *Sporotrichum* sp.

Colony Morphology. SDA, 25°C,* three days—creamy smooth; one month—creamy, glistening, waxy, soft, smooth to somewhat reticulated; old stocks wrinkled and folded with spicules.

Microscopic Morphology. 1. SDB, 25°C, three days—globose, short ovoid (5 to 7 μ), sometimes elongate yeasts (4 to 6 × 6 to 10 μ). Smaller and larger cells seen. A thin ring may form at the top of the tube. 2. CM or N PTB, 25°C, three days—mycelium and pseudomycelium formed. Masses of blastospores (~4 μ) at internodes. Terminal, thick-walled (8 to 12 μ) macroconidia (chlamydospores) are formed by most strains. Among the commonly encountered yeasts from human sources, the latter characteristic is specific for this species. On N PTB the macroconidia are blue. 3. Serum or egg albumin (Reynolds-Braude phenomenon), 37°C, two hours—sprout mycelium formed. Essentially all strains of *C. albicans* are positive in this test, and no other commonly encountered species demonstrates this reaction.

Fermentation. Glucose +, galactose +, sucrose+/−, maltose+, cellobiose−, lactose−, melibiose −, raffinose −, melizitose −, inulin −, trehalose +/w/−.

Assimilation. Glucose +, galactose +, sucrose +, maltose +, cellobiose −, trehalose +, lactose −, melibiose −, raffinose −, melezitose+/−, starch+, D-xylose+, D-ribose −, L-rhamnose−, ethanol+/−, glycerol+/−,

*SDA = Sabouraud's dextrose agar; SDB = Sabouraud's dextrose broth; CM = cornmeal; N PTB = Nickerson polysaccharide trypan blue (Nickerson-Mankowski agar); + = positive; − = negative; +/− or v = variable; w = weak.

Table 8–2. *Summary of Differential Characteristics of Candida Species Encountered in Human Disease**

	RB	Pellicle Formation	Chlamydospore	Fermentation				Assimilation							Utilization	
				Gl	*M*	*S*	*L*	*Gl*	*Gal*	*L*	*M*	*R*	*S*	*C*	*Ar*	*Et*
C. albicans	+	0	+	AG	AG	A or 0	0	+	+	0	+	0	+	0	0	v
C. guilliermondi	0	0	0	A/AG	0	A/AG	0	+	+	0	+	+	+	+	+	+
C. krusei	0	Wide film	0	AG	0	0	0	+	0	0	0	0	0	0	0	+
C. parapsilosis	0	0	0	A/AG	0	0	0	+	+	0	+	0	+	0	0	0
C. stellatoidea	0	0	Rare	AG	AG	0	0	+	+	0	+	0	0	0	0	+
C. tropicalis	0	Narrow film; bubbles	0	AG	AG	AG	0	+	+	0	+	0	+	+	v	v
C. pseudotropicalis	0	0	0	A/G	0	AG	AG	+	+	+	0	+	+	+	+	v
C. viswanathii	0	–	0	A	A	0	0	+	+	0	+	0	v	+	+	+

Gl = glucose; M = maltose; S = sucrose; L = lactose; Gal = galactose; R = raffinose; C = cellobiose; Ar = arbutin split; Et = ethanol utilized; A = acid; G = gas; v = variable; RB = Reynolds-Braude phenomenon; (+) = assimilation; (0) = no assimilation.

*Modified from Lodder.[32] For methods of determination, see Appendix.

erythritol –, ribitol +/–, D-mannitol +, L-arabinose +/–, D-arabinose –, inositol –, potassium nitrate –, L-sorbose +/–, inulin +/–.

Perfect Stage. As yet unknown. *Syringospora* with affinities to the Basidiomycetes has been proposed as the perfect stage, but this remains to be substantiated.

Candida guilliermondi (Castellani) Langeron and Guerra 1938

Synonymy. *Endomyces guilliermondi* Castellani 1912; *Monilia guilliermondi* Castellani and Chalmers 1913; *Myzeloblastanon guilliermondi* Ota 1928; *Mycotorula guilliermondi* Langeron and Guerra 1935; *Blastodendrion guilliermondi* Guerra 1935; *Castellania guilliermondi* Dodge 1935; *Monilia pseudoguilliermondi* Castellani and Chalmers 1919; *Castellania pseudoguilliermondi* Dodge 1935; *Torula fermentati* Saito 1922; *Myzeloblastanon arzti* Ota 1924; *Myzeloblastanon krausi* Ota 1924; *Blastodendrion krausi* Ciferri and Redaelli 1929; *Mycotorula krausi* Redaelli and Ciferri 1947; *Microanthomyces alpinus* Gruss 1926; *Candida melibiosi* Lodder and Kreger-Van Rij 1952; *Trichosporon appendiculare* Batista and Silveria 1959.

Colony Morphology. SDA 25°C, three days—thin, flat, glossy, cream to pinkish; one month—yellowish cream to pink, glistening, smooth, or dull, wrinkled.

Cell Morphology. 1. SDB 25°C, three days—short ovoid or ovoid cells (2 to 5 × 3 to 7 μ); also small and cylindrical cells. A ring and islets may form over the broth surface. 2. CM or N PTB 25°C, three days—variable development of pseudomycelium. Usually very fine and short with small cells, sometimes bearing ramified chains of small ovoid cells or stalagtoid, verticillated blastospores.

Fermentation. Glucose +, galactose +/w, sucrose +/w, maltose –, cellobiose –, trehalose +/w; lactose–, melibiose+/–, raffinose +/w, melezitose –, inulin +/w.

Assimilation. Glucose +, galactose +, L-sorbose +, sucrose +, maltose +, cellobiose +, trehalose +, lactose –, melibiose +, raffinose +, melezitose +, inulin +, D-xylose +, L-arabinose +, D-arabinose +, ethanol +/–, D-ribose +/w/–, soluble starch +/w/–, L-rhamnose +/w/–, ribitol +, glycerol +, erythritol –, D-mannitol +, galactitol +, salicin +, inositol –, potassium nitrate –.

The above is a description of *C. guilliermondi* var. *guilliermondi*. The var. *carpophila* differs by no fermentation of inulin, and no assimilation of melezitose, starch, L-rhamnose, or galactitol.

Perfect Stage. Mating strains are *Pichia guilliermondi*. Many strains do not mate.

The organism has been isolated from numerous human infections. It is found on normal skin and in sea water, feces of animals, fig wasps, buttermilk, leather, fish, and beer.

Candida krusei (Castellani) Berkhout 1923

Synonymy. *Saccharomyces krusei* Castellani 1910; *Endomyces krusei* Castellani 1912; *Monilia krusei* Castellani and Chalmers 1913; *Myceloblastanon krusei* Ota 1928; *Trichosporon krusei* Ciferri and Redaelli 1935; *Mycotoru-*

loides krusei Langeron and Guerra 1935; *Monilia parakrusei* Castellani and Chalmers 1919; *Castellania parakrusei* Dodge 1935; *Mycodermia chevalieri* Guilliermond 1914; *Candida chevalieri* Westerdijk 1933; *Mycoderma monosia* Anderson 1917; *Mycoderma bordetii* Kuff 1920; *Monilia inexpectata* Mazza 1930; *Mycocandida inexpectata* Talice and Mackinnon 1934; *Pseudomonilia inexpectata* Dodge 1935; *Trichosporon dendriticum* Ciferri and Redaelli 1935; *Candida dendritica* Dodge and Moore 1936; *Monilia krusoides* Castellani 1937; *Pseudomycoderma miso* Mogi 1940; *Candida castelanii* van Uden and Assis-Lopez 1953; *Candida tamarindi* Lewis and Johar 1955; *Procandida tamarindi* Novak and Zsolt 1961; *Candida lobata* Batista and Silveria 1959.

Colony Morphology. SDA 25°C, three days—flat, dull, dry; one month—greenish-yellow, dull, soft, smooth or wrinkled, dense growth of mycelium extending as a lateral fringe around the colony.

Cell Morphology. 1. SDB 25°C, three days—cylindrical and a few ovoid cells, size varies considerably (3 to 5 × 6 to 20 μ). Some become very long. A thick pellicle develops across the broth and crawls up the side of the tube. 2. CM or N PTB 25°C, three days—elongated cells in treelike arrangements or "crossed matchsticks." Blastospores are elongate and grow in verticellate branches from mycelium. In some strains few blastospores are produced.

Fermentation. Only glucose is fermented.

Assimilation. Glucose +, galactose −, L-sorbose −, sucrose −, D-ribose −, L-rhamnose −, glycerol +/−, maltose −, cellobiose −, trehalose −, melibiose −, raffinose −, melezitose −, inulin −, starch −, D-xylose −, L-arabinose −, D-arabinose −, erythritol −, ribitol −, D-mannitol −, salicin −, inositol −, ethanol +, lactose −, galactitol −, potassium nitrate −.

Pefect Stage. *Pichia kudriavezii.* Regularly associated with some forms of infant diarrhea and occasionally with systemic disease. Found in beer, milk products, skin, feces of animals and birds, and pickle brine.

Candida parapsilosis (Ashford) Langeron and Talice 1959

Synonymy. *Monilia parapsilosis* Ashford 1928; *Mycocandida parapsilosis* Dodge 1935; *Monilia onychophila* Pollicci and Nannizzi 1926; *Mycocandida onychophila* Langeron and Telice 1932; *Blastodendrion intestinale* Mattlet 1941; *Blastodendrion globosum* Zach 1933; *Blastodendrion gracile* Zach 1933; *Schizoblastosporion gracile* Dodge 1935.

Colony Morphology. SDA 25°C, three days—soft, smooth, white, sometimes lacy; one month—cream to yellowish, glistening, smooth or wrinkled.

Cell Morphology. 1. SDB 25°C, three days—short-ovoid to long-ovoid cells (2.5 to 4 × 2.5 to 9 μ). 2. CM or N PTB 25°C, three days—thin pseudomycelium, much branching, verticils of few ovoid to elongate blastospores. Thick pseudomycelium and giant cells also found.

Fermentation. Glucose +, galactose +/w/−, sucrose −, maltose −, cellobiose −, trehalose −, lactose −, melibiose −, raffinose −, melezitose −, inulin −.

Assimilation. Glucose +, galactose +, L-sorbose +/w, sucrose +, maltose +, cellobiose −, trehalose +, lactose −, melibiose −, raffinose −, melezitose +, inulin −, D-xylose +, L-arabinose +/−, D-arabinose −, D-ribose +/w/−, L-rhamnose −, ethanol +, glycerol +, erythritol −, ribitol +, D-mannitol +, D-glucitol +, salicin −, inositol −, starch −, potassium nitrate −.

Perfect Stage. *Lodderomyces elongisporus.* Regularly found in nail disease and endocarditis. Found in pickle brine, normal skin, and feces.

Candida stellatoidea (Jones and Martin) Langeron and Guerra 1939

Synonymy. *Monilia stellatoidea* Jones and Martin 1938; *Procandida stellatoidea* Novak and Zsolt 1961.

Colony Morphology. SDA 25°C, three days—slow-growing, small, creamy smooth; one month—whitish to cream color, glistening, soft, smooth or yellowish, dull, soft, verrucose or lightly folded. On blood agar it grows as small, stellate colonies, hence the name.

Cell Morphology. 1. SDB 25°C, three days—short or long ovoid (4 to 8 × 5 to 10 μ), elongate, or apiculate cells sometimes seen. 2. CM or N PTB 25°C, three days—much branched pseudomycelium in treelike arrangement composed mostly of short cells, irregular clusters of blastospores at internodes, sometimes blastospores in long chains. Chlamydospores rarely produced; may have a supporting cell.

Fermentation. Glucose +, galactose −, sucrose −, maltose +, cellobiose −, lactose −, melibiose −, raffinose −, melezitose −, inulin −, trehalose −.

Assimilation. Glucose +, galactose +, L-sorbose −, sucrose −, maltose +, cellobiose −, trehalose +, lactose −, melibiose −, raffinose −, melezitose −, inulin −, starch +, D-xylose +, L-arabinose +/−/w, D-arabinose −, D-ribose −, L-rhamnose −, ethanol +, glycerol +/−, erythritol −, ribitol −, D-mannitol +, salicin −, inositol −, potassium nitrate −.

Most isolants come from vaginal discharges.

Candida tropicalis (Castellani) Berkhout 1923

Synonymy. *Oidium tropicalis* Castellani 1910; *Endomyces tropicalis* Castellani 1911; *Monilia tropicalis* Castellani and Chalmers 1913; *Candida vulgaris* Berkhout 1923; *Mycotorula japonicum* Yamaguchi 1943; *Trichosporon lodderi* Phaff, Mrak, and William 1952; *Mycotorula dimorpha* Redaelli and Ciferri 1935; Lodder[52] lists 57 synonyms.

Colony Morphology. SDA 25°C, three days—creamy white, smooth; one month—white to cream colored, dull, soft, smooth, reticulated or wrinkled, often with overgrowth of mycelium. Old stocks become hairy and tough.

Cell Morphology. 1. SDB 25°C, three days—globose-ovoid or short-ovoid cells (4 to 8 × 5 to 11 μ). A thin film forms over the broth that may have entrapped gas bubbles. 2. CM or N PTB 25°C, three days—abundant pseudomycelium composed of elongate cells with much branching, blastospores singly, along mycelium, or in clusters. True mycelium formed.

Fermentation. Glucose +, galactose +/w, sucrose +, maltose +, cellobiose −, trehalose +/w, lactose −, melibiose −, raffinose − melezitose +/−, inulin −.

Assimilation. Glucose +, galactose +, D-ribose −, L-rhamnose −, L-sorbose +/−, sucrose +, maltose +, cellobiose +/−/w, trehalose +, lactose −, melibiose −, raffinose −, inulin −, starch +, D-xylose +, L-arabinose +/w/−, D-arabinose −, ethanol +, glycerol +/w/−, D-mannitol −, ribitol +, salicin +/−/w, inositol −, melezitose +, potassium nitrate −.

This organism appears to be closely related to *C. albicans,* and there are reports of conversion between the two. It has been isolated from feces, shrimp, and kefir.

Candida pseudotropicalis (Castellani) Basgal 1931

Synonymy. *Endomyces pseudotropicalis* Castellani 1911; *Monilia pseudotropicalis* Castellani and Chalmers 1913; *Atelosaccharomyces pseudotropicalis* de Mello, Gonzagi, and Fernandez 1918; *Myzeblastanon pseudotropicalis* Ota 1928; *Mycocandida pseudotropicalis* Ciferri and Redaelli 1935; *Castellania pseudotropicalis* Dodge 1935; *Mycotorula pseudotropicalis* Redaelli and Ciferri 1947; *Torula cremoris* Hammer and Cordes 1920; *Candida mortifera* Redaelli 1925; *Mycocandida mortifera* Langeron and Talice 1932; *Monilia mortifera* Martin, Jones, Yao, and Lee 1937; *Blastodendrion procerum* Zach 1934.

Colony Morphology. SDA 25°C, three days—creamy smooth; one month—cream to yellowish, somewhat dull, soft, smooth or slightly reticulated.

Cell Morphology. SDB 25°C, three days—short-ovoid with a few elongate cells (2.5 to 5 × 5 to 10 μ). Pseudomycelium is abundant in most strains; in rare isolants none is formed. The cells are very elongate; they fall apart and lie parallel like "logs in a stream." Blastospores not abundant. When present, they are elongated and in verticils.

Fermentation. Glucose +, galactose +, sucrose +, maltose −, cellobiose −, trahalose −, lactose +, melibiose −, raffinose +/w, inulin +, melizitose −.

Assimilation. Glucose +, galactose +, L-sorbose −, sucrose +, sorbose −, maltose −, cellobiose +, trehalose −, lactose +, melibiose −, melezitose −, inulin +, D-xylose +, L-arabinose +, D-arabinose −, L-rhamnose −, ethanol +, glycerol +/w, ribitol −, D-mannitol +/w/−, salicin +, inositol −, raffinose +, starch −, D-ribose +/w/−, potassium nitrate −.

Perfect Stage. *Kluyveromyces fragilis.* Commonly isolated from nails and lung infections. Found in cheese and dairy products.

Candida viswanathii Sandhu and Randhawa 1959

Colony Morphology. SDA 25°C, three days—cream-colored, soft, glistening; one month—creamy, soft to membranous, wrinkled, semi-dull.

Cell Morphology. 1. SDB 25°C, three

days—cells globose, ovoid to cylindrical (2.5 to 7 × 4 to 12 μ). 2. CM or N PTB, long, wavy mycelium with irregular branches at angles up to 90°, globose to ovoid blastospores in chains, verticillately arranged.

Fermentation. Glucose +, galactose +/w, sucrose −, maltose +, cellobiose −, trehalose +, lactose −, melibiose −, raffinose −, melezitose −, inulin −.

Assimilation. Glucose +, galactose +, L-sorbose +/−, sucrose +, maltose +, cellobiose +, trehalose +, lactose −, melibiose −, raffinose −, melezitose +, inulin −, starch +, D-xylose +, L-arabinose +/−, D-arabinose −, D-ribose −, L-rhamnose −, ethanol +, glycerol +, erythritol −, ribitol +, D-mannitol +, galactitol −, salicin +, inositol −, potassium nitrate −.

Isolated from spinal fluid and sputum.[73]

REFERENCES

1. Ainsworth, G. C., and P. K. C. Austwick. 1959. *Fungal Diseases of Animals*. Bucks, England, Commonwealth Agricultural Bureau.
2. Akiba, T., and K. Iwata. 1954. On the destructive invasion of a new species *Bacterium candidostruens* into Candida cells. Jap. J. Exp. Med., *24*:159–166.
3. Anderson, A., and J. Yardley. 1972. Demonstration of candida in blood smears. New Engl. J. Med., *286*:108.
4. Arthur, L. J. H. 1969. Pulmonary candidosis. Proc. Roy. Soc. Med., *62*:906–907.
5. Ashcraft, K., and L. Leape. 1970. Candida sepsis complicating parenteral feeding. J.A.M.A., *212*:454–456.
6. Bak, A. L., and A. Stenderup. 1969. Deoxyribonucleic acid homology in yeasts. Genetic relatedness within the genus *Candida*. J. Gen. Microbiol., *59*:21–30.

6a. Balandran, L., H. Rothschild, et al. 1973. A cutaneous manifestation of systemic candidiasis. Ann. Intern. Med., *78*:400–403.

7. Berye, T., and W. Kaplan. 1967. Systemic candidiasis with asteroid body formation. Sabouraudia, *5*:310–314.
8. Berkhout, C. M. 1923. De schimmelgeschlachten Monilia, Oidium, Oospora en Torula. Dissertion, University of Utrecht.
9. Block, M. B., L. M. Pachman, et al. 1971. Immunological findings in familial juvenile endocrine deficiency syndrome associated with mucocutaneous candidiasis. Am. J. Med. Sci., *261*:213–218.
10. Boeckman, C. R., and C. E. Krill. 1970. Bacterial and fungal infections complicating parenteral alimentation in infants and children. J. Pediatr. Surg., *5*:117–126.
11. Boudru, I. 1969. De la résistance des *Candida albicans* à la nystatine. J. Pharm. Belg., *1969*: 162–185.
12. Brown-Thompson, J. 1966. Reverse variations between *Candida albicans* and *Candida tropicalis*. Acta Pathol. Microbiol. Scand., *66*:143–144.
13. Buckley, H., and E. W. Lapa. 1969. The value of serological tests in diagnosis of candidiasis. Antonie van Leeuwenhoek, *35*(Section E): 19–20.
14. Buckley, R. H., Z. J. Luzas, et al. 1968. Defective cellular immunity associated with chronic mucocutaneous moniliasis and recurrent staphylococcal botryomycosis: Immunological reconstitution by allogenic bone marrow. Clin. Exp. Immunol., *3*:153–169.
15. Caroline, L., F. Rosner, et al. 1969. Elevated serum iron, low unbound transferrin and candidiasis in acute leukemia. Blood, *34*:441–451.
16. Chilgren, R. A., P. G. Quie, et al. 1969. The cellular immune defect in chronic mucocutaneous candidiasis. Lancet, *1*:1286–1288.
17. Chilgren, R. A., R. Hong, et al. 1968. Human serum interactions with *Candida albicans*. J. Immunol., *101*:128–132.
18. Cohen, A. C. 1953. Pulmonary moniliasis. Am. J. Med. Sci., *226*:16–23.
19. Curry, C. R., and P. G. Quie. 1971. Fungal septicemia in patients receiving parenteral hyperalimentation. New Engl. J. Med., *285*:1221–1225.
20. DeVita, V. T., J. Utz, et al. 1966. Candida meningitis. Arch. Intern. Med., *117*:527–535.
21. Esterly, N. B., S. R. Brammer, et al. 1967. The relationship of transferrin and iron to serum inhibition of *Candida albicans*. J. Invest. Dermatol., *49*:437–442.
22. Fetter, B. F., G. K. Klintworth, et al. 1967. *Mycoses of the Central Nervous System*. Baltimore, Williams and Wilkins, pp. 53–62.
23. Fujiwara, A., J. W. Landau, et al. 1970. Responses of thymectomized and/or bursectomized chickens to sensitization with *Candida albicans*. Sabouraudia, *8*:9–17.
24. Ghoniem, N. A., and M. Retai. 1968. Incidence of Candida species in Damieth cheese. Mykosen, *11*:295–298.
25. Giunchi, G. 1958. Candidosis following antibiotic therapy. Sci. Med. Ital., *6*:580–628.
26. Goldman, G. A., M. L. Littman, et al. 1960. Monilial cystitis effective treatment with instillations of amphotericin B. J.A.M.A., *174*:359.
27. Goldstein, E., M. H. Grieco, et al. 1965. Studies on the pathogenesis of experimental *Candida paropsilosis* and *Candida guilliermondii* infections in mice. J. Infect. Dis., *115*:293–302.
28. Gordon, M. A., J. C. Elliott, et al. 1967. Identification of *Candida albicans*, other Candida species and *Torulopsis glabrata* by means of immunofluorescence. Sabouraudia, *5*:323–328.
29. Goto, S., J. Sugiyama, et al. 1969. A taxonomic study of Antarctic yeasts. Mycologia, *61*: 748–774.
30. Granits, J. 1954. Pilzkulturen aus Reihenunter suchungen von gesunder Epidermis. Dermatol. Wochenschr., *130*:1352–1356.
31. Greshem, G. A., and C. H. Whittle. 1961. Studies on the invasive, mycelial form of *Candida albicans*. Sabouraudia, *1*:30–33.

31a. Gruby, M. 1842. Recherches anatomiques sur une plante cryptogame qui constituae le vrai muguet des enfants. C. R. Acad. Sci. (Paris), *14*:634–636.

32. Guyer, P. B., F. J. Brunton, et al. 1971. Candidiasis of the oesophagus. Br. J. Radiol., *44*:131–136.
33. Hache, J. 1966. Les candidoses aviaires. Influence de l'oxytetracycline sur la sensibilite du Poulet a l'infection par *Candida albicans*. Ecole Nat. Vet. Alfort., Thesis, *79*, 49 pp.
34. Hasenclever, H. F., and W. O. Mitchell. 1961. Pathogenicity of *C. albicans* and *C. tropicalis*. Sabouraudia, *1*:16–21.
35. Hauser, F. V., and S. Rothman. 1950. Monilial granuloma: Report of a case and review of the literature. Arch. Dermatol., *61*:297–310.
36. Hermens, P. E., J. A. Ulrich, et al. 1969. Chronic mucocutaneous candidiasis as a surface expression of deep seated abnormalities. Am. J. Med., *47*:503–519.
37. Hildrick-Smith, G., H. Blank, et al. 1964. *Fungus Diseases and Their Treatment*. Boston, Little, Brown & Co.
38. Holti, G. 1966. *Symposium on Candida Infections*. Edinburgh, E. & S. Livingstone, p. 73.
39. Hurley, R. 1966. Experimental infection with *Candida albicans* in modified hosts. J. Path. Bact., *92*:57–67.
40. Hurley, R., and V. C. Stanley. 1969. Cytopathic effects of pathogenic and nonpathogenic species of Candida on cultured mouse epithelium cells: Relation to growth rate and morphology of the fungi. J. Med. Microbiol., *2*:63–74.
41. Imperato, P. J., C. E. Buckley, et al. 1968. Candida granuloma. A clinical and immunologic study. Arch. Dermatol., *97*:139–146.
41a. Joachim, H., and S. Polayes. 1940. Subacute endocarditis and systemic mycosis *(Monilia)*. J.A.M.A., *115*:205–208.
42. John, R. W. 1968. Necrosis of oral mucosa after local application of crystal violet. Br. Med. J., *1*:157–158.
43. Kantrowitz, P. A., D. J. Fleischli, et al. 1969. Successful treatment of chronic esophageal moniliasis with a viscous suspension of nystatin. Gastroenterology, *57*:424–430.
44. Kapica, L., and A. Clifford. 1968. Quantitative determination of yeasts in sputum. Mycopathologia, *34*:27–32.
45. Kaufman, S. A., S. Scheff, et al. 1960. Esophageal moniliasis. Radiology, *75*:726–732.
46. Kim, M. H., G. E. Rodey, et al. 1969. Defective candidacidal capacity of polymorphonuclear leukocytes in chronic granulomatous disease of children. J. Pediatr., *75*:300–303.
47. Kovaks, E., B. Bucz, et al. 1969. Propagation of mammalian viruses in protista. IV. Experimental infection of *C. albicans* and *S. cerevisiae* with polyoma virus. Proc. Soc. Exp. Biol. Med., *132*:971–977.
48. Koznin, P., and C. Taschdjian. 1966. *Candida albicans:* saprophyte or pathogen? J.A.M.A., *198*:170–173.
49. Koznin, P. J., and C. J. Taschdjian. 1962. Enteric candidiasis: diagnosis and clinical considerations. Pediatrics, *30*:71–85.
50. Krause, W., H. Matheis, et al. 1969. Fungaemia and funguria after oral administration of *Candida albicans*. Lancet, *1*:598–599.
50a. Kroll, J., J. M. Einbinder, et al. 1973. Mucocutaneous candidiasis in a mother and son. Arch. Dermatol., *108*:259–262.
50b. Langenbeck, B. 1839. Auffingung von Pilzen aud der Schleimhaut der speiseröhre einen Typhus heiche. Neue Not. Geb.-Natur-u Heilk (Froriep), *12*:145–147.
51. Lehrer, R. I., and M. J. Cline. 1969. Interaction of *Candida albicans* with human leukocytes and serum. J. Bacteriol., *98*:996–1004.
51a. Lindquist, J. A., S. Rabinovich, et al. 1973. 5-Fluorocytosine in the treatment of experimental candidiasis in immunosuppressed mice. Antimicrob. Agents Chemother., *4*:58–61.
52. Lodder, J. (ed.) 1970. *The Yeasts*. 2nd ed. Amsterdam, North-Holland Publishing Company.
53. Louria D. B., D. P. Stiff, et al. 1962. Disseminated moniliasis in the adult. Medicine, *41*:307–337.
54. Louria, D. B., and R. G. Brayton. 1964. The behavior of Candida cells within leucocytes. Proc. Soc. Exp. Biol. Med., *115*:93–98.
55. Lynch, P. J., W. Minkin, et al. 1969. Ecology of *Candida albicans* in candidiasis of groin. Arch. Dermatol., *99*:154–160.
56. Maibach, H. I., and A. M. Kligman. 1962. The biology of experimental human cutaneous moniliasis. Arch. Dermatol., *85*:233–257.
57. Marples, M. J. 1965. *The Ecology of the Human Skin*. Springfield, Charles C Thomas.
58. Marples, M., and D. A. Somerville. 1968. The oral and cutaneous distribution of *Candida albicans* and other yeasts in Raratonga, Cook Islands. Trans. R. Soc. Trop. Med. Hyg., *62*:256–262.
59. Medoff, G., W. E. Dismakes, et al. 1970. Therapeutic program for Candida infection. Antimicrob. Agents Chemother., 286–290.
60. Michalska-Trenker, E. 1969. Odczyny precypitacyjne u chorych zaka zonych drozdzakeim Candida albicans. (Precipitin tests in patients infected with *C. albicans*.) Cruzlica, *37*:827–833.
61. Montes, L., and W. H. Wilborn. 1968. Ultrastructural features of host-parasite relationship in oral candidiasis. J. Bacteriol., *96*:1349–1356.
61a. Montes, L. F., R. Ceballos, et al. 1972. Chronic mucocutaneous candidiasis, myositis and thymoma. J.A.M.A., *222*:1619–1623.
62. Mourad, S., and L. Friedman. 1961. Pathogenicity of Candida. J. Bacteriol., *81*:550–556.
63. Negroni, R. 1969. Immunologia de las candidiasis. Mycopathologia, *38*:190–197.
64. Newcomer, V. D., J. W. Landau, et al. 1966. Candida granuloma. Studies of host parasite relationships. Arch. Dermatol., *93*:149–161.
65. Nickerson, W. J. 1954. An enzymatic locus participating in cellular division of yeast. J. Gen. Physiol., *37*:483–494.
66. Padhye, A. A., M. J. Thirumalachar, et al. 1967. Incidence of oral thrush in newborn infants, 1960 to 1962, at Sassoon Hospitals, Poona, and a comparative evaluation of treatment with some antifungal agents. Hindustan Antibiot. Bull., *91*:143–151.
67. Pinsloo, J. G., and P. J. Pretorius. 1966. *Candida albicans* endocarditis. Case successfully treated with amphotericin B. Am. J. Dis. Child., *111*:446–447.
68. Quillici, M. 1966. A propo de l'effet de divers serums sur la croissance de *Candida albicans*. Bull. Soc. Pathol. Exot., *59*:786–793.
68a. Rebora, A., R. R. Marples, et al. 1973. Experimental infection with *Candida albicans*. Arch. Dermatol., *108*:69–73.

68b. Rebora, A., R. R. Marples, et al. 1973. Erosio interdigitalis blastomycetica. Arch. Dermatol., *108*:66–68.
69. Record, C. O., J. M. Skinner, et al. 1971. Candida endocarditis treated with 5-fluorocytosine. Br. Med. J., *1*:262–264.
70. Reynolds, I. M., P. W. Miner, et al. 1968. Cutaneous candidiasis in swine. J. Am. Vet. Med. Assoc., *152*:182–186.
71. Reynolds, R., and A. I. Braude. 1956. The filament inducing property of blood for *Candida albicans;* its nature and significance. Clin. Res. Proc., *4*:40.
72. Roessman, U., and R. L. Friede. 1967. Candidal infection of the brain. Arch. Pathol., *84*:495–498.
73. Sandhu, R. S., and H. S. Randhawa. 1962. On the re-isolation and taxonomic study of *Candida viswanathii* Viswanathan et Randhawa 1959. Mycopathologia, *18*:179–183.
74. Scherr, G. H., and R. H. Weaver. 1953. The dimorphism phenomenon in yeasts. Bacteriol. Rev., *17*:51–92.
75. Scherr, G. H. 1955. The effect of hormones on experimental moniliasis in mice. Mycopathologia, 7:63–82.
76. Schwartzman, R. M., M. Denbler, et al. 1965. Experimentally induced moniliasis (*Candida albicans*) in the dog. J. Small Anim. Pract., *6*:327–332.
77. Sclafer, J. 1956. *Mycologie Medicale.* Paris, Expansion Scientifique Francaise, p. 191.
78. Sclafer, J., and S. Hewitt. 1960. Valeur des tests cutanés dans l'allergie à *Candida albicans.* Pathol. Biol. (Paris), *8*:323–327.
79. Shadomy, S. 1969. *In vitro* studies with 5-fluorocytosine. Appl. Microbiol., *17*:871–877.
80. Shelley, W. B. 1965. Erythema annulare centrifugum due to *Candida albicans.* Br. J. Dermatol., *77*:383–384.
81. Smiten, S. M. B. 1967. Candidiasis in animals in New Zealand. Sabouraudia, *5*:220–225.
82. Stenderup, A., and A. L. Bak. 1968. Deoxyribonucleic acid base composition of some species within genus candida. J. Gen. Microbiol., *52*:231–236.
83. Sweet, C. E., and L. Kaufman. 1970. Application of agglutinins for the rapid and accurate identification of medically important candida species. Appl. Microbiol., *19*:830–836.
84. Taschdjian, C. L., P. J. Kozinn, et al. 1969. Post mortem studies of systemic candidiasis. Sabouraudia, 7:110–117.
85. Taschdjian, C., M. B. Cuesta, et al. 1969. A modified antigen for sera diagnosis of systemic candidiasis. Am. J. Clin. Pathol., *52*:468–472.
86. Taschdjian, C. L., and P. J. Kozinn. 1957. Laboratory and clinical studies in candidiasis. J. Pediatr., *50*:426–433.
87. Taschdjian, C. L., J. J. Burchall, et al. 1960. Rapid identification of *Candida albicans* by filamentation on serum and serum substitutes. A.M.A.J. Dis. Child., *99*:212–215.
88. Thorpe, E., and H. Handley. 1929. Chronic tetany and chronic mycelial stomatitis in a child aged four and one-half years. Am. J. Dis. Child., *36*:328–338.
89. Tran Van Ky, P., T. Vaucelle, et al. 1969. Characterization of enzyme-antienzyme complexes in yeast extracts of the genus Candida following immunoelectrophoresis. Mycopathologia, *38*: 345–357.
89a. Tsuchiya, T., M. Miyasaki, et al. 1955. Studies on the classification of the genus *Candida.* Comparison of antigenic structures of standard and other strains. Jap. J. Exp. Med. *25*:15–21.
90. Watanakunakorn, C., J. Carleton, et al. 1968. Candida endocarditis surrounding a Starr Edwards prosthetic valve. Recovery of Candida in hypertonic medium during treatment. Arch. Intern. Med., *121*:243–245.
91. Winner, H. I. 1969. The transition from commensalism to parasitism. Br. J. Dermatol., *81*(Suppl. 1):62–68.
92. Winner, H. I., and R. Hurley. 1964. *Candida albicans.* Boston, Little, Brown & Co.

Chapter 9

CRYPTOCOCCOSIS AND OTHER YEAST INFECTIONS

Cryptococcosis

DEFINITION

Cryptococcosis is a chronic, subacute, or (rarely) acute pulmonary, systemic, or meningitic infection caused by the yeast *Cryptococcus neoformans.* The primary infection in man is almost always pulmonary following inhalation of the yeast; in animals it usually follows implantation or ingestion. Pulmonary infection in man is usually subclinical and transitory; however, it may arise as a complication of other diseases in debilitated patients and become rapidly systemic or even fulminant. In Europe it is known as the signal disease (malade signal), as it signals an underlying debilitating disease. It has a predilection for the central nervous system. The tissue reaction is sparse, with a few macrophages appearing in active infection, but the focus of infection may evolve into tuberculoid granulomas in chronic and healing disease. Suppuration and caseation necrosis are infrequent. The etiologic agent is unique among pathogenic fungi because of its production of a mucinous capsule in tissue and culture.

Synonymy. Torulosis, Busse-Buschke's disease, European blastomycosis, malade signal.

HISTORY

In medical literature the term "blastomycosis" has been used very loosely to designate any infection in which a budding yeast cell was found. Cryptococcosis in particular has been confused with blastomycosis. The cutaneous form of cryptococcosis, which is more often reported from Europe, was thought to be caused by a different species and came to be known as European blastomycosis. These questions were resolved in 1935 when Rhoda Benham clearly differentiated blastomycosis from cryptococcosis. She showed that the cutaneous European type of cryptococcosis was caused by the same organism that produced the meningitic form more commonly reported in America.

Busse[11] and Buschke[10] reported separately the isolation of a yeast (hefe) from the tibia of a 31-year-old woman. The lesions were described as "gumma-like" or "sarcoma-like," and the patient had lymphadenopathy and cutaneous ulcers. In 1895 Busse considered this yeast to be the cause of her malady, and designated the disease as saccharomycosis hominis. Sanfelice, in 1894, recovered an encapsulated yeast from peach juice which caused lesions in experimentally infected animals.[41] He named the organism *Saccharomyces neoformans,* thus giving taxonomic priority to this specific epithet. In 1896, Curtis described a case similar to that of Busse and Buschke. The patient had a yeast in a myxomatous tumor of the hip. He called the organism *Saccharomyces subcutaneous tumefaciens.* However, Vuillemin did not find ascospores, which are characteristic of the genus *Saccharomyces,* in the organism of

Buschke and transferred the yeast to the genus *Cryptococcus* as *C. hominis* in 1901. To confuse the taxonomy even more, the genus *Cryptococcus* had been used earlier by Kützing to describe algaelike white organisms. Sanfelice (1901) in Italy isolated an organism, similar to Buschke's yeast, from a lymph node of an ox. He named it *Saccharomyces litogenes.* Frothingham in 1902 recovered the same organism from a myxoma-like lung lesion in a horse in Massachusetts. These latter findings established the fact that the disease occurs in animals. The organisms from Sanfelice, one isolated by Klein in milk, and one isolated by Plimmer from a patient with cancer were reported to be culturally identical by Weis in 1902.

The association of the encapsulated yeast with cancers, especially myxoma-like disease, was noted early and soon led to the postulation that it was an inciter of neoplasias. Von Hansemann in 1905 appeared to be the first to see this fungus in a case of meningitis. He described the disease as "tuberculous," with yeast present in "gelatinous cysts." It was first recognized ante mortem in 1914 by Verse in a case of leptomeningitis in a 29-year-old woman.

Cutler and Stoddard (1916) delineated the clinical and pathologic differences between cryptococcosis, blastomycosis, and other mycoses. They erroneously assumed that the capsules of the organisms were cysts in the tissue caused by digestive action of the fungus ("histolytic") and named the organism *Torula histolytica.* This name has no validity or priority, and it was based on an erroneous observation. Unfortunately it became popular, and the disease was known for a long time as torulosis or *Torula* meningitis.

The original cultures of Busse, Buschke, and Curtis were included in a study of 27 isolants of pathogenic cryptococci by Benham. She concluded that there was only one species and suggested retaining the name *C. hominis.* Lodder and Kreger-Van Rij (1952) determined the priority of the name *Cryptococcus neoformans,* and this is now the accepted name. Recently the organism has been shown to have structures relating it to the Basidiomycetes, and the perfect stage name *Leukosporidium neoformans* has been proposed by Shadomy[42] (see Chapter 8).

Although the organism had been reported isolated from milk and fruit juices (including an erroneous report of isolation from the flux of mesquite), a saprobic habitat was not fully established until the work of Emmons.[16] He reported the isolation of virulent organisms from pigeon nests, roosts, and feces. Its constant association with such habitats has been amply verified and point-source infection recorded.[40]

A complete bibliography and review of the disease was published in a scholarly monograph by Littman and Zimmerman in 1956.[33]

ETIOLOGY, ECOLOGY, AND DISTRIBUTION

The pigeon, long a symbol of urban decay, appears to be the chief vector for the distribution and maintenance of *Cryptococcus neoformans.* The organism is recovered in large numbers from the accumulated filth and debris of pigeon roosts, such as attics of old buildings, cupolas, and cornices. In this desiccated environment, it is often the predominant organism present, sharing the alkaline, high-nitrogen, high-salt substrate with *Geotrichum candidum* and a few *Candida* and *Rhodotorula* species. Its prevalence in this particular ecologic niche is a situation similar to the restricted environmental habitat required by *Coccidioides immitis.* Both organisms appear unable to survive or compete with other species in other situations. Cryptococci disappear from infected debris when it is mixed with soil. Similarly *Coccidioides* cannot compete when irrigation and cultivation of its habitat allows other fungi to proliferate. In rural areas also, *Cryptococcus* is found in sites inhabited by pigeons, such as barn lofts and hay mows. The dust in such environments contains viable virulent organisms.[46] In this desiccated state the organism may be no more than 1 μ in diameter, a condition which allows it to be inhaled into alveolar spaces.

C. neoformans does not appear to infect the pigeon, whose average body temperature is 42°C, but survives passage through the gut.[32] The organisms survive but do not grow at 44°C. In moist or desiccated pigeon excreta, they remain viable for two years or more. In sheltered sites which are not in contact with the soil, the organism is found in large numbers,[15] its survival enhanced by increased relative humidity.[28] Direct exposure to sunlight, especially in summer months, soon sterilizes the habitats of *Cryptococcus;* however, it is maintained in similar locations during winter. In saprobic environments, the organisms are essentially nonencapsulated,[28] but such strains and unencapsulated mutants do become virulent following acquisition of the polysaccharide capsule.[8,9] The presence of, or the potential for, encapsulation rather than the degree of encapsulation appears to be the significant virulence factor.

Cryptococcus neoformans is the only etiologic agent of cryptococcosis. The organism is worldwide in distribution and is constantly associated with avian habitats.[37] There is no significant difference in incidence of infection which can be related to age, race, or occupation. Pigeon breeders[18] have a higher than normal occurrence of antibody but no greater rate of frank infection. This mycosis is more often reported in males than in females. This may be a function of exposure or, as has been suggested for other mycoses, a matter of hormonal differences.[37a] From time to time other species of the genus *Cryptococcus,* such as *C. albidus*[30a], have been implicated in human disease. However, such reports need to be verified.

There are essentially two types of cryptococcal disease, but their manifestation depends on host response rather than on the strain of organism. In the normal patient, infection following inhalation of the organism is usually rapidly resolved with minimal symptoms, and the disease, if any, is subclinical. Growth-inhibiting substances are normally present in body fluids.[27] If the number of organisms inhaled is considerable, an infection may be initiated in the lungs, with occasional transitory foci of infection in other anatomic sites. Even though the tissue reaction and cellular defenses are evoked slowly, they usually are adequate to contain the infection. Infrequently in normal patients, a chronic infection is established which gives rise to occasional flares of systemic, cutaneous, or meningitic involvement. Such patients require treatment. A single pulmonary lesion is often amenable to surgical intervention, followed by adequate drug therapy.

The second type of disease is associated with neoplasias, debilitating diseases, and compromised hosts, usually as a result of drug therapy. In these cases the host defenses are minimal and inadequate, and the disease readily spreads to involve almost all organs, particularly the central nervous system. The course of the infection may be protracted, with slow spread from one organ system to another, or it may be fulminant and rapidly overwhelming. This clinical type was the first recorded and for many years was the only form of the disease recognized. Presently, cryptococcosis is particularly associated with the so-called "collagen diseases," such as lupus erythematosus,[13] very frequently with sarcoidosis,[45] as well as neoplasias.[34a, 41a] Association of the disease with pregnancy[44a] and alveolar proteinosis[46a] has also been noted.

As is the case with other systemic mycoses, the subclinical or mild form of cryptococcosis in the uncompromised patient has only recently been recognized. Unlike the other fungal diseases, however, it has been impossible to estimate the frequency of occurrence of subclinical infection, for as yet no sensitive and reliable immunologic tests are available. That such a mild, subclinical form occurs quite commonly is attested to by the frequent incidental finding of old, healed, cryptococcal granulomas in routine autopsies. Current estimates of the incidence of cryptococcal disease range from 200 to 300 cases of cerebral meningitis per year in the United States to 15,000 subclinical respiratory infections in New York City alone.[2,32]

CLINICAL DISEASE

With rare exception, the portal of entry of the cryptococci is the lungs. The primary pulmonary infection may remain localized or disseminate to other organs. Involvement of other tissues and organ systems sometimes occurs in spite of resolution of lung lesions. For this reason, it is often difficult to assess the site of initial infection. Although experimental disease has been produced by feeding large numbers of yeast cells to animals, it is doubtful that human infection via the alimentary tract occurs with any frequency. Entrance of the organism through the skin and nasopharyngeal mucosa is possible but is also considered rare. The clinical types of cryptococcal disease include pulmonary, central nervous system, cutaneous and mucocutaneous, osseous, and visceral.

Pulmonary Cryptococcosis

Primary infection of the lungs has no diagnostic symptoms, and most cases are probably asymptomatic. When present, the symptoms include cough, low-grade fever, pleuritic pain, malaise, and weight loss, but these signs are seldom prominent.[12] Scanty mucoid sputum also is produced, and rarely there may be hemoptysis. Following rupture of an eroding cryptococcal focus into a bronchial branch, there is a heavy discharge of mucoid sputum containing numerous organisms. Night sweats which are prominent in tuberculosis seldom occur in cryptococcosis. In the rare, fulminant cases of disease there is pulmonary consolidation and pronounced fever. Lesions may develop in any part of the lung. They are frequently bilateral, but they may appear in only one

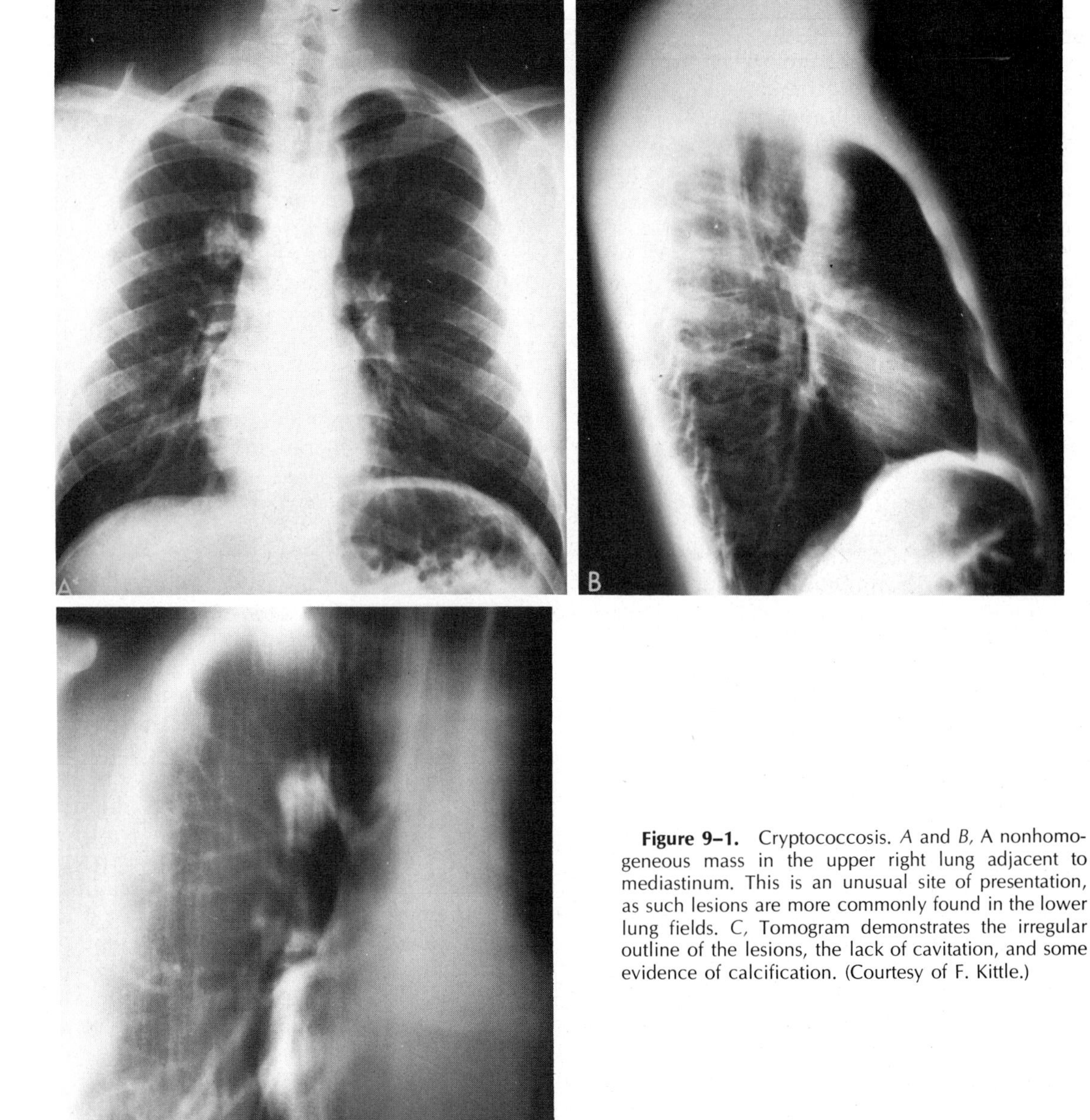

Figure 9–1. Cryptococcosis. *A* and *B*, A nonhomogeneous mass in the upper right lung adjacent to mediastinum. This is an unusual site of presentation, as such lesions are more commonly found in the lower lung fields. *C*, Tomogram demonstrates the irregular outline of the lesions, the lack of cavitation, and some evidence of calcification. (Courtesy of F. Kittle.)

lobe. Diminished breath sounds and dullness to percussion are often present, along with a pleural friction rub. In miliary disease, moist rales may be present over either the apex or base. Asymptomatic infections are detected only by x-ray. Many of the lesions are small and often go unnoticed. In addition, many lesions heal without forming granulomas (cryptococcomas) and leave no residual history of infection.

X-ray. Roentgenographic examination for pulmonary cryptococcosis reveals a variable picture, but lung changes generally fall into four categories. These depend on the recentness of the infection, the severity of the disease, and the status of host defense mechanisms.[34b, 41a]

The first and most common type is the discrete, solitary, moderately dense area of infiltration appearing in the lower lung fields. These lesions are usually from 2 to 7 cm wide and may gradually extend peripherally to simulate carcinoma, abscess, or hydatid cyst (Fig. 9–1). They are characterized by locular masses of yeasts mechanically displacing host tissue, which may rupture the bronchial wall and result in the production of mucoid, blood-tinged sputum; little or no hilar lymphadenopathy is seen. If they heal, such lesions disappear altogether or form small granulomas which infrequently calcify.

In the second type of lesion there is a broader, more diffuse infiltrate involving the lower lung areas. Vessel markings and nodular shadows are accentuated. A residual fibrosis occurs following healing (Fig. 9–2).

Extensive peribronchial infiltrates occur in the third type (Fig. 9–3). These extend widely to form woolly shadows which resemble active tuberculosis. However, fibrosis is minimal (Fig. 9–4), and caseation necrosis and cavitation are exceptional. Again calcification is rare.[41a]

Any of the above types of lesions may occur in primary infections, the degree of involvement being related to the general health of the individual and the dose of the infectious agent inhaled. Dissemination to the central nervous system may occur in any of these types, apparently with equal frequency.

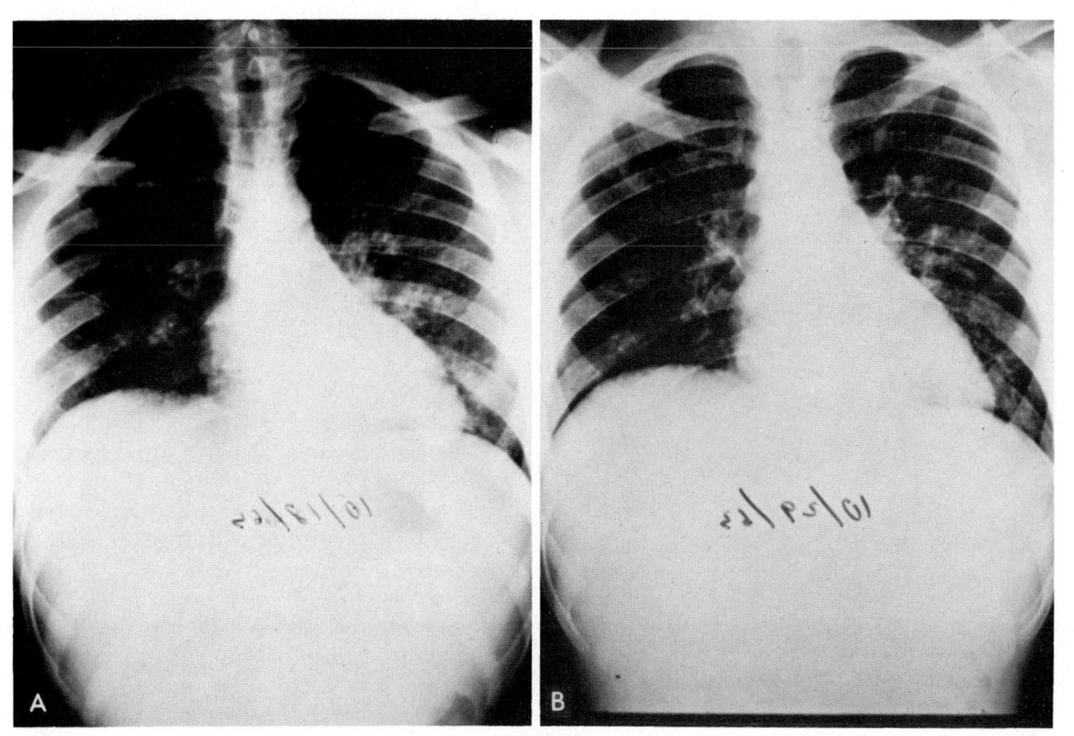

Figure 9–2. Cryptococcosis. *A,* Ill-defined interstitial infiltrate of left lung. *B,* Similar changes ten days later with some progression. The disease healed with amphotericin B treatment, and an x-ray of one year later showed a normal chest with some residual fibrosis. (Courtesy of J. Fennessey.)[40]

The fourth form of cryptococcal lung disease resembles miliary tuberculosis and usually occurs in lymphoma and leukemia patients. Lesions consisting of small gelatinous granules are present in all lung fields, but little tissue response is elicited. In time, small granulomas may be formed, or the disease becomes relentlessly progressive and disseminates.

Several points are considered important in differentiating cryptococcal disease from other mycotic and bacterial infections.[33] In cryptococcosis there is marked predilection for disease to become established in the lower lung fields. Cavitation, fibrosis or calcification, hilar lymphadenopathy, and pulmonary collapse are rare occurrences. Coin lesions, which are so common in spent histoplasmosis or coccidioidomycosis, are uncommon in cryptococcosis (Fig. 9–1). Thin-walled cavities usually do not develop. "Crab claw" shadows, so-called because they extend from a focal lesion and emulate carcinoma, are found in blastomycosis but not in cryptococcosis. Definitive diagnosis of the disease depends on the isolation of the organism. Although *C. neoformans* is not part of the resident human flora, other cryptococci may be. It is necessary to identify specifically any encapsulated yeast isolated from sputum to establish the diagnosis of cryptococcosis. In low-grade infections, however, the capsule may be absent or very small.[16a]

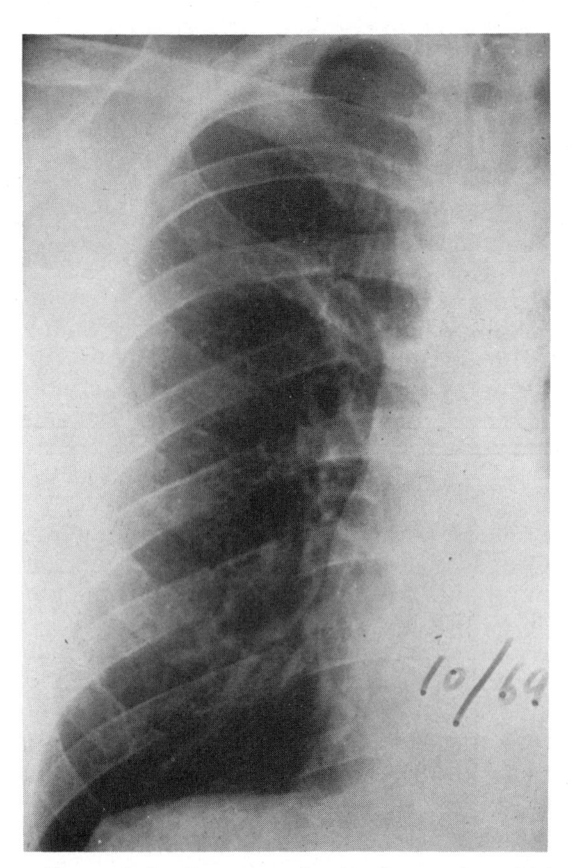

Figure 9–4. Cryptococcosis. Healing peribronchial disease. Ill-defined scarlike process superior to hilum of right lung.

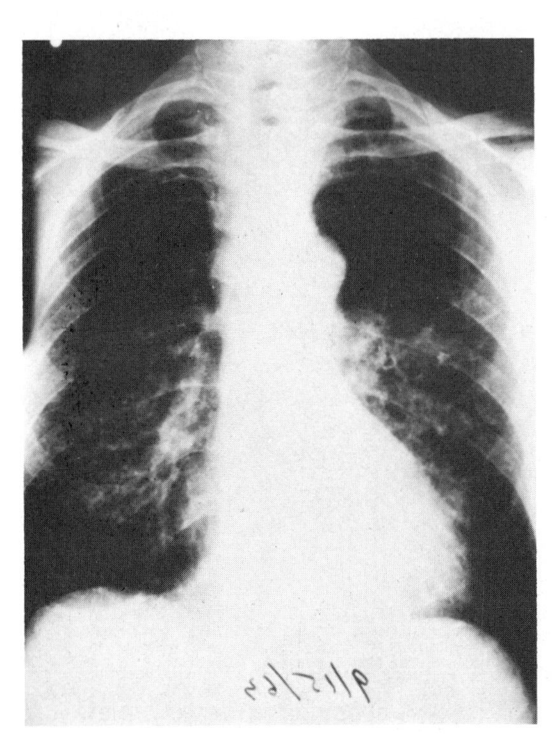

Figure 9–3. Cryptococcosis. Bilateral, interstitial infiltrate adjacent to the hilum; extension to lingula of left upper lobe.

Central Nervous System Cryptococcosis

This is the most frequently diagnosed form of cryptococcosis, though certainly not the most common site of the disease. The reason for the predilection of *Cryptococcus* for the central nervous system has not been explained. The organism probably encounters less cellular (phagocytic) response there, and it has been theorized that selective nutritional factors for the yeasts are present. Simple sources of nitrogen, such as asparagine and creatinine, are found in the spinal fluid and may stimulate growth. The absence of inhibitory factors reputed to be in serum may also play a role here, but conclusive evidence is lacking. Before drug therapy became available, this form of the disease was almost always fatal, but the mortality has been cut to about 6 per cent by the use of amphotericin B. Retreatment is often necessary, as the treatment failure rate is about 25 per cent with this drug.

The only presenting symptom found in almost all patients is headache. It is usually frontal, temporal, or retro-orbital, intermittent, and of increasing frequency and severity. Fever is also frequently present. Most patients show the remaining signs of meningitis: nuchal rigidity, tenderness of the neck, and positive Kernig's and Brudzinski's signs.[21]

A single, localized, cryptococcal granuloma of the brain (cerebritis) may produce signs only of an expanding intracranial mass. Nausea, vomiting, mental changes, coma, paralysis, and hemiparesis frequently develop. Ocular manifestations such as blurring, vertigo, diplopia, photophobia, ophthalmoplegia, nystagmus, amblyopia, and papilledema occur. The latter may require multiple lumbar punctures for control. The mental disturbances may be quite marked and include irritability, agitation, apathy, loquaciousness, confusion, defective memory, or even frank psychosis. Some patients present with epileptic seizures. Hemiplegias and hemiparesis may develop as the result of expanding granulomatous masses. Urinary incontinence and loss of patellar and Achilles tendon reflexes also occur, but some patients become hyperreflexive. Symptoms of acute infection and toxemia are absent or rare. A spinal tap usually reveals increased pressure, often up to 700 mm of water. Release of pressure during a tap may restore comatose patients to consciousness, but later they may sink back into delerium and death. The spinal fluid is clear, rarely cloudy; the cell count is elevated but usually not over 800 per ml. Lymphocytes are more numerous, but occasionally neutrophiles predominate. The protein varies from 40 to 600 mg per 100 ml. Chlorides are low (183 mEq per L or lower) and sugar is quite low (often 10 mg per 100 ml). The colloidal gold curve is "paretic" or "tabetic." Organisms may be sparse or numerous, and a spun specimen examined with India ink usually confirms the clinical diagnosis (Fig. 9–5). Frequently no organisms are seen, particularly if the lesions are restricted to the brain itself. Cryptococcal polysaccharides are usually present in the CSF in active infection, and antibodies are detectable in healing or resolved disease. The clinical and spinal fluid pictures are indistinguishable from those of tuberculous meningitis. The latter disease usually has a more rapid course and is more frequently seen in children. Differentiation of cryptococcosis from other mycoses can be made only by isolation and identification of the organism.

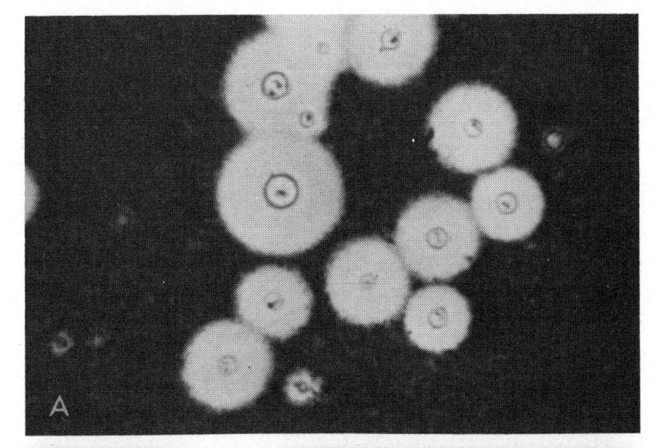

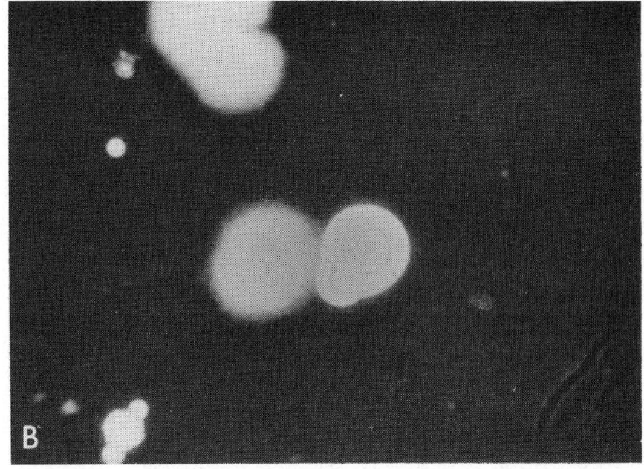

Figure 9–5. Cryptococcosis. *A,* India ink preparation of spinal fluid showing yeast cells surrounded by large capsule. *B,* Budding yeast cell within capsule. (From Rippon, J. W. *In* Burrows, W. 1973. *Textbook of Microbiology*. 20th ed. Philadelphia, W. B. Saunders Company, p. 735. Courtesy of H. Sommers.)

The course of cryptococcosis extends from a few months to 20 or more years. The usual outcome, however, is a rapidly progressive deterioration of the patient. Untreated cryptococcal meningitis is fatal, with very rare exceptions.

Cutaneous and Mucocutaneous Cryptococcosis

The first human case of cryptococcosis described by Busse and Buschke had manifestations of cutaneous lesions, bone involvement, and systemic disease, but there was no meningeal disease. Subsequently most of the cases recognized in the United States were meningitis, but in Europe cutaneous and mucocutaneous disease was considered more common. No satisfactory explanation has been given for this difference, even though extrameningitic disease is now recognized more often, while cutaneous involvement is still considered uncommon. Cutaneous and mucocutaneous disease in man is usually a manifestation of disseminated disease and occurs in about 10 to 15 per cent of cases. Transient cutaneous lesions occur in experimental infections of monkeys.[4] Such lesions are not regularly encountered when other experimental animals are used. There is seldom convincing documentation that such lesions represent the primary sites of infection.[38] Data from animal infections suggest that traumatic implantation is an important means of entry of the yeast, but the same is not true in human disease. Lesions developing at sites of injury in man may represent seeding from a small internal focus rather than an exogenous inoculation. Absence of regional lymphadenopathy particularly casts doubt on an external origin of such cutaneous lesions.

The lesions of the skin present as papules, acneform pustules, or abscesses that ulcerate with time. In the few well-documented cases of primary cutaneous inoculation, the lesions were chancreform and limited. Mucocutaneous lesions occur about one third less frequently than skin disease. They present as nodules, as granulomas, or as deep or superficial ulcers (Fig. 9–6 FS B 24).

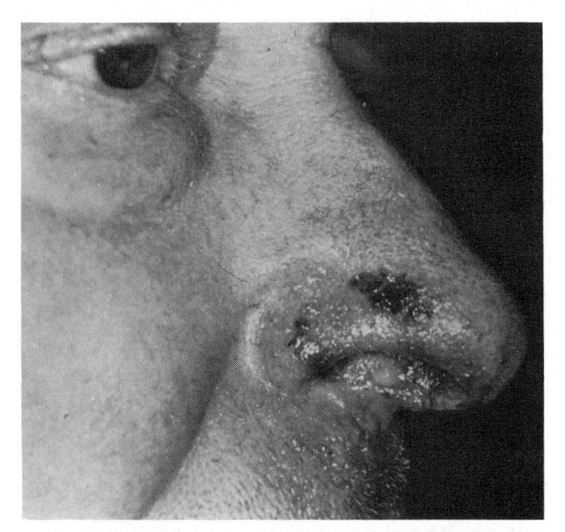

Figure 9–6. FS B 24. Cryptococcosis. Superficial ulcers involving mucocutaneous areas of nose in a case of disseminated disease.

Osseous Cryptococcosis

Bone is involved in about 5 to 10 per cent of reported cases, which is considerably less than the frequency recorded in blastomycosis, coccidioidomycosis, and actinomycosis. As in other diseases caused by fungi, the cryptococci have a predilection for bony prominences, cranial bones, and vertebrae. The joints are usually spared except by direct extension. There is no characteristic x-ray appearance, but osseous lesions are similar to those of blastomycosis and coccidioidomycosis, i.e., they are usually multiple, discrete, widely disseminated, destructive, chronic, and slowly changing (Fig. 9–7). Swelling and pain may be present. For diagnosis, the relative stationary appearance of the lesion and the lack of periosteal proliferation suggest cryptococcosis rather than other mycoses.[14] Periosteal proliferation is common in actinomycosis and regularly seen in blastomycosis and coccidioidomycosis. Synovitis without bone involvement is quite rare but has been reported. Differential diagnosis of osseous cryptococcosis must also include other diseases of the bone. Unless special stains are used, examination of biopsy or excised material may lead to a diagnosis of osteogenic sarcoma or Hodgkin's disease because the histologic picture is quite similar (Fig. 9–9,*D*).

Visceral Cryptococcosis

In disseminated disease, any organ or tissue of the body may have foci of infection. Granulomatous lesions are produced which symptomatically and even histologically resemble malignant neoplasias, particularly those of the myxomatous type. Heart, testis, prostate, and eye[30] are frequently involved, whereas kidney, adrenal, liver, spleen, and lymph nodes are usually spared. In rare cases, patients have presented with an acute abdomen with only liver involvement.[40] Rarely, massive involvement of the adrenals may give rise to signs of Addison's disease, as is sometimes seen in blastomycosis and paracoccidioidomycosis. Lesions in the gastrointestinal and genitourinary tract mimic

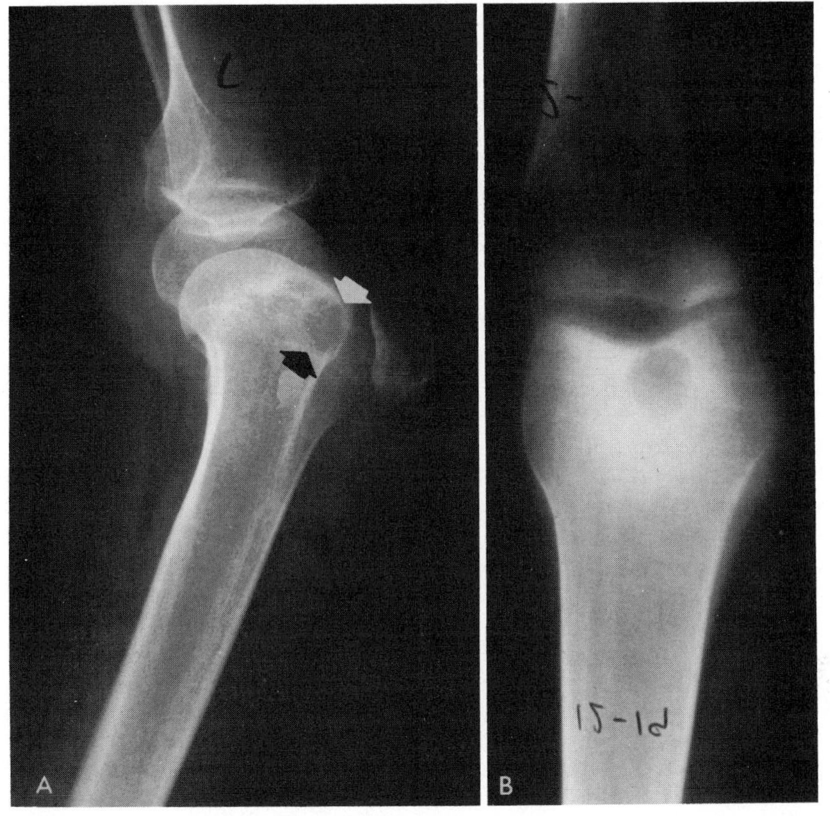

Figure 9–7. Cryptococcosis. *A,* Lateral film of knee. Spherical, well-defined lytic area in the distal end of femur (intercondylar fossa), adjacent to articular surface. Suggestive evidence of intra-articular effusion. *B,* Tomogram demonstration of sclerotic reaction about lytic lesion. The patient presented with a swollen knee from which nonencapsulated cryptococci were grown. The lung fields had areas of fibrosis. There was no other evidence of systemic disease.

tuberculosis. The eye is involved either by direct extension from the subarachnoid space or by hematogenous spread from other loci, resulting in the formation of multiple lesions in the uveal tract, retina, and vitreous (cryptococcal chorioretinitis). Endocarditis lenta is a rare form of involvement, as are cryptococcal aneurysms of the aorta.[33]

DIFFERENTIAL DIAGNOSIS

In general, cryptococcosis is most often mistaken for malignant neoplasias. This is true particularly of pulmonary and central nervous system disease. All forms of cryptococcosis mimic tuberculosis or other fungal diseases. Meningitis of cryptococcal origin is more protracted than tuberculous disease but may resemble encephalitis, dementia psychosis, dementia paralytica, or bacterial infections, especially those caused by *Brucella* and *Listeria.*

PROGNOSIS AND THERAPY

Local lesions of the lungs in normal patients have a good prognosis. They heal slowly without treatment and disappear or leave a residual scar. Rarely there is hematogenous spread to the central nervous system. When this happens the prognosis becomes very grave unless treatment is instituted. In systemic infections, the outcome is usually fatal, particularly if the patient is debilitated. Primary cutaneous or mucocutaneous lesions generally resolve spontaneously. Chronic disease of any organ system is marked by alternating intervals of remission and exacerbation, but usually the disease terminates fatally. Such periods of quiescence and recrudescence have posed great difficulties in the past in the evaluation of therapeutic modalities. At present there is only one drug available with clinically proven efficacy, amphotericin B. Whereas minor pulmonary disease or primary cutaneous involvement may cure spontaneously or by surgical excision, meningeal or systemic disease requires adequate treatment with this drug. The dose regimen is similar to that for other systemic mycoses: 1 mg per kg per day to a total of 3 to 5 g intravenously ("piggy-back" with 5 per cent dextrose given as a one- to six-hour perfusion). If the drug is not well tolerated or if the patient's BUN or creatinine level rises, the dosage can be lowered 5.0 to 7.0 mg per day, administered in conjunction with 50 mg diphenhydramine (Benadryl) and 10 grains aspirin.[5,19] Alternate-day therapy increases patient tolerance. The drug levels of sera and spinal fluid can be accurately gauged using the medium effective dose (ID_{50}) determination of Bennett.[6] Patients with underlying disease, such as lymphomas and leukemias, or in immunosuppressed states may require 5 to 8 g total dose. In the experience of most clinicians, intravenous therapy is adequate for meningeal as well as other forms of disease. Selected patients may require intrathecal administration, but arachnoiditis with severe residual damage is a not infrequent sequela to the use of amphotericin B by this method. Chronic pulmonary lesions as well as some osteal lesions have been cured by surgical excision.[12,14] However, meningeal infections following surgical procedures are not uncommon, and prophylactic amphotericin B is recommended.[12]

Five-fluorocytosine (5FC) appears to have some efficacy in the treatment of cryptococcal meningitis.[43] Since it can be given orally and is relatively nontoxic, it can be instituted immediately at full dosage levels (8 to 10 g daily to a total of 370 g). If this treatment fails, amphotericin B can be used. The failure rate of 5FC is about 60 per cent in the experience of some clinicians, whereas it is about 25 per cent for amphotericin B. Some strains of *C. neoformans* rapidly develop resistance to 5FC.

PATHOLOGY

Gross Pathology. Cryptococcal infection is accompanied by minimal inflammatory reaction; therefore, even in fatal cases of meningitis the gross appearance of the meninges and brain is almost normal. The lesions are of two principal types which correlate with the duration of the disease.[3] Early lesions are gelatinous (mucinous), whereas older lesions become granulomatous. The early lesions characteristically contain aggregates of encapsulated budding cells intermixed with a fine reticulum of connective tissue. These lesions enlarge and mechanically compress the surrounding tissue. Fibrovascular proliferation is minimal in most lesions but tends to be pronounced in healing primary infection sites, such as discrete, subpleural granulomas. Granulation tissue is seen in localized cutaneous lesions, but the pseudoepitheliomatous hyperplasia frequently observed in blastomycosis and coccidioidomycosis is absent. Fibrous encapsulation which characterizes histoplasmomas and tuberculomas is infrequently encountered in cryptococcosis. Even in well-formed granulomas containing lymphocytes and epithelioid cells, caseation necrosis and cavitation are rarely observed. Significant calcification is unusual, but the spheroidal concretions which are present in granulomas of other etiology are sometimes mistaken for the yeast cells of cryptococci. The degree of the granulomatous response is influenced by the innate resistance of the patient to infection. In cases of Hodgkin's disease, lymphoma, or leukemia in which the resistance of the patient is lowered, the inflammatory response is essentially absent.

Pulmonary Cryptococcosis. Gross examination of the lung may reveal a localized process or extensive involvement. Miliary tubercles similar to those seen in tuberculosis are present in disseminated cases. Solitary cryptococcal granulomas of the lung from uncompromised patients vary in size from 2 to 7 cm in diameter. They most frequently are observed at the periphery of the lobes, at the hilum, or in the center of a lobe. The lesions may be solid, firm, rubbery, nonspecific granulomas resembling those of sarcoid, or gelatinous, mucoid masses resembling myxomatous neoplasms. Cryptococcal lesions frequently coexist with tuberculosis, other mycoses, Hodgkin's disease, sarcoid, and lymphomas, thus confusing the diagnosis. The yeasts are sometimes seen infecting lymphoma lesions.

Cryptococcal Meningitis. As noted previously, the gross changes in the meninges, even in fatal cases of disease, may be so minimal that cryptococcal infection is not suspected. The organisms are dispersed so evenly over the subarachnoid spaces that a cloudiness is not perceived by x-ray (Fig.

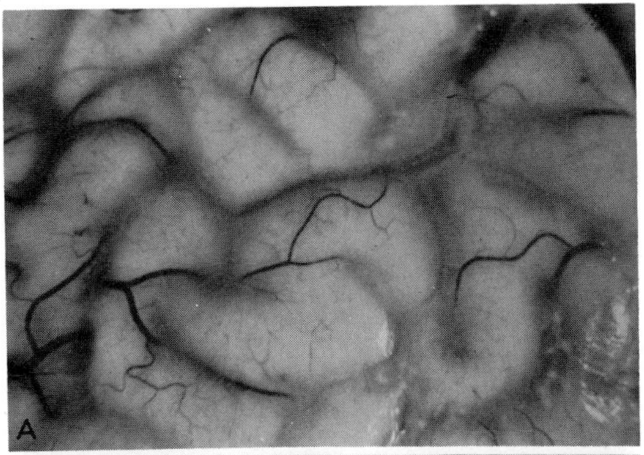

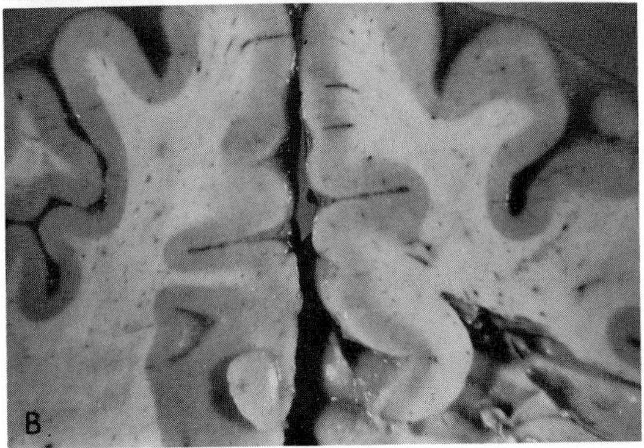

Figure 9–8. FS B 25. Cryptococcosis. *A,* Glistening surface of brain from fatal meningeal disease. The convoluted cortex is easily seen in some areas and obscured in others. There is cloudiness throughout the pia arachnoid with some small, cottony, white patches in some areas. *B,* In the sulci the pia is edematous, mucoid, and cloudy. The material within the space is gelatinous, glistening, and mucoid. There is some vascular proliferation, but no involvement of the cerebrum itself.

9–8 FS B 25). Meningeal reaction is more pronounced at the base of the brain and in the dorsal area of the cerebellum. In these areas thickening and opacity of the membranes are seen. Although pachymeningitis has been observed, leptomeningitis is seen more frequently. The meninges are hyperemic, and there is slight flattening of the convolutions of the cerebrum. The subarachnoid space is distended by a greyish, mucinous exudate, and the brain surface appears to be covered by "soap bubbles."[33] Small tubercles sometimes develop along the small vessels of the brain and meninges. The surface of the brain may appear dimpled because of the presence of cystoid lesions. When the inflammatory reaction is patchy and granulomatous, the membranes may adhere to the cortex. If the meningeal disease progresses to encephalitis, the cut surface may appear normal or contain numerous cystic spaces. Brain involvement usually results from embolization from other foci of infection. Embolic lesions are usually deep and consist of large, mucoid cysts. These cysts are found beneath the ependyma and in the periventricular and periaqueductal grey matter, the basal ganglia, the cerebral white matter, and the cerebellar dentate nucleus. Single lesions are sometimes found in the thalamus. Enlarging cerebral lesions have the radiologic appearance of a growing neoplasm or granuloma. As a consequence nearly one fourth of patients with central nervous system cryptococcosis have had a neurosurgical procedure performed before the correct diagnosis was known.

Involvement of Other Organs. In disseminated cryptococcosis, virtually any tissue may be colonized; however, localization of the infection outside of the lung and brain is infrequent. Cutaneous lesions are characterized by stretching and attenuation of the epidermis, but the hyperplasia seen in blastomycosis is absent. These lesions usually ulcerate. In the soft tissues, large cystic masses are formed that clinically and grossly mimic lipomas or myxomas. Upon excision these masses are seen to contain bright, gelatinous, shining material. Masses of organisms are present inside. In contrast, in small, hard, granulomatous lesions, the organisms are not present in large numbers, and the diagnosis may be difficult to establish. In chorioretinitis of the eye and in solitary granulomas in bone, the diagnosis of cryptococcosis may be difficult to establish histopathologically. Visceral involvement usually does not cause clinically apparent enlargement of the involved organ nor is distortion of the organ obvious upon gross examination at autopsy. The lesions are usually very small and are detected only on histologic examination.

Histopathology. Although the tissue reaction in cryptococcosis varies considerably from organ to organ and from case to case, in general the response is minimal. Frequently only tissue macrophages are found, but in rare cases there may be many giant cells and dense infiltration of plasma cells and lymphocytes. Two basic histologic patterns may occur: gelatinous and granulomatous.[3] Both types may be present in the same lesion. In the gelatinous type, masses of organisms occur, and there is mucoid degeneration of the invaded tissue (Fig. 9–9,*A* FS B 26,*A*; Fig. 9–

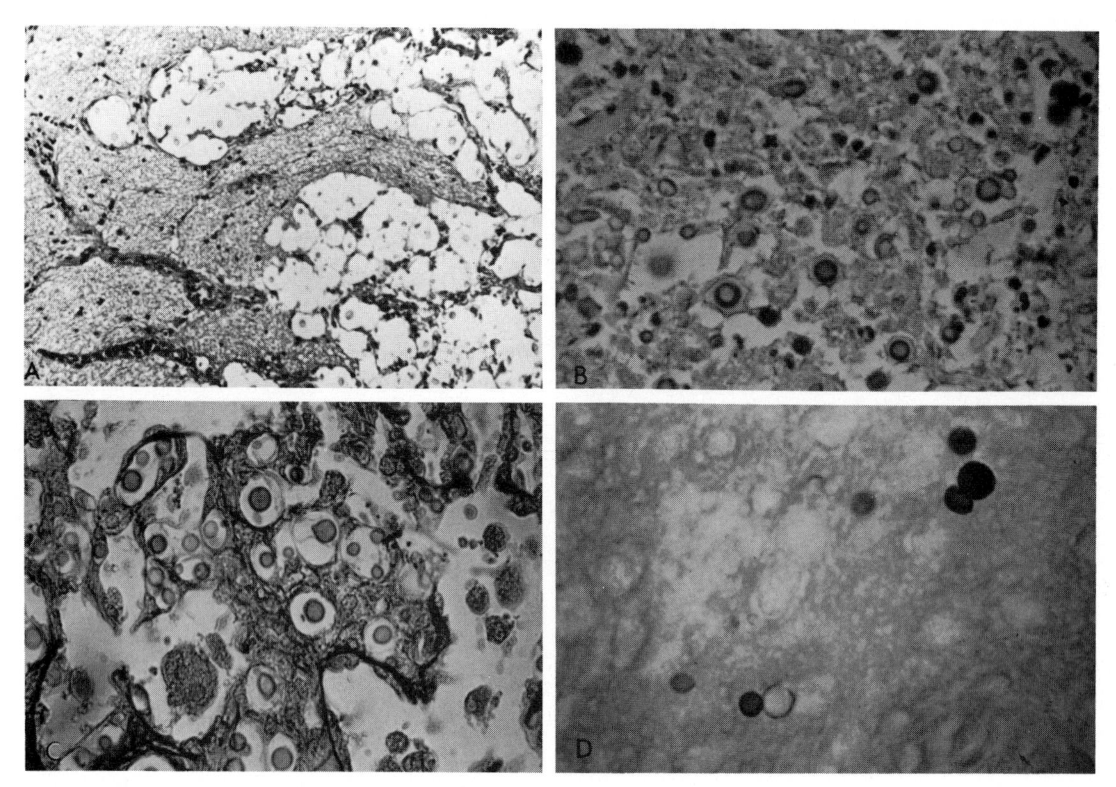

Figure 9–9. FS B 26. Cryptococcosis. *A,* Optic nerve, gelatinous-type lesion showing lack of cellular reaction. H and E. ×400. (From Rippon, J. W. *In* Burrows, W. 1973. *Textbook of Microbiology*. 20th ed. Philadelphia, W. B. Saunders Company, p. 735.) *B,* Mucicarmine stain of tissue from cryptococcosis of lung. Note "car tire" appearance of cells due to deep stain of capsule around yeast cell body. *C,* Gridley stain of lung tissue. The yeast cell body is stained, and the capsule appears as an unstained halo around it. *D,* Methenamine silver stain of synovial biopsy (same case as in Fig. 9–7), showing rare case of noncapsulated organism. ×400.

10,*A–C* FS B 27,*A–C*). Granulomatous lesions consist of histiocytes, giant cells, and lymphocytes, along with some fibroblastic activity (Fig. 9–10,*B* & *D* FS B 27,*B* & *D*). The number of yeasts in granulomatous lesions is much fewer than in gelatinous type lesions. In the latter, the yeasts are mostly free in the tissue, whereas in the former they are almost all within giant cells and histiocytes. The average size of the yeasts ranges from 5 to 10 μ, and a large, gelatinous capsule is usually but not always present. The latter does not stain in hematoxylin and eosin (H and E) preparations but can be made visible by the mucicarmine stain (Fig. 9–9,*B* FS B 26,*B*). The cells can be stained by the periodic acid–Schiff (Gridley) (Fig. 9–9,*C* FS B 26,*C*) or methenamine silver (Gomori) stain (Fig. 9–9,*D* FS B 26,*D*). Very rarely the capsule is entirely absent or so small that the mucicarmine stain is negative. In H and E slides the cells may appear as faint purple bodies surrounded by a blank space which was formerly occupied by capsular substance. The capsule is essentially immunologically inert, which accounts for the paucity of inflammatory cellular response. The yeast cells vary considerably in size, particularly in old healed lesions (Fig. 9–11). They are frequently as small as *Histoplasma capsulatum* and may be as large and thick-walled as *Blastomyces dermatitidis.* Unless budding is very apparent the yeasts also may resemble developing spherules of *Coccidioides immitis.* Candida yeast cells usually are mixed with mycelial units; however, hyphae can occur rarely in cryptococcal lesions[13] (Fig. 9–12). The capsule may be small or absent in old healed lesions, thus tending to confuse the diagnosis.

In the solitary, healed, fibrosed granuloma, the yeast are usually dead, and the capsules have mostly disintegrated. These granulomas are often encapsulated and partially calcified. They contain clefts of cholesterol and reveal considerable fibrosis. Such lesions

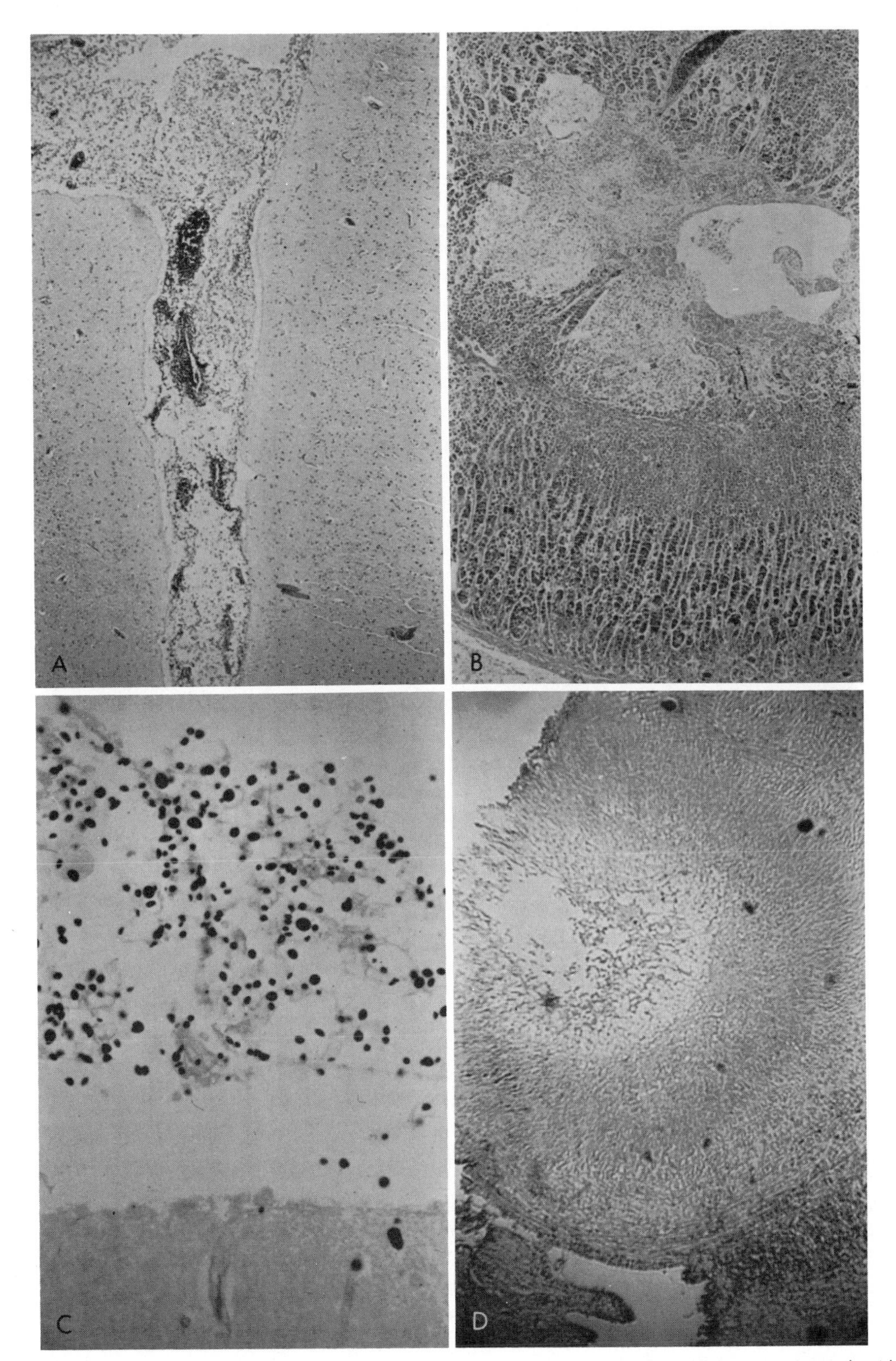

Figure 9–10. FS B 27. *A,* Histologic section from Figure 9–8. Numerous cryptococci are seen in pia arachnoid space, with focal lymphocytic aggregates around blood vessels. *B,* Adrenal gland from disseminated disease. Infiltrate and many cryptococci are seen in inner cortex, medulla, and vein walls. Note minimal cellular response, which consists of a few lymphocytes. H and E. ×100. *C,* Histologic section from Figure 9–8,*A* (methenamine silver, ×400). Numerous cryptococci in pia arachnoid space showing variation in size. *D,* Old inactive lesion in prostate. There is a central area of necrosis with margination of epithelial cells and a scattering of giant cells peripherally.

therefore resemble healed tuberculomas or histoplasmomas, both of which may also be incidental findings in a routine autopsy. Cryptococcus will not be suspected unless a methenamine silver or periodic acid–Schiff stain is done. The yeasts are essentially invisible when stained with H and E but may be extremely numerous on the specially stained preparations. The diagnosis of cryptococcosis may be verified by the following: the presence of yeasts varying considerably in size from 5 to 20 μ; no evidence of broad-based buds, but when budding is present it is thin-necked; no mature spherules with endospores; no mycelium; and a few cells with evidence of a capsule. A mucicarmine stain will demonstrate those few cells with capsules. A fluorescent antibody stain will confirm the diagnosis.

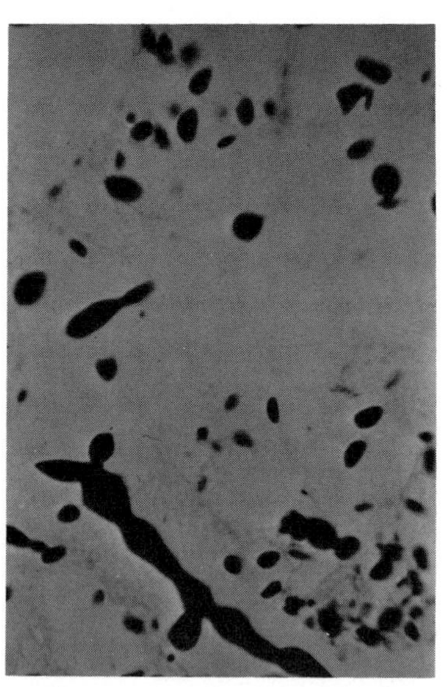

Figure 9–12. Cryptococcosis. Rare finding of mycelial-like elements in tissue of lung. Methenamine silver stain. ×600. (Courtesy of C. T. Dolan.)

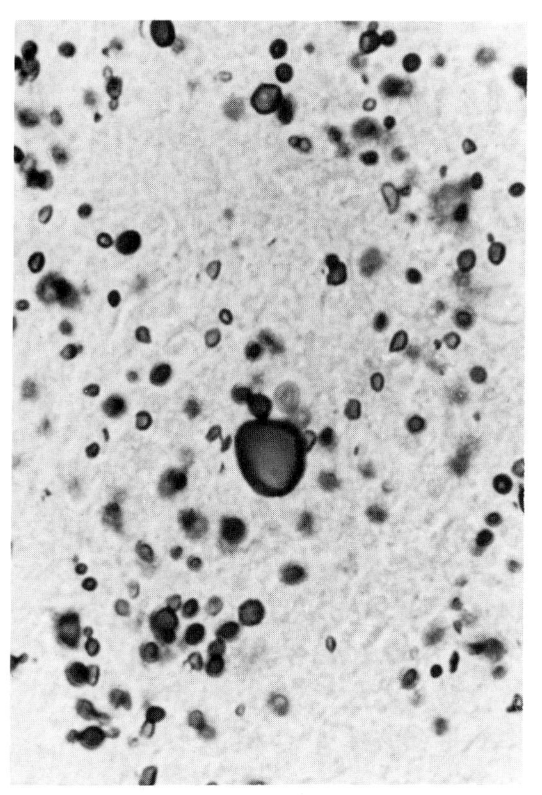

Figure 9–11. Cryptococcosis. Variation in size of yeasts present in healed lesion. This is from an old pulmonary granuloma of 1 cm in size. The lesion was calcified and fibrosed, and the organisms were dead. Note small bodies, giant yeasts, and crescent-shaped forms. Some cells retained enough capsular material to be mucicarmine-positive. Methenamine silver stain. ×400.

ANIMAL DISEASE

Natural Infection. One of the earliest recordings of cryptococcal infection was that of a myxomatous lesion of an ox.[41] Since then, spontaneous infection has been reported in horse, dog, fox, cat, cheetah, civet, monkey, guinea pig, ferret, and dairy cattle.[33] As with human infection, in animals the central nervous system and lung are most frequently involved; however, small lesions are also found on the skin. Although the lungs are probably the site of initial infection, in several reports of the disease, particularly in the horse, cat, and dog, prominent lesions were found in the facial region, nose, and hard palate. These could be interpreted as cases of direct inoculation. Massive granulomata in the stomach and intestine have been described, suggesting that the organism was swallowed, and the portal of entry was the alimentary tract. The pathology, tissue reaction, and histopathology of animal cryptococcal infection do not differ markedly from those of human disease.

Cryptococcal mastitis in dairy cows is not an uncommon disease and is worldwide in distribution. Cryptococcus was early associated with milk and dairy products. In one recent enzootic, 106 cases occurred in a herd of 235 Holstein-Friesian cows.[39] The infection was acquired through the teats from contaminated milking machines. The organisms proliferated in the udder, causing distension of the mammary, decrease in milk volume, and production of viscid secretions. In the more severe cases of infection, the regional and deeper lymph nodes were involved. Cryptococcal disease may be differentiated from bacterial mastitis by the slight elevation of temperature and absence of toxemia seen in the former.

Experimental Infection. This can be regularly produced in laboratory animals and is an important diagnostic procedure. Rabbits are more resistant than mice to infection. This may be because of the higher body temperature (39.6°C) of the rabbit. Mice are the animals of choice for most experimental work. For infection, cryptococcal organisms are obtained from a 2- to 4-day-old culture and suspended in saline. Most strains are virulent within the range of 1×10^4 to 5×10^6 cells. This count is contained in 0.02 to 0.03 ml saline and injected into the tail vein of mice. A rapid, predictably progressive, and fatal disease is produced. Mice may begin to die within a week to ten days after infection. In most animals the skull is swollen, indicating meningitis. All viable mice should be autopsied at the end of two weeks. This technique of infection is so reproducible and accurate it can be used to evaluate the efficacy of chemotherapeutic agents. It is also the most critical test used in differentiating *C. neoformans* from other *Cryptococcus* species. Intracerebral as well as intravenous inoculation is used to test the virulence of clinical isolants.

IMMUNOLOGY AND SEROLOGY

The serodiagnosis of cryptococcal infection has received much attention and is only now becoming a standardized and useful procedure. Because of the immunologically inert nature of the capsular polysaccharide, little humoral or allergic response is elicted by infection with *Cryptococcus neoformans*. For this reason no satisfactory skin test or unequivocable complement fixation or hemagglutination tests have been developed. At present, diagnostic and prognostic tests can be carried out to measure cryptococcal capsular antigen in serum or spinal fluid.[7, 22] The indirect fluorescent antibody test and the complement fixation test are used to detect the appearance of antibodies.[47] The presence of antigen is determined by the latex particle agglutination test (LCAT). In this procedure, hyperimmune rabbit antiserum is coated onto a suspension of latex particles,[7] which is then mixed with dilutions of serum or spinal fluid on a welled slide. The clumping of the latex particles is considered to be a positive reaction. This test appears to be very reliable for the detection of cryptococcal polysaccharide in body fluids, but it must be well controlled. For evaluating the progress of a patient during therapy, the latex agglutination test for antigen is combined with a method for detecting the appearance of antibody. A good prognosis is indicated by decrease in the titer of antigen concomitant with the appearance of antibody.[5] Correlation of the titers of antigen in the spinal fluid and serum has been a significant problem. Often the serum antigen titers remain positive, while those in the spinal fluid fall to undetectable levels. This is thought to indicate the presence of an extracerebral focus of infection. Persistence of either titer following a course of therapy denotes a grave prognosis.

There are several reproducible and accurate tests using immunofluorescent techniques (IFA) for the detection of antibody to cryptococci.[25, 47] In these, antigen is fixed to a slide, which is then covered with patient's serum or spinal fluid. The fixation of antibody to the antigen is detected by the subsequent addition of fluorescein-labeled, rabbit antihuman globulin to make an immunofluorescence "sandwich." In addition to these procedures, Gordon and Lapa[24] have developed a charcoal particle test to detect the presence of antibody, and immunodiffusion has also been used.[50] The value of serologic testing for following patients during treatment has been reviewed by Bardana.[5] The difficulties encountered in preparation of antigen have been reviewed by Widra.[50]

In summary, the serologic tests for cryptococcosis are useful, but it should be noted that both false-positive and false-negative results occur. The complement fixation test (CF Ag) for antigen appears to be most specific of all, since few false-positive results have been recorded. However, false-negative results run as high as 40 per cent. Latex agglutination (LCAT) for antigen is quite specific when used with proper controls,[7] but there is still a high rate of false-negative reactions. The indirect fluorescent test (IFA) for antibody is positive in 1 per cent of normal spinal fluids and sera and in up to 30 per cent of patients with blastomycosis. A tube agglutinin test (TA) is also used for the detection of antibody.[22] In the experience of some investigators, utilization of a combination of the IFA, LCAT, and CF Ag tests still results in a false-negative result rate of 15 per cent.

There have been numerous studies concerning acquired resistance to cryptococcosis.[35] and it appears that the injection of mice with capsular material or live unencapsulated yeasts affords some protection against infection.[20] By modifying the experiments which Hasenclever[26] performed with *Candida,* Abrahams has demonstrated that pertussis vaccine enhances acquired resistance of mice to cryptococcosis.[1] The exact mechanisms of this phenomenon have yet to be fully elucidated. Growth inhibiting substances have been described that occur naturally in serum, saliva, and cerebral spinal fluid.[27] This may account in part for the high degree of resistance to infection by the fungus.

The antigenic and chemical structure of *Cryptococcus neoformans,* particularly the capsule, has been the subject of much investigation. Evans in 1949 divided the pathogenic strains of *C. neoformans* into three serologic types, A, B, and C, based on the capsular material. These types comprise group III (*C. neoformans*) of the serologic scheme devised in 1935 by Rhoda Benham. Her groupings included both saprophytic and pathogenic cryptococci. The capsule of *C. neoformans* is a polysaccharide composed of xylose, mannose, and a uronic acid, probably glucuronic acid. It does not contain starch, glycogen, amino acid, protein, nucleic acids, amino sugars, or mucoitin sulfate. Several recent studies of the distribution of the serologic groups indicate that essentially all organisms from avian habitats and human infections are type A.[15,49,51] In one study most of the B and C types were from California.[51] However, several investigators have questioned whether there is a valid difference between group A and the other serotypes. They suggest that the antigenic difference may simply be a change which occurs in culture or during mouse passage and represents random mutation.[15,49]

LABORATORY IDENTIFICATION

Direct Examination. The yeasts of *Cryptococcus neoformans* are quite fragile and collapse or become crescentic in dried, fixed, or stained films. Although they are readily demonstrable in stained histopathologic slide preparations, direct examination of a wet mount should be carried out. The capsule is so distinctive that, in infected material mixed with a drop of India ink, nigrosin, or any colored colloidal mounting medium, the encapsulated organisms are outlined by negative contrast. This technique may be used to visualize the organisms in macerated biopsy material, centrifuged sediment of spinal fluid, or touch slides of autopsy material. Viewed microscopically with diminished light, the budding yeast cells can be seen within the capsule and the organisms thereby differentiated from leukocytes, myelin globules, fat droplets, and tissue cells. India ink is often contaminated with a variety of artifacts and such microorganisms as diphtheroids, spirillae, other motile bacteria, and even encapsulated yeasts. In addition, the carbon particles sometimes spontaneously agglutinate. For this reason, it is adviseable to run a saline control for comparison and regularly check the quality of the India ink. Sputum or pus can be digested in a potassium hydroxide mount, which eliminates most cells and other artifacts that may be misread as *Cryptococcus.* The capsule and organism resist this treatment and can be viewed by proper low-intensity lighting. Brain material can be crushed on a slide and examined directly or after mixing with India ink.

Culture Methods. Pathogenic strains of *C. neoformans* are not fastidious and grow well at 37°C. They are, however, variably sensitive to cycloheximide (Actidione), which is incorporated in most selective media for pathogenic fungi. This antibiotic was at one time used in the treatment of cryptococcal meningitis.[33] Material from patient sources can be streaked on SDA medium or SDA with antibacterial antibiotics. Several plates or a flask of broth (if the specimen is spinal fluid) should be used. These are incubated at 37°C. Some reports of labile forms of the yeasts (L forms?)[36] have indicated that additional procedures may be valuable, particularly if the initial cultures are negative or the patient is being treated with amphotericin B. In some cases[36] incubation of material in hypertonic media containing salt and 0.3 M sucrose followed by subculture onto blood agar or SDA has been used to successfully recover aberrant cryptococci. For specimens such as sputum where large numbers of contaminants may be present, a selective medium has been devised by Vogel.[48] It contains antibacterial antibiotics and potato dextrose and can detect urease production. *C. neoformans* produces a white colony with a pink halo, whereas *Candida albicans*, which is also white, does not produce a halo.

Growth from patient material on primary isolation media occurs in 24 to 48 hours; however, cultures should be retained for four to six weeks. Following initial isolation, colonies are streaked on heart infusion agar so that single clones can be picked and utilized for physiologic studies to confirm the identification.

Isolation of *Cryptococcus* from heavily contaminated material, such as pigeon nests and droppings, can be carried out using another selective medium. Staib[46] determined that creatinine is assimilated by *C. neoformans* but not by other members of its genus or by most species from the other common genera of the yeasts (including *Candida*). He observed also that *C. neoformans* selectively absorbs a brown pigment from the seeds of a common weed, *Guizotia abyssinica*. Shields and Ajello[44] then devised a selective medium for *C. neoformans* by combining purine and *Guizotia* seeds with diphenyl and chloramphenicol to act as mold and bacterial inhibitors. When incubated at 37°C on this medium, a few strains of *Cryptococcus laurentii* will also grow, but neither this species nor *Candida albicans* becomes pigmented, whereas almost all strains of *C. neoformans* grow well and develop a brown color. Such typical colonies can then be subcultured for further testing.

Some isolants will have very small capsules on primary isolation. Capsule production is enhanced by growing the organisms on chocolate agar at 37°C in a CO_2 incubator.

The specific identification of *C. neoformans* requires the study of a combination of factors, including physiologic characteristics, temperature tolerance, and animal pathogenicity.[21, 29] The characteristics which define the genus *Cryptococcus* are the assimilation of inositol, production of urease, and lack of mycelium on cornmeal agar. Carbon and nitrate assimilation profiles can also be carried out (see chart). The most reliable tests for the identification of *C. neoformans* are growth in culture at 37°C and pathogenicity for mice (see section on experimental animal disease).

MYCOLOGY

Cryptococcus neoformans (Sanfelice) Vuillemin 1901

Synonymy. *Saccharomyces neoformans* Sanfelice 1895; *Torula neoformans* Weiss 1902; *Blastomyces neoformans* Arzt 1924; *Torulopsis neoformans* Redaelli and Ciferri 1931; *Debaromyces neoformans* Redaelli, Ciferri, and Giordano 1937; *Saccharomyces lithogenes* Sanfelice 1895; *Cryptococcus hominis* Vuillemin 1901; *Atelosaccharomyces hominis* Verdun 1912; *Debaromyces hominis* Todd and Hermann 1936; *Torula histolytica* Stoddard and Cutler 1916; Lodder[34] lists 39 synonyms.

Perfect Stage (proposed). *Leukosporidium neoformans* Shadomy 1970.

Morphology. SDB or malt extract broth: three days, 25°C. Cells are spherical or globose, occurring singly or in pairs, sometimes groups. Budding is single or double anywhere on the cell; sometimes several buds appear. Size: 3.5 to 7.0 × 3.7 to 8 μ.

SDA or malt extract agar: three days, 25°C. Cells are globose to spheroidal, 2.5 to 8 μ in diameter. Encapsulation varies with

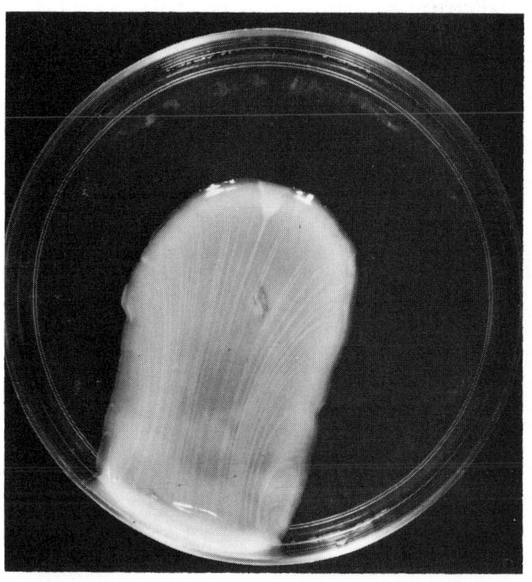

Figure 9–13. FS B 28, A. Cryptococcosis. Colony of *C. neoformans* on Sabouraud's glucose agar for one week, showing mucoid colony flowing over medium and evident sectoring.

Table 9–1. *Physiologic Differentiation of Common Cryptococcus Species*

	Nitrate Assimilation	Growth at 37°C	Animal Pathology	Carbon Assimilation						
				Gal	*Ma*	*La*	*Me*	*Er*	*Galtol*	*Su*
C. neoformans	–	+	+	+	+	–	–	v	+	+
C. laurentii and var.*	–	v	–	+	+	+	+/–	+/–	v	+
C. albidus and var.	+	v	–	+/w	+	+/–	+/v	–/w	v	+
C. terreus	+	–	–	+	v	+/w	–	–	v	–
C. leuteolus	–	–	–	+	+	–/w	+/w	+	+	+
C. melibiosum	–	–	–	+	–	w	+	–	–	–
C. flavus	–	–	–	+	+	+	+	+	w	+
C. lactativorus	–	+	–	–	–	–	–	–	–	–

*Var. of species variable in some characteristics – see Lodder.[34]

Gal = galactose, Ma = maltose, La = lactose, Me = melibiose, Er = erythritol, Galtol = galactitol, Su = sucrose, w = weak, v = variable.

the strain and the medium used. After one week, sectoring of the colony occurs in almost all species of the genus *Cryptococcus* (Fig. 9–13 FS B 28,*A*). This is because of changes in the composition and structure of the capsules. At one month, the colonies are cream-colored to yellowish or slightly pink. The texture is mucoid, and the colony may flow to the bottom of the slant. The edges are entire and without pseudomycelium. Occasional isolants produce little or no capsular material.[16a] Colonies of such organisms appear dry or glabrous. Very rare strains produce mycelium and under particular conditions are seen to form what appear to be clamp connections, teliospores, and basidiospores.[42] Organisms grown in the presence of deoxycholate tend to elongate and retain this morphology in experimental animals.[23]

Cornmeal agar: no pseudomycelium.

Fermentation. None.

Nitrogen Assimilation. Nitrate-negative; tryptone-positive.

Carbon Assimilation. Glucose +, galactose +, L-sorbose +/–, sucrose +, maltose +, cellobiose +/w, trehalose +, lactose –, melibiose –, raffinose +/w, melezitose +, inulin –, L- and D-arabinose +/w, D-ribose +/w (rare), L-rhamnose +, ethanol +/w (rare), ribitol +, galactitol +, D-mannitol +, inositol +, starch formation +/w, gelatin liquefaction –, D-xylose +.

REFERENCES

1. Abrahams, I. 1966. Further studies on acquired resistance to murine cryptococcosis: Enhancing effect of *Bordetella* pertussis. J. Immunol., *98*:914–922.
2. Ajello, L. 1969. A comparative study of the pulmonary mycoses of Canada and the United States. Public Health Rep., *84*:869–877.
3. Baker, R. D., and R. K. Haugen. 1955. Tissue changes and tissue diagnosis in cryptococcosis – a study of 26 cases. Am. J. Clin. Pathol., *25*:4–24.
4. Baker, R. D., and G. Linares. 1971. Cryptococcal dermotropism in Rhesus monkeys. Bact. Proc., *71*:118.
5. Bardana, E. J., L. Kaufman, et al. 1968. Amphotericin B and cryptococcal infection. Arch. Intern. Med., *122*:517–520.
6. Bennett, J. E. 1966. Susceptibility of *Cryptococcus neoformans* to Amphotericin B, In *Antimicrobial Agents and Chemotherapy*, 1966, Philadelphia, Am. Soc. for Microbiol., Oct., pp. 405–410.
7. Bloomfield, N., M. A. Gordon, et al. 1963. Detection of *Cryptococcus neoformans* antigen in body fluids by latex particle agglutination. Proc. Soc. Exp. Biol. Med., *114*:64–67.
8. Bulmer, G. S., M. D. Sans, et al. 1967. *Cryptococcus neoformans*. I. Nonencapsulated mutants. J. Bacteriol., *94*:1475–1479.
9. Bulmer, G. S., and M. D. Sans. 1967. *Cryptococcus neoformans*. II. Phagocytosis by human leukocytes. J. Bacteriol., *94*:1480–1483.
10. Buschke, A. 1895. Ueber eine durch coccidien hervegerufene Krankheit des menschen. Dtsch. Med. Wochenschr., *21*(No. 3):14.
11. Busse, O. 1894. Ueber parasitäre zelleinschlüsse und ihre züchtung. Zentralbl. Bakteriol., *16*:175–180.
12. Campbell, G. D. 1966. Primary pulmonary cryptococcosis. Am. Rev. Resp. Dis., *94*:236–243.
13. Collins, D. N., I. A. Oppenheim, et al. 1971. Cryptococcosis associated with systemic lupus erythematous. Arch. Pathol., *91*:78–88.
14. Daveny, J. K., and M. D. Ross. 1969. Cryptococcosis of bone. Cent. Afr. J. Med., *15*:78–79.
15. Denton, J. F., and A. F. DiSalvo. 1968. The prevalence of Cryptococcus in various natural habitats. Sabouraudia, *6*:213–217.
16. Emmons, C. W. 1955. Saprophytic sources of *Cryptococcus neoformans* associated with the pigeon (*Columba livia*). Am. J. Hyg., *62*:227–232.

16a. Farmer, S. G., and R. A. Komorowski. 1973. Histologic response to capsule-deficient *Cryptococcus neoformans*. Arch. Pathol., *96*:383–386.

17. Felton, F. G., W. E. Maldonado, et al. 1966. Experimental cryptococcal infection in rabbits. Am. Rev. Resp. Dis., *94*:589–594.
18. Fink, J. N., J. J. Barboriak, et al. 1968. Cryptococcal antibodies in pigeon breeders' disease. J. Allergy, *41*:297–301.
19. Furcolow, M. L. 1963. The use of amphotericin B in blastomycosis, cryptococcosis, and histoplasmosis. Med. Clin. North Am., *47*(5): 1119–1130.
20. Gadebusch, H. H., and A. G. Johnson. 1966. Natural host resistance to infection with *Cryptococcus neoformans*. J. Infect. Dis., *116*:551–572.
21. Goodman, J. L., L. Kaufman, et al. 1971. Diagnosis of cryptococcal meningitis. New Engl. J. Med., *285*:434–436.
22. Gordon, M. A. 1970. Practical serology of the systemic mycoses. Int. J. Dermatol., *9*:209–214.
23. Gordon, M. A., and J. Devine. 1970. Filamentation and endogenous sporulation in *Cryptococcus neoformans*. Sabouraudia, *8*:227–234.
24. Gordon, M. A., and E. Lapa. 1971. Charcoal particle agglutination test for detection of antibody to *Cryptococcus neoformans*. Am. J. Clin. Pathol., *56*:354–359.
25. Goren, M. B., and J. Warren. 1968. Immunofluorescence studies of reactions of the cryptococcal capsule. J. Infect. Dis., *118*:216–229.
26. Hasenclever, H. F., and E. J. Corley. 1968. Enhancement of acquired resistance in murine candidiasis by Bordetella pertussis vaccine. Sabouraudia, *6*:289–295.
27. Howard, J. I., and R. P. Bolande. 1966. Humoral defense mechanisms in cryptococcosis: Substances in normal human serum, saliva, and cerebral spinal fluid affecting the growth of *Cryptococcus neoformans*. J. Infect. Dis., *116*: 75–83.
28. Ishaq, C. M., G. S. Bulmer, et al. 1968. An evaluation of various environmental factors affecting the propagation of *Cryptococcus neoformans*. Mycopathologia, *35*:81–90.
29. Jennings, A., J. E. Bennett, et al. 1968. Identification of *Cryptococcus neoformans* in routine clinical laboratory. Mycopathologia, *35*:256–264.
30. Khodadoust, A. A., and J. W. Payne. 1969. Cryptococcal (torular) retinitis. Am. J. Ophthalmol., *67*:745–750.
30a. Krumholz, R. A. 1972. Pulmonary cryptococcosis. A case due to *Cryptococcus albidus*. Am. Rev. Resp. Dis., *105*:421–424.
31. Littman, M. L., and R. Borok. 1968. Relation of the pigeon to cryptococcosis: Natural carrier state, heat resistance and survival of *Cryptococcus neoformans*. Mycopathologia, *35*:329–345.
32. Littman, M. L., and J. E. Walker. 1968. Cryptococcosis: current issues. Am. J. Med., *45*:922–933.
33. Littman, J. L., and L. E. Zimmerman. 1956. *Cryptococcosis*. New York, Grune & Stratton.
34. Lodder, J. (ed.). 1970. *The Yeasts*. Amsterdam, North-Holland.
34a. Lomvardia, S., and H. I. Lurie. 1972. Epipleural cryptococcosis in a patient with Hodgkin's disease: A case report. Sabouraudia, *10*:256–259.
34b. Long, R. F., S. V. Berens, et al. 1972. An unusual manifestation of pulmonary cryptococcosis. Br. J. Radiol., *45*:757–759.
35. Louria, D. B., and T. Kaminski. 1965. Passively acquired immunity in experimental cryptococcus. Sabouraudia, *4*:80–84.
36. Louria, D. B., T. Kaminski, et al. 1969. Aberrant forms of bacteria and fungi found in blood on cerebrospinal fluid. Arch. Intern. Med., *124*: 39–48.
37. Mira, C. A., R. Anzola, et al. 1968. Aislamiento de criptococcis neoformans a portir de materiales contaminados con excreta de palomas en Medellin, Colombia. Antioquia Med., *18*:33–40.
37a. Mohr, J. A., H. Long, et al. 1972 In vitro susceptibility of *Cryptococcus neoformans* to steroids. Sabouraudia, *10*:171–172.
38. Noble, R. C., and L. F. Fajardo. 1972. Primary cutaneous cryptococcosis. Am. J. Clin. Pathol., *57*:13–22.
39. Pounden, W. D., J. M. Amberson, et al. 1952. A severe mastitis problem associated with *Cryptococcus neoformans* in a large dairy herd. Am. J. Vet. Res., *13*:121–128.
40. Procknow, J. J., J. R. Benfield, et al. 1965. Cryptococcal hepatitis presenting as a surgical emergency. J.A.M.A., *191*:269–278.
41. Sanfelice, F. 1894. Contributo alla morfologia e biologia dei blastomiceti che si sviluppano nei succhi di alcuni frutti. Ann. Igiene, *5*:239–262.
41a. Schwaz, J., and G. L. Baum. 1970. Cryptococcosis. Semin. Roentgenol., *5*:49–54.
42. Shadomy, H. J. 1971. Clamp connections in two strains of *Cryptococcus neoformans*. *Recent Trends of Yeast Research*. Atlanta Research Articles, School of Arts and Sciences, Georgia State University.
43. Shadomy, S., H. J. Shadomy, et al. 1970. *In vivo* susceptibility of *Cryptococcus neoformans* to hamycin, amphotericin B and 5-fluorocytosine. Infection and Immunity, *1*:128–134.
44. Shields, A. B., and L. Ajello. 1966. Medium for selective isolation of *Cryptococcus neoformans*. Science, *151*:208–209.
44a. Silberfarb, P. M., G. A. Saros, et al. 1972. Cryptococcosis and pregnancy. Am. J. Obstet. Gynecol., *112*:714–720.
45. Sokolowski, J. W., R. F. Schilaci, et al. 1969. Disseminated cryptococcosis complicating sarcoidosis. Am. Rev. Resp. Dis., *100*:717–722.
46. Staib, F., and G. Bethauser. 1968. Zum Nachweis von *Cryptococcus neoformans* in Stab von einem Taubenschlag. Mykosen, *11*:619–624.
46a. Sunderland, W. A., R. A. Campbell, et al. 1972. Pulmonary alveolar proteinosis and pulmonary cryptococcosis in an adolescent boy. J. Pediatr., *80*:450–456.
47. Vogel, R. A. 1966. The indirect fluorescent antibody test for the detection of antibody in human cryptococcal disease. J. Infect. Dis., *116*:573–580.
48. Vogel, R. A. 1969. Primary isolation medium for *Cryptococcus neoformans*. Appl. Microbiol., *18*:1100.
49. Walter, J. E., and E. G. Coffee. 1968. Distribution and epidemiological significance of serotypes of *Cryptococcus neoformans*. Am. J. Epidemiol., *87*:167–172.
50. Widra, A., S. McMillen, et al. 1968. Problems in serodiagnosis of *Cryptococcus*. Mycopathologia, *36*:354–358.
51. Wilson, D. E., J. E. Bennett, et al. 1968. Serological grouping of *Cryptococcus neoformans*. Proc. Soc. Exp. Biol. Med., *127*:820–823.

Torulopsosis

Torulopsosis is one of the less common yeast infections of man and animals. Most reports of disease caused by the organism, *Torulopsis glabrata*, concern patients with terminal neoplasias,[6] diabetes,[1] and prolonged intravenous therapy.[16] Rare infections occur in patients with no apparent underlying disease.[15] The yeast is a very common member of the normal flora of the oral cavity, gastrointestinal tract, and urogenital area,[1, 6, 17] as well as occurring in animals and soil[3, 19] Wickerham in 1957[20] called attention to the increasing frequency of *Torulopsis* infections in debilitated individuals, and recent literature has borne out his observation. The yeast has essentially no virulence for normal animals or man,[7, 12] but rather it represents an opportunistic invader of the compromised patient. Taschdjian[18] has reviewed the subject of opportunistic yeast infections and the factors influencing their establishment.

The early reports of torulopsosis in man and animals stressed the intracellular habitat of the yeast and the possibility of confusing it with another intracellular yeast—*Histoplasma capsulatum*[10, 13] (Fig. 9–14). More recent observations have not substantiated the early findings (Fig. 9–15). Although *T. glabrata* may be seen in great numbers within macrophages in some human lung infections, this is an exceptional presentation. For example, in reviewing histologic preparations from cases of fatal systemic infections caused by this yeast, Grimley et al.[6] found masses of fungi within tongue, esophagus, lung, and kidney, but not within the phagocytic cells of these organs. Similar observations have been made in cases with spleen and central nervous involvement. In mice altered by alloxan, cortisone, or x-ray irradiation, systemic infection was produced, but again the yeasts were not within macrophages. *In vitro* studies by Howard and Otto[8] have confirmed that the organism does not proliferate readily within mouse macrophages, an effect which is not due solely to the fungistatic effect of serum. Although the presence of serum markedly decreases growth of the organism in glucose peptone broth, maximum depression takes place only when organisms have been phagocytized in cultures of mouse macrophages.

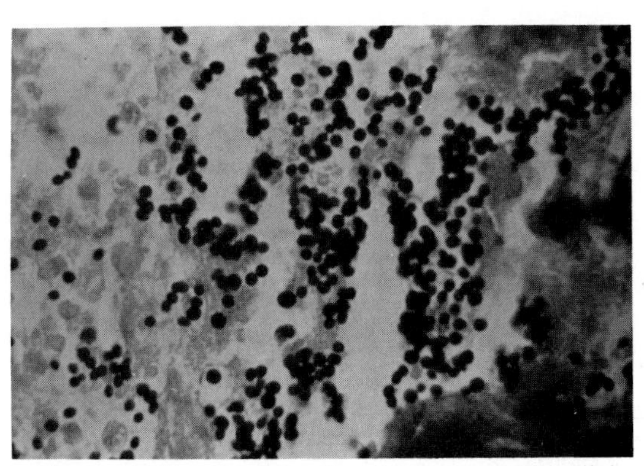

Figure 9–15. Torulopsosis. Numerous extracellular organisms in a case of endocarditis. Round, budding yeast cells are seen with no hyphal formation.

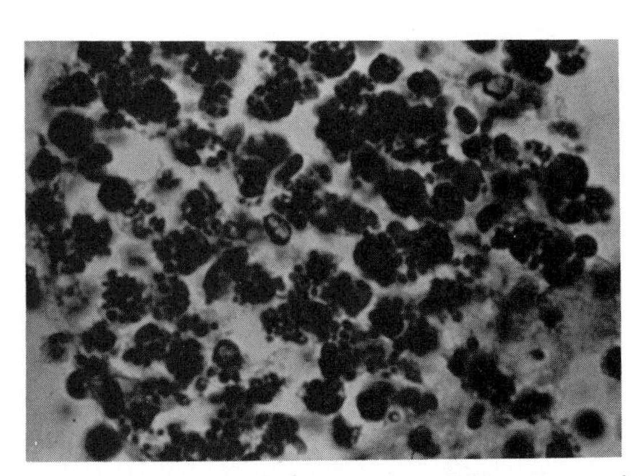

Figure 9–14. Torulopsosis. Intracellular yeasts of *Torulopsis glabrata* in an infection of the lung. Methenamine silver stain. ×400. (Courtesy of C. T. Dolan.)

CLINICAL SYMPTOMS

Sufficient case reports are lacking for the formulation of a specific clinical syndrome for torulopsosis, but a few constant features are suggestive of the disease. Almost all patients reported had received long-term therapy, such as cortisone, antibiotics, immunosuppressive agents, and irradiation, or were on hyperalimentation with intravenous catheters. The first symptoms noted were high fever (103° to 105°F), hypotension (BP 90/60), tachycardia, shaking chills, diaphoresis, and rigors. The similarity of this syndrome to the septic shock of gram-negative bacteremia has been frequently noted.[16] This form of torulopsosis is a toxic reaction to the yeast constantly being seeded into the patient, and it is not a true infection. In true torulopsosis fungemia, in which there is establishment of systemic infection, the clinical course is fulminant and rapidly fatal, although successful intervention with amphotericin B therapy has been recorded.[11] It appears, however, that many cases of shock from fungemia can be relieved by removal of the point source of the yeast, such as intravenous catheters and indwelling urinary tract catheters. Such patients usually do not require antifungal therapy.[16]

The urinary tract is often the site of *T. glabrata* infections, but the organism is also isolated with great frequency[1] from asymptomatic patients. Numbers up to 10^5 per ml have been isolated from the urine of terminal and diabetic patients, but the contribution of the yeast to local pathology remains problematic. In some patients actual colonization of the bladder, urethra, and renal pelvis may occur, which then results in a significant and sometimes fatal disease. In these cases the symptoms are those of intermittent or chronic urinary tract infection, with the repeated isolation of *T. glabrata* in significant numbers (10^5 per ml or over). Some infections have been the result of contamination during retrograde pyelograms.[14] Treatment of such cases is required and consists of a lavage of nystatin or 5-fluorocytosine to the affected area. This therapy is usually beneficial.

Colonization by the yeast and the production of disease in the vagina, uterus, and fallopian tubes has been recorded. Many such cases have occurred during pregnancy and usually clear spontaneously post partum.

Meningitis,[13] endocarditis,[9a] and pulmonary[15] and systemic infections have all been noted. Most of these cases have been fatal; however, a few responded to antimycotic therapy.[16a]

PATHOLOGY

The lesions evoked by systemic *Torulopsis* infection resemble those of candidosis. Granulomas with giant cells have been found in the lung, but the usual pathologic picture is the microabscess. These abscesses contain numerous neutrophils, some plasma cells and lymphocytes, and large numbers of yeasts. The yeasts are small and ovate in shape (2 to 3 μ by 1 to 2 μ), but some strains are larger and measure 2.5 to 3 $\times$ 4 to 5 μ. They fall in the size range between *Histoplasma* and *Candida.* Budding is easily seen in tissue sections. The bud attachment is broader in *Torulopsis* than in *Histoplasma.* Also, the latter is commonly seen within macrophages, whereas *Torulopsis* is variably intracellular. In contrast to candidosis, in torulopsosis no mycelial forms are seen in tissue in established infections.

ANIMAL DISEASE

Torulopsis was associated early with animal disease, particularly mastitis of cows. The organism has been isolated also from the liver of an elephant and from systemic infections of dogs and monkeys. Normal laboratory animals are quite resistant to infection by *Torulopsis.* Five million cells injected intravenously do not produce disease.[11] In physiologically altered mice (alloxan, x-irradiation, or cortisone), however, infections have been produced. Large abscesses containing numerous yeasts were found in them, but none of the organisms were seen intracellularly.[7] A related species, *Torulopsis pintolopesii,* is a normal inhabitant of the mouse intestine. Careless autopsy of experimentally infected animals may give erroneous results because of contamination by this organism.[5]

TREATMENT

Most cases of *T. glabrata* fungemia clear spontaneously with removal of the point source of the yeasts. In valid infection, antifungal therapy is necessary. Systemic infections respond to amphotericin B in the usual dose for fungal diseases (1 mg per kg, to a total dose of 1 to 3 g, given on alternate days as tolerated). In a case of *T. glabrata* fungemia, Louria[11] found that a 14-day course of therapy with doses of 28 to 40 mg per day cleared the infection, and the patient became afebrile. Dolan[4] has found that a regimen of 5-fluorocytosine (100 mg per kg per 24 hours for six weeks) evoked a clinical cure in cases he has treated.

Esophageal, urinary tract, and genital torulopsosis can be treated by nystatin lavage. This is preferable to amphotericin B, as the fungus is more sensitive to nystatin. A successful regimen used by Newman[14] in renal disease was continuous nystatin lavage, consisting of 800,000 units in 1000 ml of fluid every 12 hours for four days.

CULTURE METHODS

T. glabrata is resistant to cycloheximide and grows on almost all laboratory media. On blood agar, the colonies are tiny, white, raised, and nonhemolytic. On prolonged incubation they remain small.[12] Specific identification requires differentiation from *Cryptococci* and *Candida* by cultural characteristics. Lack of formation of mycelium on cornmeal agar differentiates it from the *Candida* genus, and lack of assimilation of inositol from the *Cryptococcus* sp.

MYCOLOGY

Torulopsis glabrata **(Anderson) Lodder and DeVries 1938**

Synonymy. *Cryptococcus glabratus* Anderson 1917.

Morphology. SDB or malt extract broth: three days, 25°C. Cells are ovoid to obovate, 2.5 to 4.5 × 4 to 6 μ. After one month some strains form a delicate pellicle and a ring. Most strains form sediment only.

SDA or malt extract agar: 25°C. Cells same as above. Colonies are cream-colored, soft, smooth, and glossy. They have a similar morphology on all media at 37°C.

Cornmeal agar: no mycelium.

Fermentation. Glucose and trehalose fermented; no other sugars positive. Most strains require niacin and pyridoxine; some also require biotin and thiamine.

Assimilation. Potassium nitrate–negative. Glucose +, galactose −, sucrose −, maltose −, cellibiose −, trehalose +, lactose −, inositol −, D-arabinose −, salicin −, ethanol +(w) or −, glycerol +, soluble starch −, melezitose −, D-ribose −, L-rhamnose −.

The morphology of the *T. pintolopesii* is similar to that of *T. glabrata.* It differs from the latter species in that only glucose is fermented and assimilated; no other carbon compounds are utilized.

Rhodotorulosis

Rhodotorula rubra, Rhodotorula sp., *Hansenula* sp., and *Schizosaccharomyces* sp. have on rare occasion been isolated from patients in the terminal stages of debilitating diseases. Of these yeasts, *R. rubra* is most frequently involved, and its presence has been documented in several fatal infections of the lung, kidney, and central nervous system. Frequently, *Rhodotorula* fungemia is found to be due to contaminated catheters, intravenous solutions, blood bank apparatus, and heart-lung and dialysis machines. The affected patients present with symptoms of endotoxic shock, and cultures of their blood may be positive. Removal of the source of contamination usually leads to clearing of the symptoms. Some cases require amphotericin B therapy.[9] *Rhodotorula* has been involved in a case of fatal endocarditis, in which the organism was isolated from the blood during life and from the aortic valve at autopsy.[11] Attempts to infect mice intracerebrally and intravenously with 5×10^6 cells were unsuccessful.

Rhodotorula is a common airborne contaminant of skin, lungs, urine, and feces. Its isolation from sputum is usually without significance. Its occurrence in blood cultures when contamination has been ruled out is of greater significance. It grows readily on

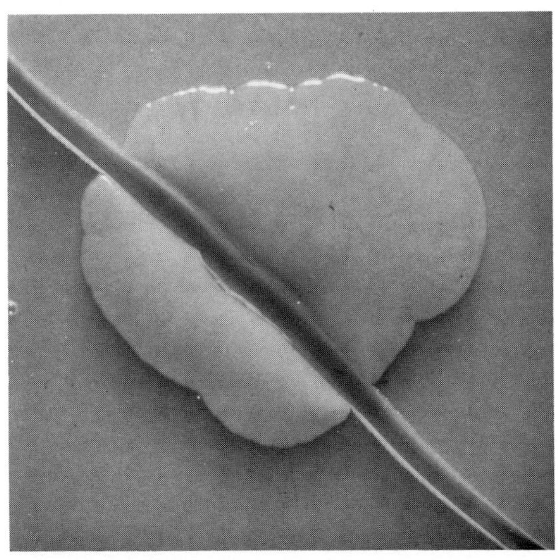

Figure 9–16. FS B 28, *B.* Rhodotorulopsis. Mucoid colony of *Rhodotorula rubra.*

almost all culture media, and its coral red, mucoid colonies are distinctive (Fig. 9–16 FS B 28,*B*). It is encapsulated and rarely forms mycelium. The organism has also been isolated from a number of food sources, cheese and milk products, air, and soil.

MYCOLOGY

Rhodotorula rubra (Demme) Lodder 1934

Synonymy. *Saccharomyces ruber* Demme 1889; *Rhodotorula mucilaginosa* (Jörgensen) Harrison 1928; *Torulopsis mena* Dodge 1935; *Torulopsis sanguinea* (Ciferri and Redaelli) 1925; *Rhodotorula sanniei* Lodder 1934; Lodder, 1970, lists 34 names in synonymy.

Perfect Stage. The perfect stage has not been described, but other species of *Rhodotorula* have a sexual stage in the class Basidiomycetes, genus *Rhodosporidium*. The chief characteristics of the genus *Rhodotorula* are the presence of carotenoid pigments, the inability to assimilate inositol, the rudimentary or absent pseudomycelium, and the lack of fermentation.

Morphology. SDB or malt extract broth: three days, 25°C. Cells are short-ovoid to elongate, single, or in short chains or clusters, 2 to 6.5 μ in diameter, with elongate cells up to 14 μ. A ring may form.

SDA or malt extract agar: morphology is similar to above. The colony is moist, smooth to mucoid, glistening, and coral-red to pink or salmon in color (Fig. 9–16 FS B 28*B*). Sectoring may be evident.

Cornmeal agar: rarely, rudimentary pseudomycelium is formed.

Fermentation. None.

Assimilation. Potassium nitrate. Glucose +, galactose +, w or −, sucrose +, maltose +, cellobiose + or −, trehalose +, raffinose +, inulin +, inositol −, soluble starch −, D-xylose +, ethanol + or −, D-mannitol + or −.

Vitamin Requirement. Thiamine.

REFERENCES

1. Ahern, D. G., J. R. Jannach, et al. 1966. Speciation and densities of yeasts in human urine specimens. Sabouraudia, *5*:110–119.
2. Anderson, H. W. 1917. Yeast-like fungi of human intestinal tract. J. Infect. Dis., *21*:341–385.
3. Cooke, W. B. 1961. Some effects of spray disposal of spent sulfite liquor on soil mold populations. Proc. Industrial Waste Conference, Purdue University, *45*:35–48.
4. Dolan, C. T. Personal communication.
5. Emmons, C. W., C. H. Binford, et al. 1970. *Medical Mycology*. Philadelphia, Lea and Febiger, p. 438.
6. Grimley, P. M., L. D. Wright, et al. 1955. *Torulopsis glabrata* infection in man. Am. J. Clin. Pathol., *43*:216–223.
7. Hasenclever, H. F., and W. O. Mitchell. 1962. Pathogenesis of *Torulopsis glabrata* in physiologically altered mice. Sabouraudia, *2*:87–95.
8. Howard, D. H., and V. Otto. 1967. The intracellular behavior of *Torulopsis glabrata*. Sabouraudia, *5*:235–239.
9. Leeber, D. A., and I. Scher. 1969. *Rhodotorula* fungemia presenting as "endotoxic" shock. Arch. Intern. Med., *123*:78–81.

9a. Lees, A. W., S. S. Rau, et al. 1971. Endocarditis due to *Torulopsis glabrata*. Lancet, *1*:943–944.

10. Lopez Fernandez, J. F. 1952. Acciopatogena experimental de la levadura *Torulopsis glabrata* (Anderson 1917) Lodder y De Vries 1938 productora de lesions histopatologicas semijantes a las de la histoplasmosis. Ann. Fac. Med. Montevideo, *37*:470–483.
11. Louria, D. B., S. M. Greenberg, et al. 1960. Fungemia caused by certain non-pathogenic strains of the family cryptococcaceae. New Engl. J. Med. *263*:1281–1284.
12. Marks, M. I., and E. O'Toole. 1970. Laboratory identification of *Torulopsis glabrata:* typical appearance on routine bacteriological media. Appl. Microbiol., *19*:184–185.
13. Minkowitz, S., D. Koffler, et al. 1963. *Torulopsis glabrata* septicemia. Am. J. Med., *34*:252–255.
14. Newman, D. M., and J. M. Hogg. 1969. *Torulopsis glabrata* pyelonephritis. J. Urol., *102*:547–548.
15. Oldfield, F. S. J., L. Kapica, et al. 1968. Pulmonary infection due to *Torulopsis glabrata*. Can. Med. Assoc. J., *98*:165–168.
16. Rodriguez, R., H. Shinya, et al. 1971. *Torulopsis glabrata* fungemia during prolonged intravenous alimentation therapy. New Engl. J. Med., *284*:540–541.

16a. Steer, P. L., M. J. Marks, et al. 1972. 5-Fluorocytosine: An oral antifungal compound. A report on clinical and laboratory experience. Ann. Intern. Med., *76*:15–22.

17. Stenderup, A., and G. T. Pederson. 1962. Yeasts of human origin. Acta Pathol. Microbiol. Scand., *54*:462–472.
18. Taschdjian, C., P. Kozinn, et al. 1970. Opportunistic yeast infections with special reference to candidiasis. Ann. N.Y. Acad. Sci., *174*:431–622.
19. Van Uden, N., L. Do Carmo Sousa, et al. 1958. On the intestinal yeast flora of horses, sheep, goats and swine. J. Gen. Microbiol., *19*:435–445.
20. Wickerham, L. J. 1957. Apparent increase in frequency of infections involving *Torulopsis glabrata*. J.A.M.A., *165*:47–48.

Miscellaneous Yeasts

Yeasts are such a constant and common part of the normal human flora, it is not surprising that many reports of their association with disease have been made. Most of these are unsubstantiated and probably represent isolation of normal flora. Indeed it is difficult to establish an unassailable diagnosis of such infections. Continuous isolation of large numbers of organisms, together with demonstrable clinical symptoms and pathologic evidence, is necessary. The latter is probably the most important factor and depends on the demonstration of the organism in tissue, with evidence of colonization and invasion. In this section some of the "miscellaneous yeast infections" recently recorded in which such evidence was present will be enumerated.

Trichosporon cutaneum is the etiologic agent of white piedra and is extensively discussed in that section. This yeast is also a common inhabitant of the respiratory tract, mouth, and pharynx. It has occasionally been seen in pulmonary and systemic infections.[8, 10] Taschdjian[8] recorded a case of systemic involvement seen by Haley and Toni. Abscesses were found in the kidney and other organs. Watson and Kallichurum[10] describe a case in which *T. cutaneum* was present in a brain abscess superimposed on a secondary metastasis from a bronchial adenocarcinoma.

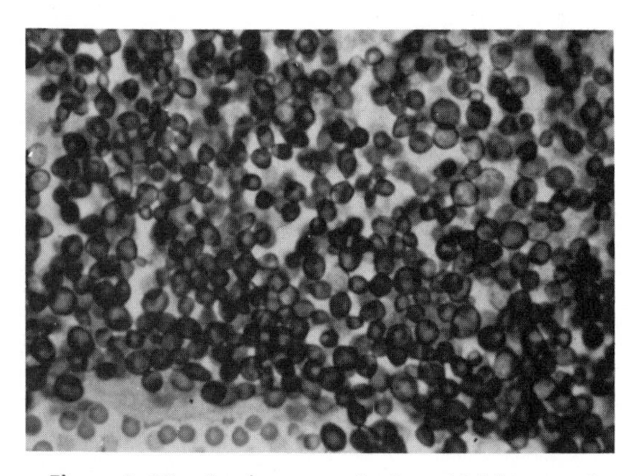

Figure 9–17. Saccharomycosis. Bronchial biopsy of a *Saccharomyces cerevisiae* infection of lung. Methenamine silver stain. ×425. (Courtesy of C. T. Dolan.)

Trichosporon capitatum is also a resident flora and has been implicated in pulmonary disease.[5]

Ascosporagenous yeasts, such as *Hansenula anomala, Pichia membranaefaciens, Saccharomyces carlebergensis, S. cerevisiae,* and *S. fragilis,* may be a small part of the normal flora or transient flora of the throat and alimentary tract.[7] In Europe, *S. cerevisiae* has been implicated in several cases of pulmonary disease, especially in brewers.[3] A well-documented case from the Mayo Clinic was reported by Dolan[4] (Fig. 9–17). The organism was seen colonizing the lung parenchyma and was repeatedly isolated from sputum. It also has been reported to cause disease in rabbits.[2] *S. carlsbergensis* and *S. pastorianus* have been implicated in mycosis of the stomach.[1] Undoubtedly as the number of debilitated patients whose lives are extended by medical means increases, the occurrence and list of opportunistic yeast infections will also increase.

REFERENCES

1. Ahnlund, H. O., B. Pallin, et al. 1969. Mycosis of the stomach. Acta Chir. Scand., *133*:555–562.
2. Devos, A. 1969. Post-mortem findings in rabbits. Vlaams Diergeneesk. Tijdschr., *38*:275–280.
3. Dodge, C. W. 1935. *Medical Mycology.* St. Louis, C. V. Mosby.
4. Dolan, C. T. 1971. Personal communication.
5. Gemeinhardt, H. 1965. The pathogenicity of *Trichosporon capitatum* in lungs of man. Zentralbl. Bakteriol (Orig.), *196*:121–133.
6. Gonyea, E. F. 1973. Cisternal puncture and cryptococcal meningitis. Arch. Neurol., *28*:200–201.
7. Mackenzie, D. W. H. 1961. Yeasts from human sources. Sabouraudia, *1*:8–14.

7a. Sleer, P. L., M. I. Marks, et al. 1972. 5-Fluorocytosine: An oral antifungal compound. A report on clinical and laboratory experience. Ann. Intern. Med., *76*:15–22.

8. Taschdjian, C., P. J. Kozinn, et al. 1970. Opportunistic yeast infections with special reference to candidosis. N.Y. Acad. Sci., *174*:431–1056.
9. Vieu, M., and G. Segretain. 1959. Contribution a l'etude de *Geotrichum* and *Trichosporum* d'origine humaire. Ann. Inst. Pasteur (Paris), *96*:421–433.
10. Watson, K. C., and K. Kallichurum. 1970. Brain abscess due to *Trichosporon cutaneum.* J. Med. Microbiol., *3*:191–193.

THE SUBCUTANEOUS MYCOSES

Chapter 10

CHROMOMYCOSIS

INTRODUCTION

The subcutaneous mycoses include a heterogeneous group of infections characterized by the development of a lesion at the site of inoculation. Unlike the systemic mycoses, whose primary mode of entry is usually pulmonary, these infections are the result of traumatic implantation of the fungus into the skin. In general, the ensuing disease remains localized to this area or slowly spreads to surrounding tissue, a picture similar to that seen with the mycetomas. In some diseases, slow extension via lymphatic channels is a frequent occurrence (sporotrichosis); in others, hematogenous and lymphatic dissemination is rarely recorded (chromomycosis).

The type of disease evoked is an interesting interplay between host reactions and defenses, and the relative virulence of the infecting agent. The species concerned are common soil saprophytes whose ability to adapt to the tissue environment and elicit disease are extremely variable. The agent of sporotrichosis is a thermodimorphic fungus similar to *Histoplasma* and *Blastomyces*, but it is of relatively limited virulence. Most soil-isolated strains of this organism are essentially nonpathogenic. The agents of chromomycosis are of even less pathogenic potential. The degree and type of dimorphism exhibited by these organisms depends on the relative resistance of the host and on the tissue in which they are growing. Some species adapt relatively easily and are the planate-dividing, yeastlike bodies of classic verrucous dermatitis. In debilitated patients, however, the same organisms are mycelial in morphology. In contrast, other species are rarely found in " chromoblastomycosis" and are restricted to opportunistic infections of the compromised host.

The entity termed "subcutaneous phycomycosis" is even more perplexing. Again, the disease is subcutaneous, with little tendency to spread, and is completely different from the other phycomycete diseases, e.g., mucormycosis. Essentially nothing is known about its frequency, its mode of transmission, or the relative pathogenicity of the infectious agent. It does not exhibit dimorphism in tissue and is grouped here only because it is "subcutaneous." Lobo's disease (keloid blastomycosis) and rhinosporidiosis are also included in this very heterogeneous group of diseases.

DEFINITION

The term "chromomycosis" includes a group of clinical entities caused by various dematiaceous (pigmented) fungi. The most common form of this disease is known as verrucous dermatitis or chromoblastomycosis. In this type of infection the fungi gain entrance through the skin by traumatic implantation. The lesions develop at the site of inoculation and are most commonly limited to the cutaneous and subcutaneous tissue. The response is one of hyperplasia, characterized by the formation of verrucoid, warty, cutaneous nodules, which may be raised 1 to 3 cm above the skin surface.

These roughened, irregular, pedunculated vegetations often resemble the florets of a cauliflower. The fungi are most often seen in tissue as planate-dividing, yeastlike bodies (sclerotic cells). The disease is usually confined to the lower legs and feet, but involvement of the hands, buttocks, ears, chest, abdomen, and other surfaces has been recorded. Some of these represent lymphatic or hematogenous spread.

The remainder of the diseases included in the chromomycosis group are localized or systemic infections in which the same species or a variety of other species are involved. The tissue reactions evoked differ from those of verrucous dermatitis and are quite variable. The morphology of the organisms is also variable. They appear much less yeastlike and are more often seen as distorted, septate, hyphál strands. These infections often occur in debilitated patients, whereas "chromoblastomycosis" occurs in normal hosts in whom infection is a function of continued trauma and exposure.

Synonymy. Chromoblastomycosis, verrucous dermatitis, phaeosporotrichosis, cladosporiosis, dermatitide verrucosa chromoparasitaria, infections by dematiaceous fungi.

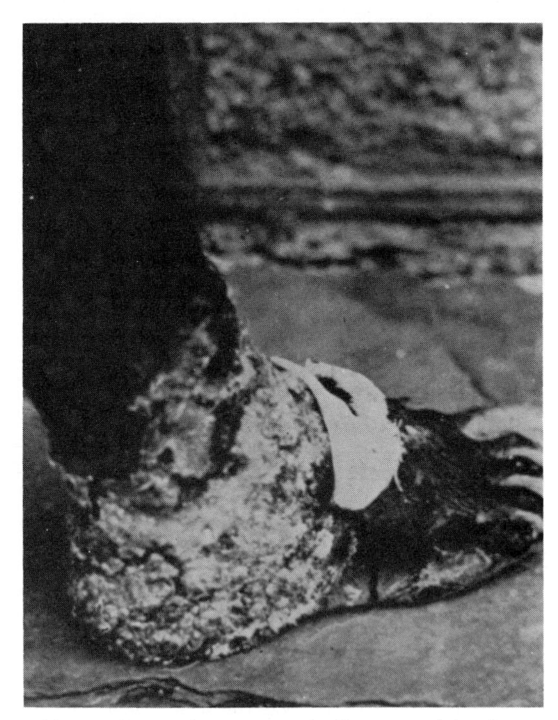

Figure 10–1. Chromomycosis. First case of the disease as described by Pedroso and Gomes.[37a]

HISTORY

Although chromomycosis is essentially a tropical and subtropical disease, the first authentic case was reported from Boston. Lane[30] and Medlar[35] in 1915 described a patient from New England with verrucous lesions on the foot. The fungus isolated was named *Phialophora verrucosa* by Thaxter, but the physician describing the case gave essentially no credit to this mycologist, and so the specific epithet is usually ascribed to Medlar. Prior to this, in 1914, Rudolph[39a] had published a report of a case of chromomycosis from Brazil, but there was no description of the fungal etiology. Pedroso and Gomes reported on four cases of the disease in 1920.[37a] One of the patients had been under observation since 1911, and thus is considered to be the first case recognized (Fig. 10–1). The fungus isolated from this patient was not similar to the one of Thaxter and was named *Hormodendrum pedrosoi* by Brumpt in 1922. In 1924 a few more cases were reported from Brazil by Terra, Torres, Fonseca, and Area Leao. Carini[10] described two additional cases and coined the term "chromoblastomycosis" for the disease. It was not until 1933 that a second case was reported by Wilson et al.[47a] from the United States. This report and another by Martin, Baker, and Conant in 1936 represented the only three cases known in the United States until recently. Since then, the disease has been recognized more frequently.[29,33]

Beurmann and Gougerot,[3a] in 1907, reported on a clinical variant of sporotrichosis in an intramuscular abscess. The organism isolated was a dematiaceous fungus named *Sporotrichum gougerotii* by Matruchot[32a] (1910), but transferred to the genus *Phialophora* by Borelli in 1955. About 50 cases of infection by this fungus have now been recorded, and Mariat et al.[34] proposed the name phaeosporotrichosis for the disease. The clinical syndromes produced and the morphology of the organism in tissue are too similar to other forms of chromomycosis to warrant separation as a special disease category. The organism is of very low virulence, and the infection is usually limited to a subcutaneous cyst.

The first case of chromomycosis of the brain was reported in 1952 by Binford et al.[4] in a patient from Maryland. Emmons named the organism isolated *Cladosporium trichoides*, but it appears to be similar to the *Torula bantiana* of Saccardo. The latter fungus was isolated from a cerebral lesion by Banti in 1911, and about a dozen subsequent cases have been recorded. The same species has been recovered from lesions of the skin and subcutaneous tissues.[15] Agents of chromomycosis belonging to other genera have been isolated from brain lesions, apparently as the result of hematogenous spread from a primary site. The morphology of the organisms in tissue is similar in all cases. For these reasons, it appears unwarranted to separate out cerebral disease by the proposed name cladosporiosis.

Reports implicating commoner agents of chromomycosis and some rarely isolated species are appearing more frequently in the literature. Some of these describe the disease in partially debilitated or compromised patients. These cases, therefore, represent opportunistic infections. The tissue reaction, as would be expected, is not the same as the one seen in verrucous dermatitis, nor is the morphology of the fungus the same. The etiologic agents are all soil-inhabiting dematiaceous fungi. They are light-brown, pigmented organisms of varied morphology in tissue, and therefore the disease they produce falls within the definition of chromomycosis.

ETIOLOGY, ECOLOGY, AND DISTRIBUTION

The etiologic agents of chromomycosis are soil-inhabiting fungi of the family Dematiaceae. Their saprophytic occurrence is well documented, and they are among the commoner fungi found in decaying vegetation, rotting wood, and forest litter. The mycelia, spores, and sclerotic cells of the organisms are pigmented in shades of light brown, yellow-brown, and brown-black. The sclerotic cells and yeastlike cells are found in "chromoblastomycosis," and the variously arranged and distorted hyphal elements occur in other forms of chromomycosis. The several species involved appear to be closely related and are very difficult to distinguish one from the other. Probably no aspect of taxonomy in medical mycology has evoked so much debate and controversy as the classification of the etiologic agents of this disease. The organisms elaborate a wide variety of spore types, depending on the strain, the substrate used, and the various physical conditions under which they are grown. The first described agent, *Phialophora verrucosa*, is fairly easily identified and remains the type species of the genus. The second agent, described as *Hormodendrum pedrosoi*, has been placed at times in several genera. The various other organisms isolated have been placed in the genera *Phialophora, Hormodendrum, Fonsecaea, Rhinocladiella*, and *Torula*. There is as yet no agreement on their taxonomy. The final determination of valid species will depend on finding either a stable, distinguishing characteristic or, hopefully, the sexual stage of the agents, a method that has become very important in the taxonomy of dermatophytes. For convenience in this chapter, we have used the arbitrary taxonomy currently and most frequently used. The various arguments concerning classification will be discussed in the section on mycology. The principal etiologic agents of chromomycosis are *Fonsecaea pedrosoi, F. compactum, Phialophora verrucosa*, and *Cladosporium carrionii*. Other organisms isolated with some frequency include *F. dermatitidis, P. gougerotii* (phaeosporotrichosis), *Cladosporium bantianum* (brain abscess), and *Phialophora jeanselmei. P. richardsiae*[45] and *P. spinifera*[37] have each been isolated once from cystic-type chromomycosis. By far the most commonly isolated agent is *F. pedrosoi.*

All of these agents have been recovered from decaying vegetation or soil habitats. Conant, in 1937, demonstrated that *P. verrucosa* was identical to *Cadophora americana*, a cause of "bluing" of lumber. *P. jeanselmei* is more frequently associated with mycetoma than chromomycosis, and has been isolated from wood pulp in Sweden and Italy under the name *Cadophora lignicola.* Several studies by Putkonen[38] and others have implicated the Finnish sauna as the point source of numerous infections with *F. pedrosoi. P. verrucosa* has also been found in such environments. *C. carrionii* has been isolated from decaying wood.

Though the first published case of chromomycosis came from Massachusetts, the disease is infrequently reported in temperate climates or among a shoe-wearing population. The organisms abound in soils in all parts of the world, but because of repeated exposure, particularly among poor rural people without shoes, the disease is more frequent in tropical and subtropical climates.

Puncture wounds are the main mode of entrance for the agents of chromomycosis. The verrucous dermatitis type of disease is always associated with a history of such a trauma or repeated trauma. In some systemic infections, particularly those involving the brain, the primary site may be pulmonary. In opportunistic infections, the organisms gain entry by various means.

More cases of the verrucous type of disease are reported among males than females, but this is probably because males have greater opportunity for soil contact and predisposition to injury while working. Patients with the disease most frequently fall within the 30- to 50-year-old range. The disease develops slowly and may be of prolonged duration. Paradoxically, the disease is rarely found in children exposed to the same environmental conditions as adults, especially in temperate climates. It has been suggested that implanted organisms may remain quiescent for long periods of time. It is possible that repeated trauma and tissue injury to the area are required before the organisms are able to incite a disease process. Good hygiene and adequate nutrition may help the individual abort a potential infection. A tabulation of reported cases of verrucous dermatitis is noted in Al-Doory.[1]

Though verrucous chromomycosis is found throughout the world, most cases come from the American tropics and subtropics.[17,29,33] It is very common in Mexico, particularly in the states of Veracruz and Tabasco. When it is looked for, numerous cases are found throughout Central America.[33] In Cuba and the Dominican Republic, there is also a high frequency of disease among the rural population. In all these areas, the most commonly isolated agent is *F. pedrosoi*, and feet and legs are the most frequent sites of infection. By contrast, in Venezuela the body sites most frequently involved are the shoulders, chest, and trunk, and the agents involved are primarily *C. carionii*, particularly in the arid Lara and Falcon states.[9] However, in the more humid areas, where climatic conditions are similar to the other countries cited, the primary agent is *F. pedrosoi*. A few cases were reported in Puerto Rico where early studies on the disease were carried out by Carrión.[11] In Colombia[17] and Ecuador, 75 per cent or more of cases involved *F. pedrosoi*, the rest being caused by *F. compactum* and *P. verrucosa*. Several hundred cases have been reviewed by Lacaz[29] in Brazil. Various clinical presentations were noted, and most cases were caused by *F. pedrosoi*. A few scattered reports come from Peru, Argentina, Rumania, Japan, Algeria, Italy, and Australia, among others. Local outbreaks have been recorded in the United States.[25]

The other clinical forms of chromomycosis, brain abscess, and phaeosporotrichosis are reported sporadically from Europe, the Americas, Africa, and India, and probably exist in all parts of the world.[4,15,28,34,48] No pattern of age or sex distribution has been noted, but underlying disease is not infrequently present.

CLINICAL DISEASE

At least four types of clinical disease are distinguishable. These are (1) the verrucous or classic chromoblastomycosis; (2) the brain abscess syndrome (cladosporiosis); (3) chromomycosis consisting of single or multiple cysts (phaeosporotrichosis); and (4) local or systemic lesions of varying pathology. The latter are most often opportunistic infections.

Verrucous Dermatitis (Chromoblastomycosis)

This type was the first form of the disease recognized, and it is still the most frequently encountered. The lesion appears at the site of some trauma or puncture wound. Initially it is a small, raised, erythematoid, nonpruriginous papule.[39] Rarely there is some pruritus.[11] The original papules or pustules are violaceous in color and histologically consist of an effusive round cell infiltrate. Often the lesions are scaly, and the organisms can be seen as distorted hyphal elements in skin scrapings (Fig. 10–2,*A* FS B 29,*A*). In time (often months or years), a new crop of lesions appears in the same area or adjacent areas. The latter follow the distribution of local lymphatic channels. Frequently these lesions become raised to 1 to 3 mm above the skin surface and are hypertrophic, with a scaly, dull, red to greyish surface. Sometimes there is peripheral spread, with healing in the center, as in cutaneous blastomycosis;[29] usually, however, the lesions tend to enlarge and become grouped. After many years they may become elevated to 1 to 3 cm, pedunculated, verrucous, and resemble florets of the cauliflower (Fig. 10–3 FS B 30). It is in

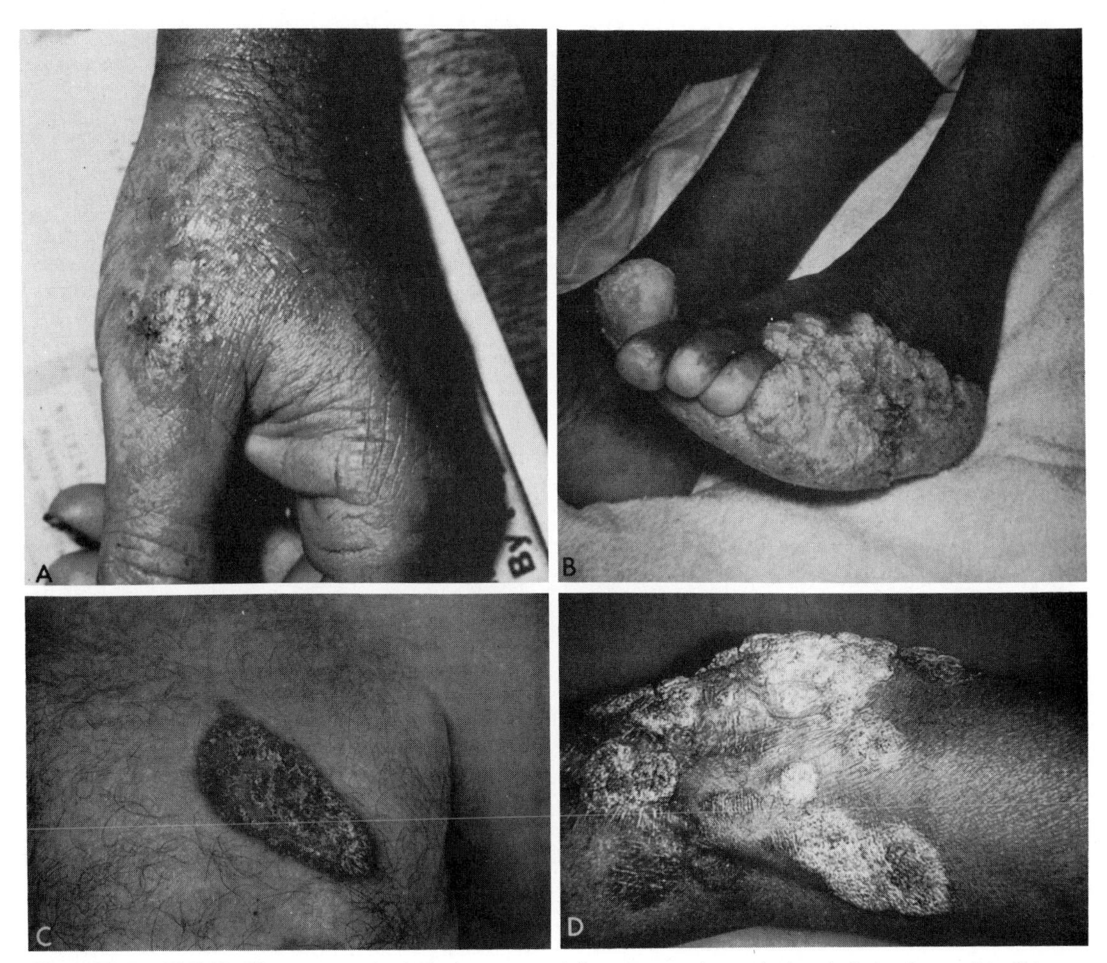

Figure 10–2. FS B 29. Chromomycosis. *A*, Erythematous, violaceous pustules on the hand of a lumber worker. Skin scraping showed distorted hyphal elements. *Phialophora verrucosa* was cultured. *B*, Chronic verrucous chromomycosis of foot due to *Fonsecaea pedrosoi*. *C*. Clinical variant seen on chest. The center is a healing atrophic scar, and an active raised border is present. *D*, Old chronic disease with some scarring and keloid formation. Many lesions were inactive.

these mature lesions of chromomycosis that the planate-dividing, yeastlike bodies (sclerotic cells) are found (Fig. 10–3 FS B 30).

Many clinical variants occur. Lesions often are traumatized, ulcerate, and become secondarily infected with bacteria. These may have a purulent exudate rather than the normal, dry, crusted appearance (Fig. 10–4). Spreading lesions with atrophic, scarred centers and raised borders are not infrequently seen[29] (Fig. 10–2,*C,D* FS B 29,*C,D*). These must be differentiated from blastomycosis. Sometimes in healing lesions, there is extensive keloid formation resulting in a fibroma-like appearance.

In general, the disease remains localized to the immediate area of the initial infection. In old cases, lesions in all stages of development may be seen. There is no apparent discomfort to the patient. Secondary infection sometimes leads to considerable lymph stasis, which may result in elephantiasis. Since chromomycosis itself is usually not debilitating, this complaint may be the initial reason for the patient to seek medical attention. There is no invasion of the bone or muscle, or fistula formation as is commonly seen in mycetoma. In rare cases there is hematogenous spread to uninvolved areas of the body, and verrucous lesions may appear on other appendages as well as on the chest (Fig. 10–2,*C* FS B 29,*C*), abdomen, or trunk.[2] Focal infection of the lungs and other internal organs have been reported.[2] About a dozen cases with dissemination to the brain are now recorded.[18] *F. pedrosoi*, *F. dermatitidis*, and *P. verrucosa* have all been isolated from brain abscesses. In this tissue they appear as long-branching, septate, brown hyphal strands indistinguishable from those seen in chromomycosis of the brain caused by *C. bantianum*. The latter organism produces disease in the absence of preexisting skin lesions. In experimental studies, all species, but especially *C. bantianum* and *F. dermatitidis*, are neurotropic.[19,27]

The early skin lesions of verrucous chromomycosis that show centrifugal spread resemble those of blastomycosis. However, in chromomycosis the lesions usually lack the sharply raised border and the multiple tiny pustules which are present in cutaneous blastomycosis. In addition to blastomycosis, the differential diagnosis must include yaws, tertiary syphilis, tuberculosis verrucosa cutis, mycetoma, leishmaniasis, candidosis, and sporotrichosis. Chromomycosis has also been confused with lupus vulgaris, lupus erythematosus, and leprosy. Lymph stasis in advanced disease along with multiple flat lesions gives the appearance of tropical "mossy foot." In all cases, demonstration of the fungus by direct examination of a potassium hydroxide mount and by culture establishes the diagnosis.

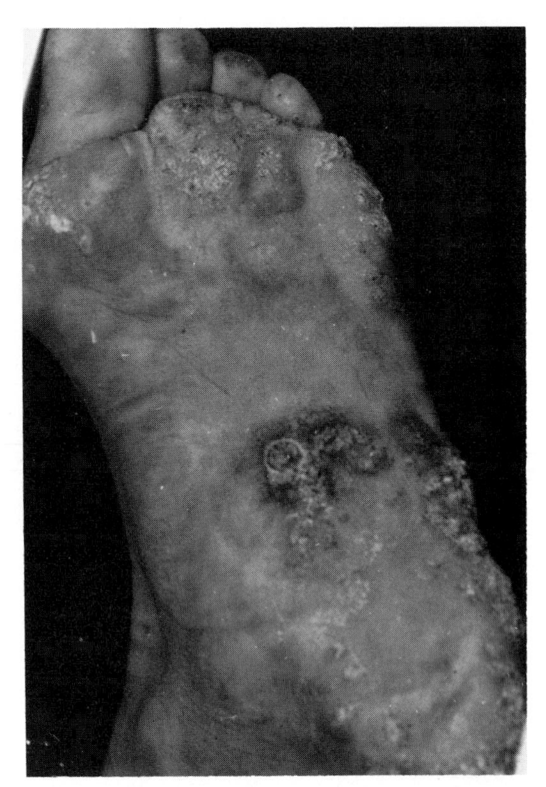

Figure 10–4. Chromomycosis. Chronic verrucous lesions and a purulent lesion resulting from bacterial superinfection. (Courtesy of F. Battistini.)

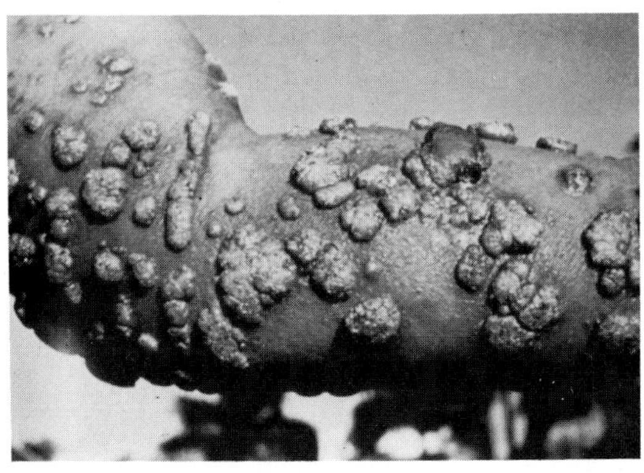

Figure 10–3. FS B 30. Chromomycosis. Pedunculated verrucous lesions resembling florets of cauliflower. (Courtesy of F. Pifano.)

Chromomycosis of the Brain (Cladosporiosis)

About two dozen cases of chromomycosis of the brain are now recorded.[4,14a,15,18,46] The etiologic agent, except as noted in the previous section, has been *Cladosporium bantianum*. The lesions may be single or multiple, the latter suggesting hematogenous spread from a focal lesion elsewhere. They consist of encapsulated abscesses, with masses of brown hyphal strands contained in the central portion. Concurrently, on occasion, small lesions have been found in the lungs and the skin of the abdomen or the ear, which possibly represent primary sites of infection. Often the affected patients are debilitated or on steroid therapy.[46]

The presenting symptoms of cerebral chromomycosis include headache, weakness or paralysis, coma, diplopia, confusion, ataxia, incoordination, and seizures. The diversity of symptoms tends to reflect the location of the lesion in the brain. Except for elevation of protein and a cell count up to 1400 per mm,[4] the spinal fluid is not grossly abnormal. In no case has the organism been recovered from spinal fluid at tap. Most patients are not diagnosed until autopsy, and in none of the remainder has there been survival beyond a few months, even with surgical intervention.

The differential diagnosis includes chronic brain abscesses of other etiology, particularly mycotic. This is especially true when focal lesions are found in the lung or other areas. The index of suspicion increases if other lesions reveal brown pigmented fungi.

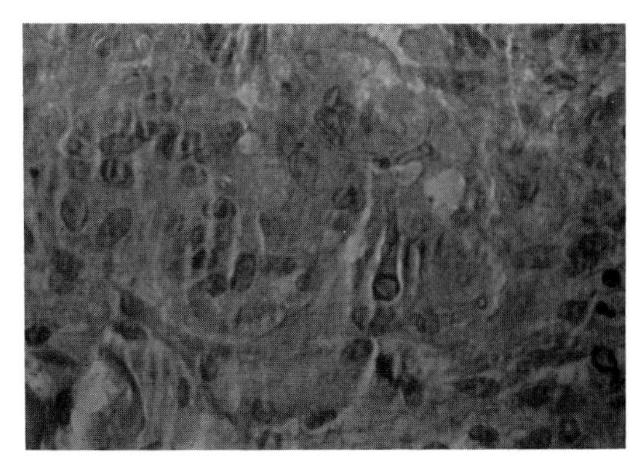

Figure 10–5. Chromomycosis. Distorted hyphae and round bodies in a lesion of chromomycosis. The subcutaneous mass extended over the thigh of the patient, who was a diabetic. (Courtesy of G. Hambrick.)

Cystic and Nonspecific Chromomycosis (Including Phaeosporotrichosis)

This group includes reports of about fifty cases of infection usually resulting from puncture wound in which there were single or multiple deep cysts containing masses of brown pigmented fungi of varied morphology.[28, 34, 48] The lesions are subcutaneous or intramuscular and tend to be stationary. From these, the most commonly isolated agent is *P. gougerotii*, but a few cases are caused by a variety of other species, such as *P. richardsiae*, *P. spinifera*, *P. jeanselmei*, and *F. dermatitidis*. The clinical designation phaeosporotrichosis has been proposed for this group, but the disease does not resemble sporotrichosis, and the etiologic agents are not related. The organisms involved are probably less virulent than the usual agents of verrucous chromomycosis, and only rarely are able to evoke a disease process.

The clinical picture most commonly recorded is an abscess which is cutaneous, subcutaneous, or intramuscular. This may evolve into a hard, cystlike process of several centimeters in diameter, covered by a raised, thickened epidermis. Occasionally the cyst may ulcerate and extrude pus containing brown pigmented hypha and sclerotic cells. This is especially true when *P. jeanselmei* or *P. gougerotii* are the etiologic agents. In most cases the regional lymph nodes are not involved, and dissemination is very rare. Aspiration of the lesions sometimes results in chronic fistula formation. In one case involving *F. dermatitidis*, the subcutaneous mass extended over most of the thigh, although it is usually contained in a fibrous cyst (Fig. 10–5). The patient was a diabetic and the lesion started at the site of an insulin injection. Several such cases of "nonspecific" chromomycotic infections have been seen, usually in debilitated patients.[16] With the increasing use of macrodisruptive medications (long-term steroids, immunosuppressives, cytotoxins), more nonspecific or opportunistic chromomycoses will be observed.

The differential diagnosis of this form of the disease includes the gumma of tertiary syphilis, sebaceous cyst, foreign body granuloma, tendon sheath ganglion, and sporotrichosis. Demonstration of pigmented fungi in material from lesions and their isolation in culture establish the diagnosis.

PROGNOSIS AND THERAPY

Verrucous chromomycosis usually remains localized and will not debilitate the patients if left untreated. Secondary infection resulting in lymph stasis and elephantiasis may interfere with normal activity. There is also a potential hazard of systemic spread, especially to the brain.

In the early stages of the disease, the most reliable therapy is surgical excision or electrodesiccation, or cryosurgery.[38a] Most cases seen in clinical practice are well advanced and require medical management. The older modes of therapy included administration of iodides (1 to 9 g IV per day for several years), iontophoresis with copper sulfate, and vitamin D injections, among others. There is a report of successful therapy using surgery and compresses soaked in the fungicide, sodium dimethyl dithiocarbamate.[31] Two new approaches to treatment with drug therapy have received considerable attention recently. The first is the use of amphotericin B either intralesionally[47] or topically as a lotion containing 30 mg per ml t.i.d.[26] Intravenous administration has also been used with success. A second promising drug for treatment is thiabendazole. This has been used in a series of studies in Central and South America.[3,20] In the first report, 2 g per day of the drug was given orally, supplemented by local application of additional drug in DMSO. In a second report, the oral dose was not well tolerated, and local applications in water-soluble base (Velvachol) only were used. Another promising drug, particularly in advanced cases of disease, is 5-fluorocytosine.[32] In a series of 12 cases, Lopes found improvement or cure in almost all patients receiving 4 doses per day of 100 mg per kg. Gonzales Ochoa[23] has successfully used this drug to treat two advanced cases of the disease. The regimen was 5 g (500 mg tablets) per day for 220 days in one case and 88 days in the other. A dose of up to 10 g per day is now recommended.

Treatment of cystic chromomycosis has usually consisted of surgical removal. Sometimes contamination of the surrounding tissue during the procedure results in recurrence and spread of the infection.

Chromomycosis of the brain has been treated surgically, but no patients have survived longer than ten months. Most cases are diagnosed at autopsy.

PATHOLOGY

Verrucous chromomycosis is generally confined to the body surfaces. The lesions range from erythematous macules to warty, hyperkeratotic, pedunculated papules. Not infrequently the surgical pathologist discovers "brown bodies" in a specimen marked epidermoid tumor or carcinoma. In histologic section, the nonulcerated lesion that is not secondarily infected shows a very high degree of pseudoepitheliomatous hyperplasia. It is not as extreme as in cutaneous blastomycosis or coccidioidal granuloma, but it is considerable and could be termed extreme acanthosis (Fig. 10–6 FS B 31). Miliary abscesses and poorly defined granulomatous nodules are

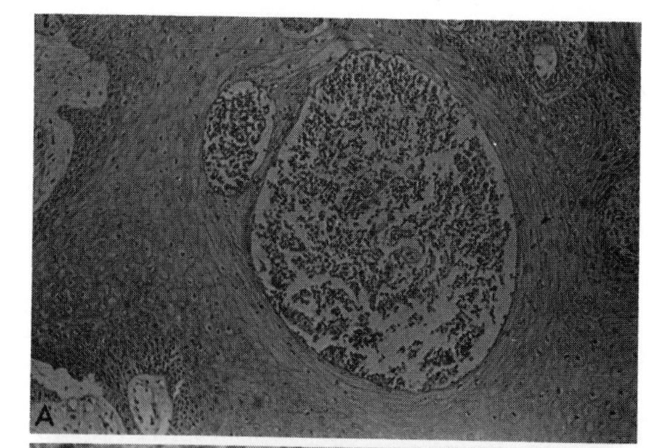

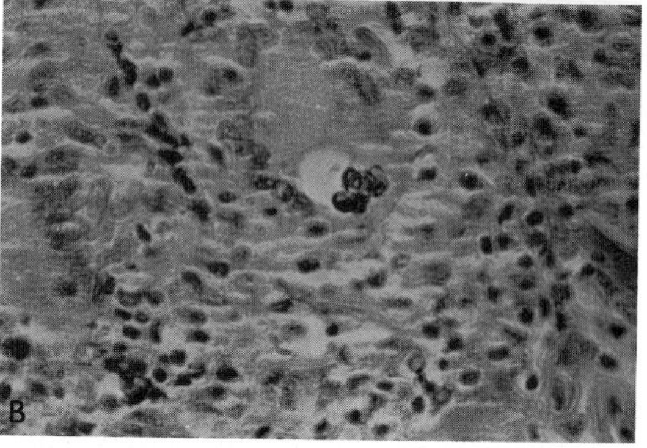

Figure 10–6. FS B 31. Chromomycosis. *A*, Massive pseudoepitheliomatous hyperplasia with small abscesses. Hematoxylin and eosin stain. ×100. *B*, High-power view of abscess showing brown bodies of fungus within giant cell. Hematoxylin and eosin stain. ×400.

seen in this disease, as in the latter two diseases. Changes consistent with chronic fibrosing inflammation are seen between the nodules. Numerous neutrophils cluster in the centers of the granulomas. Foreign body giant cells are present, as well as lymphocytes, plasma cells, and other cells of the chronic inflammatory reaction. The fungus is visible in unstained as well as stained sections. The chestnut or light yellow-brown bodies appear as rounded yeasts (5 to 15 mm in diameter), with thick, planate, septal walls, which are often grouped or in chains (Fig. 10–7). These are the sclerotic cells of the fungus.[41,42] Sometimes they lie within giant cells. In early lesions in which the organism is found mostly in the epidermis and stratum corneum, hyphal strands are seen more commonly than sclerotic cells. This is also observed when the organism metastasizes to other body sites. In Giemsa-stained sections, the fungus has a greenish-brown color. Since the organism is readily seen in hematoxylin and eosin–stained material, special fungus stains are not needed.

Ulceration and secondary bacterial inva-

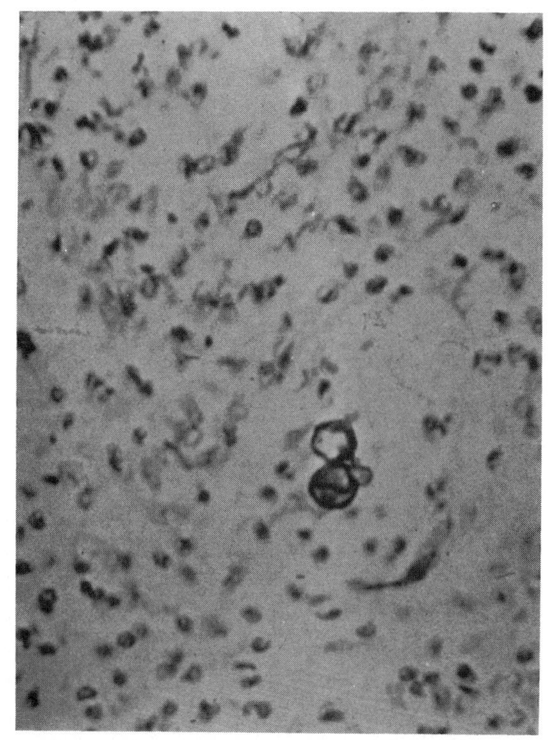

Figure 10–7. Chromomycosis. Plantate-dividing, rounded, sclerotic body of fungus in verrucoid chromomycosis. ×400. (From Rippon, J. W. *In* Burrows, W. 1973. *Textbook of Microbiology*. 20th ed. Philadelphia, W. B. Saunders Company, p. 717.)

sion alters the histologic picture of chromomycosis. Instead of the chronic inflammatory and granulomatous picture, an acute or chronic pyogenic reaction is seen.

In chromomycosis of the brain, the most frequently observed lesion is a cerebral abscess. On coronal section of the brain, this is seen as a brown to black mass within cerebral material. In most cases recorded, it is described as multilocular and well-delineated, with very thick walls. The lesion is sometimes 5 cm in diameter. Leptomeningitis with multiple granulomas occurs less frequently. When it does, the meninges are thickened, and a moderate hydrocephalus may be a complication. Within the lesions there is a central mass of neutrophils, surrounded by epithelioid and a few giant cells. Gliosis is more evident at the periphery of the abscess. Within the abscess, particularly in the center and the walls, masses of brown hyphal elements are seen between the cellular infiltrate. They are thick-walled, measuring 1 to 2 μ in diameter. Spherical cells and yeastlike bodies ranging from 9 to 20 μ are occasionally seen.

Cystic chromomycosis usually presents as a hard, defined, subcutaneous mass of several centimeters diameter. There may be marked acanthosis above the lesion. The lesion itself is a cyst limited by a thick wall of hyaline connective tissue. It is lined with granulomatous inflammatory tissue, with a center of necrotic debris. In the necrotic center and the inner walls of the cyst, abundant brown pigmented fungi are seen. Their morphology is quite variable. Hyphal elements ranging from 3 to 8 μ are common, with a few distorted bodies of up to 20 μ in diameter. Small, yeastlike rounded bodies and moniliform hyphae of gradually decreasing diameter may be present. Sometimes only compact masses of hyphae similar to those seen in chromomycosis of the brain are found.

ANIMAL DISEASE

Natural infection of animals with the organisms that cause chromomycosis occurs but is rarely reported. A few cases have been found in dogs and horses. In a case involving a horse, the lesions were described as similar to those seen in human verrucous chromomycosis. Biopsy revealed brown-pigmented, hyphal strands and yeastlike bodies. These were contained in granulomatous lesions consisting of chronic and acute inflammatory cells.[43] The organism was identified as a *Hormodendrum* species. In an interesting study of toads (*Bufo* sp.) in Colombia, C. Correa et al.[14] reported that natural infection occurred in 12 of 66 specimens examined. Granulomatous lesions were located in the liver, lungs, and kidney. Numerous sclerotic cells similar to those seen in human verrucous chromomycosis were noted. The organisms isolated were *F. pedrosoi* (three strains), *P. gougerotii* (three strains), and *C. carrionii* (seven strains).[14] There was a previous report of the disease in frogs from Brazil.[10]

Experimental infection of animals has not regularly produced the verrucous type of chromomycosis. Intracutaneous injection results in an ulcerating local lesion that heals slowly.[6b] Systemic infection involving most internal organs can be produced by intravenous inoculation of fungus, particularly *P. verrucosa.* Brain abscesses have developed following injection with *C. bantianum, F. pedrosoi,* and *F. compactum.*[1a,19] *F. dermatitidis* is also neurotropic.[27] *C. carrionii* is not neurotropic, and this characteristic is used to distinguish it from *C. bantianum.*

IMMUNOLOGY AND SEROLOGY

Serologic evaluation of patients with chromomycosis does not yet appear to be a practical adjunct to diagnosis. In one series, precipitating antibodies were found in 12 of 13 patients with chromomycosis. These appeared to be specific for the causative agent. By cross-precipitation, it was shown that *F. pedrosoi, F. compactum,* and *P. verrucosa* were closely related and distinct from *C. carrionii* and *P. jeanselmei*.[8] Conflicting results were found in another study.[13] Precipitins begin to disappear following successful treatment. Although direct microscopic demonstration of the fungus is sufficient for diagnosis, the rise and fall of precipitin titers would be of prognostic importance following the institution of therapy.

Several studies have been made of the antigenic relationships between the various etiologic agents of chromomycosis. So far, no consistent patterns have been found that firmly delineate the species involved. In an early study Conant[12] demonstrated by the complement fixation test that several strains of *P. verrucosa* were antigenically similar. Using a fluorescent antibody technique, Gordon and Al-Doory[24] were able to distinguish *C. carrionii* from *C. bantianum,* but there were many cross-reactions between conjugates prepared from the pathogenic species, *F. pedrosoi, F. compactum, F. dermatitidis,* and soil saprophytes (*Cladosporium* sp.) Cooper and Schneidau[13] demonstrated antigenic relationships between several genera using immunodiffusion and immunoelectrophoresis techniques. They found common antigens among *Cladosporium, Phialophora,* and *Fonsecaea,* but concluded that *Cladosporium* and *Phialophora* are more closely related to one another than to *Fonsecaea.*

Natural resistance to infection is high in normal individuals and disease occurs only under circumstances of repeated exposure and trauma. Hypersensitivity has been demonstrated in patients with verrucous chromomycosis. It diminishes after clinical care. The relation of this to resistance to reinfection is not known.

LABORATORY IDENTIFICATION

Direct Examination. Laboratory diagnosis of chromomycosis by direct examination of suspicious material is relatively easy, but it should always be confirmed by culture. Skin scrapings, crusts, aspirated debris, and biopsy and excised material can be examined in a potassium hydroxide mount. Brown-pigmented, branching, hyphal strands (2 to 6 μ wide) are easily seen in skin scrapings, crusts, and aspirates. In pus from cysts, very distorted hyphae (3 to 8 μ wide) and pleomorphic brown bodies (up to 20 μ in diameter), which sometimes appear to bud, may be found. In granulation tissue obtained by curettage, excision, or biopsy from verrucous chromomycosis, the sclerotic bodies predominate. They are thick-walled, brown-pigmented, exhibit planate division, and may be grouped in a chainlike formation. The size varies from 4 to 12 μ. Material from mycotic lesions of the brain can be similarly examined. Hyphal forms predominate in such cases.

Culture Methods. Since the agents of chromomycosis are not inhibited by cycloheximide (Actidione) or chloramphenicol, selective media using these antibiotics may be used. Cultures should be kept at 25°C for at least six weeks. In contrast to the saprophytic species of the soil, most of the commonly isolated agents of chromomycosis grow slowly. The specific identification of the organism is taxonomically taxing. It depends on spore types, percentage of spore types present, fine details of spore production, and perhaps the position of the planets. Part of the problem is that there are still no clearly delineated taxa for these organisms. Three general types of sporulation are found in this group (Figs. 10–8 and 10–9):

1. *Phialophora* type. In this type there is a distinct conidiophore called a phialide, which occurs terminally or along the mycelium (Fig. 10–8,*B*). This structure is generally flask-shaped and has a rounded, oval, or elongate base, a constricted neck, and an opening that may have a flaring collarette and lip. The conidia are formed at the end of the flask (semiendogenous sporulation) and are extruded through the neck. They may accumulate around the neck area, giving a picture of "flowers in a vase." The spores are oval, smooth-walled, hyaline, and have no attachment scars.

2. *Rhinocladiella* (*Acrotheca* type). The conidiophores are simple and sometimes not differentiated from the vegetative hyphae. Oval conidia are produced irregularly on the top and along the sides of the conidiophore (acropleurogenous sporulation) (Fig. 10–

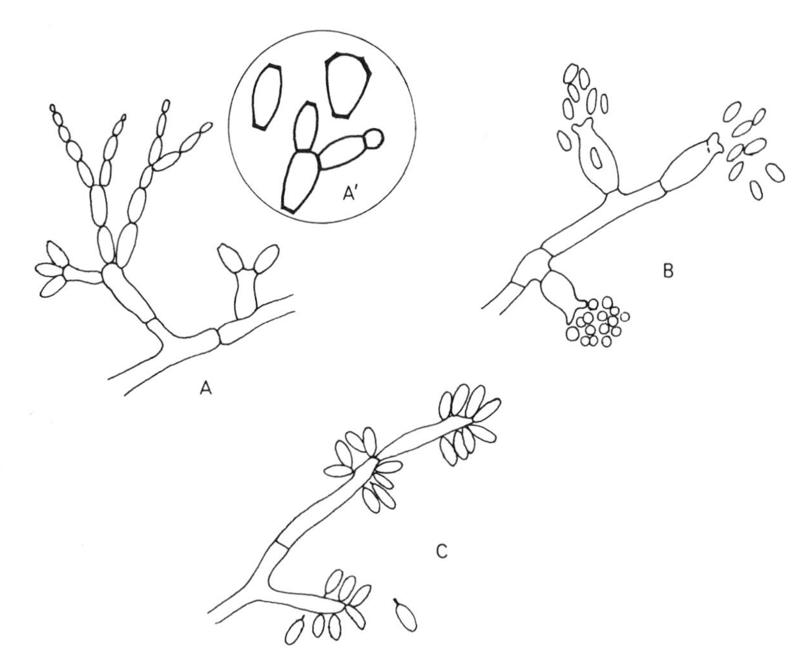

Figure 10–8. Chromomycosis. Sporulation types. *A, Cladosporium* type sporulation consists of a simple conidio-phore that buds to form spores. In *A'*, note the thickened scars or disjunctors. A "shield" cell is shown that has three disjunctors. *B, Phialophora* type sporulation. Vase-shaped phialides give rise to spores with no attachment scars. *C, Rhinocladiella* type sporulation. The conidia are borne at the ends and sides of conidiophores and have one scar of attachment.

8,*C*). They are usually single and do not bud. However, occasionally chains formed by budding and transition to the *Cladosporium* type sporulation is seen. When the conidia are detached, small bud scars can be found on the conidiophore, and there is a single scar on the spore at the attachment site. Hyphae, conidia, and conidiophores are greenish brown.

3. *Cladosporium (Hormodendrum)* type. In this type of sporulation there is a simple stalk that serves as a conidiophore (Fig. 10–8,*A* and *A'*). It is usually slightly enlarged at the distal end. Two or more spores are formed at the tip. These in turn bud and form secondary spores at their distal poles (acrepetalous sporulation). Sporulation continues, with the formation of long chains. The youngest spore is the one distal to the conidiophore. Detached spores show a thickening or scar, called a disjunctor, where they were connected to the other spore. All of the spores within the chain will have two or three (if it formed a branch) disjunctor scars, except the terminal spore which will have one. A spore having three disjunctors is described as "shield-shaped." Hyphae, conidia, and conidiophores are dark olive-aceous brown. It is thus possible, by examining the individual spores carefully, to distinguish the type of sporulation that was involved. *P. verrucosa, P. gougerotii, P. richardsiae,* and *P. spinifera* sporulate almost exclusively by the phialophora type of sporulation. Some strains, however, have been said to show *Cladosporium* and *Rhinocladiella* type sporulation.

F. dermatitidis, P. jeanselmei, P. gougerotii, and *P. spinifera* grow initially as elliptical aseptate yeasts. Mycelium gradually forms as the colony ages. *F. pedrosoi* (Fig. 10–9) and *F. compactum* produce primarily *Cladosporium* type spores, some *Rhinocladiella,* and, rarely, *Phialophora* type spores. The type and degree of sporulation is influenced by the media on which the organisms grow. In "synthetic lymph," even the sclerotic cells found in lesions can be produced *in vitro.*[41] *C. bantianum* and *C. carrionii* (Fig. 10–10) sporulate exclusively by the *Cladosporium* type of sporulation. They can be differentiated by their thermotolerance and the neurotropism exhibited by *C. bantianum* in experimentally infected mice.

Although many studies have been made of this carbon and nitrogen utilization, physiologic differentiation of the species is not yet well standardized. Fuentes[22] has shown that the pathogenic species do not liquefy gelatin, coagulate milk, or digest starch. These are useful in distinguishing the etiologic agents of chromomycosis from

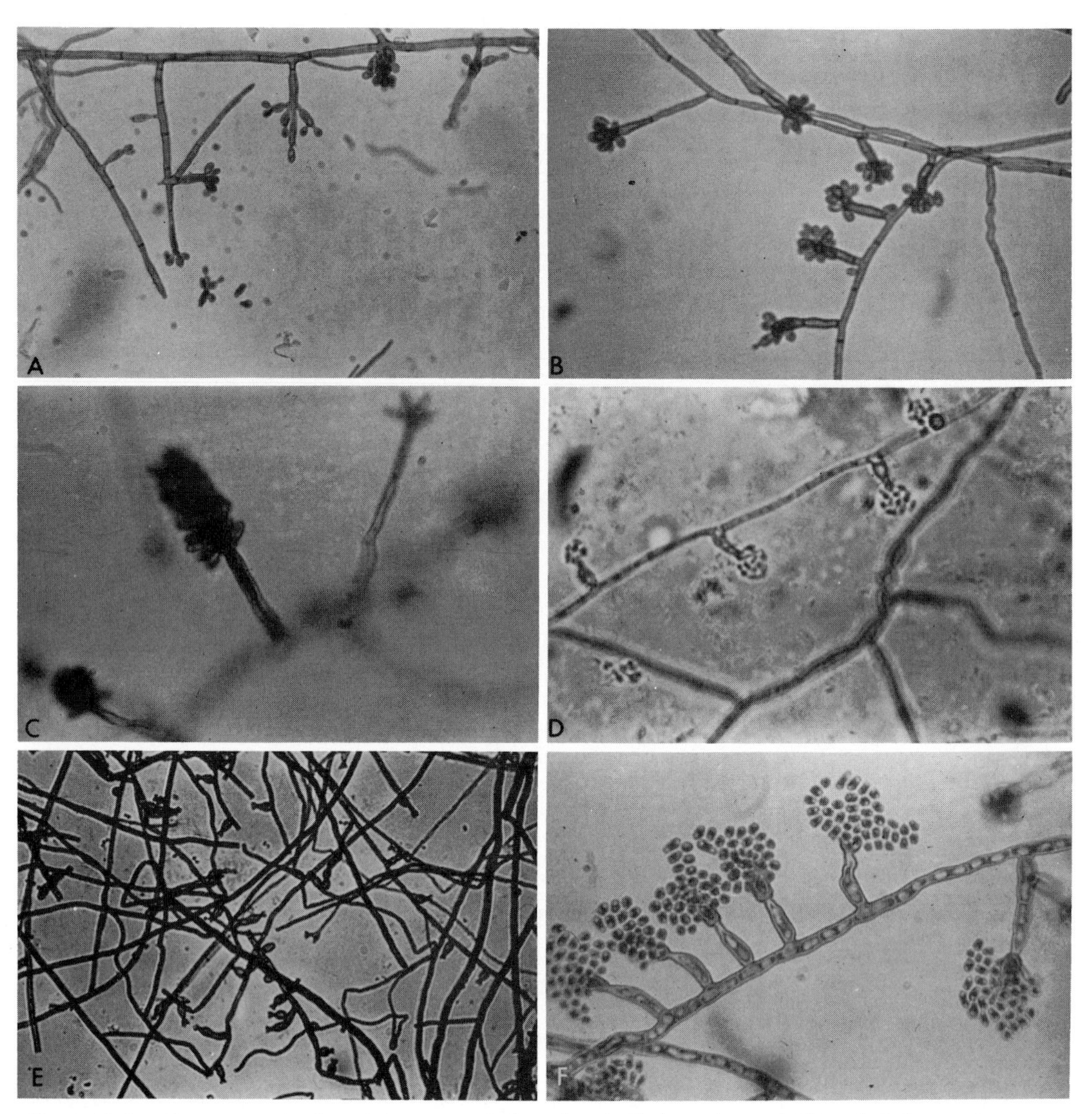

Figure 10–9. Chromomycosis. Sporulation types. *A, F. pedrosoi. Cladosporium* and *Rhinocladiella* sporulation. *B, F. pedrosoi.* Several young conidiophores, showing *Rhinocladiella* sporulation. ×400. *C,* Older conidiophores, with spores coming down sides of conidiophore. ×400. *D, F. pedrosoi.* Series of philiades showing "flowers in a vase" sporulation of the *Phialophora* type. *E, P. verrucosa.* Shows various sizes and morphology of phialidides. *F, P. verrucosa.* Accumulation of spores around flaring lips of phialide. ×400.

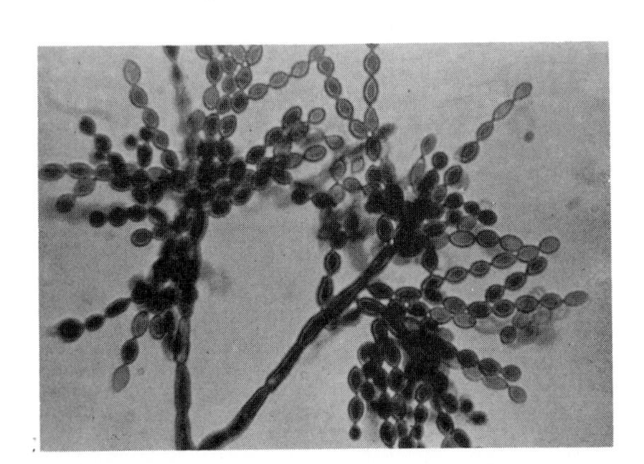

Figure 10–10. *Cladosporium carrionii.* Long, flexuose chains of elliptical spores. ×440.

soil saprophytes and air contaminants. Biondo et al. have found some consistent physiologic difference between some species.[5]

MYCOLOGY

The taxonomy of the agents of chromomycosis has been one of the most confusing in the field of medical mycology. At the present writing this problem is still unsettled, so that choosing the terminology to be used is arbitrary at best. Some of the current schools of thought will be presented and then discussed.

The type species of the genus *Phialophora* is *P. verrucosa*. Since its sporulation is exclusively of the *Phialophora* type, there appears to be no difficulty with its nomenclature.[116] This is also true of species which demonstrate only *Cladosporium* type sporulation, i.e., *C. carrionii* and *C. bantianum*. The other agents of chromomycosis have usually been described as belonging to the genus *Hormodendrum*. This genus, however, has been reduced to synonymy with the genus *Cladosporium*. Negroni in 1936 described a new genus, *Fonsecaea*, which, as redefined by Carrión, now includes *F. pedrosoi*, *F. compactum*, and *F. dermatitidis*.[11a] The inclusion of the latter organism in this genus is rather tenuous. The erection of the genus *Fonsecaea* was necessary because the species concerned produce various types of spores by various methods, a property which excludes them from the genus *Cladosporium (Hormodendrum)*, where they were originally placed. Recently Emmons[21] has proposed emending the definition of the genus *Phialophora* so that it could include the organisms now comprising the genus *Fonsecaea*. However, since these fungi only rarely produce phialides, this transfer seems unwarranted. Scholl-Schwarz[40] has stated that all the *Fonsecaea* species and a few of the *Phialophora* species should be placed in the genus *Rhinocladiella*, but there are many objections to this suggestion.[27] As it is now defined, this latter genus could not accommodate the species of *Fonsecaea* again because of the variety of spores produced. It is obvious that the organisms now included in *Fonsecaea* are closely related. Therefore, in order not to destroy already well-defined taxonomic groups by emendation, the designation *Fonsecaea* has been retained for use in this text. It is also possible that these particular species are intermediate forms be-

Table 10–1. *Characteristics of the Agents of Chromomycosis*

Species	*Dermotropic*	*Neurotropic*	*Growth at 37°C*	*Gelatin*	*Primary Culture—Yeast*	*Growth Rate*	*Spore Formation*
Fonsecaea pedrosoi	+	±	+	−	−	Slow	*Cladosporium* primarily in short chains or, less commonly, from tips and sides of conidiophore; rarely from phialides
F. compactum	+	−	+	−	−	Slow	Same as above, but *Cladosporium* spore heads reduced and compact
F. dermatitidis	+	+	+	−	+	Slow	Yeast forms early; abstrictions from tip and sides of conidiophores; semiendogenously from elongate phialides, also from peglike sterigmata
Phialophora verrucosa	+	±	+	−	−	Slow	*Phialophora* type from flaring cups on flask-shaped phialides; other sporulation types rare or absent
P. gougerotii	+	−	±	−	+	Slow	Yeast at first; abstrictions from sides and tips; semiendogenously from elongate phialides
P. richardsiae	+	−	+		−	Rapid	Phialospores semiendogenously; long phialides with saucer-shaped lips
P. spinifera	+	−	±		+	Slow	Abstrictions from tips, sides, and along mycelium; phialides spikelike, not flared; bears spores semiendogenously
Cladosporium carrionii	+	−	±	−	−	Slow	Long chains of conidia from branched conidiophore only; acrepetalous
C. bantianum	±	+	+	−	−	Slow	Same as above; differentiation by thermotolerance (43°C), neurotropism, and irregular spore size
C. species	−	−	−	+	−	Rapid	Same as *C. carrionii*; usually not thermotolerant or pathogenic for animals

tween the two genera, *Phialophora* and *Cladosporium*; thus this designation *Fonsecaea* would represent a natural phylogenetic classification. Indeed, there is serologic evidence for this progression.[13] Until such time as the perfect states of these fungi are discovered or consistent distinguishing characteristics are found, the arguments are moot.

Detailed descriptions of the etiologic agents of chromomycosis are given below and summarized in Table 10–1. *P. jeanselmei* is described in the chapter on mycetoma, a disease with which it is more frequently associated.

Fonsecaea pedrosoi (Brumpt 1922) Negroni 1936

Synonymy. *Hormodendrum pedrosoi* Brumpt 1922; *Acrotheca pedrosoi* Fonseca and Leão 1923; *Trichosporium pedrosoi* Langeron 1929; *Gomphinaria pedrosoi* Dodge 1935; *Hormodendroides pedrosoi* Moore and Almeida 1936; *Phialophora pedrosoi* Emmons in Binford et al. 1944; *Hormodendrum algeriensis* Montepellier and Catanei 1927; *Trichosporium pedrosianum* Ota 1928; *Hormodendrum rossicum* Merlin 1930; *Botyroides monophora* Moore and Almeida 1936; *Phialoconidiophora guggenheimia* Moore and Almeida 1936; *Hormodendrum japonicum* Takahashi 1937.

Colony Morphology. In culture *F. pedrosoi* grows very slowly, producing a black-brown, grey-black, olive-grey, or black colony. The texture is velvety to fluffy, and the surface varies from flat to heaped and folded. Radiations and zonations occur in some strains. The colony characteristics may vary considerably between isolants, and the organism is indistinguishable macroscopically from *P. verrucosa*. (Fig. 10–11,*A* FS B 32,*A*).

Microscopic Morphology. All three types of sporulation exist in this species, the proportion varying with the strain and the media used for growth (Figs. 10–8 and 10–9,*A–D*). The *Cladosporium* type usually predominates, with frequent admixture of the *Rhinocladiella* type and an occasional phialide. The species is quite polymorphic, and variations in these patterns are to be expected. Few spores are produced on Sabouraud's agar, but on deficient media, such as cornmeal agar, sporulation is enhanced. As distinct from species of *Cladosporium*, in *F. pedrosoi*, conidia are borne in short rather than long chains from the distal end of the conidiophore. The conidiophores are of varying lengths. The conidia are elliptical, 1.5 to 3 μ in width by 3 to 6 μ in diameter. They exhibit two dark, thick scars (disjunctors) where they were attached in chains. "Shield cells" with three scars may be found among spores on the terminal end of conidiophores if there were two fertile poles on the cell. The last cell of the chain has only one scar. *Rhinocladiella* spores are formed from bare hyphae or from knobby conidiophores. They are produced acropleurogenously (at the tip and along the side of the conidiophore) and are attached by a short apiculus. The apiculus remains on the spore when it is detached. The *Phialophora* spores are produced from phialides similar to those of *P. verrucosum,* but they may be scant or lacking. Sclerotic bodies and cleistothecia-like bodies have also been described.

Fonsecaea compactum Carrión 1940

Synonymy. *Hormodendrum compactum* Carrión 1935; *Phialoconidiophora compactum* Moore and Almeida 1936; *Phialophora compactum* Emmons in Binford 1944.

Colony Morphology. This very slow growing fungus produces a folded, heaped, brittle colony, which is dark olive-black and develops a brownish-black fuzz with age. It is this fuzz that contains most of the spores. The colony is essentially indistinguishable from that of *F. pedrosoi.*

Microscopic Morphology. The conidiophores are similar to those of *F. pedrosoi* except for the more compact form of the sporeheads. The spores are subspherical to ovoid, 1.5 × 2 × 3 μ in size, whereas the spores of *F. pedrosoi* are elliptical and larger. The other types of sporulation are the same as seen in *F. pedrosoi* and *F. compactum* and may only represent a variant of the latter species. Hundreds of isolants of *F. pedrosoi* are known, whereas only two of *F. compactum* have been found (Puerto Rico and Tennessee).

Fonsecaea dermatitidis (Kano 1937) Carrión 1950

Synonymy. *Hormiscum dermatitidis* Kano 1937; *Hormodendrum dermatitidis* Conant 1953; *Phialophora dermatitidis* Emmons 1965.

This species least comfortably fits into the genus *Fonsecaea* and has many morphologic similarities to *Phialophora jeanselmei* and *P.*

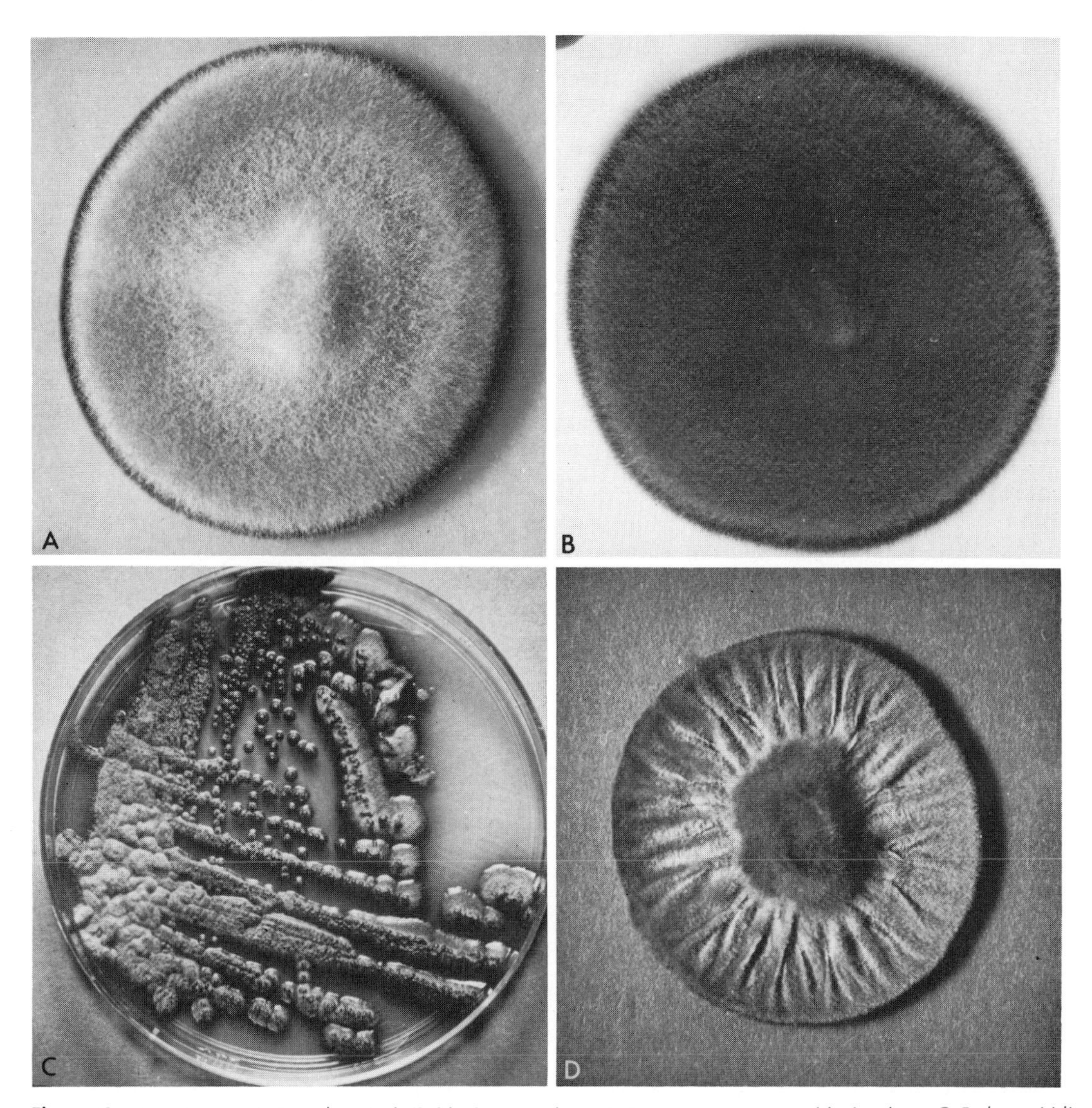

Figure 10–11. FS B 32. *A, F. pedrosoi.* Fluffy black-grey colony. *B, F. compactum.* Fuzzy black colony. *C, F. dermatitidis.* Two colony types: glistening yeastlike colonies and transition to fuzzy mycelial form. *D, P. verrucosa.* Olive-grey colony with radial grooves.

gougerotii. It is distinguished from *P. gougerotii* by its growth at 37°C.

Colony Morphology. There are two colony types. The colony which grows on initial isolation is moist, glistening, yeastlike in appearance, and olive to black in color. At this stage it is identical to *P. jeanselmei* and *P. gougerotii.* This is the "*Pullularia*" or "black yeast" stage of growth. After four to five weeks, tufts of submerged or twisted ropy strands of mycelium develop around the colony edge. These give rise to olive-grey aerial mycelium. In time, patches of the same type of mycelium appear over the surface of the colony.

Microscopic Morphology. The initial moist colony is composed of budding yeastlike cells similar to those of *Aureobasidium (Pullularia) pullulans.* When mycelia appear, phialides are produced which are similar to those seen in *P. verrucosa.* However, the lips of the phialides are usually inconspicuous. Conidiophores with pleurogenously produced spores similar to those found in *F. pedrosoi* are also seen. Sporulation of all types is usually sparse. *F. dermatitidis* has been isolated from most cases of verrucous chromomycosis in Japan, and is similar to the *Torula bergerii* reported from Canada. In these cases sclerotic cells were found in tissue, but only hyphal strands have been seen in infections of the brain and mucosa, in cystic disease, and in opportunistic infections.

Phialophora verrucosa Thaxter in Medlar 1915

Synonymy. *Cadophora americana* Nannfeldt 1927; *Phialophora macrospora* Moore and Almeida 1936; *Fonsecaea pedrosoi* var. *phialophorica* Carrión 1940.

Colony Morphology. This species grows slowly as a dark olive-grey to black colony, which is initially dome-shaped and later becomes flattened.[11b] Some strains are heaped and folded or have radial grooves (Fig. 10–11,*D* FS B 32,*D*). A grey aerial mycelium eventually covers the colony, which is compact, tough, and leathery. On cornmeal agar, growth is scant but sporulation is enhanced.

Microscopic Morphology. Well-formed, flask-shaped or vase-shaped phialides are formed along the vegetative hypha. These are 3 to 4 μ wide and 4 to 7 μ in length and have a pronounced, flaring collarette. Within their neck, small, elliptical, thin-walled spores, 1 × 3 to 2 × 4 μ in size are formed semiendogenously. The spores are produced in succession but are not attached in chains. They are surrounded by adhesive material, so they may accumulate at the lips of the phialide, forming a large ball. As each spore is released from the phialide, it leaves a residue of wall material around the lip or collarette. In this way, the size of the lip is increased as each succeeding spore is released, so that mature phialides have deeply pigmented, wide, flaring lips. Very rarely one can find examples of other types of sporulation in mature cultures. Conidia may be formed at the tip (acrogenous) or at the tip and along the sides (acropleurogenous) of a conidiophore. Sometimes these conidia may bud to produce secondary and tertiary conidia, thus forming the arborescent heads characteristic of the *Cladosporium* type of sporulation. Cleistothecia-like bodies have been found in some strains, but they did not contain ascospores. The perfect stage remains unknown.

Phialophora gougerotii (Matruchot 1910) Borelli 1955

Synonymy. *Sporotrichum gougeroti* Matruchot 1910; *Dematium gougeroti* Grigorakis 1924; *Pullularia gougerotii* Langeron 1945; *Cladosporium gougerotii* Carrión and Silva 1955.

Colony Morphology. The colony of *P. gougerotii* is initially similar to that of *Phialophora jeanselmei* and *Fonsecaea dermatitidis.* It is olive to black, moist, and yeastlike. With age, the colony becomes covered with short, olive-grey, aerial mycelium.

Microscopic Morphology. In the yeastlike colony, the cells are pleomorphic and budding, and are similar to the *Pullularia* type of black yeasts. At this stage, the cells are essentially identical to those of *P. jeanselmei* and *F. dermatitidis* with which they are often confused. On becoming mycelial, the fungus sporulates from minute projections (apiculae) along the hyphae and from elongate, sometimes branched phialophores. The phialides usually have a wider base than those of *F. dermatitidis.* The conidia are elongate to ellipsoid (2 to 2.5 × 3 to 4 μ) and often cluster around the slightly flared lip of the phialide. Peglike sterigmata are also seen along the hyphae, which bud to form blastospores that mass around the distal end.

Although this species is closely related to *P. jeanselmei* and *F. dermatitidis,* there are some differences in its morphology in tissue. *P. jeanselmei* is usually found as a crescent-shaped, poorly organized granule, and the disease produced by this organism is most frequently classed as a mycetoma. Generally, *F. dermatitidis* in tissue assumes the sclerotic cell morphology or is seen simply as hyphal strands. On the other hand, *P. gougerotii* forms moniliform hyphae and pleomorphic, budding, yeastlike bodies in tissue. *F. dermatitidis* and *P. jeanselmei* grow well at 37°C, and neither hydrolyze hypoxanthine. *P. jeanselmei* also utilizes paraffin, and *F. dermatitidis* is neurotropic. *P. gougerotii* lacks all these characteristics.

Phialophora richardsiae (Melin and Nannfeldt) Conant 1937

Synonymy. *Cadophora richardsiae* Melin and Nannfeldt 1934; *Cadophora brunnescens* Davidson 1935.

This organism was originally described as *Cadophora richardsiae* by Melin and Nannfeldt when isolated from rotting wood pulp. This species is a common lignicolous decay fungus. So far it has been isolated only once from human disease.[45] The lesion was a subcutaneous, well-encapsulated cyst, with the usual histologic morphology of a foreign body granuloma. The fungus was found in the central necrotic area and consisted of dark hyphal strands and a few budding cells.

Colony Morphology. The organism grows rapidly in cluture, producing a woolly or tufted brownish-grey colony with concentric zonation. Sporulation is almost exclusively of the *Phialophora* type.

Microscopic Morphology. A simple tapering branch (1 to 2 μ at base, 2 to 10 μ in length) from the vegetative mycelia produces semiendogenously an elliptical, hyaline conidia (1 to 2 × 2 to 4 μ). The tip of the phialide is simple, narrow, and barely discernible. As the phialide matures, the flaring saucer-shaped collar is produced, which is typical of the genus. Also as the phialide matures, the spores become thicker-walled, pigmented, and more spherical.

Phialophora spinifera Nielson and Conant 1968

This species has been recovered once from a granuloma, in which it was found to consist of spherical budding cells.[37]

Colony Morphology. The organism grows slowly, producing a moist, olive to black, yeastlike colony. It may remain yeastlike almost indefinitely on Sabouraud's agar or develop short, naplike, mycelial elements over the surface of the colony. The growth is mycelial, and sporulation is enhanced when the organism is grown on cornmeal agar.

Microscopic Morphology. Yeasts of the *Pullularia* type are found in the initial colony. Heavy sporulation occurs on the hyphal strands. Masses of spores are borne on short protuberances (apiculae) from the mycelia and from developing spinelike phialophores. The latter are simple or branched, appear rigid, and are produced at right angles to the hyphae. The tip is smooth or extended. The spores are 2.5 × 3.5 μ in size. Short, flask-shaped phialophores with flaring apical collars are also found. These are similar to those found in cultures of *P. verrucosa.* The spores from these are 1.0 × 1.5 μ in size.

Cladosporium carrionii Trejos 1954

Synonymy. *Fonsecaea pedrosoi* var. *cladosporium* Simson 1946; *Fonsecaea cladosporium* Cowell 1952.

This organism is a regular isolant from verrucous chromomycosis in Australia. It is also commonly found in Venezuela and South Africa.

Colony Morphology. The organism grows slowly, producing a small, smooth or folded, dark olive-black, compact colony (3 to 4 cm after one month). The edges are entire and are marginated by black submerged hyphae.

Microscopic Morphology. Only the *Cladosporium* type of sporulation is found. Elongate conidiophores produce long, flexuose, branching chains of conidia, 1.5 to 3.0 by 2 to 7.5 μ in size (Fig. 10–10). The organism differs from most saprophytic *Cladosporium* species by its inability to digest gelatin. The morphologically similar species, *Cladosporium bantianum,* is differentiated by (1) neurotropism in experimental animal infections, (2) growth at 43°C, (3) less regular conidia size, and (4) more rapid growth. *C. carrionii* is (1) dermo- but not neurotropic, (2) limited in growth to 35 to 36°C, (3) regular in spore size, and (4) very slow growing.

Cladosporium bantianum (Saccardo) Borelli 1960

Synonymy. *Torula bantiana* Saccardo 1912; *Cladosporium trichoides* Emmons 1952.

This organism is the most commonly isolated species from chromomycosis of the brain. Whether or not it is identical to the *Torula bantiana* of Saccardo is still a matter for debate. As pointed out by Emmons,[21] the latter species is described as having spores in unbranched chains which are 8 to 11 × 5 μ in size. Most recent isolants have spores 4 to 7 × 2 to 3 μ in size which are usually in branched chains. Some spores up to 15 and 20 μ in length have been found. The description of the lesions produced by *T. bantiana* and *C. bantianum* in the brain are similar. The latter organism is neurotropic in experimental animal infections. Intravenous inoculation of 10^5 cells regularly produces cerebral lesions in mice.

Colony Morphology. The organism grows moderately fast and produces a spreading, slightly folded, olive-grey to olive-brown colony.

Microscopic Morphology. Elongate, brown, septate conidiophores are produced which are identical to those of other species of *Cladosporium.* Spores are produced in long, branched, flexuose chains. Most spores are 2 to 3 × 4 to 7 μ in size, but some are 3 × 15 to 20 μ.

Chmelia slovaca Svoboda 1966[44] was isolated from a verrucoid lesion of the external ear, in which thick-walled, sclerotic cells were found. The fungus eventually grew as a felt-like colony, though it was quite yeasty at first. The cells were budding and yeastlike and were similar to the "Pullularia" stage of other species. Chlamydospores which were septate in many planes, blastospores, and tangled, coiled hyphae were described. The author claimed to have reproduced the disease in animals. Recent work by Borelli and Marcano did not confirm this, and they questioned whether this species was actually capable of causing disease.[7]

REFERENCES

1. Al-Doory, Y. 1972. *Chromomycosis.* Missoula, Montana, Mountain Press Publishing Company.

1a. Aravysk, R. A., and V. B. Aronson. 1968. Comparative histopathology of chromomycosis and cladosporiosis in experimental infections. Mycopathologia, *36*:322–340.

2. Azulay, R. D., and J. Serruya. 1967. Hematogenous dissemination in chromoblastomycosis. Report of a generalized case. Arch. Dermatol., *95*: 57–60.

3 Battistini, F., and R. N. Sierra. 1968. Tratamento de un caso de cromomicosis con aplicaciones locales de thiabendazole. Derm. Venez., 7:3–10.

3a. Bearmann, L., and H. Gougerot. 1907. Associations Morbides dans les sporotrichoses. Bull. Mem. Soc. Med. Hôp. Paris, *24*:591–596.

4. Binford, C. H., R. K. Thompson, et al. 1952. Mycotic brain abscess due to *Cladosporium trichoides*, a new species. Am. J. Clin. Pathol., *22*:535–542.

5. Biondo, F., K. A. Griffin, et al. 1968. Physiological and biochemical identification of dematiaceous fungi isolated from clinical materials. Bact. Proc., 1965, p. 88.

6. Borelli, D. 1960. *Torula bantiana,* agente di un granuloma cereberale. Riv. Anat. Patol. Oncol., *17*:617–622.

6a. Borelli, D. 1972. Diagnosis and Treatment of Chromomycosis. Letters to Ed. Arch. Dermatol., *106*:419.

6b. Borelli, D. 1972. A method for producing chromomycosis in mice. Trans. R. Soc. Trop. Med. Hyg., *66*:793–794.

7. Borelli, D., and C. Marcano. 1969. Observaciones sobre *Chmelia slovaca.* Derm. Venez., *8*:740–747.

7a. Brumpt, E. 1922. *Precis de Parasit.* 3rd ed. Paris. Masson et Cie, p. 1105.

8. Buckley, H., and I. G. Murray. 1966. Precipitating antibodies in chromomycosis. Sabouraudia, *5*:78–80.

9. Campins, H., and M. Scharyj. 1953. Cromoblastomicosis; Comentarios sobre 34 cases, con estudio clinico, histologico and micoligico. Gac. Med. Caracas, *61*:127–151.

10. Carini, A. 1910. Sur une moisissure qui cause une maladie spontannee du *Heptodactylus pentaduetylus.* Ann Inst. Pasteur (Paris), *24*:157–172.

11. Carrión, A. L., and M. Silva-Hutner. 1947. Chromoblastomycosis and its etiologic fungi. Ann. Cryptogamici Phytopath., *6*:20–62.

11a. Carrión, A. L., and M. Silva-Hunter. 1971. Taxonomic criteria for fungi of chromoblastomycosis with reference to *Fonsecaea pedrosoi.* Int. J. Dermatol., *10*:35–43.

11b. Cole, G. T., and B. Kendrick. 1973. Taxonomic study of *Phialophora.* Mycologia, *65*:661–688.

12. Conant, N. F., and D. S. Martin. 1937. The morphologic and serologic relationships of various fungi causing dermatitis verrucosa (chromoblastomycosis). Am. J. Trop. Med., *17*:553–578.

13. Cooper, B. H., and J. D. Schneidau. 1970. A serological comparison of *Phialophora verrucosa, Hormodendrum pedrosoi,* and *Cladosporium carionii* using immunodiffusion and immunoelectrophoresis. Sabouraudia, *8*:217–226.

14. Correa, C. R., I. Correa, et al. 1968. Lesiones micoticas (cromomicosis?) observadas en Sapos (Bufo sp.) Antioquia Med., *18*:175–184.

14a. Crichlow, D. K., F. T. Enrile, et al. 1973. Cerebellar abscess due to *Cladosporium trichoides* (*bantianum*). Am. J. Clin. Pathol. *60*:416–421.

15. Desai, S. C., M. L. Bhatikar, et al. 1966. Cerebral chromoblastomycosis due to *Clasdosporium trichoides* (Parts I and II). Neurology (Bombay), *14*:1–6, 6–18.

16. DiSalvo, A. F., and W. H. Chew. 1968. *Phialophora gougerotii:* An opportunistic fungus in a patient treated with steroids. Sabouraudia, *6*:241–245.

17. Duque, H. O. 1961. Chromoblastomicosis. Revision general y estudio de la Enferme dad en Colombia. Antioquia Med., *11*:499–521.

18. Duque, H. O. 1961. Meningo-encephalitis and brain abscess caused by *Cladosporium* and *Fonsecaea.* Am. J. Clin. Pathol., *36*:505–517.

19. Duque, H. O. 1963. Cladosporiosis cerebral experimental. Rev. Lat. Am. Anat. Patol., 7:101–110.

20. Elfren, A. 1966. Tratamiento de la cromoblastomicosis con thiabendazoles un estridio de 14 cases. Medn. Cutan., *1*:277–286.

21. Emmons, C. W., C. H. Binford, et al. 1970. *Medical Mycology.* Lea & Febiger, Philadelphia, p. 354.

22. Fuentes, C. A., and Z. E. Bosch. 1960. Biochemical differentiation of the etiologic agents of chromoblastomycosis from nonpathogenic cladosporium species. J. Invest. Dermatol., *34*:419–422.

23. Gonzalez Ochoa, A. 1970. Curation de la criptococosis y de la chromomicosis con 5-fluorocitosina. Rev. Invest. Salid Publica, *30*:63–76.

24. Gordon, M. A., and Y. Al-Doory. 1965. Application of fluorescent antibody procedures to the study of pathogenic dematiaceous fungi. J. Bacteriol., *89*:551–556.

25. Howes, J. K., C. B. Kennedy, et al. 1954. Chromoblastomycosis. Report of nine cases from a single area in Louisiana. Arch. Dermatol., *69*:83–90.

26. Hughes, W. T. 1967. Chromoblastomycosis: Successful treatment with topical amphotericin B. J. Pediatr., *71*:351–356.

27. Jotisankasa, V., H. S. Nielsen, et al. 1970. *Phialophora dermatitidis;* its morphology and biology. Sabouraudia, *8*:98–107.

28. Kempson, R. H., and W. H. Steinberg. 1963.

Chronic subcutaneous abscesses caused by pigmented fungi. A lesion distinguishable from cutaneous chromoblastomycosis. Am. J. Clin. Pathol., *39*:598–606.

29. Lacaz, C. S., J. M. Cruz Hurtado, et al. 1966. Deramitite verrucosa cromoparasitiaria. O. Hosp. (Rio), *70*:9–17.

30. Lane, C. G. 1915. A cutaneous disease caused by a new fungus *Phialophora verrucosa*. J. Cutan. Dis., *33*:840–846.

31. LeQuellec, B., A. Dodin, et al. 1966. Essais de fongiides derives du dimethyl-dithio-carbamate de zinc in vitro et in vivo dans la chromoblastomycose humaine. Bull. Soc. Pathol. Exot., *59*:192–199.

32. Lopes, C. F., E. O. Cisalpino, et al. 1971. Treatment of chromomycosis with 5-fluorocytosine. Int. J. Dermatol., *10*:182–191.

32a. Matruchotè. 1910. Sur un nouveau group de champignons pathogènes, agents des sporotrichoses. C. R. Acad. Sci. (Paris). *150*:543–545.

33. Mayorga, R. 1970. Prevalence of subcutaneous mycoses in Latin America. Proceedings International Symposium on Mycoses. Pan American Health Organ. Scientific Publ. No. *205*:18–28.

34. Mariat, F., G. Segretain, et al.: 1967. Kyste sous cutane mycosique (phaeo-sporotrichose) a *Phialophora gougerotii*. Sabouraudia, *5*:209–219.

35. Medlar, E. M. 1915. A cutaneous infection caused by a new fungus *Phialophora verrucosa* with a study of the fungus. J. Med. Res., *32*:507–522.

36. Nielsen, H. S. 1970. Dematiaceous fungi, In *Manual of Clinical Microbiology*. Blair, et al. (eds.), Am. Soc. for Microbiol., Bethesda, Md.

37. Nielsen, H. S., and N. F. Conant. 1968. A new human pathogenic *Phialophora*. Sabouraudia, *6*:228–231.

37a. Pedroso, A., and J. M. Gomes. 1920. Four cases of dermatitis verrucosa produced by *Phialophora verrucosa*. Ann. Paulistas Med. Cir., *9*:53.

38. Putkonen, T. 1961. Die chromomykose in Finnland. Der mogliche Anteil der Finnishen Sauna an ihrer Verbreitung. Hautarzt, *17*: 507–509.

38a. Ramirez, M. 1973. Treatment of chromomycosis with liquid nitrogen. Int. J. Dermatol., *12*: 250–254.

39. Roux, J., M. Fissarou, et al. 1967. Un cas de chromoblastomycose au debut. Bull. Soc. Fr. Dermatol. Syphiligr., *74*:655–656.

39a. Rudolph, M. 1914. Ueber die brasilianishe figueira Arch. Schiffs-u. Tropen. Hyg., *18*:498.

40. Scholl-Schwarz, M. B. 1968. *Rhinocladiella*, its synonym *Fonsecaea* and its relation to *Phialophora*. Antonie van Leeuwenhoek, *34*:119–152.

41. Silva, M. 1957. The parasitic phase of chromoblastomycosis: Development of sclerotic cells *in vitro* and *in vivo*. Mycologia, *49*:318–331.

42. Silva, M. 1960. Growth characteristics of the fungi of chromoblastomycosis. Ann. N.Y. Acad. Sci., *89*:17–29.

43. Simpson, J. G. 1966. A case of chromoblastomycosis in a horse. Vet. Med. Small Anim. Clin., *61*:1207–1209.

44. Svoboda, Y. 1966. *Chmelia slovaca* sp. novo. A new dematiaceous fungus pathogenic for man and animals. Biologia (Bratislava), *21*:81–88.

45. Swartz, I. S., and C. W. Emmons. 1968. Subcutaneous cystic granuloma caused by a fungus of wood pulp (*Phialophora richardsiae*). Am. J. Clin. Pathol., *49*:500–505.

46. Symmers, W. S. C. 1960. A case of cerebral chromoblastomycosis (cladosporiosis) occurring in Britain as a complication of polyarteritis treated with cortisone. Brain, *83*:37–51.

47. Whiting, D. A. 1967. Treatment of chromoblastomycosis with high local concentrations of amphotericin B. Br. J. Dermatol., *79*:345–351.

47a. Wilson, S. J., S. Hulsey, et al. 1933. Chromoblastomycosis in Texas. Arch. Dermatol., *27*: 107–122.

48. Young, J. M., and E. Ulrich. 1953. Sporotrichosis produced by *Sporotrichum gougerotii*. Arch. Dermatol., *67*:44–53.

Chapter 11

SPOROTRICHOSIS

DEFINITION

Sporotrichosis is most commonly a chronic infection characterized by nodular lesions of the cutaneous or subcutaneous tissues and adjacent lymphatics that suppurate, ulcerate, and drain. The etiologic agent, *Sporothrix schenckii,* gains entrance by traumatic implantation into the skin or, very rarely, by inhalation. Secondary spread to articular surfaces, bone, and muscle is not infrequent, and the infection may also occasionally involve the central nervous system, lungs, or genitourinary tract.

HISTORY

The first case that unquestionably presented with the clinical picture of sporotrichosis was recorded by Schenck in 1898[56] from the Johns Hopkins Hospital in Baltimore. This report described the fungus isolated from the lesion as resembling a "sporotricha." Previously, Link in 1809 and Lutz in 1889 had described cases that in all probability were sporotrichosis, but no fungus was isolated. The second recorded case, also from the United States (Chicago), was reported by Hektoen and Perkins[21] as a lesion which developed on the finger of a boy following a blow by a hammer. The lesion apparently healed spontaneously; however, 65 years later this patient still had a positive sporotrichin skin test but a negative serum agglutination titer.[48] A fungus was isolated from the lesion and referred to as *Sporothrix schenckii,* but in the title of the paper only. A very cursory but recognizable description of the organism was given in the text. Later, in 1903, the disease was described in France by de Beurmann and Ramond. The isolant from their cases was termed *Sporotrichum beurmanni* by Matruchot and Ramond in 1905. The isolant from Schenck's original case had lost its pigment by that time, and when it was examined by the French authors was considered to be a species distinct from the organism recovered by de Beurmann and Matruchot. However, Matruchot in 1910 redescribed Schenck's organism as *Sporotrichum schencki.* Curiously, he dropped the final "i" of the specific name, and until recently this spelling was commonly used. Between 1906 and 1911 de Beurmann and Gougerot identified at least ten more cases of sporotrichosis and tabulated some two hundred additional cases by 1912. This formed the basis of their excellent review of the disease,[5] which is still considered to be a classic in the field of medical literature. In addition to the cutaneous form of disease, they included for the first time descriptions of disseminated, pulmonary, osseous, and mucosal involvement.

In 1907 the first case of natural infection in animals was recorded by Lutz and Splendore in rats in Brazil. Two years later a similar organism isolated from a horse in Madagascar was named *S. equi* by Carougeau. Since that time, reports of the disease in man and animals have been published from all parts of the world but with intriguing and enigmatic shifts in geographic distribution. Though several hundred cases had been reported by 1915, after the first two decades of this century, sporotrichosis became quite rare in France and in Europe in general. The disease is still uncommon in those areas. Cawley[8] could tabulate only 200 cases in the United States before 1932. Since that time there has been a distinct rise in the number of infections recognized and increasingly frequent reports of "opportunistic" or disseminated disease.

Epidemics of sporotrichosis have been reported from time to time. One of the most famous of these was in South Africa. In a space of two years, almost 3000 cases occurred involving miners who brushed against timbers on which the fungus was growing.[59] The epidemic was terminated by treating the timbers with a fungicide.

Strains of fungi isolated from human infections in France and America were studied by Davis in 1921.[11] He concluded that they were identical and called the etiologic agent *Sporotrichum schencki.* However, Carmichael in 1962[7] pointed out the differences in sporulation between the genera *Sporothrix* and *Sporotrichum* and determined that the correct epithet for the organism was *Sporothrix schenckii.*

ETIOLOGY, ECOLOGY, AND DISTRIBUTION

All forms of sporotrichosis in man are caused by the single species *Sporothrix schenckii.* The etiologic agent of the disease in horses, which was first described by Carougeau as *S. equi,* has also been shown to be identical to *S. schenckii.* Sporotrichosis in other animals is also caused by this species.

With few exceptions, the fungus gains entry into the body through some trauma to the skin. The most common histories obtained are scratches from thorns or splinters, cuts from sedge barbs, or handling of reeds, sphagnum moss, or grasses. Sometimes brushing against infected tree bark or timber will result in disease. These modes of transmission emphasize the saprophytic association of the organism with plant life. Sporotrichosis has also occurred following parrot bites, dog bites, insect stings, injury by metal particles, hammer blows, and other traumas. In these instances, infection probably followed contamination of the wound with soil. The organism has been isolated from fleas, ants, and horse hair, which possibly represent vectors for the spores. In a few rare cases, human infection has occurred after contact with infected animals in experimental laboratory studies. The disease has also been transmitted via contaminated dressings from suppurating lesions, and there is at least one case in which the disease was transmitted by direct contact from the cheek of a mother to the cheek of a child.[61]

S. schenckii is an organism commonly found on decaying vegetation, and it has been isolated many times from soil.[23,42,47] Its distribution in the soil and the conditions that enhance its occurrence have been the subject of much speculation. In contrast to *Coccidioides immitis,* there is as yet no clear-cut explanation for the elusive and transient distribution of *Sporothrix schenckii* in soil. After the famous epidemic involving gold miners in South Africa, Findley[15] made a careful study of the conditions that favored growth of the organism in the mines. He found that the fungus grew well on untreated mine poles at a temperature of 26 to 27°C and at a relative humidity of 92 to 100 per cent. In the original report of the epidemic,[59] it has been noted that in mines where the timbers were infected with various lignicolous Basidiomycetes, such as *Poria sp.,* the timbers did not harbor *S. schenckii,* and the workers in those mines were free of infection. Findley also determined that, in areas outside the mines, the infection was most prevalent in the temperate highland plateau. Here the average relative humidity is 65 per cent, and the rainfall is 25 to 30 inches per year. Mackinnon has studied the epidemiology of the disease in an endemic area in Uruguay.[42] He found that there was a seasonal distribution of cases, with increased frequency during the hot, rainy autumn season (March to July). The conditions favoring optimal growth of the fungus were a humidity of 90 per cent and a temperature above 15°C. Yet in a similar study by González Ochoa[17–19] in Mexico, the greatest frequency of infection coincided with the dry and cooler parts of the year. The most highly endemic region of that country is the temperate plateau, where rainfall is sparse throughout the year. The seasonal distribution of cases was winter, 51.4 per cent; fall, 23 per cent; summer, 17 per cent; and spring, 9 per cent. This is in contrast to the situation in Uruguay.

Sporotrichosis is the most common subcutaneous and deep mycosis in Mexico. It outnumbers cases of mycetoma, which is also highly endemic there. In some villages in Jalisco and Michoacan, sporotrichosis is so common that lesions can be regularly spotted on inhabitants walking through the market place. Yet the area is quite dry and almost arid in contrast to other highly endemic areas of Brazil, Uruguay, and South Africa which are moist and humid. Most of the patients in Mexico give a history of working with grass—either gathering it or using it as packing material or for making baskets.[18,64] In the majority of cases in Uruguay, the infection was associated with hunting armadillos. The fungus was found not in the animals but in their nests and burrows.[42] Straw was reported to be the common source of a family epidemic in Brazil.[58]

In the United States, France, Canada, and other temperate countries, infection in soil is usually associated with gardening and contact with gardening soil.[52a] It is frequently associated with sphagnum moss,[10,29] and the disease is considered an occupational hazard of greenhouse workers and rose growers. In contrast to the situation in other endemic areas, the organism appears to thrive in moist, rich loam and humus. In a fascinating example of epidemiologic detective work, Mariat[44] traced the source of an infection acquired from a house plant to the potting soil of oak and beech litter that had been gathered fifty miles from Paris.

Although members of the genus *Sporothrix* are known to cause disease in plants, e.g., carnation bud rot, *S. schenckii* has not yet been conclusively proven to be a plant pathogen. It does appear able to grow on injured or debilitated plant tissue. This was shown experimentally by Benham, who injected spores into carnation buds which subsequently became diseased. The fungus appears to be fairly resistant to drying but is sensitive to direct exposure to sunlight[42] and severe winter weather.[47] It will grow on meat products in cold storage vaults, however.[1]

Cases of sporotrichosis have been found in all age groups. Primary infections in children as young as ten days (following a rat bite) and in adults in their seventh and eighth decades have been recorded. The apparent differences in distribution of the disease among the sexes is probably related to occupation and exposure. Though the ratio in many case studies is 3:1, male to female, others have noted the reverse distribution. In some reports from Brazil, 60 to 70 per cent of the cases occurred in females. In Mexico, González Ochoa[17] found the ratio to be about 1:1. The men acquired the disease on the legs from thorns and sedges or on the fingers and wrists from gathering grass, the women on the fingers from cultivating decorative house plants and making baskets, and the children on the face from scratches of branches. In this study, over 60 per cent of cases were found in persons less than 30 years of age. Primary lesions were most often seen in the young, whereas chronic disease was found with greater frequency in older patients. Since the number of cases appears to decrease markedly in this same population after the age of 50, it appears that many infections heal spontaneously.

Although the etiologic agent, *S. schenckii,* is found in soils all over the world, incidence of the disease varies considerably as does the type of disease contracted. In the early 1900's, sporotrichosis was very common in France, and many cases of extracutaneous dissemination were recorded. Since that time it has become rare in that country and in Europe in general. At the present time the greatest number of cases come from Mexico, Central America, and Brazil, with a scattered distribution throughout the rest of the world. De Beurmann and Gougerot[5] in 1912 were the first to suggest that sporotrichosis could be considered an opportunistic infection. Most of their patients had some underlying disease, and essentially all of the cases of extracutaneous dissemination occurred in those with severe debilitating conditions or malnutrition. This is still a common finding.[43,63,66]

Mariat,[44] citing the rural conditions in France during 1900 to 1914, has suggested that malnutrition is a significant factor in infection. He noted some experimental evidence that mice on protein-deficient diets were more susceptible to severe disease. Dietary deficiency could play a major role in the infection in rural areas of Mexico, Brazil, and Central America, where the case rates are so high.

To a certain extent all fungus infections are opportunistic, and normal, healthy, well-nourished adults seldom develop disease unless they receive an overwhelming inoculum. In the epidemiologically unique case of the miners in South Africa, the patients were in good health, and among the 3000 cases no disseminated disease was found. Frequent injury from brushing the mine poles and continual exposure to large concentrations of spores were suggested as the prime factors leading to infection in this instance. In contrast, exposure to small amounts of spores in an endemic area may gradually confer immunity. González Ochoa[17,18] has shown that a group of grass handlers who had been on the job for more than ten years had no clinical disease, but 100 per cent of them had a positive sporotrichin skin test. When disease does occur in this population, it is the fixed cutaneous type, and many cases appear to heal spontaneously.

Not only has the epidemiology of sporotrichosis changed over the years, but it appears that the pathology has also. In most of the early reports from France and the United States, numerous cigar-shaped bodies were found in lesions. More recently, lesions from cases in the United States and South Africa are described as containing very few organisms. The so-called asteroid body, which is very commonly seen in South Africa and Japan, is very rarely found in cases from the United States, Mexico, or Central America.[37,38] The asteroid body is an eosinophilic mass surrounding the organism in tissue and represents an antigen-antibody complex. It was first seen in Brazil in 1908 but was not seen again in that country until 1964. It has again become a frequent finding there.

It is apparent that there are many facets of the epidemiology and pathogenicity of sporotrichosis that are not yet clearly defined. The organism is only weakly pathogenic, and it is known that the inherent virulence of strains isolated from soil is quite variable.[23] However, this factor alone fails to explain the great variability in incidence, distribution, and severity of sporotrichosis which has been seen since its discovery in 1898.

CLINICAL DISEASE

Within a few years of the discovery of sporotrichosis, a wide spectrum of disease types was noted.[5] Most cases were the so-called "gummatous" type, involving skin, subcutaneous tissue, and local lymph channels.[63] However, disseminated disease as well as mucous membrane involvement and the primary pulmonary form were soon described. The type of disease and its pathology depend on the site of inoculation of the organism and the response of the host to it. For purposes of discussion, the clinical types are divided into five categories: lymphocutaneous, fixed cutaneous, mucocutaneous, extracutaneous and disseminated, and primary pulmonary. The first four result from traumatic implantation and the last from inhalation of spores.

Lymphocutaneous Sporotrichosis

Lymphocutaneous sporotrichosis, which is also called the gummatous type, comprises up to 75 per cent of all cases in most literature surveys.[4,8,60] The fungus gains entrance by traumatic implantation, and the first signs of infection may appear as soon as five days later. The average incubation time is three weeks, but in some patients the initial injury may have taken place six months or more before symptoms occur. The first sign of disease is the appearance of a small, hard, movable, nontender and nonattached subcutaneous nodule. This develops into a bubo, which attaches itself to the overlying skin. The lesion becomes discolored, varying from pink to purple and sometimes black. At this stage the disease may be mistaken for cutaneous anthrax. The lesion then penetrates the skin and becomes necrotic, thus forming the so-called sporotrichotic chancre. Not infrequently, the primary lesion begins as a small ulcer instead of a nodule at the site of the injury. This usually indicates location of the organisms in the upper dermis or epidermis rather than in subcutaneous tissue. The initial lesion remains for several weeks or months and tends to heal with scarring as new buboes develop in other areas. Some lesions may remain active for years. Failure to treat either type of lesion or treatment with topical agents alone results in a chronic course of infection.

In chronic sporotrichosis, the lymphatics that drain the area of the initial lesion are involved. Within a few days or weeks of the appearance of the primary lesion, multiple subcutaneous nodules develop along the loval lymphatic channels (Fig. 11–1 FS B 33,*A*). These nodules, which are similar to the primary lesion, are moveable at first but later become attached to the overlying skin. They become discolored and suppurate, and their connecting lymphatics become hard and cordlike. The clinical picture of an ulcer on the finger or wrist with an associated chain of swollen lymph nodes extending up the arm is so pathognomonic as to be an "over the telephone" diagnosis. A thin, seropurulent discharge may drain from the first few lesions; however, as larger and more distant nodes become involved, there is less tendency to necrosis and drainage. The secondary lesions are more gummatous and tend to persist for months or years. Untreated sporotrichosis usually becomes chronic, but some cases heal spontaneously. In a few instances, "gummatous" involvement may develop[63] (Fig. 11–2 FS B 33,*B*). Primary inoculation blastomycosis and coccidioidomycosis may resemble lymphocutaneous sporotrichosis.

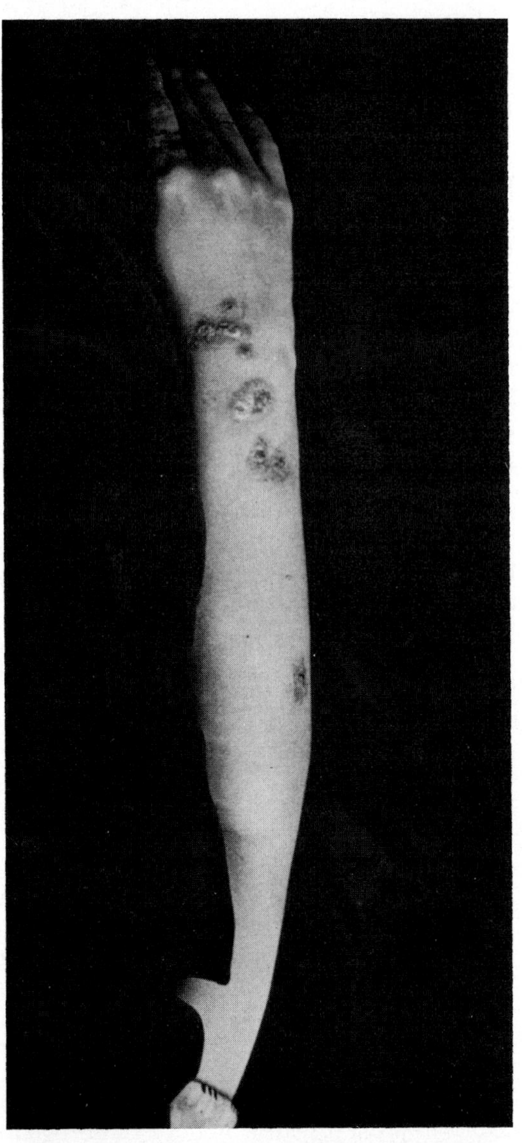

Figure 11–1. FS B 33,*A*. Lymphocutaneous sporotrichosis. The initial lesion is discolored, ulcerated, and draining. Secondary lesions are elevated but have not ulcerated.

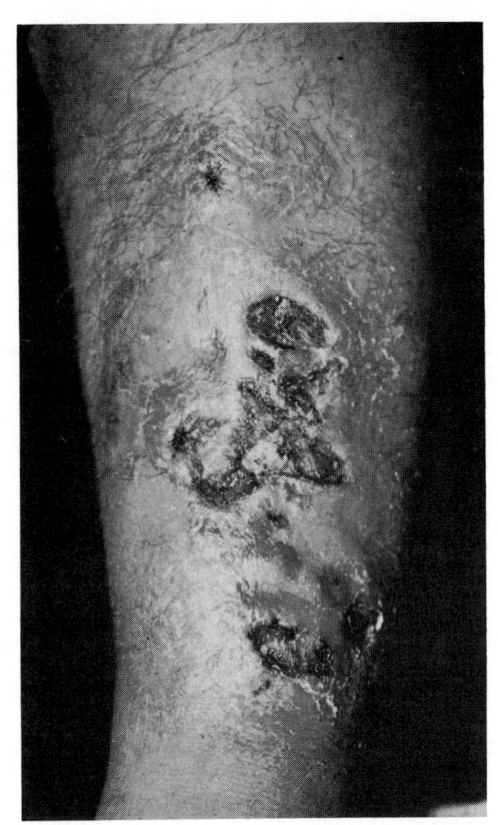

Figure 11–2. FS B 33,*B*. "Gummatous" sporotrichosis. Necrotic, ulcerating lesion of the leg. (Courtesy of S. MacMillen.)

Fixed Cutaneous Sporotrichosis

In highly endemic areas, a significant portion of the population becomes sensitized (without infection) to *S. schenckii.* This is demonstrated by reaction to the sporotrichin skin test. Primary infection in such people is quite commonly restricted to the site of inoculation and is called the fixed cutaneous type of sporotrichosis. The commonest sites of infection are the face, neck, and trunk (Fig. 11–3 FS B 34). The lesions manifest themselves as ulcerative (Fig. 11–4 FS B 35), verrucous, acneform, infiltrated, or erythematoid plaques, or as scaly, patchy, macular or papular rashes which do not involve local lymphatics and remain "fixed." Small satellite lesions are common. This type of sporotrichosis may account for 40 to 60 per cent of cases in some geographic areas. However, in most large series of cases, this type of infection usually does not exceed 20 to 25 per cent. Infections in this category very seldom evolve into systemic disease. Since

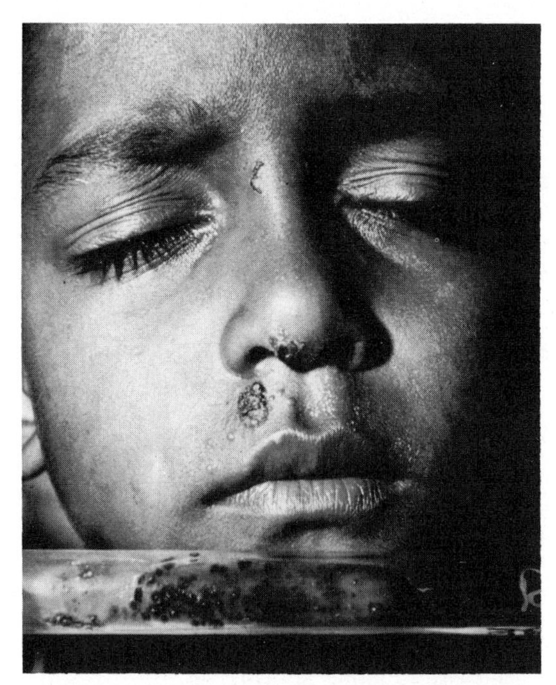

Figure 11–3. FS B 34. Fixed cutaneous sporotrichosis. Maculopapular rashes satellite lesions, and the culture of the organism. (Courtesy of F. Battistini.)

the lesions are so variable in appearance and often become crusted or weeping, they must be differentiated from many cutaneous diseases. Some of these are, in order of commonness, verrucous tuberculosis, papular necrotic tuberculosis, crustaceous syphilids, chromomycosis,[53] and cutaneous leishmaniasis. The fixed cutaneous lesions on occasion heal spontaneously but otherwise are resistant to local therapy. They may remain active for years, alternately healing and reappearing at the same location, often with a change in morphology.

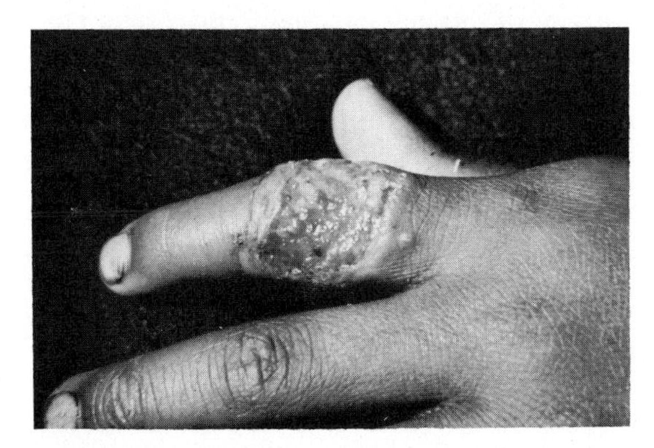

Figure 11–4. FS B 35. Fixed cutaneous sporotrichosis. Verrucous, ulcerative lesion of finger with a satellite lesion. (Courtesy of F. Pifano.)

Mucocutaneous Sporotrichosis

Infection of the mucosa without infection in other areas was described in early reports of sporotrichosis,[5] but this type of disease remains relatively rare. Involvement of the mucocutaneous areas secondary to dissemination is not infrequent, however. In either case the morphology of the lesions is reported to be similar. In the mouth, pharynx, or nose the lesion is erythematous, ulcerative, and suppurative at first and eventually becomes granulomatous, vegetative, or papillomatous. The infection is usually accompanied by pain (which is unusual in cutaneous lesions), swelling, and inflammation of the involved areas. Regional lymph nodes become hard and enlarged. Chronicity is not common, and the lesions usually heal with nondeforming scar formation. Even after healing, lesions may still harbor a small flora of *S. schenckii.* Mucous membrane sporotrichosis may resemble aphthous ulcers, oral lichen planus, or secondary cutaneous leishmaniasis.

Extracutaneous and Disseminated Sporotrichosis

This is a convenient but somewhat artificial grouping of disease types. Although it is rare, disease of the skeletal system is the most common type of infection next to the cutaneous forms of sporotrichosis. In almost all cases there was a history of the presence of cutaneous lesions prior to osseous involvement,[67] but in a few instances, direct inoculation into the knee has been recorded.[67] Rarely, dissemination involving bone, joints, and other systems may be associated with cortisone therapy.[43] It is also possible that some cases represent spread from inapparent primary pulmonary infections.

Bone, Periosteum, and Synovium. In a review of 30 cases of extracutaneous sporotrichosis, 80 per cent had lesions involving bony tissue. All 30 had had cutaneous or subcutaneous lesions also.[67] The disease is described as a destructive arthritis, with osteolytic lesions, tenosynovitis, and periosteitis. Lesions of the metacarpals and phalanges were most frequently noted, although the tibia is also commonly involved. Pathologic fracture at this latter site was noted very early in the French literature.[5] In descending frequency, other sites of bony involvement include carpal, metatarsals, tarsals, radius-ulna, femur, and ribs.

In articular sporotrichosis symptoms of pain, swelling, and chronic progressive limitation of motion are usually recorded.[67] Pain is not always a constant feature.[43] A viscous, serosanguinous joint effusion is usually present which has an elevated protein and cell count. Successful isolation of the fungus usually results from direct inoculation of such fluid onto culture media. In general the organisms are more numerous in cases of extracutaneous sporotrichosis than in cutaneous disease.

Eye and Adnexae. The eye and adnexae have been involved in about 50 cases of sporotrichosis. The lids, conjunctiva, and lacrimal apparatus are involved in about two-thirds of cases. In about 70 per cent, there were no other sites of infection. These therefore represent primary infections due to deposition of the fungus, probably on organic debris, into the conjunctival and lacrymal area. Preauricular adenopathy is reported to be absent[3] in some cases, and in only five of 48 cases were adjacent nodes enlarged.[67] The lesions are ulcerative and gummatous and run a course similar to primary cutaneous infection. Not infrequently such lesions extend to involve the orbit.

Systemic Disease. Disseminated sporotrichosis with involvement of organ systems other than the skeletal is quite rare. Sometimes there is some underlying disease, such as diabetes or sarcoidosis, or a history of long-term cortisone therapy,[43] but dissemination associated with neoplasias is very rare. In this form of sporotrichosis, the organisms are usually more numerous than in the cutaneous forms of the disease and are easily demonstrated in histologic sections by special stains. Whereas lymphatic spread is usually local and develops slowly, hematogenous dissemination may result in the sudden widespread involvement of many organ systems with multiple lesions. Most of these occur in cutaneous, osseous, or muscle tissue, but there are records of polynephritis, orchitis, epididymitis, and mastitis. Involvement of liver, spleen, pancreas, thyroid, and myocardium is very rare but has been reported.[67] Dissemination of the disease is frequently associated with a fever of 39°C or higher, anorexia, weight loss, and stiffness of joints.

Meningitis. Infection of the central nervous system is extremely rare.[30,57] In the three or four recorded cases, the primary site of infection was obscure in at least two, but the others had previous lesions in cutaneous tissue. In a report by Shoemaker,[57] disease occurred in a patient two years following myelography, culminating in neurologic decline and death. Another similar case was treated successfully with amphotericin B.[30] The presenting symptoms in all cases included dizziness, headache, confusion, and weight loss. The spinal fluid protein was elevated to about 400 mg per 100 ml, and a pleocytosis of 200 to 400 cells per ml (predominantly lymphocytes) was present, with other parameters within normal limits. *S. schenckii* was recovered from spinal fluid by culture. In the fatal cases, granulomatous microabscesses were scattered throughout the cerebral cortex.[14a]

Pulmonary Sporotrichosis

In disseminated sporotrichosis, the lung is very rarely involved when the primary infection is at some other body site. Until recently only about 30 cases of pulmonary sporotrichosis had been recorded. It now appears that most of these were primary disease, with infection resulting from the inhalation of spores, a situation analogous to primary pulmonary histoplasmosis or coccidioidomycosis.[31,50,54,67] In large urban hospitals, pulmonary sporotrichosis is not infrequent and is now considered one of the major mycoses of the lung. It appears to be particularly prevalent among chronic alcoholics.

The disease manifests itself as two general types, both of which closely resemble forms of tuberculosis. The first and most frequently reported type is chronic cavitary disease. The general course of this form is similar to that of other pulmonary mycoses. The infection begins as an acute pneumonitis or bronchitis which is sometimes minimal and may go unnoticed or be accompanied by fever, cough, and malaise. Apical areas of the lung appear to be the favored sites of infection, thus leading to an initial diagnosis of tuberculosis. The disease becomes a chronic pneumonitis, with nodular masses and development of thin-walled cavities with fibrosis and pleural effusions. If left untreated, these infections may remain stationary, but usually the disease progresses and extends. Cavitation sometimes becomes massive, with caseation necrosis often leading to a fatal outcome. Recovery has been effected with iodides, amphotericin B, or surgical excision. Spontaneous recovery is unknown.

A second type of pulmonary infection has been described which involves the lymph nodes primarily. The disease is acute and rapidly progressive,[54] but resolution of the lesions and recovery are frequent. In this type of infection, the tracheobronchial lymph nodes are involved, and the parenchyma of the lung per se is largely spared. Hilar lymphadenopathy may be of such magnitude as to cause bronchial obstruction. The disease of the lymph nodes may remain stationary for long periods of time, and spontaneous resolution is not infrequent. The syndrome of massive adenopathy is usually diagnosed as primary tuberculosis, but this can be ruled out by a negative, second-strength PPD or 1:100 O.T. In both forms of pulmonary sporotrichosis, the diseases to be considered next in the differential diagnoses are histoplasmosis, coccidioidomycosis, and sarcoidosis. The fungal infections can be eliminated by negative skin and complement fixation tests, and sarcoidosis by a negative Kviem test. Cultural and serologic confirmations are necessary to establish a diagnosis of sporotrichosis. Cultures are best done from bronchial washings or biopsy material, since sputum cultures are usually overgrown by *Candida,* and *S. schenckii* has been recovered on occasion as transient normal flora. On primary isolation from sputum, *S. schenckii* is sometimes colorless and yeastlike and resembles *Geotrichum.* For this reason it is often discarded as a contaminant.

The sporotrichin skin test using whole yeast cell antigen is usually strongly positive in cases of pulmonary sporotrichosis. After fading, this reaction sometimes reappears during therapy because of release antigen. González Ochoa has described a polysaccharide skin test which is more specific[18] than the commercially available sporotrichin test. Complement fixation titers, immunodiffusion lines, and agglutination titers appear in the serum of infected patients and can be used to follow the progress of therapy. The latter test appears to be most accurate for this purpose,[28,51,68] but immunodiffusion is quite specific and easily performed. Precipitin lines disappear with recovery from disease.

X-ray Findings. The two major forms of pulmonary sporotrichosis are reflected in the x-ray findings. The first is an acute or chronic pneumonitis, with widespread areas of miliary-type infiltration suggestive of tuberculosis, segmental involvement, or localized expanding areas of opacity simulating tumor. In any of these, extensive cavitation may develop with time (Fig. 11–5). The second form involves the tracheobronchial lymph nodes. The lesions are most frequently in the hilar areas, often with mediastinal widening (Fig. 11–6). These may remain unchanged for long periods of time. Mottling throughout several fields and atelectasis are described in some cases, which represents a mixed type of nodal and parenchymal involvement.[50,54]

DIFFERENTIAL DIAGNOSIS

Lymphocutaneous sporotrichosis is so constant in its clinical picture and in the evolution of the disease that diagnosis can be made with considerable confidence on first examination. Tularemia, anthrax, and some other bacterial infections may simulate part of the clinical picture, but these diseases are usually more acute. Sporotrichosis should be considered in any case in which multiple polymorphous eruptions or ulcers occur, particularly if they do not resolve with usual topical therapy. Other mycoses to be considered are chromomycosis, blastomycosis, mycetoma, paracoccidioidomycosis, and granulomatous trichophytosis. Gummatous syphilis, pyogenic lesions, cutaneous and lymphatic tuberculosis, staphylococcal lymphangitis, glanders, and drug eruptions such as bromoderma also must be considered. Histologic evidence of the organism is often lacking in primary sporotrichosis, and culture is the definitive procedure for diagnosis. Serologic tests may aid diagnosis, but they are usually negative in the cutaneous forms of the disease. In disseminated sporotrichosis, organisms are more numerous and usually evident in tissue sections and biopsy. Cultural confirmation is always necessary, however. Pulmonary disease is almost always diagnosed initially as tuberculosis, but this is true of all fungus infections of the lungs. Other mycoses, sarcoidosis, and tumor must also be considered.

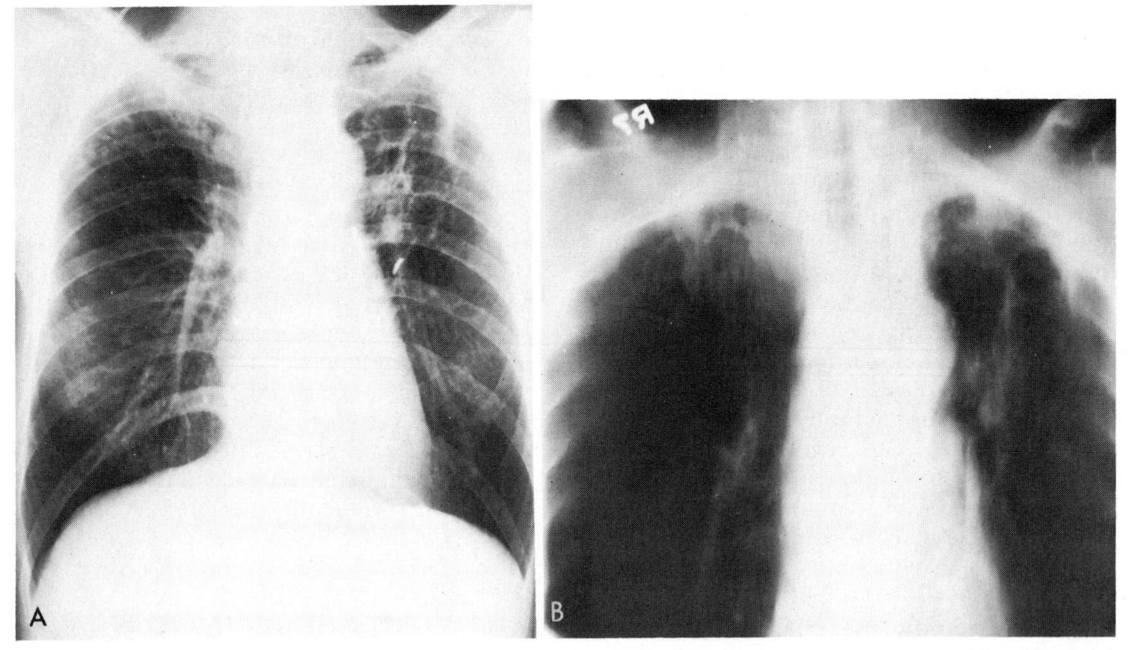

Figure 11–5. *A,* Chronic pulmonary sporotrichosis. There is a marked loss of volume in the left lobe, diffuse interstitial infiltrate, and many cystic areas in the apex. Similar changes are present in the upper right lobe, but they are less severe. (Courtesy of H. Grieble.) *B,* Tomogram showing multiple, thin-walled cavities in the apices.

PROGNOSIS AND THERAPY

Lymphocutaneous, fixed cutaneous, and mucocutaneous sporotrichosis are chronic infections that may alternately develop and regress for years in untreated patients. There is little distress or impairment of activity to the patient, and spontaneous cure occurs and is probably not infrequent, particularly in the fixed type of disease. On the other hand, the prognosis in disseminated sporotrichosis, as with all disseminated fungus diseases, is

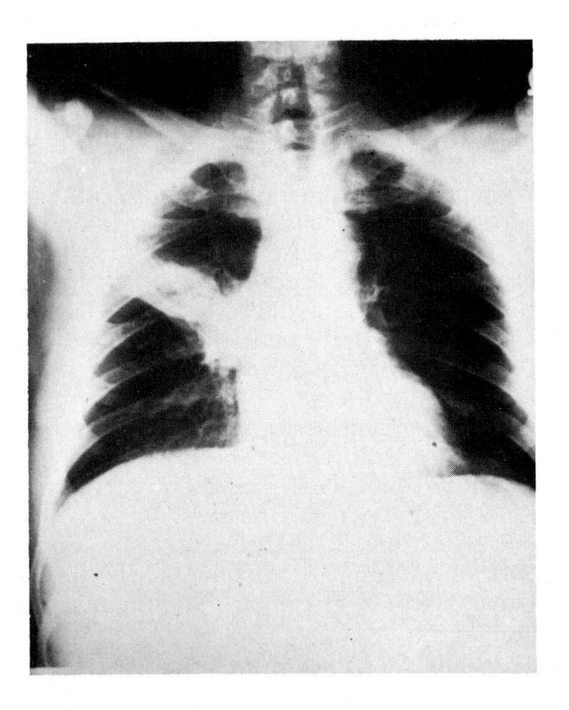

Figure 11–6. Acute pulmonary sporotrichosis. There is dense pulmonary consolidation in the middle right lung, extending from the hilum to the pleural surface. There is pronounced hilar adenopathy and some mediastinal adenopathy. The infection resolved without treatment, leaving only enlarged tracheobronchial nodes. (Courtesy of S. Kabins.)

grave. Organisms are numerous, and host defenses are inadequate. Spontaneous cure is unknown. Pulmonary disease limited to the hilar nodes may remain quiescent and sometimes resolves. In cases in which there is more parenchymal involvement or in which chronic pneumonitis has developed, progression is common, and the disease may be fatal if not treated.

Following the suggestion of Sabouraud, de Beurmann[5] was the first to employ iodides in the chemotherapy of sporotrichosis. Remarkable success was encountered, and this therapy remains the treatment of choice today.[4,14] Treatment schedules vary somewhat, but all utilize rapidly increasing doses given daily. The drug is administered orally in milk as a saturated solution of potassium iodide. The initial dose is 5 drops (1 ml) three times a day. This is increased by 3 to 5 drops each dose each day (minimum increase 1.5 ml per day) as tolerated, until 30 to 40 or more drops each dose are given (4 to 6 ml per dose, 12 to 18 ml per day). If signs of intoxication occur, such as indigestion, rash, lacrimation, cardiac problems, or swelling of salivary glands, the schedule can be tapered or the drug can be given intravenously. Treatment is continued for at least four weeks after clinical symptoms have resolved. Open lesions can be treated topically with a solution of 2 per cent KI containing 0.2 per cent iodine. Treatment failures are known, and they usually do not respond to retreatment with iodides but require other therapeutic modalities. The mode of action of potassium iodide is unknown. The fungus will grow *in vitro* on media containing a concentration of the drug up to about 10 per cent. Potassium iodide is known to cause resolution of granulomas and other abscesses. It has been suggested that this is because of enhancement of proteolysis and clearing of debris by proteolytic enzymes.[35] A direct effect on the fungus by iodine cannot be discounted.[65]

Amphotericin B is the most effective drug used for the treatment of relapsed lymphocutaneous sporotrichosis, and pulmonary and disseminated disease.[29,31,43,54,67] The two latter forms of infection may respond to iodides, and this should always be tried first. The usual dosage schedule of amphotericin B for mycotic infections (Chapter 18) is appropriate here. Infusion of the drug into the affected area has not been remarkably successful.

Relapse of the disease has occurred in a few patients treated with both iodides and amphotericin B. This is particularly true in articular involvement and systemic disease. Dihydroxystilbamidine was successfully used in one personally observed case, and others have reported similar results.[20,67] Dosage schedules vary from 50 to 225 mg per day (2 mg per kg) dissolved in 5 per cent glucose and given slowly by intravenous infusion. The total levels necessary for treatment with this drug are not so high in sporotrichosis as in blastomycosis. The course of therapy consists of ten days of the treatment followed by ten days of rest. Three such treatment regimens are usually sufficient but may be altered depending on clinical response.

In one report, griseofulvin (250 mg four times per day for three months) has been used successfully in fixed sporotrichosis.[34] Many treatment failures have occurred with this drug, and it has not gained wide popularity.[29] Some investigators are using 5-fluorocytosine in a daily dose of 100 mg per kg, but not enough experience has been gained to evaluate this mode of therapy as yet.

Ulcerated cutaneous lesions and lesions of the mucosa are often secondarily infected by bacteria. Antibacterial antibiotics are a useful adjunct in such cases.

PATHOLOGY

Gross Pathology. Because there have been so few cases of systemic sporotrichosis studied at autopsy, patterns of gross changes seen in this disease have not been described. Most cases have revealed osteomyelitis and destructive lesions of the digits and joints, widespread ulcerations of the skin, some enlargement of the spleen and liver, and a few purulent lesions of the kidney, pancreas, and other internal organs.[37,67] In the rare cases of cerebral involvement, numerous granulomatous microabscesses have been reported scattered throughout the cerebral cortex. In other cases, granulomatous basilar meningitis was present.[57]

Sufficient cases of pulmonary sporotrichosis have been studied to establish a discernible pattern of anatomic change. Once they are established in the lung, parenchyma lesions seldom resolve; they either remain fixed or expand with time, causing more and more distress to the patient. Resection of the lung has been successful in treating some of these cases.[54] Upon removal, the lesions are described as grey-white plaques, necrotic granulomata, thin-walled cavities within the lung tissue, or bronchiectatic cavities. The lesions are usually found in the upper lobes of the lung, particularly in the apical areas, although involvement of the middle and lower lobes is recorded. A thin-walled cavity is highly suggestive of fungal etiology, especially when tuberculosis has been ruled out. The lesions average 3 cm in diameter and are solitary or few in number. Rarely, close examination will reveal several small developing nodules peripheral to the main lesion. On cut section, the cavities are filled with a grey-white exudate and have a ragged necrotic lining where the organisms are found. In the bronchiectatic cavity described in one report,[54] there was a lining of necrotic material, and the space was filled with a red liquid.

Histopathology. One of the most difficult aspects of obtaining a diagnosis of sporotrichosis is the dearth of organisms present in biopsy material. Many cases of sporotrichosis are missed initially as the lesions are surgically removed because they are diagnosed as epitheliomas. Even histologically the two appear similar. It is extremely important, therefore, that cultures of biopsy material be performed whenever the diagnosis of sporotrichosis is being considered. In the typical clinical cases, culture and response to iodide treatment corroborate the correct diagnosis. The organisms are elusive in histologic sections, even when diligently looked for. In cases in which the index of suspicion is not high, they are usually not seen. Fetter[14] has modified standard fungal staining procedures to include treatment of the smear with 1:1000 malt diastase. The procedure dissolves all nonfungal polysaccharide, so that when sections are subsequently stained by PAS or methenamine silver, the organisms are much easier to find.

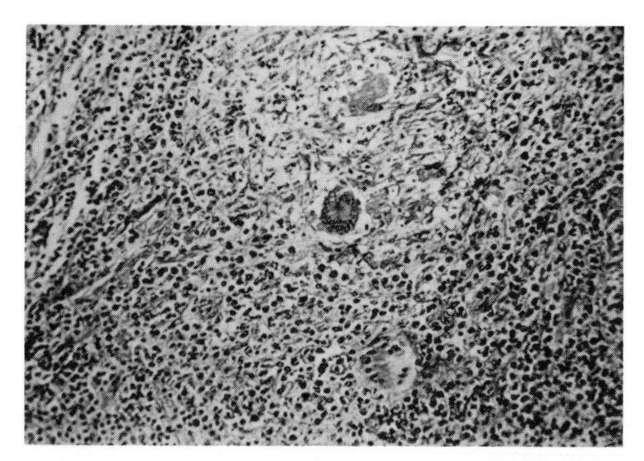

Figure 11–8. FS B 36,*B*. Sporotrichosis. High-power view of giant cells, fibroblasts, and lymphocytes. Hematoxylin and eosin stain. ×440.

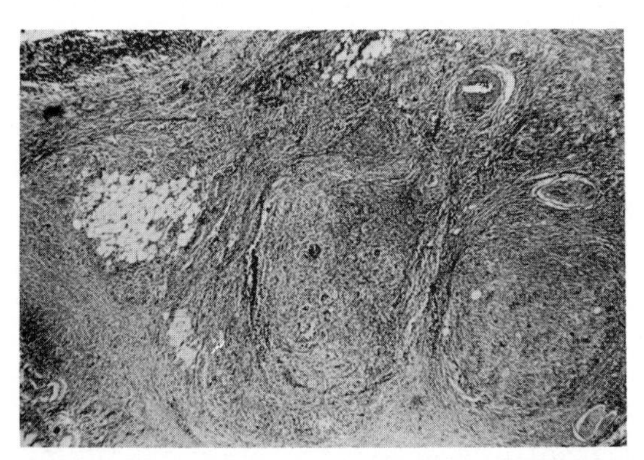

Figure 11–7. FS B 36,*A*. Sporotrichosis. Tuberculoid granuloma type lesion. Hematoxylin and eosin stain. ×100.

The histologic features of primary cutaneous sporotrichosis are a combination of granulomatous and pyogenic reactions. Lurie in 1963[37] classified the various granulomatous patterns seen as (1) sporotrichotic, (2) tuberculoid, and (3) foreign body. The reaction seen depends somewhat on the site of the lesion. The basic lesion of the sporotrichotic granulomatous reaction consists of masses of epithelioid histiocytes, which have some tendency to form concentric zones. The central area of the lesion consists of neutrophils or necrotic material surrounded by an infiltrate of neutrophils and some plasma cells and lymphocytes. In the tuberculoid reaction, this area sometimes merges into a zone of epithelioid cells mixed with fibroblasts, lymphocytes, and Langhans' giant cells (Fig. 11–7 FS B 36,*A*, and Fig. 11–8 FS B 36,*B*). Other lesions, however, may have a prominent outer layer of plasma cells, which may suggest the presence of a syphilid. In some cases, the histologic picture of a foreign body granuloma without any evidence of a pyogenic component is observed. Sometimes microabscesses without an epithelioid histiocytic component are found (Fig. 11–9 FS B 36,*C*).

In chronic sporotrichotic lesions, the pseudoepitheliomatous hyperplasia may be so extensive as to suggest a neoplasm. The rete ridges are elongated and broadened and extend into the corium, which is usually severely inflamed with some round cell infiltration into the rete itself. This combination of pseudoepitheliomatous hyperplasia and a mixed pyogenic-granulomatous-cellular reaction is highly suggestive of secondary cutaneous blastomycosis and coccidioidomycosis. In histologic sections demonstrating the above picture, sporotrichosis should be suspected if there is no evidence of other fungal etiology. Confirmation of the diagnosis often requires obtaining additional material from the patient to be used for culture.

As stated before, demonstration of the fungus in tissue section is very difficult. Application of the diastase method[14] or fluorescent antibody techniques[26, 27] (Fig. 11–9 FS B 36,*C*) aids in detection of the fungi. Even then the organisms are not numerous, and many sections may have to be examined. When they are found, the fungi are seen to be rounded, multiply budding, yeastlike cells, 3 to 5 μ in diameter. Gram stains of material from lesions show the organisms to be Gram-positive and irregularly stained (Fig. 11–10 FS B 36,*D*). In specimens in which the organisms are few in number, the cells tend to be of smaller size, and budding is infrequent. In lesions in which the fungi are more numerous, they appear cigar-shaped, 3 to 5 μ or wider in diameter, with up to three buds. These cells are most frequently seen in secondary foci of infection. Some lesions contain fungal cells with very irregular morphology, and the presence of mycelial elements in tissue has been reported on occasion.[39] Cigar-shaped cells and large spherical bodies up to 8 μ in diameter are very plentiful in experimentally infected mice (Fig. 11–11). There is no capsule around the yeast cells of *S. schenckii*; however, artifacts caused by cytoplasm shrinking away from the wall or the wall splitting and shrinking during fixation of tissue have given rise to erroneous reports of encapsulated organisms.

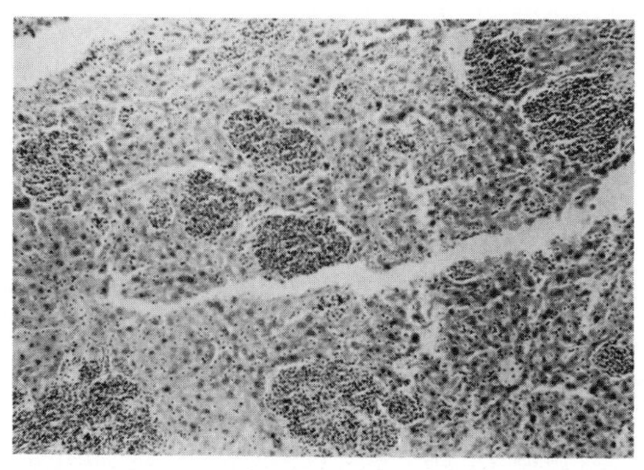

Figure 11–9. FS B 36,C. Sporotrichosis. Systemic disease. Numerous small microabscesses in a liver. Hematoxylin and eosin stain. ×110.

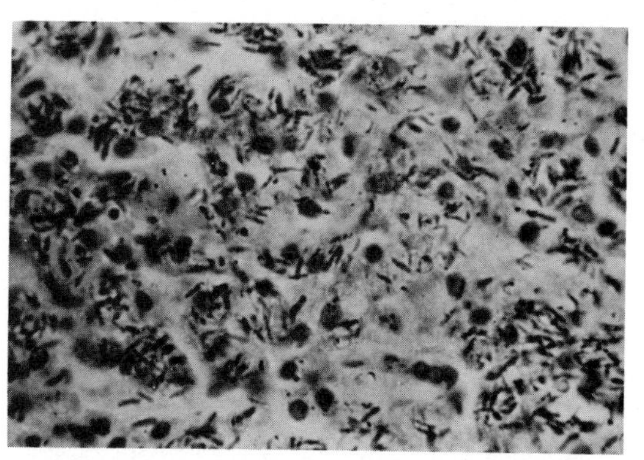

Figure 11–10. FS B 36,*D*. Sporotrichosis. Material from experimental orchitis of guinea pig. Gram stain, showing elongate, irregularly staining yeasts. ×400.

The asteroid body, when seen, is considered characteristic of sporotrichosis, but it has been found in cases of other mycoses as well. Typically it consists of a rounded or oval, basophilic, yeastlike body 3 to 5 μ in diameter. Radiating from this yeast cell are rays of eosinophilic substance up to 10 μ in thickness (Fig. 11–12 FS B 37). The asteroid phenomenon was first described in 1908 by Splendore, who coined its name. Some authors considered it to be a characteristic of certain strains which was related to their virulence, but this was later disproved. Only ten cases with asteroid bodies were recorded between 1908 and 1940. It is now regularly seen in cases reported from Japan, South America, and Africa. Careful examination of serial tissue sections is often necessary to detect it, which probably accounts for the paucity of reports of its occurrence. In meningeal involvement, such eosinophilic masses may surround several yeast cells, suggesting that they are encapsulated.[57] Although there had been much speculation as to the nature of the eosinophilic substance forming the rays of the asteroid body, it has now been established that it is an antigen-antibody complex.[38] Hoeppli observed a similar phenomenon around the ova of schistosomes in tissue, and this has been used as the basis for the development of specific serologic test for schistosomiasis.[52] Similar eosinophilic material is found around

Figure 11–11. Sporotrichosis. Impression smear from liver of experimentally infected mouse stained by flourescent antibody technique. Small yeast cells are visible.

the grains in actinomycosis and surrounding the mycelia in subcutaneous phycomycosis. Moreover, in these diseases the phenomenon is probably the result of an immune reaction.

In disseminated sporotrichosis, ulceration of cutaneous nodules occurs infrequently, and the overlying dermis and epidermis show little pathologic change. The nodule itself is not unlike the granulomata of primary lesions, and consists of an outer zone of round cells and plasma cells, an inner layer of histiocytes and giant cells, with a central area of necrosis and neutrophilic infiltrate. The presence of asteroid bodies is reported to be more common in these secondary lesions than in the primary ones.[37,38] In systemic disease of debilitated patients, the histologic picture is similar to that seen in other opportunistic fungous infections. Small microabscesses, some with necrotic centers, are seen in various organs (Fig. 11–9 FS B 36,*C*).

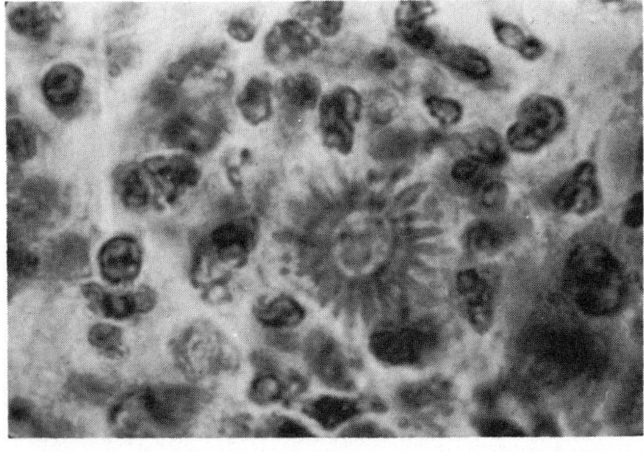

Figure 11–12. FS B 37. Sporotrichosis. Asteroid body in testicular tissue of a hamster. Hematoxylin and eosin stain. ×1000. (Courtesy of F. Mariat.)

ANIMAL DISEASE

Natural infection among animals is common, and sporotrichosis is considered a frequent and serious disease of horses. In addition, the disease has been recorded in dogs, cats, boars, rats, mules, foxes, camels, and fowl.[69] Historically infection in rats in Brazil and in mules in Madagascar was described soon after the first human cases were recorded.[16a]

In horses the pathology is similar to that seen in human disease. After introduction of the fungus, subcutaneous nodules form, followed by involvement of the local lymphatic channels and subsequent ulceration and discharge of pus. The fixed cutaneous type of disease has also been described in horses. Treatment with intravenous sodium iodide (2 doses of 2 oz each on alternate days combined with 1 oz of organic iodine in feed) has cured the infection.[16] Treatment failures following iodide therapy have been recorded. In a case reported by Davis,[12] a regimen of 10 g of KI per day orally and 6 g of NaI intravenously arrested but did not cure the infection. However, griseofulvin, 10 g per day for two weeks followed by 5 g per day for 46 days, led to clinical cure. Sporotrichosis in horses must be differentiated from equine epizootic lymphangitis, which is caused by *Histoplasma farciminosum.*

Lymphocutaneous sporotrichosis occurs in dogs also, but there is a greater tendency for the disease to disseminate in this animal than in other species. Most infections are fatal in spite of treatment. In a few cases of cutaneous disease, treatment with KI has effected cure.[36]

Experimental infections are easily produced in rats, mice, and hamsters. Mice are particularly susceptible to disease. Scrapings of the mycelial mat or suspensions of yeast may be injected intraperitoneally or intratesticularly. Within ten days a severe peritonitis or orchitis develops. Gram stain of pus from such lesions demonstrates numerous, irregularly staining, Gram-positive, cigar-shaped yeast cells (Fig. 11–11). The disease is progressive in mice, in which dissemination with destructive lesions of bones and other internal organs leads to death, usually by three weeks.[62] Most isolants from human disease regularly kill mice if given in a dose of 10^5 yeast cells. However, Howard and Orr[23] have demonstrated that few soil isolants are capable of growth at 37°C or are pathogenic for mice. This indicates that man acts as a selective agent for those strains that are able to evoke disease. This is probably true for many other "pathogenic" fungi. Serial animal passage has been shown to increase virulence. By this method, Mariat[46] has adapted strains of *Ceratocystis stenoceras* to produce a disease similar to sporotrichosis. He believes that *S. schenckii* and *C. stenoceras* may be identical.

The importance of temperature in experimental animal infection has been demonstrated by Mackinnon et al.[41] Mice were given a standardized dose of yeast cells (0.3 ml of a McFarland grade 2) and incubated at different temperatures for 15 days. Animals incubated at 2 to 5°C had miliary lesions involving several internal organs, especially the liver, muscles, and legs. At 13 to 17°C, lesions developed in muscle and legs only, with internal organs spared. Amphotericin B (0.04 mg daily) prevented disease in mice incubated at 13 to 17°C but not those at 2 to 5°C. Rats injected with yeast cells and maintained at 31°C did not develop any lesions.[40] The maximum temperature for *in vitro* growth of most strains of *S. schenckii* is 38 to 38.5°C.

IMMUNOLOGY AND SEROLOGY

Natural resistance to sporotrichosis is quite high. Since the organism is universally present in soil, factors other than exposure must be important in areas where the disease is prevalent. Mariat[44] considers nutritional state as well as repeated exposure to be of great importance in the majority of patients who contract the disease. There is some experimental evidence in support of this. Most cases of lymphatic disease are contracted by the malnourished rural poor who are in constant contact with the soil and hence the fungus. Pulmonary disease, however, is most frequently found in chronic alcoholics. Kedes[29] has coined the "alcoholic rose gardener syndrome" for patients who have constant exposure to soil and are imbibers.

In highly endemic areas, hypersensitivity develops in persons with no clinical signs of infection. In studies of grass handlers by González Ochoa,[18,64] the development of positive skin tests correlated with the length of time spent at this occupation. After ten years all workers had positive skin tests. In many individuals this hypersensitivity appears either to prevent infection or to modify the course of disease. The fixed cutaneous type of sporotrichosis is most common in this population. A similar survey was conducted in areas where clinical sporotrichosis had never been reported; there were essentially no reactors to the skin test.[24] This indicates that this procedure is a valuable epidemiologic tool for assessing the prevalence of the disease or exposure to the organism.

Strains of *S. schenckii* isolated from human and animal disease and from soil are all antigenically similar by cross-agglutination tests. Antigens prepared from these strains are used for skin testing, complement fixation tests, and immunodiffusion tests. There appear to be few cross-reactions in other fungal diseases.

The standard sporotrichin skin test consists of the intradermal injection of a 1:1000 dilution of heat-killed, packed yeast cells. A reaction of 5 mm of induration after 24 hours constitutes a positive test. As noted above, this test is useful for determining exposure to the fungus. A test using a polysaccharide derived from a crude antigen preparation has been developed by González Ochoa and Figueroa. This test appears to be more specific than the standard skin test for diagnosis of actual infection.[17,18,33,44] However, some authors have found both the crude antigen and the polysaccharide derivative equally sensitive in the diagnosis of the disease.[55]

An extensive review of the serologic procedures used in the diagnosis and prognostic evaluation of sporotrichosis has been published by Karlin and Nielsen.[28] They concluded that the yeast cell agglutination test and the latex particle agglutination test were the most sensitive serologic tests for all forms of the disease. The yeast agglutination test developed by the authors utilized lyophilized yeast cells. The material was suspended in pH 7 buffer to an O.D. of 0.12 at 420 mμ, which is equivalent to a final concentration of 0.1 mg per ml. Yeast cells that had been centrifuged for 20 minutes at 2500 rpm and then diluted 1:2400 (v/v) could also be used as antigen. In the performance of the test, serial dilutions of the patients' sera were made in phosphate buffer (pH 7) to a final volume of 0.5 ml. To each tube was added 0.5 ml of yeast antigen, followed by incubation at 37°C for one hour, then storage at 4°C overnight. After this, the tubes were centrifuged and gently agitated. The last tube containing well-formed yeast aggregates was read as the end point. A titer of 1:40 or below was considered questionable by the authors, but a higher titer was felt to be strong evidence for the presence of infection. Similar results were found by Welsh and Dolan.[68] Comparison studies were done on the same sera, utilizing the lyophilized yeast-antigen suspension in complement fixation tests and a broth culture of yeast as antigen for latex agglutination and immunodiffusion tests. They found that the yeast agglutination and latex particle agglutination tests were positive in all cases, but the immunodiffusion test was positive in 80 per cent and the complement fixation only in 39 per cent. The agglutination and immunodiffusion tests were recommended for diagnosis and prognosis of sporotrichosis. Agglutinin titers fell and precipitins disappeared with resolution of the disease.

In the immunodiffusion test there appears to be some cross-reaction with other mycotic diseases, depending on the antigen used. McMillen and Laverty[49] performed precipitin and complement fixation tests in 17 cases of sporotrichosis. Positive reactions in gel and C.F. titers of up to 1:64 were present in all cases of pulmonary and articular disease. In cutaneous disease, the gel-diffusion test was positive in 60 per cent of cases, and the maximum C.F. titer was only 1:16. In their study, a fall in titers was noted to occur with resolution of the disease. The antigen used by them was an autoclaved filtrate of the yeast phase growth that had been dried and resuspended to a concentration of 1 mg per ml. Utilizing this antigen they found no cross-reaction in sera from patients with other mycoses. Jones et al. found C.F. titers in most patients with extracutaneous disease[25] but not in those with cutaneous lesions. They found that the immunodiffusion test was unreliable as performed in their laboratory. However, in the hands of most other investigators, the agglutination and immunodiffusion tests are specific and reliable aids to the diagnosis of sporotrichosis.

LABORATORY IDENTIFICATION

Direct Examination. In the cutaneous forms of sporotrichosis, there are so few organisms present in pus, exudates, biopsy material, and aspirates that, in general, direct examination of such material is unrewarding. An exception to this is the fluorescent antibody staining technique of Kaplan and co-workers.[26,27] Utilizing this procedure, material from lesions, histologic slide preparations, and mycelium or spores from cultures can be specifically stained.

Culture Methods. *S. schenckii* grows well on almost all culture media. Aspirates from cutaneous nodules, pus, exudate, and material from curettage or swabbings from open lesions can be planted on Sabouraud's agar or blood agar and incubated at 25 or 27°C. Surgical incision and biopsy are contraindicated as methods for obtaining material for culture, since these procedures may result in spread of the disease. The fungus is resistant to cycloheximide; therefore, Mycosel agar also can be used for culture. The percentage of cultures which become positive in cases of the disease is very high, and this is the most reliable method for diagnosis. Growth usually occurs within three to five days. Cultures should be held for at least four weeks before being discarded as negative.

The colony morphology of the initial growth of *S. schenckii* is quite variable. In cultures of material from cutaneous lesions and articular aspirates, a black to brown fungal colony appears, which is shiny at first and later becomes fuzzy. Young colonies of many of these isolants and also those from cases of pulmonary sporotrichosis are white, glabrous, and yeastlike at all temperatures (Fig. 11–13 and Fig. 11–14,*A* FS B 38,*A*). In time, the colony becomes wrinkled and

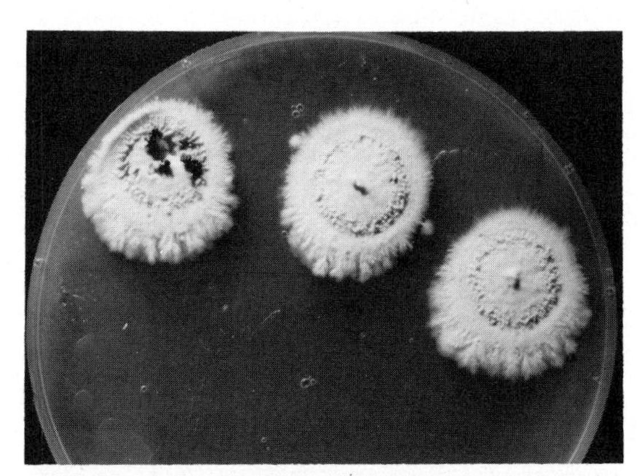

Figure 11–13. Sporotrichosis. White colonies of the initial isolant of *S. schenckii* from a case of articular disease. Subcultures of the organism became black.

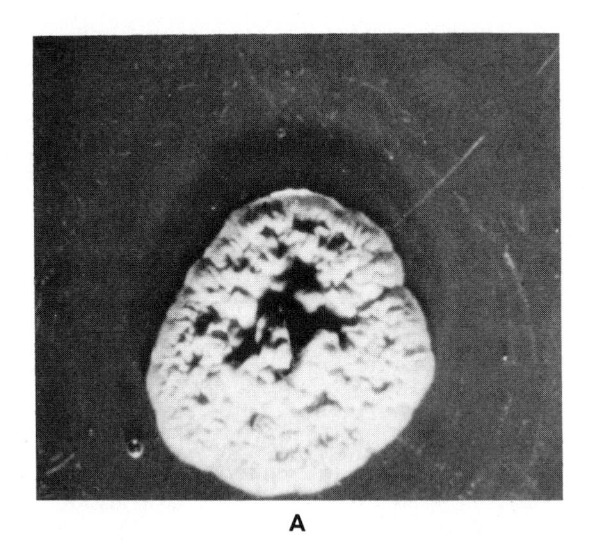

A

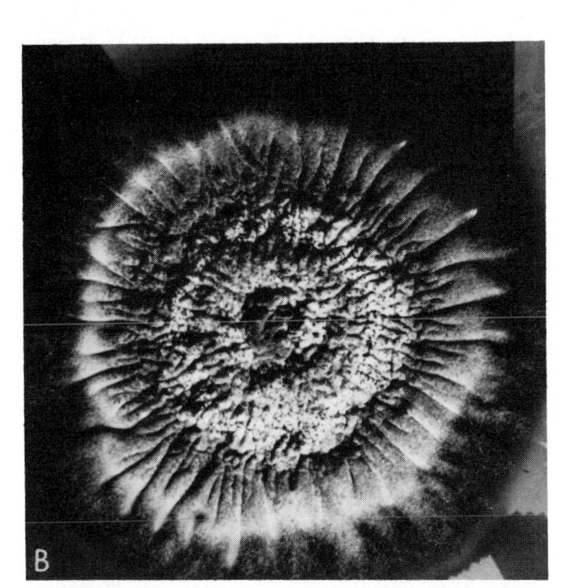

Figure 11–14. FS B 38. *A,* Sporotrichosis. White colony from pulmonary infection. This strain remained white on subculture. *B,* Sporotrichosis. Folded black-brown colony with radial grooves.

membranous, later developing areas of discoloration which normally become black and mycelial (Fig. 11–14,*B* FS B 38,*B*). Coloration is inconstant, and strains vary considerably. Even the same strain will vary from transfer to transfer. Some colonies remain white and glabrous, which may cause them to be misread as *Geotrichum* species and discarded as contaminants (Fig. 11–14,*A* FS B 38,*A*). Special media, such as cornmeal, malt, or Czapek's agar, are useful in inducing mycelium formation, sporulation, and pigmentation.

Demonstration of dimorphism is important for specific identification of *S. schenckii.* To induce mycelial to yeast transformation, the fungus is inoculated on moist blood agar tubes and incubated at 37°C. Conversion to the yeast form sometimes is restricted to the outer edge of the colony. Animal inoculation is also used for identification, especially of isolants that convert to the yeast phase with difficulty *in vitro.* There are no physiologic tests of importance for identification. *S. schenckii,* as well as many other fungal species, has a requirement for thiamine.

MYCOLOGY

Sporothrix schenckii Hektoen and Perkins 1900

Synonymy. *Sporotrichum* sp. Smith 1898; *Sporotrichum schencki* Matruchot 1910; *Sporotrichum beurmanni* Matruchot and Ramond 1905; *Sporotrichum asteroides* Splendore 1909; *Sporotrichum equi* Carougeau 1909; *Sporotrichum jeanselmei* Brumpt and Langeron 1910; *Sporotrichum councilmani* Wobach 1917; *Sporotrichum grigsby* Dodge 1935; *Sporotrichum fonsecai* Filho 1930; *Sporotrichum cracoviense* Lipinski 1928; *Rhinocladium schencki* Verdun and Mandoul 1924; *Rhinotricum schenckii* Ota 1928.

Possible Perfect Stage. *Ceratocystis stenoceras* (Robak) Moreau.[46, 50a]

The first report of *Sporothrix schenckii* infection noted that the organism was dimorphic.[50a] Though it was not seen in lesion material, the parasite grew as a mycelial fungus in culture but formed cigar-shaped yeasts when it was injected into animals.[56] Shortly thereafter, Lutz and Splendore demonstrated temperature-dependent dimorphism *in vitro. S. schenckii* grows as a budding yeast at 37°C and as a sporulating hyphal fungus at 25°C. As indicated previously, strains isolated in nature vary considerably in their ability to grow at 37°C.[23] Infection may be a selective process of those strains capable of adaptation to higher temperature and growth within animal tissue. Thermal dimorphism and infection as a result of the selective process are also exhibited by *Histoplasma capsulatum, Blastomyces dermatitidis,* and *Paracoccidioides brasiliensis.*

Early investigators concluded that the growth form seen in tissue and *in vitro* at 37°C was an extension of the conidiation process. Acrepetalous budding of conidia borne on mycelia had been observed *in vitro,* and it was thought that the conidia in tissue formed buds and thus became blastospores. However, in a detailed investigation of the process, Howard[22] demonstrated that M-Y transformation in tissue culture involves production of yeast cells directly from the mycelium itself. Conidia were seen to germinate into short mycelial units, which in turn gave rise to yeast cells. Two morphologic transformations of the mycelium were found to occur. The first involved formation of club-shaped structures at the hyphal tips or on lateral branches, which then gave rise to budding units, the blastospores (yeasts). The second method was the formation of oidia within the mycelium, followed by fragmentation. These free oidia subsequently bud and are then blastospores also. The cycle for this type of yeast cell formation was summarized as mycelium → oidia → blastospores. Previous work by the author had shown a similar mechanism in the transformation of *H. capsulatum* and *B. dermatitidis* to yeast-form growth. The budding from club-shaped structures, however, was not found in the latter two species. The yeast to mycelium transformation is effected by elongation of the parent cell, a rearrangement of the cytoplasmic membrane system, followed by septal formation.[32] The elongation continues and branching occurs. It appears that all dimorphic fungi convert from the yeast to the mycelial form of growth in a similar manner.

Colony Morphology. SDA, 25°C. The initial colony of *S. schenckii* isolated from clinical material is moist, glabrous, and yeastlike but becomes tough, wrinkled, and folded in time. The color is usually dirty white at first, although in some strains it is yellow, brown, or quite black (Fig. 11–14,*B* FS B 38,*B*). Pigmentation is extremely variable. Usually the whitish colony develops an overgrowth of fuzzy mycelium and turns darker in sectors. Old laboratory strains often lose their pigmentation altogether and are a dirty white color.

At 37°C on SDA, BHI blood, or other media containing high concentrations of glucose, the organisms grow in the yeast phase. The colony morphology varies from pasty, white, and yeastlike to a greyish yellow, sometimes glabrous, bacteria-like colony.

Microscopic Morphology. SDA, 25°C. Thin septate, branching hyphae, which generally do not exceed 1 or 2 μ in diameter, are formed at room temperature. Not infrequently they will appear as twisted ropes containing several mycelial strands (Fig. 11–15). At first, sporulation is from long, slender, tapering conidiophores rising at right angles from the hyphae. The conidiophores are erect or recumbent, 1 to 2 μ in diameter at the base, narrowing to 0.5 to 1 μ at the tip (Fig. 11–16). The length is quite variable. The apex of the conidiophore may expand to form a denticulate vesicle, and the phore elongates sympodially to form another apex. Simple, ovate conidia, 2 to 3 × 3 to 6 μ are formed at first on the apex. Their arrangement suggests a palm tree or a flower head. Sometimes the conidia are angular or obovate. With age, sporulation increases, so that spores are formed along the sides of the conidiophore and eventually along the undifferentiated hyphae as well. Dense sleeves of spores are seen in old cultures. Free spores have an apiculus or spine where they were attached to the conidiophore. Some conidia may bud once or twice to form acrepetalous clusters similar to the *Cladosporium* type of spore formation seen in *Fonsecaea pedrosoi* or related dematiaceous species. Some strains of *S. schenckii* are morphologically similar to *Ceratocystis,* and it has been suggested that it may be a conidial form of *Ceratocystis* species.[44,46] Mariat has produced a sporotrichosis-like disease in

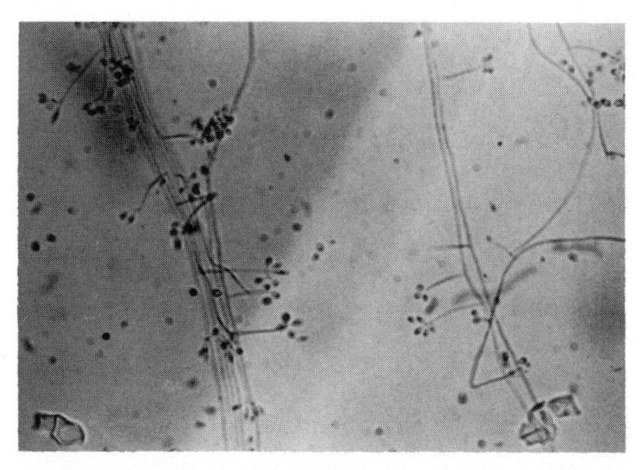

Figure 11–15. Sporotrichosis. Twisted mycelial strands bearing delicate conidiophores and aleuriospores.

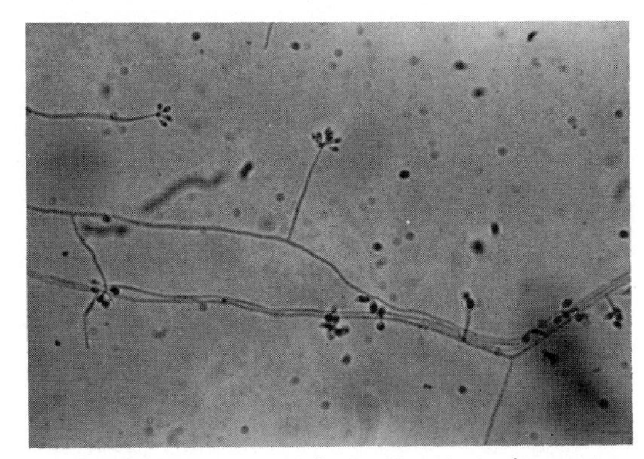

Figure 11–16. Sporotrichosis. Tapering, elongate conidiophores showing expanded denticulate vesicle at apex, bearing aleuriospores. (From Rippon, J. W. *In* Burrows, W. 1973. *Textbook of Microbiology*. 20th ed. Philadelphia, W. B. Saunders Company, p. 718.)

mice with isolants of *Ceratocystis stenoceras*.[46] The organism recovered from tissue, however, was no longer able to form perfect (sexual) stage spores. Other authors have claimed the perfect stage to be *Dolichoascus schenckii*.[61a]

At 37°C in the yeast form of growth, the cells are spherical or ovate blastospores. The size is variable, but averages 1 to 3 × 3 to 10 μ. Several buds may appear on the yeast cell and resemble the budding conidia found in culture at 25°C.

"Fungi are a mutable and treacherous tribe" is an old adage describing the only constant in the study of mycology—variation. Two variants of *S. schenckii* have been described recently which illustrate this point. Ajello and Kaplan[2] have reported on an isolant that regularly forms sclerotic bodies *in vitro*. *In vivo* this organism forms thin-walled globose cells (1.6 to 4.8 × 1.6 to 6.9 μ), which reproduce by budding, and thick-walled globose cells (11.7 to 18.7 × 14 to 23.4 μ), which reproduce by septation or by single or multiple budding. The name proposed by them for this isolant is *Sporothrix schenckii* var. *luriei*. The second variant reported by Brand and Van Niekerk[6] was isolated from 25 cases of sporotrichosis in South Africa. These isolants did not produce any pigment in culture. In experimental infections, the authors report that only a monocytic response was found in animals, whereas typical cultures induce an acute pyogenic and granulomatous response. The colonial features of these isolants were more typical of *Ceratocystis* species than of the usual strains of *S. schenckii*.

REFERENCES

1. Ahern, D. G., and W. Kaplan. 1969. Occurrence of *Sporotrichum schenckii* on cold stored meat products. Am. J. Epidemiol., *89*:116–124.
2. Ajello, L., and W. Kaplan. 1969. A new variant of *Sporothrix schenckii*. Mycosen, *12*:633–644.
3. Alvarez, G. R., and A. Lopez-Villegas. 1966. Primary ocular sporotrichosis. Am. J. Ophthalmol., *62*:150–151.
4. Becker, F. T., and H. R. Young. 1970. Sporotrichosis. A report of 21 cases. Minn. Med., *53*:851–853.
5. de Beurmann, L., and H. Gougerot. 1912. *Les Sporotrichoses*. Paris: Felix Alcan.
6. Brand, F. A., and V. Van Niekerk. 1968. An atypical strain of Sporothrix from South Africa. J. Pathol. Bacteriol., *96*:39–44.
7. Carmichael, J. W. 1962. Chrysosporium and some other aleuriosporic Hyphomycetes. Can. J. Bot., *40*:1137–1173.
8. Cawley, E. P. 1949. Sporotrichosis, a protean disease: With a report of a disseminated subcutaneous gummatous case of the disease. Ann. Intern. Med., *30*:1287–1292.
9. ContiDiaz, I. A., and E. Civila. 1969. Exposure of mice to inhalation of pigmented conidia of *Sporothrix schenckii*. Mycopathologia, *38*:1–6.
10. Crevasse, L., and P. D. Ellner. 1960. An outbreak of sporotrichosis in Florida. J.A.M.A., *173*: 29–33.
11. Davis, D. J. 1921. The identity of American and French sporotrichosis. Univ. Wisc. Studies Sci., *2*:104–130.
12. Davis, H. H., and W. E. Worthington. 1964. Equine sporotrichosis. J. Am. Vet. Med. Assoc., *145*:692–693.
13. Du Toit, C. J. 1942. Sporotrichosis on the Witwatersrand. Proc. Mine Med. Offrs.' Assoc., *22*:111–127.
14. Fetter, B. F., and J. P. Tindall. 1964. Cutaneous sporotrichosis. Clinical study of nine cases utilizing an improved technique for demonstration of organisms. Arch. Pathol., *78*: 613–617.

14a. Fetter, B. F., G. K. Klintworth, et al. 1967. *Mycoses of the Central Nervous System*. Baltimore, Williams and Wilkins, p. 198.

15. Findley, G. H. 1970. The epidemiology of sporotrichosis in the Transvaal. Sabouraudia, 7:231–236.
16. Fishburn, F., and D. C. Kelley. 1967. Sporotrichosis in a horse. J. Am. Vet. Med. Assoc., *151*:45–46.

16a. Gonçalves, A. P. 1973. Geopathology of sporotrichosis. Int. J. Dermatol, *12*:115–118.

17. González Ochoa, A. 1965. Contribuciones recientes al conaciemiento de la esporotrichosis. Gac. Med. Mex., *95*:463–474.
18. González Ochoa, A., E. Ricoy, et al. 1970. Valoracion comparative delos antigenos polisacarido y cellular de *Sporothrix schenckii*. Rev. Invest. Salud Publica, *30*:303–315.
19. Gonzalez Ochoa, A., and E. S. Figueroa. 1947. Polisaccridos del *Sporotrichum schenckii* Dato Immunologica. Rev. Inst. Salub. Enferm. Trop., *8*:143–147.
20. Harrell, E. R., F. C. Bocobo, et al. 1954. Sporotri-

chosis successfully treated with stilbamidine. Arch. Intern. Med., *93*:162–167.

21. Hektoen, L., and C. F. Perkins. 1900. Refractory subcutaneous abscesses caused by *Sporothrix schenckii.* J. Exp. Med., *5*:77–89.
22. Howard, D. H. 1961. Dimorphism in *Sporotrichum schenckii.* J. Bacteriol., *81*:464–469.
23. Howard, D. H., and G. F. Orr. 1963. Comparison of strains of *Sporotrichum schenckii* isolated from nature. J. Bacteriol., *85*:816–821.
24. Ingrish, F. M., and J. D. Schneidau. 1967. Cutaneous hypersensitivity to sporotrichin in Maricopa County, Arizona. J. Invest. Dermatol., *49*:146–149.
25. Jones, R. D., G. A. Sarosi, et al. 1969. The complement fixation in extracutaneous sporotrichosis. Ann. Intern. Med., *71*:913–918.
26. Kaplan, W., and A. González Ochoa. 1963. Application of the fluorescent antibody technique to the rapid diagnosis of sporotrichosis. J. Lab. Clin. Med., *62*:835–841.
27. Kaplan, W., and M. S. Ivens. 1961. Fluorescent antibody staining of *Sporotrichum schenckii* in cultures and clinical material. J. Invest. Dermatol., *35*:151–159.
28. Karlin, J. V., and H. S. Nielsen. 1970. Serologic aspects of sporotrichosis. J. Infect. Dis., *121*: 316–327.
29. Kedes, L. H., J. Siemski, et al. 1964. The syndrome of the alcoholic rose gardener: Sporotrichosis of radial tendon sheath. Report of a case with Amphotericin B. Ann. Intern. Med., *61*: 1139–1141.
30. Klein, R. C., M. S. Ivens, et al. 1966. Meningitis due to *Sporotrichum schenckii.* Arch. Intern. Med., *118*:145–149.
31. Kobayashi, G., and C. N. Newmark. 1969. Pulmonary sporotrichosis. J.A.M.A., *210*:1741–1744.
32. Lane, J. W., and R. G. Garrison. 1970. Electron microscopy of the yeast to mycelial phase conversion of *Sporothrix schenckii.* Can. J. Microbiol., *16*:747–749.
33. Latapi, F. 1963. [Sporotrichosis in Mexico.] Laval Med., *34*:732–738.
34. Leavell, U. W. 1965. A case of sporotrichosis cleared when given griseofulvin. J. Ky. Med. Assoc., *63*:415–416.
35. Lieberman, J., and N. B. Kurnick. 1963. Induction of proteolysis within purulent sputum by iodides. Clin. Res., *11*:81.
36. Londero, A. T., R. M. DeCastro, et al. 1964. Two cases of sporotrichosis in dogs in Brazil. Sabouraudia, *3*:273–274.
37. Lurie, H. I. 1963. Histopathology of sporotrichosis. Arch. Pathol., *75*:421–437.
38. Lurie, H. I., and W. J. S. Still. 1966. The capsule of *Sporotrichum schenckii* and the evolution of asteroid body. Sabouraudia, 7:64–70.
39. Maberry, J. D., J. F. Mullins, et al. 1966. Sporotrichosis with demonstration of hyphae in human tissue. Arch. Dermatol., *93*:65–67.
40. Mackinnon, J. E., and I. A. Conti-Diaz. 1962. The effect of temperature on sporotrichosis. Sabouraudia, *2*:56–59.
41. Mackinnon, J. E., I. A. Conti-Díaz, et al. 1964. Experimental sporotrichosis. Ambient temperature and amphotericin B. Sabouraudia, *3*:192–194.
42. Mackinnon, J. E., I. A. Conti-Díaz, et al. 1969. Isolation of *Sporothrix schenckii* from nature. Sabouraudia, *7*:38–45.
43. Manhart, J. W., J. A. Wilson, et al. 1970. Articular and cutaneous sporotrichosis. J.A.M.A., *214*: 365–367.
44. Mariat, F. 1968. The epidemiology of sporotrichosis, In *Systemic Mycoses.* Wolstenholme, G. E. W., and Porter, R. (eds.), Churchill, London, pp. 144–159.
45. Mariat, F. 1969. Variant, non-sexue de *Ceratocystis* sp. pathogenic pour le hamster. C. R. Acad. Sci. (Paris), *269*:2329–2331.
46. Mariat, F. 1971. Adaptation de ceratocystis à la vie parasitaire chez l'animal-étude de l'aquisition d'un pouvoir pathogene comparable à celui de *Sporothrix schenckii.* Sabouraudia, *9*:191–205.
47. McDonough, E. S., A. L. Lewis, et al. 1970. *Sporothrix (Sporotrichum) schenckii* in a nursery barn containing aphagnum. Public Health Rep., *85*:579–586.
48. McFarland, R. B. 1966. Sporotrichosis revisited: 65 year follow up of the second reported case. Ann. Intern. Med., *65*:363–366.
49. McMillen, S., and E. R. Laverty. 1969. Sporotrichosis: Serology as a clinical aid. Bact. Proc., p. 115.
50. Mohr, J. A., C. D. Patterson, et al. 1972. Primary pulmonary sporotrichosis. Am. Rev. Resp. Dis., *106*:260–264.

50a. Nicot, J., and F. Mariat. 1973. Caractères morphologiques et position systématique de *Sporothrix schenckii,* agent de la sporotrichose humaine. Mycopathologia, *49*:53–65.

51. Nielson, H. S. 1968. Biological properties of skin test antigens of yeast from *Sporotrichum schenckii.* J. Infect. Dis., *118*:173–180.
52. Oilver-Gonzalez, J. 1954. Anti-egg precipitins in serum of humans infected with *Schistosoma mansonii.* J. Infect. Dis., *95*:86–91.

52a. Park, C. H., C. L. Greer, et al. 1972. Cutaneous sporotrichosis: Recent appearance in northern Virginia. Am. J. Clin. Pathol., *57*:23–26.

53. Restrepo, M. A., G. Calle Velez, et al. 1967. Esporotricosis epidermica atipica. Preseutacion de das casos. Gac. Sanit., *22*:98–102.
54. Ridgeway, N. A., F. C. Whitcomb. et al. 1962. Primary pulmonary sporotrichosis. Report of two cases. Am. J. Med., *32*:153–160.
55. Rocha-Posada, H. 1968. Preuba cutanea con esporotricina. Mycopathologia, *36*:42–54.
56. Schenck, B. R. 1898. On refractory subcutaneous abscesses caused by a fungus possibly related to sporotricha. Bull. Johns Hopkins Hosp., *9*:286–290.
57. Shoemaker, E. H., H. D. Bennett, et al. 1957. Leptomeningitis due to *Sporotrichum schenckii.* Arch. Pathol., *64*:222–227.
58. Silva, Y. P., and N. A. Guimaraes. 1964. Esporotrichose familiar epidemica. Hospital (Rio), *66*:573–579.
59. Simson, F. W. 1947. Sporotrichosis infection on mines of the Witwatersrand. A symposium. Proc. Mine Med. Offrs'. Assoc., Transvaal Chamber of Mines, Johannesberg, South Africa.
60. Singer, J. J., and T. E. Muncie. 1952. Sporotrichosis: Etiological considerations and report

of additional cases from New York. N.Y. State J. Med., *52*:2147.
61. Smith, L. M. 1945. Sporotrichosis: Report of four clinically atypical cases. South. Med. J., *38*: 505–515.
61a. Thibaut, M. 1972. La forme parfaite du *Sporotrichum Schenckii* (Hektoen et Perkins 1900): *Dolichoascus schenckii*, Thibaut et Ansel 1970 nov. gen. Ann. Parasitol. Hum. Comp., *47*: 431–441.
62. Tsubura, E., and J. Swartz. 1960. Treatment of experimental sporotrichosis in mice with griseofulvin and amphotericin B. Antibiot. Chemother. (Basel), *10*:753–757.
63. Urabe, H., and T. Nagushima. 1970. Gummatous sporotrichosis. Int. J. Dermatol., *4*:301–303.
64. Velasco, O., and A. González Ochoa. 1971. Esporotricosus en individuos con esporotricina reaccion positiva previa, Rev. Invest. Saluf. Publica, *31*:53–55.
65. Wada, R. 1968. Studies on mode of action of potassium iodine upon sporotrichosis. Mycopathologia, *34*:97–107.
66. Wallk, S., and G. Bernstein. 1964. Systemic sporotrichosis with body involvement. Arch. Dermatol., *90*:355–357.
67. Wilson, D. E., J. J. Mann, et al. 1967. Clinical features of extracutaneous sporotrichosis. Medicine (Baltimore), *46*:265–280.
68. Welsh, M. S., and C. T. Dolan. 1973. Sporothrix whole yeast agglutination test. Am. Clin. Pathol., *59*:82–85.
69. Werner, R. E., B. G. Levine, et al. 1971. Sporotrichosis in a cat. J. Am. Vet. Med. Assoc. *159*:407–412.

Chapter 12

ENTOMOPHTHOROMYCOSIS

DEFINITION

Entomophthoromycosis is a chronic inflammatory or granulomatous disease which is generally restricted to the subcutaneous tissue or the nasal submucosa. It is caused by species of the order Entomophthorales[11,31] of the Zygomycetes and includes two clinically and mycologically distinct diseases. The first, rhinoentomophthoromycosis, involves the nasal submucosa and is characterized by the presence of polyps or extensive palpable subcutaneous masses that generally remain restricted to that area. The etiologic agent, *Entomophthora coronata,* is a soil organism which is pathogenic for termites, spiders, and other insects. The second disease, referred to as subcutaneous phycomycosis, is characterized by massive, palpable, indurated, nonulcerating, subcutaneous masses on the limbs, trunk, chest, back, or buttocks. The etiologic agent, *Basidiobolus haptosporus* (and possibly other species), is a soil fungus found in leaf detritus and the intestinal tract of amphibians and reptiles. Both diseases are chronic and develop slowly. Although they are seldom life-threatening, they are very disfiguring. In contrast to mucormycosis, an infection caused by species of the Mucorales which are also in the class Zygomycetes, entomophthoromycoses have not yet been associated with underlying predisposing factors and therefore do not appear to be opportunistic infections. The disease mucormycosis is considered in another chapter.

Synonymy. Subcutaneous phycomycosis, rhinophycomycosis, phycomycosis entomophthorae.

Rhinoentomophthoromycosis

Infection by *Entomophthora coronata* has been reported so far only in man and horses. In all cases, the primary infection appears to have originated in the nasal mucosa, and disease is usually restricted to the local subcutaneous tissue. Rarely it may spread to involve the paranasal sinuses, pharynx, facial muscles, and subcutaneous fat. The fungus, which was first isolated in 1897, is a common soil saprophyte found in leaf detritus and is known to be a pathogen of several arthropods, including spiders and insects. The first report of infection in higher animals was made by Emmons and Bridges,[13] who described nasal polyps in horses from Texas. These animals were abnormal in no other way, and it was assumed that the infection was the result of traumatic implantation from vegetable material or from infected insects. The first human case with substantiating mycologic evidence was reported by Bras et al. in 1965.[3] The patient, who was from the Grand Cayman Island, had an extensive swelling of the nasal area. On direct examination of scrapings and biopsy material, broad, thin-walled, aseptate hyphae were seen, and *E. coronata* was isolated in culture. Similar clinical conditions had been noted previously by Ash and Raum in 1956,[1] by Blache et al.[2] from a patient in the Cameroons in 1961, and by several other workers from Africa and Puerto Rico.[6] These were only clinical and histologic reports, however, and the identity of the etiologic agent is unknown. In a review of the subject, Clark[6] has tabulated 26 cases to the year 1967. Fifteen of these were diagnosed without cultural confirmation.

ETIOLOGY, ECOLOGY, AND DISTRIBUTION

The only etiologic agent so far isolated from the culturally confirmed cases of rhinoentomophthoromycosis is *Entomophthora coronata.* This organism has not been isolated from any other type of human disease and appears to be capable of eliciting an infection only in the nasal mucosa. The great variability in clinical response to treatment and the variety of climatic conditions in which affected patients live suggest that more than one species exists which is capable of causing this particular clinical syndrome. Almost all patients (14) have come from the tropical rain forests of Africa, particularly Nigeria. The other recorded cases have been scattered throughout the world: namely, the Caymans, India, Puerto Rico, Senegal, Congo, Brazil, Colombia,[27] and Texas. These reports are too few in number to allow speculation as to the existence or distribution of the virulent form of the organism, since *E. coronata* is found worldwide in warmer climates in moist, decaying leaves.

As is so commonly seen in other systemic and subcutaneous mycoses, the incidence of disease is higher among males than females. Over 80 per cent of the reported cases occurred in males. It is doubtful that this can be attributed to exposure alone. Unlike subcutaneous phycomycosis, in which most patients are children, rhinoentomophthoromycosis is found mainly in adults. Clark[6] notes that the youngest patient was an 8-year-old girl and the oldest a 60-year-old man, but the majority were in the second to fourth decades of life. When they have been recorded, almost all cases of disease have occurred in agricultural workers whose exposure to soil organisms is greater than that in other occupations.

CLINICAL DISEASE

Symptomatology. The usual course described in case reports is that of a nasal swelling beginning in the inferior turbinates, which slowly grows or sometimes rapidly extends to include the submucosa, sutures, ostia, foramina, and paranasal sinuses. Disease is usually bilateral but may be unilateral. The expanding mass causes disfigurement of the adjacent tissue but is usually painless (Fig. 12–1). It is palpable and not attached to the overlying skin but rather anchored to the structures beneath it. The overlying epidermis does not ulcerate, though it may become acanthotic and erythematous. The mass may be uneven and bumpy (Fig. 12–2 FS B 39,*A*), and clog the passage of the nares by pushing the turbinates against the septum. The accompanying edema may extend to include the cheeks, forehead, and lips. The eyelids may be swollen shut, resulting in leukoma of the eye.[26,27] The x-ray picture shows an opaque antrum, obliteration of the nasal air space, and mucosal thickening. The blood count is not elevated, there is no fever, and the patient is otherwise normal.

Differential Diagnosis. The clinical symptoms of rhinoentomophthoromycosis may simulate rhinomucormycosis, which also begins in the nasal mucosa. The latter disease is associated with debilitated patients and runs a rapid, often fatal, course. Entomophthoromycosis simulates other conditions of the nasal area, such as pyogenic abscess, neoplasia, pyomyositis, tuberculosis, dracunculiasis, and onchocerciasis. The correct diagnosis is easily obtained by histologic examination and culture of affected tissue.[34]

A disease which may closely simulate rhinoentomophthoromycosis was first described by Miloshev et al. in 1966.[25] A series of cases of this disease, paranasal granuloma caused by *Aspergillus flavus,* was reported by Mahgoub.[23] The condition is chronic and presents as a unilateral, nodular, painless mass which causes proptosis, destruction of the ethmoid air cells, and opacities in the maxillary antrum seen on x-ray. Nasal obstruction usually does not occur.

PROGNOSIS AND TREATMENT

Unlike rhinomucormycosis, entomophthoromycosis is relatively benign and some-

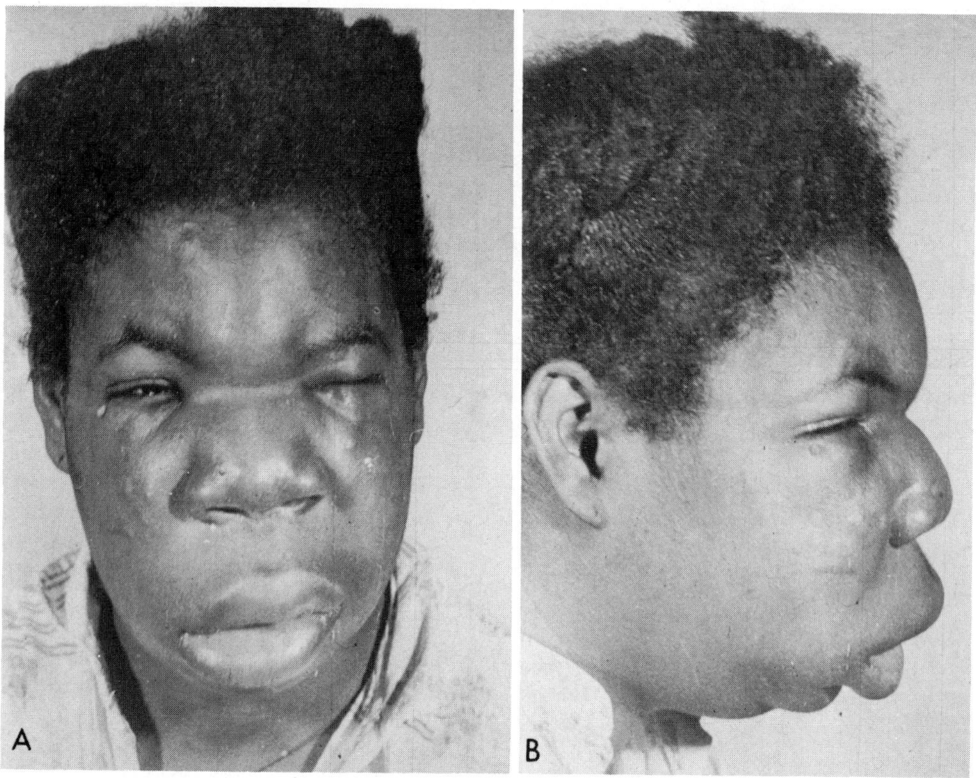

Figure 12–1. Rhinoentomophthoromycosis. Infection by *E. coronata*. *A*, Bilateral distortion of the subcutaneous tissue of the nasal region. *B*, Profile showing distortion. (From Restrepo, A., D. L. Greer, et al. 1967. Am. J. Trop. Med. Hyg., *16*:35.)

times clears spontaneously. However, most patients require treatment with potassium iodide. The lesions generally respond rapidly to this therapy, but a few have required amphotericin B treatment in addition.[24] A case reported by Restrepo et al.[27] did not respond to either potassium iodine or 2 g of amphotericin B.

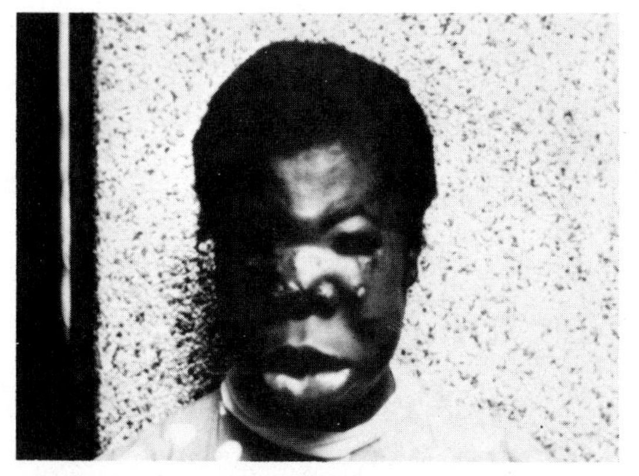

Figure 12–2. FS B 39, *A*. Rhinoentomophthoromycosis. Later stage in same patient. The lesions have regressed in some areas and become nodular in others.

PATHOLOGY

Gross Pathology. When they are removed surgically, the submucosal masses are firm, whitish in color, and have linear strands of fibrous tissue coursing through them. Focal areas of yellowish material are seen which simulate caseous necrosis. Fluid expressed from these areas contains abundant hyphal elements.

Histopathology. The histology of rhinoentomophthoromycosis and subcutaneous phycomycosis is identical and contrasts sharply to that of mucormycosis. The two major points of difference between entomophthoromycosis and mucormycosis are (1) the eosinophilic sheath (Splendore-Hoeppli phenomenon) found around the hyphae in the former disease, and (2) the lack of vascular invasion by the Entomophthorales so characteristic of the pathology found in mucormycosis. In addition, the hyphal elements of fungi causing mucormycosis are aseptate in tissue, whereas regular septation is seen in entomophthoromycosis.

In all "Phycomycete"[6] diseases, hyphal ele-

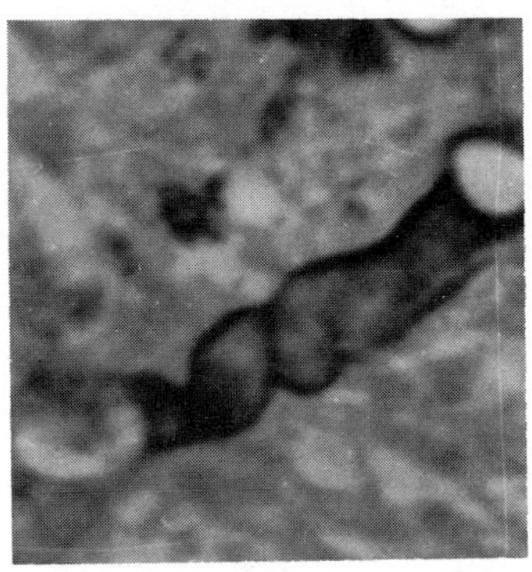

Figure 12–3. Rhinoentomophthoromycosis. Hyphae in tissue. (Courtesy of D. Greer.)

ments are readily stained in routine hematoxylin-eosin preparations. In contrast, with special fungus stains, such as Gridley and PAS, the hyphal elements are not well stained; therefore, these procedures are not of sufficient help to warrant their use. Even the very sensitive GMS stain may only stain the fungi weakly. The hyphae of *E. coronata* in tissue may lie singly or in clusters (Fig. 12–3). Their mean diameter is 8 μ, with variations between 4 μ and 10 μ and sometimes up to 22 μ. They are regularly septate but have broader hyphal units and are fewer in number than fungal elements seen in aspergillosis. The walls are thin but easily defined and surrounded by bright, radiating, granular, eosinophilic material which may be between 2 μ and 6 μ in thickness. The eosinophilic material has a fringelike or satellite arrangement and is similar to that seen in some cases of sporotrichosis (asteroid bodies), schistosomiasis, coccidioidomycosis, blastomycosis, paracoccidioidomycosis, and, rarely, in candidosis. The hyphae do not have a predilection for any particular site in infected tissues. They do not infiltrate the walls of blood vessels, are not seen in the lumen of vessels, and, in contrast to mucormycosis, there is no vascular thrombosis. Panarteritis and endarteritis are sometimes observed, however.

The cellular reaction evoked by the infection is an acute or chronic inflammatory reaction or a combination of the two. The chronic reactions tend to be of the granulomatous type. The organisms are not always found in the midst of the exudate but are seen near it. The acute reaction consists of masses of eosinophils, lymphocytes, and variable numbers of plasma cells. Neutrophils and fibroblasts are less commonly found than in chronic lesions. Vascular proliferation occurs.[34] The eosinophils may be so numerous as to form an eosinophilic abscess, as seen in some infections caused by helminths, particularly microfilariae. For this reason, cases of entomophthoromycosis were once regarded as worm infestations.

The chronic inflammatory reaction is characterized by granulomatous infiltrates. The chief cellular components are foreign body giant cells containing phagocytized hyphal elements, histiocytes, lymphocytes, and some eosinophils and plasma cells. Peripherally there is fibroblastic activity, and large collagen cords are sometimes seen. There is no caseation or coagulation necrosis. Hypha viewed in cross section may have the appearance of empty coccidioidal spherules. In hematoxylin-stained sections these elements may closely resemble old, heavily mineralized capillaries with which they sometimes have been confused.

The eosinophilic precipitate surrounding the fungal elements in tissue is probably due to an antigen-antibody reaction. Histochemical tests demonstrate the presence of phospholipids, neutral lipids, PAS-positive material, and globulins. The precipitate has a yellow autofluorescence which is useful for observing the organisms when they are sparse in tissue.

Natural animal disease has been reported only in the horse.[13, 27a] The lesions were described as nasal polyps and large granulomata of the nasal cavity. Attempts at experimental infection of several animals have so far been unsuccessful.[7]

IMMUNOLOGY AND SEROLOGY

Natural resistance to infection is high, as the etiologic agent is found worldwide, but the disease is rare. There is no satisfactory serologic procedure for diagnosis, so the true extent of subclinical disease or of cases of spontaneous remission is unknown.

LABORATORY DIAGNOSIS

Direct Examination. To provide material for direct examination soft intact vesicles are punctured, or scrapings of the affected nasal mucosa are made. In case of infection, a potassium hydroxide mount reveals broad hyphae with occasional septations (Fig. 12–4).

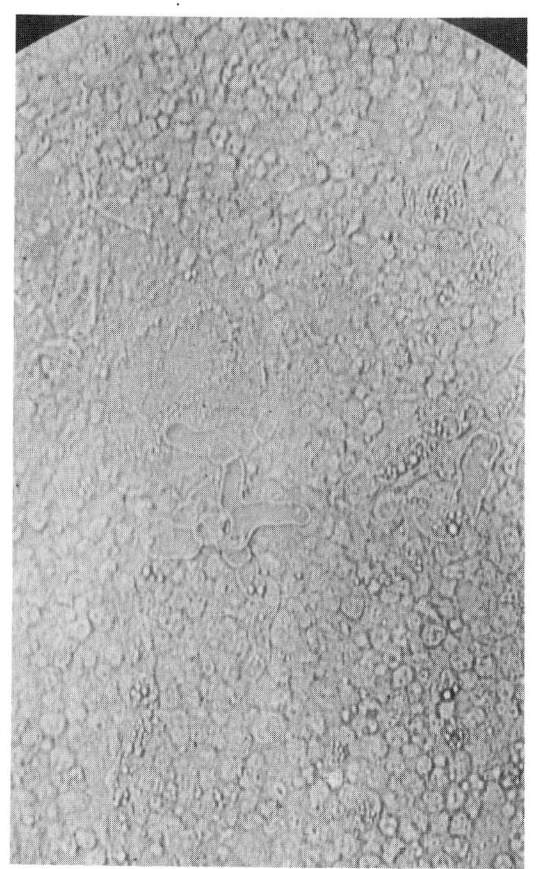

Figure 12–4. Rhinoentomophthoromycosis. Direct examination of scraping from nasal mucosa. (From Restrepo, A., D. L. Greer, et al. 1967. Am. J. Trop. Med. Hyg., *16*: 34–39.)

The hyphal walls are doubly refractile, and there is some branching. Granular inclusions are readily seen.

Culture Methods. Pathologic material for culture can be gently broken apart in a solution containing 500 mg of streptomycin and 1 megaunit of penicillin. It is then plated on Sabouraud's agar with or without antibacterial antibiotics and incubated at 25° or 37°C. Mycosel and other media containing cycloheximide cannot be used. Growth can be observed at 48 hours.[14]

To isolate the fungus from soil or leaf detritus, the material is shaken in sterile saline and allowed to stand. The supernatant is decanted and pipetted onto sterile Whatman No. 1 filter paper, which is then inserted in the cover of a Petri dish and placed over a plate containing Sabouraud's agar. The Petri dish is inverted and incubated at 25°C. Leaves can be pressed directly onto moistened filter paper inside the lid of such a plate. Conidia are propelled from the lid to the agar, and visible colonies may be observed in 24 to 72 hours.[6]

MYCOLOGY

Entomophthora coronata (Constantin) Kevorkian 1937

Synonymy. *Delacroixia coronata* Constantin 1897; *Conidiobolus coronatus* Tyrrell and MacLead 1972,[22a]

Colony Morphology. The colony grows rapidly and is glabrous and adherent at first. Furrows and folding occur, particularly when the organism is grown at 37°C. In time, the colony becomes covered with short, white, aerial mycelia and conidiophores (Fig. 12–5). The sides of the lid of the dish or tube soon become covered with conidia which are forcibly discharged by the conidiophores. The color of the colony becomes tannish to light brown with age.

Microscopic Morphology. No sporangia are formed, but large conidia (25 to 45 μ in diameter) are produced on the end of short, erect conidiophores. The conidia are ejected and travel distances up to 30 mm. If they land on nutrient medium, they will germinate and produce one or more hyphal tubes. The conidia differ from those of *Basidiobolus* sp. because they have prominent papillae on the wall which may give rise to

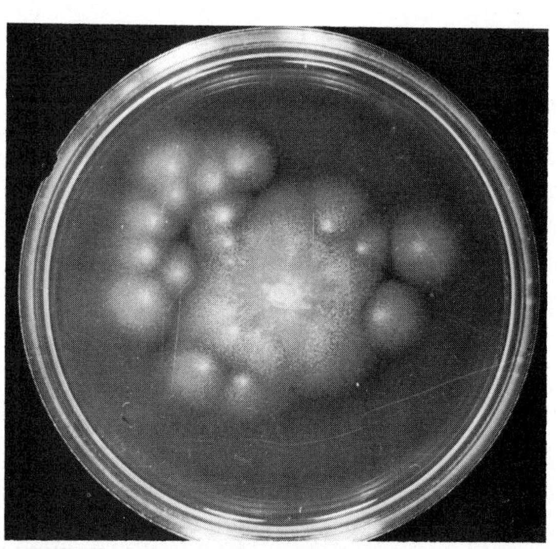

Figure 12–5. Rhinoentomophthoromycosis. Culture of *E. coronata.*

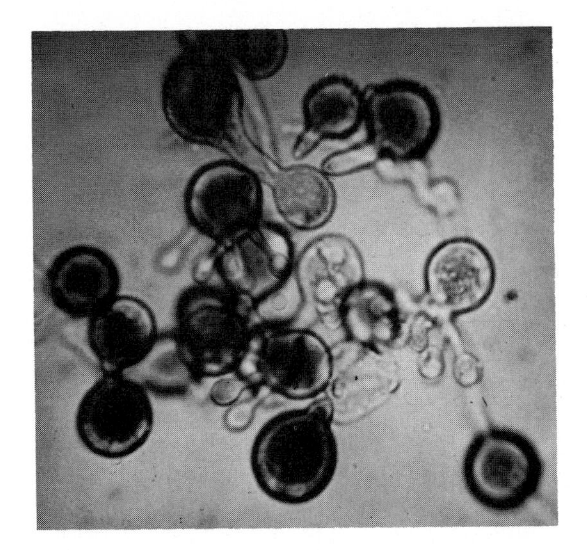

Figure 12–6. Rhinoentomophthoromycosis. Corona of papillae on conidia of *E. coronata*. (Courtesy of D. Greer.)

secondary conidia. Several papillae and conidia may be produced, giving the original spore a corona of secondary conidia (Fig. 12–6). A conidium may also produce multiple, short, hairlike appendages. Conidia falling on glass will produce a short conidiophore and eject another conidium. This conidium will in turn germinate to produce another conidium. This process repeats itself until stored nutrients are exhausted. Many chlamydospores are also present in the colonies, but zygospores are rare.[22a]

Subcutaneous Phycomycosis

The first report of this disease in which there was mycologic confirmation came from Indonesia. In 1956, Lie-Kian-Joe et al. described three cases in which there were extensive palpable masses in the subcutaneous tissue and muscle fascia.[20] The disease spread widely to involve the neck, arms, and upper chest, but eventually the lesions healed spontaneously. The etiologic agent was identified as *Basidiobolus ranarum,* but has since been reclassified as *B. haptosporus.* Previously the disease had been reported to occur in a horse[33] and possibly a man.[5] After the first report by Joe, several more cases were found in Indonesia.[21,32] These were more severe and did not heal spontaneously. Two additional patients with the disease were encountered in England;[30] one of these had lived in Indonesia. Since then many cases have been reported from Uganda[4] and India[21,23] as well. The mycosis is now known to occur in many tropical African countries and Southeast Asia. So far no cases have come from the Americas.

The disease is primarily one of the subcutaneous tissue, but on occasion deeper structures have been invaded. Two deaths have been noted; in one patient the large intestine and pelvis were involved, and in the other the neck.

The disease syndrome became well known under the name subcutaneous phycomycosis, and there is no sufficient reason to change it or group it with the entirely different clinical entities caused by other phycomycetous fungi. Therefore, the term as used here will be restricted to infections by *Basidiobolus* species.

ETIOLOGY, ECOLOGY, AND DISTRIBUTION

The original isolant from the case of Lie-Kian-Joe[20] was identified by Emmons as *B. ranarum* because some rough-walled zygospores were seen. The classification of the genus is tenuous and based on the morphology of the zygospore coat and thermotolerance of the fungus. Greer and Friedman[16,17] found only smooth-walled zygospores in human isolants and called the fungus *B. meristosporus* differentiating it from *B. haptosporus* only by the thermotolerance of the former. Drechsler,[9,10] who created the species *meristosporus,* was unsure of the correct epithet. Srinivasan and Thirumalachar[28] concluded that the organisms were *B. haptosporus.* Since there is great variability in thermotolerance of isolants from soil or from human cases, Clark and Greer have now concluded that the name *B. haptosporus* is correct and that *B. meristosporus* should be considered in synonymy with it (personal communication).

B. haptosporus is a ubiquitous species which occurs in decaying vegetation and the gastrointestinal tract of many reptiles and amphibians. It does not appear to cause any disease in these animals. It has not yet been isolated from insects, but conidia sometimes become attached to insects, and they may serve as vectors for dissemination. The fungus does not tolerate cold and dies quickly if kept in a cold environment. The organism has been isolated from tropical and subtropical environments throughout the world.[6,6a]

Subcutaneous phycomycosis is primarily a disease of children. Of the 78 cases tabulated by Clark,[6] 46 per cent of the patients were 10 years old or younger, and 82 per cent were under 20 years of age. The male-female ratio varied between 3 to 1 and 6 to 1, which tends to demonstrate a predominance of disease in male children. However, too few cases have been recorded to delineate patterns of sex distribution. Of the cases tabulated by Clark,[6] the largest number were reported from Uganda (35), Nigeria (17), and Indonesia (9). Other countries in which cases were reported include India, Burma, Senegal, Kenya, Ghana, Iraq, and Sudan. Tio[32] has since reported 12 more cases from Indonesia, and the worldwide total is now well over 100.

CLINICAL DISEASE

Symptomatology. The infection begins as a subcutaneous nodule which gradually increases in size. The portal of entry of the fungus is unknown, but in a few cases mosquito or other insect bites were noted before the onset of symptoms, which may indicate there is an arthropod vector. Introduction of the agent by an intestinal route has also been suggested. The subcutaneous swelling has a firm consistency and is well circumscribed and painless, though there may be some pruritus. The mass is palpable and attached to the overlying skin but not the underlying muscle fascia. When fingers are placed under the edge of the mass, it is found to be freely moveable. The skin tends to be atrophic and discolored or hyperpigmented, but it does not ulcerate. The mass continues to grow and sometimes involves the whole shoulder, arm and upper body, face and neck, or the entire leg and buttocks (Fig. 12–7 FS B 39, *B*). In a few cases, involvement of underlying organs, such as liver, intestine, and muscle, has been reported. These infections were probably systemic mucormycosis rather than subcutaneous phycomycosis, as no cultural confirmation was obtained. There are no reports of involvement of regional lymph nodes or hematogenous or lymphatic dissemination. A few cases regress spontaneously, but most require iodide or amphotericin B treatment. The almost exclusive restriction of disease to the subcutaneous tissue has suggested to some that the organism is lipophilic.[6] In

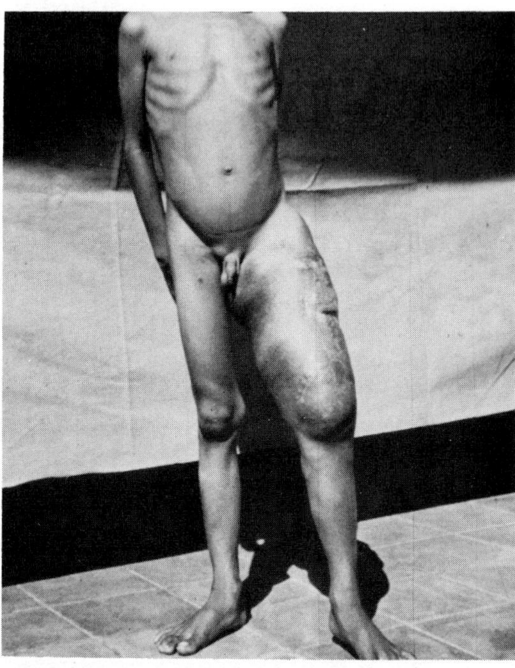

Figure 12–7. FS B 39, *B*. Subcutaneous phycomycosis. Subcutaneous lesion involving entire thigh and buttock. There was secondary bacterial infection following a biopsy. (Courtesy of C. Halde.)

the cases recorded by Tio[32] there was a marked leukocytosis (up to 29,000) and eosinophilia (up to 30 per cent); no other abnormalities or predisposing factors were noted.

Differential Diagnosis. The clinical history along with the histologic picture is so characteristic that the diagnosis is readily made. Cultural confirmation is always necessary. There are a variety of diseases which simulate this condition. These include neoplasias, pyogenic abscess, elephantiasis, and helminth infections. Another disease common in Uganda which simulates subcutaneous phycomycosis is Buruli ulcer. This is caused by *Mycobacterium ulcerans,* which also appears to be a lipophilic organism. The disease it produces also spreads slowly and involves extensive areas of subcutaneous tissue.

Histologically, the rare infections caused by *Mortierella* sp. resemble those of rhinoentomophthoromycosis and subcutaneous phycomycosis. There is a marked eosinophilia, and sparse, poorly staining hyphae are present. An eosinophilic sheath surrounding the hyphae may be seen.

PROGNOSIS AND THERAPY

The prognosis of subcutaneous phycomycosis is generally good. Among the more than one hundred cases reported, only two deaths have been directly attributable to the disease.[6] The gross disfigurement produced resolved, with little residual disease. Although a few cases regress spontaneously, most require treatment. Therapeutic trials with nystatin and griseofulvin have been unsuccessful, and the organism is insensitive to the drugs *in vitro.*[32] Iodide therapy (potassium iodide, 30 mg per kg body weight per day for a month or more) has been remarkably successful. Amphotericin B has been used in a few patients who had taken iodides for up to nine months with little or no effect. The response to this drug was disappointing in most instances.

In cases in which iodide therapy has been effective, the healing process was accompanied by diminution of the eosinophilic sheath around the hyphae and gradual disappearance of hyphae. The tuberculoid granuloma remains for some time, and necrotic tissue is replaced by fibrosis.

PATHOLOGY

Gross Pathology. In biopsy specimens the overlying skin is seen to be atrophic but rarely ulcerated. The subcutaneous mass itself is firm and thickened with fibrous tissue containing occasional yellowish foci of necrosis. Masses of mycelium in a suppurative exudate may be expressed from these areas.

Histopathology. The histologic appearance of subcutaneous phycomycosis is identical to that of rhinoentomophthoromycosis. The hyphal elements (10 to 40 μ) may be sparse in the granuloma, but the large eosinophilic sheath and its autofluorescence are helpful in locating the organism. The hyphae stain poorly with all stains, and it is only the eosinophilic sheath that indicates their presence. *Mortierella* sp. in tissue also have an eosinophilic sheath, and collection of eosinophils is found in the surrounding tissue.[30]

ANIMAL DISEASE

Natural infection in animals has been recorded only in horses.[33] The organism seems to be a natural inhabitant of the intestinal tract of reptiles but is not associated with any clinical disease. Attempts at experimental infection of several animals have not been successful.[17,32]

IMMUNOLOGY AND SEROLOGY

The rarity of infection in spite of the probable common exposure of man to this fungus in tropical and subtropical environments indicates a high natural resistance to this disease. No satisfactory serologic procedures are available as yet to allow investigation of the extent of subclinical or spontaneously resolved infections. Greer and Friedman[18] found antigenic differences between *B. meristosporus (B. haptosporus)* and *B. ranarum.* Their studies indicated that there was a common antigenic identity among several isolants of *B. haptosporus* from human disease, although there is the possibility that more than one species exists. Chemical analysis of the polysaccharides of the two species of *Basidiobolus* did not reveal significant differences.[15]

LABORATORY IDENTIFICATION

Direct Examination. Examination of biopsy material macerated in a potassium hydroxide mount demonstrates the same general findings noted for rhinoentomophthoromycosis. Broad, septate, branching hyphal elements are seen. There are numerous inclusions within the cell and doubly refractile cell walls.

Culture Methods. Biopsy material may be plated on Sabouraud's agar or any media not containing Actidione. In the experience of Tio,[32] growth is more easily obtained from large lumps of tissue than from thin slices or macerated pieces. The plates are incubated at 25° to 30°C. Growth is observable in 48 to 72 hours, and colonies attain a diameter of 7 cm in four days. Isolation of the organism from leaf detritus and the intestinal contents of reptiles is accomplished in the same manner described for *E. coronata.*

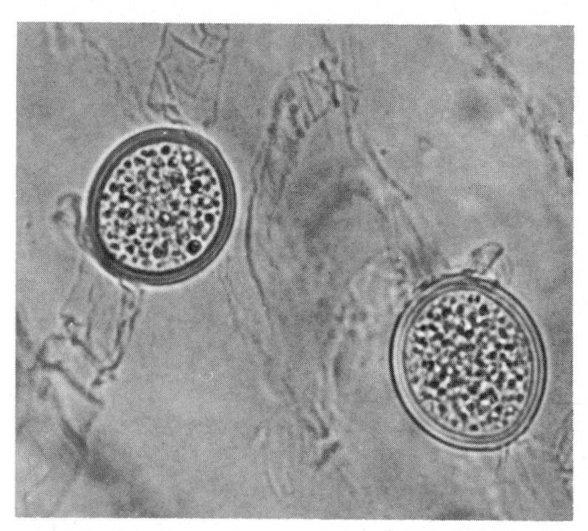

Figure 12–8. Subcutaneous phycomycosis. Zygospores of *B. haptosporus* showing "beak." (Courtesy of D. Greer.)

MYCOLOGY

Basidiobolus haptosporus Drechsler 1947

Synonymy. *B. meristosporus* Drechsler 1955;[8-10] *B. heterosporus* Srinivasan and Thirumalachar 1965;[29] *B. haptosporus* var. *minor* Srinivasan and Thirumalachar 1965.[29]

Colony Morphology. Almost all strains grow rapidly at 30°C. Ability to grow at 37°C is variable, and the same isolant will differ in its response from one plating to another.[6] The colony is flat, folded, furrowed, greyish, and waxy in consistency. Large vegetative hyphae (8 to 20 μ) are formed, which become increasingly septate as production of spores begins.

Microscopic Morphology. After seven to ten days, the colony becomes overgrown with mycelium as masses of zygospores, chlamydospores, and sporangia are formed. The zygospores are from 20 to 50 μ in diameter and have a smooth, thick wall. Sometimes slight undulations are seen in young spores. In contrast, *B. ranarum* consistently has undulating walls and produces a streptomyces-like odor. The zygospores of all *Basidiobolus* species have a prominent "beak" attached to one side, representing the remnants of the copulatory tubes (Fig. 12–8). Sporangiophores (conidiophores) are produced, each of which bears a single-celled sporangium (conidium) that is 20 to 45 μ in diameter. This, together with fragments of the sporangiophore (referred to as "basidia"), is forcibly ejected and collects on the Petri dish top.[19] Sometimes the cell within the conidium divides to form meristospores. Secondary conidia may be produced by primary conidia until stored nutrients are exhausted. After prolonged growth on laboratory medium, zygospore and conidia production diminishes, and large, thick-walled chlamydospores (20 to 40 μ) are produced in great numbers.[14]

REFERENCES

1. Ash, J. E., and M. Raum. 1956. In *An Atlas of Otolaryngic Pathology.* Washington, D.C., Armed Forces Institute of Pathology, p. 179.
2. Blache, R. P. Destombes, et al. 1961. (New subcutaneous mycoses in Southern Cameroons.) Bull. Soc. Pathol. Exot., *54*:56–63.
3. Bras, G., C. C. Gordon, et al. 1965. A case of phycomycosis observed in Jamaica; infection with *Entomophthora coronata.* Am. J. Trop. Med. Hyg., *14*(1):141–145.
4. Burkett, D. P., A. M. M. Wilson, et al. 1964. Subcutaneous phycomycosis. B. Med. J., *1*:1669–1672.
5. Casagrandi, C. 1931. Sur la presence de Basidioboles dans l'homme. Bull Soc. Interanz. Sez. Ital., *3*:399–400.
6. Clark, B. M. 1968. The epidemiology of phycomycosis, in Wolstenholme, G. E., and R. Porter (eds.), *Symposium on Systemic Mycoses.* Boston, Little, Brown and Co., pp. 179–192.

6a. Coremans-Pelsener, J. 1973. Isolation of *Basidiobolus meristosporus* from natural sources. Mycopathologia, *49*:173–176.

7. Della Torre, B., and L. Mosca. 1965. Experimental phycomycosis in rodents. Mycopathologia (Den Haag)., *26*(4):417–452.
8. Drechsler, C. 1947. A *Basidiobolus* producing elongated secondary conidia with adhesive beaks. Bull. Torrey Botan. Club, *74*:403–413.
9. Drechsler, C. 1955. A southern *Basidiobolus* forming many sporangia from globose and from elongated adhesive conidia. J. Wash. Acad. Sci., *45*:49–56.
10. Drechsler, C. 1958. Formation of sponangia from conidia and hyphal segments in an Indonesian *Basidiobolus*. Am. J. Botan., *45*:632–638.
11. Eidam, E. 1886. Basidiobolus, eine neue Gattung der Entomophthoraceen. Cohn's Beitr. Biol. Pflanzen, *4*:181–251.
12. Emmons, C. W., L.-K. Joe, et al. 1957. *Basidiobolus* and *Cercospora* from human infections. Mycologia, *49*:1–10.
13. Emmons, C. W., and C. H. Bridges. 1961. *Entomophthora coronata*, the etiological agent of a phycomycosis of horses. Mycologia, *53*:307–312.
14. Greer, D. L. 1970. Phycomycetes, in Blair, J. E. (ed.), *Manual of Clinical Microbiology*. Bethesda, Maryland, American Society for Microbiology, Chapter 45.
15. Greer, D. L., and E. Barbosa. 1967. Some chemical constituents of cell-bound and extracellular polysaccharides of *Basidiobolus ranarum*, isolated from nature, and *B. meristosporus*, isolated from subcutaneous phycomycosis. Sabouraudia, *5*:329–334.
16. Greer, D. L., and L. Friedman. 1964. Effect of temperature on growth as a differentiating characteristic between human and non-human isolants of *Basidiobolus* species. J. Bacteriol., *88*:812–813.
17. Greer, D. L., and L. Friedman. 1966. Studies on the genus *Basidiobolus* with reclassification of the species pathogenic for man. Sabouraudia, *4*:231–241.
18. Greer, D. L., and L. Friedman. 1966. Antigenic relationships between the fungus causing subcutaneous phycomycosis and saprophytic isolates of *Basidiobolus meristosporus* and *B. ranarum*. Sabouraudia, *5*:7–13.
19. Ingold, C. T. 1934. The spore discharge mechanism in *Basidiobolus ranarum*. New Phytologist, *33*:273–277.
20. Joe, L.-K., T. E. Njo-Injo, et al. 1956. *Basidiobolus ranarum* as a cause of subcutaneous phycomycosis in Indonesia. Arch. Dermatol., *74*:378–383.
21. Koshi, G., T. Kurien, et al. 1972. Subcutaneous phycomycosis caused by Basidiobolus. A report of three cases. Sabouraudia, *10*:237–243.
22. Klokke, A. H., C. K. Job, et al. 1966. Subcutaneous phycomycosis in India. Trop. Geogr. Med., *18*:20–25.
22a. MacLeod, D. M., and E. Muller-Kogler. 1973. Entomogenous fungi: Entomophthora species with pear-shaped to almost spherical conidia (Entomophthorales: Entomophtoraceae). Mycologia, *65*:823–893.
23. Mahgoub, E. S. 1971. Aspergillosis in Sudan. C. R. V[e] Congres de la Soc. Inter. de Mycol. Hum. Anim., pp. 173–174.
24. Martinson, F. D., and B. M. Clark. 1967. Rhinophycomycosis entomophthoral in Nigeria. Am. J. Trop. Med. Hyg., *16*:40–47.
25. Miloshev, B., S. Mahgoub, et al. 1966. Aspergilloma of paranasal sinuses and orbit in Northern Sudanense. Lancet, *1*:746–747.
26. Renoirte, R., J. Vandepitte, et al. 1965. Phycomycose nasofaciale (rhinophycomycose) due to an *Entomophthora coronata*. Bull. Soc. Pathol. Exot., *58*:847–862.
27. Restrepo, A., D. L. Greer, et al. 1967. Subcutaneous phycomycosis: Report of the first case observed in Colombia, South America. Am. J. Trop. Med. Hyg., *16*:34–39.
27a. Restrepo M., L., L. F. Morales, et al. 1973. Rinoficomicosis por *Entomophthora coronata* en equinos. Antioquia Medica, *23*:13–25.
28. Srinivasan, M. C., and M. J. Thirumalachar. 1965. Basidiobolus species ·pathogenic for man. Sabouraudia, *4*:32–34.
29. Srinivasan, M. C., and M. J. Thirumalachar. 1967. Studies on Basidiobolus species from India with discussion on some of the characters used in the speciation of the genus. Mycopathologia (Den Haag), *33*:56–64.
30. Symmers, W. S. C. 1960. Mucormycotic granuloma possibly due to *Basidiobolus ranarum*. B. Med. J., *1*:1331–1333.
31. Thaxter, R. 1888. The entomophthoreae of the United States. Mem. Bost. Soc. Nat. Hist., *4*(6):133–201.
32. Tio, T. H., M. Djojopranoto, et al. 1966. Subcutaneous phycomycosis. Arch. Dermatol., *93*:550–553.
33. Van Overeem, C. 1925. Beitrage zue Pilzflora von Niederlandisch Indian. 10. Ueber ein merkwuer diges Vorkommen von *Basidiobolus ranarum*. Eidana. Bull. Jardin Bot. Buitzern Borg. Ser. III, *7*:423–431.
34. Williams, A. O. 1969. Pathology of phycomycosis due to Entomophthora and Basidiobolus species. Arch. Pathol., *87*:13–20.

Chapter 13

LOBOMYCOSIS

DEFINITION

Lobomycosis is a chronic, localized, subepidermal infection characterized by the presence of keloidal, verrucoid, nodular lesions or sometimes of vegetating crusty plaques and tumors. The lesions contain masses of the spheroidal, yeastlike organism *Loboa loboi.* There is no systemic spread. The disease has been found in man and dolphins.

Synonymy. Keloid blastomycosis, Lobo's disease.

HISTORY

The disease was first described in 1931 by Jorge Lobo.[15] The patient, who was an Indian from the Amazon valley, had cutaneous keloidal lesions without lymphangitis or systemic symptoms. A fungus isolated from the patient appears to have been lost. The isolant, which is listed as J. B. 525 in the Instituto Oswald Cruz, is purported to have been isolated from the first case of the disease. It is now regarded as a mislabeled strain of *Paracoccidioides brasiliensis.*[4,10] Lobo considered the disease a mild form of paracoccidioidomycosis, but subsequent clinical, histologic, and mycologic investigation has delineated it as a separate disease.[14] The two infections may coexist in the same patient, however.[11] Since the time of its original description, lobomycosis has been reported sporadically throughout the American tropics in man and recently in dolphins from the waters of Florida and the Caribbean (Fig. 13–1). A great deal of literature has been devoted to the nomenclature of the organism,[3,12] but the controversy will not be resolved until the etiologic agent is cultured. The disease was called "keloid blastomycosis," but lesions other than the keloidal type have been found in patients.[23] Although "Lobo's disease" has been proposed as the name for the infection, the term "lobomycosis" is preferred, as it more clearly defines this entity. The final designation of the disease awaits successful isolation and study of the etiologic agent.

ETIOLOGY, ECOLOGY, AND DISTRIBUTION

The etiologic agent of lobomycosis has never been isolated in culture. Because of this, the final designation of its true taxonomic relationships cannot be made. Discussion of its nomenclature is a moot point until then. Since some term must be used for the organism, the one proposed by Ciferri,[5] *Loboa loboi,* is retained, as it has become widely adopted and implies no relationship to other organisms. Most of the other names proposed have been based on studies of contaminants.

The ecology of the organism also remains a mystery. Until recently, the disease was known only in humans. There seemed to be a common denominator of habitation in tropical rain forests and bush country, particularly the Mato Grosso in Brazil and areas of Surinam, where most of the infections have occurred. The discovery of the disease in dolphins in a marine environment makes previous assumptions as to the saprophytic nature of the fungus difficult to maintain.[17] One possibility that might account for both its inability to be cultured *in vitro* and its

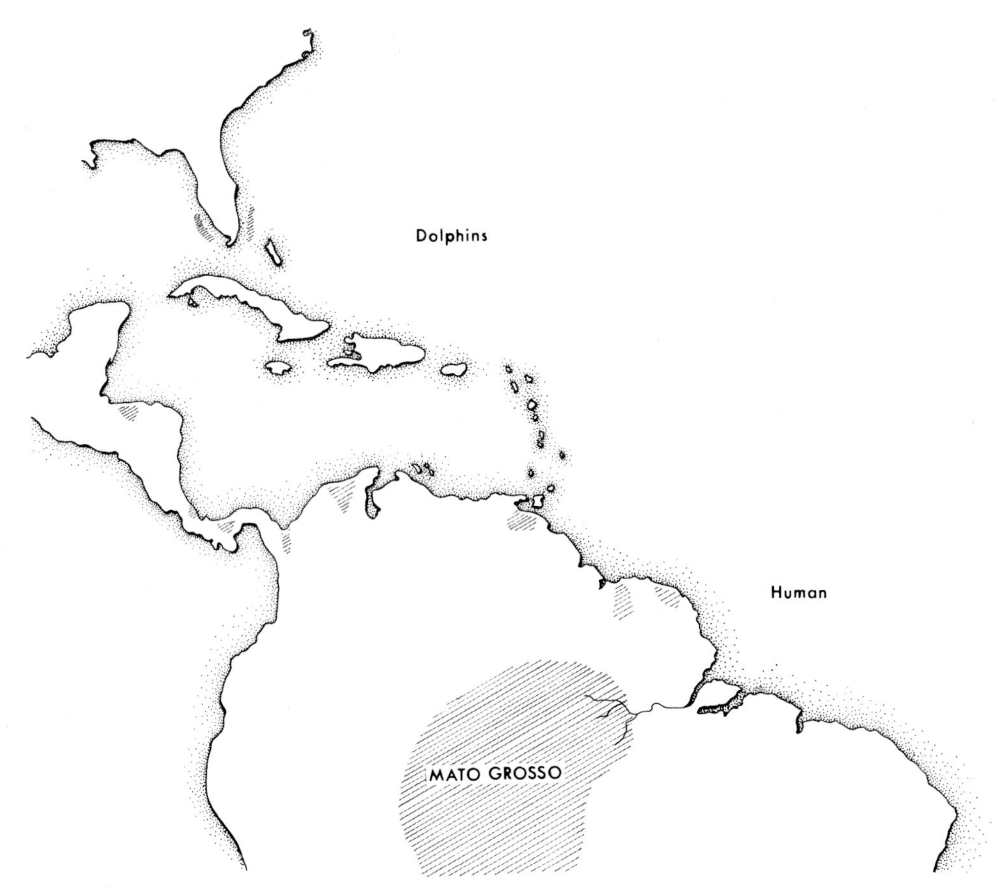

Figure 13–1. Distribution of lobomycosis based on case reports of the disease in man and dolphins.

clinical restriction to the cooler parts of the body is that the organism is an obligate parasite of some lower animal form.

Most of the human cases of lobomycosis have been in patients living in the Amazon valley of Brazil. Lacaz[10] tabulated 69 cases from this region up to 1970. The Caiabi Indians have a legend that traces the introduction of the malady into their community by diseased children captured from the Ipeni tribe.[1] The second largest series of cases is reported from Surinam. In this study, Wiersema[23] recorded 13 cases in Negroes, most of whom lived along the Saramaccaner River in the interior of the country. Scattered reports also place the disease in Colombia,[22] Venezuela,[2] French Guiana,[20] Panama, Costa Rica, and Honduras. All patients but one have been adult males, and all have been agricultural workers. The age at acquisition is difficult to assess, as the infection has a protracted and benign course.

CLINICAL DISEASE

In cases in which it has been possible to determine, the initial infection occurs at the site of some trauma to the skin.[23] The lesions begin as small, hard nodules resembling keloids that are sharply defined, freely moveable, and have a smooth surface. The color of the nodules may be slightly brownish with telangiectasia. The surrounding cutaneous area is normal, and erythema is absent. The lesions are painless or slightly pruritic. The infection is restricted to the subepidermal area and does not penetrate into the subcutaneous tissue or spread internally. The disease may be transferred to other areas of the skin by subsequent abrasions and autoinoculation,[23] in which case groups of nodules may be found on several different areas of the body (Fig. 13–2 FS B 40). The observation that groups of lesions occur on different regions of the body has led to speculation that hematogenous or lymphatic

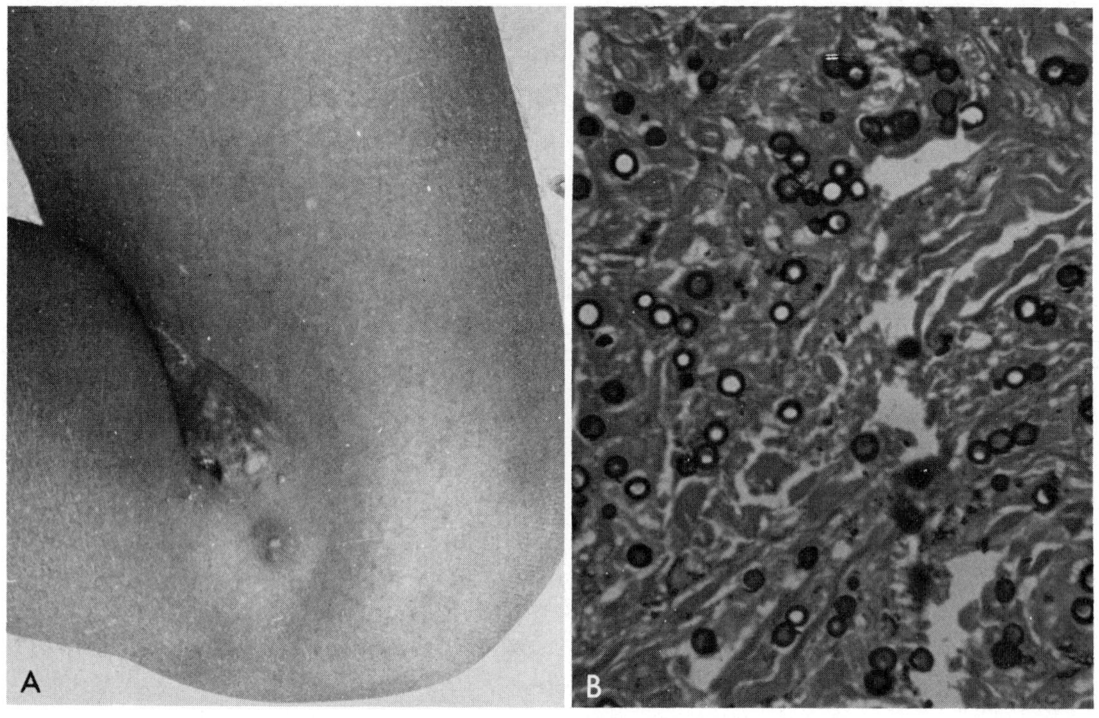

Figure 13–2. FS B 40. Lobomycosis. *A*, Old subepidermal nodules and recent satellite lesions. The latter are probably the result of autoinoculation by scratching. (Courtesy of F. Battistini.) *B*, Biopsy of lesion, demonstrating numerous yeast cells. GMS. ×400.

spread of disease may take place. There is no evidence to support this. Regional lymph nodes are not involved, and reports of this probably represent mechanical dispersion of fungi following injury. The organism is so large that penetration through uninjured skin seems improbable. Lesions spread slowly in the dermis and continue to develop over a period of many years. Older lesions become verrucoid and may ulcerate. The areas of involvement may be quite extensive, since spread by autoinoculation occurs over the years.

There appears to be no tendency for lesions to heal spontaneously. In patients with a long-term history of disease, mature verrucoid lesions are found in some areas, while young hard nodules are found in others. The favored body sites are exposed areas, such as ears, legs, arms, and face (Figs. 13–3 to 13–5).

The differential diagnosis depends on the age of the lesions. Young, sharply defined nodules resemble keloids or fibromas. The older verrucoid lesions are similar to those of chromomycosis, mycetoma, chronic pyogenic vegetations, or granulomas. Since lesions are not macroscopically distinctive, histologic examination is necessary for definitive diagnosis.

Lobomycosis causes little discomfort to the patient, and there are no generalized symptoms. In some cases, there is a history of the presence of lesions for 30 or 40 years. There is no evidence of predisposing factors, and patients appear otherwise normal.

The most successful treatment for the condition is wide surgical excision of the affected area. Care should be taken to prevent contamination of surgical wounds, as relapse is not uncommon. The sulfa drugs, especially sulfadimethoxine, have been used for therapy with varying results.

PATHOLOGY

The typical nodules consist of subepidermal granulomas that usually spare the overlying skin and do not penetrate into deep subcutaneous tissue. The epidermis is atrophic and shows neither the pseudoepitheliomatous hyperplasia nor the intraepidermal abscesses that are common in blasto-

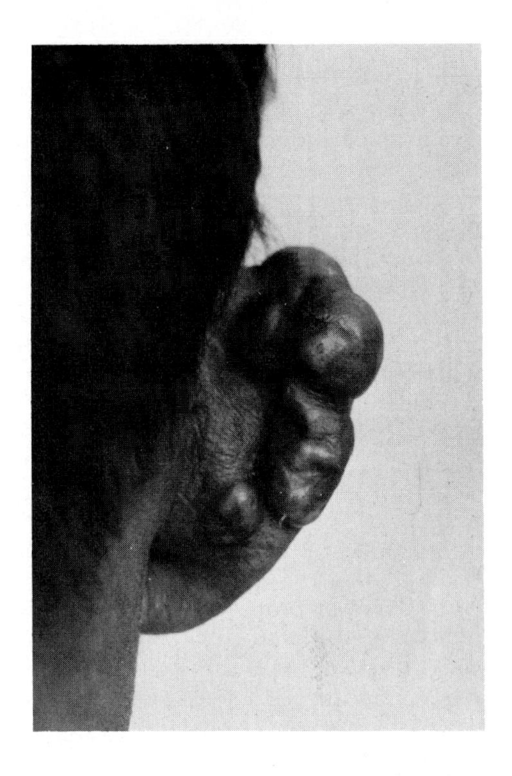

Figure 13–3. Lobomycosis. Nodular lesion on ear of 30 years' duration. (Courtesy of F. Battistini.)

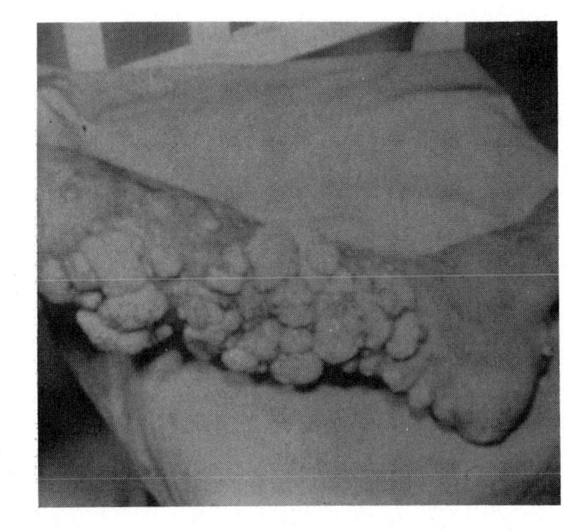

Figure 13–4. Lobomycosis. Extensive verrucoid lesions on legs.

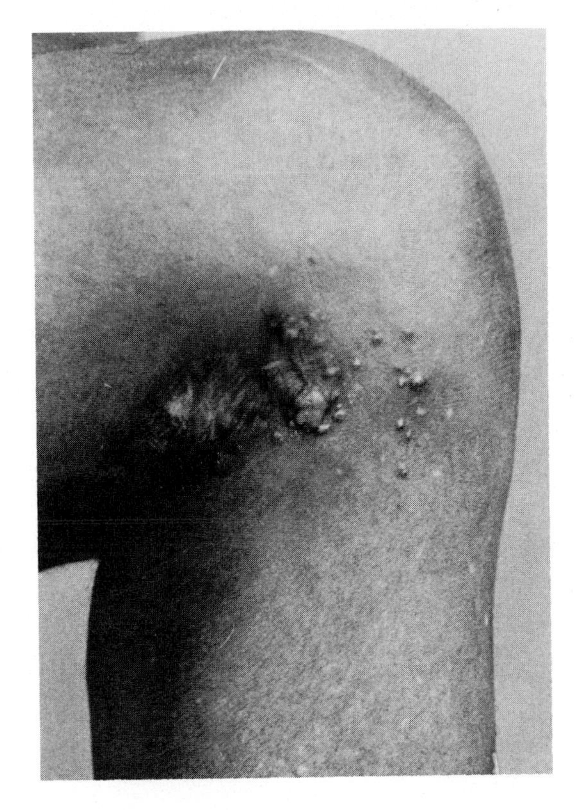

Figure 13–5. Lobomycosis. Old nodular lesions and new satellite lesions representing autoinoculation into new areas. This is the same patient as in Fig. 13–2,*A* after two more years. Note new crops of nodules.

mycosis and coccidioidomycosis. The histology of the subepidermal granuloma is variable. In the usual case, there is extensive hyaline fibrosis interspersed with masses of histiocytes and giant cells. The proportion of giant cells present increases with the age of the lesion. There is no evidence of necrosis or suppuration. The fungi are very numerous in the lesions and are mostly located within giant cells and macrophages. Giant cells with a diameter of 40 to 80 μ may contain up to ten yeast cells. The yeast cells are uniform in size (about 10 μ), and generally form a single chain. They are connected one to the other by short, bridgelike structures (Fig. 13–6). Chains of 20 or more yeast cells with several branching points are sometimes found outside the macrophages. The wall of the fungus cell is 1 μ in thickness. In hematoxylin and eosin–stained preparations, the walls are poorly colored, and the cytoplasm is usually shrunken into the center. Several dotlike nuclei may be seen. Radiating, eosinophilic,

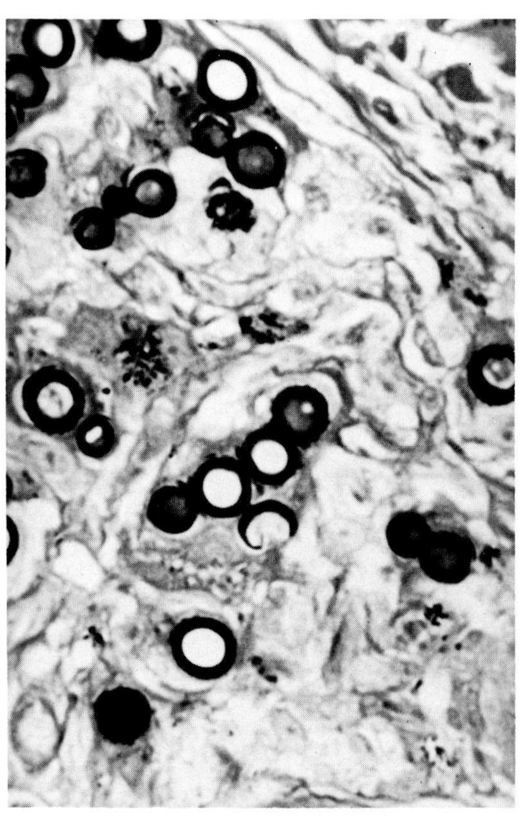

Figure 13–6. Lobomycosis. Chains of yeast cells. Note the branch and the connecting bridgelike structure. Some of the yeasts are empty of cellular contents. GMS. ×900. (From Rippon, J. W. *In* Burrows, W. 1973. *Textbook of Microbiology*. 20th ed. Philadelphia, W. B. Saunders Company, p. 720.)

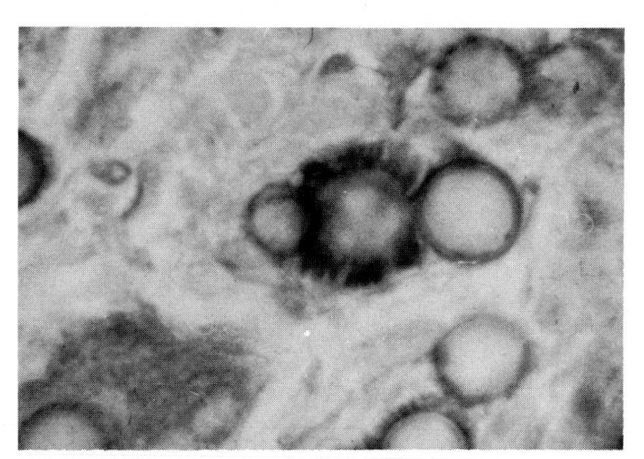

Figure 13–7. Lobomycosis. Radiating, eosinophilic, "asteroid"-like bodies around a yeast cell.

"asteroid" bodies are sometimes found (Fig. 13–7).[16] In GMS-stained material, the organism appears a very intense black color. The cell wall can be observed to narrow toward the juncture of individual cells and to be continuous from one cell to the next. Bud scars can also be found where disjunction of cells has occurred. In the Gridley stain, the yeast cells have a yellowish-brown cytoplasm, dark brown nuclei, and a pink cell wall. PAS-positive material can be seen radiating from or forming an envelope around the outside of the cell wall. Distinct, rough, spikelike formations are sometimes noted, particularly on the wall of the central cell. This material is stainable by the acid mucopolysaccharide and alcian blue methods but not by GMS.[23] Some of the yeast cells may have a bud, and a few show branching (Fig. 13–6). The numerous small buds, which are characteristic of *Paracoccidioides*, are not observed. There are no capsules present, nor is there great variation in size of yeasts as seen in cryptococcosis. In several cases examined by Wiersema,[23] there was little collagen associated with the stroma of the keloid, but a fine reticulin meshwork between the giant cells was observed by silver stain. He also observed hour glass–shaped yeast cells with a large connecting isthmus and a suggestion of pseudohyphae in some sections.

In older lesions, ulceration and some pyogenic infiltrate is seen. The ulceration appears to be mechanical rather than related to the fungus. The corneal layer in such lesions is often markedly thickened, and crusts are formed. Parakeratosis and acanthosis are present, along with empty yeast cells and giant cells which often contain much PAS-positive debris. The histopathologic appearance of lobomycosis is so characteristic that a diagnosis is readily made.

ANIMAL DISEASE

Natural infection of animals has only recently been discovered. Lobomycosis had been known only in humans until the diagnosis of the disease was made in bottle-nosed dolphins (*Tursiops truncatus*).[17] The first case occurred in a female dolphin found in the Gulf Intercoastal Waterway near Sarasota, Florida, an area quite removed from any known human cases. The migration range of this species of dolphin has not been fully determined, but it apparently does not include South or Central America. Other observers on both Florida coasts have concluded that this infection occurs with some frequency among several species of dolphins.

Grossly, the lesions on the dolphin resemble those of old lesions found in human cases of disease. They are verrucoid, crusty plaques with ulcerations. Histologically, the granulomas consist of histiocytes and giant cells. Yeasts are very numerous and uniform in size. They occur in long chains of many yeast cells and are found within giant cells.

Most reports of attempts to transfer lobomycosis to experimental animals have been negative,[18] but a few successful inoculations have been performed. Diaz[6] produced lesions in the cheek pouches of hamsters, but Wiersema[23] failed in attempts to infect mice intraperitoneally, intratesticularly, or intracutaneously. However, he was able to produce chronic disease with a typical histologic appearance in hamsters. In these experiments, biopsy material from human lesions was macerated, and yeast cells in chains were injected into the foot pads of the animals. Infection took a long time to develop, but after 8 months typical chronic nodular lesions were present. It is emphasized that lesions in experimental as well as in natural infection require much time to develop.

IMMUNOLOGY AND SEROLOGY

There have been too few cases of lobomycosis studied to detect any patterns of immunity or factors predisposing to disease. In the series of cases reported, the patients appeared to be otherwise normal and healthy. No detectable antibodies have been noted, and as yet there are no serologic procedures for diagnosis. Lacaz et al.[13] have established immunologically that paracoccidioidomycosis and lobomycosis are distinct diseases. They also discovered that the first isolant reputed to come from a case of lobomycosis was in reality a culture of *Paracoccidioides brasiliensis.* In an attempt to demonstrate common antigens between *Loboa loboi* and other fungi, Silva et al.[19] labeled with fluorescein sera from patients with lobomycosis. The sera reacted with cells of *P. brasiliensis, Candida albicans* serotypes A and B, and yeast-form *Sporothrix schenckii.* The sera did not react on biopsy tissue from a case of lobomycosis; therefore, it is doubtful that specific antibodies for the fungus of lobomycosis were present.

LABORATORY IDENTIFICATION

Direct Examination. Material obtained by curettage, surgical excision, or biopsy can be macerated and a potassium hydroxide mount made. Chains of uniform yeast cells will be abundant in case of disease. The average cell size is 9 to 10 μ, with a few ranging from 7 to 12 μ.

The etiologic agent of lobomycosis has not as yet been successfully or repeatedly grown in culture. In all instances the organisms isolated have later been found to be contaminants.[4,10,19]

MYCOLOGY

Loboa loboi (Fonseca Filho and Arêa Leão) Ciferri, Azevedo, Campos, and Siquerira Carneiro 1956

Synonymy. *Glenosporella loboi* Fonseca Filho and Area Leão 1940; *Glenosporopsis amazonica* Fonseca Filho 1943; *Paracoccidioides loboi* Almeida and Lacaz 1949; *Blastomyces loboi* Langeron and Vanbreuseghem 1952.

An isolant which was said to be obtained from the first case of lobomycosis was described by Fonseca and Leão as *Glenosporella loboi* in 1940. The fungus was used to infect experimental animals. On the basis of the pathology produced, the name *Paracoccidio-*

ides loboi was given to this organism by Almeida and Lacaz in 1949. It has since been shown to be a strain of *P. brasiliensis*.[4,10] Fonseca in 1943 isolated another fungus from a patient with lobomycosis and described it as *Glenosporopsis amazonica*. This isolant and one supplied by Borelli have since been identified by Raper[21] as *Aspergillus penicilloides*, an osmophilic *Aspergillus* of the *A. restrictus* group. Another strain of fungus isolated from a patient has now been identified as the pedunculated yeast, *Sterigmatomyces halophilus* Fell 1966.[10] All isolants with the exception of the *P. brasiliensis* have been nonpathogenic for laboratory animals and embryonated eggs.

The etiologic agent of lobomycosis remains to be cultivated *in vitro*. Its *in vivo* morphology is the same in the two mammalian species in which the disease has been recorded—man and dolphin. The organism is elliptical or lemon-shaped, has a uniform diameter of 9 to 10 μ and a doubly refractile wall, and is multinucleate. The yeasts reproduce by budding at the terminus of the cell. The buds usually remain attached by means of a tubular bridge, and chains of up to 20 yeasts are seen. More than one bud may occur, leading to branched chains of cells.

REFERENCES

1. Baruzzi, R. G., R. M. Castro, et al. 1967. Ocorrencia de blastomicose queloideana entre indios Caiabi. Rev. Inst. Med. Trop. São Paulo, *9*:135–142.
2. Battistini, F., S. G. Jover, et al. 1966. Dos casos de Blastomicosis Queloidiana o Enfirmedad de Jorge Lobo. Rev. Dermat. Venez., *5*:30–36.
3. Borelli, D. 1968. Lobomicosis: Nomenclatura de su agente. Med. Cutan., *3*:151–156.
4. Carneiro, L. S. 1952. Contribuiqao ao Estudo Microbiologica do Agente Etiologico da Doenga de Jorge Lobo. Tese. Imprensa Industrial. Recife., Pernambuco, Brasil.
5. Ciferri, R., P. C. Azevedo, et al. 1956. Taxonomy of Jorge Lobo's disease fungus. Univ. Recife Inst. Micology Publ. No. 53, pp. 1–21.
6. Diaz, L. B., M. M. Sampaio, et al. 1970. Jorge Lobo's disease. Observations on its epidemiology and some unusual morphological forms of the fungus. Rev. Inst. Med. Trop. São Paulo, *12*:8–15.
7. Destombes, P., and P. Ravisse. 1964. Etude histologique de une cas guyanais de blastomicose cheloidienne (maladie d. J. Lobo). Bull. Soc. Pathol. Exot., *57*:1018–1024.
8. Fonseca Filho, O. 1943. Doenga de Jorge Lobo. Parasitologia Medica Tomi I. Editora Guanabara, Rio de Janeiro, p. 710–714.
9. Fonseca Filho, O., and A. E. Area Leão. 1940. Contribução Para o Conhecimento das Gramunomatoses Blastomicoides. O Agente Etiologico da Doença de Jorge Lobo. Rev. Med. Cir. Brazil, *48*:147.
10. Lacaz, C. S. 1971. Keloid blastomicosis (Lobo's disease). Abstracts P.A.H.O. First Pan American Symposium on Paracoccidioidomycosis, p. 36.
11. Lacaz, C. S., R. G. Ferrí, et al. 1967. Blastomicose queloideana associada a blastomicose sul americana. Registro du um caso. Hospital (Rio), *17*:7–11.
12. Lacaz, C. S., and M. Rosa. 1969. Bibliografia sobre blastomicose sal-americana (doença de Lutz) e blastomicose queloidiforme (duença de Lobo) (1909–1968). São Paulo, Instituto de Medicina Tropical.
13. Lacaz, C. S., R. G. Ferrí, et al. 1962. Aspectos immunoquimicos na Blastomicose sul Americana e Blastomicose Queloidiana. Rev. Med. Cir. Farm., *298*:63–74.
14. Leite, J. M. 1954. Doença de Jorge Lobo. Contribuqao a seu estudo antatomopatologico. Tese. Oficinus Graficos da Revista Veterinairia. Belém. Pará, Brasil.
15. Lobo, J. 1931. Um caso de blastomicose produzida por uma especie nova, encontrada em Recife. Rev. Med. Pernambuco *1*:763–765.
16. Michalanay, J. 1963. Corpos asteroides ora blastomicose de Jorge Lobo. A proposito deum novo caso. Rev. Inst. Med. Trop. São Paulo, *5*:33–36.
17. Migaki, G., M. G. Valerio, et al. 1971. Lobo's disease in an Atlantic bottle-nosed dolphin. J. Am. Vet. Med. Assoc., *159*:578–582.
18. Nery Guamaraes, F. 1964. Inoculacoes em Hamsters da blastomicose sul americana (doença da Lutz, da blastomicose queloidiforme (doença de Lobo) e da blastomicose dos-Indios do Topajos-Zingn. O. Hospital (Rio), *66*:581–593.
19. Silva, M. E., W. Kaplan, et al. 1968. Antigenic relationship between *Paracoccidioides loboi* and other pathogenic fungi determined by immunofluorescence. Mycopathologia, *36*:98–105.
20. Silverie, C. R., P. Ravisse, et al. 1963. La blastomycose cheloidienne ou maladie de Jorge Lobo en Gyanefrancaise. Bull. Soc. Pathol. Exot., *56*:29–35.
21. Raper, K. B., and D. I. Fennell. 1965. *The Genus Aspergillus*. Baltimore, Williams and Wilkins Co., p. 234.
22. Villegas, M. R. 1965. Enfermedad de Jorge Lobo (Blastomicosis queloidiana). Presentacion de un neuro caso Colombiano. Mycopathologia, *25*:373–380.
23. Wiersema, J. P., and P. L. A. Niemel. 1965. Lobo's disease in Surinam patients. Trop. Geogr. Med., *17*:89–111.

Chapter 14

RHINOSPORIDIOSIS

DEFINITION

Rhinosporidiosis is an infection of the mucocutaneous tissue caused by *Rhinosporidium seeberi,* an as yet unclassified fungus. It is a chronic granulomatous disease characterized by the production of large polyps, tumors, papillomas, or wartlike lesions that are hyperplastic, highly vascularized, friable, and sessile or pedunculated. The nose is most commonly affected, with the conjunctiva the second most frequently involved site. Areas of infection rarely involved include the anus, penis, vagina, ears, pharynx, and larynx. The name of this clinical condition connotes an infection by the fungus *Rhinosporidium* and does not exclude sites of primary disease other than the nasal region.

HISTORY

Interestingly, both this disease and one caused by a similar etiologic agent were discovered and described by essentially the same group of investigators in a geographic area where these particular infections are rarely encountered. Coccidioidomycosis was discovered by Posadas in 1892, and rhinosporidiosis was reported by Seeber in 1900. Both men were students of Professor R. Wernicke in Buenos Aires at the same time. These diseases were at first thought to be protozoan in origin, and both are quite rare in the country of their discovery. Guillermo Seeber[27] published the first case report of rhinosporidiosis in a 19-year-old agricultural worker. The patient had a large nasal polyp that impeded breathing. He noted in his thesis that Malbran, who was also from Argentina, had seen a similar case in 1892. Independently O'Kinealy in 1903 reported a case from India under the name "localized psorospermosis" which he had first seen in 1894. Ellet in the United States also described the disease in the nose of a patient first seen in 1897. The patient was a native-born farmer in Tennessee. A full report of the case was published in 1907 by Wright.[30] In 1900 in the Programa de Zooligica Medica, Wernicke described the organism from Seeber's case as a *Coccidium,* but this paper appears to have been lost. Seeber considered it to be a *Coccidioides* related to the agent of Posadas' disease. Minchin and Fanthum in 1905[18] detailed the appearance of the organism and concluded it was a Sporozoa related to Neosporidia and Haplosporidia. Unaware of the Argentinean publication, they named it *Rhinosporidium kinealyi.* Many more reports of the disease appeared after the first published cases. Ashworth[4] made a very detailed analysis of the organism and its development in tissue and, concluding that it was a fungus, gave it the name *Rhinosporidium seeberi.* He also published the first report of the disease in Scotland. The patient was an Indian medical student with nasal polyps. Karunaratne in 1964[12] published a detailed account of the disease in man and reviewed the literature.

ETIOLOGY, ECOLOGY, AND DISTRIBUTION

The etiologic agent is a fungus called *Rhinosporidium seeberi.* Throughout the years many attempts to culture the organism have failed, so that detailed taxonomic study has been lacking. Recently it has been claimed that the organism was successfully propagated *in vitro*. Grover[9] reportedly maintained the organism in tissue culture medium 99 at 4°C and studied its replication. She found the fungus had essentially the same morphologic growth cycle as had been detailed by Ashworth from histologic sections. The organism did not produce mycelium when reproducing *in vitro*, just as it did not when growing *in vivo*. Whether this is a valid record of *in vitro* propagation remains to be substantiated. Taxonomic affinities of this organism are still lacking, but, based on his studies, Ashworth[4] concluded it to be a lower Phycomycetes. He determined that the nutritive reserve of cells in the trophic stage is fatty material, that multiple nuclear division takes place simultaneously prior to formation of spores, and that the spherule walls contain cellulose. These are all characteristics of the Olpidiaceae of the Chytridiales, a group of water molds in which Ashworth provisionally placed the organism. Rao[25] has demonstrated chitin, in addition to cellulose, in the spherule walls.

A saprobic existence for the organism has never been established, and inference from case histories is always tenuous. If the organism is indeed a *Chytridium*, an aqueous natural habitat is probable. Many case studies have noted an association of disease in the patient with frequent bathing or working in stagnant fresh water. Mandlick found a 20 per cent infection rate in a group of workers engaged in removing sand from a river bottom. The patients' co-workers on shore did not have any disease.[17] Similarities of the organism to *Ichthyosporidium*, a fungus infecting salmon and trout, have been noted, and a parasitic existence of *Rhinosporidium* in an aquatic form of life conjectured.[12] In arid countries most infections are ocular, and dust is postulated to be a vector.[13]

About 2000 cases of disease had been recorded up to 1964 when Karunaratne reviewed the world literature.[12] Of these, 88 per cent were from India and Ceylon. In one series, Allen and Dave reported seeing 60 cases within an 18-month period in India.[3] In these areas the infection is so common that many cases are not reported. After India, most published cases are reported from South America, particularly Brazil and Argentina. Forty-one cases of both animal and human infection were reported by Niño.[20,21] All occurred in a highly endemic focus in the Villa Angela region in the province of Choco in Argentina. Thirty of the infections were in horses and six in humans. All involved the nose or adjacent areas. Some 15 reports have come from Brazil,[5] with a scattering throughout Colombia, Venezuela, and other countries. About 40 reports, including one of the first, have been published from the United States. In about half these, the nasal area was involved; most of the remainder involved conjunctival disease. The disease has been recorded from almost all parts of the world, including Africa,[6,7] Iran, Russia,[11] Southeast Asia, the Near East, Mexico,[14] and Europe.[2] Many of the European cases were in individuals who were from India or had visited there for extended periods.

Though the age of patients varies from 3 years to 90 some years, most patients are between 20 and 40 years of age when diagnosed.[15,16] Since the lesions do not cause any undue discomfort to the patient and the infection is very slow in developing, it is often difficult to ascribe age at onset. One of the youngest cases on record was in a Texas girl, age 3, who had severe nosebleed as a presenting symptom.[22] In both human and equine cases of the disease, males account for 70 to 80 per cent of cases,[12,20] but this varies with age, site of infection, and geographic location. In prepubertal cases, the infection appears to occur about equally between the sexes. Eye infections seem to be more common in women, but no clear pattern is evident. When location is recorded, most patients are from rural environments, and in many there is an association with work or play in fresh water. Several report infection at the site of a previous injury.[12] Ocular infections are more frequent in arid areas and appear to occur following dust storms and injury to the eye.

CLINICAL DISEASE

The name "rhinosporidiosis" implies a sporozoan infection of the nose, which is the commonest site of infection. Of the 2000 cases tabulated, about 70 per cent involved the nasal area, 15 per cent the eye, and 8 per cent some other mucosal area or, very rarely, a cutaneous site.[12] The development

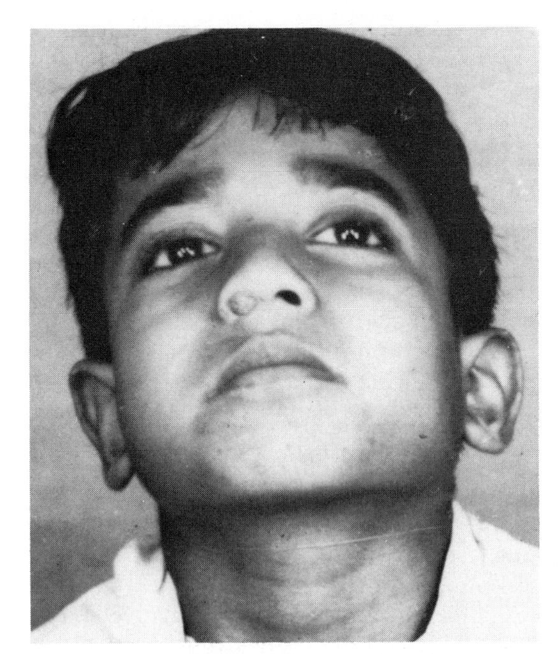

Figure 14–1. Rhinosporidiosis. Polyp developing in nasal opening. (Courtesy of S. Banerjee.)

of lesions and their gross appearance varies somewhat, depending on the anatomic site involved.

Nasal Disease. The exact mode of infection is unknown, but some initial trauma may be a predisposing factor. Karunaratne[12] conjectures that the Muslim custom of mechanically cleansing the nose before entering a mosque may predispose to infection.. Once disease is established in the nose, trauma to adjacent sites followed by autoinoculation may serve to spread the disease.

The commonest sites for initial infection are the mucous membrane of the septum, interior turbinate, and nasal floor.[26] Other areas, such as middle turbinate, middle meatus, and nasal roof, are less frequently involved. The first symptom noted by patients is the feeling of the presence of a foreign body in the nose. The infection may be accompanied by mild to intense pruritus and coryza. The lesion, which at first is sessile, develops into a pedunculated polyp (Fig. 14–1). In time the growth obstructs the air passage, and it is this symptom for which attention is sought. The nasal discharge is mucus tinged with blood and contains spores and sporangia. Mild bleeding is frequent, as the lesions are quite friable, but epistaxis may be severe. As the lesion grows it may take on a bizarre shape and appearance. Fully developed tumors are usually polypoid, globoid, and pedunculated (Fig. 14–2). The lesions may be part pedunculate and part sessile or wholly sessile. Papillary

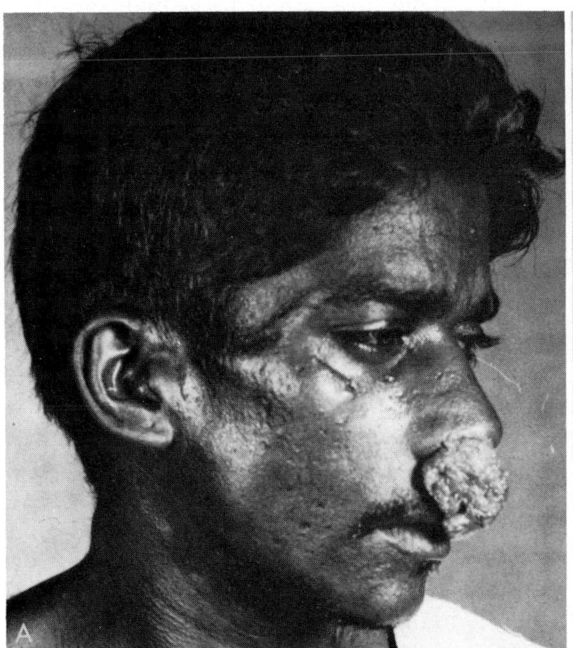

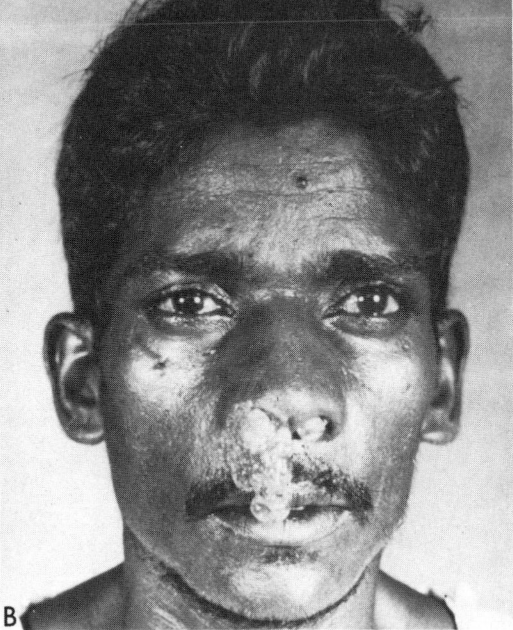

Figure 14–2. Rhinosporidiosis. *A* and *B*, Fully developed polyploid pedunculated tumor. (Courtesy of C. Satyanarayana.)

projections and lobules are sometimes seen, giving the lesion a raspberry, strawberry, or, when very corrugated, a cauliflower appearance. The color is bright pink initially, becoming deep red. The red color that develops is due partly to profuse vascularization and partly to free blood and old hemorrhagic material. The lesion is often spleenlike in appearance. On close examination, the polyps are mottled owing to the presence of numerous whitish, macroscopically visible spherules. Fibrous cords are often numerous, and the whole lesion, when cleansed of blood, may have a greyish cast. Fully mature polyps hang down through the nasal meatus and may extend beyond the lip. If they are located high up on the turbinates or nasal septum, the lesions may hang into and beyond the nasopharynx, interfering with respiration and feeding (Fig. 14–3). These lesions contain much mucus, sometimes mucinous cysts, have bulbous ends, and a general appearance of a ripe fig. They may weigh 20 g or more. Lesions in the nasopharynx alone without nasal disease are uncommon, but have been reported in about 30 patients. The polyps are similar to those of nasal origin and occasionally lead to obstruction, dyspnea, and dysphagia.

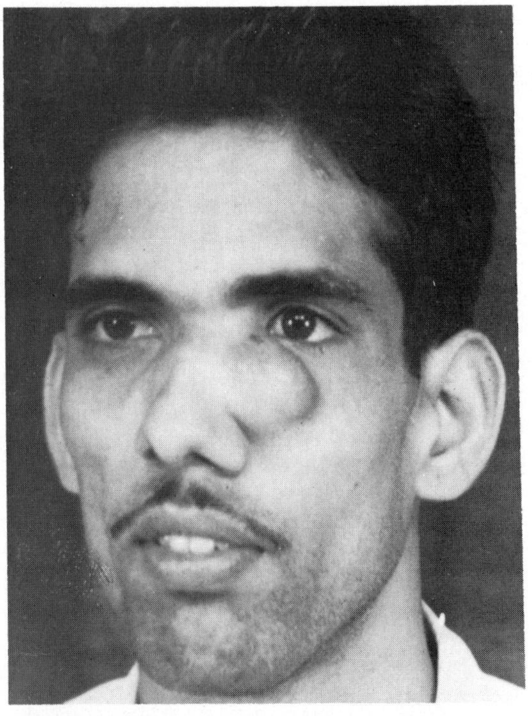

Figure 14–3. Rhinosporidiosis. Lesions developing far up in the turbinates. (Courtesy of S. Banerjee.)

Ocular Disease. The eye and adnexae account for about 15 per cent of cases in India and other moist tropical environments. In dry, dusty areas, as in Transvaal, Iran, and arid areas of the Indian subcontinent, almost all cases are ocular. Kaye,[13] commenting on the disease in South Africa, reported that infection occurred after dust storms. He theorized that dust not only caused eye injury but also transmitted the fungus, thereby leading to infection. About half the published cases of rhinosporidiosis in the United States are nasal, and most of the remaining are ocular.[22,23] The majority of these reports come from Texas, with the others scattered throughout the remaining states.

In almost 90 per cent of ocular infections, the palpebral conjunctiva is involved (Fig. 14–4). In the remaining cases, disease was of the bulb, limbus, caruncle, or canthi. Though bilateral involvement and multiple lesions are not infrequent in nasal disease, most infections of the eye are unilateral, and the lesions single.[19] The growths are sessile or stalked, the attachment being to the upper or lower fornix or tarsal conjunctiva. The lesions are often small and flat, accommodating themselves to lie between the lid and eyeball. They may cause no discomfort to the patient, who may not be aware of their presence. They are freely moveable, granular, pink to red in appearance, and with careful examination can be seen to contain whitish spherules. The patient usually does not present with any symptoms other than a growth in the eye. When lesions become large, other symptoms may occur, however. These include excessive tearing, redness of the eye, discharge, photophobia, conjunctival infection, and eversion of the lid. Some lesions are so deep blue-red and spleenlike in appearance that they suggest hemangioma, and often they are diagnosed clinically as such.

Cutaneous Disease. Lesions on the skin are infrequent, and most are associated with adjacent mucocutaneous disease. On rare occasions, skin sites at a distance from mucosa may be infected from scratching and autoinoculation.[3] Skin lesions alone, i.e., not associated with disease elsewhere, are very rare, but have been reported. These include a single lesion on the scalp, one on the abdo-

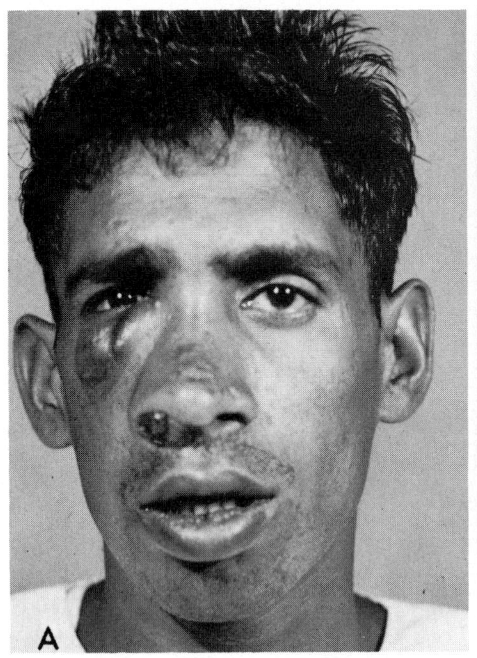

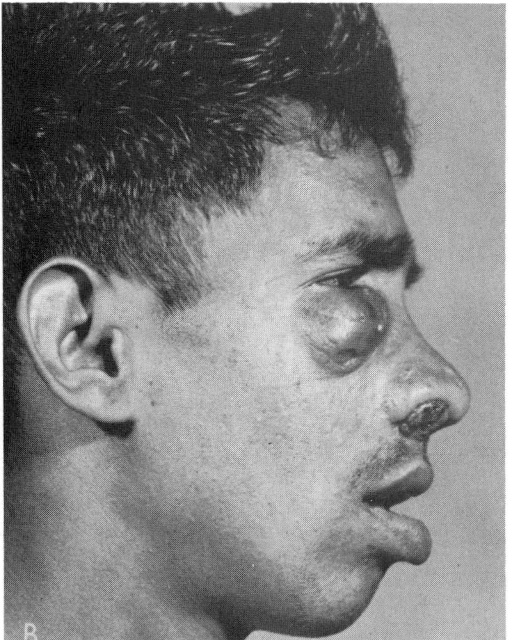

Figure 14–4. Rhinosporidiosis. *A,* Involvement of palpebral conjunctiva and nose. *B,* Profile of same patient. (Courtesy of P. Kulkavni.)

men, and one report of multiple skin lesions over several areas of the body.[12] Skin lesions begin as tiny papules which become wartlike growths. Their surface is crenated. Since they are friable, they are often ulcerated and secondarily infected with bacteria. Cutaneous lesions rarely become pedunculated. The growths that occur in the rare cases of hematogenous dissemination are described as firm, hard, subcutaneous nodules which may remain unattached to the overlying skin or invade through it. Cutaneous lesions are painless and cause no discomfort to the patient unless they are in areas where they are continually traumatized, such as the sole of the foot.

Other Areas of Involvement. Lesions occurring in other mucosal areas are infrequently encountered but reported. These include the larynx, hard palate (Fig. 14–5), epiglottis (Fig. 14–6), vagina, vulva, uvula, and anus. Vaginal and anal lesions were described as resembling condylomata, rectal polyps, or hemorrhoids. The urethra has been involved in a few cases in males. These lesions were red, pedunculated, knobby masses extending beyond the meatus of the penis. The Muslim habit of removing the last drops of urine following micturition by rubbing the meatus of the penis with a stone is cited as causing repeated trauma to the area which may predispose to infection.[25] Disease of the parotid gland, trachea, and bronchus[28] have also been reported. In the case of bronchial disease, complete obstruction of air passages occurred with a fatal outcome.

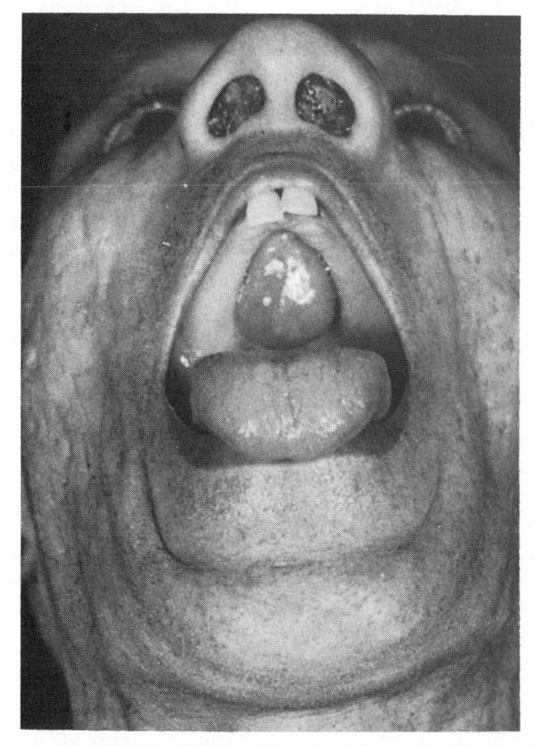

Figure 14–5. Rhinosporidiosis. Lesion on the hard palate. (Courtesy of C. Satyanarayana.)

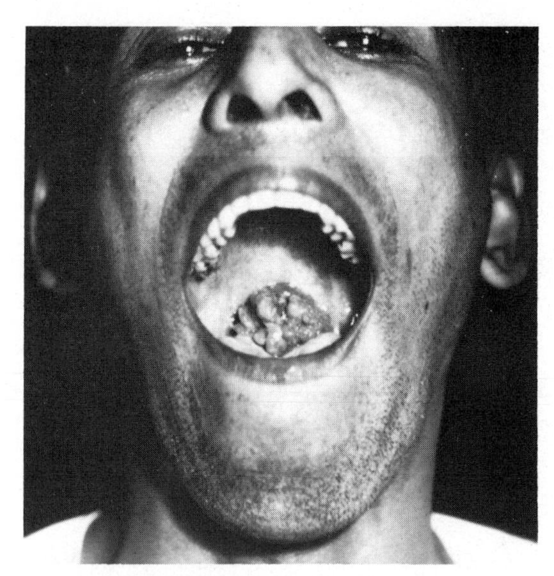

Figure 14–6. Rhinosporidiosis. Lesions involving epiglottis. (Courtesy of C. Satyanarayana.)

In the commonly encountered cases of rhinosporidiosis, the organisms are limited to the polyps; regional lymph nodes and adjacent tissue are not involved. There are a few reports of disseminated rhinosporidiosis, however. Agrawal described a case in which spherules were found infecting the spleen, liver, lung, and other viscera as well as being found in laked blood and urine.[1] In the one case of cerebral infection, lesions were found in the brain, the adjacent blood vessels, and the nose without involvement of other areas.[2] A few cases in the external ear have been recorded.[12] The lesions had the appearance of ordinary aural polyps. Only when they enlarged and caused obstruction and pressure were symptoms present. Invasion and destruction of the bone has also been recorded. Lesions occurred in osseous tissue underlying nasal, pharyngeal, or digital lesions.

DIFFERENTIAL DIAGNOSIS

The lesions of rhinosporidiosis are most often red, friable, sessile growths or polyps of the mucosal surfaces. These may resemble mucoceles, hemangiomas, condylomata, or neoplasms. Cryptococcus is the only other fungus known to elicit polypoid tumors. Direct examination of biopsy material allows the diseases to be easily differentiated.

PROGNOSIS AND THERAPY

Rhinosporidiosis is a chronic disease that usually has a long history without pain, discomfort, or debility to the patient. Disease of 30 to 40 years' duration has been noted in some reports.[12] These infections usually consist of single small lesions that cause difficulty only when they are large enough to obstruct a passage or cause pressure on vascular or neural bundles. In the rare cases of systemic disease, dissemination appears to have occurred early in the course of the infection.

Recurrence of infection is a characteristic of rhinosporidiosis, and many patients have had several surgical procedures for removal of growths. In an early case report in the United States, Wright in 1907 described a young Tennessee farmer who had had nasal polyps removed three times.[30] Ten operations in 15 years have been noted in some records. No predisposing susceptibility or immunity to reinfection has been demonstrated in rhinosporidiosis.

Treatment involves surgical removal of the affected tissue. This is accomplished by use of a hot or cold snare to avoid spreading the infection to adjacent tissue.[26] Local bacterial infection and some fatal septicemias have occurred following unskilled surgery. Copious bleeding is also a complication.[10] Local injection of amphotericin B may be used as an adjunct to surgery to prevent reinfection and spread. Other drugs generally have been without efficacy.

PATHOLOGY

Gross Pathology. Very few cases of generalized and fatal rhinosporidiosis are known. The best described case is the one reported by Agrawal.[1] The patient, a 30-year-old Hindu, had had an infection of the eyelid, skin, and palate for more than one year. At autopsy white, firm nodules were found in the lung and on the pleura, vocal cords, and epiglottis. Granulomas were also found in striated muscle and skin. Spherules were found in the sinusoids of the liver and in the spleen and kidney, but these had mostly degenerated.

Gross examination of the specimen from a usual case of rhinosporidiosis reveals it to have the appearance of an ordinary nasal polyp. In contrast to the loose, edematous, myxomatous stroma of the latter, polyps of rhinosporidiosis are rather dense, and mucinous cysts are usually absent. Opaque greyish-white granular material is apparent, which represents the mature sporangia. Cut sections show such sporangia are of varying sizes, the more mature being closer to the epithelial surface. Some of them are collapsed and assume a semilunar shape. Polyps from conjunctival infections are flattened, soft, reddish-pink to dark red, and less lobulated. Minute opaque spherules are also readily visible.

Histopathology. The layers of transitional epithelium are often invaginated to become flask-shaped and may form pseudocysts. Such areas contain spores, pus, and mucous material. The epithelium is generally hyperplastic, though it may be quite thinned in some areas. Mature sporangia often lie just beneath the thinned areas. Spores accompanied by neutrophils are sometimes found in the epithelium. The major portion of the growth consists of very vascular, fibromyxomatous connective tissue in which the parasites are found in varying stages of development. The cellular infiltrate consists of plasma cells, lymphocytes, histiocytes, and neutrophils. Occasionally eosinophils are present in large numbers, but this finding is not so constant as in ordinary nasal polyps. The cellular areas may contain mainly lymphocytes or plasma cells. Giant cells are not uncommon, especially in older lesions. Microabscesses are frequent, and there is often evidence of chronic trauma and hemorrhage. As is the case in coccidioidomycosis, freshly liberated spores from a spherule incite a polymorphonuclear response. The eosinophilic material characteristic of asteroid body formation in many fungal diseases has not been noted.

The life cycle of the organism in tissue was outlined in detail by Ashworth.[4] The infecting spore appears to be able to penetrate into the mucosal epithelium and to begin maturation in the subepithelial tissue. This spore, called the "trophic stage" by Ashworth, is 6 to 10 μ in size, has a chitinous wall, a clear protoplasm, and a vesicular nucleus with a nucleolus. He also noted the presence of a karysome-like body. As the spherule grows to a diameter of 10 to 12 μ, globular and granular material appears in the cytoplasm. This material is fatty in nature and is stained by the usual fat stains. At 50 μ the first nuclear division occurs (Fig. 14–7). Ashworth determined that there were four chromosomes. Succeeding nuclear divisions take place synchronously. At 100 μ in size, a layer of cellulose-like material is laid down on the inner surface of the wall of the spherule. The cellulose material is about 3 μ thick, and at maturity the wall itself sometimes exceeds 5 μ. The cellulose is thin in the region where a pore later develops. The pore allows the escape of mature spores. When 2000 nuclei are present, the cytoplasm is seen to condense around them. An annulus of uncondensed material, 12 μ thick, surrounds the inner wall. More nuclear divisions occur until about 16,000 young spores are present. The spherule is about 150 μ in diameter at this point. It enlarges to 250 to 350 μ, the wall becomes thinner, and the annulus begins to disappear. The mature spores migrate toward the periphery near the pore. Approximately one-third of the spores fail to mature (Fig. 14–8). Rupture of the pore occurs, possibly from internal pressure, and the spores are released into the surrounding connective tissue. The spores at maturity are 7 to 9 μ. They contain a nucleus, a basophilic karyosome-like body, and a globular cytoplasm containing eosinophilic material. An influx of foreign body giant cells is incited by rupture of the sporangia. These may be seen to invade the empty spherule together with the neutrophils. Mature sporangia usually rupture through the epithelial layer, and nasal exudate may contain numerous spores (Fig. 14–9 to 14–11). Karunaratne[12] presents evidence that freed spores lodge in the epi-

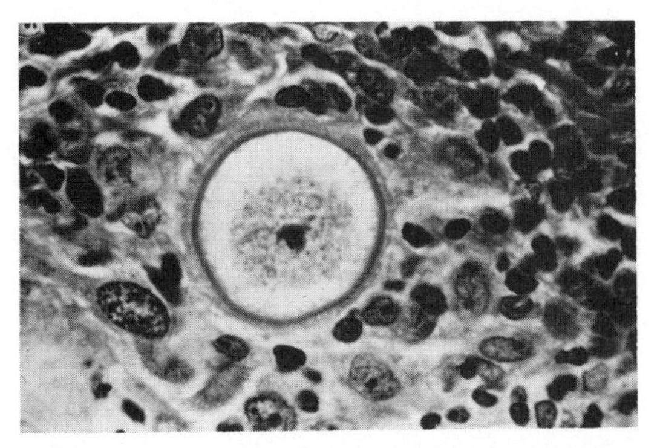

Figure 14–7. Rhinosporidiosis. Spherule just before nuclear division. Note the globular and granular material in the cytoplasm and the prominent nucleolus. Hematoxylin and eosin stain. ×400.

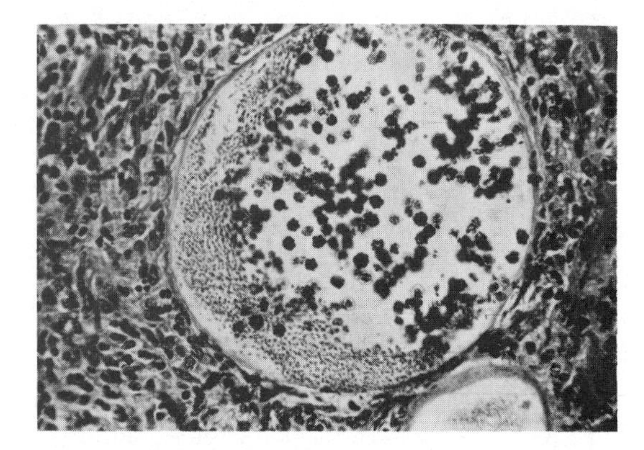

Figure 14–8. Rhinosporidiosis. Spherule with maturing endospores. Many prospores fail to develop. Hematoxylin and eosin stain. ×400.

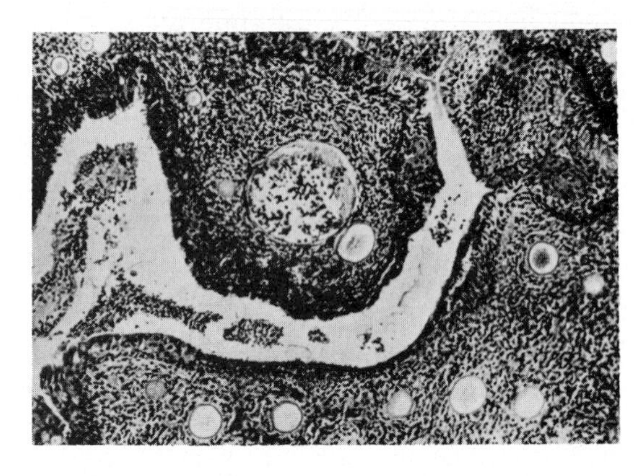

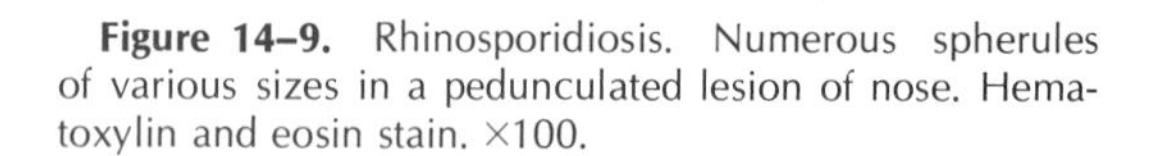

Figure 14–9. Rhinosporidiosis. Numerous spherules of various sizes in a pedunculated lesion of nose. Hematoxylin and eosin stain. ×100.

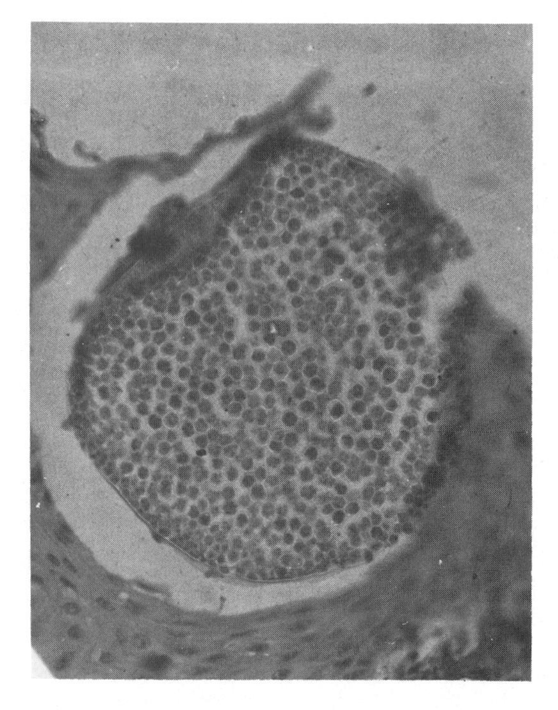

Figure 14–10. Rhinosporidiosis. Mature spherule releasing spore through pore. The spherule has broken through the lining of the epithelium. [From Satyanarayana, C. 1966. Chapter 13 in *Clinical Surgery Rhinosporidiosis.* C. Rob and R. Smith (eds.), London, Butterworth and Co.]

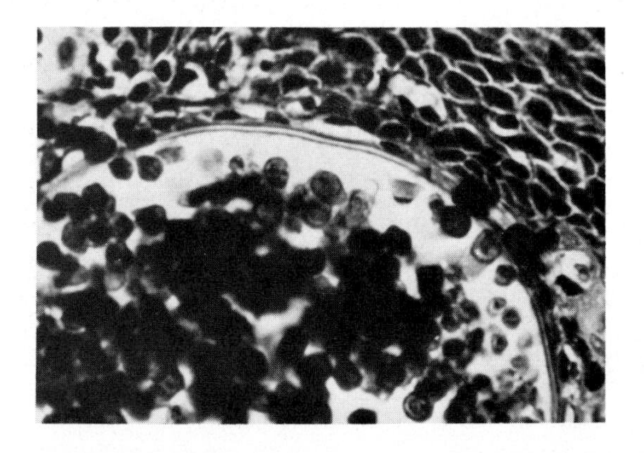

Figure 14–11. Rhinosporidiosis. Mature spherule showing thick wall and endospores near surface epithelium. Nuclei can be seen within the endospores.

thelium, become surrounded by tissue, and begin the maturation cycle over again. Controversy exists as to the fate of the spores released into tissue, as they stain differently from the so-called trophic stage. The ultrastructure of the maturing spherule has been examined by Vanbreuseghem.[29]

Rhinosporidium seeberi is easily observed in the usual hematoxylin and eosin–stained sections. Maturing sporangia and their spores are stained by the Gridley, PAS, and GMS stains. The early trophic stage, however, does not take the Gridley or PAS stain, but is readily colored by the GMS method. The inner cellulose-like material of the sporangium and the outer layer of newly released spores take the Mayer's mucicarmine stain. With the exception of the capsule of *Cryptococcus,* this is the only fungus which is colored by this method, but confusion of the two is not a problem. Small empty spherules may resemble those seen in coccidioidomycosis.

ANIMAL DISEASE

Although a prolonged and chronic disease in experimental animals has not yet been produced, natural infection in animals occurs frequently. About 80 cases in horses have been recorded, 90 per cent of which were in males. The histopathology was identical to that seen in human disease. Two dozen cases in cattle, 12 in mules, and one in a dog are also on record. Almost all of these were in male animals. The lesions in animals were all in the nose except for one in the larynx. No eye infections have been reported. Many attempts to transfer the disease to experimental animals have been attempted. In a few instances, granulomas have been produced, but sustained and progressive disease with propagation of the parasite has not occurred.[9,12]

IMMUNOLOGY AND SEROLOGY

Essentially nothing is known about immunology or resistance to the disease. The patients that have been studied have not had any debilitating diseases and infection does not appear to be opportunistic in that sense. No serologic studies are available.

LABORATORY IDENTIFICATION

Direct Examination. Examination of the lesions often reveals macroscopically visible subsurface sporangia. Dissected or excised tissue or nasal discharge can be slightly macerated and examined in a potassium hydroxide preparation. Mature sporangia up to 350 μ in diameter and spores 7 to 9 μ can be seen. Only spores are seen in nasal discharge, but a few sporangia may be present.

Culture Methods. *Rhinosporidium seeberi* has defied laboratory cultivation. Grover,[9] using tissue culture media no. 99 and incubating at 4°C, has claimed to have propagated the organism for many months and to have studied its life cycle. She was unable to infect experimental animals, however. Diagnosis of the disease can be made from histopathologic section or by direct examination.

MYCOLOGY

Rhinosporidium seeberi (Wernicke) Seeber 1912

Synonymy. *Coccidium seeberi* Wernicke 1900; *Coccidium seeberia-Wernicke* Belou 1903; *Coccidioides seeberi* Wernicke 1907; *Rhinosporidium kinealyi* Minchin and Fanthum 1905; *Rhinosporidium equi* Zschokke 1913; *Rhinosporidium ayyari* Allen and Dave 1936.

The early investigators considered the organism to be a Sporozoa. When its fungal nature was demonstrated by Ashworth,[4] it was considered to be a Phycomycetes related to the Chytridiales. Later, C. W. Dodge classified it among the Endomycetales as an ascomycete, as he considered that the internal spores produced were ascospores. Since a saprobic existence or a life cycle outside man has not been found and its cultivation *in vitro* has not yet been substantiated, it is not possible currently to delineate the true taxonomic relationships of the organism. Most mycologists feel it is a *Chytridium* related to the Olpidiaceae or, as suggested by C. W. Emmons, a *Synchytrium.* The latter group of fungi are parasites of plants, producing galls, or "plant polyps," on the host, and sporulation is very similar to that observed in *Rhinosporidium.*

REFERENCES

1. Agrawal, S., K. D. Sharma, et al. 1959. Generalized rhinosporidiosis with visceral involvement. Report of a case. Arch. Dermatol., *80*:22–26.
2. Alessandrini, O. O. 1926. Cited in Ruiz, F. R., and T. Ocana. 1930. Neuva observacione sobre *Rhinosporidium seeberi*. Rev. Med. Latino Amer., *16*:24–30.
3. Allen, F. R. W. K., and M. Dave. 1936. Treatment of rhinosporidiomycosis in man based on sixty cases. Indian Med. Gaz., *71*:376–395.
4. Ashworth, J. H. 1923. On *Rhinosporidium seeberi* (Wernicke 1903) with special reference to its sporulation and affinities. Trans. R. Soc., Edin., *53*:301–342.
5. Azevedo, P. C. 1958. *Rhinosporidium seeberi*. Tese Microbiol. Fac. Med. Univ. Parà.-Brazil, pp. 1–108.
6. Brygoo, E. R., C. Bermond, et al. 1959. First Madagascan case of rhinosporidiosis. Bull. Soc. Pathol. Exot., *52*:137–140.
7. Christian, E. C., and J. Kovi. 1966. Three cases of rhinosporidiosis in Ghana. Ghana Med. J., *5*:63–64.
8. Dube, B., and G. D. Veliath. 1964. Rhinosporidiosis in Bangalore. J. Indian Med. Assoc., *42*:59–64.
9. Grover, S. 1970. *Rhinosporidium seeberi:* A preliminary study of the morphology and life cycle. Sabouraudia, *7*:249–251.
10. Kameswaran, S. 1966. Surgery in rhinosporidiosis. Experience with 293 cases. Int. Surg., *46*:602–605.
11. Karpova, M. F. 1964. On the morphology of rhinosporidiosis. Mycopathologia, *23*:281–286.
12. Karunaratne, W. A. E. 1964. *Rhinosporidiosis in Man.* London, The Athlone Press.
13. Kaye, H. 1938. A case of rhinosporidiosis on the eye. Br. J. Ophthalmol., *22*:447–455.
14. Ketina Mora, E. 1967. Communicacion de un case de rinosporidiosis. Rev. Invest. Salub. Publica Mex., *27*:255–284.
15. Kutty, M. K., T. Sreedharan, et al. 1963. Some observations on rhinosporidiosis. Am. J. Med. Sci., *246*:695–701.
16. Kutty, M. K., and P. N. Unni. 1969. Rhinosporidiosis of the urethra. A case report. Trop. Geogr. Med., *21*:338–340.
17. Mandlick, G. S. 1937. A record of rhinosporidial polypi with some observations on the mode of infection. Indian Med. Gaz., *72*:143–147.
18. Minchin, E. A., and H. B. Fanthum. 1905. *Rhinosporidium kinealyi* n.g.n.sp. A new sporozoon from the mucous membrane of the septum masi of man. Quart. J. Microbiol. Sci., *49*:521–532.
19. Neumayr, T. G. 1964. Bilateral rhinosporidiosis of the conjunctiva. Arch. Ophthalmol., *71*:379–381.
20. Niño, F. L., and R. S. Freire. 1964. Exitenciade un foco endemico de rinosporidiosis en las provincia del Chaco. V. Estudio de neuvas observaciones y consideraciones finales. Mycopathologia, *24*:92–102.
21. Niño, F. L., and R. S. Freire. 1966. Existencia de un foco endemico de rinosporidiosis en la provincia del Chaco. Neuvas observaciones de *rhinosporidiosis equina.* Caracteres ecologicos de la region de Villa Angela. Rev. Med. Vet. Buenos Aires., *47*:421–437.
22. Norman, W. B. 1960. Rhinosporidiosis in Texas. Arch. Otolaryngol., *72*:361–363.
23. Peters, H. J., and C. G. DeBelly. 1969. Conjunctival polyp caused by *Rhinosporidium seeberi.* Report of a case. Am. J. Clin. Pathol., *51*:256–259.
24. Rao, P. N. S. 1962. Rhinosporidiosis of the conjunctiva. J. Indian Med. Assoc., *39*:601–602.
25. Rao, S. N. 1966. *Rhinosporidium seeberi:* A histochemical study. Indian J. Exp. Biol., *4*:10–14.
26. Satyanarayana, C. 1966. Chapter 13 in *Clinical Surgery Rhinosporidiosis.* C. Rob and R. Smith (eds.), London, Butterworth and Co.
27. Seeber, G. R. 1900. Un neuvo esporozuario parasito del hombre. Dos casos encontrades en polipos nasales. Tesis. Univ. Nat. de Buenos Aires.
28. Thomas, T., N. Gopinath, et al. 1956. Rhinosporidiosis of the bronchus. Br. J. Surg., *44*:316–319.
29. Vanbreuseghem, R. 1973. Ultrastructure of *Rhinosporidium seeberi.* Int. J. Dermatol., *12*:20–28.
30. Wright, J. 1907. A nasal sporozoon (*Rhinosporidium kinealyi*). N.Y. Med. J., *86*:1149–1153.

THE SYSTEMIC MYCOSES

INTRODUCTION

The systemic diseases caused by fungi fall into two very distinct categories which are delineated by the interaction of two factors: inherent virulence of the fungus and constitutional adequacy of the host. The first category includes infections caused by the true pathogenic fungi: *Histoplasma*, *Coccidioides*, *Blastomyces*, and *Paracoccidioides*. The second group of infections are termed "opportunistic" because organisms involved are inherently of low virulence, and disease production depends on diminished host resistance to infection. The common etiologic agents of opportunistic infections are *Aspergillus*, *Candida*, *Mucor*, and *Cryptococcus*.

The two categories of disease are distinct in essentially all aspects of host-parasite interaction. The true pathogenic fungi are those species that have the ability to elicit a disease process in the normal human host when the inoculum is of sufficient size. As previously discussed, pathogenicity in fungi is an accidental phenomenon and is not essential to the survival or dissemination of the species involved. In recent years it has been determined that the vast majority of such infections, usually more than 90 per cent, are either asymptomatic or of very short duration and quickly resolved. Resolution of the infection is accompanied by a strong specific resistance to reinfection that is of long duration. In the few individuals who have residual or chronic infection, the usual cellular response is a granulomatous process resembling that seen in tuberculosis.

Another characteristic of the pathogenic fungi is that they have a very restricted geographic distribution. It is also necessary for the patient to encounter the fungus when it is sporulating. Thus infection by *Coccidioides* requires a person to be present in the small ecologic areas favored by the fungus and to be there at the time of year when the fungus is growing and fruiting. The same is also true for *Histoplasma*, *Blastomyces*, and *Paracoccidioides*. The major endemic regions of these organisms are in the Americas and, with the exception of *Paracoccidioides*, primarily in the United States.

Sex, age, and race are important factors in the statistics of pathogenic fungous infections. Adult males constitute the majority of individuals with serious disease. It also appears that dark-skinned individuals have a greater risk of developing the severe disseminated forms of the disease than do Caucasians. Rural dwellers and agricultural workers are particularly susceptible to these infections because of their constant exposure to the soil-inhabiting fungi.

The pathogenic fungi exhibit a morphologic transition from the mycelial or saprophytic form to the parasitic form found in infected tissue. In most of these organisms the change is a conversion to a budding, yeastlike form which is governed principally by the temperature of incubation. Thermal dimorphism is exhibited by three of the major organisms which produce systemic infections and by one agent producing subcutaneous infections. In another important agent of systemic mycosis, transformation to the parasitic form is determined by "tissue" factors, such as carbon dioxide tension and so forth. The parasitic phases of all these fungi can be induced *in vitro*.

The second group of systemic fungous infections are those caused by opportunistic fungi. Although objections have been raised to the use of the term "opportunistic," it seems quite appropriate. The organisms involved have a very low inherent virulence, and the patient's defenses must be abrogated before infection is established. Formerly, these diseases occurred quite rarely, but in recent years they have become much more common and of great medical significance. The rise in the incidence of these opportunistic infections has paralleled the use of antibiotics, cytotoxins, immunosuppressives, steroids, and other macrodisruptive procedures that result in lowered resistance of the host. In contrast to the picture seen in infections caused by pathogenic fungi, the usual cellular response to opportunistic fungi is a suppurative necrotic process which is, at most, a poorly organized, granulomatous reaction. If the patient survives his debilitating disease or medical procedure, he is usually able to contain the fungal infection by granuloma formation, fibrosis, and calcification.

The opportunistic fungi also differ from the true pathogens in geographic distribution. A patient with lymphoma would have to travel to the San Joaquin Valley to contract coccidioidomycosis, but *Candida* and *Aspergillus* are always present in his environment. *Candida* (discussed in the section on pathogenic yeasts) is found in small numbers in the normal gut, but it proliferates in great numbers if the equilibrium of the bacterial flora is upset or if it gains entrance into the body through a "barrier break." The spores of *Aspergillus* are continually present in the air in all parts of the world. Unlike disease produced by the pathogenic fungi, establishment of opportunistic infections depends primarily upon altered host resistance rather than inoculum size. These infections are usually insidious and may not be diagnosed until autopsy. *Candida* and *Aspergillus* account for the vast majority of opportunistic fungus infections in all types of debilitated patients. Mucormycosis occurs rarely but is dramatic in its presentation and almost always is rapidly fatal.

Opportunistic infections differ from those produced by pathogenic fungi in several other ways. Recovery from the former does not establish a specific immunity, and reinfection may occur if general resistance is lowered again. There are no differences in susceptibility ascribable to age, sex, or race. The population that is afflicted with these infections is determined only by the type and severity of the underlying disease—two factors which decide the final outcome of the infection, almost independently of therapy. Lastly there is a striking difference between the tissue forms of the two groups of systemic fungous diseases. Whereas a dimorphism is exhibited by the etiologic agents of the pathogenic fungous infections, no such transition is exhibited by the organisms in opportunistic fungous diseases. *Aspergillus* growing on an agar plate, in decaying leaves, or in the lungs of a leukemic is mycelial in form. *Mucor* produces hyphal strands in moist bread or in the brain of a diabetic. The morphologic transitions exhibited by *Candida* are determined by nutrition rather than temperature or the *in vivo* environment. Although the presence of mycelial strands indicates colonization and invasion, neither yeast nor mycelium can be considered the "parasitic" form. The opportunistic fungous diseases include aspergillosis, mucormycosis, cryptococcosis, and candidosis. The latter two are discussed in the section of pathogenic yeasts. In addition to these, there are a number of rare infections caused by a variety of soil fungi. Such infections again depend on opportunity for entrance of the fungus and host debilitation, even if transient. This category also includes infections by colorless algae, fungus-like protozoa, and other organisms that are as yet unclassifiable.

The Dimorphic Pathogenic Fungi (See Table 15–1)

Chapter 15

BLASTOMYCOSIS

DEFINITION

Blastomycosis is a chronic granulomatous and suppurative disease having a primary pulmonary stage that is frequently followed by dissemination to other body sites, chiefly the skin and bone. The primary infection in the lung is often inapparent. The name blastomycosis can refer to any infection caused by a yeastlike organism, so that reference must be made in terms of specific etiology. The causative agent is the dimorphic fungus *Blastomyces dermatitidis*, which has been assumed to be a soil saprophyte in nature, but its ecologic niche has not as yet been delineated. The disease is most prevalent in men in the fourth through sixth decades of life. Formerly thought to be restricted to the North American Continent, the disease has been described from diverg-

Table 15–1. *Pathogenic Fungi**

Disease and Etiologic Agent	*Saprophytic Phase (25°C)*	*Parasitic Phase (37°C)*
	Thermal Dimorphic Fungi	
Blastomycosis *Blastomyces dermatitidis*	Septate mycelium, microaleuriospores —pyriform, globose, or double. Colonies white or beige, fluffy, or glabrous.	Budding yeast with broad based bud, 8 to 20 μ.
Histoplasmosis *Histoplasma capsulatum*	Septate mycelium, microaleuriospores, tuberculate macroaleuriospores. Colonies, white or buff, fluffy.	Small, single budded yeasts, 1 to 5 μ. 5 to 12 μ in var. *duboisii.*
Paracoccidioidomycosis *Paracoccidioides brasiliensis*	Similar to *Blastomyces dermatitidis.*	Large, multiple budding yeasts, 20 to 60 μ.
Sporotrichosis *Sporothrix schenckii*	Septate, delicate mycelium, conidia sessile on delicate conidiophores. Colonies verrucous, black, white, or grey	Fusiform, oval budding yeasts, 5 to 8 μ.
	Tissue Dimorphic Fungi	
Coccidioidomycosis *Coccidioides immitis*	Septate mycelium, fragment to arthrospores. Colonies buff or white, fluffy or "moth-eaten."	Spherules 10 to 80 μ. Endospores produced. Replicates in tissue.
Rhinosporidiomycosis *Rhinosporidium seeberi*	Not known.	Spherules 100 to 300 μ. Endospores produced. Replicates in tissue.
Adiospiromycosis *Chrysosporium parvum* and var. *crescens*	Dense, dry, fluffy tan colony. Aleuriospores on pedicles, 3 to 4 μ.	Spherules 40 μ (*C. parvum*) to 500 μ (var. *crescens*). No endospores. Does not replicate in tissue.

*From Rippon, J. W. *In* Burrows, W. 1973. *Textbook of Microbiology.* 20th ed. Philadelphia, W. B. Saunders Company, p. 722.

ent parts of Africa and the Malagasy Republic. Unlike the other common systemic mycoses, it does not appear to have a mild subclinical form, and spontaneous resolution of the infection has been documented only rarely.[72a]

Synonymy. North American blastomycosis, Gilchrist's disease, Chicago disease.

HISTORY

In 1894 T. Caspar Gilchrist described a new type of skin disease caused by a yeast (blastomycetic dermatitis) before the American Dermatologic Association. The patient's disease had been diagnosed as scrofuloderma of the hand. The attending physician (Dr. Duhring) had sent a biopsy specimen to Gilchrist for pathologic diagnosis. Gilchrist could find no tubercle bacilli but noted "curious bodies" distributed throughout all sections. He at first thought they were protozoan parasites but noted that they appeared to be budding and hence were yeastlike. The presentation of this case in May of 1894 preceded that of Busse's case of "*Saccharomyces hominis*" (cryptococcosis) by six months. At the time of Gilchrist's first publication, he thought his organism and that of Busse were identical.[31] A yeast (blastomycete) had been recovered from culture of Busse's patient, but further study of Gilchrist's blastomycete was prevented because the lesion from which it had come was completely excised before he could take material for culture.

A second case was published by Gilchrist and Stokes in July, 1896.[32] A 33-year-old male was referred to Gilchrist by a Dr. Halsted with the diagnosis of lupus vulgaris. The disease started behind the ear and gradually extended to involve much of the face. Over the next 11½ years, lesions appeared on the hand, scrotum, thigh, and back of the neck, in that order. Gilchrist noted that many lesions spontaneously healed, leaving scar tissue, and that active lesions had an elevated border with minute pustules in which the organism could be demonstrated in a potassium hydroxide mount. He also noted that the local lymph nodes were not involved. With our present understanding of the disease, we would assume that Gilchrist's patient had chronic cutaneous blastomycosis secondary to pulmonary infection rather than a primary cutaneous infection. Gilchrist was able to culture the organism from this case and noted it grew as a soft mass at first that produced "prickles" and eventually a fluffy mycelium. He thus concluded it was different from Busse's organism, which remained a yeast in all subcultures. He and Stokes named the fungus *Blastomyces dermatitidis* in 1898.[33] The patient was lost to follow-up, as he had to return home to a sick wife, but Gilchrist declared the prognosis "good" because his previous lesions had healed. Gilchrist also produced the disease in lungs and other organs in experimentally infected animals.

In the next few years many other cases of blastomycosis were recognized, especially in the Chicago area; thus the disease became known as the "Chicago disease." Montgomery and Ormsby's review of 1908 described several systemic infections in the Chicago area, and they reviewed the world literature.[55] However, they included the cases of Busse and Curtis, which were later realized to be cryptococcosis. In their own cases, however, they did describe the systemic nature of the blastomycosis and its essential pathology. For many years Gilchrist's disease was confused with cryptococcosis (under the name torulosis), coccidioidomycosis, and paracoccidioidomycosis.[82] These diseases were not fully delineated until the early 1930's. Almeida defined paracoccidioidomycosis, and Benham[8] resolved the mycologic problems of the other two.

Until 1950 it was assumed that there were two distinct forms of blastomycosis—systemic and cutaneous. The route of infection was thought to be through the skin in the cutaneous form, and by way of the lungs for the visceral type. Careful analysis of pathologic material from both types of the disease led Schwartz and Baum in 1951[73] to suggest that all cases of blastomycosis begin as primary pulmonary infections. Subsequent studies have tended to confirm this.[6,15,27,58,73,74] The confusion about what constituted a primary cutaneous lesion in this and other systemic mycoses was finally clarified by Wilson et al.[89,90] They used primary cutaneous sporotrichosis as the model infection. The lesion produced in this disease is chancriform, and there is local lymphatic involvement. When cases of cutaneous blastomycosis in the literature, including those of Gilchrist, were carefully reviewed, the lesions were noted to be of the chronic granulomatous type without apparent involvement of the local lymphatics and, therefore, secondary manifestations of primary pulmonary disease. As is the case with the other systemic mycoses, essentially all infections are acquired by inhalation of the spores. Primary cutaneous disease does occur but is extremely rare.[90]

Designating a disease by its geographic distribution is an unfortunate and often confusing practice. African histoplasmosis has been found in Japan; maduromycosis (after Madura, India) and European blastomycosis (cryptococcosis) are worldwide in distribution. Gilchrist's disease was called North American blastomycosis for many years because all cases had been found in the North American continent. In 1952 Broc and Haddad[10] described a case which was similar to Gilchrist's disease in Tunisia. The fungus isolated was named "*Scopulariopsis americana*," which is now considered a synonym of *B. dermatitidis*. Since that time autochthonous cases have been found in South Africa, Uganda, Congo, Morocco, Tanzania, and the Malagasy Republic. It is apparent that the disease is widely distributed in Africa and perhaps other places in the world as well. Infections transmitted by fomites have been recorded in Switzerland and England.

ETIOLOGY, ECOLOGY, AND DISTRIBUTION

The only etiologic agent of blastomycosis is *Blastomyces dermatitidis*. McDonough and Lewis described the perfect stage in 1967[51] as *Ajellomyces dermatitidis*. It has since been shown that most isolants from human and animal disease from various geographic locations in the United States and Canada mate with the testor strains. It therefore seems reasonable to conclude that a single species is involved in this disease in North America. Isolants from Africa form gymnothecia also but, to date, no fertile ascospores.[50] It remains to be determined whether the African strains represent a separate species. The choice of the generic name *Blastomyces* for the imperfect stage was unfortunate, as this term had been used previously in mycology and thus was illegal under the International Botanical Code. Since it is so widely accepted now, it would serve no purpose to change it, as the organism does not fit well into any of the other described genera of Hyphomycetes.

The natural habitat of *Blastomyces dermatitidis* remains an enigma. Since it now appears that essentially all infections are acquired by inhalation of spores into the lungs and follow the pattern of disease established for the other systemic mycoses, the organism should be a saprophyte in soil, producing mycelia and air-borne spores. All attempts to isolate the organism from soil in endemic areas have failed, with only a few exceptions.[20] Denton and DiSalvo[19] recovered the organism from 10 of 356 soil samples obtained near Augusta, Georgia. These were collected on only two days—March 20, 1962, and February 7, 1963. All samples taken later, even at the same time of year and at the same places, were negative for *B. dermatitidis*. The areas yielding positive samples included chicken houses, cattle loading ramps, an abandoned kitchen, a rabbit pen, and a mule stall.

McDonough[50] has studied the fate of the organism in soil. He found that yeast cells rapidly lyse when placed in soil,[52, 53a] and mycelium soon disappears under these same conditions. Conidia placed in soil survive for a few weeks only. These same conditions also eradicate *Coccidioides immitis*. *B. dermatitidis* may not be able to compete with the normal flora of soil and perhaps survives only in a very restricted ecologic environment. Several reviews of case studies[12,15,21,58,91] have noted that primary infection appears to occur during the cooler, wetter months of the year. This correlates with the reports of recovery of the organism from a saprophytic environment only in February and March. It is possible that the organism is dormant most of the time, fluorishing only rarely under particular climatic and environmental conditions of the colder seasons. It has been reported to grow on tree bark.

The geographic distribution of blastomycosis is also problematic. Unlike histoplasmosis and coccidioidomycosis, which occur within a well-established range of areas, delineation of the endemicity of blastomycosis has been hampered by the lack of two important epidemiologic tools. There is no acceptable and useful skin test available to detect subclinical and resolved infection in the population, and the organism cannot be regularly recovered from a saprophytic environment. Our knowledge of the endemic range of blastomycosis is based, therefore, only on reports of clinically apparent human and animal disease.[3]

The overall range of autochthonous cases in the American continent includes the middle western states, the southeastern states (excluding Florida), and the Appalachian states (Fig. 15–1). In general, this is the drainage pattern for the Mississippi and Ohio River basins and, to a certain extent, for the Missouri River.[27,59] From Minnesota, the endemic zone extends into the southern section of Manitoba and southwest Ontario in Canada.[41] In western New York and eastern Ontario, another endemic zone originates, which then follows the St. Lawrence River through Quebec. The highest number of Canadian cases are recorded from the latter province.[34] New England is spared the infection, as are the plains states, the two tiers of mountain states, and the west coast. All cases of blastomycosis in California have been traced to other sections of the country.[81] In analyzing 1476 human cases and 384 canine cases, Furcolow et al.[27] found Kentucky had the highest number of human infections. The highest incidence was largely to the south of the Ohio River and to the east of the Mississippi (excepting Arkansas and Louisiana). Other areas of disease concentration are North Carolina, except for the coastal areas, and the western shore of Lake Michigan, particularly the Chicago and Milwaukee regions. The prevalence of canine disease showed a distribution similar to human cases. Most infected humans and dogs came from rural environments. A notable number of human infections occurred in agricultural workers or hunters. Epidemics have been recorded[36, 72a, 80] in North Carolina, Arkansas, and Minnesota. In the former state, several people in the same community developed pleural symptoms at the same time during the winter season. No particular point source could be detected. In the latter case, an Arkansas farmer and many of his hunting hounds developed blastomycosis at about the same time following a hunting expedition in the local forest. Sarosi et al.[72a] reported on a common source outbreak of the disease that occured in 21 persons. They present evidence

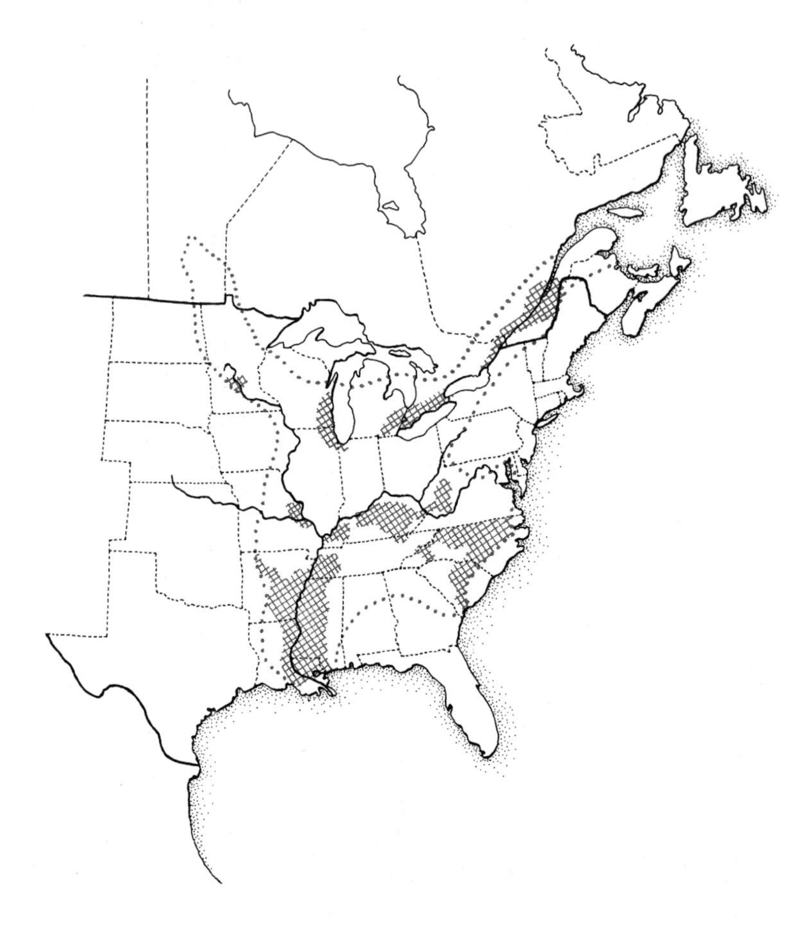

Figure 15–1. Blastomycosis. Incidence and prevalence of blastomycosis in North America. The dotted line indicates the known endemic region. The hatched areas are those with the highest incidence.

that some of the patients had mild infections that resolved spontaneously.

The discovery of blastomycosis in Africa has been too recent and the cases too few to generalize on the epidemiology of the disease there. The most interesting observation has been the great diversity of countries reporting cases over a wide geographic range. The first case found in Tunisia by Broc and Haddad[10] was reported in detail by Vermeil.[87] There is some doubt as to the authenticity of this case, however. A subsequent case has been found from the same region[61] and the identity of the culture confirmed. Since then two cases have come from the Congo,[29] one each from South Africa and Uganda,[23] four from Rhodesia,[69] one from Morocco (not confirmed),[68] and two each from Tanzania and the Malagasy Republic.[14] Segretain[75] maintains that the African form is characterized by fewer ulcerated cutaneous lesions and more gummas or subcutaneous abscesses. It has been suggested that Africa may be the original home of the fungus and the disease. The organism is postulated to have been brought to America during slave trade days.

Blastomycosis has been acquired by handling fomites. There is a case recorded in a tobacco worker in Switzerland and in a packing material handler in England.[58] Martinez-Baez et al. recorded the disease in Mexico, and it may also occur in Central America.[54] A report of the disease in South America is questionable.[62]

The sex and age distribution of blastomycosis is generally the same as that of other systemic nonopportunistic fungus infections. Of the 1114 cases in which data were available, 89 per cent were men.[26] Other studies have found variation in sex distribution.[4,21,49,91] The infection was three times as great among Negroes as Caucasians, and the former tended to be younger. Though primary infections have been recorded in children as young as six months and in a patient in the ninth decade, the peak incidence is in middle-aged adult males;[58] in over 60 per cent of cases, the patients were between 30 and 60 years of age. Twenty-one cases in children have been recorded.[85]

As has been established by analysis of numerous case reports and autopsies, blastomycosis is acquired by the inhalation of spores. Very rarely and under extraordinary circumstances, human to human transmission has apparently occurred. One case is reported to have involved conjugal transmission of yeast cells in semen.[17] The other involved two workers who both acquired blastomycosis.[64] It was surmised that there was aerosol transmission of yeasts from one patient to the other. A more acceptable explanation is that there was a common source of spores from the saprophytic mycelial phase.

CLINICAL DISEASE[15]

The clinical forms of blastomycosis as well as its possible saprobic occurrence and epidemiology are enigmatic. Beginning with Gilchrist's case in 1896 and for many years thereafter, it appeared that there were two distinct forms of the disease dependent on site of inoculation. Pulmonary disease was associated with disseminated blastomycosis and usually had a chronic, relentless, and eventually terminal outcome. The cutaneous form which was not associated with apparent pulmonary involvement waxed and waned over a period of many years. Since the work of Schwartz and Baum in 1951[73] and publication of the description of primary cutaneous disease by Wilson in 1955,[89] it has been thought that essentially all infections begin in the lung.[91] The difference between the two groups of patients is apparently a reflection of the relative ability of the host to handle the disease. If we extrapolate from the known clinical picture of coccidioidomycosis and histoplasmosis, then large numbers of self-resolving subclinical pulmonary infections of blastomycosis should occur. At present we do not know if this entity exists. Although apparent spontaneous recovery has been noted,[72a,80] most of the evidence tends to indicate that once an infection is established the disease progresses either as an occult insidious process or as a chronic expanding and eventually systemic infection.[26]

The clinical disease can be divided into four categories: (1) primary pulmonary disease, which may have severe presenting symptoms or be an inapparent infection; (2) chronic cutaneous disease, which may have occult osseous lesions as well; (3) generalized systemic disease, involving multiple organ systems and usually running a rapid course; and (4) inoculation blastomycosis, a self-limiting primary infection.

Primary Pulmonary Blastomycosis

The inhalation of the fungal spores initiates an alveolitis with invasion by macrophages. This is followed by an inflammatory reaction consisting of exudation of cells, which are predominantly polymorphonuclear leukocytes, and later granuloma formation. In this stage or at any stage in any organ, the process may be suppurative or mixed suppurative and granulomatous. The primary lesions often resemble those of tuberculosis or histoplasmosis, characterized by a "Ghon complex" with parenchymal infiltration, lymphangitis, and lymphadenitis. The symptoms are usually a mild progressive respiratory infection with a dry cough, some pleuritic pain, hoarseness, and low-grade fever. Many such lesions appear to undergo primary organization and healing, leaving a small fibrotic scar. There is little evidence of the caseation seen in tuberculosis, and frequently such lesions heal by fibrosis and absorption, leaving no re-

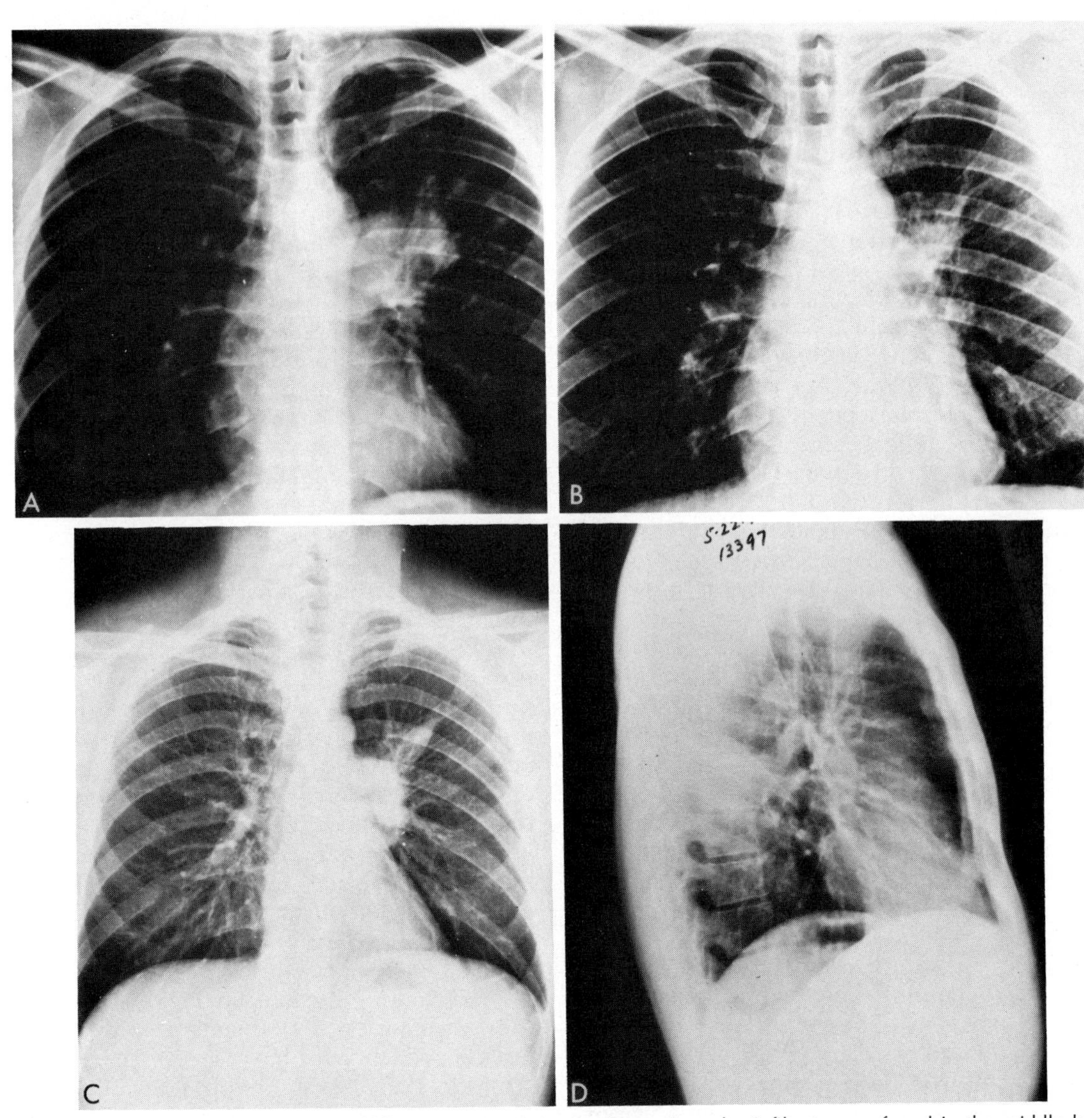

Figure 15–2. Radiologic aspects of blastomycosis. *A,* Large, dense, irregular infiltrates are found in the middle lobe abutting the hilum. Satellite densities are scattered to the main bronchi. On bronchogram the bronchi were found to be patent. (Courtesy of F. Kittle.) *B,* Same patient after two months of amphotericin B therapy. There is a marked decrease in the infiltrate. *C,* Irregular lobulated 4 × 6 cm mass in the superior segment of the lower lobe. There is some atelectasis. *D,* On lateral film the lesion is seen to be located in the posterior portion. The diagnosis of carcinoma was made and a lobectomy performed (see Fig. 15–6 FS B 44). (Courtesy of F. Kittle.)

sidual evidence of infection. However, spread by macrophages and deposition of the organism at distant sites may have already occurred. It is this type of patient that develops the chronic cutaneous or occult osseous type of disease.

An acute form of primary pulmonary disease has recently been described by Serosi et al.[72a] The clinical symptoms were of two types. The first, similar to acute histoplasmosis, consisted of fever, productive cough, arthralgia, and myalgia. The second consisted of mild to severe pleuritic pain. These cases represented a point source of infection, and all resolved without treatment so far.

If it does not resolve, the pulmonary disease evolves into an acute lobar pneumonia, acute bronchopneumonia with rapid hematogenous dissemination, or a more chronic type of infection, such as a suppurating pyogenic process or an expanding "crab claw" shadow granuloma. The patient's symptoms reflect the increasing severity of the disease. Sputum production increases and is now purulent and blood tinged; the temperature is elevated, dyspnea and weight loss occur, along with night sweats and increasing weakness. The pleura may be involved, but less commonly so than in actinomycosis. The physical signs of the disease at this stage are dullness to percussion and altered breath sounds with transitory and variable rales. Sometimes a discharging sinus or subcutaneous abscess develops over the thorax.

X-ray Examination. The roentgenographic picture of the chest during early stages of the disease is one of a widening hilar shadow or unilateral disease resembling tuberculosis or neoplasm. The latter is particularly important, as blastomycosis has been misdiagnosed as carcinoma more often than any other fungous disease.[38] In a survey of the presenting radiologic picture, Pfister et al.[60] found the infiltrative pattern was most common, followed by a nodular pattern resembling miliary tuberculosis. A high percentage of patients had x-ray findings consistent with pulmonary neoplasm (Fig. 15–2,*A-D*). As delineated by Pfister et al., a picture suggestive of blastomycosis includes dense, fibrotic pleura without either exten-

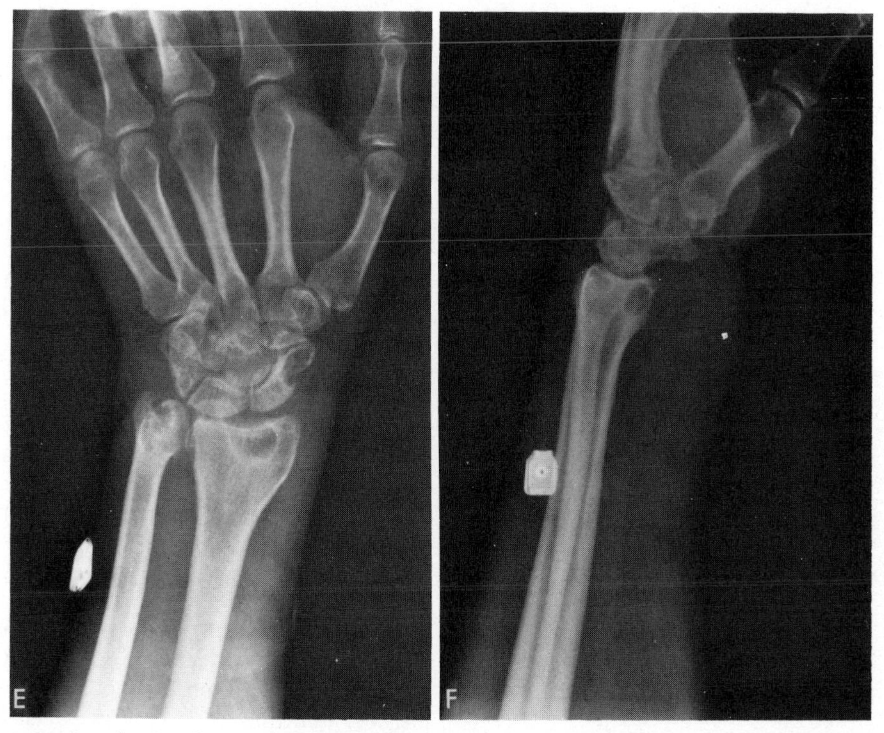

Figure 15–2 *Continued.* *E,* Left wrist. Destructive areas are seen in the carpus, the radial metaphysis, and the distal ulna. The destructive focus in the radius connects to the articular surface, and there is a slight widening of radiocarpal articular space. The foci have sclerotic margins. This was a case of monoarticular involvement diagnosed as rheumatoid arthritis. No organisms were seen in biopsy, but it was culture-positive. The lung fields were clear. *F,* Lateral view shows marked soft tissue swelling.

sive calcification or an accumulation of effusion fluid. The disease may occur anywhere in the lungs, but the posterior segments of the upper lobes are the most common sites of the infiltrative granulomatous type.[59] In most series, the apex is usually spared, but middle lobe and basilar involvement are frequent. Although they are less common than in histoplasmosis or coccidioidomycosis, in some series cavities varying in size from 1 cm to 8 cm occurred in 25 per cent of cases.[12,21,59] Hematogenous spread may produce a picture of miliary disease. If the disease has progressed this far, resolution of the pulmonary lesion does not occur, and the infection progresses to death if untreated.

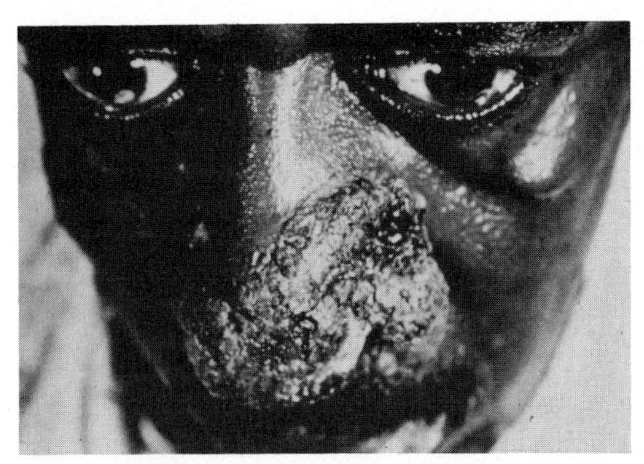

Figure 15–4. FS B 42. Blastomycosis. Evolution of the lesion into an ulcerated granuloma with a serpiginous advancing border. The central area is covered with crusts.

Chronic Cutaneous and Osseous Disease

Cutaneous blastomycosis is the most common form of extrapulmonary disease. Thus skin lesions are often the most frequent presenting symptom of the disease. For this reason many, if not most, cases of this type of disease are first seen in dermatology clinics. Up to 80 per cent of patients have skin lesions, and often the organism is first demonstrated from aspirated material from such sites.[12,21,29,58,91] The first signs to appear are subcutaneous nodules or papulopustular lesions which ulcerate. The initial lesions occur singly or in groups and are most commonly found on exposed peripheral areas, such as face, hand, wrist, and lower legs, or on mucocutaneous areas, such as the larynx[65] (Fig. 15–3 FS B 41). In time, as the disease progresses, lesions also occur on the trunk and other unexposed areas. Within weeks or months the lesions evolve into ulcerated verrucous granulomas with serpiginous, advancing borders which are raised 1 to 3 mm and have a sharp sloping edge (Fig. 15–4 FS B 42). The central area is covered with crusts and characteristically contains "black dots," representing degenerating papillary vessels.[46] This violaceous, discolored, crusty, verrucous lesion has often been misdiagnosed as basal cell carcinoma[57] (Fig. 15–5 FS B 43); however, in blastomycosis small microabscesses occur at the periphery. Aspirated material from these can be examined in a potassium hydroxide mount for demonstration of the yeast cells of the organism. The ulcers heal from the center by fibrosis and cicatrization. Biopsy from the center of the lesion is usually devoid of organisms, and only fibrosis and scar formation are seen. The yeasts are most numerous at the active edge of the lesion; therefore, biopsy and culture should be taken from this site. Lymphadenitis and lymphadenopathy are usually not present in secondary cutaneous blastomycosis. Over a period of years these lesions become deforming, thin, atrophic scars, which may cover large areas of the face, neck, or other areas (Fig. 15–6).

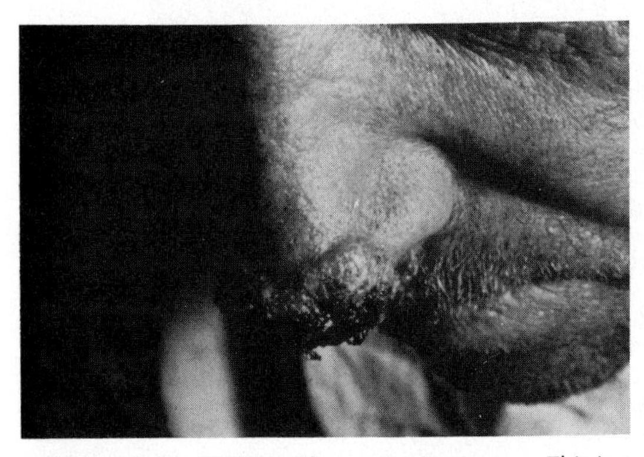

Figure 15–3. FS B 41. Blastomycosis on nose. This is a papulopustular lesion with some scarring in the center. Note small pustules on the outer edge of the lesion. The organism is easily demonstrated from aspirates of this material.

Cutaneous lesions associated with underlying bone involvement or generalized systemic disease may first appear as sinus tracts exuding purulent material containing numerous organisms. Such lesions usually lack the verrucous advancing margin seen in the usual chronic cutaneous disease.

Osseous Blastomycosis. In the several

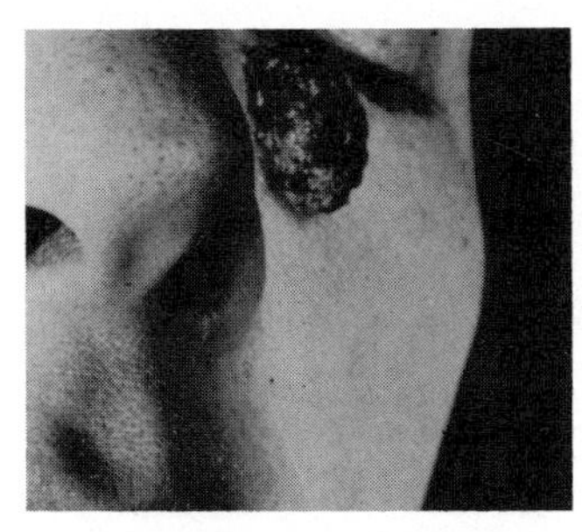

Figure 15–5. FS B 43. Blastomycosis. Dark, crusty, verrucous lesion resembling basal cell carcinoma. There were no other lesions on the patient, and the lung fields were clear. Aspiration of a small microabscess revealed the yeast cells of *B. dermatitidis*.

reviews of case reports of blastomycosis, from 25 to 50 per cent of patients have had disease of the bone.[30] Not infrequently, the only presenting symptom is an occult osteolytic lesion or monoarticular arthritis.[71] The most frequent sites of bone infection are the vertebrae, ribs, skull, long bones, and short bones. In a recent review, the long bones were found to be most commonly involved.[30] The manifestations of osseous disease are protean, and there is no distinctive radiologic picture. The lesions generally consist of a focal or diffuse suppurative osteomyelitis at the epiphyseal end of long bones. Osteolytic and osteoblastic processes may be seen by x-ray examination. The tissue reaction resembles that of granuloma formation of tuberculosis more than the cystlike lesions of coccidioidomycosis. The proliferative response in blastomycosis is less prominent than in actinomycosis (Fig. 15–2,*E*,*F*).

Involvement of the vertebrae is similar to that seen in tuberculosis. In both diseases, granulomatous lesions destroy the disk spaces, erode the vertebrae anteriorly, and produce paraspinal masses. The anterior longitudinal ligaments are dissected from beneath, but the spinal segments are spared. Compression of the spinal cord may result.

Blastomycosis of the joints often presents as a hot, swollen, septic arthritis, which is sometimes accompanied by a chronic exudative sinus tract. The periarticular tissues may be destroyed along with the synovial membrane and articular ligaments, with resultant subluxation. Since bony involvement is so frequent in blastomycosis and so often occult, complete roentgenographic examination of the entire skeleton should be made in all cases of the disease. In low-grade chronic blastomycosis, a single lesion is usually found, and the number of organisms present is small. In generalized systemic disease, however, numerous organisms are present, and several bones are involved.

Systemic Blastomycosis

Patients with extensive and unresolving pulmonary involvement almost always de-

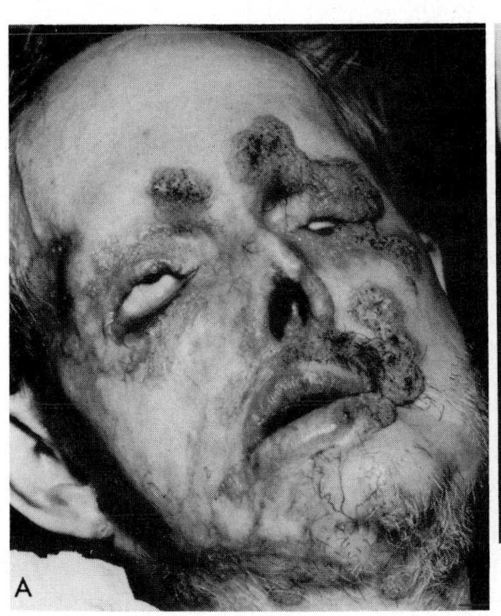

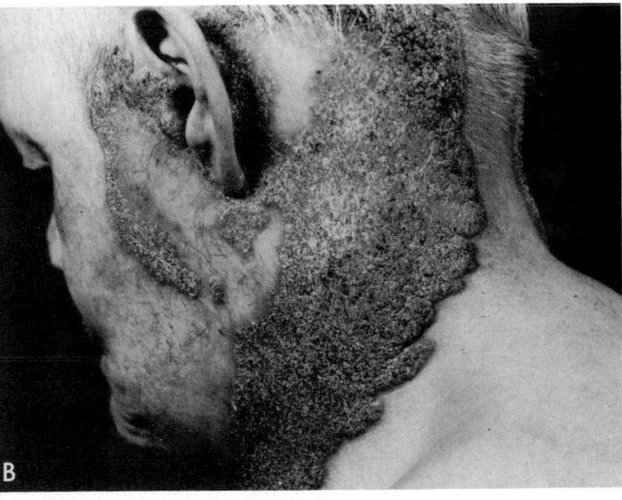

Figure 15–6. Blastomycosis. *A*, Secondary cutaneous lesions in a case of long duration. *B*, The fungus is found in the advancing border within the verrucous vegetations. The central areas clear and heal, with formation of scar tissue. (Courtesy of A. Lorincz.)

velop generalized systemic disease. These patients appear to be less able to contain the infection than those with low-grade chronic disease. Whereas few organisms are seen in biopsies from the latter, lesion material in generalized systemic disease contains numerous fungi. Showers of organisms from the pulmonary foci serve to seed other organ systems. Disseminated infections may manifest as cutaneous and osseous lesions which develop more rapidly and have less tendency to resolve than chronic disease. Usually multiple organ systems are involved, and the patient may have a rapidly deteriorating course.

The third most commonly involved site of extrapulmonary blastomycosis is the urogenital system. In published reports, the incidence varies from 5 to 22 per cent. Epididymitis appears to be the most frequently encountered condition, and it tends to be recurrent. The epididymis is swollen and tender, and testicular involvement is not infrequent. Scrotal ulcers and a draining sinus from an orchitis are sometimes seen. In our experience, the prostate is also commonly involved in disseminated blastomycosis. The prostate is enlarged, boggy, and slightly tender. Organisms can be demonstrated in material obtained after prostatic massage. The disappearance of the fungus from this material may be used as an index of therapeutic progress. Involvement of the female reproductive system is quite rare, as is disease of the kidney. Conjugal transmission of the infection has been reported.[17] As in paracoccidioidomycosis and histoplasmosis, Addison's disease may result from blastomycosis of the adrenals.[37]

The central nervous system is involved by hematogenous spread from other foci. This is not a common complication, being found in only 3 to 10 per cent of infections.[6] The most frequent neurologic symptoms are headache, convulsions, confusion, coma, paraparesis, hemiparesis, and aphasia. The cerebrospinal fluid protein and pressure are elevated. The leukocyte count is variable, and the sugar is usually normal or low. Generally the organism is not recovered in culture. Of the nine patients recorded by Buechner and Clawson,[11] there was a 67 per cent mortality rate. Recently Leers et al.[46a] described a cerebellar lesion that was not associated with meningitis or a detectable primary site of infection.

Granulomas of the liver and spleen are infrequently seen. In contrast to paracoccidioidomycosis, gastrointestinal involvement is quite rare. Endophthalmitis has been recorded in a patient who had no other evidence of disease.[24]

Inoculation Blastomycosis

This is a very rare entity, and most cases are the result of laboratory accidents or accidental implantation during autopsy examination or while embalming patients who have died of blastomycosis.[89] The skin of the fingers or the hands is usually involved. The lesions that develop are of the chancriform type. An indurated ulcer forms which is accompanied by lymphangitis and regional adenopathy. These are mild forms of the disease and heal spontaneously without peripheral extension or dissemination.

DIFFERENTIAL DIAGNOSIS

Blastomycosis must be differentiated from any chronic granulomatous or suppurative pulmonary disease. The list includes histoplasmosis, because of the overlap of endemic areas, tuberculosis, silicosis, sarcoid, and to a lesser extent actinomycosis, nocardiosis, and other bacterial diseases. High on the list of differentials is pulmonary neoplasms. Cutaneous lesions resemble scrofuloderma, lupus vulgaris, epitheliomas, bromoderma, iododerma, nodular syphilids, granuloma inguinale, swimming pool granuloma, and similar diseases.

There does not seem to be an association of blastomycosis with underlying or debilitating disease. In the many reviews of complications that occur in patients with neoplasias and leukemias, of those receiving steroids, no increase in incidence of blastomycosis was found. It is possible that subtle immunologic differences are present in the patient who develops the disease. This is particularly true in patients with the systemic form rather than the chronic cutaneous form of the disease. There is as yet no evidence to confirm this hypothesis. Blastomycosis may coexist with tuberculosis, histoplasmosis, and bronchogenic carcinoma, as well as other diseases with severe pulmonary involvement.

PROGNOSIS AND THERAPY

Prior to the advent of effective chemotherapy, the diagnosis of blastomycosis usually meant a fatal outcome. In cases of systemic disease the mortality rate was 92 per cent. Patients with chronic cutaneous disease usually had a more protracted but also fatal course. There are reports of spontaneous resolution of clinically apparent disease,[72a, 80] but it must be remembered that infections may remain quiescent for years before recurring.

The treatment of choice in all forms of blastomycosis is amphotericin B. The organism is quite sensitive to the drug (0.03 to 1.0 μg per ml *in vitro*), and relapses occur only in cases of inadequate treatment. In the reviews by Abernathy and Jansen in 1960,[2] Parker et al. in 1969,[59] Lockwood et al. in 1969,[48] and Klapman in 1970,[42] clinical cure was achieved when a total of 2 or more grams of the drug was used. There was a significant relapse rate if less than 1.5 g was used. This therapeutic regimen is the one commonly used for the other systemic mycoses. The initial dose is 10 to 20 mg in 300 to 500 ml of 5 per cent glucose given as a slow infusion over a three- to six-hour period, depending on the patient's tolerance. The dose is increased by 10 to 20 mg daily. Initially alternate day therapy may be necessary if side effects are severe. Diphenhydramine (10 to 30 mg) is used to control the nausea; 10 to 30 mg of heparin may be added to avert phlebitis, and aspirin is used for pain and headache. The dosage is increased until the standard amount of 1 mg per kg body weight per day is achieved, and blood levels of 0.5 to 3.5 μg per ml are obtained. The blood:spinal fluid concentration ratios are 30:1 to 50:1. Intrathecal injections may be necessary in cases of meningeal involvement.[58] A 5-ml quantity of spinal fluid is withdrawn, and 20 mg of hydrocortisone is mixed with it. This is slowly instilled intraspinally. After a few minutes, another 5 ml of spinal fluid is withdrawn, mixed with a solution containing 0.5 mg of amphotericin B, and reinstilled. This procedure may be repeated two or three times weekly until a total dose of 15 mg has been given. This is considered adequate for therapy of meningitis. A treatment schedule for systemic disease in children has been worked out by Turner and Wadlington.[85] Oral amphotericin B has been used, but results have been variable and serum levels unpredictable.

The aromatic diamidine, hydroxystilbamidine, had been used with some success in the treatment of blastomycosis. The drug is administered as a daily dose of 225 mg in 500 ml of saline. Total dose of 8 g is recommended. There are some severe side effects with this drug, however, and a relapse rate up to 30 per cent has been noted. In contrast to the general experience, some investigators claimed to have had excellent results using this drug.[42, 48] In patients with limited disease and preexisting kidney impairment, serious consideration may be given to the use of the drug.

Hamycin[77] has shown activity *in vitro* and in experimental infections, but clinical trials have been unrewarding. Saramycin (X5079C) has been used successfully in a few cases.[64] Subcutaneous injection of 4 mg per kg per day was efficacious in cutaneous blastomycosis. However, it appears that this drug will not be made available for general clinical use. Older modes of therapy, such as iodides, x-ray, and vaccines, have been replaced by modern therapeutic procedures. Allergic reactions may occur during therapy, but these can usually be controlled with the judicious use of steroids and do not require desensitizing procedures.

Surgical procedures are of value when large abscesses require drainage. Thoracic surgery to remove large pulmonary lesions or cavities should not be performed until the exudative phase is controlled by medical means. Larson has reviewed the long-term status of several patients treated surgically compared to those who had medical management alone.[45] Although some patients with minimal involvement were cured by resection, the authors concluded that amphotericin B should be used to treat all cases before surgical procedures are instituted.

PATHOLOGY[15]

Gross Pathology. The sharp differentiation between the two categories of blastomycosis is again evident in the gross and histologic pathology of the disease. In patients in whom there is minimal or no apparent pulmonary disease, the cutaneous, laryngeal, and mucocutaneous lesions are so verrucous and hyperplastic that they lead to a preliminary diagnosis of carcinoma. Even the cut surface of these lesions and the carcinoma-like granulomas removed from lungs give no indication of their true etiology. Accurate diagnosis depends on careful histologic examination and culture of such material. At autopsy the healed primary lesion of patients with generalized blastomycosis often appears only as a focus of pulmonary scarring or minor pleural fibrosis. In contrast to histoplasmosis, calcified fibrocaseous nodules are not a usual finding.

Whereas granuloma formation is characteristic of prolonged chronic disease, disseminated blastomycosis tends to be suppurative or mixed suppurative and granulomatous. The pulmonary foci may be scattered, small, multiple nodules or large, caseous nodules, abscesses, and enlarging granulomas (Fig. 15–7 FS B 44). Caseation occurs less commonly than in tuberculosis. Although cavitation is infrequent, it may

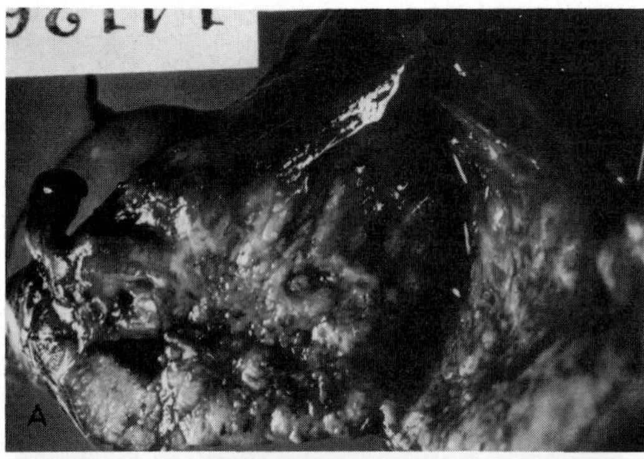

Figure 15–7. FS B 44. Blastomycosis. *A,* Cut surface of lung. The involved area is firm and granular. The texture and infiltrative pattern of the peripheral portions of lesions are indistinguishable from carcinoma of the lung. The patient had been diagnosed by x-ray as having carcinoma. The reaction was predominantly of the suppurative necrotizing granulomatous type. *B,* Same lung after perfusion. (Courtesy of S. Thomsen.)

occur at single or multiple sites. Pleurisy and pleural fibrosis are also encountered. If the disease was very active, invasion of other systems in the chest area may have occurred. Pericarditis and endocarditis, which are not usually associated with fungous infections, have been reported. Congestive heart failure resulting from pericardial adhesions and the presence of massive amounts of pericardial pus has also been noted.

The disease may extend from the pleural cavity into the vertebrae and ribs and through the chest wall, thus simulating actinomycosis. As in the latter disease, draining sinus tracts may burrow through from any osseous or subcutaneous abscess to the skin. The pus is characteristically pink-tinged because of the extravasated red blood cells. It usually contains numerous organisms which are easily demonstrated in potassium hydroxide mounts. Psoas abscesses similar to those seen in tuberculosis are also found. Lymph nodes become masses of abscesses and granulomas, with some areas of necrosis and some of fibrosis. Abscesses of the brain and meninges are found which also mimic tuberculosis. Prostatic involvement is common in systemic disease, whereas the uterus and tubes are not involved. Spleen, liver, and kidneys are usually spared, although minute abscesses are found in the late stages of extensive systemic disease.

Histopathology. The tissue response in blastomycosis varies from that of epithelioid granulomas to chronic suppuration, necrosis, and fibrosis or a combination of all of these. Again the reaction varies according to the type of disease in the patient, e.g., epitheliomatous hyperplasia, and few organisms are seen in the chronic type of infection, whereas suppuration, necrosis, and numerous organisms are more commonly encountered in the generalized systemic form.

The histologic diagnosis of blastomycosis depends on the demonstration of the yeasts of *Blastomyces dermatitidis.* The organisms are generally of uniform size, varying from 8 to 15 μ in diameter, depending on the age of the lesions (Fig. 15–8,*A* FS B 45,*A*). Some organisms may be as large as 20 to 30 μ in diameter. The wall is thick and rigid and is termed "double-contoured," or doubly refractile. The cytoplasm usually shrinks away from the wall, leaving a space, but the cytoplasm stains more prominently than the wall in hematoxylin-eosin preparations. In well-fixed, well-stained sections, the multinucleate

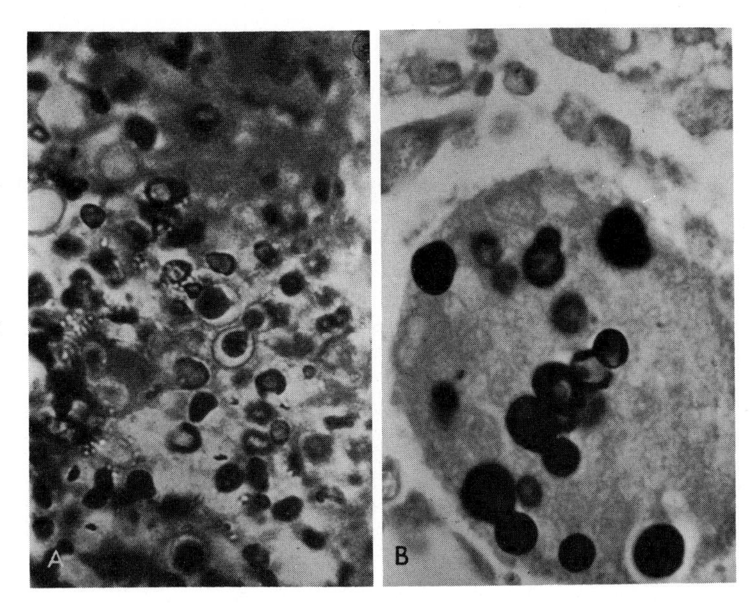

Figure 15–8. FS B 45. Blastomycosis. *A,* Budding yeast cell. The cytoplasm has shrunk from the side of the colorless wall but can be seen to be multinucleate. Hematoxylin and eosin stain. ×440. *B,* Numerous yeast cells, some of which are budding, in a giant cell. Note the broad-based buds. Methenamine silver stain. ×440.

nature of the cytoplasm is demonstrated. The most characteristic feature of *B. dermatitidis* is the broad-based bud. An unequivocal histologic diagnosis cannot be made unless this form is seen (Fig. 15–8,*B* FS B 45,*B*). Although small forms no larger than histoplasma yeast cells have been described,[6,73,74] these have broad-based buds, in contrast to *H. capsulatum,* which has a thin-necked bud on the yeast cell. Stainability of the organisms is quite variable also. Even in sections treated with special fungus stains, the yeast cells may be difficult to see. In hematoxylin and eosin preparations, the wall is colorless, and the organism may appear only as an outline in a giant cell (Fig. 15–9). The GMS or Gridley stains demarcate the wall, but its outline may be irregularly colored, especially if dead or dying organisms such as those found in chronic disease are present. In material from disseminated cases, viable organisms which were well fixed and stained are easily seen. Hyphae are very rarely present in tissue.

The finding of cells without buds necessitates the differentiation of *B. dermatitidis* from *Coccidioides immitis, Cryptococcus neoformans, Paracoccidioides brasiliensis,* and *Histoplasma capsulatum* var. *duboisii.* The yeasts of *B. dermatitidis* closely resemble those of *H. capsulatum* var. *duboisii,* and both are endemic in some of the same geographic regions. However, the latter organism lacks the broad-based bud. *H. capsulatum* is uninucleate and has a narrow-necked bud. The yeasts of *B. dermatitidis,* including the small forms, usually have several nuclei in addition to their broad-based buds. It must be remembered that both species have been isolated in culture from the same lung.[9] Groups of young spherules of *C. immitis* are especially difficult to differentiate from *B. dermatitidis.* Only with the finding of a mature spherule with endospores can one be sure of the diagnosis. The capsule of *C. neoformans* stains a brilliant pink color with Meyer's mucicarmine stain, even in old lesions in which little capsular material is present. The yeast of *B. dermatitidis* is stained only faintly

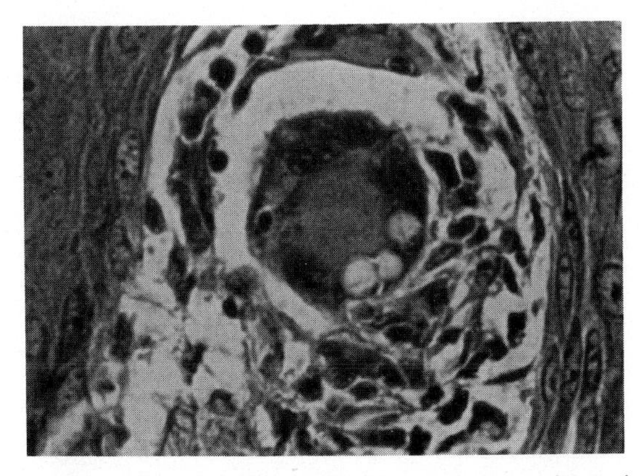

Figure 15–9. Blastomycosis. In this hematoxylin and eosin preparation, the yeast cell is outlined in a giant cell. This was from a case of cutaneous blastomycosis in which there was extensive pseudoepitheliomatous hyperplasia. Interspersed in the epidermis were a few microabscesses containing neutrophils, lymphocytes, and an occasional giant cell. Yeast cells were very rarely encountered.

by this procedure. Needless to say, in all cases isolation by culture and identification of the etiologic agent is the only unequivocal method for diagnosis of the disease.

In primary pulmonary infection, the initial cellular reaction is inflammatory. Numerous polymorphonuclear neutrophils are seen. In time, a chronic granulomatous reaction with focal suppurative areas is found—the picture of an epithelioid cell granuloma. Multiple, small abscesses are present within the granuloma. These contain leukocytes, debris, and usually a few giant cells. It is within these abscesses that the organisms are found. Not infrequently they are seen as shadows outlined within the giant cells. The hyperplasia may be so great as to suggest neoplasia, and the microabscess is often the only key to the true etiology.

In extensive disseminated blastomycosis, organization into well-formed granulomas is less pronounced, and a suppurating necrotic abscess is the histologic picture seen. As the patient's defenses diminish, the organism proliferates in great numbers, and at autopsy masses of yeasts may be found in all organ systems.

The histologic picture of chronic cutaneous disease is quite different from disseminated infection. The organisms are carried to the papillary dermis, probably by macrophages. They then incite an acanthosis that is more extensive in this than in any other fungal disease. This pseudoepitheliomatous hyperplasia has often been diagnosed as squamous cell carcinoma (Fig. 15–10). However the epithelial cells are well differentiated when the acanthosis is due to blastomycosis. If examined carefully, microabscesses are seen that contain neutrophils, some lymphocytes, and an occasional giant cell (Fig. 15–11). The fungus which is generally present in very small numbers in this form of blastomycosis is found within these abscesses. Diligent search of serial sections may be required to find the yeasts. Though the lesions begin in the dermis, prolific downgrowth of the rete ridges often results in the finding of microabscesses in the epidermal areas. Beneath the extremely irregular border

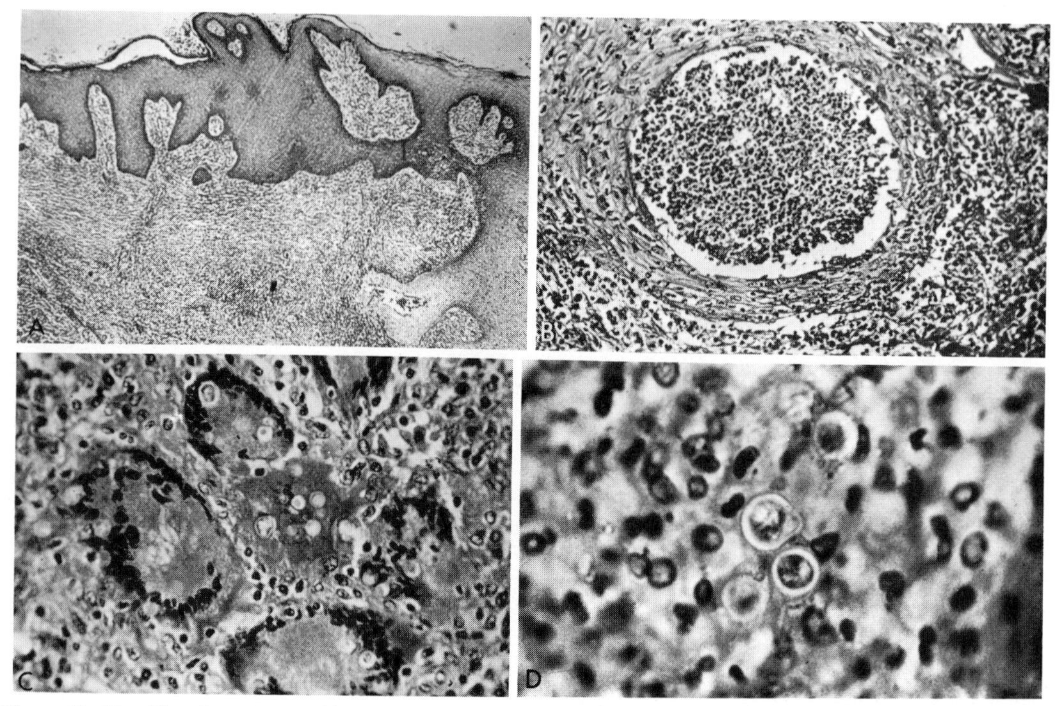

Figure 15–10. Chronic cutaneous blastomycosis. *A*, Extensive acanthosis and pseudoepitheliomatous hyperplasia. The downward growth has resulted in isolated islands in the dermis. A few microabscesses are seen in the upper epidermis. There is an infiltrate in the dermis of lymphocytes and plasma cells with a few neutrophils. ×100. *B*, A microabscess lying within the epidermis. In the neutrophilic debris were a few yeast cells. ×400. *C*, Giant cells from an organized granuloma, showing intracellular yeast cells. ×1000. *D*, Extracellular organisms that resemble the young spherules of *C. immitis*. ×1000.

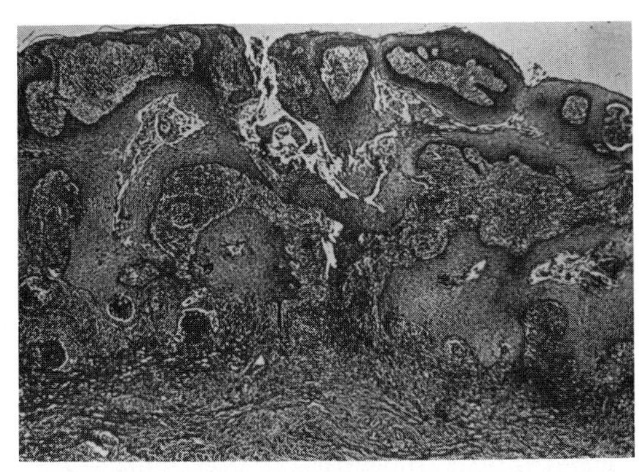

Figure 15–11. Blastomycosis. This case was diagnosed as basal cell carcinoma because of the very extensive cellular proliferation. On histologic examination, the cells were seen to be well differentiated, and a few yeast cells were found in the microabscesses. ×100.

of the epidermis, a band of infiltrate is seen. The cells include neutrophils, lymphocytes, and a few giant cells. Many dilated blood vessels are present, but fibrosis is not extensive, and connective tissue stroma is minimal. In healing lesions, the fibrosis is more apparent but not dense or firm.

ANIMAL DISEASE

Natural disease among dogs is a very common finding in endemic areas. Furcolow et al.[27] tabulated 384 canine cases, but this probably represents only a minimal indication of the true incidence. The disease in dogs was generally found to occur in the endemic areas known from records of human cases, being particularly prevalent in Arkansas, Mississippi, and Kentucky. The disease in dogs exhibits clinical symptomatology and pathology similar to human blastomycosis. Selby[76] has emphasized the occurrence of cutaneous ulcerative lesions, while others have found massive pulmonary involvement and osseous disease. Treatment with amphotericin B has been successful.[13] A total dose in the range of 3.75 mg per kg is recommended. The efficacy of serologic procedures in the diagnosis of canine disease has been reviewed by Turner et al.[84] Precipitins and complement-fixing antibodies were not dependable indicators of disease. Culture of lesion material on laboratory media was more efficient than mouse inoculation for recovery of the organism.

Disease in the horse has been recorded several times, particularly in Kentucky. Cutaneous lesions in the lower legs are the most common presenting signs, with lung involvement almost always present. An interesting case in another animal was the occurrence of blastomycosis in a northern sea lion (*Eumetopias jubata*).[88] This 8-year-old female had been housed in the Chicago Zoo for six years. She died of massive infection of the lung and meningitis from the "Chicago disease." The disease has also been recorded in the cat[78] and other animals.

Experimental disease can regularly be produced in laboratory animals.[19] In a study of comparative susceptibilities of such animals, Conti Diaz et al.[16] inoculated guinea pigs and hamsters intratesticularly and mice intravenously with 9400, 940, and 94 mycelial fragments. The organism was recovered in all three species at all doses. Soil flotation followed by injection of the supernatant intravenously into mice was used successfully by Denton and DiSalvo[19] to isolate the organism from soil. It is possible that the intratesticular inoculation of soil samples may be a better method than intravenous inoculation, as this organ is cooler than other parts of the body and most conducive to the growth of the fungus.

Sputum and contaminated clinical material mixed with an antibacterial antibiotic can be injected intraperitoneally into mice as an adjunct to cultural isolation. Mice are killed after three to four weeks and examined for the presence of pus and abscesses in the omentum and peritoneal cavity. Such lesions are usually near the pancreas, and the omentum may be adherent to other internal organs. Intravenous inoculation may result in pulmonary disease.

Landay et al.[44] have compared the relative susceptibility of the two sexes of hamster to experimental infection. As has been observed in other mycoses, males were more susceptible than females. Experimental cutaneous disease was studied by Salfelder.[70] Subcutaneous injection of *B. dermatitidis* yeast cells into animals produced an active skin lesion which did not ulcerate. Regional lymphadenitis occurred, but this was sometimes difficult to discern. The lesions healed spontaneously without hematogenous dissemination. The relative pathogenicity of the yeast and mycelial forms of the organism have been studied by Guidry and Bujard,[35] using the chorioallantoic membrane of embryonated chick eggs. At 37°C the yeast cells rapidly penetrated the mesoderm and proliferated. At the same temperature, mycelial elements did not invade until they converted into the yeast phase. At 31°C mycelial elements did not convert and were confined to the ectoderm.

Experimental disease in dogs has been studied by Ebert et al.[22] They found that injection of 10^5 cells was sufficient to produce disease. The clinical course was more severe than in experimental canine histoplasmosis. Amphotericin B therapy was found to prevent the death of some of the animals, but not all survivors were cured. All control animals died of disseminated disease. The natural occurrence in bats of several mycoses (histoplasmosis, paracoccidioidomycosis, and sporotrichosis) has been recorded, and the bat has been postulated to be a natural vector of the etiologic agents of these diseases. Tesh and Schneidau[83] produced experimental blastomycosis in bats (*Tadarida brasiliensis*) by inoculating them with 5.5×10^5 fungal cells. All animals developed severe disease, and most of them died. *B. dermatitidis* was isolated in fecal cultures obtained during the course of the disease.

BIOLOGICAL STUDIES

The morphologically distinct saprophytic and parasitic forms of *B. dermatitidis* have interested investigators for many years. It was noted early that physiologic changes occur concomitantly with morphologic changes. Levine and Ordal[47] demonstrated a severalfold increase in metabolic rate of the yeast form compared to the mycelial form of the fungus. Later work has shown that temperature is the only factor controlling dimorphism in the organism. There are some differences in enzyme production between the mycelial and yeast forms,[7,66] for in general more proteolytic enzymes are elaborated in the former than in the latter form. There are also some differences in enzyme production associated with the two sexual mating types.

As is the case with other dimorphic fungi, more chitin is present in the yeast form than in the mycelial. The yeast cell wall also contains about 95 per cent α-glucan, whereas the cell wall of the mycelial form contains only 60 per cent; the remaining 40 per cent is β-glucan.[25] Fundamental differences in membrane organization and morphology have been discerned in ultrastructural studies of the two morphologic forms. Conversion of the yeast to mycelium is accompanied by membrane reorganization and the appearance of woronin bodies where mycelial septa are going to form.[28]

The yeast form in nature is known to exist only in infected animals. Saliva containing yeast cells may contaminate the soil, but the cells quickly lyse.[52] McDonough has shown that both live and dead yeast cells lyse when placed in soil; therefore *B. dermatitidis* probably exists in nature only as a mycelial soil saprophyte. Conversion to the yeast stage and the production of disease may represent a case of accidental pathogenicity, which therefore represents a dead end as far as dissemination of the species is concerned.

Studies of the chemical composition of the yeast of *B. dermatitidis* have shown that it differs from most other pathogenic fungi in having a very high lipid content. In addition, both the yeast and mycelium forms produce an unusual metabolic product, ethylene.[56] This gas is also elaborated by ripening fruit and incites the degeneration of chlorophyll. So far the gas has been found only in a few other fungi, including *Penicillium digitatum* and the etiologic agents of histoplasmosis, blastomycosis, and paracoccidioidomycosis. A role in pathogenesis has been postulated, but no experimental evidence has been established.

IMMUNOLOGY AND SEROLOGY

The immunology and serology of blastomycosis is the least well understood of any of the systemic mycoses. No specific skin test is available, so that the existence of resolved subclinical infections which could lead to specific immunity is not known. The infection occurs in apparently normal persons, and the disease has not been associated with debilitating conditions or predisposing factors. Thus, so far as is known, it is not an opportunistic infection. Animal experiments and available patient data indicate that, once an infection is established, the disease is often chronic and persistent and does not tend to resolve. This is in sharp contrast to the other pathogenic fungi, with which subclinical infections are the rule. More research needs to be done on this point.[26,72a] There is

Figure 15–12. Blastomycosis. Immunodiffusion lines using cell sap antigen. The patient had systemic disease and was treated with amphotericin B. Wells 1, 2, 3, and 4 represent sera from various stages of the disease. Some lines are present early in the disease and disappear, others appear during convalescence. Well 1 is before treatment, well 2 is one month after initiation of therapy, well 3 is after two months of therapy when lesions are beginning to resolve. The patient was showing some allergic reactions at this point. Well 4 is seven months after completion of therapy. Some lines have disappeared.

an obvious difference in the way patients handle the infection. The general immune competence of the patient determines whether the infection will develop into the progressive pulmonary and disseminating form of disease, or the very long, chronic, but eventually progressive, cutaneous form. So far this difference in disease types has not been reflected by detectable immunologic differences between patients. Subtle defects in immune mechanisms, leukocyte function, and host defenses have been discovered in association with other infectious diseases, and this is at present an area of active investigation of mycotic infections. It is hoped that such studies may define the patient selected by the disease and the form that the disease will take.

Serologic procedures using the commercially available blastomycin are not useful in the diagnosis or prognosis of blastomycosis.[42] Skin testing with this reagent is a useless procedure. Cross-reactions to histoplasmin and coccidioidin, particularly in the early stage of the disease, are very common. The complement fixation (CF) titer for *H. capsulatum* is often higher than that for *B. dermatitidis* in patients with proven blastomycosis. Yeast phase antigen is a little more specific, but only 30 per cent of patients had a significant CF titer during active disease.[5,72] Agar gel precipitins were first described by Abernathy and Heiner,[1] but again they were not specific. Using an antigen prepared from the cell sap (see Appendix) of the organism, a specific immunodiffusion technique has been developed[67] (Fig. 15–12). There was no cross-reaction with other diseases, but only fresh sera had consistently positive results. Stored sera rapidly lost their reactivity. As has been found in the serologic evaluation of candidosis, the cell wall and cultural extracts contain many antigens in common with other fungi; the cell sap contains at least some specific antigens for the organism involved. Kaufman et al.[40a] have also developed antigens for the immunodiffusion test.

A specific fluorescent antibody has been developed by Kaplan and Kaufman.[40] When first prepared, the labeled antiglobulins stained *B. dermatitidis*, *H. capsulatum*, *P. brasiliensis*, and numerous other fungi. After adsorption with *H. capsulatum* (yeast phase) and *Geotrichum candidum*, the preparation was specific for yeast cells of *B. dermatitidis*. It does not stain the mycelial (M) phase, and there is as yet no FA test specific for the M phase. The FA antibody can be used to stain organisms in formalinized tissue and slides previously stained with hematoxylin and eosin; however, slides that have had specific fungal stains applied to them will not stain with the fluorescent-labeled antibody.[39] *Blastomyces* yeasts will be stained nonspecifically by the fluorescent dye, rhodamine. This technique is often useful for finding organisms in sputum or tissue.

In summary, there are no practical serologic procedures which may be followed.

The immunodiffusion test using cell sap antigens may be developed into a useful laboratory procedure.[67, 40a] The fluorescent antibody technique is useful in tissue sections, but all other serologic and immunologic evaluations are, at present, of no value.

LABORATORY IDENTIFICATION

Direct Examination. Because the serologic evaluation of the disease is so unreliable, the diagnosis of blastomycosis depends on histopathologic or cultural demonstration of the organism. In cutaneous disease, aspirated material from pustules at the outer edge of the lesion or pus from an open lesion can be examined in a potassium hydroxide mount (Fig. 15–13 FS B 46). This is the easiest and most rapid procedure for diagnosis. The yeast is a thick-walled spherical cell, 8 to 15 μ in diameter (Fig. 15–14). It sometimes attains a diameter of 30 μ. The bud on the parent cell has a characteristic broad base. The young bud has a thin wall which thickens as it grows. Buds may grow to the size of the parent cells and sometimes do not detach easily, resulting in a cluster of cells. The most important diagnostic aid for differentiation of *B. dermatitidis* from organisms such as *P. brasiliensis, H. capsulatum,* and other spherical forms of fungi is the width of bud attachment base (4 to 5 μ).

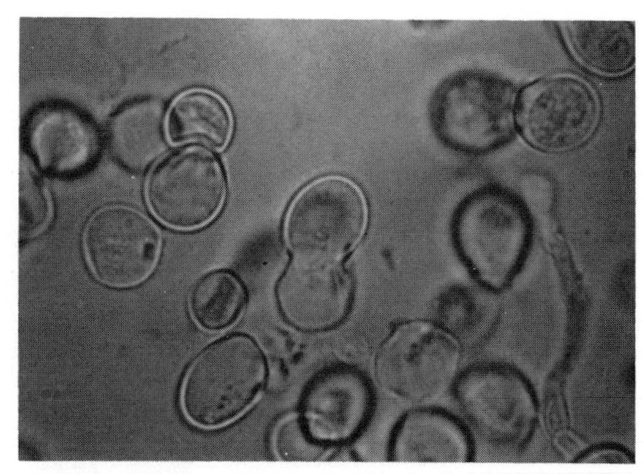

Figure 15–14. Blastomycosis. Wet mount of yeasts from culture. Note broad-based buds and retractile cell wall. ×440.

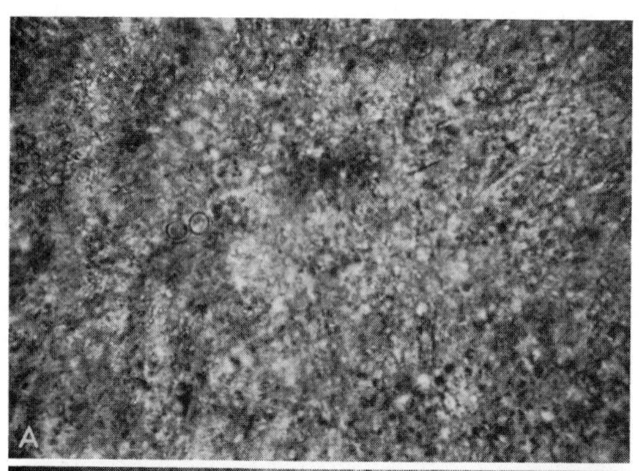

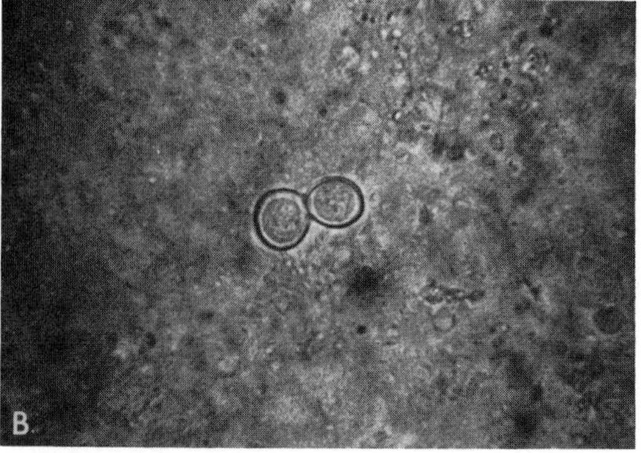

Figure 15–13. FS B 46. Blastomycosis. *A,* This is a potassium hydroxide mount of pus. Careful examination reveals a few yeast cells. ×100. *B,* Yeast cell with broad-based bud. ×440. (From Rippon, J. W. *In* Burrows, W. 1973. *Textbook of Microbiology*. 20th ed. Philadelphia, W. B. Saunders Company, p. 723.)

In pulmonary blastomycosis the first morning sputum can be examined for the presence of yeast cells. Purulent, blood-streaked portions of the material are selected for examination in a potassium hydroxide mount. Macerated tissue, aspirates from subcutaneous and osseous lesions, and pus from ulcers can be examined in a similar manner.

Culture Methods. Demonstration of the characteristic broad-based bud of the yeast cells in tissue is usually sufficient to make a diagnosis of blastomycosis. Cultures should always be taken, however, particularly when the KOH examination is negative or doubtful. The yeast form, and to a certain extent the mycelial form, is sensitive to cycloheximide, so that selective media for cultivating pathogenic fungi should be used in conjunction with antibiotic-free media.[53] Blood agar and Sabouraud's agar plates are inoculated with lesion aspirates, sputum, or pus and incubated at 25°C. Heavily contaminated material can be plated on agar containing penicillin and streptomycin or chloramphenicol. The growth rate of the organism is quite variable. Visible coremia-like spicules or "prickles"[32] may become evident within a few days to a week; however, most cultures grow very slowly, so that plates should be kept for a month or more. Since plates dry out after a few

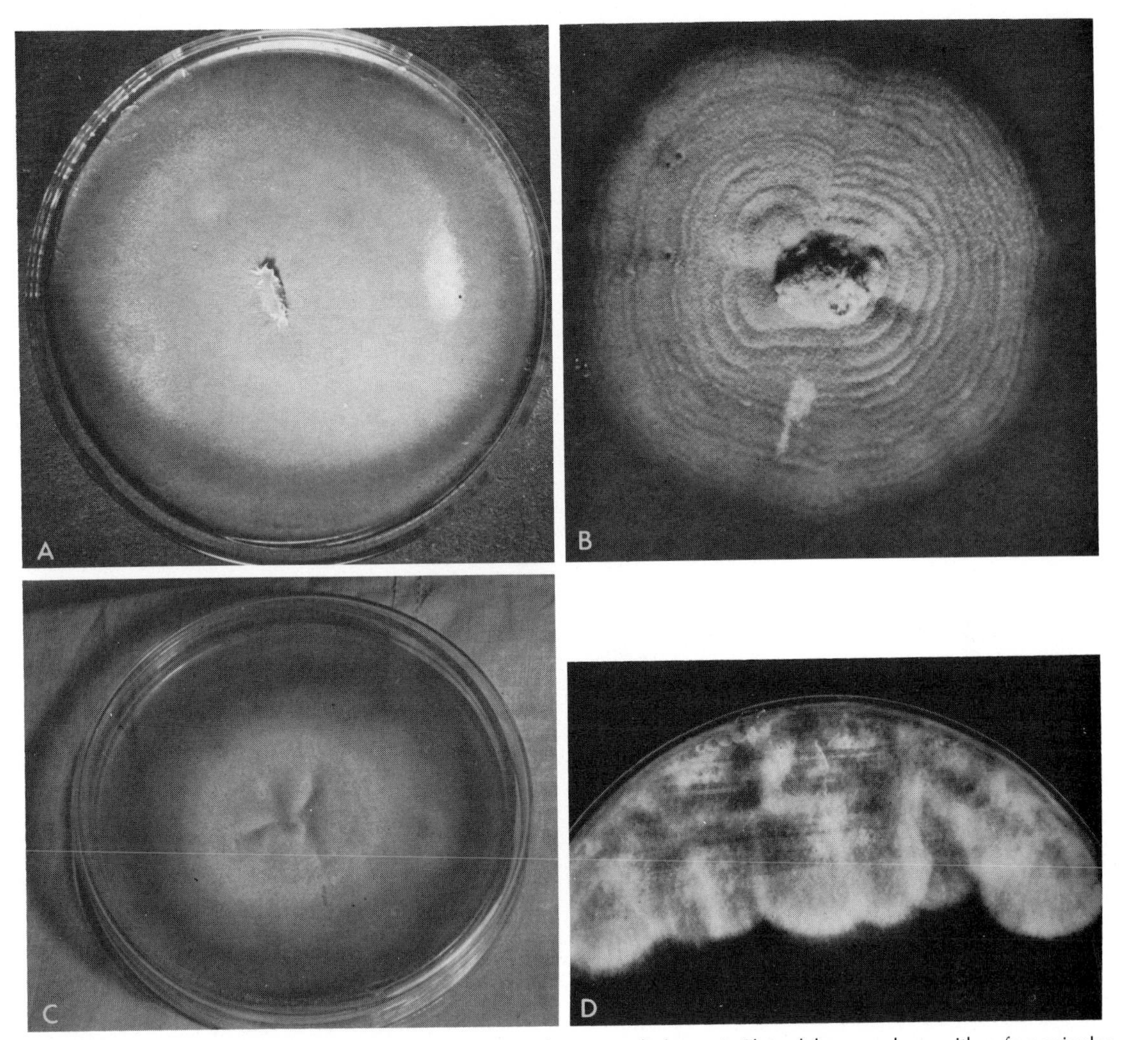

Figure 15–15. FS B 47. Blastomycosis. Variation in colony morphology. *A*, Flat, glabrous colony with a few spicules in the center and a few tufts of aerial mycelium at one edge of the periphery. *B*, Brownish-tan colony with concentric rings. *C*, White-tan colony with folds. *D*, Fluffy-white colonies from inoculum derived from a granuloma of lung.

weeks, they should be sealed with tape or kept in plastic bags during incubation.

The colony morphology of *B. dermatitidis* is quite variable and ranges from a flat, glabrous, "skin-like" consistency to a fluffy white to brownish tan colony that may have concentric rings (Fig. 15–15 FS B 47). The spores produced are the infectious elements, and cultures should be handled with caution. These spores are not characteristic and resemble those of many other fungi, especially the *Chrysosporium* species. Specific identification of *B. dermatitidis* requires demonstration of the yeast form. This is easily done by subculturing the organism on blood agar or other media and growing it at 37°C. Within a few days or weeks, the characteristic yeast forms appear. Animal inoculation can be preformed with these yeasts to determine pathogenicity. This procedure is sometimes a useful adjunct to cultural isolation, but in one series it did not yield more positives than direct culture, and in some cases it was negative when culture was positive.[84]

MYCOLOGY

Blastomyces dermatitidis Gilchrist and Stokes 1898

Synonymy. *Oidium dermatitidis* Ricketts 1901; *Cryptococcus gilchristi* Vuillemin 1902;

Zymonema gilchristi Buermann and Gougerot 1909; *Glenospora gammeli* Pollacci and Nannizzi 1927; *Blastomycoides tulanensis* Castellani 1928; *Endomyces capsulatus* Dodge and Ayers 1929; *Monosporium tulanense* Agostini 1932; *Glenospora brevis* Castellani 1933; *Endomyces capsulatus* var. *isabellinus* Moore 1933; *Zymonema dermatitidis* Dodge 1935; *Zymonema capsulatum* Dodge 1935; *Scopulariopsis americana* Broc and Haddad 1952; *Chrysosporium dermatitidis* Carmichael 1962.

Perfect State. *Ajellomyces dermatitidis* McDonough and Lewis 1968.

The generic name *Blastomyces* used by Gilchrist and Stokes to describe the etiologic agent of their newly discovered disease was illegal according to the international taxonomic code. The name had been used for an entirely different fungus, *Blastomyces luteus,* by Constantin and Rolland. However, this organism was later reduced to synonomy with *Aleurisma flavissima* by Vuillemin, thus leaving the term *Blastomyces* "open." *Zymonema,* as defined by Buermann and Gougerot in 1909, would have been an acceptable niche for *Blastomyces,* but that genus as emended by Dodge in 1935 is quite heterogeneous. Carmichael in 1962 reviewed the genus *Chyrsosporium* Corda and found that the spore production of the mycelial phase of *Blastomyces* was compatible with the description of this genus. However, this also is a very heterogenous group, and there are serious objections to including it in the *Chrysosporium* genus. Since the original designation, *Blastomyces,* is now "open," and the organism is universally known by that name, it is felt that this term should be retained. It appears to be a monotypic genus. The perfect state of the fungus is *Ajellomyces dermatitidis.*

Colony Morphology. SDA, 25°C. The growth rate and characteristics of *B. dermatitidis* are quite variable. Some strains grow rapidly, producing a fluffy white mycelium which reaches a diameter of 6 cm in two weeks time. Others grow slowly as glabrous, tan, nonsporulating colonies. Some strains sporulate heavily and have a dark brown colony consisting of concentric rings. These characteristics are heritable and segregate genetically when tested by mating.[43] Growth and sporulation have been shown to be stimulated by nitrogenous substances found in starling dung and yeast extract.[79] Most strains become pleomorphic when maintained in culture.

Colony morphology at 37°C on blood agar or Sabouraud's agar is a wrinkled and folded, glabrous, yeastlike colony. The organism quite readily converts to the yeast form when grown at 37°C. With frequent transfer, the colony becomes moist and pasty.

Microscopic Morphology. SDA, 25°C. The spores produced are aleuriospores of the *Chrysosporium* type (Fig. 15–16). They are ovoid or dumbbell-shaped, from 2 to 10 μ in diameter, and are borne on short lateral or terminal branches of the mycelium. They may resemble the microaleuriospores of *Histoplasma,* but in the latter species the small spore as well as the macroaleuriospores may be roughened with echinulations. The production of lateral globose spores intermixed with dumbbell-shaped or double spores is fairly characteristic of *B. dermatitidis.* However, sporulation of this organism is not readily differentiated from that of many other fungal species. Questions of identity can be resolved by conversion to the yeast phase and production of disease in experimentally infected mice.

When grown at an elevated temperature of 37°C, the organism produces the characteristic yeast form with broad-based buds seen in tissue. The cells again vary in size from 8 to 15 μ and up to 30 μ in diameter.

Ajellomyces dermatitidis McDonough and Lewis 1968

When mated with testor strains and grown on media containing mineral salts and oatmeal or on other deficient media, isolants of *B. dermatitidis* produce a gymnothecium and

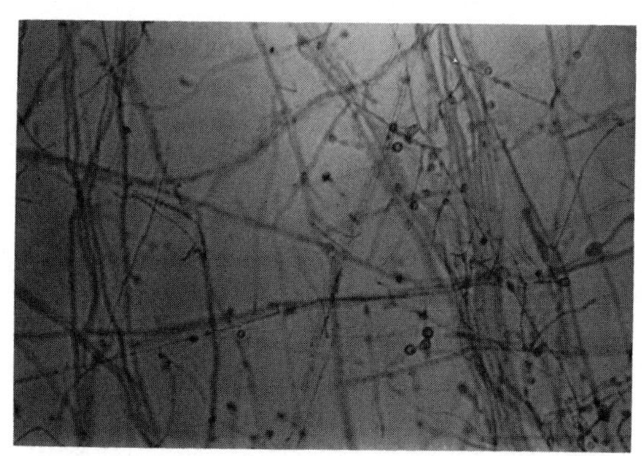

Figure 15–16. Blastomycosis. Mycelial form at 25°C. The spores are of the *Chrysosporium* type and vary from 2 to 10 μ in diameter. Some are double or dumbbell-shaped. ×100. (From Rippon, J. W. *In* Burrows, W. 1973. *Textbook of Microbiology.* 20th ed. Philadelphia, W. B. Saunders Company, p. 724.)

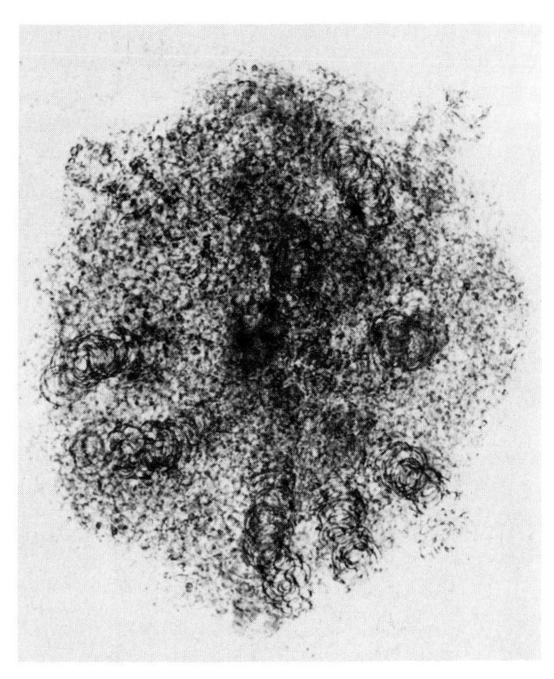

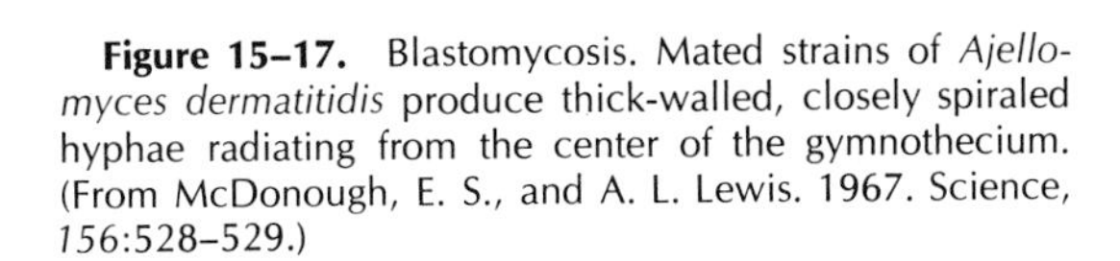

Figure 15–17. Blastomycosis. Mated strains of *Ajellomyces dermatitidis* produce thick-walled, closely spiraled hyphae radiating from the center of the gymnothecium. (From McDonough, E. S., and A. L. Lewis. 1967. Science, *156*:528–529.)

asci containing ascospores.[51] The characteristics are those of the family Gymnoascaceae, which includes the dermatophytes and the perfect stage of *Histoplasma capsulatum*. *Ajellomyces* is distinguished from other genera of the family by development of thick-walled, closely spiraled hyphae that radiate from a common center in the gymnothecium (Fig. 15–17). These spirals give rise to lateral hyphae which branch to form clusters of ascogenous hyphae (Fig. 15–18), in addition to clusters of branched hyphae constricted as the cross walls are formed. The asci are formed from the ascogenous hyphae and are globose or subglobose, 3.5 to 6 × 3.8 to 7.5 μ in diameter (Fig. 15–19). The ascus contains eight ascospores, which are smooth or roughened, 1.5 to 2 μ in diameter. Ascospore analysis shows a 1:1 ratio of the "+" and "–" mating types.[43] Mating competence

Figure 15–18. Blastomycosis. Clusters of ascogenous hyphae from lateral branches. ×100.

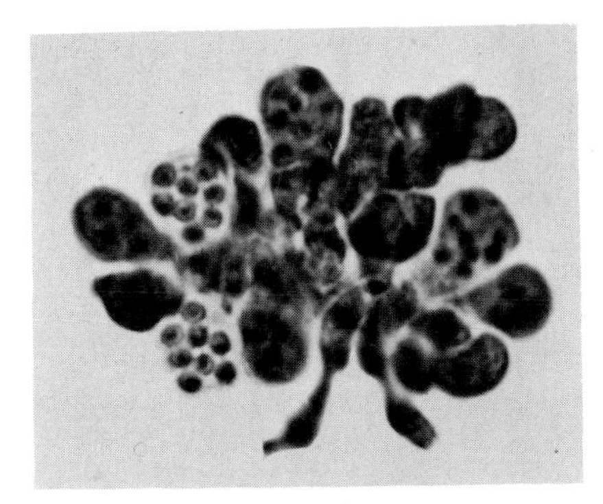

Figure 15–19. Blastomycosis. Asci in center of ascogenous hyphae containing ascospores. (From McDonough, E. S., and A. L. Lewis. 1968. Mycologia, *60*:76–83.)

falls to 11 per cent after two years of maintenance in culture. Fresh isolants produce fertile cleistothecia in 97 per cent of cases when suitably paired. The predominance of one mating type over the other in eliciting clinical disease has not been established in *A. dermatitidis* as it has for the perfect stage of *Histoplasma capsulatum.*

REFERENCES

1. Abernathy, R. S., and D. C. Heiner. 1961. Precipitation reactions in agar in North American blastomycosis. J. Lab. Clin. Med., *57*:604–611.
2. Abernathy, R. S., and G. T. Jansen. 1960. Therapy with amphotericin B in North American blastomycosis. Am. J. Int. Med., *53*:1196–1203.
3. Ajello, L. 1967. Comparative ecology of respiratory mycotic disease agents. Bact. Rev., *31*:6–24.
4. Ball, O. G., F. L. Lummus, et al. 1960. An immunologic survey for systemic fungus infections in general hospital patients of central Mississippi. Am. J. Hyg., *72* :231–243.
5. Balows, A., K. W. Deuschle, et al. 1966. Skin tests in blastomycosis. Arch. Environ. Health, *13*:86–90.
6. Baum, G. L., and J. Schwarz. 1959. North American blastomycosis. Am. J. Med. Sci., *238*:661–683.
7. Beneke, E. S., R. W. Wilson, et al. 1969. Extracellular enzymes of *Blastomyces dermatitidis.* Mycopathologia, *39*:325–328.
8. Benham, R. W. 1934. Fungi of blastomycosis and coccidioidal granuloma. Arch. Dermatol., *30*:385–400.
9. Brandsberg, J. W., F. E. Tosh, et al. 1964. Concurrent infection with *Histoplasma capsulatum* and *Blastomyces dermatitidis.* New Engl. J. Med., *270*:814–877.
10. Broc, R., and N. Haddad. 1952. Tumeur bronchique a *Scopulariopsis americana,* détermination précoce d'une maladie de Gilchrist. Bull. Mem. Soc. Med. Hôp. Paris, *68*:679–682.
11. Buechner, H. A., and C. M. Clawson. 1967. Blastomycosis of the central nervous system. II. A report of nine cases from the Veterans Administration cooperative study. Am. Rev. Resp. Dis., *95*:820–826.
12. Busey, J. F. (Chairman). 1964. Cooperative Study of the Veterans Administration. Blastomycosis. I. A review of 195 collected cases in Veterans Administration hospitals. Am. Rev. Resp. Dis., *89*:659–672.
13. Butler, W. T., and G. J. Hill. 1964. Intravenous administration of amphotericin B in the dog. J. Am. Vet. Med. Assoc., *144*:399–402.
14. Campos Magalhaes, M. J. 1968. Primier cas de blastomycose a *Blastomyces dermatitidis* observe au Mozambique. Cueriscon por l'amphotericin B. Bull Soc. Pathol. Exot., *61*:210–218.
15. Chick, E. W. 1971. North American blastomycosis, In *Handbuch der speziellen pathologischen Anatomie and Histologie. Dritter Band Fünfter Teil.* Baker, R. D. (ed.), Berlin, Springer-Verlag, pp. 465–506.
16. Conti Diaz, I. A., C. D. Smith, et al. 1970. Comparison of infection of laboratory animals with *Blastomyces dermatitidis* using different routes of inoculation. Sabouraudia, *7*:279–283.
17. Craig, M. W., W. N. Davey, et al. 1970. Conjugal blastomycosis. Am. Rev. Resp. Dis., *102*:86–90.
18. Denton, J. F., and A. F. DiSalvo. 1964. Isolation of *Blastomyces dermatitidis* from natural sites at Augusta, Georgia. Am. J. Trop. Med. Hyg., *13*:716–722.
19. Denton, J. F., and A. F. DiSalvo. 1968. Respiratory infection of laboratory animals with conidia of *Blastomyces dermatitidis.* Mycopathologia, *36*: 129–136.
20. Denton, J. F., E. S. McDonough, et al. 1961. Isolation of *Blastomyces dermatitidis* from soil. Science, *133*:1126–1127.
21. Duttera, M. J., and S. Osterhout. 1969. North American blastomycosis: A survey of 63 cases. South. Med. J., *62*:295–301.
22. Ebert, J. W., V. Jones, et al. 1971. Experimental canine histoplasmosis and blastomycosis. Mycopathologia, *45*:285–300.
23. Emmons, C. W., I. G. Murray, et al. 1964. North American blastomycosis. Two autochthonous cases from Africa. Sabouraudia, *3*:306–311.
24. Font, R. L., A. G. Spaulding, et al. 1967. Endogenous mycotic panophthalmitis caused by *Blastomyces dermatitidis.* Report of a case and review of the literature. Arch. Ophthalmol., *77*:217–222.
25. Fuminori, K., and L. Carbonell. 1971. Cell wall composition of the yeast-like and mycelial forms of *Blastomyces dermatitidis.* J. Bacteriol., *106*:946–948.
26. Furcolow, M. L., and C. D. Smith. 1973. A new interpretation of the epidemiology, pathogenesis and ecology of *Blastomyces dermatitidis* with some additional data. Trans. N.Y. Acad. Sci. (series 2) *35*:421–430.
27. Furcolow, M. L., E. W. Chick, et al. 1970. Prevalence and incidence studies of human and canine blastomycosis. Am. Rev. Resp. Dis., *102*:60–67.
28. Garrison, R. G., J. W. Lane, et al. 1970. Ultrastructural changes during the yeast-like to mycelial-phase conversion of *Blastomyces dermatitidis* and *Histoplasma capsulatum.* J. Bacteriol., *101*:628–635.
29. Gatti, F., De M. Broe, et al. 1968. *Blastomyces dermatitidis* infection in the Congo. Report of a second autochthonous case. Am. J. Trop. Med. Hyg., *17*:96–101.
30. Gehweiler, J. A., M. Paul Capp, et al. 1970. Observations on the roentgen patterns in blastomycosis of bone. Am. J. Roentgenol. Radium Ther. Nucl. Med., *108*:497–510.
31. Gilchrist, T. C. 1896. A case of blastomycetic dermatitis in man. Johns Hopkins Hosp. Report, *1*:269–298.
32. Gilchrist, T. C., and W. R. Stokes. 1896. The presence of an Oidium in the tissues of a case of pseudo-lupus vulgaris. Preliminary report. Johns Hopkins Hosp. Bull., 7:129–133.
33. Gilchrist, T. C., and W. R. Stokes. 1898. A case of pseudo-lupus vulgaris caused by a *Blastomyces.* J. Exp. Med., *3*:53–83.
34. Grandbois, J. 1963. La blasomycose nord-americaine a Canada. Laval Med., *34*:714–731.
35. Guidry, D. J., and A. J. Bujard. 1964. Comparison

of the pathogenicity of the yeast and mycelial phases of *Blastomyces dermatitidis.* Am. J. Trop. Med. Hyg., *13*:319–326.
36. Harris, J. S., J. G. Smith, et al. 1957. North American blastomycosis in an epidemic area. Public Health Rep., *72*:95–100.
37. Harville, W. E., and R. S. Abernathy. 1964. Addison's disease with unilateral blastomycosis of adrenal gland. J. Arkansas Med. Soc., *61*: 144–146.
38. Johnson, W. W., and J. Mantulli. 1970. The role of cytology in the primary diagnosis of North American blastomycosis. Acta Cytol., *17*:200–204.
39. Kaplan, W., and D. E. Kraft. 1969. Demonstration of pathogenic fungi in formalin fixed tissues by immunofluorescence. Am. J. Clin. Pathol., *52*:420–432.
40. Kaplan, W., and L. Kaufman. 1963. Specific fluorescent antiglobulins for the detection and identification of *Blastomyces dermatitidis* yeast phase cells. Mycopathologia, *19*:173–180.
40a. Kaufman, L., D. W. McLaughlin, et al. Specific immunodiffusion test for blastomycosis. Appl. Microbiol., *26*:244–247.
41. Kepron, M. W., C. B. Schaenperlen, et al. 1972. North American Blastomycosis in Central Canada. Can. Med. Assoc. J., *106*:243–246.
42. Klapman, M. H., N. P. Superfon, et al. 1970. North American blastomycosis. Arch. Dermatol., *101*:653–658.
43. Kwon-Chung, K. J. 1971. Genetic analysis on the incompatibility system of *Ajellomyces dermatitidis.* Sabouraudia, *9*:231–238.
44. Landay, M. E., E. P. Lawe, et al. 1968. Disseminated blastomycosis in hamsters after intramuscular subcutaneous and intraperitoneal injection. Sabouraudia, *6*:318–323.
45. Larson, R. E., P. E. Bernatz, et al. 1965. Results of surgical and non-operative treatment for pulmonary North American blastomycosis. J. Thoracic Cardiovasc. Surg., *51*:714–723.
46. Leavell, U. W. 1965. Cutaneous North American blastomycosis and black dots. Arch. Dermatol., *92*:155–156.
46a. Leers, W. D., N. A. Russel, et al. 1972. Cerebellar abscess due to *Blastomyces dermatitidis.* Can. Med. Assoc. J., *107*:657–660.
47. Levine, S., and Z. J. Ordal. 1946. Factors influencing the morphology of *Blastomyces dermatitidis.* J. Bacteriol., *52*:687–694.
48. Lockwood, W. R., F. Allison, et al. 1969. The treatment of North American blastomycosis. Ten years experience. Am. Rev. Resp. Dis., *100*:314–320.
49. McDonough, E. S. 1967. Epidemiology of 46 Wisconsin cases of North American blastomycosis, 1960–1964. Mycopathologia, *31*: 163–173.
50. McDonough, E. S. 1970. Blastomycosis—Epidemiology and biology of its ecologic agent *Ajellomyces dermatitidis.* Mycopathologia, *41*:195–201.
51. McDonough, E. S., and A. L. Lewis. 1967. *Blastomyces dermatitidis:* Production of the sexual stage. Science, *156*:528–529.
52. McDonough, E. S., R. Van Prooien, et al. 1965. Lysis of *Blastomyces dermatitidis* yeast phase cells in natural soil. Am. J. Epidemiol., *81*: 86–94.
53. McDonough, E. S., L. Ajello, et al. 1960. In vitro effects of antibiotics on yeast phase of *Blastomyces dermatitidis* and other fungi. J. Lab. Clin. Med., *55*:116–119.
53a. McDonough, E. S., J. J. Dubats, et al. 1973. Soil Streptomycetes and bacteria related to lysis of *Blastomyces dermatitidis.* Sabouraudia. *11*: 244–250.
54. Martinez-Batez, M., A. Reyes Mota, et al. 1954. Blastomicosis Norteamericana en Mexico. Rev. Inst. Salub. Enferm. Trop., *14*:225–232.
55. Montgomery, F. H., and O. S. Ormsby. 1908. Systemic blastomycosis. Arch. Intern. Med., *2*:1–41.
56. Nickerson, W. J. 1948. Ethylene as a metabolic product of the pathogenic fungus *Blastomyces dermatitidis.* Arch Biochem., *17*:225–233.
57. Ohlwiler, D. A., M. P. Freyes, et al. 1962. North American blastomycosis successfully treated with amphotericin B. Plast. Reconstr. Surg., *29*301–303.
58. O'Neill, R. P., and R. W. B. Penman. 1970. Clinical aspects of blastomycosis. Thorax, *25*:708–715.
59. Parker, J. D., I. L. Doto, et al. 1969. A decade of experience with blastomycosis and its treatment with amphotericin B. A National Communicable Disease Center Cooperative Mycoses Study. Ann. Rev. Resp. Dis., *99*:895–902.
60. Pfister, A. K., A. W. Goodwin, et al. 1966. Pulmonary blastomycosis: Roentgenographic clues to the diagnosis. South. Med. J., *59*:1444–1447.
61. Planques, J., L. Enjalbert, et al. 1967. Blastomycose multiviscerale a prédominance pulmonaire chez un français d'origine tunisienne guerie par l'amphotericin B. J. Fr. Med. Chir. Thorac., *21*:325–344.
62. Polo, J. F., K. Brass, et al. 1954. Enfermedad de Gilchrist en Venezuela. Rev. Sanid. Asist. Soc. (Caracas), *19*:217–235.
63. Pond, N. E., and R. J. Humphreys. 1952. Blastomycosis with cardiac involvement and peripheral embolization. Am. Heart J., *43*:615–620.
64. Procknow, J. J. 1966. Disseminated blastomycosis treated successfully with the polypeptide antifungal agent X-5079C. Evidence for human to human transmission. Am. Rev. Resp. Dis., *94*:761–772.
65. Ranier, A. 1951. Primary laryngeal blastomycosis. A review of the literature and report of a case. Am. J. Clin. Pathol., *21*:444–450.
66. Rippon, J. W. 1971. Differences between the "+" and "−" strains of keratinophilic fungi. Recent Adv. Microbiol., pp. 473–475.
67. Rippon, J. W., D. N. Anderson, et al. 1972. Blastomycosis: Specificity of antigens reflecting the mating types of *Ajellomyces dermatitidis.* Bact. Proc., p. 131.
68. Rollier, R., and M. Berrada. 1969. Un cas de blastomycose "africaine." Bull. Soc. Fr. Dermatol. Syphiligr., *76*:194–195.
69. Ross, M. D., and F. Goldring. 1966. North American blastomycosis in Rhodesia. Cent. Afr. J. Med., *12*:207–211.
70. Salfelder, K. 1965. Experimental cutaneous North American blastomycosis in hamster. J. Invest. Dermatol., *45*:409–418.
71. Sanders, L. L. 1967. Blastomycosis arthritis. Arthritis Rheum., *10*:91–98.

72. Salvin, S. B. 1963. Immunological aspects of the mycoses. Progr. Allerg., 7:213–331.
72a. Sarosi, G. A., K. J. Hammerman, et al. 1974. Clinical features of acute pulmonary blastomycosis. New Eng. J. Med., *290*:540–542.
73. Schwarz, J., and G. L. Baum. 1951. Blastomycosis. Am. J. Clin. Pathol., *21*:999–1029.
74. Schwartz, J., and G. L. Baum. 1953. North American blastomycosis. Geographic distribution, pathology and pathogenesis. Docum. Med. Geogr. Trop. (Amst.), *5*:29–41.
75. Segretain, G. 1968. La blastomycose in *Blastomyces dermatitidis.* Son extence en Afrique. Maroc Med., *48*:20–27.
76. Selby, L. A., R. T. Habermann, et al. 1964. Clinical observations on canine blastomycosis. Vet. Med., *59*:1221–1228.
77. Shadomy, S., G. M. Robertson, et al. 1969. In vitro and in vivo activity of hamycin against *Blastomyces dermatitidis.* J. Bacteriol., *97*:481–487.
78. Sheldon, W. G. 1966. Pulmonary blastomycosis in a cat. Lab. Anim. Care, *16*:280–285.
79. Smith, C. D. 1964. Evidence of the presence in yeast extract of substances which stimulate the growth of *Histoplasma capsulatum* and *Blastomyces dermatitidis* similar to that found in starling manure extracts. Mycopathologia, *22*:99–105.
80. Smith, J. G., Jr., J. S. Harris, et al. 1955. An epidemic of North American blastomycosis. J.A.M.A., *158*:641–646.
81. Sorensen, R. H., and D. E. Casad. 1969. Use of case survey technique to detect origin of Blastomyces infection. Public Health Rep., *84*:514–520.
82. Stein, R. O. 1914. Die Gilchristche Krankheit (Blastomycosis Americana) und ihre Beziehung zu den in Europa beobachteten Hefeinfektionen. Arch. Dermatol. Syph. (Berlin), *120*:889–924.
83. Tesh, R. B., and J. D. Schneidau, Jr. 1967. Experimental infection of bats (*Tadarida brasiliensis*) with *Blastomyces dermatitidis.* J. Infect. Dis., *11*:188–192.
84. Turner, C., C. D. Smith, et al. 1972. The efficiency of serologic and cultural methods in the detection of infection with Histoplasma and Blastomyces in mongrel dogs. Sabouraudia, *10*:1–10.
85. Turner, D. J., and W. B. Wadlington. 1969. Blastomycosis in childhood: Treatment with amphotericin B and a review of the literature. J. Pediatr., *75*:708–715.
86. Utz, J. P., H. J. Shadomy, et al. 1968. Clinical and laboratory studies of a new micronized preparation of hamycin in systemic mycoses in man. Antimicrob. Agents Chemother., 7:113–117.
87. Vermeil, C., A. Gordeetf, et al. 1954. Sur un case Tunisien de mycose generalisec mortelle. Ann. Inst. Pasteur (Paris), *86*:636–646.
88. Williamson, W. M., L. S. Lombard, et al. 1959. North American blastomycosis in a northern sea lion. J. Am. Vet. Med. Assoc., *139*:513–515.
89. Wilson, J. W., E. P. Cawley, et al. 1955. Primary cutaneous North American blastomycosis. Arch. Dermatol., *71*:39–45.
90. Wilson, J. W. 1963. Cutaneous (chancriform) syndrome in deep mycoses. Arch. Dermatol., *87*:81–85.
91. Witorsch, P., and J. P. Utz. 1965. North American blastomycosis: A study of 40 patients. Medicine (Baltimore), *47*:169–200.

Chapter 16

HISTOPLASMOSIS

DEFINITION

Histoplasmosis is a very common granulomatous disease of worldwide distribution caused by the dimorphic fungus *Histoplasma capsulatum.* Infection is initiated after inhalation of spores of the organism and results in a variety of clinical manifestations. Approximately 95 per cent of cases are inapparent, subclinical, or completely benign. These are diagnosed only by the x-ray finding of residual areas of pulmonary calcification and a positive histoplasmin skin test. The remaining patients may have a chronic progressive lung disease, a chronic cutaneous or systemic disease, or an acute fulminating, rapidly fatal, systemic infection. The latter form is particularly common in children. The etiologic agent has been found in practically all habitable areas of the earth in which it has been sought. Its growth is particularly associated with the presence of guano and debris of birds and bats. The fungus can survive and be transmitted from one location to another in the dermal appendages of both and also in the intestinal contents of bats, but wind is probably the most important agent of dissemination. The infection is also very common in wild and domestic animals in endemic areas. A clinically distinct form of disease, histoplasmosis duboisii, common in Africa, is caused by *H. capsulatum* var. *duboisii* (see Chapter 17). Both forms exist in Africa, and the var. *duboisii* may represent the primitive progenitor of *H. capsulatum.* The fungus is a heterothallic ascomycete whose perfect stage is *Emmonsiella capsulata,* which is classified in the family Gymnoascaceae.

Synonymy. Darling's disease, reticuloendotheliosis, reticuloendothelial cytomycosis, cave disease, Ohio Valley disease, Tingo Maria fever.

HISTORY

The most important event in the history of histoplasmosis occurred in 1944 when Amos Christie discovered that Darling's disease, which was first described in 1905, was not a rare medical curiosity, but a very common pulmonary infection. Since that time mycologists, epidemiologists, pathologists, radiologists, and clinicians have been uncovering the infection wherever they look for it. It now appears that histoplasmosis is one of the more common infectious diseases in the world. Its popularity is ever increasing, as witnessed by the publication of at least four books, three national symposia, and the voluminous case reports, research communications, reviews, and reviews of reviews on this subject which continue to appear.

In 1905 the newly appointed pathologist of the Ancon Canal Zone Hospital, Samuel Taylor Darling, did an autopsy on a Negro from Martinique. The patient had died of an overwhelming infection, with gross lesions that resembled tuberculosis. However, on microscopic examination of the lesions, round, intracellular bodies were found. The 33-year-old Darling was influenced by the recent findings of Donovan and Leishman concerning protozoan disease and by the prediction of Manson and Ross that kala-azar could exist in America. Darling believed the organisms he found in his patient to be protozoan because of their size and staining characteristics. They were found in histiocytes, resembled plasmodia, and seemed to have a capsule, so he coined the name *Histoplasma capsulatum* for them. Thousands of workers employed to build the canal were dying of various tropical diseases, so that within a few years Darling had performed some 33,000 autopsies. His first case of histoplasmosis was found on December 7, 1905,[29] and he discovered two more in January and August of the following year. The second case was also in a Negro from Martinique, and the third was in a Chinese who resided for 15 years in Panama. After studying these cases he realized that this disease was similar to kala-azar in that it also induced splenomegaly and intracellular organisms were present in lesions. He decided it was a new disease, however, since the organism lacked a blepharoplast and had the aforementioned capsule. His description and illustrations were so complete that there can be no question as to the identity of the organism in his patients. Richard P. Strong had published a description of a similar disease a few months earlier.[115] He had been working on various tropical ulcers in the Philippines and described lesions in which intracellular organisms were found on histologic examination. Although it is probable that the disease he studied was histoplasmosis, his description and drawings were less complete than those of Darling.

In 1913, Henrique da Rocha-Lima, a Brazilian studying in Hamburg, compared the microscopic sections from Darling's case with epizootic lymphangitis of horses.[30] The latter disease was known to be caused by a fungus, and on the basis of histologic similarity, he concluded histoplasmosis also was a mycotic infection. Twenty years elapsed from the first cases of Darling until the next report of the disease. In 1926 Riley and Watson described the infection in a woman resident of Minnesota. As a result, they noted that histoplasmosis was therefore not restricted to the tropics and might be the cause of obscure cases of splenomegaly with emaciation, anemia, and pyrexia. In the same year another case was reported by Phelps and Mallory. Their patient had been a banana plantation worker in Honduras.

The next significant discovery in the history of histoplasmosis occurred at Vanderbilt University. In 1929, Dr. Catherine Dodd diagnosed the first case ante mortem. The patient, a child, had parasites observable in blood smears subjected to supravital staining. Material from the child was cultured on a variety of media by William de Monbreun.[31] To his surprise, a mold grew out of the pathologic material. Thereafter, in a series of carefully executed studies he established the dimorphic nature of this fungus, the cultural characteristics of the mycelial and yeast stage, and the production of disease in experimental animals. He thus fulfilled Koch's postulates. His paper describing these findings was presented in November, 1933, at a meeting of the American Society of Tropical Medicine. As is the case with so many discoveries, there was a prior claim to the cultivation of *Histoplasma capsulatum*. A few months before de Monbreun's presentation, Hansmann and Schenken read a paper at a meeting of the American Association of Pathologists and Bacteriologists describing the growth of the organism and called it a *Sepedonium* species. Though their work was not as detailed as that of de Monbreun, all three share credit for establishing that *H. capsulatum* was a dimorphic fungus and not a protozoan.

It had been noted by many radiologists and pathologists that there were many apparently healthy people with small calcifications in their lungs who were tuberculin-negative. Review of these cases showed that a very high percentage of involved people lived in the Mississippi-Ohio Valley regions of midwestern United States. Amos Christie had had experience in California, where the mild form of coccidioidomycosis had recently been delineated. He was newly appointed to the faculty at Vanderbilt and became puzzled about the lung calcifications in tuberculin-negative people. Influenced by his California experience, he began skin testing a number of such children with an extract of the mycelial phase of the fungus, a histoplasmin. He correlated skin test reactivity with the presence of pulmonary calcifications, thus proving, in 1944, that a mild form of the disease exists.[24] A much larger skin testing series was carried out by Carroll Palmer in 1945,[83,84] which firmly established the occurrence of the subclinical form of disease and its great prevalence. Interestingly, in the same year Parsons and Zarafonetis[85] published a paper reviewing all known cases of histoplasmosis from the year 1905 to 1945. The total number was 71, and the authors concluded that the disease was widespread, but rare and always fatal.

Great interest was aroused by the work of Christie and Palmer, and the first seminar on histoplasmosis was held at the National Institutes of Health on September 13, 1948.[90] The convener was Norman Conant, and papers by Emmons, Campbell, Loosli,

Salvin, and others confirmed the efficacy of the complement fixation test and the skin test in diagnosis of the disease, and the occurrence of the disease in a wide variety of wild animals. The stage was now set for rapid development in the investigation of all phases of the disease. Negroni in 1940[81] had studied the first case in South America, thus extending the known range of the disease as far south as Argentina. He also induced the yeast form of growth *in vitro*. The organism was suspected to occur in the soil, and in 1948, Emmons[40] isolated it from rat burrows. Zeidberg et al.[126] also recovered the organism from soil and noted the very important association of fungus with bird dung. Skin testing with histoplasmin was now being done everywhere. Furcolow (1945) established a center in Kansas City to study the prevalence of the disease in this highly endemic area. Here he isolated the organism from the air, water supply, and practically every chicken house and bird roost in Missouri. The first "epidemic" of histoplasmosis occurred in 1947 in Camp Gruber, Oklahoma. Since then, famous epidemics have occurred in Milan, Michigan; Mason City, Iowa; Mexico, Missouri; Dalton, Georgia, and many other places. Ironically one such epidemic (Delaware, Ohio, 1970) was associated with Earth Day ecologic activities.

Demonstration of the organism in tissue was still difficult. Using a variant of the McManus periodic acid–Schiff stain, Hotchkiss was able to stain the polysaccharide in the yeast cell walls. Kligman et al.[67] in 1951 established the usefulness of this stain for fungi in tissue. Pathologists such as Puckett were finding that many "tuberculomas" were in fact "histoplasmomas," when they used this new method of staining.[109] The work of Schwartz, Straub, and Servianski in 1955[107,109,114] established the clinical and pathologic characteristics of the initial mild infection. They found that following inhalation of spores a Ghon complex was formed in time, along with pulmonary calcifications and, more importantly, small foci of splenic calcifications. This indicated that there was rapid hematogenous spread of the organisms early in the infection, probably within macrophages. The entire disease process could still occur without symptoms observable by the patient.

By now, the whole world was becoming aware of the disease, and histoplasmosis had become famous. Furcolow et al. in 1952 organized the first national conference, and in 1962 a second.[9] Almost everywhere investigators looked they could find the organism in pathologic material, isolate it from soil, or find reactors to the histoplasmin skin test. Darling's rare medical curiosity of 1905 had become a very common infectious disease. In the United States alone, it is estimated that 40,000,000 people have had the disease, and there are 200,000 new infections every year.[4,45]

ETIOLOGY, ECOLOGY, AND DISTRIBUTION

The etiologic agent of histoplasmosis is the dimorphic fungus *Histoplasma capsulatum*. In culture at temperatures below 35°C and on natural substrates it grows as a white to brownish mycelial fungus. The organism elaborates characteristic echinulate, oval, or pyriform macroaleuriospores (8 to 16 μ in diameter) and small (2 to 5 μ) microaleuriospores. When inhaled into the alveolar spaces, it is the latter that sprout and then transform into small budding yeasts that are 2 to 5 μ in diameter. The yeast cells are found within cells of the reticuloendothelial system. In culture at a temperature of 37°C the organism also grows in the yeastlike form. A clinically distinct form of the disease, histoplasmosis duboisii, is caused by *H. capsulatum* var. *duboisii*. Whereas the larger yeast cells of the latter are found mainly in giant cells, the small yeast forms of *H. capsulatum* are generally within histiocytes.

The ecology of *H. capsulatum* has been the subject of numerous investigations. It is firmly established that the organism grows in soil with high nitrogen content, generally associated with the guano of birds and bats.[7,10,32,41,126] The first isolation of the organism from a natural environment was from soil near a chicken house,[40,126] and since that time it has been recovered on numerous occasions from bat caves, bird roosts, chicken houses, silos inhabited by pigeons, and other such environments. In avian habitats, the organism seems to grow preferentially where the guano is rotting and mixed with soil rather than in nests or fresh deposits. In the laboratory, it has been shown to grow on shed feathers, and when the organism is injected into birds, it can be recovered from feathers *in situ*.[21, 53, 117, 119] Even in highly endemic areas, the presence of the organism is often restricted to small areas where birds congregate in large numbers. Not all guano appears to serve equally well as a substrate. The most highly endemic areas of the United States (Missouri, Kentucky, Tennessee, Southern Illinois, Indiana, and Ohio) are also the areas with the greatest concentration of starlings. Their habit of congregating in great numbers results in formation of large deposits of guano. It is possible that the Ohio-Mississippi Valley is heavily infested with *H. capsulatum* as a result of the importation of these birds from Europe. Negroni[82] suggests that comparable areas of South America escape heavy infestation because the blight of the starling has not yet been introduced. In that continent the main breeding grounds for histoplasmosis are the chicken habitats and bat caves.

Though the organism grows in great abundance only in relatively restricted environmental conditions, *H. capsulatum* is found throughout the world. In two of the most detailed studies of climatic conditions favoring growth, Furcolow[44] and Fonseca[42] found a mean temperature of 68 to 90°F (22 to 29°C), an annual precipitation of 35 to 50 inches (~1000 mm), and a relative humidity of 67 to 87 per cent or more during the growing season to be related to the greatest presence of positive skin tests and soil isolations. These conditions appear to be the most favorable for proliferation of the organism in what may be termed an "open" environment. This would include many areas in tropical, subtropical, and temperate regions of the world where there is adequate moisture. When skin testing has been done in such areas, high rates of reactivity have been found. Inexplicably some areas, such as most of Europe, seem to have escaped heavy infestation.

Outside of areas with appropriate environmental conditions, there also occur scattered areas with high endemicity. These are usually associated with existence of caves inhabited by bats or birds and may be considered "closed" environments. Though the surrounding countryside may be too dry or cold for sustained proliferation of the organism, the protected and relatively stable conditions inside the caves allow such proliferation. This probably accounts for the so-called "cave fever" in areas where otherwise there is a low incidence of the disease.[7, 10, 86] The chief vector of dissemination in the open environment is the wind. Though birds and bats may contribute to the dissemination of viable spores in this type of environment, in the closed areas they most certainly do. Klite and Diercks,[67a] Di Salvo et al.,[32] and Emmons et al.[41] have shown that bats may contaminate the environment with organisms present in their intestinal contents. They often have yeast-containing ulcers within their intestinal tract. As these animals migrate to new caves, they seed them with the fungus. Caves which are otherwise suitable for growth but are too wet remain free of infestation.[43] So far attempts to isolate organisms from the cloaca of birds have been unsuccessful, and it appears that birds do not become infected with the fungus.

Many other factors have been postulated as influencing the epidemiology and distribution of histoplasmosis. These include wind direction, flooding of riverbanks, the presence of limestone in soil, red-yellow podzolic soil, and so forth.[44] Unfortunately it appears that *H. capsulatum* will not fit as neatly into a particular environmental niche as does *Coccidioides immitis* into its lower Sonoran Life Zone.

A skin test for histoplasmosis was first used in 1941 by Van Pernis. The histoplasmin which is presently used was standardized by Emmons a few years later. It is the filtrate of an asparagine glucose broth in which the organism (mycelial phase) has been growing for 2 to 4 months at 25°C. The medium is the same "nonallergic" broth used for making old tuberculin (OT). Histoplasmin is injected intradermally as a 1:100 or 1:1000 dilution. An area of induration of at least 5 mm after 48 hours is considered a positive test. Cross-reactions occur in patients with blastomycosis and, to a lesser extent, coccidioidomycosis and possibly other fungal diseases. Since the early 1950's hundreds of thousands of people have been skin tested in all parts of the world. From such studies it appears that the largest area with high prevalence (80 to 90 per cent) is the middle section of the North American continent (Fig. 16–1). However, scattered, often isolated, areas with high prevalence of the disease exist in all parts of the world.[25] These include southern Mexico, northern Panama, Honduras, Guatemala, Nicaragua, Venezuela, Colombia, Peru, Brazil, Surinam,

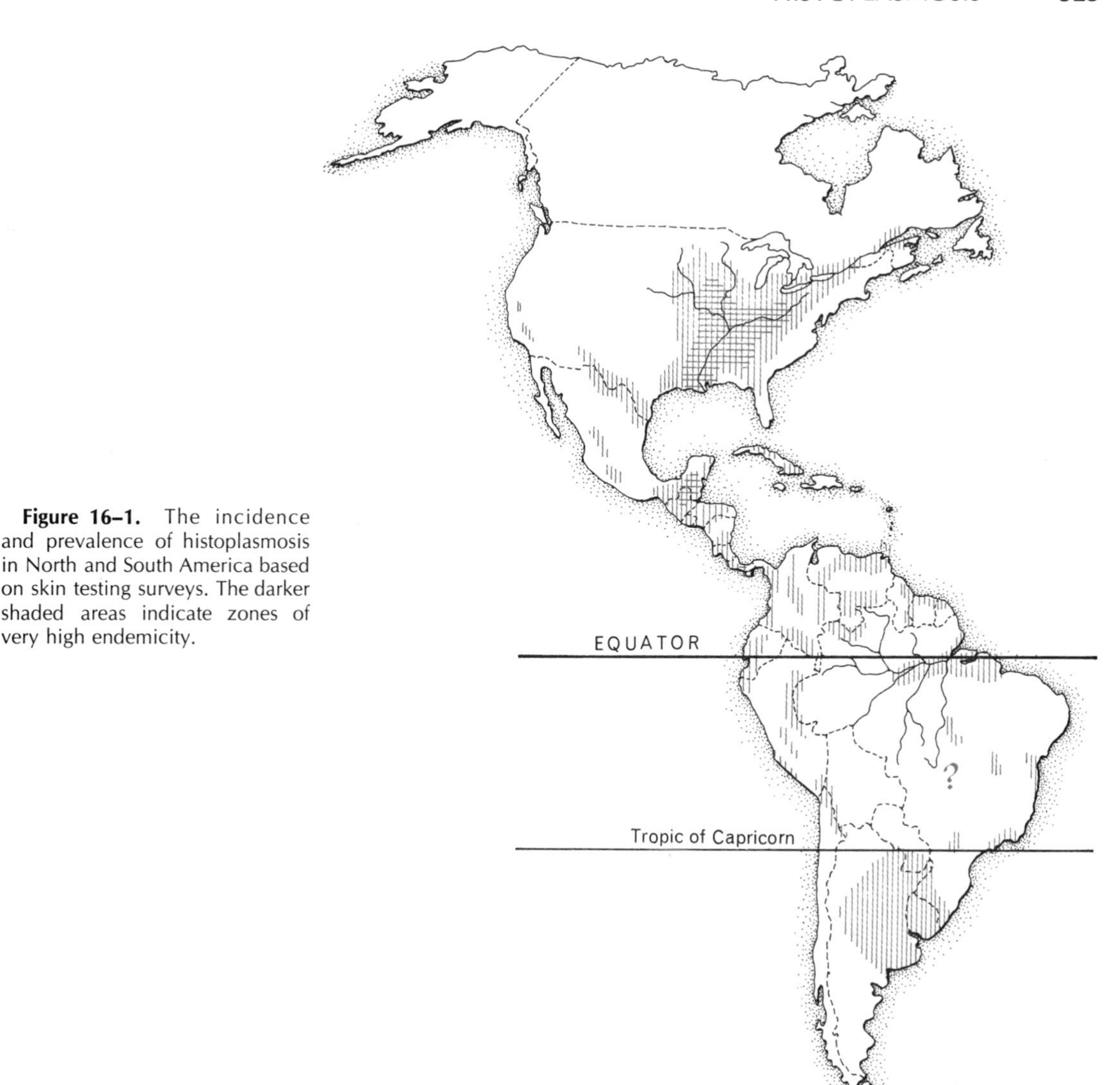

Figure 16–1. The incidence and prevalence of histoplasmosis in North and South America based on skin testing surveys. The darker shaded areas indicate zones of very high endemicity.

Burma, Indonesia, Philippines, Turkey, and others.[5, 37, 38, 49, 86, 92, 112, 113]

In a series of studies conducted in the highly endemic area of Kansas City, Furcolow[46] found that, by age 20, between 80 and 90 per cent of the population had a positive histoplasmin skin test. The same is true in the Cincinnati–southern Ohio region, southern Illinois, central Missouri, and areas of Kentucky, Tennessee, and Arkansas. Focal areas of high endemicity also occur in Michigan, Minnesota, Georgia, and Louisiana.[37,38] At present the histoplasmin skin test merely indicates that one has probably lived in the central United States at one time, and the test itself has essentially no diagnostic value.

Epidemics of acute respiratory histoplasmosis are frequently recorded. These occur when there is exposure to an aerosol containing numerous spores, as with the family that cleaned a silo in Indiana, the Boy Scouts that cleared a park in Mexico, Missouri, and children in a school where the air conditioner intake was located over a pile of starling droppings in Milan, Michigan. About fifty such epidemics have been recorded,[110,124] which are particularly interesting for the epidemiologist and public health workers. One of the more excellent studies is the one by Dodge et al.,[33] demonstrating the association of a bird roosting site with infection of school children. It serves as a model of epidemiologic detective work.

All ages and sexes are subject to primary pulmonary infection, but the sex distribution and form of disease change with increasing age of the patient. Certain select children acquire a fulminating, rapidly fatal disease in which there is massive proliferation of histiocytes and the *Histoplasma* organism within them. Both sexes are equally affected by all forms of the disease until the age of puberty. Then, chronic progressive disease is at least three times more frequent in adult males than in females. There is no documented racial difference in susceptibility. Point exposure is much more important than occupation. A rapidly progressive opportunistic infection occurs in some patients with the lymphoma-leukemia-Hodgkin's group of diseases, but the rarity of such cases does not justify the inclusion of *Histoplasma* as an important fungal "opportunist."

CLINICAL DISEASE

Histoplasma capsulatum is a pathogenic fungus. Inhalation of a sufficient quantity of spores will cause an infection in the lungs of the normal healthy patient. This is followed by a rapid, transient, hematogenous and systemic spread of the organism. In the vast majority of patients, the infection is aborted, leaving only residual calcifications in the lung and sometimes the spleen. Resolution of the disease confers a certain degree of immunity to reinfection. In a few patients the disease becomes chronic and progressive and follows a course similar to chronic tuberculosis. Massive reinfection may result in a fatal acute allergic reaction in highly sensitized lungs. In two instances, the organism can be considered an opportunistic fungus. The first is in cases of the mysterious histiocytosis that accompanies the fulminant disease of childhood, and the second is the very uncommon systemic disease associated with the above mentioned lymphomatous diseases.

Symptomatology. With the exception of rare, accidental, cutaneous inoculation, infection by *H. capsulatum* occurs by way of the lungs. The resulting disease may be divided into two major categories – pulmonary and disseminated – each of which has acute and chronic subcategories. Presumed *Histoplasma* choroiditis (uveitis) is discussed in Chapter 22 under mycotic infections of the eye.

Clinical Forms of Histoplasmosis

- I. Pulmonary histoplasmosis
  - A. Acute
    1. Asymptomatic
    2. Mild
    3. Moderately severe
    4. Severe, epidemic
  - B. Chronic progressive
- II. Disseminated histoplasmosis
  - A. Acute
    1. Benign
    2. Progressive adult
    3. Fulminant of childhood
  - B. Chronic progressive and mucocutaneous

Pulmonary Histoplasmosis

ACUTE ASYMPTOMATIC DISEASE

From the studies of Christie, Peterson, Palmer, Furcolow, and others, it is estimated that at least 95 per cent of all primary cases of histoplasmosis are not referable to specific symptomatology. Multiple calcifications may be seen in the lungs of patients living in highly endemic areas, and such patients cannot give a history of relevant symptoms.

MILD DISEASE

Mild disease may present with a flulike syndrome that includes nonproductive cough, pleuritic pain, shortness of breath, and hoarseness. Histoplasmosis is a common cause of "summer fever" in children.[72]

MODERATELY SEVERE DISEASE

In moderately severe disease, the above signs are exaggerated and also include fever, night sweats, weight loss, some cyanosis, and occasional hemoptysis. In this latter group of patients, the organism can sometimes be cultured from sputum, and, when fever is present, it has been recovered from bone marrow obtained by sternal puncture.

The roentgenographic picture of all these types of primary disease is similar. There is the appearance of multiple lesions scattered in all lung fields. Initially they appear as disseminated infiltrates or discrete nodular foci of activity (Fig. 16–2,*A*). Hilar lymphoadenopathy and a Ghon-like complex are almost always present. This may lead to an initial diagnosis of tuberculosis, or an expanding lesion may appear to be a lymphoblastoma or other neoplasm. At all stages of the disease the appearance is essentially identical to tuberculosis and can only be differentiated from it by proper serologic and cultural procedures.[111]

The lesions resolve slowly. A few undergo complete resolution or may leave small areas of fibrosis; other lesions heal by fibrosis, and, characteristically, these calcify (Fig. 16–2,*B*,*C*). As pointed out by Schwartz and Straub,[107,114] the resolved primary complex of histoplasmosis has a consistent x-ray pattern. The lymph nodes are large and the tubercles small, and there are halos around the primary calcifying nodules. They are usually numerous, round, oval, and of uniform size (Fig. 16–2,*D*). The healing lesions of tuberculosis tend to be more irregular. In histoplasmosis, calcified lesions can also be seen in the spleen and occasionally in the liver. The radiologic appearance of "popcorn"-type calcifications is also commonly seen in healed disease (Fig. 16–3).

There are other residual manifestations of primary histoplasmosis. One very common one is frequently picked up on routine x-ray (Fig. 16–2,*E*). This is the presence of a solitary lesion that has not yet calcified. Such "coin lesions" are difficult to distinguish from various neoplasms and are often surgically removed to establish a diagnosis (Fig. 16–2,*F* FS B 48).

Another manifestation of histoplasmosis that mimics tuberculosis is the histoplasmoma. These are found in the lungs of adults and are usually 2 to 3 cm in size. The central area is necrotic, and there is a fibrotic capsule. Calcification proceeds in the central portion, but the lesion appears to enlarge with alternating concentric rings of fibrosis and calcification. Again they may have the appearance of neoplasms. Calcification is even more common in histoplasmosis than in tuberculosis and is also frequently encountered in coccidioidomycosis. It is unusual in cryptococcosis and blastomycosis.

EPIDEMIC HISTOPLASMOSIS

The fourth type of primary histoplasmosis is the epidemic form. This has been described under a variety of names as cave fever, acute miliary pneumonitis, primary atypical pneumonia, and many more. The syndrome occurs when individuals are exposed to large quantities of spores. The incubation period is from seven to fourteen days. The symptoms are a high fever, severe pleuritic pain, dyspnea, weakness, and severe pneumonitis. These are the same symptoms found in acute pulmonary histoplasmosis, but they are more exaggerated. The skin test becomes positive more rapidly, and it has been suggested that some of the lung pathology is allergic in nature. In about 50 such epidemics involving approximately 500 people, only seven deaths have been recorded.[110]

CHRONIC PULMONARY HISTOPLASMOSIS

Chronic cavitary histoplasmosis occurs most commonly in adult males. This disease is essentially indistinguishable from chronic cavitary tuberculosis, but the correct diagnosis must be made since the treatment is different. Cavitary histoplasmosis may develop immediately following the acute stage or after years of apparent quiescence of disease. The symptoms include productive cough, occasional hemoptysis, low-grade fever, progressive weakness, and roentgenographic picture of pulmonary cavitation (Fig. 16–2,*G*). Rales are more common in reactivated disease of older patients than in recently acquired disease in younger patients. This cavitary form of the disease does not resolve, and eventual dissemination with systemic manifestations is the rule.

Reactivation histoplasmosis has come to be a more commonly recognized syndrome in highly endemic areas. As is seen in tuberculosis, the mortality peaks of histoplasmosis are in infancy and in the fifth or sixth decades. With endogenous reactivation, lesions are in one or both upper lobes and are chronic and fibrocaseous in nature. Recent surveys conducted at Veterans Administration Hospitals and at state tuberculosis sanitoria have emphasized the high frequency and the greater morbidity and mortality of reactivated histoplasmosis as compared with reactivated tuberculosis. Many instances of the coexistence of the two diseases have been noted, especially when marked hemoptysis is present.[99] Reinfection from exogenous sources is extremely rare, as healed primary infection renders the patient essentially immune to recurrence of the disease. Exacerbation of old lesions is attributed to endogenous reactivation.[116]

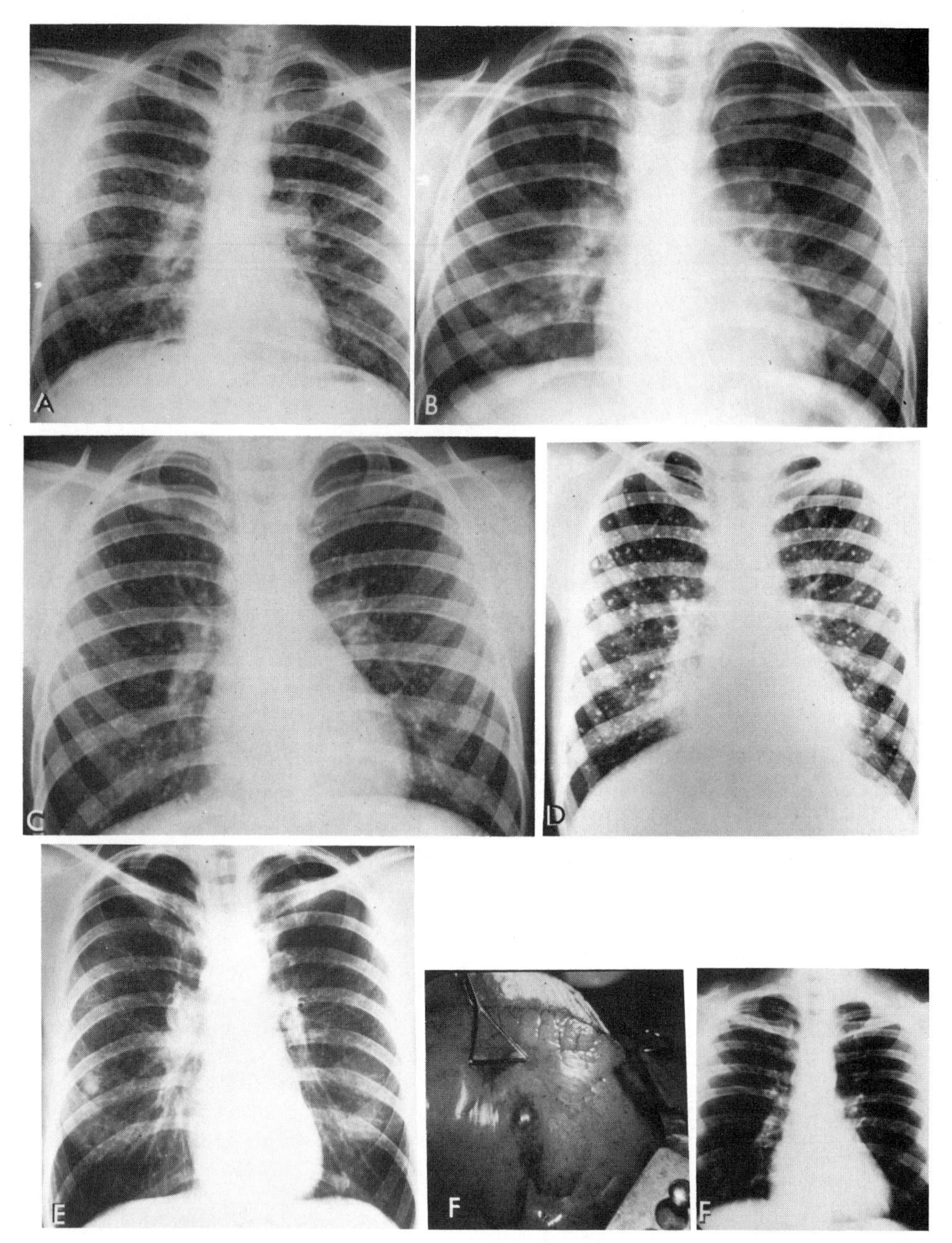

Figure 16–2. Radiologic aspects of histoplasmosis. *A,* Diffuse interstitial and nodular infiltrates are seen in both lungs and are associated with hilar and mediastinal involvement. *B,* Multiple miliary nodules in both lungs of a child associated with hilar and mediastinal adenopathy. *C,* Two years later there are multiple calcified densities throughout both lung fields. *D,* Multiple, small, calcified areas in lung parenchyma. Bilateral involvement. (Courtesy of J. Fennessy.) *E,* A well-circumscribed peripheral nodule in right middle lobe associated with enlarged hilar nodes. *F* (FS B 48), Well-circumscribed peripheral nodule in right lower lobe and appearance of lesion at surgery. (Courtesy of F. Kittle.)

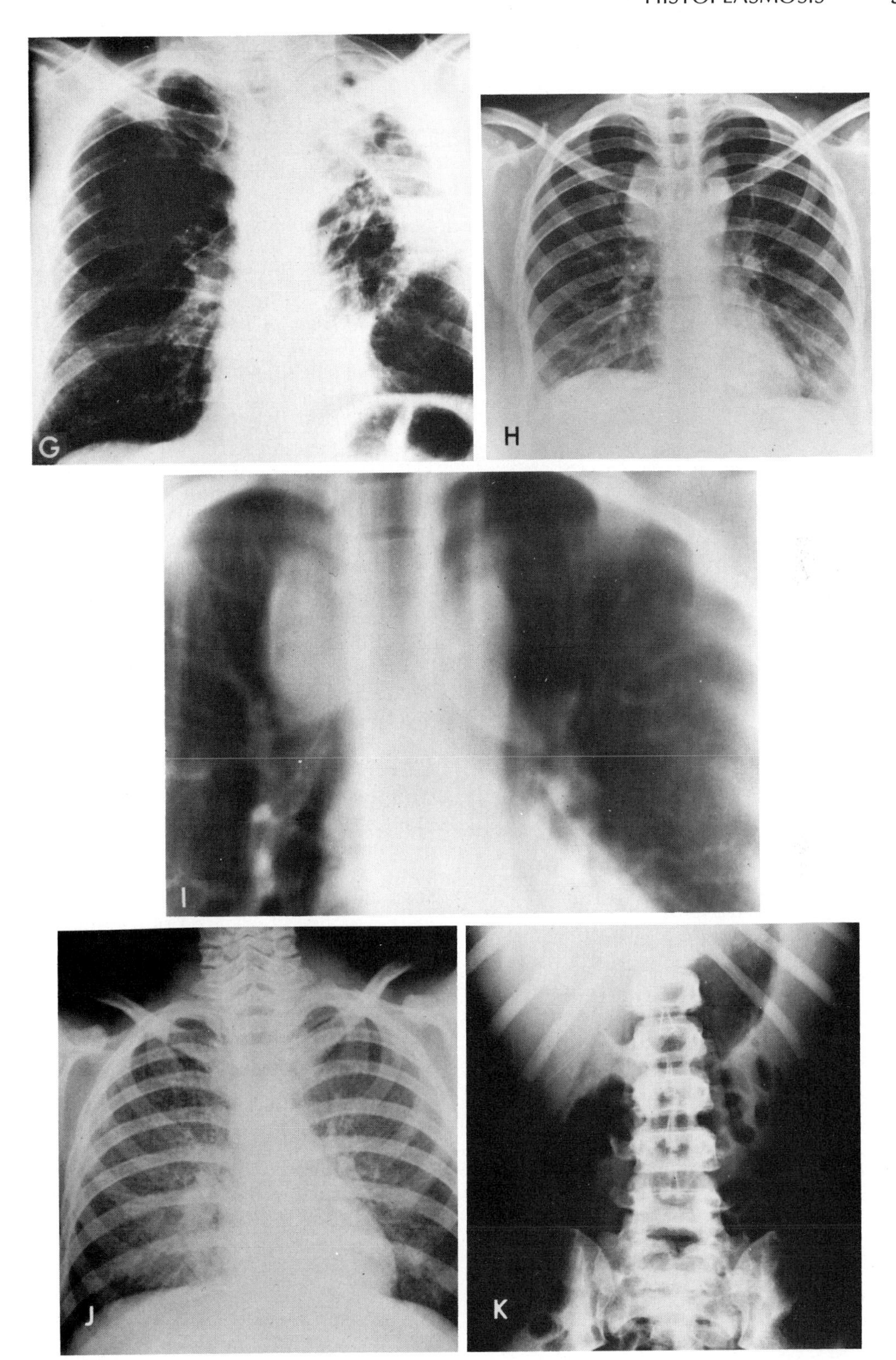

Figure 16–2 *Continued.* *G,* Cavitary histoplasmosis. There is a reduction in volume of the left lung, with multiple cavities in upper lobe. A very large lateral cavity has an air-fluid level. There is also marked pleural thickening. *H,* Superior mediastinal syndrome. There is a large right-sided mediastinal mass in the region of the tracheobronchial nodes which is smooth in outline and shows no calcification. *I,* A tomogram confirms the location in tracheobronchial node and shows displacement of upper lobe bronchus downward. *J,* Fatal disseminated histoplasmosis in a 10-year-old child with leukemia. There are diffuse interstitial and pulmonary infiltrates of both lungs associated with hilar adenopathy and hepatosplenomegaly. *K,* Plane film of abdomen showing splenomegaly in same case.

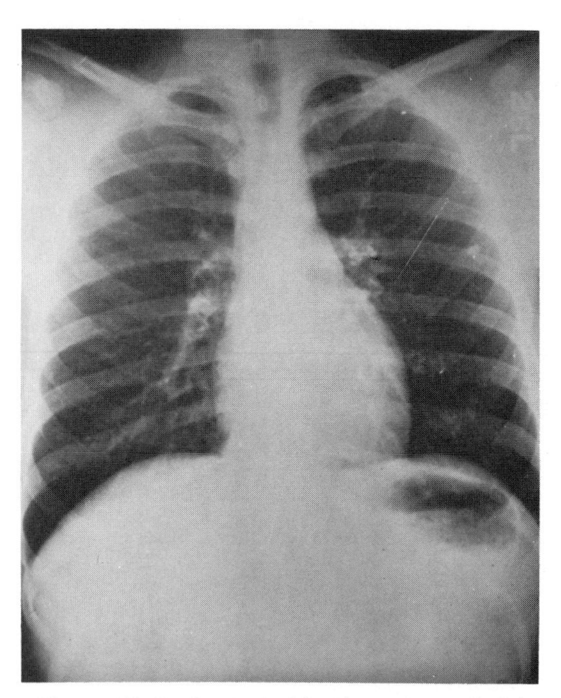

Figure 16–3. Resolved histoplasmosis. Left hilar lymph nodes contain dense and irregular "popcorn"-type calcifications. There is also a small calcified lesion in the periphery of the left lung. (Courtesy of H. Grieble.)

A rare type of chronic pulmonary histoplasmosis has been termed the *superior mediastinal syndrome.* There is mechanical obstruction of the vena cava, a condition which is usually due to neoplasias, tuberculomas, or other enlarging granulomas (Fig. 16–2,*H,I* and 16–11 FS C 6). As is usual in all forms of the disease, the hilar nodes are involved first, but the infection extends to include the mediastinal nodes. The description of this entity as a sequela of histoplasmosis notes there is massive fibrosis of the mediastinum, apparently as a result of the rupture of a caseous lymph node. Empyema and systemic dissemination may occur if the lesion breaks down.[109] Because of this and the danger of obstruction to the vena cava, surgical removal of such masses is warranted. Adequate chemotherapeutic coverage must be instituted during and after the procedure to prevent reinfection. Another unusual form of histoplasmosis is pericarditis and constrictive pericarditis resulting from chronic pulmonary disease.[52, 125] Bronchopleurocutaneous fistula as a complication of disease has also been reported.[61b]

Disseminated Histoplasmosis

BENIGN DISEASE

As has been amply demonstrated by the presence of miliary calcification in the spleen, liver, and other reticuloendothelial areas, dissemination is a common occurrence in primary pulmonary disease.[109] Small tubercles are formed which subsequently heal, leaving only residual areas of calcification of the type seen in healed pulmonary disease.

PROGRESSIVE ADULT DISEASE

Progressive dissemination in the adult was the form of the disease first described by Darling and the only form known for many years. Pulmonary symptoms are usually not prominent, but hepatosplenomegaly, a characteristic of leishmanial diseases, is very apparent. Other symptoms include anemia, leukopenia, and weight loss. Sometimes, however, there is associated massive consolidation of the lung. This form of disease is often rapidly fatal, and there are many variations in its clinical presentation.

Progressive histoplasmosis with a fulminant course also occurs in some patients with lymphomatous disease[45] (Fig. 16–2,*K*). The original infection probably occurred many years before and lay dormant until the neoplasia developed. *Histoplasma* in such cases is acting as an opportunist. The disease disseminates and the fungus infects many organs which are not usually involved in systemic disease. These include endocardium, adrenals, meninges, and bone.[11,106] Histoplasma causing chronic ulcerative enteritis[55] and peritonitis[94] has also been reported.

FULMINANT DISEASE OF CHILDHOOD

In children less than one year old, fulminating, acute disseminated histoplasmosis is not an uncommon entity in highly endemic areas. A few survive with or without therapy, but in others the course is rapidly fatal. In the latter group there is a phenomenal proliferation of histiocytes, most of which contain numerous *Histoplasma* yeast cells. The number and size of these histiocytes often cause obliteration of the normal architecture of affected organs. Endothelial tissues are particularly involved, especially those of the spleen, bone marrow, lymph nodes, liver, and intestinal mucosa (see Fig. 16–10). It has been suggested that such patients have an endothelial defect but there is as yet no evidence to substantiate this view.

CHRONIC PROGRESSIVE AND MUCOCUTANEOUS HISTOPLASMOSIS

This is a rare form of the disease with some similarities to cutaneous blastomycosis and paracoccidioidomycosis. Pulmonary involvement is seen to be minimal by x-ray examination, but solitary or multiple ulcers occur on the mouth, pharynx, larynx, stomach, and mucosa of the large and small intestine. Patients have also presented with a small lesion on the tongue (Fig. 16–4 FS B 49), with an exfoliative erythroderma,[101] or with peritonitis[94] or balanitis (Fig. 16–5 FS B 50). Granulomatous lesions of the skin resembling secondary blastomycosis have also been recorded.[109] Histoplasmosis rarely involves bones or joints; however, a few cases have presented as polyarthritis.[24a]

Primary cutaneous histoplasmosis is a very rare entity. It is usually the result of accidental injection of contaminated material in a laboratory or, in one personally observed case, the result of implantation of the fungus with a thorn while the patient was collecting bat guano from which the organism was isolated (Fig. 16–6 FS C 1). In this latter case, the lesion, which was chancriform, was accompanied by local lymphadenopathy and resolved without treatment.

DIFFERENTIAL DIAGNOSIS

As stated previously, at all stages of its pathogenesis, histoplasmosis mimics tuberculosis. The differentiation of these diseases is always difficult. Only culture and adequate serologic evidence provide the correct diagnosis. Histoplasmosis may coexist with any number of granulomatous diseases of the lung, including tuberculosis, sarcoidosis, actinomycosis, or other mycotic infections.[99]

Primary histoplasmosis in its acute stage closely resembles the acute infection produced by other mycoses, viral and bacterial pneumonias, lipoid pneumonia, and Hamman-Rich syndrome (diffuse interstitial pulmonary fibrosis).

Acute disseminated histoplasmosis with its accompanying hepatosplenomegaly, lymphopathy, anemia, and leukopenia resembles the acute stage of visceral leishmaniasis as well as many of the lymphomatous diseases. In most forms of histoplasmosis, thick blood smears or material from sternal punctures often reveal the organism more readily than

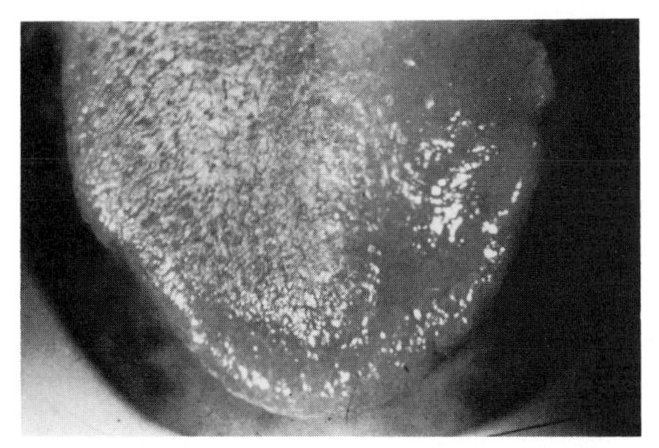

Figure 16–4. FS B 49. Disseminated histoplasmosis. Denuded erythematous lesion on tongue.

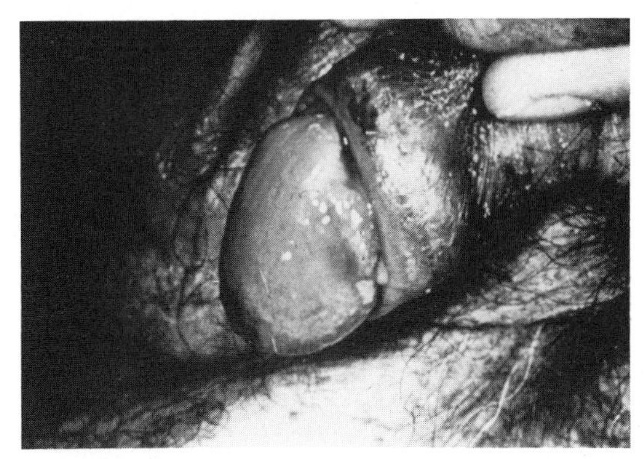

Figure 16–5. FS B 50. Disseminated histoplasmosis. Granulomatous lesion of penis.

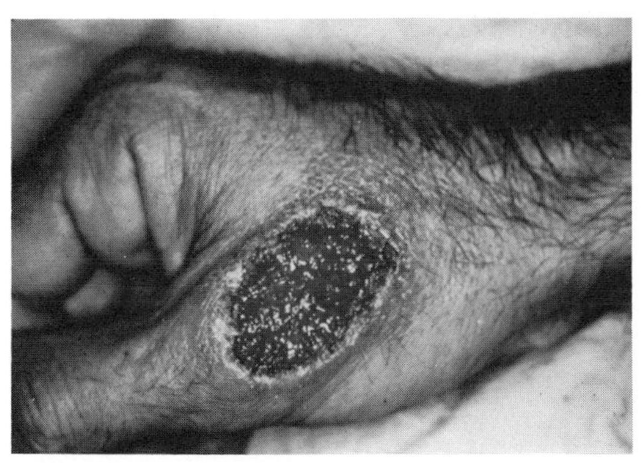

Figure 16–6. FS C 1. Primary cutaneous histoplasmosis. Well-demarcated ulcer with raised border. The lesion was the result of the inoculation of spores by a thorn. The hand is that of the late explorer Francis Brenton. (Courtesy of S. McMillen.)

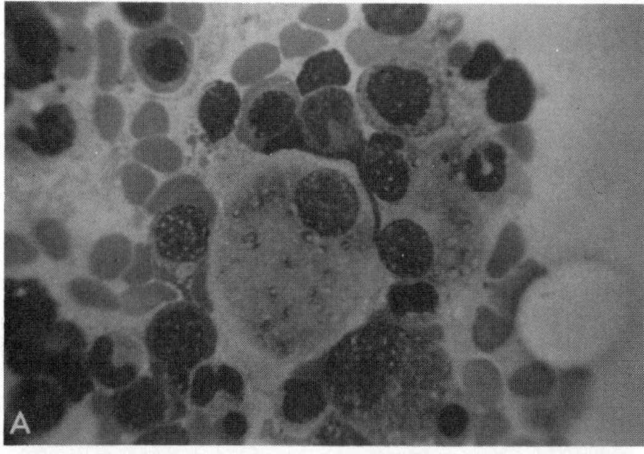

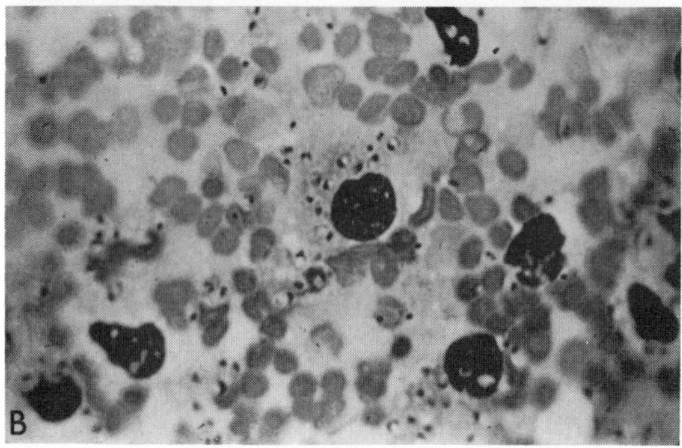

Figure 16–7. FS C 2. Histoplasmosis. *A*, Sternal puncture showing intracellular yeast cells. Wright's stain. ×400. *B* (FS C 3), Similar preparation. GMS. ×400.

culture or direct examination of sputum (Fig. 16–7 FS C 2, 3). The blood smears can be made from the buffy coat of centrifuged citrated blood or from the bottom of the tube, where sedimented, heavily infested cells are found. Other diseases to be considered in the differential diagnosis of histoplasmosis are infectious mononucleosis, brucellosis, malaria, and Gaucher's disease. When cutaneous or mucocutaneous lesions are present, they may suggest neoplasias, sporotrichosis, syphilis, toxoplasmosis, bacterial cellulitis, tuberculosis cutis, or other systemic mycotic infection.

The histoplasmin skin test is of no diagnostic value, as it only connotes past or present exposure to the organism. Serial complement fixation tests, especially when there is a rise in titer, are very significant. It must be remembered that the histoplasmin skin test may cause a one- to two-tube rise in the complement fixation test.

PROGNOSIS AND THERAPY

In essentially all cases of histoplasmosis, when the patient does not have an underlying disease or the inoculum is not overwhelming, the disease resolves uneventfully. Even in moderately severe primary infections, bed rest and supportive measures are sufficient to effect resolution of the disease. However, disseminated disease, chronic cavitary, mucocutaneous, or systemic infections require antimycotic therapy.

The treatment to be used is the same as that for all the systemic mycoses: amphotericin B.[73,100] The dosage schedule is also the same, as are its side effects and measures to ameliorate these side effects (see Chapter 18). The total course of therapy is 0.5 g in severe primary disease, and 2 to 3 g in chronic disease. In contrast to coccidioidomycosis and blastomycosis, recovery is rapid, and relapse is essentially unknown. In the rare case of cerebral involvement, slow (over 6 hours) intrathecal administration of 0.2 to 0.3 mg of amphotericin B administered by means of an Ommaya reservoir has been recommended.[120]

Ethyl vanillate, MDR-12, nystatin, 2-hydroxystilbamidine, sulfadiazine, 5-fluorocytosine, and antibacterial or antitubercular antibiotics are ineffective in histoplasmosis. Sulfadiazine shows some effect in the treatment of disease in experimental animals, and before amphotericin B was available appears to have been efficacious in a few human cases.[82] Saramycetin (X5079C) has been shown to be effective and essentially nontoxic, but is unavailable for either clinical trial or purchase.

After medical management has brought about clinical arrest or the disease spontaneously resolves, surgical excision of large cavities or granulomatous masses may be considered.[61a, 89a] Prophylactic use of amphotericin B is advisable to preclude reactivation or dissemination during the surgical procedure.

PATHOLOGY

Gross Pathology. The gross pathology induced by acute primary disease is essentially unknown, as this form of infection is a mild, uncomplicated disorder, and few patients come to autopsy. To extrapolate from the animal experiments of Proknow,[91] following infection there is an initial alveolitis, and macrophages engulf the organisms. Subsequently there is an invasion by neutrophils and lymphocytes, and a progressive inflammatory reaction of the usual pyogenic type occurs. The macrophages probably carry the organisms to other body parts very early in the course of infection. The pyogenic response in the lung is followed by the formation of epithelioid cell tubercles indistinguishable from those of tuberculosis. In the few fatal cases of acute histoplasmosis due to overwhelming exposure to spores, there was massive pleural and alveolar effusion, causing death by asphyxiation within a few days. When the course was prolonged, numerous yellowish nodular masses (forming tubercles) and confluent nodular masses were found in the lungs, along with great enlargement of the hilar and mediastinal lymph nodes. In less severe disease, where there was a need to remove a nodule for diagnosis, the lesion was an epithelioid tubercle with Langhans' giant cells that would have been called, if caseating, "histologically compatible with tuberculosis, bacilli not seen" or, if not caseating, sarcoid. Healed lesions from subclinical or mild disease completely calcify after several years. They are small, 15 mm or less, and may be scattered throughout all lung fields. Similar nodules are present in the spleen (Figs. 16–8 FS C 4 and 16–9 FS C 5) and less frequently in the liver. The organisms in tubercles are usually dead and demonstrable only by the Grocott-Gomori methenamine stain or other special fungal stains.

In chronic progressive cavitary disease, the lesions are identical to those of cavitary tuberculosis. Caseation, necrosis, and layers of fibrosis and calcification are present. The

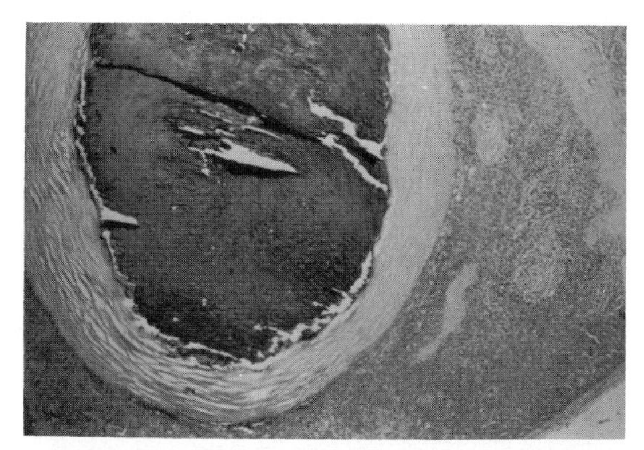

Figure 16–8. FS C 4. Histoplasmosis. Fibrocaseous lesion of spleen completely calcified. This represents a residual lesion from an infection acquired many years previously. The lesion was inactive and the organism dead. Yeasts were visible on GMS stain. Hematoxylin and eosin stain. ×100. (Courtesy of S. Thomsen.)

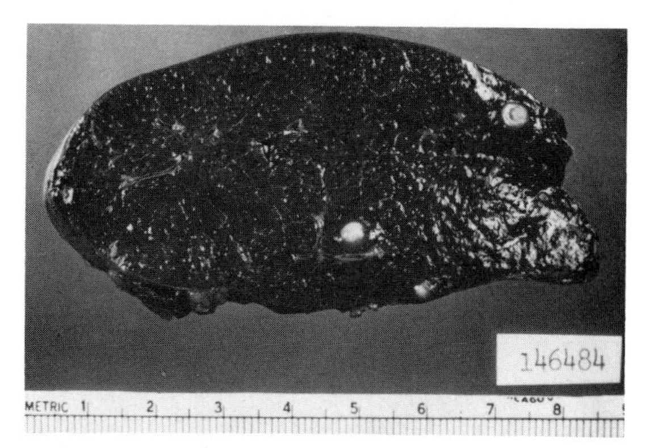

Figure 16–9. FS C 5. Histoplasmosis. Spleen from a patient who had primary histoplasmosis many years previously. Well-circumscribed, partially calcified, light yellow, raised nodules are present in cut surface of spleen. They are from 0.2 to 0.4 cm in diameter. The lesions are typical of healed fibrocalcific granulomata of old resolved histoplasmosis. (Courtesy of S. Thomsen.)

cavity may be 8 or more cm in diameter, and a few have been noted to contain *Aspergillus* fungus balls. *H. capsulatum* yeast cells can be demonstrated in the walls, and the hyphae of the *Aspergillus* are in the lumen.[99,109]

The pulmonary "coin" lesion, which is usually solitary, may vary from 0.5 to several cm in size (Fig. 16–2,*F* FS B 48). It is usually located in one of the lower lobes and generally attached to the pleura. There is a fibrous thickening of the overlying pleura. The lesion itself has a fibrous capsule with some calcific material and a caseous center. The organisms present are usually dead and are in the central caseous part. The histoplasmoma has essentially the same characteristics except that it is larger and may be seen to expand with time.[50]

In the rapidly fatal progressive form of histoplasmosis, there is, as first noted by Darling in 1905, gross enlargement of the liver and spleen. This may be the only abnormality apparent at autopsy. Histologically, phagocytized yeast cells are seen in the areas rich in reticuloendothelial cells, such as liver, spleen, bone marrow (Fig. 16–10), lymph nodes (Fig. 16–11 FS C 6), and sometimes the adrenals. If the disease had a more protracted course, tubercles are present in essentially all organs, lymph nodes, and tissue. In certain organs such as the adrenals, there is evident caseous necrosis, and such patients may have signs of Addison's disease.[27]

In chronic disease, all of the above mentioned pathology is present but exagger-

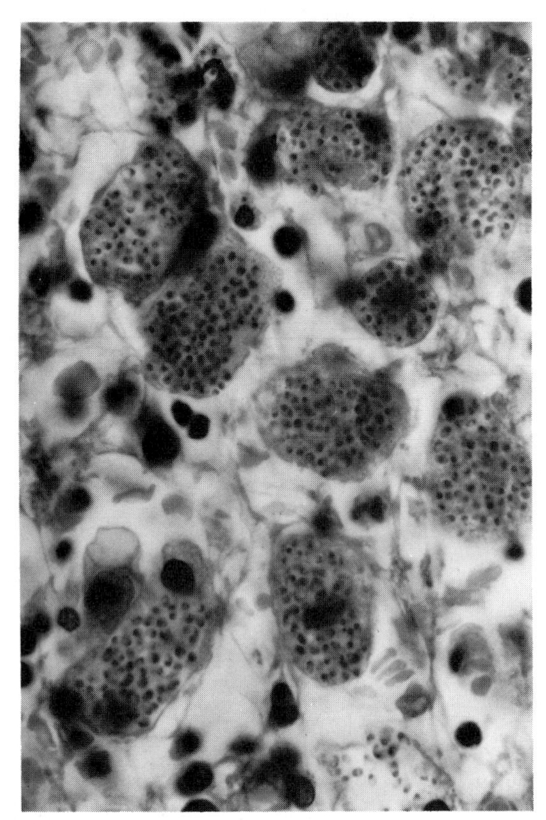

Figure 16–10. Histoplasmosis. Histiocytes containing numerous yeast cells. This was from a case of fulminant disease of childhood. Hematoxylin and eosin stain. ×4000. (Courtesy of E. Humphreys.)

ated.[11] In addition, ulcerating lesions may be found in the intestine, larynx, pharynx, genitals, tongue, meninges, and endocardium. In the latter instance, extensive vegetations resembling those of *Candida* endocarditis are present. A complete analysis of gross and histopathologic appearance of organ systems in all forms of histoplasmosis has been compiled by Schwarz.[109]

Histopathology. The histologic appearance of the lesions depends on the recentness of the infection and severity and form of the disease. In the acute, rapidly fatal form (histiocytomycetic disease), numerous yeast cells are seen only within the histiocytes. Neutrophils, plasma cells, and lymphocytes are not abundant and do not contain yeast cells. The yeast cells within the histiocytes are budding, of uniform size (about 3 μ), and are visible in Gram, Giemsa, Wright's, or hematoxylin-eosin stained smears. They resemble *Leishmania donovani* but lack the rodlike parabasal body (kinetoplast) (Figs. 16–7 FS C 2, 3 and 16–12 FS C 7). The internal morphology is best seen with Giemsa stains. *Histoplasma* yeast cells are stained by any of the special fungus stains, whereas *Leishmania* is not. *Toxoplasma* also must be ruled out in such cases. It is smaller and not found within histiocytes and does not take the special fungus stains. *Torulopsis glabrata* also may be intracellular. Only culture or the use of the immunofluorescent technique on tissue sections will differentiate them.

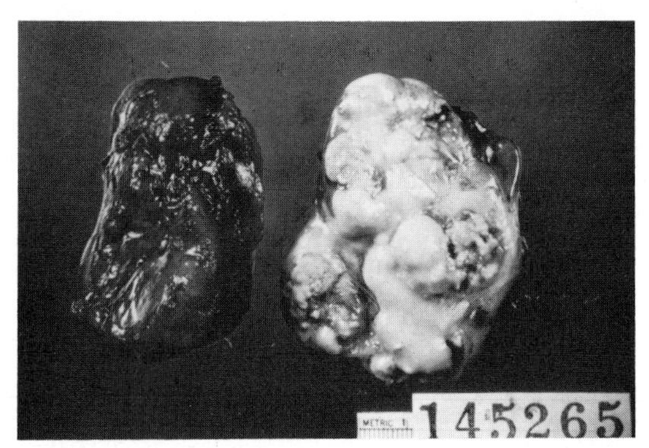

Figure 16–11. FS C 6. Histoplasmosis. Superior mediastinal syndrome. The mass removed was well encapsulated and nodular. On cut surface it could be seen that the lymph nodes were nearly all destroyed by necrotizing and caseous granulomas. The necrotic center contained a thick, creamy, light-green fluid (see Fig. 16–2,*H,I*). (Courtesy of S. Thomsen.)

In the less fulminating forms of the disease, epithelioid granulomas are formed that contain plasma cells, lymphocytes, macrophages, neutrophils, and giant cells. The organisms are seen in any of the phagocytic

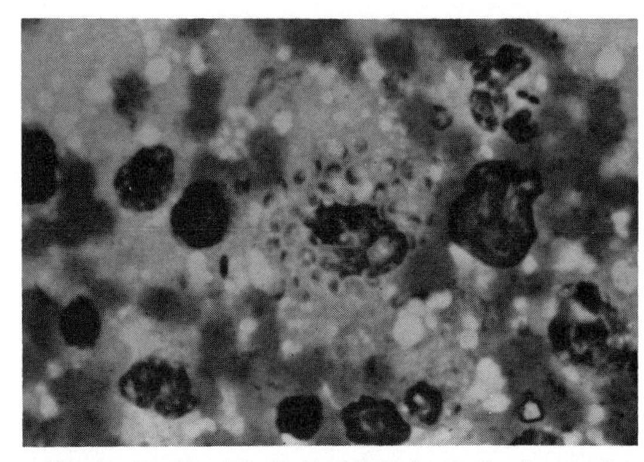

Figure 16–12. FS C 7. Histoplasmosis. Impression smear of liver from a case of disseminated disease. Intracellular organisms are seen, and their internal morphology is easily discerned. Wright's stain. ×400.

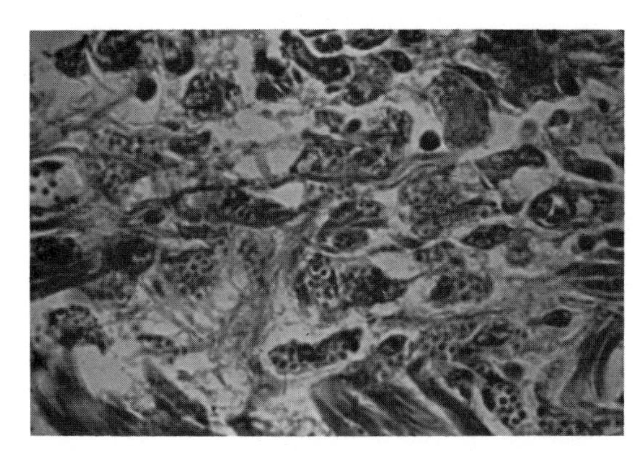

Figure 16–13. Histoplasmosis. Biopsy specimen from chronic cutaneous disease. Intracellular organisms are seen, but there is little budding.

cells, but in smaller numbers than in fulminant disease. Budding is less prominent, and the uniformity of size and staining characteristics are the basis for presumptive diagnosis. This is also true in biopsy material from cutaneous and mucocutaneous lesions (Fig. 16–13). The stain of choice in such cases is the methenamine silver stain. In chronic cavitary disease, the organisms are even fewer in number, and very rarely are seen to bud.

In old lesions, coin lesions, histoplasmomas, and calcified nodules, the organisms are usually dead and more difficult to find in histologic sections. In some areas they may be abundant but are very irregular in size and morphology. Organisms up to 20 μ have been found. In many lesions of chronic or resolved pulmonary disease, the yeast cells are very rare, and few if any will show budding. In such cases, cultures are usually negative, and a diagnosis must be made on histologic evidence alone after extensive search of many sections. The yeast cells of *H. capsulatum* must be differentiated from newly released spores and young spherules of *Coccidioides immitis*, from *Cryptococcus neoformans*, and from the small cell form of *Blastomyces dermatitidis*. There is usually enough capsular material around even dead cryptococci to stain with Mayer's mucicarmine stain. The so-called "capsule" of *H. capsulatum* represents unstained cell wall, and it does not take the mucicarmine stain. If the lesion in question is caused by *C. immitis*, spherules or parts of spherules are found after intensive search. Even the small forms of *B. dermatitidis* show a broad-based bud and can be seen to be multinucleate in a hematoxylin and eosin stain. *H. capsulatum* is uninucleate and has a narrow-necked bud. Small hyphae have been reported to be present in a case of *Histoplasma* endocarditis, but such a finding is rare. In such a case, differentiation of the disease from candidosis is necessary. In lesions in which differentiation of *Histoplasma* from other etiologic agents cannot be made, specific fluorescent antibody staining can be attempted.[62]

ANIMAL DISEASE

Natural disease in wild and domestic animals is very common in endemic areas. Dogs seem to be particularly susceptible, and the infection has been reported in cats, swine, cattle, and horses. Among wild animals, histoplasmosis has been reported in numerous species, ranging from the house mouse to the Kodiak bear (the latter housed in an Ohio zoo). Indeed the only recorded case of histoplasmosis in Switzerland is that of a badger.[16] This animal was found along the roadside, and the fungus was demonstrated in its submandibular lymph nodes. The only animal species considered to be an important vector for dissemination of the fungus is the bat.[32]

The type and severity of infection in animals and the pathology elicited are as protean as those found in human cases. They vary from small calcified nodules found incidental to other diseases to an acute, disseminating, rapidly fatal form. There is no substantiated evidence that natural infection occurs in avian species. Menges[77] has reviewed the animal species in which infection has been reported and the varieties of pathology encountered. Epizootic lymphangitis caused by *H. farciminosum* is discussed in Chapter 23.

The mouse is the most susceptible laboratory animal to experimental infection. Rowley[97] demonstrated that from 1 to 10 yeast cells will infect a mouse, and Ajello and Runyon[6] found that one macroaleuriospore was sufficient to evoke disease. For this reason the mouse can be used as an adjunct to other procedures for isolation of the yeast from patient material, for determining the presence of the organism in soil samples, and for testing the efficacy of chemotherapeutic agents in experimental infections.

The procedures for use of the mouse as an

aid to isolation of the fungus are well standardized. Material obtained at biopsy can be macerated and injected intraperitoneally. Sputum, however, must be mixed with antibiotics before injection. Usually one ml of a mixture of penicillin (7 mg per ml) and streptomycin (10 mg per ml) is added to 5 ml of sputum. Chloramphenicol (0.5 mg) may also be used. Since sublethal doses of *H. capsulatum* are known to immunize mice and abort infections, mice should be autopsied from two to four weeks after inoculation. The spleen and liver are removed and minced with a scissors and smeared on Sabouraud's or blood agar plates. These are sealed with tape and incubated at room temperature for at least four weeks. The initial growth may be glabrous but is usually a white fuzz. For identification, a portion of the colony is transferred to blood agar slants and incubated at 37°C to demonstrate dimorphism. The mycelial growth occurring at 25°C is examined for the presence of tuberculate macroaleuriospores. Care should be exercised in handling the mycelial phase.

The procedure for isolation of the organism from soil is similarly executed.[71] A saline suspension of the soil is mixed with antibiotics in the same concentrations as described for sputum. The material is then injected intraperitoneally. The animals are usually held for one month before autopsy. Mice may also be infected by housing them with contaminated soil.

For the purpose of testing therapeutic agents, it is necessary to assay the virulence of the particular strain of *Histoplasma* to be used. Though most strains will kill mice when a dose of 10^6 or 10^7 cells is injected intraperitoneally, some strains are less virulent.[28] In antibiotic testing it is preferable to determine the LD_{50} and inoculate this dose by the intravenous route. The tail vein is suitable for this procedure, and with experience few technical problems are encountered. Accurate counts of the number of viable units injected can be determined by the Janus green method of Berliner and Reca.[14]

Dogs are quite susceptible to experimental histoplasmosis. The early sequence of events following infection has been studied by administering aerosols of spores to dogs.[91] Other laboratory animals, such as rabbits, hamsters, and rats, vary considerably in their susceptibility to infection. Experimental disease has also been produced in monkeys.[62a]

Experimental disease has been produced in poikilothermic animals also. In one experiment lizards and frogs were injected with either the yeast or mycelial form of the fungus. Regardless of which form was injected, lesions in animals incubated at 25°C contained mycelium, and yeast cells were found in those incubated at 37°C.[104]

Pigeons and chickens have been injected with *H. capsulatum*. Infection is produced, but it is of short duration. The body temperature of birds is thought to be too high for a progressive disease to develop.[108]

BIOLOGICAL STUDIES

One of the most intriguing aspects of its biology is the dimorphism exhibited by *H. capsulatum*. As with the other pathogenic fungi and in contrast to opportunistic agents of fungous disease, a radical morphologic and physiologic change accompanies *in vivo* existence. In *H. capsulatum* the form found in its natural saprophytic habitat is a mycelium producing numerous spores, but when this fungus is inhaled by susceptible animal species, it grows as a small, delicate, budding yeast. The factors that govern this dimorphism have been the subject of much investigation. Initially it was thought that substances from serum or tissue within the host were necessary for conversion to the yeast. *In vitro* conversion to the yeast form of growth could be accomplished on blood agar but not on nutrient agar incubated at 37°C, even though nutrient agar incubated at this temperature was known to be sufficient for the growth of the yeast form of *B. dermatitidis*. Since *H. capsulatum* grew as a mycelium at 25°C on blood agar, it appeared that a factor in addition to temperature affected the mycelial to yeast conversion. Investigations by Pine,[87,88] Scherr,[103] and others established that it was free -SH groups provided by the blood in blood agar which effected the transformation. They demonstrated that, in a salt medium containing glucose, mycelial to yeast conversion could be accomplished if cysteine was included to provide the free -SH groups. It was later shown that this conversion could be effected by electrolytically lowering the oxidation-reduction potential of liquid media to that of living tissue.[96] Cysteine is the most effective -SH-containing compound for converting and maintaining the yeast form of growth. In a study of a series of low molecular weight, sulfur-containing compounds, Garrison et al.[48] have shown that cysteine has the greatest stimulatory effect on yeast respiration, and that it stimulated growth of the yeast form to a greater degree than the mycelial form. The reduced environment appears to enhance cell wall production as a consequence of its stimulation of energy metabolism.

It has now been established that temperature is the single most important factor regulating the mycelial to yeast conversion of *H. capsulatum*, and therefore it is a thermal dimorphic fungus. Pine[88] and Scherr[103] demonstrated this temperature dependence *in vitro*, and Howard[59] found that phagocytized intracellular yeast cells in tissue culture replicated as yeasts when maintained at 30°C. However, if the phagocytized yeasts were incubated at 25°C, they sprouted to form mycelium. A similar temperature dependence of *in vivo* growth form has been demonstrated in poikilothermic animals. The morphogenesis of the yeast form and its growth in tissue culture have been investigated and reviewed by Howard.[57, 58, 59a] Nutritional and physiologic aspects of both forms have been reviewed by Bauman.[12]

Domer et al. have studied the differences in cell wall composition accompanying morphogenesis.[34,35] They found quantitatively more chitin and less mannose and amino acids in the walls of the yeast form than in the mycelial form. The composition of the mycelial form cell wall closely resembled that of the yeast *Saccharomyces cerevisiae*. They found that there are at least two chemotypes of the cell walls of the yeast form. Chemotype II was similar in its monosaccharide and chitin content to the yeast form of *B. dermatitidis*.[34] These chemical differences in cell wall composition are undoubtedly reflected in the different antigenic patterns of the yeast form of *H. capsulatum*. Other biological studies have included the standardization of growth curves for production of antigens. It was found by Reca and Campbell[18,93] that different conditions of growth will markedly affect the antigens produced by the organism. Protoplasts of the yeast phase can be produced by a number of agents.[15] The presence of "L" forms in infected tissue has been postulated but not established. "L" forms have been found in other fungous infections. Electron microscopic studies of the organism have also been carried out.[36, 48a]

IMMUNOLOGY AND SEROLOGY

The presence of a high natural resistance to infection by *H. capsulatum* is attested to by the fact that in the vast majority of infected persons there are few, if any, symptoms, and the disease resolves rapidly. In cases of overwhelming exposure to fungus, disease is established in all patients and also occurs in a few patients who have had only minimal exposure. Subtle, as yet undefined, immune differences or slight, even transient, debilitation may account for these latter cases. As noted previously, *H. capsulatum* is rarely an opportunistic fungus invader in the usual sense, which tends to make the problem of the patient population that develops serious disease more enigmatic. It now appears that recovery from mild infection renders the patient firmly resistant to reinfection. Circulating antibodies are considered of little consequence in the immunity established following resolution of infection. In experiments using tissue culture it was shown that the phagocytosis and intracellular growth of *H. capsulatum* yeast cells was not affected by the presence of serum containing antibodies to the organism.[58] However, Howard et al.[61] have demonstrated that mononuclear phagocytes from immunized animals restricted the intracellular growth of the yeast cells as compared with phagocytes from nonimmune animals. It appears, then, that immunity to the infection is concerned with cellular rather than humoral defenses.

Serologic Tests

The Skin Test. As has been noted previously, past or present exposure to the fungus usually confers a lifelong reactivity to the histoplasmin skin test. Thus this test is of little diagnostic or prognostic value except in cases of disseminated disease, in which absence of reactivity denotes anergy,[46] or in very early infections, in which the test may also be negative. The skin test usually becomes positive within two weeks after exposure.

The Complement Fixation Test. Serial samples of serum used in the complement fixation test are of great value in establishing the diagnosis as well as delineating the prognosis of the disease. This was first demonstrated by Salvin in experimentally infected animals. Complement fixing antibodies may appear as early as two weeks following infection, and almost all patients will have a demonstrable titer by four weeks. Following resolution of the disease, the titer gradually falls and usually disappears by the ninth month. Reaction to yeast phase antigen generally remains longer than to that prepared from the mycelial phase. In some recorded cases, complement fixing titers after clinical recovery have remained positive for up to four years, when the histoplasmin (mycelial) antigen was used, and up to nine years with the yeast cell antigen. However, a high titer (1:32 or over) that remains at that level or rises is usually considered to be indicative of active or progressive disease. The serologic pattern and its interpretation is not as well established in histoplasmosis as it is in coccidioidomycosis.

There are many pitfalls in the use of complement fixation tests for diagnostic and prognostic evaluation. The careful analysis of Campbell[17–19] has shown that high antibody titers to histoplasmin may be present in cases of cryptococcosis, blastomycosis, and more rarely coccidioidomycosis. Cross-reactions occur in the sera from experimentally infected animals also.[98] In addition, it has been noted that administration of the histoplasmin skin test may cause a one- to two-fold increase in complement fixation titers.[66] Other workers, however, could not corroborate this.[102] Campbell has also demonstrated that the reactivity of the antigen varies from batch to batch, with age of the cells, with composition of the media, and with other as yet unknown factors.[18, 39, 60] This is not surprising, considering the multiple antigenic sites present on the organism. Efforts to fractionate and isolate a specific antigenic component to be used in the complement fixation test have been reviewed by Larsh and Bartels.[70] Of the many fractions isolated, the most promising from the point of view of specificity is a polysaccharide derived from histoplasmin. At present, therefore, the complement fixation test is useful but cannot be said to give unequivocal evidence for the presence or absence of disease.

The immunodiffusion test using concentrated histoplasmin is a useful adjunct to the complement fixation test. It appears to become positive by the third or fourth week following infection. As first described by Heiner,[54] the antigen and patient's sera were allowed to diffuse toward each other from wells in an Ouchterlony plate. He noted that two lines were of significance. An m line was present in persons who had recovered from the disease or in cases of early infection. An h line closer to the well containing the serum corresponded with the presence of active infection (Fig. 16–14). It disappears with resolution of disease, but may be present for up to two years following clinical cure. A c band, which is sometimes found, appeared to represent a common antigen between *H. capsulatum* and *B. dermatitidis*. Other bands designated x and z have appeared in some patients' sera, but their significance is unknown. Further evaluation of what appeared to be an extremely impor-

Figure 16–14. Histoplasmosis. Immunodiffusion plate with a known positive serum in one outer well and an unknown one in an adjacent well. The antigen is in the center well. Note the lines of identity in the two tests. The h line is the finer line near the serum-containing well. The m line is near the antigen-containing well.

tant diagnostic tool has shown variations and cross-reactions in this test also,[63,105] but it remains one of the most important serologic procedures for diagnosis and prognostic evaluation of histoplasmosis.

The latex agglutination test, in which serum is mixed with latex particles coated with histoplasmin, has been extensively investigated by Hill and Campbell.[56] It may be positive before the complement fixation test and is also useful in cases in which the patient's serum is anticomplementary.

The use of a battery of serologic tests tends to rule out both false-positive, false-negative, and cross-reactions. Their interpretation is always difficult, and the only unequivocal evidence for the diagnosis of histoplasmosis is isolation of the fungus *H. capsulatum.* Only when a single specific antigen has been isolated, purified, and standardized will the serologic testing in this and other fungous diseases be of absolute diagnostic value. It is possible, however, that developing a test for the presence of antigens derived from the organism which may be present in the patient's blood would be a more useful procedure than trying to evaluate the presence of the various antibodies. This is already a practical procedure in the serodiagnosis of cryptococcosis.

Cell-fixed antibodies have been demonstrated in experimental disease.[70] In this study, spleen cells from infected animals are mixed with *Histoplasma* yeast cells which stick to the sensitized cells, forming rosettes.

Skin testing for histoplasmosis has been discussed at length in other sections of this chapter. The recommended procedure is to inject 0.1 mg of a 1:100 dilution of a standardized histoplasmin intracutaneously. The test is read after 48 and 72 hours, though it may not become maximal for five days. An area of induration of 5 mm or more is considered positive. Care should be exercised in administration of the test. Hypersensitive individuals may develop a severe reaction, with fever, nausea, and necrosis at the area where the test was administered. A positive test indicates past or present exposure to the organism. A negative test in a patient with known histoplasmosis indicates anergy and suggests a grave prognosis. A new tine test which allows easier application of antigen has recently been developed.[23] Since skin testing has little diagnostic value and may interfere with the complement fixation tests for histoplasmosis and coccidioidomycosis, the general use of this procedure is to be discouraged.

Several studies have been made of the appearance of particular species of globulins in patients infected with *H. capsulatum.* Results initially indicated that, in new infections, IgM globulins were the first to increase in quantity. This was followed by increases in IgG and IgA. In chronic disease, the IgM and IgG levels were normal, but there was an absolute increase in IgA.[121, 122] More recent and extensive work reviewed by Larsh and Bartels[70] has shown that there is wide variation among patients in the types and quantities of the immunoglobulins in any stage of the disease.

Antigenic analysis of *H. capsulatum* has revealed that there are several serotypes. Kaufman and Blumer[65] have shown that there are four antigenic components and several antigenic patterns. The 1,4 type is identical serologically to *H. capsulatum* var. *duboisii,* indicating that the latter organism is probably a variety of *H. capsulatum* and not a separate species. Pine and Boone[89] have demonstrated that some differences in serologic types are reflected by differences in cell wall composition.

A fluorescent antibody specific for the yeast cells of *H. capsulatum* has been developed by Kaufman and Kaplan.[64] The usefulness of this procedure in making or confirming a diagnosis in the absence of cultures has been reviewed by Kaplan.[62]

Summary of Practical Serologic Testing in Histoplasmosis

Screening Tests. The latex agglutination becomes positive early in the disease. Immunodiffusion (ID) can differentiate active from inactive disease. It is positive about the same time as the latex agglutination (two to five weeks after development of symptoms). The fluorescent antibody test is also very useful where available.

Confirmatory Tests and Tests for Prognostic Evaluation. The complement fixation (CF) test is positive later in disease (six weeks or more) than most other tests. A titer of 1:32 is considered significant, but some patients with active disease have titers of only 1:8 or 1:16. Serial titers should be done. A rise in titer indicates dissemination. At this writing it appears that the CF and ID are the most important serologic procedures for diagnosis and evaluation of prognosis.

LABORATORY DIAGNOSIS

The diagnosis of histoplasmosis is made by the isolation and identification of *Histoplasma capsulatum.* Serologic techniques may be useful but cannot be considered absolute. Demonstration of the organism in stained sputum and preparations of biopsy or necropsy material, especially if confirmed by immunofluorescence, can be reliable affirmation of the diagnosis in the absence of culture.

Direct Examination. Detection of the organism in sputum by direct examination is difficult. The KOH procedure, which is useful in other fungous diseases, is usually negative in histoplasmosis. Sputum is a more useful specimen when cultured. Sputum, the buffy coat of centrifugal blood specimens, the sediment of such specimens, biopsy material, and sternal puncture material are best examined following staining. The material is spread on a slide and fixed for ten minutes with methyl alcohol. The fixed material is then stained either by the Wright or Giemsa method. From personal experience, it is felt that the Giemsa method is superior. Overstaining tends to emphasize the yeast cells. The organism is seen within macrophages and is 2 to 4 μ in diameter. It is usually ovoid, with the bud at the smaller end. The large-form cells of the var. *duboisii* are described in Chapter 17. There is a halo of unstained material around the cell, representing the cell wall. Within the cell there is a large vacuole, and a crescent-shaped, red-staining mass of protoplasm is usually present at the larger terminus of the cell. The buds are delicate and usually detach during the staining procedure. Though they are predominantly in macrophages or monocytes, yeast cells may appear extracellular because of destruction of the phagocytic cell. Many other yeasts, foreign bodies, artifacts, and parasites may simulate the appearance of *H. capsulatum* yeast cells.

Culture. In cases of primary pulmonary histoplasmosis, the most valuable specimen for isolation of the organism is sputum. The first morning sputum is collected after the patient has rinsed out his mouth. Needless to say, the organism is found in sputum; saliva will not be useful for culture. Gastric lavage done early in the morning before feeding the patient may also be rewarding. Sputum is planted directly on blood agar and Sabouraud agar plates and incubated at 25°C. Areas of sputum that are purulent or blood-streaked are selected for culture. Gastric lavage is centrifuged, and the sediment is planted on the same media as sputum. From personal experience and that communicated by Larsh, blood agar without antibiotics is the medium of choice. With diligent and frequent examination of such plates, the tiny colonies of *H. capsulatum* can be subcultured, even when there is abundant growth of contaminating bacteria and *C. albicans.* Other pathologic material, such as biopsy specimens, sternal punctures, aspirations, and so forth, are also minced and spread over blood agar and Sabouraud's agar plates. Plates should be sealed with tape, or stacks of them enclosed in plastic bags to prevent drying out. All plates should be kept for from four to six weeks. Any colony showing hyphal growth is subcultured for further identification. The initial colonies on blood agar are usually glabrous, cerebriform, and pinkish to reddish brown. In time they become mycelial and white to brownish. The colony is indistinguishable from that of *B. dermatitidis* and many other fungi. Demonstration of the characteristic tuberculate macroaleuriospores and conversion to the yeast phase are necessary for identification. The latter is accomplished by growth on blood agar slants at 37°C.

MYCOLOGY

Histoplasma capsulatum Darling 1906

Synonymy. *Cryptococcus capsulatus* Castellani and Chalmers 1919; *Torulopsis capsulatus* Almeida 1933; *Posadasia capsulata* Moore 1934; *Posadasia pyriforme* Moore 1934; *Histoplasma pyriforme* Dodge 1935.

Perfect State. *Emmonsiella capsulata* Kwon-Chung 1972.

Colony Morphology. SDA, 25°C. The colony that develops on Sabouraud's agar is initially white or buff-brown. These two morphologic types have been termed A (albino) and B (brown) by Berliner[13] (Fig. 16–15 FS C 8). Both types may be isolated from the same patient, and with subculture the brown type shows sectoring and eventual conversion to the albino type. As reported by Berliner, these types remain constant when the organism is kept in the yeast form. The

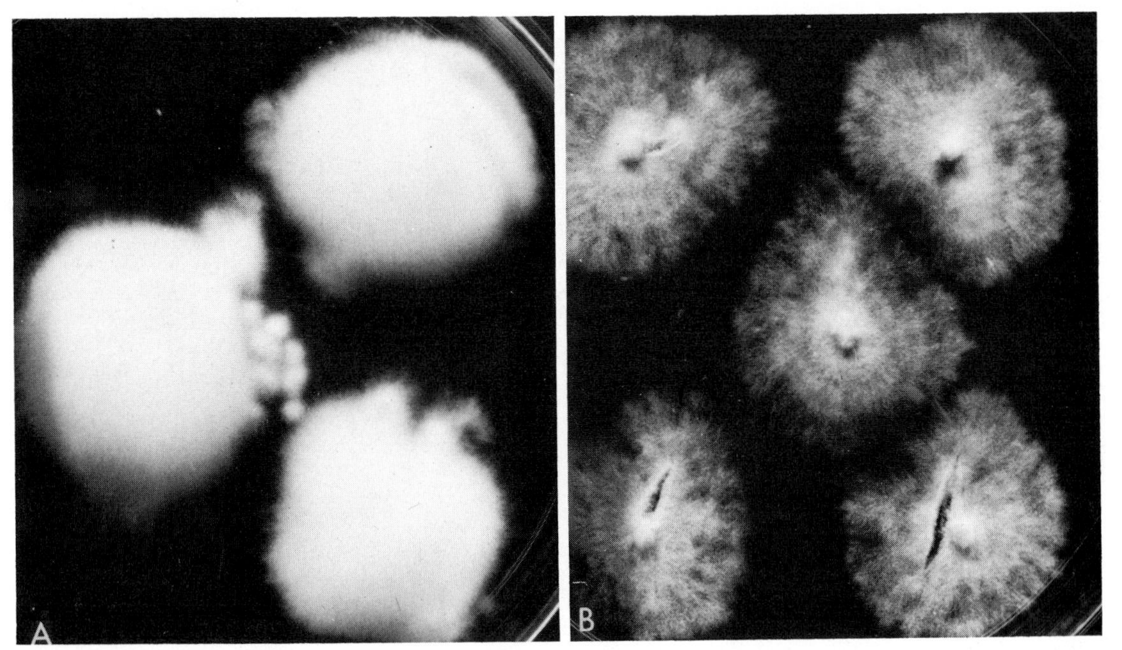

Figure 16–15. FS C 8. Histoplasmosis. *A*, Type A (albino) colony form. *B*, Type B (brown) colony form of *H. capsulatum*. There is sectoring of B to A. (Courtesy of M. Berliner.)

brown type has more numerous tuberculate macroaleuriospores than the white.

Microscopic Morphology. SDA, 25°C. The characteristic tuberculate macroaleuriospores are seen on examination of hyphae (Fig. 16–16 FS C 9). They are borne on narrow tubular aleuriophores, 8 to 14 μ in diameter, and are round or pyriform. The tubercles or fingerlike projections are quite variable in size and morphology, and in subculture only about 30 per cent of the macroaleuriospores show tubercles (Fig. 16–17). With special stains they appear to be mucopolysaccharide in nature, and they do not contain cytoplasm. The saprophytic fungus *Sepedonium* also produces tuberculate macroaleuriospores but is not dimorphic or pathogenic for animals. The latter organism is found in soil and parasitizes fleshy fungi. Microaleuriospores are also produced and are particularly abundant in fresh isolants of *H. capsulatum*. These spores are borne at the tips of short, narrow phores and at right angles to the vegetative hyphae. They are spherical, 2 to 4 μ in diameter, and occasionally there is a secondary lateral spore which sometimes appears to bud, giving the spore a dumbbell shape. The walls are smooth but often somewhat roughened. After many subcultures, the white mycelial growth produces few spores of either type.

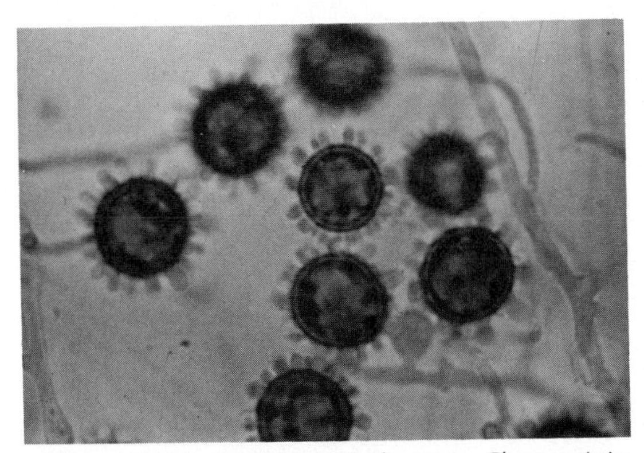

Figure 16–16. FS C 9. Histoplasmosis. Characteristic macroaleuriospores of *H. capsulatum*. (From Rippon, J. W. *In* Burrows, W. 1973. *Textbook of Microbiology*. 20th ed. Philadelphia, W. B. Saunders Company, p. 732.)

Colony Morphology and Microscopic Morphology. Blood agar, 37°C. The yeast form of *H. capsulatum* does not develop directly from either type of aleuriospore when transferred to an incubation temperature of 37°C, but arises from within the mycelium itself.[57] Spores first germinate and then convert to the yeast phase. The colony produced is at first glabrous and tenacious. Examination shows it to be a mixture of pseudohyphae, hyphae in the process of conversion, and freely budding yeast cells. On subsequent transfer, the colony becomes white, smooth, and yeastlike. The cells are now oval, bud-

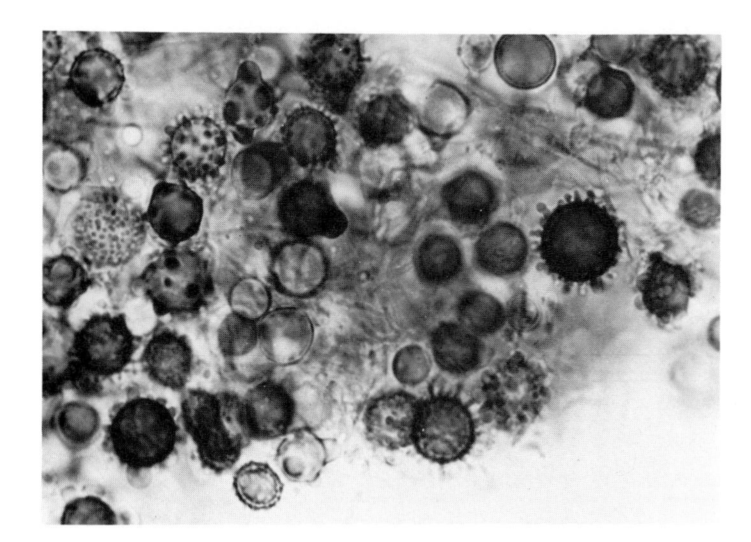

Figure 16–17. Histoplasmosis. Variation in size and morphology of macroaleuriospores. (From Berliner, M. 1968. Sabouraudia, 6:111–118.)

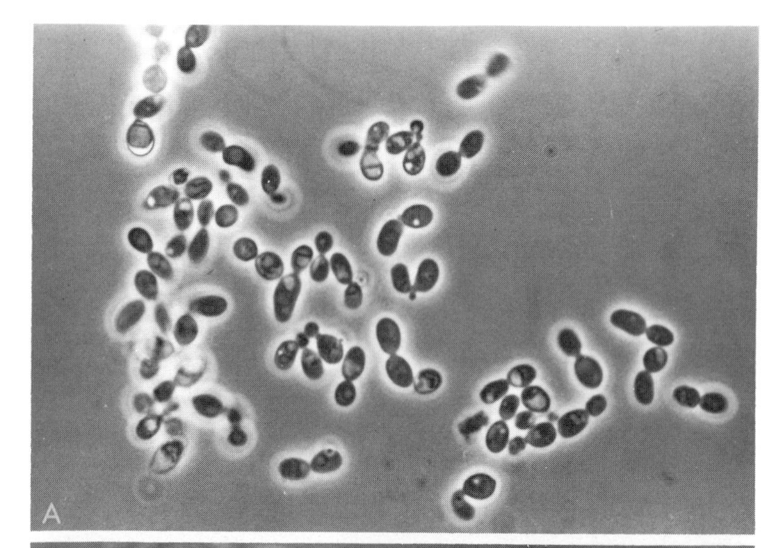

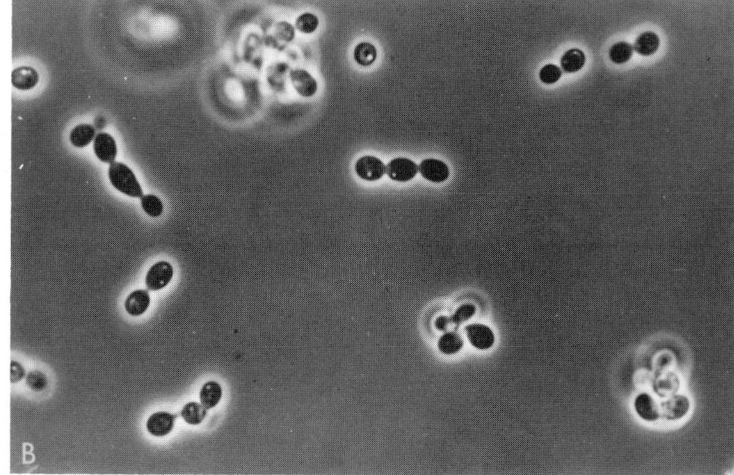

Figure 16–18. Histoplasmosis. Microscopic morphology of yeast form cells from culture. Courtesy of M. Berliner. ×800.

ding yeasts. The average size is 2 to 3 × 3 to 4 μ (Fig. 16–18). Under certain conditions, such as growth in tissue explants, a size of 20 μ may be attained.[109] The buds occur at the narrow end of the oval yeast cell. They have a narrow neck (0.2 to 0.3 μ), and the attachment is often drawn out as a narrow thread. This characteristic is useful in differentiating it from small forms of *B. dermatitidis*, which have a broad-based bud. Giemsa stained cells show that a *H. capsulatum* yeast cell contains a single nucleus with a densely staining nucleolus. The yeast cells of *B. dermatitidis* contain from six to eight nuclei. As reported by Howard,[58–60] the generation time within mouse histiocytes is 11 hours. Comparable times *in vitro* are evident by the growth curve studies of Reca and Campbell.[93]

Emmonsiella capsulata Kwon-Chung 1972[69]

The perfect stage of *H. capsulatum* was first reported to be *Gymnoascus demonbreunii*, a homothallic fungus, by Ajello and Cheng.[8] It appears that this was a soil keratinophilic contaminant. The work of Kwon-Chung[68] has demonstrated that *H. capsulatum* has a perfect stage, now named *Emmonsiella capsulata*. The perfect stage is heterothallic and morphologically consistent with fungi classed in the family Gymnoascaceae of the Ascomycetes. In mated pairs, tightly coiled hyphae radiate from a common source at the base of a naked young ascocarp (Fig. 16–19). The appearance is similar to the perfect stage of *B. dermatitidis (Ajellomyces dermatitidis)*, but differs from it in that the highly branched hyphae arising from such coils are irregularly curved and are not constricted at the cross walls.[69] The mature gymnothecia are globose, have a buffy pigment, and range in size from 80 to 250 μ in diameter (Fig. 16–20). The asci are pyriform. The ascospores are globose, eight in number, hyaline, smooth, and 1.5 μ in diameter. Whether these characteristics are sufficient to create a separate genus for the

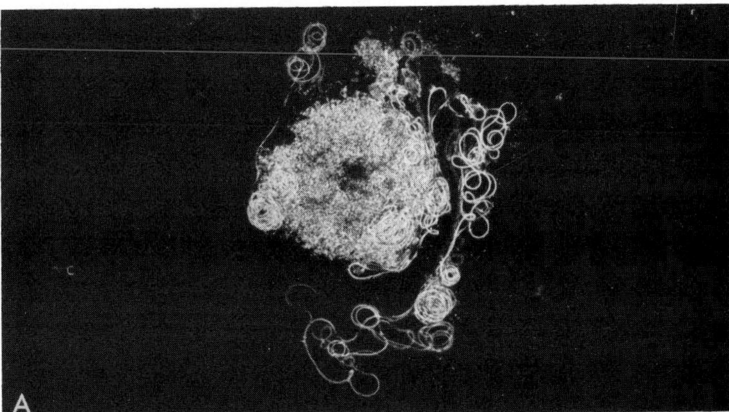

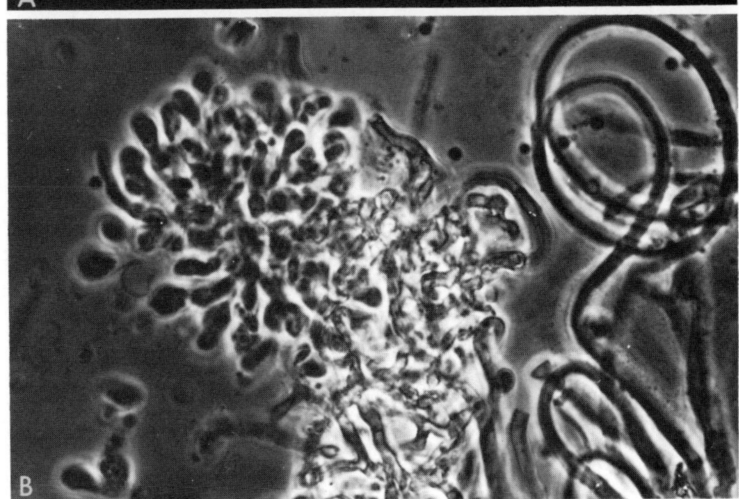

Figure 16–19. Histoplasmosis. Coiled hyphae of mated strains. (Courtesy of M. Berliner.)

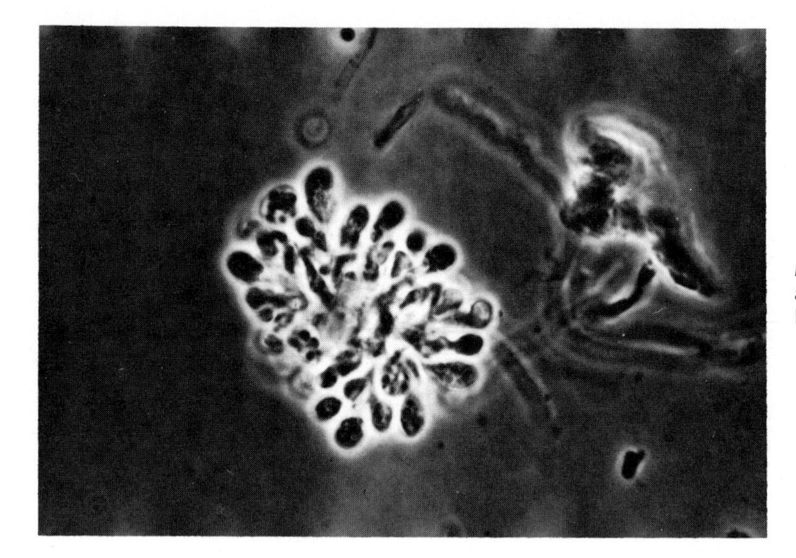

Figure 16–20. Histoplasmosis. Mature gymnothecium showing asci and ascospores. (Courtesy of M. Berliner.)

perfect stage of *H. capsulatum* and not include it in the genus of the perfect stage of *B. dermatitidis* remains to be determined. The sex distribution from soil isolants is approximately 1:1 for the + and − strains. Isolants from clinical disease, however, are predominantly of the (−) minus strain.[69] This may indicate a difference in virulence between the sexes of *E. capsulata*. Gymnothecia are also produced when tester strains of *E. capsulata* are paired with strains of *H. capsulatum* var. *duboisii*. Aberrant ascospores are produced, and their viability has at this writing not been ascertained.

REFERENCES

1. Ajello, L. 1958. Geographic distribution of *Histoplasma capsulatum*. Mykosen, *1*:147–155.
2. Ajello, L. 1964. Relationship of *Histoplasma capsulatum* to avian habitats. Public Health Rep., *79*:266–270.
3. Ajello, L. 1968. Comparative morphology and immunology of members of the genus *Histoplasma*. Mykosen, *11*:507–514.
4. Ajello, L. 1970. *The Medical Mycological Iceberg*. Proc. Int. Sympos. Mycoses, Sci. Pub. PAHO No. *205*:3–10.
5. Ajello, L. 1971. Coccidioidomycosis and histoplasmosis—A review of their epidemiology and geographic distribution. Mycopathologia, *45*:221–230.
6. Ajello, L., and L. C. Runyon. 1953. Infection of mice with single spores of *Histoplasma capsulatum*. J. Bacteriol., *66*:34–40.
7. Ajello, L., P. E. C. Manson-Bahr, et al. 1960. Amboni caves, Tanganyika. A new endemic area for *Histoplasma capsulatum*. Am. J. Trop. Med. Hyg., *9*:633–637.
8. Ajello, L., and S.L. Cheng. 1967. Sexual reproduction in *Histoplasma capsulatum*. Mycologia, *59*:689–697.
9. Ajello, L., E. W. Chick, et al. 1971. *Histoplasmosis*. Proceedings of Second National Conference. Springfield, Ill., Charles C Thomas.
10. Ajello, L., T. Briceno-Maaz, et al. 1959. Isolation of *Histoplasma capsulatum* from an oil bird (*Steatornis caripensis*) cave in Venezuela. Mycopathologia, *12*:199–206.
11. Baker, R. D. 1964. Histoplasmosis in routine autopsies. Am. J. Clin. Pathol., *41*:457–470.
12. Bauman, D. S. 1971. Physiology of *Histoplasma capsulatum*. In *Histoplasmosis*. Proceedings of Second National Conference. Ajello, L., E. W. Chick, et al. (eds.), Springfield, Ill., Charles C Thomas, Chapter 12.
13. Berliner, M. 1968. Primary subcultures of *Histoplasma capsulatum*. 1. Macro- and micromorphology of the mycelial phase. Sabouraudia, *6*:111–118.
14. Berliner, M. D., and M. E. Reca. 1966. Vital staining of *Histoplasma capsulatum* with Janus Green B. Sabouraudia, *5*:26–29.
15. Berliner, M. D., and M. Reca. 1969. Protoplasts of systemic dimorphic fungal pathogens. Mycopathologia, *37*:81–85.
16. Burgisser, von H., R. Fankhauser, et al. 1961. Mykose bei einem Dachs in der Schweiz histologisch Histoplasmose. Pathol. Microbiol., *24*: 794–802.
17. Campbell, C. C. 1960. The accuracy of serologic methods in diagnosis. Ann. N.Y. Acad. Sci., *89*:163–177.
18. Campbell, C. C. 1965. Problems associated with antigenic analysis of *Histoplasma capsulatum* and other mycotic agents. Am. Rev. Resp. Dis., *92*:113–118.
19. Campbell, C. C. 1967. Serology in the respiratory mycoses. Sabouraudia, *5*:240–259.
20. Campbell, C. C. 1971. History of the development of serologic tests for histoplasmosis, In *Histoplasmosis*. Proceedings of Second National Conference. Ajello, L., E. W. Chick, et al. (eds.), Springfield, Ill., Charles C Thomas, Chapter 42.

21. Campbell, C. C., G. B. Hill, et al. 1962. *Histoplasma capsulatum* isolated from a feather pillow associated with histoplasmosis in an infant. Science, *136*:1050–1051.

22. Chandler, J. W., T. K. Smith, et al. 1969. Immunology of the mycoses. II. Characterization of immunoglobulin and antibody responses in histoplasmosis. J. Infect. Dis., *119*:247–254.

23. Chase, H. V., P. J. Kadull, et al. 1968. Comparison of histoplasmin tine test with histoplasmin mantoux. Am. Rev. Resp. Dis., *9*:1058–1059.

24. Christie, A., and J. C. Peterson. 1945. Pulmonary calcification in negative reactors to tuberculin. Am. J. Public Health, *35*:1131–1147.

24a. Class, R. N., and F. S. Cascio. 1972. Histoplasmosis presenting as acute polyarthritis. New Engl. J. Med., *287*:1133–1134.

25. Comstock, G. W. 1959. Histoplasmin sensitivity in Alaskan natives. Am. Rev. Resp. Dis., *79*:542.

26. Comstock, G. W., C. N. Vicens, et al. 1968. Differences in distribution of sensitivity to histoplasmin and isolations of *Histoplasma capsulatum.* Am. J. Epidemiol., *18*:195–209.

27. Crispell, K. R., W. Parson, et al. 1956. Addison's disease associated with histoplasmosis. Report of four cases and review of the literature. Am. J. Med., *20*:23.

28. Daniels, L. S., M. D. Berliner, et al. 1968. Varying virulence in rabbits infected with different filamentous types of *Histoplasma capsulatum.* J. Bacteriol., *96*:1535–1539.

29. Darling, S. T. A. 1906. A protozoan general infection producing pseudotubercles in the lungs and focal necrosis in the liver, spleen and lymph nodes. J.A.M.A., *46*:1283–1285.

30. da Rocha-Lima, H. 1912–1913. Beitrag zur kenntnis der Blastomykosen Lymphangitis epizootica und Histoplasmosia. Zentralbl. Bakteriol., *67*:233–249.

31. De Monbreun, W. A. 1934. The cultivation and cultural characteristics of Darling's *Histoplasma capsulatum.* Am. J. Trop. Med., *14*:93–125.

32. Di Salvo, A. F., L. Ajello, et al. 1969. Isolation of *Histoplasma capsulatum* from Arizona bats. Am. J. Epidemiol., *89*:606–614.

33. Dodge, H. J., L. Ajello, et al. 1965. The association of a bird roosting site with infection of school children by *Histoplasma capsulatum.* Am. J. Public Health, *55*:1203–1211.

34. Domer, J. E. 1971. Monosaccharide and chitin content of cell wall of *Histoplasma capsulatum* and *Blastomyces dermatitidis.* J. Bacteriol., *107*:870–877.

35. Domer, J. E., J. G. Hamilton, et al. 1967. Comparative study of the cell walls of the yeastlike and mycelial phase of *Histoplasma capsulatum.* J. Bacteriol., *94*:466–474.

36. Dumont, A., and C. Piche. 1969. Electron microscopic study of human histoplasmosis. Arch. Pathol., *87*:168–178.

37. Edwards, L. B., F. A. Acquavita, et al. 1969. An atlas of sensitivity to tuberculin PPD-B and histoplasmin in the United States. Am. Rev. Resp. Dis. (Suppl.), *99*:1–132.

38. Edwards, P. Q. 1971. Histoplasmin sensitivity patterns around the world, In *Histoplasmosis.* Proceedings of Second National Conference. Ajello, L., E. W. Chick, et al. (eds.), Springfield, Ill., Charles C Thomas, Chapter 14.

39. Ehrhard, H.-B., and L. Pine. 1972. Factors influencing the production of H and M antigens by *Histoplasma capsulatum:* Effect of physical factors and composition of medium. Appl. Microbiol., *23*:250–261.

40. Emmons, C. W. 1949. Isolation of *Histoplasma capsulatum* from soil. Public Health Rep., *64*: 892–896.

41. Emmons, C. W., P. D. Klite, et al. 1966. Isolation of *Histoplasma capsulatum* from bats in the United States. Am. J. Epidemiol., *84*:103–109.

42. Fonseca, J. C. 1971. Analisis estadistico y ecologio-epidemiologico de la sensibilidad a la histoplasmina en Colombia, 1950–1968. Antioquia Med., *21*:109–154.

43. Fonseca, J. C., and M. Victoria Cadivid. 1971. Busqueda de *Histoplasma capsulatum* en la Caverna de Nus (Antioquia-Colombia) y en murcielagos residentes en ella. Antioquia Med., *21*:23–30.

44. Furcolow, M. L. 1958. Recent studies on the epidemiology of histoplasmosis. Ann. N.Y. Acad. Sci., *72*:127–164.

45. Furcolow, M. L. 1962. Opportunism in histoplasmosis. Lab. Invest., *11*:1134–1139.

46. Furcolow, M. L. 1963. Tests of immunity in histoplasmosis. New Engl. J. Med., *268*:357–361.

47. Furcolow, M. L., M. A. Doto, et al. 1961. Course and prognosis of untreated histoplasmosis. J.A.M.A., *177*:292–296.

48. Garrison, R. G. 1970. The uptake of low molecular weight sulfur containing compounds by *Histoplasma capsulatum* and related dimorphic fungi. Mycopathologia, *40*:171–180.

48a. Garrison, R. G., and J. W. Lane. 1973. Scanning-beam electron microscopy of the conidia of brown and albino filamentous varieties of *Histoplasma capsulatum.* Mycopathologia, *49*: 185–191.

49. González Ochoa, A. 1959. Histoplasmosis primaria pulmonar aguda en la republica Mexicana. Rev. Inst. Salub. Enferm. Trop., *19*:341–350.

50. Goodwin, R. A., and J. D. Snell. 1969. The enlarging histoplasmona. Am. Rev. Resp. Dis., *100*:1–12.

51. Greene, C. H., L. S. DiLalla, et al. 1960. Separation of specific antigens of *Histoplasma capsulatum* by ion-exchange chromatography. Proc. Soc. Exp. Biol. Med., *105*:140–141.

52. Gregoriades, D. G., H. L. Landeluttig, et al. 1961. Pericarditis with massive effusion due to histoplasmosis. J.A.M.A., *178*:331–334.

53. Guidry, D. J., and H. A. Spence. 1967. Deposition of *Histoplasma capsulatum* in the subcutaneous tissues and feathers of inoculated chick embryos. Sabouraudia, *5*:288–292.

54. Heiner, D. C. 1958. Diagnosis of histoplasmosis using precipitin reactions in agar gel. Pediatrics, *22*:616–629.

55. Henderson, R. G., H. Pinkerton, et al. 1942. *Histoplasma capsulatum* as a cause of chronic ulcerative enteritis. J.A.M.A., *118*:885–889.

56. Hill, G. B., and C. C. Campbell. 1962. Commercially available histoplasmin sensitized latex

particles in an agglutination test for histoplasmosis. Mycopathologia, *18*:169–179.
57. Howard, D. H. 1962. The morphogenesis of the parasitic forms of dimorphic fungi. Mycopathologia, *18*:127–139.
58. Howard, D. H. 1965. Intracellular growth of *Histoplasma capsulatum*. J. Bacteriol., *89*:518–523.
59. Howard, D. H. 1967. Effect of temperature on intracellular growth of *Histoplasma capsulatum*. J. Bacteriol., *93*:438–444.
59a. Howard, D. 1973. Fate of *Histoplasma capsulatum* in guinea pig polymorphonuclear leukocytes. Infect. Immun., *8*:412–419.
60. Howard, D. H., and V. Otto. 1969. Protein synthesis by phagocytized yeast cells of *Histoplasma capsulatum*. Sabouraudia, *7*:186–194.
61. Howard, D. H., V. Otto, et al. 1971. Lymphocyte mediated cellular immunity in histoplasmosis. Infect. Immun., *4*:605–610.
61a. Hughes, F. A., C. E. Eastridge, et al. 1971. Surgical treatment of pulmonary histoplasmosis. In *Histoplasmosis*. Proceedings of Second National Conference. Ajello, L., E. W. Chick, et al. (eds.), Springfield, Ill., Charles C Thomas, Chapter 54.
61b. Johns, L. E., R. G. Garrison, et al. 1973. Bronchopleurocutaneous fistula due to infection with *Histoplasma capsulatum*. Chest, *63*:638–641.
62. Kaplan, W. 1971. Application of the fluorescent antibody technique to the diagnosis and study of histoplasmosis, In *Histoplasmosis*. Proceedings of Second National Conference, Ajello, L., E. W. Chick, et al. (eds.), Springfield, Ill., Charles C Thomas, Chapter 41.
62a. Kaplan, W., L. Kaufman, et al. 1972. Pathogenesis and immunological aspects of experimental histoplasmosis in Cynomolgus monkeys (*Macaca fascicularis*). Infect. Immun., *5*:847–853.
63. Kaufman, L. 1971. Serological tests for histoplasmosis: Their use and interpretation, In *Histoplasmosis*. Proceedings of Second National Conference. Ajello, L., E. W. Chick, et al. (eds.), Springfield, Ill., Charles C Thomas, Chapter 40.
64. Kaufman, L., and W. Kaplan. 1961. Preparation of a fluorescent antibody specific for the yeast phase of *Histoplasma capsulatum*. J. Bacteriol., *82*:729–735.
65. Kaufman, L., and S. Blumer. 1966. Occurrence of serotypes among *Histoplasma capsulatum* strains. J. Bacteriol., *91*:1434–1439.
66. Kaufman, L., R. T. Terry, et al. 1967. Effect of a single histoplasmin skin test on serologic diagnosis of histoplasmosis. J. Bacteriol., *94*:798–803.
67. Kligman, A. M., H. Mescon, et al. 1951. The Hotchkiss.McManus stain for the histopathologic diagnosis of fungus disease. Am. J. Clin. Pathol., *21*:86–91.
67a. Klite, P. D., and F. H. Diercks. 1965. *Histoplasma capsulatum* in fecal contents and organs of bats in the Canal Zone. Am. J. Trop. Med. Hyg., *14*:433–439.
68. Kwon-Chung, K. J. 1972. Sexual stage of *Histoplasma capsulatum*. Science, *175*:326.
69. Kwon-Chung, K. J. 1973. Studies on *Emmonsiella capsulata*. I. Heterothallism and development of the ascocarp. Mycologia, *65*:109–121.
70. Larsh, H. W., and P. A. Bartels. 1970. Serology of histoplasmosis. Mycopathologia, *41*:115–131.
71. Larsh, H. W., A. Hinton, et al. 1953. Efficacy of the flotation method in isolation of *Histoplasma capsulatum* from soil. J. Lab. Clin. Med., *41*:478–485.
72. Little, J. A. 1960. Benign primary pulmonary histoplasmosis a common cause of unexplained fever in children. South. Med. J., *53*:1238–1240.
73. Lehan, P. H. 1957. Experience with the therapy of sixty cases of deep mycotic infection. Dis. Chest, *32*:597–614.
74. Mantovani, A., A. Mazzoni, et al. 1968. Histoplasmosis in Italy. 1. Isolation of *Histoplasma capsulatum* from dogs in the province of Bologna. Sabouraudia, *6*:163–164.
75. McDearmann, S. 1971. An interpretation of serologic tests in *Histoplasma capsulatum* infections, In *Histoplasmosis*. Proceedings of Second National Conference. Ajello, L., E. W. Chick, et al. (eds.), Springfield, Ill., Charles C Thomas, Chapter 43.
76. McVeigh, I., and K. Morton. 1965. Nutritional studies on *Histoplasma capsulatum*. Mycopathologia, *25*:294–308.
77. Menges, R. W. 1971. Clinical manifestation on animal histoplasmosis, In *Histoplasmosis*. Proceedings of Second National Conference. Ajello, L., E. W. Chick, et al. (eds.), Springfield, Ill., Charles C Thomas, Chapter 20.
78. Miller, H. E., F. M. Keddie, et al. 1947. Histoplasmosis. Cutaneous and mucomembranous lesions, mycologic and pathologic observations. Arch. Dermatol., *56*:715–739.
79. Mochi, A., and P. Q. Edwards. 1952. Geographical distribution of histoplasmosis and histoplasmin sensitivity. Bull. W.H.O., *5*:259–291.
80. Nicholas, W. M., J. A. Wier, et al. 1961. Serologic effects of histoplasmin skin testing. Am. Rev. Resp. Dis., *83*:276–279.
81. Negroni, P. 1940. Estudio micologico del primer caso Argentino de histoplasmosis. Rev. Inst. Bacteriol. Malbrán, *9*:239–294.
82. Negroni, P. 1965. *Histoplasmosis*. Translated by S. McMillen. Springfield, Ill., Charles C Thomas.
83. Palmer, C. E. 1946. Geographic differences in sensitivity to histoplasmin among student nurses. Public Health Rep., *61*:475–487.
84. Palmer, C. E. 1945. Nontuberculous pulmonary calcification and sensitivity to histoplasmin. Public Health Rep., *60*:513–520.
85. Parsons, R. J., and C. Zarafonetis. 1945. Histoplasmosis in man. A report of 7 cases and a review of 71 cases. Arch. Intern. Med., *75*:1–23.
86. Pifano, F., H. Campins, et al. 1962. El estado actual de la histoplasmosis sistemica en venezuela presentacion de un neuvo caso comprobado en el paiz. Arch. Venez. Med. Trop., *4*:157–164.
87. Pine, L. 1960. Morphological and physiological characteristics of *Histoplasma capsulatum*, In *Histoplasmosis*. Sweany, H. D. (ed.), Springfield, Ill., Charles C Thomas, pp. 40–75.
88. Pine, L. 1970. Growth of histoplasma capsulatum. VI. Maintenance of the mycelial phase. Appl. Microbiol., *19*:413–420.
89. Pine, L., and C. J. Boone. 1968. Cell wall composi-

tion and serologic reactivity of histoplasma serotypes and related species. J. Bacteriol., *96*:789–798.

89a. Polk, J. W. 1971. Surgery for cavitary histoplasmosis. In *Histoplasmosis*. Proceedings of Second National Conference. Ajello, L., E. W. Chick, et al. (eds.), Springfield, Ill., Charles C Thomas, Chapter 55.

90. Proceedings of the Seminar on Histoplasmosis. September 13, 1948. Bethesda, Md., National Institutes of Health.

91. Procknow, J. J., M. I. Page, et al. 1960. Early pathogenesis of experimental histoplasmosis. Arch. Pathol., *69*:413–426.

92. Randhawa, H. S. 1970. Occurrence of histoplasmosis in Asia. Mycopathologia, *41*:75–89.

93. Reca, M. E., and C. C. Campbell. 1967. Growth curves with yeast phase of *Histoplasma capsulatum*. Sabouraudia, *5*:267–273.

94. Reddy, P. A., P. A. Brasher, et al. 1970. Peritonitis due to histoplasmosis. Ann. Intern. Med., *72*:79–81.

95. Ridley, M. F., and T. A. Nowell. 1959. Another Australian case of histoplasmosis. Med. J. Aust., *2*:640–641.

96. Rippon, J. W. 1968. Monitored environment system to control growth, morphology and metabolic rates in fungi by oxidation-reduction potential. Appl. Microbiol., *16*:114–121.

97. Rowley, D. A., and M. Huber. 1955. Pathogenesis of experimental histoplasmosis in mice. I. Measurement of infecting dosages of the yeast phase of *Histoplasma capsulatum*. J. Infect. Dis., *96*:174–183.

98. Salfelder, K., and J. Schwartz. 1964. Cross reaction to *Histoplasma capsulatum* in mice. Sabouraudia, *3*:164–166.

99. Salfelder, K., et al. 1973. Multiple deep fungus infections. Curr. Top. Pathol., *57*:123–177.

100. Saliba, N. 1971. Amphotericin B, basic techniques and dosage for histoplasmosis, In *Histoplasmosis*. Proceedings of Second National Conference. Ajello, L., E. W. Chick, et al. (eds.), Springfield, Ill., Charles C Thomas, Chapter 48.

101. Samovitz, M., and T. Dillon. 1970. Disseminated histoplasmosis presenting as exfoliative erythroderma. Arch. Dermatol., *101*:216–219.

102. Saslaw, S., R. L. Perkins, et al. 1967. Histoplasmosis. To skin test or not to skin test. Proc. Soc. Exp. Biol. Med., *125*:1274–1277.

103. Scherr, G. H. 1957. Studies on the dimorphism of *Histoplasma capsulatum*. Exp. Cell Res., *12*:92–107.

104. Scherr, G. H., and J. W. Rippon. 1959. Experimental histoplasmosis in cold blooded animals. Mycopathologia, *11*:241–249.

105. Schubert, J. H., H. J. Lynch, et al. 1961. Evaluation of the agar plate precipitin test for histoplasmosis. Am. Rev. Resp. Dis., *84*:845–849.

106. Schulz, D. M. 1954. Histoplasmosis. A statistical morphologic study. Am. J. Clin. Pathol., *24*:11–26.

107. Schwarz, J., F. N. Silverman, et al. 1955. The relation of splenic calcification to histoplasmosis. New Engl. J. Med., *252*:887–891.

108. Schwarz, J., G. L. Baum, et al. 1957. Successful infection of pigeons and chickens with *Histoplasma capsulatum*. Mycopathologia. *8*:189–193.

109. Schwarz, J. 1971. Histoplasmosis, In *Handbuch der Speziellen Pathologischen Anatomie und Histologie Dritter Band Funfter Teil*. Baker, R. D. (ed.), Berlin, Springer-Verlag.

110. Serosi, G. A., J. D. Parker, et al. 1971. Histoplasmosis outbreaks, their patterns, In *Histoplasmosis*. Proceedings of Second National Conference. Ajello, L., E. W. Chick, et al. (eds.), Springfield, Ill., Charles C Thomas, Chapter 16.

111. Silverman, F. N. 1960. Roentgenographic aspects of histoplasmosis, In *Histoplasmosis*. Sweeny, H. C. (ed.), Springfield, Ill., Charles C Thomas, pp. 337–381.

112. Solanke, T. F., O. O. Akinuemi, et al. 1969. A case of histoplasmosis in a Nigerian. J. Trop. Med. Hyg., *72*:101–104.

113. Sotgiu, G., A. Montavani, et al. 1970. Histoplasmosis in Europe. Mycopathologia, *41*:53–74.

114. Straub, M., and J. Schwarz. 1960. General pathology of human and canine histoplasmosis. Am. Rev. Resp. Dis., *82*:528–541.

115. Strong, R. P. 1906. Study of some tropical ulcerations of skin with particular reference to their etiology. Philippine J. Sci., *7*:91–116.

116. Sutliff, W. D., and L. Ajello. 1970. *Histoplasma capsulatum* in the environment of sporadic histoplasmosis cases. Mycopathologia, *40*: 45–51.

117. Suthill, L. C., and C. C. Campbell. 1965. Feathers as substrate for *Histoplasma capsulatum* in its filamentous phase of growth. Sabouraudia, *4*:1–2.

118. Tesh, R. B., A. A. Arafa, et al. 1968. Histoplasmosis in Colombian bats. Am. J. Trop. Med. Hyg., *17*:102–106.

119. Tewari, R. P., and C. C. Campbell. 1965. Isolation of *Histoplasma capsulatum* from feathers of chickens inoculated intravenously and subcutaneously with the yeast phase of the organism. Sabouraudia, *4*:17–22.

120. Utz, J. P. 1971. Amphotericin B. Review of recent pharmacological developments, In *Histoplasmosis*. Proceedings of Second National Conference. Ajello, L., E. W. Chick, et al. (eds.), Springfield, Ill., Charles C Thomas, Chapter 49.

121. Walter, J. E. 1969. The significance of antibodies in chronic histoplasmosis by immunoelectrophoretic and complement fixation test. Am. Rev. Resp. Dis., *99*:50–58.

122. Walter, J. E., and G. B. Price. 1968. Chemical, serologic and dermal hypersensitivity activities of two fractions of histoplasmin. Am. Rev. Resp. Dis., *98*:474–479.

123. White, F. C. 1955. Chronic pulmonary disease in histoplasmin reactors: A review of 229 cases discovered through chest clinic examinations. Am. Rev. Tuberc., *72*:274–296.

124. Wilcox, K. R. 1958. The Walworth, Wisconsin, epidemic of histoplasmosis. Ann. Intern. Med., *49*:388–418.

125. Wooley, C. F., and D. M. Hosier. 1961. Constrictive pericarditis due to *Histoplasma capsulatum*. New Engl. J. Med., *264*:1230–1232.

126. Zeidberg, L. D., L. Ajello, et al. 1952. Isolation of *Histoplasma capsulatum* from soil. Am. J. Public Health, *42*:930–935.

Chapter 17

HISTOPLASMOSIS DUBOISII

DEFINITION

Histoplasmosis duboisii, or African histoplasmosis, is a clinically distinct form of histoplasmosis. It is characterized by the presence of granulomatous and suppurative lesions, primarily of the cutaneous, subcutaneous, and osseous tissues. There is little evident involvement of the lung. The histologic picture is that of masses of large yeast forms within numerous giant cells. These points differ from the *capsulatum* type histoplasmosis, in which the lung is always involved and the yeast forms are small and intracellular in histiocytes. The etiologic agent is called *Histoplasma duboisii,* although whether it is a separate species or variety of *H. capsulatum* remains to be determined.

It is unwise and often misleading to give a geographic designation to a disease. Therefore the term "histoplasmosis duboisii" is preferable to the name "African histoplasmosis," which is sometimes used. In fact, the disease has recently been recorded in Japan.[30]

HISTORY

In 1943, J. T. Duncan first observed a large yeast form of *Histoplasma* in the biopsy of an annular, papulocircinate skin lesion. The patient was an English mining engineer who had worked in Ghana.[12,13] Duncan found no cultural differences between the organism isolated and *H. capsulatum* but noted that the large yeast cells in tissue resembled *Blastomyces.* He did not choose to give the organism a special designation and did not publish his results until 1947.[12] In 1945 Catanei and Kervran described a large-celled type of histoplasmosis in a patient from the Sudan.[7] In retrospective studies it may be that histoplasmosis duboisii was the disease reported by Blanchard and Lefrou in 1922 as a saccharomycete infection in a Congolese patient. It also appears to be the "levures" found in a Senegalese by Brumpt in 1936.[8,28] In 1952, Dubois isolated a fungus from a patient with cutaneous lesions. Also in 1952, Vanbreuseghem studied this fungus from material sent him by Dubois and designated it a new species *Histoplasma duboisii.*[14,15] However, in 1957, after study of several strains, Drouet named the etiologic agent *Histoplasma capsulatum* var. *duboisii.*[11]

ETIOLOGY, ECOLOGY, AND DISTRIBUTION

Whether the etiologic agent is a stable variant of *H. capsulatum* or a separate species has been the subject of much discussion and debate. Now that the perfect stage of *H. capsulatum* has been found (*Emmonsiella capsulata*), and since it is heterothallic, it will be possible to mate the *duboisii* strains and resolve the issue. Taschdjian[27] has already shown that hyphal fusion of the two strains occurs, which may indicate they are the same species or closely related. In addition, preliminary studies by Kwon-Chung indicate that matings between the two are fertile.

Vanbreuseghem has summarized the differences between the *duboisii* type and the *capsulatum* type of histoplasmosis.[23,28] He stressed that the characteristic giant forms, increased cell wall thickness, and lipoidal bodies of the yeast phase of *H. duboisii* were distinctly different from those of *H. capsulatum.* If experimental animals such as guinea pigs and hamsters are given large inocula of either type of yeast, they rapidly die, and small-form yeasts are seen in histologic sections. If small doses are used, by four to eight months only giant cells are found in animals inoculated with the *duboisii* strain, whereas small yeasts are present if *H. capsulatum* is injected.

In 1967 Al-Doory and Kalter[2] reported the isolation of a *duboisii* type *Histoplasma* from pooled soil samples obtained from a village near Darajani in Kenya, but the validity of their isolant is questioned by other investigators. So far this is the only record of the recovery of the organism from soil. Some human infections have been associated with chicken runs and bat-infested caves. These are the same ecologic environments associated with the *capsulatum* type histoplasmosis. In one case, a schoolgirl contracted the disease after sweeping out a schoolroom contaminated with bat guano.[4] A lesion developed at the site of an abrasion. This may be a case of either direct inoculation or hematogenous localization following inhalation of spores.

So far almost all cases of *H. duboisii* (about a hundred) have been found in Africa between the two major deserts: the Sahara to the north and the Kalahari to the south. As pointed out by Vanbreuseghem,[28] the area is encompassed by 15° latitude north and 10° latitude south of the equator, and from the Senegal on the Atlantic coast to Uganda in the east (Fig. 17–1). This is almost identical to the endemic area of African trypanosomiasis, and is a region with a high average rainfall, high humidity, and little variation in diurnal temperature. A single possible case of the *duboisii* type has recently been reported from Japan.[30]

In the reviews by Cockshott and Lucas[8] and Vanbreuseghem,[28] the age range of patients was quite variable. Patients were found to be from 2 years to 70 years of age, but there was a cluster of cases in the second decade. The male-female ratio was 2:1 or more. All races were affected, and, as in paracoccidioidomycosis, the possibility that there are many subclinical cases of disease exists. The endemic areas of histoplasmosis delineated by the few skin test surveys made in Africa with histoplasmin (*capsulatum*) do not necessarily correspond to areas in which *duboisii* type histoplasmosis has been found. Clinically identifiable *capsulatum* type histoplasmosis is not uncommon in many regions of Africa. Histoplasmin prepared from *duboisii* strains is sometimes negative in patients with the *duboisii* type infection, as is

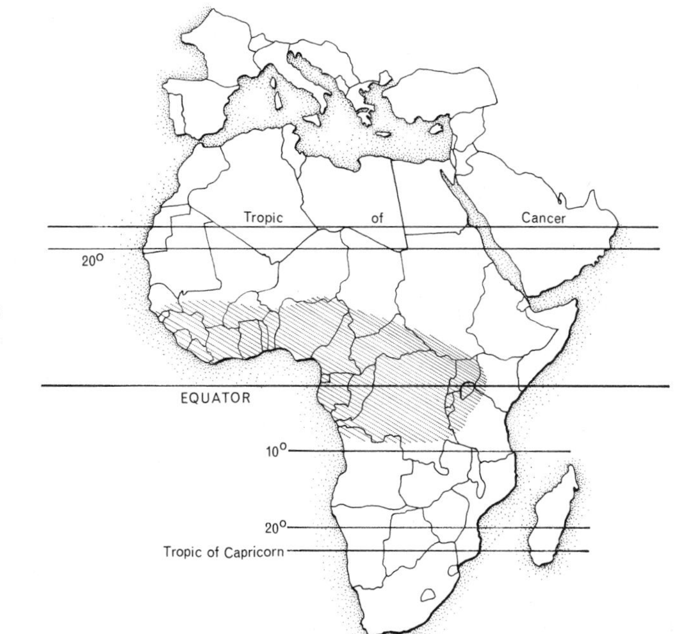

Figure 17–1. Incidence of histoplasmosis duboisii in Africa.

histoplasmin prepared from *capsulatum* strains. Early infection or anergy cannot always be ruled out in such cases.

CLINICAL DISEASE

Histoplasmosis duboisii is still virtually an undefined disease. Cockshott and Lucas have tabulated about one hundred cases and these have presented a somewhat similar disease pattern.[8,18] The cases fall roughly into two categories: (1) solitary lesions of the skin, subcutaneous tissue, or bone, and (2) disseminated or multiple lesions of the above tissues and other organs. The questions of portal of entry and the presence of subclinical disease have been much discussed. If the pattern known for the other systemic mycoses holds true with this disease, the primary infection is pulmonary. This is the portal of entry in coccidioidomycosis, blastomycosis, paracoccidioidomycosis, and *capsulatum* type histoplasmosis, even though very rare primary cutaneous disease is known for all these entities. *Duboisii* yeast cells have been recovered several times from sputum and in one case in which there was no clinical or radiologic suggestion of a lung lesion. However, in the several necropsy examinations performed on infected patients, pathologists have been unable to discover any pulmonary foci of infection. It has been demonstrated in animal studies that the *duboisii* organism is not highly virulent. As is now presumed in paracoccidioidomycosis, the *duboisii* organism may be inhaled and then transported hematogenously to a favorable site for proliferation.

The clinical categories delineated by Cockshott and Lucas are presented here.

1. Localized type. In these cases there is a single lesion which may be represented by a circumscribed skin lesion (as in Duncan's first case), a subcutaneous granuloma, or a single lesion of bone. There are no signs of systemic disturbance, such as fever, anemia, or weight loss. The disease is chronic, and the lesions go through phases of quiescence and recrudescence. Often they heal spontaneously.

2. Disseminated type. As the term indicates, multiple lesions occur that may involve the skin, lymph nodes, bones, and abdominal viscera. Systemic *duboisii* histoplasmosis often runs a rapidly progressive and fatal course, particularly in patients with hepatic and splenic involvement. The lesions occur randomly throughout the body. Of particular interest is the involvement of bone. In one series, skeletal lesions demonstrated by x-ray were present in two-thirds of all cases.[9] They are often multiple and resemble the cranial defects seen in multiple myeloma or those seen in the phalanges and carpals in sarcoidosis. Lymphadenopathy may be very prominent, as is seen in paracoccidioidomycosis. This may be confined to one area, such as the groin, several areas, or be generalized.

As would be expected in a systemic disease, fever, weight loss, and other signs of debilitation are present. Anemia may be quite severe, as there is often localization of the organisms in the marrow and loss of blood from skin ulcers. *Capsulatum* histoplasmosis involves the reticuloendothelial system and can be seen within histiocytes in material obtained by sternal puncture and elsewhere. The organism rarely deposits in bone. The *duboisii* type also involves the reticuloendothelial system but is found within giant cells and causes a destructive granulomatous process in the bones. Some writers have commented on the presence in infected patients of dysproteinemia, a low albumin/globulin ratio. However, in contrast to Europeans, this low a/g ratio is normal for Nigerians and other Africans. The white cell count is usually normal.

Cutaneous lesions are papular, nodular, ulcerative, circinate, eczematoid, or psoriasiform. The verrucoid or chancriform lesion characteristic of primary inoculation mycosis is seldom described. In disseminated disease, numerous skin granulomata form and are seen to resolve as new ones evolve. In one case observed, the time from the first appearance of the lesions to their resolution was from two to four months. The lesions appeared first as flat papules that grew into dome-shaped nodules, which were paler than the dark Negro skin. The lesions were usually sessile, but a few became penduncu-lated. As they enlarged, pits were formed, and the central area became denuded, forming an ulcer with a rolled border. Some small lesions healed without ulceration, leaving a depigmented or hyperpigmented macule. Spontaneous healing in the center with annular expansion of the active border may mimic lesions seen in blastomycosis.

Subcutaneous lesions may arise from foci in the superficial flat bones, ribs, or skull, or they may develop without underlying bony involvement. They may be abscesses that suppurate and burrow to the skin or become freely moveable granulomata. The acute phase is one of swelling, tenderness, and pain, followed by evolution into a soft, fluctuant, cold abscess. This is similar to the development of subcutaneous lesions in disseminated blastomycosis.

Mucosal involvement occurs in a few cases. Lesions at this site are described as dome-shaped papules about 1 cm in diameter, which are found on the glossal or buccal mucosa.

Hepatic and splenic involvement has also been described in a few instances. Most postmortem cases have shown infiltration of the liver and spleen, with typical granulation tissue containing numerous giant cells and large yeasts. The intestine has only rarely been involved. One patient died as the result of a perforation of an ileocecal granuloma.

Osseous involvement is a very common feature of disseminated *duboisii* histoplasmosis.[8,9,18] The marrow elements of the skeletal system are involved with particular frequency. Granulomatous lesions develop which destroy bone trabeculae and expand to erode cortical bone. If the periosteum is involved, it is at first lifted off the surface, and there is production of new extracortical bone. The granulomata often are not contained with the bony process but extend to adjacent soft tissue. In a survey of 56 cases, the skull was involved in 12, the jaw in 4, the scapula in 3, the clavicle in 3, the sternum in 2, the spinal column in 5, the arm in 3, the forearm in 10, the hand in 3, the thigh in 8, the leg in 6, and the foot in 4. Multiple lesions of the skull are the rule, and the ribs also may have several foci. Involvement of the spine may result in paraplegia. A "spinal block" occurred in one patient following destruction of the pedicle of the seventh dorsal vertebra with extension to the nerve trunk. The peculiar predilection of *duboisii* histoplasmosis for the bone is a characteristic of the disease.

Summary of the Differences Between capsulatum Histoplasmosis and duboisii Histoplasmosis. Histoplasmosis of the *capsulatum* type has two stages: (1) the primary, often subclinical, type is pulmonary and usually heals, leaving calcified opacities in the lung; (2) the second stage, in the rare patient that has generalized disease, is systemic spread to involve, in addition to the lungs, the liver, spleen, lymph nodes, and oro- and nasopharynx. Bone involvement is very rare. The organism is most commonly seen within histiocytes. The primary portal of entry for histoplasmosis duboisii is not known. Although it is presumed to be the lung, the organism does not remain there long enough to cause discernible pathology. The first clinical signs of disease are single granulomata involving skin, lymph nodes, or bone. The disseminated disease involves multiple lesions in bone, skin, lymph nodes, liver, and spleen. The bone, especially the marrow elements, is most commonly involved. The organisms are found within giant cells. The course of both forms of disseminated histoplasmosis, if untreated, is progressive and fatal.

DIFFERENTIAL DIAGNOSIS

The single lesion of histoplasmosis duboisii resembles many other dermatologic conditions. These include eczema, lichen planus, psoriasis, the tineas, annular tuberculosis, Majocchi erythema annulare centrifugum, molluscum contagiosum, rodent ulcer, and warts. In disseminated disease, a tentative diagnosis is made by the appearance of multiple abscesses in bone, subcutaneous tissue, and pleomorphic skin lesions. Blastomycosis also exists in the endemic region and may present with a similar clinical picture. The yeast cells of both appear similar on direct examination of lesion material.

PROGNOSIS AND THERAPY

The isolated lesion of *duboisii* histoplasmosis may heal spontaneously but is amenable to surgical removal. Disseminated disease is quite grave and is generally fatal when liver and spleen are involved. Amphotericin B has been used successfully for treatment in many disseminated cases. The recommended dose is 0.25 mg to 1 mg per kg diluted in 5 per cent glucose. This is given as a slow infusion over six hours. Alternate-day therapy can be employed. A total of 1 to 3 g is necessary for a full course of treatment. The side effects are frequent, and consist of fever, rigors, and headache, as well as nausea and vomiting. Salicylates, barbiturates, and chlorpromazine are usually sufficient to control the side effects. Cockshott and Lucas[8] recommend administration of 200 mg sodium amytal and 100 mg chlorpromazine by mouth to adults one-half hour before infusion of amphotericin B. This may be repeated after four hours. Dramatic results are seen within one week of therapy. A full curative course may require from 6 to 22 weeks.

PATHOLOGY

Gross Pathology. In the necropsy studies that have been done on patients who have died of histoplasmosis duboisii, granulomatous and suppurative lesions have been found. Both may occur in the same individual. The lesions involve the cutaneous and subcutaneous tissues, bones, joints, and abdominal viscera. Parenchymal lung lesions are rare, and the alimentary tract is spared except in one recorded case. The central nervous system is not involved, except when a vertebral lesion causes compression of the cord. Lesions in the skull do not penetrate the dura mater.

Histopathology. The histologic picture is quite characteristic. In contrast to the granuloma of *capsulatum* type histoplasmosis, where giant cells are infrequent, the *duboisii* type lesion consists largely of aggregates of giant cells. These may be very large (up to 200 μ or more) and contain numerous ovoid, double contoured, walled yeast cells, with a diameter of 12 μ to 15 μ. The yeasts are sometimes present in chains of four or five cells (Fig. 17–2 FS C 10). They resemble the yeast cells seen in blastomycosis but lack the broad-based bud. Other types of inflammatory cells, especially neutrophils, may also be present. In caseating lesions, extracellular degenerating yeasts may be found. These vary in size from very small bodies to large spheroidal empty cells that may be crenated and crescent-shaped. The fungus is visible with the usual hematoxylin-eosin stain, but is better demonstrated by the methenamine silver, Gridley, or periodic acid–Schiff stains (Hotchkiss-McManus). Organisms may be rare in healing fibrotic lesions.

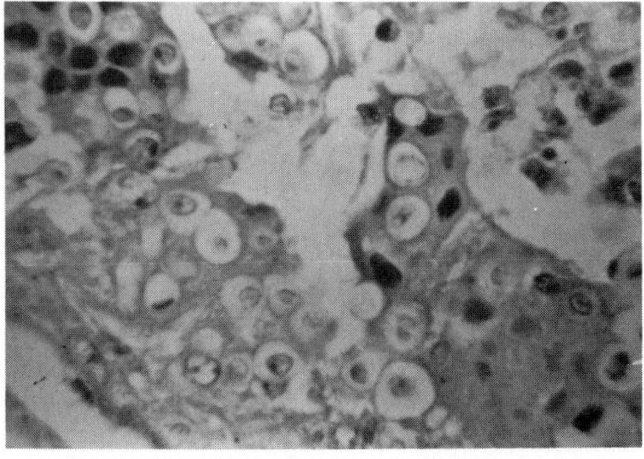

Figure 17–2. FS C 10. Histoplasmosis duboisii. Large yeast cells. The cytoplasm has shrunk away from the wall. Nuclei are visible. Hematoxylin and eosin stain. ×450. (From Rippon, J. W. *In* Burrows, W. 1973. *Textbook of Microbiology,* 20th ed. Philadelphia, W. B. Saunders Company, p. 731.)

ANIMAL INFECTION

Natural infection in animals has been recorded in baboons of the species *Papio papio* and *Papio cyanocephalus.*[19,29] Both had granulomatous lesions of the skin, subcutaneous tissue, and lymph nodes, and in one invasion of the bone occurred. In neither were the lungs or internal viscera involved. The granulomatous reaction with many giant cells observed in histopathologic sections was similar to that seen in human cases. In one case the animal had been in captivity for five years in England,[29] and in the report by Mariat and Segretain[19] the five animals with the disease had been at the Pasteur Institute in Paris for 18 months to two years. This indicates the long duration of the infection before clinical symptoms are observable. These animals had been captured in Gambia and Guinea.

Experimental infection in animals has been studied by a number of authors,[20,28] who have shown the *duboisii* organism to be of relatively low virulence. Whereas *H. capsulatum* uniformly produces fatal infection in hamsters, this was rarely achieved by inoculation of the *duboisii* strains. Infection which spontaneously resolved was produced in guinea pigs, rabbits, and pigeons. When the fungus is first injected, small-cell yeasts (*capsulatum* type) are formed which are 2 to 5 μ. These are gradually replaced by the large cells of the *duboisii* type. Once the transition has occurred, no small cells are seen. Small coccoid bodies (0.5 to 1.5 μ) representing degenerating yeasts are visible in Kupffer cells. The initial cellular reaction is histiocytic but is replaced by giant cells, particularly in infections of hamsters and mice. Some of the giant cells are between 220 and 500 μ and contain huge numbers of organisms.

Biological Studies. Many physiologic studies have been carried out in an attempt to distinguish the *capsulatum* and *duboisii* forms of *Histoplasma*.[1] Montemartini and Ciferri reported that histidine or ammonium sulfate could serve as a sole nitrogen source for *H. capsulatum,* but not for the *duboisii* strain. Urease production by *H. capsulatum* and lack of it by *duboisii* was reported by Coremans.[10] Rosenthal and Sokolsky[24] found tyrosine but not gelatin to be hydrolyzed by *H. capsulatum,* while the *duboisii* strains hydrolyzed gelatin but not tyrosine. Berliner[5] also found *H. capsulatum* was unable to hydrolyze gelatin, while *duboisii* strains were able to liquefy it within 24 to 96 hours. Blumer and Kaufman[6] reevaluated the work of Coremans, Rosenthal and Sokolsky, and Berliner. They could not verify the findings of these investigators and concluded that strain variations were too great to allow differentiation of *duboisii* and *capsulatum* biochemically.

Transition of the two types to the yeastlike phase *in vitro* was extensively studied by Pine et al.[21] They found that both types of *Histoplasma* initially grow as small yeasts. However, the *duboisii* strains gradually transform into large, thick-walled yeasts, whereas the *capsulatum* strains remain typically small (2 to 5 μ). Once the *duboisii* strains transform to the larger size, their daughter cells are always of the large *duboisii* cell type. Both the *capsulatum* type and *duboisii* type cells may produce spheroplasts, which are degenerating forms, often devoid of cytoplasm, and subject to crenation. Such spheroplasts have been found in pus and necrotic tissues from patients with both types of infection.

IMMUNOLOGY AND SEROLOGY

Inherent immunity or immunity resulting from resolution of clinical or subclinical infections has not been established in Histoplasmosis duboisii. Skin testing using histoplasmin (*capsulatum*) has been done extensively in Africa, and there are high reaction rates in some areas.[3,25,28] These rates often do not correspond with the known cases of histoplasmosis duboisii. In the few clinical cases in which both *duboisii* histoplasmin and *capsulatum* histoplasmin have been used, the results were equivocal.[9,12,13,22]

Fluorescent antibody studies have shown a close relationship between the two types of *Histoplasma.* Originally it was thought that the two could be separated by "specific" antisera. However, Kaufman and Blumer[16,17] through reciprocal cross-staining and adsorption procedures defined five serotypes of *H. capsulatum*—1,2; 1,4; 1,2,3; 1,2,4; and 1,2,3,4. This led to a reinvestigation of the *duboisii* strains. All *duboisii* strains were antigenically indistinguishable from the 1,4 serotypes of *H. capsulatum.*

Ajello,[1] in reviewing the genus *Histoplasma,* was in agreement with Duncan,[13] who in 1958 said, "The species *H. duboisii* rests on a very slender basis, merely its morphology under certain conditions of parasitic life and to some extent an analogous tendency to develop large cells in yeast phase in culture in response to unfavorable environmental factors, which seem to be more marked in this species than in *H. capsulatum.*" Both authors concluded that the identity of the organism as a type or as a separate species should be preserved. The clinical disease produced by *duboisii* strains is distinctly different from that produced by *H. capsulatum.* Since the sexual mating types of *Histoplasma capsulatum* have been discovered, the matter of species separation will be settled presently.

LABORATORY IDENTIFICATION

Direct Examination. The organisms are usually numerous in pus from skin lesions, abscesses, draining sinuses from bone lesions, or biopsy material. In potassium hydroxide mounts, the yeasts appear as large (12 to 15 μ in diameter), thick-walled yeasts. They may have fat droplets within the cells, and the presence of broad-based buds has been reported in a few instances. The latter is a characteristic of the yeast cells of *Blastomyces dermatitidis,* which have been reported to cause disease in several parts of Africa also. It is therefore necessary to isolate the organism in culture for correct diagnosis. The *duboisii* organism can be seen occasionally in sputum when no clinical disease is evident.

Culture Methods. Material for culture can be placed on Mycosel agar or other antibacterial, antibiotic-containing media and incubated at 25°C. It grows slowly, producing a flat, brown, glabrous colony or a fluffy white to beige colony. Growth sometimes is not initiated for four to six weeks. In a manner similar to *H. capsulatum,* the *duboisii*

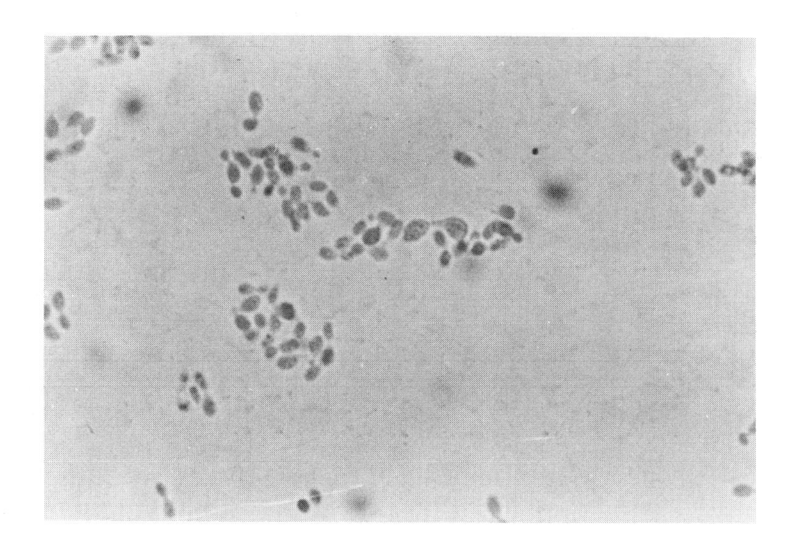

Figure 17–3. Histoplasmosis duboisii. Variation in morphology of cells grown at 37°C on blood agar.

strains grow on almost all laboratory media and can be converted to the yeast phase by inoculation on blood agar or KY agar[21] and then incubated at 37°C.

MYCOLOGY

Histoplasma capsulatum var. *duboisii* Drouhet 1957[11]

Synonymy. *Histoplasma duboisii* Vanbreuseghem 1952.[14]

Colony Morphology. The colony morphology of cultures grown at 25°C is identical to that of *H. capsulatum.* On microscopic examination, microaleuriospores and tuberculate macroaleuriospores are produced which are identical to those of normal strains of *H. capsulatum.*

Morphology at 37°C. The organism converts easily to the yeast phase. Initially on blood agar the small yeast forms are seen, but in time large, thick-walled cells (12 to 15 μ) are produced (Fig. 17–3). Great morphologic variation is encountered when KY agar is used. Once they are converted, the *duboisii* cells reproduce by the production of large *duboisii* yeast cells, and small *capsulatum* type cells are no longer seen.

REFERENCES

1. Ajello, L. 1968. Comparative morphology and immunology of members of the genus *Histoplasma.* Mykosen, *11*:507–514.
2. Al-Doory, Y., and S. S. Kalter. 1967. The isolation of *Histoplasma duboisii* and keratinophilic fungi from soils of East Africa. Mycopathologia, *31*:289–295.
3. Ball, J. D., and P. R. Evans. 1954. Histoplasmin sensitivity in Uganda. Br. Med. J., *2*:848–849.
4. Basset, A., M. Basset, et al. 1963. Formes cutanées de l'histoplasmose Afrique. Bull. Soc. Fr. Dermatol. Syphiligr., *70*:61–64.
5. Berliner, M. D. 1967. Gelatin hydrolysis for identification of the filamentous phases of Histoplasma, Blastomyces, and Chrysosporium. Sabouraudia, *5*:274–277.
6. Blumer, S., and L. Kaufman. 1968. Variations in enzymatic activities among isolates of *Histoplasma capsulatum* and *Histoplasma duboisii.* Sabouraudia, *6*:203–206.
7. Catanei, A., and P. Kervran. 1945. Nouvelle mycose humaine observée au Soudan français. Arch. Inst. Pasteur Alger., *23*:169–172.
8. Cockshott, W. P., and A. O. Lucas. 1964. Histoplasmosis duboisii. Quart. J. Med., *33*:223–238.
9. Cockshott, W. P., and A. O. Lucas. 1964. Radiological findings in *Histoplasma duboisii* infections. Br. J. Radiol., *37*:653–660.
10. Coremans, J. 1963. Un test biochimique de différentiation de *Histoplasma duboisii* Vanbreuseghem 1952 avec *Histoplasma capsulatum* Darling 1906. C. R. Soc. Biol. (Paris), *157*:1130–1132.
11. Drouhet, E. 1957. Quelques aspect biologiques et mycologiques de l'histoplasmose. Pathol. Biol. (Paris), *33*:439–461.
12. Duncan, J. T. 1947. A unique form of Histoplasma. Trans. R. Soc. Trop. Med. Hyg., *40*:364–365.
13. Duncan, J. T. 1958. Tropical African histoplasmosis. Trans. R. Soc. Trop. Med. Hyg., *52*:468–474.
14. Dubois, A., P. G. Janssens, et al. 1952. Un cas d'histoplasmose africaine avec une nota mycologique sur *Histoplasma duboisii* n.sp. par R. Vanbreuseghem. Ann. Soc. Belge Med. Trop., *32*:569–584.
15. Dubois, A., and R. Vanbreuseghem. 1952. L'histoplasmose africaine. Bull. Acad. Roy. Med. Belg., *17*:551–564.

16. Kaufman, L., and S. Blumer. 1966. Occurrence of serotypes among *Histoplasma capsulatum* strains. J. Bacteriol., *91*:1434–1439.
17. Kaufman, L., and S. Blumer. 1968. Development and use of a polyvalent conjugate to differentiate *Histoplasma capsulatum* and *Histoplasma duboisii* from other pathogens. J. Bacteriol., *95*:1243–1246.
18. Lucas, A. O. 1967. The clinical features of some of the deep mycoses in West Africa, In *Systemic Mycoses.* CIBA Foundation Symposium. G. E. W. Wolstenholme and R. Porter (eds.), Boston, Little, Brown and Company, pp. 96–112.
19. Mariat, F., and G. Segretain. 1956. Etude mycologique d'une histoplasmose spontanée de singe africain (*Cynocephalus babuin*). Ann. Inst. Pasteur (Paris), *91*:874–891.
20. Okudaira, M., and J. Swartz. 1961. Infection with *Histoplasma duboisii* in different experimental animals. Mycologia, *53*:53–63.
21. Pine, L., E. Drouhet, et al. 1964. A comparative morphological study of the yeast phases of *Histoplasma capsulatum* and *Histoplasma duboisii.* Sabouraudia, *3*:211–224.
22. Pine, L., L. Kaufman, et al. 1965. Comparative fluorescent antibody staining of *Histoplasma capsulatum* and *Histoplasma duboisii* with a specific anti-yeast phase. *H. capsulatum* conjugate. Mycopathologia, *24*:315–326.
23. Renoirte, R., J. L. Michaux, et al. 1967. Nouveaux cas d'histoplasmose africaine et de cryptococcose observés en République démocratique du Congo. Bull. Acad. Roy. Med. Belg., 7:465–527.
24. Rosenthal, S. A., and H. Sokolsky. 1965. Enzymatic studies with pathogenic fungi. Dermat. Internat. *4*:72–79.
25. Stott, H. 1954. Histoplasmin sensitivity and pulmonary calcifications in Kenya. Br. Med. J., *1*:22–25.
26. Swartz, J. 1953. Giant forms of *Histoplasma capsulatum* in tissue explants. Am. J. Clin. Pathol., *23*:898–903.
27. Taschdjian, C. L. 1952. Hyphal fusion studies on *Histoplasma capsulatum* and *Histoplasma duboisii.* Mykosen, *2*:1–6.
28. Vanbreuseghem, R. 1964. L'histoplasmose africaine un histoplasmose causes por *Histoplasma* duboisii Vanbreuseghem 1952. Bull. Acad. Roy. Med. Belg., *4*:543–585.
29. Walker, J., and E. T. C. Spooner. 1960. Natural infection of the african baboon *Papio papio* with the large cell form of *Histoplasma.* J. Pathol. Bacteriol., *80*:436–438.
30. Yamato, H., H. Hitomi, et al. 1957. A case of histoplasmosis. Acta Med. Okayama, *11*:347–364.

Chapter 18
COCCIDIOIDOMYCOSIS

DEFINITION

Coccidioidomycosis is a benign, inapparent, or mildly severe upper respiratory infection which usually resolves rapidly. Rarely, the disease is an acute or chronic, severe disseminating, fatal mycosis. Recovery from the mild forms of the disease usually results in lifelong immunity to reinfection. If infection is established, the disease may progress as a chronic pulmonary condition or as a systemic disease involving the meninges, bones, joints, and subcutaneous and cutaneous tissues. Such involvement is characterized by the formation of burrowing abscesses. The initial tissue response and that found in rapidly disseminating disease is suppuration. However, in established chronic and slowly advancing infection, a granulomatous reaction is found, with some areas showing a mixed-type cellular infiltrate. The presence of *Coccidioides immitis*, the etiologic agent of coccidioidomycosis, is associated with a hot, semi-arid environment. It is probably the most virulent of the fungal pathogens. The highly endemic areas of disease include the southwestern United States and northern Mexico, with endemic foci in Central America, Venezuela, Colombia, Paraguay, and Argentina. The disease has been reported to occur in Russia, central Asia, Nigeria, and Pakistan, but these reports still await confirmation.

Synonymy. Posadas' disease, coccidioidal granuloma, Valley fever, desert rheumatism, Valley bumps, California disease.

HISTORY

Coccidioidomycosis was the first of the severe fatal mycoses in which an inapparent or mild form of disease was found to occur commonly in inhabitants of its endemic region. The pathogenesis of the common, mild disease rather than the rare, severe infection was delineated by Dickson and Gifford in the late 1930's, just as it was in the 1940's for histoplasmosis by Christie and Palmer. This pattern of disease is now being discovered for other systemic mycoses. Coccidioidomycosis is probably the most studied and best understood of the systemic human mycotic infections, although many unknown aspects of the disease remain to be investigated.

As pointed out by Fiese,[22] human coccidioidomycosis is probably a relatively new disease. The endemic areas where it is found were very sparsely populated until the advent of European explorers and the subsequent settling of agricultural and ranching populations in these regions. The indigenous populations that existed there, such as the Yokuts of the San Joaquin valley, were soon eliminated by exposure to the European diseases of influenza, cholera, syphilis, and smallpox.

As was the case for several of the mycoses, the discoverers of coccidioidomycosis initially described the organism as a protozoan. The patient in whom the disease was first found was a soldier from the Argentine pampas. He had had recurrent tumors of the skin for four years before entering the University hospital in Buenos Aires in 1891. There, his disease was studied by Alejandro Posadas, a student of the famous pathologist Robert Wernike. The patient lived another seven years, during which time the investigators noted the progress and pathologic development of the disease. They considered the disease to be a neoplasia, a form of mycosis fungoides (tumors resembling a mushroom), but recognized the presence within the lesions of an as yet undescribed parasite. This organism was likened to the protozoa of the order Coccidia. In 1892 Posadas[61] published the preliminary reports of the case in Argentina, and Wernicke described the same patient[90] in a paper printed in Germany. Posadas was able to reproduce the disease in various animals by inoculating them with lesion material from the patient. The disease was considered to be rare in Argentina, and the second case from that country was not reported until 35 years later.[53] When the known cases of coccidioidomycosis in Argentina were reviewed by Negroni in 1967,[54] only 27 had been reported. Thus the clinically evident form of coccidioidomycosis is rare in the nation where it was first discovered, as is another mycosis, rhinosporidiosis, which was first discovered by the same investigators, Wernicke and Posadas. An interesting sidelight concerns the mortal remains of the first patient with coccidioidomycosis. In 1948 Dr. Flavio Niño found an unidentified head resembling that of the patient described by Posadas in the anatomy museum of the medical school. Further study confirmed this was so, and therefore the specimens from the first case of coccidioidomycosis were rediscovered after being lost for half a century. This head and other appendages of the patient are now a featured exhibit of the medical school museum (Fig. 18–1).

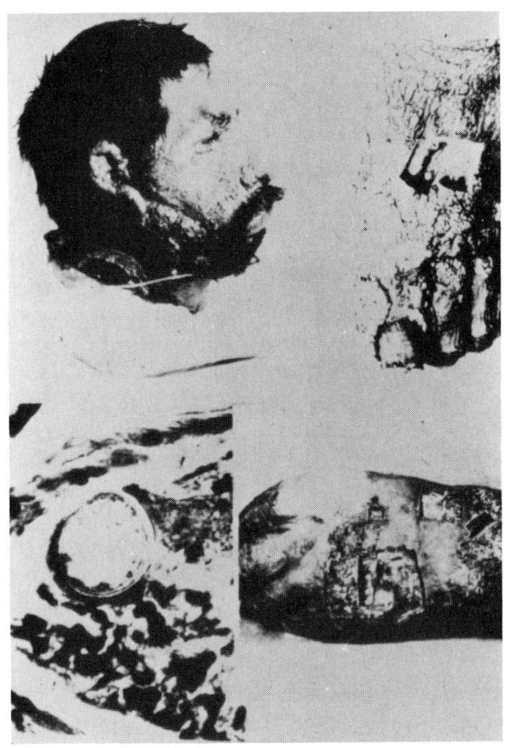

Figure 18–1. Coccidioidomycosis. Museum specimens of the first case of Posadas and Wernike, showing the recurrent verrucous granulomas of skin and a tissue section containing a spherule. These specimens were recovered by Dr. Flavio Niño.

At almost the same time that Posadas was describing the disease in Argentina, Emmet Rixford was studying the first case in California. The patient, Joas Furtado Silverra, a Portuguese from the Azores, had emigrated to the San Joaquin Valley in 1886. He soon developed recurrent nodular cutaneous lesions similar to those observed in Posadas' patient. This particular type of presentation is now known to be fairly rare when compared with other forms of the disease (see Fig. 18–7,*B* FS C 13,*B*). Rixford and Gilchrist[66] studied the parasite, which they observed in the lesion material. At the suggestion of C. W. Stiles, they named the organism *Coccidioides* (Coccidia-like) *immitis* (*im* = not, *mitis* = mild) and described it as a protozoan of the class Sporozoa. The organism appeared to them to be related to the members of the order Coccidia, which includes the causative agent of coccidiosis of chickens. Bacteriologic studies were performed on biopsy material from the patient, but the agar plates were noted to be overgrown by a "mould" and discarded as "contaminated." A second fatal case of disease was found by them in 1894, again in a Portuguese from the Azores. In 1896, Rixford and Gilchrist published a detailed account of these two cases. They recognized the similarity of the organism seen in their patients to that found in Wernicke's case, but considered the parasite to be the etiologic agent of a distinct and new infectious disease which was not related in any way to neoplasia.

The true nature of the etiologic agent of coccidioidomycosis was elucidated by Ophüls. In 1900, the third American case was discovered, again in a Portuguese from the Azores. Bacteriologic investigation revealed growth only of a mold that was considered, at least at first, to be a contaminant. The regularity of its isolation, however, eventually led to the association of the mold with the disease. Thus Ophüls and Moffitt in 1900[57] described the etiologic agent as being a fungus. In 1905 Ophüls published his works on the life cycle of *C. immitis*.[56]

In 1915, ten years after the work of Ophüls, Ernest Dickson reviewed the 40 known cases of the disease and stressed the importance of its occurrence in the southern California region. The experience of Ophüls and others indicated that the lung was the portal of entry for the fungus, which later disseminated to the skin and other organs. The discovery of healed pulmonary lesions in patients that did not have disseminated disease began to suggest that there was a milder form of the infection. Subsequently, a laboratory accident dramatically proved this point. A student named Chope, in Ophüls' laboratory, inadvertently inhaled a quantity of spores of *C. immitis*. When he became ill, he was declared to have coccidioidal granuloma, a universally fatal infection. Though it was severe at first, the disease resolved instead of progressing toward a fatal outcome. After recovery, the student decided to leave California and moved to Baltimore. In 1932 Stewart and Meyer[82] isolated *C. immitis* from soil in the San Joaquin Valley near a site where four Filipinos had contracted their severe or fatal infections. Thus, they established a soil reservoir for the organism in the area. The stage was now set for the next important discovery in the history of the disease.

Dickson continued his work at Stanford University, compiling statistics on the incidence and pathology of coccidioidomycosis. At the same time Myrnie A. Gifford, who practiced medicine in Bakersfield, located in the San Joaquin Valley, had examined a patient with fever, pleurisy and pneumonia who later developed erythema nodosum. This condition was referred to as Valley bumps and Valley fever. A few months later she was able to recover *C. immitis* from the sputum of another patient with the same syndrome. In 1936 she noted that a bout of the bumps (erythema nodosum) had preceded development of coccidioidal granuloma in three of 15 patients. The association of bumps and infection was emphasized by Dickson in 1937 in a publication that named the disease "coccidioidomycosis" and delineated a primary and secondary phase. By 1938, the cooperative studies of Dickson and Gifford had resulted in establishing that Valley fever and its synonyms were in fact mild forms of coccidioidomycosis and that the disease was much more common than previously suspected.[15]

With a grant from the Rosenberg Foundation, Dickson, Gifford, and a new member of the team, C. E. Smith, began an extensive study of coccidioidomycosis in the San Joaquin Valley. Smith criss-crossed the desert in an old Ford named the "Flying Chamydospore." He gathered data concerning the incidence, severity, and epidemiology of the disease. He developed a precipitin test and standardized the coccidioidin skin test.[79] With such data he and his associates were able to define the various forms which the initial phase of the disease may take, establish the duration of the incubation period, and study epidemics resulting from simultaneous exposure of many people to areas heavily contaminated with fungal spores.

In the 1940's airfields were built in endemic areas in the Valley.[77] This brought about the exposure of hundreds of nonimmune men to the spore-containing dust of the area, and large numbers of infections resulted. Smith formulated dust control methods, such as oiling roads, planting grass, and using swimming pools rather than dusty athletic fields for recreation. Such methods reduced the infection rate by 65 per cent.[78] Meanwhile the disease was discovered to be a problem in other areas of the Southwest. Its endemicity was established in Arizona, Nevada, New Mexico, Texas, Utah, and northern Mexico.[28] The great risk of disease in nonimmune individuals forced the closing of a prisoner of war camp in Florence, Arizona. Recently, new areas of endemicity have been established in other countries. In 1948, Campins et al.[7,8] diagnosed the first case in Venezuela. Subsequently, cases and endemic foci of disease have been found in Honduras in 1950, in Guatemala in 1960,[52] and in Colombia in 1967.[69]

Since the discovery of the benign form of coccidioidomycosis, the literature produced on this subject has been voluminous. Studies and reviews of the clinical and pathologic forms of the disease, its epidemiology, ecology, serology, and therapy, have made it one of the most "famous" of the mycotic infections. An excellent review of the subject was published by Marshall Fiese in 1958.[22] Coccidioidomycosis has been the subject of two international symposia, the latest held in 1965, which summarized the knowledge of the disease to that time.[2] The latest bibliography compiled for the disease is that by Al-Doory in 1972.[4]

ETIOLOGY, ECOLOGY, AND DISTRIBUTION

Coccidioides immitis is the only etiologic agent of coccidioidomycosis. Most strains isolated from infections or the soil are fairly uniform in regard to morphology, physiology, and pathogenicity. As shown by Huppert, Sun, and Bailey,[33] however, there may be great variability in morphology among some isolants of the organisms, and confusion with keratinophilic fungi and other soil saprophytes has occurred frequently.[20] The relationship of this organism to other fungi has been the subject of much debate. To the first investigators the morphology of the endosporulating spherule in tissue suggested a kinship to the protozoa. When the fungal nature of the organism was established, the same endospore formation within spherules implied an affinity to the Zygomycetes, a group in which progressive cleavage and sporangiospore formation occurs. Many mycologists presently place *C. immitis* in this class of fungi. However, it has been shown to be keratinophilic,[54] and there have been sporadic attempts to relate it to other fungal groups. As yet no satisfactory classification of the organism exists, and the relationship of *C. immitis* to other fungi is still unsettled. Although the fungus can regularly be isolated from soil, many erroneous reports of its occurrence have been made.

The most highly endemic regions of coccidioidomycosis are the southwestern United States and northern Mexico (Fig. 18–2). As delineated by Maddy[47–49] and later confirmed by others, this area conforms to the ecologic classification termed the Lower Sonoran Life Zone. The zone is characterized by the presence of such plants as *Larrea tridentata* (the creosote bush), mesquites, opuntias and other cacti, yuccas, and agaves (Fig. 18–3). Among the animals found there are species of *Perognathus* (pocket mice), *Dipodomys* (kangaroo rats), *Citellus* (ground squirrel), the long-eared desert fox, the big-eared and white-haired bats, and a few birds, including some owls which are predatory on rodents, e.g., *Otus asio* and *Speotyto cunicularia*. The climatic conditions of this region are semi-arid rather than completely dry; the rainfall averages 10 inches per year (254 mm) and occurs all in one season. Temperatures average 100°F in the summer, then fall in winter to a mild 33 to 38°F. These conditions describe the west side of the southern San Joaquin valley, which appears to be the more highly endemic part. As was shown by Egeberg and others,[17] within this region the growth of the fungus is restricted to a few small areas. *C. immitis* survives at depths of 20 cm and is usually absent on the surface during periods of hot, dry weather. The favored soils usually have a high content of carbonized organic material and a high salt concentration, particularly of $CaSO_4$ and borates. The sterile desert surface is reinvaded after the rainy season, possibly when the capillary action of evaporation reestablishes a favorable concentration of salts. Survival of the organism also occurs in rodent burrows, for, as was shown by Emmons,[19] native rodents become infected, particularly lactating and pregnant females. Distribution of the fungus to new areas by such infected animals has been postulated. Maddy[49] reported that the burial of infected rodent carcasses in an area previously free of *C. immitis* was followed by recovery of positive soil cultures for the duration of the experiment (seven years). Predators (owls, foxes) of such infected rodents do not become infected, however.[84]

The unique ecologic habitat favored by *C. immitis* has been the subject of many laboratory investigations. The fungus proliferates readily on almost any type of sterile soil at the extremes of naturally occurring pH and temperature. Growth is more luxuriant on rich soil than on poor. However, survival of the organism in soil with normal bacterial and fungal flora is greatly decreased. It appears that *C. immitis* is ill fitted to survive in competition with other soil microorganisms and indeed is inhibited by some. Egeberg et al.[17] found that, in the soils particularly favored by *C. immitis*, the most important inhibitory organisms were *Bacillus subtilis* and *Penicillium janthinellum*. These species proliferate during the rainy season, but as the temperature increases and evaporation increases salinity of the soil, their growth is inhibited. *B. subtilis* is sensitive to high salt concentration, and *P. janthinellum* to a temperature of 100°F. *C. immitis* is very tolerant of a wide range of salt concentrations and almost uniquely

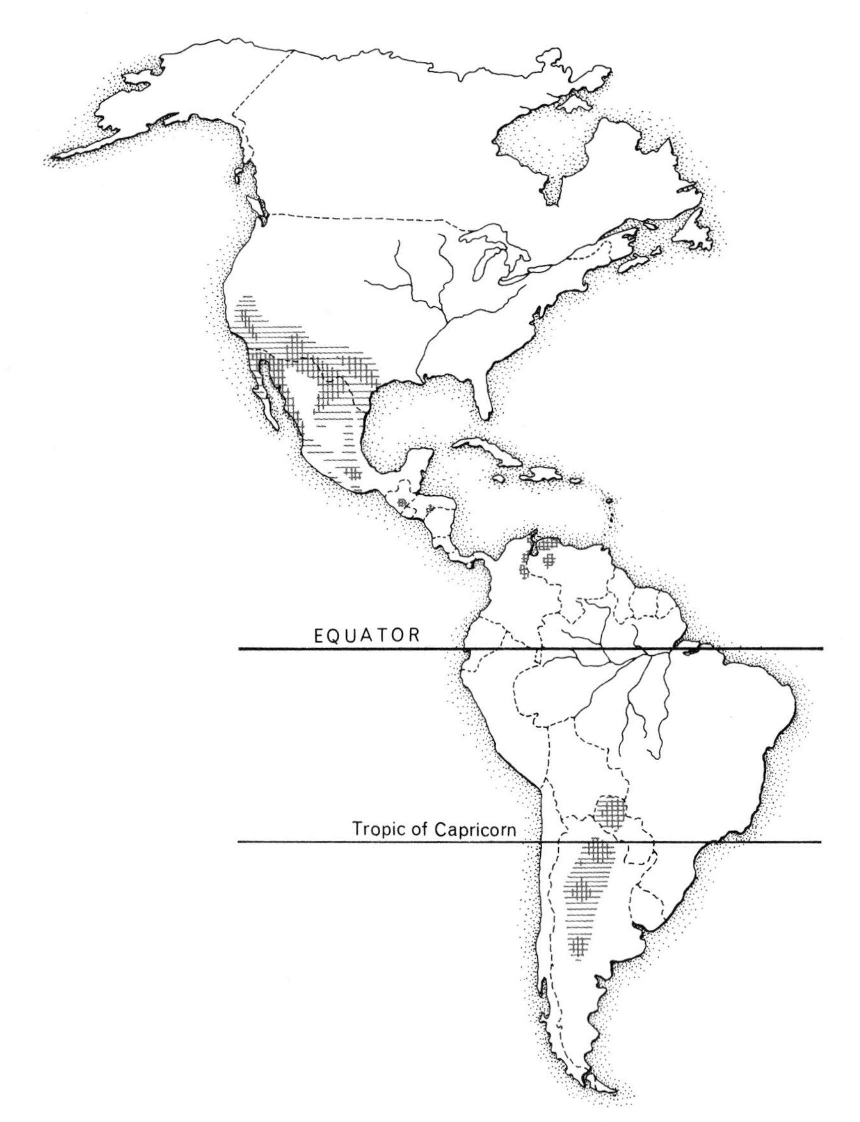

Figure 18–2. Incidence and prevalence of coccidioidomycosis.

tolerant to boron-containing salts.[80] Growth also occurs up to a temperature of 130°F. Both spherules and arthrospores survive for long periods under adverse conditions,[24] although the arthrospores survive extreme conditions much longer. It appears, then, that the factors which delineate the ecologic areas inhabited by *C. immitis* are dependent upon the tolerance of the fungus to adverse soil composition and high temperatures, which are inhibitory to competing organisms.

C. immitis presumably is disseminated primarily by dust aerosols that are prevalent in early summer and continue until the first rains of winter. The spores may be distributed naturally by wind storms, by man-made disruptions during construction work, and by farming, or in "digs" for archeological or biological specimens. The latter have been the origin of many small "epidemics."[89] Even the digging by desert rodents will form infectious aerosols. Swatek et al.[83] studied translocation of the organism which was not associated with rodent burrows. They found the fungus in the upper two inches of virgin desert soil and concluded that the wash of a water-soil slurry during the rainy season would also spread the organism to new areas.

Sensitivity to the intradermal injection of coccidioidin has been used to delineate the endemic areas of the disease. The most comprehensive survey and review of skin

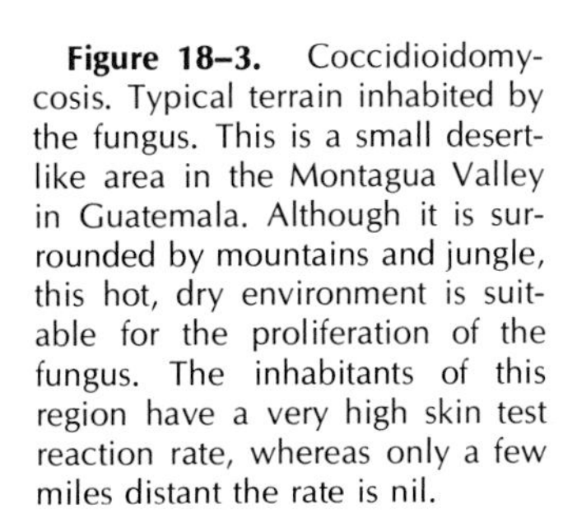

Figure 18–3. Coccidioidomycosis. Typical terrain inhabited by the fungus. This is a small desert-like area in the Montagua Valley in Guatemala. Although it is surrounded by mountains and jungle, this hot, dry environment is suitable for the proliferation of the fungus. The inhabitants of this region have a very high skin test reaction rate, whereas only a few miles distant the rate is nil.

test sensitivity in the United States is that of Edwards and Palmer.[16] The most highly endemic area in the world is Kern County, California. In some areas within this region, the positive skin test rate is 90 per cent or more, the average being 50 to 70 per cent. In the San Joachin Valley, the prevalence of positive skin tests gradually tapers off north of this area. The northernmost foci of infestation so far recorded are Red Bluff and Chico, California.[89] Outside of California, very high rates of skin test sensitivity (50 per cent or more) are found in Maricopa (Phoenix) and Pima (Tucson) counties in Arizona and in several west and southwest counties of Texas near the Mexican border. Rates of reactivity vary considerably over the remaining areas of the southwest. In regions where histoplasmosis is endemic, there is a low incidence of positive reactors, which reflects cross-reactions between the antigens of coccidioidin and histoplasmin rather than true presence of coccidioidomycosis.

There are three endemic areas of northern Mexico which are ecologically comparable to the San Joaquin Valley.[28] The first is termed the "northern zone," which is a continuation of endemic areas in the southwestern United States. Skin reactivity rates of 50 per cent or more are found in northern Baja California, the Sonora, and the Chihuahua states. The rate is less in the other northern states of Coahuila, Nuevo Leon, and Tamaulipas. The second zone of endemicity is the Pacific littoral zone. This encompasses the Sonora and Sinoloa states west of the Sierra Madre Occidental Mountains. The reactivity rate here is over 50 per cent, but it diminishes in the coastal states to the south. The northern zone descends toward the central states of Mexico but is separated from them by the Sierra Madre Oriental Mountains. This area constitutes the central zone of endemicity. Coccidioidin reaction rates are lower than in the other regions, but may reach 50 per cent in some scattered areas. These endemic areas correspond roughly to the limits of the Lower Sonoran Life Zone in Mexico. In addition, however, there are two highly endemic areas that are in tropical regions. The first, in the states of Colima and Michoacua, has a rate of reaction to coccidioidin of up to 30 per cent. The annual rainfall in this area is 717 mm. A second area in the state of Guerrero has a reactivity rate of 10 per cent and a rainfall of 1309 mm. Endemic foci are also found in other countries of central and South America that are not defined ecologically as Lower Sonoral Life Zones. It appears, therefore, that *C. immitis* may survive under a variety of ecologic conditions.

The highest endemic area of coccidioidomycosis outside the southwestern United States–northern Mexico region appears to be in Venezuela. In the dry, arid, northwestern states of Falcon, Lara, and Zulia, skin test sensitivities of 50 per cent or more have been found. The average temperature varies from 24 to 27°C, with a rainfall of 600 mm or less.[7]

In Central America, two endemic areas have so far been discovered. One is the

Montagua Valley in Guatemala (Fig. 18–3), and the other is the Comayagua Valley of Honduras. Skin reaction rates up to 30 per cent have been found there. These regions are semi-arid, with xerophilic vegetation and a rainfall of 30 inches (800 mm). This is higher by far than the Sonoran Life Zone, but similar to the endemic region in central Paraguay.

The other endemic regions for the disease are found in Paraguay, Argentina, and Colombia. The area in Paraguay includes the north and the western open plains of the Gran Chaco. The climate is hot and dry, with an average rainfall of up to 20 to 28 inches per year in certain parts. Among the Indians of the western Chaco, skin test sensitivities of 43 per cent have been found.[3] In Argentina the endemic areas are found in the central arid areas of the Patagonia from the 40th parallel north to the Salta-Hondo-Dulce Rivers on the 27th parallel.[54] The highest skin test reactivity (20 per cent) was found at the northern end of the zone near the Rio Hondo. A recently discovered endemic focus in Colombia is in an area adjacent to Venezuela. A part of the region is in the departments of Guajira and Magdalena, where there is rainfall of 500 mm per year, and the remainder is in the department of Cesar that has 1000 to 2000 mm of precipitation yearly.[69] It appears that an important factor in all these areas where rainfall is higher than normal for the Sonoran Life Zone is that the rains occur all in one short season and are followed by hot, dry, dusty weather.

Cases have been reported outside the American endemic zones. Most of these have been traced to either acquisition of the infection while visiting an endemic area or transmission by fomite.[2] A recent case occurring in Australia was acquired after a trip to the arid interior of that country, but so far *C. immitis* has not been found to occur there.[85] Reports of the disease in Russia have been published for some time. Stepanishtcheva et al.[81] record a total of 35 cases up to 1971. The validity of these cases must be examined further. The disease has also been recorded in buffalo in West Pakistan.[36] Corroboration of this report is also lacking. There is also a recent report of canine disease in Nigeria.[54a]

There are considerable differences in the severity of coccidioidomycosis which are related to sex and race. Skin test surveys show there is no difference between the sexes in acquisition of primary infections, and, until the age of puberty, the dissemination rates are about equal. In adult males, however, the risk of dissemination is 265 per 100,000, compared to 74 per 100,000 for females.[22] Following initial infection, erythema nodosum, a manifestation usually associated with high resistance to development of severe disease, is found in 25 per cent of women but only in 5 per cent of men. This difference in susceptibility of adult males holds true for all pathogenic fungous infections but not for opportunistic infections. The exception is in pregnant women, in whom the dissemination rate about equals or exceeds that of men.[76] The explanation for these observations obviously involves differences in hormones and hormonal balance, but this subject has not yet been adequately investigated.

An essentially unique feature of coccidioidomycosis is the increased susceptibility of persons with pigmented skin. Most all the cases of disseminated disease in the early literature were among Portuguese and Filipino farm laborers. In the many studies that have been made since, differences in rates of occurrence of disseminated disease have consistently been noted. Equalizing as much as possible for occupational exposure and socioeconomic conditions, Filipinos and Negroes run the highest risk of dissemination, with the rate considerably less for Indians and Indian mixtures (Mexicans). Caucasians have the lowest incidence of dissemination. A little more than 1 per cent of all primary infections develop into serious disease, and of these, dissemination is ten times more likely to occur in Negroes than Caucasians.

Primary infection with *C. immitis* is dependent on exposure to the fungus and can occur at any age. Agricultural workers and construction crews are particularly prone to infection because of exposure to spore-bearing dust. Considering the number of "epidemics" recorded among them, archeology students could be considered involved in a "high-risk occupation."

Coccidioides immitis is a pathogenic fungus and is probably the most virulent of all of the etiologic agents of the human mycoses. Inhalation of a few spores will produce infection in the normal human host, and a sufficient quantity of spores will produce an overwhelming disease. In the great majority of patients, primary disease is followed by recovery (without treatment) and development of a strong specific immunity to reinfection. The few patients who develop serious disease probably have some as yet undefined difference in their defense mechanisms from those that recover. These differences may be genetic or transient, but they influence the ultimate outcome of the host-parasite interaction. The same is true to a certain extent in all of the pathogenic fungous infections. These differences in host susceptibility are the areas most in need of imaginative investigation. In a few rare instances, coccidioidomycosis may be an opportunistic infection in patients with debilitating disease.[67] This again is related to exposure. A leukemic patient would have to travel to an endemic area to acquire a *Coccidioides* infection, whereas *Candida* and *Aspergillus* are always with him. It must be remembered also that an old, residual, coccidioidal granuloma may reactivate years later if the patient becomes debilitated (see Fig. 18–5,*C*). In such cases the infection could be considered opportunistic.

CLINICAL DISEASE

Infection by *Coccidioides* results after inhalation of dust containing arthrospores. Initial infection is followed by the production of two vastly different clinical diseases. The first is a completely inapparent or moderately severe disease followed by complete resolution of the infection and establishment of strong immunity to reinfection. The second is the rare form in which the infection becomes established and is followed by a chronic progressive disease or an acute, rapidly fatal dissemination. Such diversity in the disease syndrome following inhalation of infectious units is also seen in histoplasmosis, paracoccidioidomycosis, and other fungous infections. The extensive studies by Smith, Fiese, and others on coccidioidomycosis have shown that in about 60 per cent of patients primary infections are asymptomatic, and 40 per cent have mild to severe acute pulmonary disease. However, only in about 0.5 per cent (one or two per thousand) is a serious disease established. This is known as secondary coccidioidomycosis. The clinical classification presented here reflects these two major forms of the disease.

I. Primary coccidioidomycosis
   A. Pulmonary
      1. Asymptomatic
      2. Symptomatic
   B. Cutaneous (rare)
II. Secondary coccidioidomycosis
   A. Pulmonary
      1. Benign chronic
      2. Progressive
   B. Single or multisystem dissemination
      1. Meningeal
      2. Chronic cutaneous
      3. Generalized

Primary Coccidioidomycosis

PULMONARY

Asymptomatic Pulmonary Disease. This classification includes the majority of patients who become infected. There are no symptoms or at least none significant enough to have been remembered by the patient. Such patients are the "skin test convertors" who become reactive to intradermal injections of coccidioidin in the absence of demonstrable disease. There is no residual scar or lesion in the lungs, and it is only the conversion to coccidioidin sensitivity that indicates an infection has occurred. Such a clinical course does not always indicate a particularly strong resistance to the infection, however. In some patients with extrapulmonary involvement (e.g., meningitis), no demonstrable pulmonary lesions could be found, and there was no history of symptoms coincident with exposure.[10]

The asymptomatic group of patients who have no residual pathology merges gradually into the groups with some indication of past infection in addition to a positive skin test. In these there may be a history of a mild, undiagnosed, flulike disease. On roentgenographic examination such patients usually show small, healed areas of fibrosis and, commonly, calcifications representing old hilar, parenchymal, or pleuritic lesions.

Symptomatic Pulmonary Disease. Symptoms vary considerably in this group of patients. Following exposure and an average incubation period of 10 to 16 days, they become mildly ill with a "cold" or may have severe respiratory disease. The extremes of the incubation period are seven days to four weeks, and the duration of the primary disease is from a few days to several weeks. The incubation period and duration and severity of symptoms are usually a function of the magnitude of the exposure. Inhalation of a few spores is associated with an illness of short duration. Sometimes no respiratory disease is noted, and symptoms are confined to allergic manifestations, such as toxic erythemas, arthralgias, episcleritis, and so forth.

Patients have, in varying degrees, one or more of the following symptoms and signs.

1. *Fever.* Most patients are febrile for some period during their illness. There is nothing characteristic about its pattern. The fever usually rises to a peak with diurnal variation and then diminishes. It may last a few hours to a few months, may reach 40.5°C, and can be accompanied by night sweats. Persistence or recrudescence of fever following initial recovery often heralds dissemination. Spread of disease may occur in the absence of and without increase of fever, however.

2. *Pain.* Chest pain is very common and is usually the first sign of the disease. It occurs in 70 to 90 per cent of symptomatic patients. Although it may manifest only as a dull discomfiture, pleuritic pain may be so severe and sudden in onset as to emulate rib fracture, myocardial infarction, or cholecystitis. Pain is often accompanied by friction rub and, sometimes, demonstrable pleural fluid levels. Interference with respiration often occurs as the pleuritic pain is made worse by coughing or deep inspiration. Interference with swallowing may result from substernal pain caused by enlargement of the mediastinal nodes.

3. *Respiratory embarrassment.* This symptom is uncommon but may result from the presence of pleural effusion or pneumonic spread of the disease. Although quite rare, shortness of breath may also be due to spontaneous pneumothorax.

4. *Cough.* Cough is less common and less severe in coccidioidomycosis than in other pulmonary infections. It is often nonproductive, but with severe pulmonary involvement sputum may be white or purulent and sometimes blood-streaked. In such cases, the organisms can often be seen on direct examination of a potassium hydroxide mount of the sputum, but their absence does not rule out the diagnosis.

5. *Anorexia.* This symptom frequently occurs even in mild disease. The duration is usually short, but some patients lose 20 to 30 pounds within a three-week period. In disseminated disease, anorexia may persist and lead to profound cachexia.

Other symptoms are present in varying degrees. Generalized aching, malaise, myalgia, and backaches are often present. In mild cases malaise and lassitude may be the only complaints. Headache may be quite severe in acute, uncomplicated disease. It may subside in a week or persist for a much longer period of time. Spinal fluid is normal at this stage. Meningitis is an infrequent sequela of primary disease and is accompanied by abnormal spinal fluid findings.

X-ray. The roentgenographic findings of symptomatic disease vary from a normal chest picture to extensive pulmonary infiltration, lymphadenopathy, or massive pleural effusion. In the mildest infections, only superficial inflammation of the bronchial mucosa or the alveolar spaces is present, and there is nothing detectable in the thoracic shadow. However, about 80 per cent of patients show some pulmonary changes on x-ray. Infection may involve any area of the lung, including the apices, but involvement of the lower lobes or the bases of the upper and middle lobes is more common. Thus primary coccidioidomycosis more closely resembles primary atypical rather than bacterial pneumonia. There is no single x-ray pattern noted, and individual roentgenograms often defy classification. Probably the most common finding is that of a pneumonic infiltrate. There are soft, hazy, uniform lines which radiate from the hila toward the base or periphery. Although it may persist for months, this type of infiltrate usually resolves about a week to ten days after cessation of clinical symptoms. In other cases only a soft, fuzzy, hilar thickening is found which usually resolves within two weeks.

The best characterized lesions are well circumscribed nodules found in the lung parenchyma. These are most often basilar in the middle or lower lobes. They usually have a diameter of 2 to 3 cm and are single or, rarely, multiple. Such lesions resemble the nodules of primary or metastatic tuberculosis. They usually resolve, leaving little residue, or become thin-walled cavities. Such cavities fibrose and later calcify. Large inactive cavities may persist for years.

An infrequently seen pattern is that of hilar and mediastinal lymphadenopathy. Although hilar nodes are frequently involved, mediastinal nodes appear to be secondarily invaded. Such a picture is often found in patients who later have disseminated disease. When hilar nodes alone are involved, they regress in a few days to a week, but mediastinal nodes may take months to heal, even if dissemination does not occur. Enlarged nodes most often accompany parenchymal infiltrates in acutely ill patients. Rarely they are the only sign present and can be confused with neoplasias, such as Hodgkin's disease. Another pattern observed in about one-fifth of patients consists of small pleural effusions. They may suffice only to obscure the costophrenic sulcus and usually resolve rapidly.

Primary infection that is becoming rapidly progressive and disseminated is indicated by progression of the above findings on serial x-rays. Pneumonia and pleural effusion become extensive, and there is widening of the mediastinum. Fuzzy miliary foci that become confluent are seen, and consolidation of one or more lobes usually indicates an imminent fatal terminus.

The other laboratory findings in primary coccidioidomycosis include an elevated leukocyte count, often with an eosinophilia; an elevated sedimentation rate; and conversion of the coccidioidin skin test to positive. The latter occurs from 2 to 21 days following initial symptoms. Humoral antibodies develop later. In very mild infections precipitins and complement fixing antibodies do not appear. In more severe primary infections they are present for varying lengths of time, but disappear with resolution of the disease. Persistence of a complement fixation titer is associated with establishment of serious disease and probable dissemination. Another test that can gauge progress of the disease is the erythrocyte sedimentation rate. It is markedly elevated, even in mild or moderate primary disease. It gradually falls with recovery. Persistence of, or a steady increase in, an elevated sedimentation rate denotes development of progressive disease and possible dissemination.

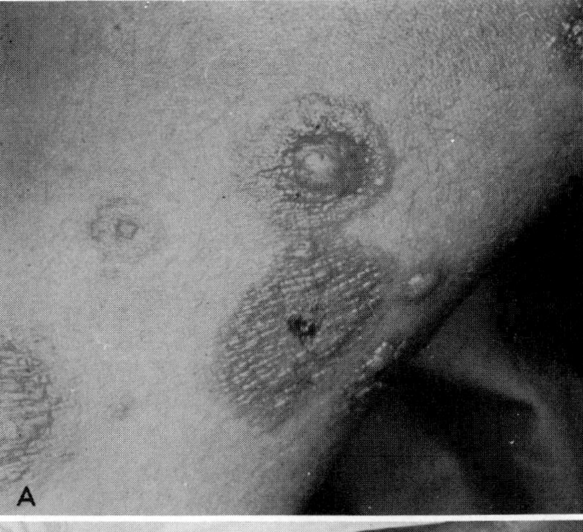

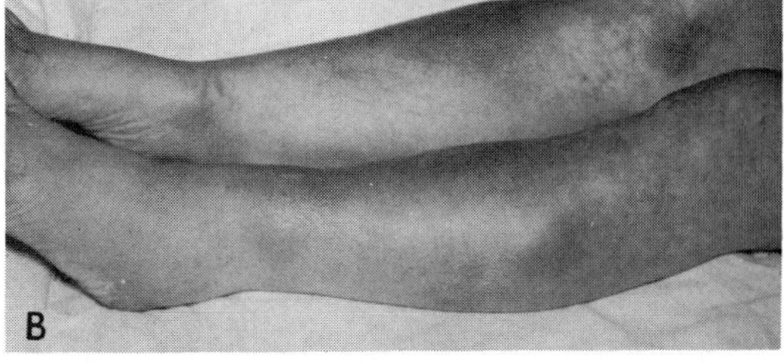

Figure 18–4. FS C 11. Allergic manifestations of coccidioidomycosis. *A*, Erythema multiforme on extensor surface of arm. Intact bullous eruption on erythematous base. Target or "bulls-eye" lesion. *B*, Erythema nodosum. Erythematous tender nodules with a bluish cast on the anterior aspect of lower extremities.

Allergic Manifestations of Primary Disease. In about 5 to 10 per cent of cases some form of allergic manifestation occurs. An erythema nodosum or multiforme-like toxic eruption is found (Fig. 18–4 FS C 11) in 25 per cent of adult white females and 4 per cent of white males. The earliest signs of cutaneous lesions are small, generalized, erythematous, macular or papular rashes. They appear within a day or two after the onset of the primary symptoms and resolve within a week. Such toxic erythemas are found in many acute febrile illnesses. They precede development of the skin test hypersensitivity. Though they are present in only a small percentage of patients, in an endemic area they are almost pathognomonic for coccidioidomycosis. Development of any form of "desert bumps" or "desert rheumatism" usually denotes strong resistance to the infection. Very rarely, however, the disease may progress in patients with such symptoms, most often to pulmonary cavitation or meningitis. In the latter case, the patients are usually skin test negative.

The lesions of erythema nodosum are most often restricted to the lower extremities. Usually they are found on the anterior tibial areas, are more numerous about the knees, and may extend to the thighs. Typically they appear as a crop of bright red, tender, itching, or painful raised nodules. They are firm and elastic and deeply imbedded in the skin (Fig. 18–4,*B* FS C 11,*B*). They range in size from a few millimeters to several centimeters. The acute lesions regress within a few days. Although the color is initially red to purple, after resolution, the areas may show some degree of brownish postinflammatory hyperpigmentation, which lasts for several weeks or months.

The erythema multiforme–like toxic eruption develops on the upper half of the body. The favored sites are the neck, face, collar, upper back, thorax, and occasionally the arms. They appear as reddish to purple nodules, papules, macules, or vesicles and fade in time to a violaceous or brownish hue (Fig. 18–4,*A* FS C 11,*A*). Sometimes they are accompanied by erythema nodosum on the legs. One crop of lesions is the usual pattern, but several crops may appear. Reappearance of such lesions often follows fatigue from physical exertion.

"Desert rheumatism" is the name for the arthritis that sometimes accompanies primary disease. The symptoms vary from vague arthralgias to periarticular swelling, stiffness, erythema, and heat. Joint effusion is not found. Other less common allergic phenomena involve the eye. These include episcleritis, phlyctenular conjunctivitis, and keratitis, all of which resolve usually within a few days. When they occur they are often overshadowed by the more spectacular skin rashes and joint pains that may be present in the same patients.

PRIMARY CUTANEOUS COCCIDIOIDOMYCOSIS

This is the rarest form of infection, and it has been adequately documented only in about three cases. The first case was reported in 1927 and involved inoculation of spores by a prick of a cactus needle, as described by Guy and Jacobs.[28a] Introduction of spherules into the abraded skin of his finger by a mortician embalming a patient who had died of the disease was the second.[92] The third case involved a laboratory accident in which a graduate student introduced spores into her thumb.[86] In all three cases a chancre-like lesion developed, there was regional adenitis, and the lesions healed uneventfully within a few weeks. Wilson et al.[92] have set up rigid criteria for documentation of primary cutaneous disease. These include no history of pulmonary disease immediately preceding onset, clear evidence of traumatic inoculation, regional adenitis in nodes that drain the area, and a chancriform lesion. The lesion is similar to the primary lesion seen in sporotrichosis. It is a painless, firmly indurated nodule or nodular plaque with central ulceration. Reported cases of primary cutaneous disease, in which the lesions were described as being multiple or verrucous or were abscessed, torpid, cutaneous ulcers, in reality represent secondary lesions following an inapparent primary pulmonary infection. Descriptions of primary infection occurring through the oral and nasal mucosa have not been adequately documented. A case of primary endophthalmitis has been reported, however.[29]

Secondary Coccidioidomycosis[30]

PULMONARY

Benign Chronic Pulmonary Disease. In about 2 to 8 per cent of symptomatic infections, residual chronic cavitation occurs. It is considered benign because dissemination is a feature of the primary infection and rarely develops from chronic cavitary disease. As a complication of primary disease, cavitation occurs more frequently in Caucasians, a group which is much less prone to disseminated disease. Development of cavitations is not necessarily related to the severity of the primary infection.

The commonest presentation of this form of disease is a single, chronic, thin-walled cavity which often has a fluid level and little surrounding tissue reaction. It is usually symptomless and discovered only on routine X-ray (Fig. 18–5,*A*). If symptoms occur the most common indication of the presence of the lesion is hemoptysis, which is also the most frequent and serious complication. In a few patients, cough, production of sputum, and malaise are also present. The distribution of the lesions in the lung fields is the same as that of the primary pulmonary disease from which it develops.

Hemoptysis is always a danger in the cavitary form of the disease. In addition, if the lesions are peripheral they may rupture, causing pleural effusion, empyema, spontaneous pneumothorax, or bronchopleural fistula. Resolution of these cavities is often resistant to chemotherapy, and they must be removed surgically. On surgical removal, the cavity is found to have a tough fibrotic wall, with a granular interior containing spherules and sometimes hyphae of *C. immitis.* Enlarging

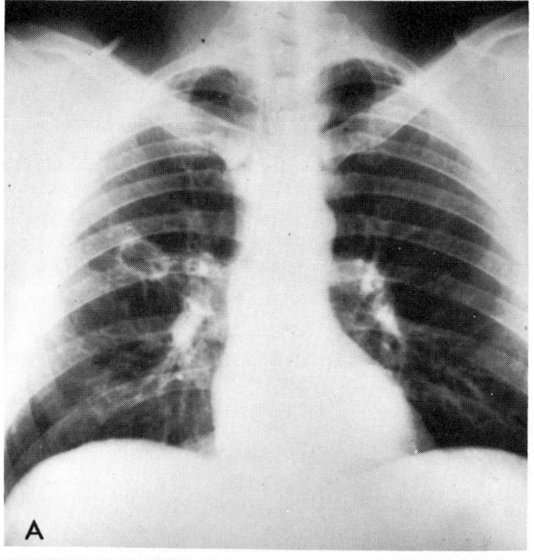

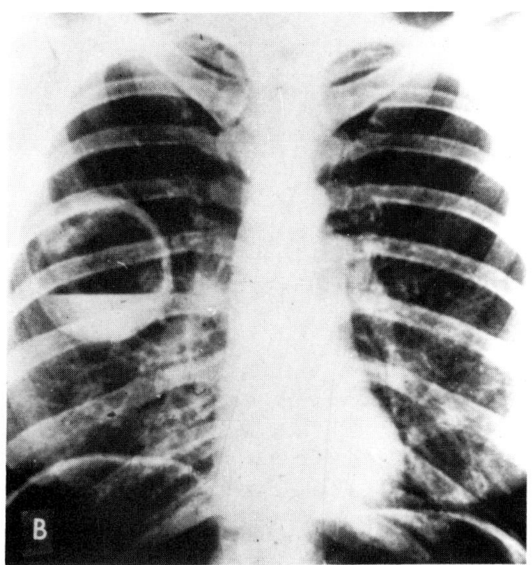

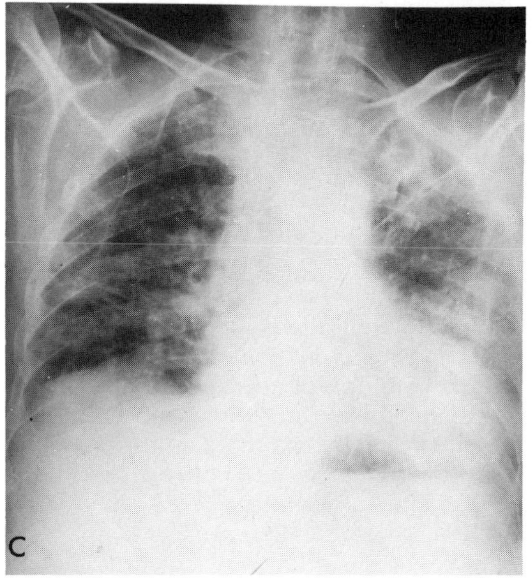

Figure 18–5. Radiologic aspects of secondary coccidioidomycosis. *A,* "Coin lesion." In the superior segment of the right lower lobe, there is a thin-walled, sharply defined cavity. This lesion has been inactive for years, and essentially no reaction is seen around it in this x-ray. (Courtesy of F. Kittle.) *B,* Chronic coccidioidoma. There is a large, sharply defined cavity which has an air-fluid level in the middle right lung, lower lobe. *C,* Disseminated disease from an old cavity lesion. The lungs are underinflated. Both lung fields show diffuse interstitial infiltrates, particularly evident on the right. In addition, there is a thin-walled cavity on the left apically located that is surrounded by intense tissue reaction. In the left base there is a confluence of shadows consistent with alveolar disease. The picture on the left is consistent with a diagnosis of pneumonia, whereas that on the right would probably be interpreted as hematogenous spread. (Courtesy of H. Grieble.)

cavities are usually the result of bacterial superinfection.

Though they are typically thin-walled and cystlike, there are many variations of coccidioidal cavities. The commonest is one with a thick wall, presenting as a moderately dense pulmonary shadow. Such lesions are difficult to differentiate from tuberculous cavities or bacterial lung abscesses. The size of all types of cavities varies from a few millimeters to several centimeters.

Another form of chronic residual disease is the coccidioidoma, which is comparable in morphology and development to the histoplasmoma or tuberculoma. It is formed as the result of a resolved or arrested pneumonitis or granuloma. Though these do not represent a hazard to the patient, since reactivation of such lesions is very rare, they do represent a diagnostic difficulty because of their resemblance to neoplasias. Their size varies from a few millimeters to 4 or 5 centimeters (Fig. 18–5,*B*). The outline is often lobulated or irregular, reflecting the coalescence of several sites of pneumonitis. Most frequently they are single, but multiple lesions may be seen. Although they are usually stable for long periods of time, in some excavation occurs followed by refilling. This is most often associated with superinfection by bacteria.

Some calcification of the lesion is common. On surgical removal, this type of lesion is seen to have a thick, fibrous wall with a center which is soft, yellow-grey or yellowish, necrotic, and caseous. On histologic examination, spherules and sometimes hyphae are seen. The spherules are often atypical, and may be wrinkled and distorted empty ghosts. A few viable organisms are usually present. Satellite tubercles from the main cavity are common. The complement fixation titers in both chronic cavity disease and coccidioidoma are low or absent.

Less common complications of primary disease include bronchiectasis, pulmonary fibrosis, empyema, pneumothorax, hydropneumothorax, and chronic pericarditis. Since primary disease is frequently endobronchial, chronic or slowly progressive bronchiectasis may result. There is chronic productive cough, hemoptysis, fever, and frequent bacterial or viral superinfection. The complement fixation titers are usually higher in such patients than in those with cavitary disease or coccidioidoma. Persistent pulmonary fibrosis of itself is not a serious complication, but on x-ray examination it causes diagnostic problems. Empyema is an infrequent but more serious complication which results from rupture of a chronic cavity or follows the pleural effusion of the primary disease. Together with pneumothorax and hydropneumothorax, it is associated with a brief spread of the infection to other pleural areas. Occasionally more serious consequences develop, such as multiple sinuses draining through the chest wall. Chronic empyema can be a very debilitating condition and requires chemotherapy. In contrast, pericarditis is a benign residual complication of primary disease. Mechanical construction of the heart, however, may require surgical correction.

Progressive Pulmonary Disease.[13a] Although residual pulmonary disease remains stable, though chronic, it may also become progressive. This sometimes begins immediately after the primary disease has begun to resolve or after long periods of time following stabilization. Moderate debilitation, malnutrition, old age, and chronic pulmonary disease are factors which predispose to reactivation and progression of residual coccidioidomycosis. This progressive disease may take several forms. One is a simple, locally extending lesion. Others include enlarging or multiplying cavities and nodules, abscessing nodules, extending infiltrates, or lobar consolidation. Such activation and progression are associated with increasing symptoms of severe chest disease, and the patient should be followed by serial x-ray examinations. Progressive chest disease, particularly following previously stabilized and dormant disease, is a very serious condition (Fig. 18–5,*C*). The infection may stabilize again, but there is usually extrapulmonary dissemination or a relentless pulmonary course with a fatal outcome. It is therefore necessary to initiate specific chemotherapy in such circumstances.[70]

SINGLE OR MULTISYSTEM DISSEMINATION[22,70,93,95]

Dissemination of the disease is dependent on several factors. Overwhelming exposure to the fungus may result in an almost immediate dissemination and a severe, rapidly fatal disease. In Negroes and Filipinos dissemination usually follows closely the onset of the primary infection. It is the general rule that, if dissemination is going to occur, it will do so within the first few weeks of primary disease. The signs that herald dissemination are a persistent elevated sedimentation rate and complement fixation titer, along with persistant and increasing fever and malaise. In patients with these signs dissemination is often rapidly fatal. It results in acute meningitis, involvement of multiple organ systems, and frequently cutaneous and subcutaneous abscesses.

Other patients may contain their infections to a greater extent and have apparent remissions. They appear more resistant to the infection and generally have a marked eosinophilia along with their pneumonitis and pleural effusion, but their defenses are in some way less than adequate to contain the disease. Symptoms such as anorexia and fever reappear, and a rise in the complement fixation titer at this point indicates imminent progress and dissemination of the infection. The resulting disease more frequently has a long, chronic, protracted course than that of patients with a rapidly fatal infection. Lung lesions and the involvement in other organs go through periods of quiescence and recrudescence.

A third category of patients is those in whom dissemination occurs late in the course of the disease (Fig. 18–5,*C*). This form of infection is particularly prevalent in white males. Late dissemination probably represents the emergence of active infection from inapparent disease. The course may be quite insidious, as is found in cases of chronic meningitis which may become symptomatic long after the primary infection has abated.

In patients receiving steroid therapy inactive lesions of coccidioidomycosis may be reactivated, and dissemination may occur.[67] In such cases, the organism is considered to be an opportunist. Opportunistic dissemination also may occur in some cases of leukemia, lymphoma, or other neoplasias.

Meningitis. Involvement of the meninges may be acute, subacute, or chronic. It is most often acute and rapidly fatal in Negroes and Filipinos. The disease in Caucasians, however, is more frequently a subacute or long, chronic process. Meningitis is found in about 25 per cent of cases of secondary coccidioidomycosis. After tuberculosis and cryptococcosis, it is the third most common form of chronic granulomatous meningitis. The disease is uniformly fatal if not treated, but may have a protracted course with periods of remission.

The acute form of coccidioidal meningitis develops as a sequela of primary pulmonary disease. There is no remission of symptoms associated with the primary infection, and fever, lassitude, anorexia, and weight loss continue. Neurologic signs may be minimal at first. The spinal fluid is turbid and yellow, and there is a marked pleocytosis. The complement fixing titers of serum and spinal fluid are often quite high.

Chronic meningitis may be very insidious in its course. In a recent series of cases reported by Caudill et al.,[10] over half the patients had had no previous history of coccidioidal disease. In all patients, the skin test was negative, but significant complement fixing titers were present in serum and spinal fluid. The initial spinal fluid examination was grossly abnormal, and a pleocytosis, elevated protein, and decreased glucose were found. The authors conclude that early examination of spinal fluid is necessary to exclude this and other treatable granulomatous meningitides in patients presenting vague, protracted, neurologic signs and psychiatric syndromes. In other series, the complement fixing antibody was present in the spinal fiuid in only 75 per cent of cases. Unfortunately, the correct diagnosis in some patients is not made until autopsy.[9] Complications of meningitis, usually types of obstructive hydrocephalus, are frequent. Chemotherapy is required, and intrathecal combined with intravenous amphotericin B has been used successfully for treatment.[10, 94, 96]

Chronic Cutaneous Disease. This was the form of coccidioidomycosis seen in the first case of Posadas and that of Rixford. Both patients had long courses of disease with recurrent lesions until they died of multisystem involvement, 11 and 8 years after the appearance of the first cutaneous lesions. This form of disseminated disease is seen most often in Negroes and other dark-skinned individuals and is quite rare in Caucasians. Lesions often appear first on the nasolabial folds, face, scalp, or neck. At first the lesions are epidermal thickenings without inflammation. Gradually they become larger and wartlike to form the characteristic type lesions termed "verrucous granuloma," which are similar to those of chronic cutaneous blastomycosis. The lesions may grow and spread and resemble fungating epitheliomas. Other lesions remain small and eventually heal, leaving atrophic scars (Fig. 18–6 FS C 12). As disseminated disease progresses to involve other tissues, skin lesions begin to appear as indolent ulcers (Fig. 18–7 FS C 13). These are frequently seen over joints and represent sinus tracts from subcutaneous and osseous foci of infection (Fig. 18–8 FS C 14).

Generalized Disease. Dissemination often is followed by the establishment of miliary lesions in many organ systems. Commonly involved are lymph nodes, which sometimes heal by fibrosis. Splenic lesions are granulomatous, tumorlike, or suppurative, depending on which stage the disease was in when the organ became involved. Osteomyelitis is found in about half the cases of disseminated infection. The bones most commonly involved are the ribs, vertebrae, and all the extremities (Fig. 18–9). In long bones, lesions are commonly found in the distal portions and within the prominences. This is similar to the distribution of lesions found in blastomycosis. Lesions in the flat bones are usually centrally located. The reaction is most often osteolytic, but in chronic and solitary lesions osteoblastic activity is found. Sometimes chronic osseous disease is the only extrapulmonary lesion noted. Joints are the sites of lesions, with accompanying arthritis in about a third of disseminated cases.

Involvement of other organs varies considerably in cases of disseminated coccidioidomycosis.[30, 31] The skeletal muscles are usually spared, but a psoas abscess may be found.

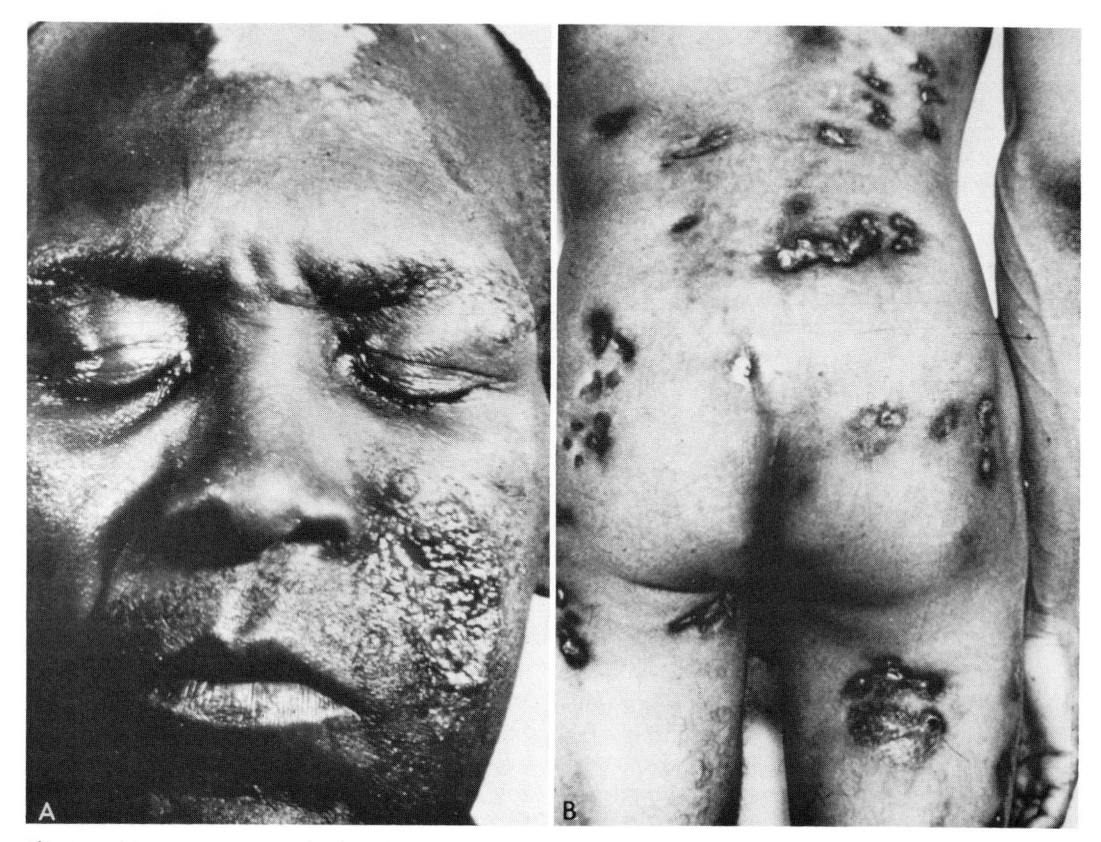

Figure 18–6. FS C 12. Coccidioidomycosis. *A,* Chronic lesions of the face. Active lesions are seen on the cheek. An atrophic, depigmented scar representing a healed lesion is on the forehead. *B,* Chronic granulomatous lesion and suppurating lesions in an advanced case of disseminated coccidioidomycosis.

The adrenals are rarely foci of infection. Thus the complication of Addison's disease is much less frequently encountered than in blastomycosis and histoplasmosis. Pericardial granulomas are present in about one-fourth of cases. Involvement of heart muscle is extremely rare. In contrast to the findings in aspergillosis and paracoccidioidomycosis, the gastrointestinal tract is usually spared. Miliary granulomas or abscesses may be found in the renal parenchyma, but the pelvis is generally without lesions. The male genital system is very rarely involved, but when it is, the prostate is the usual site of infection. Disease in this organ is much less frequent than in systemic blastomycosis. Lesions in the female genitalia are also uncommon.

DIFFERENTIAL DIAGNOSIS

In the endemic areas, coccidioidomycosis should be considered in the differential diagnosis of any nonspecific illness. This is also true with persons who have visited these areas at any time prior to onset of symptoms, particularly when the signs and symptoms are vague and nonspecific.

Primary disease is often confused with other acute pulmonary infections, such as influenza, primary atypical pneumonia, bronchitis, bronchial pneumonia, or simply a "cold." Since this form of the disease most often resolves uneventfully and without treatment, the patient is none the worse. Secondary coccidioidomycosis, however, requires therapy so that it is imperative to make a specific diagnosis. The disease must be differentiated from tuberculosis, neoplasias, other mycotic infections, syphilis, tularemia, glanders (melioidosis), and osteomyelitis of bacterial origin.

PROGNOSIS AND THERAPY

Recovery from primary disease occurs in almost all symptomatic and asymptomatic

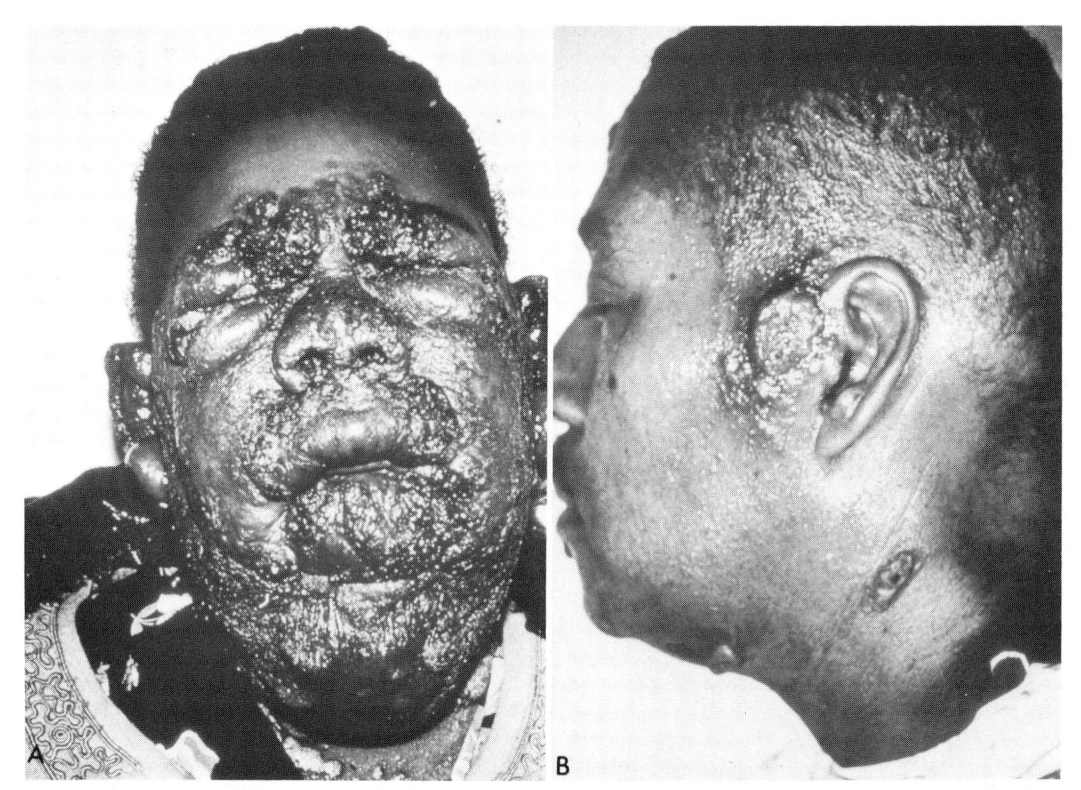

Figure 18–7. FS C 13. Chronic cutaneous coccidioidomycosis. *A*, Granulomatous lesions on face, neck, and chin. *B*, Advanced chronic cutaneous disease. This is the rare nodular form and similar to the morphology present in the first cases of Posadas and Wernicke and of Rixford. (Courtesy of M. Gifford.)

cases. Therefore primary disease is treated with bed rest and restriction of activity. Steroids are sometimes used to control severe allergic manifestations, but patients so treated should be closely watched as steroids may promote dissemination. Surgical intervention should be considered in benign residual disease of the lung because of the danger of recurrent hemoptysis. Cavities sometimes close spontaneously, however. All other forms of secondary coccidioidomycosis have a grave prognosis unless specific chemotherapy is instituted.

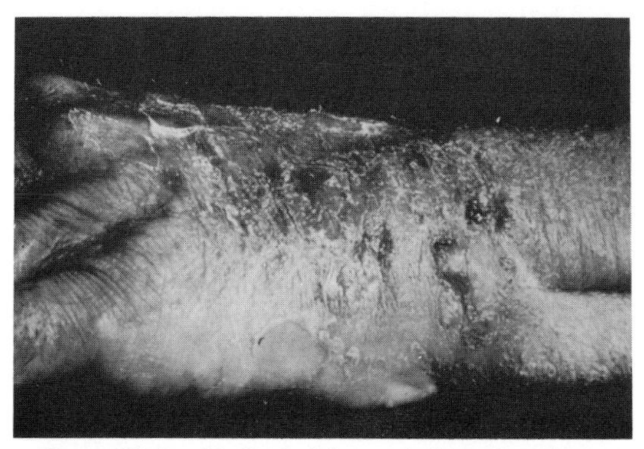

Figure 18–8. FS C 14. Advanced disseminated coccidioidomycosis. The nodular lesions are now mixed with suppurative draining lesions. Subcutaneous and osseous lesions were present in the arm and wrist.

The drug of choice in coccidioidomycosis is amphotericin B. The late William Winn had extensive experience in treating disseminated disease, and he formulated the generally accepted dosage schedule.[93,94,98] The regimen is essentially the same as that used for other systemic mycoses. Based on body weight, it is 1 mg per kg per day for the adult and from 1.25 to 1.50 mg per kg per day for children. The total dose is 1 to 3 g, but persistent disease may require more. Patients are started on a 5.0-mg schedule initially, which is gradually increased as tolerated. Alternate day therapy may be required. The drug is diluted in 500 ml of 5 per cent dextrose solution. It must be made up fresh, as the drug is unstable in solution over long periods of time. Storage in brown bottles or covering the bottle is not neces-

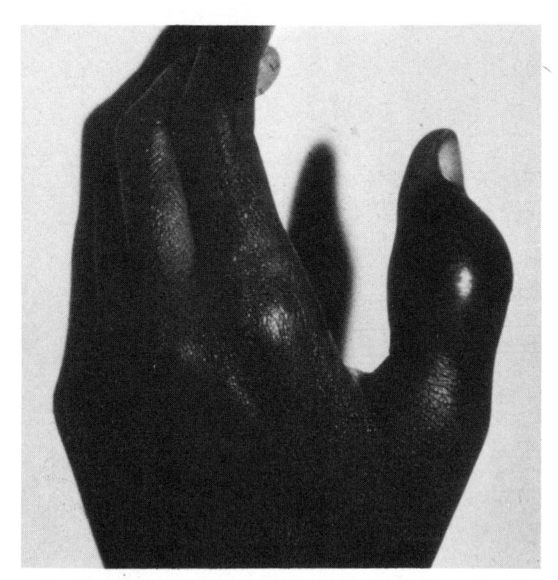

Figure 18–9. Coccidioidomycosis. Dactylitis. The lesion is swollen and hot but has not eroded through the skin.

sary, as the drug is stable exposed to light for the six-hour infusion time.[4a] Additives are used to control side effects when necessary. These include 10 to 30 mg of diphenhydramine hydrochloride (Benadryl) or 10 mg of chloroprophenpyridamine maleate for nausea. Phlebitis can be controlled by the addition of 10 to 30 mg of heparin. The solution is administered intravenously as a slow infusion over 4 to 6 hours. A 22- to 24-gauge needle is recommended. An *in vitro* method for monitoring therapy has been described by Borchardt et al.[4b]

Renal toxicity is the most significant side effect of drug therapy. It is mandatory to follow the blood urea nitrogen level and the creatinine level during therapy. The latter is now considered the most sensitive index of renal damage. Most nephrotoxicity is reversible, but irreversible changes occur when more than an average of 3 g of amphotericin B is given. Pretreatment evaluation of kidney function will often forewarn the possibility of difficulties and can be used as a guide to modify the dosage regimen.

Meningitis usually requires the use of intrathecal, in addition to intravenous, amphotericin B therapy. The treatment regimen may involve the use of the Ommaya reservoir, and often the Pudez ventriculoatrial shunt when hydrocephalus is present. Details of such treatments are given by Winn[93,96] and Zealear and Winn.[98]

Hundreds of other drugs have been tried for the treatment of coccidioidal disease. A partial list of these is given on p. 182 of Fiese.[22] None of these, however, has been found to be efficacious. It is hoped that, for coccidioidomycosis as well as the other systemic mycoses, a better and less toxic drug than amphotericin B will be found in the future.

PATHOLOGY[30,31]

Gross Pathology. Death in the primary stage of coccidioidomycosis is very infrequent, so that the early gross and histopathologic changes that occur have not been well described. From the few natural cases that have been studied and from data obtained in experimental disease in animals, the following sequence has been delineated. Spores that land on the bronchi evoke an endobronchial reaction, such as a bronchitis or a bronchopneumonia. Spores reaching the alveolar space induce an alveolitis with an initial cellular invasion of macrophages. This is rapidly followed by an acute pyogenic reaction. The course varies considerably at this point. Necrosis, caseation, and granuloma formation may follow in any combination and in any degree. Thus the lesions seen in gross specimens show localized or patchy pneumonitis, a caseous pneumonitis, or involvement of whole lobes with consolidation and necrotic cavitation (Fig. 18–10 FS C 15). In the milder forms of disease, microabscesses and tubercles are found, and in severe infections numerous necrotic granulomatous nodules are present. Frequently, small areas of necrosis lead to cavity formation. The peribronchial and peritracheal lymph nodes are enlarged, as they become infected very early in acute primary disease. With resolution of primary infection, nonnecrotic granulomatous areas hyalinize and appear later as small bands of scar tissue or as firm nodules several centimeters in diameter. Cavities are formed in areas of necrosis. These become surrounded by a granulomatous reaction, thus forming thick-walled, well-organized nodules or cavities with central necrosis or thin-walled cavities with central caseation. The former may develop into coccidioidomas, and the latter are seen as the diagnostically trouble-

Figure 18–10. FS C 15. Coccidioidomycosis. Cut section of lung. There is focal consolidation in upper and lower lobes with generalized consolidation and focal abscess formation in the middle lobe. Fibrosis is evident in hilar lymph nodes.

some, thin-walled "coin lesions" (Fig. 18–5,*A*). Such lesions most often are solitary, but may be multiple in either one lobe or several lobes. They very closely resemble the circumscribed nodules known as histoplasmomas and tuberculomas. In endemic areas such lesions can be given a presumptive diagnosis of coccidioidal granuloma; in nonendemic areas the true etiology usually must await surgical removal and pathologic examination.

The gross pathology of progressive and disseminated disease has been the subject of a number of reviews. The types of lesions and their distribution are essentially identical to those of blastomycosis. In both diseases pulmonary involvement may be minimal or extensive lung disease may be present, but invariably there is extensive cutaneous and subcutaneous involvement. In addition, both diseases frequently cause lesions in bone with the same favored sites. It is easy to appreciate why these two diseases were often confused in their early history. The pulmonary findings are similar to those seen in resolved primary or residual disease. In disseminated disease, however, miliary abscesses and nodules caused by hematogenous spread of the organism may be found. It must be remembered that, although the lung is the portal of entry for the organism, many cases of fatal dissemination have been recorded where no pulmonary pathology was found.

One of the commonest causes of death in coccidioidomycosis is meningitis. Of 181 such cases tabulated by Huntington,[30,31] leptomeningitis was present in 100. The characteristic lesion is a diffuse granulomatous meningitis which envelops the brain. Such lesions often cause an obstructive hydrocephalus. The brain matter itself is very rarely the site of disease. In the few cases in which infection of the brain occurred, the lesions varied from small, soft granulomas to large abscesses with ragged borders.[22]

Based on autopsy studies, statistically the spleen and liver are second only to the lungs as sites of disease. Both organs are usually infected late in the course of the disease, so there is no clinical evidence of involvement. Often spherules and endospores are seen with little or no cellular reaction around them, indicating an anergic patient. Other lesions vary from acute pyogenic tumors to well-formed granulomas. Sites of disease in other organs in descending frequency are kidney, skin, bone, and adrenals. Clinically important disease is produced in the lung, meninges, skin, and bone (in that order) during the course of the infection. In these areas well-formed granulomas as well as areas of necrosis and abscess formation are to be expected. Other organs are involved late in the disease during the terminal "fungemia," and often such lesions are usually nothing more than small areas of necrosis and abscess formation.

Histopathology. The cellular reaction evoked in infection by *C. immitis* is of three basic types. These are pyogenic (purulent), granulomatous, and mixed. The pyogenic reaction occurs around the initial infecting spores and also within granulomas at the time of the rupture of the spherule and release of endospores (Fig. 18–11). This is thought to be the result of the chemotactic response of neutrophils elicited by the substances released during spherule rupture. Granuloma formation occurs around the developing spherule (Fig. 18–12). Spherules are frequently found within histiocytes during the early development of the granuloma, and in later stages within foreign body giant cells. The granulomatous response demonstrates the sequential invasion of lymphocytes, plasma cells, monocytes, histiocytes, epithelioid cells, and giant cells. After organization, these lesions exhibit fibrosis, caseation, and calcification. Calcification occurs as frequently or to a greater extent

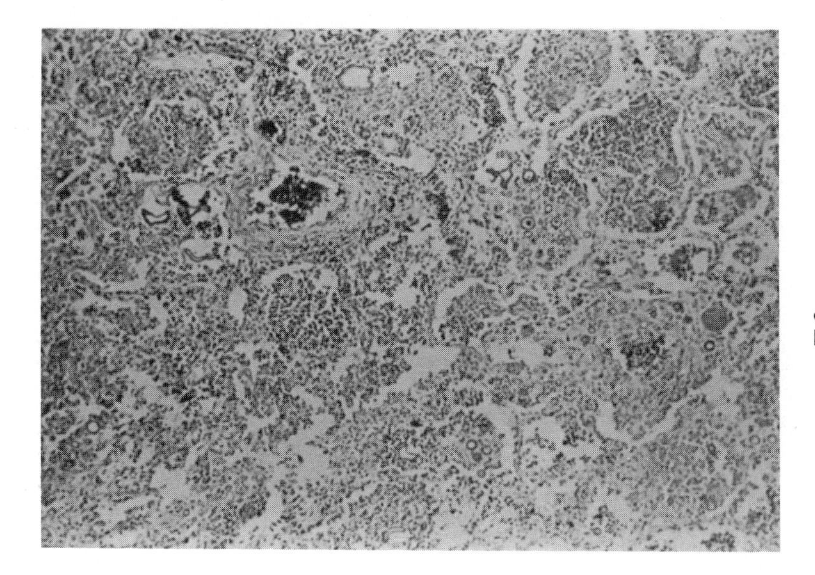

Figure 18–11. Coccidioidomycosis. Section of lung with purulent bronchopneumonia. ×100.

in coccidioidomycosis as in histoplasmosis. In lesions in which the organisms are growing and reproducing, a mixed-type cellular reaction is found. The intermittent pyogenic reaction occasioned by release of spherules is responsible for this histologic picture. In old, inactive lesions in which the organisms are dead, one is more likely to see a "pure" granuloma. As the resistance of the patient diminishes in disseminated disease purulent lesions become more prominent.

The character of the cutaneous lesions of chronic coccidioidomycosis is particularly noteworthy. The pseudoepitheliomatous hyperplasia may be as pronounced as that seen in cutaneous blastomycosis (Fig. 18–13). More than once skin biopsies from patients with this form of disease have been read as neoplasia. One can understand why Posadas thought his case was a type of mycosis fungoides. The spherules of *C. immitis* are much more easily seen than the yeast cells of *Blastomyces*, so that misdiagnosis is less frequent in coccidioidomycosis than in blastomycosis. As the patient's defenses diminish in progressive disseminated disease, suppurative necrotic lesions of the cutaneous and subcutaneous tissue are found.

Cutaneous lesions not associated with the presence of organisms are found in primary disease. These include erythema nodosum and multiforme-like eruptions. These are toxic erythemas and consist of panniculitis and angiitis.

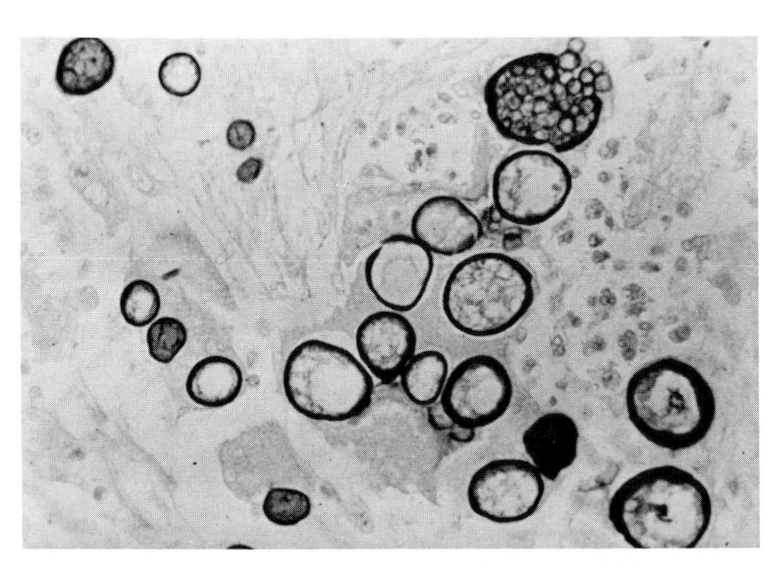

Figure 18–12. Coccidioidomycosis. Gridley stain of section from lymph node showing various stages in the development of spherules. ×400. (From Rippon, J. W. *In* Burrows, W. 1973. *Textbook of Microbiology*. 20th ed. Philadelphia, W. B. Saunders Company, p. 726.)

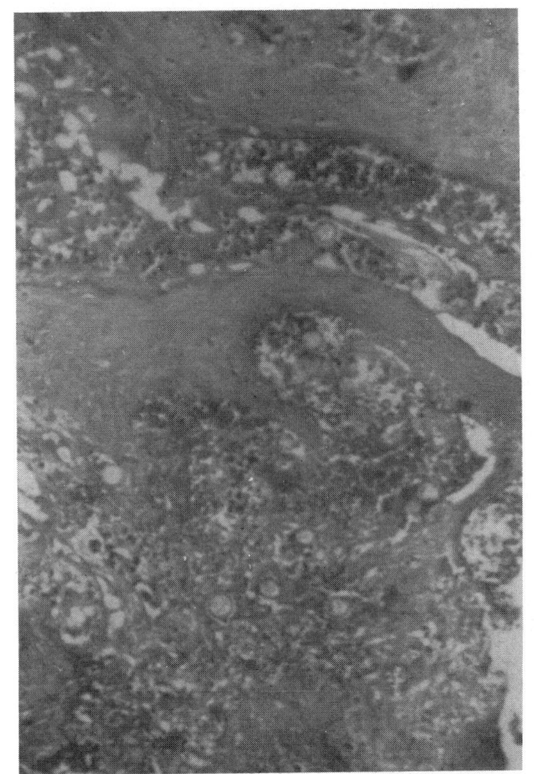

Figure 18–13. Coccidioidomycosis. Psuedoepitheliomatous hyperplasia of skin. Note the spherules. Hermatoxylin and eosin stain. ×100.

There are also other histologic findings not associated with the presence of organisms. One is the diffuse myocarditis that is not infrequently found in disseminated disease. The reaction is a nonspecific inflammatory response, and its etiology is not known. Another is the renal lesion associated with amphotericin B. toxicity.[65] At first the lesions are mainly confined to the tubules and are reversible with cessation of chemotherapy. Later, severe glomerular necrosis occurs and is usually considered irreversible. The drug remains in the body at low levels following cessation of therapy. Thus the glomerular necrosis may occur after the end of the treatment regimen.

Coccidioides immitis is easily seen in routine hematoxylin and eosin stained tissue sections. Following inhalation the arthrospore develops into a spherule within a few hours or days. Arthrospores are barrel-shaped and average 2.5 to 4 by 3 to 6 μ in size. They become more rounded as they transform into spherules (Fig. 18–12). At maturity, these are 30 to 60 μ in diameter (Figs. 18–14 and 18–15 FS C 16). The wall is thick (to 2 μ) and quite prominent. The cytoplasm is eosinophilic and contains many nuclei. As the spherules near maturity, endospore production begins by a process called "progressive cleavage." Furrows form and divide the protoplasm into multinucleate masses called "protospores." Often there is a central vacuole in the spherule. Secondary cleavage lines are formed which divide the contents of the spherule into uninucleate endospores that are 2 to 5 μ in diameter. At maturation, the spherule wall breaks and the endospores are released. It is at this time that neutrophils can be seen invading the spherule and surrounding the newly released spores. It is probable that this is the

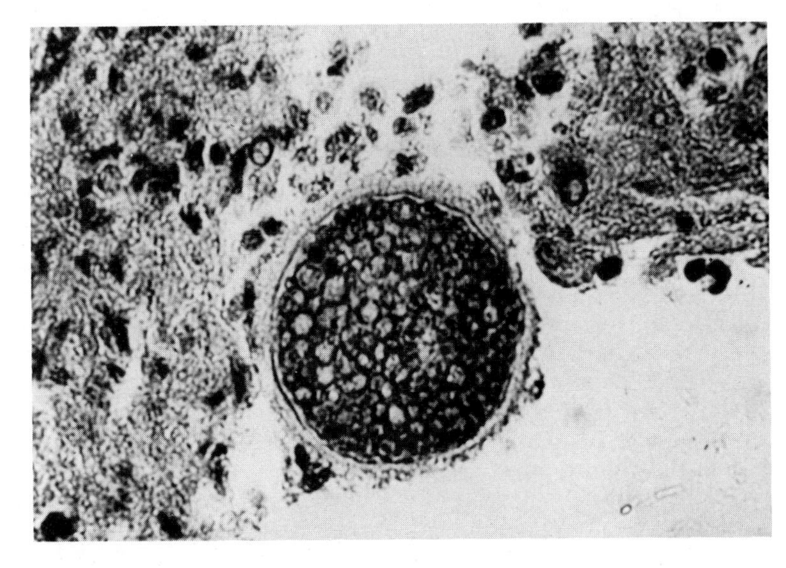

Figure 18–14. Coccidioidomycosis. Mature spherule with endospores. Hematoxylin and eosin stain. ×500.

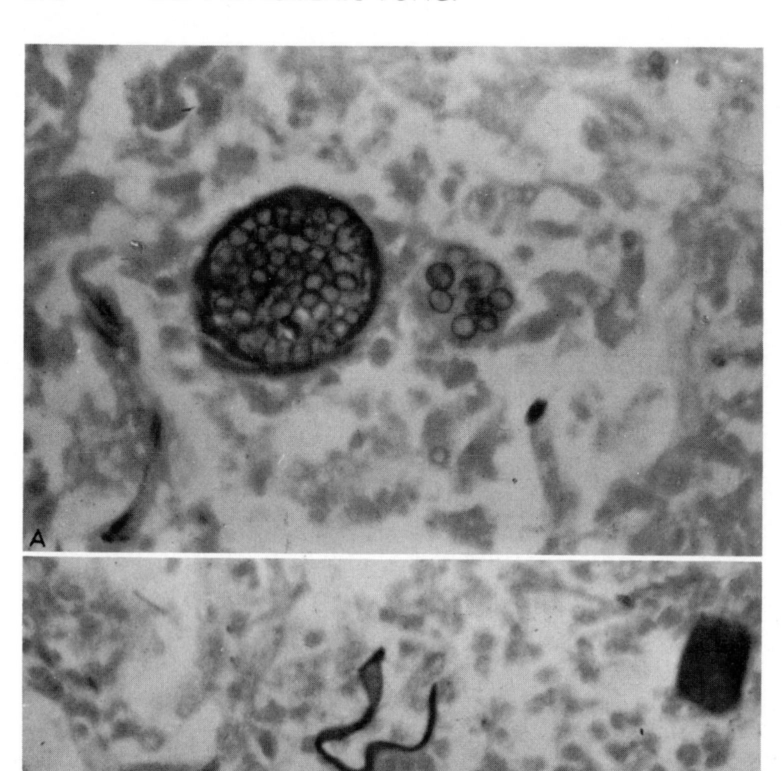

Figure 18–15. FS C 16. Coccidioidomycosis. *A,* Spherule with endospores and released endospores. Note that the endospores somewhat resemble budding yeast cells of *Blastomyces* or other fungi. *B,* Endospores and cell wall of empty spherule. Gridley stain. ×450. (Courtesy of P. Graff.)

time when the patient's defenses are most effective in killing these newly released spores and eliminating the infection. The surviving spores gradually evolve into spherules, and the process is repeated. In some tissue sections, spherules may be seen to have eosinophilic radiations emanating from the wall (Fig. 18–16). This is probably an example of the Splendore-Hoeppli phenomenon, the eosinophilic material representing antigen-antibody complexes.

In sections from old inactive lesions, the spherules may appear very distorted, broken, crescent-shaped, and empty (Fig. 18–15 FS C 16). Endospores which approximate the size of neutrophils are often difficult to distinguish from artifacts and leukocytic debris in hematoxylin and eosin stains. The special stains for fungi, such as Gridley, methenamine silver, and Hotchkiss-McManus (PAS) are useful in such cases. Since the organisms are usually dead, or the specimen was immersed in formaldehyde before mycotic disease was suspected, cultures cannot be performed, and it is often difficult to make a specific diagnosis on histopathologic grounds alone. Small, empty spherules may approximate the size of the yeast cells of *B. dermatitidis,* and persistent endospores may be confused with the cells of *Cryptococcus neoformans, Histoplasma capsulatum,* and *Paracoccidioides brasiliensis.* The appearance of all these organisms in old lesions is atypical. Cryptococcosis can be diagnosed histopathologically by the use of Mayer's mucicarmine stain. The close approximation of two small cells of *Coccidioides* may be misinterpreted as budding, so that this cannot be relied upon

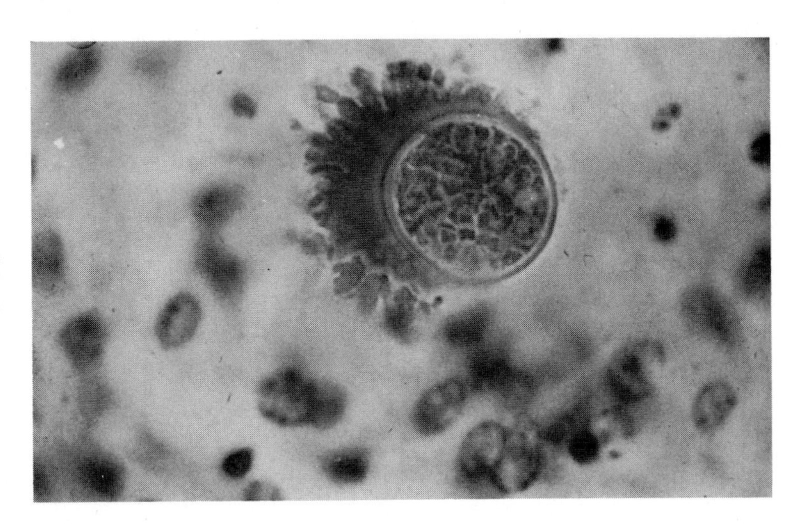

Figure 18–16. Coccidioidomycosis. Spherule with "asteroid" formation. Hematoxylin and eosin stain. ×450.

for discerning the correct etiology. Search of many fields is often necessary to find a spherule with some endospores. Use of specific fluorescent antibody is an aid in these cases.[37]

In old cavitary lesions of the lung and in some granulomas, hyphae of *C. immitis* are found (Fig. 18–17). In one series the incidence was 73 per cent in cavities and 30 per cent in granulomatous lesions.[63] Hyphae have also been described in meningeal lesions. The hyphae observed do not contain arthrospores; therefore, unless spherules are present, a specific etiology cannot be assigned. With careful search, however, a few endospore-containing spherules can be seen. It must be remembered that hyphae of *Aspergillus* may also be found in old cavities of coccidioidal granuloma.

ANIMAL DISEASE

Natural infection among animals is very common in the zones which are highly endemic for coccidioidomycosis. Emmons[19] in 1942 was the first to note coccidioidal lesions in the lungs and lymph nodes of native rodents in the southwestern deserts of the United States. The species included deer mice, pocket mice, kangaroo rats, and ground squirrels. It has been postulated that such infected animals may act as vectors of dissemination. This is as yet an unsettled question. The disease in such animals seems to be minimal, and predators feeding on them do not usually acquire the infection.[84] However, it was found that the burial of carcasses of animals that had died of experimental disease rendered the surrounding

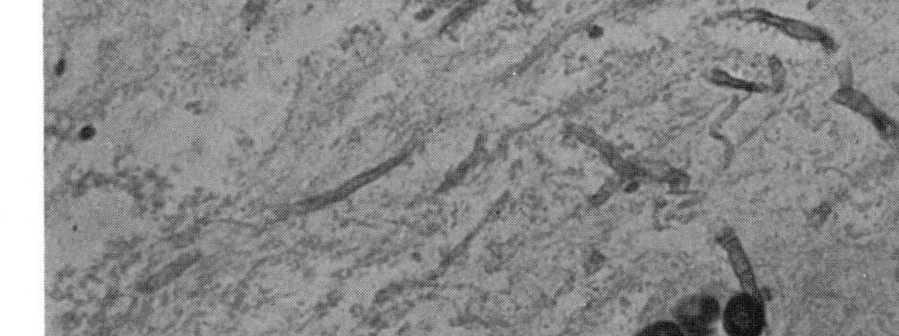

Figure 18–17. Coccidioidomycosis. Spherules and mycelium in fibrocaseous nodule. ×400.

soil areas positive for the growth of *C. immitis* for many years.[49] There are very few species of native mammals that occupy the endemic zones of the *C. immitis*, and none of these seem to be susceptible to the serious forms of coccidioidomycosis.

The first report of natural infection in domestic animals was that of Giltner[27] in 1918. He described lesions in the bronchial and mediastinal lymph nodes of cattle. He reproduced the disease in swine by injection of lesion material. Since then, coccidioidomycosis has been reported in horses, cattle, swine, llamas, burros, rabbits, dogs, sheep, tigers,[29a] and primates. The disease in most of these animals is minimal; the disseminated form is regularly found only in primates and dogs.

Several investigators have determined the susceptibility of monkeys to the disease and its transmission to cage mates.[38] As few as ten arthrospores were able to cause infection in this animal. When infected monkeys were housed with noninfected monkeys, transmission of the disease occurred in a few cases.

The first case of disseminated disease in dogs was noted in 1940 by Farness. This occurred in his Great Dane, Sarah.[21] Since that time severe disease in dogs has been reported frequently. Maddy[47] reviewed several hundred canine cases and concluded that dissemination in such breeds as the boxer and Doberman pinscher is even more common than in man. The systemic disease is quite similar to that seen in human cases,[54a] except that osseous lesions are proliferative rather than lytic. Benign disease akin to Valley fever has also been noted in some dogs. Infection in the cat has not been reported, and this species appears to be quite resistant to infection.

From the first report in 1918 until the present, there have been many records of the disease in cattle. Essentially all of these have been noted in abattoirs where a few lesions are found in the lymph nodes or lungs. Disseminated disease is extremely rare, so that the presence of such lesions in lymph nodes does not require condemnation of the carcass. Maddy has skin tested some 12,000 cattle from different areas of Arizona. He found the reaction rate varied from 0 to 84 per cent. The overall average was 24.6 per cent.[50] In endemic areas, horses and sheep also have high rates of skin test sensitivity. However, disseminated disease is also very rare in these animals, though it does occur.

Experimental disease can be produced in most laboratory animals. The intravenous inoculation of ten arthrospores or less evokes a chronic infection in mice. At a dosage of 100 to 500 arthrospores, a rapidly fatal disease is produced. The virulence of different strains is variable, however.[24] Inoculation of spores into poikilothermic animals does not elicit disease.

Biological Studies. Because of the great importance of coccodioidomycosis in its endemic areas, many investigations of the biology of *C. immitis* have been carried out. These have been concerned with practical aspects, such as vaccines, soil fungicides, and so forth, and many basic studies on the pathogenesis, metabolism, physiology, and development of the fungus have been done.

No serious work on a vaccine has been done with any of the other mycoses. Because it is known that strong immunity follows resolution of primary infection and because of the significant susceptibility of certain racial groups and occupations in endemic areas, much effort has gone into the development of a vaccine. Both living and nonliving vaccines have been tried. Attenuated spherules have been produced that markedly increase the resistance of mice to challenge infections.[42,44] Such attentuation was found to be an unstable characteristic, however, and thus the use of such vaccines was deemed dangerous. Scalarone and Levine[44,72] found that mice were more resistant to challenge infections following intravenous or intramuscular injections of formalin-killed spherules. A stronger immunity was acquired using the intramuscular route as compared to the intravenous route. The intravenous route induced higher precipitin titers but weaker delayed hypersensitivity. This route of injection also did not induce the leukocytolytic and macrophage aggregation response of white blood cells in the presence of *Coccidioides* antigen. These responses were conferred by intramuscular injection of spherule vaccine. The use of these vaccines in humans has conferred skin test reactivity on some,[43] but in other subjects it was lacking or at most weak.[59] Controlled clinical trials of the vaccine have not been carried out. Other types of vaccines prepared from arthrospores and cellular components have also been investigated.[11,74] At this writing, a safe, effective vaccine does not appear to be forthcoming.

Another practical aspect of coccidioidomycosis that has been investigated is the possibility of soil disinfection. This would be particularly appropriate in highly endemic areas where construction or agricultural work is contemplated. So far, one partially effective fungicide has been found. Used as a soil spray, 1-chloro-2-nitropropane leaves an area noninfectious for 24 hours after use.[18,73] Results of field tests have not as yet been reported.

Studies on the virulence of *C. immitis* have been carried out by several investigators. Friedman et al.[25] found considerable variation among strains of *C. immitis* in their ability to cause disease in mice. Gale et al.[26] have determined that the dissemination rate was higher in pigmented than in albino mice. In addition, they demonstrated the toxicity of culture filtrates of the organism. Mice surviving toxic challenge were more resistant to subsequent infection when challenged with an infection by injection of spores.

The effect of amphotericin B on *C. immitis* has also been investigated. The growth inhibiting property is associated with injury to the cell membrane, particularly at loci containing sterols. Exposure to the drug results in a lesion which causes loss of intracellular metabolites, particularly potassium, inorganic phosphate, and sugars. However, even after treatment of endospores with 4 μg per ml of the antibiotic for five days, capacity to grow was restored by subsequent treatment with cysteine or glutathione.[75] The organisms thus treated do not become spherules but rather grow as mycelial units. The effect is thought to be owing to the lowering of the oxidation-reduction potential of the growth medium, which now permits growth of antibiotic-treated cells, but only in the mycelial form.

The phenomenon of dimorphism in *C. immitis* has been investigated by a number of workers. Unlike the thermal dimorphic fungi, such as *Histoplasma capsulatum* or *Blastomyces dermatitidis,* morphogenesis in *C. immitis* is not governed by temperature but is regulated by several factors, including CO_2 tension. Increased CO_2 and growth in a liquid medium are sufficient to produce "culture spherules." Such "culture spherules" may not be altogether comparable to "tissue spherules," however.[5] It has been postulated that other factors affect the transition of the mycelium to spherule *in vivo.*

All the enzymes of the Embden-Meyerhof pathway and the pentose shunt have been found in *C. immitis.* The enzymes of the Krebs cycle are present except for α-ketoglutaric dehydrogenase. However, the finding of isocitric lyase suggests the operation of the glyoxylate shunt[46] in the organism. This metabolic pathway is common among many fungi. In *C. immitis,* the enzyme levels were always higher in the mycelial as compared to the "culture spherule" stage.

The cell walls of all growth forms of *C. immitis* contain an abundant chitinous "core." In addition, there are β-glucans and some other polysaccharides. The inner wall of the spherule is lined with phospholipids. Among the interesting components of the cell is the rare sugar 3-O-methylmannose,[60] known previously only in three species of bacteria.

IMMUNOLOGY AND SEROLOGY

The innate human immunity to infection by *C. immitis* is extremely high. At the time of initial inhalation of spores, this is a general resistance or nonspecific immunity to the infection. As discussed in previous sections, in fully 60 per cent of people infected, the entire disease process and its resolution is completely asymptomatic. Most of the remaining 40 per cent have a mild febrile to moderately severe respiratory disease. Following recovery from the infection, the patient is left with a strong specific immunity to reinfection. This immunity can be gauged to a certain extent by responsiveness to the coccidioidin skin test. It is also reflected by the allergic manifestations present during the initial infection. If dissemination is going to occur, it does so at an early stage of primary infection. Recovery from primary infection essentially rules out dissemination, even in those with residual cavitary disease. In patients demonstrating toxic erythemas, dissemination is even less likely. As has been noted, these various responses are related to sex and race. Thus the white female is most likely to demonstrate allergic manifestations and least likely to have disseminated disease. On the other hand, dark-skinned men very rarely have allergic lesions and most often have disseminated disease. There are, of course, exceptions to this general pattern. In a few documented cases, dissemination has occurred in the presence of erythema nodosum. This may have been related to the infectious dose. However, cases of so-called "late" dissemination are more likely to be "late" clinical appearance of what was an occult, quiescent, disseminated infection.

Our knowledge of the various serologic and immunologic patterns and their relation to clinical disease is greater in coccidioidomycosis than in any other of the mycoses. Beginning with Smith,[77–79] who correlated the results of 39,500 sera with the various clinical forms of the infection, and adding the work of Huppert, Pappagianis, Levine, and others, we now have valuable serologic aids for the diagnostic and prognostic evaluation of this disease. These include the well-standardized skin test reagent, coccidioidin, the complement fixation test, the precipitin tests, and the more recently developed latex agglutination, immunodiffusion, and fluorescent antibody tests.

Coccidioidin Skin Test. The standard procedures for the manufacture and use of coccidioidin were developed by C. E. Smith and associates.[79] These were modified from those used in the production of the tuberculin skin test reagent. The coccidioidin skin test consists of the intradermal injection of 0.1 ml of a 1:100 dilution of standardized mycelial coccidioidin. The reaction is read at 48 and 72 hours. The induration must be 5 mm or more to be considered positive. There may be an immediate response, but this is ignored. Usually there is erythema accompanying the induration, which may consist only of a faint blush. The following precautions and sources of error are to be noted. The needle used should be a disposable one or one that has not previously been used for injection of biological materials. Tuberculin is not inactivated by autoclaving and may evoke a false-positive reaction. Subcutaneous injection of the material will also lead to false results. The maximum reaction is reached at an average of 36 hours. Reading too early or too late may lead to false interpretations. A few patients demonstrate reactivity to other fungal skin test antigens, though usually to a lesser degree than to coccidioidin. Patients with any allergic manifestations may have a severe reaction to the standard dose of coccidioidin, and dilutions should be used. Levine et al.[45] have prepared a skin test antigen from the spherules of *C. immitis*. It appears to be much more sensitive than standard coccidioidin, and therefore may be more useful in diagnosis of the disease and in epidemiologic studies.

Erythema nodosum, erythema multiforme, and desert rheumatism are accompanied by a strong and often severe reaction to the intracutaneous injection of coccidioidin. The sequelae include headache, fever, worsening of the toxic lesions, swelling, and possible necrosis at the test site. For this reason, in patients manifesting any allergic reactions, a 1:1000 or 1:10,000 dilution of coccidioidin is used. Allergy to coccidioidin can be passively transferred by the intradermal injection of leukocytes from skin test–positive patients into the skin of coccidioidin test–negative patients.[64] This cannot be accomplished by injecting serum. Repeated skin testing does not appear to induce a positive skin response or a detectable antibody titer. In contrast to histoplasmin, the use of the coccidioidin skin test reagent does not cause a false rise in the complement fixation titer. However, patients originally negative to dilute coccidioidin (1:100) may be positive following injection of more concentrated material.[64]

Patients with primary coccidioidomycosis develop skin test sensitivity quite rapidly. The reaction is positive in 87 per cent of cases during the first week and in 99 per cent after the second week of clinical symptoms. The latex agglutination test is sometimes positive before this. The extremes recorded for development of coccidioidin sensitivity are 2 to 21 days after onset of symptoms. A negative skin test may be the result of testing too early in the course of the disease or an indication of impending dissemination. The latter is particularly true when the test was previously positive and then converts to negative. Such anergy is associated with a grave prognosis.

According to the method of Smith, coccidioidin is produced by growing the fungus in a modification of the asparagine, salts, glucose medium used for preparation of PPD tuberculin. Ten strains of *C. immitis* are used. These were isolated from patients with various clinical forms of the disease and from various geographic areas. No significant differences in the coccidioidin produced by the various strains have been observed, although fine analysis has shown some antigenic differences between some isolants.[32]

The fungus is grown for 6 to 10 weeks in static broth culture, after which aqueous merthiolate is added, and the culture filtrate is separated out by passage through a millipore or Berkefeld filter. The material is tested for sterility, potency, and specificity. Many lots must be discarded because of the lack of specificity and potency. Coccidioidin is also used in the complement fixation and precipitin tests. Batches are standardized for each test separately, and a single batch may be excellent for one test and unusable for the others. The material appears to be stable indefinitely. The complement fixing antigens are destroyed by autoclaving, but those responsible for the skin test and the precipitin test are not.

A tine test has been developed using coccidioidin.[88] In comparison to the standard intracutaneous injection of 0.1 ml of a 1:100 dilution, it was found to have certain advantages for large scale surveys, but lacked the flexibility of differing dosages needed in clinical evaluation.

Tube Precipitin Test (TP). In essentially all symptomatic and asymptomatic cases of primary coccidioidomycosis, the coccidioidin skin test will be positive and is usually the first serologic procedure to become so. Other serologic procedures, such as the tube precipitin and complement fixation tests, are positive for varying lengths of time in most patients that have symptomatic disease. Precipitins usually appear within the first week of clinical illness and persist for only a few days or weeks, even if dissemination occurs. Within the first few weeks of clinical disease, 90 per cent of symptomatic patients will have a positive tube precipitin test, but by the fourth month only 10 per cent do. Eventually patients with all forms of the disease become negative by this test. Therefore, a positive tube precipitin test indicates early active disease. The precipitin test is highly specific for coccidioidomycosis, and cross-reactions are very rare.

The test is carried out by the usual method for detecting precipitating antibodies. Into each of four tubes is placed 0.2 ml of undiluted serum. This is followed by the addition of 0.2 ml of undiluted coccidioidin in the first tube, 0.2 ml of a 1:10 dilution in the second, and 0.2 ml of a 1:40 dilution in the third. A control is run using plain asparagine broth. The solutions are mixed and read daily for five days. A positive result is indicated by the appearance of a flocculus or precipitate at the bottom of the tube.

Complement Fixation Test (CF). Complement fixing antibodies are the last to appear, but they have the greatest prognostic significance. Sometimes their presence is not detected until three months after the appearance of clinical symptoms. They, like the precipitins, appear in 90 per cent of symptomatic patients, but in only 10 per cent of asymptomatic cases. Though they usually disappear by six to eight months, they may persist at low titers for years in patients with resolved, uncomplicated disease. In disseminated infection, they rise to very high titers and persist through the fatal terminus of generalized disease. For this reason a persistent or gradually rising CF titer combined with clinical symptoms indicates present or imminent dissemination.

The CF test is based on the quantitative Kolmer complement fixation technique for syphilis. Serial dilutions of sera are made to 1:512. Constant amounts of the other reagents are added. After proper incubation the test is read by noting the presence or absence of hemolysis. Details are given in the Appendix. A single positive complement fixation test may indicate either early active primary disease, a residual disease, such as coccidioidoma, or progressive disease. Therefore, serial samples are necessary for proper evaluation.

The CF test for coccidioidomycosis is quite specific, and cross-reactions with other mycotic infections are rare. In cases of acute or disseminated histoplasmosis, however, the test may be weakly positive. Though the coccidioidin CF test is not usually positive in blastomycosis, the blastomycin CF test may be positive in dilutions of serum of 1:128 or higher in cases of coccidioidomycosis. This antigen (blastomycin) is of absolutely no value in any serologic procedure for any disease.

Immunodiffusion and Latex Particle agglutination. The tube precipitin (TP) and complement fixation tests (CF) have proven to be very useful in the diagnosis and prognosis of coccidioidomycosis. Unfortunately, as routine screening procedures or in small diagnostic laboratories they have many drawbacks. Chief among these are time, the necessity for special equipment and technical skill, and expense. For this reason two new procedures have largely replaced the standard methods for initial screening. These are the immunodiffusion test (ID) and the latex particle agglutination test (LPA). Both have been developed and tested in large measure by Huppert and associates.[32,34,35] As a single test, the LPA test was positive in 70.6 per cent of cases of disease, compared to 13.5 per cent for the tube precipitin (TP) test. The former had a 3 per cent false-positive rate. Again, as single tests, the CF and immunodiffusion (ID) tests were of equal sensitivity, being positive in about 80 per cent of cases. However, as paired procedures, the LPA and ID detected 93 per cent of cases, compared to 84 per cent for TP with CF. Thus as an initial diagnostic screen, the combination of the two tests (LPA and ID) is superior on several counts. If these two are positive, a CF should be done for confirmation and setting of a base line for later serial examinations.

Using culture filtrates, toluene extracts of mycelium, and other procedures, Huppert et al.[32,34] have found several antigenic types and combinations in different strains of *C. immitis*. This raises the following interesting possibility. What we think of as the limited distribution of the fungus is in large part the result of skin testing with coccidioidin prepared from strains found in essentially only one geographic area. The finding by Huppert et al. of strains with no antigens in common to other strains suggests that, in diverse areas such as Russia and tropical America, infections by strains of the fungus may exist that would not be detected by the standard serologic procedures using our presently available reagents. Skin testing in such regions should be done using antigens prepared from indigenous strains. This situation has recently been established for histoplasmosis, when it was found that great antigenic differences exist among isolants from diverse geographic areas. At this writing four important antigens have been described from strains of *C. immitis*. They are detectable as independent lines on a double diffusion ID plate. Different strains may contain all, none, or any combination of these antigens. The antigen 1 of Huppert et al.[32,34] is involved in the tube precipitin test and the latex particle test. It is prepared from toluene lysates of mycelium. Many strains lack this antigen. Antigen 2 is the specific antigen involved in the complement fixation (CF) and immunodiffusion tests (ID) and is the standard antigen for these tests. Two other antigens have also been noted in ID tests by these investigators. The identity and significance of these other lines have not been determined.

The ID test becomes positive at about the same time as the CF test (Fig. 18–18). As far as has been determined, the duration of antibodies detectable by the two tests is roughly parallel. The ID test for coccidioidomycosis is performed, as are other double diffusion tests. Serum from the patient is placed into one well in an agar plate, and antigen is placed in an adjacent well. They are allowed to diffuse toward each other. The plates are read at 24 hours and each day for five days. Lines appear which represent specific antigen-antibody combinations. Quantitation of the test can be done using serum dilutions. The prognostic significance of the appearance and disappearance of particular lines has yet to be determined. Multiple lines are usually seen in active disease, whereas a single line is frequently associated with stable chronic infection. Spherules and arthrospores also contain a number of different antigens,[40,40a,41] but their diagnostic significance has not yet been delineated.

The latex particle agglutination test[35] is performed by placing two drops of inactivated serum and one drop of sensitized latex particles in a small tube. This is rotated at 180 rpm for four minutes. The tubes are read for the presence of agglutinated par-

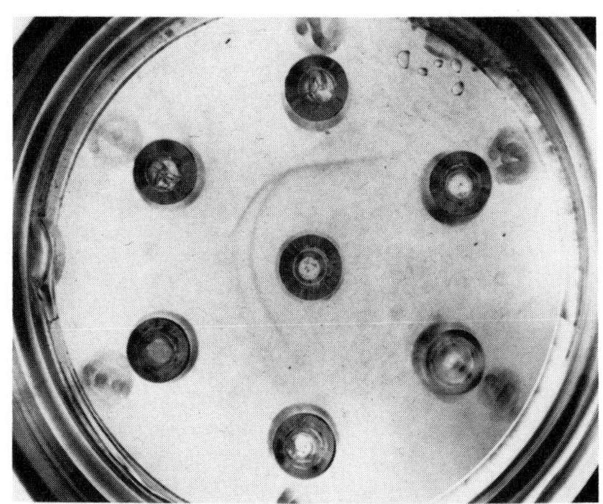

Figure 18–18. Coccidioidomycosis. Serology. Immunodiffusion plate. The center well contains antigen, and three wells contain sera from patients with different types of the disease. The wells with two precipitin lines are from cases of active disease. The serum in the well showing one line is from a patient with a "coin lesion."

ticles. The latex particles are sensitized by being coated with coccidioidin antigen.

Specific fluorescent antibody techniques have been developed for detection of spherules in tissue and arthrospores in soil. In addition, FA inhibition tests are sometimes positive in sera that is anticomplementary or that is negative by the tube precipitin test.[37]

The detection of antigen in serum would probably be the most specific diagnostic and prognostic test for the disease. So far the development of this procedure has not been completed.

Summary of Practical Serologic Procedures. For initial diagnosis and screening of sera, the LPA and ID are positive in 90 per cent of cases with clinical symptoms. The complement fixation (CF) and tube precipitin (TP) tests are used for diagnosis as well as for confirmation. Serial studies of the CF titer are used in prognostic evaluation. The skin test is widely used for epidemiologic studies; a single test is of value in diagnosis only in patients with a history of a previous negative reaction. A negative skin test in a patient that previously had a positive test indicates anergy associated with advanced disseminated disease. The tube precipitin test, when positive, indicates early active disease.

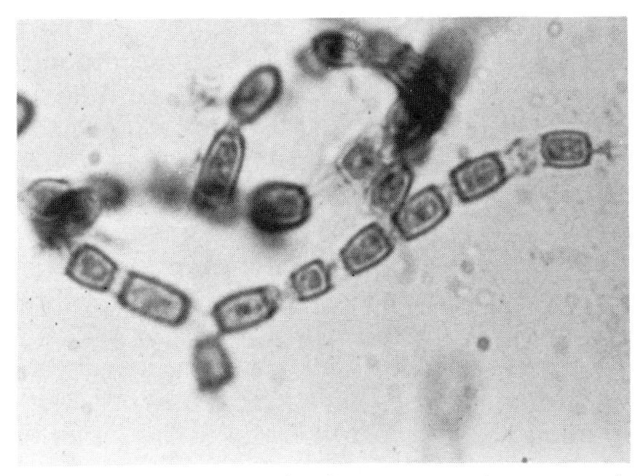

Figure 18–19. Coccidioidomycosis. Arthrospores alternating with empty cells in mycelial form of *C. immitis.* ×450.

LABORATORY IDENTIFICATION

Direct Examination. Because of the hazards involved in the handling of cultures of *C. immitis,* direct examination of clinical materials and histopathologic studies are of great importance in the diagnosis of this disease. Mature spherules of the organism are large and, when present, are easily seen on direct microscopic examination. Young spherules and newly released endospores are difficult to differentiate from host cells and other artifacts, however. Sputum, pus, gastric washings, and exudates from cutaneous lesions can be examined using a 10 per cent potassium hydroxide mount and subdued lighting. The spherules containing endospores vary in size from 10 to 80 μ, with an average diameter of 30 to 60 μ (in experimentally infected mice, they may reach 200 μ). The wall of the spherule is doubly refractile and up to 2 μ in thickness (Fig. 18–20). The endospores have an average diameter of 2 to 5 microns. Their size and numbers are quite variable. Imma-

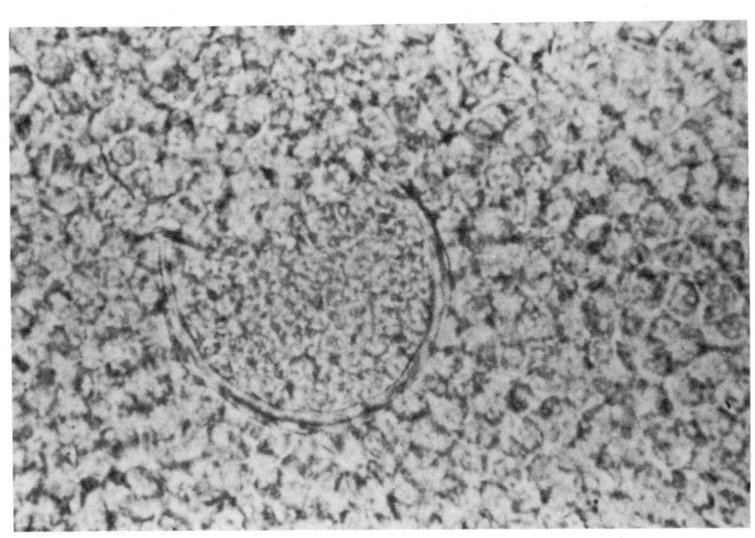

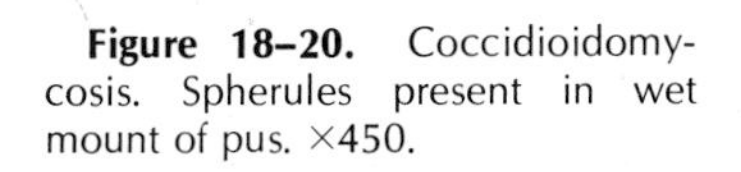

Figure 18–20. Coccidioidomycosis. Spherules present in wet mount of pus. ×450.

ture sporangia and endospores are also variable in size, and they resemble nonbudding yeast cells of *Blastomyces dermatitidis;* therefore, a careful search should be made for spherules containing endospores. Once they are seen, this is sufficient corroborative evidence for the diagnosis of coccidioidomycosis. In sputum or other samples held for periods of time, the spherules may form sprout mycelium and germ tubes.

Culture Methods. In cases in which the organism is not demonstrable in direct examination, or to confirm the identity of the organism, culture of the material may be performed. The fungus grows on almost all routine laboratory media with or without antibiotics. Because of the hazards inherent in culture work, the use of Petri dishes is not advised without special precautions. Material to be examined can be spread on slants or bottles of Sabouraud's agar and incubated at room temperature. Growth usually occurs by the third to fourth day, and sporulation by the tenth to 14th day. If subcultures are to be done, they should be made in a gloved isolation hood. When the culture is ready for examination, formaldehyde may be poured over the slants. Teased mounts of the mycelium are made with lactophenol cotton blue. Identification depends on finding the thin-walled rectangular, ellipsoidal, or barrel-shaped arthrospores that are 2.5 to 4 by 3 to 6 μ in size. Confirmation of identification, however, requires the production of spherules in experimentally infected animals.

MYCOLOGY

Coccidioides immitis Stiles in Rixford and Gilchrist 1896[66]

Synonymy. *Posadasia esferiformis* Cantón 1898; *Blastomycoides immitis* Castellani 1928; *Pseudococcidioides mazzai* da Fonseca 1928; *Geotrichum immite* Agostini 1932; *Coccidioides esferiformis* Moore 1932; *Glenospora metaeuropea* Castellani 1933; *Glenospora louisianoideum* Castellani 1933; *Trichosporon proteolyticum* Negroni and de Villafane Lastra 1938.

C. immitis is a dimorphic fungus. It grows as a mold with a septate mycelium in soil and culture media, and as an endosporulating spherule in animal tissue and under certain *in vitro* growth conditions. Unlike most of the other pathogenic fungi, this dimorphism is not controlled by temperature, and the organism does not have a budding, yeastlike form.

Colony Morphology. SDA, 25°C. Growth of *C. immitis* is apparent within three to four days after inoculation. At first, it is moist, glabrous, and greyish, but the colony rapidly develops abundant, floccose, aerial mycelium that soon covers the slant (Fig. 18–21). The mycelium is initially white, but usually becomes tan to brown with age.

As demonstrated by Huppert et al.,[33] there is considerable variation in the colonial morphology of *C. immitis* (Fig. 18–22 FS C 17). Some of the strains studied by them were so different in their morphology that they would not have been correctly identified by routine laboratory procedures. A few strains produced colonies that had no aerial mycelium, others had radial grooves, and in others the color varied from grey to pale lavender, pink, buff, cinnamon, yellow, and brown. Some strains produced a diffusible brown pigment. All of the isolants, however, regardless of their morphologic variation, were pathogenic for mice and produced spherules in tissue.

Microscopic Morphology. Sporulation begins within a few days after the initiation of growth. The spores usually appear first on side branches of the vegetative hyphae. The hyphae themselves are thin and septate, but

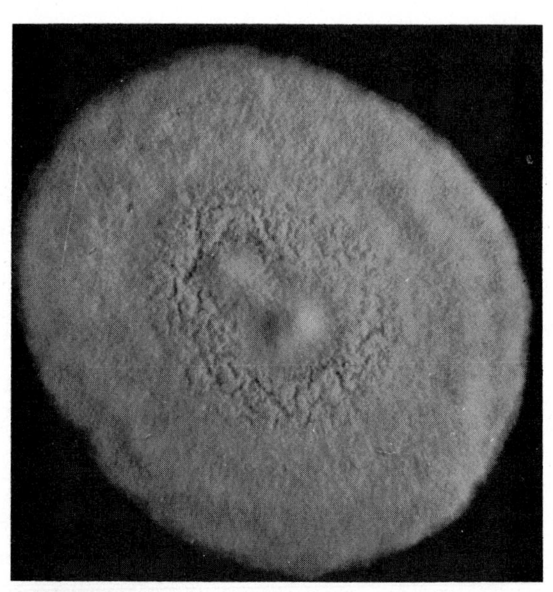

Figure 18–21. Coccidioidomycosis. Typical colonial morphology of *C. immitis*. The cracked, powdery appearance of the center indicates fragmentation to form arthrospores.

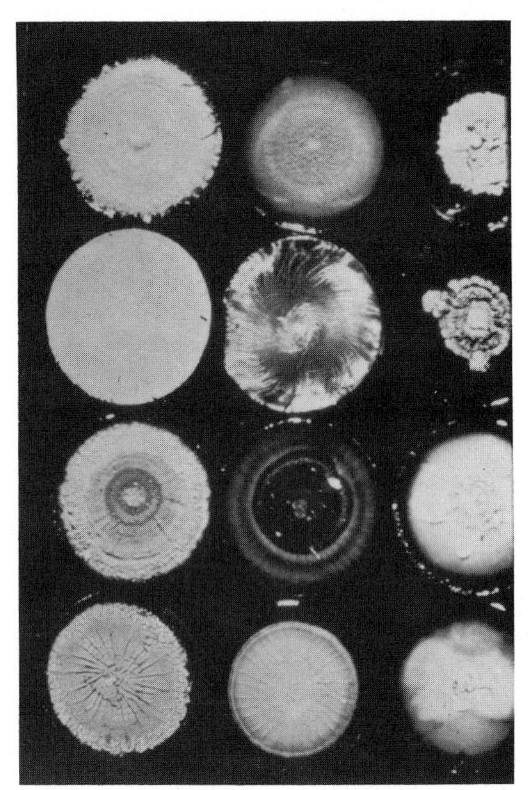

Figure 18–22. FS C 17. Coccidioidomycosis. Variation in morphology, texture, and color in isolants of *Coccidioides immitis*. [From Huppert M., S. H. Sun, et al. Copyright 1967. Natural variability in *Coccidioides immitis*, In *Coccidioidomycosis*. Proceedings of Second Coccidioidomycosis Symposium. Ajello, L. (ed.), Tucson, University of Arizona Press, p. 323, by permission.]

the side branches are almost twice as thick and have numerous septations. Thick-walled arthrospores are then produced, which alternate with thin-walled empty cells (disjunctors). The arthrospores are barrel-shaped, 2.5 to 4 × 3 to 6 μ in size, and are released by fragmentation of the mycelium. They retain portions of the walls of the disjunctor cells as ornaments on the walls of either end. This characteristic is often helpful in distinguishing the species. The cytology and nuclear cycle of arthrospore formation have been carefully detailed by Kwon-Chung.[39] As the culture ages, the vegetative hyphae also fragment into arthrospores. No other type of spore formation is seen in culture of this fungus, but racquet mycelium may be found.

Huppert et al. also found a wide variation in sporulation. This variation occurred within the same strain or was constant within a strain. A few straight or flexuose cells ranged in size from 1.5 to 2.3 by 1.5 to 15 μ. More commonly, the size range was 3 to 4.5 by 3 to 12 μ. Some spores were thin-walled and elongated, and some had rounded ends. In some, spore-bearing hyphae occurred in clusters and branched at acute angles from vegetative hyphae, and in others, the sporulating hyphae tended to lie in parallel rows.

A type of dimorphism with the production of culture spherules is produced *in vitro* using particular media and increased CO_2 tension (see Appendix) (Fig. 18–23). Whether or not these spherules are comparable to the ones found in lesion material is debatable.[46] Details of the development of the spherules in tissue are discussed in the section on histopathology.

The taxonomic relationship of *C. immitis* to other fungi is one of the most baffling in medical mycology. The organism was at first thought to be a protozoan. When the fungal nature of the organism was discovered, its spherules were considered to be asci, and it was classed as an Ascomycetes.[56] However, the number of endospores produced was irregular, and they were formed by progressive cleavage—characteristics not observed in fungi of the class Ascomycetes. In the 1920's Castellani grouped *C. immitis* with the Hyphomycetes or Fung. Imperfecti, and many mycologists agreed with that classification at the time. Endosporulation by progressive cleavage is a characteristic of the zygomycetous fungi in their formation of sporangia and sporangiospores. Thus in

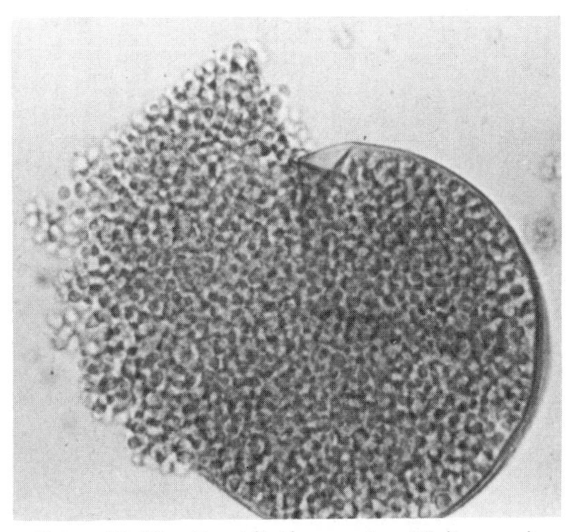

Figure 18–23. Coccidioidomycosis. "Culture spherule." ×400.

1926 Ciferri and Redelli suggested that *C. immitis* was a Zygomycetes, but the problem of its taxonomy remains unsettled. The soil or saprophytic form of the organism, with its septate mycelium and production of alternately spaced arthrospores, is similar to several other soil fungi,[20] including haploid Basidiomycetes and Hyphomycetes. However, it most closely resembles the imperfect stage of some of the Gymnoascaceae, which include members of the genus *Auxarthron* and the group designated "gymnoascus imperfecti." *C. immitis* has been reported to be keratinophilic, and in soil it is frequently found in association with other keratinophilic species of the genera *Arthroderma, Trichophyton,* and *Microsporum*.[58] None of these organisms has as yet been shown to form an endosporulating spherule. The only other pathogenic fungus demonstrating this characteristic in tissue is *Rhinosporidium seeberi,* but the two organisms do not appear to be closely related. The Zygomycetes in general do not have septate mycelium, do not form arthrospores, and contain some cellulose in their cell wall. Thus, we have a fungus (*C. immitis*) that has characteristics of a Zygomycetes in its parasitic stage and is gymnoascaceous in its saprophytic stage. The problem will be fully settled only when the perfect stage of *C. immitis* is found. If, in fact, *C. immitis* were found to be in the Gymnoascaceae, then essentially all the true pathogenic fungi, both dermatophytes and systemic infecting organisms, would be classed in that small family of the Ascomycetes.

REFERENCES

1. Ajello, L., K. Maddy, et al. 1965. Recovery of *Coccidioides immitis* from the air. Sabouraudia, *4*:92–95.
2. Ajello, L. (ed.) 1967. *Coccidioidomycosis.* Proceedings of Second Coccidioidomycosis Symposium. Tucson, University of Arizona Press.
3. Ajello, L. 1971. Coccidioidomycosis and histoplasmosis. A review of their epidemiology and geographical distribution. Mycopathologia, *45*:221–230.
4. Al-Doory, J. 1972. A bibliography of coccidioidomycosis. Mycopathologia, *46*:113–188.
4a. Block, E. R., and J. E. Bennett. 1973. Stability of amphotericin B in infusion bottles. Antimicrob. Agents Chemother., *4*:648–649.
4b. Borchardt, K., K. Litwack, et al. 1973. *In vitro* monitoring of amphotericin B therapy in disseminated coccidioidomycosis. Arch. Dermatol., *108*:119–120.
5. Burke, R. C. 1951. *In vitro* cultivation of the parasitic phase of *Coccidioides immitis.* Proc. Soc. Exp. Biol. Med., *76*:332–335.
6. Campbell, C. C., M. Reca, et al. 1967. Reactions of *Histoplasma capsulatum* antigens in sera from mammalian species including man infected with *Coccidioides immitis.* Ref. 2, pp. 243–248.
7. Campins, H. 1967. Coccidioidomycosis in Venezuela. Ref. 2, pp. 279–286.
8. Campins, H., M. Sharyj, et al. 1949. Coccidioidomicosis (Enfermedad de Posadas). Su comprobacion en Venezuela. Arch. Venez. Pat. Trop., *1*:215–234.
9. Case Records of the Massachusetts General Hospital. 1971. Coccidioidal meningitis. New Engl. J. Med., *11*:621–630.
10. Caudill, R. G. 1970. Coccidioidal meningitis. Am. J. Med., *49*:360–365.
11. Converse, J. 1965. The effect of nonviable and viable vaccines in experimental coccidiomycosis. J. Bacteriol., *74*:106–107.
12. Converse, J. L., and R. E. Reed. 1966. Experimental epidemiology of coccidioidomycosis. Bacteriol. Rev., *30*:679–694.
13. DeMartini, J. C., and W. E. Riddle. 1969. Disseminated coccidioidomycosis in two horses and a pony. J. Am. Vet. Med. Assoc., *155*: 149–156.
13a. Deppisch, L. M., and E. M. Donowho. 1972. Pulmonary coccidioidomycosis. Am. J. Clin. Pathol., *58*:489–500.
14. Dickson, E. C. 1937. "Valley fever" of the San Joaquin Valley and fungus Coccidioides. California West. Med., *47*:151–155.
15. Dickson, E. C., and M. A. Gifford. 1938. Coccidioides infection (coccidioidomycosis); the primary type of infection. Arch. Intern. Med., *62*:853–871.
16. Edwards, L. B., and C. Palmer. 1957. Prevalence of sensitivity to coccidioidin, with special reference to specific and nonspecific reactions to coccidioidin and histoplasmin. Dis. Chest, *31*:35–60.
17. Egeberg, R. O., A. E. Elconin, et al. 1964. Effect of salinity and temperature on *Coccidioides immitis* and three antagonistic soil saprophytes. J. Bacteriol., *88*:473–476.
18. Elconin, A. F., M. C. Egeberg, et al. A fungicide effective against *Coccidioides immitis* in the soil. Ref. 2, pp. 319–322.
19. Emmons, C. W. 1942. Isolation of Coccidioides from soil and rodents. Public Health Rep., *57*:109–111.
20. Emmons, C. W. 1967. Fungi which resemble *Coccidioides immitis.* Ref. 2, pp. 333–338.
21. Farness, O. J. 1940. Coccidioidal infection in a dog. J. Am. Vet. Med. Assoc., *97*:263–264.
22. Fiese, M. J. 1958. *Coccidioidomycosis.* Springfield, Ill., Charles C Thomas.
23. Fosburg, R. G., B. F. Baisch, et al. 1969. Limited pulmonary resection for coccidioidomycosis. Ann. Thoracic Surg., *7*:420–427.
24. Friedman, L., C. E. Smith, et al. 1962. Studies of the survival characteristics of the parasitic phase of *Coccidioides immitis* with comments on contagion. Ann. Rev. Resp. Dis., *85*:224–231.
25. Friedman, L., W. G. Roessler, et al. 1956. The virulence and infectivity of twenty-seven strains of *Coccidioides immitis.* Am. J. Hyg., *64*:198–210.
26. Gale, D., E. A. Lockhart, et al. 1967. Studies of

Coccidioides immitis. I. Virulence factors of *C. immitis.* Sabouraudia *6*:29–36.

27. Giltner, L. T. 1918. Occurrence of coccidioidal granuloma (oidiomycosis) in cattle. J. Agricult. Res., *14*:533–542.
28. González Ochoa, A. 1967. Coccidioidomycosis in Mexico. Ref. 2, pp. 293–300.
28a. Guy, W. H., and F. M. Jacobs. 1927. Granuloma coccidioides. Arch. Dermatol., *16*:308–311.
29. Hagele, A. J., D. J. Evans, et al. 1967. Primary endophthalmic coccidioidomycosis: Report of a case of exogenous, primary coccidioidomycosis of the eye diagnosed prior to enucleation. Ref. 2, pp. 37–40.
29a. Henrickson, R. V., and E. L. Biberstein. 1971. Coccidioidomycosis accompanying hepatic disease in two Bengal tigers. J. Am. Vet. Med. Assoc., *161*:674–677.
30. Huntington, R. W. 1971. Coccidioidomycosis, In *Handbuch der speziellen pathologischen anatomie and histologie. Dritter Band/Funfter teil.* R. D. Baker (ed.), Berlin, Springer-Verlag, pp. 147–210.
31. Huntington, R. W., W. J. Waldmann, et al. 1967. Pathologic and clinical observations on 142 cases of fatal coccidioidomycosis with necropsy. Ref. 2, pp. 143–168.
32. Huppert, M. 1970. Serology of coccidioidomycosis. Mycopathologia, *41*:107–113.
33. Huppert, M., S. H. Sun, et al. 1967. Natural variation in *Coccidioides immitis.* Ref. 2, pp. 323–330.
34. Huppert, M., J. W. Bailey, et al. 1967. Immunodiffusion as a substitute for complement fixation and tube precipitin tests in coccidioidomycosis. Ref. 2, pp. 221–226.
35. Huppert, M., E. T. Peterson, et al. 1968. Evaluation of a latex particle agglutination test for coccidioidomycosis. Am. J. Clin. Pathol., *49*:96–102.
36. Ilahi, A., H. Afzal, et al. 1966. Coccidioidomycosis in buffaloes in West Pakistan. Pakist. J. Anim. Sci., *3*:14–20.
37. Kaplan, W. 1967. Application of the fluorescent antibody technique to the diagnosis and study of coccidioidomycosis. Ref. 2, pp. 227–231.
38. Kruse, R. H., T. D. Green, et al. 1967. Infection of control monkeys with *Coccidioides immitis* by caging with inoculated monkey. Ref. 2, pp. 387–396.
39. Kwon-Chung, K. J. 1969. *Coccidioides immitis:* Cytological study on the formation of the arthrospores. Can. J. Genet. Cytol., *9*:43–53.
40. Landay, M. E., R. W. Wheat, et al. 1967. Serological comparison of the three morphological phases of *Coccidioides immitis* by agar gel diffusion method. J. Bacteriol., *93*:1–6.
40a. Landay, M. E. 1973. Spherules in the serology of *Coccidioides immitis.* II. Complement fixation tests with human sera. Mycopathologia, *49*:45–52.
41. Landay, M. E., R. W. Wheat, et al. 1967. Studies on the comparative serology of spherule and arthropsore growth forms of *C. immitis.* Ref. 2, pp. 233–236.
42. Levine, H. B. 1970. Development of vaccines for coccidioidomycosis. Mycopathologia, *41*:177–185.
43. Levine, H. B., and C. E. Smith. 1967. The reactions of eight volunteers injected with *Coccidioides immitis* spherule vaccine; first human trials. Ref. 2, pp. 197–200.
44. Levine, H. B., and G. M. Scalarone. 1971. Deficient resistance to *C. immitis* following intravenous vaccination. III. Humoral and cellular responses to intravenous and intramuscular doses. Sabouraudia, *9*:97–108.
45. Levine, H. B., A. González Ochoa, et al. 1973. Dermal sensitivity to *Coccidioides immitis:* A comparison of response elicited by spherulin and coccidioidin. Am. Rev. Resp. Dis., *107*:379–386.
46. Lones, C. 1967. Studies of the intermediary metabolism of *Coccidioides immitis.* Ref. 2, pp. 349–353.
47. Maddy, K. 1958. Disseminated coccidioidomycosis of the dog. J. Am. Vet. Med. Assoc., *132*:483–489.
48. Maddy, K. T. 1965. Observations on *Coccidioides immitis* found growing naturally in soil. Arizona Med., *22*:281–288.
49. Maddy, K. T., and H. G. Crecelius. 1967. Establishment of *Coccidioides immitis* in negative soils following burial of infected animals and animal tissues. Ref. 2, pp. 309–312.
50. Maddy, K. T., H. G. Crecelius, et al. 1960. Distribution of *Coccidioides immitis* determined by testing cattle. Public Health Rep., *75*:955–962.
51. McCullough, D. C., and J. C. Harbert. 1969. Isotope demonstration of CSF pathways. Guide to antifungal therapy in coccidioidal meningitis. J.A.M.A., *209*:558–560.
52. Mayorga, R. 1967. Coccidioidomycosis in Central America. Ref. 2, pp. 287–292.
53. Mazza, S., and S. Parodi. 1929. Micosis chaquena producida por el *Pseudococcidioides mazzai.* Prensa Med. Argent., *16*:268–272.
54. Negroni, P. 1967. Coccidioidomycosis in Argentina. Ref. 2, pp. 273–278.
54a. Oduye, O. O. 1972. Canine disseminated coccidioidomycosis in Nigeria. Bull. Epizootic Dis. Africa, *20*:113–120.
55. Omieczynski, D. J., and F. E. Swatek. 1967. The comparison of two methods for direct isolation of *Coccidioides immitis* from soil using three different media. Ref. 2, pp. 265–272.
56. Ophüls, W. 1905. Further observations on a pathogenic mould formerly described as a protozoan (*Coccidioides immitis,, Coccidioides pyogenes*). J. Exp. Med., *6*:443–485.
57. Ophüls, W., and H. C. Moffitt. 1900. A new pathogenic mould (formerly described as a protozoan: *Coccidioides immitis*). Preliminary report. Phila. Med. J., *5*:1471–1472.
58. Orr, G. F. 1968. Some fungi isolated with *Coccidioides immitis* from soils of endemic areas in California. Bull. Torrey Bot. Club, *95*:424–431.
59. Pappagianis, D., C. E. Smith, et al. 1967. Serologic status after positive coccidioidin skin reactions. Am. Rev. Resp. Dis., *96*:520–523.
60. Porter, J. F., E. R. Scheer, et al. 1971. Characterization of 3-O-methyl mannose from *Coccidioides immitis.* Infect. Immun., *4*:660–661.
61. Posadas, A. 1892. Un neuvo caso de micosis fungoidea con psorospermias. Circulo Med. Argentino, *15*:585–597.
62. Posadas, A. 1900. Psorospermiose infectante generalise. Rev. Chir. (Paris), *21*:277–282.
63. Puckett, T. F. 1954. Hyphae of *Coccidioides immitis* in tissues of the human host. Am. Rev. Tuberc., *70*:320–327.

64. Rapaport, F. T., H. S. Lawrence, et al. 1960. Transfer of delayed hypersensitivity to coccidioidin in man. J. Immunol., *84*:358–367.
65. Rhodes, E. R., J. J. McPhaul, et al. 1967. The interpretation of renal changes associated with administration of amphotericin B.: Importance of pre-treatment studies, Ref. 2, pp. 137–140.
66. Rixford, E., and T. C. Gilchrist. 1896. Two cases of protozoan (coccidioidal) infection of the skin and other organs. Johns Hopkins Hosp. Rep., *1*:209–269.
67. Roberts, P. L., J. H. Knepshield, et al. 1968. Coccidioides as an opportunist. Arch. Intern. Med., *121*:568–570.
68. Roberts, J., J. M. Counts, et al. 1970. Production *in vitro* of *Coccidioides immitis* as a diagnostic aid. Am. Rev. Resp. Dis., *102*:811–813.
69. Robledo, V. M., A. Restrepo M., et al. 1968. Encuesta epidemiologica sobre coccidioidomicosis en algunas zonas aridas de Colombia. Antioquia Med., *18*:503–522.
70. Salkin, D. 1967. Clinical examples of reinfection coccidioidomycosis. Am. Rev. Resp. Dis., *95*:603–611.
71. Sawaki, Y., M. Huppert, et al. 1966. Patterns of human antibody reactions in coccidioidomycosis. J. Bacteriol., *91*:422–427.
72. Scalarone, G. M., and H. B. Levine. 1969. Attributes of deficient immunity in mice receiving *Coccidioides immitis* spherule vaccine by the intravenous route. Sabouraudia, *7*:169–177.
73. Schmidt, R. J., and D. H. Howard. 1968. Possibility of *C. immitis* infection in museum personnel. Public Health Rep., *83*:882–888.
74. Sinski, J. T., E. P. Lowe, et al. 1965. Immunization against experimental lethal Simian coccidioidomycosis using whole killed arthrospores and cell fractions. Mycopathologia, *57*:431–441.
75. Sippel, J. E., and H. B. Levine. 1969. Annulment of amphotericin B inhibition of *Coccidioides immitis* endospores: Effects on growth respiration and morphogenesis. Sabouraudia, 7: 159–168.
76. Smale, L. E., and K. G. Waechter. 1970. Disseminated coccidioidomycosis in pregnancy. Am. J. Obstet. Gynecol., *107*:356–361.
77. Smith, C. E. 1940. Epidemiology of acute coccidioidomycosis with erythema nodosum. Am. J. Public Health, *30*:600–611.
78. Smith, C. E., R. R. Beard, et al. 1946. Effect of season and dust control on coccidioidomycosis. J.A.M.A., *132*:833–838.
79. Smith, C. E., M. T. Saito, et al. 1956. Pattern of 39,500 serologic tests in coccidioidomycosis. J.A.M.A., *160*:546–552.
80. Sorensen, R. H. 1967. Survival characteristics of diphasic *Coccidioides immitis* exposed to the rigors of a simulated natural environment. Ref. 2, pp. 313–318.
81. Stepanishtcheva, Z. G., A. M. Arievitch, et al. 1972. Investigation of deep mycoses in the USSR. Int. J. Dermatol., *11*:181–183.
82. Stewart, R. A., and K. F. Meyer. 1932. Isolation of *Coccidioides immitis* (Stiles) from soil. Proc. Soc. Exp. Biol. Med., *29*:937–938.
83. Swatek, F. E. 1970. Ecology of *Coccidioides immitis*. Mycopathologia, *40*:3–12.
84. Swatek, F. E., D. T. Omieczinski, et al. 1967. *Coccidioides immitis* in California. Ref. 2, pp. 255–264.
85. Symmers, W. S. C. 1971. An Australian case of coccidioidomycosis. Pathology, *3*:1–8.
86. Trimble, J. R., and J. Doucette. 1956. Primary cutaneous coccidioidomycosis. Report of a case of laboratory infection. Arch. Dermatol., *74*:405–410.
87. Utz, J. P. 1967. Recent experience in the chemotherapy of the systemic mycoses. Ref. 2, pp. 113–118.
88. Wallraff, E. B., and P. R. O'Bar. 1969. Evaluation of a coccidioidin tine test as a skin testing technique. Am. Rev. Resp. Dis., *99*:943–945.
89. Werner, S. B., D. Papagianis, et al. 1972. An epidemic of coccidioidomycosis among archeology students in northern California. New Engl. J. Med., *286*:507–512.
90. Wernicke, R. 1892. Ueber einen Protozoenbefund bei mycosis fungoides (?). Zentralbl. Bakteriol., *12*:859–861.
91. Wilson, J. W., and W. R. Bartok. 1970. Immunoallergic aspects of coccidioidomycosis. Int. J. Dermatol., *9*:253–358.
92. Wilson, J. W., C. E. Smith, et al. 1953. Primary cutaneous coccidioidomycosis: the criteria for diagnosis. California Med., *79*:233–239.
93. Winn, W. A. 1962. The diagnosis and treatment of coccidioidomycosis. Arizona Med., *19*:211–217.
94. Winn, W. A. 1964. The treatment of coccidioidal meningitis. The use of amphotericin B in a group of 25 patients. California Med., *101*: 78–89.
95. Winn, W. A. 1967. A working classification of coccidioidomycosis and its application to therapy. Ref. 2, pp. 3–10.
96. Winn, W. A. 1967. Coccidioidal meningitis, a follow-up report. Ref. 2, pp. 55–62.
97. Wright, E. T., and L. H. Winer. 1971. The natural history of experimental *Coccidioides immitis* infection. Int. J. Dermatol., *10*:17–23.
98. Zealear, D. S., and W. Winn. 1967. The neurosurgical approach in the treatment of coccidioidal meningitis: Report of ten cases. Ref. 2, pp. 43–54.

Chapter 19

PARACOCCIDIOIDOMYCOSIS

DEFINITION

Paracoccidioidomycosis is a chronic granulomatous disease which characteristically produces a primary pulmonary infection, often inapparent, and then disseminates to form ulcerative granulomas of the buccal, nasal, and occasionally the gastrointestinal mucosa. The lymph nodes are very commonly involved, and sometimes there is extension to cutaneous tissue or systemic involvement of multiple organ systems. The disease in its inception and development is similar to blastomycosis and coccidioidomycosis. The infection is geographically restricted to areas of South and Central America. The only etiologic agent is *Paracoccidioides brasiliensis.* The name of the disease, paracoccidioidomycosis, is particularly unwieldy. Because of its historical development, this term means a fungous disease, like a fungous disease, like a protozoan disease.

Synonymy. South American blastomycosis, Brazilian blastomycosis, Lutz-Splendore-Almeida's disease, paracoccidioidal granuloma.

HISTORY

Adolfo Lutz in 1908[29] described two patients who had extensive granulomatous lesions in the nasopharyngeal region and intense cervical adenopathy. In material from the lesions, he observed a fungus which reproduced in tissue by multiple budding but otherwise resembled the parasitic form of *Coccidioides immitis.* He called the disease "pseudococcidioidal granuloma," but did not name the organism. The Italian microbiologists Carini and Splendore published several studies of the organism between 1908 and 1912.[41,55] Splendore classified it as an ascomycetous yeast in the genus *Zymonema,* a group in which he also included *Candida albicans* and all dimorphic fungi.[55] In many reports that followed, the disease was confused with coccidioidomycosis and other mycoses. In the former disease, the fungus forms endospores in tissue, which are released by fracture of the spherule wall. It was thought that in reproduction by *P. brasiliensis* the endospores passed through pores in the cell wall to the outside and remained attached to the surface of the parent cell. This process was termed "cryptosporulation," based on misinterpretation of the budding process. The excellent work of Almeida published between 1927 and 1929 finally clarified the subject and separated the two diseases.[3] Negroni studied the organism and in 1921 demonstrated dimorphism *in vitro.* After reviewing the characteristics of the fungus *in vitro* and *in vivo,* Almeida and Lacaz in 1928 suggested the genus name *Paracoccidioides.* The similarity of sporulation of the mycelial form of this organism and that of *Blastomyces dermatitidis* led Conant and Howell to rename it *Blastomyces brasiliensis* in 1942.[15] Later studies have shown that the two fungi are distinct, and the original designation of Almeida and Lacaz is retained.

In Lutz's two cases and in the many cases reviewed and tabulated by Splendore,[55] the most evident lesions were the mucocutaneous ones about the mouth. The presence of lesions in the alimentary tract and apparent systemic dissemination from this area led to the long-held conclusion that the disease began by either the inoculation of the fungus into the oral mucosa or the ingestion of vegetable material containing the organism, followed by infection of the gastrointestinal tract. Only recently, because of the careful investigations of several authors, has it become apparent that the primary infection is pulmonary. These studies have been aided greatly by skin test surveys. Just as histoplasmosis was considered to be a rare disease until 1946, since only a few cases of the advanced fatal form were known, paracoccidioidomycosis was also thought to be rare. It now appears that subclinical and resolved infections are common in the endemic areas.

In 1914, Stein in Austria developed a skin test for blastomycosis. Patients with the Brazilian disease did not react consistently with this test, indicating that the diseases were indeed separate entities. Fonseca was the first to devise a specific skin test for patients with paracoccidioidomycosis. A specific polysaccharide developed by Fava-Netto in 1959 has been used more recently. This has demonstrated that subclinical infection is quite frequent, and that the disease is much more common than formerly appreciated.

Several excellent reviews have been published on the disease, particularly by the Brazilian group of physicians and mycologists. Notable among these are the reviews of Lacaz[26,27] and Fonseca.[19] Recently the first Pan American Symposium on Paracoccidioidomycosis was held, and the proceedings of this meeting summarize the present state of our understanding of the disease and its etiology.[41]

ETIOLOGY, ECOLOGY, AND DISTRIBUTION

The names of the disease and its etiologic agent give some indication of the confusion that has existed about these subjects. As mentioned above, the organism was initially thought to be a type of *Coccidioides* that exhibited cryptosporulation. The disease was also confused with lobomycosis. However, early in its history it was appreciated that almost all isolants were similar in morphology, and that probably only one agent was responsible for the specific disease entity, paracoccidioidomycosis. Unlike the etiologic agents of other fungal diseases, long lists of synonymous names were not forthcoming. The similarity of sporulation of the mycelial form and to some extent the budding of the yeast form have more recently been used to relate the agent of Lutz's disease to that of Gilchrist's disease. The inclusion of both species in the single genus *Blastomyces* had been proposed.[15] *B. dermatitidis* has a single, large, broad-based bud in the yeast phase, and *P. brasiliensis* has several small, narrow-necked buds. The work of Carbonell[10] has clearly demonstrated that the budding processes of the two species are fundamentally different. Therefore, the accepted designation for the etiologic agent of paracoccidioidomycosis is *Paracoccidioides brasiliensis.*

The recent skin test studies of large populations and the recovery of the organism from soil have begun to clarify the ecology and epidemiology of the disease. Chaves-Batista et al.[14] recovered an organism from soil which they considered to be the agent of paracoccidioidomycosis. This was later identified as *Aspergillus penicilloides.* Negroni claims to have isolated it once from soil in Argentina.[39] More recently, in a study in a highly endemic region in Venezuela, Albornoz[1] recovered the organism several times from the area of Paracotos. This area has a population with a high rate of skin test sensitivity (up to 53 per cent).[2] The region is a coffee growing area and is characterized by "humid mountain forests" (Holdridge plant classification zone). The median temperature is 23°C, the altitude 648 M, and the rainfall is 1400 mm annually. Restrepo et al.[51,53] have done retrospective studies on patients in Colombia. They found that at the time of diagnosis of the disease up to 89 per cent of patients had been born or lived in the subtropical selvatic areas. These forests are included within a few of the Holdridge Life Zones characterized by particular climatic and biological associations. The mean temperature is 18 to 23°C, there is a rainfall of 800 to 2000 mm annually and an elevation of 500 to 1800 M. Laboratory studies by Restrepo et al.[50] had indicated that the fungus survives longer in acid soil than in other soil types. Albornoz isolated the fungus from acidic surface soil, the "A horizon," an area of great biological activity and turnover. The organism has also been recovered from the intestinal contents of bats,[23] but there was no evidence of disease in the animals. Borelli has suggested that the organism is a commensal or pathogen of poikilothermic animals.[6] At present, the

Figure 19–1. Paracoccidioidomycosis. Approximate distribution of the disease in North, Central, and South America based on case reports. (After several authors, particularly the work of Dante Borelli.)

saprophytic habitat and ecologic associations of the fungus are not fully elucidated.

Over 5000 clinically diagnosed cases of paracoccidioidomycosis have been recorded in the literature (Fig. 19–1).[7,26] These represent only the serious, clinically apparent forms of the infection. The increased incidence of disease recorded in recent years probably reflects a greater degree of diagnostic accuracy. In Brazil, where over 3000 cases have been reported, the highly endemic area are in the states of São Paulo, Rio de Janeiro, Rio Grande do Sul, Mato Grosso, and Minas Gerais. The disease is rare in the tropical rain areas of Manaus, Amazonas, and in the dry northeastern states of Brazil.[33] About 600 cases have been reported in Colombia, 600 in Venezuela, 100 from Argentina, 100 from Peru, 50 from Ecuador, 50 from Uruguay, 30 from Paraguay, 20 from Central American countries, and 20 from Mexico.[7,21] It is estimated that 225 new, serious cases or 0.8 per 100,000 population occur annually. This does not include the subclinical or mild forms of the disease. Imported and long-latent cases have been recorded in the United States and Europe.[20,36a]

Using the paracoccidioidin skin test, Greer[22] studied the distribution of sensitivity in patients' families and in control groups in urban and rural environments. He found that in rural environments the number of reactors was the same as in families of patients and control families. The reactivity rate, especially in the controls, was less in urban environments. Although Albornoz and Albornoz[2] had found a higher rate of skin sensitivity in males, Greer[22] and Restrepo[52] found no difference in sex distribution. Although members of a family would appear to have equal risk of exposure, no epidemics of clinical disease were noted, even when more than one person was skin test–positive. These studies on skin test reactivity indicate that many subclinical infections go unrecognized.

The finding of 50:50 distribution between the sexes in skin test sensitivity contrasts sharply with the rate recorded for clinically significant disease. The sex distribution of cases in several series varies from 7:1 to 70:1, or from a general average of 90 per cent males to 10 per cent females. This is similar to the sex distribution noted in other systemic mycoses. Some of this unequal distribution of disease is attributable to exposure but some is conjectured to be related to hormonal differences. Prepuberal children of either sex rarely have clinically apparent paracoccidioidomycosis. Although the age range of patients is 10 to 70, the majority of clinically diagnosed cases is found in males between 20 and 50 years of age. In trying to associate the disease with other factors, it is noted that most patients are rural workers—many are tree cutters or work at similar occupations. In addition, many have other diseases, such as malaria, Chagas' disease, and tuberculosis, and essentially all are at least moderately malnourished. This suggests that physiologic, hormonal, and immunologic factors play a role in the appearance and distribution of the clinically severe form of the disease.

CLINICAL DISEASE

As noted earlier, the most commonly encountered form of clinical paracoccidioidomycosis is the presence of lesions on the oropharynx and gingivae. This, together with the less frequent occurrence of disease in the alimentary tract and anorectal region, gave the impression that primary infection occurred by implantation of the organism into the mucosa or by ingestion with food. It was suggested that cleaning the teeth with grass caused minor traumas which allowed invasion of the fungus. Recent careful observations by many investigators make it apparent that primary infection is pulmonary, with subsequent dissemination of the fungus to other regions. This is similar to the clinical course observed in essentially all other systemic mycoses. The disease falls into three clinical categories: pulmonary, mucocutaneous-lymphangitic, and disseminated.

Pulmonary Paracoccidioidomycosis

Benign pulmonary disease has only recently been recognized with regularity.[8] This is the inapparent subclinical infection that had previously been seen in Brazil in European immigrants residing in endemic areas.[30] Now a syndrome is developing consisting of minor lung changes and skin test reactivity in otherwise healthy individuals. The lung changes appear commonly as bilateral lesions. The initial reaction is an alveolitis, with an influx of macrophages and a few giant cells, which is sometimes followed by caseation in small areas. At this stage of the disease, the macrophage is the probable mode of dissemination of the fungus to other areas of the body. A pyogenic reaction with numerous neutrophils then occurs.[31] Interstitial changes accompany the intra-alveolar lesions. The lesions develop slowly over a period of time, losing their initial soft appearance and attaining the characteristics of interstitial fibrosis. A marked hilar adenopathy may be seen by x-ray examination (Fig. 19–2,*C*). In time, small calcifications may appear, but this finding is less common than in other mycoses or tuberculosis. While studying calcifications of the nodes of the hili pulmonis in routine autopsies, Angulo[4] found histologically identifiable yeast cells of *P. brasiliensis* in many patients who had died of other causes.

Progressive pulmonary disease is rare. Clinically it is accompanied by respiratory insufficiency, dyspnea, fever, rales, chest pain, and cough productive of sputum which is sometimes blood-tinged. All lobes of the lung may be affected, but the involvement of the lower lobes seems to occur more frequently than of the upper (Fig. 19–2,*A*). Infiltration, striation, and the presence of small nodules are seen on x-ray. Such areas

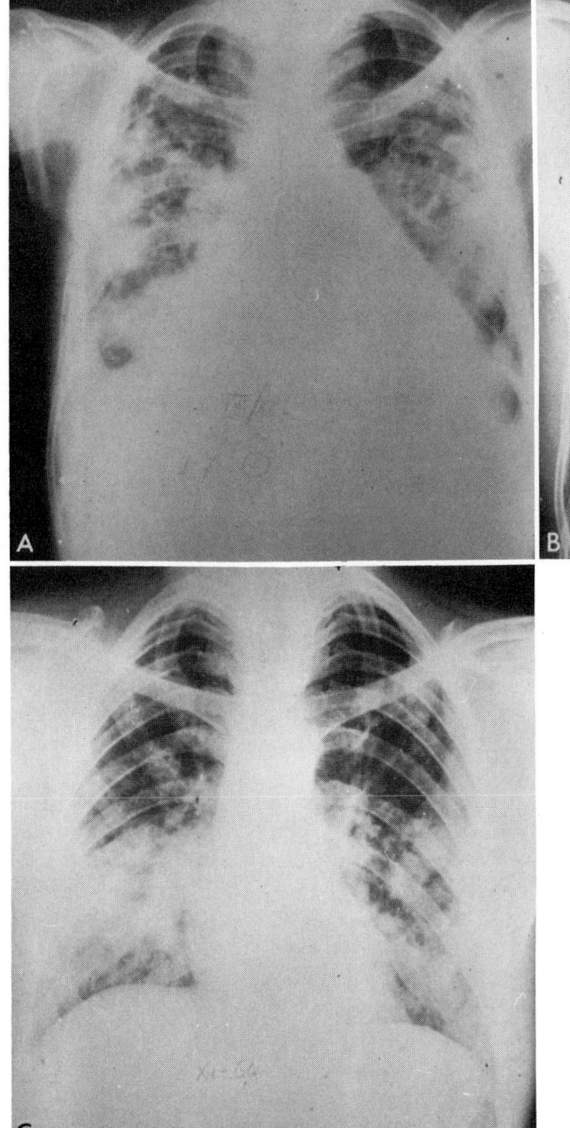

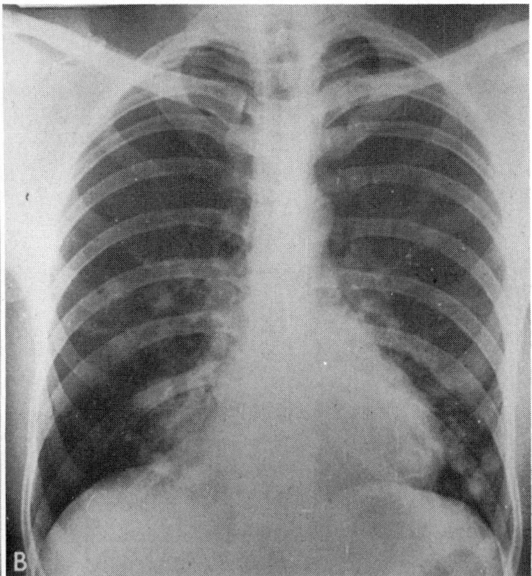

Figure 19–2. Radiologic aspects of paracoccidioidomycosis. *A,* Diffuse cavitary and alveolar disease in both lungs, with a right pleural effusion and prominent hilar nodes. There is massive enlargement of the cardiac silhouette. *B,* There are many sharply defined nodules in this old established disease; none appear to have cavitated. *C,* Diffuse, patchy, alveolar infiltrate in lower lobes of both lungs and prominent hilar adenopathy on right.

may become confluent. Chronic infections are associated with slight to moderate suppuration, but extensive fibrosis and deterioration of respiratory function occur, which may result in cor pulmonale. Cavities are not common and are seen in only about one-third of cases (Fig. 19–2,*B*).

Extensive mucocutaneous and lymphangitic disease may be present with only minor pulmonary changes. When the lungs are involved by secondary dissemination through hematogenous spread, numerous miliary granulomatous lesions are found throughout all areas.

Mucocutaneous-Lymphangitic Paracoccidioidomycosis

The primary pulmonary infection with *P. brasiliensis* is usually subclinical, and the secondary involvement of the oropharyngeal mucosa may be the most apparent presenting symptom. The lesions in the mouth begin as papules or vesicles which then ulcerate. This condition is referred to as "moriform stomatitis." In the personally observed cases, the ulcerative lesions had a rolled border, and a blanched white exudative base studded with small hemorrhagic dots (Fig. 19–3 FS

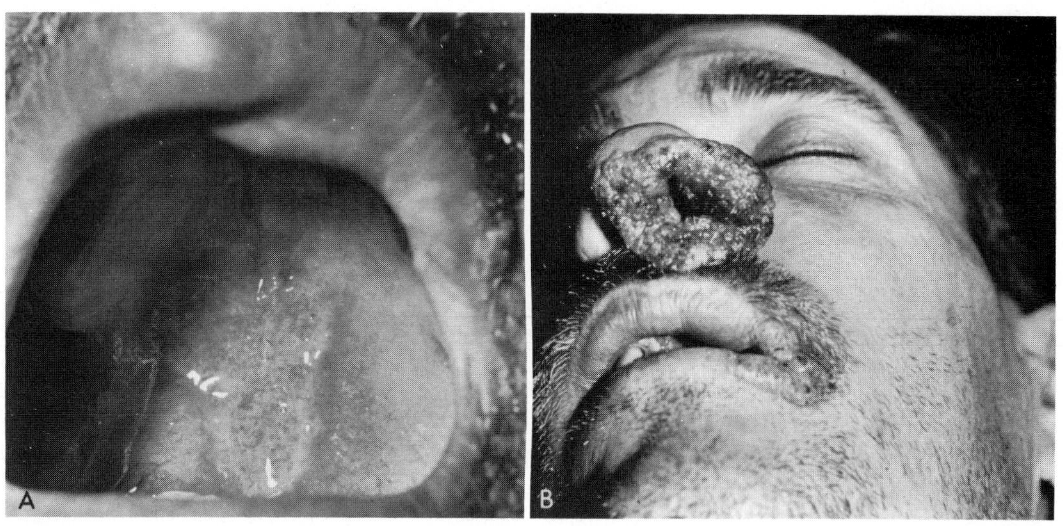

Figure 19–3. FS C 18. Mucocutaneous paracoccidioidomycosis. *A,* There is a papular lesion on the pharyngeal mucosa that has a blanched white exudative base, a rolled border, and small hemorrhagic dots in the center. *B,* Granulomatous lesions involving the nose. At this stage, the disease resembles secondary cutaneous leishmaniasis. (Courtesy of F. Pifano.)

C 18). Though they are initially painless, extensive lesions cause distress on ingestion and mastication, and are painful. This may lead to cachexia. The lesions have a granulomatous appearance and spread slowly to sometimes form extensive vegetations. In advanced cases, the involvement becomes deeper and may be accompanied by destruction of the epiglottis and uvula, perforation of the hard palate, and extension to the lip and adjacent cutaneous areas of the nose (Figs. 19–4 and 19–5). Involvement of the lip may cause a hard edema and is followed by destruction of the area, with extensive scarring and fibrosis (Fig. 19–6 FS C 19). Gingival involvement, which is sometimes

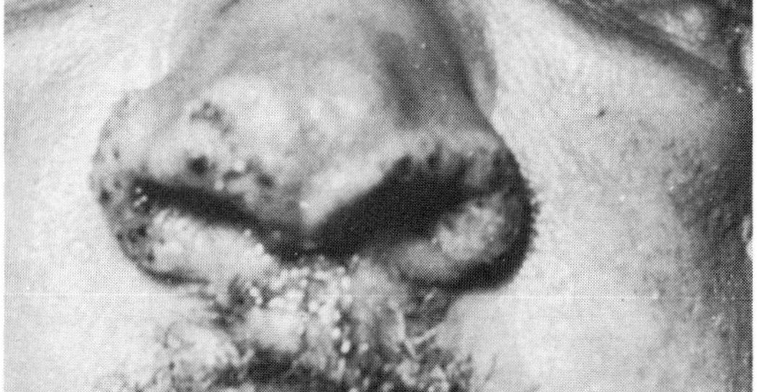

Figure 19–4. Mucocutaneous paracoccidioidomycosis. Chronic granulomatous lesions involving lips and nose.

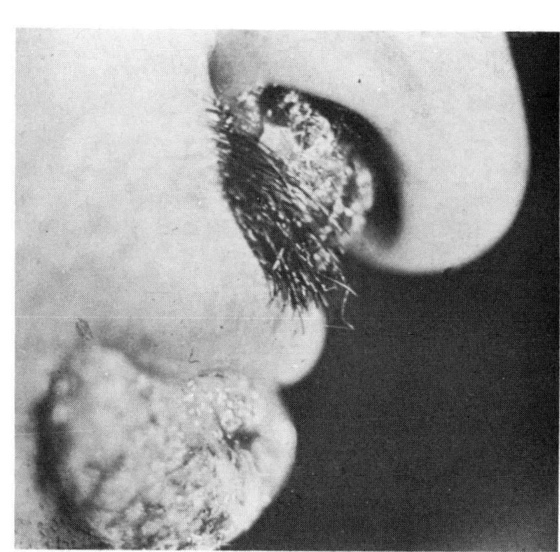

Figure 19–5. Mucocutaneous paracoccidioidomycosis. Granulomatous lesions of nose and lower lip. (Courtesy of A. Restrepo.)

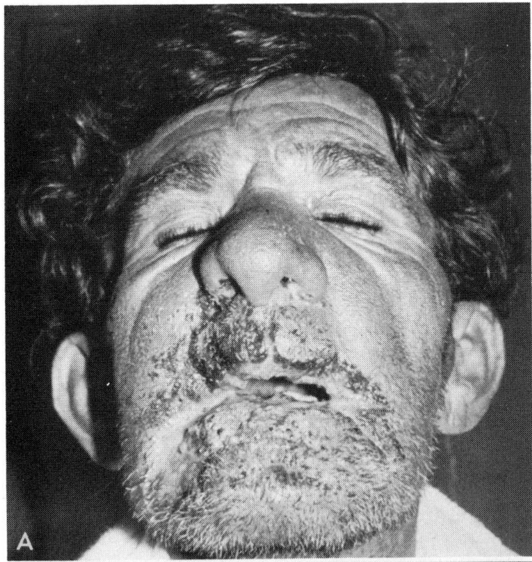

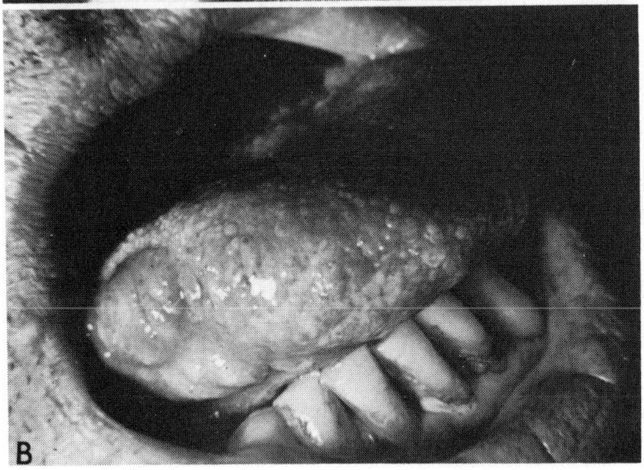

Figure 19–6. FS C 19. Mucocutaneous paracoccidioidomycosis. *A,* Extensive destruction of facial feature. *B,* Granulomatous lesions of the tongue. (Courtesy of F. Battistini.)

the presenting symptom, causes loss of teeth by spontaneous shedding. Lesions may appear to occur in waves, so that many stages of their development may be seen in the same patient. Involvement of the ocular mucosa is infrequent but has been recorded.

Regional lymph nodes that drain the oronasal areas are involved soon after the appearance of initial lesions. Lymphadenopathy is a common and characteristic feature of this disease (Fig. 19–7). The cervical nodes are usually involved first and are often adherent to the overlying skin. Frequently, the nodes suppurate and drain, thereby forming sinus tracts. The drainage material contains abundant fungi. Massive nodes in the neck may be an early sign of the disease. Visceral node involvement may be so extensive as to suggest malignant neoplasia.[34]

Cutaneous lesions of various types occur in this disease.[9] Usually they begin at a mucocutaneous border or by extension from mucosal lesions. They are typically ulcerative and crusted granulomas. As a result of hematogenous or lymphatic dissemination, satellite lesions occur and in advanced cases not only are grouped on the face and neck but also develop on any area of the body (Figs. 19–8 and 19–9). Chancriform lesions which are an indication of primary inoculation disease are extremely rare. The solitary skin lesion present may be considered to be primary infection following subcutaneous implantation only when there is no evidence of lung involvement or disease elsewhere. The skin test and immunodiffusion test are negative in such cases, as they are in the early stages of the more classic infection.

The predilection of the dissiminating organism for the mucocutaneous areas has been the subject of much conjecture and research. *In vitro* experiments indicate that growth of the organism occurs within a narrow range of temperature.[32] It had long been maintained by Mackinnon that the predilection for development of oropharyngeal and nasal lesions in man is owing to the cooling effect of air on these areas, and there is mounting experimental evidence[31,56] to indicate this is so. With increased accommodation of the parasite to *in vivo* existence and diminution of host defenses, spread occurs involving other areas.

Systemic Paracoccidioidomycosis

Disseminated visceral disease similar to that seen in other mycoses is infrequent. When dissemination occurs, the lymphatic system, spleen, intestine, adrenals, and liver are the common sites of systemic disease; other organs, such as the testes, brain, meninges, heart, large arteries, and bones, are rarely involved.[4, 36a]

Intestinal lesions are not infrequent and begin in the submucosal lymphoidal tissue. These erode into the lumen of the intestine, forming ragged ulcers. The conjectured occurrence of primary infection in the alimentary or anorectal areas has not been supported by experimental animal infection, and disease in such areas must be considered secondary.[28] The stomach is seldom involved except by extension from abdominal lymph nodes.

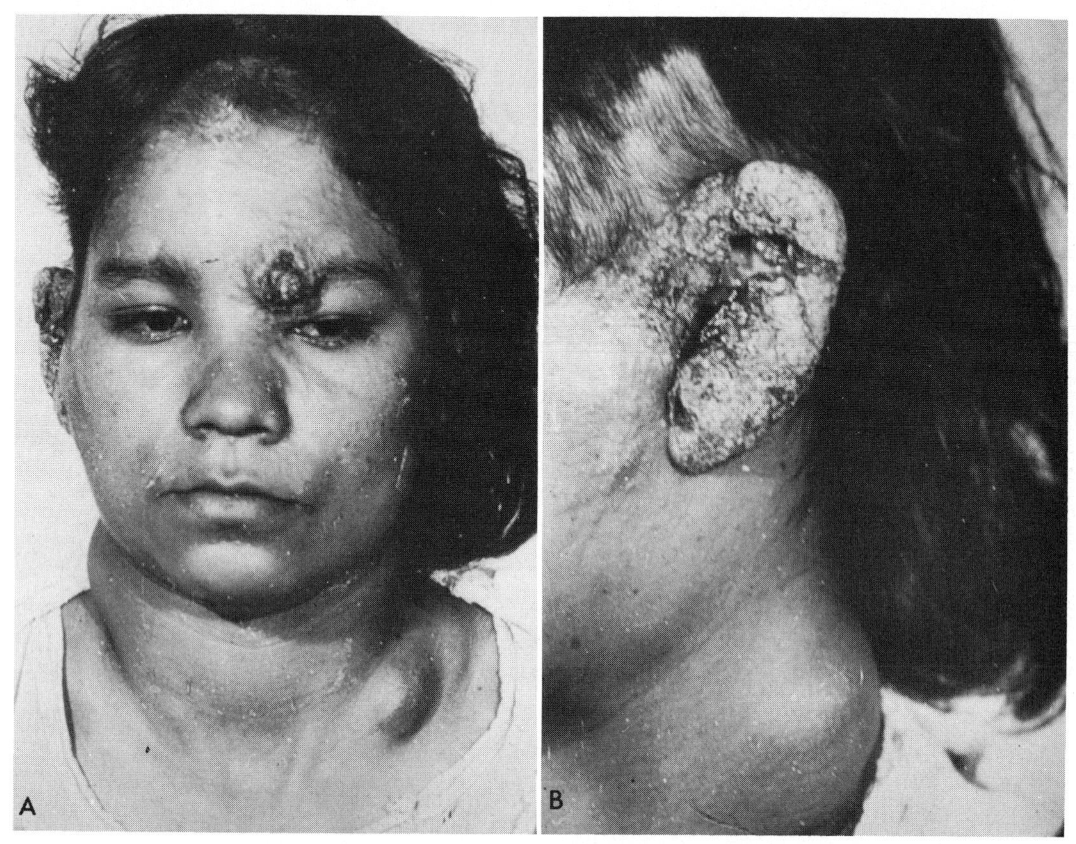

Figure 19–7. Cutaneous paracoccidioidomycosis. *A,* In this advanced case, lesions are found on the ear and forehead. Note the enlarged lymph nodes. *B,* Granulomatous lesion of the ear. (Courtesy of A. Restrepo.)

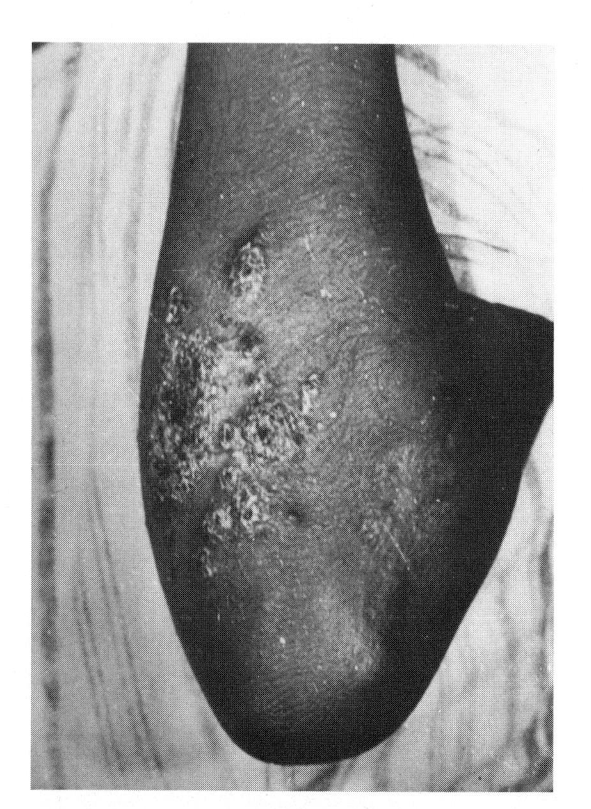

Figure 19–8. Cutaneous paracoccidioidomycosis. Crusted lesions on the forearm.

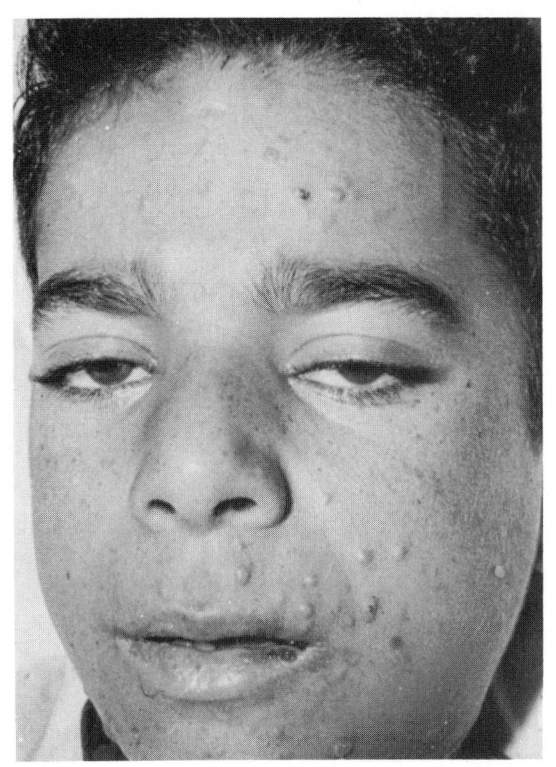

Figure 19–9. Cutaneous paracoccidioidomycosis. In this disseminated case, acneform lesions were found on all body surfaces. (Courtesy of A. Restrepo.)

Symptoms of alimentary involvement are abdominal pain, anorexia, vomiting, and fever. Enlargement of the mesenteric lymph nodes is also a prominent feature. The liver, spleen, and other organs are soon involved. In some series of reports, the spleen was found to be involved in 98 per cent of cases of disseminated disease. Ascites may be prominent. Invasion of the adrenals, with necrosis and destruction, is common and was present in 50 per cent of cases in one series.[37] Such patients often have Addison's disease as a further complication. The particular susceptibility of the adrenal glands to infection has been demonstrated in experimental animals.[31] Osseous lesions are relatively rare and, when found, are usually present in areas of bone which have a good blood supply. They usually consist of isolated osteolytic foci or, rarely, diffuse osteomyelitis. Bone involvement may be more common than previously appreciated.[4]

Central nervous system involvement has been recorded in about 40 cases (5 per cent) in several series tabulating the pathology of disseminated systemic disease. On rare occasion, meningitis was the presenting symptom. Papilledema and other evidence of intracranial pressure, as well as headache, vomiting, seizures, and hemiplegia, were present. Cerebrospinal fluid is within normal limits as far as pressure, color, glucose, and chlorides are concerned. The cell count may be increased, with a predominance of mononuclear leukocytes. The protein ranges from normal to 200 mg per 100 ml. The lesions are granulomatous and predominantly basilar in location. They may be single and massive or numerous and small. Tubercles composed of giant cells, macrophages, plasma cells, and phagocytized yeast cells are present and often show evidence of necrosis.

DIFFERENTIAL DIAGNOSIS

Pulmonary disease emulates other mycoses and tuberculosis. Cases in which all three (tuberculosis, paracoccidioidomycosis, and histoplasmosis) were present have been recorded. Mucocutaneous disease simulates cutaneous leishmaniasis. Both are endemic in the same geographic regions, and both cause secondary lesions leading to destruction of nasal septum and other areas of the oropharyngeal mucosa. Lymphatic enlargement simulates Hodgkin's disease and other malignant diseases, as well as scrofuloderma. The ulcerative or vegetative wartlike lesions suggest chromomycosis and sporotrichosis. Advanced cases may simulate yaws, syphilis, systemic tuberculosis, and visceral leishmaniasis. With abdominal involvement, there are many features similar to abdominal actinomycosis, but no sinus tracts form.

As previously mentioned, paracoccidioidomycosis is frequently found in patients with other diseases. High on this list are malnutrition, ascariasis, and other helminth infections. Concomitant diseases also include Chagas' disease, tuberculosis, and schistosomiasis.

PROGNOSIS AND THERAPY

In a manner similar to histoplasmosis, paracoccidioidomycosis was formerly recognized only in its advanced, often fatal, form. A benign, self-limiting form of paracoccidioidomycosis also occurs, though probably not to such an extent as in the former disease.

Diagnosis of late-stage disease was universally fatal prior to the advent of the sulfas. In 1940, Oliveira Ribeiro demonstrated the efficacy of sulfonamides. Blood levels of 5 mg per 100 ml of sulfadiazine or sulfamerazine (4 g per day) were able to control the infection, but with cessation of therapy clinical remission occurred. With the slowly excreted sulfas, such as sulfamethoxypyridazine and sulfadimethoxine, patients could be maintained on 0.5 g per day, following an initial dose of 1 g per day for a week. Again cessation of therapy was usually followed by relapse of disease.

In 1959 amphotericin B was introduced by Sampaio and Lacaz as a therapeutic regimen for paracoccidioidomycosis. The yeast phase of most strains is sensitive to 0.6 μg per ml of this drug. It is generally felt that a serum level twice the minimal inhibitory dose *in vitro* is necessary for effective treatment. Doses ranging from 0.25 to 1.2 mg per kg per day are given, every day or on alternate days. The initial dose is 25 mg each day, increased to 50 mg, to 75 mg, and finally to the required regimen. It is given as the deoxycholate salt dissolved in 5 per cent glucose and administered as a slow infusion over several hours. The course of treatment is essentially the same as that used for other systemic mycoses. A total dose of 1 to 4 g is recommended, depending on the severity of disease and the degree of nephrotoxicity encountered.[40]

Sulfonamides can be effective in clearing the mildest forms of disease, but even in these cases it is recommended that a course of 1 g of amphotericin B be completed.[40] Dihydroxystilbamidine has usually failed when used alone. Iodides are contraindicated, as they tend to disseminate the infection.

PATHOLOGY

Gross Pathology. A picture similar to that seen in coccidioidomycosis and blastomycosis has been noted in autopsy examination of patients with disseminated paracoccidioidomycosis. The main differences are the extensive involvement of lymphoid tissues in paracoccidioidomycosis and the occasional widespread involvement of the intestinal tract. Ulcers occur in the intestinal mucosa. The lesions appear to arise in the lymphoid tissue beneath the mucous membrane and then extend into it. This is also seen in nodes that drain from mucocutaneous lesions of other areas. The entire node may be involved, showing focal or diffuse necrosis. This predilection for lymphoid tissue also explains the almost universal presence of lesions in the spleen in disseminated cases. Nodules in the liver and, less frequently, the kidney are also found. Granulomatous lesions are also found in other organ systems with about the same frequency as is seen in blastomycosis. In contrast to the case in the latter disease, bone involvement is relatively rare in paracoccidioidomycosis.

Histopathology. The histopathologic reaction seen in cases of paracoccidioidomycosis is essentially identical to that of blastomycosis. There is a granulomatous reaction interspersed with pyogenic abscess formation. Langhans' and foreign body giant cells are numerous and may contain organisms. Focal necrotic and caseous areas are seen with zones of macrophages, histiocytes, lymphocytes, plasma cells, and fibroblasts.[56] Microabscesses with a polymorphonuclear reaction are interspersed between areas that are purely granulomatous. Epithelioid granulomas which are sharply outlined and without necrosis resemble Boeck's sarcoid. In older lesions, fibrosis may be very extensive.

Cutaneous lesions and those of the buccal mucosa characteristically show a pseudoepitheliomatous hyperplasia together with a marked pyogenic and granulomatous response. Intraepithelial microabscesses are frequent. Clusters of epithelioid cells, plasma cells, and lymphocytes are present. The same types of lesions are seen in coccidioidomycosis and particularly in blastomycosis. Differentiation from these diseases by examination of histopathologic material alone requires observation of the yeast cell of the fungus with its pathognomonic "pilot wheel" peripheral budding. The organism may be evident in hematoxylin-eosin stained material, but is more easily observed in methenamine silver stains. The typical budding cell varies from 12 to 40 μ in diameter. The periphery of the spherule is studded with small buds that communicate with the mother cell by small necks. The buds are visible when they are 2 μ in diameter, and they grow to be 5 μ or more. The parent cells may appear empty. Buds of many sizes or only a few large buds may occur on the

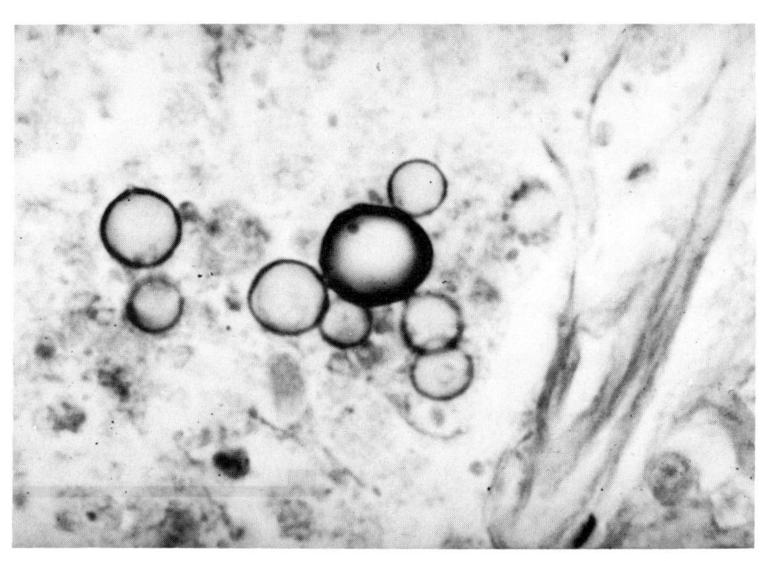

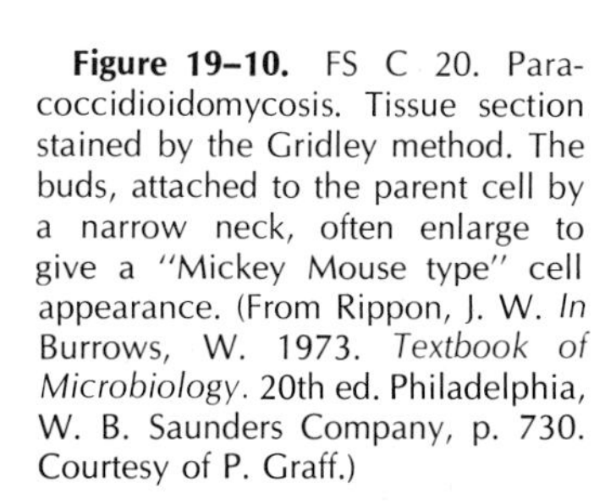

Figure 19–10. FS C 20. Paracoccidioidomycosis. Tissue section stained by the Gridley method. The buds, attached to the parent cell by a narrow neck, often enlarge to give a "Mickey Mouse type" cell appearance. (From Rippon, J. W. *In* Burrows, W. 1973. *Textbook of Microbiology*. 20th ed. Philadelphia, W. B. Saunders Company, p. 730. Courtesy of P. Graff.)

same cell, giving a "Mickey Mouse cap" appearance to the yeast (Fig. 19–10 FS C 20). Young cells in hematoxylin and eosin preparations have chromatin dots representing nuclei in a centripetal or even distribution.

The diagnosis of paracoccidioidomycosis cannot be made histopathologically without the presence of the typical budding cell. Spherules from 5 to 15 μ in size can be seen in many lesions which appear identical to young spherules of coccidioidomycosis. Single budding cells resemble *Blastomyces dermatitidis*, and small cells (2 to 3 μ) lying free or within macrophages or giant cells may be misidentified as the yeasts of *Histoplasma capsulatum*. In some necrotic areas, enormous numbers of cells which are sometimes crescent-shaped, distorted, and crumpled are seen.

In disseminated disease, numerous epithelioid cell granulomas occur. These often have necrotic centers and resemble the lesions of miliary tuberculosis. Granulomatous osteomyelitis and meningitis have been reported. Caseous necrosis of the adrenals is common and resembles that seen in histoplasmosis.[37]

ANIMAL DISEASE

Natural infection in animals is unknown. In the laboratory *P. brasiliensis* is not highly virulent for most experimental animals. Pus from lesions can be inoculated intratesticularly into guinea pigs, which then develop an orchitis. Examination of material from such lesions shows an abundance of yeast cells. The mountain rat (*Proechimys guayanensis*) is particularly susceptible to infection. Use of the animal has been suggested as a diagnostic aid when few organisms are present.[5] Systemic infections have been produced in mice, rabbits, guinea pigs, and hamsters, but with great difficulty. Mice inoculated by the pulmonary route developed an alveolitis initially. In some, this was followed by spread to the regional lymph nodes, with subsequent hematogenous dissemination.[31] If the mice were kept at an ambient temperature of 18°C, the infection ran a fulminant course with fatal dissemination in all animals. When kept at an ambient temperature of 35°C, few animals developed disease.[56] Mackinnon[32] has demonstrated that growth of the organism in liquid media decreases sharply at temperatures above 35°C.

BIOLOGICAL STUDIES

As with other dimorphic fungi, the factors that affect mycelial to yeast transformation and the physiologic alterations that accompany this change have been the subject of much research. Yeast cells are seen to develop directly from mycelia by a budding process not unlike the production of aleuriospores.[11] This process appears fundamentally different from that observed in *Histoplasma* or *Sporothrix*. As with other dimorphic fungi, the yeast form has a higher chitin content than the mycelium. The buds of yeast cells placed at 25°C elongate directly into hyphae to form a corona of many hyphal units from a single cell.[43] This transformation is accompanied by a change from α-glucan in the yeast cell wall to the β-glucan found in the mycelium.[13]

The effect of drying and temperature on the viability of the organism was studied by Conti-Diaz et al.[16] The fungus was not able to survive in dry soils, though it retained viability in humid samples for long periods of time. Maintenance in humid or dry soils at 2°C was deleterious. Viability is retained for longer periods of time in acid soils than in neutral or alkaline soils.[50] These studies may explain the geographic limitations of the organism to the warm humid areas and acidic soils of South and Central America.

IMMUNOLOGY AND SEROLOGY

The recent finding by surveys of skin test sensitivity that exposure to the fungus is much more prevalent than previously thought has modified several concepts of this disease.[22,52] As has been established in histoplasmosis, it appears that a sizeable population encounters the etiologic agent of paracoccidioidomycosis, and a subclinical, self-limiting disease develops. This indicates that there is a high natural resistance to infection for both diseases. It appears that the patients who develop severe disease initially or from a latent infection long after exposure are physiologically or immunologically deficient, possibly only transiently. Patients with severe disease have an acquired deficiency of cell-mediated immunity. They are usually anergic and show no reactivity to a number of skin tests. In addition, there is a reduction of lymphocyte blast transformation after stimulation, and there is an increased tolerance to heterologous skin grafts.[35,36] Whether initial resolution of infection confers a strong immunity, as is seen in coccidioidomycosis and histoplasmosis, is not known.

The first satisfactory and reasonably specific serologic test for paracoccidioidomycosis was developed by Fava-Netto in 1955.[17,18] Using a polysaccharide antigen, he found that there were few cross-reactions with other mycotic diseases in either the complement fixation test or the precipitin test. Precipitins (by the tube technique) appear early in the disease and disappear after therapy. Complement fixing antibodies appear later and are present for several months following clinical cure. The reappearance of positive serologic tests without clinical signs has been interpreted as indication of a smoldering infection. The complement fixation test is positive in 84 to 95 per cent of patients with active lesions. Restrepo has developed an immunodiffusion test reported in a series of papers.[45,48,52] She found that antigen prepared from the yeast phase was more reliable than that from the mycelial phase. There was a good correlation with the complement fixation test, but the immunodiffusion test was positive in 16 per cent of patients who were negative by complement fixation or whose sera were anticomplementary. In an extensive study, Kaufman[25] found low titer cross-reactions in the complement fixation test with sera from patients with aspergillosis and candidosis. However, the immunodiffusion test of Restrepo was specific for the disease. This test usually becomes negative following clinical cure.

A number of common antigens between *P. brasiliensis* and *B. dermatitidis*, *H. capsulatum* and *H. capsulatum* var. *duboisii*, have been found in that order of relatedness. By immunoelectrophoresis of cell fractions of *P. brasiliensis*, the specific antigen was determined to be a cathodic fraction identified as an alkaline phosphatase.[57] A fluorescent antibody test for use with tissue sections and cultures has been described.[24, 54, 54a]

Skin testing with antigens prepared from yeast cultures or tissue of infected guinea pigs has been used to evaluate the immune state of the patient. Patients recovering from disease have a strong delayed hypersensitivity exhibited by an area of induration of 10 mm or more. Patients in terminal stages of the disease are usually anergic. However, both types of patients will have a positive immunodiffusion test. Skin test reactivity along with a negative immunodiffusion test indicates the presence of a subclinical or mild, self-limited infection. A skin test antigen termed "paracoccidioidin" that was prepared by Mackinnon has been found to be specific for the disease in a 1:100 dilution. At a dilution of 1:10, however, there was cross-reaction with sera from cases of histoplasmosis and coccidioidomycosis. The antigen of Restrepo is even more specific,[48] but a few instances of cross-reaction in cases of histoplasmosis still occur.

LABORATORY IDENTIFICATION

Direct Examination. Sputum, biopsy material, crusts, material from the granulomatous bases or the outer edge of ulcers, and pus from suppurative draining lymph nodes contain numerous fungal elements. The material is placed on a slide with a drop of 10 per cent potassium hydroxide, then heated to clear the specimen (Figs. 19–11 and 19–12). The yeasts are readily observed. They vary from young, recently separated buds 2 to 10 μ in size to mature cells up to 30 μ or more. Some yeasts may be 60 μ in diameter. The cells have a double refractile wall (0.2 to 1 μ thick); they are spherical, oval, or elliptical, and may occur in chains of four or more. From one to a dozen narrow-necked buds of uniform or variable size arise from the mother cells.

Culture Methods. Material from pus, lesions, biopsies, and sputum can be plated on media containing antibacterial antibiotics and cycloheximide and incubated at 25°C. It can also be cultured at 37°C on blood agar with antibacterial antibiotics. The yeast phase of this organism, like that of *B. dermatitidis* and *H. capsulatum*, is sensitive to cycloheximide. At 25°C the fungus grows very slowly, and a colony usually does not appear for 15 to 25 days. It may require ten days more to reach a diameter of 1 cm. Growth of stock cultures is more rapid. Transfer to blood agar incubated at 37°C converts the organism to the yeast form. As no characteristic spores are produced at 25°C, conversion to the typical morphology of the yeast form is essential for diagnosis. Restrepo and Correa[46a] have found that yeast extract agar was superior to other media for the initial isolation of the fungus.

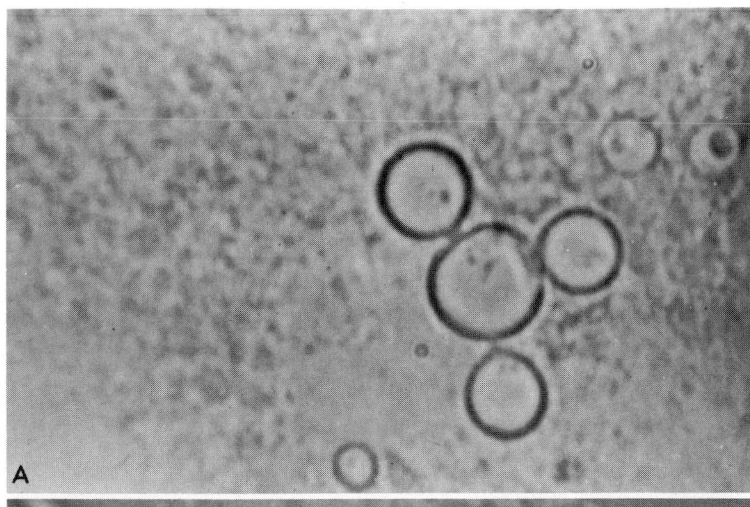

Figure 19–11. Paracoccidioidomycosis. Direct examination of sputum by KOH mount. *A*, Parent cell with large buds. *B*, Buds in chains. ×940. (Courtesy of A. Restrepo.)

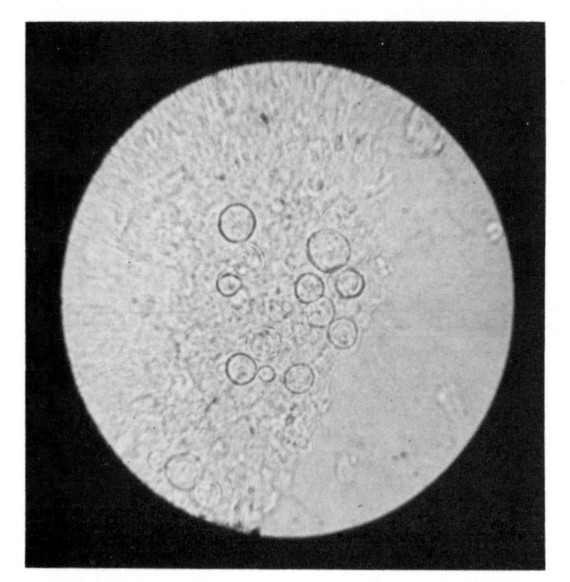

Figure 19–12. Paracoccidioidomycosis. Direct examination by KOH mount of scraping from lesion on oral mucosa. Most of the yeasts appear as single cells or with one bud. They may be mistaken for several other fungi. ×400. (Courtesy of E. Belfort.)

MYCOLOGY

Paracoccidioides brasiliensis (Splendore) Almeida 1930

Synonymy. *Zymonema brasiliense* Splendore 1912; *Zymonema histosporocellularis* Haberfield 1919; *Mycoderma brasiliensis* Brumpt 1912; *Mycoderma histosporocellularis* Neveu-Lemaire 1921; *Monilia brasiliensis* Vuillemin 1922; *Coccidioides brasiliensis* Almeida 1929; *Coccidioides histosporocellularis* Fonseca 1932; *Paracoccidioides cerebriformis* Moore 1935; *Proteomyces faverae* Dodge 1935; *Paracoccidioides tenuis* Moore 1935; *Lutziomyces histoporocellularis* Fonseca Filho 1939; *Blastomyces brasiliensis* Conant and Howell 1941; *Aleurisma brasiliensis* Neves and Boglioli 1951.

Colony Morphology. SDA, 25°C. The fungus grows slowly, producing a variety of colonial forms. These vary from a glabrous leathery, brownish, flat colony with a few tufts of aerial mycelium to a wrinkled, folded, floccose to velvety, white, and later beige form. The colonial forms are indistinguishable from *B. dermatitidis* (Fig. 19–13,*A* FS C 21,*A*). The color of the reverse is yellowish brown to brown.

SDA, Kelley's, BHI blood agar, 37°C. The mycelium converts to the yeast phase with ease at 37°C (Fig. 19–14,*B*). The organism grows slowly, producing a wrinkled, folded, glabrous, whitish colony.

Microscopic Morphology. SDA, 25°C. Microscopic examination of sporulating strains shows a variety of spores, none of which are characteristic of the species (Fig. 19–15). Most strains grow for long periods of time (up to ten weeks) without the production of spores. All cultures produce intercalary chlamydospores and coiled hyphae. In media deficient in glucose, some strains produce intercalary cells which become rectangular, or triangular, thick-walled arthrospores. They may be in an alternate pattern within the mycelium. Pear-shaped aleuriospores and arthroaleuriospores are also seen. Isolants vary in the proportion, timing, and variety of spores produced. Sporulation is enhanced on yeast extract agar.[46]

A

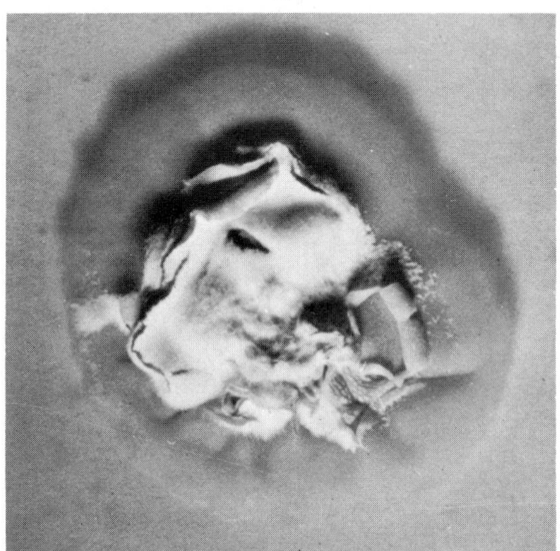

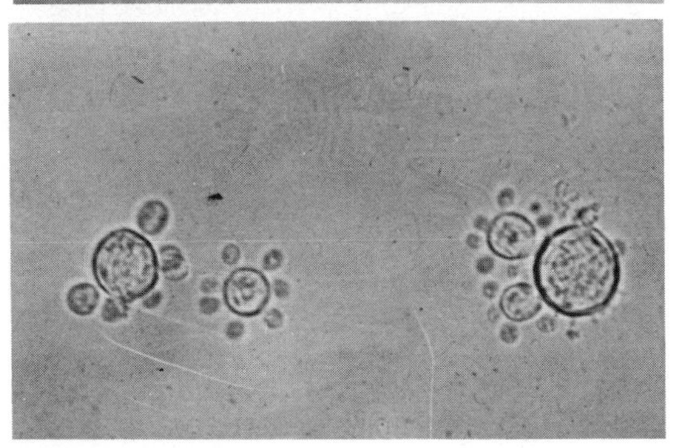

B

Figure 19–13. FS C 21. Paracoccidioidomycosis. *A*, Colony morphology. The glabrous, leathery, cracked colony has some areas of aerial mycelium. *B*, Yeast form at 37°C. Note the "pilot wheel" and "Mickey Mouse type" forms. ×440.

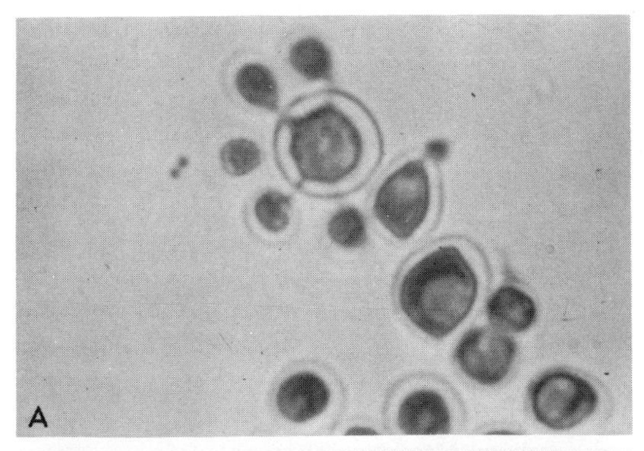

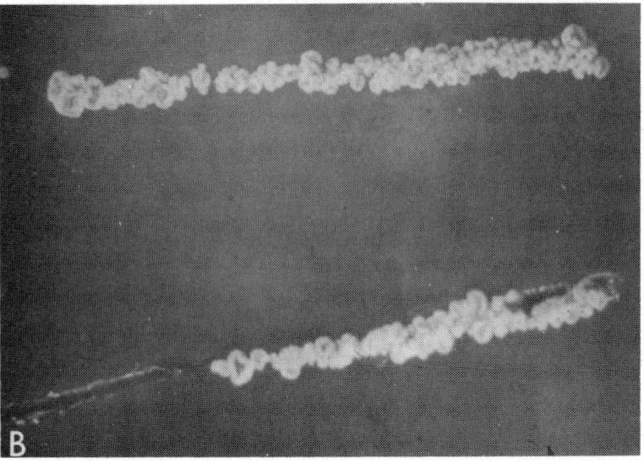

Figure 19–14. Paracoccidioidomycosis. *A,* Lactophenol mount of yeast form cells. Note the cytoplasmic connection from parent cell to bud. ×450. (Courtesy of A. Restrepo.) *B,* Yeast form growth at 37°C on Kelly's medium.

When cultures are transferred to 37°C, distorted mycelial elements of varying lengths are seen to be intermixed with yeast cells. The yeasts are 2 to 30 μ or more, and are oval or irregular in shape (Fig. 19–13,*B*

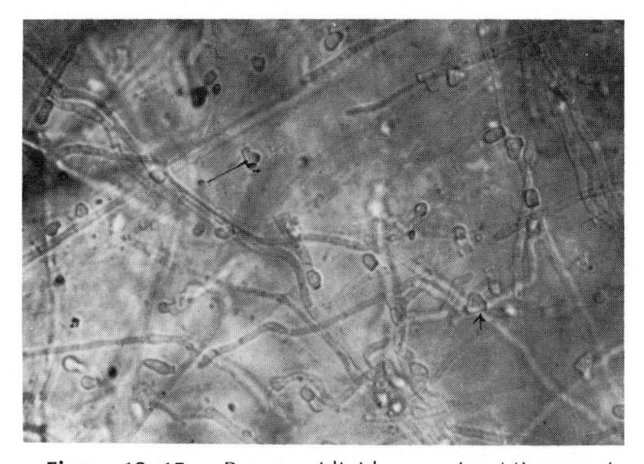

Figure 19–15. Paracoccidioidomycosis. Microscopic morphology of mycelial form at 25°C. There are aleuriospores and intercalary chlamydospores, rectangular and bulging arthrospores (at arrows), and arthroaleuriospores. ×100. (From Restrepo, M. A. 1970. A reappraisal of the microscopical appearance of the mycelial phase of *Paracoccidioides brasiliensis*. Sabouraudia, *8*:141–144.)

FS C 21,*B* and Fig. 19–14,*A*). They have from one to several thin-necked, round buds of uniform or varying size, which develop from all areas of the parent cell. The walls of *P. brasiliensis* are thinner than those of *B. dermatitidis*, and the buds are easily dislodged. The bud of the latter species has a broad base. *P. brasiliensis* has no capsule. Biochemical properties are, so far, not useful in its identification, and a perfect stage has not as yet been described.

REFERENCES

1. Albornoz, M. B. 1971. Isolation of *Paracoccidioides* from rural soil in Venezuela. Sabouraudia, *9*:248–253.
2. Albornoz, M. B., and R. Albornoz. 1971. Estudio de la sensibilidad específica en residentes de un area endémica a la paracoccidioidamicosis en Venezuela. Mycopathologia, *45*:65–75.
3. Almeida, R. P. 1931. Estudos comparativos do granuloma coccidioidica nos Estados Unidos e no Brasil. Novo genero para a parasita brasiliero. An. Fac. Med. Univ. São Paulo, *5*:3–19.
4. Angulo, A. 1971. Anatomo-clinical aspects of paracoccidioidomycosis. Ref. 41, pp. 129–133.
5. Belfort, A. E. 1967. Paracoccidioidomycosis: Diagnostico mediante inoculacion a *Proechimys guayensis*. Rev. Dermatol. Venezol., *3*:91–97.
6. Borelli, D. 1961–1962. Hipótesis sobre ecologia de paracoccidioides. Derm. Venez., *3*:130–132.
7. Borelli, D. 1970. Prevalence of systemic mycoses in Latin America, In *International Symposium on Mycoses.* PAHO Scientific Publications No. 205. Washington, D.C., World Health Organization, pp. 28–38.
8. Borrero, R. J., A. Restrepo, et al. 1965. Blastomicosis sur americana de forma pulmonar pura. Antioquia Med., *15*:503–516.
9. Calle Velez, G. 1971. Dermatological aspects of paracoccidioidomycosis. Ref. 41, pp. 118–121.
10. Carbonell, L. M. 1967. Cell wall changes during the budding process of *Paracoccidioides brasiliensis* and *Blastomyces dermatitidis.* An electron microscopic study. J. Bacteriol., *94*:213–223.
11. Carbonell, L. M., and J. Rodriguez. 1965. Transformation of mycelial and yeast forms of *Paracoccidioides brasiliensis* in cultures and in experimental inoculations. J. Bacteriol., *90*: 504–510.
12. Carbonell, L. M., and J. Rodriguez. 1968. Mycelial phase of *Paracoccidioides brasiliensis* and *Blastomyces dermatitidis.* An electron microscopic study. J. Bacteriol., *96*:533–534.
13. Carbonell, L. M., F. Kanetsuna, et al. 1970. Chemical morphology of glucan and chitin in the cell wall of the yeast phase of *Paracoccidioides brasiliensis.* J. Bacteriol., *101*:636–642.
14. Chaves-Batista, A., S. K. Shome, et al. 1962. Pathogenicity of *Blastomyces brasiliensis* isolated from soil Publicação 373. Inst. de Micrologia, Universidade Recife, Brasil.

15. Conant, N. F., and A. Howell. 1942. The similarity of the fungi causing South American blastomycosis (paracoccidioidal granuloma) and North American blastomycosis (Gilchrist's disease). J. Invest. Dermatol., *5*:353–370.
16. Conti-Diaz, I. A., J. E. Mackinnon, et al. 1971. Effect of drying on *Paracoccidioides brasiliensis*. Sabouraudia, *9*:69–78.
17. Fava-Netto, C. 1965. The immunology of South American blastomicosis. Mycopathologia, *26*: 349–358.
18. Fava-Netto, C., and A. Raphael. 1961. Intradermal reaction with *Paracoccidioides brasiliensis* polysaccharide in South American blastomycosis. Dermatol. Trop., *2*:27.
19. Fonseca, J. B. 1957. Blastomycosis Sul-Americana Estudo das lesões dentárias a paradentárias sob a ponto de vista clínico e histopathológico. Tese Fac. Farmacia e Odontologia da Univ. São Paulo, pp. 1–182.
20. Fountain, F. F., and W. D. Sutliff. 1969. Paracoccidioidomycosis in the United States. Am. Rev. Resp. Dis., *99*:89–93.
21. González Ochoa, A., and L. D. Soto. 1957. Blastomicosis sulamericana. Casos mexicanos. Rev. Inst. Salub. Enferm. Trop., *17*:97–101.
22. Greer, D. L., D. D'Costa de Estrade, et al. 1971. Dermal reactions to paracoccidioidin among family members of patients with Paracoccidioidomycosis. Ref. 41, pp. 76–84.
23. Groose, E., and J. R. Tamsitt. 1965. *Paracoccidioides brasiliensis* recovered from intestinal tract of three bats (*Artibeus literatus*) in Columbia, S.A. Sabouraudia, *4*:121–125.
24. Kaplan, W. 1971. Application of immunofluorescence to the diagnosis of paracoccidioidomycosis. Ref. 41, pp. 224–226.
25. Kaufman, L. 1971. Evaluation of serological tests for paracoccidioidomycosis. Preliminary report. Ref. 41, pp. 221–224.
26. Lacaz, C. S. 1967. Compêndio de Micologia Medicar. Universidade de São Paulo. (Ed. Sarvier-Xavier).
27. Lacaz, C. S. 1955–1956. South American blastomycosis. An. Fac. Med. Univ. São Paulo, *29*:1–120.
28. Linares, L., and L. Friedman. 1971. Pathogenesis of paracoccidioidomycosis in experimental animals. Ref. 41, pp. 287–291.
29. Lutz, A. 1908. Uma mycose peudoscoccidica localizada na boca e observada no Brasil. Contribuição ao conhecimenta das hyfoblastomycoses americanas. Brasil-méd., *22*:121.
30. Machado, J., and S. L. Miranda. 1960. Conciderações relativas a blastomicose sul-americana. Da participação pulmonar entre 338 cases consecutivos. O Hospital (Rio), *58*:431–449.
31. Mackinnon, J. E. 1959. Pathogenesis in South American blastomycosis. Trans. R. Soc. Trop. Med. Hyg., *53*:487–494.
32. Mackinnon, J. 1966. Algo mas sobre blastomicosis y temperatura ambiental. Tórax, *15*:127–129.
33. Mackinnon, J. 1970. On the importance of South American blastomycosis. Mycopathologia, *41*: 187–193.
34. Martin, J. 1971. Blastomycose sul-americaine aigue evoluant sous le masque d'une lymphopathie maligne. Int. J. Dermatol., *10*:246–250.
35. Mendes, N. F. 1971. Lymphocyte cultures and skin allograft survival in patients with S.A. blastomycosis. J. Allergy Clin. Imm., *48*:40–45.
36. Mendes, M., and A. Raphael. 1971. Impaired delayed hypersensitivity in patients with S. A. blastomycosis. J. Allergy, *47*:17–22.
36a. Murray, H. W., M. L. Littman, et al. 1974. Disseminated Paracoccidioidomycosis (South American Blastomycosis) in the United States. Am. J. Med., *56*:209–220.
37. Negro, G. 1961. Localização supra-renal da blastomicose sul-americana. Thesis, Univ. São Paulo, pp. 1–144. São Paulo, Of. Graficas Saraiva, S. A.
38. Negroni, P. 1966. El *Paracoccidioides brasiliensis* vive saprofitacamanta en el suelo Argentino. Pren. Med. *53*:2831–2832.
39. Negroni, P., and R. Negroni. 1968. Aspectos clinicao e inmunologicos de la blastomicosis sudamericana en la Argentina. Tórax, *17*:63–66.
40. Negroni, P. 1971. Prolonged therapy for paracoccidioidomycosis: Approaches, complications and risks. Ref. 41, pp. 147–155.
41. Pan American Health Organization. 1972. First Pan American Symposium on Paracoccidioidomycosis. Medellin Colombia 25–27, October, 1971. Scientific Publication No. 254. Washington, D.C., World Health Organization.
42. Pollak, L., and A. Angulo-Ortega. 1967. Las Micosis broncopulmonares en Venezuela. Tórax, *16*:135–145.
43. Ramirez-Martinez, J. R. 1971. *Paracoccidioides brasiliensis*. Conversion of yeastlike forms into mycelia in submerged culture. J. Bacteriol., *105*:523–526.
44. Ramirez-Martinez, J. R. 1971. A comparative study of ribonucleic acid species on the yeast-like and mycelial forms of *Paracoccidioides brasiliensis*. Sabouraudia, *9*:157–163.
45. Restrepo M., A. 1966. La prueba de immunodiffusion en el diagnostica de la paracoccidioidomicosis. Sabouraudia, *4*:223–230.
46. Restrepo M., A. 1970. A reappraisal of the microscopical appearance of the mycelial phase of *Paracoccidioides brasiliensis*. Sabouraudia, *8*: 141–144.
46a. Restrepo M., A., and I. Correa. 1973. Comparison of two culture media for primary isolation of *Paracoccidioides brasiliensis* from sputum. Sabouraudia, *10*:260–265.
47. Restrepo M., A., and S. L. Espinal. 1968. Algunas considerciones ecologicas sobre la paracoccidioidomicosis en Colombia. Antioquia Med., *18*:433–446.
48. Restrepo M., A., and L. H. Moncada. F. 1970. Serological procedures in the diagnosis of paracoccidioidomycosis. Proc. Int. Symp. on Mycoses. PAHO Sci. Pub. No. 205, pp. 101–110. Washington, D.C., World Health Organization.
49. Restrepo M., A., and J. D. Schneidau. 1967. Nature of the skin reactive principle in culture filtrates prepared from *P. brasiliensis*. J. Bacteriol. *91*: 1741–1748.
50. Restrepo M., A., H. Moncado, et al. 1969. Effect of hydrogen ion concentration and of temperature on the growth of *Paracoccidioides brasiliensis* in soil extract. Sabouraudia, *7*:207–215.
51. Restrepo M., A., D. L. Greer, et al. 1971. Relation-

ship between the environment and paracoccidioidomycosis. Ref. 41, pp. 84–91.

52. Restrepo M., A., M. Robledo, et al. 1968. Distribution of paracoccidioides sensitivity in Colombia. Am. J. Trop. Med. Hyg., *17*:25–37.

53. Restrepo M., A., F. Gutierrez, et al. 1970. Paracoccidioicomycosis (South American blastomycosis). A study of 39 cases observed in Medellin Colombia. Am. J. Trop. Med. Hyg., *19*:68–76.

54. Silva, M. E., and W. Kaplan. 1965. Estudios adicionais sobre a preparaçao de antiglobulinas marcadas pela fluoresceina para a fase cerebriforme do *Paracoccidioides brasiliensis*. Rev. Inst. Med. Trop. São Paulo, 7:289–295.

54a. Silva, M. E., and W. Kaplan. 1965. Specific fluorescein labelled antiglobulins for the yeast form of *Paracoccidioides brasiliensis*. Am. J. Trop. Med., *14*:290–294.

55. Splendore, A. 1912. Una attezione micotica con localizazione nella mucosa della boca osservata in Brasile, determinate da funghi apparteneti alla tribú degli Exiascei (*Zymonema brasiliense* n. sp.). Estraido do volume in honore del Prof. Angello Celli nel 25° anno di insegnamento. Roma, Tipografia Nazional di G. Bertero, and Bull. Soc. Pathol. Exot., *5*:313–319.

56. Yarzabal, L. A. 1971. Pathogenesis of paracoccidioidomycosis in man. Ref. 41, pp. 261–270.

57. Yarzabal, L., J. M. Torres, et al. 1971. Antigenic mosaic of *Paracoccidioides brasiliensis*. Ref. 41, pp. 239–244.

THE SYSTEMIC MYCOSES / The Opportunistic Fungi

Chapter 20

ASPERGILLOSIS

DEFINITION

Aspergillosis broadly defined is a group of diseases in which members of the genus *Aspergillus* are involved. The gamut of pathologic processes include (1) toxicity due to ingestion of contaminated foods; (2) allergy and sequelae to the presence of spores or transient growth of the organism in body orifices; (3) colonization without extension in preformed cavities and debilitated tissues; (4) invasive, inflammatory, granulomatous, necrotizing disease of lungs and other organs; and rarely (5) systemic and fatal disseminated disease. The type of disease evoked depends on the local or general physiologic state of the host, as the etiologic agents are ubiquitous and opportunistic. The same agents are involved in many important animal diseases, such as mycotic abortion of sheep and cattle, pulmonary infections of birds, and toxicosis due to ingestion of grain containing fungal byproducts. The spectrum of human infection also includes otomycosis, mycotic keratitis, and rarely mycetoma. The diversity of involvement attests to the adaptive capacity of the organisms, so that "aspergillosis" in reality is a spectrum of diseases.

HISTORY

Aspergillosis was one of the first fungal diseases of animals recognized. Mayer and Emmert[33] in 1815 described an infection in in the lungs of a jay (*Corvus glandarius*). The term "aspergillosis" was coined by Fresenius in 1850 from work on fungous infections of the air sacs of birds. He named the isolant *Aspergillus fumigatus*.[14] Human infection was recognized by Bennett in 1842, and *Aspergillus* pneumomycosis was described by Sluyter in 1847.[53a] By description, the fungus in Bennett's case does not appear to be an *Aspergillus*, so that the report of Sluyter probably represents the first case of human aspergillosis. Bronchial and pulmonary disease, together with such an accurate description of the etiologic agent as to be recognizable as *A. fumigatus*, was published by Virchow in 1856[59] This presentation of four cases is a classic paper of descriptive pathology. Except for the association of the pulmonary form with "pigeon feeder's" disease by Virchow and later by Dieulafoy, Changemesse, and Widal in 1897, in the following years very few cases of the disease were reported. In 1897 Renon[47] published a book containing an excellent review of the field and also the association of the disease with certain occupations, such as wig cleaner, in addition to pigeon handlers and concluded the moldy grain was the source of the spores. By 1910 almost all forms of aspergillosis in animals had been delineated, including mycotic abortion and fatal respiratory disease.[1]

The association of *Aspergillus* infection with other diseases or with debilitation has occurred relatively recently. Cleland in 1924 found aspergillosis in necrotic lung tissue that resulted from a gunshot. Lapham in 1926 described aspergillosis as an infection secondary to tuberculosis. "Bronchiectasiant aspergilloma" (fungus ball) was defined by Deve in 1938.[9] However, it was not until 1952 that the diversity and importance of aspergillosis was realized. Monod et al.[38] expanded the work of Deve on aspergilloma, and Hinson et al.[21] illustrated various clinical manifestations of the disease. The allergic form of aspergillosis and the colonizing type aspergillosis are now realized to be common and frequently encountered diseases.[15,17–19,42] Since that time a gamut of disease processes have been recognized.

The importance of *Aspergillus* as an agent of opportunistic infections has only recently been recognized. Systemic disseminated disease is essentially a product of the antibiotic era and may be termed a "disease of medical progress." Fortunately this form of aspergillosis is still relatively rare[56] The debilitated patient, however, offers a special milieu for opportunistic fungi. In one large series of 454 leukemia cases, 14 per cent of deaths were attributable to fungous infections.[4] Aspergillosis was second only to candidosis in this series. Several recent reviews have emphasized the increasing occurrence of aspergillosis in patients with leukemias.[34,36] In one series there was a greater increase of aspergillosis complicating leukemias as compared to lymphomatous disease.[34] Aspergillosis is also a significant problem in organ transplant recipients. In the review by Bach et al.,[2] there was a significant rise in the incidence of aspergillosis associated with the extent of antirejection therapy necessary, and particularly with the use of antilymphocyte serum. The increased use of cytotoxin, steroids, and transplantation for prolongation of life will doubtless result in an increased frequency of disseminated aspergillosis.

ETIOLOGY, ECOLOGY, AND DISTRIBUTION

There are about 300 species of the genus *Aspergillus*. They are among the most common fungi of all environments. They are common in the soil and decaying vegetation throughout the world,[12] but are also found on all types of organic debris: spilled food, wet paint, cracked dialysis bags, opened medications, refrigerator walls, "sanitizing" fluids, used dressings, and so forth. They are the "bottle imp" of fluffy growth that is found in chemical reagents and distilled water bottles that have remained undisturbed for some time. *Aspergillus* organisms have been recovered from sulfuric acid-copper sulfate plating baths and formalinized pathology museum specimens. *Aspergillus* spores are airborne and constantly inhaled. After exposure to a cloud of *Aspergillus fumigatus* spores, the fungus can be recovered from sputum for many days afterward. Spores have been recovered by weather balloons in the upper atmosphere, from snow in the Antarctic, and from winds over the Sahara. The omnipresence and ubiquity of *Aspergillus* organisms are a major factor in designating this planet as "our moldy earth."[8b]

Although aspergilli are constant in the human environment, only about eight species have been consistently and authentically involved in human infectious disease. It is important to appreciate that any species could be involved in asthmatic allergy because of inhaled spores, and many species transiently germinate and grow on bronchial secretions,[50a] wounds, and diseased and normal skin. These are the common "contaminants" found in cultures of such material, but in the past were often mistaken as etiologic agents for numerous diseases and found their way into medical literature. However, the species that have been recovered from authentic cases of aspergillosis are few in number. *Aspergillus fumigatus* accounts for almost all diseases, both allergic and invasive. Allergic aspergillosis has also been caused by *A. flavus*, *A. nidulans*, *A. niger*, *A. terreus*, and *A. clavatus*. Aspergillomas are frequently reported to be caused by *A. niger* in the United States and more commonly by *A. fumigatus* in England and Europe. Aspergillosis of the lungs may be caused by *A. fumigatus*, *A. terreus*, *A. flavus*, *A. nidulans*, and *A. niveus*. Three different species in pure culture (*A. flavus*, *A. fumigatus*, and *A. nidulans*) have been isolated from different lesions in the lungs and heart of a single case.[61] Disseminated aspergillosis has involved *A. fumigatus*, *A. flavus*, and *A. restrictus*.[56,64] *A. niger* is frequently encountered in otomycosis and has been the cause of primary cutaneous disease.[6] *A. amsteloidami*, among others, has been recovered from cases of mycetoma. *A. flavus* is the cause of many primary infections of the nasal sinuses,[35] and *A. terreus* is a rare etiology of primary cerebral aspergillosis. All these species have thermotolerance in common, and many soil isolants have been shown to be pathogenic for experimental animals.[13,44] Fungi, especially aspergilli, are remarkably adaptable and can be "trained" to become pathogenic.[48]

The first necropsy of a case of human aspergillosis was described by Virchow in 1856. Thereafter, most cases of the disease were recorded from Europe. The infection has now been recognized in all age groups, both sexes, all races, and in all parts of the world. The rare primary disease is more common in adult males than in children or females. The mean age is about 40. Secondary aspergillosis is a terminal event in the debilitated patient and depends on the availability of such subjects, not on sex or age. Small foci of primary infection are not infrequently found in routine lung necropsies, but these are incidental and not involved with the demise of the patient.

In the early French literature, the disease was associated with moldy barns, grain, pigeon fanciers (*gaveurs des pigeons*), wig dusters, and so forth.[21, 59] These patients were exposed to clouds of spores, and many were malnourished and tubercular. Point source of heavy contamination with spores in primary disease is always possible. This was evident in the case of two sisters, both of whom died with acute pulmonary aspergillosis,[54] and in several case reports of farmers who inhaled masses of spores.[58] Disease has also resulted from accidental injection of contaminated dialysis fluid, antibiotics, and so forth.[49] Primary disseminated aspergillosis not associated with underlying disease or heavy exposure is quite rare but does occur.[7, 54] Secondary aspergillosis has been associated with antibiotics, steroids, radiation,[56] cytotoxins and neoplasias,[4, 22, 34, 36, 51] tuberculosis,[40] eclampsia and dialysis,[49] "barrier breaks" during surgery,[7a] aspergillomas with tuberculosis cavities,[51, 52] congenital heart disease,[41] sarcoid,[27] and histoplasmosis.[45]

There have been several good reviews of aspergillosis that have contributed to the emerging understanding of the disease; disseminated aspergillosis by Tan,[56] allergic aspergillosis by Henderson,[17–19] the French literature by Hinson et al.,[21] the English literature by Pepys et al.,[42] and the current European literature by Orie et al.[40]

CLINICAL DISEASE

It is now recognized that there are basically three categories of disease involving aspergilli. These are (1) allergic aspergillosis, (2) colonizing aspergillosis, and (3) invasive disease. In addition, aspergilli can exist as saprobes in bronchi or on body surfaces without eliciting any pathology.

It is now appreciated that invasive or disseminated disease is the result of a combination of three factors: (1) lowered resistance due to debilitating disease or drugs, (2) local point of entry for the fungus, and (3) disruption of normal flora and inflammatory response by antibiotics and steroids. It must also be remembered that aspergillosis can be both primary and secondary. Allergic and colonizing aspergillosis are often primary conditions, whereas invasive and especially disseminated aspergillosis are secondary to other diseases. This division of primary and secondary is often overemphasized.[51] Obvious underlying disease presents no problem in classification, but to call aspergillosis primary when nothing else is found speaks more of our lack of diagnostic ability than the inherent pathogenicity of the fungus. Impaired defenses can be generalized, regional, local, or transient. The ubiquitous opportunist, *Aspergillus* would take advantage of any such situation.

The pathogenetic spectrum of aspergillosis is complicated, confusing, and, as stated earlier, represents a group of diseases. The clinical diseases will be discussed as categories that, most simply, allow separation and characterization of the particular syndrome. They are:

A. Pulmonary aspergillosis
   1. Allergic aspergillosis
      a. Asthma
      b. Bronchopulmonary aspergillosis
   2. Colonizing aspergillosis (aspergilloma)
   3. Invasive aspergillosis
B. Disseminated aspergillosis
C. Central nervous system aspergillosis
D. Cutaneous aspergillosis
E. Nasal orbital aspergillosis
F. Iatrogenic aspergillosis

Allergic aspergillosis is a disease that has only recently been defined, and its frequency appreciated.[10, 15, 17, 42] The other forms of disease are still relatively rare. Aspergillus infections of the ear and eye are discussed in Chapter 23 and toxicosis due to ingestion of metabolic products in Chapter 24.

Pulmonary Disease

ALLERGIC ASPERGILLOSIS

Asthma. Asthmatic allergy to spores of the various *Aspergillus* species is a well-known and defined disease. It does not vary significantly from allergy to other types of dander and pollen or spores. The spores seldom germinate in the bronchial passages. The clinical picture is that of segmental shadowing, most frequently in the upper lobes. It has a tendency to recur in the same segment.[17] Transient infiltrates characterize the immediate reaction, and linear fibrosis is found in chronic disease. Intermittent fever, cough, wheezing, chills, malaise, aches, pains, and peripheral blood and sputum eosinophilia are seen. Pulmonary function studies demonstrate a restrictive defect and impairment of gas exchange. Reaginic antibodies (IgE) are present and can be transferred into donors or monkeys.[15] These patients usually do not have precipitating antibodies or demonstrate an Arthus type reaction to injected antigen.[10]

Bronchopulmonary Aspergillosis. Also called mucomembranous *Aspergillus* bronchitis, this form of aspergillosis may develop as an exaggeration of the above disease, in late-onset asthma, or as some other manifestation of atopic diathesis. The clinical symptoms are similar, but more chronic and severe. Bronchial plugging occurs. The peripheral shadows are consistent with areas of collapse distal to bronchial occlusion. Bronchoscopy reveals the fungus to be growing in patches of various sizes perpendicularly to the bronchi. Fruiting heads and spores are sometimes produced. The underlying mucus membrane is red and congested but not invaded. Coughing produces sputum in which the fungus may be seen by direct smear (Fig. 20–1) and from which it may be cultured. In the more than 200 patients reviewed, positive sputum cultures were a constant finding, and they remained so during the length of the disease.[10, 15, 17] A delayed type skin reaction, as well as immediate skin reactions, is demonstrated by these patients. In most of these cases, a type 3 or Arthus hypersensitivity can be elicited. The pathology of the disease is thought to be owing to both the immediate and delayed types of hypersensitivity. The most frequently isolated organism, *Aspergillus fumigatus,* is known to produce powerful endotoxins[57] and a C substance.[31] The former may

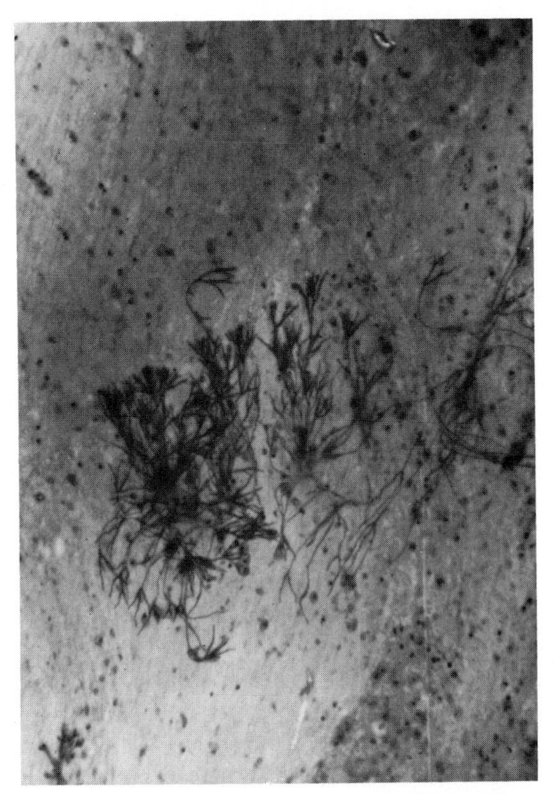

Figure 20–1. Bronchopulmonary aspergillosis. Direct smear of sputum. Clumps of aspergilli are seen that demonstrate fingerlike dichotomous branching. ×100. (From Rippon, J. W. *In* Burrows, W. 1973. *Textbook of Microbiology.* 20th ed. Philadelphia, W. B. Saunders Company, p. 737.)

also contribute to the disease. An absolute eosinophilia is present. Precipitins are demonstrable in the sera of almost all patients.[17, 42] It is now considered that allergic aspergillosis is one of the more common causes of pulmonary eosinophilia.[17] The other diagnostic criteria are episodic airway obstruction, a one-second forced expiratory volume less than 70 per cent of vital capacity, and the finding of abundant fungi in culture. This latter criterion consists of culturing the fungus from more than one specimen or growing more than two colonies from one specimen. The sputum produced is gelatinous and sometimes bloody. The radiologic picture is that of segmental shadowing and a migrating type of recurrent pneumonitis, usually in the upper lobes and especially in the apical segments. The other lobes and lingula are not frequently involved. The shadows of mucoid impaction are continuous, with the hilum proximally and distally blunted. Dilated bronchi sometimes appear opaque and at other times lucent. Cylindrical bronchiectasis of the upper lobes is seen on

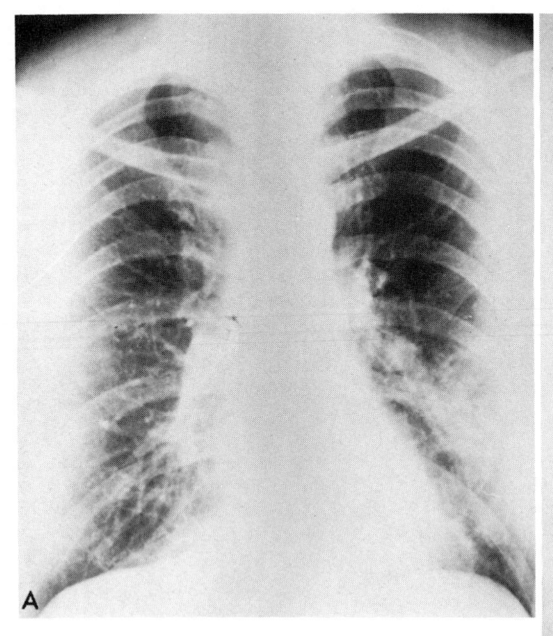

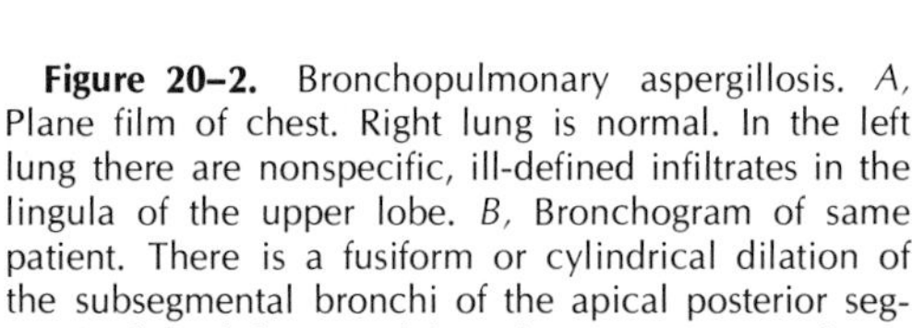

Figure 20–2. Bronchopulmonary aspergillosis. *A*, Plane film of chest. Right lung is normal. In the left lung there are nonspecific, ill-defined infiltrates in the lingula of the upper lobe. *B*, Bronchogram of same patient. There is a fusiform or cylindrical dilation of the subsegmental bronchi of the apical posterior segment of the left upper lobe. There is tapering of posterior bronchi to close the normal diameter. Elsewhere in the lung, chronic bronchitic changes are seen. (Courtesy of J. Fennessy.)

bronchograms (Fig. 20–2).[17,33a] A striking feature of this disease is the marked change in x-ray patterns associated with relatively mild clinical symptoms.

COLONIZING ASPERGILLOSIS (ASPERGILLOMA)

This condition, with the development of "fungus balls," may result from chronic allergic aspergillosis[9] or from colonization of preformed cavities caused by other diseases. The former condition is associated with "eosinophilic" pneumonia and bronchiectasis. The patients usually have a history consistent with chronic allergic aspergillosis, and an aspergilloma forms in an ectatic bronchus (Fig. 20–3 FS C 22). Clinically the symptoms are those of the allergic disease, but bouts of severe hemoptysis are more regularly seen. These patients have transferable reaginic (IgE) antibodies and precipitins (IgG), and demonstrate immediate and late skin sensitivity.

Another type of primary aspergilloma is seen in patients without allergic disease. Over a period of time, localized areas of infiltrate are seen to develop fuzzy edges that gradually become rounded and form cavities. These may show internal opacity with a radiolucent crest (Monod's sign). If resected at this time, masses of live fungal elements are found. Some cavities may excavate or increase in size, and others may disappear with residual fibrosis. Resection of old unexcavated lesions show deliquescent masses of dead hyphae.

Aspergilloma secondary to other cavitary disease is now considered to be a frequently encountered disease. According to a large survey conducted by the British Tuberculosis Association,[5] it occurred in 12 per cent of healed tuberculosis patients on radiologic evidence, and an additional 5 per cent had precipitating antibodies suggesting colonization. These patients lack reaginic antibodies, but IgG antibodies are universally present (Fig. 20–4). The immediate and delayed skin reactions are usually lacking. The second

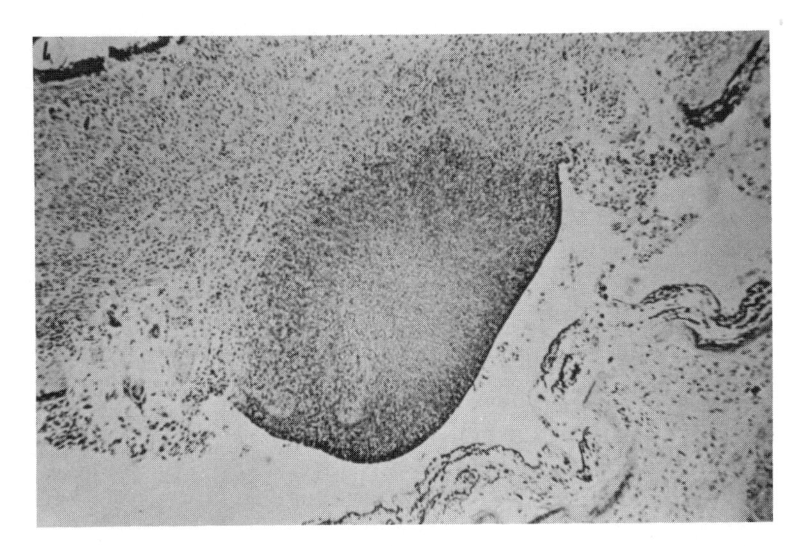

Figure 20–3. FS C 22. Aspergilloma. Small aspergilloma formed in an ectatic bronchus. There is erosion of underlying epithelium. ×100.

most common predisposing cavity is that resulting from healed lesions of sarcoidosis.[10,27] It is also found on bronchogenic cysts, histoplasmosis cavities,[45] and in areas of old radiation fibrosis.[28]

The fungus grows as one or more brownish balls within the cavity. These consist of tangled masses of mycelium, which are often found sporulating if there is an air space above them. The cavity is usually not filled, and both cavity and fungus ball may be seen to enlarge over a period of time.[38,52] The smooth wall of the cavity is lined by metaplastic epithelium. Inflammatory reaction around the cavity may be severe, and fibrosis occurs. Two or more bronchi may be connected to the same cavity. Where they converge into the cavity they are both dilated, giving an x-ray picture essentially pathognomonic for the disease.[16]

The characteristic x-ray picture is that of a uniform opacity, not dense, which is round or oval in shape and has a radiolucent crescent (Monod's sign) over the upper portion of an internal mass (Fig. 20–5). Sometimes a complete circle of air is seen around the mass.[38] Changing the position of the patient causes the density to shift. The most frequent location of aspergillomas is the apex of the lungs. Both right and left sides may be affected,[25] but there appears to be some predilection for the right.

The principal clinical feature of aspergilloma is recurrent hemoptysis (Fig. 20–6 FS C 23). This varies from small episodes of blood-tinged sputa to fatal exsanguinating hemorrhage (Fig. 20–7 FS C 24). In a recent series, the mortality rate due to hemoptysis was considerably higher than had previously been noted, indicating that aspergilloma

Figure 20–4. Aspergilloma. Immunodiffusion test of a patient with aspergilloma. The patient's serum is in the center well, and an *Aspergillus fumigatus* antigen is in the peripheral well. At least 18 bands were detected. More bands are present in cases of aspergilloma than in any other form of aspergillosis.

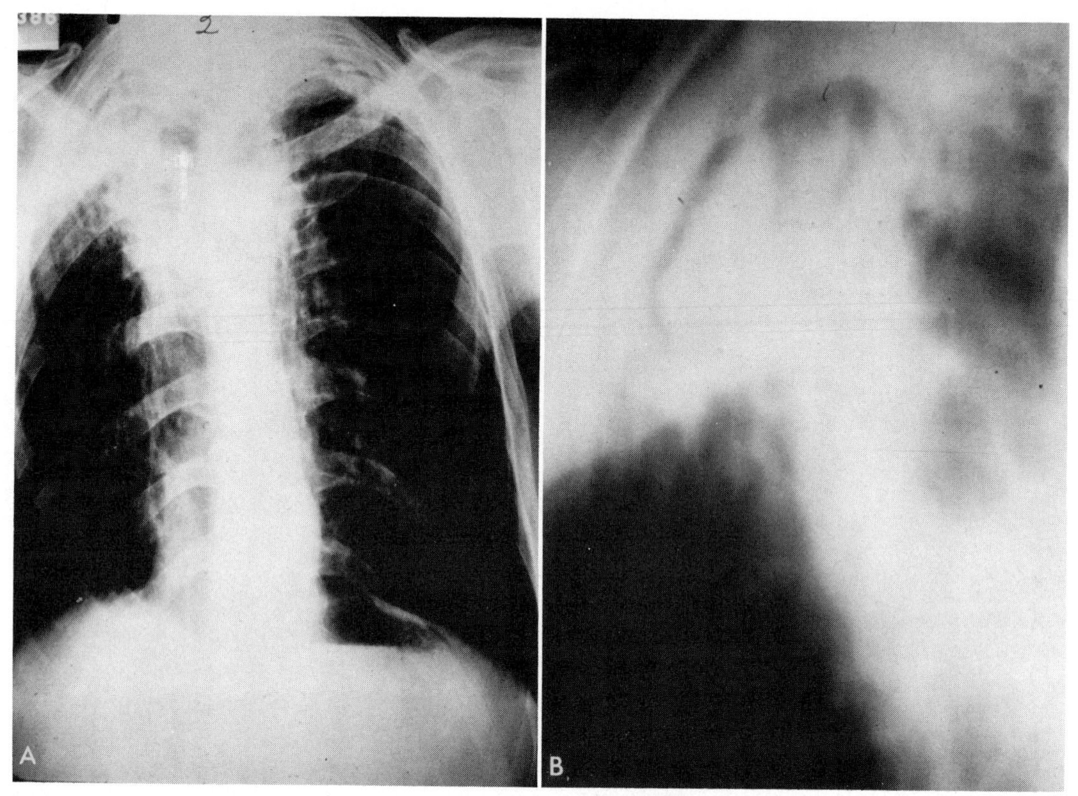

Figure 20–5. Aspergilloma. *A,* Plane film of chest. The lungs are over-inflated. There is scarring in upper lobes and bilateral, apical, pleural thickening, which is most marked on the right. There is opacification of the right apex. A density partly surrounded by air is seen in the center. There is also reduction in volume of the upper lobe. *B,* Tomogram of same patient. There is a large cavity containing a large solid density partly surrounded by a crescent of air (Monod's sign). (Courtesy of J. Kasik.)

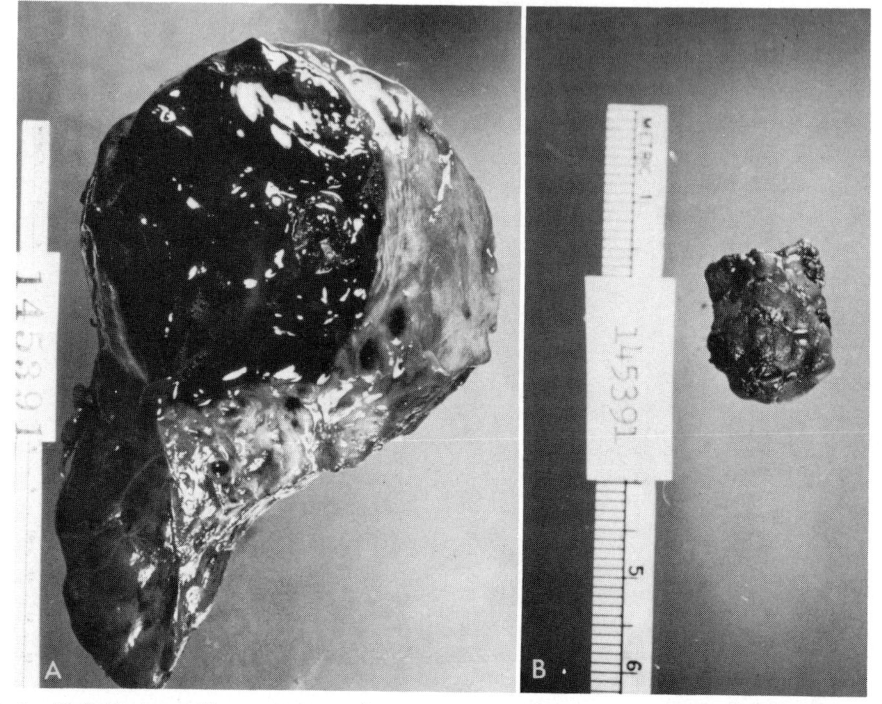

Figure 20–6. FS C 23. Aspergilloma. *A,* Aspergilloma in an old sarcoid cavity. The patient had recurrent hemoptysis. In the excised lobe, a large clot fills the cavity. The aspergilloma is in the upper central area. *B,* Fungus ball removed from cavity. (Courtesy of S. Thomsen.)

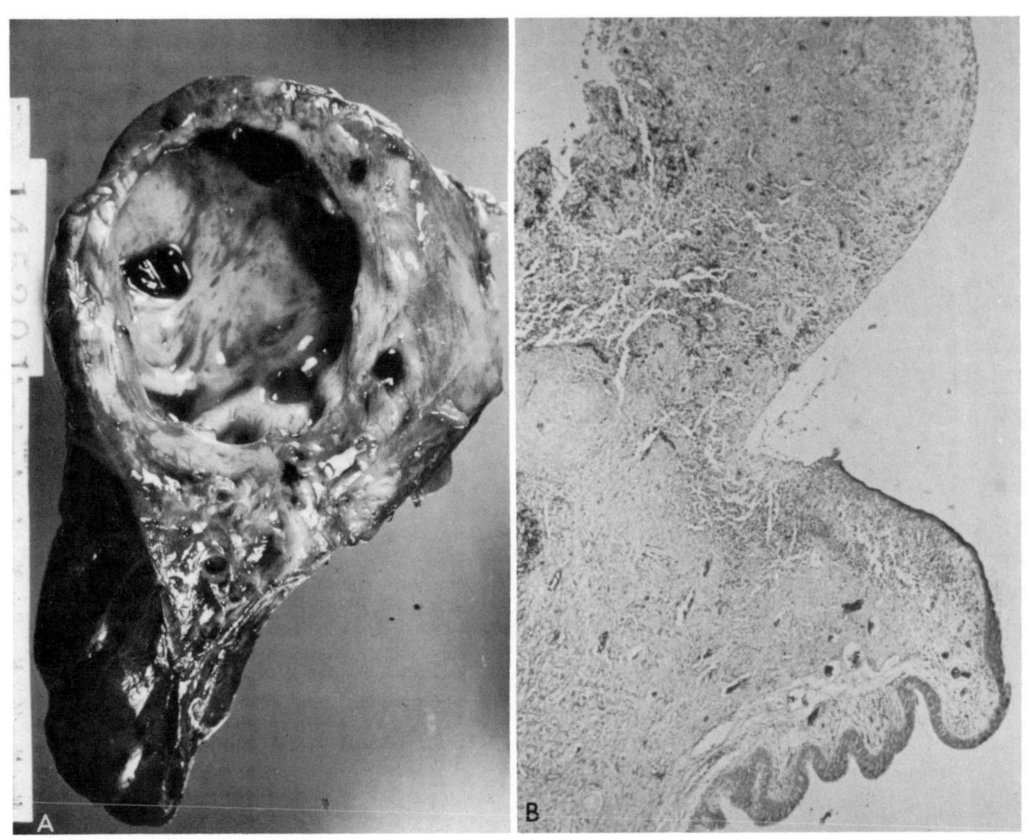

Figure 20–7. FS C 24. Aspergilloma. *A,* Cavity cleared of clot and fungus ball. The wall is thick, fibrous tissue. There are several expanded bronchi leading from the central cavity. *B,* Section of cavity wall. The epithelial lining has been eroded in one area, exposing granulation tissue. This necrosis and erosion may be due to endotoxins released by the organism. ×100.

should not be considered a benign condition.[10] Confirmation of the diagnosis by culture is often lacking, because the fungus ball may not be connected with a patent bronchus. In such cases, the etiology may not be established until after surgical removal and culture of the lesion (Fig. 20–8 FS C 25). Increase in size of the lesions has been noted in many series.[9,52] It has been postulated that the fungus ball acts as a valve, permitting entry of air during inspiration and preventing its escape at exhalation. This necessitates that the open end be downward and may infer the reason for the frequency of apical location. Lesions may show lobation of the fungal mass, with excavation of the surrounding tissue, indicating that an invasive process is possible. Zonation of the mycelial elements with layers of pyknotic host cellular debris verifies a dynamic process in the evolution of the lesion (Fig. 20–9). It has been postulated that some of the pathologic disease seen in the tissue is owing to endotoxin released by the fungus.

Resolution or regression of cavities has been noted,[26] but this appears to be an uncommon occurrence. When the epithelial lining is eroded, the fungal mass comes into contact with formed granulation tissue, and invasion of the cavity by granulation tissue may occur. Old cavities may contain deliquescing masses of dead hyphal elements and show resorption of the debris by giant cells.[25–27]

Although aspergillomas are associated with old tuberculosis cavities, the coexistence of the two diseases in the same area is controversial. *A. fumigatus* produces antibiotic substances, and tubercle bacilli are usually absent from lesions in which the fungus is present. Rarely, however, acid-fast bacilli have been found among hyphal masses in an aspergilloma.[52]

The saprophytic infestation of the bronchial tree by *Aspergillus* is also known.[10] These patients show no evidence of hypersensitivity or clinically significant pathology, but aspergilli, such as *A. fumigatus,* are constantly

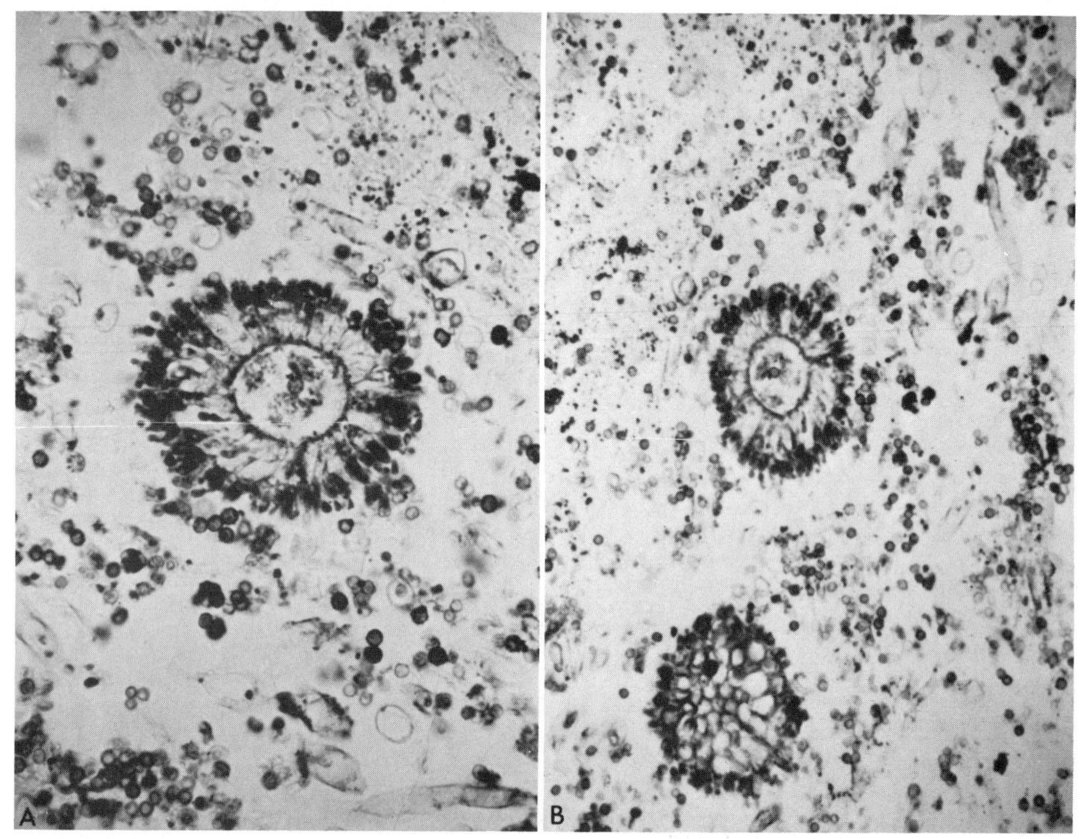

Figure 20–8. FS C 25. Aspergilloma. *A,* Unstained section from a case of aspergilloma caused by *Aspergillus niger.* Since there is an air space above the ball, fruiting heads of the aspergillus are formed. This does not occur in any other form of aspergillosis. *B,* By the color and architecture of the spores and the presence of long primary and short secondary series of sterigmata, one can essentially identify the organism as *A. niger* on histologic section. ×440. (Courtesy of S. Thomsen.)

excreted and can be cultured from their sputum.

INVASIVE ASPERGILLOSIS

In this uncommon form of the disease, the mycelium is present in the lung tissue. This condition can be seen to develop from any of the preceding diseases, or *de novo.* It is chronic or acute, sometimes with a rapid fulminating, fatal outcome. The usual predisposing factors for opportunistic infection are present, but the disease may occur in an apparently normal patient as well.[56,58] Rarely, the disease has developed after severe coccal pneumonia.[10] Clinically the disease presents as a pneumonia with fever, cough, leukocytosis, and other signs of respiratory distress. The x-ray findings may show diffuse involvement or consolidation with a single mass, resembling the picture of lung cancer. Sometimes the lesions round up, excavate, and become aspergillomas (Figs. 20–10 and 20–11).

Several recent reviews have emphasized the increasing frequency of invasive aspergillosis among patients with leukemias and lymphomas.[34,36] The former group seem to be particularly prone to developing this complication.[34] In one review of patients with renal transplants, aspergillosis was more frequently observed in those receiving antilymphocyte serum therapy.[2]

The sinuous hyphae radiate from a central focus, branching freely and giving a characteristic picture histologically (Fig. 20–12 FS C 26). The surrounding tissue necrosis may be quite extensive. This has suggested to some that part of the pathologic disease is owing to an endotoxin derived from the fungus.[11,51] The characteristic radiating fungal masses are sometimes seen in routine autopsy studies. They apparently have no causal relationship to the demise of the

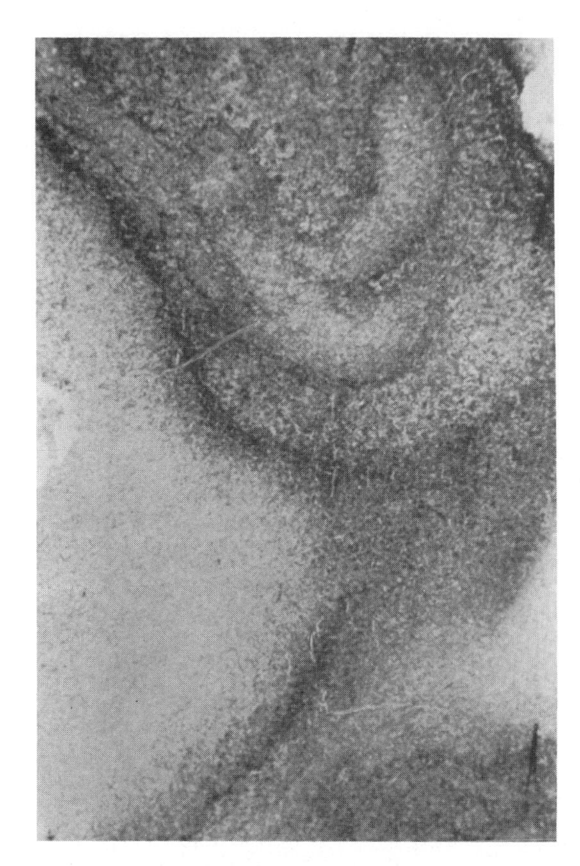

Figure 20–9. Aspergilloma. Cut section of aspergilloma, showing zonation of hyphal growth and deliquescence of central hyphal strands. GMS. ×100. (Courtesy of S. Thomsen.)

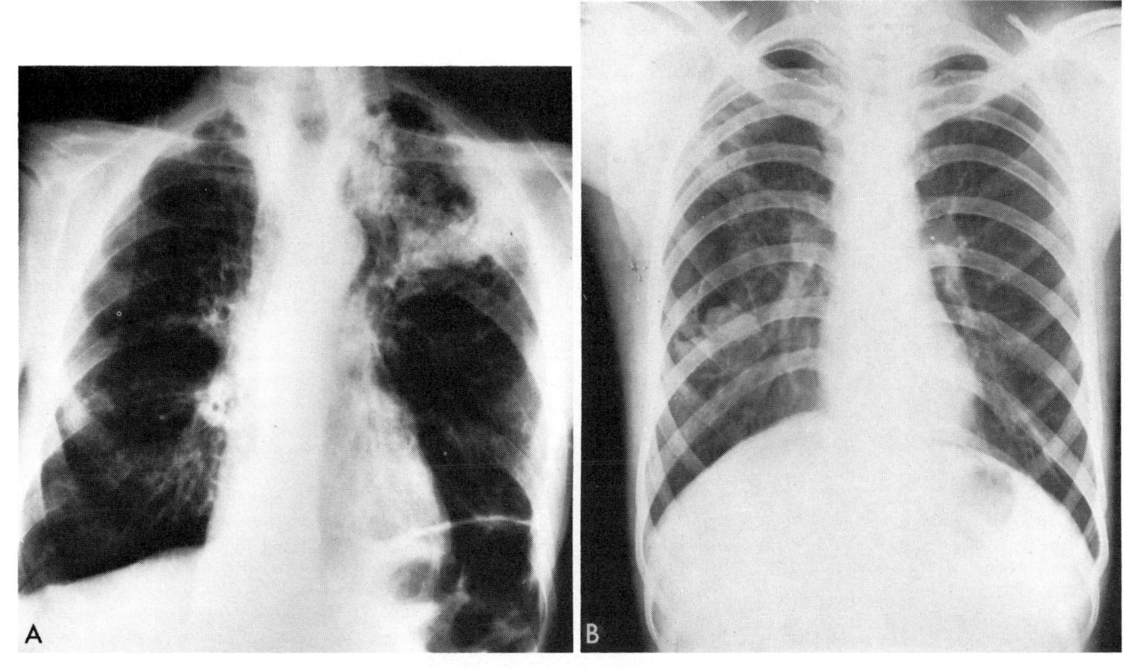

Figure 20–10. Invasive aspergillosis. *A,* A large, irregular cavity may be seen in the left upper lobe, surrounded by an infiltration that extends toward the hilum. Elsewhere there is evidence of obstructive lung disease. The patient had leukemia. *B,* Numerous thin-walled cavities are seen in the lungs. The one in the right lower lobe contains an air-fluid level. Irregular alveolar infiltrates containing small cavities are seen in the right upper lobe. The patient had leukemia.

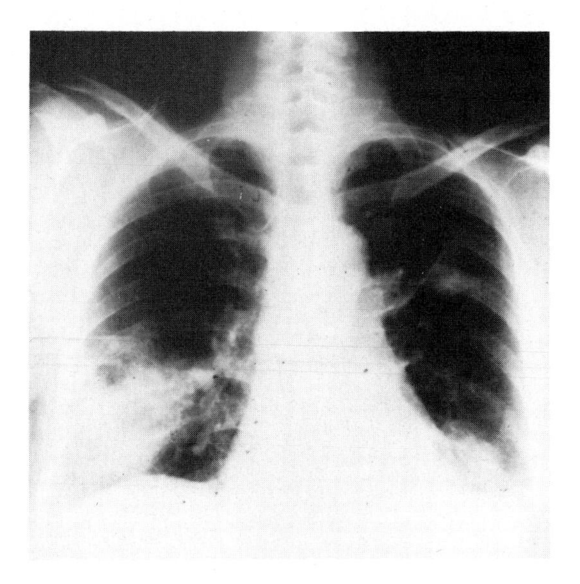

Figure 20–11. Invasive aspergillosis. There is a dense infiltrate in the right lower lobe, containing an irregularly outlined cavity. Nonspecific zones of alveolar infiltrate are present in the opposite lung. Subsequently these zones increased in size, and cavities appeared. The patient had hepatitis and was treated with steroids. (Courtesy of J. Fennessy.)

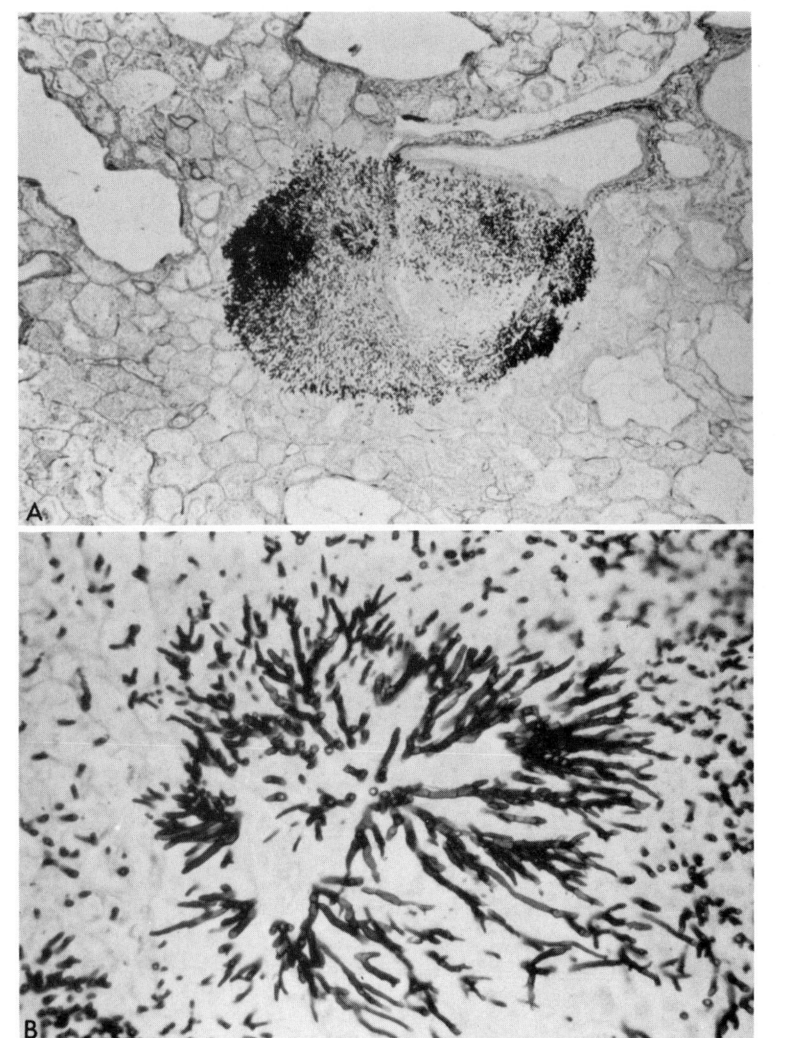

Figure 20–12. FS C 26. Invasive aspergillosis. *A,* Characteristic radiating pattern of invasive *Aspergillus.* ×100. *B,* Enlarged view of central area. GMS. ×440. (Courtesy of S. Thomsen.)

patients, but represent an incidental or terminal colonization. This situation is especially common in farming areas, where frequent contact with spores is possible. That these restricted invading colonies are common in healthy domestic animals has been demonstrated by Austwick.[1]

Disseminated Aspergillosis

A new and as yet rare disease, this appears to be a product of the antibiotic-steroid-cytotoxin era. The first case reported by Linck in 1939 and almost all subsequent cases have implicated either antibiotics or steroids or both. Many had underlying disease or associated disease, including pneumonia, diarrhea, leukemia, eclampsias, and hepatitis.[49,56] A few patients did not have any apparent abnormality, however. The presenting symptoms are those of an acutely ill patient. Dissemination to various organs causes associated symptoms, such as neurologic signs,[49] the Budd-Chiari syndrome,[62] and so forth. In the review by Tan et al.,[56] the lungs were the most frequently involved organ, followed by the brain and kidney. Only rarely has dissemination occurred that does not involve the lung. Other portals of entry aside from the pulmonary are possible, however. A "mycotic aneurysm" of the aorta was reported by Meyerwitz et al.[34a] In this immunosuppressed patient, the primary focus appeared to be the lung. Involvement of bone is rare, but has been recorded.[46a,52a]

Central Nervous System Aspergillosis

This disorder has been found as the result of iatrogenic procedures, in drug addicts following injection of contaminated material, and sometimes without known portal of entry or primary focus. The presenting symptoms are those of acute meningitis, and the disease is rapidly fatal. Extensive necrotic lesions of brain, meninges, and vessels are seen at autopsy. Very few clinical cures have been recorded following intrathecal treatment with amphotericin B. Some cases of pulmonary disease have presented as spinal cord compression.[52a]

Cutaneous Aspergillosis

Although it has been reported, this is a rare disease. It is usually secondary to dissemination, but may be primary. In the latter, the lesions consist of multiple nodules, and the skin is thick, edematous, and has a purplish discoloration. One such case was mistaken for lepromatous leprosy and was so treated for over a decade. The lesions consisted of multiple granulomata containing *Aspergillus* in the superficial middle and deep dermis. Numerous giant cells were present, and there was central necrosis. The etiologic agent was *A. niger*. The disease responded to topical nystatin.[6]

The skin lesions in disseminated cases start as small red discrete papules which become pustular. Biopsy shows them to consist of small microabscesses and granulomata with central necrosis. Small radiating colonies of *Aspergillus* are seen.[12a,58]

Nasal-Orbital Aspergillosis

This disorder consists of an aspergilloma in the nasal sinuses that may eventually involve the orbit of the eye. Throughout the world this is a rare disease, but it is not uncommon in the Sudan. Seventeen cases have been reviewed by Milosev et al.[35] The lesions most often involved the ethmoid sinuses, although disease in all cavities was recorded. The most common presenting sign was unilateral proptosis and swelling of surrounding tissue of varying severity. The etiologic agent in all cases was *A. flavus*. Other species have also been reported. Treatment consisted of surgical removal of granulomata and draining of cavities. An association of cure with the relief of partial anaerobiosis was hypothesized by the authors.

Iatrogenic Aspergillosis

Aspergilli are constantly present in the environment and grow on all types of organic debris. For this reason they may often contaminate hospital rooms and supplies. They may thus, inadvertently gain entrance to susceptible patients by many portals. *A. fumigatus* was growing in the procedure room and may have been growing on leaking dialysis bags in a case of fatal disseminated disease following dialysis.[49] Meningeal involvement has occurred following intrathecal penicillin therapy. In a personally observed case, severe meningitis occurred in an otherwise healthy man three weeks following a disk fusion. In this case *Aspergillus terreus* grew out from many spinal fluid cultures. Treatment with amphotericin B was successful in arresting the infection, but the patient is at present paraplegic due to the severe arachnoiditis that accompanied therapy.

DIFFERENTIAL DIAGNOSIS

The diagnosis of pulmonary aspergillosis can be made only with demonstration of fungal elements in pathologic material and repeated isolation of the fungus in culture—the latter in the so-called "abundant fungus" method (see Laboratory Identification). Hyphal strands are easily demonstrated in sputum by direct examination, and the organisms are easily grown on standard laboratory medium. In a typical case, the diagnosis is readily made. The mere presence of *Aspergillus* spores or a colony in culture is not sufficient evidence for this diagnosis. In some aspergillomas and invasive pulmonary aspergillosis, fungal elements are not present in sputum. The former may be diagnosed by typical radiologic picture only; however, a recent blood clot within a cavity gave a similar x-ray picture. In the latter disease, the diagnosis may not be made until surgical removal or at autopsy.

Aspergillosis is so frequently a secondary infection in tuberculosis, carcinoma, sarcoid, bronchiectasis and other fungal diseases that an underlying malady should be sought. Mucormycosis and aspergillosis as well as other opportunistic infectious agents are not uncommonly seen in the same patient.

PROGNOSIS AND THERAPY

The prognosis of aspergillosis depends entirely on the type of disease evoked and the physiologic state of the patient. Allergic aspergillosis tends to become more severe and the duration of attacks longer and more chronic. The resulting pathology is that of increasing respiratory distress with bronchiectasis, segmental collapse, and fibrosis. Colonizing aspergillosis (aspergillomas) may remain chronic for years without much distress except for bouts of hemoptysis and associated lung pathology. Such colonization (fungous balls) are known to become invasive, and a high mortality is noted in some series.[5,10]

Invasive aspergillosis is an extremely serious disease requiring intensive therapy, the success of which depends in large part on the patients' underlying illness and immune competence. Few survive.

In terms of treatment, allergic aspergillosis is approached both as a hypersensitivity disease and as an infection. In some cases of short duration, prednisone has cleared the condition without relapse.[15] A dose of 25 mg daily was initiated, which led to improvement in one week, and the drug was gradually decreased and discontinued over the next few months. Some investigators caution against the use of prolonged steroids in chronic cases.[10] The main object of therapy is to minimize the periods of bronchial plugging by mucus containing the fungus. Vigorous physiotherapy, bronchoscopic aspiration, and lavage are recommended. Steroids are used only during the acute phase. Disodium chromoglycate, an antiallergic agent without the side effects of steroids, has been suggested.[10] Inhalation of nystatin or natamycin (pimaricin) has been used to clear the bronchial walls of the organism by some investigators.[18]

Aspergillomas have been considered benign conditions which sometimes resolve spontaneously. However, several recent series have stressed the danger of death from massive exsanguinating hemoptysis during the development of invasive aspergillosis.[10,28] Therefore, therapy of aspergilloma is advised. The lesions are large and avascular, so that systemically administered antifungal agents are not usually helpful. Medical management has been used, especially in cases in which surgical resection was not possible. Ramirez introduced the method of infusing the lesion with amphotericin B through an indwelling catheter.[45a] Ikemoto has used intrabronchial instillation of amphotericin B. This led to cessation of growth and expectoration of the ball. The drug (200 mg dissolved in 10 ml of water) was instilled by tracheal puncture using a lumbar puncture (7 cm, 22 gauge) needle.[24,25] In a six-year follow-up, there had been no recurrence of the disease.[25] Intracavitary needling for the instillation of a paste containing amphotericin B or nystatin has been described by Krakowka et al.[29] Nystatin had a small edge as far as effectiveness was concerned. Aerosol amphotericin B was not useful in one series.[25]

The treatment of choice for aspergilloma is surgical resection whenever it is feasible. In a review of 70 cases, with 14 new cases, Kilman et al.[28] concluded that lobectomy was followed by the least number of complications. Local surgical evacuation followed by irrigation with natamycin (pimaricin) has been used by Henderson and Pearson.[18] Natamycin (pimaricin) given as an aerosol, 0.25 per cent suspension in Alevaire, 1 ml four times daily, has also been used.[10] The old standard treatment for all fungous infections was potassium iodide. It has also been used in aspergillosis with some success.

Invasive aspergillosis is often rapidly fatal and its treatment difficult. Chronic aspergillosis in an immunologically competent patient has been successfully treated, however. The aspergilli are not particularly sensitive to amphotericin B, but they appear to be quite sensitive to nystatin and natamycin (pimaricin). Natamycin has been given as an aerosol in the dosage mentioned above. The drug did not appear to be toxic when administered by this method. A dramatic success was achieved by Vetter and Schorr[58] using aerosol nystatin. Their patient, a child, had a diffuse, extensive, pulmonary disease with metastatic skin nodules. Nystatin orally did not give detectable blood levels. However, when dissolved in 10 per cent propylene glycol (1.5 M units in 500 ml) and vaporized in a mist tent (8:00 P.M. to 6:00 A.M.), blood levels could be detected. The patient required 2½ months of treatment, after which she was put on oral nystatin. She subsequently developed new skin lesions and required more inhalation treatment. She now appears healthy.

Surgical resection of the involved lung has been used when feasible. This procedure has varied from segmental lobectomy to pneumonectomy. Systemic and meningitic aspergillosis are generally fatal in spite of treatment.

PATHOLOGY

Gross Pathology. In routine autopsies of patients who have died of other diseases, it is not uncommon to find colonies of aspergilli in the bronchi, in ectatic cavities, and in other stagnant air spaces. The mycelial mat growing on such areas may be quite extensive, but there is usually little or no evidence of inflammation in the tissue.

The gross pathology of the several forms of aspergillosis is quite variable. In allergic aspergillosis, there is abundant hyphal growth in mucus plugs and a prominent inflammatory reaction in bronchi. In acute invasive disease, hyphae can be seen invading the wall of the bronchi and the surrounding parenchyma. The pathologic picture here is that of an acute, necrotizing, pyogenic, pneumonitic process. The pulmonary lesions are usually necrotizing abscesses or infarcts (Fig. 20–13 FS C 27). Similar to the etiologic agents of mucormycosis, aspergilli have a marked tendency to invade blood vessels, causing thrombosis. Such invasion may also result in dissemination to other organs. The pulmonary abscesses may be as massive as those seen in mucormycosis, or they may be small, multiple, yellowish lesions. In the rare chronic granulomatous aspergillosis, consolidation of the entire lung may occur. In contrast, there is little tissue destruction resulting from aspergillomas, as these develop in preformed stationary cavities. Cut sections of such cavities show brownish hyphal masses in the center and a fibrous or epithelial lining of the cavity wall (Fig. 20–6 FS C 23).

When dissemination to brain, kidney, and myocardium occurs, the metastatic lesions are acute necrotizing pyogenic abscesses. Infarcts surrounded by extravasated blood and ischemic tissue are also commonly seen (Fig. 20–14 FS C 28). Such reddish brown areas seen in cut sections of brain, heart, or kidney are almost pathognomonic for disseminated aspergillosis.

Histopathology. If the *Aspergillus* is growing in a cavity with an airspace (fungus ball), sporulating conidiophores are frequently observed (Fig. 20–8 FS C 25). This does not occur if the fungus is invading tissues. The hyphal units of aspergilli are stained reasonably well with hematoxylin and eosin, but unless searched for they will be missed. The Gridley and methenamine silver stains readily stain the fungus (Fig. 20–9). The PAS stain has sometimes been disappointing when used to demonstrate the presence of aspergilli in pathologic specimens.

The diameter of the hyphae ranges from

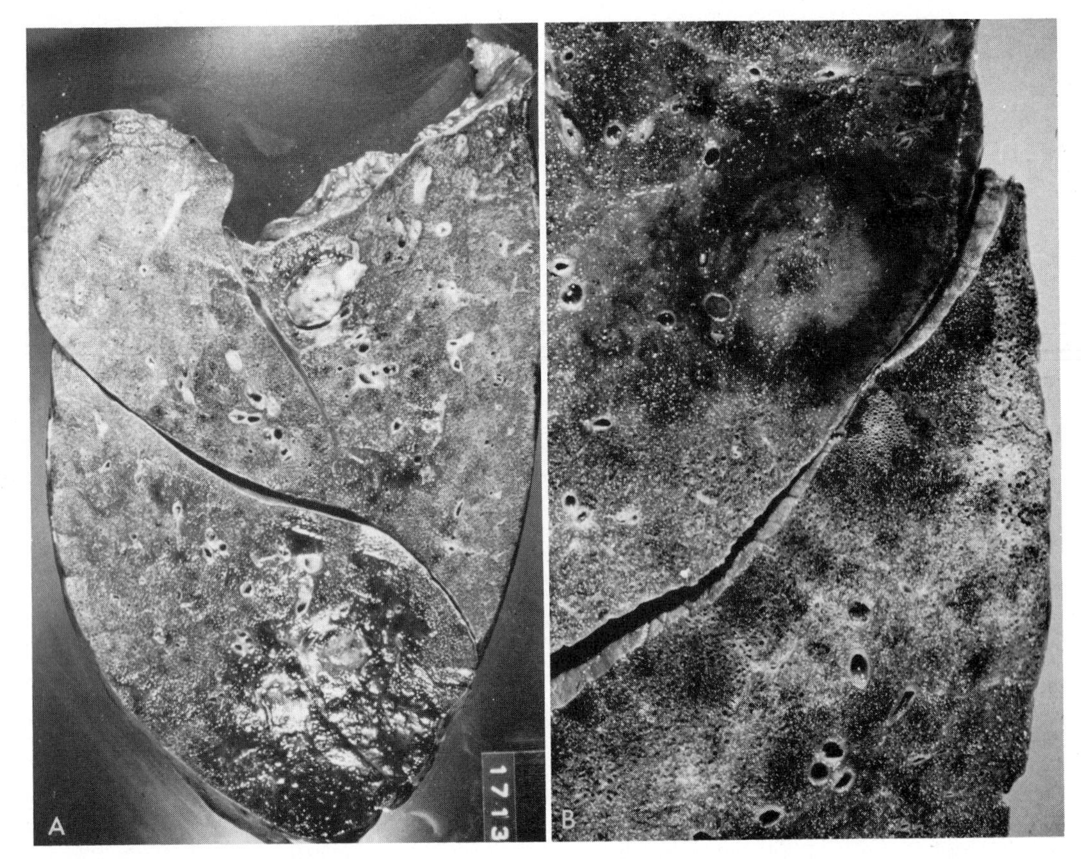

Figure 20–13. FS C 27. Invasive aspergillosis. *A*, Massive hemorrhagic consolidation and focal abscess formation of apical area of lungs. There is abscess formation in the lower lobe. *B*, Infarcted area of lung, showing hemorrhage and necrosis due to thrombosis in a mycotic abscess.

2.5 to 4.5 μ, with an average of 3 μ.[39] Hyphal strands characteristically demonstrate repeated dichotomous branching, with several hyphae oriented in the same direction (Fig. 20–15). This gives a brushlike or extended fingers-like appearance (Fig. 20–16). The hyphae are often seen to radiate from a focal point. The branches arise regularly at about a 45° angle. Multiple septation is usually present and demonstrated by the special fungal stains. If septae are not readily apparent, the hypha may resemble *Mucor* species. Conidia are not seen in tissue sections, but cross sections of hyphae may look like spores or yeasts. In the center of hyphal aggregates there may be swollen cells and cystlike and distorted hyphae. These may measure up to 8 μ in diameter. Members of the terreus-flavipes group produce aleuriospores in culture and in tissue.[44] Eosinophilic halos are sometimes seen around hyphae[39] (Fig. 20–17 FS C 29). Aspergillosis is differentiated from candidosis histologically by the following: (1) dichotonously branching hyphae and lack of budding yeasts are found in the former; (2) in candidosis there is always a mixture of budding yeasts; pseudomycelium, and irregularly branching septate mycelium.

The histologic reaction evoked by the invading aspergillus is generally that of the acute pyogenic type accompanied by necrosis. The leukocytic debris and perifocal necrosis may be extensive and is sometimes referred to as "diffusion necrosis." It is postulated that some of this necrosis is owing to the release of endotoxins from the fungus.[11,52,55]

Since chronic aspergillosis is an uncommon disease, granulomatous lesions of the lungs, nasal area, or other organs are rarely observed. Granulomata from nasal sinus aspergillosis are very hard and fibrosed.[35] Histologically they show aggregates of foreign body giant cells and well-formed fibrous tissue. The hyphae are usually sparse and distorted, and do not show the regular branching pattern seen in acute invasive aspergillosis.

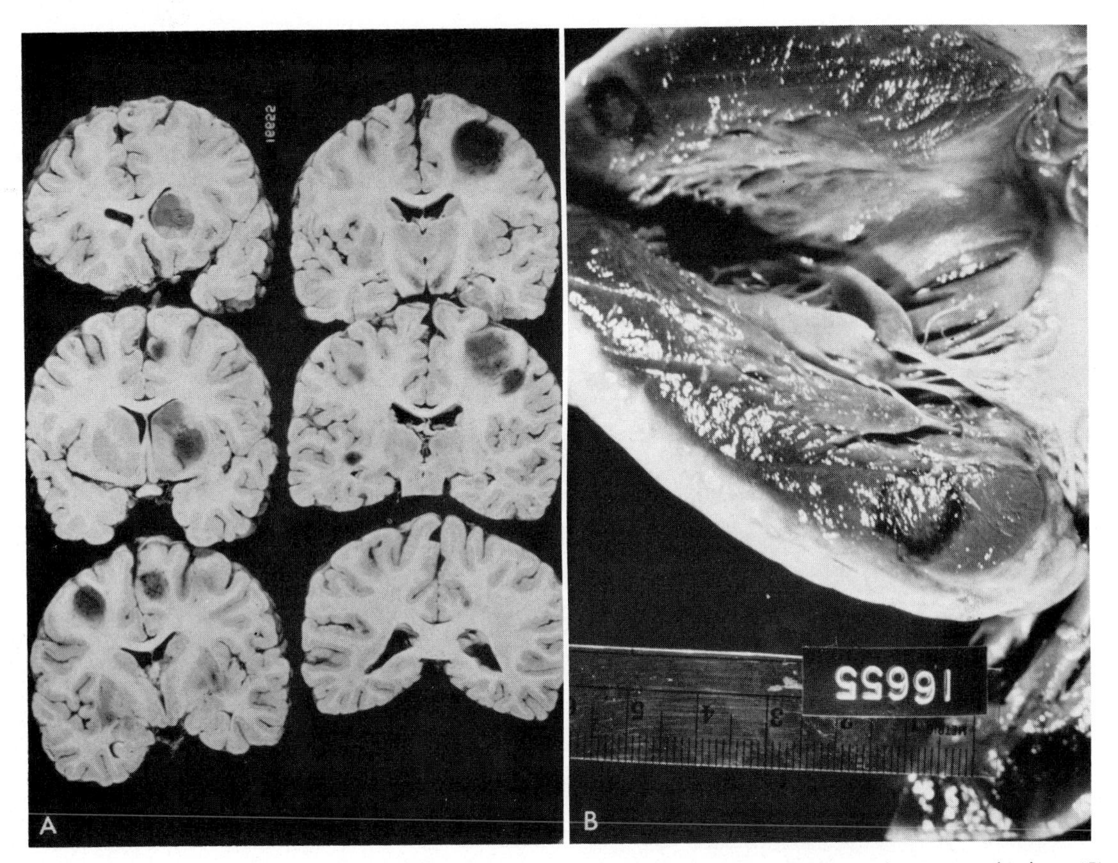

Figure 20–14. FS C 28. Invasive aspergillosis. Infarcted areas in disseminated aspergillosis in brain (*A*) and in heart (*B*). Aspergilli are one of the few fungi that invade the myocardium. (Courtesy of M. Warnock.)

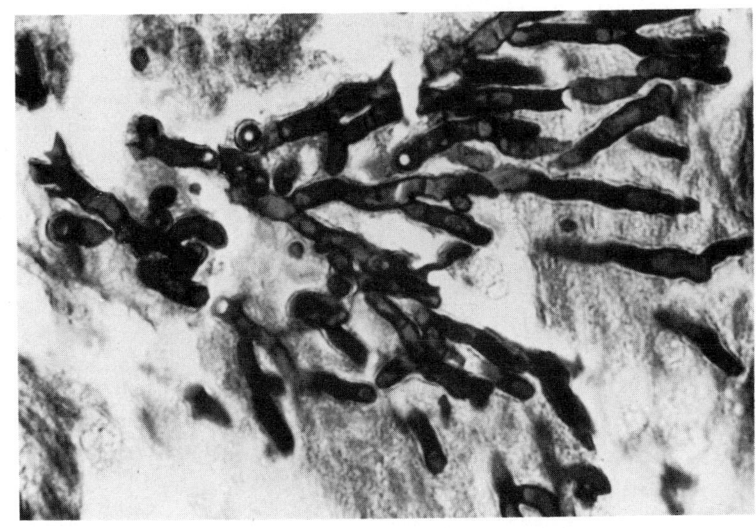

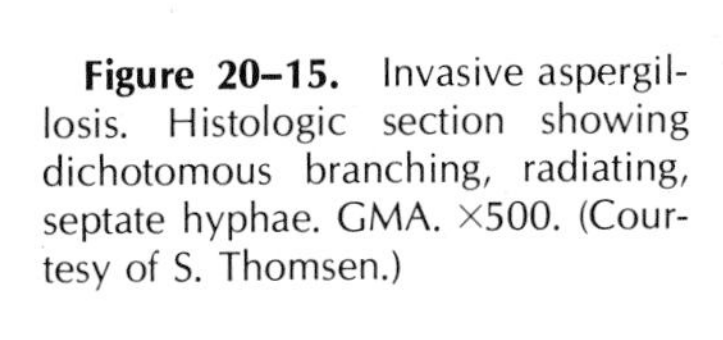

Figure 20–15. Invasive aspergillosis. Histologic section showing dichotomous branching, radiating, septate hyphae. GMA. ×500. (Courtesy of S. Thomsen.)

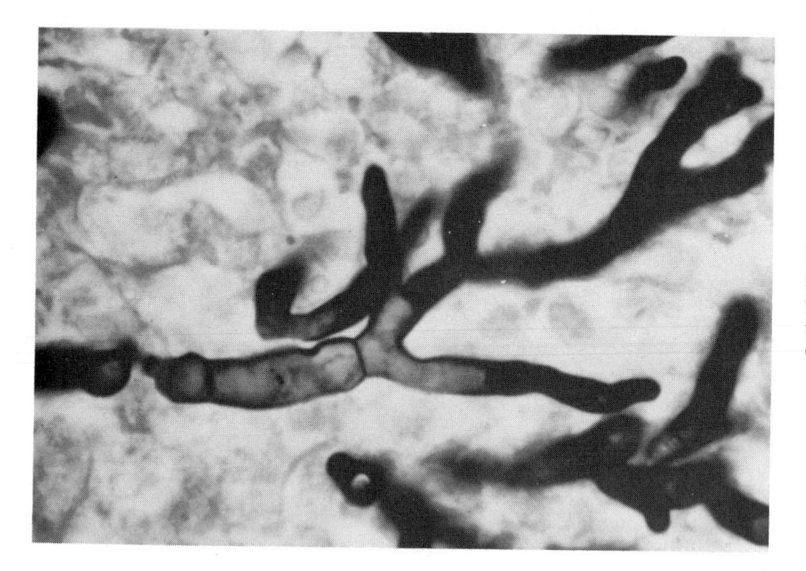

Figure 20–16. Invasive aspergillosis. Dichotomous branching and septation of hyphae. Sometimes there is an indentation of the hyphal wall at the point of septal formation. GMS. ×500.

ANIMAL DISEASE

Natural infection with aspergilli in birds has been known for many years and was the first form of the disease described. It is the most important fungal disease and one of the most important infectious agents of birds. Pleural, air sac, alimentary, meningeal, and systemic infections of chickens, turkeys, penguins,[6a] and other birds have been described. Such infections sometimes occur in epidemics, resulting in severe economic loss. Many epidemics have been traced to moldy grain. Acute aspergillosis is particularly common in young chickens, and it accounts for 10 per cent of all deaths in this age group. Acute illness and diarrhea are followed by convulsions and death in 24 to 48 hours. The pathology seen in infected tissue is essentially the same as noted in human disease.

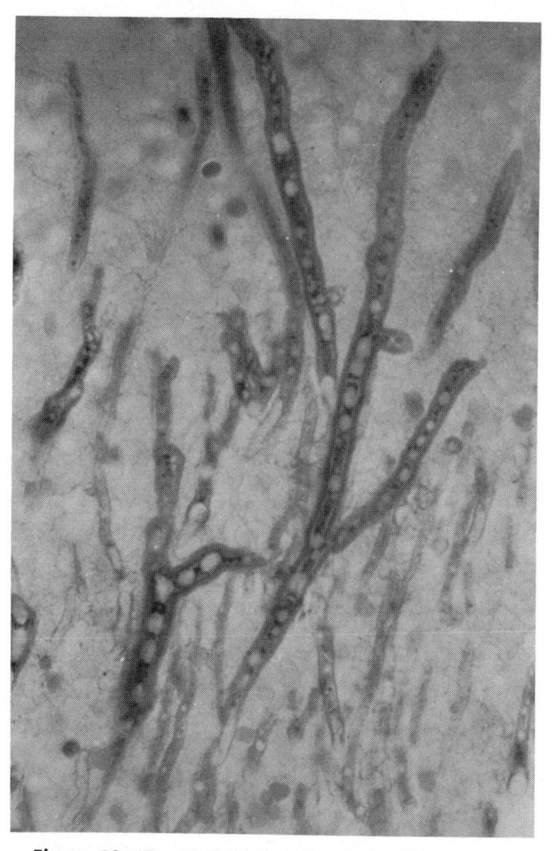

Figure 20–17. FS C 29. Invasive aspergillosis. This was from a case of chronic disease. There are eosinophilic sheaths around the hyphae, the Splendore-Hoeppli phenomenon. Hematoxylin and eosin stain. ×400.

Aspergillosis in mammals is not as commonly encountered as in birds. It has been reported sporadically in the veterinary literature. The most important and well-defined form of animal aspergillosis is called "mycotic abortion." In this condition, there is an infection of the placenta of the cow, with subsequent loss of the calf. Radiating mycelial elements are seen in tissue section. Invasive aspergillosis has been reported in horses, sheep, pigs, and several other species. Foci of radiating hyphae of aspergilli have also been found in the tissues of normal healthy cattle.[1]

Experimental Disease. Injection of the spores of *A. fumigatus* or other thermotolerant species into mice regularly results in the production of disease. Strains vary markedly in virulence, however; Sidransky and Friedman obtained only a transient pneumonitis

Figure 20–18. Aspergillosis. Torticollis of the mouse produced by infection of the central nervous system by *A. terreus*.

when mice treated with antibiotics were exposed to *Aspergillus* spores.[53] A fulminating disseminated aspergillosis developed if the mice received steroids also. Similar findings were described by Sandhu.[50] Chick observed that experimental leukemia of chickens produced by injection of myoblastosis virus rendered the birds more susceptible to aspergillosis than normal controls.[8] In comparing the relative virulence of several *Aspergillus* species, Ford and Friedman[13] found that members of the *A. flavus* group consistently killed mice in doses as low as 10^4 spores. In studies by Pore and Larsh,[44] torticollis, lateral and truncal ataxia, and multiple acute lesions of the cerebellum and cerebrum were produced by members of the terreus-flavipes group (Fig. 20–18). Aleuriospores which are formed by the species of this group in culture were also found in histologic sections of cerebral lesions. Rippon et al.,[48] by manipulation of oxidation-reduction potentials and temperature, induced *A. sydowi* to grow as a budding yeastlike form. These cells caused a systemic mycosis when injected into mice. Smith[53b] has found some improved resistance in experimental infection in mice when the animals had been given a live vaccine of spores prior to challange. A new species, *A. bisporus*, is pathogenic for mice but has not as yet been recovered from human infection.[29a]

A disease similar to allergic aspergillosis has been produced in rhesus monkeys by Golbert and Patterson.[15] Both the immediate (due to reaginic antibody) and toxic complex (due to precipitating antibody) types of reactivity were transferred to the animals. This was accomplished by infusion of the patient's serum into the animals (passive systemic transfer), followed by aerosol challenge with *Aspergillus* antigen. Cutaneous reactivity and pulmonary lesions consistent with the donor's illness were found.

BIOLOGICAL INVESTIGATIONS

The effect of steroids, amphotericin B, and nystatin on the growth of *Aspergillus* species in culture was studied by Mohr et al.[37] Of the steroids tested, estradiol in concentrations of 0.05 μg per ml inhibited growth, but testosterone, progesterone, and potassium iodide (1 part per 100) did not. Amphotericin B at 5 μg inhibited one strain but not the others. Nystatin in a concentration of 100 units per ml inhibited all strains. A propensity for *A. fumigatus* to grow in microaerobic cavities and necrotic tissue was suggested by Okudaira and Schwartz.[39] However, they found the growth rates equal under aerobic or microaerobic environments, but greater at 37°C than at 25°C.

Endotoxins from *A. fumigatus* were first described by Henrici.[20] Acute hemorrhagic necrosis was produced when endotoxin derived from *Aspergillus* was inoculated into rabbits. This may explain the acute toxic reaction produced by inhalation of masses of spores. Tilden[57] described and characterized a hemolysin and several endotoxins from cultures of *A. fumigatus* and *A. flavus*. The endotoxins were extremely potent when tested in rabbits. A neurotoxin was isolated by Ceni and Besta in 1902, but its identity is unknown. *A. flavus* produces a potent hepatotoxin and carcinogenic agent, aflatoxin, when growing on certain grains. It has been suggested that some human disease may be owing to this agent following the ingestion of contaminated food.[11] The production of toxins by pathogenic fungi is, in general, rare.

IMMUNOLOGY AND SEROLOGY

Aspergillosis is an opportunistic infection, and the normal individual is not susceptible to the disease. However, inhalation of large masses of spores of some species, particularly *A. fumigatus*, can cause an acute toxic reaction or fulminating infection or both. There is no clinical or experimental evidence that an immunity develops if the patient survives the disease. Hypersensitivity reactions are developed by some atopic patients as a result of continued exposure to spores, mycelium, or antigen. Reaginic antibodies are present in patients with asthma, and both reaginic and precipitating antibodies are found in cases of bronchopulmonary aspergillosis.

Serologic evaluation of the various forms of aspergillosis has seen significant advances in the past several years. It had long been known that atopic patients had a positive skin test to *Aspergillus* extracts and that some patients with the invasive disease had either an immediate and delayed reaction or only a delayed reaction. The studies by Pepys, Longbottom, and others have correlated the various clinical types of disease with skin reactions and immunodiffusion (I.D.) tests.[10,17,19,31,42] Serum precipitins demonstrated by the immunodiffusion test are present in a few (9 per cent) patients who have asthma but who do not have pulmonary eosinophilia. Precipitin lines occur in patients with pulmonary eosinophilia, allergic aspergillosis, and aspergillomas. In some studies, a stronger reaction and a larger number of bands occurred in patients with allergic aspergillosis who also had fungi growing in the lumen of the bronchial tree.[19] Precipitins are always present in aspergilloma patients. This is especially useful for diagnosis when sputa is culturally negative due to lack of connection of the cavity with a patent bronchiole. The precipitin bands disappear after surgical removal of the lesion. Few bands or none are found in cases of invasive aspergillosis.[63] The latter patients are usually anergic.

In our experience certain patterns of bands are present which correspond to the clinical type of disease. A single weak band may be found in asthmatic patients. Four bands or more are present in allergic aspergillosis with colonization of bronchi, and up to 18 bands have been found in cases of aspergilloma. Invasive disease, if chronic, may result in the detection of a few precipitin lines. However, the usual acute, fulminating, invasive aspergillosis occurs in patients with a defective or abrogated immune system who are unable to produce antibodies and whose I.D. test is therefore negative. It is important to note that sera may be negative to a single *Aspergillus* antigen, so that a battery of antigens prepared from various strains and species should be used.[8a] Sometimes concentrating the sera is also necessary. Complement fixation tests generally agree with immunodiffusion results, and may be positive earlier in the disease.[60] Ky et al.[30] and others have shown that some of the antigenic components are enzymes of fungal origin, and that patients' sera contain antibodies specific for them. Bardana et al.[3a] have studied the antigens and the types of antibody elicited during infection.

LABORATORY IDENTIFICATION

Direct Examination. The procedures and appearance of material for direct examination will vary somewhat, depending on the form of the disease. Sputum from allergic aspergillosis is usually gelatinous and thick. Therefore, the use of mucolytic agents is sometimes required before examination can be made. Long, branching, hyphal strands may be seen either on direct examination or, more easily, after digestion with 10 per cent potassium hydroxide tinted with ink. In aspergillomas in which the cavity is connected to a patent bronchiole, numerous tangled masses of hyphae may be present in bloody sputum. Such preparation often show the typical conidiophores topped by an expanded spore-bearing vesicle. This finding indicates that the *Aspergillus* is colonizing the surface of a bronchus, cavity, or surface in contact with air. The vesicle and conidiophore of *A. fumigatus* are so distinctive as to make possible specific identification from such preparations. If the "fungus ball" of the aspergilloma is in a cavity that does not connect with a patent bronchiole, both culture and direct examination will be negative.

Biopsy material, surgical specimens, and tissue debris from cases of invasive aspergillosis can be digested with potassium hydroxide to render visible the mycelium of fungus. In tissue, conidiophores are usually absent, but characteristically dichotomously branching, septate hyphae are found.

Aspergillus species are common airborne contaminants of all surfaces, including skin, mouth, lungs, wounds, and so forth. For this reason it is necessary to be cautious in evaluating the isolation of an *Aspergillus* organism from patient materials. For the diagnosis of aspergillosis from cultures of sputa, Henderson[17] has defined as necessary what he terms the presence of "abundant fungus." This consists of growing the *Aspergillus* from several specimens or several colonies from a single specimen. The presence of mycelial elements on direct examination, in addition to positive culture, is a more reliable guide to correct diagnosis than simply the isolation of an *Aspergillus* organism. Similarly, a positive culture from surgically removed tissue is considered more significant if the organism can be demonstrated in histologic sections also.

The etiologic agents of aspergillosis grow readily on almost all laboratory media. They are usually sensitive to cycloheximide, so that only antibacterial antibiotics should be present in the isolation medium. Most pathogenic aspergilli have an optimum growth temperature at 37°C, which inhibits some fungal contaminants. *A. fumigatus*, a very thermotolerant species, grows to a temperature of 45°C. Conidiophores characteristic of the organism are present within 48 hours when plated on agar and incubated at 37°C. Species identification is aided by preparation of slide cultures for the examination of conidiophores, conidiospores, vesicles, and spore mass *in situ*, in addition to the observation of colony morphology, pigmentation, and growth rate at 25°C and 37°C. There are about 300 recognized species and varieties of *Aspergillus*. For identification of unusual species, reference should be made to Raper and Fennell's *The Genus Aspergillus*.[46] The commonly encountered species from pathologic material, such as *A. fumigatus, A. terreus, A. flavus*, and *A. niger*, are fairly easy to identify and are discussed in the following section.

MYCOLOGY

Aspergillus fumigatus Fresenius 1850[14]

Synonymy. *Aspergillus aviarius* Peck 1891; *Aspergillus bronchialis* Blumentritt 1901; *Aspergillus calyptratus* Gudemans 1901; *Aspergillus cellulosae* Hopffe 1919; *Aspergillus glaucoides* Spring 1852; *Aspergillus pulmonumhominis* Welcker 1855; *Aspergillus ramosus* Hallier 1870; *Aspergillus nigrescens* Robin 1853.

Colony Morphology. The organism grows rapidly on SDA or Czapek solution agar (25°C or 37°C), producing a flat, white colony that quickly becomes grey-green with the production of conidia (Fig. 20–19,*A* FS C 30,*A*). The texture may vary from strictly velvety to deep felt, floccose, or somewhat folded. The reverse is generally colorless. The spore mass of the conidial heads are columnar, compact, and often crowded. They range in size from 200 to 400 μ by 50 μ.

Microscopic Morphology. The conidiophore is short, smooth, and up to 300 μ in length and 5 to 8 μ in diameter (Fig. 20–20). It may have a slightly green or brownish coloration, especially toward the upper part near the vesicle. The conidiophore gradually enlarges, passing imperceptibly to form the expanded flask-shaped vesicle. The vesicle is 20 to 30 μ in diameter, producing, on the upper half only, a single series of sterigmata (6 to 8 μ long). The sterigmata bend upward, paralleling the axis of the conidiophore. The conidia are green in mass, echinulate, globose to subglobose, and 2 to 3 μ in diameter. Elliptical conidia are found in the var. *ellipticus*. Sclerotia or cleistothecia are not found. *A. fisheri* is morphologically similar to *A. fumigatus*, but is homothallic, producing numerous cleistothecia. It has never been associated with human disease, although it is extremely thermotolerant.

Aspergillus flavus Link 1809

Synonymy. *A. fasciculatum* Batista and Maia 1957; *A. humus* Abbott 1926; *A. luteus* (Van Teigh) Dodge 1935; *A. nolting* Hallier 1870; *A. pollinis* Howard 1896; *A. sojae* Sakaguchi and Yamada 1933; *A. wehmeri* Costantin and Lucet 1905.

Colony Morphology. SDA or Czapek's at 25°C. Growth is rapid (6 to 7 cm in ten days) or slow (3 to 4 cm), and consists of a close-textured basal mycelium, which is flat or radially furrowed or wrinkled (Fig. 20–19,*B* FS C 30,*B*). Conidial heads are abundant and of intense yellow to yellow-green in color. The reverse is colorless to pinkish drab or darker. Sclerotia are prominent in some isolants. They become red-brown and may be 400 to 700 μ in diameter. Spore heads are radiate, splitting to form loose columns. They have an average diameter of 300 to 400 μ. In the var. *columnaris*, the spore heads are columnar, and sterigmata are in one series.

Microscopic Morphology. The conidiophores are thick-walled, unpigmented, coarsely roughened, long (up to 1 mm in

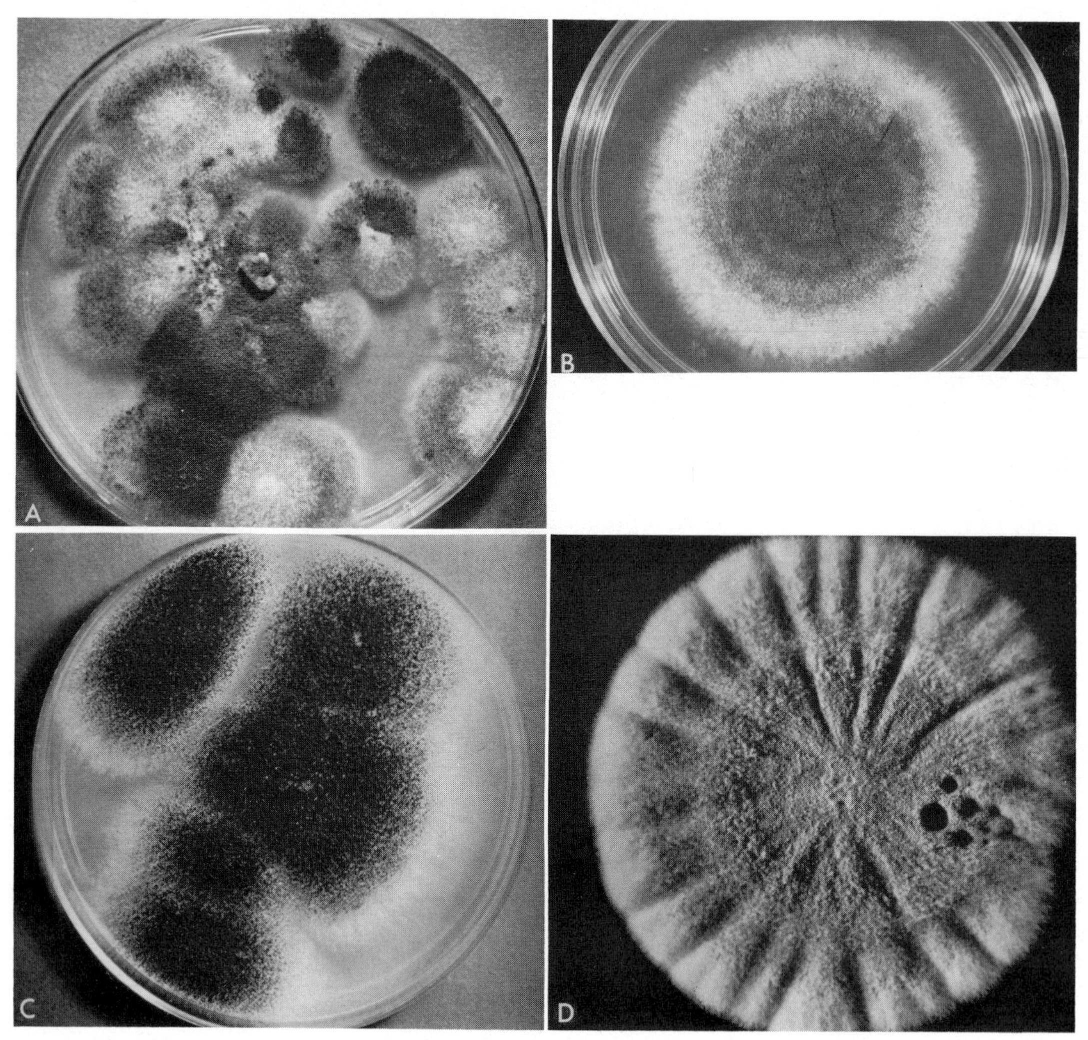

Figure 20–19. FS C 30. Colonial morphology. *A, A. fumigatus. B, A. flavus. C, A. niger. D, A. terreus.*

length or more), and 10 to 20 μm in diameter below the vesicles (Fig. 20–20). Vesicles are globose to subglobose, 10 to 65 μ in diameter, and fertile (produce sterigmata) over almost the entire area. Sterigmata are biserate or uniserate. The primary sterigmata are up to 10 μ in length, and the secondary branches are about 5 μ in length. The conidia are elliptical at first, but later they are mostly globose (3.5 to 4.5 μ in diameter) and conspicuously echinulate.

Aspergillus niger Van Tieghem 1867

Synonymy. *Sterigmatocystis antacustica* Cramer 1859; *Aspergillus fuliginosis* Peck 1934; *Aspergillus fumaricus* (Wehmer) Thom and Church 1926; *Aspergillus longobesidia* Bainer 1922; *Aspergillus nigricans* Wreden 1867; *Aspergillus nigraceps* Berk 1888; *Aspergillus pyri* English 1940; *Aspergillus welwitschiae* (Bresadola) Hennings 1907.

Colony Morphology. On SDA or Czapek solution agar (25°C) it grows to form a restricted colony with a diameter of 2.5 to 3.0 cm in 10 days (Fig. 20–19,*C* FS C 30,*C*). The compact basal mycelium is white to yellow and soon bears abundant conidial structures which are black. The conidial spore heads are large, black, and globose at first, becoming radiate or splitting to loosely columnar. There is a distinct "moldy" odor produced by the fungus.

Microscopic Morphology. Conidiophores are 1.5 to 3.0 mm by 15 to 20 μ, smooth, and colorless or turning dark towards the vesicle

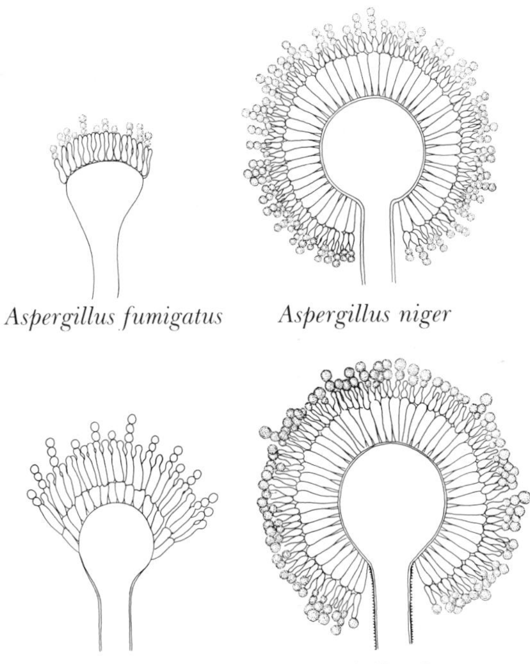

Figure 20–20. Conidiophores of *A. fumigatus, A. niger, A. flavus,* and *A. terreus.*

(Fig. 20–20). The vesicle is globose, about 60 μ in diameter, and bears sterigmata over all the surface. The sterigmata are in two series (biserate). The primaries are long (30 μ or more by 6 μ), brownish in color, and may be septate. The secondaries are short, 8 μ by 3 μ, and bud off conidia. These are globose, 4 to 5 μ in diameter, brown to black, and very rough.

A. niger is one of the most easily identifiable aspergilli, with its white to yellow mat later bearing black conidia. This species is very commonly found in aspergillomas and is the most frequently encountered agent of otomycosis.

Aspergillus terreus Thom 1918

Synonymy. *Sterigmatocystis hortai* Langeron 1922; *Aspergillus hortai* Dodge 1935; *Aspergillus galeritus* Blochwitz 1929; *Aspergillus terreus* var. *boedijni* Thom and Raper 1945.

Colony Morphology. SDA or Czapek's at 25°C. The growth of the colony is rapid, reaching a diameter of 3 to 5 cm by 10 days (Fig. 20–19,*D* FS C 30,*D*). The consistency is floccose to velvety, sometimes furrowed or tufted, and the colony sporulates profusely. The massed spore heads are columnar and range in color from cinnamon-buff to wood brown. The heads are 30 to 50 μ in diameter and 150 to 500 μ or more in length.

Microscopic Morphology. The conidiophores are long and slender (100 to 250 μ by 5 to 6 μ), smooth, and of uniform diameter throughout (Fig. 20–20). The vesicles are hemispherical or domelike, 10 to 16 μ in diameter, and merge imperceptibly with the conidiophore. The sterigmata are in two series. The primaries are 5 to 7 μ in diameter. The secondaries are 5 to 8 μ in length and 1 to 2 μ in diameter. The conidia are elliptical, smooth, and 2 to 2.5 μ in diameter. Aleuriospores are produced by the mycelial mat submerged in the agar. They are 6 to 7 μ in diameter. Such cells are also produced in tissue during infection. *A. terreus* is another thermotolerant species that has been isolated from several cases of invasive aspergillosis, particularly meningitis.

REFERENCES

1. Austwick, P. K. 1962. The presence of *Aspergillus fumigatus* in the lungs of dairy cows. Lab. Invest., *11*:1065–1072.
2. Bach, M. C., J. L. Adler, et al. 1973. Influence of rejection therapy on fungal and nocardial infections in renal transplant recipients. Lancet, *27*:180–184.
3. Baker, R. D. 1962. Leukopenia and therapy in leukemia as factors predisposing to fatal mycoses. Am. J. Clin. Pathol., *37*:358–373.
3a. Bardana, E. J., J. K. McClatchy, et al. 1972. The primary interaction of antibody to component of aspergilli. I. Immunological chemical characterization of nonprecipitating antigen. J. Allerg. Clin. Immunol., *50*:208–221.
4. Bodey, G. P. 1966. Fungal infections complicating acute leukemia. J. Chronic Dis., *19*:667–687.
5. British Thoracic and Tuberculosis Association Report. 1970. Aspergilloma and residual tuberculous cavities—The results of a re-survey. Tubercle, *51*:227–245.
6. Cahill, K. M., A. M. El Mofty, et al. 1967. Primary cutaneous aspergillosis. Arch. Dermatol., *96*:545–546.
6a. Campbell, G. 1972. Aspergillosis in captive penguins. Vet. Serv. Bull. Dept. Agricul. and Fisher. Dublin, *2*:39–41.
7. Case Records of the Massachusetts General Hospital (Case 44). 1971. New Engl. J. Med., *285*:337–346.
7a. Case Records of the Massachusetts General Hospital (Case 24). 1973. New Engl. J. Med. *288*:1290–1296.
8. Chick, E. W., and N. C. Durham. 1963. Enhancement of aspergillosis in leukemic chickens. Arch. Pathol., *75*:81–84.
8a. Coleman, R. M., and L. Kaufman. 1972. Use of Immunodiffusion test in the serodiagnosis of aspergillosis. Appl. Microbiol., *23*:301–308.

8b. Cooke, W. B. 1971. *Our Mouldy Earth.* Cincinnati, Ohio, U.S. Dept. Of Interior. Fed. Water Pollution Control Admin. Robert A. Taft. Water Research Center Advanced Waste Treatment Research Laboratory.
9. Deve, F. 1938. Une nouvelle forme anatomoradiologique de mycose pulmonaire primitive. Le méga-mycétome intrabronchetasique. Arch. Med. Chir. Appl. Resp., *13*:337–361.
10. Edge, J. R., D. Stansfeld, et al. 1971. Pulmonary aspergillosis in an unselected hospital population. Chest, *59*:407–413.
11. Eisenberg, H. W. 1970. Aspergillosis with aflatoxicosis. New Engl. J. Med., *283*:1348.
12. Emmons, C. W. 1962. Natural occurrence of opportunistic fungi. Lab. Invest., *11*:1026–1032.
12a. Findley, G. H., and H. F. Roux. 1971. Skin manifestations in disseminated aspergillosis. Br. J. Dermatol., *85*:94–97.
13. Ford, S., and L. Friedman. 1967. Experimental study of the pathogenicity of aspergilli for mice. J. Bacteriol., *94*:928–933.
14. Fresenius, G. 1850–1863. In *Beitrage zur Mycologie.* Frankfort, H. L. Bronner, p. 81.
15. Golbert, T. M., and R. Patterson. 1970. Pulmonary allergic aspergillosis. Ann. Intern. Med., *72*:395–403.
16. Goldberg, B. 1962. Radiological appearance in pulmonary aspergillosis. Clin. Radiol., *13*: 106–114.
17. Henderson, A. H. 1968. Allergic aspergillosis. Thorax, *23*:501–512.
18. Henderson, A. H., and J. E. G. Pearson. 1968. Treatment of bronchopulmonary aspergillosis with observations on the use of natamycin. Thorax, *23*:519–523.
19. Henderson, A. H., M. P. English, et al. 1968. Pulmonary aspergillosis. Thorax, *23*:513–518.
20. Henrici, A. T. 1939. An endotoxin from *Aspergillus fumigatus.* J. Immunol., *36*:319–338.
21. Hinson, K. E. W., A. J. Moon, et al. 1952. Bronchopulmonary aspergillosis: A review and report of eight new cases. Thorax, 7:317–333.
22. Hutter, R. V. P., and H. S. Collins. 1962. Occurrence of opportunistic fungus infections in a cancer hospital. Lab. Invest., *11*:1035–1045.
23. Ikemoto, H. 1965. Treatment of pulmonary aspergilloma with amphotericin B. Arch. Intern. Med., *115*:598–601.
24. Ikemoto, H. 1971. Six years follow up of a patient in whom Aspergillus fungus ball disappeared after amphotericin B therapy. Jap. J. Med. Mycol., *12*:9–10.
25. Ikemoto, H., K. Watanabe, et al. 1971. Pulmonary aspergilloma. Sabouraudia, *9*:30–35.
26. Ikemoto, H., and K. Watanabe. 1971. Spontaneous disintegration of Aspergillus fungus ball of the lung. Jap. J. Med. Mycol., *12*:11–12.
27. Israel, H. L., and A. Ostrow. 1969. Sarcoidosis and aspergilloma. Am. J. Med., *47*:243–250.
28. Kilman, J. W., N. C. Andrews, et al. 1969. Surgery for pulmonary aspergillosis. J. Thorac. Cardiovasc. Surg., *57*:642–647.
29. Krakówka, P., K. Traczyk, et al. 1970. Local treatment of aspergilloma of the lung with a paste containing nystatin or amphotericin B. Tubercle, *51*:184–191.
29a. Kwon-Chung, K. J. 1971. A new pathogenic species of Aspergillus. Mycologia, *63*:478–489.
30. Ky, P. T. V., J. Biguet, et al. 1971. Immunoelectrophoretic analysis and characterization of the enzymic activities of *Aspergillus flavus* antigenic extracts. The significance for the differential diagnosis of human aspergillosis. Sabouraudia, *9*:210–220.
31. Longbottom, J. L., and J. Pepys. 1964. Pulmonary aspergillosis. Diagnostic and immunological significance of antigens and C-reactive substance in *Aspergillus fumigatus.* J. Pathol. Bacteriol. *88*:141–151.
32. Marsalek, E., Z. Žižka, et al. 1960. Plicni aspergillozas generlizaci vyvolana druhem *Aspergillus restrictus.* Cas. Lek. Cesk., *41*:1285–1292.
33. Mayer, A. C., and Emmert. 1815. Vershimmelung (Mucedo) in lebenden Korper. Dtsch. Arch. Anat. Physiol. (Meckel.), *1*:310.
33a. McCarthy, D. S., G. Simon, et al. 1970. The radiological appearances in allergic bronchopulmonary aspergillosis. Clin. Radiol., *21*: 366–375.
34. Meyer, R. D., L. S. Young, et al. 1973. Aspergillosis complicating neoplastic disease. Am. J. Med., *54*:6–15.
34a. Meyerwitz, R. L., R. Friedman, et al. 1971. Mycotic "mycotic aneurysm" of the aorta due to *Aspergillus fumigatus.* Am. J. Clin. Pathol., *55*:241–246.
35. Milosev, B., S. Mahgoub, et al. 1969. Primary aspergilloma of paranasal sinuses in the Sudan. Br. J. Surg., *56*:132–137.
36. Mirsky, H. S., and J. Cuttner. 1972. Fungal infection in acute leukemia. Cancer, *30*:1348–1352.
37. Mohr, J. A., B. A. McKown, et al. 1971. Susceptibility of Aspergillus to steroids, amphotericin B and nystatin. Am. Rev. Resp. Dis., *103*:283–284.
38. Monod, O., G. Pesle, et al. 1957. L'aspergillome brochestasiant. Sem. Hôp. (Paris), *33*:3588–3602.
39. Okudaira, M., and J. Schwartz. 1962. Tracheobronchopulmonary mycoses caused by opportunistic fungi. Lab. Invest., *11*:1053–1064.
40. Orie, N. G. M., G. A. de Vries, et al. 1960. Growth of Aspergillus in the human lung. Am. Rev. Resp. Dis., *82*:649–662.
41. Paulk, E. A., R. C. Schlant, et al. 1965. Aspergilloma associated with congenital heart disease. Dis. Chest, *47*:113–117.
42. Pepys, J., et al. 1959. Clinical and immunologic significance of *Aspergillus fumigatus* in the sputum. Am. Rev. Resp. Dis., *80*:167–180.
43. Pimental, J. C. 1959. Further morphological aspects of "pulmonary aspergilloma:" its probable initial residual stages. Gaz. Med. Port., *12*:195.
44. Pore, R. S., and H. W. Larsh. 1968. Experimental pathogenicity of *Aspergillus terreus–flavipes* group species. Sabouraudia, *6*:89–93.
45. Procknow, J., and D. Loewen. 1960. Pulmonary aspergillosis with cavitation secondary to histoplasmosis. Am. Rev. Resp. Dis., *82*: 101–111.
45a. Ramirez, J. 1964. Pulmonary aspergilloma: Endobronchial treatment. New Engl. J. Med., *271*: 1281–1285.
46. Raper, K. B., and D. I. Fennel. 1965. *The Genus Aspergillus.* Baltimore, Williams and Wilkins Company.
46a. Redmond, A., I. J. Carré, et al. 1965. Aspergillosis

(Aspergillus nidulans) involving bone. J. Pathol. Bacteriol., *89*:391–316.
47. Renon, L. 1897. *Etude Sur les Aspergilloses Chez les Animaux et Chez l'Homme.* Paris, Masson et Cie.
48. Rippon, J. E., T. P. Conway, et al. 1965. Pathogenic potential of Aspergillus and Penicillium species. J. Infect. Dis., *115*:27–32.
49. Ross, D. A., MacNaughton, M. C., et al. 1968. Fulminating disseminated aspergillosis complicating peritoneal dialysis in eclampsia. Arch. Intern. Med., *121*:183–188.
50. Sandhu, D. K., R. S. Sandhu, et al. 1970. Effect of cortisone on bronchopulmonary aspergillosis in mice exposed to spores of various Aspergillus species. Sabouraudia, *8*:32–38.
50a. Sandhu, D. K., and R. S. Sandhu. 1973. Survey of Aspergillus species associated with the human respiratory tract. Mycopathologia, *49*:77–87.
51. Seabury, J. H., and M. Samuels. 1963. The pathogenic spectrum of aspergillosis. Am. J. Clin. Pathol., *40*:21–33.
52. Segretain, G. 1962. Pulmonary aspergillosis. Lab. Invest., *11*:1046–1052.
52a. Seres, J. L., H. Ono, et al. 1972. Aspergillosis presenting as spinal cord compression. J. Neurosurg., *36*:221–224.
53. Sidransky, H., and L. Friedman. 1959. The effect of cortisone and antibiotic agents on experimental pulmonary aspergillosis. Am. J. Pathol., *35*:169–183.
53a. Sluyter, T. 1847. De Vegetalibus organismi animalis parasitis. Diss. Inaug. Berlini, p. 14.
53b. Smith, G. R. 1972. Experimental aspergillosis in mice. Aspects of resistance. J. Hygiene, *70*: 741–754.
54. Strelling, M. K., K. Rhaney, et al. 1968. Fatal acute pulmonary aspergillosis in two children of one family. Arch. Dis. Child., *41*:34–43.
55. Symmers, W. S. C. 1962. Histopathologic aspects of the pathogenesis of some opportunistic fungal infections as exemplified in the pathology of Aspergillosis and the Phycomycoses. Lab. Invest., *11*:1073–1090.
56. Tan, K. K., K. Sugai, et al. 1966. Disseminated aspergillosis. Case report and review of world literature. Am. J. Clin. Pathol., *45*:697–703.
57. Tilden, E. B., E. H. Hatton, et al. 1961. Preparation and properties of the endotoxins of *Aspergillus fumigatus* and *A. flavus.* Mycopathologia, *14*: 325–346.
58. Vedder, J. S., and W. F. Schorr. 1969. Primary disseminated pulmonary aspergillosis with metastatic skin nodules. J.A.M.A., *209*: 1191–1195.
59. Virchow, R. 1856. Beiträge zur Lehre von den beim Menschen vorkommenden pflanzlichen Parasiten. Virchows Arch. Pathol. Anat., *9*:557–593.
60. Walker, J. E., and R. D. Jones. 1968. Serologic tests in diagnosis of aspergillosis. Dis. Chest, *53*:729–735.
61. Welsh, R. A., and J. M. Buchness. 1955. Aspergillus endocarditis, myocarditis and lung abscesses. Report of a case. Am. J. Clin. Pathol., *25*:782–786.
62. Young, R. C. 1969. The Budd-Chiari syndrome caused by aspergillus. Arch. Intern. Med., *124*:754–757.
63. Young, R. C., and J. E. Bennett. 1971. Invasive aspergillosis. Absence of detectable antibody response. Am. Rev. Resp. Dis., *104*:710–716.
64. Young, R. C., A. Jennings, et al. 1972. Species identification of invasive aspergillosis in man. Am. J. Clin. Pathol. *58*:554–557.

Chapter 21

MUCORMYCOSIS

DEFINITION

Mucormycosis is generally an acute and rapidly developing, less commonly chronic, infection of a debilitated patient caused by various species of the order Mucorales. Depending on the portal of entry, the disease involves the rhino-facial-cranial area, lungs, gastrointestinal tract, skin, or less commonly other organ systems. The infecting fungi have a predilection for invading vessels of the arterial system, causing embolization and subsequent necrosis of surrounding tissue. A suppurative, pyogenic reaction is elicited; granuloma formation is not frequently encountered. The infection is most commonly associated with the acidotic diabetic in whom the disease runs a rapid, fatal course, usually within ten days, or with malnourished children, severely burned patients, or those with leukemia, lymphomas, and other debilitating diseases. Infection also occurs as a sequela to immunosuppressive therapy, use of cytotoxins and corticosteroids, or any procedure such as surgery which results in tissue debilitation. Several species of the genera *Rhizopus, Mucor,* and *Absidia* are involved, but the clinical syndrome and pathologic findings are similar regardless of the specific etiology. The organisms are members of the order Mucorales of the Zygomycetes, quite distinct taxonomically and in the type of disease evoked from other pathogenic phycomycetous fungi. For this reason the term "mucormycosis" is preferred to others that have been suggested to describe this infection. The disease is also encountered with some frequency in domestic animals and less commonly in wild animals. In addition to the aforementioned organisms, other genera, such as *Mortierella* and *Hyphomyces,* have been isolated from such infections. In animals, the disease has not been associated with any specific predisposing factors.

Synonymy. Phycomycosis, hyphomycosis, zygomycosis.

HISTORY

Even though the first reports of this disease are 100 to 150 years old, mucormycosis has been a rare finding in pathologic processes until recently. Most early records of infection are from animals, such as dogs, pigs, horses, and cows, with only rare reports of the disease occurring in man. The first published account of a phycomycetous disease was that of Mayer[52] in 1815. He described hyphae (Mucedo) in the lungs of a blue jay, but scrutiny of his drawings reveals that the fungi were, in fact, aspergilli. Furthermore, avian mucormycosis has not been reported since. The earliest record of human involvement appears to be that described by Sluyter in 1847. He described *Mucor* in a pulmonary cavity, but Virchow examined the material and considered it to be *Aspergillus.* Fürbringer[30] in 1876 presented excellent data on the pulmonary disease–producing capacity of *Mucor* and described several cases of the infection. He saw nonseptate hyphae and sporangia in pathologic specimens, so that his report is probably valid,[6] although some authors have questioned it.[23] Paltauf in 1885[58] recorded the first disseminated case. The detail and accuracy of his description leaves no question as to the acceptability of this as a case of mucormycosis. Experimental infections were produced by Lichtheim,[48] who described two species, *Mucor (Absidia) corymbifer* and *Mucor (Rhizopus) rhizopodiformis.* These species, when injected into rabbits, evoked widespread abscesses, and death occurred. In 1886 Lindt[48a] described two more species, *Mucor pusillus* and *Mucor ramosus,* from human infections which were also pathogenic by animal inoculation.

The most commonly seen form of the disease in the United States is the rhino-facial-cranial syndrome.[2,46] Yet until 1943 this syndrome had only been recorded once, in 1885, by Paltauf.[58] A report of the disease in 1929 by Christensen and Nielson is considered dubious, but the three cases recorded by Gregory et al.[36] in 1943 delineated the syndrome, its history, symptomatology, and development so accurately that this paper is considered a classic in descriptive medicine. About 200 case reports have been recorded since their paper,[2, 34, 46, 60] and this form of the disease is now regularly encountered.

Until recently there was a similar paucity of case reports describing mucormycosis of the digestive tract. In Paltauf's review of disseminated mucormycosis, a few patients had involvement of the gastric mucosa, and Gregory[36] found only five cases in the literature up to the year 1943. Only six case reports were accepted as valid up to 1960 by Neame and Rayner,[57] who tabulated 20 more. Gastrointestinal mucormycosis appears to be an infrequent form of the disease and is associated most frequently with malnutrition. Infection or dissemination to other organs or body areas has been documented in a few scattered reports in the literature.[32, 42, 46, 61]

Pulmonary disease was a very rare medical curiosity after the first few case reports. At present it is becoming increasingly common as a complication of lymphoma and leukemia[54] and occasionally diabetes.

ETIOLOGY, ECOLOGY, AND DISTRIBUTION

The early case reports in the 1880's of Lindt, Paltauf, and others contained descriptions of a few species of the Mucorales as the etiologic agents of this disease. Since then many additional species and genera have been reported, but most of these have been placed in synonymy with the originally recorded species. The most frequently encountered agents of human mucormycosis are *Rhizopus arrhizus, R. oryzae,* and *Absidia corymbifera,* with some cases caused by *A. ramosa* and *Mucor pusillus.* It is very difficult to assess the variety of species that may evoke the infection, as cultural information is present in less than 10 per cent of all recorded cases of mucormycosis. Historically *A. corymbifera* under its various synonyms was the most frequently recorded agent in the early literature. However, in more recent reports essentially all human infections have been caused by the two species of *Rhizopus. A. corymbifera* is still commonly encountered in animal disease, however.[3] Various other species of *Mucor, Rhizopus, Absidia, Mortierella, Cunninghamella,* among others have been sporadically reported, but the authenticity of most of these is in doubt.[23] The first isolation of *Mucor ramosissimus* has recently been recorded from a chronic destructive lesion quite different from the usual forms of mucormycosis.[72] In this case the patient had no detectable underlying disease. In experimental infections, other species of the genus *Rhizopus* (all of which were thermotolerant) have been shown to have pathogenic potential,[63] but so far have not been recorded from natural infection.

The spectrum of disease and etiologic agents of mucormycosis in animals is somewhat different from that associated with human infections. No set of predisposing factors has been established and chronic granulomatous disease and cutaneous lesions are more commonly encountered. *Absidia corymbifera* is the most frequently recorded agent from systemic infections, and a species unique to animal disease, *Hyphomyces destruens,* is found in cutaneous and subcutaneous lesions.

The species of the Mucorales isolated from cases of mucormycosis are ubiquitous, a constant component of decaying organic debris, and thermotolerant. The human infecting species grow rapidly on any carbohydrate substrate and usually produce numerous sporangia and sporangiospores which are airborne. This, together with their thermotolerance, rapid growth, and saccharolytic ability, may account for the predominance of the *Rhizopus* species causing human infections, especially rhinocerebral and pulmonary disease in diabetics. It has been postulated that some animal and human gastric infections probably are the result of ingestion of contaminated food, however.[57] The *Hyphomyces* isolated from hoof disease of animals does not sporulate and probably is acquired from contact with decaying vegetation and dung.

The distribution of the various clinical

forms of mucormycosis is not based on age, sex, geography, or race, but on factors predisposing the patient to infection. Rhinocerebral disease is most often associated with acidosis of the uncontrolled diabetic;[34,46] pulmonary infection is primarily a disease of patients with leukemia and lymphoma,[41,54] and gastric involvement is encountered most frequently in malnourished individuals, particularly children with kwashiorkor.[34,46,57] Focal or general tissue debility, such as burns or surgical procedures, also accounts for some infections.[29,62,70] Hepatic disease, renal failure, tuberculosis, corticosteroids,[68] cytotoxins, antibiotics, and antineoplastic chemotherapeutic agents have also been cited as predisposing conditions and agents. The recent increase in mucormycosis is attributable in part to the use of new drugs that, although life-saving at the time, prolong the existence of the patient without curing the basic malady. This allows opportunistic infectious agents to proliferate.

CLINICAL DISEASE

Mucormycosis is the most acute and fulminant fungous infection known. Rhinocerebral disease in the acidotic patient terminates in death, often within seven days or less. The particular predilection of the organism for the acidotic patient, whether due to diabetes,[2] diarrhea,[18] uremia,[46] or aspirin intake, has been the subject of much investigation. Although significant, delayed cellular migration and diminished phagocytic function of the ketoacidotic patient alone cannot account for the rapid growth of the organism in tissue. Growth *in vitro* of *Rhizopus* is favored by the metabolic conditions encountered in the ketoacidotic diabetic. The organism has an optimum growth rate at an acid pH and a temperature of 39°C; it thrives in a medium of high glucose content and has an active ketoreductase.[59] Fortunately the disease is comparatively rare, even in such predisposed patients. This indicates the probable occurrence of other, as yet undefined, determinants in the host and in the prospective parasite.

Mucormycosis presents a spectra of disease types dependent upon the type of previous debility of the patient and the portal of entry of the organism. The ensuing pathology following infection is similar, regardless of the anatomic site or species involved. The marked predilection of the organism for invading major vessels and the resultant emboli cause ischemia and necrosis of adjacent tissue. Landau and Newcomber,[46] Abramson et al.,[2] and others have classified the clinical forms of human mucormycosis according to the major anatomic areas involved. These are rhinocerebral, thoracic, abdominal-pelvic and gastric, and cutaneous mucormycosis.

Rhinocerebral Mucormycosis

This is the acute, rapidly progressive infection most commonly encountered in the North American and European literature. The patient is generally an uncontrolled diabetic who is acidotic. This form of mucormycosis is the exclusive domain of the genus *Rhizopus*. When cultures have been done, essentially all cases were caused by *R. oryzae* or *R. arrhizus*. The age range of patients is from 16 days to 75 years, though about half are patients under 19 years old. The infection begins in the upper turbinates or paranasal sinuses,[18,26] or less commonly in the palate[70] and pharynx, causing severe cellulitis. The acidotic patient may respond to treatment of the diabetes, but if an infection has been established, the course may be an insidious and relentless development of the mucormycosis. Sometimes, however, the disease may regress and remain quiescent when the diabetes is controlled, only to be reactivated by another onset of acidosis. The usual presenting symptoms involve first the nose and then the eye, brain, and possibly the meninges. The nasal area has a thick discharge that is dark and blood-tinged, and there are reddish black necrotic areas in the turbinates and septum. A cloudy sinus with a discernible fluid level is seen on x-ray (Fig. 21–1). Also present are paranasal sinusitus or pansinusitus, fistula, and sloughs of the hard palate (Fig. 21–2 FS C 31), loss of function of the fifth and seventh cranial nerve, and occasionally a hard, somewhat palpable, demarcated swelling on either side of the nasal bridge. Severe orbital cellulitis at this stage denotes a poor prognosis, as it heralds impending or actual invasion by the fungus of the orbit and central nervous system. The findings in the eye, when it is involved, include orbital pain, with ophthalmoplegia, ptosis, localized anesthesia, proptosis, limitation of movement of the

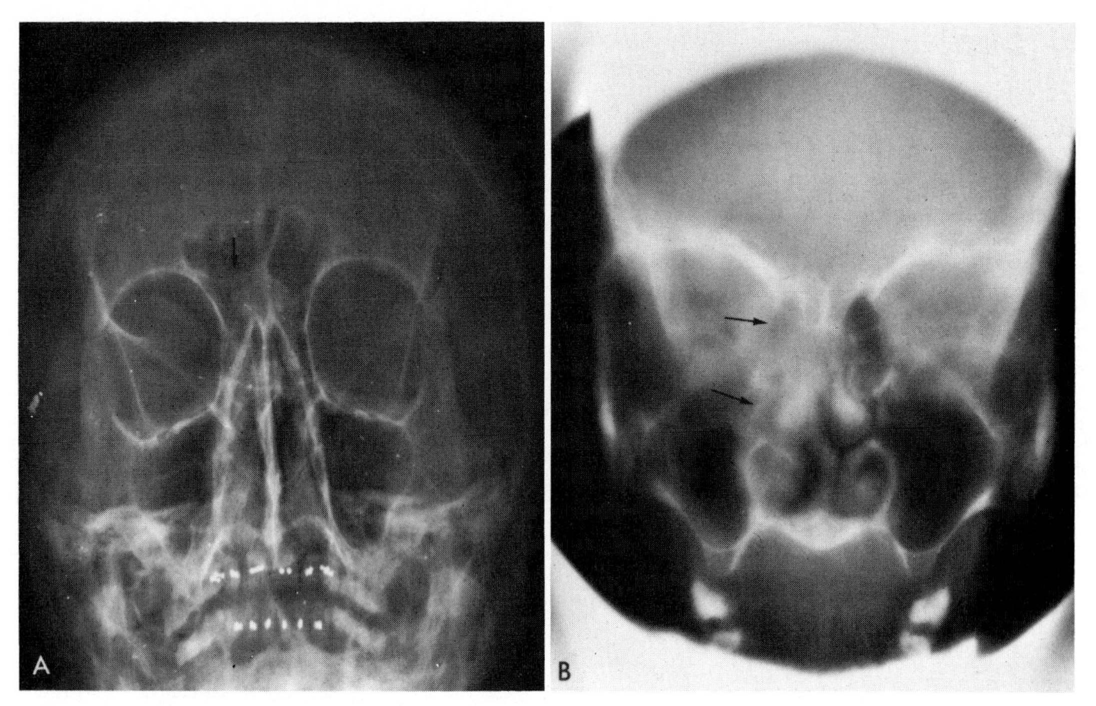

Figure 21–1. Rhinocerebral mucormycosis. *A,* Plane film sinus view. There is a well-defined air-fluid level in right ethmoid (arrow). *B,* Tomogram of same patient. There is total opacification of right ethmoid sinuses with demineralization of their lateral walls (arrows) and medial wall. Soft tissue of right nasal turbinates is enlarged, and the air passage is almost completely occluded. (Courtesy of J. Fennessy.)

bulbus oculi, fixation of the pupil, and loss of vision[28] (Fig. 21–3 FS C 32). Funduscopic examination may reveal normal findings, dilatation of the retinal veins, or thrombosis of retinal arteries, and hyphae may be seen coursing through the vitreous. Anesthesia of the areas supplied by the first and second branches of the trigeminal nerve occurs, along with homolateral facial nerve palsy due to involvement of the seventh cranial nerve (Fig. 21–4 FS C 33).

As the disease progresses, the fungus shows a predilection for invading large blood vessels. There is subsequent infarction and necrotic sequelae in the brain, which may be accompanied by softening of the frontal

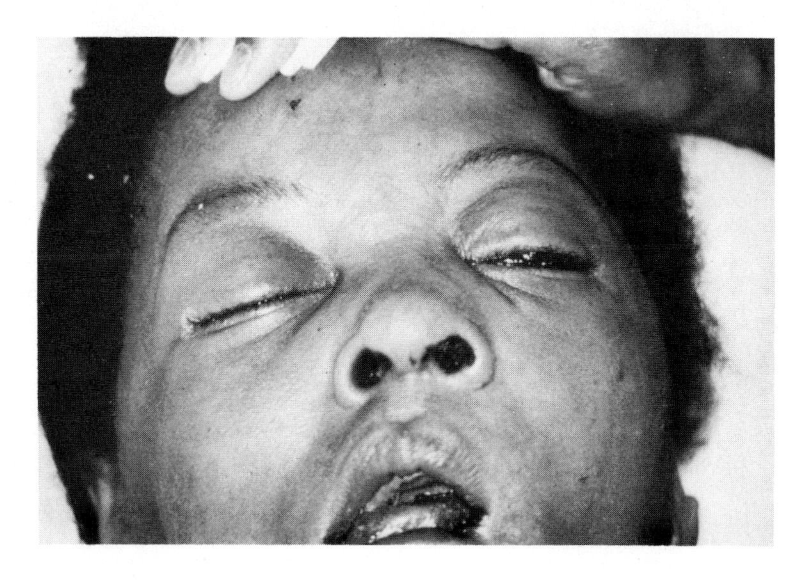

Figure 21–2. FS C 31. Rhinocerebral mucormycosis. The patient is an acidotic diabetic. There is a dark discharge in the nares, necrosis of the hard palate, a palpable swelling extending from the nasal bridge, facial nerve palsy and ptosis.

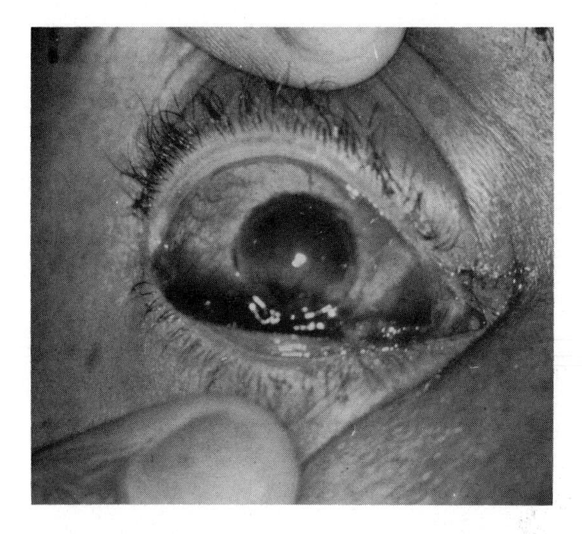

Figure 21–3. FS C 32. Rhinocerebral mucormycosis. The eye shows ophthalmoplegia, fixation of pupil, and haziness in the anterior chamber.

lobes. Lethargy progressing to coma ensues, and the patient succumbs in seven to ten days. Demise as early as two days and as late as two months has been recorded. Additional symptoms are retro-orbital headaches and intracranial pressure, which may give rise to seizures and convulsions. Fever is not prominent. Laboratory findings include a white count of 10,000 to 30,000 cells per ml and up to 41,000, a high initial blood sugar, and a CO_2 of about 13 mEq per liter. Spinal fluid findings are variable and nondiagnostic. Spread of the infection to the lungs and other organs is recorded. The overall mortality rate was 80 to 90 per cent, but many survivors have been recorded following treatment with amphotericin B.[2,6]

There are rare records of chronic nasal mucormycosis in otherwise healthy patients. These have consisted of tumors containing hypha, surrounded by a granulomatous reaction. One such condition was successfully treated with amphotericin B.[9] There is also a report of a single granulomatous lesion of the brain that contained hyphae consistent with mucormycosis in an otherwise healthy individual. The lesion was excised owing to increased intracranial pressure, and the patient recovered.[56]

Thoracic Mucormycosis

Pulmonary involvement may be primary, particularly in leukemia and lymphoma patients,[54] following inhalation of spores, or secondary to aspiration of infectious material from rhinofacial disease. The presenting symptoms are those of a progressive, non-

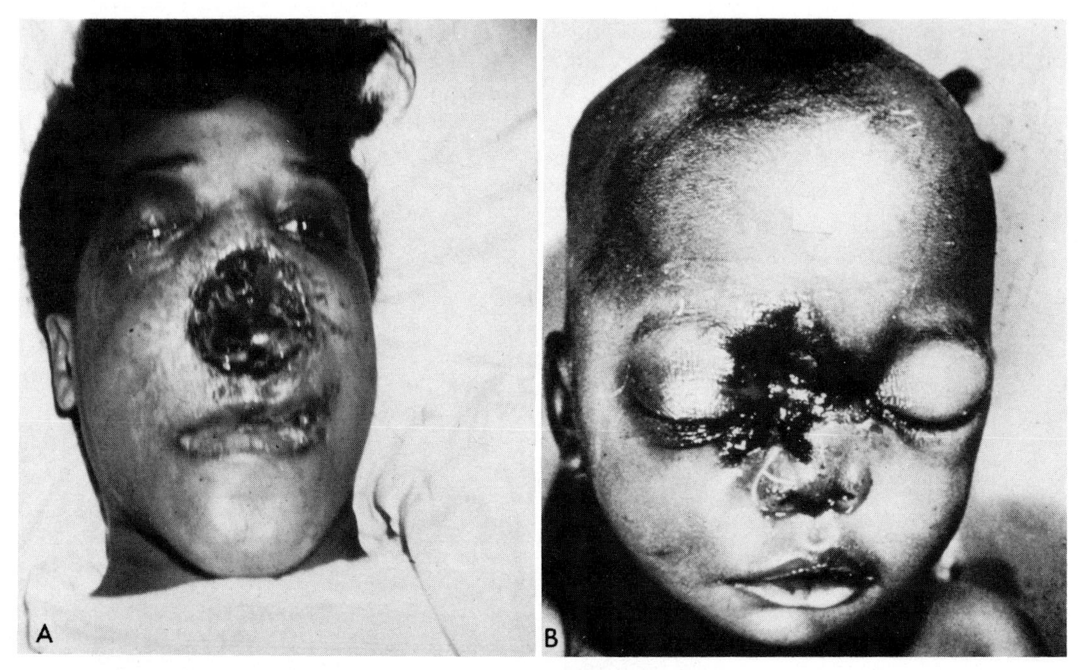

Figure 21–4. FS C 33. Rhinocerebral mucormycosis in acidotic diabetics. *A*, Adult with sloughing of nasal area. *B*, Child with sloughing of nasal area and frontal area of cranial bones. (Courtesy of L. Calkins.)

specific bronchitis and pneumonia, with superimposed signs of thrombosis and infarction.[4,14] A severe and sudden onset may be accompanied by pain, a pleural friction rub, and bloody sputum. X-ray examination demonstrates signs of nonspecific pneumonia and infarction.[7a] The disease is generally progressive to a fatal outcome in as few as three or as long as 30 days. Some patients, even with acute disease, have survived with amphotericin B therapy.[53] It has been pointed out that pulmonary mucormycosis is more prevalent among patients with leukemia or lymphoma than those with other types of cancer,[54] but primary pulmonary mucormycosis may occur in diabetics as well. Cardiac involvement has been recorded, and myocardial infarction due to direct invasion of coronary blood vessels has occurred.[41]

Localized chronic pulmonary lesions have been recorded.[32] Some have been successfully treated by lobectomy. Pulmonary lesions have also been incidental findings at autopsy when death was due to other causes.

Abdominal-Pelvic and Gastric Disease

Mucormycosis of the gastrointestinal tract as a primary disease is most often associated with undernourished patients, particularly children. In a review of twenty cases, Neame[57] found severe underlying disorders of the gastrointestinal tract to be predisposing factors. These included kwashiorkor, amebic colitis, typhoid, and in one case pellagra. All patients were undernourished or suffering from severe malnutrition with the associated physiologic imbalances. Primary gastric disease has been reported in diabetics also[44] and occasionally where no predisposing illness could be found.[1]

The symptoms of abdominal mucormycosis vary considerably and depend on the site and extent of involvement. Nonspecific abdominal pain, atypical peptic ulcer, pain, diarrhea, "coffee ground" hematemesis, and bloody stools are recorded. Ulceration of the gastric mucosa with thrombosis of associated vessels was most commonly observed at autopsy in one series.[57] No particular area is favored, as reports list involvement of the colon, stomach, ileum, and, by extension, gall bladder, liver, pancreas, and spleen. Signs of peritonitis may become evident as lesions commonly perforate the gastrointestinal wall. The course of the disease is usually 70 days. The immediate cause of death is shock from hemorrhage of bowel, resulting in peritonitis and bowel infarction.[27]

Primary mucormycosis of other organs of the pelvic-abdominal area is quite rare. Unilateral renal involvement has been recorded. This patient presented with symptoms of renal infarction and was cured following nephrectomy and amphotericin B treatment.[61] Histologic examination showed focal necrosis of parenchyma and adipose tissue. Other cases of renal involvement have recently been reported.[46a] Nonseptate hyphae unassociated with inflammatory cells were found. Infection limited solely to the bladder wall, endometrium of the uterus, arteriosclerotic thrombus of the abdominal aorta,[46] and saphenous vein graft[71] have also been recorded.

Cutaneous Lesions

Involvement of skin, particularly otitis externa, was frequently reported in the older literature, but more rigid criteria for establishing this diagnosis have invalidated most of these reports. Species of the Mucorales are quite frequently isolated from normal as well as diseased skin, ear canals, and so forth. Transient colonization of injured areas is not infrequent, and colonization with rapidly fatal disease is a particular danger to burn patients.[29] Such patients may be protected from bacterial disease by the use of antibiotics, only to succumb to mucormycosis. A few cases of cutaneous mucormycosis based on histologic evidence alone are recorded. A lesion on a chest wall of a 6-year-old boy revealed nonseptate hyphae. The lesion consisted of a reticulated, atrophic, central area with a slowly progressive border of lichenified papules.[24] In another case, indolent chronic ulcers yielded a *Mortierella* species on culture.[23] Granulomatous and abscessed areas contained nonseptate hyphae. The disease appeared to be similar to *Hyphomyces* infection of horses. Cutaneous lesions over the malleolus of a diabetic has been reported,[46] and another cutaneous lesion is reported to have developed following incision of a carbuncle.[7] Mycetoma-like lesions have also been reported from which *Rhizopus* and *Mucor* have been isolated.[24,42] Disease of the oral mucosa is also known.[70] Cutaneous disease as a manifestation of disseminated infection has been documented.[54a]

DIFFERENTIAL DIAGNOSIS

In a diabetic patient with acute and rapidly spreading sinusitis, cellulitis of orbital tissues, bronchitis, or bronchial or lobar pneumonia, a diagnosis of mucormycosis should be seriously entertained. Demonstration of the fungus by direct examination and isolation in culture should be diligently sought. Because of the fulminant course of the disease, early diagnosis is imperative for administration of appropriate therapy. Other opportunistic mycoses may simulate the condition, but usually are not so rapidly overwhelming. Fulminating bacterial or viral infections also must be considered in the differential diagnosis.

In tissue the hyphal strands seen in mucormycosis resemble those present in entomophthoromycosis morphologically, but the pronounced eosinophilia and deposition of Splendore-Hoeppli material around the organism found in the latter disease are usually missing. *Mortierella* and *Hyphomyces*, however, incite a chronic eosinophilic reaction similar to that seen in infections caused by *Basidiobolus* and *Entomophthora*.

PROGNOSIS AND THERAPY

The overall mortality of rhinocerebral mucormycosis in the diabetic formerly was 80 to 90 per cent. The prime consideration is control of the diabetes, but even when antifungal therapy is instituted, the prognosis is grave. In recent series in which the disease is diagnosed ante mortem and adequate therapeutic measures instituted, the survival rate is about 50 per cent.[2,8,9,46] Based upon the experimental results reported by Chick et al.[11,12] concerning the efficacy of amphotericin B in experimental mucormycosis in rabbits, this drug has been used in the successful treatment of the disease. Battock et al.[8] reported survival following institution of the following measures: alternate day IV amphotericin B administration at a rate of 1.2 mg per kg, control of diabetes, and local surgical debridement of involved nasal tissue. Mean 48-hour serum levels of the drug were 0.34 to 0.45 μg per ml. This does not exceed the maximal tolerated levels in humans found by other investigators (0.5 to 1.5 μg per ml). The *in vitro* sensitivity of the isolated strains varies from 0.03 μ to more than 1000 μg per ml. In one of Battock's cases, the minimal inhibitory concentration was 1000 μg per ml. This emphasizes that the drug may be of some efficacy *in vivo* in spite of lack of action *in vitro*. This discrepancy may be partly due to the testing procedure. Medoff and Kobayashi suggest use of germinated spores in broth with increments of the drug as being more physiologically relevant.[53] Abramson et al.[2] suggest a cumulative dose of 3 g. Prockop and Silva-Hutner[60] did not have success with this drug but claim cycloheximide to have aided recovery. No good evidence exists for the efficacy of iodides or desensitization. Both regimens have now been abandoned.

Most cases of gastric and pelvic disease are diagnosed at autopsy, so that there is not sufficient accumulated experience to comment on efficacy of treatment. In renal disease and in the case of saphenous graft infection, amphotericin B has been used with clinical cure of the disease.[61,71] Though most cases of pulmonary mucormycosis have been fatal or required much surgical manipulation, a few have survived. Medoff and Kobayashi report on a leukemic patient with acute pulmonary mucormycosis successfully treated with amphotericin B.[53] Alternate-day therapy and surveillance of blood levels of the drug and sensitivity of the organism to the drug were employed.

PATHOLOGY

Gross Pathology. Infarction and necrosis of tissue are the main findings in autopsy specimens from mucormycosis.[6] The findings are similar in all anatomic sites: brain, lung, stomach, intestine, and liver. Gangrenous degeneration is a common apellation of such involvement. The infected areas of brain show softening and sloughing of adjacent tissue with punctate hemorrhage. There is usually gross consolidation in the lungs when involved. Cut surfaces show massive hemorrhage, with recent emboli and gray "puttylike" material in old infected areas. In gastric involvement, engorged vessels and ulcers up to 3 to 4 cm are seen that have black necrotic central areas. In old established chronic lesions, some granulomatous changes may be found.

Histopathology. The tissue reaction to infection by species of the Mucorales is

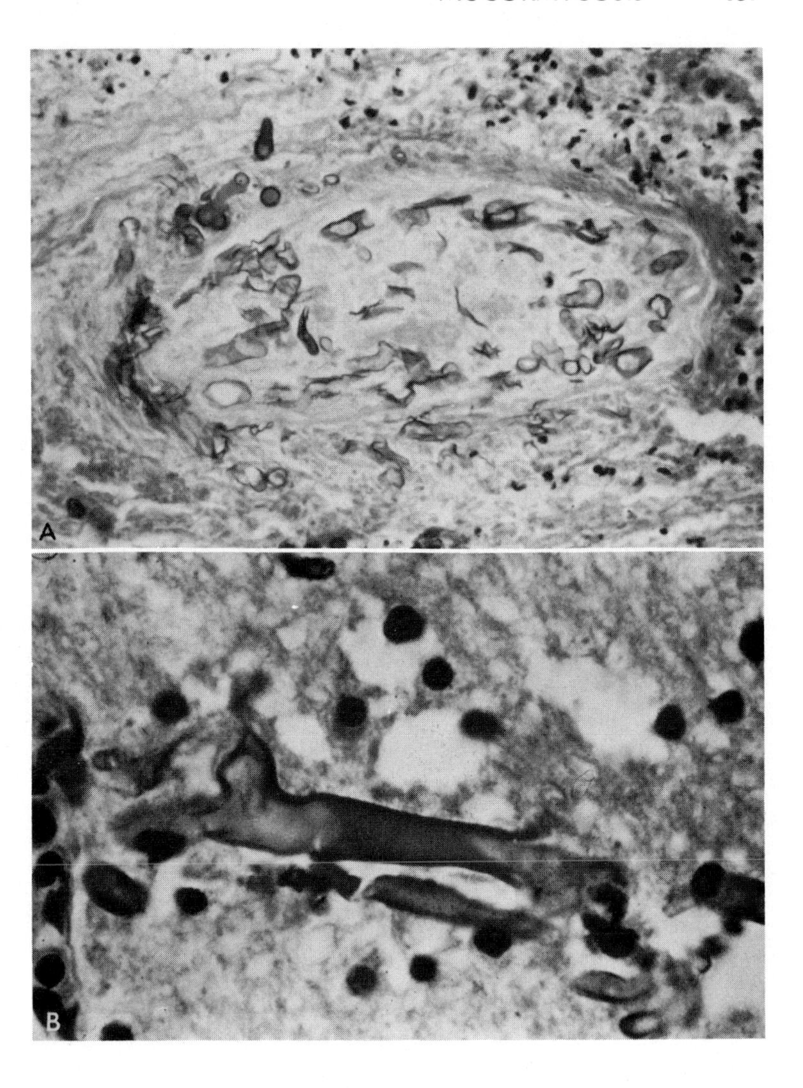

Figure 21–5. FS C 34. Mucormycosis. *A,* Invasion of arterial wall and thrombus formation which resulted in necrosis of adjacent tissue. Note the broad, irregular, nonseptate hyphae. Hematoxylin and eosin stain. ×150. *B,* Broad, irregular, nonseptate hyphae in necrotic debris. ×500. (From Rippon, J. W. *In* Burrows, W. 1973. *Textbook of Microbiology.* 20th ed. Philadelphia, W. B. Saunders Company, p. 738.)

variable. Sections with marked invasion of hyphal elements sometimes show little or no cellular response.[6,61,69] More commonly, there are varying degrees of edema, necrosis, and accumulations of neutrophils, plasma cells, and sometimes giant cells. Eosinophils are not as frequently seen as in entomophthoromycosis. Often the tissue reaction is suppurative, but it may show some granulomatous changes.[47]

The most characteristic feature of mucormycosis is involvement of blood vessels. The fungi show a marked predilection to invade directly the walls of large and small arteries (Fig. 21–5 FS C 34), and less frequently into veins. This invasion is followed by thromboses, which in turn cause infarction and necrosis of adjacent tissue. This sequence of events is the major contributing factor to symptomatology of the patient and the gross and histopathologic findings. In most series of rhinocerebral disease, thrombosis of the internal carotid artery was present in 33 per cent of patients, whereas cavernous sinus thrombosis was infrequent.[6,46] The fungus also invades bones, nerves, and fatty tissues; muscles are usually spared. In one personally observed case, all major cranial vessels were affected, and the sella turcica and pituitary were obliterated.

In contrast to most fungi, the etiologic agents of mucormycosis are readily seen in hematoxylin and eosin preparations of tissue. The special fungous stains, such as PAS and Gridley, do not demonstrate the organisms well, although the Gomori methenamine silver stain is usually quite adequate (Fig. 21–6,*A* FS C 35,*A*). The hyphae are typically broad (10 to 20 μ), sparsely septate, and haphazardly branched. Widths varying from

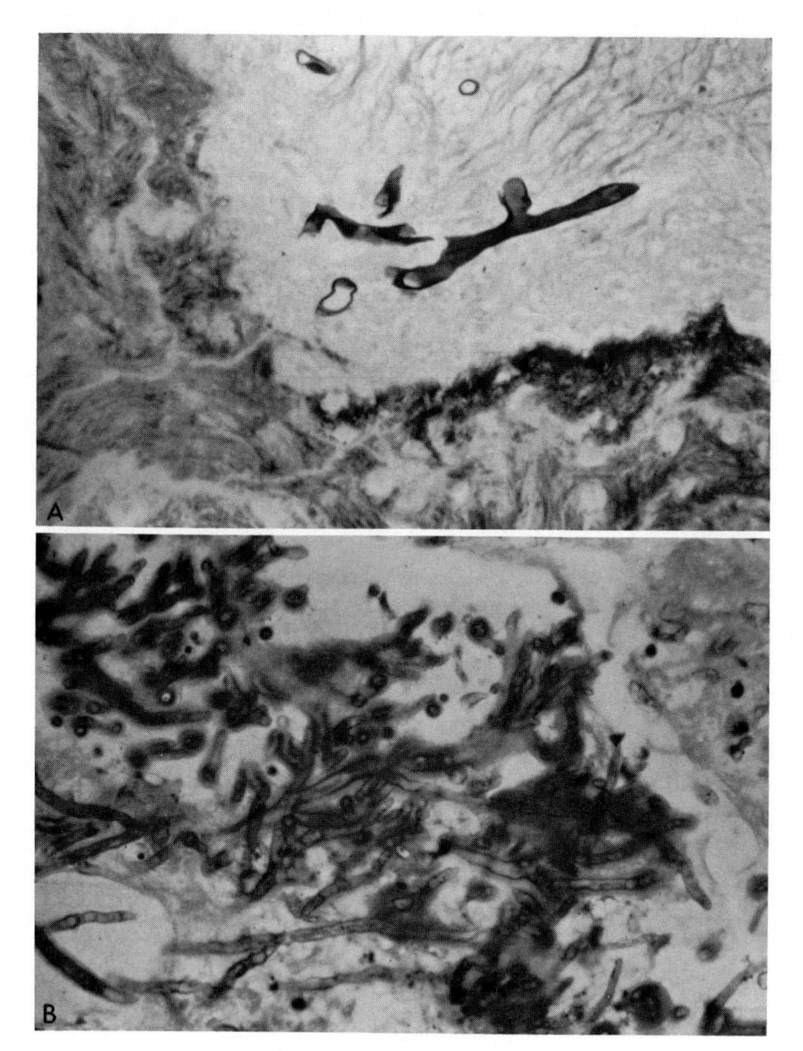

Figure 21-6. FS C 35. Mucormycosis. *A,* Irregular, broad, nonseptate hyphae in GMS stain. ×400. *B,* Unusual deposition of eosinophilic halo around hyphae in lesion from chronic lung disease. The hyphae are thinner, and there are more septations than are usually found. The organism was not cultured. Hematoxylin and eosin stain. ×400. (Courtesy of S. McMillen.)

6 to 50 μ have been recorded. The mycelium is sometimes quite distorted, and the walls vary considerably in thickness. Cross sections of hyphae have sometimes been mistaken for nematodes or empty spherules of *Coccidioides immitis.* Thick-walled hyphal strands may also resemble sclerosed capillaries or arterioles. Mucormycosis is easily differentiated from *Aspergillus* infection, in which the thinner width, regular septation, and brushlike or fingerlike branching are characteristic. True septa in mucormycosis are rarely seen. Hyphal folds may resemble a septum, but examination under oil immersion reveals them to be patent. The eosinophilic halo (Splendore-Hoeppli phenomenon) is usually absent, although "asteroid" or "club" formation has been reported in animal disease[15] (Fig. 21-6,*B* FS C 35,*B*).

ANIMAL DISEASE

Natural infection with the Mucorales is widespread among wild animals and is a significant veterinary problem in domesticated livestock.[3,15] Among wild animals, disease has been reported in monkeys,[37,51] mice,[3] minks, okapi,[39] and others.[3] Both gastric and rhinofacial types of the disease are found in monkeys. In the rhinofacial form, involvement included the paranasal sinuses, orbit, facial skin, and lymph nodes; in addition, osteomyelitis, cellulitis, vasculitis, lymphadenitis, and sialadenitis were also observed in this diabetic animal. An unidentified *Mucor* was isolated. The okapi infection was in a neonatal animal and involved the gastrointestinal tract. As is usual in animal infection, *Absidia corymbifera* was isolated which was thought to have been the result of ingestion of moldy grain.

Mucormycosis of domestic animals is chiefly known as a granulomatous disease involving lymph nodes. The disease falls into two categories. The commoner form is an acute or chronic involvement of mesenteric, bronchial, mediastinal, or submaxillary lymph nodes and sometimes liver, lung, and kidney. The granulomata are usually not discovered until the animal wastes away, or in routine inspection of abattoirs.[15] Cattle and pigs are most frequently involved.[3,19,50,55,66] Fatal infection in adult animals has occurred following abortion[19] and in neonatal calves both as terminal systemic infection and incidental to slaughter.[13,40] Most all internal organs are involved. Culture is usually not obtained, but *A. corymbifera* is the most commonly isolated species when fungal isolation has been done. In one series, *Mucor circinelloides,* a rarely encountered species, was repeatedly cultured from diseased cattle.[55] Parrots fed millet infected with the fungus died of "toxic manifestations of the brain;" however, the organism did not elicit a disease when experimental infection of animals or birds was attempted. *Mortierella* sp. have also caused disease in cattle.[53a]

The second common form of animal disease is that of ulceration of the stomach and intestinal tract. Any age may be affected, but most rapid fatal infections occur in young animals. Scouring of calves is the first symptom, and the disease progresses rapidly. This form accounts for practically all cases of abdominal ulcerations. The early lesions are small, raised, inflamed foci which become ulcers with raised hemorrhagic margins and grey, depressed, central necrotic areas. *Absidia corymbifera, A. ramosa, Rhizopus microsporus,* and various other Mucorales have been reported. Eating moldy grain is the probable source. Bovine mycotic abortion, though usually caused by *Aspergillus* sp., can also be the result of invasion by members of the Mucorales.

Horses and dogs are infrequently affected. Lesions involving the nose only, the nose and other areas, cutaneous lesions, and fatal systemic infection have been recorded.[16,25,49] In one case, a sinus tract formed at the site of a bite by another dog, and systemic involvement followed.[25] The histopathology is similar to human infection. *Mucor pusillus* and other Mucorales have been identified.

A special form of mucormycosis in horses is caused by *Hyphomyces destruens.*[10,40a] This disease was first noted in 1884 in India under the name of bursautee and later in France in 1896. In 1903 it was described in Indonesia as hyphomycosis destruens. Since then, it has frequently been confused with infection by nematode larvae under such names as Florida horse leeches and bursatti. A disease with identical symptomatology is caused by larvae of *Habronema.* Only histologic examination will differentiate them. About 40 cases of hyphomycosis have been recorded in Texas and several in Florida. Infection usually starts at the site of a cut on the hoof, hock, or fetlock. Necrosis, granulation tissue, and the formation of fistulous tracts ensues. This is followed by formation of yellow necrotic lesions which contain masses of the fungus. Histopathologic examination reveals the yellow masses to be foci of coagulative necrosis, containing numerous neutrophils, eosinophils, and occasional giant cells. Masses of branching hyphae with occasional septations are seen. The fungi vary in width from 5 to 10 μ. Infection of the mucosa, legs, abdomen, mammary gland, neck, head, and lips has also been recorded. The disease is chronic and progressive; spontaneous remission is unknown. Fatal systemic involvement so far has not been reported. Radical surgery or euthanasia is the usual management. The organism *Hyphomyces destruens* does not sporulate, but appears to be related to the genus *Mortierella.* It is probably contracted from the soil. Natural infection of birds has not been reported, but spontaneous disease of guinea pigs and mice used in cancer experiments has occurred.[3] *Hyphomyces* infection of the dog is occasionally recorded.[36a]

Experimental Disease. Mucormycosis has been the subject of experimental investigation for some time. Many of these studies have been concerned with the particular propensity of acidotic diabetics to acquire infection. One of the earliest records of experimental disease is that of Lichtheim, who produced fatal infections in rabbits with various species.[8] He also noted that the species that were pathogenic were able to grow at elevated temperatures. This work was the basis of Vuillemin's separation, in 1903, of the *Mucor* and *Absidia* species into a new genus, *Lichtheimia,* which was to contain all species demonstrated to be pathogenic. This artificial separation is no longer recognized.

Duffy[20] and others[63,63a] showed that alloxan

diabetes in rabbits was similar to human diabetes and had an acute acidotic form, usually fatal, and a chronic form. Later a number of investigators demonstrated that intranasal or intravenous injection of various species of the Mucorales could produce cerebral or pulmonary mucormycosis. Elder and Baker[21] found that intrathecal injection of spores into rabbits with acute acidotic diabetes produced extensive or fatal infections, whereas normal rabbits or those with chronic diabetes did not succumb to disease. In normal rabbits, the spores did not germinate. These results have been duplicated in mice and rats similarly treated. Sheldon et al.[67] demonstrated that in experimental infection most mast cells in normal rats quickly degranulated, and a rapid inflammatory reaction occurred, with limitation and abortion of the disease process. Previous degranulation of mast cells by the histamine liberator 48/80 only slightly delayed the reaction, but in acute alloxan diabetic animals, degranulation was impaired. The inflammatory reaction was greatly delayed and severe sometimes fatal infection occurred. Gale and Welch[31] demonstrated that normal human serum markedly inhibited the growth of *Rhizopus oryzae*, whereas sera from an acidotic diabetic did not. Later, when the diabetic patient had recovered clinically, inhibition of growth by his sera was similar to that of normal controls. The serum factor inhibitory to the growth of *Mucor* is apparently independent from the anti-*Candida* and anti-dermatophyte complexes described in previous chapters.

Josefiak and Smith-Foushee[43] have shown that a limited disease could be produced in a Selye pouch (pneumoderm) in normal rats. Chick et al.[11,12] using this technique demonstrated the efficacy of amphotericin B in alloxan diabetic rabbits. They noted that the drug inhibited hyphal formation when spores were introduced into the pneumoderm. This important finding led to the use of this drug in therapy of the disease.

The extraordinary and extensive tissue damage, necrosis, and thrombosis (even in the presence of thrombocytopenia) that is seen in cases of mucormycosis has stimulated a search for toxins and thromboplastins. The presence of thromboplastic substances or a substance of the host activated by the invading organism has been hypothesized, but as yet there is no laboratory evidence.[41] An endotoxin has been described, but little research has been done on it.[64] Potent endotoxins, such as those of *Aspergillus fumigatus*, have been found in some fungi, and it is possible a similar substance exists in the pathogenic Mucorales. A hemolysin, which was shown to be a lipid, has been recovered from some pathogenic Mucorales.[45] The biological significance remains to be determined, as hemolysis is not a prominent feature of the natural disease in man or animals.

SEROLOGY AND IMMUNOLOGY

The rarity of mucormycosis indicates a high natural resistance to the disease. This has been amply verified by laboratory investigation. Mucormycosis is an almost perfect example of opportunism—a host with a very specific set of predisposing factors and a parasite, essentially avirulent normally, which has the potential for growth in that particular tissue environment.

There are no serologic procedures of consequence for the diagnosis of mucormycosis. Patients with the disease, as well as many normal patients, will react to intradermal injections of culture extracts. Such preparations are of no practical value and are remnants of the time when desensitization and autologous vaccines were in vogue. The Mucorales are ubiquitous members of the environment, and their airborne spores are omnipresent. The spores apparently are potent allergens, as are the spores of *Aspergillus* sp. Up to 60 per cent of patients with allergic bronchitis or asthma react positively to scratch tests or intradermal injections of culture extracts of various Mucorales.

LABORATORY IDENTIFICATION

Direct Examination. Since mucormycosis is the most acutely fulminant fungal disease of man, rapid diagnosis is extremely important if management and therapy are to be successful. Unfortunately fungal elements are usually not numerous in discharges, so that diagnosis may have to rest on clinical evidence alone. For the "classic" disease in the "classic" patient, this is sufficient. Since spores of the Mucorales are common contaminants of the environment and sputa, direct examination with demonstration of the organism is more meaningful for diagnosis than is culture. Fungal elements can be found in scrapings from the upper turbinates or in aspirated material from sinuses in rhinocerebral disease, and in sputum in pulmonary disease. Broad, sparsely septate, branching hyphae are seen in well-prepared potassium hydroxide mounts. They are thick-walled and refractile, with an average diameter of 10 to 15 μ. The size may range from 3 to 30 μ. Swollen cells (up to 50 μ) and distorted hyphae are sometimes seen.[25,49] Biopsy material, whenever practical, is usually an excellent source for observing and culturing the organism.

Culture Methods. The Mucorales that are involved in human or animal disease all grow on standard laboratory media without cycloheximide. The growth is rapid and usually noticeable 12 to 18 hours after planting of the specimen. Establishing a diagnosis on cultural evidence alone is difficult. The pathogenic species of the Mucorales are constant inhabitants of the environment, contaminants of skin, discharges, and sputa, and grow on almost all moist organic substrates. Bronchiectatic patients as well as normal individuals may cough up spores of these organisms for days following exposure to a spore-ridden environment. Since the organisms overgrow the Petri dish so rapidly, colony counts that help in establishing the diagnosis of aspergillosis are difficult to carry out. An uncritical diagnosis of mucormycosis cannot be given on cultural evidence alone, and conversely the isolation of a Mucorales from a patient cannot be discarded as transient flora or a contaminant. All forms and sources of evidence must be marshalled and critically judged for an accurate diagnosis.

Discharges, scrapings, and biopsy material can be planted on Sabouraud's agar and incubated at 37°C. Almost all the pathogenic Mucorales are easily isolated from such material. The media may contain antibacterial antibiotics, but most isolants are sensitive to cycloheximide. Organisms from animal disease and from gastric mucormycosis are often difficult to culture. Sometimes the pronounced saccharolytic abilities of the organisms are used to advantage by placing the specimen on a section of sterile bread which is on the surface of an agar plate.[57]

Once isolated, the specific identification of the organism is often difficult. There are many species of Mucorales in the environment, and many of these are thermotolerant. Even to an expert, species separation is a difficult task. Physiologic reactions as a basis for identification have not been established, although some have been reported.[65] In the heterothallic species, identification can be made by sexual crosses with known testor strains.

Sensitivity to amphotericin B varies considerably among isolants of the Mucorales.[73] For this reason, it is important to determine the sensitivity of the organism isolated to the drug. The usual procedure is to disperse known concentrations of amphotericin B into tubes and add known amounts of melted Sabouraud's agar (see Appendix). The tubes are then planted with some mycelial strands or spores of the isolant. Sensitivities of isolants may vary from 0.1 μg per ml to a 1000 μg per ml. The drug appears to be fungistatic, as spores washed free of the drug from tubes showing inhibition are still viable. *In vivo* efficacy is not always able to be correlated with results obtained by *in vitro* testing. It has been observed that the drug is efficacious in treating some cases of mucormycosis in spite of the high concentration necessary for inhibition of growth determined by *in vitro* tests. Medoff and Kobayashi[53] devised a more "physiologic" method for testing. In their procedure, the spores are allowed to germinate in a liquid medium, and then amphotericin B is added. Growth is determined turbidometrically.

MYCOLOGY

The commonly encountered genera of the Mucorales can usually be identified by using Gilman's key.[33] For species identification of *Rhizopus* and other genera, the Zycha monograph should be consulted.[74,75] Ellis and Hesseltine[22] have extensively reviewed the genus *Absidia.*

Rhizopus Ehrenberg 1820

This genus is characterized by simple or branched brown sporangiophores which arise singly or in groups from nodes directly above the rhizoids (hold fasts) (Fig. 21–7,*A*). The nodes are connected by a stolon. An evagination of the sporangiophore called a columella extends into the sporangium. The mycelium is tenacious, woolly, and coarse.

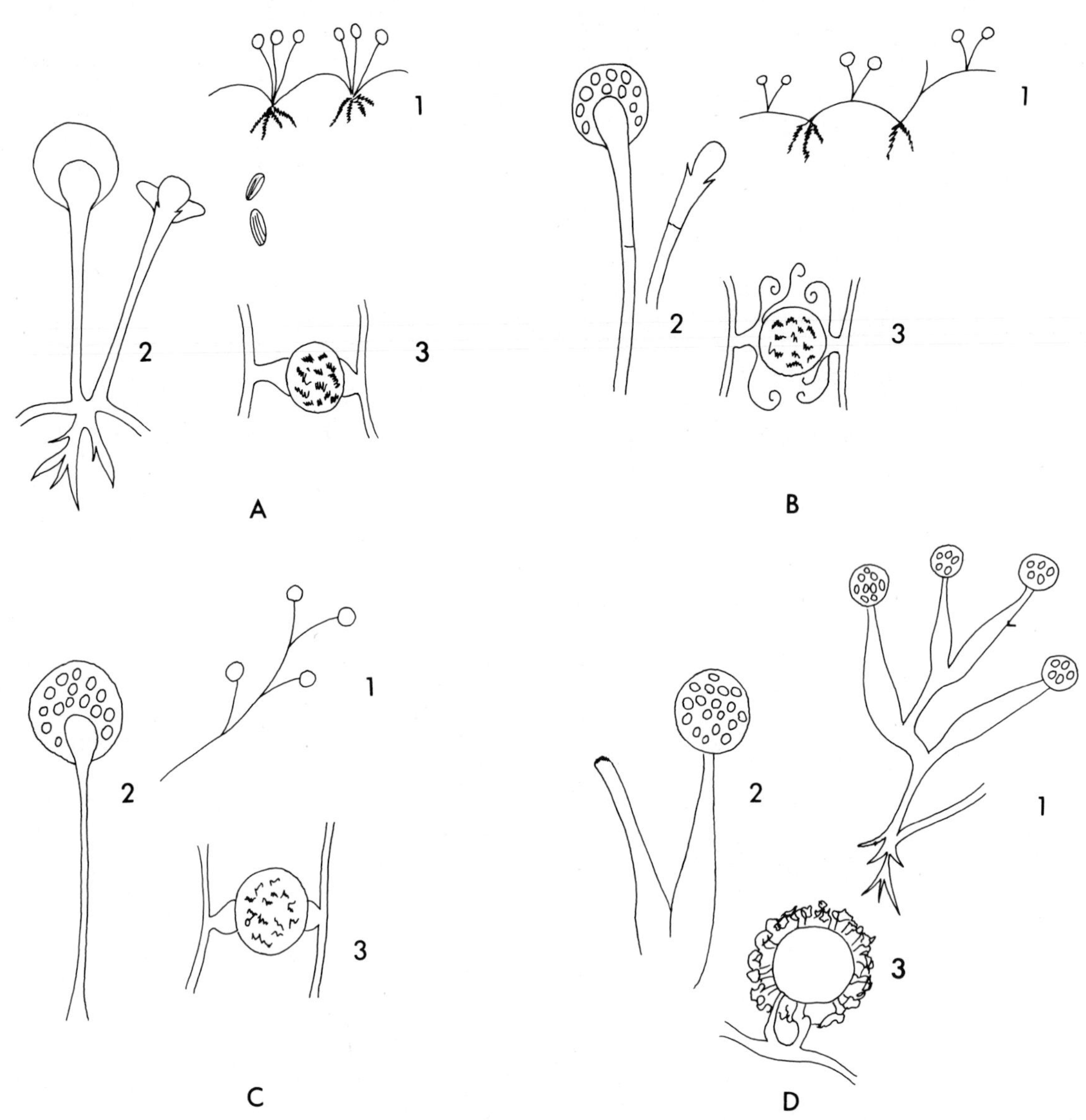

Figure 21–7. Characteristics of the four genera of the Mucorales associated with human infection. *A, Rhizopus.* (1) Sporangiophores arise in groups directly above the rhizoids. (2) A columella extends into the sporangium. (3) Naked round crenated, or roughened zygospores formed by perfect species. *B, Absidia.* (1) Sporangiophores arise between nodes; rhizoids present. (2) Columella extends into sporangium. (3) Zygospores are enveloped by circinate hyphae. *C, Mucor.* (1) Branched sporangiophores arise randomly along aerial mycelium. There are no rhizoids. (2) Columella extends into sporangium. (3) Zygospores in perfect species resembles those of the genus *Rhizopus. D, Mortierella.* (1) Sporangiophores are branched. (2) Sporangiophores taper toward the attachment of sporangium and lack columella. (3) Zygospore covered by thick case of mycelium.

Sporangia are dark-walled, spherical, and filled with hyaline or colored spores. Of the many reported species from human disease, *R. oryzae* and *R. arrhizus* are most often implicated, and are perhaps the only *Rhizopus* species truly pathogenic.

Rhizopus oryzae Went and Prinsen Geerlings 1895

Synonymy. *Rhizopus achlamydosporus* Takeda (fide Hesseltine) 1965; *Rhizopus formosaensis* Nakazawa 1913.

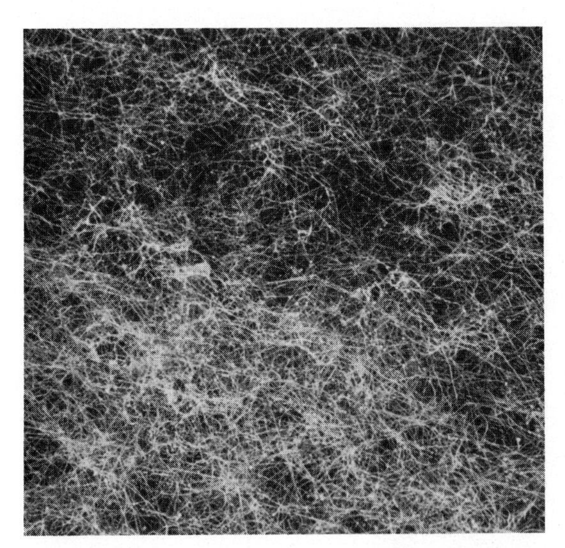

Figure 21–8. FS C 36. *Rhizopus oryzae.* Loose, cottony colony. The light yellow sporangia are visible above the mycelium. ×5.

This is the most frequently reported isolant from mucormycosis.

Colony Morphology. The colony is fast growing, white at first, becoming yellowish brown with age (Fig. 21–8 FS C 36). Sporangia are dark brown.

Microscopic Morphology. The sporangiophores are simple or branched, up to 4 mm in length. Rhizoids are yellow brown (Fig. 21–9). Sporangia are 100 to 350 μ in diameter. The sporangiospores are light brown, striated, irregularly shaped, and 6 to 8 μ by 7 to 9 μ in size. Numerous gemmae may be present. Zygospores are not known. The organism grows at a temperature of 40°C. This species is used in Japan for making koji.

Rhizopus arrhizus Fisher 1892

Synonymy. *Rhizopus nodosis* Namyslowski 1906; *Rhizopus ramosis* Moreau 1913; *Rhizopus maydis* Brudeslein 1917; *Mucor arrhizus* Hagem 1908; *Mucor nurveqicus* Hagem 1908; *Rhizopus bovinus* van Beyma 1931; *Rhizopus chinensis* Nakazawa 1913.

This species is also frequently isolated from mucormycosis.

Colony Morphology. The general morphology is similar to that of *R. oryzae,* but the colony less often becomes brown.

Microscopic Morphology. Sporangiophores are similar to those of *R. oryzae,* but the spores are oval to flattened and angular with a ridged surface. They are 4.5 to 9 μ by 4.5 to 5.5 μ in size. Rhizoids are not as well developed in this species as in *R. stolonifer.* Growth at temperatures of 40°C is recorded. Gemmae are rarely seen.

Rhizopus nigricans Ehrenberg 1820

Synonymy. *Mucor stolonifer* Ehrenberg 1818; *Rhizopus niger* Ciaglinski and Hewelke 1896; *Rhizopus artocarpi* Raciborski 1900; *Mucor niger* Gedoelst 1902.

Zygospores are present (160 to 220 μ in diameter) (Fig. 21–10). This is a very common contaminant and has not been isolated from true infections.[63] The general morphology is similar to *R. arrhizus,* but the rhizoids are well developed. Spores are large, 7 to 15 μ in diameter.

Rhizopus cohnii Berlese and de Toni 1888, *R. microsporus* Van Tieghem 1875, and *R. equinus* Costantin and Lucet 1903 have also been reported from infections. *R. rhizopodiformis* and *R. oligosporus* Saito 1905 have been shown to be pathogenic for animals, but so far they have not been isolated from natural disease.[63, 63a]

Mucor (Micheli) Link 1824

Members of this genus bear simple or branched sporangiophores which arise randomly along the aerial mycelium (Fig. 21–7,*C*). Stolons and rhizoids are absent. The colony is rapidly growing and generally grey or yellowish grey. The sporangium is large and spherical, and the wall may dissolve (deliquesce) at maturity. Few species have been recovered from well-documented cases of mucormycosis, and infection due to members of this genus is rare.

Mucor ramosissimus Sanutsevitsch 1927 was isolated from a chronic destructive facial lesion of 24 years' duration.[72] The colony is cinnamon-buff to grey-olive, with sporangia 15 to 70 μ in diameter. Small sporangia lack a columella. Spores are ovoid to globose, 3.3 to 5.5 μ by 3.5 to 8 μ.[38]

Mucor pusillus Lindt 1886 has been reported from a disseminated infection,[54a] and *Mucor circinelloides* Van Tieghem 1875[23] and *Cunninghamella elegans* Lendner 1908 have been isolated from documented cases. *M. javanicus* Wehmer 1900, *M. racemosus* Fresenius 1850, and *M. spinosus* Van Tieghem 1876 have also been recorded as being isolated from disease processes, but the authenticity of the cases is in doubt.

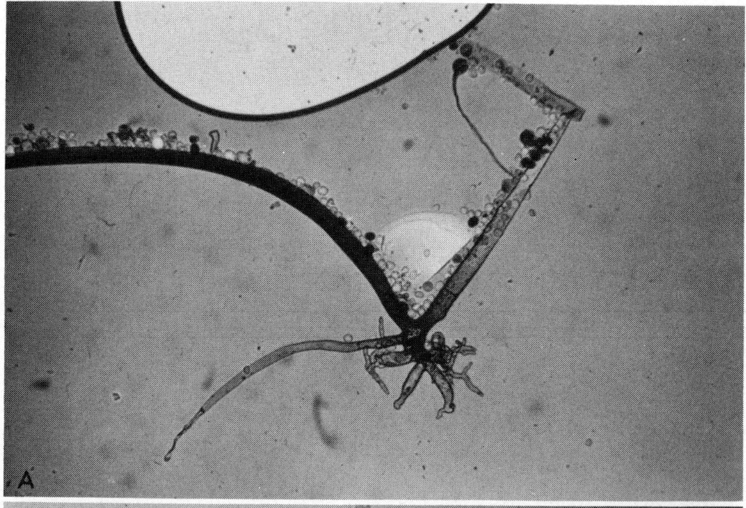

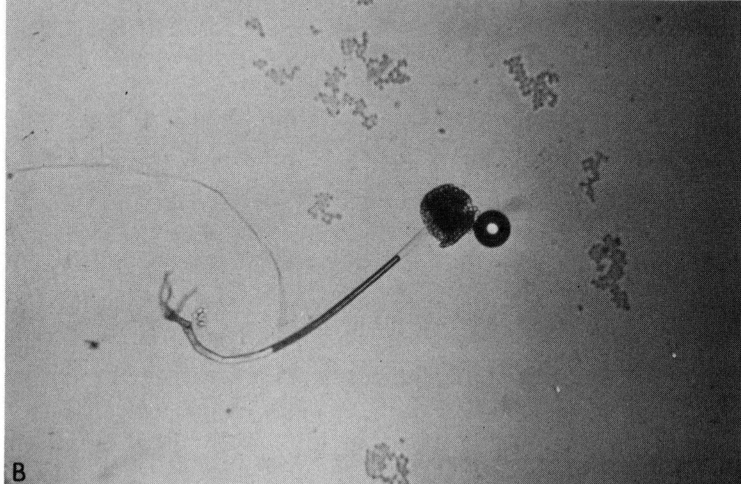

Figure 21–9. *Rhizopus* sp. *A,* Light yellow-brown rhizoids. ×400. *B,* Rhizoids and a small sporangiophore and sporangium. ×100.

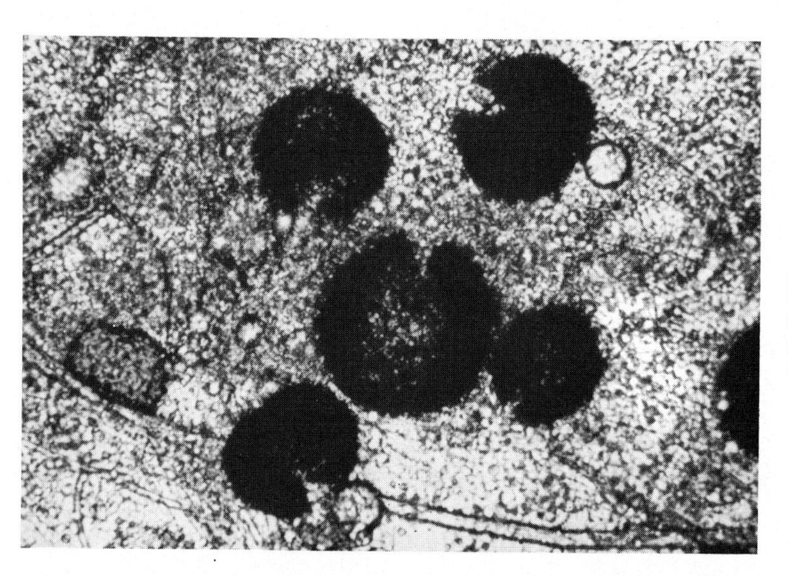

Figure 21–10. Zygospores of *Rhizopus nigricans.* Some have been broken open. One developing zygote shows attachment of suspensor cell. ×450. (Courtesy of S. McMillen.)

Absidia Van Tieghem 1876

This genus is characterized by the presence of rhizoids and by branching sporangiophores which arise between the nodes (Fig. 21-7,*B*). This latter characteristic is in contrast to the location of sporangiophores in the genus *Rhizopus*. The sporangia are pyriform. Two species of this genus are frequently isolated from animal infections and sometimes from human disease. These are *A. corymbifera* (Cohn) Saccurdo and Trotter 1912 and *A. ramosa* (Lindt) Lendner 1909.[22] Scholer[65] could not distinguish them biochemically and indicates the species may in fact be identical.

Absidia corymbifera (Cohn) Saccardo and Trotter 1912

Synonymy. *Mucor corymbifer* Cohn 1884. Ellis and Hesseltine[22] list 16 other synonyms.

Colony Morphology. Growth is rapid and produces a floccose, light olive-grey colony. The color fades in time. Sporangiophores are very long, up to 450 μ, and 4 to 8 μ in diameter. They arise from stolons and are light grey in color. Numerous smaller and irregularly shaped sporangiophores are also produced. The sporangiophores branch repeatedly to form corymbs. Sporangia are 20 to 35 μ in diameter and greyish. Columellae are 16 to 27 μ in diameter, ovoid to spatulate, and may have a few spines. The spores are formed by internal cleavage in a manner similar to that found in *Coccidioides*. Mature spores are globose to oval, 2 to 3 × 3 to 4 μ. Rhizoids arise from swollen areas (nodes) along a stolon. They are 12 μ in diameter, up to 370 μ in length, and hyaline. Zygospores are produced. They are 40 to 80 μ in diameter, thick-walled, and roughened with equitorial ridges. The brown suspensor cells develop circinate filaments that almost completely encircle the zygospores. Giant cells are numerous. The fungus grows well at 37°C, and growth is obtained up to 42°C.

Absidia ramosa (Lindt) Lendner 1908 differs from *A. corymbifera* chiefly in having uniformly oval spores, and when mated it produces fewer zygospores. Scholer[65] has developed a carbohydrate utilization pattern to distinguish *A. corymbifera* from several other species of *Absidia*. However, the physiologic pattern of *A. ramosa* is identical to that of *A. corymbifera*.

Mortierella Coemans 1863

This genus is characterized by a sporangium lacking a columella. The sporangiophore tapers to the attachment of the sporangium (Fig. 21-7,*D*). The spores are small, 1 to 2 μ by 2 to 4 μ. Small one-celled conidia called stylospores are also present. They are spiny or echinulate. The colonies are grey to yellowish grey and do not develop much aerial mycelium. *Hyphomyces destruens* isolated from infection in horses has many of the colonial characteristics of *Mortierella* but does not sporulate. In tissue, *Mortierella* and *Hyphomyces* evoke a marked eosinophilia similar to that seen in *Entomophthora* infections. *M. wolfii* Mehrotra and Baijal 1963 has been isolated from cases of abortion and fatal pneumonia of cattle. It was isolated from ensilage.[53a]

REFERENCES

1. Abramowitz, I. 1964. Fatal perforations of the stomach due to mucormycosis of the gastrointestinal tract. S. Afr. med. J., *38*:93–94.
2. Abramson, E., D. Wilson, et al. 1967. Rhinocerebral phycomycosis in association with diabetic ketoacidosis. Report of two cases and a review of clinical and experimental experience with amphotericin B therapy. Ann. Intern. Med., *66*:735–742.
3. Ainsworth, G. C., and P. K. C. Austwick. 1973. In *Fungal Diseases of Animals*. 2nd ed. Bucks, England, Commonwealth Agricultural Bureau.
4. Baker, R. D. 1956. Pulmonary mucormycosis. Am. J. Clin. Pathol., *26*:1235.
5. Baker, R. D. 1960. Diabetes and mucormycosis. Editorial. Diabetes, *9*:143–145.
6. Baker, R. D. 1971. Mucormycosis, In *The Pathologic Anatomy of Mycoses*. Dritterband, Funfter teil. Berlin, Springer-Verlag, Ch. 21.
7. Baker, R. D., J. H. Seabury, et al. 1962. Subcutaneous and cutaneous mucormycosis and subcutaneous phycomycosis. Lab. Invest., *11*:1091–1102.

7a. Bartram, R. J., M. Watnick, et al. 1973. Roentgenographic findings in pulmonary mucormycosis. Am. J. Roentgenol., *117*:810–815.

8. Battock, D. J., H. Grausz, et al. 1968. Alternate day amphotericin B therapy in the treatment of rhinocerebral phycomycocis (mucormycosis). Ann. Intern. Med., *68*:122–137.
9. Baum, J. L. 1967. Rhinoorbital mucormycosis, occurring in an otherwise apparently healthy individual. Am. J. Ophthalmol., *63*:335–339.
10. Bridges, C. H., and C. W. Emmons. 1961. A

phycomycosis of horses caused by *Hyphomyces destruens*. J. Am. Vet. Med. Assoc., *138*: 579–589.
11. Chick, E. W., J. Evans, et al. 1958. The inhibitory effect of amphotericin B on localized *Rhizopus oryzae* (mucormycosis) utilizing the pneumoderm pouch of rats. Antibiot. Chemother., *8*:506–510.
12. Chick, E. W., J. Evans, et al. 1958. Treatment of experimental mucormycosis (*Rhizopus oryzae* infection) in rabbits with amphotericin B. Antibiot. Chemother., *8*:394–399.
13. Cordes, D. O., W. A. Royal, et al. 1967. Systemic mycosis in neonatal calves. New Zealand Vet., *15*:143–149.
14. Darja, M., and M. Davy. 1963. Pulmonary mucormycosis with cultural identification. Can. Med. Assoc. J., *89*:1235–1238.
15. Davis, C. L., W. A. Anderson, et al. 1955. Mucormycosis in food producing animals. J. Am. Vet. Med. Assoc., *126*:261–267.
16. Dawson, C., N. G. Wright, et al. 1969. Canine phycomycosis—A case report. Vet. Rec., *84*:633–634.
17. Deal, W. B., and J. E. Johnson. 1962. Gastric phycomycosis. Report of a case and review of the literature. Gastroenterology, *57*:579–586.
18. Deweese, D. D., A. J. Schleuning, et al. 1965. Mucormycosis of the nose and paranasal sinuses. Laryngoscope, *75*:1398–1407.
19. Donnelly, W. J. C. 1967. Systemic bovine phycomycosis: A report of two cases. Irish Vet. J., *21*:82–87.
20. Duffy, E. 1945. Alloxan diabetes in rabbits. J. Pathol. Bacteriol., *57*:199–212.
21. Elder, T. D., and R. D. Baker. 1956. Pulmonary mucormycosis in rabbits with alloxan diabetes. Arch. Pathol., *61*:159–168.
22. Ellis, J. J., and C. W. Hesseltine. 1966. Species of *Absidia* with ovoid sporangiospores. Sabouraudia, *5*:59–77.
23. Emmons, C. W. 1964. Phycomycosis in man and animals. Rev. Patol. Vegetale *4*:329–337.
24. Englander, G. S. 1953. Mycetoma caused by *Rhizopus*. Arch. Dermatol., *68*:741.
25. English, M. P., and N. M. Lucke. 1970. Phycomycosis in a dog caused by unusual strains of *Absidia corymbifera*. Sabouraudia, *8*:126–132.
26. Faillo, P. S., H. P. Sube, et al. 1959. Mucormycosis of the paranasal sinuses and the maxilla. Oral Surg., *12*:304–309.
27. Feinberg, R., and T. S. Risley. 1959. Mucormycotic infection of an arteriosclerotic thrombus of the abdominal aorta. New Engl. J. Med., *260*: 626–629.
28. Fleckner, R. A., and J. H. Goldstein. 1969. Mucormycosis. Br. J. Ophthalmol., *53*:542–548.
29. Foley, F. D., and J. M. Shuck. 1968. Burn wound infection with phycomycetes requiring amputation of the hand. J.A.M.A., *208*:596.
30. Fürbringer, P. 1876. Beovbachtungen über Lungenmycose beim Menschen. Virchows Arch. Pathol. Anat., *66*:330.
31. Gale, R. G., and A. M. Welch. 1961. Studies on opportunistic fungi. I. Inhibition of *Rhizopus oryzae* by human serum. Am. J. Med. Sci., N.S., *241*:604–612.
32. Gale, A. M., and W. P. Kleitsch. 1972. Solitary pulmonary nodule due to phycomycosis (mucormycosis). Chest, *62*:752–755.
33. Gilman, J. C. 1957. *A Manual of Soil Fungi*. 2nd ed. Ames, The Iowa State College Press.
34. Ginsberg, J., A. G. Spanding, et al. 1966. Cerebral phycomycosis (mucormycosis) with ocular involvement. Am. J. Ophthalmol., *62*:900–906.
35. Green, W. H., H. I. Goldberg, et al. 1967. Mucormycosis infection of the craniofacial structures. Am. J. Roentgenol., *101*:802–806.
36. Gregory, J. E., A. Golden, et al. 1943. Mucormycosis of the central nervous system. Report of three cases. Bull. Johns Hopkins Hosp., *73*:405–419.
36a. Heller, R. A., H. P. Hobson, et al. 1971. Three cases of phycomycosis in dogs. Vet. Med. Small Anim. Clin., *66*:472–476.
37. Hessler, J. R., J. C. Woodard, et al. 1967. Mucormycosis in a rhesus monkey. J. Am. Vet. Med. Assoc., *151*:909–913.
38. Hesseltine, C. W., and J. J. Ellis. 1964. An interesting species of Mucor, *M. ramosissimus*. Sabouraudia, *3*:151–154.
39. Hewer, T. F., H. Pearson, et al. 1968. Aspergillosis and mucormycosis in two newborn *Okapi johnstoni* (Sclater). Br. Vet. J., *124*:282–286.
40. Hogben, B. R. 1967. Systemic mycotic lesions in apparently normal bobby calves: Incidence and appearance. New Zealand Vet. J., *15*:30–32.
40a. Hutchins, D. R., and K. G. Johnston. 1972. Phycomycosis in the horse. Aust. Vet. J., *48*:269–278.
41. Hutter, R. V. P. 1959. Phycomycetous infection (mucormycosis) in cancer patients, a complication of therapy. Cancer, *12*:330–350.
42. Jonquieres, E. D., C. A. Castello, et al. 1972. Seudomicetoma cutaneo ficomicotico. Int. J. Dermatol., *11*:89–95.
43. Josefiak, E. J., and J. H. Smith-Foushee. 1958. Experimental mucormycosis in the healthy rat. Science, *127*:1442.
44. Kahn, L. B. 1963. Gastric mucormycosis: Report of a case with a review of the literature. S. Afr. Med. J., *37*:1265–1269.
45. Kujiwara, A., J. W. Landau, et al. 1970. Hemolytic activity of *Rhizopus nigricans* and *Rhizopus arrhizus* (et seq.). Mycopathologia, *40*:131–139, 139–144.
46. Landau, J. W., and V. D. Newcomber. 1962. Acute cerebral phycomycosis (mucormycosis). J. Pediatr., *61*:363–383.
46a. Langston, C., D. A. Roberts, et al. 1973. Renal phycomycosis. J. Urol., *109*:941–944.
47. LaTouche, C. J., T. W. Sutherland, et al. 1964. Histopathological and mycological features of a case of rhinocerebral mucormycosis (phycomycosis) in Britain. Sabouraudia, *3*: 148–150.
48. Lichtheim, L. 1884. Ueber pathogene Mucorineen und die durch sie erzeugten Mykosen der Kaninchens. Z. Klin. Med., *7*:140.
48a. Lindt, W. 1886. Mitteilungen über einige neue pathogene Schimmelpilze. Arch. Exp. Pathol. Pharmacol., *21*:269.
49. Lucke, V. M., D. G. Morgan, et al. 1969. Phycomycosis in a dog. Vet. Res., *84*:645–646.
50. MacKenzie, A. 1969. The pathology of respiratory infections in pigs. Br. Vet. J., *125*:294–303.
51. Martin, J. E., D. J. Kroe, et al. 1969. Rhino-orbital phycomycosis in a rhesus monkey (*Macaca mulatta*). J. Am. Vet. Med. Assoc., *155*:1253–1257.

52. Mayer, A. F., and J. C. Emmert. 1815. Verschimmelung (Mucedo) in lebender Körper. Dtsch. Arch., Anat. Physiol. (Meckel), *1*:310–318.
53. Medoff, G., and G. S. Kobayashi. 1972. Pulmonary mucormycosis. New Engl. J. Med., *286*:86–87.
53a. Menna, M. E. di, M. E. Carter, et al. 1972. The identification of *Mortierella wolfii* isolated from cases of abortion and pneumonia in cattle and a search for its infection source. Res. Vet. Sci., *13*:439–442.
54. Meyer, R. D., P. Rosen, et al. 1972. Phycomycosis complicating leukemia and lymphoma. Ann. Intern. Med., *77*:871–879.
54a. Meyer, R. D., M. H. Kaplan, et al. 1973. Cutaneous lesions in disseminated mucormycosis. J.A.M.A., *225*:737–738.
55. Morquer, R., C. Lombard, et al. 1965. Pathologénie de quelques Mucorales pour les animaux. Une nouvelle mucormycose chez les Bovides. Bull. Trim. Soc. Mycol. Fr., *81*:421–449.
56. Muresan, A. 1960. A case of cerebral mucormycosis diagnosed in life with eventual recovery. J. Clin. Pathol., *13*:34–36.
57. Neame, P., and D. Rayner. 1960. Mucormycosis—A report of twenty-two cases. Arch. Pathol., *70*:261–268.
58. Paltauf, A. 1885. Mycosis mucorina: ein Beitrag zur Kenntnis der menschlichen Fadenpilzer-Krankungen. Virchows Arch. Pathol. Anat., *102*:543.
59. Polli, C. 1965. (On the distribution of ketoreductase in microorganisms.) Pathol. Microbiol. (Basel), *28*:93–98.
60. Prockop, L. D., and M. Silva-Hutner. 1967. Cephalic mucormycosis (phycomycosis). A case with survival. Arch. Neurol., *17*:379–386.
61. Prout, G. R., and A. R. Goddard. 1960. Renal mucormycosis. Survival after nephrectomy and amphotericin B therapy. New Engl. J. Med., *263*:1246.
62. Rabin, E. R., G. D. Lundberg, et al. 1961. Mucormycosis in severely burned patients. New Engl. J. Med., *264*:1286–1288.
63. Reinhardt, D. J., W. Kaplan, et al. 1970. Experimental cerebral zygomycosis in alloxan-diabetic rabbits. I. Relationship of temperature tolerance of selected zygomycetes to pathogenicity. Infect. Immun., *2*:404–413.
63a. Reinhardt, D. J., and J. J. Licota. 1974. *Rhizopus, Mucor* and *Absidia* as agents of cerebral zygomycosis. Abs. Am. Soc. Microbiol., Mm 21.
64. Salvin, S. B. 1952. Endotoxin in pathogenic fungi. J. Immunol., *69*:89–99.
65. Scholer, H. J., and E. Müller. 1966. Beziehungen zwischen biochemischer Leistung und Morphologie bei Pilzen aus der Familie der Mucoraceen. Pathol. Microbiol., *29*:730–741.
66. Scholz, H. D., and L. Meyer. 1965. *Mortierella polycephala* as a cause of pulmonary mycosis in cattle. Berl. Münch. Tierärztl. Wschr., *78*:27–30.
67. Sheldon, W. H., and H. Bauer. 1960. Tissue mast cells and acute inflammation in experimental cutaneous mucormycosis of normal, 48/80 treated, and diabetic rats. J. Exp. Med., *112*:1069–1083.
68. Simon, R., G. G. Hoffman, et al. 1964. Phycomycosis. Aerospace Med., *35*:668–675.
69. Symmers, W. S. 1962. Histopathological aspects of the pathogenesis of some opportunistic fungal infections, as exemplified in the pathology of Aspergillosis and the Phycomycetoses. Lab. Invest., *11*:1073–1090.
70. Taylor, R., G. Shklar, et al. 1964. Mucormycosis of oral mucosa. Arch. Dermatol., *89*:419–425.
71. Thomford, N. R., T. H. Dee, et al. 1970. Mucormycosis of a saphenous vein autograph. Arch. Surg., *101*:518–519.
72. Vignale, R., J. E. Mackinnon, et al. 1964. Chronic destructive mucocutaneous phycomycosis in man. Sabouraudia, *3*:143–147.
73. Watson, K. C., and P. B. Neame. 1960. *In vitro* sensitivity of amphotericin B on strains of Mucoraceae pathogenic to man. J. Lab. Clin. Med., *56*:251–257.
74. Zycha, H. 1935. *Mucorineae*. Kryptogamen-flora. Lepzig Mark Brandenberg. *6A*:1–264. Berlin, Gebrueder Borntraeger.
75. Zycha, H., R. Siepmann, et al. 1969. *Mucorales*. Weinheim, J. Cramer.

Chapter 22

MISCELLANEOUS AND RARE MYCOSES, ALGOSIS, AND PNEUMOCYSTOSIS

From time to time numerous soil and airborne fungi have been reported as being responsible for a variety of disease processes. The majority of these reports are invalid or at least questionable. Pathogenic fungi are often difficult to grow, whereas contaminants are frequently isolated and described as the etiology of a particular disease process. However, there are a few organisms not normally considered virulent that may adapt to and take advantage of a situation and establish an infection. This is particularly true of the so-called "opportunistic infections"—candidosis, aspergillosis, mucormycosis, and so forth, covered in previous chapters. Such organisms as *Cephalosporium, Trichosporon, Penicillium, Rhodotorula,* and so forth have been isolated from such situations also and their etiologic roles substantiated in some cases. The circumstances usually involve a barrier break, such as surgery or indwelling catheter, steroid therapy, or the use of immunosuppressive drugs or cytotoxins. The increased use of such modalities will vastly increase the opportunities for such superinfections, and the list of organisms involved will also increase.

The use of steroids, antibiotics, and cytotoxins has also brought to light the presence of a peculiar organism, *Pneumocystis carinii,* that appears to be of common occurrence in rodents and man, but until recently was not recognized. It is now known to be of major importance as an opportunistic infectious agent. The taxonomic affinities of this organism to other microorganisms are as yet unsettled. It had previously been grouped with the protozoans, but its cell wall contains chitin which, along with certain ultramicroscopic characteristics, indicates a relation to the fungi. In addition to the true fungi and actinomycetes, there are a few valid cases of infection caused by the colorless alga, *Prototheca,* and even a report of disease due to blue-green algae. In this chapter some of the well-established but infrequent mycoses will be discussed, together with a résumé of unique or very rare infections. Reviews of the rare mycoses have recently been published by English[6] and Parker.[12]

Geotrichosis

DEFINITION

Geotrichosis is an infection caused by the ubiquitous fungus, *Geotrichum candidum.* Lesions are bronchopulmonary, bronchial, oral, cutaneous, and, very rarely, alimentary.

HISTORY

Geotrichum was first isolated from decaying leaves by Link in 1809. Bennett in 1842 described the organism as causing a superinfection in an old tuberculous cavity.[1] He differentiated it from "*Monilia*" (*Candida*), and if this report is valid, it represents the first case of geotrichosis. Confusion with infections caused by *Candida* or isolation of the organism from normal flora invalidates most of the early records of *Geotrichum* as a fungal pathogen. The reports of pulmonary geotrichosis made by Linossier in 1916 and Martin in 1928 are probably authentic.[9] The description of the organism in eczematous dermatitis in 1935 by Ciferri and Redaelli, however, probably represents isolation of normal flora from a pathologic process of other etiology. Geotrichosis, when substantiated as valid, is usually secondary to some other debilitation, such as tuberculosis, or is a complication of steroid therapy.

ETIOLOGY AND DISTRIBUTION

Though several species of the genus have been described, only *Geotrichum candidum* has been associated with human infections. Schnoor[16] isolated the fungus from 29 per cent of 314 stool specimens obtained from healthy people. In a recent survey of sputum, feces, urine, and vaginal secretions from 2643 subjects, both healthy and with some other disease, the organism was present in 18 to 31 per cent of specimens.[14] However, it was not associated with any specific illness. *G. candidum* has also been recovered from cottage cheese, dairy products, plant material, and healthy skin. It is therefore endogenous as normal flora in man and appears to be ubiquitous in nature.

CLINICAL DISEASE

Pulmonary involvement is the most frequently reported form of the disease but bronchial, oral, cutaneous, and alimentary infections have also been noted.

Pulmonary Geotrichosis

Pulmonary disease simulates tuberculosis and is usually secondary to it. Clinically there is a light grey, thick, and mucoid sputum which in some cases is purulent and rarely blood-tinged. Fine to medium rales are heard. The condition is most often chronic, and there is little debilitation or fever present. On x-ray examination, smooth, dense, patchy infiltrations and some cavities are seen. These are present in the areas of the lung commonly associated with tuberculosis, such as the hilar and apical regions. About 24 cases of pulmonary geotrichosis are recognized in the literature.[9] Although usually chronic, the disease may run a fulminant, sometimes fatal, course if patients are put on steroids.[13] To establish the diagnosis of geotrichosis, it is necessary to see the organisms in quantity in sputum and to obtain multiple positive cultures.[11a] Other etiologies of lung disease must be excluded. Specific treatment for the disease includes iodides, aerosol nystatin, and amphotericin B.[9]

Bronchopulmonary geotrichosis, which is similar to the allergic type of aspergillosis, is another form of the disease. *Geotrichum* is seen growing in the lumen of the bronchae, and the patient has symptoms of severe asthma.[15] Colistin, methanesulfonate, desensitization, and prednisolone have relieved the symptoms, and patients recover without specific antifungal therapy.

Bronchial Geotrichosis

In this form, the lung is not involved, and the disease consists of an endobronchial infection. Symptoms include a prominent chronic cough, gelatinous sputum, lack of fever, and medium to coarse rales. On x-ray, diffuse peribronchial thickenings are seen. A fine mottling may be present on the middle and basilar pulmonary fields. About nine cases have been recorded.[9] In the early literature the allergic form of chronic bronchopulmonary geotrichosis was not differentiated from the infectious or invasive forms.[14,17] On bronchoscopy, fine white patches are seen in the bronchial tree. They are similar in appearance to the lesions of thrush. The sputum contains large numbers of the organism. Specific antifungal therapy is necessary for control of this form of geotrichosis.

Oral Geotrichosis

This disorder is identical in appearance to thrush and formerly was often confused with it. The two can only be differentiated by direct microscopic examination and culture. Few authentic cases of this disease exist.[9]

Gastrointestinal Geotrichosis

This disorder has been recorded a few times. The first case was that of Almeida and Lacaz.[9] Recently, Neagoe reported on enterocolitis associated with glutamic acid therapy.[11] Symptoms disappeared when the therapy was discontinued. The establishment of an etiologic relationship of the organism to the disease is very difficult, as *Geotrichum*, like *Candida*, is part of the normal flora.[5, 11a] Intestinal geotrichosis has been described in animals.[8]

Cutaneous Geotrichosis

A few cases of cutaneous infection have been recorded. In a recent case the lesion, a cystic mass, appeared in the soft tissue of the hand following a skin grafting procedure. Drainage was followed by healing.[7]

PATHOLOGY

A case of disseminated geotrichosis was reported by Chang and Buerger in 1964.[5] The patient had carcinoma of the ascending colon and had received fluorouracil as a treatment. Lesions containing *Geotrichum* were extensive and generalized throughout the body. Infection of the colonic mucosa adjacent to the carcinoma was seen. This was the probable point of entry for the organism. The gross lesions appeared as necrotic foci and were observed in heart, lungs, and spleen. Microscopic examination revealed suppuration and necrosis. The fungus was seen as a mixture of yeastlike cells and septate mycelium with oval and spherical arthrospores. Some areas contained the rectangular arthrospores characteristic of *Geotrichum*. In addition, some of the lesions had *Candida*-like organisms in them. If this truly was a case of geotrichosis, it represents an opportunistic infection. Unlike the chronic bronchial forms, systemic infection is rapidly fatal.

Establishing a diagnosis of geotrichosis is very difficult. All other etiologies must be eliminated. Similar to *Candida*, the organism is part of the normal flora of the mouth and intestine but is a much less aggressive opportunist. The two cannot be differentiated on histologic appearance alone; only identification by culture is adequate. In the terminal stages of debilitating disease, it is possible that both organisms may invade and set up focal lesions.

ANIMAL DISEASE

Natural disease in animals is also quite rare. The organism has caused an adenitis in pigs, intestinal disease in the ocelot,[4] and disseminated disease in the dog.[8] Early reports of mastitis in cows and infection in other domestic animals probably reflect confusion with *Candida* sp. In the case of the disseminated disease in the dog, lesions consisted of areas of coagulative necrosis and macrophage invasion in the lung, lymph nodes, and kidney. Small granulomas were seen in liver, spleen, bone marrow, and eye.

Experimental Infection. In laboratory animals, this is very difficult to establish.[10] In most of the recorded literature, the attempts have been unsuccessful or the identity of the organism questionable. The fungus is easily confused with other yeastlike organisms, especially *Trichosporon capitatum*. This latter yeast, when injected into rabbits, regularly produces abscesses.

LABORATORY IDENTIFICATION

There are no serologic procedures useful for the diagnosis of the disease, and the skin test is of no value. The latter and the oidiomycin skin test only indicate host sensitization by normal flora. Direct and histologic examination may strongly suggest the presence of *Geotrichum*, but cultural confirmation is always necessary.

Direct Examination. Sputum, pus, or lesion material is examined as a potassium hydroxide mount. As in examination for other fungi, sputum may be digested first (by the methods used for isolation of tubercle bacilli), the specimen concentrated, spun, and the sediment examined. Gram stains are also useful. The characteristic morphology is an oblong or rectangular arthrospore, 4 to 8 μ in size. The ends are square or rounded. Spherical cells, 4 to 10 μ, are also seen.

Culture Methods. Material to be cultured can be spread over media containing antibiotics and incubated at 25°C.

MYCOLOGY

Geotrichum candidum Link 1909

Synonymy. *Oidium pulmoneum* Bennett 1842.

Perfect State. *Endomyces geotrichum* Butler and Petersen 1972. Carmichael[3] lists 100 synonyms of *Geotrichum*. They are usually in the genera *Oidium, Oospora,* and *Mycoderma.*

Morphology. SDA, 25°C. The organism grows as a dry, mealy, white to cream colony (Fig. 22–1 FS C 37). At 37°C, growth is very slow and subsurface. Microscopically the hyphae are seen to fragment into arthrospores, which are quite variable in size. The arthrospores are seen to germinate at one end, giving the appearance of a bud. The latter develops into a septate mycelium, however. True blastospore production is not found in the genus. Carbohydrates are not fermented. Auxanograms usually show assimilation of glucose and galactose. Lactose, maltose, erythritol, cellibiose, melibiose, and sucrose are not assimilated. It is urease-negative. The identification of *G. candidum* is very difficult, especially its differentiation from *Trichosporon* species.

Recently Bulter and Petersen[2] described the perfect state of the organism as *Endomyces geotrichum* in the class Ascomycetes.

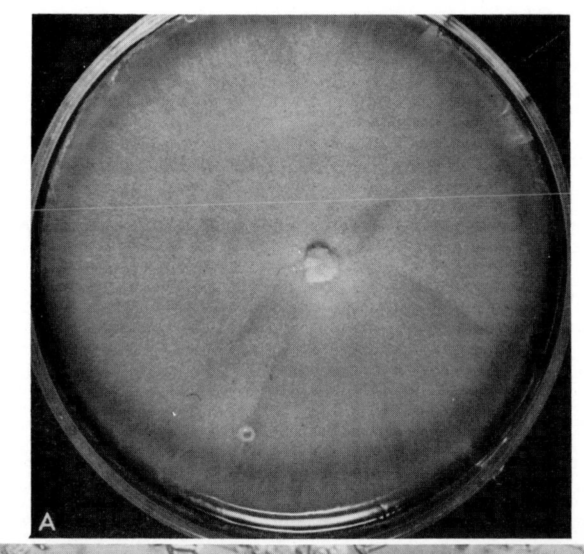

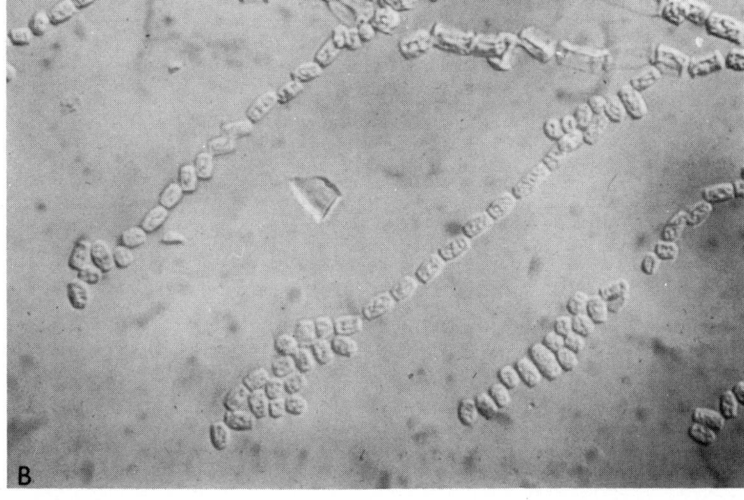

Figure 22–1. FS C 37. Geotrichosis. *A,* Colony of *Geotrichum candidum.* It is off-white, mealy, and glabrous. *B,* Fragmentation of mycelium to form typical arthrospores. ×400. (Courtesy of S. McMillen.)

REFERENCES

1. Bennett, J. H. 1842. On the parasitic fungi found growing in living animals. Trans. R. Soc. Edinb., *15*:277–294.
2. Bulter, E. E., and L. J. Petersen. 1972. *Endomyces geotrichum*, a perfect state of *Geotrichum candidum*. Mycologia, *64*:365–375.
3. Carmichael, J. W. 1957. *Geotrichum candidum*. Mycologia, *49*:820–829.
4. Carroll, J. M., et al. 1968. Intestinal geotrichosis (*Geotrichum candidum* in the ocelot [*Felis pardalis*]). Am. J. Vet. Clin. Pathol., *2*:257–261.
5. Chang, W. W. L., and I. L. Buerger. 1964. Disseminated geotrichosis. Case report. Arch. Intern. Med., *113*:356–360.
6. English, M. P. 1967. Some unusual mycoses. Rev. Med. Vet. Mycol., *6*:103–108.
7. Goldman, S., R. R. Lipscomb, et al. 1969. *Geotrichum* tumefaction of the hand. J. Bone Joint Surg., *51*:587–590.
8. Lincoln, S. D., and J. L. Adcock. 1968. Disseminated geotrichosis in a dog. Pathol. Vet., *5*:282–289.
9. Morenz, J. 1971. Geotrichosis, In *Handbuch der Speziellen Pathologischen Anatomie*. Vol. 3. Baker, R. D. (ed.), Berlin, Springer-Verlag.
10. Morquer, R., C. Lombard, et al. 1955. Pouvoir pathogène de quelques espèces de *Geotrichum*. C. R. Acad. Sci. (Paris), *240*:378–380.
11. Neagoe, G., and M. Neagoe. 1967. Enterocolitis durch *Geotrichum candidum* nach Therapie mit Glutaminsaüre. Dtsch. Z. Verdau.-u. Stoffwechsdkr., *27*:205–208.

11a. Olin, R. 1971. Pulmonary geotrichosis and candidiasis. Minnesota Med., *54*:881–886.

12. Parker, J. C., and G. K. Klintworth. 1971. Miscellaneous and uncommon disease attributed to fungi and actinomycetes. In *Handbuch der Speziellen Pathologischen Anatomie*. Vol. 3. Baker, R. D. (ed.), Berlin, Springer-Verlag.
13. Peninou-Castaing, S., R. Collas, et al. 1964. Geotrichose pulmonaire devolution fatale associée à une tuberculose pulmonaire. Poumon, *20*:287–296.
14. Peter, M., G. Horváth, et al. 1967. Date referitoare é le frecventa genului *Geotrichium* in diferite produse biologice umane. Medna. Interna. (Bucur.). *19*:875–878.
15. Ross, J. D., K. D. G. Reid, et al. 1966. Bronchopulmonary geotrichosis with severe asthma. Br. Med. J., *1*:1400–1402.
16. Schnoor, T. G. 1939. The occurrence of Monilia in normal stools. Am. J. Trop. Med., *19*:163–169.

16a. Vlassopoulos, K., and S. Bartsokas. 1972. [Intestinal geotrichosis—Case report.] Acta Microbiol. Hellenica, *17*:197–206.

17. Webster, B. H. 1959. Bronchopulmonary geotrichosis: A review of four cases. Dis. Chest, *35*:273–281.

Adiaspiromycosis

DEFINITION

Adiaspiromycosis refers to the *in vivo* development, without replication, of adiaspores from inhaled aleuriospores of the fungal genus *Chrysosporium (Emmonsia)*. The condition is commonly found in the lungs of rodents and other small animals that live or burrow in the soil. The environment inhabited by the animals and the fungus ranges from hot, dry desert and semiarid habitats to banks of streams and ponds in tropical jungles or semiarctic, coniferous forests. The distribution is worldwide.[12,13,19] The finding of adiaspores in human tissue is quite rare. Such spores in sufficient numbers in human or animal tissue to cause distress or disease is also uncommon. The name "adiaspore" signifies the enlargement, without replication, of a fungal spore under the influence of elevated temperature.[11] Adiaspiromycosis is the term used when such development occurs within an animal body.

Synonymy. Haplomycosis, adiasporosis.

HISTORY

Many of the human and animal mycoses were initially described as protozoan diseases. This was usually done on pathologic evidence alone. It is only chance observation and appreciation that fungi may have different morphologic forms under different environmental conditions of growth that has led to the discovery of the true nature of the etiologic agent. Adiaspiromycosis was discovered in 1942 by Emmons while examining rodents from Arizona for infection by *Coccidioides immitis*.[10] Since that time the former disease has been found to be extremely common in rodents and small animals from all parts of the world. The disease is not new, as it now has been recognized in specimens of preserved lungs from animals trapped in Sweden in 1845. It is likely that the organism had been seen in tissue previously, but described as a protozoan. The fungal nature of the condition was appreciated by Kirschenblatt[14] in 1939. He recorded cysts in the lungs of several rodents from Georgia (*Transcaucasia*). He described the etiology as a fungus, *Rhinosporidium pulmonale,* but without culture or a Latin description.

In a search for an animal reservoir of coccidioidomycosis, Emmons and Ashburn[10] in 1942 examined 303 rodents trapped in southern Arizona. *Coccidioides immitis* was isolated from 25 of these, but a new fungus was present in 101 animals. Both fungi were found in eight. Beside the spherules of *C. immitis,* another type of fungal spore was found. This had a pale-staining cytoplasm and very thick laminated walls, was uninucleate, and did not form endospores. Later work showed that the spores from the cultures of the new fungus grew to these large bodies (<40 μ) when injected into mice or grown at 37°C. The new fungus was named *Haplosporangium parvum* because of the purported resemblance to the phycomycete, *Haplosporangium bisporale.* Basing their toxonomy on critical mycologic observation, Ciferri and Montemartini[4] considered the organism to represent a new hyphomycete genus and named it *Emmonsia parvum.* Isolants of *E. parvum* from the desert areas of the southwest United States produced adiaspores 20 to 40 μ in diameter. However, larger adiaspores were being discovered in material from Canada and the northern United States. The strains that produced spores with a diameter up to 400 μ were designated a new species, *Emmonsia crescens,* in 1960.[11] Other workers have concluded that there is not sufficient evidence to separate the two species, and the latter may be regarded as a variety of the former.[3,17] They conclude that the fungus is best regarded as a species of the genus *Chrysosporium.*

ETIOLOGY, ECOLOGY, AND DISTRIBUTION

The only agents so far identified as producing adiaspores in the lungs of animals are *Chrysosporium parvum* and *C. parvum* var. *crescens.* The organisms identified as *E. brasiliensis* and *E. ciferrina* were later shown to be *chrysosporium pruinosum.*[17] This latter species produces large spores in culture. When these are injected into animals, they remain viable and can be recultured, but they do not enlarge to form adiaspores.

Adiospiromycosis has been recorded in a number of animal species. The disease is very common in mice, moles, rats, rabbits, ground squirrels, and other rodents. It has also been found in carnivores, such as skunks, weasels, martens, and minks, and in other animals, such as armadillos, wallabies, and opposums.[12,13] What was probably the first human case of the disease was described by Doby-Dubois[6] in 1964. A single adiaspore was found in a nodule in the parenchyma near a bronchus in a patient who had pulmonary aspergillosis and cystic disease of the lung. Another case has been discovered in Honduras, again represented by a single adiaspore in a nodule.[5] A few cases have been recorded in Czechoslovakia[15] and in Venezuela.[18] In a recent review by Salfelder at al.,[18] the authors accept ten human infections as valid. In one report, enormous numbers of adiaspores of the var. *crescens* were found in the lung biopsy of an 11-year-old boy. It appeared that functional impairment of the lung was present.[15] When large numbers of spores are present in natural or experimentally produced disease, severe respiratory distress may ensue. Sometimes the infection occurs as an epidemic among animals and can cause severe pneumonia and death.[16]

A spring peak in the incidence of the disease in animals has been reported by Dvořák et al.[8] These authors also speculate that adiaspores may serve in dissemination of the species to other locales through the mediation of the lungs of infected animals.[7] Following death of the animals, the mycelial stage of the organism grows in the burrows of the rodents or in surrounding soil.

CLINICAL DISEASE AND PATHOLOGY

Naturally occurring adiaspiromycosis is restricted to the lungs. The spores do not migrate, replicate, or disseminate, so that lesions are found only in the endobronchial or alveolar spaces. When few spores are present, there is little cellular response. A few mononuclear cells may surround the developing spore. In cases of heavy exposure, granulomas may develop. In some of the original animals examined by Emmons,[10] adiaspores and the spherules of *C. immitis* were found in the same granuloma. Very rarely spores may be found in other organs.[15]

The size of the adiaspores produced by *C. parvum* is quite variable. In natural disease, the average size is 10 to 13 μ. In experimental disease using white mice, the adiaspores reach a diameter of 40 μ. This is an increase in volume of 10^4. *C. parvum* var. *crescens* is much more common throughout the world. The adiaspores produced average 200 to 400 μ in diameter. Such spores are seen surrounded by a few histiocytes.[2] Injection of the fungus into white mice may be followed by the production of adiaspores up to 600 μ in diameter and of a wall with a thickness of 70 μ. This is a volume increase of 10.[6]

Natural disease is limited to the lung, but systemic disease can be produced if aleuriospores from a mycelial culture are injected into mice, dogs, rabbits, and rats. The spores enlarge and produce adiaspores in any organ, but they do not replicate. If spores are given in sufficient numbers, death may ensue from mechanical obstruction.

On staining, the fungus shows a laminate wall and often a vacuolate cytoplasm. The inner layers of the wall take the PAS stain, but the outer layers are variable in reaction (Fig. 22–2 FS C 38). The cytoplasm is pale-staining and in *C. parvum* has one nucleus. *C. parvum* var. *crescens* is multinucleate.

Culture. The organism grows on most laboratory media at 25°C as a mycelial colony. At 37°C, adiaspores are produced, but they do not proliferate at this temperature.

MYCOLOGY

The taxonomy of this organism is still unsettled. The original epithet used by Emmons, *Haplosporangium*, cannot accommodate the organism, and a new monotypic genus was created to include it. Later Emmons divided the known strains into two species. *E. parvum* produced adiaspores that were uninucleate and reached a diameter of 40 μ. This was the type commonly isolated

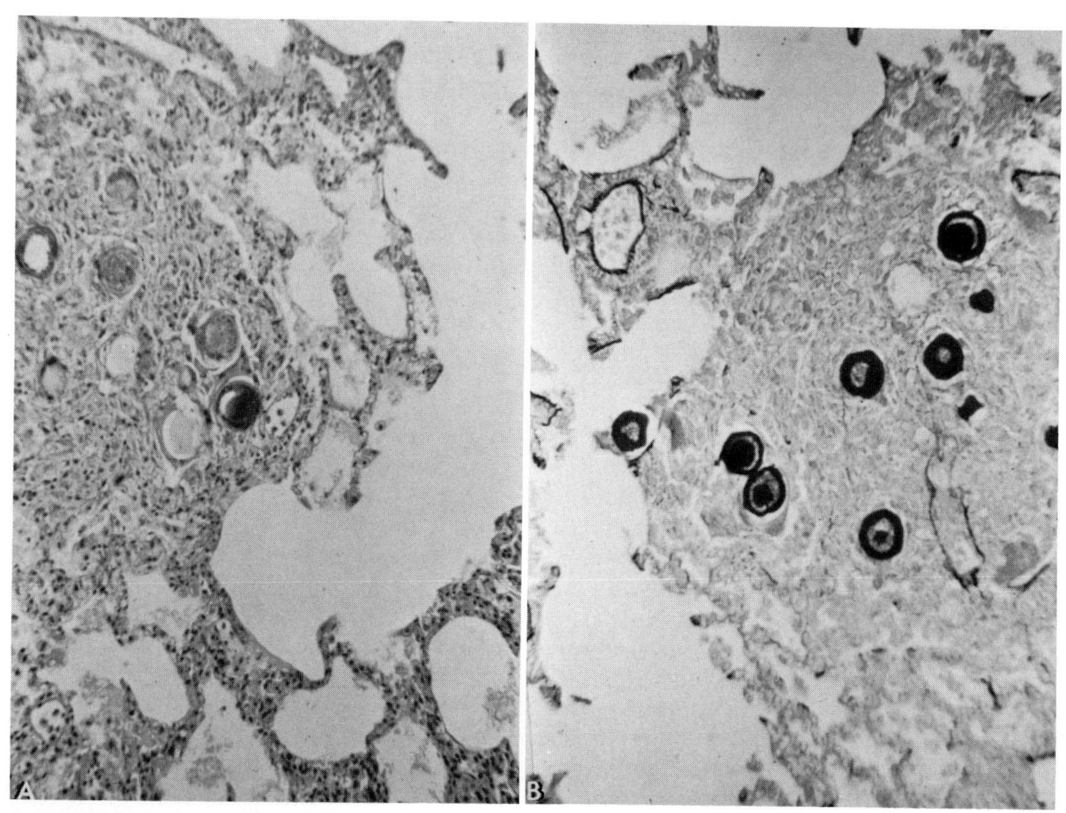

Figure 22–2. FS C 38. Adiaspiromycosis. *A*, Section of rat lung showing large vacuolated spherules and granuloma formation. Hematoxylin and eosin stain. ×60. *B*, Section of lung stained by Gridley method showing thick laminate walls of spherules. ×60.

from the southwestern United States and perhaps once in Kenya. In the thousands of lesions and isolants found in other parts of the world, the adiaspores were larger and multinucleate. Emmons[11] considered this sufficient evidence to create a new species, *E. crescens*, for them. However, in culture the colonial characteristics, sporulation, and mycelial morphology of the two organisms are essentially identical. For this reason other workers consider them to be varieties of the same species. Furthermore, Carmichael[3] and Padhye and Carmichael[17] discuss the similarity of the *Chrysosporium*, *Emmonsia*, *Blastomyces*, and *Paracoccidioides* genera. Critical mycologic investigation is needed before natural relationships among these organisms can be made.

Chrysosporium parvum Carmichael 1962

Synonymy. *Haplosporangium parvum* Emmons and Ashburn 1942; *Emmonsia crescens* Emmons and Jellison 1960; *Emmonsia parva* Ciferri and Montemartini 1959.

The organism grows initially as a glabrous, colorless colony, which in time produces white aerial mycelium. It reaches a diameter of 5 cm in two weeks. The colony often has alternate areas of tufted mycelium or mycelium in coremia, with areas of a glabrous consistency.

On microscopic examination the hyphae are seen to be septate and branching, with a diameter of 0.5 to 2 μ. Numerous aleuriospores are produced on conidiophores that branch at right angles from the vegetative hyphae. These spores are spherical, though slightly flattened in the vertical axis, 3 to 3.5 μ in diameter, and may have fine to coarse spinulation. Secondary spores may be formed from the primary aleuriospores. The var. *crescens* is reported to have slightly larger spores. These may reach a diameter of 4.5 μ and are more ovoid in shape.

When the aleuriospores and hyphae of *C. parvum* are placed on media at 40°C, the mycelium degenerates but the spores enlarge. The size reached by the enlarging spores will depend on how crowded they are. They average 15 to 25 μ, but some will grow to 40 μ or more, with a wall 2 μ thick. They remain uninucleate and do not replicate. Emmons cautions that a crowded inoculum when transferred will show apparent multiplication, but this is owing only to the further enlarging of spores in a less crowded environment. The aleuriospores of the var. *crescens* when grown at 40°C develop into adiaspores from 200 to 700 μ in diameter. These may have walls up to 70 μ thick and contain several hundred nuclei. If adiaspores are placed at 25°C, they sprout, producing numerous mycelical strands. A type of "budding" of the adiaspores can be produced that resembles *Paracoccidioides brasiliensis*.[9] If aleuriospores of var. *crescens* are incubated at 40°C for 12 days (attaining a diameter of 200 μ) and then incubated at temperature of 25°C, they will sprout, with the formation of many hyphae. If the culture is reincubated at 40°C within 4 to 8 hours, the hyphae initially will form secondary adiaspores, giving the appearance of the multiple budding yeast cells of *P. brasiliensis*. In some culture preparations, several secondary adiaspores may be formed within the parent spore, giving the impression of endosporulation. The process is quite different, however, and no true multiplication occurs.

REFERENCES

1. Bakerspigel, A. 1968. Canadian species of *Sorex*, *Microtus*, *Peromyscus* infected with *Emmonaia*. Mycopathologia, *37*:273–279.
2. Boisseau-Lebreuil, M. T. 1972. Evolution d'*Emmonsia crescens* agent de l'adiaspiromycose en adiaspores dans la poumon de souris de laboratoire expérimentalement infestées avec la phase saprophytic du champignon. Mycopathologia, *43*:267–281.
3. Carmichael, J. W. 1962. Chrysosporium and some other aleuriosporic hyphomycetes. Can. J. Bot., *40*:1137–1173.
4. Ciferri, R., and A. Montemartini. 1959. Taxonomy of *Haplosporangium parvum*. Mycopathologia, *10*:303–316.
5. Cueva, J. A., and M. D. Little. 1971. *Emmonsia crescens* infection adiaspiromycosis in man in Honduras. Am. J. Trop. Med., *20*:282–287.
6. Doby-Dubois, M., M. L. Chevrel, et al. 1964. Premier cas humain d'adiaspiromycose par *Emmonsia crescens*. Bull. Soc. Pathol. Exot., *57*:240–244.
7. Dvořák, J., M. Otčenášek, et al. 1966. [Conception on the circulation of *Emmonsia crescens* Emmon et Jellison 1960 in nature.] Folia Microbiol. (Praha), *13*:150–157.
8. Dvořák, J., Otčenášek, et al. 1969. The spring peak of adiaspiromycosis due to *Emmonsia crescens* Emmons and Jellison 1960. Sabouraudia, 7:12–14.
9. Emmons, C. W. 1964. Budding in *Emmonsia crescens*. Mycologia, *56*:415–419.
10. Emmons, C. W., and L. L. Ashburn. 1942. The isolation of *Haplosporangium parvum* n. sp. and

Coccidioides immitis from wild rodents. Public Health Rep., *57*:1715–1727.
11. Emmons, C. W., and W. L. Jellison. 1960. *Emmons crescens* sp. n. and adiaspiromycosis (haplomycosis in mammals). Ann. N.Y. Acad. Sci., *89*:91–101.
12. Jellison, W. L. 1969. *Adiaspiromycosis: Haplomycosis.* Missoula, Montana, Mountain Press.
13. Jellison, W. L., and J. W. Vinson. 1961. The distribution of *Emmonsia crescens* in Europe. Mycologia, *53*:524–535.
14. Kirschenblatt, J. D. 1939. A new parasite of the lungs in rodents. C. R. (Doklady) Acad. Sci. (USSR), *23*:406–408.
15. Kodousek, R., V. Vojtek, et al. 1970. Systemic pulmonary adiaspiromycosis (caused by *E. crescens*). Čas. Lék. Čes., *109*:923–924.
16. McDiarmid, A., and P. K. C. Austwick. 1954. Occurrence of *Haplosporangium parvum* in the lungs of the mole *Talpa europea.* Nature (Lond.), *174*:843–844.
17. Padhye, A. A., and J. W. Carmichael. 1968. *Emmonsia brasiliensis* and *Emmonsia ciferrina* are *Chrysosporium pruinosum.* Mycologia, *60*:445–447.
18. Salfelder, K., A. Fingerland, et al. 1973. Two cases of adiaspiromycosis. Beitr. Pathol., *145*:94–160.
19. Taylor, R. L., D. C. Cavanaugh, et al. 1968. Adiaspiromycosis in small mammals of Viet Nam. Mycologia, *60*:450–451.

Penicilliosis

The common blue-green molds, *Penicillium* sp., are ubiquitous in nature and constantly present in the human environment. Whereas species of the related genus *Aspergillus* are regularly associated with invasive human disease, documented infection by members of the genus *Penicillium* is very rare. The fungus is regularly isolated from sputum, bronchial secretions, and other body surfaces,[5] but even when repeatedly isolated and in quantity, its etiologic role is equivocal unless verified by histopathologic examination. There are well over 300 species of the genus, and some are known to produce toxins while growing on decaying vegetation. Sometimes these products produce severe and fatal toxicosis in animals and gastrointestinal disturbances in man. *Penicillium* sp. are also well established as agents of external ear infections. The former disease is discussed in the chapter on mycotoxicoses and the latter in infections of the eye and ear.

The ubiquity of *Penicillium* species and their ease of isolation from a variety of human sources have led to a long list of diseases in which these organisms have been the purported etiologic agents. In the vast majority of cases, the mold isolated was a contaminant and not associated with the pathology. One of the earliest records of human infection by *Penicillium* was that of Chute in 1911.[3] The organism *P. glaucum* was isolated from urine and said to be the etiologic agent of a bladder infection. Details are lacking. In 1915 Mantrelli and Negri[8] isolated a *Penicillium* described as *P. mycetogenum* from black-grained mycetoma. It has not been found since.

Bronchopulmonary penicilliosis has been the most frequently reported form of the disease. There are ten records in the literature, starting with Castellani in 1918, and the disease has been reported sporadically since. Delore reported a case in 1955[4] and Huang and Harris in 1962.[7] The latter was a systemic infection. In most of these cases, adequate substantiation of the etiologic role of *Penicillium* in the disease process is lacking. In some, a matted mycelial mass was seen in a pulmonary cavity or cyst. There was no cellular reaction. It is probable that the fungus was growing saprophytically in a preformed cavity, possibly just before or immediately after death of the patient. *P. crustaceum* is the organism most often cited. What might be termed an allergic bronchopulmonary disease was described by Delore et al.[4] The patient improved on cortisone therapy. Although a mycotic septicemia was alleged, owing to "finding mycelium in a calf muscle biopsy," the evidence is inadequate. An authenticated case of *Penicillium* infection of the urinary tract was reported by Gillium and Vest[6] in 1951. The patient had sporadic attacks of fever and right flank pain. Several boli of pink-tinged mycelial mats were voided in the urine. By catheterization, it was found that these came only from the right side calyx. The organism was identified as *P. citrinum.* No treatment was instituted

The only verified case of systemic penicilliosis was reported in 1963 by Huang and Harris.[7] The patient had acute leukemia and was treated with antibiotics and steroids. At autopsy, the patient had disseminated cerebral and pulmonary penicilliosis and gastrointestinal candidosis. Vascular invasion, thrombosis, and infarction were the basic pathologic features. Masses of mycelium were seen in the thrombi. Thus the lesions were identical to those found in invasive aspergillosis. The organism was identified as *P. commune,* an infrequently encountered soil saprophyte.

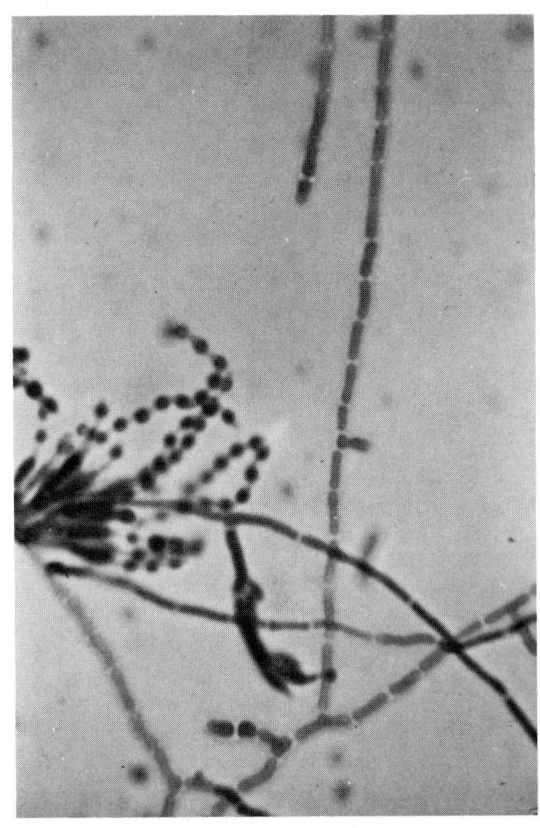

Figure 22–4. Culture mount of *Penicillium marneffei.* ×400. (Courtesy of A. DiSalvo.)

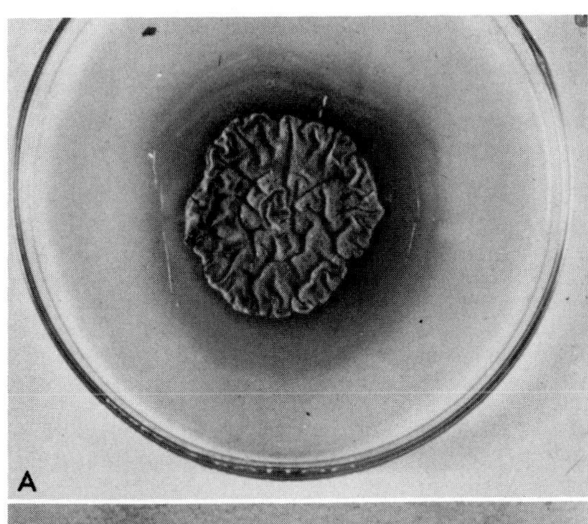

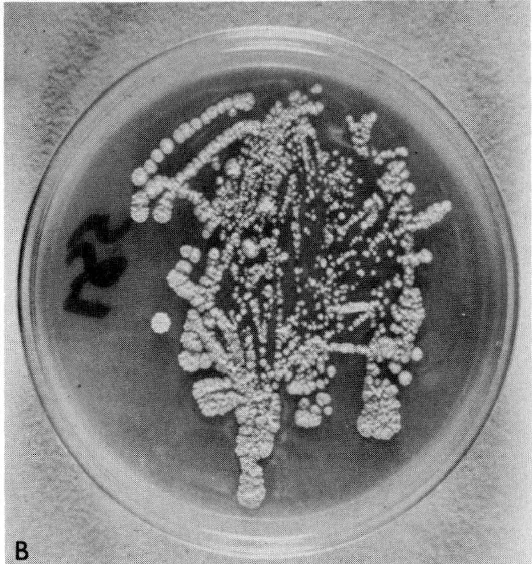

Figure 22–3. FS C 39. *Penicillium marneffei. A,* Colony at 25°C in Sabouraud's glucose agar. A defusable red pigment is produced. The colony is wrinkled, folded, and has red-stained mycelium and greenish conidia. *B,* At 37°C the organism grows as a yeastlike colony. (Courtesy of A. DiSalvo.)

An unusual form of penicilliosis has been described in rodents and once in man. The disease was first reported in the bamboo rat, which is found throughout Southeast Asia in the high plateaus. Several animals were noted as having a disease characterized by ascites, splenomegaly, hepatomegaly, and lesions in the Peyer's patches. The organism *P. marneffei* was found as small bodies (4.5 μ), somewhat resembling *Histoplasma* yeasts, within macrophages. These bodies reproduced by planate division, however, and not by blastospore formation. Mycelial elements (15 to 20 μ) have also been found.[1,9,10] The human case reported by Di Salvo et al.[5] was in a 67-year-old male with Hodgkin's disease. The organism was demonstrated in and cultured from the spleen (Figs. 22–3 FS C 39, 22–4, and 22–5 FS C 40). The case was recorded from South Carolina, indicating that this organism may have a worldwide distribution.

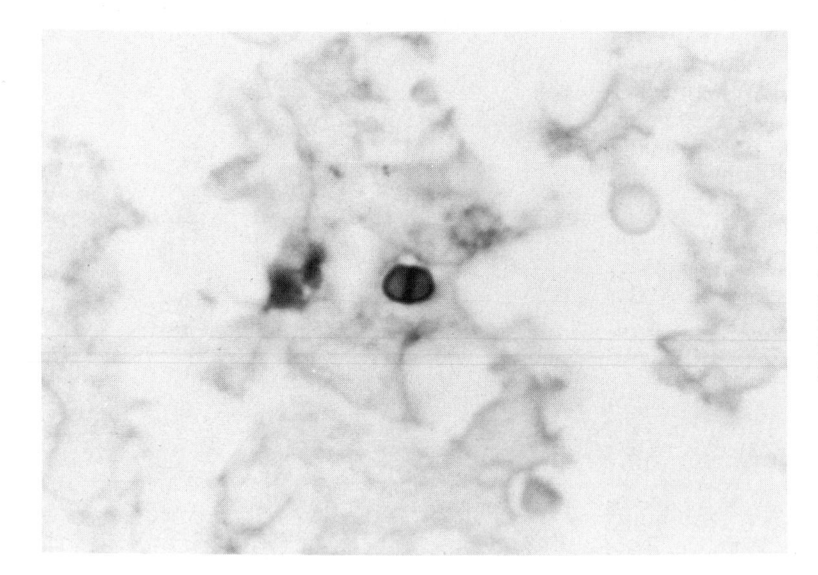

Figure 22–5. FS C 40. Planate dividing yeast cell of *P. marneffei* from liver biopsy of patient. ×440. (From DiSalvo, A. F., A. Fickling, et al. 1973. *Penicillium marneffei* infection in man. Description of first natural infection. Am. J. Clin. Pathol., *59*:259–263.)

REFERENCES

1. Capponi, M., P. Sureau, et al. 1956. Penicillose de *Rhizomys sinensis*. Bull. Soc. Pathol. Exot., *49*:418.
2. Castellani, A. 1920. The higher fungi in relation to human pathology. Lancet, *1*:895–901.
3. Chute, A. L. 1911. An infection of the bladder with *Penicillium glaucum*. Boston Med. Surg. J., *164*:420–422.
4. Delore, P., J. Coudert, et al. 1955. Un case de mycose bronchique avec localisations musculaires septicemiques. Prensa Med., *63*:1580–1582.
5. Di Salvo, A. F., A. Fickling, et al. 1973. *Penicillium marneffei* infection in man. Description of first natural infection. Am. J. Clin. Pathol., *59*:259–263.
6. Gillium, J., and S. A. Vest. 1951. Penicillium infection of the urinary tract. J. Urol., *65*:484–489.

6a. Garrison, R. G., and K. S. Boyd. 1973. Dimorphism of *Penicillium marneffei* as observed by electron microscopy. Can. J. Microbiol., *19*:1305–1309.

7. Huang, S. N., and L. S. Harris. 1963. Acute disseminated penicillosis. Am. J. Clin. Pathol., *39*:167–174.
8. Mantrelli, C., and G. Negri. 1955. Ricerche sperimentali sull' agente eziologico di' un micetoma a grani negri (*Penicillium mycetogenum*) n.f. G. Acad. Med. Torino, *21*:161–167.
9. Segretain, G. 1959. *Penicillium marneffei* n. sp. agent d'une mycose du systeme reticuloendothelial. Mycopathologia, *11*:327–353.
10. Segretain, G. 1962. Some new or infrequent fungous pathogens. In *Fungi and Fungous Diseases*. Dalldorf, G. (ed.), Springfield, Ill., Charles C Thomas.

Basidiomycosis

Basidiomycetes, though very common in nature, are very rarely associated with human disease. Allergic reactions to inhalation of the spores have been frequently cited, but actual infection is very rare. The possible exception to this is the recent description of the perfect stage of *Cryptococcus*, which appears to be in the basidiomycete genera *Leukosporidium* (see Chapter 9).

Emmons in 1954[3] reported on the repeated isolation of the mushroom *Coprinus micaceus* from the sputa of one patient. Mycelial strands were found on direct examination. Ciferri et al.[2] reported that *Schizophyllum commune* was isolated and mycelial elements were seen in sputum from another patient with chronic lung disease.

Meningitis was attributed to an *Ustilago* sp. by Moore et al.[4] Lesions contained multinucleate giant cells and macrophages. Within these were seen structures that resembled the sprout mycelium and echinulate spores of

Ustilago zeae. No cultures were obtained, however. In another case of meningitis, the organism *Schizophyllum commune* was isolated on 16 cultures from the first spinal tap and on 19 cultures from a second tap obtained four months later. The patient recovered uneventfully.[1] Restrepo et al.[4a] also recovered *Schizophyllum commune* from human disease. Their patient, a 4-year-old girl, had ulcerations in the mouth and a perforated palate. No underlying disease was noted, and the lesion responded to amphotericin B therapy.

Probably the only well-documented case of a systemic basidiomycete infection is that of Speller and MacIver in 1971.[5] In a patient who had had a mitral valve prosthesis, endocarditis of the aortic valve developed. When this was removed, it was found to contain much septate mycelium in association with an acute inflammatory process. On culture, an *Oidium* producing fungus was isolated. At autopsy, involvement of other organs was not noted. It was determined by mating experiments that the organism was the haploid stage of *Coprinus cinereus*, one of the inky cap mushrooms. The mycelium of several *Coprinus* sp. grows on plaster, and this may have been the source of exposure during the original surgical procedure.

REFERENCES

1. Chavez Batista, A., J. A. Maia, et al. 1955. Basidioneuromycosis in man. Sociedade de Biologia de Pernambuco Anals, *13*:52–60.
2. Ciferri, R., A. Chavez Batista, et al. 1956. Isolation of *Schizophyllum commune* from sputum. Atti Inst. Bontanico Laboratorio Crittogamico Univ. di Pavia, *14*(1–3):118–120.
3. Emmons, C. 1954. Isolation of myxotrichum and gymnoascus from lungs of animals. Mycologia, *46*:334–338.
4. Moore, M., W. O. Russel, et al. 1946. Chronic leptomeningitis and ependymitis caused by *Ustilago*, probably *U. zeae*: Ustilagomycosis, the second reported instance of human infection. Am. J. Pathol., *22*:761–773.

4a. Restrepo, A., D. L. Greer, et al. 1971. Ulceration of the palate caused by a Basidiomycete *Schizophyllum commune*. Sabouraudia, *9*:201–204.

5. Speller, D. C. E., and A. G. MacIver. 1971. Endocarditis caused by a *Coprinus* species. A fungus of the toadstool group. J. Med. Microbiol., *4*:370–374.

Protothecosis

DEFINITION

Protothecosis is an infection caused by members of the genus *Prototheca*. These organisms are generally considered to be achloric algae. The disease has been recorded in man and several domestic and wild animals. The lesions have varied from cutaneous and subcutaneous infections to systemic invasion involving several internal organs. The organisms are ubiquitous in nature and can be isolated from human and animal skin, feces, and sputum in the absence of disease.[7,18]

HISTORY

Prototheca was a genus erected by Krüger in 1894 to encompass some nonpigmented unicellular organisms found in the mucinous flux of trees. They were considered to be yeasts and therefore included in the fungi. West in 1916 reclassified them as algae because their spores are produced internally in a manner identical to that of the green alga *Chlorella*. As late as 1957, however, Ciferri[6] reclassified them as saccharomycetes. They have the general appearance of achlorophyllic mutants of *Chlorella*,[5,16] but there are differences in cell wall composition,[9] physiology, and ability to survive environmental stress. It is now generally considered that the genus developed from *Chlorella* at some point in evolution.

ETIOLOGY, ECOLOGY, AND DISTRIBUTION

Algae are autotrophic organisms. Their importance in medicine has generally been restricted to toxicity to animals and man following ingestion of contaminated water or to imparting an unpleasant taste and odor to drinking water.[16] No infection due to a chlorophyll-containing alga has been substantiated. The achloric *Prototheca* are heterotrophic and require external sources of organic carbon and nitrogen. There are a few authentic records of various types of infection being caused by the members of this genus. Members of the genus are commonly isolated throughout the world.

Early reports of the association of *Prototheca* and disease are for the great part invalid. *Prototheca ciferri (P. zopfii)* was isolated from feces in patients with tropical sprue, and it was suggested that there was a relationship.[4] Almeida et al.[2] found *Prototheca* in cases of actinomycosis and paracoccidioidomycosis. These were probably contaminants. The subject of human infection by algae was reviewed in 1940 by Redaelli.[15]

CLINICAL DISEASE

Only a few valid cases of human infection by *Prototheca* have been recorded. The first was described by Davies et al. in 1964.[9] The patient was a rice farmer in Sierra Leone. The lesion began on the inner side of the right foot as a depigmented area. Within three years it had become a rugose papule with a raised edge covering two-thirds of the foot. In tissue sections and culture the grouped, rounded bodies of *Prototheca* were observed. A skin biopsy showed hyperkeratosis and pseudoepitheliomatous hyperplasia. The organism was seen in the epidermis and in the papillary and reticular dermis. *Prototheca* had been noted to be sensitive to pentamidine. A total of 4.9 g was given to the patient but without benefit. At last report,[8] the lesion was advancing, and organisms were found in the lower femoral lymph nodes. This isolant was named *P. segbwema* after the location of the hospital in which the organism was isolated, but the name has been reduced to synonymy with *P. zopfii.*

What may be termed an opportunistic infection by *Prototheca* was reported by Klintworth et al.[12] in 1968. The patient was a diabetic and had widespread metastases of breast cancer. *Prototheca wickerhamii* was isolated from several ulcerating papulopustular lesions on the leg. Neutrophils and macrophages were present, but the cellular response was minimal. The patient died of the carcinoma, but no autopsy was performed. The authors mention other cases, as yet unreported, of algosis. Nosanchuk and Greenberg[14] reported a case and reviewed the literature. A total of eight human infections have been recorded, and five of these have involved the olecranon bursa.[12a, 18a]

Several cases of infection in wild and domestic animals have been reported. Ainsworth and Austwick recovered the organism from the inflamed udder of a cow.[1] *Prototheca* was cultured and seen in intestinal tissues and several organs in a case of enterocolitis reported by Van Kruiningen in 1970.[19] Three cases of the disease in cows have been recorded by Migaki.[13] Frese, Gedek, and Shiefer[17] found the organism in lymph nodes, bone, and subcutaneous tissue of the limbs of a two year old deer in Germany. Primary cutaneous disease in the dog has also been reported.[18b]

PATHOLOGY

Direct examination, when recorded, has shown that the organism is easily mistaken for cells of various fungi, particularly the etiologic agents of chromomycosis and histoplasmosis duboisii. In *Prototheca* one can observe a number of daughter cells (autospores) forming within the theca. No budding (blastospore formation) is found.

In tissue sections containing cells of *Prototheca,* little cellular reaction is observed. Depending on the species, the size of the cells varies from 8 to 20 μ in diameter. The morphology of the cells also varies among the several species described.

Identification in Tissue. *Prototheca* is a large nonbudding cell readily seen in tissue. It is spheroid, ovoid, or elliptical, with a prominent wall. The round cell (theca) contains several thick-walled autospores. The organisms are not readily apparent in hematoxylin and eosin stained smears, but stain well with Grocott's modification of the Gomori methenamine silver or the PAS stain.

PROGNOSIS AND THERAPY

Davies et al.[8] have shown that *Prototheca* are moderately sensitive to several therapeutic agents, particularly pentamidine. This drug was used in the treatment of one case, but had little effect. The organism was inhibited by 10 μg of amphotericin B and 100 units per ml of nystatin. Experimental disease in animals has been tried by several authors, but the results have been equivocal. Most human infections have resolved following surgical excision.[14]

ORGANISMS INVOLVED

The classification of the genus *Prototheca* has been examined by several authors.[3,7] Arnold and Ahearn[3] have investigated the carbohydrate and alcohol assimilation patterns of several isolants. They conclude that there are five species in the genus: *Prototheca filamenta, P. stagnora, P. zopfii, P. moriformis,* and *P. wickerhamii* (Table 22–1). So far only *P. zopfii* and *P. wickerhamii* have been involved in human or animal infection. Pore[14a] has devised a selective medium for the isolation of *Prototheca*. The authenticity of *P. filamenta* as a *Prototheca* is now in doubt.

PHYCOLOGY

Prototheca zopfii Kruger 1894

Synonymy. *Prototheca portoricensis* var. *trispora* Ashford, Ciferri, and Salman 1930; *Prototheca ciferii* Negroni and Blaisten 1940; *Prototheca segbwema* Davies, Spencer, and Wakelin 1964.

Morphology. SDA, 25°C. The colony is dull white and yeastlike in consistency. It does not grow on medium containing cycloheximide. The cells are variable in size and shape, being 8.1 to 24.2 × 10.8 to 26.9 μ in diameter. The autospores are spherical and 9 to 11 μ in diameter (Fig. 22–6). The morphology varies, depending on the medium employed for growth.

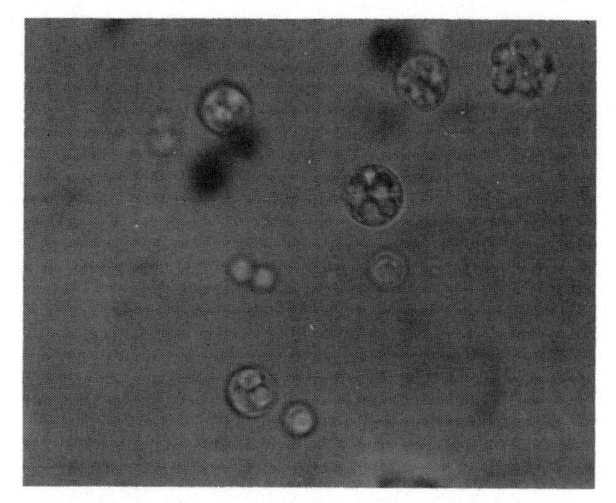

Figure 22–6. Protothecosis. *Prototheca zopfii.* Single cells and thecae with autospores. ×60. (Courtesy of M. Feo.)

Assimilation. Glucose +, sucrose −, maltose −, lactose −, galactose +, cellobiose −, xylose −, raffinose −, trehalose −, methanol −, ethanol −, *n*-propanol +, *n*-butanol +, glycerol +, sorbitol −, inositol −, dulcitol −, erythritol −.

Prototheca wickerhamii Tubaki and Soneda 1959

Morphology. SDA, 25°C. The colony is moist and cream-colored. Growth is optimal at 30°C. The cells are similar to those of *P. zopfii* in shape but somewhat smaller. The cells vary from 8.1 to 13.4 × 10.8 to 16.1 μ when grown on glucose-containing media. The autospores are smaller (4 to 5 μ in diameter) and more numerous (up to 50 per theca).

Assimilation. Same as for *P. zopfii,* with the following exceptions: trehalose +, and *n*-propanol −.

Table 22–1. *Key to the Genus Prototheca**

1. Sucrose assimilated	2
Sucrose not assimilated	3
2. Trehalase and lactose assimilated	*P. filamenta*
Trehalose and lactose not assimilated	*P. stagnora*
3. Trehalose assimilated	4
Trehalose not assimilated	*P. zopfii*
4. *n*-Propanol assimilated	*P. moriformis*
n-Propanol not assimilated	*P. wickerhamii*

*From Arnold, P., and D. G. Ahearn. 1972. The systemics of the genus Prototheca with a description of a new species *P. filamenta.* Mycologia, *64*:265–276.

REFERENCES

1. Ainsworth, G. C., and P. K. C. Austwick. 1955. A survey of animal mycoses in Britain. Trans. Br. Mycol. Soc., *38*:369–386.
2. Almeida, F., C. S. Lacaz, et al. 1946. Consideracoes sobre tres cajos di micoses humanas. An. Fac. Med. Univ. São Paulo, *22*:295–299.
3. Arnold, P., and D. G. Ahearn. 1972. The systemics of the genus Prototheca with a description of a new species *P. filamenta.* Mycologia, *64*:265–276.
4. Ashforth, B. K., R. Ciferri, et al. 1930. A new species of *Prototheca* and a variety of the same isolated from the human intestine. Arch. Protisteuk., *70*:619–638.

5. Butler, E. E. 1954. Radiation induced chlorophyll-less mutants of *Chlorella*. Science, *120*:274–275.
6. Ciferri, O. 1957. Metabolismo comparativo delle Protothecae e delle mutanti achloriche di Chlorella. G. Microbiol., *3*:97–108.
7. Cooke, W. B. 1968. Studies in the genus *Prototheca*. II. Taxonomy. J. Elisha Mitchell Scientific Soc., *84*:217–220.
8. Davies, R. R., and J. L. Wilkinson. 1967. Human protothecosis. Supplementary studies. Ann. Trop. Med. Parasitol., *61*:112–115.
9. Davies, R. R., H. Spencer, et al. 1964. A case of human protothecosis. Trans. R. Soc. Trop. Med. Hyg., *58*:448–451.
10. Fetter, B. F., G. K. Klintworth, et al. 1971. Protothecosis—Algal infection. In *Handbuch der Speziellen Pathologischen Anatomie und Histologie*. Baker, R. D. (ed.), Berlin, Springer-Verlag.
11. Frese, K., and B. Gedek. 1968. Ein Fall von Protothecosis beim Reh. Berlin Münch. Tierärztl. Wochenschr., *81*:174–178.
12. Klintworth, G. K., B. F. Fetter, et al. Protothecosis, an algal infection. J. Med. Microbiol., *1*:211–216.
12a. Mars, P. W., A. R. Rabson, et al. 1971. Cutaneous protothecosis. Br. J. Dermatol., *85*:76–84.
13. Migaki, G., F. M. Garner, et al. 1969. Bovine protothecosis. Pathol. Vet., *6*:444–453.
14. Nosanchuk, J. S., and R. D. Greenberg. 1973. Protothecosis of the olecranon bursa caused by achloric algae. Am. J. Clin. Pathol., *59*:567–573.
14a. Pore, R. S. 1973. Selective medium for the isolation of Protochecca. Appl. Microbiol., *29*:648–649.
15. Redaelli, P. 1940. As algoses Resenha. Clin. Cient. São Paulo, *9*:443–447.
16. Schwimmer, M., and D. Schwimmer. 1955. *The Role of Algae and Plankton in Medicine*. New York, Grune & Stratton.
17. Shiefer, B., and B. Gedek. 1968. Zum Verhalten von Prototheca species im Gewebe von Saugetieren. Berlin Münch. Tierärztl. Wochenschr., *81*: 485–490.
18. Sonck, C. E., and Y. Koch. 1971. Vertreter der Gattung Prototheca als Schmarotzer auf der Haut. Mykosen, *14*:475–482.
18a. Sudman, M. S. 1974. Protothecosis. A critical review. Am. J. Clin. Pathol., *61*:10–19.
18b. Sudman, M. S., J. A. Majka, et al. 1973. Primary mucocutaneous protothecosis in a dog. J. Am. Vet. Med. Assoc., *163*:1372–1374.
19. Van Kruiningen, H. J. 1970. Protothecal enterocolitis in a dog. J. Am. Vet. Med. Assoc., *157*: 56–63.

Pneumocystosis

DEFINITION

Pneumocytosis is an infection of the alveolar areas of the lung, leading to an interstitial pneumonia. The etiologic agent is *Pneumocystis carinii*. This organism is generally considered to be a protozoan, but is probably allied to the fungi. It is of worldwide distribution and appears to be part of the normal flora of rodents and perhaps other animals, including man. The disease is most often encountered in infants, in whom it may reach epidemic proportions, and in patients debilitated by a number of disease states or iatrogenic procedures. Pneumocystosis is becoming increasingly more significant and prevalent in this latter group.

Synonymy. Chronic interstitial plasma cell pneumonia.

HISTORY

Pneumocystosis was first observed in the lungs of guinea pigs by Chagas in 1909,[5] who was working on experimental infection of *Trypanosoma cruzi*. He considered the structures to be a part of the life cycle of that organism. Carini[3,4] in 1910 found the organism in rats that had not been infected with other parasites. After studying the various structures seen in tissue sections, Delanöe and Delanöe[7] in 1912 named the organism *Pneumocystis carinii*. Following careful examination and comparison with other parasites, Wenyon in 1926[25] classified the organism as a protozoan in the class Sporozoa, subclass Coccidiomorpha.

Many fungous diseases were first considered to have protozoan etiologies because the initial descriptions were made on histologic evidence alone. Only later by use of several lines of investigation did their true biological affinity become clear. Several enzymes, including chitinase, attack the cell wall of *Pneumocystis*,[21] as well as the walls of true fungi. This, in addition to observations on its ultrastructure, suggests a relation of the organism to the fungi. This evidence is at most tenuous, so that the phylogenetic affinities of the organism are as yet still in doubt. *Pneumocystis* infection is included in this chapter because the disease is assuming greater importance and must be considered in the differential diagnosis of several fungal and actinomycotic diseases, particularly those termed "opportunistic infections." Rossle in 1923[19] described a chronic interstitial pneumonia of infants. Rossle's work evoked a great deal of interest and led to the discovery of the disease in several parts of central Europe. This type of pneumonia was not yet associated with the *Pneumocystis* organism, but in retrospect these were probably cases of pneumocystosis. Van der Meer and Burg[23] in 1942 were probably the first investigators to associate *Pneumocystis* with the form of chronic interstitial pneumonia that was characterized by large numbers of plasma cells. Rossle's work and that of Van der Meer stimulated investigation of this form of pneumonia, and it was realized that the disease was much more common than previously appreciated. Between 1941 and 1949 some 700 cases were reported in Switzerland alone, and the disease was found in many other countries thereafter. Essentially all these cases involved infections of prematurely born infants or "sickly" neonates.

The first cases in adults were noted by McMillan and Hamperl in Europe,[9a] and by the mid 1960's some 19 cases had been tabulated from the United States.[8,11] Of these, 14 were complications of neoplastic disease, including leukemias, Hodgkin's disease, lymphosarcoma, and multiple myeloma. In a retrospective study, Esterly and Warner reviewed 12 cases at the University of Chicago Medical Center.[8] Six were from children with lymphoreticular malignancies, and six were from adults with similar diseases. All but one had had steroid therapy, and all had received antibiotics. At various times and in various reviews, the disease has been associated with cytomegalic inclusion pneumonia, renal transplants,[6] steroid therapy, particularly after cessation of the regimen, and hyaline membrane disease.[8] It now appears that any debilitation, particularly when complicated by cytotoxin or steroid therapy, may result in *Pneumocystis* pneumonia.

CLINICAL DISEASE

The clinical course of the pulmonary disease is insidious.[18] There is gradually progressive dyspnea, tachypnea, tachycardia, and cyanosis, with little or no fever unless there is complication by other infections. Chest x-ray reveals gradual spreading of perihilar haziness, with some granular components or formation of indistinct nodules.[1a] There is an alveolar or interstitial infiltrate but no pulmonary infiltrate (Fig. 22–7). Very frequently the course is complicated by superinfection with bacterial or other fungal opportunists. Only careful analysis of the several possible causes of the infection results in adequate therapy and resolution of the disease.[6] Walzer et al.[24a] have reviewed 194 cases from the United States.

PATHOLOGY

Gross Pathology. At autopsy, greyish yellow or pink-tinged consolidations of the lung are seen[22] (Fig. 22–8). In outward appearance and consistency, these areas often resemble lobular pancreatic tissue. Septal tissue is prominent, but the pleura is usually normal. In most of the cases of Esterly and Warner,[8] prominent septal cells and hyaline membranes were found.

Histopathology. The most characteristic finding in severely involved tissue is pink, honeycombed, foamy, vacuolated, intra-alveolar material (Fig. 22–9 FS C 41). This material is PAS-positive, usually separated from the alveolar wall, and the cysts of *P. carinii* are easily seen within it. Plasma cells and macrophages may or may not be present,[8] probably reflecting the immune competency of the patient. The organism lives freely in the intra-alveolar airspace and within the alveolar wall. This accounts for the accompanying cyanosis resulting from interference with blood gas exchange.

The most useful stain for observing the organisms is the Grocott-Gomori methenamine silver method.[8] The cysts of *P. carinii* measure about 4 μ in diameter. Many show "parentheses" within the round cysts. Wrinkled, crescent-shaped and cup-shaped cells are also observed (Figs. 22–9,*B* FS C 41,*B* and 22–10). Campbell,[2] from electron micrograph studies, has suggested a life cycle. Mature cysts (up to 5 μ in diameter) contain intracystic bodies. The crescent-shaped cells represent cysts from which small cells (1.2 to 2 μ in diameter) have escaped. These are at first thin-walled, but

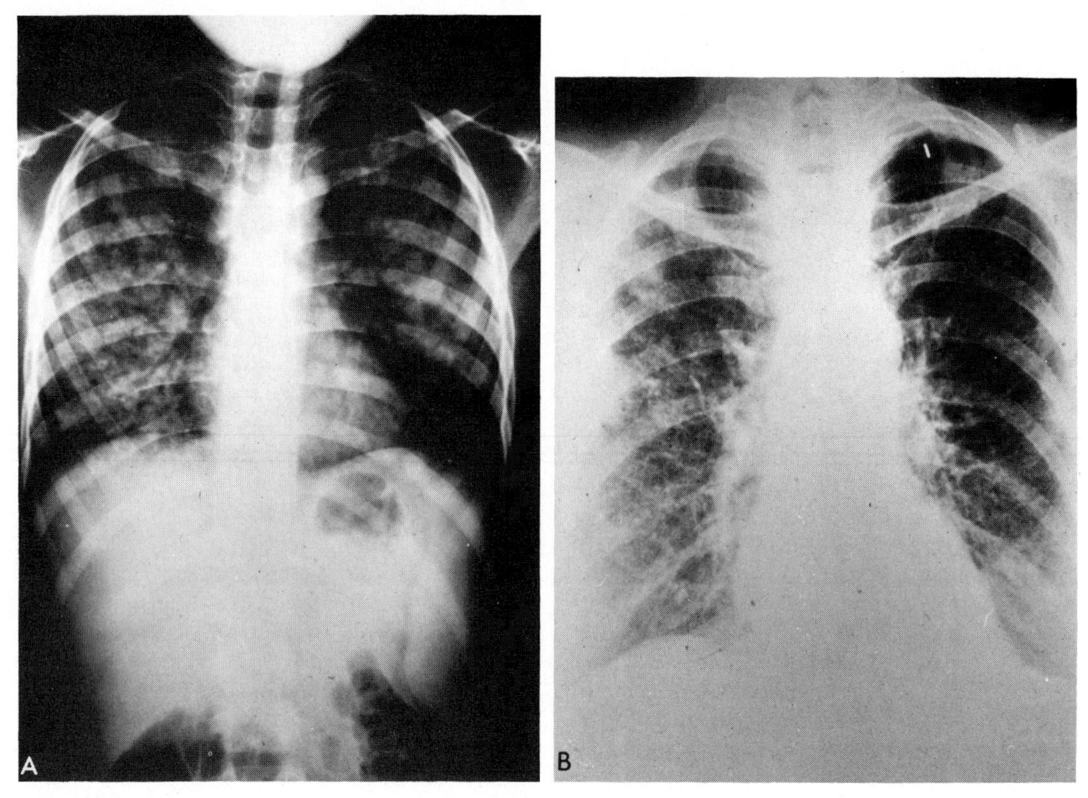

Figure 22–7. Pneumocystosis. The two most frequently seen x-ray patterns. *A,* Diffuse alveolar infiltrate throughout both lungs. Air bronchograms are seen. The liver and spleen are enlarged. Child with leukemia. *B,* Diffuse interstitial infiltrates in both lungs, but more evident in right lung. There are small areas of alveolar infiltrate as well. (Courtesy of J. Fennessy.)

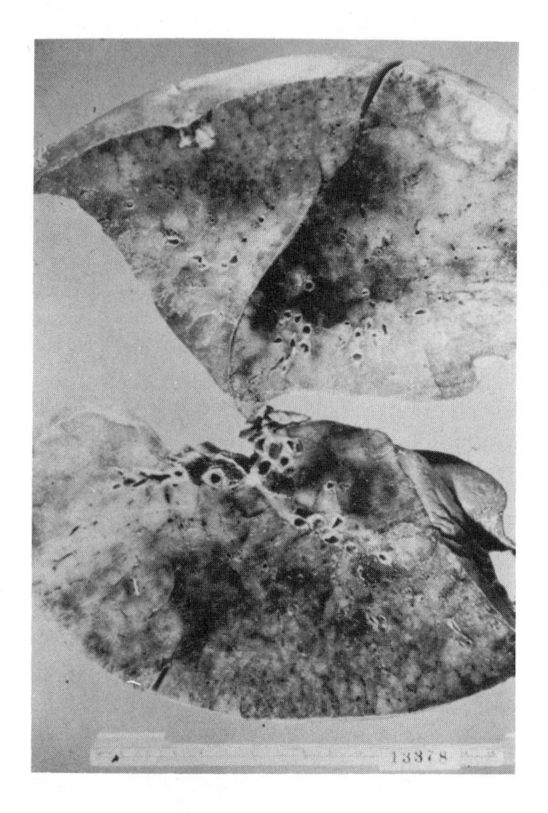

Figure 22–8. Pneumocystosis. Lobular "pancreas-like" consolidation of lung.

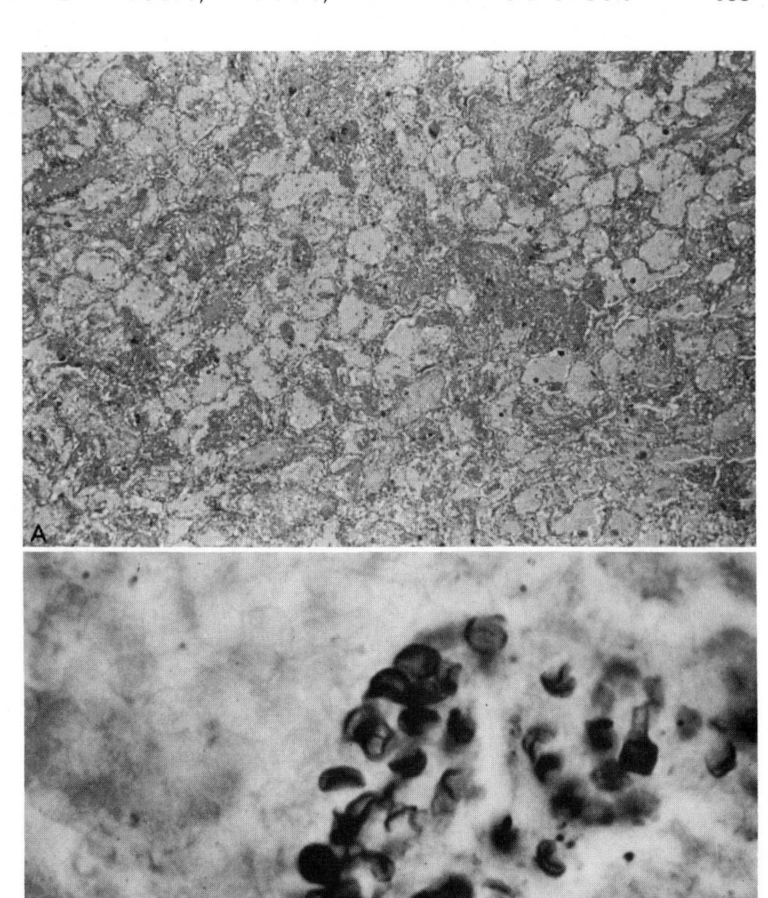

Figure 22–9. FS C 41. Pneumocystosis. *A,* Section of lung showing pink, honeycombed, foamy, vacuolated, intra-alveolar material. There are plasma cells, lymphocytes, histiocytes, and some giant cells. ×60. *B,* Wrinkled, crescent-shaped, cup-shaped, "parentheses" and round cells are seen within a giant cell. ×400. (Courtesy of S. Thomsen.)

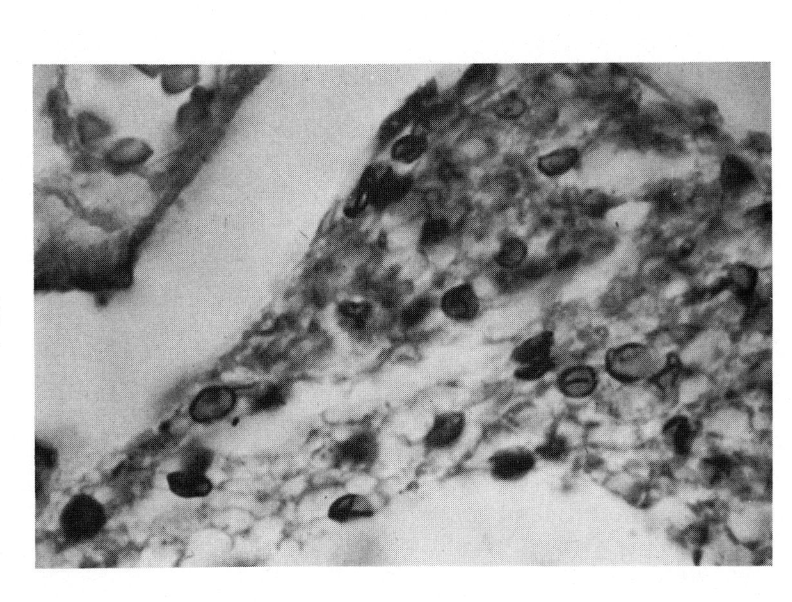

Figure 22–10. Pneumocystosis. Free-living cells in foamy material within alveolar space. GMS. ×400.

grow and possibly fuse to form thick-walled, mature cysts again containing intracystic bodies. In lung material studied after pentamidine therapy, the cysts were found to be abnormal, and did not contain intracystic bodies. Such "ghosts" were seen within macrophages, a finding not associated with active disease.

PROGNOSIS AND THERAPY

The disease is still too new to assess clinical response to a variety of therapeutic modalities. The most commonly used agent in successfully treated cases is pentamidine isothionate.[13,16] The dosage recommended by Cohen et al.[6] is 4 mg per kg for 14 days. Recurrences are noted in some series[12] and perhaps can be related to inadequate therapy. Kirby et al.[14] have successfully treated the disease with pyrimethamine and sulfadiazine. Trimethoprin and sulfamethoxazole have also been found effective and may become the treatment of choice.[12a]

DIAGNOSIS

The etiologic agent of *Pneumocystis* pneumonia has not as yet been cultured. Therefore, diagnosis depends on demonstration of the organism in tissue. Using the bronchial brush technique of Fennessy,[9] Repsher et al.[17] were able to confirm the presumptive diagnosis in 11 of 19 patients. They conclude that the procedure is well tolerated, has few serious complications, and is the most sensitive method presently available. The Gomori method was used to stain the smears. Smith et al.[20] have modified this stain using DMSO and increased its sensitivity for the demonstration of organisms. An indirect fluorescent stain for the detection of antibodies in sera is being developed.[15] So far the test is positive in 50 per cent of provisionally diagnosed cases of pneumocystic pneumonia. Antibodies are present in 14 per cent of healthy contacts of such patients, but none were detected in 50 sera of a healthy control population. The former may represent transient experience and sensitization to the organism. A direct fluorescent antibody test has also been developed.[14a]

Barton and Campbell[1] have reproduced the disease in rats by simply giving them cortisone. This indicates that the organism is probably widespread in nature, and perhaps in small numbers is a part of the normal flora. Hendley[10] was able to effect activation of the disease in rats and demonstrated transmission of the disease among cage mates.

REFERENCES

1. Barton, E. G., Jr., and W. G. Campbell, Jr. 1969. *Pneumocystis carinii* in lungs of rats treated with cortisone acetate: Ultrastructural observations relating to the life cycle. Am. J. Pathol., *54*:209–236.

1a. Bragg, D. G., and B. Janis. 1973. The roentgenographic manifestations of pulmonary opportunistic infections. Am. J. Roentgenol., *117*:798–801.

2. Campbell, W. G. 1972. Ultrastructure of Pneumocystis in human lung. Life cycle in human pneumocystosis. Arch. Pathol., *93*:312–324.

3. Carini, A. 1910. [On protozoan parasites found in rodents.] Bol. Soc. Med. Cirug. São Paulo, *18*:204.

4. Carini, A., and J. Maciel. 1916. Ueber *Pneumocystis carinii.* Zentralbl. Bakteriol. (Orig.), 77:46–50.

5. Chagas, C. 1909. Nova tripanozomiaze cicloevolutiro do schizo trypanum n. gen. n. sp. agente etiolojio de nova entidade morbica do homen. Mem. Inst. Osw. Cruz, *1*:159–218.

6. Cohen, M. L., E. B. Weiss, et al. 1971. Successful treatment of *Pneumocystis carinii* and *Nocardia asteroides* in a renal transplant patient. Am. J. Med., *50*:269–276.

7. Delanöe, P., and Mme. P. Delanöe. 1912. Sur les rapports de cystes de carini du poumon des rats avec le *Trypanosoma lewisi.* C. R. Acad. Sci. (Paris), *155*:658–660.

8. Esterly, J. A., and N. E. Warner. 1965. *Pneumocystis carinii* pneumonia. Arch. Pathol., *80*:433–441.

9. Fennessy, J. J. 1966. A technique for the selective catheterization of segmental bronchi using arterial catheters. Am. J. Roentgenol. Radium Ther. Nucl. Med., *96*:936–943.

9a. Hamperl, H. 1957. Variants of *Pneumocystis* pneumonia. J. Pathol. Bacteriol., *74*:353–356.

10. Hendley, J. D. 1971. Activation and transmission in rats of infection with Pneumocystis. Proc. Soc. Exp. Biol. Med., *137*:1401–1404.

11. Hendry, W. S., and R. Patrick. 1962. Observations on thirteen cases of *Pneumocystis carinii* pneumonia. Am. J. Clin. Pathol., *38*:401–405.

12. Hughes, W., W. W. Johnson, et al. 1971. Recurrent *Pneumocystis carinii* pneumonia following apparent recovery. J. Pediatr., *79*:755–759.

12a. Hughes, W. T., P. C. McNabb, et al. 1974. Efficacy of trimethoprim and sulfamethoxazole in the prevention and treatment of *Pneumocystis carinii* pneumonitis. Antimicrob. Agents Chemother., *5*:289–293.

13. Ivady, G. 1962. Further experience in the treatment of interstitial plasma cell pneumonia with pentamidine. Wochenschr. Kinderheilkd., *111*:297–299.

14. Kirby, H. B., B. Kenamore, et al. 1971. *Pneumocystis carinii* pneumonia treated with pyrimetha-

mine and sulfadiazine. Ann. Intern. Med., *75*:505–509.
14a. Lim, S. K., W. C. Eveland, et al. 1973. Development and evaluation of a direct fluorescent antibody method for the diagnosis of *Pneumocystis carinii* infections in experimental animals. Appl. Microbiol., *26*:666–671.
15. Norman, L., and I. G. Kagan. 1972. A preliminary report of an indirect fluorescent antibody test for detecting antibodies to cysts of *Pneumocystis carinii* in human sera. Am. J. Clin. Pathol., *58*:170–176.
16. Patterson, J. H. 1966. *Pneumocystis carinii* pneumonia and altered host resistance: Treatment of one patient with pentamidine isothionate. Pediatrics, *38*:388–397.
17. Repsher, L. H., G. Schröter, et al. 1972. Diagnosis of *Pneumocystis carinii* pneumonitis by means of endobronchial brush biopsy. New Engl. J. Med., *287*:340–341.
18. Rifkind, D., T. D. Faris, et al. 1966. *Pneumocystis carinii* pneumonia: Studies on the diagnosis and treatment. Ann. Intern. Med., *65*:943–956.
19. Rossle, R. 1923. Referat uber Ehtzundung. Verh. Dtsch. Pathol. Ges., *19*:18–68.
20. Smith, J. W., and W. T. Hughes. 1972. A rapid staining method for *Pneumocystis carinii.* J. Clin. Pathol., *25*:269–271.
21. Smith, J. W., W. T. Hughes, et al. 1971. Characterization of *Pneumocystis carinii* by biophysical and enzymatic methods. Bacteriol. Proc., *71*:121.
22. Spencer, H. 1968. *Pathology of the Lung.* Oxford, Pergamon Press, pp. 337–341.
23. Van der Meer, G., and S. L. Burg. 1942. Infection à Pneumocystis chez l'homme et chez les animâux. Ann. Soc. Belge Trop. Med., *22*:301–307.
24. Vanek, J., O. Jirovec, et al. 1953. Interstitial plasma cell pneumonia in infants. Ann. Paediatr., *180*:1–21.
24a. Walzer, P. D., D. P. Perl, et al. 1974. *Pneumocystis carinii* pneumonia in the United States. Ann. Intern. Med., *80*:83–93.
25. Wenyon, C. M. 1926. In *Protozoology. Vol. 2.* London, Balliere, Tindall and Cox.

Beauveriosis

The organism *Beauveria bassiana* is of great historical interest in microbiology. Bassi in 1835[1] showed it to be responsible for the disease known as muscardine of silkworms. This was the first demonstration of a microorganism specifically evoking a disease process, and it preceded the classic works of Koch and Pasteur by many years. *Beauveria* has since been found to be a commonly encountered soil fungus pathogenic for many insects, particularly beetles.

Georg et al.[3] isolated the organisms from pulmonary disease in the giant tortoise. A human infection of the lung was recorded by Lahourcade in 1966.[2] The patient was a female (22 years old) who had lived in North Africa. For many years she had manifested ulcerative cervical lymphadenopathy and was diagnosed as having tuberculosis. A chest x-ray revealed a thick-walled pulmonary cavity plus other small pulmonary lesions. *B. bassiana* was isolated from surgically excised material and bronchial aspirates. Hyphae were visible in tissue section. She was treated with amphotericin B, and at last report her condition was stable.

REFERENCES

1. Bassi, A. 1835. Del male del segno calcinaccio o muscordino malattia che attigge: Bachi da set a. Patel. Teorica Tip. Orcesi, Lodi.
2. Freour, P., M. Lahourcade, et al. 1966. Une mycose nouvelle: Etude clinique et mycologique d'une localisation pulmonaire de "Beauveria." Bull. Soc. Med. Hôp. Paris, *117*:197–206.
3. Georg, L. K., W. M. Williamson, et al. 1962. Mycotic pulmonary disease of captive giant tortoises. Sabouraudia, *2*:80–82.

Cercosporomycosis

Among the more strange cases of human fungous disease is the one recorded by Lie-Kien-Joe et al.[3] and Emmons.[2] The patient was a 12-year-old boy from Indonesia. The disease appeared many years previously as a small nodule on the cheek. At the time of examination, there were extensive indurated, verrucous, and ulcerated cutaneous and subcutaneous lesions of the face. The nasal mucosa was also involved. There was no lymphadenopathy or fever, and no underlying disease was detected (Fig. 22–11).

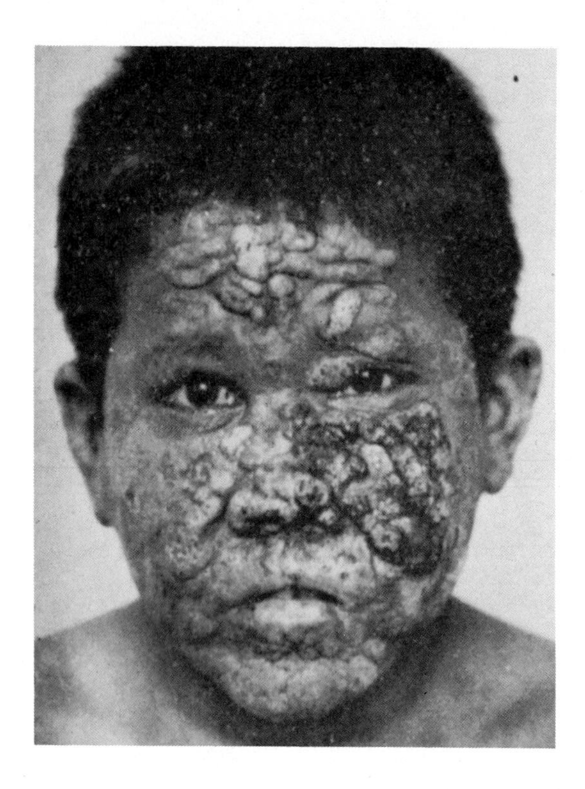

Figure 22–11. Cercosporomycosis. Numerous granulomatous lesions around face. (From Lie-Kien-Joe, et al. 1957. Arch. Dermatol., *75*:864–870.)

Eleven biopsies were taken during a period of several years. Epithelial hypertrophy was seen, along with hyperkeratosis and intraepithelial abscesses. Granulomas were found in subcutaneous tissue. Brown septate mycelium, 4 to 8 μ in diameter, were seen in all areas affected. Cultures grew at 25°C but not at 37°C. The same fungus grew from all specimens. It produced an olive-grey colony. Conidiophores grew irregularly and could be seen to have terminal scars from which conidia were released. These were unicellular when young (4 to 6 μ) and multicellular when mature (26 to 120 μ). Experimental disease in animals was attempted but unsuccessful; however, lesions were produced in many plants. The organism was identified as *Cercospora apii* by Emmons,[2] but Chupp[1] does not agree with the specific epithet. The patient died years later with extensive involvement of the face, but an autopsy was not obtained.

REFERENCES

1. Chupp, C. 1957. The possible infection of the human body with *Cercospora apii*. Mycologia, *49*:773–774.
2. Emmons, C. W., Lie-Kien-Joe, et al. 1957. *Basidiobolus* and *Cercospora* from human infections. Mycologia, *49*:1–10.
3. Lie-Kien-Joe, Njo-Injo Tjoei, et al. 1957. A new verrucous mycosis caused by *Cercospora apii*. Arch. Dermatol., *75*:864–870.

Histoplasmosis farciminosum (Epizootic Lymphangitis)

Epizootic lymphangitis is an infection of horses and mules caused by *Histoplasma farciminosum*. The disease was first described by Rivolta in 1873.[2] He noted the budding yeast cells within lesions from horses and named the organism *Cryptococcus farciminosum*. In 1934, Redaelli and Ciferri transferred it to the genus *Histoplasma* because of the similarity in life cycle and the appearance of the macroaleuriospores. Ajello[2] has reviewed the genus *Histoplasma* and contends that the etiologic agent of epizootic lymphangitis should not be included in it.

The disease is widespread throughout Scandinavia, Russia, central and southern Europe, northern Africa, India, and southern Asia.[1] It is particularly prevalent in the northern sections of Egypt and in India.[7,9] The horse is most commonly involved and constituted 89 per cent of the cases in one survey.[9] Mules and sometimes donkeys are also involved. Male animals have a higher rate of infection (61.6 per cent) than females. The incidence of infection is highest in January in Egypt, and some 724 cases were reported between 1960 and 1970.

The type of clinical disease most commonly seen involves subcutaneous and ulcerated lesions of the skin. The local lymphatics are involved, and thus this is a primary inoculation mycosis. Frequently both front and hind legs are involved, as is the neck, other areas rubbed by a harness, or an area which is the site of repeated injury. A mixed necrotic, pyogenic, and granulomatous process is seen in tissue section. The yeast cells are seen free and intracellularly within giant cells. The yeasts are 3 to 5 $\times$ 2.5 to 3.5 μ in diameter.

Fawi[6] has recently reported on the primary pulmonary form of the disease, which in time caused death of the animals. The horses had multiple soft gray granulomata throughout the parenchyma of the lung, which varied in size from 0.5 to 20 cm in diameter. The lymph nodes were not involved, nor were other organs of the body. Previously Bennett had recorded "cryptococcal" pneumonia in 1931.[3] Disseminated disease involving all organ systems has been recorded sporadically.[10] Experimental disease has been produced by intradermal or intraperitoneal injection of mice and rabbits.[11]

Treatment of the infection has included amphotericin B, hamycin, and several other modalities. The yeast form of the organism is quite sensitive to hamycin *in vitro*.[8]

Diagnosis of the disease depends on isolation of the organism. Lesion material can be planted on Sabouraud's agar plates containing antibiotics and grown at 25°C. The yeast form is maintained by incubation at 37°C, and growth is enhanced if Hartley horse blood agar is used in an atmosphere of 20 per cent CO_2. Fawi[5] has devised a fluorescent antibody test. Sera of the horse are placed on fixed smears of the organisms. Forty-seven of 50 sera from proven cases were positive and controls negative.

Histoplasma farciminosum (Rivolta) Redaelli and Ciferri 1934

Synonymy. *Cryptococcus farciminosum* Rivolta 1873.

Morphology. SDA, 25°C. The fungus grows slowly and produces a greyish white colony. Microscopically septate mycelium are seen. A variety of spores are found, including arthrospores, chlamydospores, and some blastospores. Round, smooth macroaleuriospores are produced that somewhat resemble the macroaleuriospores of *H. capsulatum* but lack the tubercles. At 37°C a yeast resembling *H. capsulatum* is seen.

REFERENCES

1. Ainsworth, G. C., and P. K. C. Austwick. 1959. *Fungal Diseases of Animals*. Kew, England, Commonwealth Mycological Institute, p. 22.
2. Ajello, L. 1968. Comparative morphology and immunology of members of the genus *Histoplasma*. Mycosen, *11*:507–514.
3. Bennett, S. C. J. 1931. Cryptococcus pneumonia in Equidae. J. Comp. Pathol., *44*:85–105.
4. Bullen, J. J. 1949. The yeastlike form of *Cryptococcus farciminosum*. J. Pathol. Bacteriol., *6*:117–120.
5. Fawi, M. T. 1969. Fluorescent antibody test for the sero diagnosis of *Histoplasma farciminosum* infections in Equidae. Br. Vet. J., *125*:231–234.
6. Fawi, M. T. 1971. *Histoplasma farciminosum*, the etiologic agent of equine cryptococcal pneumonia. Sabouraudia, *9*:123–125.
7. Mohan, R. N., K. N. Sharma, et al. 1966. A note on an outbreak of epizootic lymphangitis in equines. Indian Vet. J., *43*:338–339.
8. Padhye, A. A. 1969. *In vitro* antifungal activity of hamycin against *Histoplasma farciminosum*. Mykosen, *12*:203–205.
9. Refai, M, and A. Loot. 1970. Incidence of epizootic lymphangitis in Egypt. Mykosen, *13*:247–252.
10. Singh, T. 1966. Studies on epizootic lymphangitis. Study of clinical cases and experimental transmission. Indian J. Vet. Sci., *36*:45–59.
11. Singh, T., and B. M. L. Varmani. 1966. Studies on epizootic lymphangitis. A note on pathogenicity of *Histoplasma farciminosum* (Rivolta) for laboratory animals. Indian J. Vet. Sci., *36*:164–167.

Allescheriosis

Allescheria boydii (Monosporium apiospermum) is a common etiologic agent of mycetoma and is discussed in detail in Chapter 3. The organism has also been shown to cause pulmonary infection. The fungus was isolated from sputa and diseased lung. Granulomas were seen, and granules similar to those found in mycetoma of the foot were present. The disease has been associated with underlying respiratory conditions, such as sarcoidosis, chronic bronchitis, emphysema, and bronchogenic cysts.[3,6] About seven cases have been reported.

Allescheria boydii has been isolated from a meningitis reported by Benham and Georg[2] and Aronson et al.[1] The patient had had a spinal anesthetic, and thereafter began to develop signs of meningitis. *A. boydii* was grown from several spinal taps. After eight months the patient died. At autopsy a granulomatous meningitis was present covering the cord, brain stem, and cerebellum. A systemic case of allescheriosis was reported by Rosen et al.[5] Lesions containing the organism were found in the brain and thyroid. The patient had received steroids and immunosuppressive therapy for subacute glomerulonephritis. Chronic prostatitis due to *A. boydii* has also been reported.[4]

REFERENCES

1. Aronson, S. M., R. Benham, et al. 1953. Maduromycosis of the central nervous system. J. Neuropathol. Exp. Neurol., *12*:158–168.
2. Benham, R. W., and L. K. Georg. 1948. *Allescheria boydii,* causative agent of a meningitis. J. Invest. Dermatol., *10*:99–110.
3. Louria, D. B., P. H. Lieberman, et al. 1966. Pulmonary mycetoma due to *Allescheria boydii.* Arch. Intern. Med., *117*:748–751.
4. Meyer, E., and R. D. Herrold. 1961. *Allescheria boydii* isolated from a patient with chronic prostatis. Am. J. Clin. Pathol., *35*:155–159.
5. Rosen, F., J. H. Deck, et al. 1965. *Allescheria boydii* unique systemic dissemination to thyroid and brain. Can. Med. Assoc. J., *93*:1125–1127.
6. Travis, R. E., E. W. Ulrich, et al. 1961. Pulmonary allescheriosis. Ann. Intern. Med., *54*:141–152.

Aureobasidiomycosis

Aureobasidium pullulans has been isolated from a skin lesion by Vermeil et al.[1] The plaques were on the inferior aspect of the thigh, feet, and calves. The lesions were verrucous and had a black, studded appearance. There was no edema or induration. On histologic examination, the lesion resembled a keloid. There was extensive fibrosis and a granulomatous reaction, and numerous black, yeastlike bodies were seen free and within macrophages.

REFERENCES

1. Vermeil, C., A. Gordeff, et al. 1971. Blastomyaose cheloidienne a *Aureobasidium pullulans.* Mycopathologia, *43*:35–39.

Phomamycosis

Phoma hibernica has been found to colonize previously injured skin. Bakerspigel's report[1] notes that the patient had had recurring lesions on the lower leg that had been treated with steroids, coal tar, and grenz rays. A granuloma developed in the treated area, from which *Phoma hibernica* was isolated. On histologic examination, hyphal elements were seen in tissue.

REFERENCES

1. Bakerspigel, A. 1970. The isolation of *Phoma hibernica* from a lesion on a leg. Sabouraudia, 7:261–264.

Cephalosporiomycosis

Cephalosporium sp. have been recovered from many cases of mycetoma (Chapter 3), onychomycosis (Chapter 7), and mycotic keratitis (Chapter 23). *Cephalosporium* sp. are abundant in soil, sewage, and so forth, and represent the imperfect stage of several perfect fungi. Only rarely have they been related unquestionably to other types of human infection. In one case a midline granuloma with eroding destruction of the hard palate and involvement of the maxilla and mandible was described. In biopsy material from this case, hyphae were seen in tissue sections, and the organism was cultured on several occasions.[1] An authenticated case of meningitis was reported by Drouhet et al.[2] In this case, a 33-year-old woman developed sciatica following a spinal anesthetic given for a cesarean operation. She was treated with steroids. In time, after a second pregnancy, she developed neurologic and psychiatric problems. A surgical exploration revealed a granulomatous meningitis from which *Cephalosporium* sp. was cultured, and hyphal units were seen in tissue section. The patient died 15 months after onset of symptoms in spite of amphotericin B therapy.

There are many reports in the literature of *Cephalosporium* being isolated from skin, bile, blood, gastric juice, pleural fluid, and so forth. Most of these lack verification.

The genus *Cephalosporium* is now considered in synonymy with the genus *Acremonium*.

REFERENCES

1. Cowen, D. E., D. E. Dines, et al. 1965. *Cephalosporium* midline granuloma. Ann. Intern. Med., *62*: 791–795.
2. Drouhet, E., L. Martin, et al. 1965. Mycose meningo-cerebrale a *Cephalosporium*. Presse Med., *31*: 1809–1814.

Fusariomycosis

Fusarium species are very common soil organisms and plant pathogens. A few are important agents of mycotic keratitis (Chapter 23). Probably the only authentic cases of human infection caused by *Fusarium* are those of colonization of burned skin. There have been a few recent reports of this form of disease. Holzegel and Kempf[1] reported such a case in 1964. Previously Peterson and Baker had isolated *Fusarium roseum* from several cultures of burned skin.[2] In one case observed personally, *Fusarium oxysporium* was isolated in large numbers from 21 consecutive cultures of a severely burned man. At autopsy, some mycelium was seen in the crusts and debris of the burned cutaneous tissue. Invasion of surrounding tissue was not observed, however. What, if any, contribution to the pathology was attributable to the fungus is unknown.

REFERENCES

1. Holzegel, K., and H. F. Kempf. 1964. *Fusarium* mykose und der Haut eines Verbrannten. (*Fusarium* mycosis of the skin of a burned patient.) Dermatol. Wocherschr., *150*:651–658.
2. Peterson, J. E., and J. J. Baker. 1959. An isolate of *Fusarium roseum* from human burns. Mycologia, *51*:453–456.

Paeciliomycosis

Paecilomyces was isolated from the blood, from a thrombus overlying the mitral valve, and from an embolus in the iliac artery following the death of a patient who had had a valve replacement.[1] Following the surgery the patient had signs of cardiac failure and embolization along with pyrexia and emaciation. Fungal elements (1.5 to 3 μ) were demonstrated within thrombi of the mitral valve and in the iliac embolus. Rounded bodies were also seen. A tuberculoma-like lesion was found in the vessel wall near the iliac embolus. Caseous necrosis, epithelioid cells, and giant cells were present. *Paecilomyces varioti* was cultured.[1] Georg et al.[2] isolated *Paecilomyces fumoso-roseus* from pulmonary lesions in a giant tortoise. Mycelium was present in many small abscesses throughout the lung.

REFERENCES

1. Ays, C. J., P. A. Don, et al. 1963. Endocarditis following cardiac surgery due to the fungus *Paecilomyces*. S. Afr. Med. J., *37*:1276–1280.
2. Georg, L. K., W. M. Williamson, et al. 1962. Mycotil pulmonary disease of captive giant tortoises due to *Beauveria bassiana* and *Paecilomyces fumoso-roseus*. Sabouraudia, *2*:80–86.

Phaeohyphomycosis (Chromomycosis?)

Curvularia geniculata is another common soil saprophyte that has recently been isolated from a case of endocarditis. The patient had received a Starr-Edwards aortic valve prosthesis. Four months later, the patient was readmitted to the hospital because of elevated temperatures and signs of endocarditis. The chest was reopened and vegetations were found adhering to the valve. A replacement was made, but the patient died of renal failure shortly thereafter. Septate hyphae were found in the vegetations, and at autopsy mycotic emboli were found in the left renal and internal iliac arteries. Mycotic infarcts were seen in the heart, spleen, left kidney, and right cerebral hemisphere, as were mycotic microabscesses in the thyroid. *Curvularia geniculata* was grown from all infected material.

C. geniculata has also been isolated from cases of keratitis (Chapter 23).

REFERENCES

1. Kaufman, S. M. 1971. Curvularia endocarditis following cardiac surgery. Am. J. Clin. Pathol., *56*: 466–470.

Alternaria sp. is a very common soil saprophyte and plant pathogen, and its spores are frequently involved in human asthma. It is seldom as yet an established cause of human infectious disease. It may, however, colonize denuded, macerated, or previously injured skin. It has been isolated repeatedly from such areas. Whether the organism is contributing to the pathology is doubtful.

REFERENCES

1. Botticher, W. W. 1966. *Alternaria* as a possible human pathogen. Sabouraudia, *4*:256–258.
2. Higashi, N., and Y. Asada. 1973. Cutaneous alternariosis with mixed infection of *Candida albicans*. Arch. Dermatol., *108*:558–560.

Helminthosporium is another common dematiacious fungus of the soil. It is frequently isolated from sputum and skin, but it is rarely recorded as responsible for infectious disease. Dolan, Weed, and Dines reported two such cases. Both patients had chronic pulmonary disease and presented with purulent sputum, hemoptysis, and fever. Multiple cavities and abscesses were found in the lung. These contained histiocytes, lymphocytes, plasma cells, and granulocytes. Light brown, septate, branching mycelium was seen in the lesion tissue and found in bronchi, bronchioles, and alveoli. The fungus grown was a *Helminthosporium* sp.

REFERENCES

1. Dolan, C. T., L. A. Weed, et al. 1970. Bronchopulmonary helminthosporiosis. Am. J. Clin. Pathol., *53*:235–242.

Chaetoconidium sp. was isolated from a lesion on an ankle and later from subcutaneous nodules on the knee. The patient was a 16-year-old male on immuno supressive therapy following a renal allograph. The lesions developed at the site of an injury. He also had cryptococcosis. The fungus *Chaetoconidium* is a common soil saprophyte.

REFERENCES

1. Lomvardias, S., and G. E. Madge. 1972. Chaetocladium and atypical acid-fast bacilli in skin ulcers. Arch. Dermatol., *106*:875–876.

Drechslera hawaiiensis is a soil dematiaceous fungus of wide geographic distribution. It has been isolated from a fatal case of meningoencephalitis in a patient who had unsuspected lymphocytic lymphosarcoma.[1] Examination of autopsy material revealed a severe granulomatous and suppurative leptomeningitis and vasculitis. Brownish colored mycelial fragments were seen in tissue sections from the base of the brain. The involved areas showed massive infiltration of lymphocytes and plasma cells with granulomatous reaction. In areas of vasculitis there were numerous giant cells, and the mycelial elements were greatly distorted, swollen, and bizarre. The fungus grew as a cottony blackish brown mold. The spores resembled those of *Helminthosporium*, but it was classified by Ellis[2] as a *Drechslera* because of the acropetalous succession of porospores produced on a sympodially extending conidiophore.

REFERENCES

1. Fuste, F. J., L. Ajello, et al. 1973. *Drechslera hawaiiensis:* Causative agent of a fatal fungal meningoencephalitis. Sabouraudia, *11*:59–63.
2. Ellis, M. B. 1971. *Dematiaceous Hyphomycetes.* Kew, England, Commonwealth Mycological Institute, p. 415.

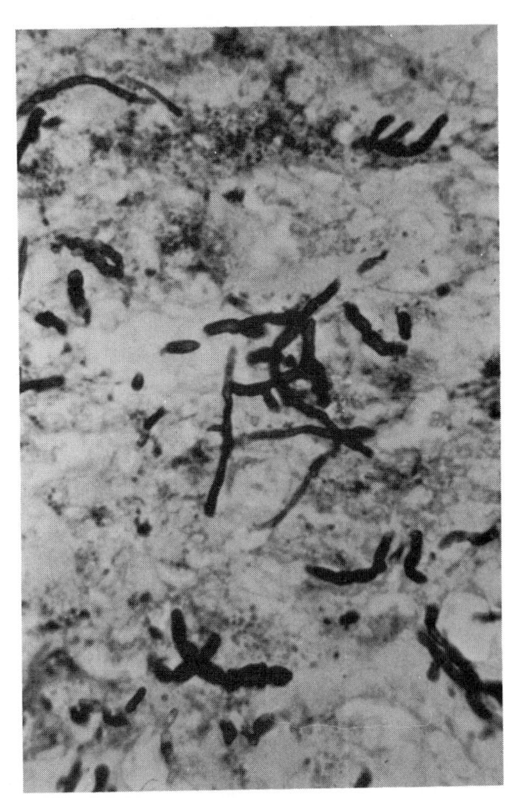

Figure 22–12. Phaeohyphomycosis. Tissue section showing irregular, branching hyphae. *Phialophora parasitica.* GMS. ×500.

Phialophora parasitica was isolated from the subcutaneous tissue of a kidney transplant patient on immunosuppressive therapy. This is another example of the ever increasing list of soil fungi that can be isolated from opportunistic infections. The common denominator in all these cases is a patient with abrogated defenses. The fungus grows as mycelium in tissue as it does in soil, and there is no transformation to a "parasitic" form.

The fungus isolated in this case was *P. parasitica.* In culture it is characterized by formation of collarless phialides that extrude ovate to elliptical phialospores. In tissue section the fungus appears as septate, branching, brown-colored mycelial elements of varied morphology (Fig. 22–12). Since the hyphae in tissue are brown, infection by this organism as well as that by *Curvularia,* could be called chromomycosis or opportunistic infection caused by dematiaceous fungi. The latter would be the more meaningful designation, as the word "chromomycosis" is associated with a particular set of clinical entities (Chapter 10). The authors of the paper describing this case suggested the term "phaeohyphomycosis" to include cases of opportunistic infection caused by dematiaceous fungi. Since the agents of these infections are taxonomically diverse but all appear as brownish mycelial units in tissue, this appears to be a useful designation.

REFERENCES

1. Ajello, L., L. K. Georg, et al. 1973. A case of phaeohyphomycosis caused by a new species of *Phialophora.* Abstracts, American Society for Microbiology.

Scopulariopsosis

Scopulariopsis brevicaulis is a common soil saprophyte of wide distribution. González Ochoa and Dallal[1] have isolated it from soil of caves and mines inhabited by *Histoplasma.* They demonstrated pathogenicity for mice of several strains. *In vivo* yeastlike forms were noted. Sekhon et al.[2] described a subcutaneous infection involving tendon sheath and muscle of the right ankle. The fungus isolated was sensitive to griseofulvin and hamycin.

REFERENCES

1. González Ochoa, A., and E. Dallal y Castillo. 1960. Frequencia de *Scopulariopsis brevicaulis* en muestras de suelos en cuevas y minas del pais. Rev. Inst. Salubr. En ferm. Trop. (Mex.), *20*:247–252.
2. Sekhon, A. S., D. J. Williams, et al. 1974. Deep scopulariopsosis: A case and sensitivity studies. Abs. Am. Soc. Microbiol., *74*:Mm 18.

Chapter 23

MYCOTIC INFECTIONS OF THE EYE AND EAR

The eye and the ear are continually subjected to challenge by a variety of fungi, bacteria, yeasts, and other microorganisms that are present in the external environment. Through evolution these organs have developed several means of preventing colonization and infection from airborne organisms. Any breach in the normal defenses preventing infection may lead to colonization and disease by the omnipresent potential invaders. Injury to the cornea of the eye is the most common predisposing factor leading to infection of this organ, whereas the accumulation of debris, particularly in damp tropical environments, allows colonization and infection of the external ear. The majority of the fungal organisms involved are soil saprophytes whose airborne spores find the injured tissue a suitable environment for growth. Therefore, the mycology involved is quite different from that encountered in systemic or cutaneous infections. The list of fungal species associated with such infections is very long and may include essentially every fungus that inhabits the earth. However, a few species seem to be more aggressive opportunists and account for the majority of infections recorded. *Fusarium* sp., *Aspergillus fumigatus*, and *A. niger* are most frequently involved in external mycotic keratitis, whereas *A. niger* and *A. fumigatus* most commonly cause infections of the external ear. The characteristics of these particular organisms that enhance their ability to colonize these organs are as yet undetermined.

MYCOTIC INFECTIONS OF THE EYE

Mycotic infections of the eye have assumed increasing importance in ophthalmology. The several reviews that have appeared on this subject since 1960[10,23,29,46,51,71] attest to the increased incidence and awareness of this disease. These reviews also emphasize the difficulties involved in predicting its occurrence, defining its morbidity, and instituting effective treatment.

There are three categories of eye disease in which fungi are involved. Each has a particular set of predisposing factors, type of pathology, group of etiologic agents, and specific therapeutic procedures. These categories are (1) mycotic keratitis, which is an infection of the cornea following trauma or superficial disease by fungi from an external source; (2) endogenous oculomycosis, which is ocular infection by dissemination (usually hematogenous) of systemic fungal disease; and (3) extension oculomycosis, which is the extension into the orbit of fungal disease from adjacent tissue. In addition, there are miscellaneous infections of the adnexa, tear ducts, eye lids, and conjunctiva.

History

Recorded cases of fungal infections of the eye have appeared sporadically in the literature. What was probably canaliculitis was described by Cesoni in 1670,[63] and the diagnosis of the disease and its etiologic agent (an actinomycete) was made in 1854 by Graefe.[28] The first report of mycotic keratitis was made by Leber in 1879.[39] He described an infection of the cornea by an *Aspergillus.* However, mycotic keratitis was considered to be a very rare entity, and by 1951 only 63 cases were recorded.[10] In a review of the pathology of fungus infections of the eye made in 1958, only 31 cases had been reported in which the organism had been demonstrated in tissue.[46] Other reviews emphasized the rarity of this type of infection until recently. The first instance of keratomycosis recorded at the Armed Forces Institute of Pathology was in 1933, and only two other cases had been noted by 1952. In the next four years, however, twelve more patients were recorded.[72] By 1963 a total of 150 reports of mycotic infections of the eye had appeared in the literature, and of these 85 had occurred since 1951. This increase coincides in time with the general use of corticosteroids in ophthalmology. At present, this disease is a frequently encountered sequela of injury to the eye.

The endogenous dissemination of systemic mycosis to involve the orbit, retina, optic nerve, sclera, conjunctiva, and adjacent tissues has also been reported sporadically in the literature.[16,21,31] In most of these reports, ocular involvement has been incidental to primary pathology elsewhere in the body, and the eye does not appear to be a target organ following dissemination of these diseases.

The third category of disease is the special case of rhinocerebral mucormycosis. This disease also was considered rare,[6] and by 1966 only 55 cases were recorded in the world literature. Of these, most had occurred between 1960 and 1966.[26] This reflects both a real increase in the number of cases and improved methods of diagnosis for the recognition of the disease. Rhinocerebral mucormycosis is now a well-defined syndrome that has a particular set of predisposing factors.[26]

Mycotic Keratitis

Clinical Features. Mycotic ulcers of the eye usually occur subsequent to trauma to the cornea by vegetative matter, soil, or surgery. Colonization of debilitated tissue may occur as a sequela to exposure keratitis, congenital defects, and ulcers initiated by other causes. Usually there is a severe inflammatory reaction, with vascularization, ciliary flush, flare of the anterior chamber, and folding of Descemet's membrane. The initial tissue reaction may be minimal if masked by corticosteroid treatment. However, the ulcer itself is characterized by a raised epithelium and often a white shaggy border (Fig. 23–1 FS C 42). Frequently there is a distinct radiating margin with penetration of fungal elements into the corneal stroma. This fuzzy, hyphate border extends beyond the ulcer edge and may form satellite lesions (Fig. 23–2 FS C 43). This picture, seen in conjunction with white endothelial plaques in the center of the cornea, strongly suggests fungal infection.[20,35,72] Eventually a sterile

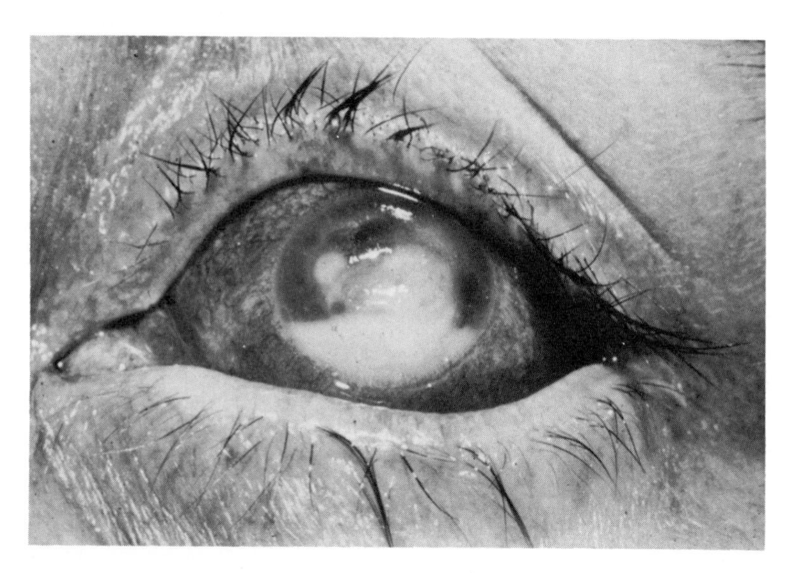

Figure 23–1. FS C 42. Mycotic keratitis. Central, shaggy-edged ulcer and satellite lesion with marked hypopyon. *Allescheria boydii.* (From Ernest, J. T., and J. W. Rippon. 1966. Am. J. Ophthalmol., *62:* 1202–1204.)

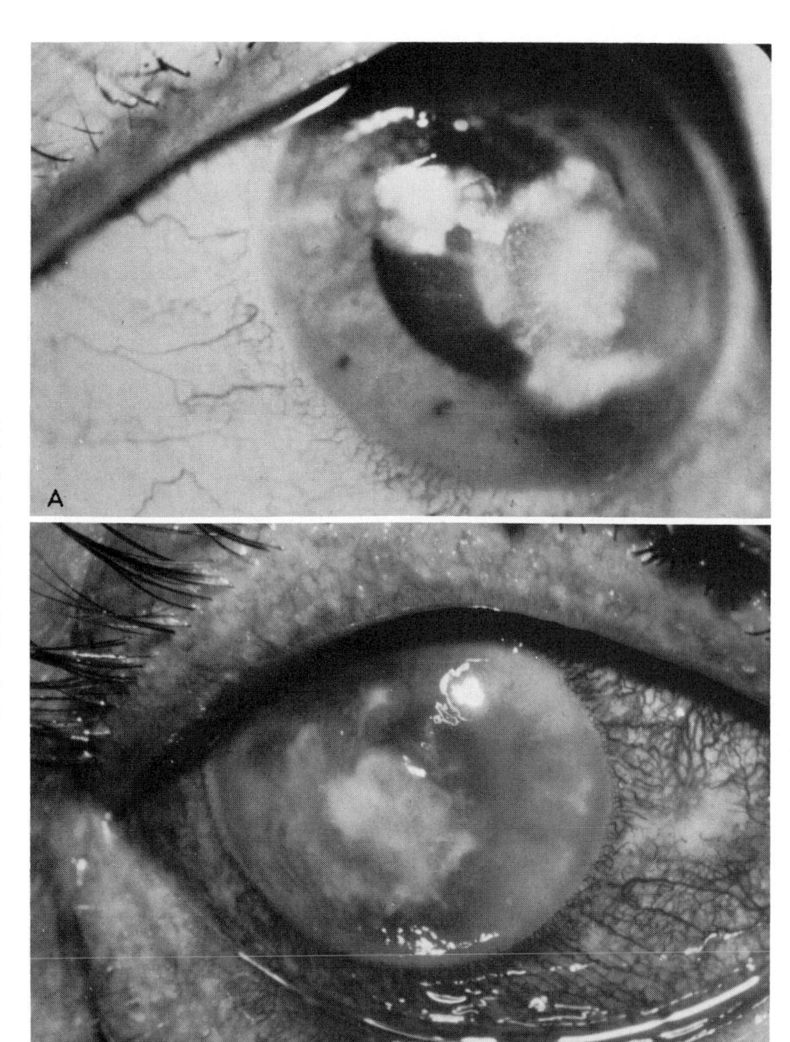

Figure 23–2. FS C 43. Mycotic keratitis. *A,* Fuzzy, hyphate border and satellite lesions. There is little inflammatory reaction owing to use of steroids. *Cephalosporium* sp. *B,* Typical fungal ulcer with necrotic, hyphate infiltrate within cornea underlying ulcer. Satellite lesions are present, along with a small hypopyon. This is a case from southern Florida caused by *Fusarium solani.* The infection was cured by use of pimaricin. (Courtesy of R. K. Forster.)

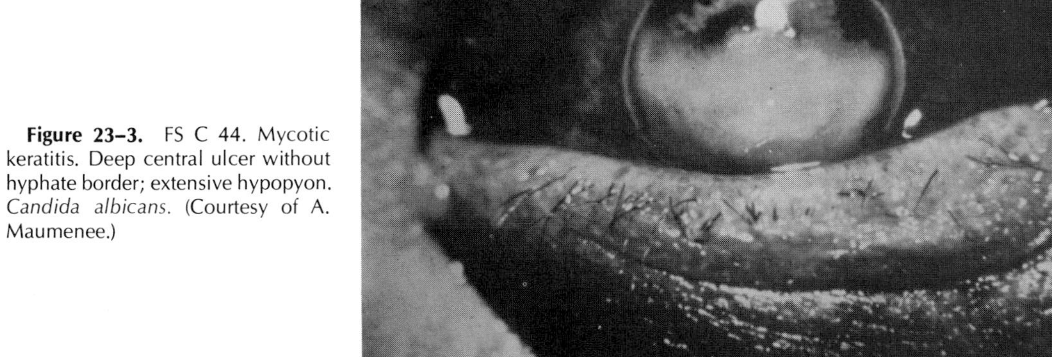

Figure 23–3. FS C 44. Mycotic keratitis. Deep central ulcer without hyphate border; extensive hypopyon. *Candida albicans.* (Courtesy of A. Maumenee.)

hypopyon develops (Fig. 23–3 FS C 44), and there may be a persistent "corneal ring" beyond the edge of the ulcer, composed of neutrophils, eosinophils, and plasma cells.

Another clinical syndrome, endophthalmitis, may be a complication of surgery[19] (Fig. 23–6), the result of deep penetrating injury or progression into the globe of the above described ulcerative lesions. Francois[23] has tabulated 74 cases up to 1968 in which the infection followed surgery. All but two cases followed cataract operations. Starch from glove powder has been implicated as the vector carrying the fungi.[56,57] The lesion may not be apparent for several days or up to six months following surgery. A developing fuzzy-edged white mass forms behind the iris and ciliary body in the vitreous, and a grayish infiltrate occurs which gradually becomes more extensive. Vision is not appreciably impaired for some time, however. Redness and pain may develop, along with a hypopyon. The latter clinical features may recede in time, but the infection itself runs a relentless progressive course. Almost all cases have been resistant to treatment and have required enucleation.

Pathology. Unless eradicated rapidly, the lesions of mycotic keratitis advance into the deep stroma of the cornea. It has been repeatedly emphasized that hyphae were present not only throughout the thickness of the cornea but also frequently only in the middle and deeper layers. Histologically an acute suppurative inflammatory process is seen early in the disease, which may be accompanied by coagulative necrosis. This response often subsides, and the reaction becomes minimal. Fungal hyphae are usually aligned parallel to the lamellae of the cornea, and Descemet's membrane appears to act as a barrier to penetration of the globe proper (Fig. 23–4 FS C 45). However, in some patients, especially if steroids were used, perpendicular penetration through the membrane leading to perforation of the cornea and invasion of the internal orbit has been reported[36,46] (Figs. 23–5 FS C 46 and 23–6).

Predisposing Factors. As has been pointed out by several authors, the recent rise in the frequency of keratomycosis corresponds to the general use of corticosteroids in the treatment of inflammatory eye disease. In experimental fungal infections of the cornea in rabbits, infection was noted in 80 per cent of test animals when cortisone was instilled along with fungal spores. By comparison, there was only a 20 per cent incidence of infection in controls that had had spores instilled but no steroids.[41] In addition, external corneal trauma (especially caused by foreign bodies containing vegetative material) and many other conditions predispose the eye to colonization by fungi.[19,20,64] These other conditions include previous corneal disease, exposure keratitis, radiation keratitis, herpetic lesions, serpiginous lesions, and functional disorders, such as facial palsy and bullous keratopathy.

Jones et al.,[32–34] in reviewing 39 cases that had occurred in southern Florida, detected a seasonal distribution. The majority of cases

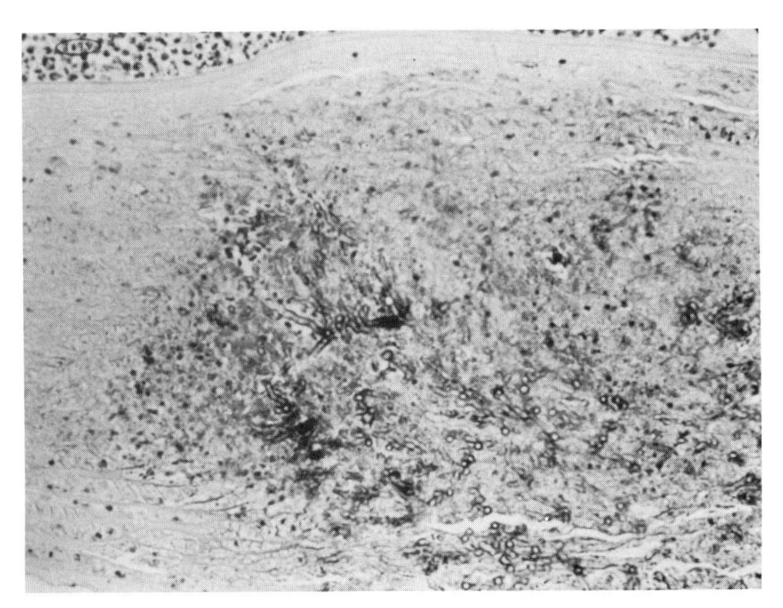

Figure 23–4. FS C 45. Mycotic keratitis. Hyphal strands of *Aspergillus* in cornea. ×100.

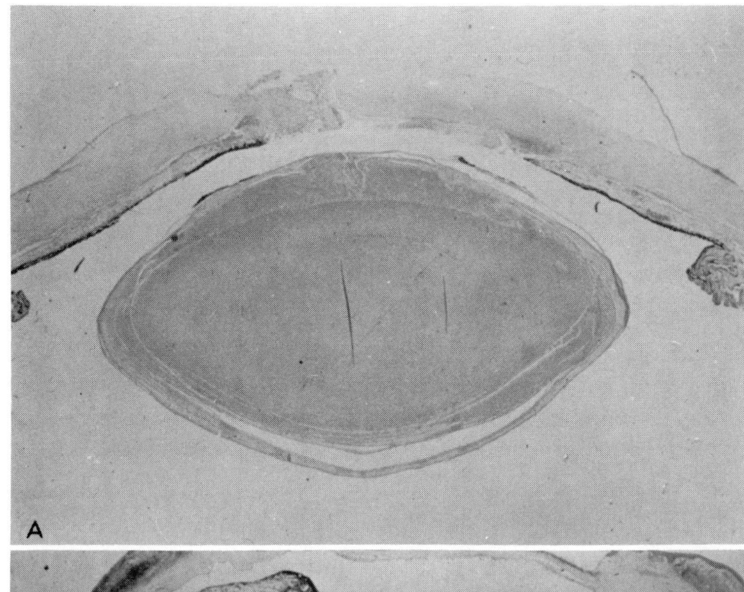

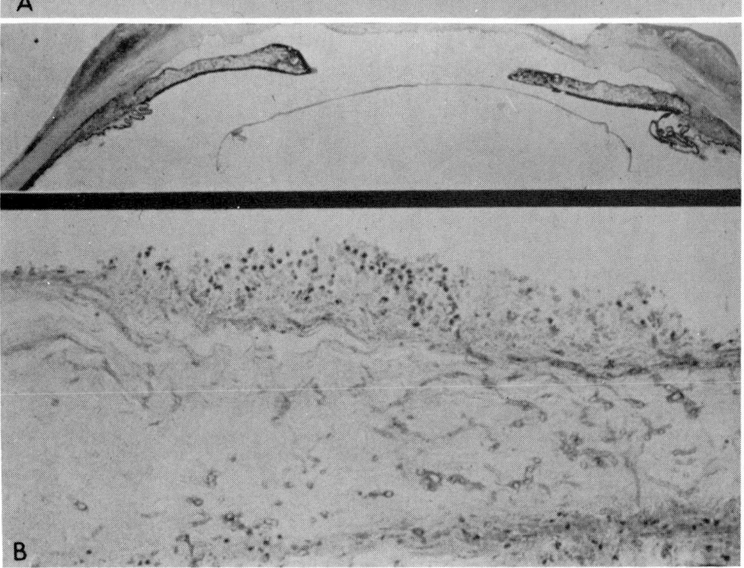

Figure 23–5. FS C 46. Mycotic keratitis. *A,* Ulceration and perforation of cornea. *B,* Ulceration of cornea without perforation. Mycelial elements are seen in high-power view. The fungus was not cultured. ×100.

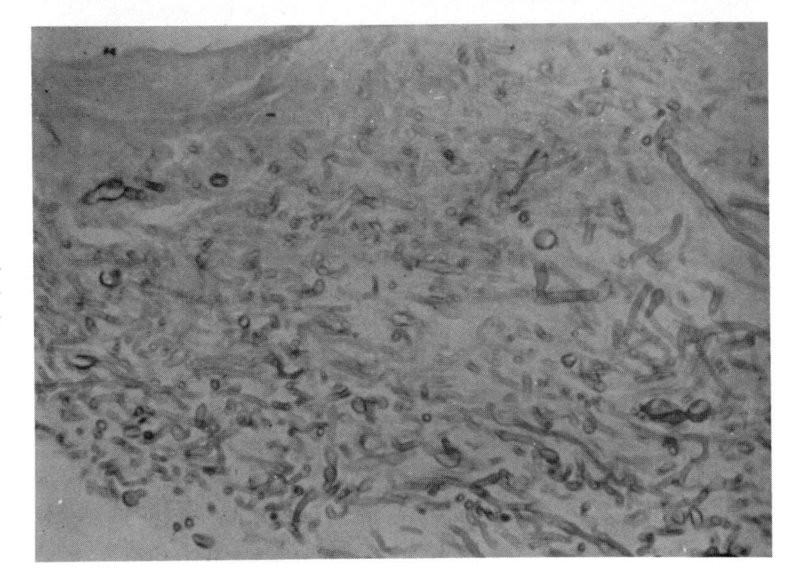

Figure 23–6. Mycotic keratitis. Mycelium of *Cephalosporium* sp. in a sloughed corneal transplant. ×400.

occurred in the rainy winter months in that geographic area. Twenty-nine of the cases were caused by *Fusarium solani,* and almost all patients were fruit pickers or agricultural workers. In the 1962 review by Chick and Conant,[10] trauma was noted in 97 per cent, previous use of steroids in 37 per cent and of antibiotics in 39 per cent of cases.

Etiology. The list of fungi isolated from patients with keratitis is long and varied. There is a considerable fungal flora in the normal eye[1,30,54,55] and an even greater flora in the diseased eye. For this reason many of the reports of oculomycosis in the literature are invalid, as the organisms isolated probably represent normal fungal flora. In patients in whom infection is substantiated by demonstration of fungal elements in debrided material or on histologic examination of tissue, organisms belonging to at least 20 genera have been isolated.[17,23,64] Essentially all of these are saprophytic soil fungi not usually associated with infections in man. Fungi are notorious opportunists. It is not surprising, therefore, to find fungal colonization when natural defenses of the eye are abrogated. Even in cases of keratitis in which the etiologic agent was one of the pathogenic fungi, the organism was observed in its saprophytic phase in the diseased cornea.[13]

Depending on the series,[29,72] in up to one-half the cases, members of the genus *Aspergillus* were involved, particularly *A. fumigatus, A. flavus,* and *A. niger.* This is not surprising, as these are particularly aggressive opportunists and are frequently involved in pulmonary and systemic disease in patients with a variety of debilitations. As might be expected, another common opportunistic organism, *Candida albicans,* also accounts for a large percentage of cases. The third most commonly isolated agent, *Fusarium solani,* is uniquely associated with mycotic keratitis, as it is does not cause any other form of human disease.[3,23,33] The factors that account for its particular affinity for colonizing injured corneal tissue is not known, but there is some suggestive evidence. It appears to be collagenolytic and grows rapidly in corneal tissue. It also is somewhat heat-tolerant.

Some of the other reported causes of keratitis include *Allescheria* (about five cases),[4,13,27,29] *Curvularia lunata*[69] and *C. geniculata,*[49] *Tetraploa,*[47] *Fusidium, Gibberella, Glenospora, Penicillium, Neurospora, Volutella, Phialophora, Macrophoma, Trichosporon, Ustilago, Sporothrix schenckii, Scopulariopsis, Sterygmatocystis nigra, Periconia keratitidis, Verticillium* sp.,[17,23,51] and so forth. How many of these are valid infections cannot be determined. Organisms that have been isolated from culture and the infection confirmed by demonstration of hyphal elements in the involved tissue include *Fusarium* sp.[63a] (Figs.

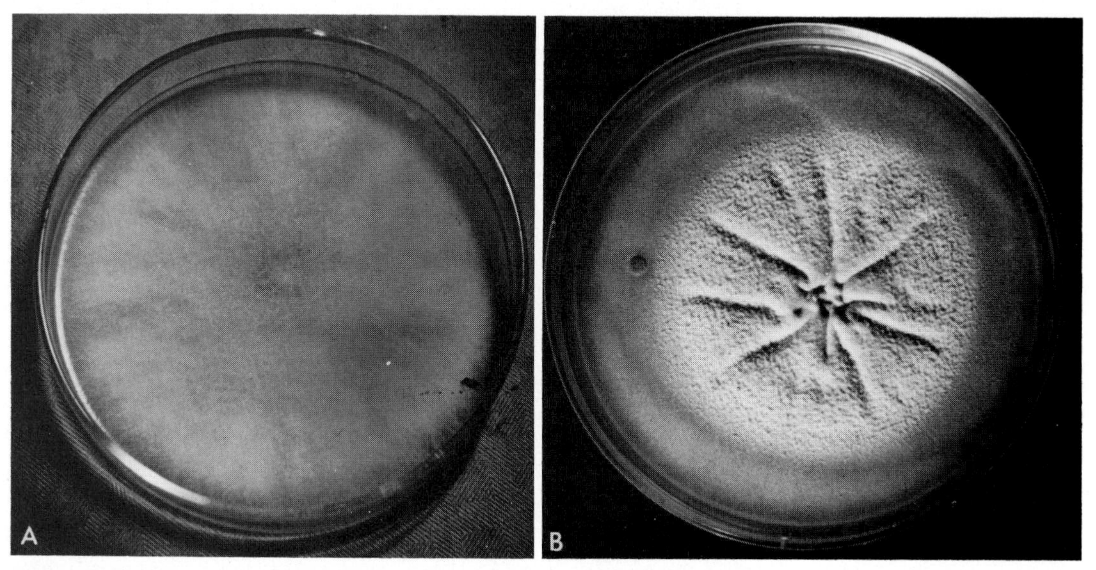

Figure 23–7. FS C 47. *A,* Colony of *Fusarium* sp. It is rapidly growing, cottony, and flat. At first white, it soon develops a rose-mauve or other color, depending on the species. *B,* Colony of *Cephalosporium* species. It is often glabrous and rusty brick-colored at first, but becomes fluffy (often folded) and cream-white with age.

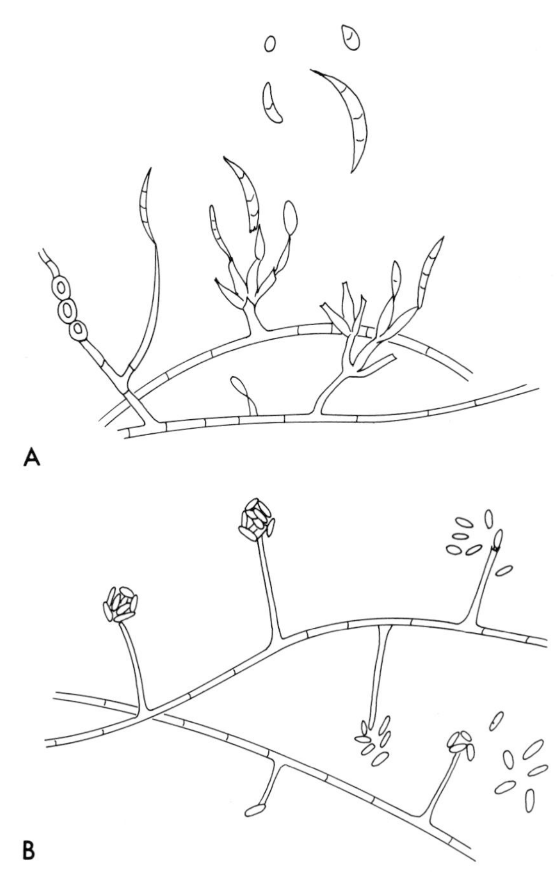

Figure 23–8. Microscopic morphology of (*A*) *Fusarium,* with characteristic crescent-shaped, septate, macroconidia and of (*B*) *Cephalosporium,* with the conidia in a spherical cluster atop a slender tapering conidiophore.

23–7,*A* FS C 47,*A* and 23–8,*A*), *Aspergillus* sp., *Candida* sp.,[21,42,45] *Curvularia, Tetraploa, Graphium,*[2a] *Cylindrocarpon, Botryodiplodia,*[38a] *Allescheria,* and *Cephalosporium*[8] (Figs. 23–7,*B* FS C 47,*B* and 23–8,*B*). In addition to these, other organisms have been cultured from intraocular mycosis resulting from cataract surgery. These include *Sporothrix, Cladosporium,* and *Scopulariopsis.*[23] The actinomycete, *Actinomyces bovis (israelii)* has also been recovered from corneal ulcers.[25]

Diagnosis.[61, 68] In making the diagnosis of mycotic keratitis, it is important to remember two factors. Fungal elements are difficult to find, and the etiologic agents are saprophytic soil organisms. In keratitis, the fungi are usually found deep within the corneal structure and are often absent on its surface. Therefore, a superficial swabbing of the affected area may not be sufficient for demonstration of the etiologic agent. It is often necessary to use extensive debridement in order to obtain viable fungal material. A direct smear is valuable for rapid diagnosis. In potassium hydroxide mount, fungal elements are easily seen (Fig. 23–9). If the organisms are present in small numbers, a Gram stain is useful, especially if a yeast is the etiologic agent. The Gridley stain may also be employed.

The second point of importance in the mycology of eye disease is that the fungi are saprophytic soil organisms. Therefore, they require different cultural procedures from those used for systemic or cutaneous fungal pathogens. Multiple cultures should be taken. A single colony on one culture may represent a contaminant. However, multiple colonies of the same organism usually indicate fungal etiology of the disease, especially if there is histologic evidence also. The preferred medium for culture is Sabouraud's agar slants or plates that do not contain antibiotics. The culture media usually used for the isolation of pathogenic fungi contains cyclohexamide (Actidione), which inhibits growth of saprophytic fungi. For this reason its use is contraindicated for the isolation of the etiologic agents of mycotic keratitis. Fungi in general are not fastidious and thus will grow on almost all antibiotic-free laboratory media. Most of the organisms involved in mycotic keratitis are inhibited by temperatures of 37°C, so that the cultures should be kept at room temperature. Growth is usually apparent in about three days, sometimes within 24 hours. Cultures should be kept for several weeks before being discarded, however, as some species grow very slowly. If growth occurs, slide cultures can be made. The fungus is identified by the morphology and color of its colony, and by the production type and arrangement of its spores. Most of the organisms are species rarely encountered in a medical mycology laboratory, so that identification is often difficult. Such cultures should be sent to reference laboratories and other medical mycologists for verification of identification.

Therapy. In the past many agents have been used for the treatment of mycotic keratitis, and essentially all have failed. There are few specific drugs for the treatment of any form of mycotic disease, and this is particularly true of mycotic keratitis. Of those available griseofulvin is without benefit, nystatin is of limited effective range, and amphotericin B is quite irritating to the

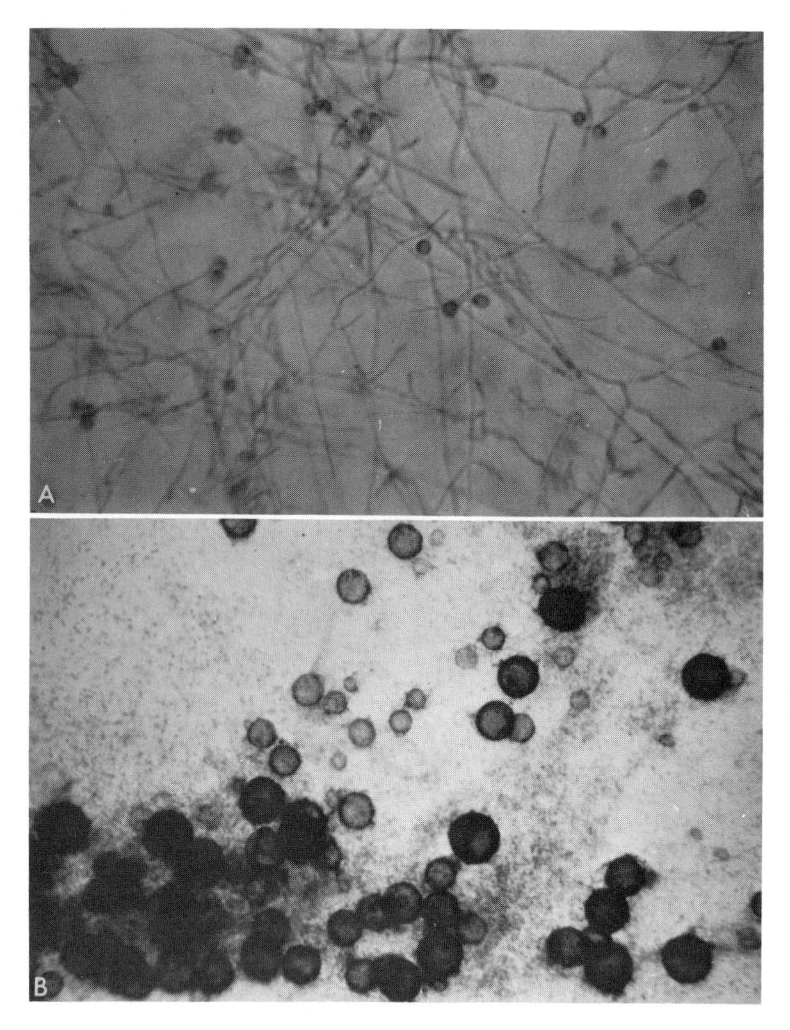

Figure 23–9. Mycotic keratitis due to *Allescheria boydii. A,* KOH mount of corneal scraping showing mycelium and conidia characteristic of *A. boydii.* This fungus is associated with mycetoma and demonstrates a type of tissue-induced morphologic dimorphism which forms granules in tissue. In this eye infection (Fig. 23–1), it is growing in its "saprophytic" form. ×300. *B,* In culture, the fungus produced cleistothecia filled with ascospores in addition to the conidia and mycelium. ×5. (From Ernest, J. T., and J. W. Rippon. 1966. Am. J. Ophthalmol., *62*:1202–1204.)

involved tissue. A new drug, pimaricin (natamycin) appears to be very efficacious. Unfortunately it is as yet unavailable for general use. However, when released it will be the drug of choice for most infections. Sensitivity testing of the isolated organism should be done, as the response of this group of fungi to available drugs is unpredictable. Several studies have been done on the efficacy of these drugs in experimental mycotic keratitis.[18,59]

Amphotericin B. Many of the organisms isolated from eye infections are resistant to this drug at pharmacologic levels. However, higher concentrations may be used with benefit as drops or a wash. The standard regimen is instillation of one or two drops in the conjunctival sac at intervals of one-half to two hours. The concentration is usually 2 to 4 mg per ml aqueous solution. If the infection is severe, deep, or systemic, intravenous therapy is used as an adjunct. Administered either as drops or by subconjunctival injection, the drug is severely irritating, and treatment may be painful. An exaggerated inflammatory response, scarring, and permanent damage may be associated with the use of this drug. How much of this is caused by the drug itself and how much is due to the bile salts surfactant included in the preparation is difficult to assess. Unfortunately the drug is unavailable without additives. Concerning this latter point, studies have shown that, as an orally administered preparation used in conjunction with antibacterial antibiotics and without surfactants, it does not seem to cause untoward distress. It is necessary to conclude, therefore, that if other treatments are available use of amphotericin B is contraindicated.

In the review of experience with amphotericin B, Jones et al.[33,34] noted treatment failures in 13 of 20 cases when the drug was used alone. These were all infections caused by *Fusarium*. Some of their patients required conjunctival flaps or penetrating grafts in addition to drug therapy. They had greater success with the drug when the disease was caused by fungi other than *Fusarium*. This was also noted in the review of cases by Anderson and Chick.[2] These authors stressed extensive mechanical debridement as a part of the therapeutic procedure.

Nystatin. This agent has been used successfully, even in some cases in which the fungus was resistant to amphotericin B[13,44] (Fig. 23–10). Treatment consists of instillation of one drop of the antibiotic every hour in a concentration of 8000 to 20,000 (and sometimes 100,000) units per ml in saline. It is somewhat irritating, but much less so than amphotericin B. Again it is necessary to emphasize the necessity for determining the sensitivity of the etiologic agent to the drug used.

Pimaricin (Natamycin). This drug was derived from the actinomycete, *Streptomyces nataliense* in 1955. It is another polyene antibiotic in the group of the tetraenes. This group also includes nystatin and Rimocidin. Its clinical usefulness has been studied since 1958, but it has not gained wide acceptance in the treatment of systemic infections. Although pimaricin is insoluble and moderately toxic when administered systemically, it is essentially without irritation or discomfort when used topically.[33,34] Several series have recently been published in which this drug was used successfully in mycotic keratitis.

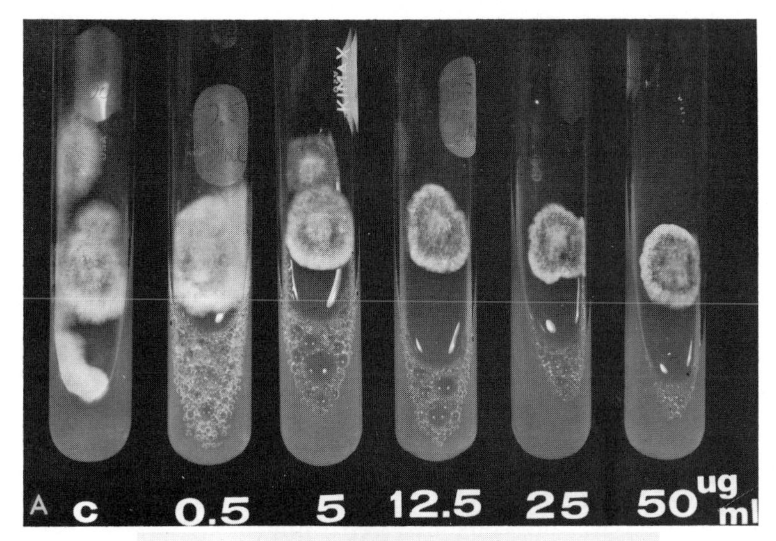

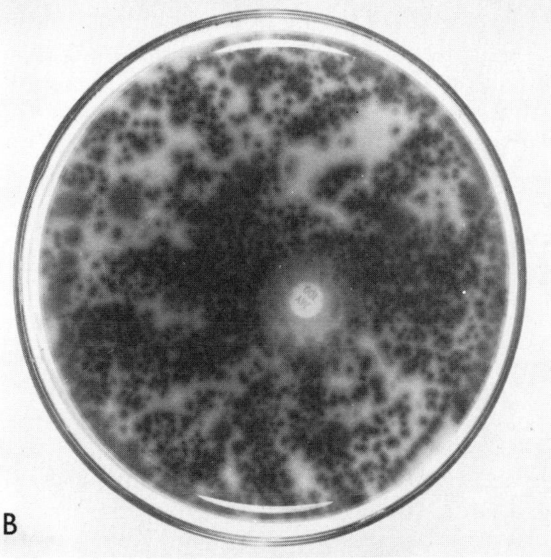

Figure 23–10. Antibiotic testing in a case of mycotic keratitis. *A*, The etiologic agent (*A. boydii*) was resistant to amphotericin B, but there was a zone of inhibition around a disk containing 100 units of nystatin (*B*). (From Ernest, J. T., and J. W. Rippon. 1966. Am. J. Ophthalmol., *62*:1202–1204.)

It is particularly effective in *Fusarium* infections.[33,35,48,55] The treatment schedule has now been standardized.[33] The drug is administered as a 5 per cent solution. This is prepared by dissolving the dry antibiotic in distilled water with 4 N sodium hydroxide. This is then quickly neutralized to pH 7 with 4 N hydrochloric acid. It is stirred for two hours until it becomes a lotionlike suspension. After adjusting the pH to 6.5 or 7.0, it is autoclaved for 20 minutes at 110°C, 15 psi, and then dispensed in plastic dropper bottles. It should be stored in the dark at 4°C. The suspension is given as drops on the lower tarsal conjunctiva every one to three hours until the infection subsides. There were only two treatment failures noted by Jones et al.[34] in 18 consecutive cases. In previous studies there had been more failures when a 1 per cent suspension was used.

Following the resolution of the infection, a variety of surgical procedures may be necessary, depending on the extent of damage to the eye. A conjunctival flap or penetrating keratoplasty is contraindicated as a treatment of the infection without previous medical management.[32] Prior to the establishment of specific drug therapy, these procedures were the only form of treatment available, but their success rate was very low.

The new drug 5-fluorocytosine (5-FC) was used in one case in which there was endogenous disease and mycotic keratitis due to *Candida albicans.* Treatment was successful.[60] This patient had had several courses of intravenous and intraocular amphotericin B and intraocular 1 per cent pimaricin without resolution of the disease. The patient was subsequently given 5-fluorocytosine, 150 to 200 mg per kg per day orally, and hourly drops of a 1 per cent solution, and the infection was brought under control. There was a relapse, however, that required more treatment. So far, 5-FC has not been shown to have *in vivo* activity against any fungus other than a few yeasts, such as *Candida* and *Cryptococcus.* Fishman,[21] however, has found that some strains of *Candida* are resistant to the drug.

Endogenous Oculomycosis

This type of disease results from dissemination of fungus from other loci in the body. In almost all the systemic mycoses, eye involvement has been reported but is considered a rare complication of such diseases. Often involvement of the eye is a terminal event in which there is widespread dissemination of the disease to all organ systems. The diseases noted in order of frequency are candidosis, cryptococcosis, coccidioidomycosis, blastomycosis, sporotrichosis, paracoccidioidomycosis, and histoplasmosis. About 18 cases of ocular involvement as a sequela of cryptococcosis have been recorded.[23] Most of these involved a chorioretinitis. Approximately nine cases of eye disease in systemic coccidioidomycosis have been found,[16] and about six cases of blastomycosis have been recorded.[9] The latter oculomycosis was first described in 1914 by Churchill and Stober.[11] In another case described by Font,[22] this was the only manifestation, and the diagnosis of blastomycosis was made only after finding the organisms in the enucleated eye. There were retrospective chest findings.

The most common cause of endogenous oculomycosis is *Candida albicans.* More than 30 cases of ocular involvement in disseminated candidosis have been described in the literature.[21,45,60] In a recent series of six, Fishman[21] described *Candida* endophthalmitis as a complication of candidemia. All patients had received antibiotics, and all had had indwelling catheters for prolonged periods of time. Only one patient had a neoplasm, and none had received steroid or immunosuppressive therapy. Symptoms of ocular distress occurred 3 to 15 days after positive blood cultures were obtained. The symptoms included painful eye, pain around the eye, blurring of vision, and spots before the eye. However, in two patients there were no symptoms, though lesions of the eye were found on examination. Intravitreous extension occurred in three cases, and anterior uveitis was found in the other three. The treatment consisted of intravenous amphotericin B. The dosage varied, depending on response. As little as 300 mg was used with success in one case, but others required up to 1.4 g. The minimum inhibitory concentration of amphotericin B for the *Candida* isolated from these cases varied between 0.04 and 0.078 μg per ml. Two isolants were resistant to 5-fluorocytosine, and one had a minimum inhibitory concentration (MIC) of 0.05.

The usual presenting symptoms in all

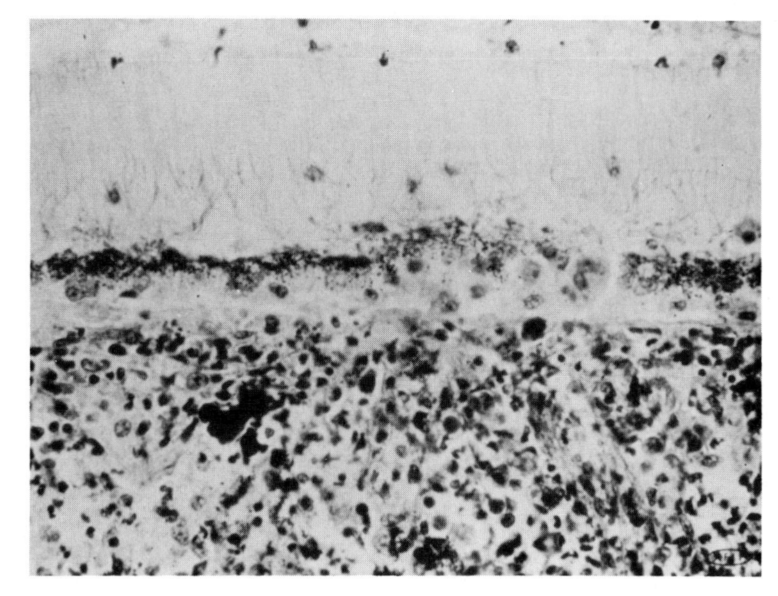

Figure 23–11. FS C 48. Endogenous oculomycosis. Granulomatous uveitis in a case of disseminated blastomycosis. Some yeast cells can be seen among the cellular infiltrate within choroid. ×400.

types of disseminated mycosis involving the eye are a progressive granulomatous uveitis (Fig. 23–11 FS C 48), diffuse retinitis, and deep vitreous abscess. Glaucoma may also be present. The inflammatory response is noted to be minimal in ocular cryptococcosis. The diagnosis of endophthalmic disease usually depends on isolation of the organism from other loci of infection, and treatment involves systemically administered amphotericin B.

Presumed *Histoplasma* choroiditis[37, 38, 43, 44, 62, 66] (benign histoplasmosis, *Histoplasma* granulomatous uveitis, or multifocal choroiditis) appears to be an allergic disease rather than a direct infection. Its association with histoplasmosis is tenuous, and substantiated evidence of a causal relationship is lacking. In true disseminated histoplasmosis, the eye is rarely involved. In such cases the pathology resembles that of tuberculosis and other granulomatous infections and does not resemble the clinical or pathologic features seen in multifocal choroiditis. In the latter disease, the primary lesion in the acute stage is a yellowish white, slightly elevated choroidal infiltrate that is usually multifocal. Fluid leakage may occur from these macular lesions. During periods of chronic serous fluid leakage, new vessels from the choroid may be formed. Their rupture leads to hemorrhage.[37] Eventually, hypertrophic scars form in such areas, with severe visual loss.

Presumed *Histoplasma* uveitis was first described by Woods and Wahlen in 1959.[70] They reported that a preexisting uveitis worsened following a histoplasmin skin test and concluded there was a causal relationship. Since that time several hundred cases of a similar (though possibly not identical) disease have been diagnosed.[37, 38, 43, 62, 66] In no case has the organism been identified in tissue or has specific fluorescent staining been positive for debris of the *Histoplasma* yeast cell. Furthermore, there is no well-documented evidence that treatment with amphotericin B improved the patient's condition, although such has been reported.[15] However, steroids and photocoagulation have been efficacious in controlling the disease.[37]

Although it has been reported that 90 per cent or more of patients with multifocal choroiditis have a positive histoplasmin skin test,[37, 66] it is difficult to assess or correlate the association of skin test reactivity with the disease. In some areas of the United States, between 80 and 90 per cent of the general population will demonstrate a skin test reactivity, and cross-reactions with other fungal antigens are common.[12, 24] It has also been shown that patients with presumed histoplasmin choroiditis are hyperergic to several antigens other than histoplasmin.[66] In this same study the curve expressing frequency of uveitis at different ages paralleled the appearance of skin testing reactivity in different age groups. In recent studies concerned with the disease, serum antibodies to double stranded RNA and DNA have been found in patients who had the idiopathic and secondary types of uveitis.[7, 14] In France a type of uveitis similar to multifocal choroiditis was associated with sensitivity to *Candida* antigen.[65]

Extension Oculomycosis

This disease is usually associated with acute rhinocerebral mucormycosis of the diabetic, but it is also seen in cases of rhinosporidiosis and meningeal cryptococcosis. These latter infections are discussed in other chapters. In acute mucormycosis the infection starts in the upper portion of the nasal septum and extends into the orbit of the eye, the frontal sinuses, the major cerebral vessels, and subsequently into the central nervous system. A complete discussion of this disease is given in Chapter 21. In this section a summary of the associated ocular manifestations will be given.

Clinical Symptoms. The presenting symptoms of mucormycosis involve first the nose and then the eye. The nasal area shows a thick discharge, dark and blood-tinged, as well as reddish-black necrotic areas of the turbinates and septum. A cloudy sinus sometimes with a discernible fluid level can be seen on x-ray examination. In addition, there may be paranasal inflammation or pansinusitis; fistula formation and sloughs of the hard palate; loss of function of fifth and seventh cranial nerves; and often a hard, somewhat palpable, demarcated swelling on either side of the nasal bridge. Severe orbital cellulitis at this stage denotes a poor prognosis, as it heralds impending or actual invasion by the fungus of the eye and usually the central nervous system. The findings of the eye include orbital pain, with ophthalmoplegia, ptosis, localized anesthesia, proptosis, limitation of movement of the eyeball, fixation of the pupil, and loss of vision. Hyphae may be seen coursing through the vitreous. The disease is unilateral or bilateral.[6, 26]

As the infection progresses, the fungus shows a predilection for blood vessels, with subsequent infarction and necrotic sequelae in the brain and softening of the frontal bones. Lethargy progressing to coma ensues, and the patient succumbs in seven to ten days. In some personally observed cases of fulminating disease, death occurred as early as two to three days after initial symptoms of infection. The overall mortality rate of this disease is between 80 and 90 per cent. Most of the patients who survive have received antifungal therapy in addition to control measures for their diabetes.

Predisposing Factors. Acute mucormycosis has classically been associated with diabetes. There is experimental evidence, however, indicating that the acidotic state is the prime factor predisposing to infection. Rats or rabbits made diabetic by alloxan treatment were resistant to *Mucor* infection, but, when made acidotic as well, they rapidly succumbed to infection by the organism. Acute mucormycosis has been recorded in patients who were acidotic due to other causes, such as diarrhea or salicylate ingestion.

Treatment. Control of diabetes appears to be the most important factor in halting the infection. In the few reports of survivors before antifungal therapy became available, the infection receded when the acidotic state of the patient was corrected. However, the disease sometimes relapsed during another bout of uncontrolled diabetes with acidosis. At present, use of systemic and local amphotericin B has controlled the infection in several patients. It is important to do sensitivity studies on the fungus isolated from the infection, as the organisms vary considerably in their response to the drug.

Etiology and Diagnosis. The fungus most commonly isolated from this form of the disease is *Rhizopus oryzae* and *R. arrhizus*. A few other species in the genera *Mucor* and *Rhizopus* have also been reported as etiologic agents of the infection. All organisms are common saprophytic soil fungi, often termed the "black bread molds." In general, the organisms are readily cultivated from patient material. A scraping of the upper nares is used for preparation of a direct mount (KOH mount is preferred), and this same material can be used for culture. The fungi grow on any media without antibiotics. On direct mount, the fungus appears as wide (4 to 15 μ), nonseptate, branching hyphae. The culture is incubated at room temperature, and growth usually occurs within a few days or as early as 18 hours.

Infection of the Lacrymal Ducts and Conjunctiva

Cases of canaliculitis (dacryocanaliculitis) have been described for many years, and the disorder is a well-recognized syndrome.

The patients present with a swelling at the opening of the duct, tearing, pruritus, and inflammation of the conjunctiva. A hard, knobby concretion is felt within the duct, causing stasis and pressure. Most often only one duct is involved, and it is usually the inferior canal. In a review of 146 cases, the superior duct was involved in only 41.[23] The treatment is simple and effective and consists of removal of the plug. This is generally accompanied by administration of antibiotics, such as sulfas and penicillin.

Examination of the excised material usually demonstrates intertwined hyphal elements and bacilliform bodies that are 1 μ in diameter, Gram-positive, and usually acid-fast–negative. The etiologic agent was called *Streptothrix* in the older literature and the disease named streptotricosis. Culture of material from canaliculitis has yielded a variety of organisms, including *Actinomyces israelii*,[53] *Nocardia asteroides*[52] (acid-fast–positive), and various *Corynebacterium*.

Another form of the disease is dacryocystitis. This is an inflammatory disease of the lacrymal duct caused by a variety of bacteria and fungi. Among these are *Actinomyces, Candida, Aspergillus, Blastomyces* (?), *Paecilomyces*,[31a] *Sporothrix, Rhinosporidium, Cephalosporium,* and *Trichophyton*. Probably only a few such reports are valid cases of disease in which there was direct involvement of a fungal agent.

Although the conjunctiva may be involved in a number of systemic mycoses, conjunctivitis is rarely a presenting symptom of a mycotic disease. Infection of the conjunctiva in the absence of other foci of disease is known in rhinosporidiosis but in essentially no other fungous infection. Infection of the eyelid alone, however, has been recorded on several occasions. This is usually due to dermatophytes[50] or *Pityrosporum* sp.[23]

REFERENCES

1. Albesi, E. J., and R. Zapater. 1972. Flora fungica de la conjunctiva en ojos sanes. Arch. Oftal. B. Aires, *47*:329–334.
2. Anderson, B., and E. W. Chick. 1963. Mycokeratitis: Treatment of fungal corneal ulcers with amphotericin B and mechanical debridement. South. Med. J., *56*:270–273.

2a. Apostol, J. G., and S. L. Meyer. 1972. *Graphium* endophthalmitis. Am. J. Ophthalmol., *73*: 566–569.

3. Arrechea, A., R. Zapater, et al. 1971. Queratomicosis por *Fusarium solani*. Arch. Oftal. B. Aires, *46*:123–127.
4. Bakerspigel, A. 1971. Fungi isolated from keratomycosis in Ontario, Canada. I. *Monosporium apiospermum*. Sabouraudia, *9*:109–112.
5. Birge, H. L. 1962. Reclassification of mycotic disease in clinical ophthalmology. Am. J. Ophthalmol., *53*:630–635.
6. Borland, D. S. 1959. Mucormycosis of central nervous system. Am. J. Dis. Child., *97*:852–856.
7. Burns, R. M., M. S. Rheins, et al. 1967. Anti-DNA in the sera of patients with uveitis. Arch. Ophthalmol., *77*:777–779.
8. Byers, J. L., M. G. Holland, et al. 1960. Cephalosporium keratitis. Am. J. Ophthalmol., *49*: 267–269.
9. Cassady, J. V. 1946. Uveal blastomycosis. Arch. Ophthalmol., *35*:84.
10. Chick, E. W., and N. F. Conant. 1962. Mycotic ulcerative keratitis. A review of 148 cases from the literature (abstract). Invest. Ophthalmol., *1*:419.
11. Churchill, T., and A. M. Stober, 1914. A case of systemic blastomycosis. Arch. Intern. Med., *13*:568–574.
12. Edwards, P. W., and J. H. Klaer. 1956. Worldwide geographic distribution of histoplasmosis and histoplasmin sensitivity. Am. J. Trop. Med., *5*:235–257.
13. Ernest, J. T., and J. W. Rippon. 1966. Corneal ulcer due to *Allescheria boydii*. Am. J. Ophthalmol., *62*:1202–1204.
14. Epstein, W. V., M. Tan, et al. 1971. Serum antibody to double stranded RNA and DNA in patients with idiopathic and secondary uveitis. New Engl. J. Med., *285*:1502–1506.
15. Falls, H. F., and C. L. Giles. 1960. The use of amphotericin B in selected cases of chorioretinitis. Am. J. Ophthalmol., *49*:1288–1298.
16. Faulkner, R. F. 1962. Ocular coccidioidomycosis. Am. J. Ophthalmol., *53*:822–827.
17. Fazakas, A. 1955. Mycotic infections of the eye and their treatment. Klin. Mbl. Augenheilk., *127*:701–721.
18. Fine, B. S., and L. E. Zimmerman. 1960. Therapy of experimental intraocular Aspergillus infection. Arch. Ophthalmol., *64*:849–861.
19. Fine, B. S., and L. E. Zimmerman. 1959. Post operative mycotic endophthalmitis diagnosed clinically and verified histopathologically. Br. J. Ophthalmol., *43*:753–758.
20. Fine, B. S., and L. E. Zimmerman. 1959. Exogenous intraocular fungus infections (with particular reference to complications of intraocular surgery). Am. J. Ophthalmol., *48*: 151–165.
21. Fishman, L. S., J. R. Griffin, et al. 1972. Hematogenous Candida endophthalmitis. A complication of candidemia. New Engl. J. Med., *286*:675–681.
22. Font, R. L., A. G. Spaulding, et al. 1967. Endogenous mycotic panophthalmitis caused by *Blastomyces dermatitidis*. Arch. Ophthalmol., *77*:217–222.
23. François, J. 1968. *Les Mycoses Oculaires*. Paris. Masson and Cie.
24. Furcolow, M. L. 1963. Tests of immunity in histoplasmosis. New Engl. J. Med., *268*:357–361.

25. Gingrich, W. D., and M. E. Pinkerton, 1952. Anaerobic *Actinomyces bovis* corneal ulcer. Arch. Ophthalmol., *67*:549–553.
26. Ginsberg, J., A. G. Spaulding, et al. 1966. Cerebral phycomycosis (mucormycosis) with ocular involvement. Am. J. Ophthalmol., *62*:900–906.
27. Gordon, M. A., et al. 1959. Corneal allescheriosis. Arch. Ophthalmol., *62*:758–763.
28. Graefe, A. von. 1854. Polypen des Tränenschlauchs. Albrecht v. Graefes Arch. Ophthalmol., *1*:283–284.
29. Halde, C., and J. Okumoto. 1966. Ocular mycoses. A study of 82 cases. Amsterdam, Excerpta Medical International Congress, 1966, pp. 705–712.
30. Hammeke, J. C., and P. P. Ellis. 1960. Mycotic flora of the conjunctiva. Am. J. Ophthalmol., *49*:1174–1178.
31. Harley, R. D., and S. E. Mishler. 1959. Endogenous intraocular fungus infection. Trans. Am. Acad. Ophthalmol. Otolaryngol., *63*:264–268.
31a. Henig, F. E., N. Lehrer, et al. 1973. Paecilomycosis of the lacrimal sac. Mykosen, *16*:25–28.
32. Jones, B. R., D. B. Jones, et al. 1970. Surgery in the management of keratomycosis. Trans. Ophthalmol. Soc. U.K., *89*:887–897.
33. Jones, D. B., R. K. Forster, et al. 1972. *Fusarium solani* keratitis treated with natamycin (pimaricin). Arch. Ophthalmol., *88*:147–154.
34. Jones, D. B., R. Sexton, et al. 1970. Mycotic keratitis in South Florida. A review of 39 cases. Trans. Ophthalmol. Soc. U.K., *89*:781–787.
35. Kaufman, H. E., and R. M. Wood. 1965. Mycotic keratitis. Am. J. Ophthalmol., *59*:993–1000.
36. Kaufman, L., R. T. Terry, et al. 1967. Effects of a single histoplasmin skin test on the serologic diagnosis of histoplasmosis. J. Bacteriol., *94*:798–803.
37. Krill, A. E., and D. Archer. 1970. Choroidal neovascularization in multifocal (presumed histoplasmin) choroiditis. Arch. Ophthalmol., *84*:595–604.
38. Krill, A. E., M. Chrishti, et al. 1969. Multifocal inner choroiditis. Trans. Am. Acad. Ophthalmol. Otolaryng., *73*:722–742.
38a. Laverde, S., L. H. Moncada, et al. 1973. Mycotic keratitis; 5 cases caused by unusual fungi. Sabouraudia, *11*:119–123.
39. Leber, T. 1879. Keratomykosis aspergillina als Ursache von Hypopion ketaritis. Albrecht v. Graefes Arch. Ophthalmol., *25*:285–301.
40. Levitt, J. M., and J. Goldstein. 1971. Keratomycosis due to *Allescheria boydii.* Am. J. Ophthalmol., *71*:1190–1191.
41. Ley, A. P. 1956. Experimental fungus infections of the cornea. Am. J. Ophthalmol., *42*:59–71.
42. Manchester, P. T., and L. K. Georg. 1959. Corneal ulcer due to *Candida parapsilosis (C. parakrusei).* J.A.M.A., *171*:1339–1341.
43. Maumenee, A. E., and A. M. Silverstein (eds.). 1964. *Immunopathology of Uveitis.* Baltimore, Williams & Wilkins Co.
44. McGrand, J. C. 1970. Symposium on direct fungal infection of the eye. Keratomycosis due to *Aspergillus fumigatus* cured by nystatin. Trans. Ophthalmol. Soc. U.K., *89*:799–802.
45. Mendelblatt, D. L. 1953. A review and a report of the first case demonstrating the *Candida albicans* in the cornea. Am. J. Ophthalmol., *36*:379.
46. Naumann, G., W. R. Green, et al. 1967. Mycotic keratitis. A histopathologic study of 73 cases. Am. J. Ophthalmol., *64*:668–682.
47. Newmark, E., and F. M. Polack. 1970. *Tetraploa* keratomycosis. Am. J. Ophthalmol., *70*: 1013–1015.
48. Newmark, E., A. C. Ellison, et al. 1970. Pimaricin therapy of *Cephalosporium* and *Fusarium keratitis.* Am. J. Ophthalmol., *69*:458–466.
49. Nityananda, K., P. Sivasubramanam, et al. 1964. A case of mycotic keratitis caused by *Curvularia geniculara.* Arch. Ophthalmol., *71*:456–458.
50. Ostler, H. B., M. Okumoto, et al. 1971. Dermatophytosis affecting the periorbital region. Am. J. Ophthalmol., *72*:934–938.
51. Parker, J. C., G. Klentworth, et al. 1971. Miscellaneous uncommon diseases. In *Handbuch der Speziellen Pathologischen Anatomie.* Vol. 3. Baker, R. D. (ed.). Berlin, Springer-Verlag, pp. 986–987.
52. Penikett, E. J. K., and D. L. Rees. 1962. *Nocardia asteroides* infection of the nasal lacrimal system. Am. J. Ophthalmol., *53*:1006–1008.
53. Pine, L., H. Hardin, et al. 1960. Actinomycotic lactimal canaliculitis. A report of two cases with a review of characteristics which identify the causal organism *Actinomyces israeli.* Am. J. Ophthalmol., *49*:1278–1288.
54. Pine, L., W. A. Shearin, et al. 1961. Mycotic flora of the lacrimal duct. Am. J. Ophthalmol., *52*:619–625.
55. Polack, F. M., H. E. Kaufman, et al. 1971. Keratomycosis, medical and surgical treatment. Arch. Ophthalmol., *85*:410–416.
56. Posner, A. 1960. Treatment and prevention of fungus endophthalmitis. Eye, Ear, Nose Thr. Monthly, *39*:71–72.
57. Posner, A. 1960. The role of starch derivative glove powder in hospital infection. Eye, Ear, Nose Thr. Monthly, *39*:175.
58. Prabhakar, H., et al. 1969. Mycotic and bacterial flora of the conjunctival sacs in healthy and diseased eyes. Indian J. Pathol. Bacteriol., *12*:158–161.
59. Richards, A. B., Y. M. Clayton, et al. 1970. Antifungal drugs for oculomycosis. IV. The evaluation of antifungal drugs in the living cornea. Trans. Ophthalmol. Soc. U.K., *89*: 847–861.
60. Richards, A. B., B. R. Jones, et al. 1970. Corneal and intraocular infection by *Candida albicans* treated with 5-fluorocytosine. Trans. Ophthalmol. Soc. U.K., *89*:867–885.
61. Rippon, J. W. 1972. Mycotic infections of the eye. Diagnosis and treatment. Ophthalmol. Digest., *34*:18–25.
62. Schlaegel, T. F., J. C. Weber, et al. 1967. Presumed histoplasmic choroiditis. Am. J. Ophthalmol., *69*:919–925.
63. Segelken, Von. 1902. Ein kasuistischer Beitrag zur Aeriologic der konkremente in den Tränenröhrchen. Klin. Mbl. Augenheilk., *40*(II):134–143.
63a. Snyder, W. C., and T. A. Tousson. 1965. Current status of taxonomy of *Fusarium* species and their perfect stages. Phytopathology, *55*: 833–837.
64. Theodore, F. H. 1962. The role of so-called saprophytic fungi in eye infections. In *Fungi and Fungus Diseases.* Daldorf, F. (ed.), Springfield, Ill., Charles C Thomas, Publisher, pp. 22–32.

65. Vallery-Radot, C. 1966. Uveite et allergie a *Candida albicans*. Rev. Fr. Allerg., *6*:27–32.
66. Van Metre, T. E., and A. E. Maunseuee. 1964. Specific ocular uveal lesions in patients with evidence of histoplasmosis. Arch. Ophthalmol., *71*:314–324.
67. Weber, J. C., and T. F. Schlaegel. 1969. Delayed skin test reactivity of uveitis patients. Am. J. Ophthalmol., *62*:732–744.
68. Wilson, L. A., and R. R. Sexton. 1968. Laboratory diagnosis in fungal keratitis. Am. J. Ophthalmol., *66*:647–653.
69. Wind, C. A., and F. M. Polack. 1970. Keratomycosis due to *Curvularia lunata*. Arch. Ophthalmol. *84*:694–696.
70. Woods, A. C., and H. E. Wahlen. 1960. The probable role of benign histoplasmosis in the etiology of granulomatous uveitis. Am. J. Ophthalmol., *49*:205–220.
71. Zapater, R. C., A. de Arrochea, et al. 1972. Queratomicosis por *Fusarium dimerum*. Sabouraudia, *10*:274–275.
72. Zimmerman, L. E. 1963. Keratomycosis. Survey Ophthalmol., *8*:1–25.

Otomycosis

Otomycosis, or otitis externa of fungal etiology, is a chronic or subacute infection of the pinna, the external auditory meatus, and the ear canal. Various fungi may be involved, along with several bacteria.[5,6] The presenting symptoms include scaling, pruritus, and pain. The canal is seen to be crusted, edematous, and erythematous, and there is a collection of cerumen. A feeling of fullness of the ear is often present, along with impairment of hearing. Suppuration and a foul odor may also be found and are caused by bacterial invasion of the subepithelial layers. In disease in which there is little bacterial involvement, the lesions are dry and eczematized. In severe acute infection, the cartilagenous structures of the ear may be involved. In almost all cases, the tympanic membrane is spared, however. In 80 to 90 per cent of cases of otitis externa, a variety of bacteria are found, including *Pseudomonas, Proteus, Micrococcus, Streptococcus, Escherichia,* and *Corynebacterium.*[9] Frequently the condition responds to antibacterial therapy alone. A fungal etiology in such cases is doubtful, even if it is positive by culture. The establishment of a true fungal etiology for otitis externa requires demonstration of mycelial elements in scrapings, as well as a positive culture.[6–8]

External ear infection due to invasion by fungi is a well-established syndrome which is frequently encountered. The fungi most commonly involved are *Aspergillus fumigatus* and *A. niger.*[11] Otoscopic examination sometimes reveals the fruiting heads of the organisms lining the wall of the canal (Fig. 23–12 FS C 49). Other species of fungi have also been documented as causing this type of infection, but only rarely. These include *Scopulariopsis,*[12] *Aspergillus flavus,* other *Aspergillus* species, *Penicillium, Mucor,* and *Candida* sp. *Peyronellaea* has been reported in animal infections.[3] Numerous other species have been listed, but few have been unequivocally documented as the etiologic agent. Dermatophytosis has been reported in infections of the pinna proper. Such infections usually are extensions from disease elsewhere on the skin.

Treatment of the disease has included many agents and modalities. In a recent series of six cases in which *A. niger* was the etiologic agent, application of nystatin ointment for three to four weeks cleared the infection,[2] and this drug has also been successfully used in infections caused by a variety of other fungi.[5] Phenylmercuric borate ointment has been used with good results in a series of 34 cases.[6] The usual treatment involves cleaning the ear and removing the macerated debris. Swabs with Burow's solution, 5 per cent aluminum acetate solution, or urea–acetic acid solution accomplish debridement of this material. Often this is sufficient to effect a clinical cure. As there is usually a bacterial component, antibacterial therapy is often necessary. The preparations most generally utilized are chloramphenicol, bacitracin, polymyxin, neomycin, and Aureomycin. In chronic ear infection in which the bacterial component is minimal, metacresol acetate (Cresatin), thymol (1 per cent) in metacresyl acetate, and phenylmercuric acetate (0.02 per cent) in water[12] or as a borate ointment (0.005 and 0.04 per cent) have effected clinical cure. Amphotericin B ointment also has been tried, but the results have been equivocal.

The fungi responsible for the disease are generally airborne saprophytes, so that cultures should be made on Sabouraud's agar

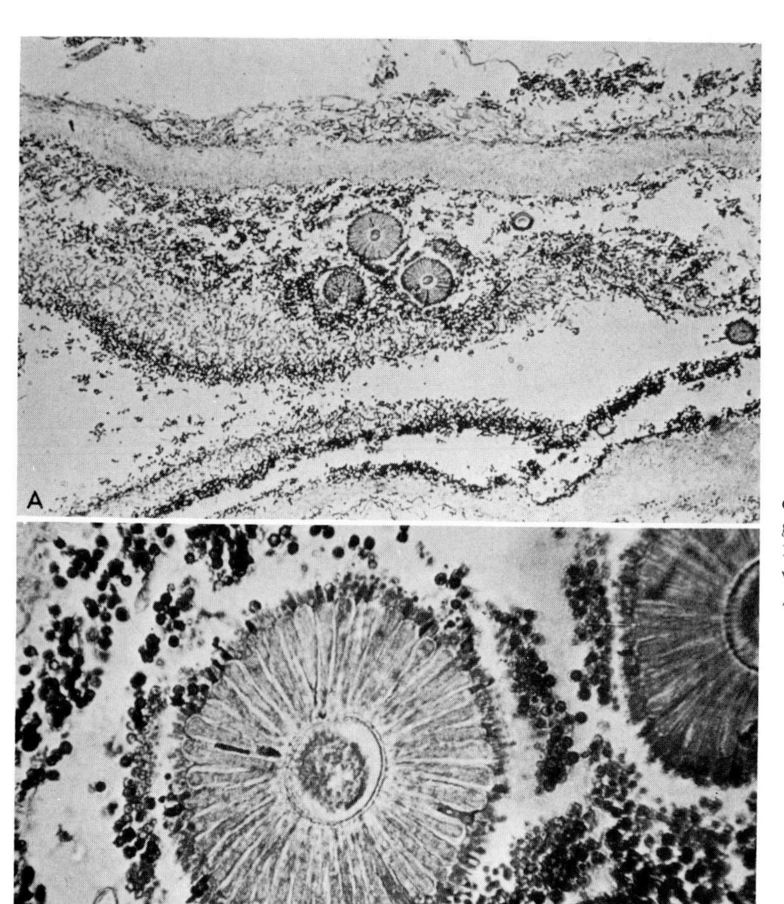

Figure 23–12. FS C 49. Otomycosis. *A,* Fruiting heads of *Aspergillus niger* seen in the ear canal. ×100. *B,* High-power view of the vesicle sterigmata and conidia of *A. niger.* ×400.

without cycloheximide. Optimal growth is obtained at room temperature. Isolation of several colonies of the same fungus, in addition to evidence by direct examination, is necessary for a specific diagnosis. Otomycosis has been reproduced experimentally in animals.[10]

REFERENCES

1. Beaney, G. P. E., and A. Broughton. 1967. Tropical otomycosis. J. Laryngol. Otol., *81*:987–997.
2. Bezjak, V., and O. P. Arya. 1970. Otomycosis due to *Aspergillus niger.* E. Afr. Med. J., *47*:247–253.
3. Dawson, C. D., and A. W. D. Lopper. 1970. *Peyronellaea glomerata* infection of the ear pinna in goats. Sabouraudia, *8*:145–148.
4. English, M. P. 1957. Otomycosis caused by a ringworm fungus. J. Laryngol. Otol., *71*:207–208.
5. Gregson, A. E. W., and C. J. LaTouche. 1961. The significance of mycotic infection in the aetiology of otitis externa. J. Laryngol. Otol., *75*: 167–170.
6. Grigoriu, D., and N. Font. 1970. Les otomycoses. Dermatologica, *141*:138–142.
7. Kingery, F. A. 1965. The myth of otomycosis. J.A.M.A., *191*:129.
8. McGonigla, J. J., and O. Jilson. 1967. Otomycosis, an entity. Arch. Dermatol., *95*:45–46.
9. Singer, D. E., et al. 1952. Otitis externa. Bacteriological studies. Ann. Otol. (St. Louis), *61*:317–330.
10. Sood, V. P., A. Sinha, et al. 1967. Otomycosis; A clinical entity–Clinical and experimental study. J. Laryngol. Otol., *81*:999–1012.
11. Stuart, E. A., and F. Blank. 1955. Aspergillosis of the ear. A report of twenty-nine cases. Can. Med. Assoc. J., *72*:334–337.
12. Yamashita, K., and T. Yamashita. 1972. *Polypaecilum insolitum (Scopulariopsis divaricata)* isolated from cases of otomycosis. Sabouraudia, *10*:128–131.

Part Three

Allergic Diseases, Mycetismus, and Mycotoxicosis

Chapter 24

In addition to actual infection by a fungus, there are several diseases that are caused by the inhalation of fungal debris, eliciting an allergic reaction; by the ingestion of organisms, usually mushrooms, containing toxins (mycetismus); or by the ingestion of products that have been altered by previous growth of fungi that contain toxic substances (mycotoxicosis). An overview of these several types of disease will be given in this section. Each of these disease types has been the subject of a great amount of investigation, and several books have been devoted to them. The present chapter is intended to be an introduction to these specialized subjects and a guide to the major reference sources.

Fungal toxins and allergies have been known to affect man and animals since antiquity. Fatal mushroom poisoning was the subject of much Greek, Roman, and Hindu literature, and interest in the subject has continued through the middle ages and renaissance to the present. Formulas for the practical or political use of such poisons, as well as means of alleviating the distress if one, perchance, is the victim, are extant in these sources. On a wider scale of economic importance, the entire population and probably the survival of western civilization has been dramatically affected by such intoxications as ergot poisoning. Most of these accounts were relegated to curious sidelights of history until recently. In 1960 hundreds of turkey poults were killed by a strange disease in England termed "turkey X disease." This disease, caused by a toxin produced by *Aspergillus flavus* growing on grain, sparked a renewed interest in the mycotoxicoses. It is surmised that many previously unexplained diseases of man and animals may be caused by as yet unknown toxins that result from the growth of fungi on foodstuffs. At present, aflatoxicosis, as well as other similar syndromes, is being investigated, and a voluminous literature is evolving.

Distress and death mixed with much superstition and religous folklore have been associated with the ingestion of mushrooms and other fleshy fungi during all recorded history. Even in eastern European countries where mycophagy has been practiced for centuries, there are still significant numbers of fatalities each year due to misidentification of mushrooms used for human consumption. The intoxication and hallucination following ingestion of certain species in certain Indian tribal rites is the subject of much faddist, anthropologic, and scientific "lore." Many mushrooms are superior items on a gourmet's menu; others, however, may lead to a variety of gastrointestinal, neurologic, or other types of disturbances or may ultimately be fatal. Thus the study of the various effects of ingesting fungi on man has popular as well as scientific and medical interest.

Although asthma and general allergy to the inhalation of fungal spores has long been recognized as a clinical entity, only recently have specific occupational diseases, some causing irreversible damage or death, been recognized. Many cases of "idiopathic" chronic lung disease are now being recognized as manifestations of hypersensitivity to fungal spores, fungal products, or fungal debris present in a particular environment.

New syndromes of all three types of mycotic disease are continually being recognized. These diseases will be grouped for convenience into three major subheadings: (1) mycetismus – disease caused by the ingestion of fungi (mushroom poisoning); (2) mycotoxicosis – ingestion of preformed fungal products or products formed under the influence of fungi; and (3) allergies or hypersensitivities manifest by particular individuals to the presence of fungal allergens in the environment.

Mycetismus (Mushroom Poisoning)

Mycophagy, the consumption of fleshy fungi, has been practiced by man since before recorded history. Experience over the centuries has implicated certain species as having undesirable or prized side effects. These range from minor gastrointestinal distress to hallucination, delirium, coma, and death. Needless to say, such a variety of responses has held a measure of fascination for the "mushroom hunter" equal to or surpassing the "hunter" of other forms of dangerous game. Despite these inherent dangers, the quest of a determined mycophagist or the naive amateur continues in experimentation with new forms, hopefully without unpleasant sequelae.

The author is a devoted mycophagist and has practiced this study for some years. One learns, for example, that *Hydnum coralloides,* found in the fall, is sweet and has a definite shrimplike flavor when sautéed. *Collybia radicata* has a chickenlike quality but may give the impression of a lingering pepperlike aftertaste, and *Clavaria aurea,* slightly fried in butter, is nutlike and superb as an appetizer or dessert. The gourmet discerns that white truffles are more delicate in flavor than the black, and in spring the advent of morels is to be savored with much joy. On the other hand, folklore and scientific investigation delineate certain species as forbidden fruit. There is no way to determine which is an edible and which is a poisonous species except by careful and expert identification. Peeling of the cap (pileus) to avoid possible toxins, turning dark silver when the fungus is placed in water, or changing color when the mushroom is bruised do not constitute accurate delineation of a "poisonous" from a "safe" mushroom. Even the most experienced mycophagist must be cautious in picking the prey. Many books are available on the subject, but it must be appreciated that there are thousands of species of mushrooms, many difficult to identify and quite variable in morphologic characteristics, and that idiosyncrasies of the consumer may determine the pleasure or distress of the feast.[3, 13a, 15] A "toadstool" is the vernacular for a toxic mushroom. This has no basis on scientific grounds. In the final analysis, a toadstool is a mushroom on which a toad is sitting.

Mushroom poisoning is not as common among inhabitants of the United States as it is among those of eastern Europe. In the latter area, the gathering and cooking of fungi is a centuries-old practice. Even though tradition and folklore have determined what are good mushrooms and what are "toadstools" (poisonous forms), the highest rate of mycetismus is still among peoples who regularly collect and consume known "safe" species. Each year there are many cases of poisoning and death, especially among children. In the western hemisphere, the popularity of mushroom gathering is not so great, so that there is less experience in these forms of toxicity. However, there are a significant number of cases each year, so that attention and interest should be maintained. In Canada there are approximately 150 cases per year of mycetismus, with the most occurring in Ontario (75 per cent).[13] Similar to the situation in Europe, the majority (<80 per cent) are in children. In the United States the rate of cases of intoxication is low but constant from year to year. Inexperience in identification is the main reason for the hazards, and thankfully most cases are not serious.

Probably the first report of fatal mycetismus in the United States was recorded in 1838. This is noted on a tombstone in Trinity Episcopal Church, Fishkill, New York. The inscription reads: William Gould, wife, and son were "poisoned by eating fungi (toadstools)." Since that time numbers of small, often family-centered, epidemics have occurred.[2, 5, 8] There have been many attempts to classify the particular distresses associated with the various fungi involved. The one proposed by Ford in 1923[4] and modified first by Alder and later by Tyler[16] in 1963 appears to be the most useful. Ford divided the effects of toxic mushroom ingestion into five categories based on the primary organ system involved or the most important set of symptoms. There are (1) *mycetismus gastrointestinalis,* (2) *mycetismus choleriformis,* (3) *mycetismus nervosus,* (4) *mycetismus sanguinareus,* and (5) *mycetismus cerebralis.* Tyler classified the effects based on pharmacology as: protoplasmic poisons, neurologic effects, gastrointestinal irritants, and disulfiram-like compounds. In this section the two groupings will be integrated and expanded.

Mycetismus Gastrointestinalis

This is usually a mild to severe reaction to the ingestion of fleshy fungi. The symptoms occur within a few minutes to hours after ingestion and usually disappear uneventfully within 36 to 72 hours. There is usually nausea, vomiting, and diarrhea. The symptoms vary greatly in severity. There may be mild discomfort or violent retching and diarrhea accompanied by cramps. The water loss may be great and cause subsequent electrolyte imbalance. Sometimes there are cerebral manifestations, with dizziness and slight hallucination. These are not nearly so prevalent as is seen in other types of mycetismus. Treatment is usually not necessary, but gastric lavage may be required.

Species that produce mild mycetismus include *Russula emetica, Boletus satanas, Boletus miniato-olivacius, Lactarius torminosus, Entoloma lividum,* and *Lepiota morgani.* Individuals may have idiosyncratic reactions to several species that do not affect others. Some species taste bad, such as *Lactarius vellereus.* This species contains two sesquiterpenes ($C_{15}H_{20}O_2$), which are responsible for the disagreeable peppery flavor.[12] Other species may have a pleasant flavor but leave a disagreeable after taste. There are a few species that produce moderate to severe distress, which nevertheless resolves without serious sequelae. Sometimes fatalities have occurred, however, especially among children and the debilitated, when the nausea, vomiting, and diarrhea are quite severe. Normally the patients recover suddenly about 72 hours after ingestion. This is the case in the not uncommon "jack-o-lantern" (*Clitocybe illudens,* now called *Omphalalotus olearius*) poisoning of children (Fig. 24–1,*C* FS C 50,*C*). Other species that may cause this type of poisoning include *Lactarius torminosus, Rhodophyllus sinuatus,* and *Tricholoma pardinum.*

Figure 24–1. FS C 50. *A, Amanita virosa* showing white volva and annulus. The cap is off-white and without scales. *B, A. muscaria* showing white gills, volva, and annulus. This is the yellow-capped variety. Note the white scales on cap. *C, Clitocybe illudens.* The jack-o'-lantern. This brilliant orange mushroom glows in the dark. *D, Inocybe fastigiata.* The gills are purplish brown. The cap is smooth, brownish, and split.

Mycetismus Choleriformis

This is a severe and frequently fatal illness. The chief characteristic of this form of disease is the long symptomless period following ingestion. The latent period may extend from 6 to 24 hours. The symptoms begin suddenly and are characterized by violent vomiting, continuous diarrhea, dehydration, and muscle cramps. The fluid loss often amounts to 2 to 3 liters per day. There is also a feeble rapid pulse. No sensory loss is noted at this stage, but the patient may be listless and apathetic. The poisoning is distinctly diphasic. Symptomatic treatment of the distress results in remission; however, there is often a fatal relapse. The terminal symptoms include a sudden onset of confusion, coma, sweating, miosis, muscular twitching, lacrimation, and salivation. This is due to the action of the chief toxins, amanitine and muscarine, on smooth muscles and exocrine glands innervated by the cholinergic nerves. The poison has a cholinergic or more specifically a parasympathomimetic action. At this point large doses of atropine, often as much as 0.02 mg per kg, are used in treatment.[2] It is noted that patients with muscarine poisoning tolerate doses that would ordinarily be toxic. Renal and hepatic symptoms appear and may result in hepatic coma. The death rate as noted in several series varies from 35 to 90 per cent.[2, 5, 16] At autopsy, the patient is usually severely jaundiced as a result of an acute hepatic necrosis. Fat deposition in the liver is prominent, and the stroma of the enlarged organ may almost be completely replaced by adipose tissue. There is also fatty infiltration of myocardium, kidney, and central nervous system, which gives rise to various clinical signs as terminal events.

The principal species responsible for this form of poisoning are *Amanita phalloides* (death angel) and *A. virosa* (destroying angel) (Fig. 24–1,*A* FS C 50,*A*). Other members of the genus contain varying amounts of the amanitine toxins.

Muscarine poisoning may also produce some of the above symptoms. This compound is found in *Amanita muscaria, A. pantherina,* and some *Inocybe* species. It is also reported to be found in small amounts in members of the genera *Clitocybe, Russula,* and *Boletus.* These latter species are seldom involved in fatal poisoning unless large amounts are consumed.

Mycetismus Nervosus

The more common symptoms of muscarine poisoning are evinced by this disorder than by mycetismus choleriformis. It involves both the parasympathetic nervous system and the gastrointestinal tract. The symptoms appear early, about one to two hours after ingestion. There is violent gastrointestinal involvement, profuse sweating, convulsions, contracted pupils, and salivation. Depending on the dose of the toxin, the patient develops delirium, hallucination, or coma. Death may result if sufficient toxin is present and is due to cardiac and respiratory failure. An acutely dilated heart is seen at autopsy in addition to necrosis of various organs. Atropine is also used in the treatment of this form of toxicosis. In addition to *A. muscaria* (Fig. 24–1,*B* FS C 50,*B*) and *A. pantherina, Inocybe infelix* and *I. fastigiata* (Fig. 24–1,*D* FS C 50,*D*) have caused fatal mycetismus nervosus. These latter species contain higher quantities of muscarine than do most of the *Amanita* species.

Mycetismus Sanguinarius

This is the special category for *Helvella (Gyromitra)* poisoning. The fungus responsible, *Helvella (Gyromitra) esculata,* resembles the highly prized *Morchella* species (morels). *H. esculata* contains several heat-stable and heat-labile substances. The so called "helvellic acid" does not exist as such, and the primary agent has been found to be gyromitrin.[11] This compound, monomethyl hydrazine ($C_4H_8N_2O$), has a chocolate-like odor and is volatile. Poisoning may occur from inhalation of the fumes during cooking.

The symptoms occur six to eight hours after ingestion of the raw or undercooked fungus. There is hemoglobinuria and transient jaundice, in addition to milder symptoms of amanitine poisoning. Hemolysis may be massive. In Europe the death rate is 2 to 4 per cent of patients ingesting the mushroom, but fatalities in America are very rare.

Mycetismus Cerebralis

This is the syndrome that results from the ingestion of mushrooms containing the hallucinogen, psilocybin. The symptoms begin within 30 to 60 minutes of ingestion.

They resemble the psychomimetic symptoms produced by the ergot-derived compound, lysergic acid diethylamide (LSD). At first they consist of anxiety and difficulty of concentration. There are changes in sensory perception, with distortion of tactile sensations. Visual acuity is also distorted. Perception of color, depth, size, and shape are altered, and "kaleidoscopic changes" in observed objects are commonly experienced. Psychic elevation may alternate with depression, and hallucinations occur. A fascinating account of such a "trip" is related by Stein in Singer et al.[14] Symptoms resolve in five to ten hours, and the patient is generally back to normal by 24 hours. A psychic "hangover" may last for a few days or longer however. The sacred mushroom of the Mexican Indians, *Psilocybe cubensis,* contains psilocybin in some quantity. Other species that contain this toxin include several *Psilocybe* species, *Paneolus sphinctrinus, P. fimicola, P. pupilionaceous, P. campanulatus,* and it is reported to be present in *Stropharia* species.

Since the publication of Ford there have been many studies of the toxic principles of these mushrooms, and new syndromes of mushroom poisoning caused by different species have been described. A few of these new syndromes are listed. *Cortinarious (Dermocybe) orellanus* Fr. is the cause of orellana syndrome. It consists of an incubation time of 2 to 17 days, with onset of high fever and renal failure. At autopsy there is acute renal necrosis. Numerous intoxications and fatalities of this type have been recorded around Warsaw and Posen.[9] *Coprinus atrementarius* was reported to have had a disulfiram-like [tetraethylthiuram disulfide (Antabuse)] effect following consumption with alcohol. This has been discredited by List and Reith.[10] The authors found no TETD or similar compounds in the fungus. The active ingredients isolated caused nausea, and these effects were not modified by consumption of alcohol. *Galerina veneta* is a common small mushroom that is extremely toxic. The symptoms are similar to amanitine poisoning. The active ingredients have not been isolated, but the fungus may contain amanitine or related toxins.

The toxins of the *Amanita* genus have been extensively studied.[19, 20] Chemically there are five toxins: phalloidine, phalloin, β- α-, and γ-amanitine. They are all cyclopeptides composed of a few amino acids with a molecular weight around 1000 (Fig. 24–2). They are heat-stable. Each contains an S atom from a cysteine residue, but it is neither in the SH or SS form.

Phalloidine is the best characterized of the toxins (Fig. 24–2). The molecular formula is $C_{35}H_{43}O_{10}N_8S \cdot 6H_2O$ and consists of cysteine, alanine, hydroxyproline, threonine, oxindolylalanine, and delta-oxyleucine. Its toxicity (LD_{50}) is 2 mg per kg in mice. Phalloin is closely related chemically to phalloidine. It was the last of the five amanities to be isolated. The only difference between phalloin and phalloidine is the lack of an O in a branched amino acid of the former. The LD_{50} is 1 mg per kg. α- and β-Amanitine also are closely related chemically. They differ by the substitution of an OH group in β-amanitine and by an NH_2 in the α form. The LD_{50} of α-amanitine is 0.1 mg per kg—one of the most potent poisons known. β-Amanitine has an LD_{50} of 0.5 mg per kg. The chemical structure of γ-amanitine is not worked out as yet. Its LD_{50} is 0.8 mg per kg. A 100-g quantity of *A. phalloides* contains 10 mg phalloidine, 8 mg of α-aminitine, 5 mg β-amanitine, and 0.5 mg γ-amanitine. "Only one little cap placed in a stew pot . . ." is a traditional quote

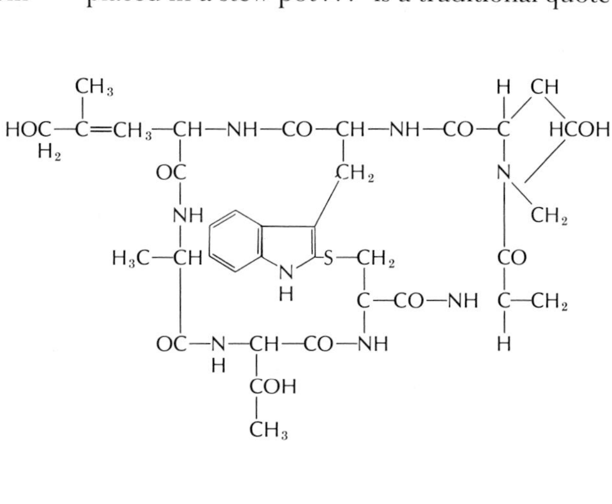

Figure 24–2. Chemical structure of phalloidine.

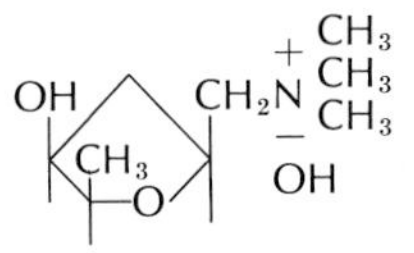

Figure 24–3. Chemical structure of muscarine.

from ancient formulas for the elimination of rivals, kinsmen, or political figures.

Pharmacologically the first effect of amanitine poisoning is hepatic glycogen mobilization. Enzymes of lipid, carbohydrate, and protein metabolism are inhibited. One target effect known is the rupture of the SS bridge within cyclopeptide-containing enzymes. Fatty degeneration of organs follows rapidly. The toxin can be detected in minute quantities using paper chromatographic techniques.[16]

Muscarine is the chief toxin responsible for mycetismus nervosus (Fig. 24–3). It is derived from *Amanita muscaria,* and small amounts are found in *Boletus, Clitocybe, Lepiota, Hebeloma,* and *Russula* species. *Inocybe* species contain the highest quantity of the compound, which may represent as much as 0.73 per cent of the dry weight. A popular account of its folklore and history, *Soma: Divine Mushroom of Immortality,* is given by Wasson.[18] Muscarine is a small molecule and has been determined to be 5-aminomethyltetrahydro-3-hydroxy-2-methylfuran. It has a potent effect on postganglionic parasympathetic effector sites.

In addition to muscarine, there are so-called "pilzatropines" present in *A. muscaria* and *Inocybe* sp. These have an anticholinergic atropine-like effect, as well as a hallucinogenic effect. They have not been isolated or purified.

Psilocybin induces mycetismus cerebralis (Fig. 24–4). There are two compounds with equal physiologic activity: psilocybin $C_{12}H_{17}N_2O_4P$ (*O*-phosphoryl-4-hydroxy-*N,N*-dimethyltryptamine) and psilocin, which is rapidly formed in the body by dephosphorylation of psilocybin. The ingestion of as little as 20 to 25 μg produces a hallucinogenic effect equal to the same amount of LSD, to which it is related structurally.

REFERENCES

1. Buck, R. W. 1969. Mycetism. New Engl. J. Med., *280*:1363.
2. Cann, H. M., and H. L. Verhulst. 1961. Mushroom poisoning. Am. J. Dis. Child., *101*:128–131.
3. Christensen, C. M. 1965. *Common Fleshy Fungi.* Minneapolis, Burgess Publishing Co.
4. Ford, W. W. 1923. A new classification of mycetismus (mushroom poisoning). Trans. Assoc. Am. Physicians, *38*:225–229.
5. Grossman, C. M., and B. Malbin. 1954. Mushroom poisoning: A review of the literature and report of 2 cases caused by a previously undescribed species. Ann. Intern. Med., *40*:249–259.
6. Heim, R. 1963. *Les Champignons Toxiques et Hallucinogenes.* Paris, Editions N. Boubee et Cie.
7. Heim, R., and R. G. Wasson. 1958. *Les Champignons Hallucinogenes du Mexique.* Paris, Mus. Natl. Histoire Natural.
8. Kempton, P. E., and V. L. Wells. 1969. Mushroom poisoning in Alaska: Helvella. Alaska Med., *10*:24–32
9. Kuschinsky, G. (ed.). 1970. *Taschenbuch der modernie Arzneimittle behandlung.* 5th ed. Stuttgart, Georg Thieme Verlag, p. 664.
10. List, P. H., and H. Reith. 1960. Der faltentintling *Coprinus atramentarius* Bull., und seine dem tetraäthylthiuram disulfid ähnliche Wirkung. Arzneimittelforsch, *10*:34–40.
11. List, P. H., and P. Luft. Gyromitrin das Gift der Früjahrslorchel *Helvela (Gyromitra) esculata* Pers. ex Fr. Z. Pilzkunde, *34*:1–8.
12. List, P. H., and H. Hackenberg. 1969. Velleral und iso Velleral, scharf schmeckende Stoffe aus *Lacterius vellereus* Fries. Arch. Pharm. (Weinheim), *302*:125–143.
13. Lough, J., and D. G. Kinnear. 1970. Mushroom poisoning in Canada: A report of a fatal case. Can. Med. Assoc. J., *102*:858–860.

13a. Miller, O. K. 1972. *Mushrooms of North America.* New York, E. P. Dutton & Co.

14. Singer, R. 1958. Observations on agarics causing cerebral mycetismus. Mycopathologia, *9*:261–284.

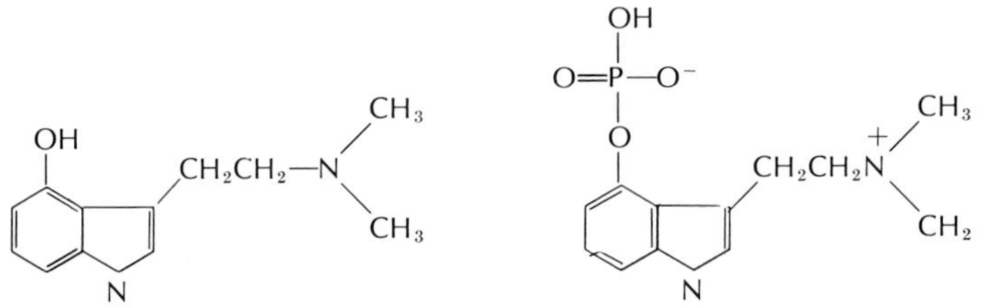

Figure 24–4. Chemical structure of psilocin (a) and psilocybin (b).

15. Smith, A. H. 1958. *The Mushroom Hunter's Field Guide.* Ann Arbor, University of Michigan Press.
16. Tyler, V. E. 1963. Poisonous mushrooms. Progr. Chem. Toxicology, *1*:339–384.
17. Wasson, R. G. 1961. The hallucinogenic fungi of Mexico. Bot. Mus. Leaflets (Harvard Univ.), *19*:137–162.
18. Wasson, R. G. 1971. *Soma: Divine Mushroom of Immortality.* New York, Harcourt Brace Jovanovich.
19. Wieland, T., and O. Wieland. 1959. Chemistry and toxicology of *Amanita phalloides.* Pharmacol. Rev., *11*:87–107.
20. Wieland, T. 1968. Poisonous principles of mushrooms of the genus *Amanita.* Science, *159*:946–952.

Mycotoxicosis

Mycotoxicosis is an intoxication due to the ingestion of preformed substances produced by the action of certain molds on particular foodstuffs. It is analogous to *Clostridium botulinum* or staphylococcal food poisoning in which the organism has produced the offending agent in the food before consumption. In the case of mycotoxins the evidence indicates that the toxins are produced only in particular environments, on particular substrates, under certain conditions, and only by particular strains of the species involved. Examples include the following:

1. The common soil contaminant *Cladosporium cladosporoides,* growing on tall fescue (grass), produces a skin erythema–causing toxin which is not formed when the fungus grows on other substrates.[18]
2. *Claviceps purpurea* elaborates a great variety of toxins (ergot alkaloids) while growing on the fruiting heads of rye, but few or none when growing on other substrates.[4]
3. Only a few strains of *Fusarium moniliforme* or other molds produce the toxin responsible for human and bovine toxic alimentary aleukia, and these only under conditions of cold.[16]
4. Some strains of *Aspergillus flavus* produce the hepatoma-inducing toxin, aflatoxin, when growing on peanuts or wheat, but not during the formation of fermented soya sauce from soy beans.[14] This indicates that particular chemical moieties must be present in the substrate in order for the particular toxic metabolites to be formed.

Throughout the ages epidemics of poisoning due to the ingestion of contaminated foods have been recorded. Probably the most famous of these were the waves of ergot poisoning that occurred during the middle ages. The symptoms of this toxicosis—hallucination, black leg (due to vascular necrosis), and other distresses—were well known and the etiology established long before the science of mycology was founded.[4] Picking only the rye heads free of diseased grain served to prevent the disease. After this, interest in mycotoxins diminished, and little was made of such diseases up to and including the first half of the twentieth century. Ergot is now grown commercially for extraction of pharmacologically active alkaloids. Some of these are used to control bleeding, induce uterine contractions, and treat some vascular diseases. Mycotoxicosis, in general, remained of some interest to veterinary medicine due to the economic loss which occurred when domestic animals ate certain moldy forage.

All this was changed in 1960 with the investigations surrounding turkey X disease.[3,19] The symptoms in turkey poults, ducklings, calves, and pigs were loss of weight, followed by ataxia, convulsions, and death, and the animals often showed signs of jaundice. Hepatic necrosis, fibrosis, and sometimes neoplasia were found in affected animals, particularly in ducklings. The common factor was found to be Brazilian ground nut (peanut meal) in the feed. Further investigation revealed that a fluorescent toxic substance was present in the peanuts and was attributable to growth on the nuts of *Aspergillus flavus.* A large amount of investigation has occurred since, and it has been realized that mycotoxins are present in many foodstuffs contaminated by a number of different organisms and can cause a wide variety of diseases.[1,18a,30]

Aflatoxin was isolated from grains infected by *Aspergillus flavus* and by experimentation defined as the cause of the strange manifestations of turkey X disease. Later investigations indicated that there were a series of com-

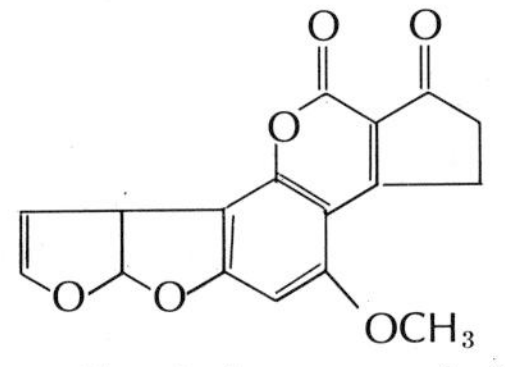

Figure 24–5. Chemical structure of aflatoxin B.

pounds with similar activity. These were designated aflatoxin B_1, B_2, $B_{2\alpha}$, G_1, G_2, and $G_{2\alpha}$. The best described is aflatoxin B, which has a molecular weight of 312 and a formula of $C_{17}H_{12}O_6$ (Fig. 24–5). It has a highly unsaturated structure and is chemically related to the coumarins. In quantities of 0.3 ppm it regularly produces hepatomas and hepatic necrosis in ducklings, trout, and other animals. After a simple procedure for its determination had been devised,[6, 16a] it was found to contaminate many human foods. In one example it was found in grain used in the making of beer in a Chicago brewery. Since the entire supply of its grain was affected, the company went bankrupt. Aflatoxin has been found in the ground nuts (peanuts) used as a major staple of food of some African tribes and has been associated with the high rate of liver and renal disease (including cancer) of these people.[17] It was determined, however, that stains used for the production of koji and shoyu in the Orient do not produce the active ingredients.[14] At present, such products as peanut butter, beer, stored grains, nuts, meal, and animal feed can be monitored to prevent intoxication by aflatoxin. The toxin is produced by *Aspergillus flavus, A. parasitius,* and *Penicillium puberulum.* Interestingly, cancer production in animals given aflatoxin is prevented by hypophysectomy.[10]

Toxic alimentary aleukia in animals is a well-established syndrome of forage herds, particularly in Russia. During World War II, conditions necessitated the storage of grain for human consumption in the open field during the winter. Thereafter there were epidemics of disease and death among the population that consumed the grain. The disease was characterized by necrotic rashes on the skin, leukopenia, agranulocytosis, necrotic and hemorrhagic diathesis, vertigo, and ulcerative and gangrenous lesions of the pharynx that led to aphasia and death. It was found that particular strains of *Fusarium sporotrichoides, Cladosporium, Alternaria, Mucor,* and *Penicillium* produced toxins responsible for the condition. When stored under normal conditions, the infestation of grain and production of toxin was not observed. Three compounds with toxic properties have been isolated and identified. They are fusariogenin, epicladosporic acid, and fagicladosporic acid. Another fungus, *Fusarium roseum,* produces a mildly intoxicating substance when growing on grain used for flour. When bread or other staples are made from it, a syndrome known as "drunken bread eater" is produced.

One recent epidemic of atypical interstitial pneumonia of cattle was found to be the result of feeding on moldy sweet potatoes. Several toxins termed "ipomeanols" were found to be responsible.[8, 28] They were produced by the sweet potato in response to infection by *Fusarium javanicum.* After extensive investigation into the problem, Wilson et al.[28] found that there were several toxins elaborated. Two were the hepatotoxins, ipomeamarone (Fig. 24–6) and ipomeamaronal, and a third, referred to as "lung oedema (LO) factor," was shown to be structurally related to others and called 4-ipomeanol. It has an empirical formula of $C_9H_{12}O_3$ and a molecular weight of 168. A dose of 1 mg IV in mice causes death in five to eight hours owing to severe pulmonary interstitial edema, but it is without effect on the liver.

Fusarium moniliforme when growing on corn or other grain produces a potent toxin. This toxin causes a disease called leukoencephalomalacia in donkeys and horses that consume the infected grain. The affected animals show increasing ataxia, and at autopsy large areas of necrosis in the brain are seen. This disease has been known since the nineteenth century.[27] Other *Fusarium* species produce a variety of toxins when growing on fodder, forage, or grain.

Stachybotryotoxicosis has been extensively studied in the Soviet Union. The disease syndrome was first described in 1931. It was known as MZ (*massavie zabolivanie,* massive

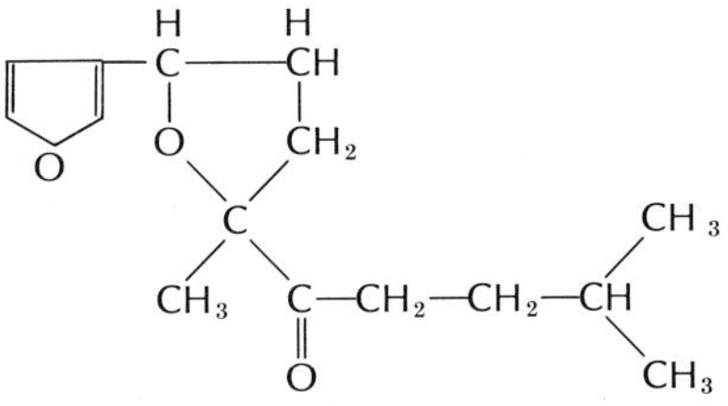

Figure 24–6. Ipomeamarone.

illness), because it was enzootic in that country, causing the death of thousands of horses, cattle, sheep and swine. The first manifestations are epithelial desquamation and swelling of lymph nodes. A leukopenia develops, and death follows within five days of onset of symptoms. At autopsy profuse hemorrhage is found, along with necrosis of many organs. The organism responsible is *Stachybotrys alterans (altra)* which grows on hay and forage.[8a, 9, 29] Farm workers inhaling the toxin or absorbing it percutaneously have been affected to a milder degree. They demonstrate cutaneous rashes, pharyngitis, and leukopenia.

A rather strange disease, first described in New Zealand, is called facial eczema. It appears in sheep and cattle who eat forage infected with *Pithomyces chartarum.* When growing on grass, this organism elaborates a powerful toxin called "sporidesmin," or "sporidesmolides," which has the empirical formula of $C_{33}H_{58}N_4O_8$ (Fig. 24–7). The toxin causes liver damage characterized by acute cholangitis, which in time evolves into fibrotic obstruction of the involved portals. In addition, vasculitis and lymphangitis occur. Severely affected animals also become photosensitized and develop a severe sloughing eczema on exposed skin sites. The face and nose desquamate in sheep, as does the udder in cows. The photosensitization is due to retention of phyllo-erythrin, a normal breakdown product of chlorophyll usually excreted in bile. *Pithomyces* is now known to be of worldwide distribution.[20]

The endotoxins have been known since the pioneering work of Henrici.[12] As early as

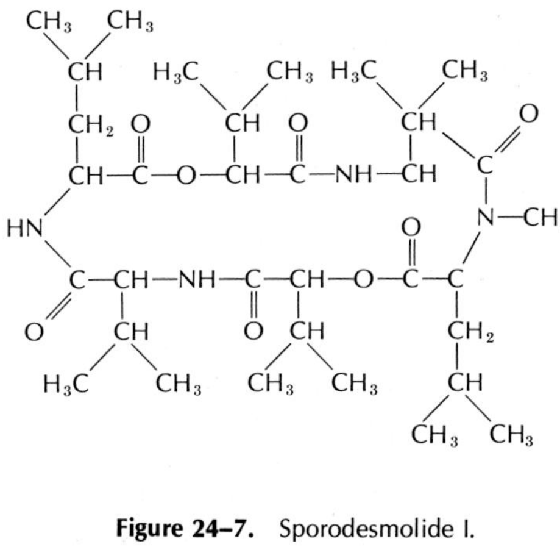

Figure 24–7. Sporodesmolide I.

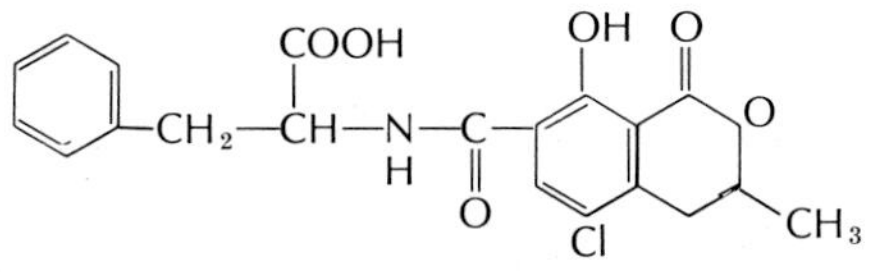

Figure 24–8. Ochratoxin A.

1902 Ceni and Besta correlated certain pellagra-like syndromes with the eating of corn infected by this fungus. Red clover infested by *Rhizoctonia leguminacata* causes excessive salivation of foraging animals, the active principle of a parasympathomimetic alkaloid,[7] but the chemical structure has not yet been elucidated.

Some other examples of mycotoxicosis include the following: *Penicillium cyclopium* produces a tremogenic toxin when growing on hay that kills cattle, horses, and sheep. The animals suffer severe convulsions and death. A chronic and eventually fatal respiratory disease of certain natives of New Guinea is associated with moldy sweet potatoes. The etiology is as yet unknown, but this may be similar to the *Fusarium javanicum* toxicosis described by Wilson et al.[27, 28]

Aspergillus ochraceus produces ochratoxins A and B when growing on feed grains (Fig. 24–8). The disease produced in chickens and other animals is similar to several other mycotoxicoses. There is suppression of hemopoiesis, acute nephrosis, hepatic necrosis, and enteritis.[23] Nephropathy and edema are evoked in swine fed corn on which *Penicillium viridicatum* has grown.[5] The edema produced in these animals is so massive that ascites, hydrothorax, and hydropericardium develop, in addition to subcutaneous edema and mesenteric edema. Affected animals die within a few hours. Lupinosis is a disease of sheep in South Africa and other countries that is characterized by hepatic necrosis. The disease was thought to be caused by eating the stems and pods of various legumes, particularly *Lupinus albus.* It has now been shown to be the result of a toxin produced by the fungus, *Phomopsis leptostromiformus,* while growing the plants.[26] *Penicillium rubrum* also produces several potent agents called rubritoxins.[11]

It is probable that in the future many more diseases of animals and man, both toxic and neoplastic, will be found to be caused by the chronic ingestion of toxins elaborated by fungi growing on articles of consumption.[21, 22, 23a]

REFERENCES

1. Allcroft, R. 1965. Aspects of aflatoxicosis in farm animals. Ref. 29, pp. 153–162.
2. Bell, D. K., and B. Doupnik. 1971. Infection of ground nut pods by isolates of *Aspergillus flavus* with different aflatoxin producing potentials. Trans. Br. Mycol. Soc., *57*:166–169.
3. Blount, W. P. 1961. Turkey x disease. Turkey (J. Br. Turkey Fed.), *9*(2):52, 55–58, 61, 77.
4. Bove, F. J. 1970. *The Story of Ergot.* Basel, S. Karger.
5. Carlton, W. W., and J. Tuite. 1970. Neuropathological edema syndrome induced in miniature swine by corn cultures of *Penicillium viridicatum.* Pathol. Vet., 7:68–80.
6. Coomes, T. J., and J. C. Sander. 1963. The detection and estimation of aflatoxin in ground nuts and ground nut materials. Part I. Paper chromatographic procedure. Analyst, *88*:209–213.
7. Crump, M. H., E. B. Smalley, et al. 1963. Mycotoxicosis in animals fed legume hay infested with *Rhizoctonia leguminocola.* J. Am. Vet. Med. Assoc., *143*:996–997.
8. Doupnik, B., O. H. Jones, et al. 1971. Toxic fusaria isolated from mouldy sweet potatoes involved in an epizootic of atypical interstitial pneumonia in cattle. Phytopathology, *6*:890.

8a. Eppley, R. M., and W. J. Bailey. 1973. 12,13-Epoxy-Δ9-trichothecenes as the probable mycotoxins responsible for stachybotryotoxicosis. Science, *181*:758–760.

9. Forgacs, J., and W. T. Carll. 1962. Mycotoxicoses. Adv. Vet. Sci., *7*:273–282.
10. Goodall, C. M., and W. H. Butler. 1969. Aflatoxin carcinogenesis: Inhibition of liver cancer induction in hypophysectomized rats. Int. J. Cancer, *4*:422–429.
11. Hayes, A. W., and B. J. Wilson. 1970. Effects of rubratoxin B on liver composition and metabolism in the mouse. Toxic. Appl. Pharmacol., *17*:481–493.
12. Henrici, A. T. 1939. An endotoxin from *Aspergillus fumigatus.* J. Immunol., *36*:319–338.
13. Herzberg, M. (ed.). 1969. Proc. of UNJR Conference on Toxic Microorganisms (Honolulu). Washington, U.S. Dept. of Agriculture.
14. Hesseltine, C. W., O. L. Shotwell, et al. 1966. Aflatoxin formation by *Aspergillus flavus.* Bacteriol. Rev., *30*:795–805.
15. Hodges, F. A., J. R. Zust, et al. 1964. Mycotoxins: Aflatoxins isolated from *Penicillium puberulum.* Science, *145*:1439.
16. Joffe, A. 1965. Toxin production in cereal fungi causing toxic alimentary aleukia in man, In Ref. 29.

16a. Jones, B. D. 1972. Methods of aflatoxin analysis. Trop. Prod. Inst. Reports No. 670, 58 pp.

17. Keen, P., and P. Martin. 1970. Is aflatoxin carcinogenic in man? The evidence in Swaziland. Trop. Geogr. Med., *23*:44–53.
18. Keyl, A. C., J. C. Lewis, et al. 1966. Toxic fungi isolated from tall fescue. Mycopathologia, *31*:327–331.

18a. Kirksey, J. W., and R. J. Cole. 1973. New toxin from *Aspergillus flavus.* Appl. Microbiol., *26*: 827–828.

19. Lancaster, M. C., F. P. Jenkins, et al. 1961. Toxicity associated with certain samples of ground nuts. Nature (Lond.), *192*:1095–1096.
20. Leach, C. M., and M. Tulloch. 1971. *Pithomyces chartarum,* a mycotoxin producing fungus, isolated from seed and fruit in Oregon. Mycologia, *63*:1086–1089.
21. Louria, D. B. 1967. Deep seated mycotic infections, allergy to fungi and mycotoxins. New Engl. J. Med., *227*:1065–1071; 1126–1134.
22. Louria, D. B., J. K. Smith, et al. 1970. Mycotoxins other than aflatoxins: Tumor-producing potential and possible relation to human disease. Ann. N.Y. Acad. Sci., *174*:583–591.
23. Peckham, J. C., B. Doupnik, et al. 1971. Acute toxicity of ochratoxins A and B in chicks. Appl. Microbiol., *21*:492–494.

23a. Purchase, I. F. H. 1971. Mycotoxins in human health. London, Macmillan.

24. Slifkin, M., and J. Spaulding. 1970. Studies of the toxicity of *Alternaria mali.* Toxic. Appl. Pharmacol., *17*:375–386.
25. Trenk, H. L., M. E. Butz, et al. 1971. Production of ochratoxins in different cereal products by *Aspergillus ochraceus.* Appl. Microbiol., *21*:1032–1035.
26. Van Warmelo, W. F. O. Marasas, et al. 1970. Experimental evidence that lupinosis of sheep is a mycotoxicosis caused by the fungus *Phomopsis leptostromiformis* (Kuhn) Bubak. J. S. Afr. Vet. Med. Assoc., *41*:235–247.
27. Wilson, B. J., and P. R. Manonpot. 1971. Causative fungus agent of leucoencephalomalacia in equine animals. Vet. Rec., *88*:484–486.
28. Wilson, B. J., M. R. Boyd, et al. 1971. A lung oedema factor from mouldy sweet potatoes (*Ipomoea batatas*). Nature (Lond.), *231*:52–53.
29. Wogan, G. N. 1965. *Mycotoxins in Foodstuffs.* Cambridge, MIT Press.
30. Wogan, G. N., and R. S. Pony. 1970. Aflatoxins. Ann. N.Y. Acad. Sci., *174*:623–635.
31. Wright, D. E. 1968. Toxins produced by fungi. Ann. Rev. Microbiol., *22*:269–282.

Allergic Diseases

Hypersensitivity to the inhalation of fungal spores or fungal products manifests itself in many ways. In certain atopic individuals, the sensitivity presents as bronchial asthma, and the allergens are usually the airborne spores of common soil fungi. Sensitivity may also develop in normal individuals who are chronically exposed to the spores of certain fungi in the course of their occupation. These include farmer's lung, teapicker's disease, maple bark stripper's disease, bagassosis, and so forth. In the first group, asthma, the disease is due to an idiosyncracy of immunologic response, and a long list of fungal spores have been involved along with dust, animal dander, pollens, and so forth. In the second group, the allergies are produced in normal people, and the disease is a consequence of chronic contact with quantities of specific spores that are particularly allergenic. A few of the well-documented allergic diseases will be briefly discussed.

Asthma[6a]

Fungal spores have long been incriminated as a common allergen in provoking bronchial asthma. *Alternaria, Helminthosporium, Penicillium,* and *Aspergillus* spores are most frequently involved. There is a definite seasonal incidence, with a high count of spores in the fall and a lesser peak in the spring. This corresponds to the growth peaks of these fungi.[6,19]

There are two general types of asthma. The childhood form reaches a peak in early adolescence and gradually subsides, and the second form begins after forty and may become more severe with increasing age. The latter runs a more rapid and fatal course if unremitting asthma (status asthmaticus) develops. The patients have reaginic antibodies (IgE) to the offending substances. On challenge, the antigen-antibody complex causes activation of mast cells, release of histamine, and there is subsequent bronchospasm and formation of mucus plugs in bronchioles. A delayed Arthus-like reaction (IgG-mediated) may complicate the disease. In fatal cases, the lungs are overaerated and show extensive bronchial plugging; however, there is little evidence of emphysematous changes. Both large and small bronchi are thickened and filled with semisolid yellowish plugs of mucus. The mucus contains large numbers of layered eosinophils, Charcot-Leyden crystals, and clusters of epithelial cells (the epithelial *zellballen* described by Vierordt in 1883). Histologically the bronchial mucus glands are thickened and in greater than normal numbers. The lumen of the glands contains the thick mucinous material. Edema of the epithelium is pronounced, and areas denuded of epithelium are present. The basement membrane is thickened and is sometimes ruptured, thus allowing extravasations of underlying tissue into the bronchial lumen. There is a large invasion of eosinophils and some neutrophils. The bronchi as well as bronchioles down to 0.2 cm are involved and often completely obstructed. A proteinaceous edema fluid is present in many alveolar spaces. Fungal spores (unsprouted) are sometimes observed in the plugs. Patients with bronchial asthma do not have precipitating antibodies in their sera. They have an immediate wheal and flare to the intracutaneous injection of the allergen, but do not have a delayed reaction. A delayed reaction in addition to the immediate reaction indicates the asthma is complicated by IgG-mediated Arthus-like reactions in the lung as well. This occurs in bronchopulmonary aspergillosis (see Chapter 19). In this disease there are sprouted spores and mycelium in the bronchial lumen, and the patients have IgG antibodies as well as IgE.

Farmer's Lung

This is an acute or chronic sometimes fatal disease of persons continually exposed to moldy hay or similar material.[7,20,21] It is a manifestation of sensitization to several allergens, principally the spores of the actinomycetes *Micromonospora vulgaris, Micromonospora (Micropolyspora) faeni,* and *Thermopolyspora polyspora.* Several fungi also cause the identical syndrome.[11] *Aspergillus,* especially *A. niger, A. fumigatus,* and *A. flavus,* and *Penicillium,* most commonly *P. simplicissimum, P. herquei, P. rubrum, P. italicum,* and *P. caseicolum,* are frequently involved. In particular areas or in particular occupations, other genera and species may be the predominant allergens.

The syndrome in which farm workers showed acute symptoms and chronic lung changes was first delineated in Canada by Cadham in 1924.[3] Subsequently Campbell[4] described the "farmer's lung" in England, and it has now been found to be a frequent and serious occupational disease of worldwide distribution.[7,20] It occurs more often in the temperate than tropical zones, and is more frequent in areas of high rainfall where the conditions for mold growth on hay are optimum. Office workers have suffered the same disease owing to the growth of the organism in air conditioning vents.[1]

Three stages of clinical disease are described in the several series on farmer's lung.[16] They are (1) the acute stage, found most frequently in harvesters and thrashers, in whom there is an overwhelming initial exposure to the spores; (2) the subacute stage, found sometimes in harvesters and animal keepers who have long-spaced exposure periods, but this form occurs more frequently in silo and grain mill workers; and (3) the chronic form, found in silo and mill workers who have low-level but constant exposure to the allergens. A similar disease occurs in horses (the "heaves"), in cows ("fog fever"), and among bird fanciers.[8]

On x-ray examination, the acute stage shows diffuse mottling of the middle and lower lobes. The patients have chills, fever, and general malaise. This form usually lasts only a few days and may leave no residual, or only slight fibrotic, changes. The subacute or subchronic form shows areas of infiltrated foci which are variable in size. Small tubercle-like granulomas are sometimes seen, and there is segmental atelectasis. The condition may resolve without major consequence if exposure challenge is prevented.

Airway obstruction is the major feature of chronic farmer's lung, but gas exchange studies are usually normal. The patients suffer from acute dyspnea following exertion. They have livid lips, shallow breathing, chronic cough, and some weakness. Radiographically fine to gross fibrotic changes and interstitial infiltrates are seen, and sometimes scattered miliary nodules (Fig. 24–9). Chronic disease is debilitating in varying degrees and often progressive to a fatal terminus.

The histopathology of the disease has been described by several investigators.[21,22] In the acute phase there is an interstitial edema and some lymphocytic infiltration. This may resolve but leave some intralobular fibrosis. In progressive forms, granulomatous lesions develop. These resemble sarcoid. The lesions are in the parenchyma near the bronchioles, in the septa, or in subpleural areas. These contain histocytes which develop into layers of epithelioid cells, which are in turn surrounded by a ring of lymphocytes and plasma cells. Large foreign-body giant cells are interspersed in the lesions. These may contain nonbirefringent fibrous material. There are also bronchiolitis and vasculitis.[21] As the disease becomes chronic, more severe and permanent changes occur. The granulomas may calcify and sometimes enlarge. Pulmonary fibrosis becomes massive; there are cystic changes and changes associated with pulmonary hypertension.

Whereas asthma was associated with an immediate type 1 reaction mediated by IgE,

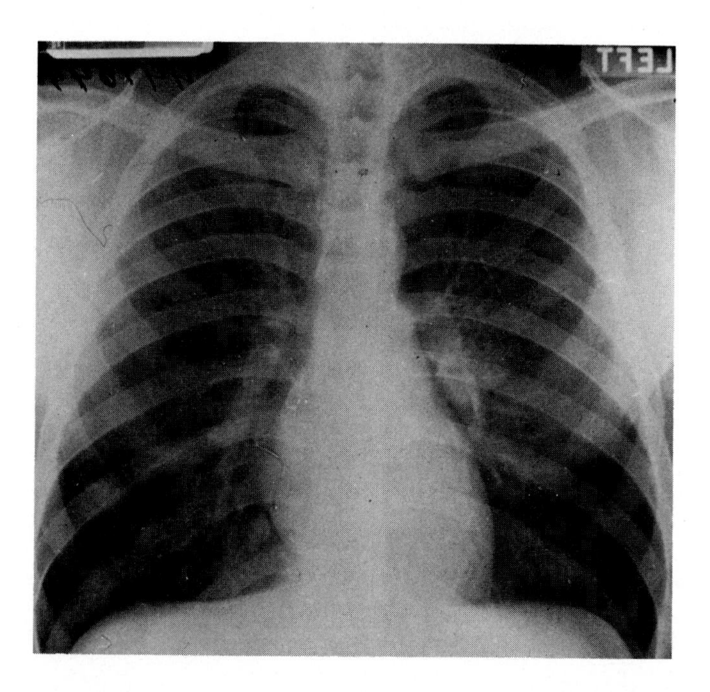

Figure 24–9. Farmer's lung of several years duration. There are diffuse, reticulated, interstitial infiltrates in the bases of both lungs. (Courtesy of N. Gross.)

the changes in farmer's lung are compatible with an Arthus (type III) reaction, and IgG precipitating antibodies have been demonstrated in sera. Later in the disease, there are numerous epithelioid tubercles in the lung, indicating that a delayed hypersensitivity (type IV) reaction of the tuberculin type is also a part of the syndrome. In addition, the presence of foreign-body giant cells within the granulomas suggests a nonallergic foreign-body, inflammatory component to the disease. It can be appreciated, therefore, that farmer's lung is a complicated disease syndrome involving several allergic and nonallergic components.

A serologic procedure for diagnosis has been devised by Kobayashi et al.[15] and Pepys and Jenkins.[20] This is an immunodiffusion test in which patients' sera is placed in one well, and either a decoction of moldy hay or specific actinomycete antigen is placed in the other. Precipitin lines appear if specific antibody is present. Jameson[13] has described a rapid immuno-osmophoresis technique to detect the same antibodies. Positive tests for the detection of precipitins have been found to correlate well with clinical disease, and this test together with a positive delayed reaction to the skin test is considered sufficient for diagnosis of disease.[5a, 26]

The organisms responsible for the disease are thermophilic fungi and actinomycetes that grow on rotting hay or other material. The hay does not appear to contribute to the pathology. The hay flora thrives at temperatures of 40 to 60°C. The usual etiology is a mixture of *Thermopolyspora*, *Micromonospora*, (*Micropolyspora*) and other actinomycetes and various fungi. Exposure to *Aspergillus fumigatus* alone has caused acute fatal disease.[10]

Maple Bark Stripper's Disease (Coniosporosis or Cryptostromosis)

This disease was first described by Towey in 1932.[25] It occurred in 35 bark peelers in a lumbering community in northern Michigan. The logs had been stored over the winter. The symptoms included cough, dyspnea, fever, night sweats, and substernal pain. The disease resolved uneventfully. More recently it has been described in Wisconsin and other places.[5, 24] In the series by Emmanuel, lung biopsy showed small cream-colored granulomas. Microscopically there were chronic granulomatous changes that consisted of histiocytes, fibroblast, giant cells, lymphocytes, and a few neutrophils. Small spores 4 to 5 μ in diameter were sometimes present within the phagocytic cells. These were brownish red in color, and no budding was evident. Initially there was confusion with the yeast cells of *Histoplasma capsulatum*, but the spores were shown to be from the fungus *Cryptostroma corticale*, which grows on the bark of maple trees. The inhaled spores do not grow in the lung but remain dormant. When lung tissue containing spores is planted on Sabouraud's agar and incubated at 25°C, the fungus will grow. A white pleomorphic colony develops. The hyphae are septate and profusely branched. The spores are oval, 4 to 5 μ, and brownish red in color. Patients with the disease have both positive skin tests and precipitins by the immunodiffusion test.[24] An allergic pneumonitis caused by *Alternaria* sp. growing on logs has recently been described by Schlueter et al.[21a]

Scopulariopsosis

A similar disease to cryptostromosis was brought to my attention by Dr. Hans Grieble of the Veterans Administration Hospital, Hines, Illinois. The patient had been in Viet Nam, where he had injected himself over a period of a month with a brown fluid sold to him by the natives as a crude opium preparation. In time he developed severe debilitating respiratory distress and was shipped back to the United States. On chest x-ray, several nodular lesions could be seen in the lower lung fields (Fig. 24–10,*A*). One of these was removed surgically for diagnosis. Histologic examination revealed a forming granulomatous reaction (Fig. 24–10,*B*) and the presence of numerous brown bodies that were 4 to 7 μ in diameter (Fig. 24–10,*C*). The material was cultured on a variety of media. Numerous colonies of a greyish white fungus grew on all media (Fig. 24–10,*C*). The organism was identified as *Scopulariopsis brumptii*. Antibodies to an antigen prepared from the fungus were present in the patient's sera. The disease in the patient resolved uneventfully and without treatment.

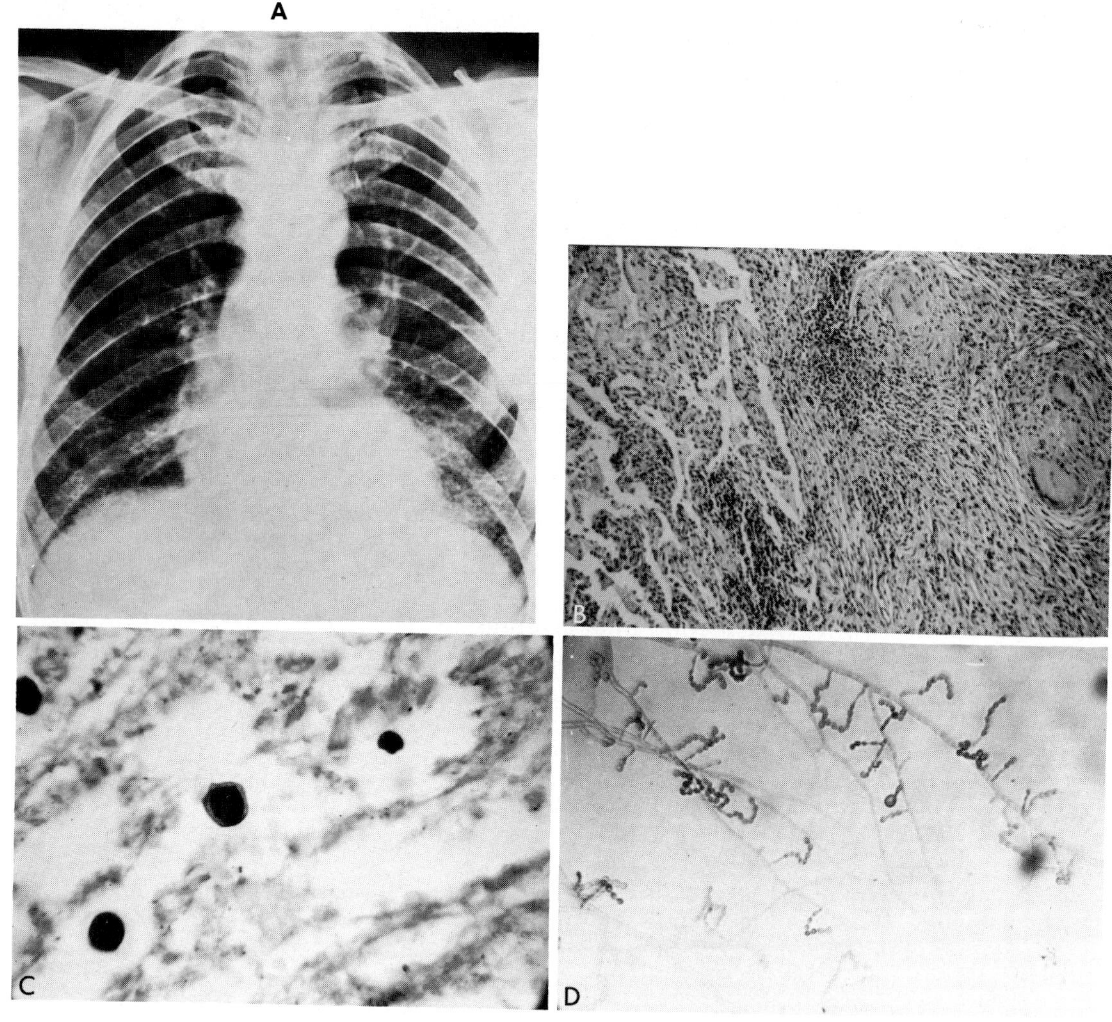

Figure 24–10. Phaeoscopulariopsosis. *A,* X-ray of chest. The lung is overinflated. There are two discrete, well-defined nodules in the lower lobe of the right lung. In the left lung there is a solitary, discrete, well-defined nodule in the superior segment of the lower lobe. (Courtesy of H. Grieble.) *B,* Well-formed granuloma with foreign body giant cells. ×100. *C,* High-power magnification showing various size spores of fungus. Note lack of budding. ×500. *D,* Culture mount of *Phaeoscopulariopsis (Scopulariopsis) brumptii* (Salvanet-Duval, 1935) isolated from lesion. ×100.

Bagassosis

This disease is the result of inhaling dust from bagasse, the material left after the juice has been expressed from sugar cane. Until 1921 the material was burned, but it is now used in making wallboard and similar products. In addition to fiber, silica, and quartz, the dust contains numerous fungal spores.[12] The fungi represent many species and genera. Actinomycetes are also present and may be involved in the disease.[18] The syndrome was first described in 1941 by Jamison and Hopkins.[14] The symptoms are cough, dyspnea, some fever, and production of blood-streaked sputum. Mottling of the lung fields is seen on x-ray examination. Histologically there is an interstitial inflammatory reaction, with giant cells and intra-alveolar foam cells present in the early stages of the disease. The giant cells are seen to contain birefringent material. As the disease progresses, interstitial fibrosis occurs, along with chronic nonspecific changes in the lung. As is seen in farmer's lung, continued exposure may result in fatal pulmonary disease.[2] Patients have precipitating antibodies in their sera.[9]

Byssinosis

This disease is similar to bagassosis but involves inhalation of dust from cotton processing plants. The disease was first described by Kay in 1831 among cotton workers, carders, and sorters in Lancashire cotton mills. Nichols et al.[17] demonstrated that the pericarp of the cotton seed pods was the active material and caused histamine release. Bronchoconstriction results, and, in cases of chronic exposure, inflammatory, fibrotic, and nonspecific changes occur. Fungal spores are found in the dust, but their contribution to this particular allergy is probably not great.

Lycoperdonosis

This disease is caused by the inhalation of the spores of the fleshy basidiomycete, *Lycoperdon* (puffball). In folk medicine this material is used to treat epistaxis. Inhalation of large quantities of spores results in an allergic pulmonary disease. The symptoms include rales, fever, and cough. Spores can be seen in sputum and will sprout if placed on culture media. Chest x-ray reveals fine miliary nodules throughout several lung fields.[23]

REFERENCES

1. Banaszak, E. F., W. H. Thiede, et al. 1970. Hypersensitivity pneumonitis due to contamination of an air conditioner. New Engl. J. Med., *283*:271–276.
2. Boonpucknavig, V., N. Bhamarapravati, et al. 1973. Bagassosis. Am. J. Clin. Pathol., *59*:461–474.
3. Cadham, F. T. 1924. Asthma due to grain dust. J.A.M.A., *83*:27.
4. Campbell, J. M. 1932. Acute symptoms following work with hay. Br. Med. J., *2*:1143–1144.
5. Emmanuel, D. A., F. J. Wenzel, et al. 1966. Pneumonitis due to *Cryptostroma corticles* (maple bark disease). New Engl. J. Med., *244*:1413–1418.

5a. Edwards, J. H. 1972. The isolation of antigens associated with farmer's lung. Clin. Exp. Immunol., *11*:341–355.

6. Frankland, A. W. 1971. Seasonal allergic rhinitis. Proc. R. Soc. Med., *64*:447–450.

6a. Gross, N. 1974. Bronchial Asthma: Current Immunologic, Pathophysiologic, and Management Concepts. New York, Harper and Row.

7. Hapke, E. J., M. E. Seal, et al. 1968. Farmer's lung. Thorax, *23*:451–468.
8. Hargreave, F. E., J. Pepys, et al. 1966. Bird breeders (fancier's) lung. Lancet, *1*:445–449.
9. Hearn, C. E. D., and V. Halford-Strevens. 1968. Immunological aspects of bagassosis. Br. J. Ind. Med., *25*:283–292.
10. Höer, P. W., L. Horbach, et al. 1964. Das Krankheitsbild der Farmer lunge und seine Beziehung zu den Pilzinfektionen. Z. Klin. Med., *158*:1–21.
11. Hořejší, M., J. Šach, et al. 1960. A syndrome resembling farmer's lung in workers inhaling spores of *Aspergillus* and Penicillia moulds. Thorax, *15*: 212–217.
12. Hunter, D., and K. M. A. Perry. 1946. Bronchiolitis resulting from handling of bagasse. Br. J. Ind. Med., *3*:64–74.
13. Jameson, J. E. 1968. Rapid and sensitive precipitin test for diagnosis of farmer's lung using immunoosmophoresis. J. Clin. Pathol., *21*:376–382.
14. Jamison, S. C., and J. Hopkins. 1941. Bagassosis: Fungus disease of the lung. New Orleans Med. Surg. J., *93*:580–582.
15. Kobayashi, M., M. A. Stahmann, et al. 1963. Antigens in mouldy hay as the cause of farmer's lung. Proc. Soc. Exp. Biol. Med., *113*:472–476.
16. Kovats, F., and B. Bufy. 1968. *Occupational Mycotic Diseases of the Lung*. Budapest, Akademiai Kiado.
17. Nicholls, P. J., G. R. Nicholls, et al. 1967. Inhaled particles and vapors. 2nd ed. Davies, C. N. (ed.), Oxford, Pergamon Press, pp. 93–100.
18. Lacey, J. 1971. *Thermoactinomyces sacchari* sp. nov. A thermophilic actinomycete causing bagassosis. J. Gen. Microbiol., *66*:327–338.
19. Ordman, D. 1970. Seasonal respiratory allergy in Windhoek: The pollen and fungus factors. S. Afr. Med. J., *44*:250–253.
20. Pepys, J., and P. A. Jenkins. 1965. Precipitin (F.L.H.) test in farmer's lung. Thorax, *20*:21–35.
21. Seal, R. M. E., E. J. Hapke, et al. 1968. The pathology of acute and chronic stages of farmer's lung. Thorax, *23*:469–489.

21a. Schlueter, D. P., J. N. Fink, et al. 1972. Woodpulp workers' disease: A hypersensitivity pneumonitis caused by *Alternaria*. Ann. Intern. Med., 77:907–914.

22. Spencer, H. 1968. *Pathology of the Lung*. Oxford, Pergamon Press, pp. 449–453.
23. Strand, R. D., E. D. Neuhauser, et al. 1967. Lycoperdonosis. New Engl. J. Med., *277*:89–91.
24. Tewksbury, D. A., F. J. Wenzel, et al. 1968. An immunological study of maple bark disease. Clin. Exp. Immunol., *3*:857–863.
25. Towey, J. W., H. C. Sweany, et al. 1932. Severe bronchial asthma apparently due to fungus spores found in maple bark. J.A.M.A., *99*:453–459.
26. Wenzel, F. J., D. A. Emmanuel, et al. 1972. A simplified hemagglutination test for farmer's lung. Am. J. Clin. Pathol., *57*:206–208.

Part Four

Taxonomy and Fungal Genetics

Chapter 25

E. D. Garber

Professor of Biology, University of Chicago

TAXONOMY

Taxonomists agree that a species is a natural unit, but the delineation of a species may be subjective and subject to change. Although taxonomists prefer to evaluate morphologic characters to determine their stability and worth in defining species, geographic, ecologic, cytologic, and, whenever possible, physiologic and biochemical criteria are also assessed to support decisions based on morphologic characters. In general, organisms with relatively few, but often plastic, morphologic characteristics present difficult taxonomic problems. The fungi are unquestionably a challenge to taxonomists.

Heterotrophy, the hyphal organization of cells, and the chemical composition of the cell walls characterize all groups of the fungi. The major dichotomy distinguishing the lower and higher fungi involves the presence or absence of a cross wall in the hyphae, but exceptions have been found even at this level. A second important dichotomy in fungal systematics concerns the presence or absence of a sexual stage. The Fungi Imperfecti lack a demonstrable sexual stage, whereas the ascomycetous and basidiomycetous fungi are distinguished by the morphology of their sexual organs and the products of meiosis. Although the perfect higher fungi present the usual challenges to fungal taxonomists, the taxonomy of the imperfect species is confused and controversial.[23] Unfortunately, many isolants and strains of animal and human fungal pathogens are imperfect. Medical mycologists, however, have not been deterred in their efforts to assign specific epithets to these imperfect isolants and strains. Consequently, such criteria as biochemical characteristics, the spectrum of susceptible hosts, and the sites and appearance of lesions in infected susceptible hosts must serve as additional criteria in delineating imperfect species. This approach to the taxonomy of the imperfect species satisfies the need to conform to Linnean protocol. "Whatever other function might be attributed to taxonomy—such, for example, as investigating phylogeny—this one, the creation of a data storage and retrieval agency, is surely the inescapable one."[20]

By 1934, the imperfect dermatophtyes, for example, were assigned to 40 genera containing 300 species. To restore some semblance of taxonomic reality, Emmons[12] used macroconidial characters to construct three genera and 13 species. The obvious need for the sexual stage to obtain additional morphologic characters for taxonomic purposes stimulated a search for appropriate inducer substrates yielding fruiting bodies for pairs of imperfect strains. In many cases, the search was successful. Species of the imperfect genus *Trichophyton* have been identified as species of *Arthroderma,* and species of the imperfect genus *Microsporum* as species of *Nannizzia,* thereby resolving several taxonomic problems in these genera. Moreover, the discovery of the sexual stage for a number of fungal pathogens provided suitable material for genetic studies in these species.

Chromosome morphology and number have been used in plant and animal taxonomy, but fungal cytology is notoriously difficult and would be an impractical means of characterizing the imperfect strains and species. In a number of genera, hyphal anastomosis does not occur between members of different species in one genus. Cytologic observations could be used to detect anastomosis. Although genetic and environmental factors may inhibit or prevent anastomosis between strains of the same species, positive evidence for anastomosis indicates a close genetic and presumably taxonomic relationship. Genetic techniques provide an alternate means of detecting anastomosis.

Heterokaryons are readily synthesized by mixing conidia or mycelial fragments from strains with different nutritional requirements in a liquid or on a solid complex medium to obtain a mycelial mat. Fragments from the washed and shredded mycelium are added to the minimal medium and incubated until hyphal growth is detected well beyond the periphery of the colony. Hyphal tips capable of growth on the minimal medium contain intermingled nuclei from each strain as the result of anastomosis of hyphae from the two strains.[33] Morphologic and color mutant strains provide visual markers to demonstrate heterokaryosis for colonies originating from conidia, mycelial fragments, or hyphal tips taken from the presumptive heterokaryotic colonies.[32] While these techniques may not assist the medical mycologist in assigning imperfect isolants or strains to the appropriate species, the experimental fungal taxonomist cannot afford to overlook any means of delineating imperfect species.

The production of fertile fruiting bodies by a pair of imperfect strains constitutes acceptable evidence that the two strains belong to the same taxon. The converse situation, however, does not indicate that the strains belong to different taxa. For heterothallic species, crosses fail when the two strains have the same mating type allele. When testor strains with a known mating type allele are available for a species, the imperfect strains are paired with each testor strain using the appropriate regime to obtain fertile fruiting bodies. This procedure has been successfully used as a taxonomic tool.

Stockdale[46] obtained fertile fruiting bodies for a number of strains of the imperfect species *Microsporum gypseum* which were assigned to the perfect species *Nannizzia incurvata*. Five strains included in the imperfect species were assumed to belong to other species of *Nannizzia* because they did not yield fertile fruiting bodies in the different combinations of crosses with testor strains of *N. incurvata*. Stockdale[47] extended this effort to include crosses involving one strain of *Gymnoascus gypseum*,[18] one strain of *Epidermophyton radiosulcatum* var. *flavum*,[49] and the five unassigned strains of *M. gypseum*. Three strains of *M. gypseum* yielded fertile gymnothecia only in matings with *G. gypseum*, and two strains gave fertile gymnothecia only in crosses with *E. radiosulcatum*. These observations were responsible for recognizing two additional species of *Nannizzia*: *G. gypseum* = *N. gypsea*, and *E. radiosulcatum* = *N. fulva*. A successful cross between one strain assigned to *M. fulvum* and the strain of *E. radiosulcatum* supported the decision to recognize the perfect species *N. fulva*. With the discovery of the sexual stage for an increasing number of dermatophytes and other fungal pathogens, medical mycologists now have a more stable and natural taxonomic treatment of these fungi.

Maintaining a collection of testor strains for different perfect species and making a series of crosses with the imperfect strains are time-consuming and tedious tasks. Biochemical systematics offered few meaningful alternatives to the conventional criteria of fungal taxonomy until electrophoresis became a flexible and routine laboratory technique. Mycelial or cellular extracts and culture filtrates can be used to obtain protein profiles and zymograms by this technique to determine whether members of the same taxon yield characteristic patterns of sites of protein or of enzymatic activity, respectively. The case for electrophoretic patterns as adjuncts to the conventional taxonomic criteria has been presented by Manwell and Kerst.[28]

> The more difficult a taxonomic problem is, the more likely that an incomplete solution will be obtained, regardless of whether the method is morphological or molecular. The advantage of using protein characteristics is that any given protein is the ultimate product of only one or two cistrons; accordingly, the protein systematist is close to the genotype and yet still concerned with the fundamental basis of phenotypes.

In assessing the available literature on the value of electrophoretic patterns for fungal taxonomy, Clare et al.[8] and Garber and Rippon[16] concluded that electrophoresis represents a valuable tool in delineating fungal species.

The procedures for preparing cellular or mycelial crude extracts for electrophoresis do not differ essentially from those used for enzyme assays. Starch or acrylamide gels and cellulose acetate strips have been used as the supporting media. Acrylamide gels are generally more versatile than starch gels, because more extract can be added to the wells in the acrylamide gel or to the surface of an acrylamide column, and the resolution of protein sites or enzyme sites is more easily controlled. The relatively small volume of extract used for cellulose acetate strip electrophoresis is less likely to yield a large number of sites. The number of different enzymes detectable in gels after electrophoresis is relatively large and increasing each year.[3,29]

To have taxonomic value, electrophoretic patterns must be reproducible and exhibit one or more invariant sites characterizing all members of one species and distinguishing different species of one genus.[31] In some cases, it may be necessary to use two different enzyme systems to accomplish this purpose. Zymograms are more extensively used than protein profiles for taxonomic purposes, but no claims for the superiority of zymograms have been reported. Such cultural regimes as the composition of the medium, the duration of growth, and the extraction procedures must be standardized to ensure reproducible patterns. Culture filtrates with sufficient protein to yield profiles or sufficient enzymatic activity to give detectable sites can also be used for electrophoresis.[30]

Shechter et al.[44] compared protein profiles for two species of *Microsporum*, four species of *Trichophyton*, and one species of *Epidermophyton*. Although different protein profiles were obtained for mycelial extracts from one strain grown in different complex media, each species had a characteristic protein profile when all the strains were grown in the same medium. Culture filtrates from these strains also yielded characteristic protein profiles for each species. Schechter (personal communication) obtained characteristic protein profiles for seven species of *Microsporum*, 12 species of *Trichophyton*, one species of *Keratinomyces*, and one species of *Epidermophyton* in a more extensive electrophoretic study of dermatophytes. Moreover, no significant intraspecific variation was noted for four to six isolants of *M. gypseum*, *M. canis*, *T. mentagrophytes*, *T. tonsurans*, *T. rubrum*, and *E. floccosum*. These observations indicate that electrophoresis has taxonomic value for the dermatophytes and should be extended to other fungal pathogens. Clare et al.[8] have reported that phytopathogenic and saprophytic species in a number of fungal genera can be characterized by their protein profiles.

Zymograms generally yield fewer sites than protein profiles, so that invariant sites are more easily detected. Unfortunately, electrophoretic surveys usually have not included reasonably large collections of strains for each species. Nevertheless, the available literature on zymograms indicates that they have taxonomic value for many imperfect and perfect species.[8,16] The growing interest in zymograms as taxonomic tools for mycologists suggests that electrophoresis should be considered in delineating imperfect pathogenic fungal species. Additional criteria, however, should be used to determine whether protein profiles and zymograms can be used not only to assign strains to the appropriate species but also to detect errors of classification or of culture transfers.

Kurzeja and Garber[24] compared amylase zymograms from 90 strains assigned to *Aspergillus nidulans*, using culture filtrates with amylolytic activity. Nine patterns accommodated all the strains (Fig. 25–1). Patterns 1 and 2 were exhibited by 36 and 45 strains, respectively, and each of the remaining seven patterns by one to two strains. Assuming that only patterns 1 and 2 characterized members of *A. nidulans*, the strains with the other patterns represent cryptic or sibling species. Alpha-esterase, beta-esterase, and phosphatase zymograms also characterized the strains with different amylase patterns. Only the cross between strains with amylase patterns 1 or 2 gave hybrid cleistothecia. Consequently, the zymograms distinguished genetically isolated taxa whose morphologic characteristics should be evaluated to provide a basis for reorganizing the taxonomy of the *A. nidulans* group.

Zymograms have been used to determine whether morphologic characters in strains of the *A. niger* group warranted the recognition of certain strains as species. Raper and Fennell[36] viewed certain morphologic characteristics of strains NRRL 2317 (*A. tubingensis*), NRRL 346 (*A. carbonarius*), and NRRL 337 and NRRL 341 (*A. foetidus*) as sufficiently distinctive to warrant assigning

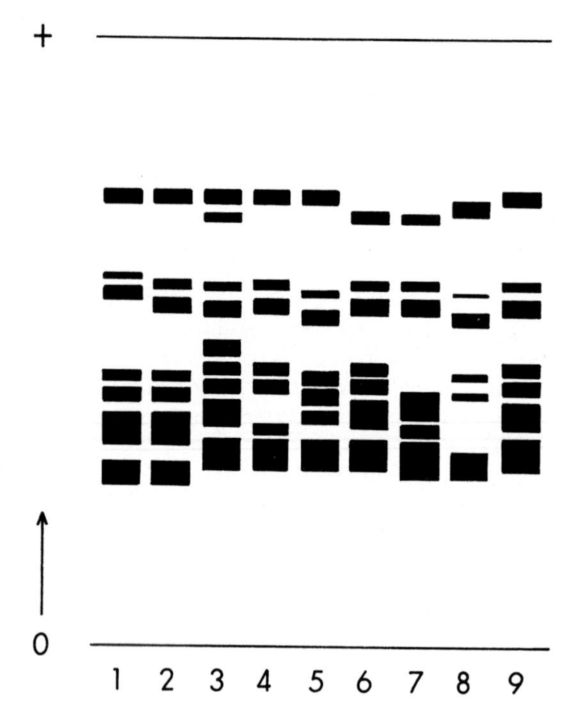

Figure 25–1. Nine patterns of sites of amylolytic activity in zymograms of culture filtrates for 90 strains assigned to *Aspergillus nidulans*. Strains with patterns 1 and 2 belong to this species, and strains with one of the remaining seven patterns are cryptic or sibling species. (From Kurzeja, K. C., and E. D. Garber. 1973. A genetic study of electrophoretically variant extracellular amylolytic enzymes of wild-type strains of *Aspergillus nidulans*. Can. J. Genet. Cytol., *15*:275–287.)

specific epithets to these strains. Amylase zymograms for these and for typical strains of *A. niger* were not significantly different to distinguish any of the tested strains.[1] Furthermore, heterokaryons or vegetative diploid strains were obtained for all combinations of these strains.[19] While heterokaryosis and diploidization are not orthodox taxonomic criteria, they do merit consideration in defining taxonomic relationships. An infraspecific category such as subspecies might be sufficient to emphasize certain morphologic characteristics distinguishing the four strains of *A. niger*.

When strains assigned to the same species have been demonstrated to yield the same characteristic protein profile or zymogram, it may be possible to explain an atypical profile or zymogram for one or more strains by assuming an error in classification or in the transfer of cultures. Nealson and Garber[31] obtained atypical zymograms for one strain assigned to *A. rugulosus* which did not yield fertile cleistothecia in crosses with genetically marked typical strains of this species. Reddy and Threlkeld[37] observed atypical zymograms for one strain assigned to *Neurospora sitophila* which were typical, however, for strains of *N. crassa*. Appropriate crosses indicated that the atypical strain did not yield hybrid perithecia in a cross with a typical strain of *N. sitophila* and gave fertile perithecia in a cross with a typical strain of *N. crassa*. These observations suggest that an atlas of zymograms for species in one genus might aid the fungal taxonomist in assigning isolants or strains to the appropriate species when distinguishing morphologic characters are subtle.

Although fungal taxonomists will continue to emphasize morphologic and, to some extent, ecologic characteristics in defining species, other criteria will have to be considered in handling imperfect isolants or strains of fungal pathogens. While heterokaryosis, the production of fertile fruiting bodies by appropriate crosses with testor strains, and electrophoretic patterns may be impractical for the diagnostic purposes of medical mycologists, the experimental taxonomist may profit by applying these currently unorthodox criteria in constructing definitive and natural taxonomic treatments of fungal pathogens and recognizing species, particularly of the imperfect pathogens.

GENETICS

The extraordinary diversity of the fungi bewildered the early mycologists seeking stable morphologic characters to distinguish species and to assign species to genera. This problem required detailed study of the vegetative and reproductive structures and cells, as well as the different modes of sexual reproduction. Eventually, fungal taxonomists appreciated the value and significance of the life cycles in establishing relationships among the fungi. Although the classification and systematics of the fungi are not yet satisfactorily resolved, mycologists have accepted four major groupings: Phycomycetes, Ascomycetes, Basidiomycetes, and Fungi Imperfecti. The imperfect fungi occupy a tenuous position because the lack of a demonstrable sexual stage is a significant factor in maintaining the nomenclatural convenience of Fungi Imperfecti. The tactics and strategy for a genetic study of a fungal species are directly influenced by the nuclear composition of the vegetative and reproductive cells and the nature of its life cycle. Consequently, the assignment of a species to one of the major groups constitutes the first step in understanding the sources of genetic variation and determines the experimental designs in producing mutants, making crosses, and analyzing the segregant progeny.

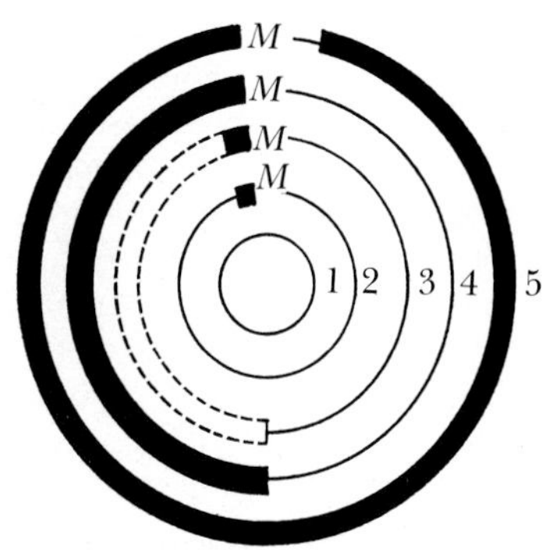

Figure 25–2. Diagrammatic illustration of five basic life cycles in the fungi. Each circle represents a different life cycle, in which "M" identifies the time of meiosis, and the sequence of events follows clockwise. The thin line represents the haplophase, the interrupted line the dikaryotic stage, and the thick line the diplophase. The life cycles are numbered: (1) asexual, (2) haploid, (3) haploid-dikaryotic, (4) haploid-diploid, and (5) diploid. (After Burnett, 1968,[4] and based on Raper, 1954.[34])

Life Cycles

The life cycle of a perfect fungus comprises the alternation of a haplophase (N) with a diplophase (2N). The vegetative and reproductive structures produced during the haplophase, the production of gametic nuclei or cells, the nature of fertilization and formation of the diploid nucleus or cell, the duration of the diplophase, the timing of meiosis, and the type of meiotic products are the data needed to construct the life cycle. For genetic studies, the most desirable material to initiate a strain or for mutagenesis is a haploid cell with a single nucleus. Conidia are commonly used for this purpose. In species lacking conidia, the products of meiosis, such as ascospores or basidiospores, serve the same purpose. All the products of one meiotic event can be removed from one ascus or basidium with the aid of a micromanipulator, or a random sample of meiotic products can be collected for genetic analysis.

Raper[34] proposed seven basic life cycles, which Burnett[4] reduced to five by recognizing broader categories (Fig. 25–2): asexual, haploid, haploid-dikaryotic, haploid-diploid, and diploid. The asexual, haploid, and haploid-dikaryotic life cycles are commonly encountered among the fungal pathogens.

The absence of a detectable sexual stage characterizes the species with an asexual

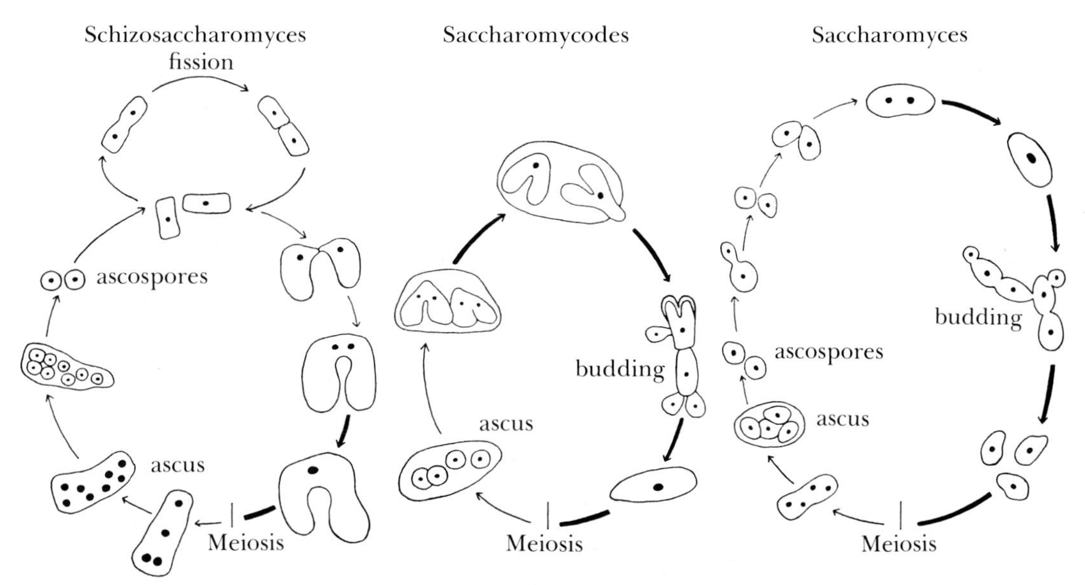

Figure 25–3. Life cycles of three yeast genera: *Schizosaccharomyces, Saccharomycodes*, and *Saccharomyces*. In *Schizosaccharomyces*, the life cycle is haploid, and vegetative cells multiply by fission. The life cycle for *Saccharomycodes* is diploid, and for *Saccharomyces*, haploid-diploid; vegetative cells multiply by budding. (After Burnett, 1968.[4])

life cycle. A second name is applied to an asexual species yielding a sexual stage. Prior to the discovery of the parasexual cycle,[33] genetic studies in the asexual species were restricted to the detection of spontaneous mutants or to the production of mutants by different mutagens. Genetic recombination, however, can be demonstrated for asexual species with the parasexual cycle.

In many groups of Phycomycetes and some Ascomycetes, nuclear fusion is immediately followed by meiosis and the dispersal of the meiotic products. Consequently, the species with haploid life cycles are distinguished by their short diplophase (Fig. 25–3). In *Saccharomyces cerevisiae*, for example, the diplophase can be extended.

The diplophase in species with haploid-dikaryotic life cycles is very short. The hyphal cells include at least two nuclei with different genotypes, which divide more or less synchronously for variable lengths of time, depending on the species. At the appropriate time in the life cycle, two genotypically different nuclei fuse to yield the zygotic nucleus. In many Discomycetes, binucleate ascogenous hyphae form immediately before the development of the ascus. In many of the Basidiomycetes, a monokaryotic mycelium originates from one

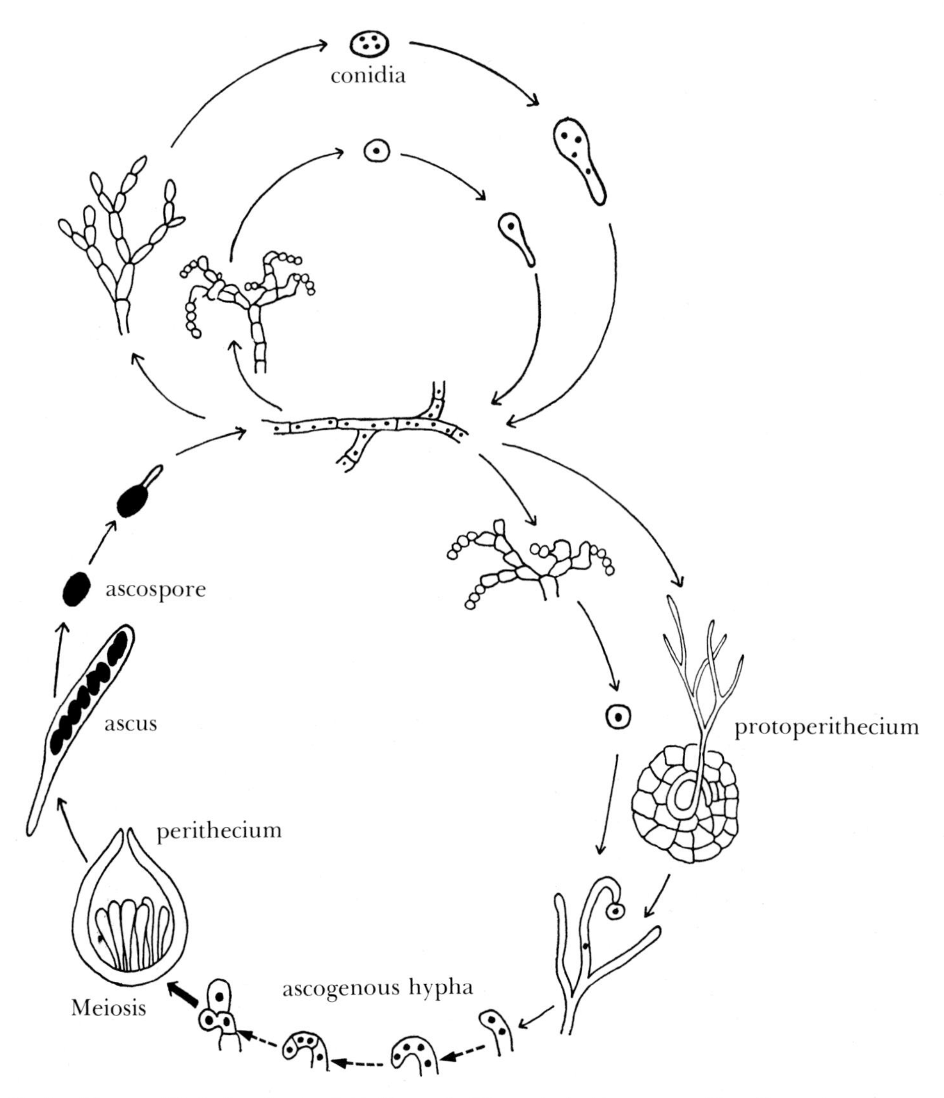

Figure 25–4. Diagram illustrating the haploid-dikaryotic life cycle in *Neurospora*. The dikaryotic stage occurs in the ascogenous hyphae, and the mating type systems involve two alleles at one locus. (After Burnett, 1968.[4])

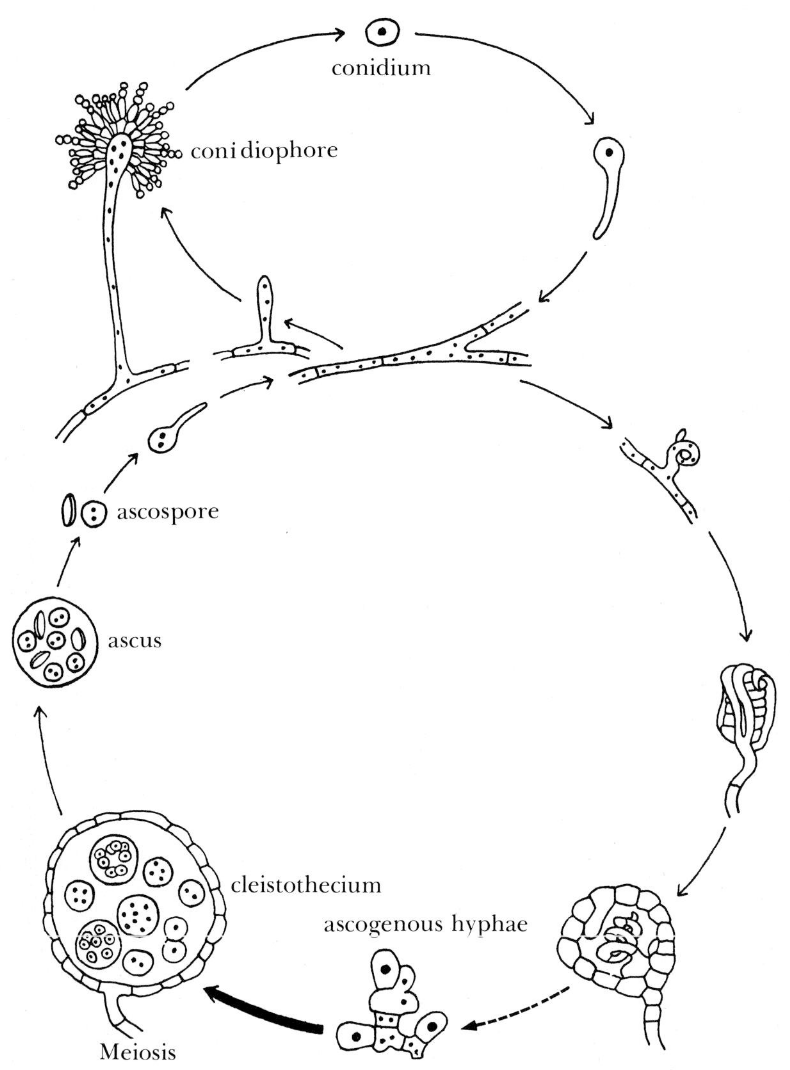

Figure 25–5. Diagram illustrating the haploid-dikaryotic life cycle in perfect homothallic species of *Aspergillus*. The brief dikaryotic stage occurs in the binucleate ascogenous hyphae. (After Burnett, 1968.[4])

basidiospore. Once hyphal anastomosis occurs for two compatible strains, the dikaryotic (N + N) mycelium persists until the sexual stage is produced. In the smuts, fleshy Basidiomycetes, and some yeasts, the dikaryotic mycelium persists for most of the life cycle. The haploid-dikaryotic life cycle is found in four genera containing species of particular interest to geneticists: *Neurospora* (Fig. 25–4), *Aspergillus* (Fig. 25–5), *Coprinus* (Fig. 25–6), and *Ustilago* (Fig. 25–7).

Only a few species of aquatic Oomycetes display the haploid-diploid life cycle in which there is a regular alternation of haplophase and diplophase. The diploid life cycle found in a number of species of Oomycetes is characterized by the restriction of the haplophase to gametes or gametangia.

Mating Type Systems

Normal sexual reproduction in fungi can be described in terms of the proper development of the sexual structures, which may or may not be related to the differentiation of the sexes, gametic nuclei or gametes, fertilization, meiosis, and the dispersal of the meiotic products. Genic mutations, however, can interfere with the normal sequence of

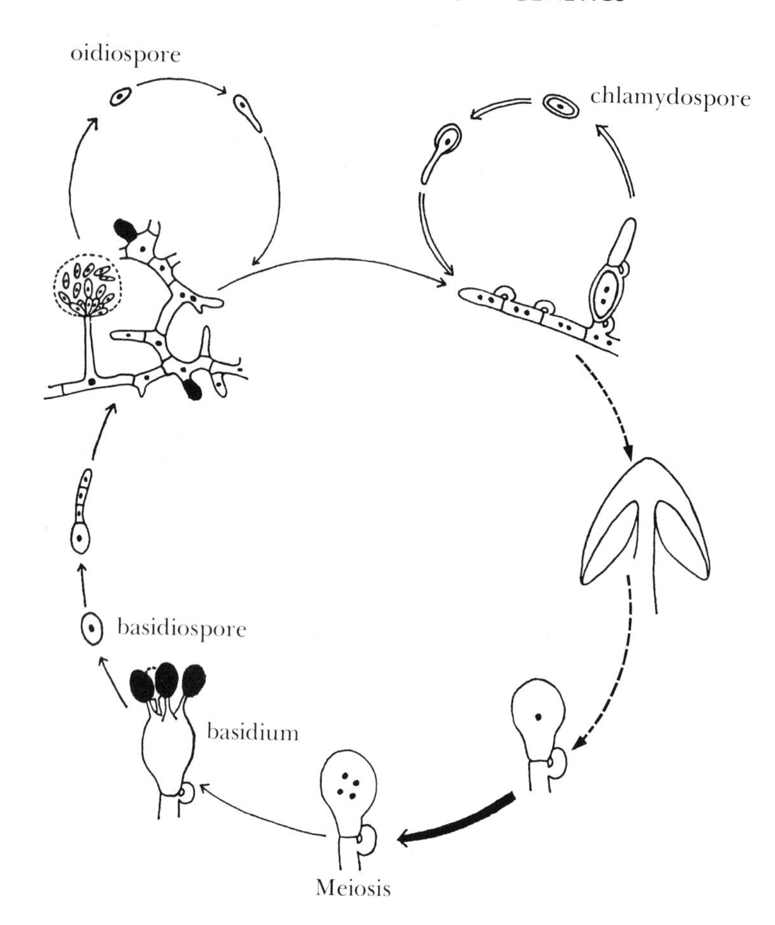

Figure 25–6. Diagram illustrating the haploid-dikaryotic life cycle in *Coprinus*. The dikaryotic stage is binucleate, of long duration, and often characterized by clamp connections at the cross walls in the hyphae. Vegetative multiplication involves oidiospores or chlamydospores. When oidiospores are produced, they arise from the monokaryotic mycelium prior to hyphal anastomosis. (After Burnett, 1968.[4])

events at any step. Two basic mating systems occur in the perfect species: homothallism and heterothallism. In heterothallism, thalli from different meiotic products must provide the gametic nuclei or gametes for fertilization and completion of the life cycle. In homothallism, one meiotic product yields the thallus, which produces the fruiting bodies to complete the life cycle. Therefore, homothallism favors inbreeding and genetic homogeneity, while heterothallism results in outbreeding and genetic heterogeneity. By using genetic markers, Carr and Olive[5] and Pontecorvo et al.[33] demonstrated outcrossing in the homothallic species *Sordaria fimicola* and *Aspergillus nidulans*, respectively.

Genetic heterogeneity from outcrossing can be accomplished by sexual dimorphism, which occurs in the aquatic Phycomyetes. A common means to accomplish outcrossing in other fungi involves the development of heterothallism. In many heterothallic Phycomycetes, Ascomycetes, and the rusts, each of the two mating types normally produces male and female sex organs or the equivalent structures. The higher Basidiomycetes have no sex organs per se, but usually have more than two mating types.

While mycologists do not agree on a satisfactory nomenclature for the diverse mating type systems,[35, 53] one approach to this problem has considered the genetic basis of the mating types. In the most straightforward and usually encountered system, there are two alleles at one locus, which have been termed *a-1* and *a-2*, + and −, *A* and *a*, or *α* and *a*, depending on the species. For example, in *Neurospora crassa*, the thallus with one mating allele produces protoperithecia, and fertilization eventually follows the addition of conidia or hyphae from the thallus to the other allele. Thalli with different alleles, however, are not distinguishable by detectable morphologic characteristics.

In the class Basidiomycetes, the Hymenomycetes and Gasteromycetes contain bipolar species, that is, species with multiple alleles at the mating type locus. Only thalli with different alleles are compatible, and the result of a mating is a dikaryotic mycelium

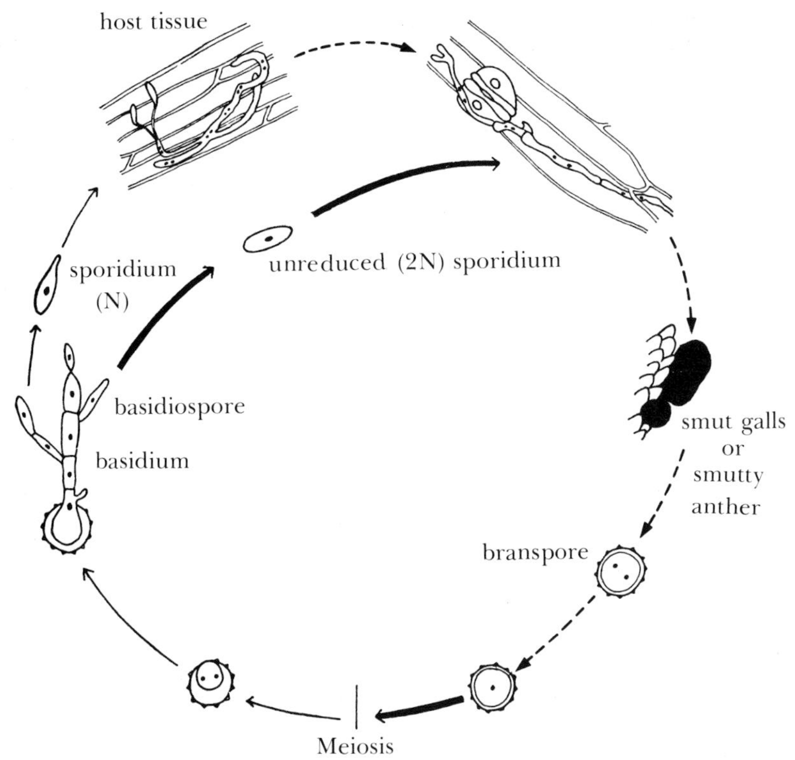

Figure 25–7. Diagram illustrating the haploid-dikaryotic life cycle of *Ustilago*. The occasional diploid sporidium is solopathogenic and produces an infectious mycelium in the susceptible host without the hyphal anastomosis required for infection by the haploid sporidia with compatible mating type alleles. (After Burnett, 1968.[4])

which produces the sexual structures in the appropriate physicochemical environment. The meiotic products from a cross between two dikaryotic strains with different pairs of mating type alleles appear in Table 25–1.

The compatibility of strains of a tetrapolar species is determined by alleles at two independently segregating mating type loci. The meiotic products of compatible matings for a tetrapolar species are presented in Table 25–2. The mating type loci may have two or more alleles. Although strains with the same allele at one locus are incompatible, they will form unstable heterokarytic hyphae.

Table 25–1. *Outcome of Matings Between Haploid Strains Derived from Meiotic Products of Strains With Multiple Alleles at One Mating Locus: a–1 a–2 and a–3 a–4*

	Haploid Strains			
	a–1	a–2	a–3	a–4
a–1	–	+	+	+
a–2	+	–	+	+
a–3	+	+	–	+
a–4	+	+	+	–

When a perfect species is used for genetic studies, it must be established whether the species is homothallic or heterothallic. For heterothallic species, the number of loci determining the mating type system and the number of alleles at each locus must be determined. Without this information, successful crosses cannot be made, and the mating allele or alleles in each meiotic product cannot be identified. Although hybrid fruiting bodies always result from crosses in heterothallic species, genetic markers are needed to detect hybrid fruiting bodies for homothallic species.

Saprophytes and phytopathogens provided the experimental material for the genetic study of fungi.[9,13,14] The methodology developed for these species is readily adapted to the human and animal fungal pathogens. Mutants are absolutely necessary in genetic studies. Although isolants from wild populations or laboratory strains frequently provide such visually altered phenotypes as color

Table 25–2. *Types of Tetrads Produced by a Tetrapolar Dikaryon (A1B1 + A2B2) and the Outcome of Sib and Cross Mating With Haploid Strains Derived From the Meiotic Products From Another Dikaryon (A1B1 + A3B3) in a Species With a Multiple Allelic Series at Two Independently Segregating Loci For Mating Type*

		Haploid Strains						
		Sib Mating				*Cross Mating*		
Tetrad Type		A1B1	A1B2	A2B1	A2B2	A1B3	A3B1	A3B3
PD*	A1B1	−	−	−	+	−	−	+
	A1B1	−	−	−	+	−	−	+
	A2B2	+	−	−	−	+	+	+
	A2B2	+	−	−	−	+	+	+
NPD	A1B2	−	−	+	−	−	+	+
	A1B2	−	−	+	−	−	+	+
	A2B1	−	+	−	−	+	−	+
	A2B1	−	+	−	−	+	−	+
T	A1B1	−	−	−	+	−	−	+
	A1B2	−	−	+	−	−	+	+
	A2B1	−	+	−	−	+	−	+
	A2B2	+	−	−	−	+	+	+

*PD = parental ditype; NPD = nonparental ditype; T = tetratype.

and colonial variants, mutagenic agents are commonly used to produce a spectrum of different types of mutations. Ultraviolet light and such chemical mutagens as ethylmethanesulfonate (EMS) or nitrosoguanidine (NTG) are frequently employed to increase the yield of mutant phenotypes. Vegetative or sexual spores are treated with the mutagen and plated on the appropriate medium, and the survivors are screened as colonies to detect the mutants.

Fungal mutations can be assigned to three broad categories of altered phenotypes: (a) visually altered phenotypes, particularly conidial color or colonial morphology; (b) nutritional deficiency; and (c) resistance to inhibitory or lethal compounds, such as fungicides, antimetabolites, antibiotics, and a number of inorganic or organic compounds. Mutations leading to reduced virulence or the loss of virulence are particularly interesting to medical mycologists. Mutants screened for one phenotype may also be altered for another phenotype. For example, a mutant resistant to an inhibitor may also have an altered color or colonial morphology.

Mutants with an altered spore color or colonial morphology are readily detected and easily scored in progeny from a cross. The color mutants usually grow as well as the wild type, and are equally compatible in making crosses. Colonial mutants may grow poorly and not cross with compatible wild type strains or with another colonial mutant. Colonial mutants, on the other hand, are more frequently encountered than color mutants in screening wild type populations (Table 25–3). Weitzman and Silva[52] crossed a number of colonial mutants in *Nannizzia incurvata* in constructing a linkage map in this species. Mutant conidial colors were used to demonstrate heterokaryosis and the parasexual cycle in *Aspergillus fumigatus*.[48]

Nutritionally deficient mutants grow on the complex medium but cannot grow on the minimal medium lacking the appropriate supplement. The usual supplements for the auxotrophic mutants are amino acids, purines, pyrimidines, vitamins, or growth factors. While auxotrophic mutants are

Table 25–3. *Frequencies of Spontaneous Morphologic Mutants From 174 Single Ascospore Cultures of* Nannizzia incurvata*

Mutant Phenotypes	***Isolation Frequency***
Pale (pa)	46
Microconidial (mi)	32
Fluffy (flu)	21
Spectrum (spec)	13
Pink (pi)	18
Orange (or)	5
Bizarre (bz)	2
Miscellaneous mutants	10

*From Weitzman, I., and M. Silva. 1966. Variation in the *Microsporum gypseum* complex. II. A genetic study of spontaneous mutation in *Nannizzia incurvata*. Mycologia, *58*:570–579.

excellent genetic markers in their own right, they are commonly used to synthesize heterokaryons and to detect diploid vegetative spores which are prototrophic, that is, capable of growing on the minimal medium. The auxotrophic markers are generally recessive and are heterozygous in the diploid strains. Auxotrophic mutants of pathogens may be avirulent when the host does not furnish an adequate concentration of the required compound in an available form to support the *in vivo* growth of the mutant.[15]

Mutants resistant to inhibitory or lethal compounds are obtained by plating large populations of spores on medium containing the proper concentrations of these compounds, so that the wild type spores are obviously inhibited, and the mutant spores are clearly uninhibited in terms of the diameter of the colonies. Resistant mutants may arise in the course of treatment of patients with a fungicide or antimetabolite. It is possible to estimate the probability of such resistant types occurring by mutation by *in vitro* tests. Furthermore, mutants resistant to one inhibitory compound may also be resistant to other inhibitory compounds.[2]

Pathogenic fungi have not been used for many genetic studies. This situation should change because the perfect stage has now been demonstrated for a number of species. Weitzman[51] determined the compatibility of 23 strains of the imperfect species *Microsporum gypseum* by demonstrating fruiting bodies for a number of combinations of strains on a complex medium containing hair as the inducer substrate (Fig. 25–8). By comparing patterns of compatibility and incompatibility, 22 strains were assigned to *Nannizzia incurvata* or *Gymnoascus gypseum* (= *N. gypsea*). Isolated ascospores from one ascus were used to establish that each species is heterothallic, with one locus and two alleles responsible for compatibility.

The spontaneous appearance of white aerial mycelium on the surface of colonies has been termed "pleiomorphism." Colonies from the aberrant mycelium have a fluffy surface, loss of pigmentation, and either a reduced conidiation or no conidia. Pleiomorphism has been attributed to environmental influences, cytoplasmic effects, or genic mutation. Weitzman[50] obtained monogenic segregation for crosses involving a number of spontaneous pleiomorphic strains

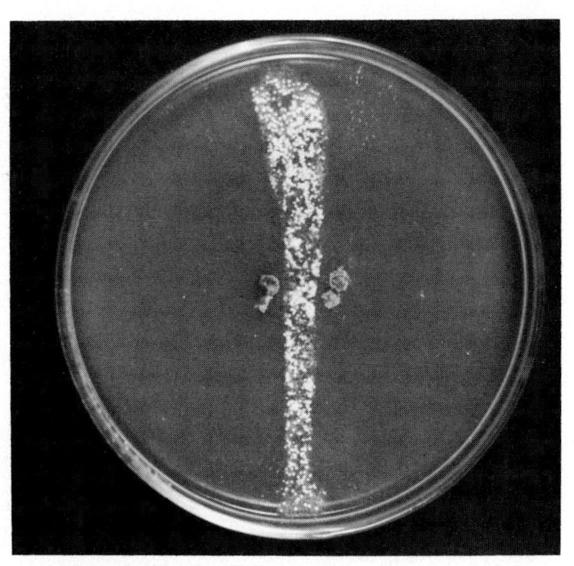

Figure 25–8. Gymnothecia along the streak of powdered hair for two compatible strains of *Nannizzia incurvata*. (Courtesy of Dr. Irene Weitzman.[51])

of *Nannizzia incurvata* and *N. gypsea* (Fig. 25–9). Weitzman and Silva[52] extended the study of pleimorphism in comparing the spontaneous mutants with those produced by ultraviolet light in *N. incurvata*. Most of the spontaneous mutants were in one linkage group, which had seven distinct phenotypes (Fig. 25–10). Furthermore, three of the 13 tested UV-produced mutants were in the same linkage group with the spontaneous mutants.

The classic approaches to establishing the determinants of virulence in the pathogenic fungi have not yielded much information of value to medical mycologists. Biochemical and genetic methods have provided considerable information about the virulence determinants of pathogenic bacteria.[45] These investigations offer appropriate models for similar ventures with the pathogenic fungi. Cheung and Maniotis[7] used a biochemical-genetic approach to implicate high elastase production as a virulence determinant in *Arthroderma benhamiae*.

The Parasexual Cycle

Genetic studies with imperfect fungi were once restricted to the study of spontaneous or induced mutants and of spontaneous or synthesized heterokaryons. With the discovery of the parasexual cycle in *Aspergillus*

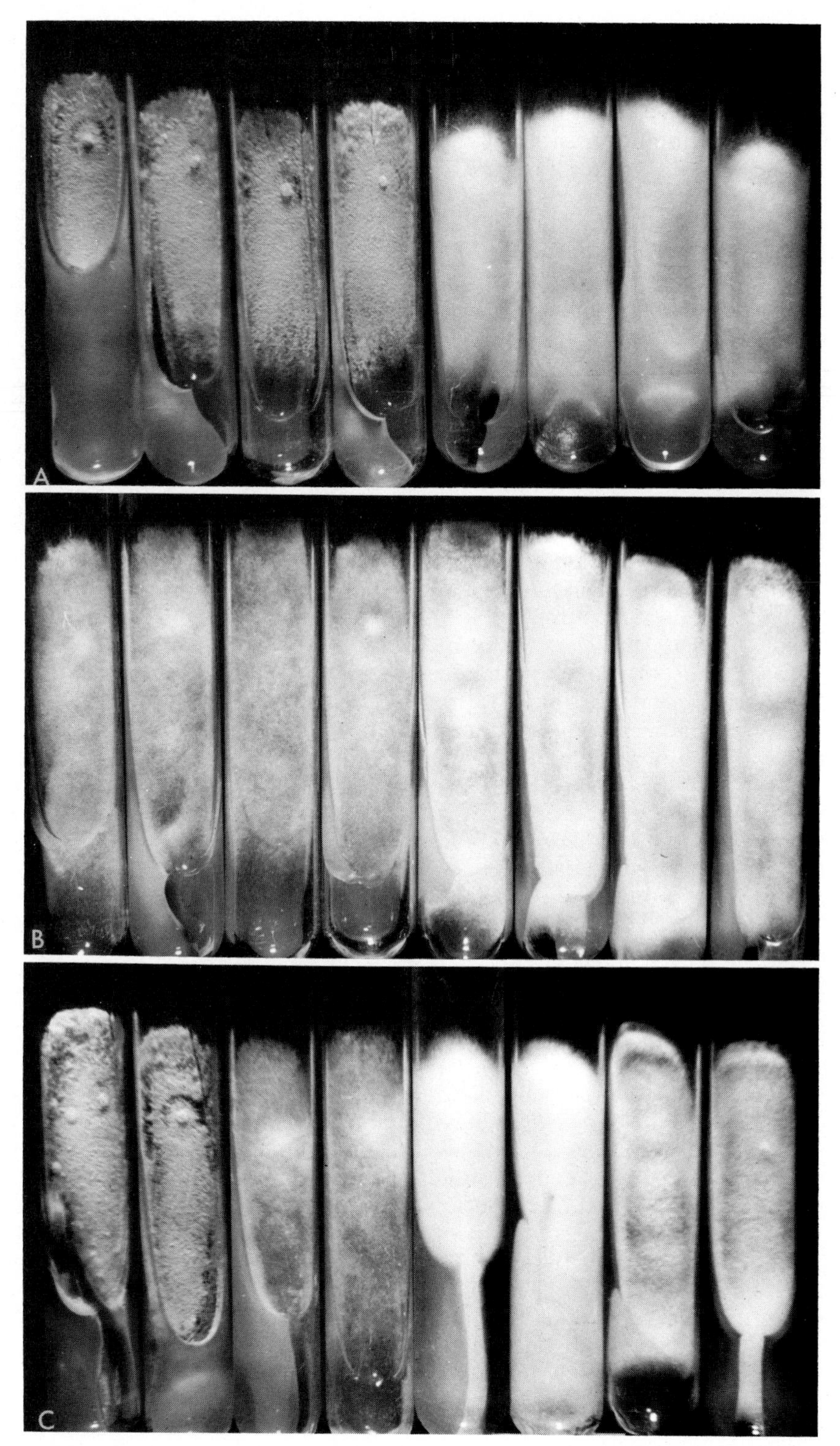

Figure 25–9. Segregation patterns from a cross between a double mutant fl_3 wo_4 strain and a wild type strain of *Nannizzia gypsea*. *A*, Parental ditype ascus: four wild type cultures on the left and four double mutant cultures on the right. *B*, Nonparental ditype ascus: 4 $\pm$ wo_4 cultures on the left and 4 fl_3 $\pm$ cultures on the right. *C*, Tetratype ascus: wild type cultures on the left, followed by $\pm$ wo_4, fl_3 wo_4, and fl_3 $\pm$ (Courtesy of Dr. Irene Weitzman.[50])

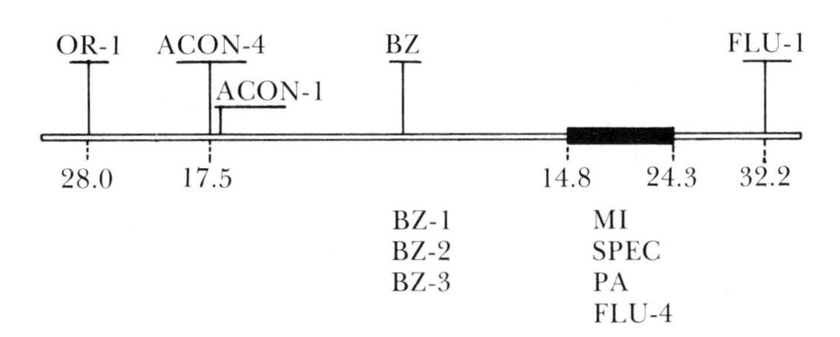

Figure 25–10. Location of the most frequent spontaneous morphologic mutants in linkage group I in *Nannizzia incurvata*. The mutable region is indicated by the black segment. Linkage values were calculated in terms of the *bz* locus. (Courtesy of Dr. Irene Weitzman.[52])

nidulans,[33] genetic recombination can be demonstrated without resorting to the sexual stage. A demonstration of the parasexual cycle requires three observations: (a) the synthesis of a heterokaryon using strains with different nutritional requirements and, whenever possible, different morphologic or color markers; (b) the isolation and detection of diploid vegetative spores; and (c) the recovery of haploid or diploid segregant vegetative spores (Fig. 25–11). The nutritional deficiencies serve to detect the heterokaryotic mycelium by their ability to grow on the minimal medium which does not support the growth of the parental strains. The detection of diploid spores on heterokaryotic colonies is facilitated by using conidial color markers. Patches or sectors of conidia with the wild type color provide visual evidence for diploidy, that is, heterozygosity for the recessive color markers. Diploid conidia can be detected by plating conidia from the heterokaryotic colonies on the minimal medium and observing colonies which yield the wild type conidia. Patches or sectors of conidia with one or the other mutant color on diploid colonies provide visual evidence for recombinant conidia, which can then be isolated and scored for the presence or absence of each of the nutritional markers. The parasexual cycle has been demonstrated for a number of perfect and imperfect ascomycetous filamentous species, including the pathogen *Aspergillus fumigatus*[48] and two species of *Ustilago*.[10, 21] Heterokaryons in a species do not necessarily indicate that the species will have a parasexual cycle.[11] Furthermore, not all diploid strains in a species produce segregant spores.[48]

The frequency of diploid conidia from heterokaryons is relatively low and differs among species: *Aspergillus nidulans*, 1 to 10 per 10^6 conidia, *A. niger*, 3 to 4 per 10^5 conidia, and *Penicillium chrysogenum*, 2 to 3 per 10^8 conidia. Strømnaes and Garber[48] obtained diploid conidia from approximately 70 per cent of heterokaryons with different genotypes in *A. fumigatus* and a broad range of frequencies for diploid conidia for those heterokaryons which gave diploid conidia.

Although diploid conidia can be differentiated, by their greater volume and DNA content, from the haploid conidia, a genetic study of presumptive diploids can also provide supporting evidence. For example, diploid strains of *A. nidulans* heterozygous for white and yellow conidia have the wild type (green) conidia. Such strains will produce patches or sectors of white or yellow conidia on a background of green conidia. The recombinant conidia, however, may be diploid or haploid, and a decision can be made on the basis of genetic tests.

Mitotic crossing-over in a heterozygous diploid strain produces homozygosity for all of the loci in one chromosome arm distal from the site of crossing-over (Fig. 25–12). The frequency of mitotic crossing-over in many species is sufficiently low to warrant the assumption that only one crossing-over will occur in a single nucleus. It should be noted that mitotic crossing occurs with a much reduced frequency as compared with meiotic-crossing over. When a color marker is located at the end of the chromosome arm involved in a mitotic crossing-over, other mutant genes in the same chromosome arm may also be homozygous. Consequently, a linkage map sequencing these genes can be constructed. In perfect species with a parasexual cycle, the linkage maps from mitotic crossing-over conform to those constructed from data obtained by meiotic crossing-over.

In haploidization, a process presumably initiated by the spontaneous nondisjunction of chromosomes during mitosis, recombinant conidia have the haploid chromosome number. Käfer[22] estimates that one haploidization in *A. nidulans* occurs in approximately 50 mitotic divisions in the mycelium. As the result of mitotic nondisjunction, a trisomic (2N+1) nucleus is produced, which then

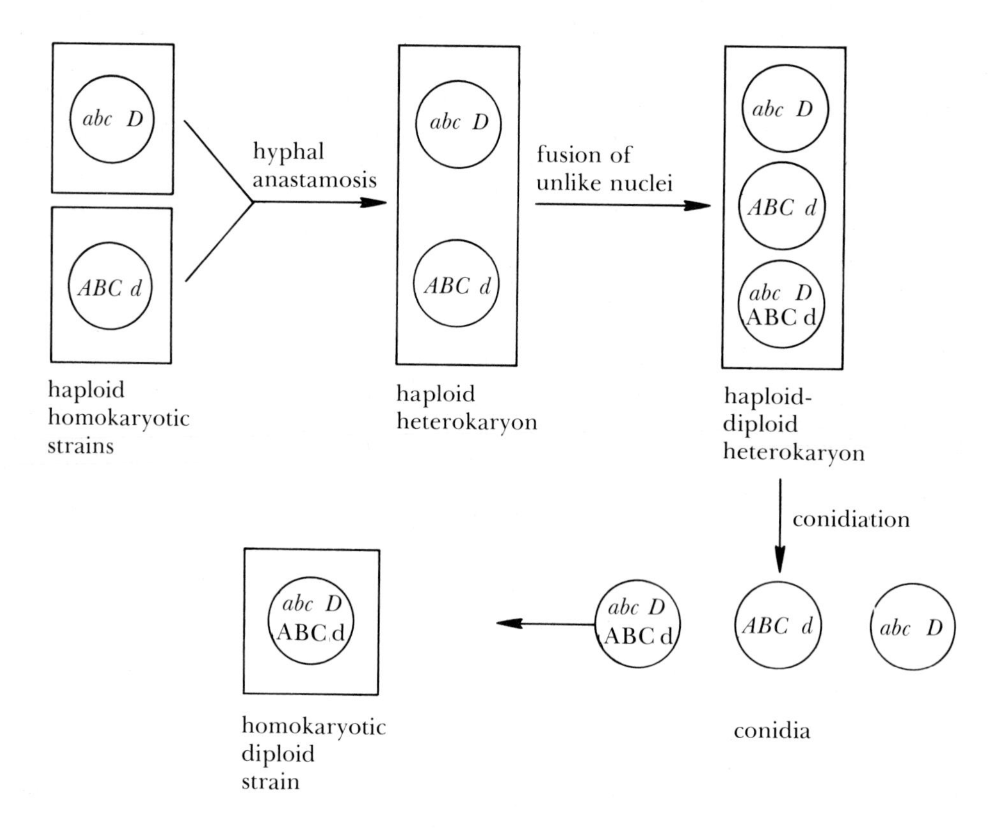

Segregant Conidia	Phenotype	Origin
abc D	abc	haploidization
ABC d	d	
abc d	abc d	
ABC D	wild type	
abc D	c	mitotic crossing over
ABc d		
abc D	bc	
Abc d		
abc D		
abc d	abc	

Figure 25–11. The parasexual cycle in *Aspergillus nidulans,* illustrated with three linked mutant genes and one independently segregating mutant gene.

reverts to a diploid (2N) nucleus which may be homozygous for all the genetic markers on the extra chromosome. Another product of mitotic nondisjunction is a monosomic (2N−1) nucleus which continues to lose one of each pair of chromosomes during successive mitotic divisions, eventually yielding a haploid (N) nucleus. In *A. nidulans*, haploidization has provided genetic data indicating eight linkage groups, which correspond to the haploid number of chromosomes. The different haploid nuclei resulting from haploidization in the same or different heterozygous colonies of *A. nidulans* include random combinations of the eight chromosomes. Because mitotic crossing-over and haploidization occur with low frequencies, recombinant nuclei from heterozygous dip-

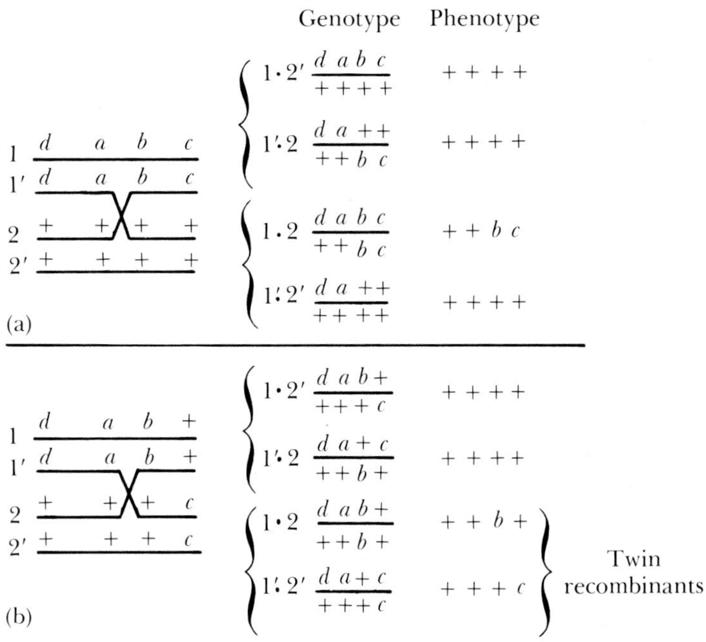

(a)

1	*d*	*a*	*b*	*c*
1′	*d*	*a*	*b*	*c*
2	+	+	+	+
2′	+	+	+	+

	Genotype	Phenotype
1·2′	*d a b c* / + + + +	+ + + +
1′·2	*d a* + + / + + *b c*	+ + + +
1·2	*d a b c* / + + *b c*	+ + *b c*
1′·2′	*d a* + + / + + + +	+ + + +

(b)

1	*d*	*a*	*b*	+
1′	*d*	*a*	*b*	+
2	+	+	+	*c*
2′	+	+	+	*c*

	Genotype	Phenotype	
1·2′	*d a b* + / + + + *c*	+ + + +	
1′·2	*d a* + *c* / + + *b* +	+ + + +	
1·2	*d a b* + / + + *b* +	+ + *b* +	Twin recombinants
1′·2′	*d a* + *c* / + + + *c*	+ + + *c*	

Figure 25–12. Diagram illustrating the consequences of mitotic crossing-over in a heterozygous vegetative nucleus of a diploid strain. After recombination, the homologous chromosomes proceed to opposite poles during anaphase, and only certain combinations of nonsister chromatids yield chromosomes homozygous for those mutant genes distal from the site of crossing-over. When the distal pair of linked genes are in repulsion, twin or reciprocal recombinant conidia are recovered from a single crossing-over.

loid colonies are presumed to have resulted from one or the other processes, but not from a combination of these processes.

In perfect species with a parasexual cycle, it may be more efficient to determine linkage groups by haploidization than by the random combinations of mutant genes in strains to be used for linkage studies by the sexual stage. Once mutant genes have been assigned to the same linkage group, meiotic crossing-over can be used to produce conventional linkage maps. Mitotic crossing-over, however, can be exploited to assign mutant loci to one or the other arm of the linkage group, that is, the chromosome associated with the linkage group. Linkage groups in imperfect species can be detected only by the parasexual cycle. The discovery of a potent selection agent in the antimetabolite *para*-fluorophenylalanine for haploid hyphae has greatly facilitated the detection of segregant haploid conidia produced by diploid strains grown on medium containing this antimetabolite.[27]

THE BIOCHEMICAL-GENETIC APPROACH TO PATHOGENICITY

Pathogenicity is distinguished from saprophytism and symbiosis in that the host is adversely affected by the parasite and displays a characteristic response, a disease. Furthermore, pathogenicity must be defined in terms of a specific dynamic interaction between host and parasite, in which the host provides the environment determining whether the parasite will or will not multiply and metabolize. Virulence, reduced virulence, or avirulence are established only by bioassay, that is, by the infection of a susceptible host to cause disease. Certain morphologic or biochemical characteristics may be correlated with virulence, and the loss of these characteristics may or may not be found in avirulent strains or mutants. Consequently, strains or mutants with reduced virulence and avirulent strains or mutants can be valuable tools to probe the nature of the mechanisms responsible for the determinants of virulence. An appreciation of the diverse facets of virulence can lead to an understanding of the nature of resistance and susceptibility.

Bacterial pathogens have been extensively studied to determine the nature of their pathogenicity. Relatively little information is available for the pathogenic fungi. In a comprehensive review of the pertinent literature, Smith[45] considered such virulence determinants in bacterial pathogens as the ability to initiate infection, *in vivo* growth and multiplication, aggressive activity, production of "toxins" and the mechanism of "toxicity," and the biochemical bases of host and tissue specificity. When virulence involves a number of determinants, an avirulent strain or mutant may lack only one determinant, but avirulence can result from the loss of any one of a number of determinants. Comparative studies of virulent and avirulent strains or mutants often indicate concomitant changes in morphologic or biochemical characteristics, but such changes must be viewed with caution. These altered characteristics may be indirectly associated with the virulence determinants. Furthermore, avirulent strains from repeated subculturing may have accumulated several mutations responsible for altering different determinants. Consequently, spontaneous or induced avirulent mutants obtained from virulent strains provide suitable candidates for comparative studies which could indicate specific determinants.

The dimorphic fungal pathogens occur *in vivo* as the yeast stage and *in vitro* as the mycelial stage.[42] Strains in the mycelial stage must be converted into the yeast stage to be virulent. Although the *in vitro* environment can be manipulated by altering temperature or the oxidation/reduction potential to convert the yeast stage to the mycelial stage, the regulatory controls and the genetic basis for such control have not yet been established. Mutants incapable of producing the yeast stage should be avirulent. It may be possible to devise a screening procedure to detect such mutants *in vitro* and then to determine their virulence by infecting susceptible hosts.

Kwon-Chung[25] demonstrated two alleles at one locus for the compatibility of strains of *Histoplasma capsulatum.* In a survey of isolants from soil and from human infections determining the mating type alleles in these isolants, approximately equal numbers of strains with one or the other allele were found for the soil isolants, but approximately 80 per cent of the isolants from human infections had the minus allele[26] These observations suggest that minus allele strains may have a selective advantage in the host. A detailed comparison of the enzymologic and biochemical characteristics of strains with different mating type alleles might yield clues to the virulence determinants of this species.

Rippon[40] compared ten enzymes from strains with different mating type alleles in five species of pathogenic fungi and in the yeast and mycelial phase for one of these species, *Blastomyces dermatitidis.* Several enzymes were detected in strains with one but not the other allele, suggesting that the loci determining the activity of these enzymes might be closely linked with the locus for the mating type alleles in these species. Furthermore, several extracellular proteolytic enzymes were produced by the mycelial but not the yeast phase of *B. dermatitidis.* These observations, however, do not necessarily implicate the proteolytic enzymes in the virulence of this species. They may play a role in hyphal anastomosis and, therefore, in the production of the sexual stage.

Extracellular proteolytic enzymes have been viewed as virulence determinants for pathogenic fungi and have been investigated to determine their involvement in causing disease.[6, 43] While extracellular enzymes indeed have a role in establishing the pathogen in the host-environment, not all manifestations of the syndrome need be associated with the activity of all these enzymes. For example, dermatophytes produce extracellular keratinases, but not all fungi producing these enzymes are pathogenic. If the production of extracellular keratinase were a significant aspect of virulence, mutants producing relatively small amounts or undetectable amounts of keratinases should have reduced virulence or be avirulent. Consequently, the saprophytic fungi producing extracellular keratinase presumably lack the virulence determinants present in the dermatophytes.

With the discovery of the sexual stage for a number of fungal pathogens and of the parasexual cycle for those fungi still lacking a sexual stage, it is now possible to determine the genetic basis for the production of extracellular enzymes presumed to be involved with virulence. Mutants or strains lacking activity for specific enzymes can be crossed with the wild type strains to establish a relationship between enzymatic activity and virulence by assaying segregants for activity and virulence.

Rippon[38] investigated the production of elastase, a protease degrading the scleroprotein elastin, in strains of species of *Trichophyton, Microsporum,* and *Epidermophyton.* Three strains of *M. fulvum* (= *Nannizzia fulva*) were isolated from patients and assayed for elastolytic activity. Two strains from inflammatory infections were elastase-positive, and one strain from a noninflammatory infection was elastase-negative. Tests of segregating progeny from a cross between an elastase-positive and an elastase-negative strain indicated that these phenotypes were determined by two alleles. Although these observations suggest that elastase is involved in the production of inflammatory infections caused by *M. fulvum*, elastase-negative mutants from an elastase-positive strain are needed to establish that the mutants do not produce an inflammatory infection.

Strains of *Trichophyton mentagrophytes* (= *Arthroderma benhamiae*) with a downy colonial morphology are usually isolated from chronic low-grade infection, and those with a granular colonial morphology from human and animal infections with suppurating lesions.[17] Furthermore, downy strains are less virulent than granular strains in that the former strains produce less severe lesions in infected guinea pigs. Cheung and Maniotis[7] demonstrated a correlation between high levels of elastase activity in strains with the granular colonies and low levels of elastase activity in strains with the downy colonies. Crosses between strains with different colonial morphology and elastase levels indicated that two alleles at one locus determine colonial morphology, and two alleles at another locus determine the level of elastase activity. Furthermore, the loci for these pairs of contrasting phenotypes were closely linked, thereby accounting for the correlation between colonial morphology and elastase activity or virulence. According to these observations, the level of elastase activity can be viewed as a virulence determinant for this species.

Rippon[39] reported that *Trichophyton schoenleinii* produces an extracellular proteolytic enzyme degrading collagen. Although such an enzyme appears to be a likely candidate for a virulence determinant in this species, no evidence was presented to associate the production of the collagenase with virulence. Rippon and Peck[41] demonstrated the significance of extracellular collagenase as a virulence determinant in the bacterial pathogen *Streptomyces (Actinomadura) madurae* by obtaining collagenase-negative mutants which were avirulent. Collagenase-positive revertants from the collagenase-negative strain were virulent. Finally, one active mutant from the parental virulent strain produced approximately threefold greater collagenase-activity and was obviously more virulent than the parental strain.

Fungal geneticists have devised efficient screening methods to produce different types of mutants by treating spores with mutagens, and to score progeny from crosses to determine their genotype. These methods can be adapted to the study of the virulence determinants of fungal pathogens. A survey of mammalian laboratory species and of methods to produce infections may be needed to find the appropriate susceptible strains and route of infection for virulence assays. The investigation of the relation between colonial morphology and levels of elastase in strains of *Arthroderma benhamiae* represents an excellent example of the biochemical-genetic approach to studying the virulence determinants in a fungal pathogen.

REFERENCES

1. Beebe, J. D. 1970. Amylases of selected species of *Aspergillus*. Ph.D. Thesis, University of Chicago.
2. Beraha, L., and E. D. Garber. 1966. Genetics of phytopathogenic fungi. XV. A genetic study of resistance to ortho-phenylphenate and sodium dehydroacetate, In *Penicillium expansum*. Am. J. Bot., *53*:1041–1047.
3. Brewer, G. J. 1970. *Introduction to Isozyme Techniques*. New York, Academic Press.
4. Burnett, J. H. 1968. *Fundamentals of Mycology*. New York, St. Martin's Press.
5. Carr, A. J. H., and L. S. Olive. 1959. Genetics of *Sordaria fimicola*. III. Cross compatibility among self-fertile and self-sterile cultures. Am. J. Bot., *46*:81–91.
6. Chattaway, F. W., D. A. Ellis, et al. 1963. Peptidases of dermatophytes. J. Invest. Dermatol., *41*:31–37.
7. Cheung, S. S., and J. Maniotis, 1973. A genetic study of an extracellular elastin-hydrolyzing protease in the ringworm fungus *Arthroderma benhamiae*. J. Gen. Microbiol., *74*:299–304.
8. Clare, B. G., N. T. Flentje, et al. 1968. Electrophoretic patterns of oxidoreductases and other proteins as criteria in fungal taxonomy. Aust. J. Biol. Sci., *21*:275–295.
9. Clowes, R. C., and W. Hayes. 1968. *Experiments in Microbial Genetics*. New York, John Wiley and Sons.
10. Day, A. W., and J. K. Jones. 1969. Sexual and parasexual analysis in *Ustilago violacea*. Genet. Res., *14*:195–221.
11. Dutta, S. K., and E. D. Garber. 1960. Genetics of phytopathogenic fungi. III. An attempt to demonstrate the parasexual cycle in *Colletotrichum lagenarium*. Bot. Gaz., *122*:118–121.
12. Emmons, C. W. 1934. Dermatophytes. Natural grouping based on the form of the spores and accessory organs. Arch. Dermatol. Syph., *30*: 337–362.
13. Esser, K., and R. Kuenen. 1967. *Genetics of Fungi*. New York, Springer-Verlag.
14. Fincham, J. R. S., and P. R. Day. 1971. *Fungal Genetics*. 3rd ed. Philadelphia, F. A. Davis Company.
15. Garber, E. D. 1956. A nutrition-inhibition hypothesis of pathogenicity. Am. Nat., *90*:183–194.
16. Garber, E. D., and J. W. Rippon. 1968. Proteins and enzymes as taxonomic tools. Adv. Appl. Microbiol., *10*:137–154.
17. Georg, L. K. 1954. The relationship between the

downy and granular forms of *Trichophyton mentagrophytes.* J. Invest. Dermatol., *23*:123–141.
18. Griffin, D. M. 1960. The re-discovery of *Gymnoascus gypseum,* the perfect state of *Microsporum gypseum,* and a note on *Trichophyton terrestre.* Trans. Br. Mycol. Soc., *43*:637–642.
19. Hayhome, B. A. 1970. Genetic studies on the regulation of the amylolytic enzymes of *Aspergillus niger.* Ph.D. Thesis, University of Chicago.
20. Heslop-Harrison, J. 1962. In *Microbial Classification.* Ainsworth, G. C., and Sneath, P. H. A. (eds.), London, Cambridge University Press, pp. 14–36.
21. Holliday, R. 1961. The genetics of *Ustilago maydis.* Genet. Res., *2*:204–230.
22. Käfer, E. 1961. The processes of spontaneous recombination in vegetative nuclei of *Aspergillus nidulans.* Genetics, *46*:1581–1609.
23. Kendrick, B. (ed.). 1971). *Taxonomy of Imperfect Fungi.* Toronto, University of Toronto Press.
24. Kurzeja, K. C., and E. D. Garber. 1973. A genetic study of electrophoretically variant extracellular amylolytic enzymes of wild-type strains of *Aspergillus nidulans.* Can. J. Genet. Cytol., *15*:275–287.
25. Kwon-Chung, K. J. 1972. Sexual stage of *Histoplasma capsulatum.* Science, *175*:326.
26. Kwon-Chung, K. J. 1973. Studies on *Emmonsiella capsulata.* I. Heterothallism and development of the ascocarp. Mycologia, *65*:109–121.
27. Lhoas, P. 1961. Mitotic haploidization by treatment of *Aspergillus niger* diploids with para-fluorophenylalanine. Nature (Lond.), *190*:744.
28. Manwell, C., and K. V. Kerst. 1965. Possibilities of biochemical taxonomy of bats using hemoglobin, lactate dehydrogenase esterases and other proteins. Comp. Biochem. Physiol., *17*:741–754.
29. Maurer, H. R. 1971. *Disc Electrophoresis.* New York, Walter de Gruyter.
30. Meyer, J. A., E. D. Garber, et al. 1964. Genetics of phytopathogenic fungi. XII. Detection of esterases and phosphatases in culture filtrates of *Fusarium oxysporum* and *F. xylarioides* by starch-gel zone electrophoresis. Bot. Gaz., *125*:298–300.
31. Nealson, K. H., and E. D. Garber. 1967. An electrophoretic survey of esterases, phosphatases, and leucine aminopeptidases in mycelial extracts of *Aspergillus.* Mycologia, *59*:330–336.
32. Pontecorvo, G. 1946. Genetic systems based on heterokaryosis. Cold Spring Harbor Symp. Quant. Biol., *11*:193–201.
33. Pontecorvo, G., J. A. Raper, et al.: 1953. The genetics of *Aspergillus nidulans.* Adv. Genet., *5*:142–238.
34. Raper, J. R. 1954. Life cycles, sexuality and sexual reproduction, In *Sex in Microorganisms.* Wenrich, D. H., et al. (eds.), Washington, D.C., American Association for the Advancement of Science, pp. 42–81.
35. Raper, J. R. 1966. *Genetics of Sexuality in Higher Fungi.* New York, Ronald Press.
36. Raper, K. B., and D. I. Fennell, 1965. *The Genus Aspergillus.* Baltimore, Williams and Wilkins.
37. Reddy, M. M., and S. F. H. Threlkeld. 1971. Genetic studies of isozymes in *Neurospora.* I. A study of eight species. Can. J. Genet. Cytol., *13*:298–305.
38. Rippon, J. W. 1967. Elastase: Production by ringworm fungi. Science, *157*:947.
39. Rippon, J. W. 1968. Extracellular collagenase from *Trichophyton schoenleinii.* J. Bacteriol., *95*:43–46.
40. Rippon, J. W. 1971. Differences between "+" and "−" strains of keratinophilic fungi, In *Recent Advances in Microbiology* (10th Internatl. Congr. Microbiol.), Mexico D. F., Libreria, Internacional S. F., pp. 473–475.
41. Rippon, J. W., and G. L. Peck. 1967. Experimental infection with *Streptomyces madurae* as a function of collagenase. J. Invest. Dermatol., *49*:371–378.
42. Romano, A. H. 1966. Dimorphism, In *The Fungi.* Vol. 2. Ainsworth, G. C., and Sussman, A. S. (eds.). New York, Academic Press, pp. 181–209.
43. Rosenthal, S. A., and H. Sokolsky. 1965. Enzymatic studies with pathogenic fungi. Derm. Int., *4*:72–79.
44. Shechter, Y., J. W. Landau, et al. 1966. Comparative disc electrophoretic studies of proteins from dermatophytes. Sabouraudia, *5*:144–149.
45. Smith, H. 1968. Biochemical challenge of microbial pathogenicity. Bacteriol. Rev., *32*:164–184.
46. Stockdale, P. M. 1961. *Nannizzia incurvata* gen. nov., sp. nov., a perfect state of *Microsporum gypseum* (Bodin) Guiart et Grigorakis. Sabouraudia, *1*:114–126.
47. Stockdale, P. M. 1963. The *Microsporum gypseum* complex (*Nannizzia incurvata* Stockd., *N. gypsea* (Nann.), comb. nov., *N. fulva* sp. nov.). Sabouraudia, *3*:114–126.
48. Strømnaes, Ø., and E. D. Garber. 1963. Heterocaryosis and the parasexual cycle in *Aspergillus fumigatus.* Genetics, *48*:653–662.
49. Szathmary, S., and Z. Herpay. 1960. Perithecium-formation of *Microsporon gypseum* and its cognate *Epidermophyton radiosulcatum* var. *flavum* Szathmary 1940 on soil. Mycopathologia, *13*:1–14.
50. Weitzman, I. 1964. Variation in *Microsporum gypseum.* I. A genetic study of pleiomorphism. Sabouraudia, *3*:195–204.
51. Weitzman, I. 1964. Incompatibility in the *Microsporum gypseum* complex. Mycologia, *56*:425–435.
52. Weitzman, I., and M. Silva. 1966. Variation in the *Microsporum gypseum* complex. II. A genetic study of spontaneous mutation in *Nannizzia incurvata.* Mycologia, *58*:570–579.
53. Whitehouse, H. L. K. 1949. Multiple allelomorph heterothallism in the fungi. New Phytol., *48*: 212–244.

Part Five

Pharmacology of Antimycotic Drugs

Chapter 26

There is a wide range of antibiotics available for the treatment of bacterial infections. This is in part due to vast differences between the physiology of the prokaryotic bacteria and the eukaryotic host. Because of these differences certain substances (antibiotics) will interfere with the metabolism of the infecting agent but have little or no effect on the host. Though some of the antibiotics used in the treatment of bacterial infections may have moderate to serious side effects, one has the prerogative of changing from one to another antibiotic agent if the response to treatment is not adequate or if side effects occur. This is not true in infections caused by fungi. As a consequence of an eukaryotic organism parasitizing a eukaryotic host, the range of physiologic differences between host and parasite is much smaller.

So far the search for antifungal agents has been disappointing. We have at this point three well-established drugs, one for each of the three major categories of infection, together with a few other drugs that are of limited use and only in particular types of infections. Amphotericin B, a very toxic agent, is the standard drug for the treatment of systemic fungous infections. Nystatin, being essentially insoluble, is restricted to the topical treatment of yeast infections. Griseofulvin is the most efficacious agent for treating all forms of dermatophyte infections, but it is not useful for other types of fungous diseases. In addition to these, there are some newer agents being investigated that have been found useful in some types of disease. These include tolnaftate and haloprogin for some forms of dermatophyte infections, and hydroxystilbamidine and saramycetin in a few systemic infections. A new compound, 5-fluorocytosine is being investigated as a therapeutic agent for several forms of fungous infections. Again it holds promise of being useful in some cases and has the advantage of being relatively nontoxic. Its overall value remains to be established, and experience is needed to determine standardization of therapeutic regimen and side effects. Other new drugs include miconazole and clotrimazole.

POLYENE ANTIBIOTICS

The first antifungal antibiotic to be isolated was nystatin in 1949.[33] The compound was found to contain a series of aliphatic unsaturated chains (—CH═CH—CH═CH—) also present in carotene or vitamin A and was designated a "polyene." Since that time numerous polyene antibiotics have been described.[43] Of these, only a few have received much attention as useful antifungal agents owing to the high degree of toxicity inherent in all of them.[44]

The number of conjugated double bonds varies in the various polyene antibiotics and allows them to be classified according to that number. The basic structure of all of them is a macrolide ring containing a number of conjugated double bonds. Those that have been investigated as antifungal agents are listed in Table 26–1. The macrolide ring may also contain a number of substitutes, such as amino sugar, carboxyl groups, and so forth. The addition of some of these confers amphoteric properties on the molecule. Thus amphotericin A and B, nystatin, pimaricin, and etruscomycin are amphoteric, whereas filipin, lacking an ionizable radical, is a neutral polyene. The presence or absence of such qualities does not appear to affect the mode of action of the antibiotic or its toxicity.

The biological effects of the polyene antibiotics have been the subject of much investigation.[25, 27, 30] It appears that in low concentrations all the polyenes affect only those cells

Table 26–1. *Polyene Antibiotics of Interest as Antimycotic Agents and Date of Discovery*

Tetraenes	
Nystatin	Hazen and Brown, 1949
Amphotericin A	Vanderputte et al., 1956; Gold et al., 1956
Pimaricin (Tennecetin)	Struyk et al., 1958; Divekar et al., 1961
Etruscomycin	
Pentaenes	
Filipin	Gottlieb, 1955
Fungichromin	Tytell, 1955
Hexaenes	
Flavacidin	Albert, 1947
Mediocidin	Utahara et al., 1954
Heptaenes	
Amphotericin B	Vanderputte et al., 1956; Gold et al., 1956
Trichomycin	Nakano, 1960
Candicidin	Kligman and Lewis, 1953
Candidin	Taber et al., 1954

that contain sterols in their cell membrane.[45] This is also the basis for their selective toxicity and application as therapeutic agents. In cells that contain rigid cell walls, such as fungi and helminths, they interfere with selective membrane functions that may result in stasis or death of the organism. In cells without rigid external walls (fungal protoplasts, protozoa, mammalian erythrocytes), the alteration of permeability of the membrane may result in lysis of the cell. In organisms that do not contain sterols in their cell membranes and thus are not sensitive to polyenes (i.e., bacteria), it has been possible to confer sensitivity by growing them on media containing cholesterol.[79] Mutants of yeasts that do not contain cholesterol in their membranes are also resistant to nystatin.[25, 81] Fungi can be protected from the effect of polyene antibiotics by including sterols in the growth medium.[30] The amount of damage to a given membrane appears to be a function of the number of carbon atoms in the antibiotic molecule. The larger antibiotics, nystatin and amphotericin B, are less damaging than the smaller molecules, filipin and etruscomycin.

The major effect of polyene antibiotics appears to be on the selective permeability of biological membranes. It has been demonstrated that certain substances (hydrogen ions) normally excluded gain entrance, and that certain essential cytoplasmic constituents are lost. The effect and degree of interference with normal membrane function are then reflected by other cellular abnormalities, such as cessation of growth, abnormal growth, decrease in metabolic rate, ballooning of cells, lysis, and so forth. In some cases the effect may be reversed by removal of the antibiotic from the environment. Thus *Coccidioides* spherules inhibited by the presence of amphotericin B will resume activity and growth if transferred within a certain amount of time to antibiotic-free media.

The extent of damage to the membrane and the nature of the cytoplasmic constituents released are influenced by several factors. Among these are the particular polyene molecule investigated, the concentration of the substance in the environment, and the time of exposure. One of the first effects noted in experimental studies on the mode of action of polyene antibiotics is the loss of intracellular potassium. This is followed by the appearance in the media of other low-molecular-weight cellular constituents. As time of exposure increases, larger compounds, including proteins and nucleic acids, leak from the cell. Also the toxicity of the polyene antibiotics for mammalian tissues may be more pronounced on particular membrane systems. Thus amphotericin B seems to have a special affinity or toxicity for cells of the renal tubules.

The two polyene antibiotics most commonly employed in antimycotic therapy, amphotericin B and nystatin, will be discussed in detail. In addition, some compounds having promise as useful agents of the future or established as effective treatments for certain specific diseases will also be included.

An excellent review of the subject of antifungal antibiotics up to 1964 is found in Chapters 41 to 57 of the text by Hildick-Smith et al.[36] and in the recent review by Hildick-Smith in 1968.[35]

Nystatin

Nystatin (Mycostatin, Moronal, Nystan, Fungicidin) $C_{46-47}H_{73-75}O_{18}N$ (tentative), yellow powder, m.p. dec. above 160°C. No definite m.p., $[\alpha]_D^{25}$ − 10° (glacial acetic acid). An amphoteric tetraene with a sugar moiety mycosamine (Fig. 26–1). U.V. max., 290, 307, 322. Essentially insoluble in water 4.0 mg per ml, methanol (11.2 mg per ml), or ethanol (1.2 mg per ml). Quite soluble in dimethylformamide, propylene glycol, dimethylacetamide, glacial acetic acid, 0.5 N

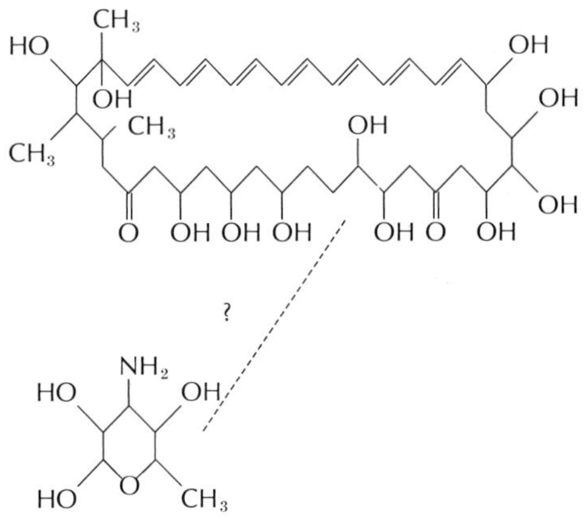

Figure 26–1. Nystatin.

methanolic hydrochloric acid, and sodium hydroxide. It inactivates rapidly in the latter three. All solutions begin to lose activity after preparation. Sensitive to heat, light, and oxygen. LD_{50}IP in mice, 200 mg per kg; IV, 3 mg per kg. Activity not diminished by blood or serum. Extracted from mycelium of *Streptomyces noursei* by MeOH.

History. A search for antimycotic agents by Hazen and Brown led to the discovery of nystatin in 1949.[33] The first of the polyene antibiotics was isolated from an actinomycete recovered from a pasture of the Nourse dairy farm in Fouquier County, Virginia. The organism was named *Streptomyces noursei* and was shown to produce two antifungal antibiotics. One was soluble and identified as Actidione (cycloheximide). The other was insoluble and was later named "nystatin" after the New York State laboratory where the developers were employed. Most of the common fungal pathogens are inhibited *in vitro* by 1.5 to 13 μg per ml of nystatin. However, the lack of solubility of the compound in water is paralleled by its lack of absorption following ingestion. This together with its significant toxicity when administered parenterally prevents the use of this drug in the treatment of systemic fungous infection.[51] It has been found to be of great value in the topical treatment of cutaneous and mucocutaneous infections, especially those caused by the *Candida* species.

Range of Antimicrobial Activity. When first isolated a milligram of nystatin was arbitrarily assigned an activity value of 1000 units. In bioassays, the activity of the drug in dilution series was expressed as "units," depending on the fraction of a milligram per milliliter needed to inhibit growth. In time, purification procedures allowed the production of the drug with a range of up to 6000 units per mg. The antibiotic now generally available for clinical use has a range of 2500 to 3500 units per mg. For bioassay, nystatin is dissolved in dimethylformamide to give 1000 units per ml, then diluted with phosphate buffer pH 6 to 30 units per ml. Aliquots are placed in liquid or solid medium and test organisms planted.[14] The compound is more active between pH 4.4 to 6.6 on solid media, but in liquid media the optimum pH is 7. The inhibitory range of nystatin is given in Table 26–2. A simple laboratory procedure for sensitivity testing uses a disk containing 100 units. This is done by inoculating a plate with organism and placing a disk of anti-

Table 26–2. *Antifungal Range of Nystatin**

Fungi	*Least Amount Inhibiting Growth (units/ml agar)*†
Candida albicans, C. stellatoidea *C. parapsiosis, C. krusei* *C. tropicalis* *Histoplasma capsulatum* (yeast) *Paracoccidoides brasiliensis*	7.8–7.9
Cryptococcus neoformans	3.5–3.9
Blastomyces dermatitidis (yeast)	1.9
Sporothrix schenckii (yeast)	31.2
Rhodotorula sp., *Torulopsis* sp., *Saccharomyces* sp.	7.9
Penicillium sp.	7.9
Fusarium sp.	31.0
Allescheria boydii	500.0
Trichophyton tonsurans, Microsporum audouinii, M. canis	16.0

*Derived from several authors and personal experience.

†Average of several experiments. It must be noted that strains of species vary greatly in sensitivity, particularly among myceliate fungi. Most of the yeasts, particularly *Candida*, are rather constant in the range of sensitivity.

biotic on it. Zones of inhibition can be read within 24 to 72 hours.

The antibiotic has no effect on bacteria, and in concentrations of 100,000 units per ml it is not toxic to ECHO, coxsackie, or polio virus. In tissue culture it is used in a concentration of 30 units per ml to control fungal contaminants. In addition to antifungal activity, nystatin inhibits *Trichomonas vaginalis* at dilutions of 1:60,000. It also has leishmanicidal activity.[26] Nystatin resistance in *Candida* sp. has been produced in the laboratory.[25,49] Mutants of yeasts with low or no ergosterol in their cell membranes are also resistant to nystatin.[81] Such *in vitro*–produced resistant strains are nonpathogenic in experimental infection, and there is usually cross-resistance to other polyenes.[82] Back mutation to ergosterol- and cholesterol-containing membrane systems has been accomplished. Such cells are again pathogenic and sensitive to nystatin. Significant natural resistance to the drug in organisms recovered from patients following treatment has not been substantiated or is quite rare.[46]

Pharmacology. Even after massive oral doses, levels of no more than 1 to 2.5 mg per ml can be attained in serum. This indicates the compound is not absorbed to any degree from the intestinal tract. The use of intravenous therapy is precluded by the high toxicity, as demonstrated in several clinical trials.

Indications and Dosage. Nystatin is available as oral tablets containing 500,000 units. These can be given as two tablets t.i.d. in cases of intestinal overgrowth of *Candida* and continued until stool or oral cultures show normal numbers. As a prophylactic measure, it has been given at a dose of one tablet t.i.d. for seven to ten days. There is no evidence that this is efficacious in preventing colonization by *Candida* following abdominal surgery.

A nystatin suspension is used in the treatment of oral candidosis of children. It is available commercially in the form of a powder. After addition of water, each ml contains 100,000 units. It is stable for seven days. For children with recalcitrant thrush, 1 to 2 ml of the suspension is placed directly in the mouth q.i.d. Treatment is continued for 7 to 14 days or until clinical response. In adults this suspension may be used for oral candidosis, and as a lavage both for esophageal candidosis and in the treatment of mycotic keratitis. For oral disease, the suspension is given at a dose of 4 ml q.i.d. until clinical response is achieved. As a lavage in the treatment of mycotic keratitis, it is given as conjunctival drops every one to two hours or as necessary. This is done only after demonstration of sensitivity to the antibiotic by the etiologic agent. It is particularly useful in *Candida* infections.

Cutaneous Candidosis. Nystatin ointment at a concentration of 100,000 units per g is applied to the affected site twice daily. It is commonly combined with antibacterial antibiotics and a steroid, and contained in a cream or ointment base. One popular preparation includes neomycin, 2.5 mg, gramicidin, 0.25 mg, and triamcinolone acetonide, 1 mg, in addition to 100,000 units per g of nystatin.

For diaper disease, a nystatin topical powder is available. Each gram of talc contains 100,000 units of nystatin. It is applied liberally two or three times a day. *Candida* vaginitis is treated with vaginal suppositories containing 100,000 units of nystatin dispersed in lactose, ethyl cellulose, stearic acid, and starch. One to two tablets are inserted daily for 14 days. Recalcitrant infections may require some oral nystatin to lower the anal contamination.

Nystatin has been used as an aerosol in pulmonary candidosis (Chapter 8) and aspergillosis (Chapter 20). It was administered in a nebulizer containing 25,000 units suspended in 5 ml of saline twice daily for six weeks.[53]

Side Effects. Rare but occasional hypersensitivity has been reported in the topical use of nystatin. Usually as a cutaneous, vaginal, or oral preparation, it generally produces no serious side effects or irritations. Some records of diarrhea, nausea, and vomiting exist, and allergic contact dermatitis has been recorded.[16,78]

Amphotericin B (Fungizone)

$C_{46}H_{73}NO_{20}$, MW 960.10; deep yellow prisms or needles from dimethylformamide; m.p. gradually dec. above 170°C:$[\alpha]_D^{24}$ + 333° (in acidic dimethylformamide); $[\alpha]_D^{24}$ − 33.6° (in 0.1 N methanolic HCl). Amphoteric heptaene with mycosamine moiety. Molecular structure unknown. U.V. max., 364,383,408. Insoluble in water at pH 7, pH 2, or pH 11 ~ 0.1 mg per ml. Solubility in water increased by sodium deoxycholate. Solubility in dimethylformamide, 2 to 4 mg per ml; dimethylformamide + HCl, 60 to 80 mg per ml; dimethyl sulfoxide, 30 to 40 mg per ml. Solid and solutions stable between pH 4 and 10 for long periods at room temperature if protected from air and light. LD_{50} IP mice, 280 mg per kg; IV, 4.5 mg per kg; LD_2 IV, 2.2 mg per kg. Blood level of human tolerance 0.5 to 1.5 μ per ml. Extracted from broth cultures of *Streptomyces nodosus* by *iso*-PrOH at pH 10.5.

History. Although the first specific antifungal agent isolated was the polyene nystatin, the great toxicity of the substance when used intravenously and the lack of absorption from the gut precluded its use for treating systemic fungus infection. Gold et al.[27] in 1956 described a culture of streptomyces (known as M-4575) that was inhibitory to the *in vitro* growth of *Saccharomyces cerevisiae, Rhodotorula glutinis, Candida albicans,* and *Aspergillus niger.* Vanderputte et al.[75] isolated and characterized two antifungal substances from this streptomycete now known as amphotericin A and B. Amphotericin B is the more effective antifungal agent. Commercial amphotericin B preparations contain about 1 to 2 per cent amphotericin A. The streptomycete is now called *Streptomyces nodosus* and was isolated from a soil sample collected from near Tembladoro on the Orinoco River in Venezuela.[73] The usefulness of the antibiotic was at first restricted by its complete insolubility in water and its unpredictable absorption from the gut. It was found, however, that the wetting agent desoxycholate greatly increased the solubility of the drug. Presently, the marketed powder contains 50 mg of amphotericin B, 41 mg of sodium deoxycholate, and a phosphate buffer. When needed, it is dissolved in 10 ml of sterile water. It can then be added to more sterile water or to 5 per cent glucose up to an optimum dilution of 1 mg per 10 ml water. Though it appears to be a solution, the mixture is a fine colloidal dispersion. Any contamination with electrolytes, viz., sodium chloride, will cause the formation of flocs, making the mixture unsuitable for clinical use. A solution of amphotericin B in water or 5 per cent glucose is stable for about 24 hours.

The advent of a soluble form of the drug allowed its investigation as an antifungal agent for systemic infections.[48] Even though it has numerous serious side effects, particularly nephrotoxicity, it is the only proven standardized agent for the treatment of systemic mycoses. In addition, it has been used in the treatment of Chagas' disease, cutaneous leishmaniasis[63] and leprosy,[24] and schistosomiasis.[29] The drug is particularly efficacious in advanced mucocutaneous leishmaniasis and has been effective in the treatment of visceral leishmaniasis and Chagas' disease.

Range of Antimicrobial Activity. Laboratory tests have shown that the growth of a wide variety of fungi is inhibited by amphotericin B. Most of the fungi causing systemic infections in man are sensitive to the drug and at fairly constant levels.[27, 35, 37, 47, 72a] Tests on numbers of strains of several species by a number of investigators show little variation in level of drug causing inhibition. Therefore, the drug is always useful in treating disease caused by these organisms. These fungi are included in Group I of Table 26–3. There are some fungi that are quite variable in sensitivity, and strains may

Table 26–3. *Minimal Inhibitory Concentrations (μg/ml) of Amphotericin B for Various Fungi* and Protozoans*

Blood Level of Human Tolerance: 0.5–1.5 μg/ml

Group	*Organism*	*Average MIC*	*Range MIC*
I	*Candida albicans*	1.9	0.05–3.25
	Histoplasma capsulatum (Y)	0.04	
	Blastomyces dermatitidis (Y)	0.1	
	Coccidioides immitis	0.5	
	Paracoccidioides brasiliensis	0.1	
	Sporothrix schenckii (Y)	0.07	
	Cryptococcus neoformans	0.2	
	Leishmania brasiliensis	0.01	
	Torulopsis glabrata	0.25	
	Rhodotorula mucilaginosa	0.5	
II	*Candida parapsilosis*		3.0–>40
	Candida tropicalis	3.5	0.2–25
	Mucor heimalis		1.9–>40
	Rhizopus oryzae		0.9–>40
	Allescheria boydii	>40	
	Aspergillus fumigatus		1.0–>40
III	*Microsporum audouinii*	1.0	
	Microsporum canis	3.7	
	Trichophyton mentagrophytes	2.5	
	Trichophyton rubrum	7.3	
	Sporothrix schenckii (M)	>40	
	Geotricum candidum		3.0–14.0
	Phialophora verrucosa	>40	
	Fonsecaea pedrosoi	>40	
	Fonsecaea compaetum	>40	
	Acremonium kiliense	>40	
	Fusarium sp.		3–>40

*Results read at 24°C 72 hours post inoculation either broth or agar.

differ significantly in *in vivo* and *in vitro* response to the drug. Group II of Table 26–3 includes some important pathogens and opportunists, such as *Mucor* and *Rhizopus* sp. isolated from mucormycosis, *Allescheria boydii* from both eye infections and mycetoma, and some of the *Candida* species, such as *C. parapsilosis* and *C. tropicalis.* The drug may or may not be of benefit in diseases caused by these organisms. It has been the author's experience that results of sensitivity testing of organisms in Group I vary little from the figures given, and routine sensitivity testing is not necessary. The organisms in Group II are totally unpredictable, and each isolant must be tested for its level of sensitivity. It should be cautioned, however, that organisms relatively resistant by *in vitro* testing have been isolated from disease in which the drug was at least somewhat efficacious. The reverse is also true and probably involves host factors as well as inherent sensitivity of the fungus to the drug. Details of the sensitivity test are given in the Appendix. Group III includes the *in vitro* sensitivities of various other fungi with which the drug is ineffective, of questionable value, or not used. Resistance to amphotericin B has been produced *in vitro*[49, 69] but not authenticated *in vivo*. Cross-resistance with nystatin may or may not develop in *in vitro* studies.[82]

Absorption, Distribution, and Excretion.[50] Amphotericin B is poorly and unpredictably absorbed from the gastrointestinal tract following oral administration. Intake of 3 g per day produces plasma levels of about 0.1 to 0.5 μg per ml. In another study, 5 g was given during one day. Following this, 100 to 300 μg were detected in the 24-hour urine specimen. A daily dose of 1.6 to 5.6 g was followed by blood levels of up to 0.18 μg per ml. Some studies have shown no detectable blood level following ingestion of as much as 7.2 g.[48]

Oral amphotericin B is very useful in the treatment of chronic cutaneous candidosis. In the protocol outlined by Montes et al.,[56] patients received 1 to 1.8 g per day for six months. This was followed by resolution of the disease in most cases.

The intravenous injection of 1 to 5 mg of amphotericin B per day initially, followed by gradually increasing the dose to the desired level of 1 mg per kg per day, will lead to a peak serum level between 0.5 to 3.5 μg per ml, with an average of 1 mg per ml. This peak is maintained for about six to eight hours, with the half-life usually 24 hours. It is estimated that only 2 to 5 per cent of the drug given is excreted in a biologically active form in the urine. There are several molecular alterations that occur, particularly in the liver. Other studies on blood levels have shown a single 1 mg per kg dose was followed by levels of 0.36 μg per ml at 24 hours, 0.3 to 5.5 mg per ml at 48 hours, and demonstrable levels persisting at 72 hours.[13] For this reason alternate-day therapy is the most frequently used modality today.[3] Cerebrospinal fluid levels are about 1/40 that of blood levels.

The drug is slowly excreted in the urine. Detectable levels are frequently present at 72 and 96 hours after cessation of treatment. It has been detected in the kidney one year after the last intravenous dose.[59] For this reason, manifestation of nephrotoxicity may develop long after cessation of the therapy. Renal insufficiency does not appear to influence excretion or plasma levels.[22a] There is some evidence that the drug is initially bound to lipoproteins and then gradually released over a period of time. Several studies have shown that the drug is able to penetrate cells such as erythrocytes and leukocytes. In the latter case, it inhibited the growth of intracellular yeasts of *Histoplasma capsulatum.*

Side Effects.[37] Essentially no one receiving a full dose of amphotericin B therapy escapes without some side effects.[55] These vary from mild headache, chills, and fever to severe hemolytic anemia and acute nephritis. The list of reported side effects following intravenous administration includes fever, chills, headache, nausea, vomiting, anorexia, anaphylaxis, malaise, severe pain, phlebitis, hemolytic anemia, hypokalemia, hypotension, arrhythmias, diarrhea, melena, myalgia, arthralgia, and various manifestations of nephrotoxicity. The latter is the most constant and serious problem. Intrathecal administration is complicated by arachnoiditis, vertigo, diplopia, peripheral neuropathy, and convulsions. Arachnoiditis is the most serious and may be followed by paraplegia or other residua.

In most series almost all patients suffer some degree of chills and fever; anorexia is found in about half, headache in 40 per cent, and nausea and vomiting in about 20 per cent. Usually patients develop a degree of tolerance to some of these effects, and the dosage may be increased. However, chills and fever accompany almost every injection of the drug. If administration of the drug has been interrupted by seven days or more, it is necessary to gradually build up tolerance again.

Liver. Most reports have indicated there is little demonstrable damage to the liver in the majority of patients. In a few patients, particularly infants with widely disseminated disease, hepatic failure and jaundice have been seen. Focal necrosis was observed at autopsy.

Kidney. Renal toxicity is the most important single side effect of amphotericin B therapy. Some degree of azotemia is almost always seen during a course of therapy. The severity of this side effect may necessitate interruption of treatment. Blood urea nitrogen, and particularly, creatinine clearance must be monitored as an index of this complication. It is therefore necessary to have a complete renal function profile to act as a baseline before beginning treatment.

From studies in experimental animals, it appears that renal blood flow is reduced to about 10 per cent following injection of the drug.[12] This vasoconstriction is almost immediate, and no tolerance develops. At first there is a tubular leak, manifested by a few cells in the urine. Later, distal tubular necrosis occurs. Hypokalemia and hyposthenuria precede the development of azotemia. Nephrocalcinosis and hypokalemia are attributed to tubular acidosis, and alkali therapy during a prolonged treatment regimen may be necessary.[52] The urine from patients with moderate to severe renal damage contains the usual signs of nephritis: red and white blood cells, albumin, and granular and cellular casts. Though the response of patients varies, most of the damage is reversible. Permanent impairment or progressive fatal degeneration may occur, however. It is generally held that no permanent damage occurs when the regimen is less than 3 g and little if less than 5 g, but permanent damage is almost universally found if more than 5 g has been administered. The high doses (10 to 20 g) formerly used in treatment protocols led to a number of deaths due to progressive renal failure.[37] The methyl ester of amphotericin B is about 1.8 times less toxic,[39] but at this writing it has not received clinical evaluation.

Hematopoietic System. When high blood levels of amphotericin B are reached (0.5 to 5 μg per ml), there may develop a moderate to severe hemolytic anemia[40,41] After long-term treatment at lower doses, some depression of the erythropoietic system may occur, but without effect on the production of leukocytes or thrombocytes. The drug causes a normochromic or normocytic anemia accompanied by hypoferremia. The decrease in hemoglobin may be as much as 2.0 mg per 100 ml. This is almost always reversible.

The cardiac effects of drug therapy are rare and usually not severe. However, increase in heart size[15] and decrease in cardiac force and frequency have been noted. These symptoms usually disappear after therapy. However, irreversible cardiac arrest has occurred.[35] Tissue culture studies indicate cytotoxic effects occur at concentrations of 0.5 to 1 μg per ml. Impairment of prostate and testicular function has been found in animal studies,[72] but so far it has not been documented in human studies. Many patients have a transient hypocholesterolemia during treatment.[64] This reverses with the termination of therapy.

Therapeutic Uses. The drug was formerly available as oral tablets containing 250 mg and used for enteric candidosis. Considering the success obtained by Montes et al.[56] in the treatment of chronic cutaneous candidosis it is hoped that an oral preparation will again be marketed. A 3 per cent lotion is available for topical application. The most commonly used form for intravenous or intrathecal therapy is a lyophilized powder which contains 50 mg of amphotericin B, 41 mg sodium deoxycholate, buffers, and diluent. The contents are dissolved in 10 ml of sterile water and added to 5 per cent dextrose to give a concentration of 0.1 mg per ml. It is given as a slow infusion over six hours.[2a]

Several treatment schedules are available.[2,3,13,23] One involves the administration of 0.25 mg per kg the first day, followed by increments of 0.25 mg per kg each day until the 1.5 mg per kg level is reached. Alternate-day double dose therapy is recommended to increase patient tolerance.[3] Intrathecal amphotericin B has been used in cases of meningitis, but it is of questionable benefit to the patient owing to the high percentage of severe complications.[1a, 19a] The usual method is to dissolve 0.5 mg in 5 ml of spinal fluid and to reinject the spinal fluid. This is done two to three times a week until a total dose of 15 mg is reached. This is considered adequate by most investigators (see chapters on specific diseases). Intrathecal steroids are often used. Some of the side effects engendered by intravenous therapy can be ameliorated by additives. Commonly aspirin, cortisone, and heparin are used (Chapter 17). The total course required varies with the severity of the disease, the immune state of the patient, and the etiologic agent.

The total experience with amphotericin B has shown it to be efficacious in most systemic fungal infections. The series vary, but in general the following data have been accumulated. Essentially all cases of pulmonary and disseminated blastomycosis have been cured with adequate treatment. About 50 per cent of acute pulmonary or disseminated coccidioidomycosis responded to therapy. Many of the treatment failures were associated with pregnancy. Whereas the mortality rate of disseminated cryptococcosis formerly was 83 per cent and that of cryptococcal meningitis 100 per cent, amphotericin B has decreased this to about 30 per cent.[23] About 75 per cent of patients with disseminated or chronic cavitary histoplasmosis are improved or cured following therapy. Pulmonary and disseminated candidosis usually respond to a schedule of 1 to 3 g. Iatrogenic candidemia (from catheters and such) has responded to as little as 10 to 300 mg total dose. In cases of *Candida* cystitis, the urinary bladder was irrigated with 15 mg per day, which cleared the infection. *Candida* endocarditis does not usually respond to amphotericin B therapy. American cutaneous leishmaniasis has been cured following a total dose of 200 to 1450 mg of the drug.

Laboratory studies have indicated some synergistic effects with other antibiotics, such as rifampicin[42, 42a, 54] and 5-fluorocytosine. It may be possible in the future to augment the efficacy of amphotericin B by using combinations of drugs. This may allow the administration of the drug at lower levels, hopefully lessening the toxic side effects. The newly described methyl ester of amphotericin B is much less toxic than the parent compound.[39] Its efficacy in clinical trial remains to be determined.

Pimaricin (Tennecetin, Natamycin, Pimifucin, Myprozine)

$C_{33}H_{47}O_{13}N$ (Fig. 26–2), MW 665.75, amphoteric colorless crystal insoluble in H_2O, acetone, alcohol, chloroform; soluble in dilute acid or alkali; m.p. ca. 200°C (dec.), $[\alpha]_D^{25}$ + 250° (MeOH), U.V. max. (in MeOH), 279, 290, 303, 318 mμ, similar to other tetraenes (nystatin, amphotericin A, and Rimocidin). Extracted from mycelium of *Streptomyces natalensis* by lower alcohols, glycols, or formamides. Original isolant from Pietermaritzburg in Natal, South Africa. LD in mice, 650 mg per kg IP, 1500 mg per kg per os.

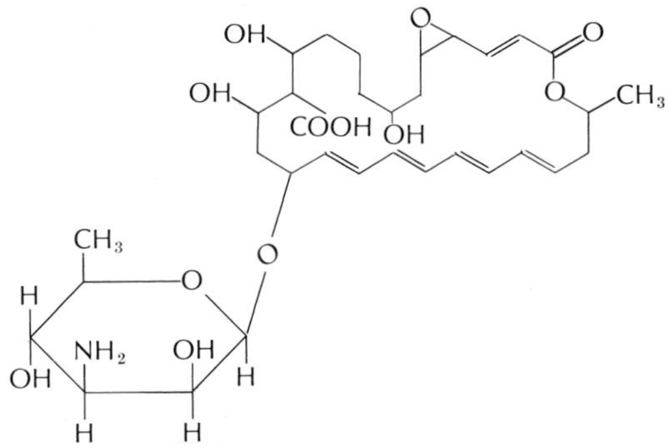

Figure 26–2. Pimaricin.

This drug has been under investigation since 1959.[1] It has been advocated for use as an aerosol in pulmonary candidosis and aspergillosis[34] and appears to be popular as a topical treatment for various cutaneous fungous infections in Europe and some other countries.[58] To date, its most important and almost unique place as an antifungal agent is in mycotic keratitis (Chapter 23).

Griseofulvin (Fulcin, Grifulvin, Grisovin, Lamoryl, Grisactin, Sporostatin, Spirofulvin, Likuden, Poncyl)

$C_{17}H_{17}ClO_6$, MW 352.77 (Fig. 26–3). Antibiotic from *Penicillium griseofulvum*[11, 57] and *P. janczewskii* Zal. [= *P. nigricans* (Banier) Thom], octrahedra from benzene, m.p. 220°C, $[\alpha]_D^{17}$ + 370° absorption max., 286,325. Practically insoluble in water; slightly soluble in ethanol, methanol, acetone, benzene, chloroform, ethyl acetate, and acetic acid; soluble in dimethylformamide

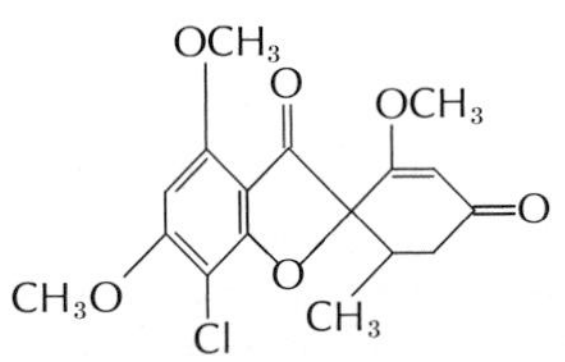

Figure 26–3. Griseofulvin.

(25°C) 12 to 14 g per 100 ml. Also soluble in dimethyl sulfoxide. It is thermostable and unaltered by autoclaving. Several spectrophotometric procedures for its determination have been devised.[8, 22, 80]

History. In a series of works on the biochemistry of microorganisms, Oxford, Raistrick, et al. discovered griseofulvin.[57] In 1939 they reported on its isolation from *Penicillium griseofulvum* (Dierk.) and its chemical and physical properties, but did not study its biological effects. Sometime later (1956) a compound was described by Brian, Curtis, and Hemming from a culture of *P. janczewskii*.[10] This substance caused the mycelium of *Botrytis allii* to become stunted and form spirals, and the compound became known as curling factor. Later Brian identified the "curling factor" as identical to the griseofulvin of Oxford et al. It was proposed by Brian that griseofulvin would be useful as an antimycotic agent of plants.[9] It was not until the work of Gentles in 1958 that it was shown to be useful in the treatment of animal ringworm infections. Since that time, it has become the standard treatment for all forms of human and animal dermatophytosis.

It is now known that griseofulvin is produced by a number of *Penicillium* sp. The antibiotic is extracted from the mycelial mat following growth on broth. Several chemical alterations have been tried. Dechlorogriseofulvin and bromogriseofulvin are less active on assay (curling of *B. allii* test) than the parent substance. However, the *n*-butyl and *n*-propyl homologues are twenty times more active. The biosynthesis of griseofulvin was described by Rhodes et al.[60]

Range of Antimicrobial Activity. *In vitro* griseofulvin is active against almost all the dermatophytes at concentrations less than 1 μg per ml when tests are read at 72 hours (Table 26–4). In higher concentrations, it inhibits many other fungi to varying degrees.[36] Therapeutically it is useful against dermatophyte infections, but has little or no efficacy in other types of fungous infections.[28] It is not active against bacteria, most yeasts, Phycomycetes, viruses, or rickettsiae.[9, 35, 36, 62]

In vitro resistance can be produced by growing some strains of *Trichophyton, Epidermophyton*, and *Microsporum* on increasing concentrations of the drug. This resistance does not disappear with subsequent transfer to antibiotic-free media. When the fungus is used to infect animals, the disease produced

Table 26–4. *Minimal Inhibitory Concentrations of Griseofulvin for Common Dermatophytes**

Organism	*Average MIC*
Microsproum audouinii	0.4 –0.6
M. canis	0.2 –0.24
M. gypseum	0.4 –0.46
Trichophyton mentagrophytes	0.38–0.43
T. rubrum	0.14–0.18
T. tonsurans	0.30–0.34
T. gallinae	0.40–0.44
T. schoenleinii	0.34–0.38
T. violaceum	0.36–0.40
T. megninii	0.30–0.34
T. verrucosum	0.28–0.30
T. concentricum	0.26–0.30
Epidermophyton floccosum	0.38–0.42

**In vitro* agar plates 25°C read at 72 hours; at 160 hours the MIC are double, on the average.

From Roth, F. J., B. Sallman, et al. *In vitro* studies of the antifungal antibiotic griseofulvin. J. Invest. Dermatol., *33*:403–418. © 1959 The Williams & Wilkins Co., Baltimore.

is the same as that elicited by parent antibiotic sensitive strains. Natural resistance of strains from human infections in which the drug has been used in therapy appears to occur occasionally. The antibiotic is fungistatic and not fungicidal. Actively growing mycelia are killed by the drug, but it does not affect dormant hyphae. In time, the mycelium destroys griseofulvin by demethylation.[7]

Mode of Action. Large quantities of the antibiotic are taken up by actively growing mycelium. There appears to be a two-stage binding of the drug. The first step is an immediate reaction not requiring energy, whereas the second stage is prolonged and requires active metabolism by the organism. Griseofulvin is bound to lipids within the cell but not to RNA or DNA. DNA metabolism is enhanced, and there is a greater content of this substance in drug-treated cells than in normal cells.

The drug appears to have a colchicine-like action and has been used in the treatment of gout. It also has some anti-inflammatory activity[18] and has been used in a number of diseases, such as gout,[76] angina pectoris,[19] and shoulder-hand syndrome.[35]

Therapeutic Uses.[83] The drug is useful in all forms of dermatophyte infection (Chapter 7). The usual oral daily dose is 10 mg per kg for children between 30 and 50 pounds, and 125 to 250 mg for children over 50 pounds. The usual adult dose is 0.5 to 1.0 g per day. In severe infection, up to 2 g per day can be given. Absorption is enhanced if the drug is taken after a fatty meal.[17] It is rapidly deposited in the epidermis and subsequently in the stratum corneum.[22]

Side Effects. Usually there are no serious side effects to therapy. About 55 per cent of patients have headache at the onset of therapy. This usually disappears as therapy is continued. Some patients complain of gastrointestinal distress, nausea, vomiting, and diarrhea. Rare neurologic symptoms include neuritis, lethargy, mental confusion, syncope, vertigo, blurred vision, and transient macular edema. Other rare side effects include urticaria, stomatitis, glossodynia, albuminuria, cylinduria, and renal insufficiency. All these resolve after cessation of treatment. Estrogen-like effects have been noted in some children. In most studies, topical application of griseofulvin is without significant benefit.

A serious consequence of griseofulvin therapy has been porphyria cutanea tarda,[83] as well as other porphyrias.[61] Protoporphyrin deposition in the liver, liver cirrhosis, and hepatomas have been produced in the mouse[38] but not in rats, guinea pigs, or rabbits. Rats have received up to 2000 mg per kg for three months without notable effects, and guinea pigs maintained on 500 mg per kg for 12 months have shown no ill effects.

Tolnaftate (Focusan, Hi-Alazin, Sporiline, Tinactin, Tinaderm, Tonoftal)

$C_{19}H_{17}NOS$, *m, N*-dimethylthiocarbanilic acid *O*-2 naphthyl ester (Fig. 26–4). Crystals from alc., m.p. 110.5–111.5°. Soluble in chloroform; sparingly soluble in ether; slightly soluble in alcohol; essentially insoluble in water.

This compound was one of the first synthetic chemical agents to have topical antifungal activity. *In vitro* studies indicate that it inhibits the growth of *T. mentagrophytes* at concentrations of 0.0075 to 0.075 μ per ml. The compound is without effect on yeasts (*C. albicans*) or bacteria. As a 1 per cent compound in polyethylene glycol or vanishing cream base, it is effective against infections of the cutaneous regions caused by most dermatophytes. It is not effective in tinea capitis or tinea unguium. The relapse rate for this drug in cutaneous infections is about the same as that for griseofulvin. The main advantage of this preparation is that it can be applied topically. Toxic or allergic reactions have not yet been recorded with frequency. Tolnaftate USP (Tinactin) is available as a cream, powder, or 1 per cent solution. It is usually applied twice daily until symptoms disappear. Pruritus is usually relieved within 72 hours, but treatment should be continued for 14 to 21 days to prevent relapse of infection.[31]

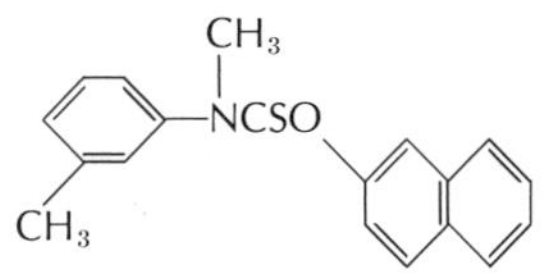

CH₃ / NCSO / CH₃

Figure 26–4. Tolnaftate.

Haloprogin (Halotex, Polik)

$C_9H_4Cl_3IO$, MW 361.41, 3-iodo-2-propynyl-2,4,5-trichlorophenyl ether (Fig. 26–5). Crystals with m.p. at 113–114°C.

This compound under the name of Halotex has been marketed as a topical antifungal agent. In *in vitro* studies,[65] it has a recorded MIC of 0.78 μg per ml for *T. rubrum* and 0.48 μg per ml for *T. tonsurans*. It has approximately equal efficacy[32] against most of the etiologic agents of dermatophyte infection. In *in vivo* studies of its metabolism, the compound was rapidly absorbed percutaneously (in rats) and appeared unchanged in urine. No tissue storage was detected. As a 1 per cent compound per gram in various bases, it appears to be well tolerated and efficacious in most dermatophyte infections and has an activity about equal to that of tolnaftate as presently used. It is not effective in tinea capitis and tinea unguium. No reports of significant irritation or contact allergy are as yet available. It has the advantage of being effective against *Candida* sp. also.

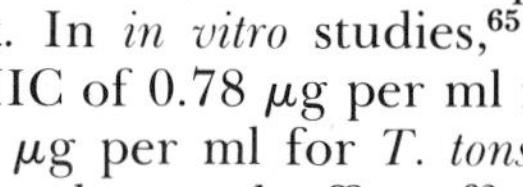

Cl / Cl / $OCH_2C{\equiv}CI$ / Cl

Figure 26–5. Haloprogin.

5-Fluorocytosine (Ancobon, Flucytosine)

This compound has recently been introduced as an antifungal agent[2a, 32a, 64a, 72a, 74, 77]

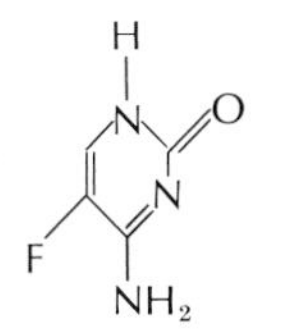

Figure 26–6. 5-Fluorocytosine.

(Fig. 26–6). It was first shown to have antifungal activity in 1964 and has since been demonstrated to have efficacy in cryptococcal meningitis and *Candida* sepsis.[71] Unfortunately relapses have occurred in both. It has the advantage of being relatively nontoxic by oral dose[74] of 150 mg per kg per day. Treatment must be continued until clinical cure is assured. *C. neoformans* appears to develop resistance only after treatment with lower doses.[5a] However, almost 50 per cent of *Candida albicans* strains are resistant without previous exposure at the prescribed human tolerance levels.[64a, 66, 68] Because it can be given orally and appears to have few side effects, it may be considered as a first choice in cryptococcal meningitis or as an adjunct to IV amphotericin B. It has also been used successfully in infections due to *Torulopsis glabrata.* Some cases of pancytopenia after its use have been recorded (Chapter 8). Hepatic insufficiency does not influence serum levels.[4] Investigators in Mexico and Latin America have been using the drug at maximal levels for the treatment of mycotic mycetoma, sporotrichosis, chromomycosis, and other fungal diseases. At this writing, it is too soon to evaluate its efficacy in these diseases. In experimental infections, it has been found ineffective in sporotrichosis but of benefit in cladosporiosis.[5]

It should be noted that *in vitro* studies must be carried out in media free of cytosine or uridine. Thus media containing beef extract, peptone, or yeast extract will negate the effect of the compound. It appears that the range of activity of 5-fluorocytosine is restricted to a few yeasts. It has, however, been demonstrated to potentiate the effect of amphotericin B on yeasts *in vitro.*[54]

Miconazole

R 14889 Janssen Lab. MW 479.16, $C_{18}H_{14}Cl_4N_2O$, 2,4-dichloro-β(2,4-dichlorobenzyl oxylphenethyl) imidazole nitrate (Fig. 26–7). Slightly soluble in water (0.03 per cent) or organic solvents. Pharmacology: administration of 40 mg per kg per day orally in rats produced no signs of toxicity after three months. Necropsy normal. Acute toxicity orally 600 mg per kg in rats. *In vitro* range of activity is similar to that of tolnaftate. Most dermatophytes are sensitive in a range of 1–10 μg per ml. *Nocardia* sp., *Streptomyces* sp., *Histoplasma capsulatum* (Y), and *Blastomyces dermatitidis* (Y) are sensitive in a range of 0.1–1 μg per ml. Other agents of deep mycoses vary from 1–100 μg per ml.

This product is new, and at this writing there has been little clinical experience with it. Most studies have involved its use as a topical agent. As a 2 per cent cream applied one or two times per day for ten days, it is about as effective as tolnaftate for the treatment of dermatophytosis and more effective than topical nystatin for cutaneous candidosis. Infections with *Trichophyton verrucosum* were found resistant. One investigator has reported complete cure of tinea unguium after eight months therapy. Its efficacy in systemic and subcutaneous infections is being studied by a number of investigators.

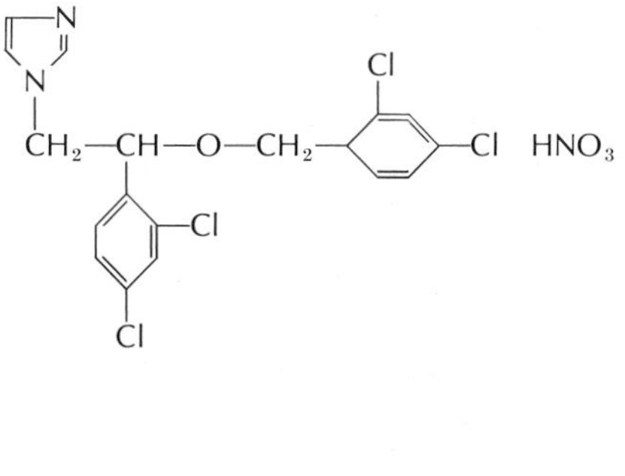

Figure 26–7. Miconazole.

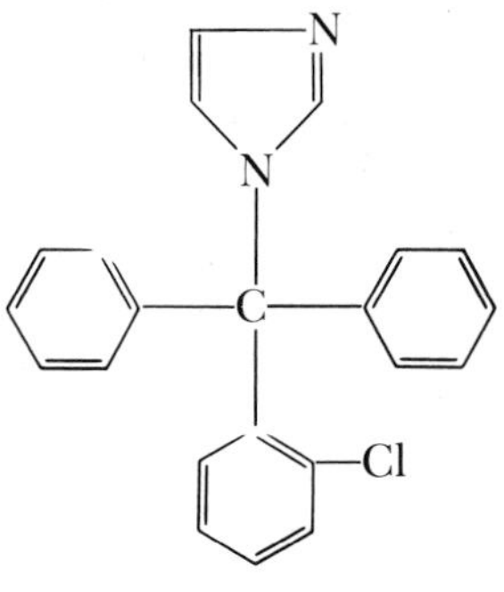

Figure 26–8. Clotrimazole.

Clotrimazole

Clotrimazole (Bay b 5097) bis-phenyl(2-chlorophenyl)-1-imidazole methane (Fig. 26–8) (Delbay Pharmaceuticals) is a chlorinated trityl imidazolyl, which has been shown to have *in vitro* activity against a number of pathogenic fungi at a concentration of 4 μg per ml or less.[67] *Histoplasma capsulatum, Blastomyces dermatitidis, Sporothrix schenckii, Cryptococcus neoformans,* and *Coccidioides immitis* were found to be inhibited in a range of 0.20 to 3.13 μg per ml. This is comparable to the range of amphotericin B, which is 0.10 to 6.25 μg per ml. The activity against *Candida albicans, Aspergillus,* and *Allescheria* was quite variable. Against dermatophytes, it was found that 0.78 μg per ml would suppress the growth of most species. Presently it is being marketed as a cream and an ointment for the treatment of dermatophyte and *Candida* infections of the skin and as suppositories for the treatment of vaginal candidosis. Investigations are presently under way in regard to its efficacy in systemic mycoses.[2a, 32a, 37a, 50a]

REFERENCES

1. Alteras, I., and I. Cojocaru. 1969. Apercu critique sur l'action antifongique de la pimaricine. Mykosen, *121*:139–149.

1a. Alazraki, N. P., J. Fierer, et al. 1974. Use of hyperbaric solution for administration of intrathecal amphotericin B. New Engl. J. Med., *290*:641–646.

2. Andriole, V. T., and H. M. Kavetz. 1962. The use of amphotericin B in man. J.A.M.A., *180*:269–272.

2a. Bennett, J. E. 1974. Chemotherapy of systemic mycoses. New Engl. J. Med., *289*:30–32, and *290*: 320–324.

3. Bindschadler, D. D., and J. E. Bennett. 1969. A pharmacologic guide to the clinical use of amphotericin B. J. Infect. Dis., *20*:427–436.

4. Block, E. R. 1973. Effect of hepatic insufficiency on 5-fluorocytosine concentrations in serum. Antimicrob. Agents Chemother., *3*:141–142.

5. Block, E. R., A. E. Jennings, et al. 1973. Experimental therapy of cladosporiosis and sporotrichosis with 5-fluorocytosine. Antimicrob. Agents and Chemother., *3*:95–98.

5a. Block, E. R., A. E. Jennings, et al. 1973. 5-Flurocytosine resistance in *Cryptococcus neoformans.* Antimicrob. Agents Chemother., *3*:649–656.

6. Blum, S. F., S. B. Shohet, et al. 1969. The effect of amphotericin B on erythrocyte membrane cation permeability; its relation to in vivo erythrocyte survival. J. Lab. Clin. Med., *73*:980–987.

7. Boothroyd, B., E. J. Napier, et al. 1961. The demethylation of griseofulvin by fungi. Biochem. J., *80*:34–37.

8. Bedford, C., D. Busfield, et al. 1959. Spectrophotofluorometric assay of griseofulvin. Nature (Lond.), *184*:364–365.

9. Brian, P. W. 1960. Control of fungal diseases of plants with griseofulvin. Trans. St. John's Hosp. Dermatol. Soc. *45*:4–6.

10. Brian, P. W., P. J. Curtis, et al. 1946. A substance causing abnormal development of fungal hyphae produced by *Penicillium janczewskii* Zal. I. Biological assay, production and isolation of curling factor. Trans. Br. Mycol. Soc., *29*:173–187.

11. Brain, P. W., P. J. Curtis, et al. 1955. Production of griseofulvin by *Penicillium raistrickii.* Trans. Br. Mycol. Soc., *38*:305–308.

12. Butler, W. T., G. J. Hill, et al. 1964. Amphotericin B renal toxicity in dog. J. Pharmacol. Exp. Therap., *143*:47–56.

13. Campbell, G. D., H. E. Einstein, et al. 1969. Indications for chemotherapy in the pulmonary mycoses. A report of the committee on fungus diseases, subcommittee on therapy – American College of Chest Physicians. Dis. Chest, *55*:160–162.

14. Carlson, J. R., and J. W. Snyder. 1959. *Candida albicans* plate assay of nystatin. Antibiot. Chemother. (Basel), *9*:139–144.

15. Chung, D. K., and M. G. Koenig. 1971. Reversible cardiac enlargement during treatment with amphotericin B and hydrocortisone. Report of three cases. Am. Rev. Resp. Dis., *103*:831–841.

16. Cosey, R. J. 1971. Contact dermatitis due to nystatin. Arch. Dermatol., *103*:228.

17. Crounse, R. G. 1961. Human pharmacology of griseofulvin – The effect of fat intake on gastrointestinal absorption. J. Invest. Dermatol., *37*:529–533.

18. D'Ary, F., E. M. Howard, et al. 1960. The antiinflammatory action of griseofulvin in experimental animals. J. Pharm. Pharmacol., *12*:659–665.

19. DePasquale, N. P., J. W. Burks, et al. 1963. The treatment of angina pectoris with griseofulvin. J.A.M.A., *184*:421–422.

19a. Diamond, R. D., and J. E. Bennett. 1973. A subcutaneous reservoir for intrathecal therapy of fungal meningitis. New Engl. J. Med., *288*: 186–188.

20. Douglas, J. B., and J. K. Healy. 1969. Nephrotoxic effects of amphotericin B including renal tubular acidosis. Am. J. Med., *46*:154–162.

21. Eisenberger, R. S., and W. H. Oatway. 1971. Nebu-

lization of amphotericin B. Am. Rev. Resp. Dis., *103*:289–292.

22. Epstein, W. L., V. P. Shah, et al., 1972. Griseofulvin levels in stratum corneum. Arch. Dermatol., *106*:344–348.

22a. Feldman, H. A., J. D. Hamilton, et al. 1973. Amphotericin B therapy in an anephric patient. Antimicrob. Agents Chemother., *4*:302–305.

23. Furcolow, M. L. 1963. The use of amphotericin B in blastomycosis, crytpococcosis and histoplasmosis. Med. Clin. North Am., *47*:1119–1130.

24. Furtardo, T. A., E. Cisalpino, et al. 1962. Amphotericin B in the treatment of lepromatous leprosy. O. Hospital (Rio), *62*:1099–1109.

25. Gale, G. R. 1963. Cytology of *Candida albicans* as influenced by drugs acting on the cytoplasmic membrane. J. Bacteriol., *86*:151–157.

26. Ghosh, B. K., D. Haldar, et al. 1961. Leishmanicidal property of nystatin and its clinical application. Ann. Biochem. Exp. Med., *21*:25–28.

27. Gold, W., H. A. Stout, et al. 1956. Amphotericins A and B: Antifungal antibiotics produced by a streptomycete. I. *In vitro* studies. *Antibiotics Annual*, 1955–1956. New York, Medical Encyclopedia, pp. 579–586.

28. González Ochoa, A. 1960. Griseofulvin in deep mycoses. Ann. N.Y. Acad. Sci., *89*:254–257.

29. Gordon, B. L., P. A. S. St. John, et al. 1963. Chemotherapy of schistosomiasis in mice with amphotericin B. Bacteriol. Proc., *63*:95.

30. Gottlieb, D., H. E. Carter, et al. 1958. Protection of fungi against polyene antibiotics by sterols. Science, *128*:361.

31. Gould, A. H. 1964. Tolnaphtate in dermatology. A new potent topical fungicide. Dermatol. Trop., *3*:255–253.

32a. Hamilton-Miller, J. M. T. 1972. A comparative *in vitro* study of amphotericin B, clotrimazole and 5-fluorocytosine against clinically isolated yeasts. Sabouraudia, *10*:276–283.

32. Harrison, E. F., P. Zwadyk, et al. 1970. A topical antifungal agent. Appl. Microbiol., *19*:746–750.

33. Hazen, E., and R. Brown. 1950. Two antifungal agents produced by a soil actinomycete. Science, *112*:423.

34. Henderson, A. H., and S. E. G. Pearson. 1968. Treatment of bronchopulmonary aspergillosis with observations of the use of natamycin. Thorax, *25*:519–523.

35. Hildick-Smith, G. 1968. Antifungal antibiotics. Pediat. Clin. North Am., *15*:107–118.

36. Hildick-Smith, G., H. Blank, et al. 1964. *Fungus Diseases and Their Treatment.* Boston, Little, Brown & Co., Chapters 41–57.

37. Holeman, C. W., and H. Einstein. 1963. The toxic effects of amphotericin B in man. California Med., *99*:90–93.

37a. Holt, R. J., and L. Newman. 1972. Laboratory assessment of the antimycotic drug clotrimazole. J. Clin. Pathol., *25*:1089–1097.

38. Hurst, E. M., and G. E. Paget. 1963. Protoporphyrin cirrhosis and hepatoma in the livers of mice given griseofulvin. Br. J. Dermatol., *75*:105–112.

39. Keim, G. R., J. W. Poutsiaka, et al. 1973. Amphotericin B methyl ester hydrochloride and amphotericin B: Comparative acute toxicity. Science, *179*:584–585.

40. Kinsky, S. C., J. Avruch, et al. 1962. The lytic effect of polyene antifungal antibiotics on mammalian erythrocytes. Biochem. Biophys. Res. Commun., *9*:503–507.

41. Kinsky, S. C. 1970. Antibiotic interaction with model membranes. Ann. Rev. Pharmacol., Palo Alto, Annual Reviews, pp. 119–142.

42. Kobayashi, G., G. Medoff, et al. 1972. Amphotericin B potentiation of rifampicin on an antifungal agent against the yeast phase of *Histoplasma capsulatum.* Science, *177*:709–710.

42a. Kobayashi, G., C. Sze-Schuen, et al. 1974. Effects of rifamycin derivatives, alone and in combination with amphotericin B against *Histoplasma capsulatum.* Antimicrobiol. Agents Chemother., *5*:16–18.

43. Korzybski, T., Z. Kowszk-Gindifer, et al. 1967. Antifungal antibiotics, in *Antibiotics.* Vol. I. Oxford, Pergamon Press, pp. 769–846.

44. Lampen, J. O. 1969. Amphotericin B and other polyenic antifungal antibiotics. Am. J. Clin. Pathol., *52*:138–146.

45. Lane, J. W., R. G. Garrison, et al. 1972. Drug-induced alterations in the ultrastructure of *Histoplasma capsulatum* and *Blastomyces dermatitidis.* Mycopathologia, *48*:289–296.

46. Larsh, H. W. 1962. The prevalence of the drug resistance of *Histoplasma capsulatum* and *Candida albicans* in patients with pulmonary histoplasmosis. Lab. Invest., *11*:1140–1145.

47. Lepper, M. H., J. Lockwood, et al. 1959. Studies on the epidemiology of strains of *Candida* in hospital wards. *Antibiotics Annual*, 1958–1959. New York, Medical Encyclopedia, pp. 666–671.

48. Littman, M. L., and P. L. Horowitz. 1958. Coccidioidomycosis and its treatment with amphotericin B in man. Am. J. Med., *24*:568–592.

49. Littman, M. L., M. A. Pisano, et al. 1958. Induced resistance of *Candida* species to nystatin and amphotericin B. *Antibiotics Annual*, 1957–1958. New York, Medical Encyclopedia, pp. 981–987.

50. Louria, D. B. 1958. Some aspects of absorption distribution and excretion of amphotericin B. Antibiot. Med. Clin. Ther., *5*:295–301.

50a. Mahgoub, E. S. 1972. Laboratory and clinical experience with clotrimazole (Bay b 5097). Sabouraudia, *10*:210–217.

51. McCoy, E., and J. S. Kiser. 1959. An evaluation of nystatin and candicidin against a standardized systemic *Candida albicans* infection in mice. *Antibiotics Annual*, 1958–1959. New York, Medical Encyclopedia, pp. 903–909.

52. McCurdy, D. K., M. Frederic, et al. 1968. Renal tubular acidosis due to amphotericin B. New Engl. J. Med., *278*:124–130.

53. McKendrick, G. D. W., and J. M. Medlock. 1958. Pulmonary moniliasis treated with nystatin aerosol. Lancet, *1*:631.

54. Medoff, G., G. S. Kobayashi, et al. 1972. Potentiation of rifampicin and 5-fluorocytosine as antifungal antibiotics by amphotericin B. Proc. Natl. Acad. Sci. USA *69*:196–199.

55. Miller, R. P., and J. H. Bates. 1969. Amphotericin B toxicity. A follow-up report of 53 patients. Ann. Intern. Med., *71*:1089–1095.

56. Montes, L., M. D. Cooper, et al. 1971. Prolonged oral treatment of chronic mucocutaneous candidiasis with amphotericin B. Arch. Dermatol., *104*:45–55.

57. Oxford, A. E., H. Raistrick, et al. 1939. XXIX. Studies on the biochemistry of microorganisms. LX. Griseofulvin $C_{18}H_{17}O_6Cl$, a metabolic product of *Penicillium griseofulvum* Dierck. Biochem. J., *33*:240–248.
58. Polay, A. 1969. Erfahrungen mit einem neuen antimykotikum dem Pimaricin. Mykosen, *12*: 677–686.
59. Renolds, B. S., Z. M. Tomkiewicz, et al. 1963. The renal lesions related to amphotericin B treatment of coccidioidomycosis. Med. Clin. N. Amer., *47*:1149–54.
60. Rhodes, A., G. A. Sommerfield, et al. 1963. Biosynthesis of griseofulvin. Biochem. J., *88*:349–356.
61. Rimington, C., P. N. Morgan, et al. 1963. Griseofulvin administration and porphyrin metabolism. A survey. Lancet, *2*:318–322.
62. Roth, F. J., B. Sallman, et al. 1959. *In vitro* studies of the antifungal antibiotic griseofulvin. J. Invest. Dermatol., *33*:403–418.
63. Sampaio, S. A., J. T. Godoy, et al. 1960. Treatment of American mucocutaneous leishmaniasis with amphotericin B. Arch. Dermatol., *82*:627–635.
64. Schaffner, C. P., and H. W. Gordon. 1968. The hypocholesterolemic activity of orally administered polyene macrolides. Proc. Natl. Acad. Sci. USA, *61*:36–41.
64a. Schönbeck, J., and S. Ansehu. 1973. 5-Fluorocytosine resistance in *Candida* sp. and *Torulopsis glabrata*. Sabouraudia, *11*:10–20.
65. Seki, S., B. Nomiya, et al. 1963. Laboratory evaluation of M-1028 (2,4,5-trichlorophenyl γ-iodopropargyl ether), a new antimicrobial agent. In *Antimicrobial Agents and Chemotherapy*. Sylvester, J. C. (ed.), Ann Arbor, American Society of Microbiology, pp. 569–572.
66. Shadomy, S. 1969. *In vitro* studies with 5-fluorocytosine. *Appl. Microbiol., 17*:871–877.
67. Shadomy, S. 1971. *In vitro* and fungal activity of clotrimazole (Bay b5097). Infect. Immun. *4*: 143–148.
68. Shadomy, S., C. B. Kirchoff, et al. 1973. *In vitro* activity of 5-fluorocytosine against *Candida* and *Torulopsis* species. Antimicrob. Agents Chemother., *3*:9–14.
69. Sorensen, L. J., E. G. McNall, et al. 1959. The development of strains of *Candida albicans* and *Coccidioides immitis* which are resistant to amphotericin B. *Antibiotics Annual*, 1958–59. New York, Medical Encyclopedia, pp. 920–923.
70. Struyk, A. P., I. Hoette, et al. 1958. Pimaricin, a new antifungal antibiotic. *Antibiotics Annual*, 1957–1958. New York, Medical Encyclopedia, pp. 878–885.
71. Tassel, D., and M. A. Medoff. 1968. Treatment of Candida sepsis and Cryptococcus meningitis with 5-fluorocytosine: A new antifungal agent. J.A.M.A., *206*:830–832.
72. Texter, J. H., and D. X. Coffey. 1969. The effects of amphotericin B on prostatic and testicular function in the dog. Trans. Am. Assoc. Genitourin. Surg., *61*:101–114.
72a. Titsworth, E., and E. Grunberg. 1973. Chemotherapeutic activity of 5-fluorocytosine and amphotericin B against *Candida albicans* in mice. Antimicrob. Agents Chemother., *4*: 306–308.
73. Trejo, W. H., and R. E. Bennett. 1963. *Streptomyces nodosus* sp. nov. The amphotericin producing organism. J. Bacteriol., *85*:436–439.
74. Utz, J. 1972. Flucytosine. Editorial. New Engl. J. Med., *286*:777–778.
75. Vandeputte, J., J. L. Wachtel, et al. 1956. Amphotericin A and B, antifungal antibiotics produced by a streptomycete. The isolation and properties of crystalline amphotericins. *Antibiotics Annual*, 1955–1956. New York, Medical Encyclopedia, pp. 587–595.
76. Wallace, S. L., and A. W. Nissen. 1962. Griseofulvin in acute gout. New Engl. J. Med., *266*:1099–1101.
77. Warner, J. F., R. F. McGee, et al. 1971. 5-Fluorocytosine in human candidiasis. *Antimicrobial Agents and Chemotherapy*. II. Bethesda, American Society of Microbiology, pp. 473–475.
78. Wasilewski, C. 1970. Allergic contact dermatitis from nystatin. Arch. Dermatol., *102*:216–217.
79. Weber, M. M., and S. C. Kinsky. 1965. Effect of cholesterol on the sensitivity of *Mycoplasma laidlawii* to the polyene antibiotic filipin. J. Bacteriol., *89*:306–312.
80. Weinstein, G. D., and H. Blank. 1960. Quantitative determination of griseofulvin by a spectrophotometric assay. Arch. Dermatol., *81*:746–749.
81. Woods, R. A. 1971. Nystatin resistant mutants of yeast: Alterations in sterol content. J. Bacteriol., *108*:69–73.
82. Woods, R. A., and K. A. Ahmed. 1968. Genetically controlled cross resistance to polyene antibiotics in *Saccharomyces cerevisiae*. Nature (Lond.), *218*:1080–1084.
83. Zaias, N., D. Taplin, et al. 1966. Evaluation of microcrystallin griseofulvin in tinea capitis. J.A.M.A., *198*:805–807.
84. Ziprkowski, L., A. Zeinberg, et al. 1966. The effect of griseofulvin in hereditary porphyria cutanes tarda. Investigations of porphyrins and blood lipids. Arch. Dermatol., *93*:21–22.

APPENDIX

CULTURE METHODS, MEDIA, STAINS, AND SEROLOGIC PROCEDURES

As one gains experience and, hopefully, expertise in medical mycology, one develops a set of procedures for the isolation of fungi, their identification and serodiagnosis, and histopathologic methods. These are usually derived from the work of others and are modified for the particular laboratory situation in which one is working. In this section will be listed procedures and methods that have been found useful in the author's experience. As newer methods are described, the older ones are modified or dropped, so that any procedure listed here may be altered to fit a particular need.

Culture Methods

As has been noted in the text, the identification of fungi depends primarily on morphologic criteria. Morphology, color, spore production, and so forth are all affected by the particular medium on which the organism is growing, as well as by other environmental factors. A fungus growing on blood agar or an eosin methylene blue plate would hardly be recognized as the same organism when grown on Czapek-Dox agar. Since the identification of the fungus is based on observed morphology, there have evolved standard descriptions of organisms growing on particular media. The standard most universally used in medical mycology is the morphology of the organism growing on Sabouraud's agar. Raymond Sabouraud developed a peptone and honey or sugar medium in the late 1800's and described the morphology of many species of dermatophytes. Today the medium bearing his name, Sabouraud's agar. is composed of glucose and peptone at a pH of 5.6 and is still the standard one used in the medical mycology laboratory. This is not because the organisms grow best on it or produce the most spores; it is simply because of tradition. The standard description of the dermatophytes and most of the systemic and subcutaneous infecting organisms is based on the morphology of the organism when grown on this medium. The *Aspergillus, Penicillium*, and other opportunists or contaminants are identified by their morphology when growing on Czapek-Dox or similar "deficient" media. Morphologic description of the yeasts is often based on the growth observed on cornmeal or a similar medium. Thus the medical mycology laboratory requires only three media, basically: Sabouraud's agar, Czapek-Dox agar, and cornmeal agar. This is sufficient for the identification of essentially all the fungi one will encounter.

To facilitate the isolation and identification of fungi, some modifications of the above media have been made. Sabouraud's formula plus antibiotics becomes a selective medium for the isolation of dermatophytes and most systemic pathogens, and is called Mycosel or Mycobiotic agar. However, in our experience and that of others, we find that some respiratory pathogens are more easily isolated from blood agar than from Sabouraud's agar with antibiotics. Therefore, a blood agar plate is often added to the regimen of primary media used in respiratory infections.

The culture procedures for the isolation of fungi used in our laboratory are as follows: For dermatologic specimens a wide slant (25 × 100-mm tube) containing Mycosel or Mycobiotic agar is used. Skin scrapings or hairs are furrowed into the slant, which is then incubated at room temperature for four weeks. When mycelium is visible on the slant, a transfer is made to a slide culture. This consists of 1 per cent water agar in a Petri dish to act as a moist chamber, and a bent glass rod to act as support for the slide culture. The slide culture itself has a small amount (a glob) of Sabouraud's agar (without antibiotics) on one side and cornmeal 1 per cent glucose agar on the other. Both are inoculated with the growth to be identified, and a sterile cover slip is placed on top of each of the media. The slide culture is then placed in the moist chamber, covered, and allowed to grow for ten days. The Sabouraud's agar is for observation of the morphology of spores and hyphal structures of the organism. The cornmeal 1 per cent glucose agar stimulates pigment production in *T. rubrum* and generally suppresses it in other dermatophyte species (see Fig. 7–41). After growth has occurred, the cover slip is removed from the slide, and by using a razor blade the glob of medium is gently moved. This is done so that the strands of mycelium adherent to the slide remain undisturbed. A drop of lactophenol cotton blue is placed on the slide where the growth occurred, and the cover slip is replaced. It can now be examined with a microscope to observe the spore morphology, size and arrangement, and mycelial appendages.

For the culture of material containing pathogens causing subcutaneous and systemic infections, two media are recommended. The first is a Mycosel or Mycobiotic agar (or the equivalent) plate, and the second is a brain-heart infusion (BHI) sheep blood (5 per cent) agar plate. If sputum is to be cultured, a viscous blood-tinged area is selected. A KOH mount of a sample from the area is made for direct examination, and an aliquot of the specimen is streaked thoroughly on the two agar plates. These are incubated at room temperature for four to six weeks. To prevent drying, plates can be stored in a plastic bag. The plates should be checked every day for five days for the appearance of any mycelial growth. If present, it should be transferred immediately to a Sabouraud's slant. After five days, the plates are checked biweekly until discarded. After the organism has grown on Sabouraud's agar, it can be processed for identification.

Media

Most media for general diagnostic work are available commercially. The preparation and use of these media and some special media are described.

WATER AGAR

Agar	20 g
Water	1000 ml

Preparation. Heat to dissolve and then autoclave at 121°C for 15 minutes.

Use. This is used as the base of Petri plates for slide cultures. It is a deficient media and is sometimes useful in suppressing mycelial growth and stimulating asexual sporulation or cleistothecium production of some fungi. *Allescheria boydii* often produces asci and ascospores when grown on water agar.

SABOURAUD'S DEXTROSE AGAR

Glucose (dextrose)	40 g
Peptone	10 g
Agar	20 g
Water	1000 ml

Preparation. Heat to dissolve and then autocalve at 121°C for 15 minutes. Dispense in plates or tubes.

Use. This is the most commonly used medium in medical mycology and the basis of most morphologic descriptions. It is not good for maintaining stock cultures. For this, Emmons' modification (2 per cent glucose), sugar-free media, or cornmeal agar should be used. The latter, however, do not stimulate typical pigment production, and thus growth on such agars cannot be used for comparison with the standard morphologic descriptions of most species.

SABOURAUD'S CYCLOHEXIMIDE-CHLORAMPHENICOL AGAR*

Glucose	10.0 g
Peptone	10.0 g
Agar	15.5 g
Cycloheximide	0.4 g†
Chloramphenicol	0.05 g‡
Water	1000 ml

Preparation. Heat to dissolve and then autocalve at 121°C for 15 minutes. Dispense in tubes or plates.

Use. This preparation is the selective

*Formulations available as Mycobiotic agar (Difco Laboratories, Detroit, Michigan) and Mycosel (Baltimore Biological Laboratory, Inc., Baltimore, Maryland).

†Actidione (Upjohn).

‡Chloromycetin (Parke, Davis).

medium most commonly used in medical mycology. The cycloheximide inhibits the growth of most saprophytic fungi. Some pathogenic fungi are also sensitive to it to some degree. Among these are *Cryptococcus neoformans; Candida tropicalis* and a few other *Candida* species; *Trichosporon cutaneum*; and, at 37°C, the yeast forms of *Blastomyces dermatitidis* and *Histoplasma capsulatum*. The medium is used only as an isolation medium at room temperature. The chloramphenicol inhibits the growth of bacteria, so that *Actinomyces, Nocardia,* and *Streptomyces* will not grow.

CORNMEAL AGAR

Cornmeal	125 g
Water	3000 ml

Preparation. Heat to 60°C with stirring for one hour. Filter. Then add sufficient water to make 3 liters and add 50 g agar. Heat to dissolve, divide to manageable portions, autoclave at 121°C for 15 minutes, and dispense in tubes or plates.

Use. This is a deficient medium useful for stimulating mycelium production in *Candida* and other yeasts. It suppresses vegetative growth of many fungi and stimulates sporulation. It can also be used as a stock culture medium.

When made up with glucose (1 per cent), it is used as a stimulant for pigment production of *T. rubrum.*

CASEIN MEDIUM

Prepare separately:

1. Skim milk (dehydrated nonfat milk)	10 g
Water	100 ml
2. Water	100 ml
Agar	2 g

Autoclave both at 121°C for 15 minutes.

Preparation. Cool broth to approximately 45°C, mix, and pour in small Petri dishes.

Use. *N. asteroides* does not hydrolyze casein; *N. brasiliensis* and *Streptomyces* do. The medium is also used for differentiation of some agents of mycotic mycetoma.

GELATIN TEST MEDIA

Gelatin	4.0 g
Water	1000 ml

Preparation. Dissolve and adjust to pH 7. Dispense in tubes; autoclave at 121°C for five minutes.

Use. Inoculate with small fragments of suspected organisms. *N. asteroides* grows poorly; *Streptomyces* produces poor to good growth (stringy or flaky); *N. brasiliensis* produces compact, round colonies.

XANTHINE OR TYROSINE AGAR

Nutrient agar	23 g
Tyrosine	5 g
or	
Xanthine	4 g
Water	1000 ml

Preparation. Dissolve nutrient agar in 1000 ml of water. Add tyrosine or xanthine crystals and shake. Autoclave at 121°C for 15 minutes. Shake to disperse cystals evenly and dispense into Petri plates.

Use. For the differentiation of *Nocardia* and *Streptomyces.* See Chapters 2 and 4.

STARCH HYDROLYSIS TEST MEDIUM

Heart infusion broth	25 g
Casitone	4 g
Yeast extract	5 g
Soluble starch	5 g
Agar	15 g
Water	1000 ml

Preparation. Heat to dissolve and adjust to pH 7. Autoclave at 121°C for 15 minutes. Dispense in tubes for *Actinomyces* or on plates for *Nocardia* and fungi.

Use. To test for starch hydrolysis. Inoculate and incubate for several days to two weeks. Add Gram's iodine to plate. If the starch has been hydrolyzed, there will be a clear halo around the colony.

STARCH PRODUCTION AGAR

$(NH_4)_2SO_4$	1 g
$MgSO_4$	0.5 g
KH_2PO_4	1 g
Glucose	10 g
Thiamine	200 μg
Agar	25 g
Water	1000 ml

Preparation. Dissolve and adjust to pH 4.5. Autoclave at 121°C for 15 minutes. Dispense in plates.

Use. For detecting extracellular starchlike substances of the cryptococci. Inoculate plate and incubate at room temperature. After three weeks, pour Lugol's iodine over plate. If starch is present, there will be a blue halo around the colony.

Media for the Stimulation of Ascospore Formation of Perfect Yeasts and Fungi

ALPHACEL AGAR

Alphacel*	20 g
$MgSO_4 \cdot 7H_2O$	1 g
KH_2PO_4	1 g
$NaNO_2$	1 g
Tomato paste (Hunt's)	10 g
Oatmeal, baby (Beech Nut)	10 g
Agar	18 g
Water	1000 ml

Preparation. Dissolve and adjust to pH 5.6. Autoclave at 121°C for 15 minutes. Dispense in plates for molds or on slants for yeasts.

Use. For mating test cultures for ascospore formation.

MALT AGAR FOR YEAST ASCOSPORE FORMATION (Wickerham)

Malt extract	50 g
Agar	30 g
Water	1000 ml

Preparation. Dissolve in water, autoclave at 121°C for 15 minutes, and dispense in plates.

Use. Streak yeasts on agar slants. Incubate at room temperature for 24 to 48 hours. To observe ascospores of yeasts, make slides of growth and stain by Dorner method.

Stain for Ascospores (Dorner)

1. Prepare film on slide, dry, and heat fix.
2. Flood slide with carbolfuchsin; heat over flame for four minutes.
3. Cool. Wash with water; decolorize with acid alcohol (1 per cent HCl in ethanol).
4. Drip down 1 per cent nigrosin and allow to dry.

USE. Ascospores red, ascus colorless, yeast cells pale pink, background black.

*Nutritional Biochemical Co. Cleveland, Ohio.

Other Media for Special Purposes

POTATO-CARROT MEDIA

Carrots	20 g
Potatoes	20 g

Add some water and purée in blender; q.s. water to 1000 ml. Add 20 g agar.

Preparation. Heat to dissolve agar and autoclave 15 minutes at 121°C. Dispense in tubes and slant.

Use. Stock culture medium for many hyphomycetes.

HYPERTONIC MEDIA FOR GROWTH OF L FORMS OF FUNGI

Sabouraud's broth plus 30 per cent sucrose. Sterilize and pour over previously made Sabouraud's slants.

BIRD SEED AGAR FOR CRYPTOCOCCI (STAIB)

Guizotia abyssinica seeds (niger or thistle seeds)	70.00 g
Creatinine	0.78 g
Chloramphenicol	0.05 g
Glucose	10.00 g
Agar	20.00 g
Water	1000 ml
Diphenyl	100.00 mg

Preparation

1. Suspend seed powder in 350 ml of water and autocalve 10 minutes at 121°C.
2. Filter and bring volume to 1000 ml.
3. Dissolve other ingredients except diphenyl and autoclave 121°C for 15 minutes.
4. Cool to 45°C. Add diphenyl (dissolved in 10 ml 95 per cent alcohol). Stir and pour in plates.

Use. *Cryptococcus neoformans* forms dark brown gelatinous colonies; other yeasts are colorless or light brown.

ASPARAGINE BROTH

L-Asparagine	7.00 g
K_2HPO_4 c.p. anhydrous	1.31 g
Ammonium chloride	7.00 g
$Na_3C_6H_5O_7 \cdot 5H_2O$ (sodium citrate, c.p.)	0.90 g

OUTLINE OF DRUG DILUTION

Tube Number	1	2	3	4	5	6	7	8	9
Amphotericin B (5 mg/ml solution)	Undil soln.	1 ml of no. 1	1 ml of no. 2	0.5 ml of no. 1	0.5 ml of no. 4	1 ml of no. 5	1 ml of no. 5	1 ml of no. 7	1 ml of no. 8
Water		1 ml	1 ml	4.5 ml	4.5 ml	1 ml	4 ml	1 ml	1 ml
Concentration of drug in 0.1 ml	500 μg	250 μg	125 μg	50 μg	5 μg	2.5 μg	1 μg	0.5 μg	0.25 μg

$MgSO_4 \cdot 7H_2O$	1.50 g
Ferric citrate USP	0.30 g
Glucose USP	10.00 g
Glycerine c.p.	25.00 g
Water	1000 ml

Preparation. The asparagine is dissolved in 300 ml of hot (50°C) distilled water. Each organic salt is dissolved in separate aliquots of 25 ml of distilled water. Hot distilled water is used for the ferric chloride. Each salt is added in the order given above to the hot asparagine, mixing well after each addition. The glucose and glycerine are added last. The solution is brought up to volume by addition of distilled water. It is then dispensed in a wide-bottom flask, such as the Fernback flask, to give a depth of 1 to 1½ inches. The flasks are autoclaved at 121°C for 15 minutes.

Use. For the preparation of histoplasmin, coccidioidin, or other antigens. Spores are spread over the surface of the medium, and it is incubated (stationary) at room temperature for four weeks. A mycelial mat should cover the surface. At the end of four weeks the mat is separated from the culture broth by filtration. The antigens are prepared as described for the *Aspergillus* antigen preparations in a later section.

AMPHOTERICIN B SENSITIVITY

Amphotericin B (Fungizone, intravenous), 50 mg, Sabouraud's agar or broth.

Preparation of Medium. Prepare 18 tubes of Sabouraud's broth or agar (the latter must be kept molten) containing 9.9 ml in each tube. Two control tubes containing 10 ml are also needed.

Preparation of Drug. Into the phial containing 50 mg of Amphotericin B, place 10 ml of distilled water (sterile). This solution contains 5 mg per ml.

Preparation of Drug Dilutions. Set up eight empty tubes labeled 2 to 9 and handle as shown above:

Preparation of Drug Dilutions in Medium. Into the tube of medium labeled no. 1, place 0.1 ml of the original solution of amphotericin B. Into the tube of medium labeled no. 2, place 0.1 ml of dilution tube no. 2. Repeat with the remaining tubes 3 to 9. Mix thoroughly and, if agar is used, slant the tube.

Final Concentration of Amphotericin B in Medium. See Below:

When cool, the medium is inoculated with a standardized suspension of spores of the fungus or yeast cells. Control tubes containing no antibiotics are inoculated. The test should be done in duplicate. The test is incubated at 37°C and read at 48 hours.

SERUM LEVELS OF AMPHOTERICIN B

To test serum levels of amphotericin B, use the standard strain of *Candida tropicalis* ATCC #13803. Dilutions of serum are made in molten Sabouraud's agar. The medium is slanted and inoculated with a standard inoculum (usually 10^6) of *C. tropicalis.* It is incubated at 37°C. The test can also be done using spores of *Paecilomyces varioti.* An amphotericin B sensitivity series is done at the same time to ascertain the sensitivity of the organism. All tests are read at 48 hours. By comparison of end points, the level of amphotericin B in the serum can be ascertained.

CONCENTRATION IN MEDIUM

Tube Number	1	2	3	4	5	6	7	8	9
Concentration in μg/ml	50	25	12.5	5.0	0.5	0.25	0.1	0.05	0.025

PROCEDURE FOR DETERMINING 5-FC MIC (METHOD OF E. R. BLOCK*)

1. Each organism to be tested is subcultured on yeast-nitrogen base agar slants, incubated at 30°C for 48 hours, and then harvested in sterile normal saline.
2. The saline suspension of each organism is adjusted to obtain an optical density of 0.300 at a wavelength of 600 nm (using a spectrophotometer tube with an internal diameter of 13 mm). Normal saline is used as a blank.
3. This latter suspension (O.D. = 0.300) is then diluted 1:10,000 (10^{-4}) in sterile distilled water and is ready to be used in the test procedure.
4. Preparation of the test media is as follows:
   a. Nine tubes are required for each organism to be tested.
   b. This includes a control tube which has organisms but no 5-FC and eight tubes with twofold serial dilutions of 5-FC extending from 2.5 to 320 μg per ml.
   c. For each set of nine tubes, 35 ml of medium is prepared as below:

Yeast-nitrogen base (YNB)	5 ml	(each 100 ml contains 6.7 g YNB and 10.0 g dextrose in distilled H_2O); to use, dilute 1:10.
50% Glucose	2 ml	
Distilled water	28 ml	

   d. Add 5 ml of the final suspension of the test organism (from step 3) to 35 ml of medium to make a final volume of 40 ml. Mix well and then add 4 ml of this mixture to each of the nine test tubes. This will leave 4 ml extra. This step will be repeated with each organism to be tested.
   e. To prepare the 5-FC concentrations, 160 mg of powdered 5-FC (obtained as bulk powder from Hoffman-La Roche, Inc.) is dissolved in 100 ml of distilled water (with help of KOH) and filter sterilized. Serial twofold dilutions of this original stock concentration are made as below.
   f. If 1.0 ml of A is added to the first of the nine tubes containing 4 ml of medium and test organism, the final concentration of 5-FC will be 320 μg per ml. Similarly, if 1.0 ml of B through H is added to the next seven tubes, twofold dilutions of 5-FC, ranging from 160 to 2.5 μg per ml, will be obtained. In the ninth tube (control tube), 1.0 ml of sterile distilled water is added.
5. All tubes are then incubated on a rotating drum (2 rpm) at 32°C and read at 48 hours. The minimum inhibitory concentration is defined as the lowest concentration of drug in which no visible growth is observed.

*Modified by Block from Block, E. R., and J. E. Bennett. 1972. Pharmacological studies with 5-flourocytosine. Antimicrob. Agents Chemother., *1*:476–482.

GRISEOFULVIN SENSITIVITY TEST

The contents of one capsule containing 125 mg of griseofulvin are dissolved in 62.5 ml of acetone (concentration 2 mg per ml). Also prepare six tubes containing 9.9 ml of melted Sabouraud's agar. Six dilutions of griseofulvin are then prepared as follows:

1. Original dilution – griseofulvin in acetone
2. 3 ml of tube 1 and 1 ml H_2O
3. 1 ml of tube 1 and 1 ml H_2O
4. 1 ml of tube 1 and 3 ml H_2O
5. 1 ml of tube 2 and 4 ml H_2O
6. 1 ml of tube 3 and 9 ml H_2O

Add 0.1 ml of dilution 1 to agar tube A, then 0.1 ml of dilution 2 to tube B, and so forth.

Tube	Concentration of Drug in 0.1 ml	Final Concentration in Tube
A	200 μg	20 μg

OUTLINE OF DRUG DILUTION (5-FC)

A = 160 mg/100 ml 5-FC
B = 20 ml of 160 mg/100 ml stock + 20 ml distilled water = 80 mg/100 ml 5-FC
C = 20 ml of 80 mg/100 ml dilution + 20 ml distilled water = 40 mg/100 ml 5-FC
D = 20 ml of 40 mg/100 ml dilution + 20 ml distilled water = 20 mg/100 ml 5-FC
E = 20 ml of 20 mg/100 ml dilution + 20 ml distilled water = 10 mg/100 ml 5-FC
F = 20 ml of 10 mg/100 ml dilution + 20 ml distilled water = 5 mg/100 ml 5-FC
G = 20 ml of 5 mg/100 ml dilution + 20 ml distilled water = 2.5 mg/100 ml 5-FC
H = 20 ml of 2.5 mg/100 ml dilution + 20 ml distilled water = 1.25 mg/100 ml 5-FC

B	150 μg	15 μg
C	100 μg	10 μg
D	50 μg	5 μg
E	30 μg	3 μg
F	10 μg	1 μg

Slant tubes and inoculate each and a control without the drug with standard inoculum of the organism to be tested.

AGAR FOR MATING FUNGI AND MAINTENANCE OF STOCK CULTURES

Glucose	5.0 g
Yeast extract	0.9 g
Soil	20.0 g
Agar	20.0 g
Hair (human or horse)	As needed

Preparation. Dissolve and autoclave at 121°C for 15 minutes. Dispense into Petri plates.

Use. Mating of *Aspergillus heterothallicus, A. fennelli,* and many other fungi can be carried out by simply planting the two mating types on the agar near each other and allowing them to grow toward each other. At the boundary of the intermingling of the two thalli, a line of cleistothecia will be formed. For mating *Arthroderma* and *Nannizzia* species, it is often necessary to add keratin. This can be done by placing a few strands of hair in the center of the plate and then inoculating. Experience varies, but it appears that horse tail hair is the most satisfactory; however, sterile human hair can also be used. Some authors use chopped or ground hair. If mating is not successful, other additives may be utilized to enhance the romantic environment. These include beetle wings, oatmeal, tomato paste, chopped chicken feathers, and so forth.

Homothallic organisms are often stimulated to produce their perfect stage by being grown in deficient media. Soil agar is excellent for most of these. Some will require salts, and some produce ascospores best on plain 2 per cent water agar. *Allescheria* particularly is stimulated by this. Deficient medium is also the best for stock cultures. Various combinations of ingredients have been formulated and modified. These include potato-carrot, tomato-oatmeal, cornmeal, lima bean, and so forth. Many sound quite appetizing. This is probably because most mycologists of my acquaintance are also gourmets and excellent cooks.

Different formulations for the maintenance of stock culture have been devised over the years. Many species require special media or special care. The most useful and simplest techniques that have been successful in the author's experience will be outlined here.

Stock cultures of all organisms are kept on Sabouraud's agar (Emmons' formula) and cornmeal agar. Both media are available commercially. The slants are stored at room temperature. An additional set is flooded with sterile litmus milk (or reconstituted dry milk) and frozen. It is stored in the minus 20°C deep freeze. This combination of four tubes in two temperatures and one set protected by a colloid has been found to maintain most species satisfactorily, both alive and with cultural and spore characteristics intact. The room temperature stocks are transferred every six to eight months, and the deep freeze cultures every two to three years. If available, the lyophilization of spores in a colloid medium is the superior method of preserving fungi.

MEDIUM FOR PRODUCTION OF CULTURE SPHERULES OF *COCCIDIOIDES IMMITIS* FROM COUNTS AND CRECELIUS (1970)*

Basic Medium (tenfold concentration)

KH_2PO_4	6.260 g
$ZnSO_4$	0.036 g
$MgSO_4$	3.944 g
$NaHCO_3$	1.176 g
$CaCl_2$	0.029 g
NH_4 acetate	12.334 g
Glucose	39.635 g
$KHPO_4$	6.846 g
NaCl	0.014 g
Casein hydrolysate	20.000 g
Distilled water	1 liter

Preparation. The pH is adjusted to 6.0 and the medium autoclaved at 121°C for ten minutes. The medium is then diluted 1:10 with sterile distilled water and dispensed in 10-ml aliquots into a 50-ml sterile Erlenmeyer flask stoppered with a cotton plug and the cap covered with aluminum foil. The basic medium can be stored this way for some time without deterioration. Final medium is prepared when needed and cannot be stored.

*Roberts, J. A., J. M. Counts et al. 1970. Production in vitro of *Coccidioides immitis* spherules and endospores as a diagnostic aid. Am. Rev. Resp. Dis. *102*:811–813.

Additives for Final Medium

Biotin	1:10,000
Glutathione	0.005 M

These are prepared fresh and sterilized by filtration. The final medium is prepared by the addition of 0.5 ml of the glutathione solution and 0.5 ml of the biotin solution to each flask of basic medium to be used.

Use. Mycelium and arthrospores of *C. immitis* are inoculated and the cultures incubated without shaking at 40°C. Within three weeks spherules are produced.

Serologic Techniques

Serologic techniques used in the diagnosis of fungous infections have been the subject of an immense amount of literature and controversy. A few of these techniques have proved to be of great value and are standard procedures used in the diagnosis and prognostic evaluation of fungous infections. These procedures include complement fixation tests, precipitin tests, agglutination tests, and various immunofluorescent procedures. The complement fixation test is technically difficult, in addition to being time consuming, and requires experience and skill in its methodology and interpretation. Many of the immunofluorescent procedures are very useful but require expensive equipment and special training. Therefore, these tests are usually restricted to the large facilities of state and federal laboratories.

For the small laboratory, the most promising serologic procedures which are both rapid and inexpensive are the several latex agglutination tests and the immunodiffusion techniques. The latex agglutination tests that are commercially available at this writing are screening tests for histoplasmosis and coccidioidomycosis and a test for the presence of antigen in cryptococcosis. These all appear to be quite specific and, if properly carried out, are of great value in the diagnosis of disease. When these screening tests are positive, sera can be sent to reference laboratories for evaluation by complement fixation and other serologic procedures. The immunodiffusion test is also simple to carry out and can be used as a screening procedure. Double diffusion plates are available commercially that require only small amounts of both serum and antigen. In our laboratory the routine test consists of the patient's serum in the center well and a battery of six antigens in the outer wells. These are antigens prepared from *Histoplasma, Blastomyces, Candida, Aspergillus, Coccidioides,* and *Sporothrix schenckii.* The tests are read at 48 and 72 hours after incubation at room temperature and then placed in the cold and read after five days. If lines appear, a drop of 1.5 per cent EDTA (ethylenediaminetetraacetic acid) is placed in the center well and allowed to diffuse for one hour. EDTA will cause the disappearance of soluble lines. These are generally owing to C-reactive substance in the antigen preparation. If lines are present, they are considered a probable positive result, and the serum is tested with specialized serologic procedures, such as the complement fixation test. The interpretation of such lines is discussed in the various chapters concerned with specific diseases.

Production of good antigens is, at best, an art. So far, consistency of antigen preparation has not been achievable. In this section will be presented the methods that have been developed or found to be useful by S. McMillen, myself, and several other investigators. As a final note, it is hoped that better and more specific tests will be developed in the future, as most of these methods are quite crude. In some patients, especially those with severe debilitation and those suffering from candidosis or aspergillosis, the tests are often negative due to diminished or absent antibody production by the patient. A better test in these as in all cases would be the detection of antigen produced by the fungus in the infected patient. One such test is presently available and is quite useful—the antigen test in cryptococcosis. It is hoped that in the future more such procedures will be developed and made available.

ASPERGILLUS ANTIGEN

Medium. Sabouraud's broth.

Inoculum. *Aspergillus fumigatus* var. *ellipticus.* In our experience the antigen from this strain is the most sensitive for all forms of aspergillosis caused by most species of *Aspergillus.* In our laboratory we have also prepared antigen from *A. fumigatus* (five strains), *A. niger* (three strains), *A. terreus,* and *A. flavus.* On rare occasions the test is negative using the var. *ellipticus* antigen and is picked up by one of the other antigens when it is the etiologic agent.

Method. Dispense broth to a depth of 1 to 1½ inches in a large flask in order that you have a wide surface for the growth of the mycelial mat. Autoclave. Inoculate with a quantity of spores laid on the surface. Incubate (stationary) in the light at room temperature. A large surface mat will be produced.

At four weeks, aseptically remove 5 to 10 ml of the broth and test as is with positive control serum. If there is no activity, add 2 volumes cold acetone to the remainder of the sample and refrigerate for 24 to 48 hours. Centrifuge and collect the sediment. This is redissolved in 0.5 ml of distilled water. Retest the concentrate with known sera. If lines appear, challenge with a drop of 1.5 per cent EDTA or 5 per cent sodium citrate. Soluble lines representing C-reactive substance are commonly found in *Aspergillus* antigen preparations.

In general, most culture filtrates of *Aspergillus* have the greatest antigen content after four weeks' incubation. Those that grow more slowly may require six weeks. There is little increase in antigen content after that time.

To process the broth filtrate, decant from the flask and filter. Precipitate with 2 volumes cold acetone and refrigerate for 48 hours. The material is now stable and may be stored until convenient for further processing. Decant acetone and collect precipitate. Redissolve in water, pool the solutions from several flasks, and reprecipitate with equal amounts of (v/v) cold acetone. Decant and collect filtrate. Redissolve in water. Serial dilutions are necessary to establish potency and proper working dilution. In general, this is double the point of solubility of the precipitate. It can be stored frozen and does not lose potency. Most antigen is inactivated by heat.

BLASTOMYCES CELL SAP ANTIGEN

Medium. Brain-heart infusion broth. 100 ml in 250-ml cotton-stoppered flask.

Inoculum. *Blastomyces dermatitidis* yeast cells grown on BHI blood agar slants at 37°C for two weeks.

Method. Wash down two slants of *Blastomyces* yeast cells with 5 ml of sterile saline. Pipette 1 ml of saline suspension of yeast cells to each of five flasks of BHI broth. Incubate in a shaking water bath at 37°C for two weeks or until the cell mass is large. Harvest cells by filtration and wash with several changes of distilled water. Collect cells and freeze in liquid nitrogen. Break cells in a French press immersed in liquid nitrogen. It is important that the French press be immersed in liquid nitrogen at least one hour before using. Place the French press on a hydrolic press and bring diaphragm pressure to 2000 lb per in^2. Collect debris in stainless steel test tube. Check for per cent breakage of cells by examining with a microscope some debris suspended in water. If the breakage is less than 90 per cent, cells must be reprocessed. The debris is then suspended in sufficient saline to make an opalescent slurry. This is spun at 3000 rpm for 30 minutes to separate cell sap from cell walls. The cell sap may be frozen and stored at −95°C. When it is to be used, the sap is diluted with saline to give a reading of 2.5 g protein per 100 ml, using a Hitachi refractometer (see method for *Candida* cell sap). Culture filtrate and cell walls contain antigens that cross-react with many fungi. In our laboratory, the cell sap has been found to be the most specific antigen for use in *Blastomyces* immunodiffusion and complement fixation tests. Serum to be tested must be fresh, as many false-negatives have been found using stored sera. In our experience the French press and liquid nitrogen are the only way to obtain breakage of *Blastomyces* yeast cells.

CANDIDA CELL SAP ANTIGEN

Preparation. *Candida albicans* is grown in Sabouraud's broth for 24 to 48 hours at 37°C in a shaking water bath. The cells are harvested by centrifugation, washed three times in saline, and resuspended in sterile distilled water. The latter should be a very turbid suspension. To 20 ml of such a suspension, add 5 ml of a 5-micron glass bead suspension. A 10-ml lot is then sonicated in a stainless steel "cold shoulder" cell in a sonifier (Branson High Intensity). The exposure needed is 90 to 95 watts for 40 minutes. This should yield 80 to 90 per cent breakage of cells. The material is spun at 3000 rpm for 30 minutes in order to separate the cell walls from the cell sap. The cell sap is diluted to give a reading about 1.015 specific gravity on a Hitachi refractometer or equivalent apparatus. This is equivalent to

1.5 mg protein per 100 ml. This constitutes the "s" antigen of Taschdjian et al. Merthiolate, to a final concentration of 1:10,000, may be added as a preservative. The antigen remains stable for months in the refrigerator or indefinitely in the freezer. About equal results can be obtained using the commercially available Hollister-Stier *Monilia albicans* (Oidiomycin) antigen.

SPOROTHRIX (SCHENCKII) ANTIGEN

There are several methods described for the preparation of this antigen. A good M form antigen useful as a screening test can be made using the technique described for *Aspergillus* antigen. A yeast form antigen (Y) which is more specific and comparable to yeast form histoplasmin antigen can also be prepared. Several media may be used, such as nutrient broth, Sabouraud's broth, or asparagine broth. Some authors prefer *Trichophyton* #3 (Difco) broth. The formula is as follows.

Bacto–Vitamin-free Casamino Acids	2.5 g
Bacto-Inositol	50 mg
Thiamin	200 μg
Glucose	40 g
KH_2PO_4	1.8 g
$MgSO_4 \cdot 7H_2O$	0.1 g
Distilled water	1000 ml

Dissolve and dispense in 250-ml flasks. Autoclave at 121°C for 15 minutes. Inoculate with a heavy suspension of yeasts. Incubate at 37°C (rotating) for 30 days. Harvest cells and save both filtrate and yeasts. The filtrate antigen is prepared as described for *Aspergillus* antigen using cold acetone. A cell sap antigen "s" can be prepared by sonicating yeast cells or putting them through a French press (method for *Blastomyces* antigen). The cell sap antigen is the most specific in our hands.

An M form broth antigen can be prepared using *Trichophyton* #3 broth and the method for *Aspergillus* antigen.

HISTOPLASMA CAPSULATUM ANTIGEN

There are many formulas for the preparation of this antigen. The most successful in our experience is the simple "filtrate" antigen. This is prepared by growing the organism as a mycelial mat in asparagine broth medium (stationary) at room temperature for four weeks. The procedure for concentration of the antigen by cold acetone precipitation is the same as noted above for *Aspergillus* antigen. The *Histoplasma* antigen prepared from the mycelial stage (M) has been found to be superior to that prepared from the yeast stage (Y) in the immunodiffusion test.

COCCIDIOIDES ANTIGEN

A good *Coccidioides* antigen can be prepared from mycelial mat growth in asparagine broth. The culture is kept stationary at room temperature for four weeks. The antigen is concentrated by twofold acetone precipitation, as noted above in the section on *Aspergillus* antigen. This is the immunodiffusion antigen and complement fixation antigen (IDCF) of Huppert and Bailey.

Huppert and Bailey have prepared another antigen from the mycelial mat of *Coccidioides*. After the mat is separated from the culture filtrate, a measured volume of water is added to make a thick slurry. Toluene at a rate of 3 ml per 100 ml of water is added to the slurry. The flask is stoppered and incubated at 37°C for three days. The residue is filtered off through a Buchner filter. The lysate is then used as the antigen. This is termed the IDTP (immunodiffusion, tube precipitation) antigen, as it corresponds to that used in the old tube precipitin tests. (Huppert, M., and J. W. Bailey. 1965. The use of immunodiffusion tests in coccidioidomy cases. Am. J. Clin. Pathol., *44*:369–373.)

COMPLEMENT FIXATION TEST USING FUNGAL ANTIGENS (AFTER C. E. SMITH)

The serum is inactivated for 30 minutes at 56°C. Standard controls are run simultaneously. Serial dilutions are made of the serum with saline – 1:2, 1:4, 1:8, 1:16, and so forth, up to 1:512. To each 0.25 ml of diluted serum is added 0.25 ml of coccidioidin (or other antigen) and two units of complement contained in 0.5 ml. The mixture is incubated for two hours in a 37°C water bath. Then 0.5 ml of 2 per cent sheep red blood cells sensitized with two units of antisheep ambocepter is added. The mixture is agitated and incubated at 37°C in a water bath for one hour. Readings are made before and after overnight refrigeration. Complete hemolysis, representing no previous fixation of complement, is read as negative. No hemolysis (complete fixation of complement) is read at 4+. Intermediate degrees are recorded as +, ++, or +++. If complement is fixed, serum is retained for comparison with future samples. A modification of this procedure used in some laboratories is to mix sera, complement, and coccidioidin (or other antigen) and refrigerate (4°C) overnight. The sheep erythrocytes (sensitized) are then added and the mixture incubated at 37°C for two hours. This method is reputed to be more sensitive and to give a one-tube dilution higher than the method employing two-hour binding at 37°C.

Procedures for Direct and Histopathologic Examination for Fungi

DIRECT EXAMINATION

The Potassium Hydroxide Mount. The most useful technique for direct examination of all types of specimens for fungi is the potassium hydroxide mount. One drop of 10 per cent potassium hydroxide is placed on the slide and mixed with the specimen. It is cleared by warming the slide. Do not boil, as this produces KOH crystals. The walls of the fungi are tough and not harmed by the alkali. Most tissue cells and many artifacts are dissolved.

India Ink Mount. India ink slides are a traditional method used for the detection of cryptococci in spinal fluid. The spun sediment of spinal fluid is mixed with one small drop of India ink and examined. The preparation should be thin and hazy brown, not thick and black. It is examined with the high dry lens in the microscope. There are numerous artifacts that resemble cryptococcal yeasts. Most are red and white blood cells, glial cells, macrophages, and so forth. These are destroyed by KOH, so that this is a necessary control to rule out artifact. Only when the typical budding yeast cell with its halolike capsule is seen is the examination termed positive (see Fig. 9–5). Airborne cryptococci as well as other fungi have been found growing in bottles of India ink, so the reagent must be checked from time to time.

Cryptococci in tissue can be seen by using touch slide. The slide is touched to the suspected formalinized tissue or the tissue contents squeezed onto a slide and India ink added. A KOH control must be performed. Macerated brain can also be examined as an India ink preparation.

Scrape Mounts. These are prepared by scraping the cultures of fungi to obtain mycelium and spores. An inoculating needle bent to form a hook is useful. The mycelium and spores are teased apart in a drop of lactophenol cotton blue on a slide. The teasing is performed with the corner of a cover slip. After sufficient separation of mycelial strands, the cover slip is placed on it and the preparation examined. The use of slide cultures is superior to scrape mounts for the identification of fungi.

Lactophenol Cotton Blue (LCB)

Phenol crystals	20 g	(melt in water bath before weighing)
Lactic acid (85%)	20 g	
Glycerin	40 g	
Cotton Blue (C4B) (Poirriers's blue)	0.05 g	
Distilled water	20 ml	

Permanent mounts of slide cultures or scrape mounts can be made by sealing the edge of the coverslip with nail polish. A superior method is to dip the preparation in collodion before staining. After drying the slide is stained with LCB for 10 to 15 minutes. Then the following dehydration procedure is followed: (From Rivalier, E., and S. Seydel. 1932. Cultures mincés sur lamés gelasées calorées et ecaminées in situ in preparatinas definitines paur etude des cryptogámes microscopiques. C. R. Soc. Biol., *110*:181–184.)

1. Wash in 70% ethanol rapidly.
2. 95% ethanol 10 minutes.
3. Absolute alcohol 10 munutes or acetone.
4. Acetone-xylene 10 minutes.
5. Xylene.
6. Immediately (before drying) add mounting media (histology lab as Permount) and coverslip.

HISTOPATHOLOGIC EXAMINATION

Tissue specimens for examination of fungi are handled in the same manner as routine tissue. They are fixed in 10 per cent formalin solution and processed by the usual procedures for tissue sectioning. Few fungi are visible in routine hematoxylin and eosin stained preparations, so that special staining procedures are often necessary to visualize them. The various modifications of the periodic acid–Schiff stain are the most frequently employed or the Grocott variation of the Gomori methenamine silver stain. All have general acceptance and are excellent aids for diagnosis. They also have many pitfalls in the staining technique and the presence of artifacts. The author prefers the Gridley stain and the Grocott-Gomori methenamine stain. In this section will be described some of the more useful stains, the procedures used, and the difficulties encountered.

The Periodic Acid–Schiff Stains. The periodic acid–Schiff stains are based on the Feulgen reaction and have found wide use in biology. The Feulgen reaction depends on the formation of aldehyde groups in a polysaccharide chain and the *in situ* recoloration of the Schiff base (leukofuchsin). The Schiff base is produced by the action of sulfurous acid on the dye, basic fuchsin. This involves the addition of sulfur dioxide to the two amino groups of pararosaniline (basic fuchsin). The dye is then colorless.

The cell walls of fungi contain chitin and other complex polysaccharide chains. When these are treated with periodic or chromic acid, the carbon to carbon chain is broken in places, and pairs of aldehydes are formed. When Schiff's base comes in contact with the properly paired aldehydes, a reaction takes place in which the leukosulfonic acid–Schiff base is oxidized and the aldehyde reduced. The dye is attached to the polysaccharide molecule at the site of the reaction. This is termed the Feulgen reaction (Fig. A–1). The color produced is a magenta. The depth of the color produced depends on the number of available aldehyde groups produced by hydrolysis. Since the cell walls of fungi have a high concentration of complex polysaccharides, they stain an intense purple-red. Both the cell walls and capsular material of *Cryptococcus neoformans* stain by the PAS

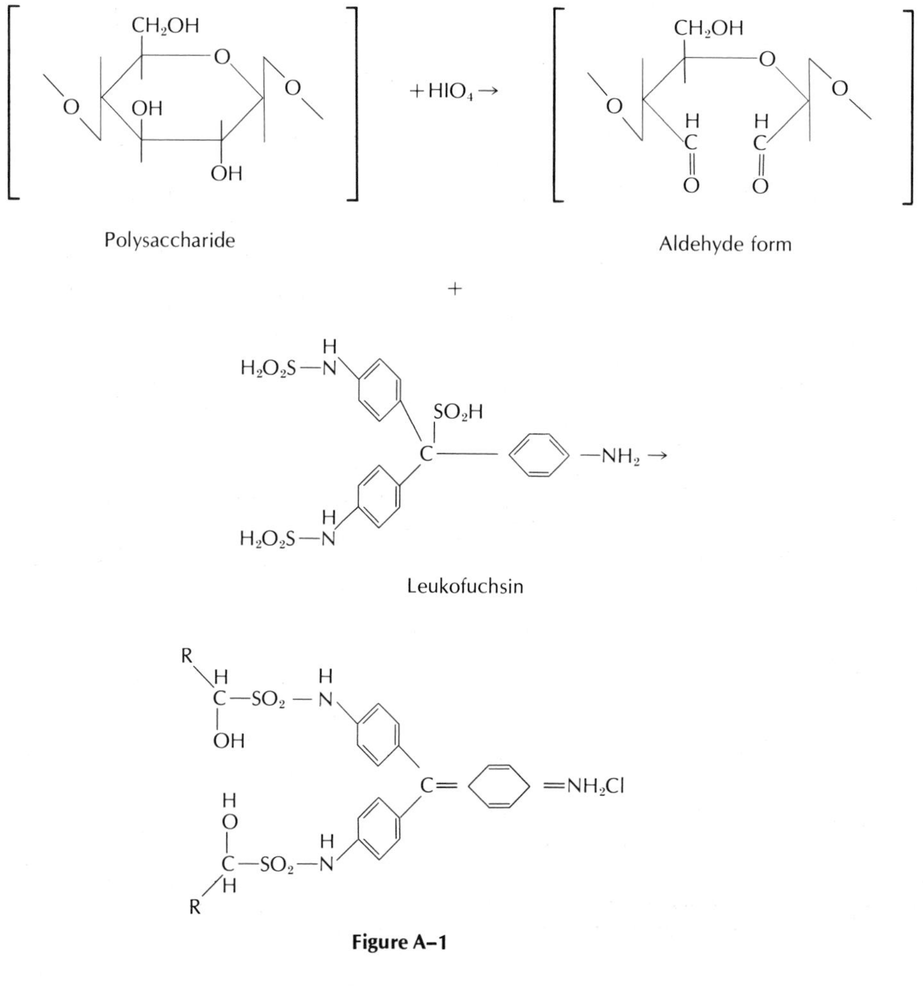

Figure A–1

technique. The cell wall will be dark, and a halo of deep, red-staining, concentrated capsular material will surround the cell. As the capsule becomes thinner, the color becomes progressively paler.

Advantages. This stain is very useful for observing fungi in tissue. The fungus is usually stained much darker than the surrounding tissue. The morphology of the fungus is easily discernible, and even a few organisms in a field will be observable, especially if the background is stained by metanil yellow or fast green. Some pathologists prefer to have the slides stained by the hematoxylin-eosin method as a counter stain after the PAS has been completed. This makes individual fungus cells difficult to see, as there are many red or magenta objects present. The author prefers a slide stained by the Gridley method (metanil yellow), a second slide stained by the methenamine silver stain, and a third slide stained by hematoxylin and eosin to observe the tissue reaction and morphology.

Disadvantages. Many components of tissue that are carbohydrate in composition are stained by the PAS method. These include glycogen, starch, cellulose, glycolipids, mucin, fibrin threads, and elastic tissue. They are not usually stained as intensely as the fungi, however. Bacteria such as *Nocardia* or *Actinomyces* are not stained by the PAS method but are by the methenamine stains.

Gridley Stain for Fungi

Solutions

A. Chromic acid solution (this must be prepared fresh)

Chromic acid	4.0 g
Distilled water	100.0 ml

B. Feulgen reagent (Coleman)

Bring to boil 200 ml distilled water. Remove from flame and add 1 g basic fuchsin. When dissolved, cool and filter. To the solution add 2 g sodium metabisulfite (hypo—photographic) and 10 ml hydrochloric acid (normal). Permit the solution to bleach for 24 hours. Add 0.5 g activated charcoal powder (Norit or equivalent), shake for one minute, and filter using coarse filter paper. The solution should be colorless. The finished reagent may be stored in the refrigerator. It is usable until it becomes pale pink.

C. Normal hydrochloric acid

HCl, specific gravity 1.19	83.5 ml
Distilled water	916.5 ml

D. Sodium metabisulfite solution (10 per cent)

Sodium metabisulfite	10.0 g
Distilled water	100.0 ml

E. Sulfurous acid rinse

Sodium metabisulfite, 10 per cent	6.0 ml
Normal hydrochloric acid	5.0 ml
Distilled water	100.0 ml

F. Aldehyde-fuchsin solution

Basic fuchsin	1.0 g
Alcohol, 70 per cent	200.0 ml
Paraldehyde	2.0 ml
Hydrochloric acid (conc.)	2.0 ml

The solution is allowed to stand for three days at room temperature until it has turned a deep blue. It may be stored in the refrigerator. Before use, filter and bring to room temperature.

G. Metanil yellow solution

Metanil yellow	0.25 g
Distilled water	100 ml
Glacial acetic acid	2 drops

Procedure. Paraffin blocks are cut at 6 microns and fixed to the slide. The slides are passed through the following steps.

1. Deparaffinize through xylene, absolute alcohol, and 95 per cent alcohols to distilled water.
2. Place in 4 per cent chromic acid (oxidizer) for one hour.
3. Wash in running water for five minutes.
4. Place in Coleman's Feulgen reagent for 15 minutes.
5. Put slides through three changes of sulfurous acid rinse.
6. Wash in running tap water for 15 to 30 minutes.
7. Place slides in aldehyde-fuchsin solution for 15 to 30 minutes (the time allowed depends on personal experience and quality of reagents).
8. Rinse off excess stain in 95 per cent alcohol.
9. Rinse in water.
10. Lightly counterstain with metanil yellow for one minute.
11. Rinse in water.
12. Dehydrate through the usual series of alcohols to absolute alcohol.
13. Clear with xylene.

Results. Mycelium and yeasts, deep blue or deep rose; conidia, deep rose to purple; background, yellow; elastic tissue, mucin, and so forth, deep blue.

Gridley, H. F. 1953. A stain for fungi in tissue sections. Am. J. Clin. Pathol., *23*:303–307.

Grocott Modification of Gomori Methenamine Silver Stain

Solutions

A. Chromic acid (5 per cent)
  Chromic acid (CrO_3) 5.0 g
  Distilled water 100.0 ml

B. Methenamine (3 per cent)
  Hexamethylene tetramine USP $(CH_2)_6N_4$ 3.0 g
  Distilled water 100.0 ml

C. Methenamine silver nitrate stock solution
  Silver nitrate, 5 per cent solution 5.0 ml
  Methenamine, 3 per cent solution 100.0 ml

A white precipitate will form, but it disappears on shaking. The solution is stable for months when refrigerated.

D. Sodium bisulfite (1 per cent)
  Sodium bisulfite ($NaHSO_3$) 1.0 g
  Distilled water 100.0 ml

E. Light green stock
  Light green SF (yellow) 0.2 g
  Distilled water 100.0 ml
  Glacial acetic acid 0.2 ml

F. Working light green solution
  Light green (stock) 10.0 ml
  Distilled water 50.0 ml

G. Silver nitrate (5 per cent)
  Silver nitrate ($AgNO_3$) 5.0 g
  Distilled water 100.0 ml

H. Borax solution (5 per cent)
  Borax (USP) $Na_2B_4O_7 \cdot 10H_2O$) 5.0 g
  Distilled water 25.0 ml

I. Methenamine silver nitrate working solution
  Borax, 5 per cent solution 2.0 ml
  Distilled water 25.0 ml

Mix and add 25 ml of methenamine silver nitrate stock solution.

J. Gold chloride (0.1 per cent)
  Gold chloride, 1 per cent solution ($AuCl_3 \cdot HCl \cdot 3H_2O$) 10 ml
  Distilled water 90 ml

This solution may be used repeatedly.

Procedure. Cut paraffin blocks at 6 microns, place on slides, and fix. Deparaffinize through xylene (two changes), absolute alcohol, and 95 per cent alcohols to distilled water. Previously stained sections may also be used, as chromic acid will remove previous stains.

1. Place slides in chromic acid solution for one hour.
2. Wash in running tap water for 15 seconds.
3. Rinse in 1 per cent sodium bisulfite for one minute to remove residual chromic acid.
4. Wash in tap water for five to ten minutes.
5. Wash in four changes distilled water.
6. Place in working methenamine silver nitrate solution, and then place container in oven at 58 to 60°C for 30 to 60 minutes (until sections are a light yellowish brown). Most fungi require 60 minutes. To remove slides use paraffin-coated forceps. Dip slides in distilled water. Check control slide with microscope for silver impregnation. Fungi should be a dark brown at this stage.
7. Rinse in six changes of distilled water.
8. Place in 0.1 per cent gold chloride to tone for two to five minutes.
9. Rinse in distilled water.
10. Remove unreduced silver with 2 per cent sodium bisulfite for two to five minutes.
11. Wash thoroughly in running tap water.
12. Counterstain in light green for 30 to 45 seconds.
13. Dehydrate with two changes 95 per cent alcohol, absolute alcohol, and clear with two changes xylene. Mount slides in Permount.

Results. Fungi, sharply delineated in black; bacteria (*Nocardia, Actinomyces*), black; *Nocardia* may require up to 90 minutes staining in the methenamine silver solution; mucin, taupe to dark grey; inner parts of mycelium, old rose; background, pale green. In addition, depending on staining procedure, reticulum fibers, elastin fibers, and so forth are stained black. Red blood cells and some other cells are old rose to greyish rose.

Grocott, R. G. 1955. A stain for tissue section and smears using Gomori's methenamine-silver nitrate technic. Am. J. Clin. Pathol., *25*(no. 7):975–979.

Acid Retention Stain for *Nocardia*

1. Heat-fix slide.
2. Flood slide with kinyoun carbolfuchsin and heat for three to five minutes. Wash thoroughly.
3. Decolorize with 1 per cent H_2SO_4 for 1½ minutes.
4. Counterstain with methylene blue for five to ten seconds.

Results. Nocardia, red; background, blue.

Mayer's Mucicarmine Histologic Stain for *Cryptococcus neoformans*

Solutions

A. Weigert's iron hematoxylin

1. Solution A
  Hematoxylin 1.0 g
  Alcohol, 95 per cent 100.0 ml

2. Solution B
   Ferric chloride, 29 per cent aqueous solution 4.0 ml
   Distilled water 95.0 ml
   HCl (conc.) 1.0 ml
3. Working solution
   Equal parts of solution A and B Prepare fresh

B. Metanil yellow solution
   Metanil yellow 0.25 g
   Distilled water 100.0 ml
   Glacial acetic acid 0.25 ml

C. Mucicarmine stain
   Carmine 1.0 g
   Aluminum chloride, anhydrous 0.5 g
   Distilled water 2.0 ml

Mix stain in small test tube. Heat over small flame for two minutes. Liquid will become almost black and syrupy. Dilute with 100 ml of 50 per cent alcohol and allow to stand for 24 hours. Filter and dilute 1:4 with tap water for use.

Staining Procedure

1. Cut paraffin sections at 6 microns and fix to slide. Deparaffinize and rehydrate through xylene, absolute alcohol, and 95 per cent alcohols in the usual manner to distilled water. Remove mercury precipitates through iodine and hypo (photographic) solutions if necessary.
2. Stain for seven minutes in working solution of Weigert's hematoxylin.
3. Wash in tap water for five to ten minutes.
4. Rinse in diluted mucicarmine solution for 30 to 60 minutes or longer. (Check control slide after 30 minutes.)
5. Rinse in distilled water.
6. Stain in metanil yellow solution for one minute.
7. Rinse quickly in distilled water.
8. Rinse quickly in 95 per cent alcohol.
9. Dehydrate in two changes of absolute alcohol; clear with two to three changes of xylene and mount in Permount.

Results. Mucin, deep red to rose; nuclei, black; background, yellow. Care should be used not to overstain with metanil yellow.

Mallory, F. B. 1942. *Pathologic Techniques.* Philadelphia, W. B. Saunders Company, p. 130.

Brown and Brenn Modification of Gram's Stain for *Nocardia* and *Actinomyces* in Tissue

Solutions

A. Crystal violet solution
   Crystal violet 1.0 g
   Distilled water 100.0 ml

B. Sodium bicarbonate solution
   Sodium bicarbonate 5.0 ml
   Distilled water 100.0 ml

C. Gram's iodine
   Iodine 1.0 g
   Potassium iodide 2.0 g
   Distilled water 300.0 ml

D. Ether-acetone mixture
   Equal parts of acetone and ether (v/v)

E. Basic fuchsin solution
   Saturated aqueous solution of basic fuchsin 0.1 ml
   Distilled water 100.0 ml

F. Picric acid–acetone
   Picric acid 0.1 g
   Acetone 100.0 ml

G. Acetone-xylene mixture
   Equal parts of acetone and xylene (v/v)

Procedure

1. Cut paraffin sections 6 microns and fix to slides. Deparaffinize and rehydrate with xylene, absolute alcohol, and 95 per cent alcohol to distilled water in usual manner.
2. Place slides on rack and pour on approximately 1 ml of 1 per cent crystal violet and 5 drops of 5 per cent sodium bicarbonate. Allow to stand for one minute while agitating slowly.
3. Wash in water.
4. Flood slide with Gram's iodine solution for one minute.
5. Rinse in water. Gently blot dry with filter paper.
6. Decolorize in ether-acetone mixture by dropping on slide until no more color comes off.
7. Stain with basic fuchsin solution for one minute.
8. Wash in water. Blot dry.
9. Transfer slide to acetone.
10. Decolorize in picric acid–acetone solution until sections are yellowish pink.
11. Rinse in acetone.
12. Rinse in acetone-xylene mixture.
13. Clear in several changes of xylene.
14. Mount in Permount or equivalent.

Results. Gram-positive bacteria, blue; Gram-negative bacteria, red; nuclei, red; background, yellow.

Brown, J. H., and L. Brenn. 1931. Methods for differentiational staining of Gram-positive and Gram-negative bacteria in tissue sections. Bull. Johns Hopkins. Hosp., *48*: 69–73.

AUTHOR INDEX

INDEX

Note: Page numbers in italics refer to illustrations. Page numbers followed by the letter "t" refer to tables.